CORROSION RESISTANCE TABLES

CORROSION RESISTANCE TABLES

METALS, PLASTICS, NONMETALLICS, AND RUBBERS

Philip A. Schweitzer
CHEM PRO Equipment Corporation
Fairfield, New Jersey

MARCEL DEKKER, INC. New York and Basel

MARCEL DEKKER, INC.

270 Madison Avenue, New York, New York 10016

LIBRARY OF CONGRESS CATALOG CARD NUMBER: 76-46718

ISBN: 0-8247-6488-9

Current printing (last digit):
10 9 8 7 6 5 4 3 2 1

PRINTED IN THE UNITED STATES OF AMERICA

INTRODUCTION

For years the Chemical Engineer has been faced with the problem of selecting the proper material of construction to be used for the construction of processing vessels, storage tanks, valves, piping systems, and whatever else the chemicals being handled come into contact with. During the early days of the chemical engineering profession the list of materials from which to choose was relatively small, consisting primarily of various metals and their alloys.

As technology made advances new alloys were developed as well as synthetic materials such as plastics. Simultaneously new chemicals came into being, many of which required the newly developed materials of construction for their safe handling.

However the problem of material selection is not limited to the chemical engineer. Many industries use chemicals in their everyday operations which can not be classified as members of the Chemical Processing Industry, yet they have the same problems of selecting the proper materials of construction to handle these products.

This is particularly true in the area of pollution control. Many processes for the elimination of air and/or water pollution involve chemical changes in the pollutants which result in problems of corrosion which were not present before.

The dictionary defines corrosion as "the eating or wearing away by slow degrees," a definition which is rather broad and which does not restrict the application of the word to the destruction of metals. This is as it should be because of the availability of so many nonmetallic materials of construction.

Most reference sources for the selection of proper materials of construction are devoted to either a specific group of metals or alloys; plastics; rubbers (synthetic and natural); gasketting materials; or packing materials. An individual with the problem of selecting materials of construction usually must make a selection in more than one of the preceding categories which necessitates his having to go to several sources for the answers. It is the purpose of this book to provide one source from which all of the components of a system may be selected including processing vessels, tanks, pumps, piping, valves, gaskets, packing, etc.

These corrosion tables have been prepared with the intention of assisting not only the practicing engineer but whoever else may be charged with the responsibility of selecting the proper materials of construction for a specific application. Any table of this type should only be used as a guide since it is extremely difficult and at times impossible to duplicate actual operating conditions. To fully guarantee the suitability of a particular material of construction corrosion tests should be conducted under actual operating conditions.

In the charts which follow all of the chemicals listed are assumed to be in the pure state or in concentrated or saturated aqueous solutions unless otherwise indicated. Concentration percentages used are weight percents.

Two types of notations have been employed to indicate the suitability of a specific material of construction - one for metallic materials and one for nonmetallic materials. This is necessary because of the differing mechanisms of corrosion.

The four most common forms of corrosion of metals are:

1. General Corrosion
2. Galvanic Corrosion
3. Stress Corrosion Cracking
4. Intergranular Corrosion

General Corrosion

Metals resist corrosion by the formation of a passive film on the surface, which in a sense is also corrosion, but once formed it prevents further degradation of the base metal. Most metals will form this film after a period of time upon exposure to the air. Passivation is the name applied to the chemical treatment procedure which helps to form the film more rapidly. For example, exposure of austenitic stainless steels to nitric acid will produce this protective film.

As long as these films remain intact they will protect the base metal from further attack. Certain chemicals attack various films at varying rates. When this occurs the metal is said to corrode. Such interaction between a chemical and a metal can be determined from the corrosion tables. This type of corrosion is known as general corrosion.

Galvanic Corrosion

This type of corrosion occurs when two or more dissimilar metals are in contact, or when metals having the same analysis have different surface conditions and an electrolyte is present.

Under these conditions an electolytic cell is formed. All metals or alloys have certain "built-in" properties which cause it to react as an anode or a cathode when in contact with dissimilar metals or alloys. Whether a particular material will react as a cathode or an anode can be determined from the relative positions of the materials in the galvanic series. The further apart two materials are from each other in the galvanic series, with all other factors being equal, the greater the rate of corrosion. The material closest to the anodic end will be the one to corrode. For example if tin and zinc were in contact the zinc would corrode whereas if tin and copper were in contact the tin would corrode.

The rate of attack is also affected by the relative size of the materials and the specific electrolyte present. A small anode area in contact with a large cathode area will result in a rapid severe attack. Conversely, a large anode area in contact with a small cathode area will lessen the rate of galvanic attack since the same total e.m.f. driving force of corrosion will be spread out over a larger area. The higher the degree of ionization of the electolyte the greater the rate of attack.

Galvanic corrosion can be prevented by judicious use of dissimilar metals.

GALVANIC SERIES

Anodic End

Magnesium
Magnesium Alloys
Zinc
Aluminum 5052
Aluminum 6061
Cadmium
Aluminum AA 2017
Iron & Carbon Steel
Copper Steel
4-6% Chromium Steel
Ferretic Stainless (active)
    400 Series
Austenitic Stainless (active)
    18-8 Series
Lead-Tin Solder
Lead
Tin
Nickel (active)
Inconcl (active)

Hastelloy C (active)
Brasses
Copper
Bronzes
Cupro-Nickel Alloys
Monel
Silver Solder
Nickel (passive)
Inconel (passive)
Ferretic Stainless (passive)
Austenitic Stainless (passive)

Titanium

Hastelloy C (passive)
Silver
Graphite
Gold
Platinum

Cathodic End

Stress Corrosion Cracking

All metals are subject to stress corrosion cracking which results from residual or applied stress in the metal. Residual stresses can occur as a result of cold forming or quenching after heat treating. Annealing after fabrication will minimize these stresses.

Applied stresses can result from faulty design, vibrations, flexing and excessive expansion and contraction due to thermal changes.

The particular corrodent will determine the degree of effect. Halogen salts are among the materials that will attack metals most vigorously when stresses are present.

INTERGRANULAR CORROSION

This form of corrosive attack will occur only at the grain boundaries of austenitic stainless steels. All austenitic stainless steels contain carbon in small amounts. When this material is heated to the sensitizing range (850°F to 1650°F), such as in welding, carbon is precipitated out at the grain boundaries in the form of chromium carbide. The chromium has been taken from the grain boundaries, changing the composition and making this area even more susceptible to corrosive attack.

This effect can be minimized by annealing the stainless steel after it has been sensitized. This is done by heating the alloy to 1800°F or higher, depending upon the specific alloy, and then quickly cooling it through the sensitizing range to prevent the carbon from precipitating out.

Another way to reduce this effect is to limit the carbon content to less than 0.03 per cent. This extra low carbon (ELC) stainless steel can be welded without the danger of carbide precipitation.

Corrosion of Nonmetallics

Plastic materials are attacked by solvation or a chemical reaction. Solvation is the penetration of the plastic by a corrosive element which causes softening, swelling, and ultimate failure. Plastics in contrast to metals do not exhibit a corrosion rate, usually they completely resist attack or they deteriorate rapidly. Because of this difference in corrosion mechanism the two types of notation were established.

Since corrosion is a function of temperature the tables reflect this by indicating the suitability of each material at varying temperatures. When a material is indicated to be unsatisfactory at a specific temperature it is also unsatisfactory at all temperatures above the one shown. A blank in the chart indicates that no data is available.

There are many plastic formulations which may vary from manufacturer to manufacturer. An indication in the chart that a specific material is suitable for use with a specific corrodent does not mean that the formulation used by all manufacturers would be suitable. This must be verified with the specific manufacturer.

A word of caution should also be given regarding temperatures. The table only shows the resistivity to corrosive attack at various temperatures. Before the material is specified for use the physical properties at the operating temperature must be examined to insure that the material has the proper mechanical strength for the application.

USING THE TABLES

The tables are arranged alphabetically according to corrodent. Unless otherwise noted, the corrodent is considered pure, in the case of liquids, and a saturated aqueous solution in the case of solids. All percentages shown are weight percents.

There are two pages for each corrodent: one containing the resistance to metals, glass, and carbon and the other containing the resistance to plastics and rubbers.

Corrosion is shown as a function of temperature. Symbols used to designate specific corrosion rates are shown on the bottom of the even numbered pages. The use of the temperature scale is explained by the following example:

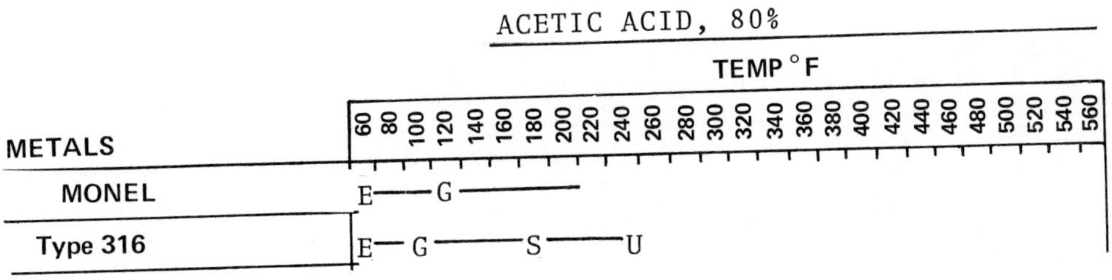

From the above it is seen that monel has a corrosion rate varying with temperature as follows:

(E) <2 mils penetration/year between 60 and 120°F
(G) <20 mils penetration/year between 120 and 210°F
No data beyond 210°F

Type 316 stainless steel has a corrosion rate varying with temperature as follows:

(E) <2 mils penetration/year between 60 and 100°F
(G) <20 mils penetration/year between 100 and 180°F
(S) <50 mils penetration/year between 180 and 245°F
(U) >50 mils penetration/year above 245°F

Notations

1. Material is subject to pitting.
2. Material is subject to stress cracking.
3. Material is subject to crevice attack.
4. Not for use with carbonated beverages.
5. Material should be passivated.
6. Alkaline.
7. Acid free.
8. Acid free and passivated.
9. Material is subject to intergranular corrosion.
10. Dynel surfacing veil should be used.
11. This material will cause stress cracking in many polyethylenes. A stress crack resistant high molecular weight polyethelene must be used.
12. Material is subject to stress cracking when wet.

all Data applies to all concentrations.
elc Material is low carbon grade.
sat Solution is saturated.

CONTENTS

CORROSION RESISTANCE TABLES

READING THE TEMPERATURE SCALE

When using the tables in this volume, note that the tick marks refer to temperatures midway between the temperatures cited (see the sample table, below).

Thus, both the alloy 20Cb3 and brass are corroded by pyridine at a rate which is less than 20 mils/year between 60 and 210°F. Aluminum and bronze have the same resistance up to 350°F. Carbon steel has a good resistance to corrosion up to approximately 128°F, and copper, only to 80°F.

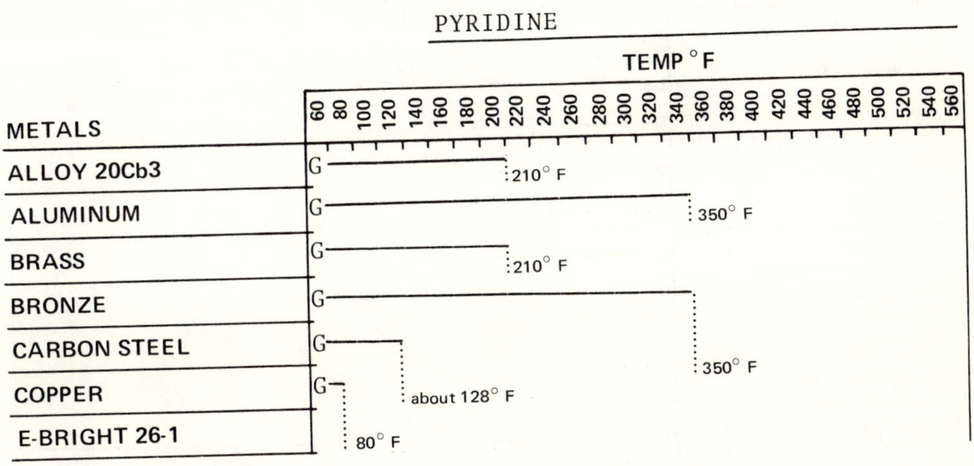

ACETALDEHYDE

TEMP °F: 60 80 100 120 140 160 180 200 220 240 260 280 300 320 340 360 380 400 420 440 460 480 500 520 540 560

METALS	Rating
ALLOY 20Cb3	E — G
ALUMINUM	G
BRASS	U
BRONZE	G
CARBON STEEL	G
COPPER	U
E-BRIGHT 26-1	
HASTELLOY-B	
HASTELLOY-C	E
HASTELLOY-D	
HIGH SILICON IRON	G
INCONEL	G
LEAD	G
MONEL	G
NICKEL	G
STAINLESS STEELS	
Type 304/307	E — G
Type 316	E — G
TANTALUM	G
TITANIUM	E
ZIRCONIUM	E

NONMETALLICS	Rating
BOROSILICATE GLASS	R
CARBON	R

TEMP °F: 60 80 100 120 140 160 180 200 220 240 260 280 300 320 340 360 380 400 420 440 460 480 500 520 540 560

FOR METALS

E = < 2 Mils Penetration/Year; G = < 20 Mils Penetration/Year
S = < 50 Mils Penetration/Year; U = > 50 Mils Penetration/Year

FOR NONMETALLICS

R = Resistant
U = Unsatisfactory

Please see page 1 for notes on calibration.

ACETALDEHYDE

TEMP °F: 60 80 100 120 140 160 180 200 220 240 260 280 300 320 340 360 380 400 420 440 460

PLASTICS	
ABS	U
ACRYLICS	U
ASBESTOS REINF., EPOXY	
ASBESTOS REINF., PHENOLIC	
CHLORINATED POLYETHER	R——U
CPVC	U
EPOXY	R——
FLUOROCARBONS FEP	R——
FLUOROCARBONS TFE	R——
FURFURYL ALCOHOL	R——
NORYL	
PHENOLIC	
POLYAMIDES–NYLON	R—
POLYESTERS	
BISPHENOL A-FUMURATE	U
HYDROGENATED BISPHENOL-A-BISPHENOL-A	
ISOPHTHALIC	
CHLORINATED POLYESTERS	
POLYSTYRENES	
POLYETHYLENE	U
POLYPROPYLENE	R——U
PVC–TYPE 1	U
PVC–TYPE 2	U
POLYVINYLIDENE CHLORIDE	R——
VINYLIDENE FLUORIDE	R——
VINYL ESTER	U

RUBBERS	
BUTYL GR-1	R—
FLUORO ELASTOMERS VITON A	U
FLUORO ELASTOMERS KEL-F 3700	
HARD RUBBER	
HYPALON	U
KOROSEAL	U
NATURAL RUBBER (GRS)	U
NEOPRENE GR-M (CR)	R——
NITRILE BUNA N (NBR)	U
NORDEL	
POLYBUTADIENE	
SILICONE	
POLYURETHANE	U

TEMP °F: 60 80 100 120 140 160 180 200 220 240 260 280 300 320 340 360 380 400 420 440 460

ACETAMIDE

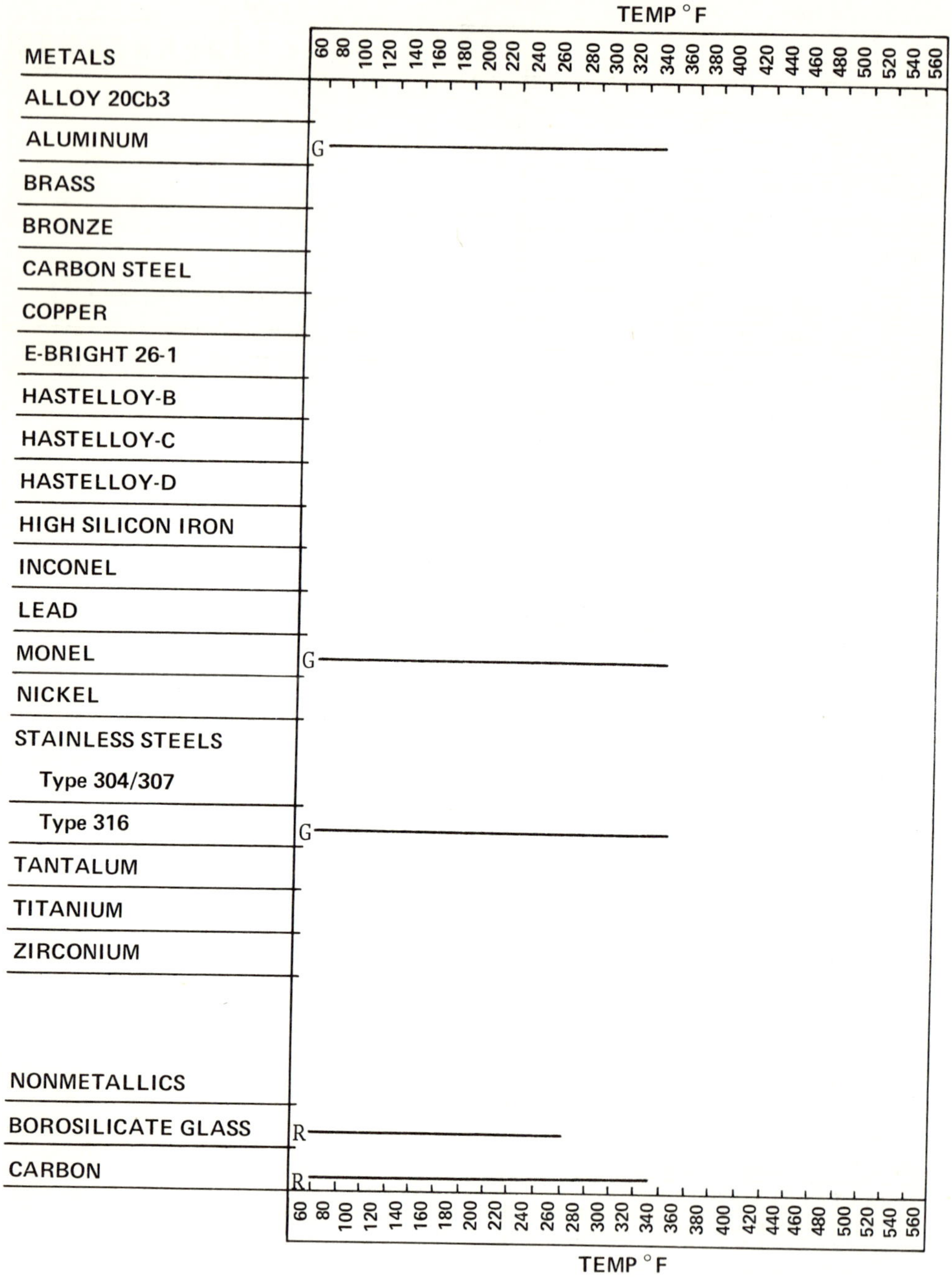

FOR METALS

E = < 2 Mils Penetration/Year; G = < 20 Mils Penetration/Year

S = < 50 Mils Penetration/Year; U = > 50 Mils Penetration/Year

FOR NONMETALLICS

R = Resistant

U = Unsatisfactory

ACETAMIDE

TEMP °F (60–460)

PLASTICS	
ABS	
ACRYLICS	
ASBESTOS REINF., EPOXY	
ASBESTOS REINF., PHENOLIC	
CHLORINATED POLYETHER	R
CPVC	
EPOXY	
FLUOROCARBONS FEP	R
FLUOROCARBONS TFE	R
FURFURYL ALCOHOL	
NORYL	
PHENOLIC	
POLYAMIDES–NYLON	
POLYESTERS	
BISPHENOL A-FUMURATE	
HYDROGENATED BISPHENOL-A-BISPHENOL-A	
ISOPHTHALIC	
CHLORINATED POLYESTERS	
POLYSTYRENES	
POLYETHYLENE	
POLYPROPYLENE	R
PVC–TYPE 1	
PVC–TYPE 2	
POLYVINYLIDENE CHLORIDE	
VINYLIDENE FLUORIDE	R
VINYL ESTER	

RUBBERS	
BUTYL GR-1	
FLUORO ELASTOMERS VITON A	R
FLUORO ELASTOMERS KEL-F 3700	
HARD RUBBER	
HYPALON	
KOROSEAL	
NATURAL RUBBER (GRS)	
NEOPRENE GR-M (CR)	R
NITRILE BUNA N (NBR)	R
NORDEL	R
POLYBUTADIENE	
SILICONE	
POLYURETHANE	

TEMP °F (60–460)

ACETATE SOLVENTS, CRUDE

TEMP °F: 60 80 100 120 140 160 180 200 220 240 260 280 300 320 340 360 380 400 420 440 460 480 500 520 540 560

METALS	Rating
ALLOY 20Cb3	E
ALUMINUM	E
BRASS	S
BRONZE	S
CARBON STEEL	G
COPPER	G
E-BRIGHT 26-1	
HASTELLOY-B	
HASTELLOY C	
HASTELLOY-D	
HIGH SILICON IRON	G
INCONEL	
LEAD	
MONEL	G
NICKEL	E
STAINLESS STEELS	
Type 304/307	E
Type 316	E
TANTALUM	
TITANIUM	
ZIRCONIUM	

NONMETALLICS	
BOROSILICATE GLASS	
CARBON	

TEMP °F: 60 80 100 120 140 160 180 200 220 240 260 280 300 320 340 360 380 400 420 440 460 480 500 520 540 560

FOR METALS

E = < 2 Mils Penetration/Year; G = < 20 Mils Penetration/Year
S = < 50 Mils Penetration/Year; U = > 50 Mils Penetration/Year

FOR NONMETALLICS

R = Resistant
U = Unsatisfactory

ACETATE SOLVENTS, CRUDE

TEMP °F

PLASTICS	Rating (from 60 °F)	Extent of bar (°F, read from chart)
ABS	U	
ACRYLICS		
ASBESTOS REINF., EPOXY		
ASBESTOS REINF., PHENOLIC		
CHLORINATED POLYETHER		
CPVC		
EPOXY		
FLUOROCARBONS FEP	R	60–400
FLUOROCARBONS TFE	R	60–460
FURFURYL ALCOHOL		
NORYL		
PHENOLIC		
POLYAMIDES—NYLON		
POLYESTERS		
BISPHENOL A-FUMURATE		
HYDROGENATED BISPHENOL-A-BISPHENOL-A		
ISOPHTHALIC		
CHLORINATED POLYESTERS		
POLYSTYRENES		
POLYETHYLENE	R	60–80
POLYPROPYLENE	R	60–80
PVC—TYPE 1	U	
PVC—TYPE 2	U	
POLYVINYLIDENE CHLORIDE		
VINYLIDENE FLUORIDE		
VINYL ESTER		

RUBBERS		
BUTYL GR-1		
FLUORO ELASTOMERS VITON A		
FLUORO ELASTOMERS KEL-F 3700		
HARD RUBBER		
HYPALON		
KOROSEAL		
NATURAL RUBBER (GRS)		
NEOPRENE GR-M (CR)		
NITRILE BUNA N (NBR)		
NORDEL		
POLYBUTADIENE		
SILICONE		
POLYURETHANE		

TEMP °F: 60 80 100 120 140 160 180 200 220 240 260 280 300 320 340 360 380 400 420 440 460

ACETATE SOLVENTS, PURE

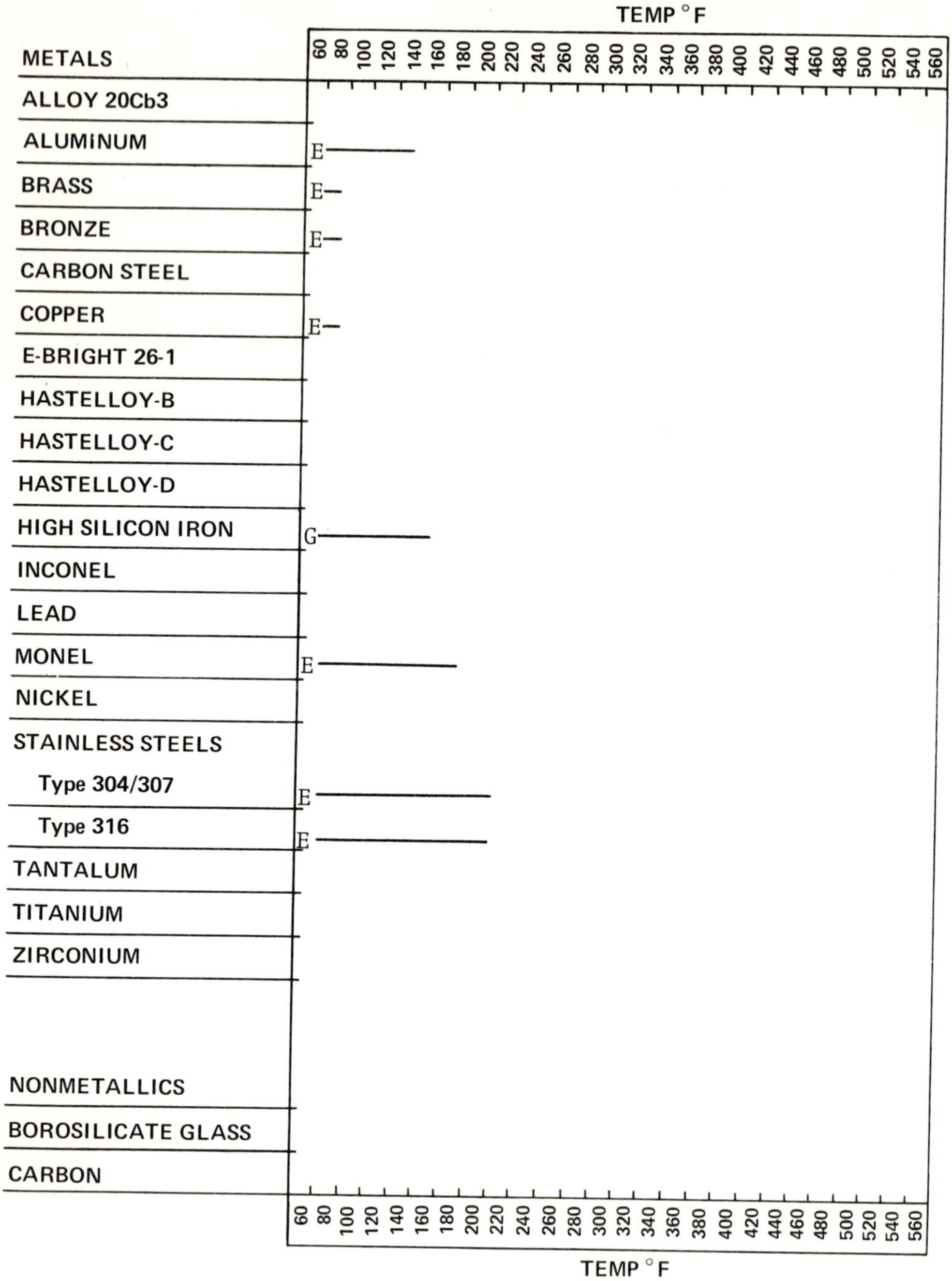

FOR METALS

E = < 2 Mils Penetration/Year; G = < 20 Mils Penetration/Year

S = < 50 Mils Penetration/Year; U = > 50 Mils Penetration/Year

FOR NONMETALLICS

R = Resistant

U = Unsatisfactory

ACETATE SOLVENTS, PURE

TEMP °F: 60, 80, 100, 120, 140, 160, 180, 200, 220, 240, 260, 280, 300, 320, 340, 360, 380, 400, 420, 440, 460

PLASTICS	
ABS	U
ACRYLICS	
ASBESTOS REINF., EPOXY	
ASBESTOS REINF., PHENOLIC	
CHLORINATED POLYETHER	
CPVC	
EPOXY	
FLUOROCARBONS FEP	R
FLUOROCARBONS TFE	R
FURFURYL ALCOHOL	
NORYL	
PHENOLIC	
POLYAMIDES–NYLON	
POLYESTERS	
BISPHENOL A-FUMURATE	
HYDROGENATED BISPHENOL-A-BISPHENOL-A	
ISOPHTHALIC	
CHLORINATED POLYESTERS	
POLYSTYRENES	
POLYETHYLENE	R
POLYPROPYLENE	R
PVC–TYPE 1	U
PVC–TYPE 2	U
POLYVINYLIDENE CHLORIDE	
VINYLIDENE FLUORIDE	
VINYL ESTER	

RUBBERS	
BUTYL GR-1	
FLUORO ELASTOMERS VITON A	
FLUORO ELASTOMERS KEL-F 3700	
HARD RUBBER	
HYPALON	
KOROSEAL	
NATURAL RUBBER (GRS)	
NEOPRENE GR-M (CR)	
NITRILE BUNA N (NBR)	
NORDEL	
POLYBUTADIENE	
SILICONE	
POLYURETHANE	

TEMP °F: 60, 80, 100, 120, 140, 160, 180, 200, 220, 240, 260, 280, 300, 320, 340, 360, 380, 400, 420, 440, 460

ACETIC ACID, GLACIAL

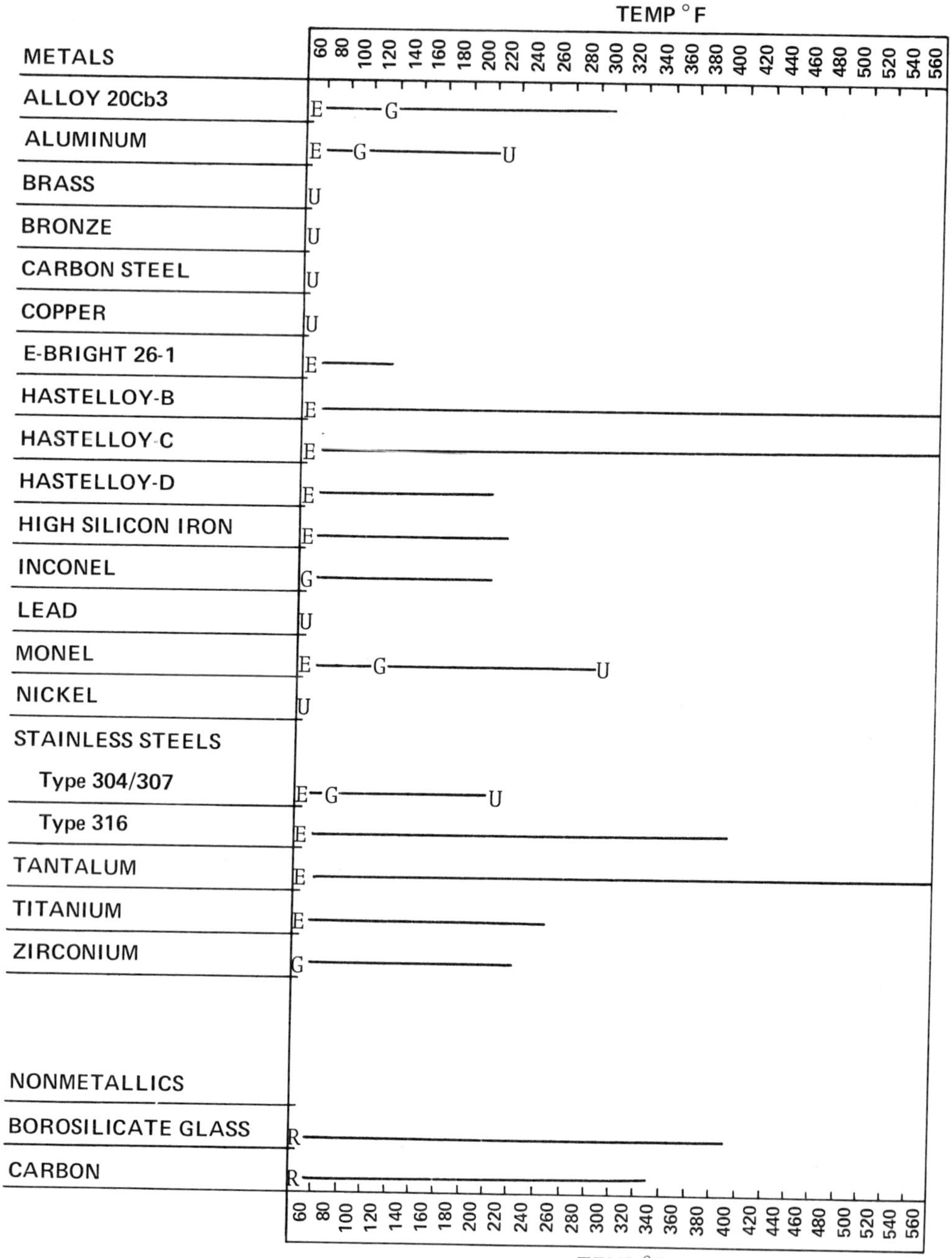

FOR METALS

E = < 2 Mils Penetration/Year; G = < 20 Mils Penetration/Year
S = < 50 Mils Penetration/Year; U = > 50 Mils Penetration/Year

FOR NONMETALLICS

R = Resistant
U = Unsatisfactory

ACETIC ACID, GLACIAL

TEMP °F (60–460)

PLASTICS	Rating
ABS	U
ACRYLICS	U
ASBESTOS REINF., EPOXY	U
ASBESTOS REINF., PHENOLIC	R—
CHLORINATED POLYETHER	R—
CPVC	U
EPOXY	R—U
FLUOROCARBONS FEP	R—
FLUOROCARBONS TFE	R—
FURFURYL ALCOHOL	R—
NORYL	R—
PHENOLIC	
POLYAMIDES–NYLON	R—
POLYESTERS	
BISPHENOL A-FUMURATE	U
HYDROGENATED BISPHENOL-A–BISPHENOL-A	
ISOPHTHALIC	U
CHLORINATED POLYESTERS	R—
POLYSTYRENES	U
POLYETHYLENE	U
POLYPROPYLENE	R—U
PVC–TYPE 1	R—U
PVC–TYPE 2	U
POLYVINYLIDENE CHLORIDE	R—
VINYLIDENE FLUORIDE	R—U
VINYL ESTER	R—

RUBBERS	Rating
BUTYL GR-1	R–U
FLUORO ELASTOMERS VITON A	U
FLUORO ELASTOMERS KEL-F 3700	
HARD RUBBER	R—
HYPALON	U
KOROSEAL	U
NATURAL RUBBER (GRS)	U
NEOPRENE GR-M (CR)	U
NITRILE BUNA N (NBR)	R—U
NORDEL	U
POLYBUTADIENE	
SILICONE	R—
POLYURETHANE	U

TEMP °F

ACETIC ACID, 80%

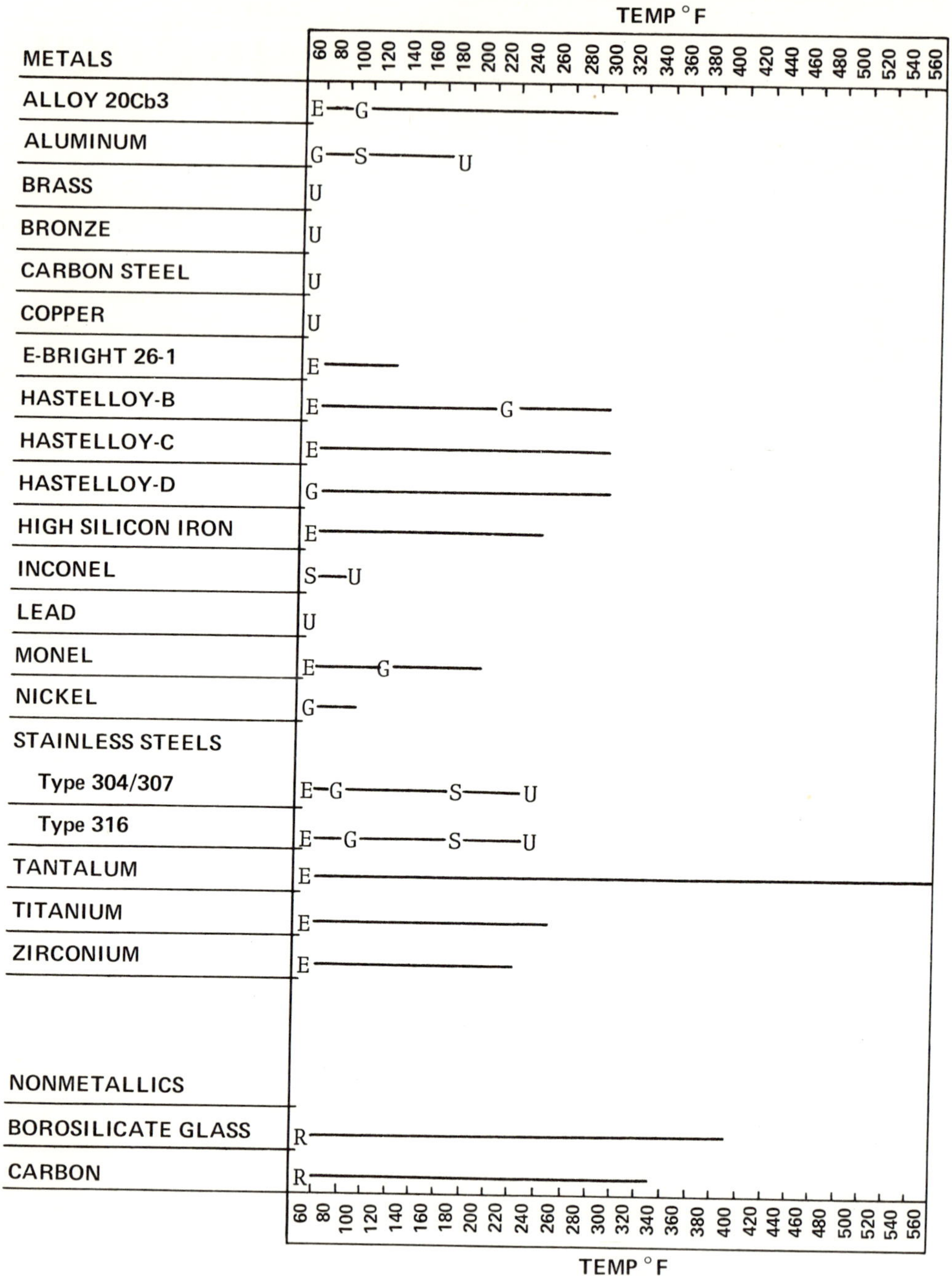

FOR METALS

E = < 2 Mils Penetration/Year; G = < 20 Mils Penetration/Year

S = < 50 Mils Penetration/Year; U = > 50 Mils Penetration/Year

FOR NONMETALLICS

R = Resistant

U = Unsatisfactory

ACETIC ACID, 80%

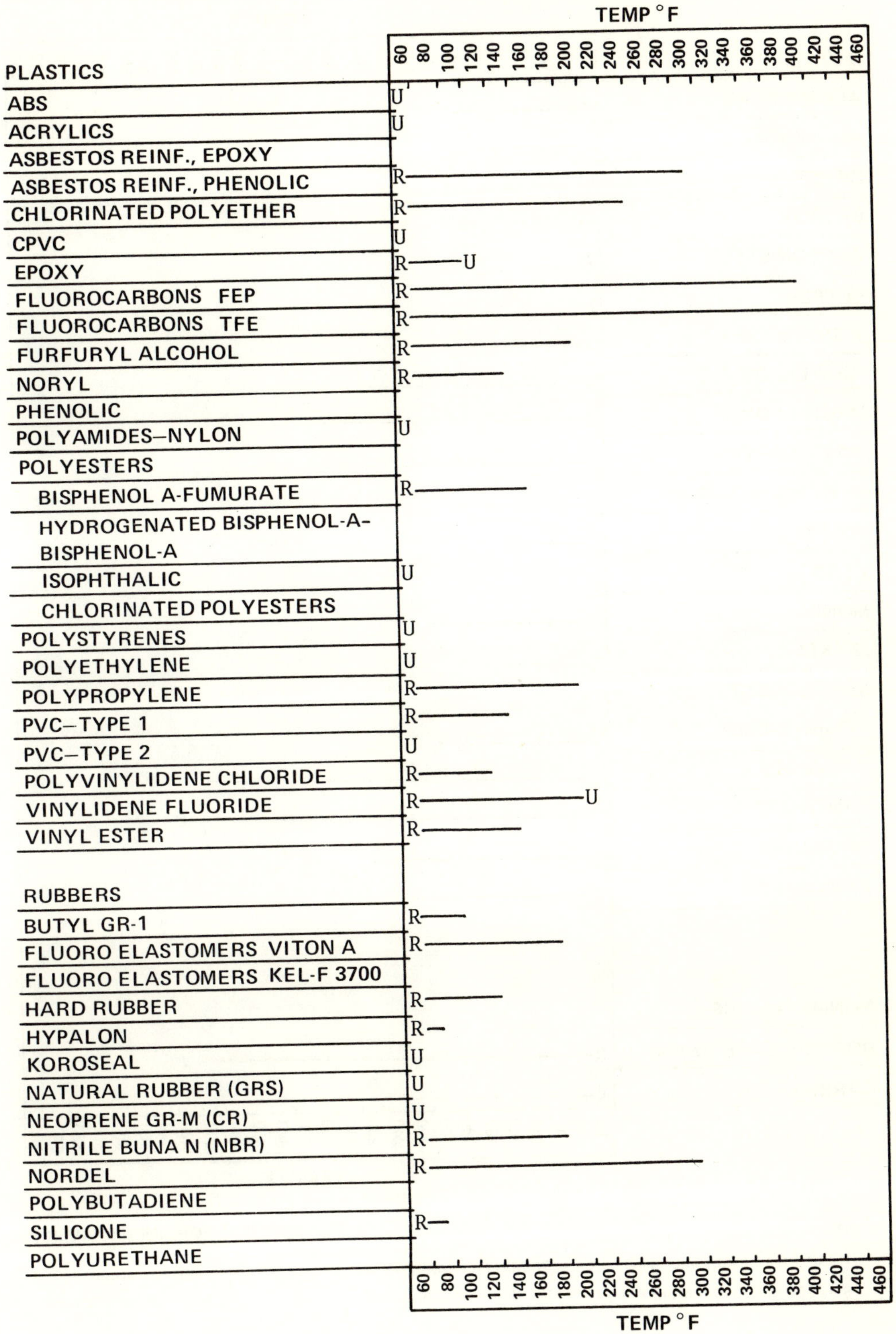

ACETIC ACID, 50%

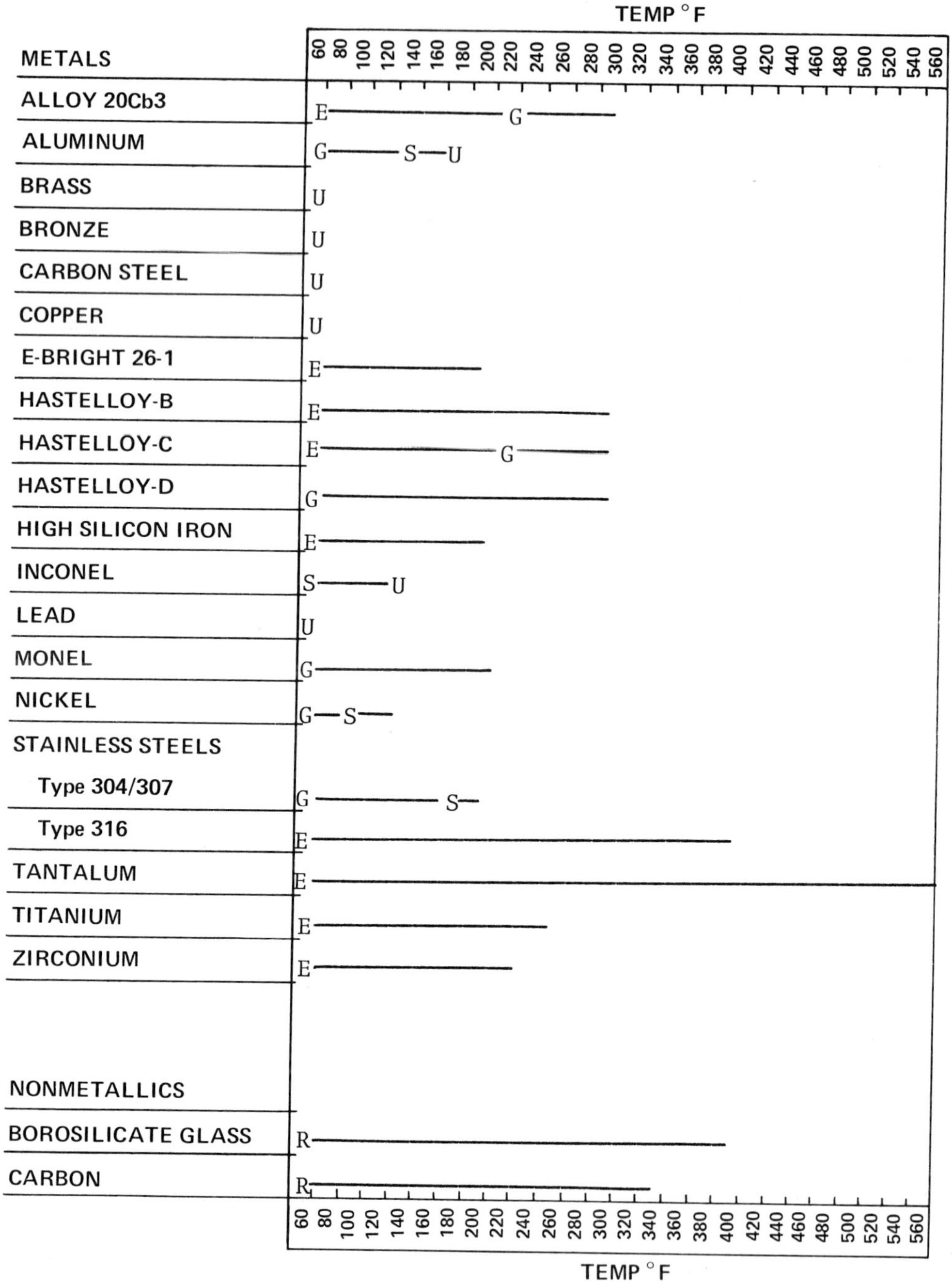

FOR METALS

E = < 2 Mils Penetration/Year; G = < 20 Mils Penetration/Year

S = < 50 Mils Penetration/Year; U = > 50 Mils Penetration/Year

FOR NONMETALLICS

R = Resistant

U = Unsatisfactory

ACETIC ACID, 50%

TEMP °F (60 – 460)

PLASTICS	
ABS	R———U
ACRYLICS	U
ASBESTOS REINF., EPOXY	
ASBESTOS REINF., PHENOLIC	R———
CHLORINATED POLYETHER	R———
CPVC	U
EPOXY	R———U
FLUOROCARBONS FEP	R———
FLUOROCARBONS TFE	R———
FURFURYL ALCOHOL	R———
NORYL	R———
PHENOLIC	
POLYAMIDES—NYLON	U
POLYESTERS	
BISPHENOL A-FUMURATE	R———
HYDROGENATED BISPHENOL-A-BISPHENOL-A	R———
ISOPHTHALIC	R———U
CHLORINATED POLYESTERS	R—
POLYSTYRENES	U
POLYETHYLENE	
POLYPROPYLENE	R———
PVC—TYPE 1	R———
PVC—TYPE 2	R—
POLYVINYLIDENE CHLORIDE	R———
VINYLIDENE FLUORIDE	R———
VINYL ESTER	R———

RUBBERS	
BUTYL GR-1	R———
FLUORO ELASTOMERS VITON A	R———
FLUORO ELASTOMERS KEL-F 3700	
HARD RUBBER	R———
HYPALON	R—
KOROSEAL	U
NATURAL RUBBER (GRS)	U
NEOPRENE GR-M (CR)	U
NITRILE BUNA N (NBR)	R———
NORDEL	R—
POLYBUTADIENE	
SILICONE	R—
POLYURETHANE	

TEMP °F (60 – 460)

ACETIC ACID, 20%

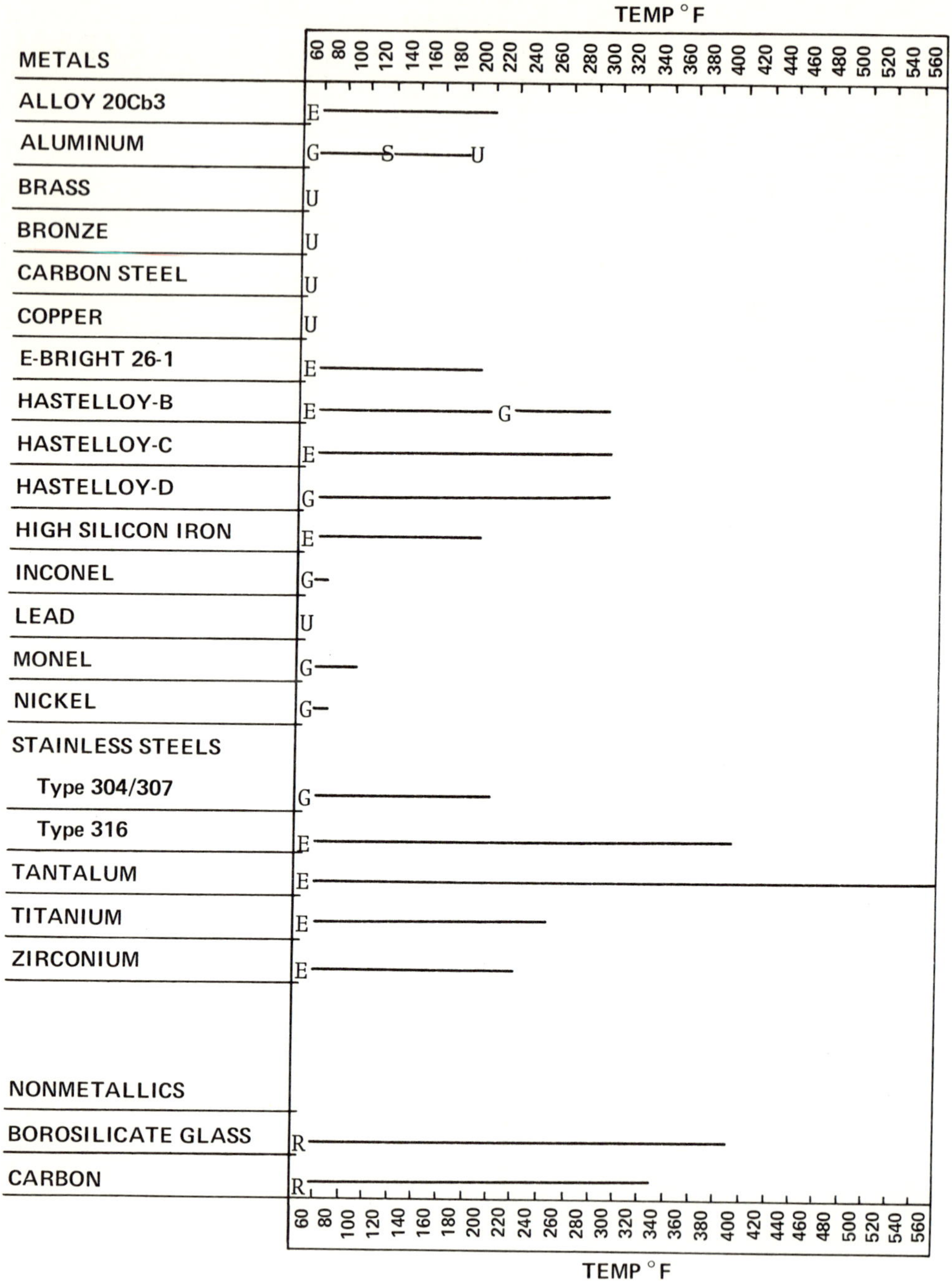

FOR METALS

E = < 2 Mils Penetration/Year; G = < 20 Mils Penetration/Year

S = < 50 Mils Penetration/Year; U = > 50 Mils Penetration/Year

FOR NONMETALLICS

R = Resistant

U = Unsatisfactory

ACETIC ACID, 20%

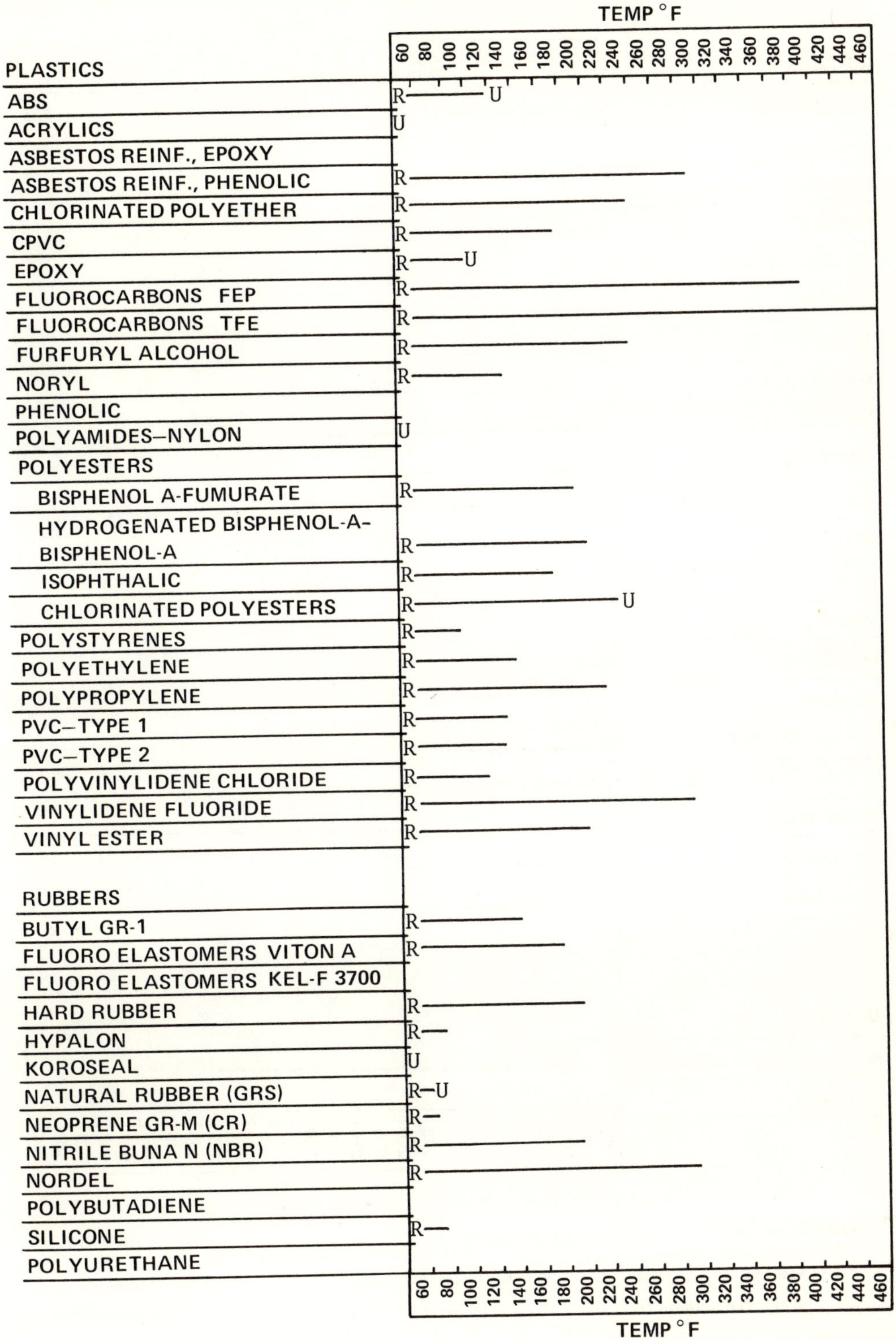

ACETIC ACID, 10%

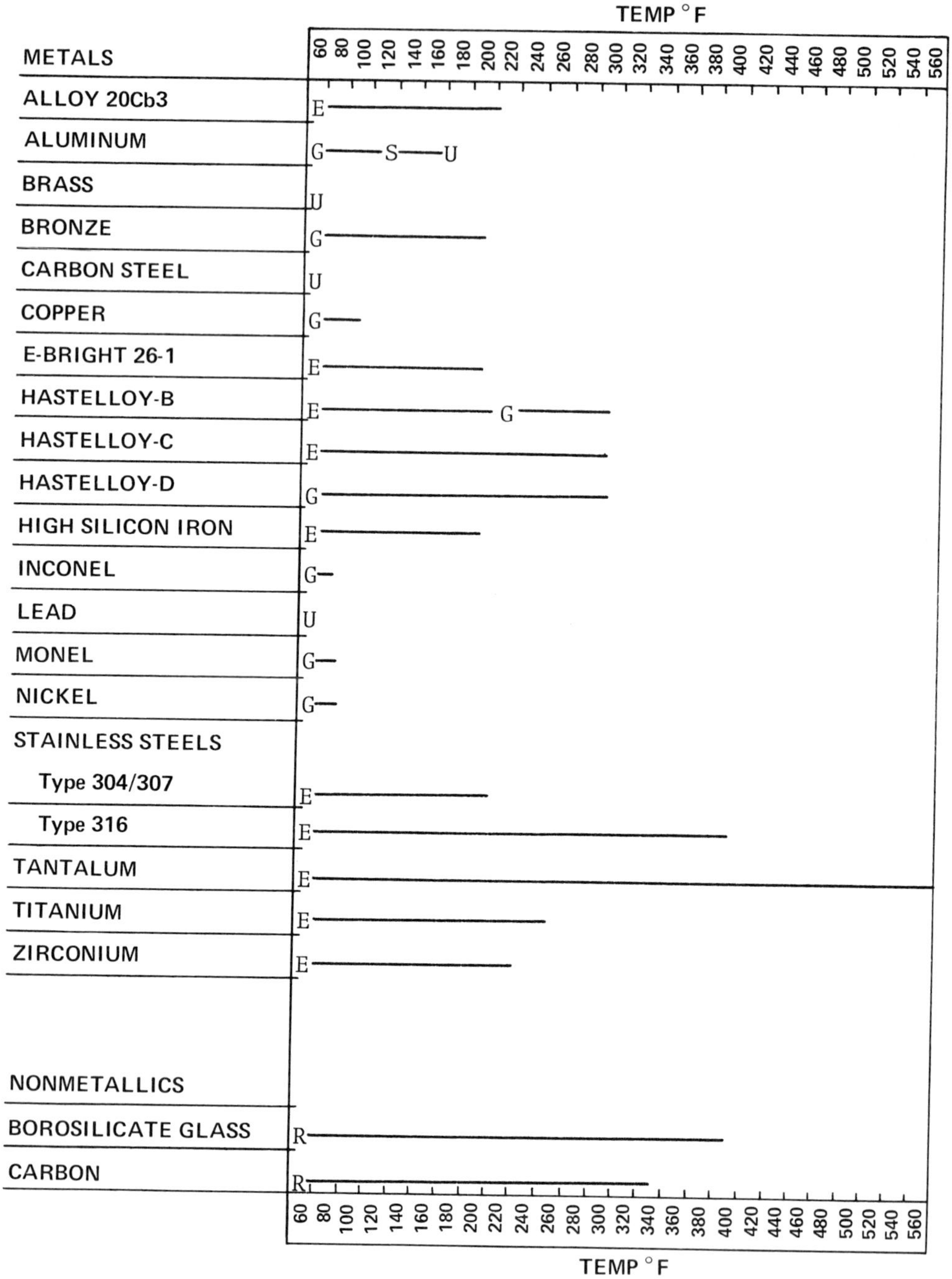

FOR METALS

E = < 2 Mils Penetration/Year; G = < 20 Mils Penetration/Year

S = < 50 Mils Penetration/Year; U = > 50 Mils Penetration/Year

FOR NONMETALLICS

R = Resistant

U = Unsatisfactory

ACETIC ACID, 10%

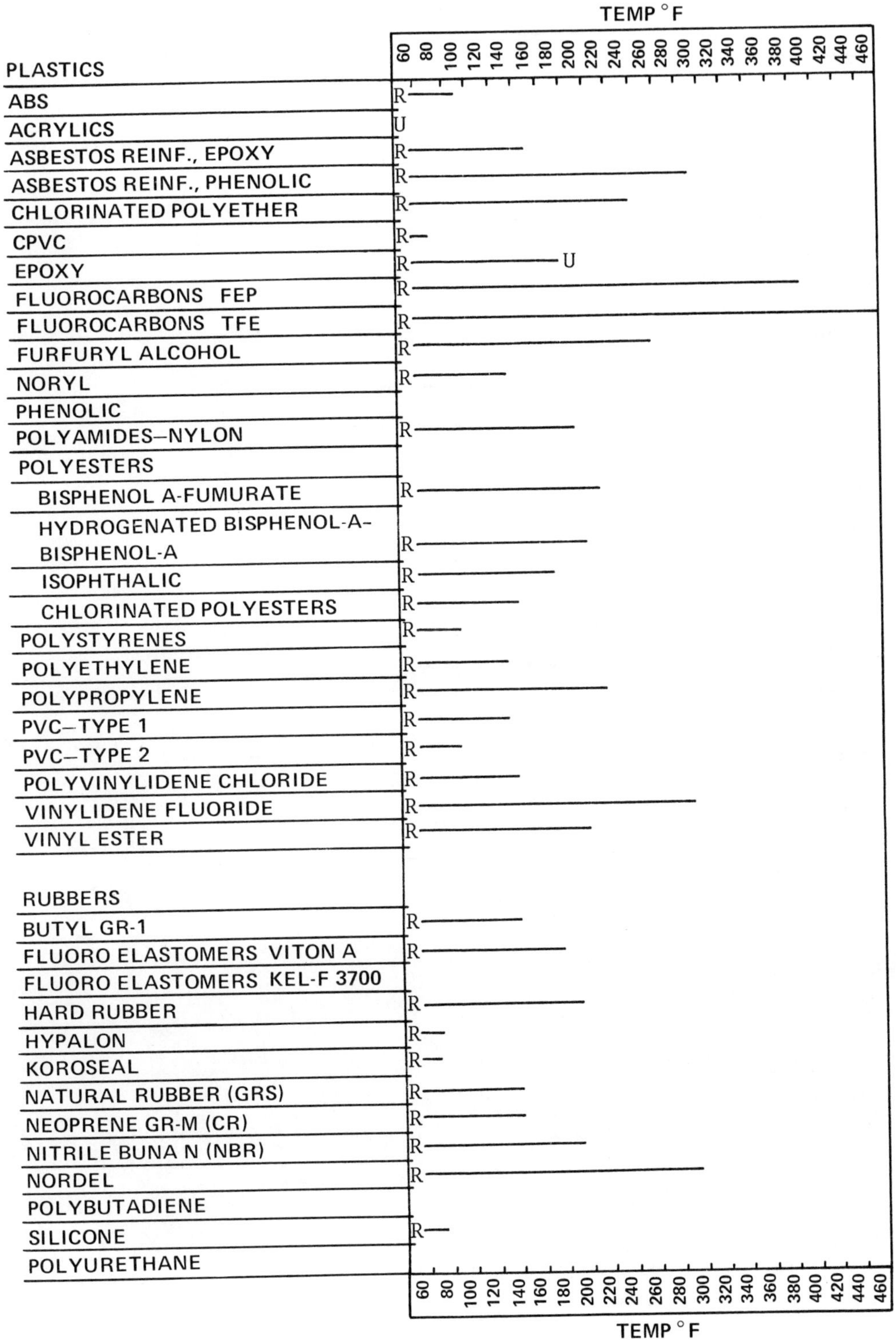

ACETIC ANHYDRIDE

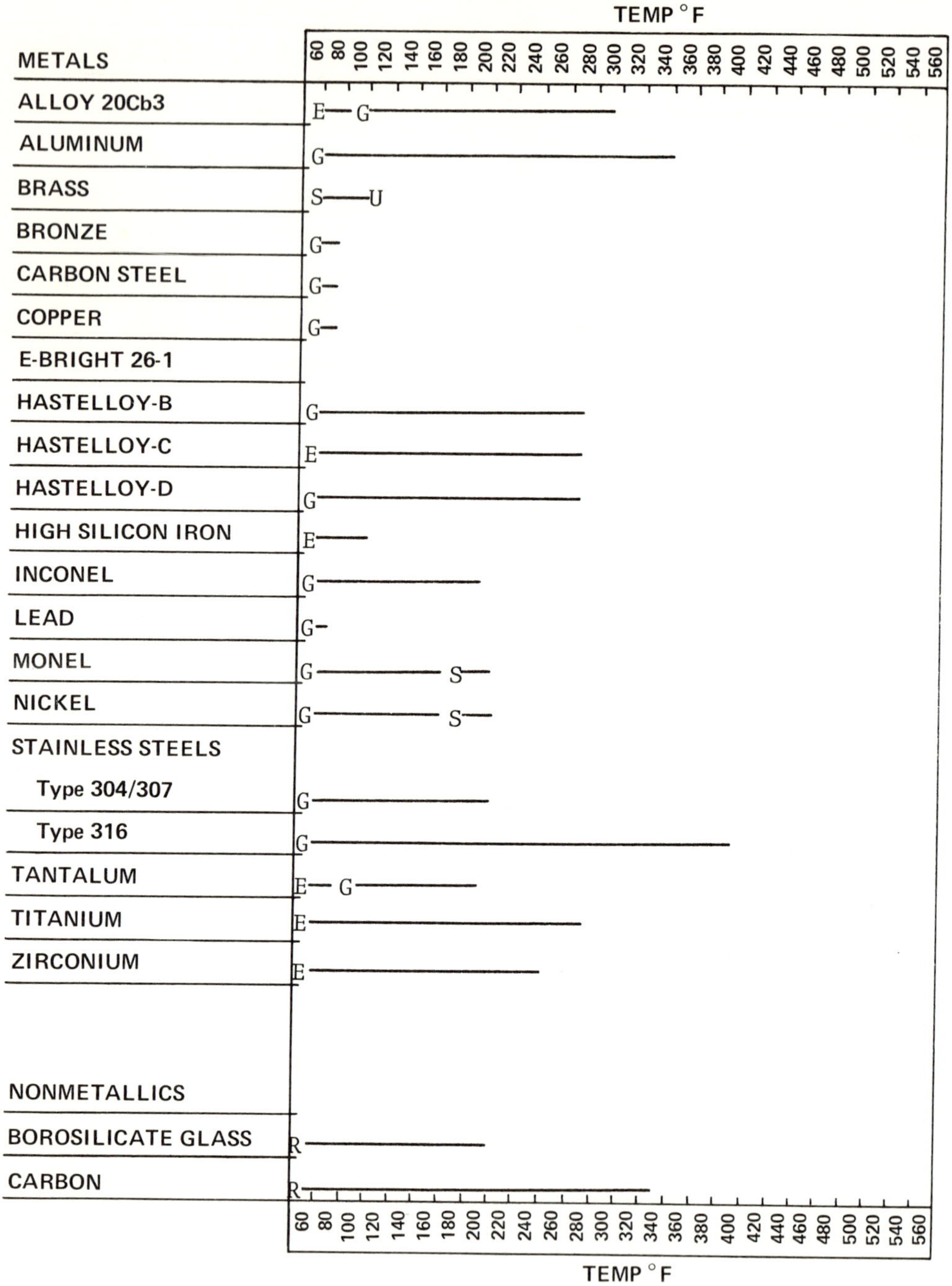

FOR METALS

E = < 2 Mils Penetration/Year; G = < 20 Mils Penetration/Year

S = < 50 Mils Penetration/Year; U = > 50 Mils Penetration/Year

FOR NONMETALLICS

R = Resistant

U = Unsatisfactory

ACETIC ANHYDRIDE

TEMP °F: 60 80 100 120 140 160 180 200 220 240 260 280 300 320 340 360 380 400 420 440 460

PLASTICS	
ABS	U
ACRYLICS	U
ASBESTOS REINF., EPOXY	R
ASBESTOS REINF., PHENOLIC	R
CHLORINATED POLYETHER	R
CPVC	U
EPOXY	U
FLUOROCARBONS FEP	R
FLUOROCARBONS TFE	R
FURFURYL ALCOHOL	R
NORYL	
PHENOLIC	
POLYAMIDES—NYLON	R
POLYESTERS	
BISPHENOL A-FUMURATE	R
HYDROGENATED BISPHENOL-A-BISPHENOL-A	U
ISOPHTHALIC	U
CHLORINATED POLYESTERS	R
POLYSTYRENES	U
POLYETHYLENE	U
POLYPROPYLENE	R
PVC—TYPE 1	U
PVC—TYPE 2	U
POLYVINYLIDENE CHLORIDE	R
VINYLIDENE FLUORIDE	R
VINYL ESTER	R
RUBBERS	
BUTYL GR-1	R
FLUORO ELASTOMERS VITON A	U
FLUORO ELASTOMERS KEL-F 3700	
HARD RUBBER	R
HYPALON	R
KOROSEAL	
NATURAL RUBBER (GRS)	U
NEOPRENE GR-M (CR)	R
NITRILE BUNA N (NBR)	U
NORDEL	
POLYBUTADIENE	
SILICONE	
POLYURETHANE	U

TEMP °F: 60 80 100 120 140 160 180 200 220 240 260 280 300 320 340 360 380 400 420 440 460

ACETONE

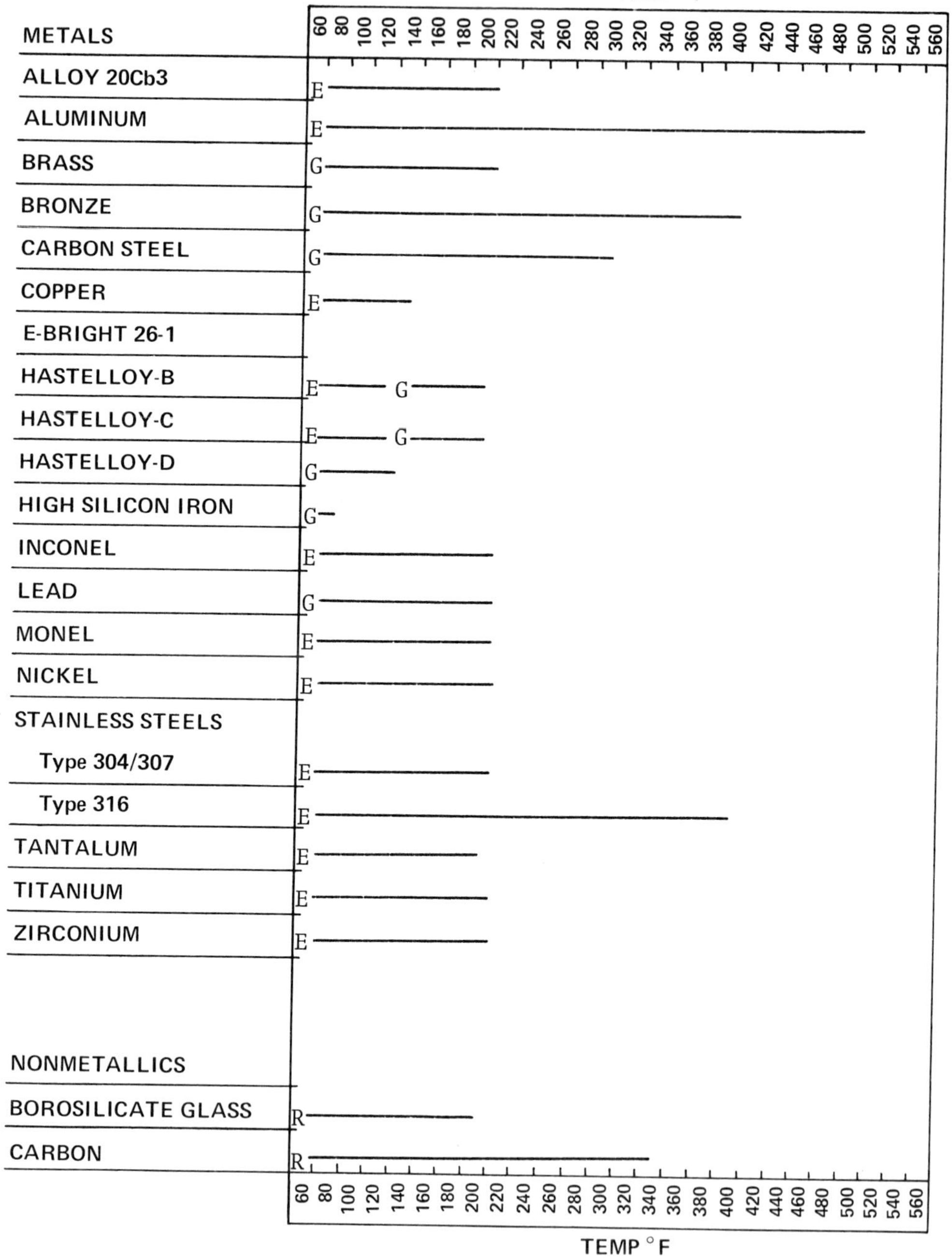

FOR METALS

E = $<$ 2 Mils Penetration/Year; G = $<$ 20 Mils Penetration/Year

S = $<$ 50 Mils Penetration/Year; U = $>$ 50 Mils Penetration/Year

FOR NONMETALLICS

R = Resistant

U = Unsatisfactory

ACETONE

TEMP °F: 60 80 100 120 140 160 180 200 220 240 260 280 300 320 340 360 380 400 420 440 460

PLASTICS	TEMP °F
ABS	U
ACRYLICS	U
ASBESTOS REINF., EPOXY	U
ASBESTOS REINF., PHENOLIC	U
CHLORINATED POLYETHER	R—U
CPVC	U
EPOXY	R———U
FLUOROCARBONS FEP	R—————————————————
FLUOROCARBONS TFE	R————————————————————
FURFURYL ALCOHOL	R———————
NORYL	
PHENOLIC	
POLYAMIDES—NYLON	R—
POLYESTERS	
BISPHENOL A-FUMURATE	U
HYDROGENATED BISPHENOL-A-BISPHENOL-A	
ISOPHTHALIC	U
CHLORINATED POLYESTERS	U
POLYSTYRENES	
POLYETHYLENE	R———U
POLYPROPYLENE	R————————
PVC—TYPE 1	U
PVC—TYPE 2	U
POLYVINYLIDENE CHLORIDE	R—
VINYLIDENE FLUORIDE	U
VINYL ESTER	U
RUBBERS	
BUTYL GR-1	R————
FLUORO ELASTOMERS VITON A	U
FLUORO ELASTOMERS KEL-F 3700	
HARD RUBBER	U
HYPALON	U
KOROSEAL	
NATURAL RUBBER (GRS)	U
NEOPRENE GR-M (CR)	U
NITRILE BUNA N (NBR)	U
NORDEL	R————————————
POLYBUTADIENE	
SILICONE	R—
POLYURETHANE	U

TEMP °F: 60 80 100 120 140 160 180 200 220 240 260 280 300 320 340 360 380 400 420 440 460

ACETONE, 50% WATER

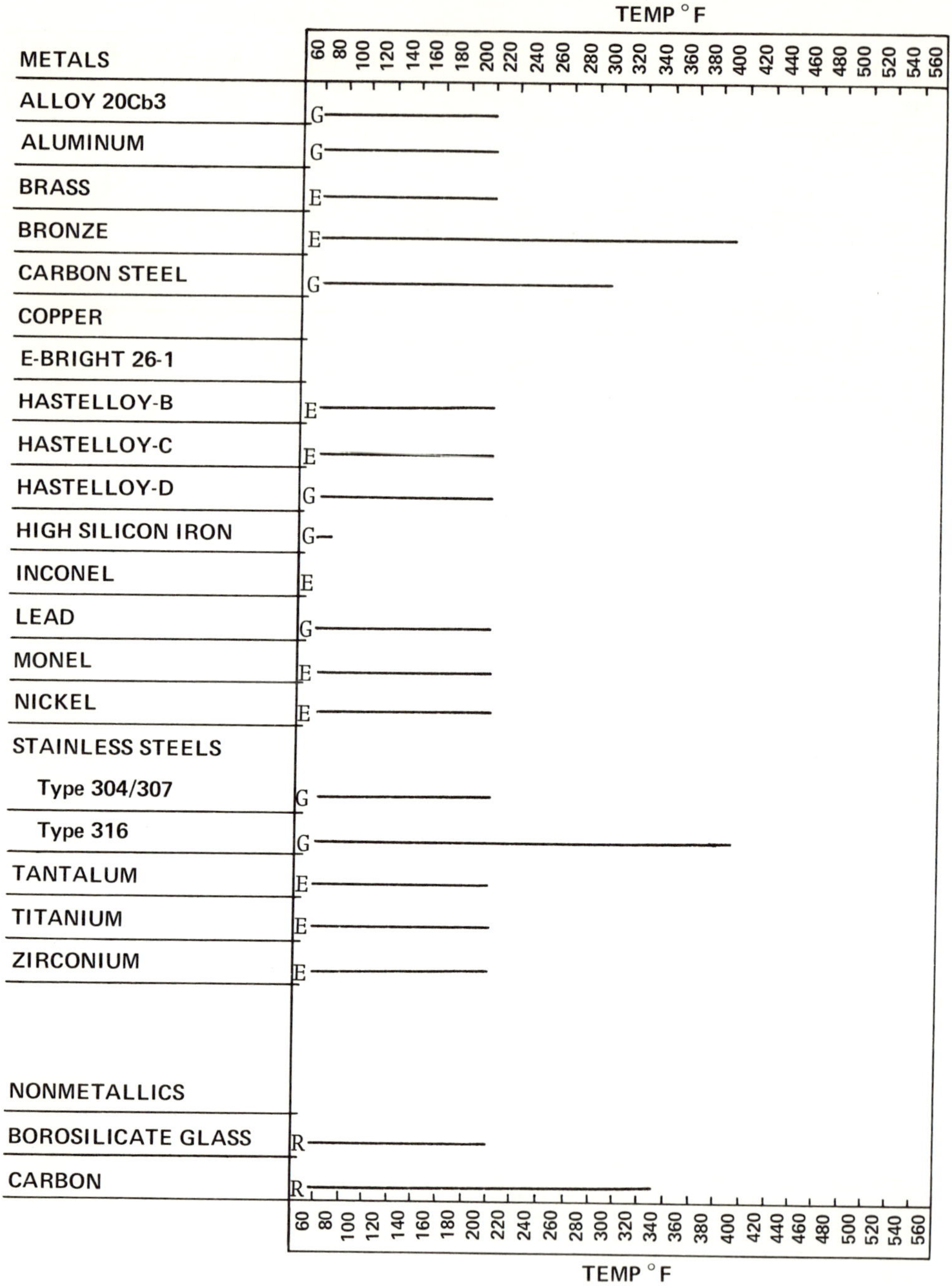

FOR METALS

E = < 2 Mils Penetration/Year; G = < 20 Mils Penetration/Year

S = < 50 Mils Penetration/Year; U = > 50 Mils Penetration/Year

FOR NONMETALLICS

R = Resistant

U = Unsatisfactory

ACETONE, 50% WATER

TEMP °F: 60 80 100 120 140 160 180 200 220 240 260 280 300 320 340 360 380 400 420 440 460

PLASTICS	
ABS	
ACRYLICS	
ASBESTOS REINF., EPOXY	U
ASBESTOS REINF., PHENOLIC	
CHLORINATED POLYETHER	
CPVC	
EPOXY	
FLUOROCARBONS FEP	R—
FLUOROCARBONS TFE	R—
FURFURYL ALCOHOL	R—
NORYL	
PHENOLIC	
POLYAMIDES–NYLON	
POLYESTERS	
BISPHENOL A-FUMURATE	
HYDROGENATED BISPHENOL-A-BISPHENOL-A	
ISOPHTHALIC	U
CHLORINATED POLYESTERS	
POLYSTYRENES	
POLYETHYLENE	
POLYPROPYLENE	R—
PVC–TYPE 1	
PVC–TYPE 2	
POLYVINYLIDENE CHLORIDE	
VINYLIDENE FLUORIDE	R—
VINYL ESTER	

RUBBERS	
BUTYL GR-1	R—
FLUORO ELASTOMERS VITON A	
FLUORO ELASTOMERS KEL-F 3700	
HARD RUBBER	U
HYPALON	
KOROSEAL	
NATURAL RUBBER (GRS)	U
NEOPRENE GR-M (CR)	U
NITRILE BUNA N (NBR)	U
NORDEL	
POLYBUTADIENE	
SILICONE	R—
POLYURETHANE	

TEMP °F: 60 80 100 120 140 160 180 200 220 240 260 280 300 320 340 360 380 400 420 440 460

ACETONITRILE

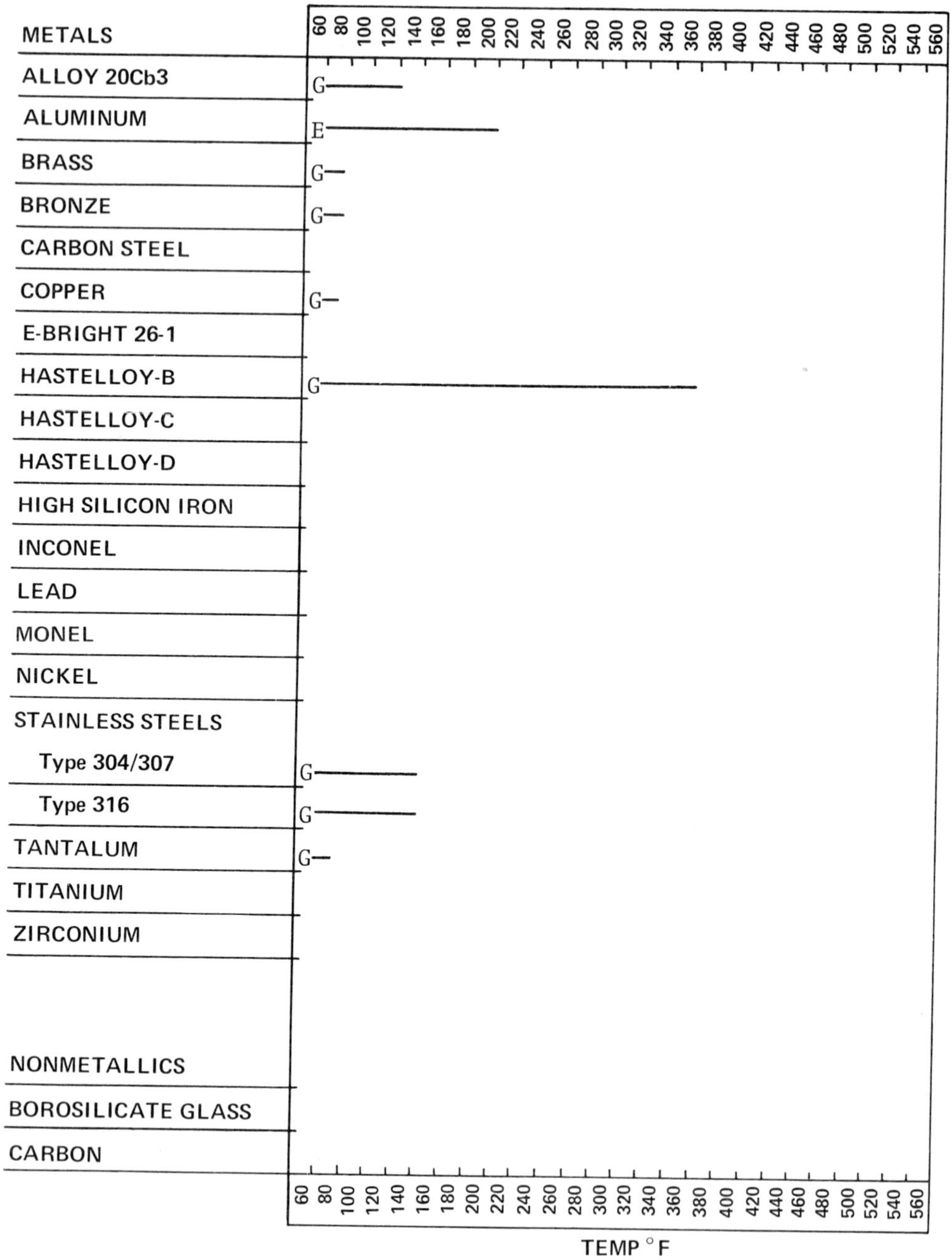

FOR METALS

E = < 2 Mils Penetration/Year; G = < 20 Mils Penetration/Year

S = < 50 Mils Penetration/Year; U = > 50 Mils Penetration/Year

FOR NONMETALLICS

R = Resistant

U = Unsatisfactory

ACETONITRILE

TEMP °F: 60, 80, 100, 120, 140, 160, 180, 200, 220, 240, 260, 280, 300, 320, 340, 360, 380, 400, 420, 440, 460

PLASTICS	
ABS	U
ACRYLICS	
ASBESTOS REINF., EPOXY	R
ASBESTOS REINF., PHENOLIC	U
CHLORINATED POLYETHER	R
CPVC	
EPOXY	
FLUOROCARBONS FEP	R
FLUOROCARBONS TFE	R
FURFURYL ALCOHOL	
NORYL	
PHENOLIC	
POLYAMIDES—NYLON	R
POLYESTERS	
BISPHENOL A-FUMURATE	U
HYDROGENATED BISPHENOL-A-BISPHENOL-A	U
ISOPHTHALIC	U
CHLORINATED POLYESTERS	
POLYSTYRENES	
POLYETHYLENE	
POLYPROPYLENE	R
PVC—TYPE 1	
PVC—TYPE 2	
POLYVINYLIDENE CHLORIDE	
VINYLIDENE FLUORIDE	R
VINYL ESTER	

RUBBERS	
BUTYL GR-1	
FLUORO ELASTOMERS VITON A	
FLUORO ELASTOMERS KEL-F 3700	
HARD RUBBER	
HYPALON	
KOROSEAL	
NATURAL RUBBER (GRS)	
NEOPRENE GR-M (CR)	
NITRILE BUNA N (NBR)	
NORDEL	
POLYBUTADIENE	
SILICONE	
POLYURETHANE	

TEMP °F: 60, 80, 100, 120, 140, 160, 180, 200, 220, 240, 260, 280, 300, 320, 340, 360, 380, 400, 420, 440, 460

ACETOPHENONE

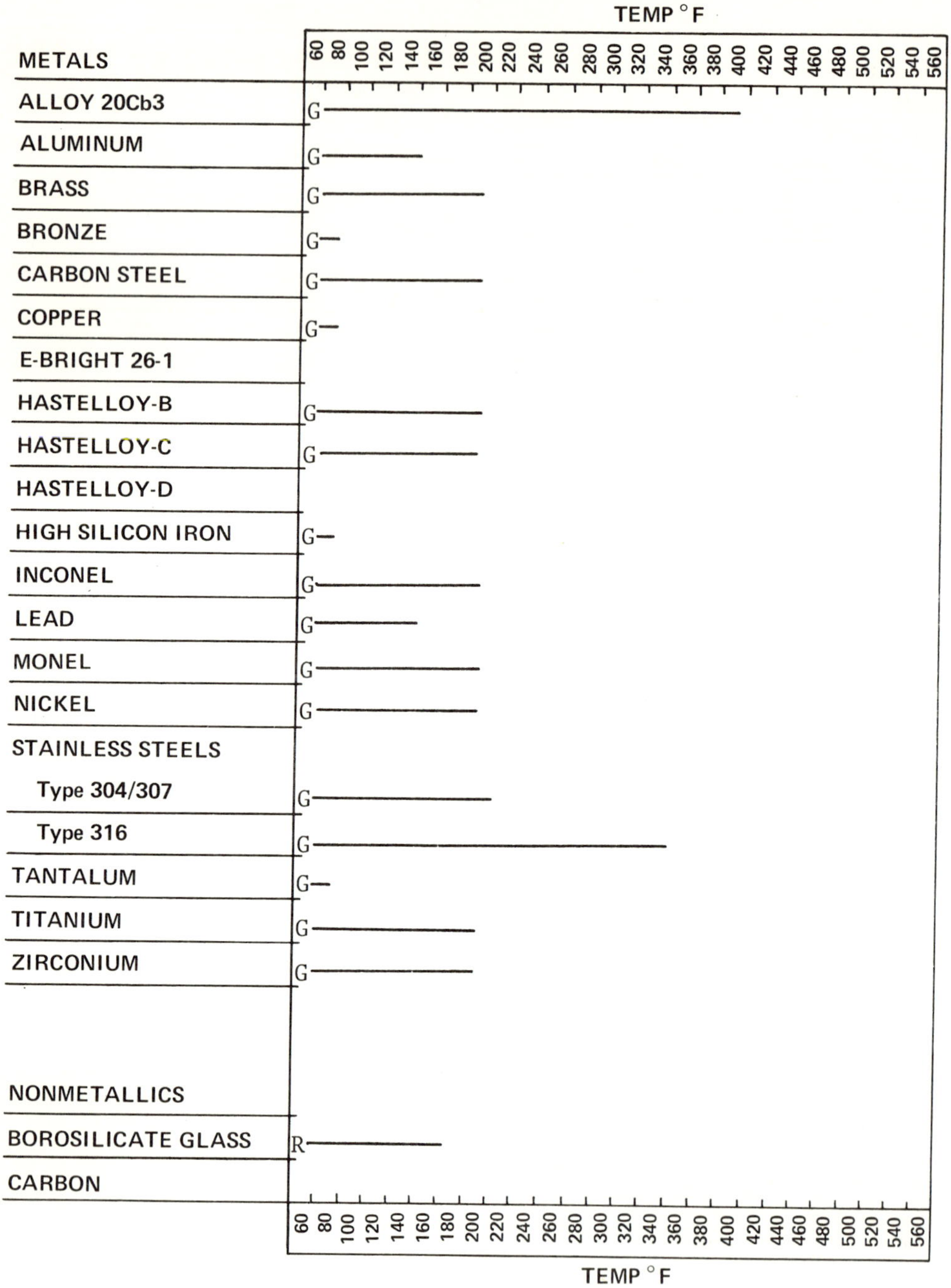

FOR METALS

E = < 2 Mils Penetration/Year; G = < 20 Mils Penetration/Year

S = < 50 Mils Penetration/Year; U = > 50 Mils Penetration/Year

FOR NONMETALLICS

R = Resistant

U = Unsatisfactory

ACETOPHENONE

TEMP °F: 60 80 100 120 140 160 180 200 220 240 260 280 300 320 340 360 380 400 420 440 460

PLASTICS	
ABS	U
ACRYLICS	
ASBESTOS REINF., EPOXY	
ASBESTOS REINF., PHENOLIC	
CHLORINATED POLYETHER	
CPVC	
EPOXY	
FLUOROCARBONS FEP	R——
FLUOROCARBONS TFE	R——
FURFURYL ALCOHOL	
NORYL	
PHENOLIC	
POLYAMIDES–NYLON	R——
POLYESTERS	
BISPHENOL A-FUMURATE	U
HYDROGENATED BISPHENOL-A-BISPHENOL-A	
ISOPHTHALIC	U
CHLORINATED POLYESTERS	R—
POLYSTYRENES	
POLYETHYLENE	
POLYPROPYLENE	R——U
PVC–TYPE 1	
PVC–TYPE 2	
POLYVINYLIDENE CHLORIDE	
VINYLIDENE FLUORIDE	R——U
VINYL ESTER	U

RUBBERS	
BUTYL GR-1	R—
FLUORO ELASTOMERS VITON A	U
FLUORO ELASTOMERS KEL-F 3700	
HARD RUBBER	
HYPALON	
KOROSEAL	
NATURAL RUBBER (GRS)	U
NEOPRENE GR-M (CR)	U
NITRILE BUNA N (NBR)	U
NORDEL	
POLYBUTADIENE	
SILICONE	
POLYURETHANE	

TEMP °F: 60 80 100 120 140 160 180 200 220 240 260 280 300 320 340 360 380 400 420 440 460

ACETYL CHLORIDE

TEMP °F: 60 80 100 120 140 160 180 200 220 240 260 280 300 320 340 360 380 400 420 440 460 480 500 520 540 560

METALS	
ALLOY 20Cb3	G
ALUMINUM	
BRASS	U
BRONZE	U
CARBON STEEL	G
COPPER	U
E-BRIGHT 26-1	
HASTELLOY-B	E
HASTELLOY-C	
HASTELLOY-D	
HIGH SILICON IRON	G
INCONEL	G
LEAD	
MONEL	G
NICKEL	G
STAINLESS STEELS	
Type 304/307	G
Type 316	G
TANTALUM	G
TITANIUM	
ZIRCONIUM	G

NONMETALLICS	
BOROSILICATE GLASS	
CARBON	R

TEMP °F: 60 80 100 120 140 160 180 200 220 240 260 280 300 320 340 360 380 400 420 440 460 480 500 520 540 560

FOR METALS

E = < 2 Mils Penetration/Year; G = < 20 Mils Penetration/Year

S = < 50 Mils Penetration/Year; U = > 50 Mils Penetration/Year

FOR NONMETALLICS

R = Resistant

U = Unsatisfactory

ACETYL CHLORIDE

TEMP °F: 60 80 100 120 140 160 180 200 220 240 260 280 300 320 340 360 380 400 420 440 460

PLASTICS	
ABS	U
ACRYLICS	
ASBESTOS REINF., EPOXY	
ASBESTOS REINF., PHENOLIC	
CHLORINATED POLYETHER	U
CPVC	
EPOXY	
FLUOROCARBONS FEP	R
FLUOROCARBONS TFE	R
FURFURYL ALCOHOL	R
NORYL	
PHENOLIC	
POLYAMIDES–NYLON	R
POLYESTERS	
BISPHENOL A-FUMURATE	U
HYDROGENATED BISPHENOL-A-BISPHENOL-A	
ISOPHTHALIC	
CHLORINATED POLYESTERS	U
POLYSTYRENES	
POLYETHYLENE	U
POLYPROPYLENE	
PVC–TYPE 1	
PVC–TYPE 2	
POLYVINYLIDENE CHLORIDE	R
VINYLIDENE FLUORIDE	R
VINYL ESTER	

RUBBERS	
BUTYL GR-1	
FLUORO ELASTOMERS VITON A	R
FLUORO ELASTOMERS KEL-F 3700	
HARD RUBBER	
HYPALON	U
KOROSEAL	
NATURAL RUBBER (GRS)	
NEOPRENE GR-M (CR)	
NITRILE BUNA N (NBR)	U
NORDEL	
POLYBUTADIENE	
SILICONE	
POLYURETHANE	U

TEMP °F: 60 80 100 120 140 160 180 200 220 240 260 280 300 320 340 360 380 400 420 440 460

ACETYLENE

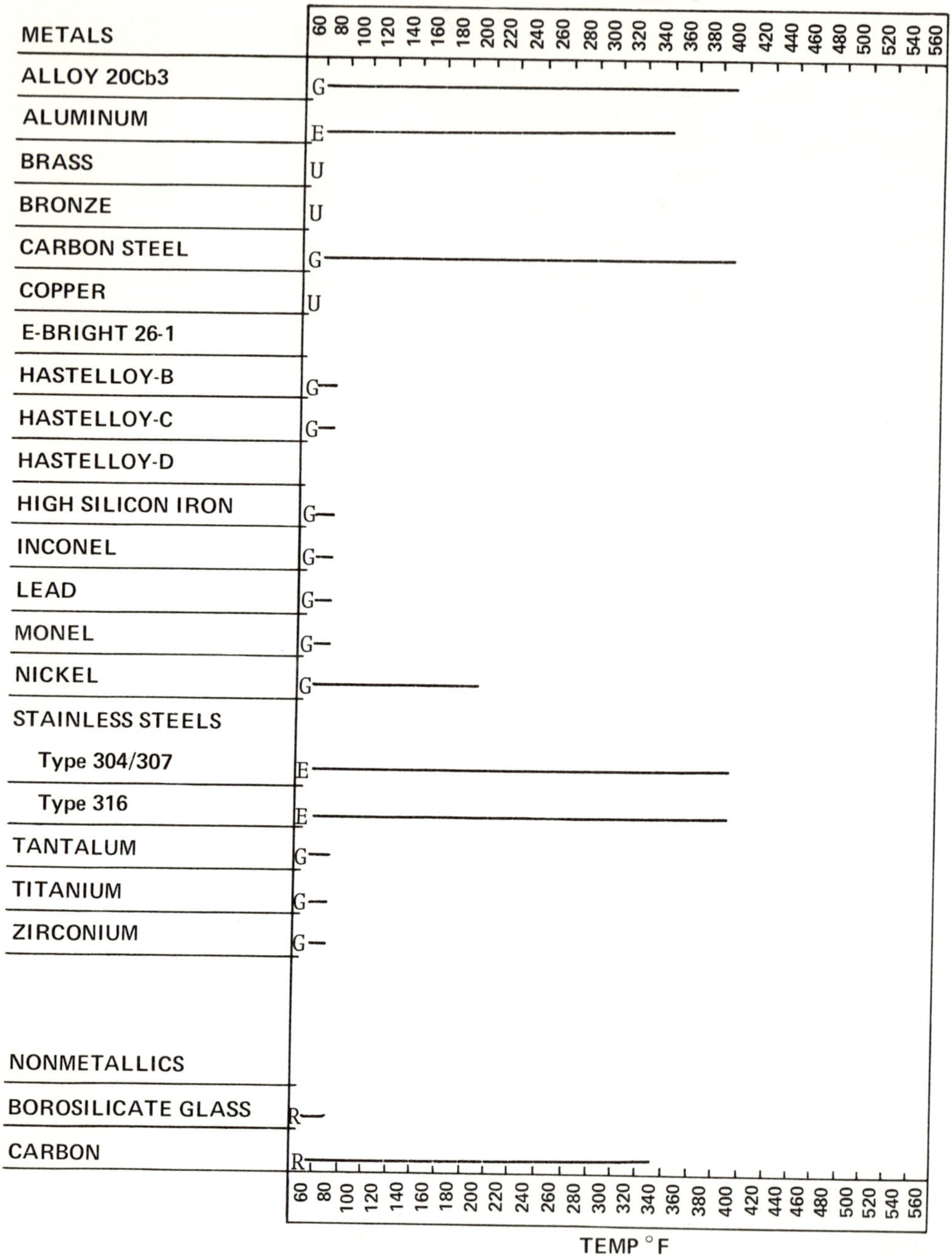

FOR METALS

E = < 2 Mils Penetration/Year; G = < 20 Mils Penetration/Year

S = < 50 Mils Penetration/Year; U = > 50 Mils Penetration/Year

FOR NONMETALLICS

R = Resistant

U = Unsatisfactory

ACETYLENE

TEMP °F: 60 80 100 120 140 160 180 200 220 240 260 280 300 320 340 360 380 400 420 440 460

PLASTICS	
ABS	R
ACRYLICS	
ASBESTOS REINF., EPOXY	
ASBESTOS REINF., PHENOLIC	
CHLORINATED POLYETHER	
CPVC	R
EPOXY	
FLUOROCARBONS FEP	R
FLUOROCARBONS TFE	R
FURFURYL ALCOHOL	
NORYL	
PHENOLIC	
POLYAMIDES–NYLON	R
POLYESTERS	
BISPHENOL A-FUMURATE	
HYDROGENATED BISPHENOL-A-BISPHENOL-A	
ISOPHTHALIC	
CHLORINATED POLYESTERS	
POLYSTYRENES	
POLYETHYLENE	U
POLYPROPYLENE	R–U
PVC–TYPE 1	R
PVC–TYPE 2	R
POLYVINYLIDENE CHLORIDE	R
VINYLIDENE FLUORIDE	
VINYL ESTER	

RUBBERS	
BUTYL GR-1	R
FLUORO ELASTOMERS VITON A	R
FLUORO ELASTOMERS KEL-F 3700	
HARD RUBBER	R
HYPALON	R
KOROSEAL	
NATURAL RUBBER (GRS)	R
NEOPRENE GR-M (CR)	R
NITRILE BUNA N (NBR)	R
NORDEL	R
POLYBUTADIENE	
SILICONE	
POLYURETHANE	

TEMP °F: 60 80 100 120 140 160 180 200 220 240 260 280 300 320 340 360 380 400 420 440 460

ACRYLONITRILE

TEMP °F

METALS	
ALLOY 20Cb3	G
ALUMINUM	E
BRASS	G
BRONZE	G
CARBON STEEL	G
COPPER 1	G
E-BRIGHT 26-1	
HASTELLOY-B	G
HASTELLOY-C	G
HASTELLOY-D	
HIGH SILICON IRON	G
INCONEL	G
LEAD	
MONEL	G
NICKEL	G
STAINLESS STEELS	
Type 304/307	G
Type 316	G
TANTALUM	G
TITANIUM	G
ZIRCONIUM	G

NONMETALLICS	
BOROSILICATE GLASS	
CARBON	R

TEMP °F: 60 80 100 120 140 160 180 200 220 240 260 280 300 320 340 360 380 400 420 440 460 480 500 520 540 560

FOR METALS

E = < 2 Mils Penetration/Year; G = < 20 Mils Penetration/Year

S = < 50 Mils Penetration/Year; U = > 50 Mils Penetration/Year

FOR NONMETALLICS

R = Resistant

U = Unsatisfactory

ACRYLONITRILE

TEMP °F (60–460)

PLASTICS	
ABS	
ACRYLICS	
ASBESTOS REINF., EPOXY	
ASBESTOS REINF., PHENOLIC	
CHLORINATED POLYETHER	R
CPVC	
EPOXY	
FLUOROCARBONS FEP	R
FLUOROCARBONS TFE	R
FURFURYL ALCOHOL	R
NORYL	
PHENOLIC	
POLYAMIDES—NYLON	
POLYESTERS	
BISPHENOL A-FUMURATE	U
HYDROGENATED BISPHENOL-A-BISPHENOL-A	U
ISOPHTHALIC	U
CHLORINATED POLYESTERS	U
POLYSTYRENES	
POLYETHYLENE	R
POLYPROPYLENE	R
PVC—TYPE 1	
PVC—TYPE 2	
POLYVINYLIDENE CHLORIDE	R
VINYLIDENE FLUORIDE	R
VINYL ESTER	

RUBBERS	
BUTYL GR-1	U
FLUORO ELASTOMERS VITON A	U
FLUORO ELASTOMERS KEL-F 3700	
HARD RUBBER	R
HYPALON	
KOROSEAL	
NATURAL RUBBER (GRS)	R
NEOPRENE GR-M (CR)	U
NITRILE BUNA N (NBR)	U
NORDEL	
POLYBUTADIENE	
SILICONE	
POLYURETHANE	

TEMP °F

ADIPIC ACID

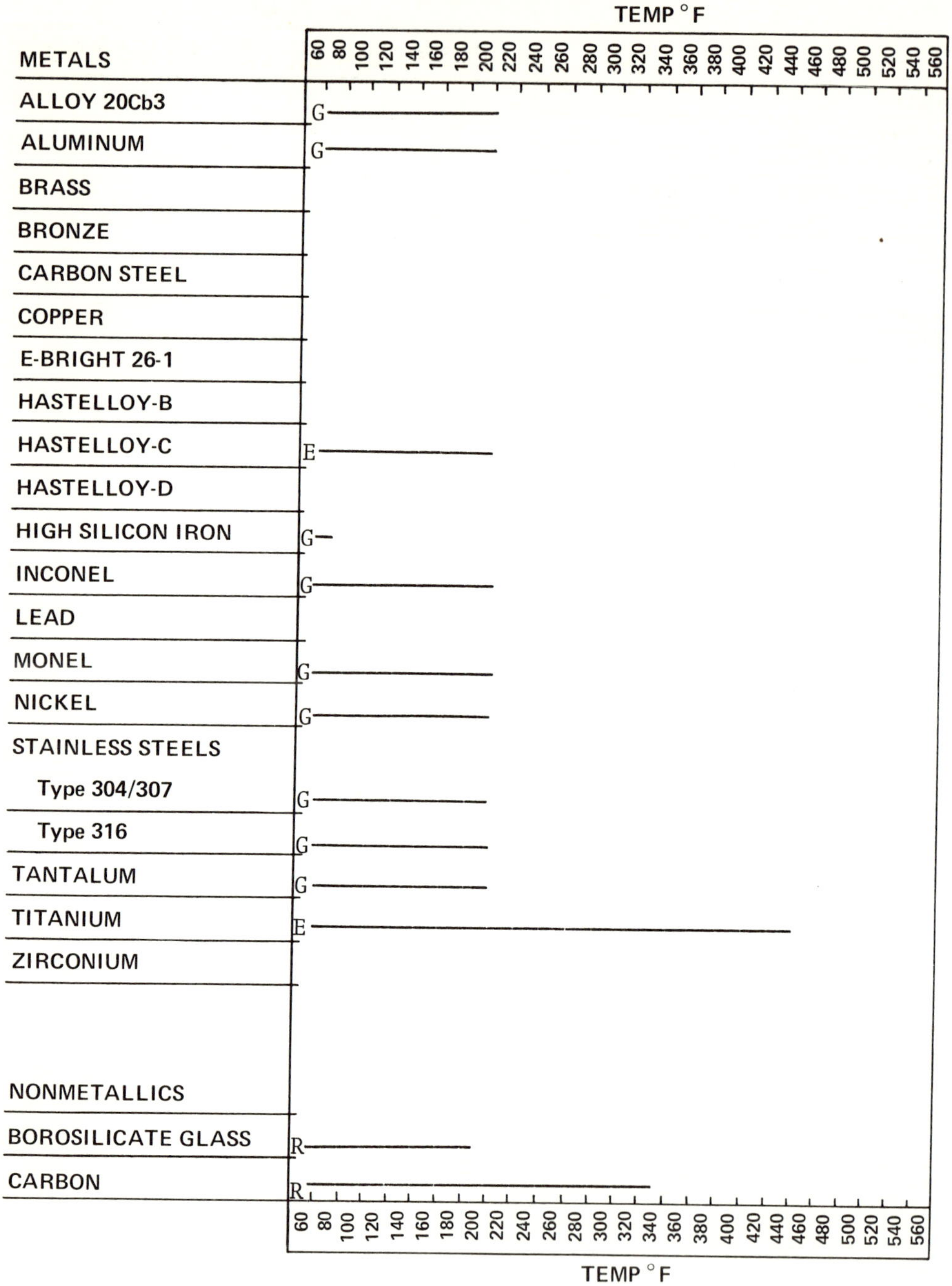

FOR METALS

E = < 2 Mils Penetration/Year; G = < 20 Mils Penetration/Year

S = < 50 Mils Penetration/Year; U = > 50 Mils Penetration/Year

FOR NONMETALLICS

R = Resistant

U = Unsatisfactory

ADIPIC ACID

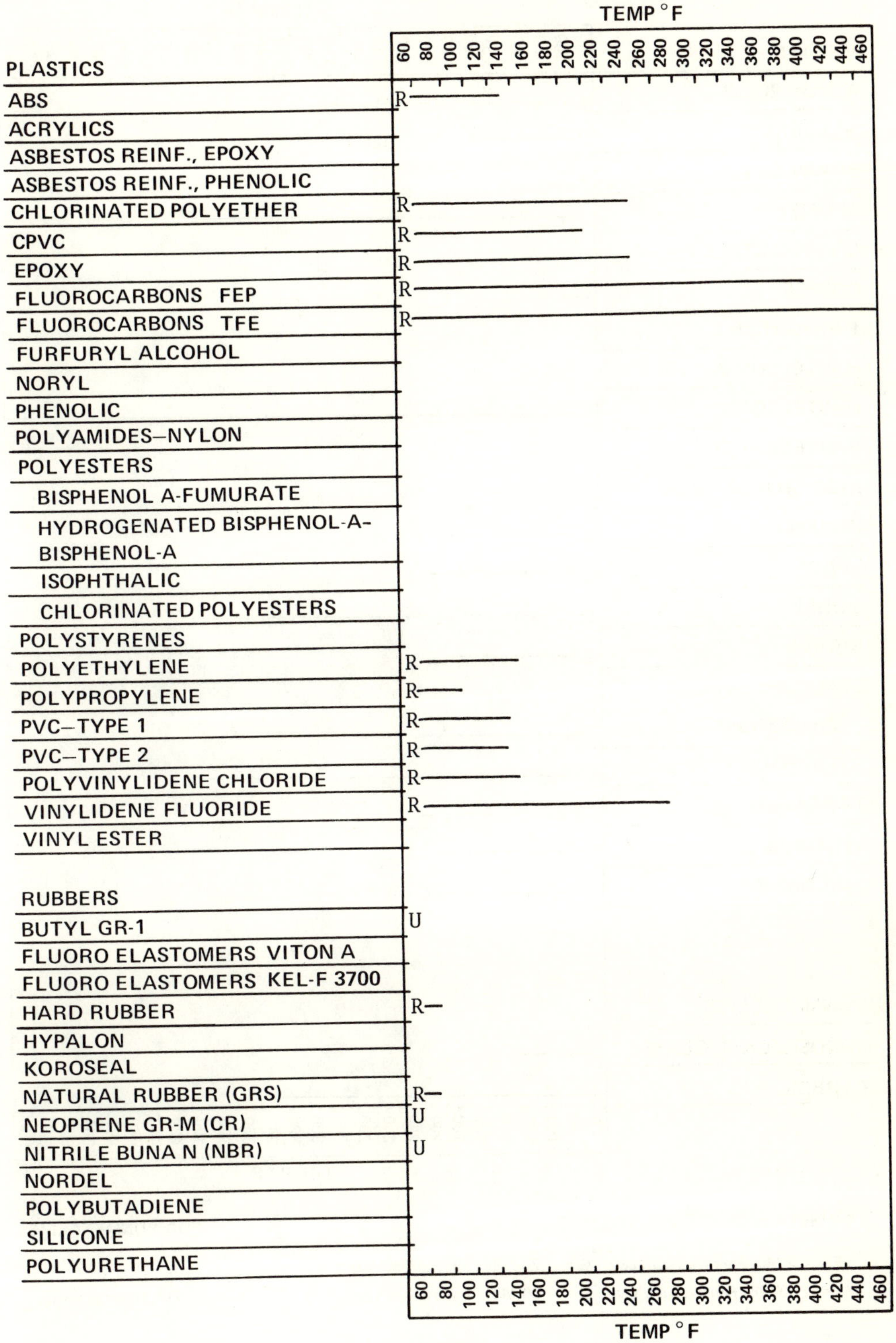
TEMP °F
60 80 100 120 140 160 180 200 220 240 260 280 300 320 340 360 380 400 420 440 460
PLASTICS
ABS R
ACRYLICS
ASBESTOS REINF., EPOXY
ASBESTOS REINF., PHENOLIC
CHLORINATED POLYETHER R
CPVC R
EPOXY R
FLUOROCARBONS FEP R
FLUOROCARBONS TFE R
FURFURYL ALCOHOL
NORYL
PHENOLIC
POLYAMIDES–NYLON
POLYESTERS
BISPHENOL A-FUMURATE
HYDROGENATED BISPHENOL-A-BISPHENOL-A
ISOPHTHALIC
CHLORINATED POLYESTERS
POLYSTYRENES
POLYETHYLENE R
POLYPROPYLENE R
PVC–TYPE 1 R
PVC–TYPE 2 R
POLYVINYLIDENE CHLORIDE R
VINYLIDENE FLUORIDE R
VINYL ESTER
RUBBERS
BUTYL GR-1 U
FLUORO ELASTOMERS VITON A
FLUORO ELASTOMERS KEL-F 3700
HARD RUBBER R
HYPALON
KOROSEAL
NATURAL RUBBER (GRS) R
NEOPRENE GR-M (CR) U
NITRILE BUNA N (NBR) U
NORDEL
POLYBUTADIENE
SILICONE
POLYURETHANE
60 80 100 120 140 160 180 200 220 240 260 280 300 320 340 360 380 400 420 440 460
TEMP °F

ALLYL ALCOHOL

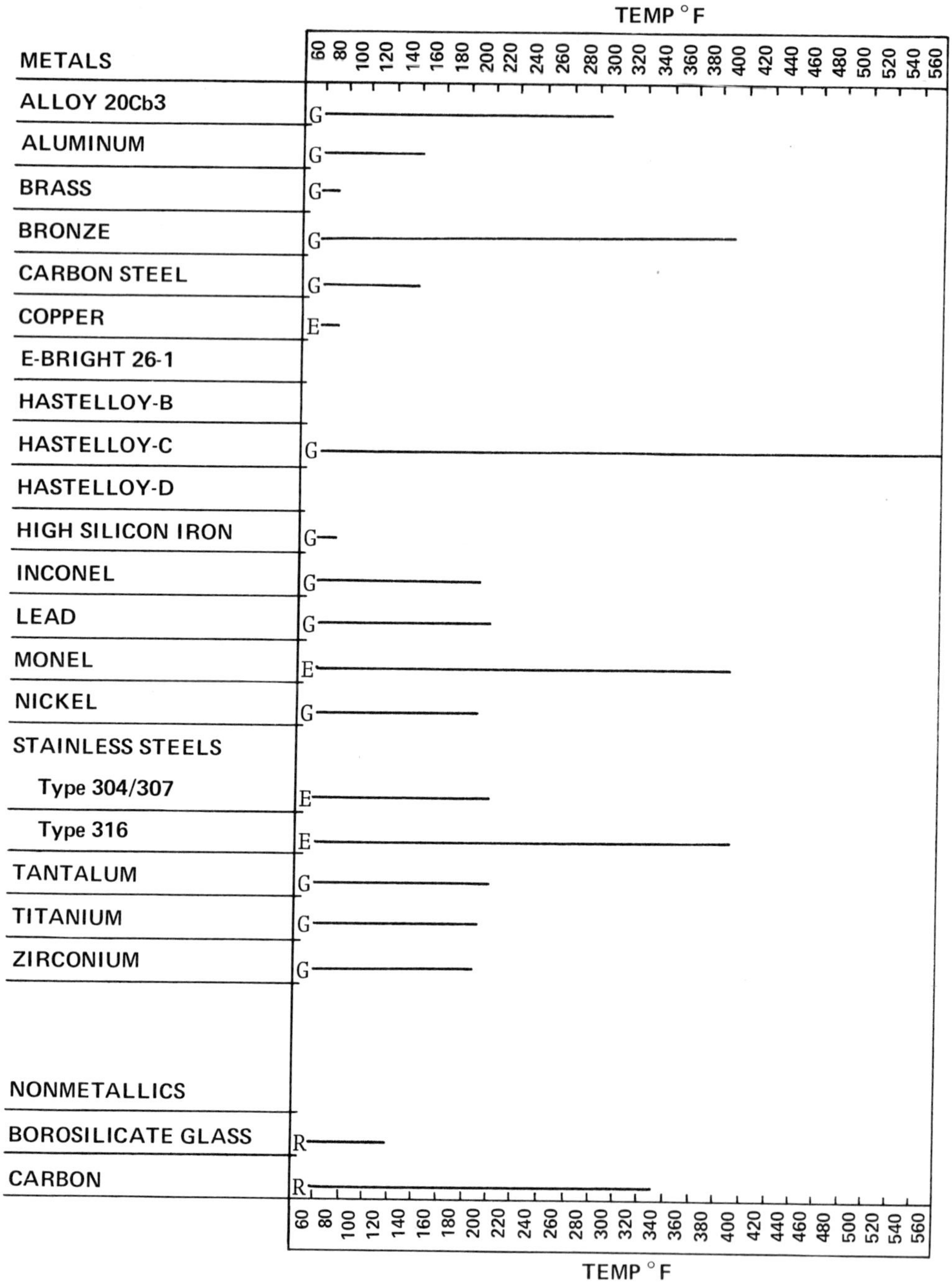

FOR METALS

E = < 2 Mils Penetration/Year; G = < 20 Mils Penetration/Year

S = < 50 Mils Penetration/Year; U = > 50 Mils Penetration/Year

FOR NONMETALLICS

R = Resistant

U = Unsatisfactory

ALLYL ALCOHOL

TEMP °F: 60 80 100 120 140 160 180 200 220 240 260 280 300 320 340 360 380 400 420 440 460

PLASTICS		
ABS		U
ACRYLICS		
ASBESTOS REINF., EPOXY		
ASBESTOS REINF., PHENOLIC		
CHLORINATED POLYETHER		R—
CPVC	96%	R—
EPOXY		
FLUOROCARBONS FEP		R—
FLUOROCARBONS TFE		R—
FURFURYL ALCOHOL		
NORYL		
PHENOLIC		
POLYAMIDES—NYLON		
POLYESTERS		
BISPHENOL A-FUMURATE		
HYDROGENATED BISPHENOL-A-BISPHENOL-A		
ISOPHTHALIC		
CHLORINATED POLYESTERS		
POLYSTYRENES		
POLYETHYLENE		R—U
POLYPROPYLENE		R—
PVC—TYPE 1		R—U
PVC—TYPE 2		R—U
POLYVINYLIDENE CHLORIDE		
VINYLIDENE FLUORIDE		R—
VINYL ESTER		

RUBBERS	
BUTYL GR-1	R—
FLUORO ELASTOMERS VITON A	R—
FLUORO ELASTOMERS KEL-F 3700	
HARD RUBBER	U
HYPALON	
KOROSEAL	
NATURAL RUBBER (GRS)	R—
NEOPRENE GR-M (CR)	R—
NITRILE BUNA N (NBR)	R—
NORDEL	R—
POLYBUTADIENE	
SILICONE	
POLYURETHANE	

TEMP °F: 60 80 100 120 140 160 180 200 220 240 260 280 300 320 340 360 380 400 420 440 460

ALLYL CHLORIDE

METALS	TEMP °F: 60 80 100 120 140 160 180 200 220 240 260 280 300 320 340 360 380 400 420 440 460 480 500 520 540 560
ALLOY 20Cb3	G————————
ALUMINUM	U
BRASS	
BRONZE	
CARBON STEEL	U
COPPER	
E-BRIGHT 26-1	
HASTELLOY-B	E————————
HASTELLOY-C	
HASTELLOY-D	
HIGH SILICON IRON	G—
INCONEL	G————U
LEAD	U
MONEL	G————————
NICKEL	G————————
STAINLESS STEELS	
Type 304/307	G———
Type 316	
TANTALUM	
TITANIUM	
ZIRCONIUM	G————————
NONMETALLICS	
BOROSILICATE GLASS	
CARBON	

TEMP °F: 60 80 100 120 140 160 180 200 220 240 260 280 300 320 340 360 380 400 420 440 460 480 500 520 540 560

FOR METALS

E = < 2 Mils Penetration/Year; G = < 20 Mils Penetration/Year

S = < 50 Mils Penetration/Year; U = > 50 Mils Penetration/Year

FOR NONMETALLICS

R = Resistant

U = Unsatisfactory

ALLYL CHLORIDE

TEMP °F: 60, 80, 100, 120, 140, 160, 180, 200, 220, 240, 260, 280, 300, 320, 340, 360, 380, 400, 420, 440, 460

PLASTICS	
ABS	U
ACRYLICS	
ASBESTOS REINF., EPOXY	
ASBESTOS REINF., PHENOLIC	
CHLORINATED POLYETHER	R—
CPVC	U
EPOXY	R—
FLUOROCARBONS FEP	R—
FLUOROCARBONS TFE	R—
FURFURYL ALCOHOL	
NORYL	
PHENOLIC	
POLYAMIDES–NYLON	
POLYESTERS	
BISPHENOL A-FUMURATE	U
HYDROGENATED BISPHENOL-A-BISPHENOL-A	
ISOPHTHALIC	
CHLORINATED POLYESTERS	
POLYSTYRENES	
POLYETHYLENE	R—
POLYPROPYLENE	R—
PVC–TYPE 1	U
PVC–TYPE 2	U
POLYVINYLIDENE CHLORIDE	
VINYLIDENE FLUORIDE	R—
VINYL ESTER	R—

RUBBERS	
BUTYL GR-1	U
FLUORO ELASTOMERS VITON A	R—U
FLUORO ELASTOMERS KEL-F 3700	
HARD RUBBER	
HYPALON	
KOROSEAL	
NATURAL RUBBER (GRS)	U
NEOPRENE GR-M (CR)	U
NITRILE BUNA N (NBR)	U
NORDEL	
POLYBUTADIENE	
SILICONE	
POLYURETHANE	

TEMP °F: 60, 80, 100, 120, 140, 160, 180, 200, 220, 240, 260, 280, 300, 320, 340, 360, 380, 400, 420, 440, 460

ALUM (ALUMINUM POTASSIUM SULFATE)

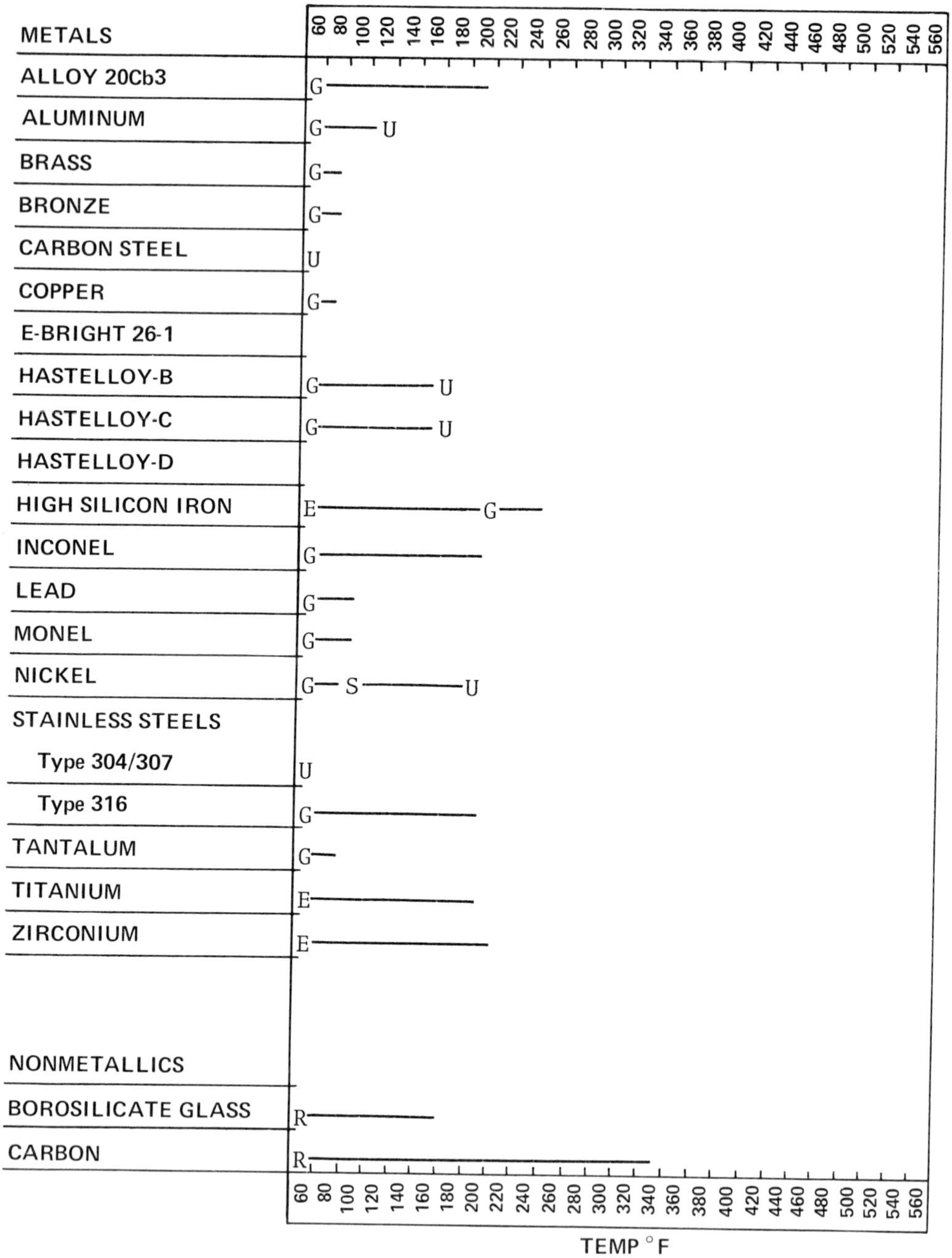

FOR METALS

E = < 2 Mils Penetration/Year; G = < 20 Mils Penetration/Year

S = < 50 Mils Penetration/Year; U = > 50 Mils Penetration/Year

FOR NONMETALLICS

R = Resistant

U = Unsatisfactory

ALUM (ALUMINUM POTASSIUM SULFATE)

PLASTICS	
ABS	R
ACRYLICS	R
ASBESTOS REINF., EPOXY	
ASBESTOS REINF., PHENOLIC	
CHLORINATED POLYETHER	R
CPVC	R
EPOXY	R
FLUOROCARBONS FEP	R
FLUOROCARBONS TFE	R
FURFURYL ALCOHOL	
NORYL	R
PHENOLIC	
POLYAMIDES—NYLON	U
POLYESTERS	
BISPHENOL A-FUMURATE	R
HYDROGENATED BISPHENOL-A-BISPHENOL-A	
ISOPHTHALIC	R
CHLORINATED POLYESTERS 10%	R
POLYSTYRENES	
POLYETHYLENE	R
POLYPROPYLENE	R
PVC—TYPE 1	R
PVC—TYPE 2	R
POLYVINYLIDENE CHLORIDE	R
VINYLIDENE FLUORIDE	
VINYL ESTER	R
RUBBERS	
BUTYL GR-1	R
FLUORO ELASTOMERS VITON A	R
FLUORO ELASTOMERS KEL-F 3700	R
HARD RUBBER	R
HYPALON	R
KOROSEAL	R
NATURAL RUBBER (GRS)	R
NEOPRENE GR-M (CR)	R
NITRILE BUNA N (NBR)	R
NORDEL	
POLYBUTADIENE	R
SILICONE	
POLYURETHANE	

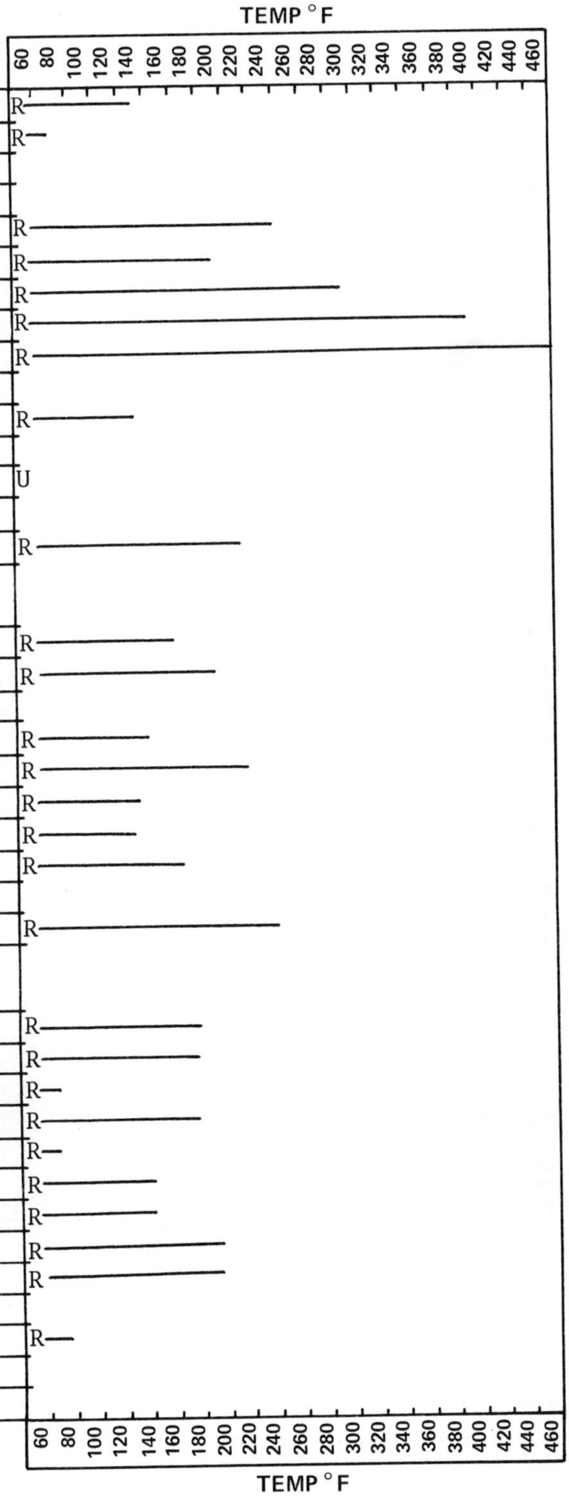

ALUM AMMONIUM

METALS		TEMP °F
ALLOY 20Cb3		
ALUMINUM	E	60
BRASS	U	60
BRONZE	G	60
CARBON STEEL	U	60
COPPER	G	60
E-BRIGHT 26-1		
HASTELLOY-B		
HASTELLOY-C		
HASTELLOY-D		
HIGH SILICON IRON		
INCONEL		
LEAD		
MONEL	G	60
NICKEL		
STAINLESS STEELS Type 304/307		
Type 316	G	60
TANTALUM	E	60
TITANIUM		
ZIRCONIUM		

NONMETALLICS	Rating	TEMP °F
BOROSILICATE GLASS		
CARBON	R	60–340

TEMP °F: 60, 80, 100, 120, 140, 160, 180, 200, 220, 240, 260, 280, 300, 320, 340, 360, 380, 400, 420, 440, 460, 480, 500, 520, 540, 560

FOR METALS

E = < 2 Mils Penetration/Year; G = < 20 Mils Penetration/Year

S = < 50 Mils Penetration/Year; U = > 50 Mils Penetration/Year

FOR NONMETALLICS

R = Resistant

U = Unsatisfactory

ALUM AMMONIUM

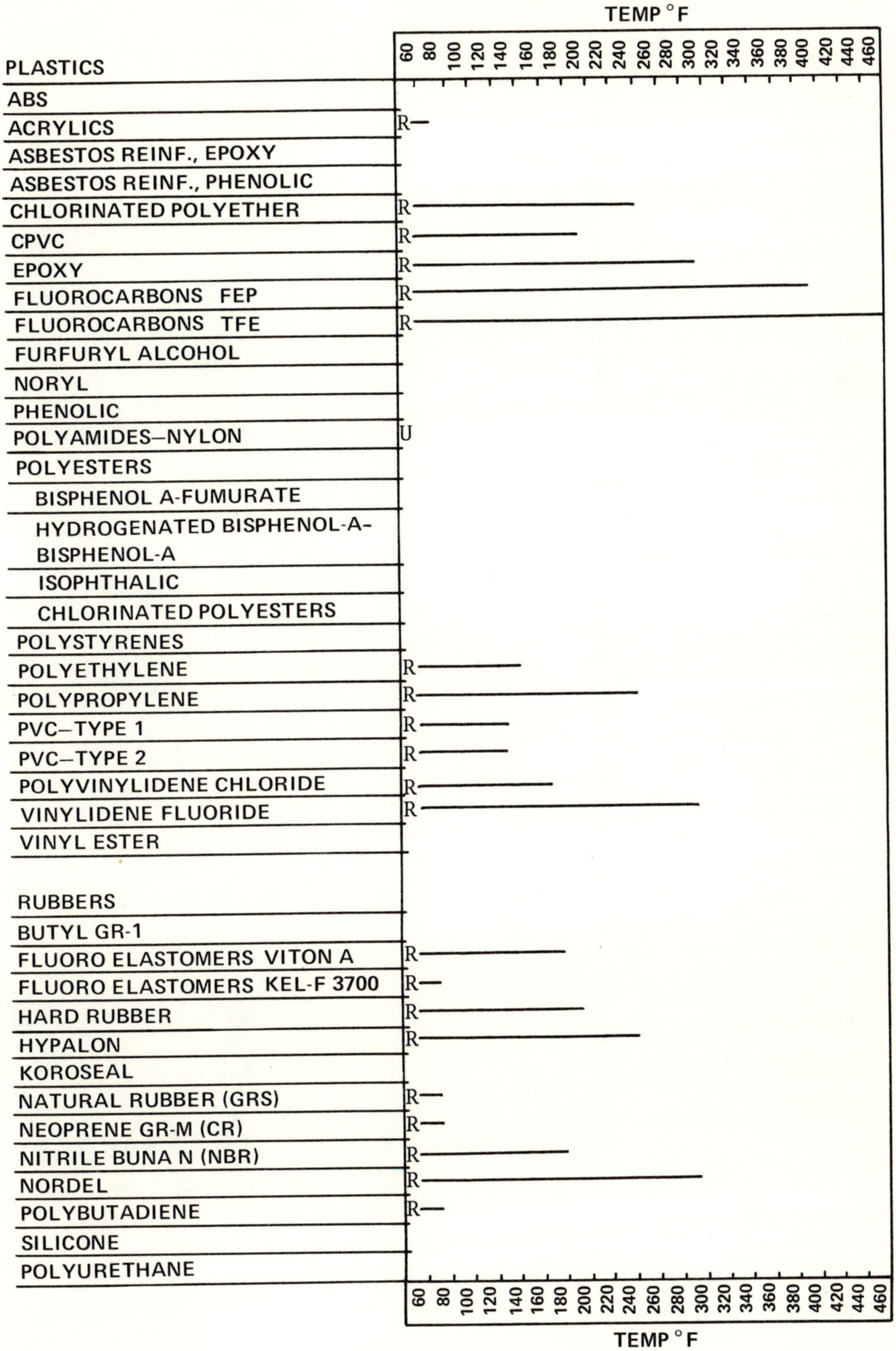

ALUM AMMONIUM SULFATE

TEMP °F (60–560)

METALS	Rating
ALLOY 20Cb3	
ALUMINUM	E—
BRASS	U
BRONZE	G—
CARBON STEEL	U
COPPER	G—
E-BRIGHT 26-1	
HASTELLOY-B	
HASTELLOY-C	
HASTELLOY-D	
HIGH SILICON IRON	
INCONEL	
LEAD	
MONEL	G—
NICKEL	
STAINLESS STEELS	
Type 304/307	
Type 316	G—————
TANTALUM	
TITANIUM	
ZIRCONIUM	

NONMETALLICS	
BOROSILICATE GLASS	
CARBON	

TEMP °F: 60 80 100 120 140 160 180 200 220 240 260 280 300 320 340 360 380 400 420 440 460 480 500 520 540 560

FOR METALS

E = < 2 Mils Penetration/Year; G = < 20 Mils Penetration/Year
S = < 50 Mils Penetration/Year; U = > 50 Mils Penetration/Year

FOR NONMETALLICS

R = Resistant
U = Unsatisfactory

ALUM AMMONIUM SULFATE

TEMP °F: 60 80 100 120 140 160 180 200 220 240 260 280 300 320 340 360 380 400 420 440 460

PLASTICS	
ABS	
ACRYLICS	R
ASBESTOS REINF., EPOXY	
ASBESTOS REINF., PHENOLIC	
CHLORINATED POLYETHER	
CPVC	
EPOXY	
FLUOROCARBONS FEP	R
FLUOROCARBONS TFE	R
FURFURYL ALCOHOL	
NORYL	
PHENOLIC	
POLYAMIDES–NYLON	U
POLYESTERS	
BISPHENOL A-FUMURATE	
HYDROGENATED BISPHENOL-A-BISPHENOL-A	
ISOPHTHALIC	
CHLORINATED POLYESTERS	
POLYSTYRENES	
POLYETHYLENE	
POLYPROPYLENE	R
PVC–TYPE 1	
PVC–TYPE 2	
POLYVINYLIDENE CHLORIDE	
VINYLIDENE FLUORIDE	R
VINYL ESTER	

RUBBERS	
BUTYL GR-1	R
FLUORO ELASTOMERS VITON A	R
FLUORO ELASTOMERS KEL-F 3700	R
HARD RUBBER	R
HYPALON	R
KOROSEAL	
NATURAL RUBBER (GRS)	R
NEOPRENE GR-M (CR)	R
NITRILE BUNA N (NBR)	R
NORDEL	
POLYBUTADIENE	R
SILICONE	
POLYURETHANE	

TEMP °F: 60 80 100 120 140 160 180 200 220 240 260 280 300 320 340 360 380 400 420 440 460

ALUM CHROME

TEMP °F (60, 80, 100, 120, 140, 160, 180, 200, 220, 240, 260, 280, 300, 320, 340, 360, 380, 400, 420, 440, 460, 480, 500, 520, 540, 560)

METALS	
ALLOY 20Cb3	
ALUMINUM	E
BRASS	U
BRONZE	G
CARBON STEEL	U
COPPER	G
E-BRIGHT 26-1	
HASTELLOY-B	
HASTELLOY-C	
HASTELLOY-D	
HIGH SILICON IRON	
INCONEL	
LEAD	
MONEL	G
NICKEL	
STAINLESS STEELS	
Type 304/307	G
Type 316	G
TANTALUM	
TITANIUM	
ZIRCONIUM	
NONMETALLICS	
BOROSILICATE GLASS	
CARBON	R

TEMP °F

FOR METALS

E = < 2 Mils Penetration/Year; G = < 20 Mils Penetration/Year

S = < 50 Mils Penetration/Year; U = > 50 Mils Penetration/Year

FOR NONMETALLICS

R = Resistant

U = Unsatisfactory

ALUM CHROME

TEMP °F (60–460)

PLASTICS	ALUM CHROME
ABS	
ACRYLICS	R
ASBESTOS REINF., EPOXY	
ASBESTOS REINF., PHENOLIC	
CHLORINATED POLYETHER	R
CPVC	R
EPOXY	
FLUOROCARBONS FEP	R
FLUOROCARBONS TFE	R
FURFURYL ALCOHOL	
NORYL	
PHENOLIC	
POLYAMIDES—NYLON	U
POLYESTERS	
BISPHENOL A-FUMURATE	
HYDROGENATED BISPHENOL-A-BISPHENOL-A	
ISOPHTHALIC	
CHLORINATED POLYESTERS	
POLYSTYRENES	
POLYETHYLENE	R
POLYPROPYLENE	R
PVC—TYPE 1	R
PVC—TYPE 2	R
POLYVINYLIDENE CHLORIDE	R
VINYLIDENE FLUORIDE	R
VINYL ESTER	

RUBBERS	ALUM CHROME
BUTYL GR-1	
FLUORO ELASTOMERS VITON A	R
FLUORO ELASTOMERS KEL-F 3700	R
HARD RUBBER	R
HYPALON	R
KOROSEAL	
NATURAL RUBBER (GRS)	R
NEOPRENE GR-M (CR)	R
NITRILE BUNA N (NBR)	R
NORDEL	R
POLYBUTADIENE	R
SILICONE	
POLYURETHANE	

TEMP °F (60–460)

ALUM POTASSIUM

TEMP °F: 60 80 100 120 140 160 180 200 220 240 260 280 300 320 340 360 380 400 420 440 460 480 500 520 540 560

METALS	
ALLOY 20Cb3	
ALUMINUM	E
BRASS	U
BRONZE	G
CARBON STEEL	U
COPPER	G
E-BRIGHT 26-1	
HASTELLOY-B	
HASTELLOY-C	
HASTELLOY-D	
HIGH SILICON IRON	
INCONEL	
LEAD	
MONEL	G
NICKEL	
STAINLESS STEELS Type 304/307	
Type 316	G
TANTALUM	
TITANIUM	
ZIRCONIUM	

NONMETALLICS	
BOROSILICATE GLASS	
CARBON	R

TEMP °F: 60 80 100 120 140 160 180 200 220 240 260 280 300 320 340 360 380 400 420 440 460 480 500 520 540 560

FOR METALS

E = < 2 Mils Penetration/Year; G = < 20 Mils Penetration/Year

S = < 50 Mils Penetration/Year; U = > 50 Mils Penetration/Year

FOR NONMETALLICS

R = Resistant

U = Unsatisfactory

ALUM POTASSIUM

TEMP °F (60–460)

PLASTICS	Rating
ABS	
ACRYLICS	R
ASBESTOS REINF., EPOXY	
ASBESTOS REINF., PHENOLIC	
CHLORINATED POLYETHER	R
CPVC	R
EPOXY	
FLUOROCARBONS FEP	R
FLUOROCARBONS TFE	R
FURFURYL ALCOHOL	
NORYL	
PHENOLIC	
POLYAMIDES—NYLON	U
POLYESTERS	
BISPHENOL A-FUMURATE	
HYDROGENATED BISPHENOL-A-BISPHENOL-A	
ISOPHTHALIC	
CHLORINATED POLYESTERS 5%	R
POLYSTYRENES	
POLYETHYLENE	R
POLYPROPYLENE	R
PVC—TYPE 1	R
PVC—TYPE 2	R
POLYVINYLIDENE CHLORIDE	R
VINYLIDENE FLUORIDE	R
VINYL ESTER	

RUBBERS	Rating
BUTYL GR-1	
FLUORO ELASTOMERS VITON A	R
FLUORO ELASTOMERS KEL-F 3700	R
HARD RUBBER	R
HYPALON	R
KOROSEAL	
NATURAL RUBBER (GRS)	R
NEOPRENE GR-M (CR)	R
NITRILE BUNA N (NBR)	R
NORDEL	R
POLYBUTADIENE	R
SILICONE	
POLYURETHANE	

TEMP °F (60–460)

ALUMINUM CHLORIDE

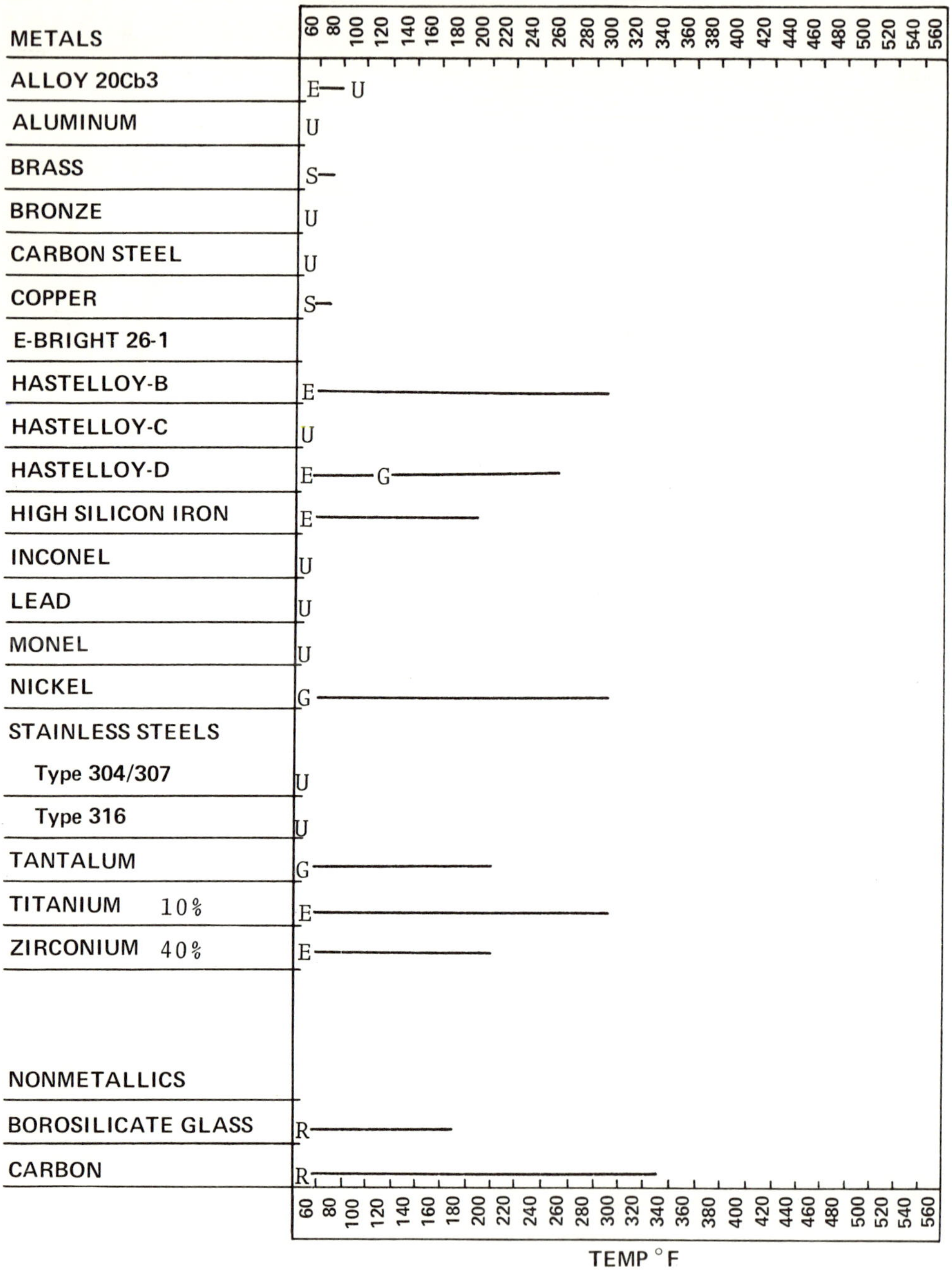

FOR METALS

E = < 2 Mils Penetration/Year; G = < 20 Mils Penetration/Year

S = < 50 Mils Penetration/Year; U = > 50 Mils Penetration/Year

FOR NONMETALLICS

R = Resistant

U = Unsatisfactory

ALUMINUM CHLORIDE

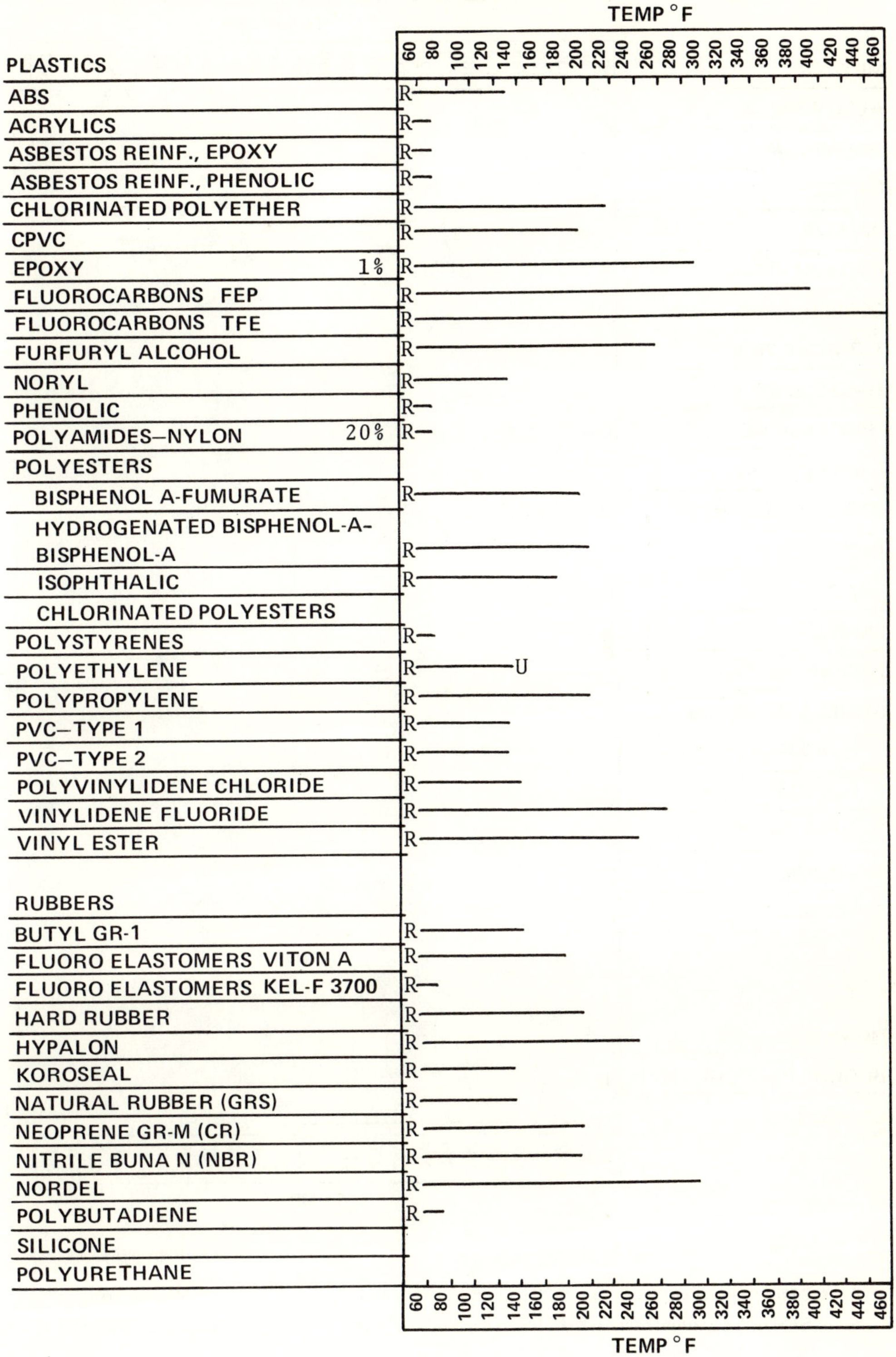

ALUMINUM FLUORIDE (SAT)

TEMP °F: 60 80 100 120 140 160 180 200 220 240 260 280 300 320 340 360 380 400 420 440 460 480 500 520 540 560

METALS	
ALLOY 20Cb3	U
ALUMINUM	G——
BRASS	
BRONZE	
CARBON STEEL	U
COPPER	
E-BRIGHT 26-1	
HASTELLOY-B	
HASTELLOY-C 10%	G—
HASTELLOY-D	
HIGH SILICON IRON	U
INCONEL	
LEAD	G—
MONEL	G—
NICKEL	G—
STAINLESS STEELS	
Type 304/307	U
Type 316	G—
TANTALUM	U
TITANIUM	E—
ZIRCONIUM	U

NONMETALLICS	
BOROSILICATE GLASS	U
CARBON	R———————————— (to 340)

TEMP °F: 60 80 100 120 140 160 180 200 220 240 260 280 300 320 340 360 380 400 420 440 460 480 500 520 540 560

FOR METALS

E = < 2 Mils Penetration/Year; G = < 20 Mils Penetration/Year

S = < 50 Mils Penetration/Year; U = > 50 Mils Penetration/Year

FOR NONMETALLICS

R = Resistant

U = Unsatisfactory

ALUMINUM FLUORIDE (SAT)

TEMP °F: 60, 80, 100, 120, 140, 160, 180, 200, 220, 240, 260, 280, 300, 320, 340, 360, 380, 400, 420, 440, 460

PLASTICS		Rating
ABS		R
ACRYLICS		
ASBESTOS REINF., EPOXY		
ASBESTOS REINF., PHENOLIC		
CHLORINATED POLYETHER		R
CPVC		R
EPOXY		R
FLUOROCARBONS FEP		R
FLUOROCARBONS TFE		R
FURFURYL ALCOHOL		R
NORYL		
PHENOLIC		
POLYAMIDES–NYLON		R
POLYESTERS		
BISPHENOL A-FUMURATE		R
HYDROGENATED BISPHENOL-A-BISPHENOL-A		U
ISOPHTHALIC	10	R
CHLORINATED POLYESTERS	10	R
POLYSTYRENES		
POLYETHYLENE		R
POLYPROPYLENE		R
PVC–TYPE 1		R
PVC–TYPE 2		R
POLYVINYLIDENE CHLORIDE		R
VINYLIDENE FLUORIDE		R
VINYL ESTER		R
RUBBERS		
BUTYL GR-1		R
FLUORO ELASTOMERS VITON A		R
FLUORO ELASTOMERS KEL-F 3700		
HARD RUBBER		R
HYPALON		R
KOROSEAL		
NATURAL RUBBER (GRS)		R
NEOPRENE GR-M (CR)		R
NITRILE BUNA N (NBR)		R
NORDEL		R
POLYBUTADIENE		
SILICONE		
POLYURETHANE		

TEMP °F: 60, 80, 100, 120, 140, 160, 180, 200, 220, 240, 260, 280, 300, 320, 340, 360, 380, 400, 420, 440, 460

ALUMINUM HYDROXIDE

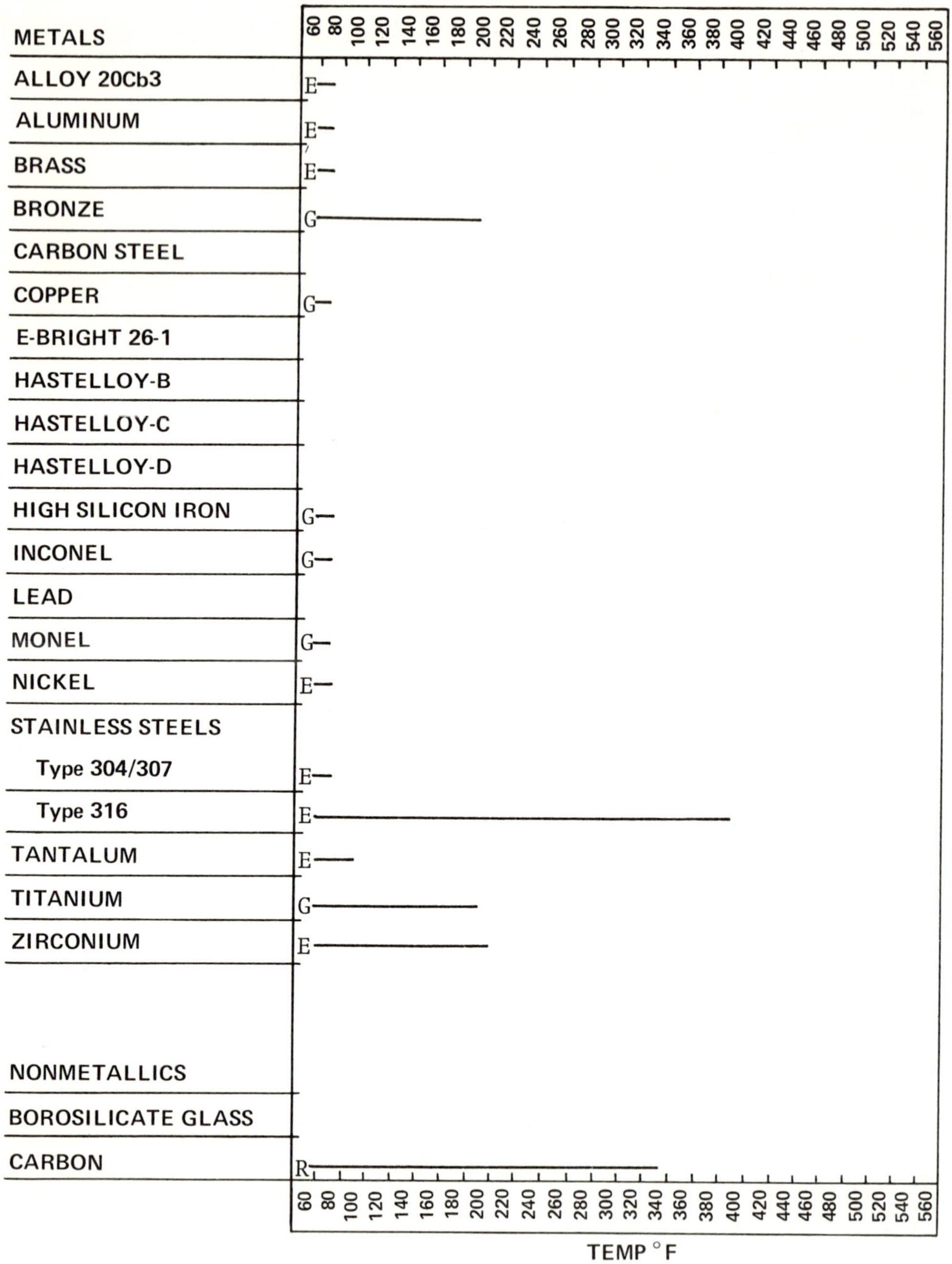

FOR METALS

E = < 2 Mils Penetration/Year; G = < 20 Mils Penetration/Year

S = < 50 Mils Penetration/Year; U = > 50 Mils Penetration/Year

FOR NONMETALLICS

R = Resistant

U = Unsatisfactory

ALUMINUM HYDROXIDE

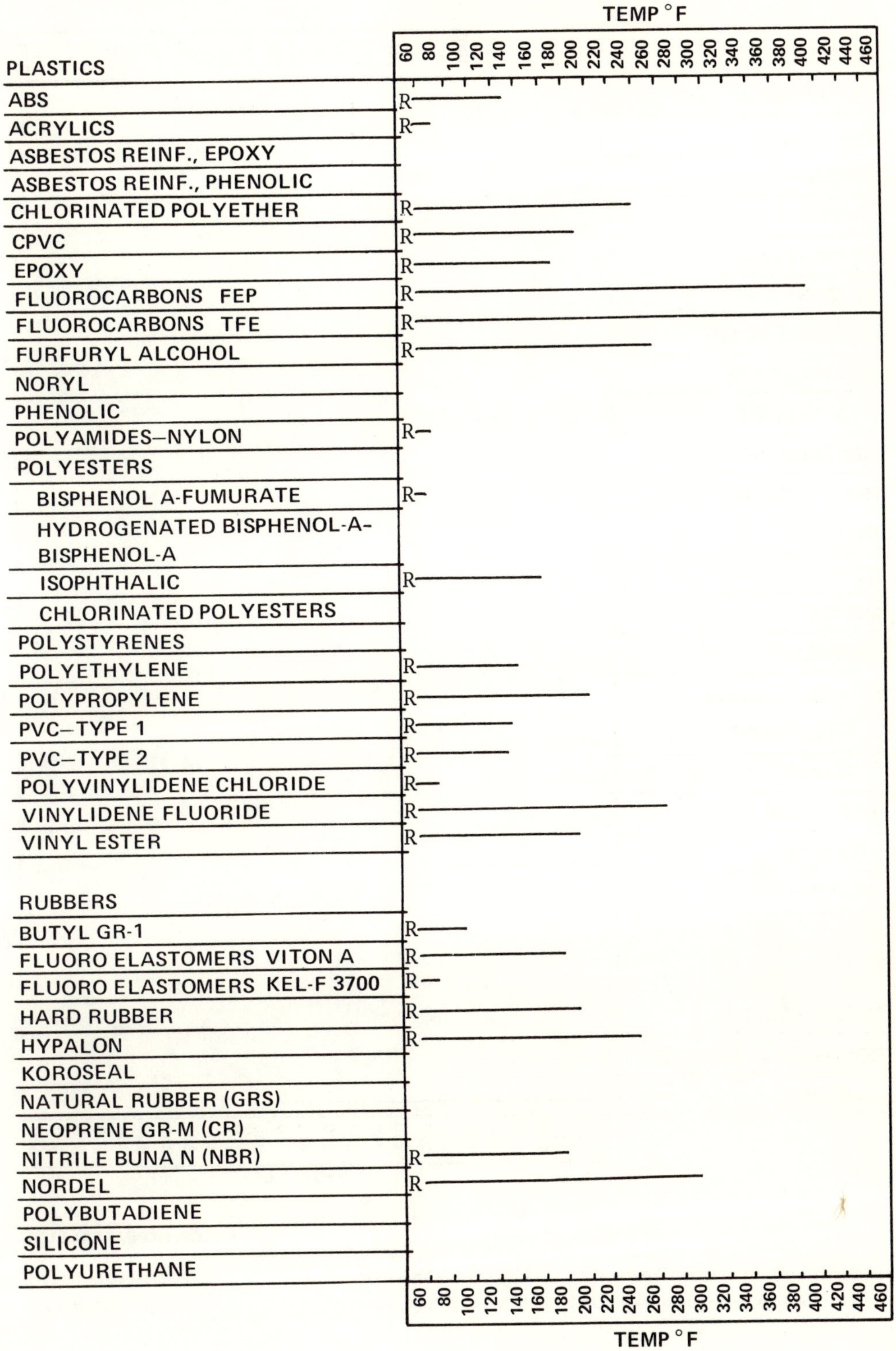

ALUMINUM NITRATE (SAT)

TEMP °F: 60 80 100 120 140 160 180 200 220 240 260 280 300 320 340 360 380 400 420 440 460 480 500 520 540 560

METALS	Rating
ALLOY 20Cb3	G–
ALUMINUM	G—U
BRASS	
BRONZE	
CARBON STEEL	U
COPPER	
E-BRIGHT 26-1	
HASTELLOY-B	
HASTELLOY-C	
HASTELLOY-D	
HIGH SILICON IRON	G–
INCONEL	
LEAD	
MONEL	
NICKEL	
STAINLESS STEELS	
Type 304/307	G—
Type 316	G—
TANTALUM	G—
TITANIUM	E————————
ZIRCONIUM	

NONMETALLICS	Rating
BOROSILICATE GLASS	R——
CARBON	R————————————————

TEMP °F: 60 80 100 120 140 160 180 200 220 240 260 280 300 320 340 360 380 400 420 440 460 480 500 520 540 560

FOR METALS

E = < 2 Mils Penetration/Year; G = < 20 Mils Penetration/Year

S = < 50 Mils Penetration/Year; U = > 50 Mils Penetration/Year

FOR NONMETALLICS

R = Resistant

U = Unsatisfactory

ALUMINUM NITRATE (SAT)

PLASTICS	Rating	TEMP °F (max)
ABS		
ACRYLICS		
ASBESTOS REINF., EPOXY		
ASBESTOS REINF., PHENOLIC		
CHLORINATED POLYETHER	R	250
CPVC	R	200
EPOXY	R	250
FLUOROCARBONS FEP	R	400
FLUOROCARBONS TFE	R	460+
FURFURYL ALCOHOL		
NORYL		
PHENOLIC		
POLYAMIDES—NYLON		
POLYESTERS		
BISPHENOL A-FUMURATE	R	200
HYDROGENATED BISPHENOL-A-BISPHENOL-A		
ISOPHTHALIC	R	160
CHLORINATED POLYESTERS		
POLYSTYRENES		
POLYETHYLENE	R	140
POLYPROPYLENE	R	220
PVC—TYPE 1	R	140
PVC—TYPE 2	R	140
POLYVINYLIDENE CHLORIDE	R	180
VINYLIDENE FLUORIDE	R	280
VINYL ESTER	R	200

RUBBERS	Rating	TEMP °F (max)
BUTYL GR-1	R	180
FLUORO ELASTOMERS VITON A	R	180
FLUORO ELASTOMERS KEL-F 3700		
HARD RUBBER	R	180
HYPALON	R	250
KOROSEAL	R	140
NATURAL RUBBER (GRS)	R	140
NEOPRENE GR-M (CR)	R	200
NITRILE BUNA N (NBR)	R	200
NORDEL	R	300
POLYBUTADIENE		
SILICONE		
POLYURETHANE		

ALUMINUM OXYCHLORIDE

METALS	TEMP °F (60 80 100 120 140 160 180 200 220 240 260 280 300 320 340 360 380 400 420 440 460 480 500 520 540 560)
ALLOY 20Cb3	
ALUMINUM	
BRASS	
BRONZE	
CARBON STEEL	
COPPER	
E-BRIGHT 26-1	
HASTELLOY-B	
HASTELLOY-C	
HASTELLOY-D	
HIGH SILICON IRON	
INCONEL	
LEAD	
MONEL	
NICKEL	
STAINLESS STEELS	
Type 304/307	
Type 316	
TANTALUM	
TITANIUM	
ZIRCONIUM	

NONMETALLICS	TEMP °F (60 80 100 120 140 160 180 200 220 240 260 280 300 320 340 360 380 400 420 440 460 480 500 520 540 560)
BOROSILICATE GLASS	R (60–160)
CARBON	

FOR METALS

E = < 2 Mils Penetration/Year; G = < 20 Mils Penetration/Year

S = < 50 Mils Penetration/Year; U = > 50 Mils Penetration/Year

FOR NONMETALLICS

R = Resistant

U = Unsatisfactory

ALUMINUM OXYCHLORIDE

TEMP °F: 60 80 100 120 140 160 180 200 220 240 260 280 300 320 340 360 380 400 420 440 460

PLASTICS	
ABS	R
ACRYLICS	
ASBESTOS REINF., EPOXY	
ASBESTOS REINF., PHENOLIC	
CHLORINATED POLYETHER	R
CPVC	R
EPOXY	
FLUOROCARBONS FEP	R
FLUOROCARBONS TFE	R
FURFURYL ALCOHOL	
NORYL	
PHENOLIC	
POLYAMIDES—NYLON	
POLYESTERS	
BISPHENOL A-FUMURATE	
HYDROGENATED BISPHENOL-A-BISPHENOL-A	
ISOPHTHALIC	
CHLORINATED POLYESTERS	
POLYSTYRENES	
POLYETHYLENE	R
POLYPROPYLENE	R
PVC—TYPE 1	R
PVC—TYPE 2	R
POLYVINYLIDENE CHLORIDE	R
VINYLIDENE FLUORIDE	R
VINYL ESTER	

RUBBERS	
BUTYL GR-1	
FLUORO ELASTOMERS VITON A	U
FLUORO ELASTOMERS KEL-F 3700	
HARD RUBBER	
HYPALON	
KOROSEAL	
NATURAL RUBBER (GRS)	
NEOPRENE GR-M (CR)	
NITRILE BUNA N (NBR)	
NORDEL	
POLYBUTADIENE	
SILICONE	
POLYURETHANE	

TEMP °F: 60 80 100 120 140 160 180 200 220 240 260 280 300 320 340 360 380 400 420 440 460

ALUMINUM POTASSIUM SULFATE

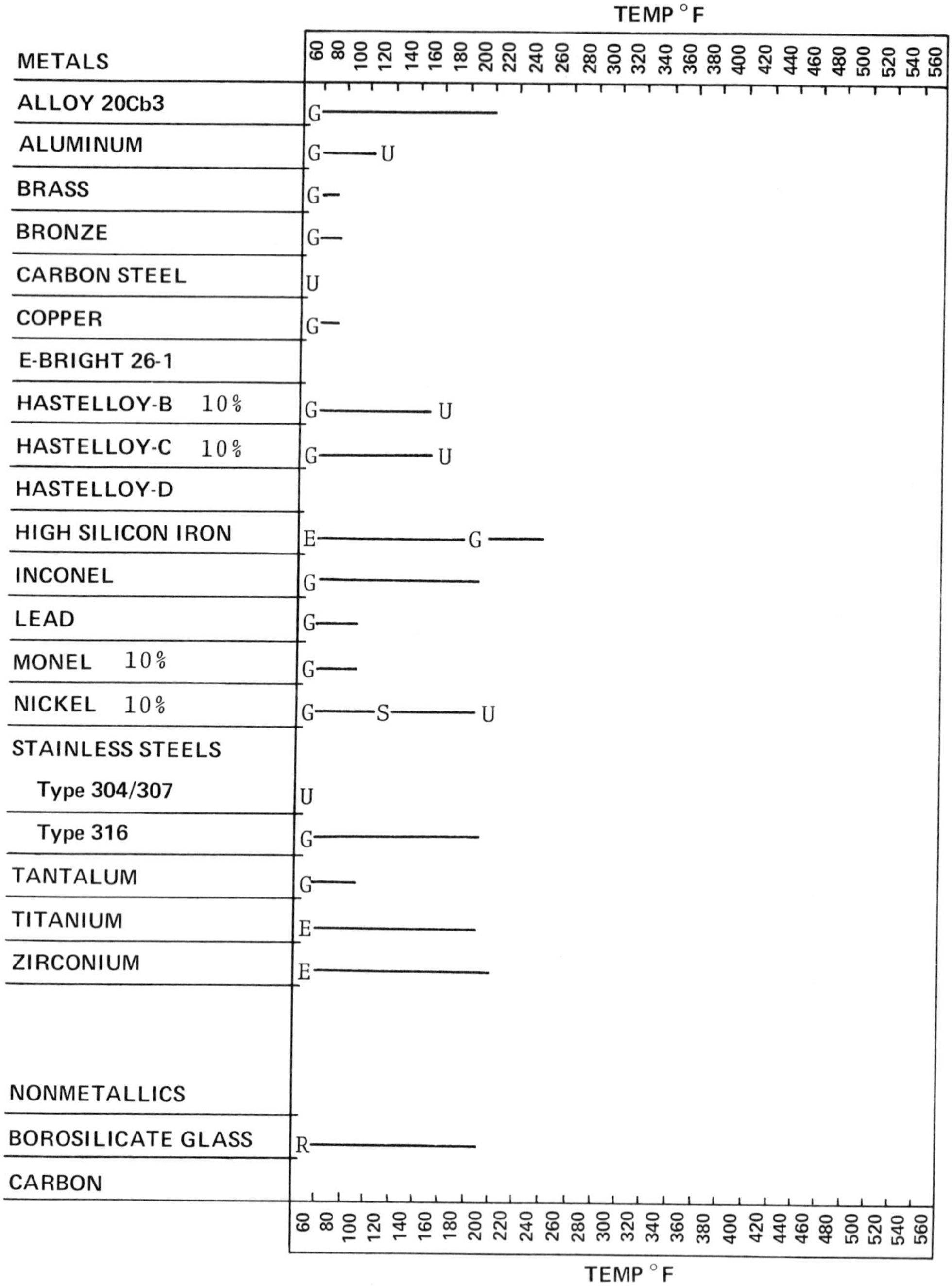

FOR METALS

E = < 2 Mils Penetration/Year; G = < 20 Mils Penetration/Year
S = < 50 Mils Penetration/Year; U = > 50 Mils Penetration/Year

FOR NONMETALLICS

R = Resistant
U = Unsatisfactory

ALUMINUM POTASSIUM SULFATE

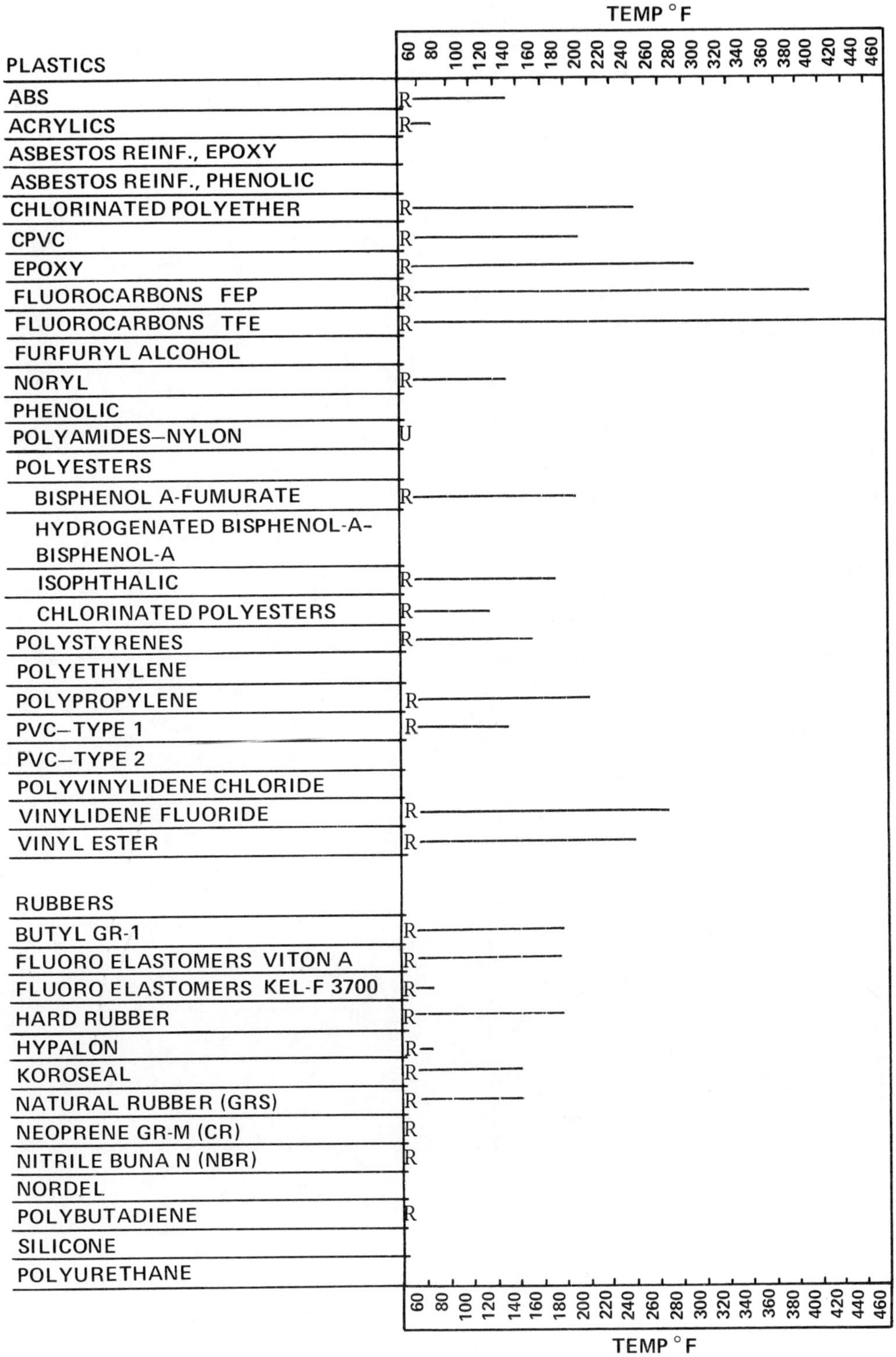

ALUMINUM SULFATE (SAT)

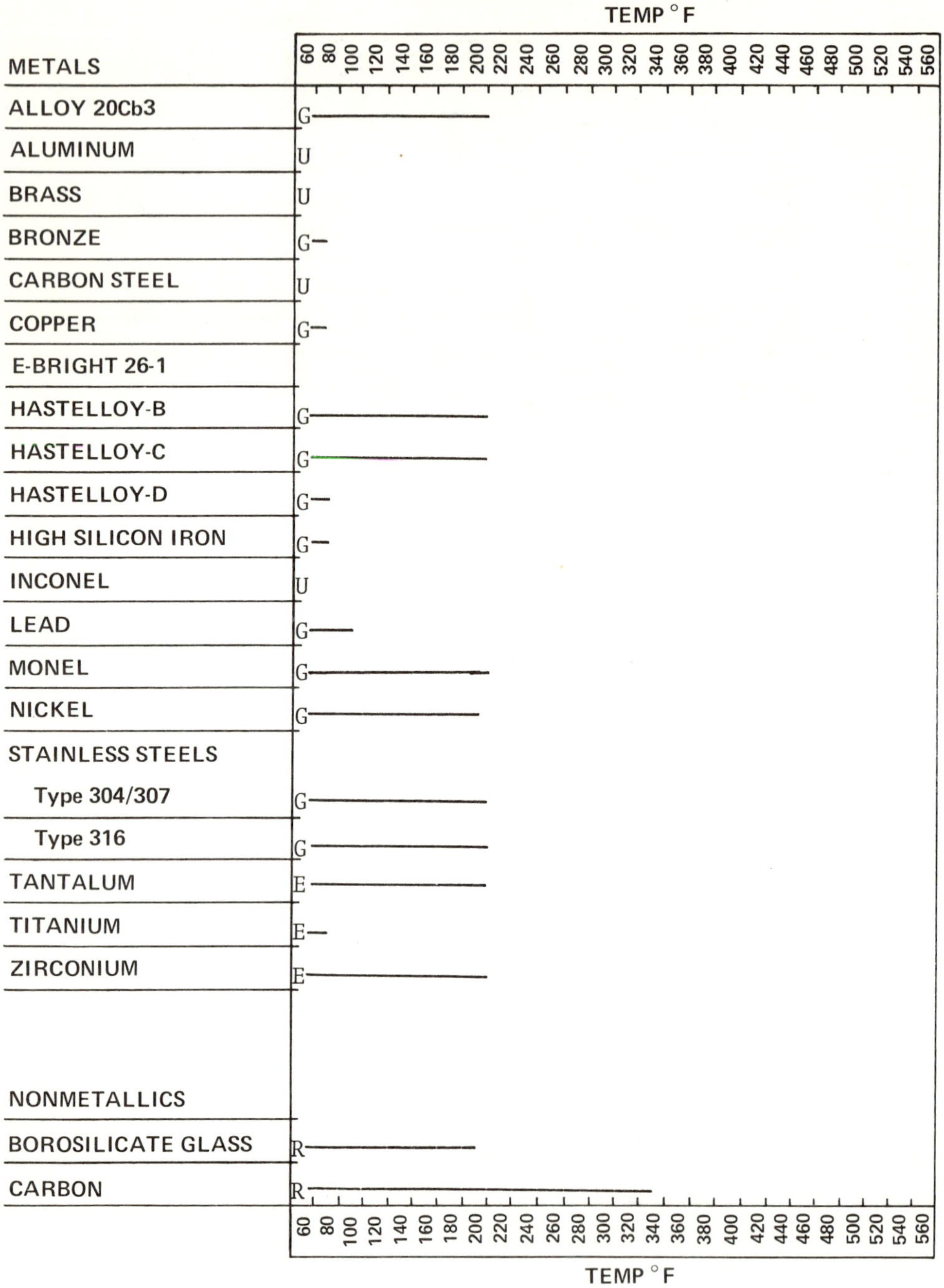

FOR METALS

E = < 2 Mils Penetration/Year; G = < 20 Mils Penetration/Year

S = < 50 Mils Penetration/Year; U = > 50 Mils Penetration/Year

FOR NONMETALLICS

R = Resistant

U = Unsatisfactory

ALUMINUM SULFATE (SAT)

TEMP °F: 60, 80, 100, 120, 140, 160, 180, 200, 220, 240, 260, 280, 300, 320, 340, 360, 380, 400, 420, 440, 460

PLASTICS		Rating
ABS		R
ACRYLICS		R
ASBESTOS REINF., EPOXY		R
ASBESTOS REINF., PHENOLIC		R
CHLORINATED POLYETHER		R
CPVC		R
EPOXY		R
FLUOROCARBONS FEP		R
FLUOROCARBONS TFE		R
FURFURYL ALCOHOL		R
NORYL		
PHENOLIC		R
POLYAMIDES–NYLON	10%	R
POLYESTERS		
BISPHENOL A-FUMURATE		R
HYDROGENATED BISPHENOL-A-BISPHENOL-A		R
ISOPHTHALIC		R
CHLORINATED POLYESTERS		R
POLYSTYRENES		
POLYETHYLENE		R—U
POLYPROPYLENE		R
PVC–TYPE 1		R
PVC–TYPE 2		R
POLYVINYLIDENE CHLORIDE		R
VINYLIDENE FLUORIDE		R
VINYL ESTER		R

RUBBERS	Rating
BUTYL GR-1	R
FLUORO ELASTOMERS VITON A	R
FLUORO ELASTOMERS KEL-F 3700	R
HARD RUBBER	R
HYPALON	R
KOROSEAL	R
NATURAL RUBBER (GRS)	R
NEOPRENE GR-M (CR)	R
NITRILE BUNA N (NBR)	R
NORDEL	R
POLYBUTADIENE	R
SILICONE	
POLYURETHANE	

TEMP °F: 60, 80, 100, 120, 140, 160, 180, 200, 220, 240, 260, 280, 300, 320, 340, 360, 380, 400, 420, 440, 460

AMMONIA (ANHYDROUS)

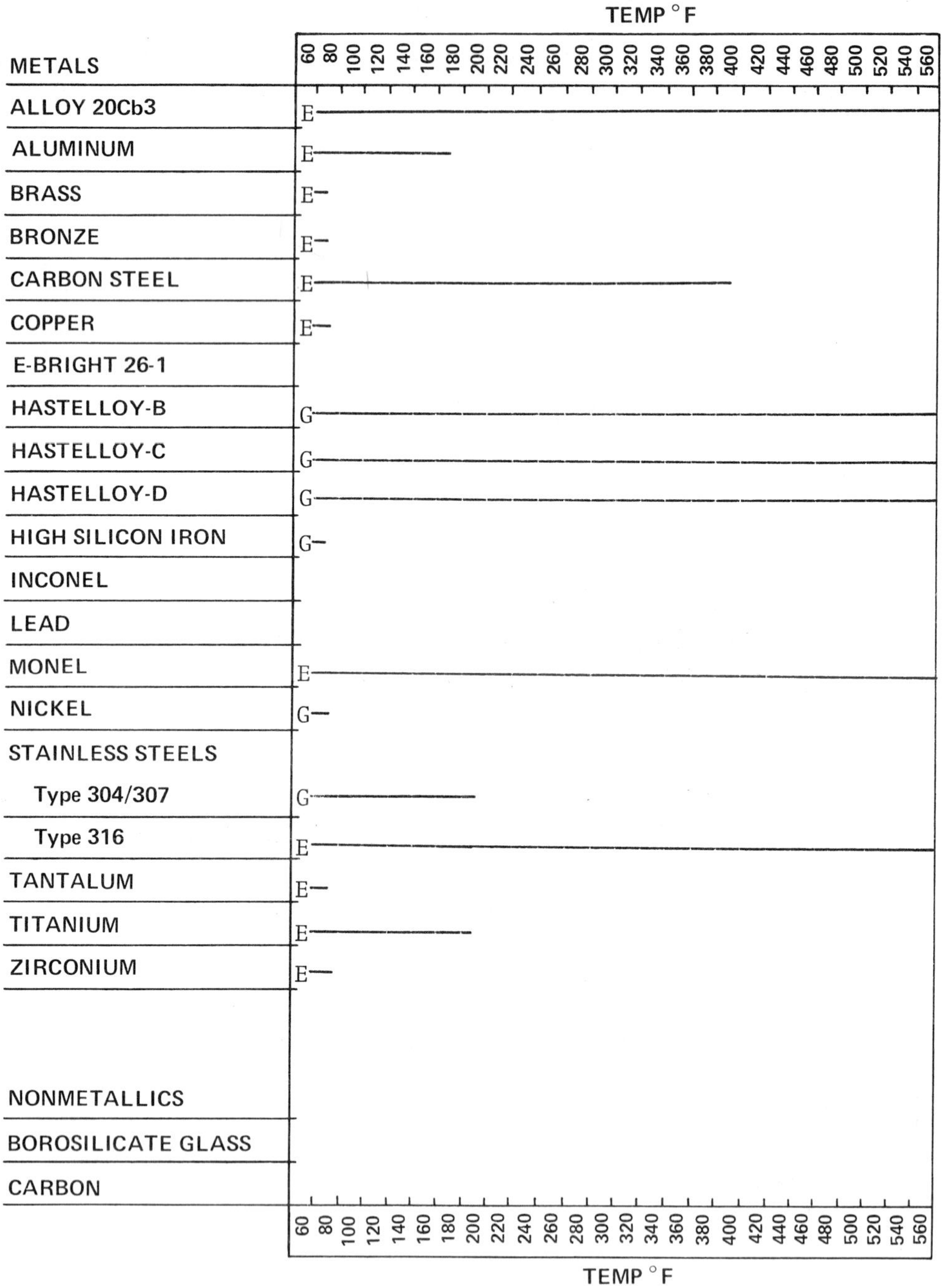

FOR METALS

E = < 2 Mils Penetration/Year; G = < 20 Mils Penetration/Year

S = < 50 Mils Penetration/Year; U = > 50 Mils Penetration/Year

FOR NONMETALLICS

R = Resistant

U = Unsatisfactory

AMMONIA (ANHYDROUS)

TEMP °F: 60 80 100 120 140 160 180 200 220 240 260 280 300 320 340 360 380 400 420 440 460

PLASTICS	
ABS	U
ACRYLICS	R
ASBESTOS REINF., EPOXY	
ASBESTOS REINF., PHENOLIC	
CHLORINATED POLYETHER	R
CPVC	R
EPOXY	U
FLUOROCARBONS FEP	R
FLUOROCARBONS TFE	R
FURFURYL ALCOHOL	R
NORYL	
PHENOLIC	
POLYAMIDES—NYLON	R
POLYESTERS	
BISPHENOL A-FUMURATE	
HYDROGENATED BISPHENOL-A-BISPHENOL-A	
ISOPHTHALIC	U
CHLORINATED POLYESTERS	
POLYSTYRENES	U
POLYETHYLENE	R
POLYPROPYLENE	R
PVC—TYPE 1	R
PVC—TYPE 2	R
POLYVINYLIDENE CHLORIDE	
VINYLIDENE FLUORIDE	R
VINYL ESTER	
RUBBERS	
BUTYL GR-1	U
FLUORO ELASTOMERS VITON A	U
FLUORO ELASTOMERS KEL-F 3700	
HARD RUBBER	U
HYPALON	R
KOROSEAL	
NATURAL RUBBER (GRS)	U
NEOPRENE GR-M (CR)	R
NITRILE BUNA N (NBR)	R
NORDEL	R
POLYBUTADIENE	
SILICONE	
POLYURETHANE	

TEMP °F: 60 80 100 120 140 160 180 200 220 240 260 280 300 320 340 360 380 400 420 440 460

AMMONIA GAS

TEMP °F (60, 80, 100, 120, 140, 160, 180, 200, 220, 240, 260, 280, 300, 320, 340, 360, 380, 400, 420, 440, 460, 480, 500, 520, 540, 560)

METALS	TEMP °F
ALLOY 20Cb3	E–
ALUMINUM	
BRASS	U
BRONZE	U
CARBON STEEL	E——
COPPER	U
E-BRIGHT 26-1	
HASTELLOY-B	E————————
HASTELLOY-C	E————————
HASTELLOY-D	E—
HIGH SILICON IRON	
INCONEL	
LEAD	
MONEL	U
NICKEL	E—
STAINLESS STEELS	
Type 304/307	E—
Type 316	E—
TANTALUM	
TITANIUM	
ZIRCONIUM	G——
NONMETALLICS	
BOROSILICATE GLASS	
CARBON	R————————————————

TEMP °F (60, 80, 100, 120, 140, 160, 180, 200, 220, 240, 260, 280, 300, 320, 340, 360, 380, 400, 420, 440, 460, 480, 500, 520, 540, 560)

FOR METALS

E = < 2 Mils Penetration/Year; G = < 20 Mils Penetration/Year

S = < 50 Mils Penetration/Year; U = > 50 Mils Penetration/Year

FOR NONMETALLICS

R = Resistant

U = Unsatisfactory

AMMONIA GAS

TEMP °F: 60 80 100 120 140 160 180 200 220 240 260 280 300 320 340 360 380 400 420 440 460

PLASTICS		
ABS	dry	R—
ACRYLICS		R—
ASBESTOS REINF., EPOXY		R—
ASBESTOS REINF., PHENOLIC		R—
CHLORINATED POLYETHER		R—
CPVC	dry	R—
EPOXY	dry	R—U
FLUOROCARBONS FEP		R—
FLUOROCARBONS TFE		R—
FURFURYL ALCOHOL		
NORYL		
PHENOLIC		R—
POLYAMIDES—NYLON		R—
POLYESTERS		
BISPHENOL A-FUMURATE		R—
HYDROGENATED BISPHENOL-A-BISPHENOL-A		
ISOPHTHALIC		R—
CHLORINATED POLYESTERS		
POLYSTYRENES		U
POLYETHYLENE		R—
POLYPROPYLENE		R—
PVC—TYPE 1		R—
PVC—TYPE 2		R—
POLYVINYLIDENE CHLORIDE		U
VINYLIDENE FLUORIDE		R—
VINYL ESTER		R—

RUBBERS	
BUTYL GR-1	
FLUORO ELASTOMERS VITON A	U
FLUORO ELASTOMERS KEL-F 3700	R—
HARD RUBBER	
HYPALON	R—
KOROSEAL	
NATURAL RUBBER (GRS)	U
NEOPRENE GR-M (CR)	R
NITRILE BUNA N (NBR)	R—
NORDEL	
POLYBUTADIENE	R—
SILICONE	
POLYURETHANE	

TEMP °F: 60 80 100 120 140 160 180 200 220 240 260 280 300 320 340 360 380 400 420 440 460

AMMONIUM ACETATE

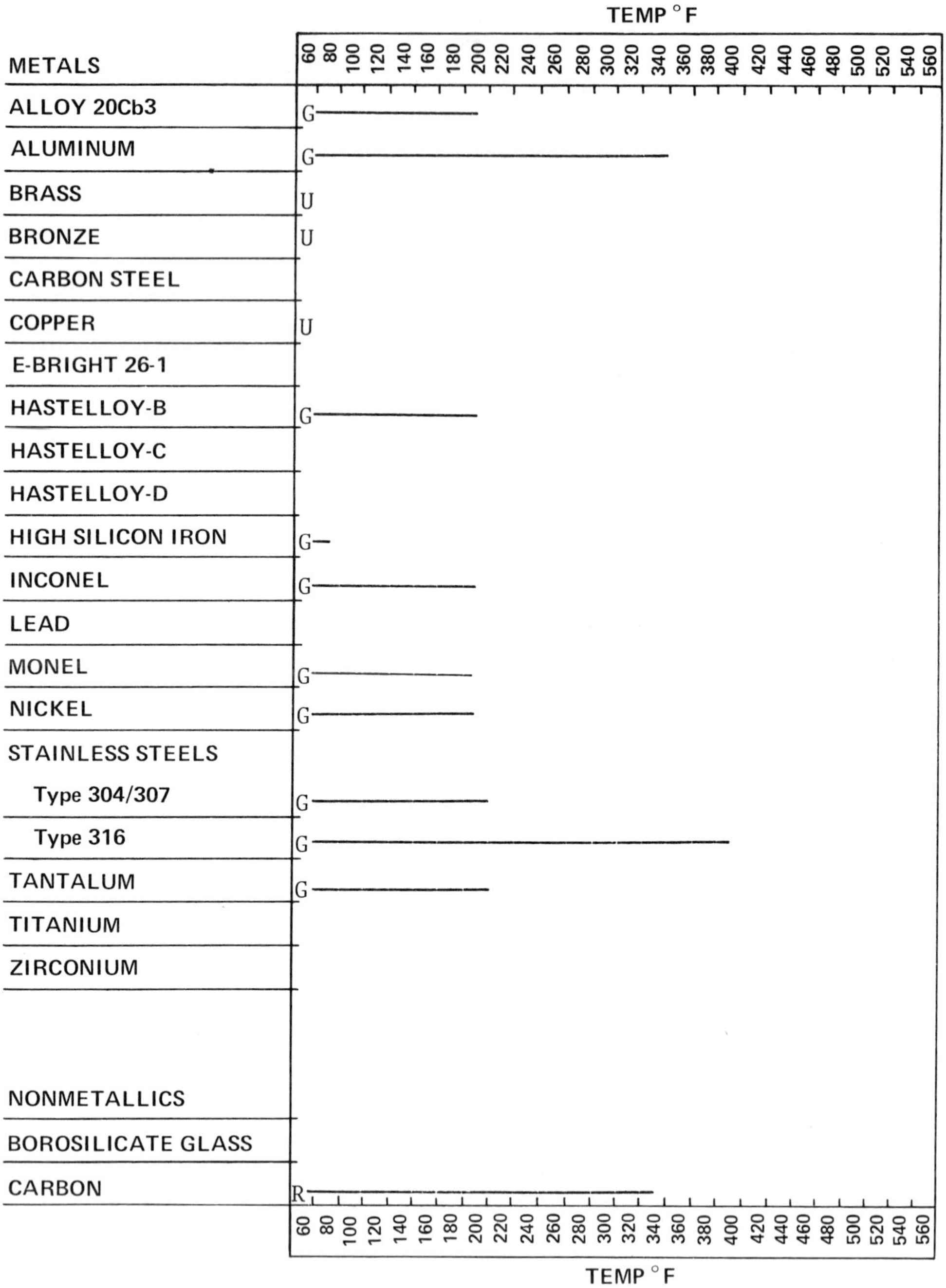

FOR METALS

E = < 2 Mils Penetration/Year; G = < 20 Mils Penetration/Year

S = < 50 Mils Penetration/Year; U = > 50 Mils Penetration/Year

FOR NONMETALLICS

R = Resistant

U = Unsatisfactory

AMMONIUM ACETATE

TEMP °F (60–460)

PLASTICS		Rating
ABS		
ACRYLICS		
ASBESTOS REINF., EPOXY		
ASBESTOS REINF., PHENOLIC		
CHLORINATED POLYETHER		R
CPVC		R
EPOXY		
FLUOROCARBONS FEP		R
FLUOROCARBONS TFE		R
FURFURYL ALCOHOL		
NORYL		
PHENOLIC		
POLYAMIDES–NYLON		
POLYESTERS		
BISPHENOL A-FUMURATE	65%	R
HYDROGENATED BISPHENOL-A-BISPHENOL-A		
ISOPHTHALIC		U
CHLORINATED POLYESTERS		
POLYSTYRENES		
POLYETHYLENE		R
POLYPROPYLENE		R
PVC–TYPE 1		R
PVC–TYPE 2		R
POLYVINYLIDENE CHLORIDE		R
VINYLIDENE FLUORIDE		R
VINYL ESTER	65%	R

RUBBERS	Rating
BUTYL GR-1	
FLUORO ELASTOMERS VITON A	RU
FLUORO ELASTOMERS KEL-F 3700	
HARD RUBBER	
HYPALON	
KOROSEAL	
NATURAL RUBBER (GRS)	
NEOPRENE GR-M (CR)	
NITRILE BUNA N (NBR)	U
NORDEL	R
POLYBUTADIENE	
SILICONE	
POLYURETHANE	

TEMP °F (60–460)

AMMONIUM BICARBONATE

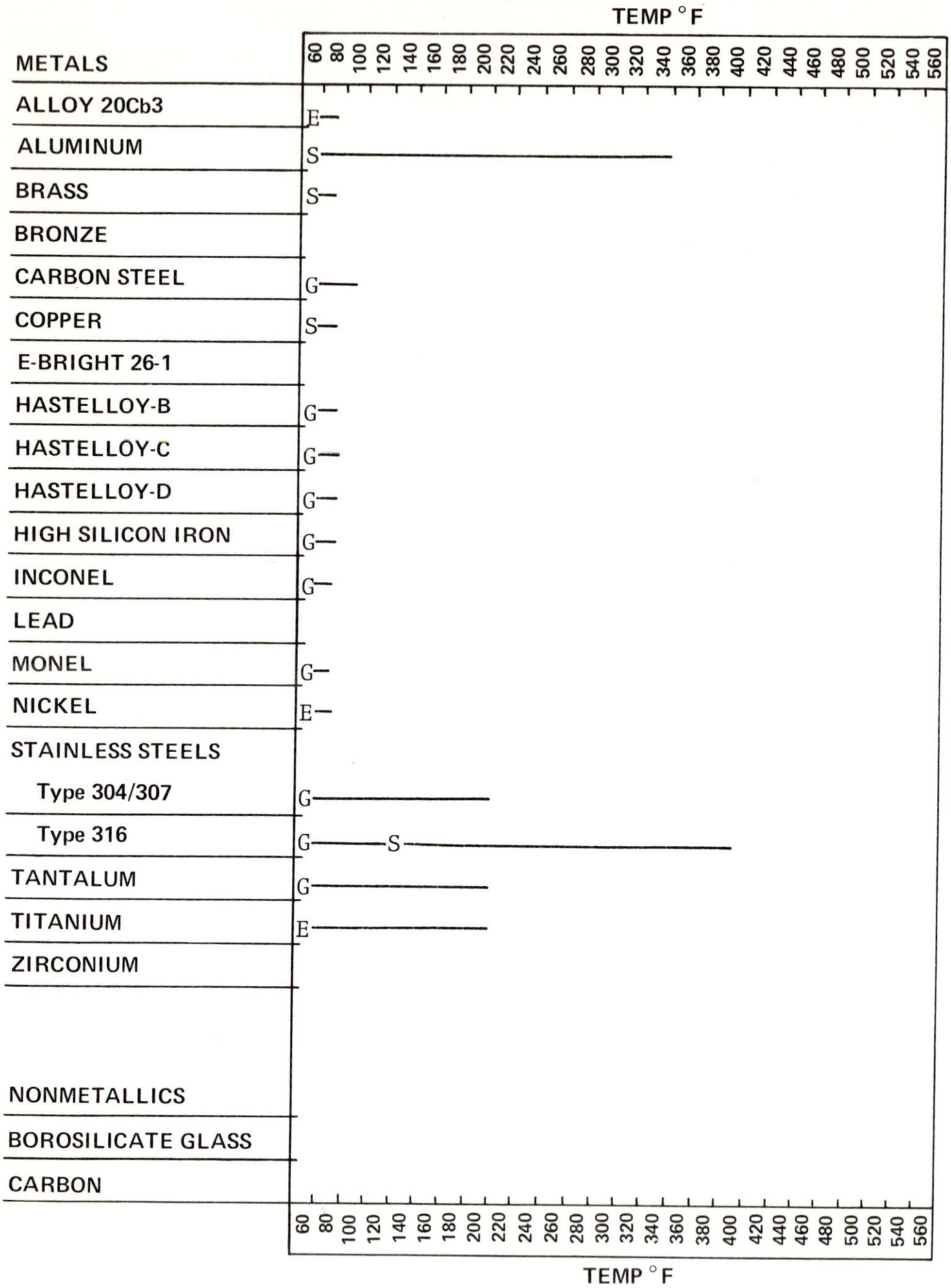

FOR METALS

E = < 2 Mils Penetration/Year; G = < 20 Mils Penetration/Year

S = < 50 Mils Penetration/Year; U = > 50 Mils Penetration/Year

FOR NONMETALLICS

R = Resistant

U = Unsatisfactory

AMMONIUM BICARBONATE

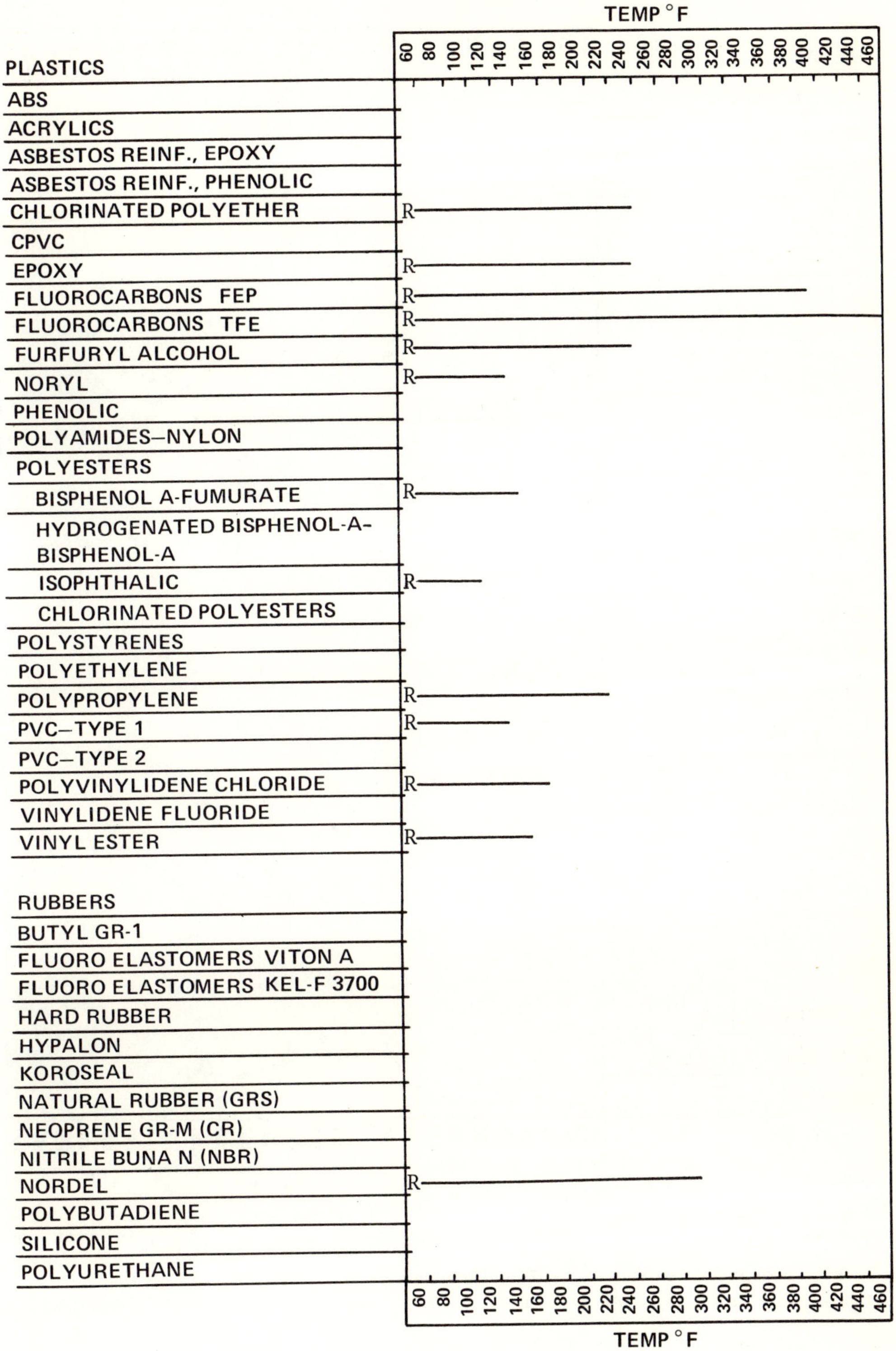

AMMONIUM BIFLUORIDE

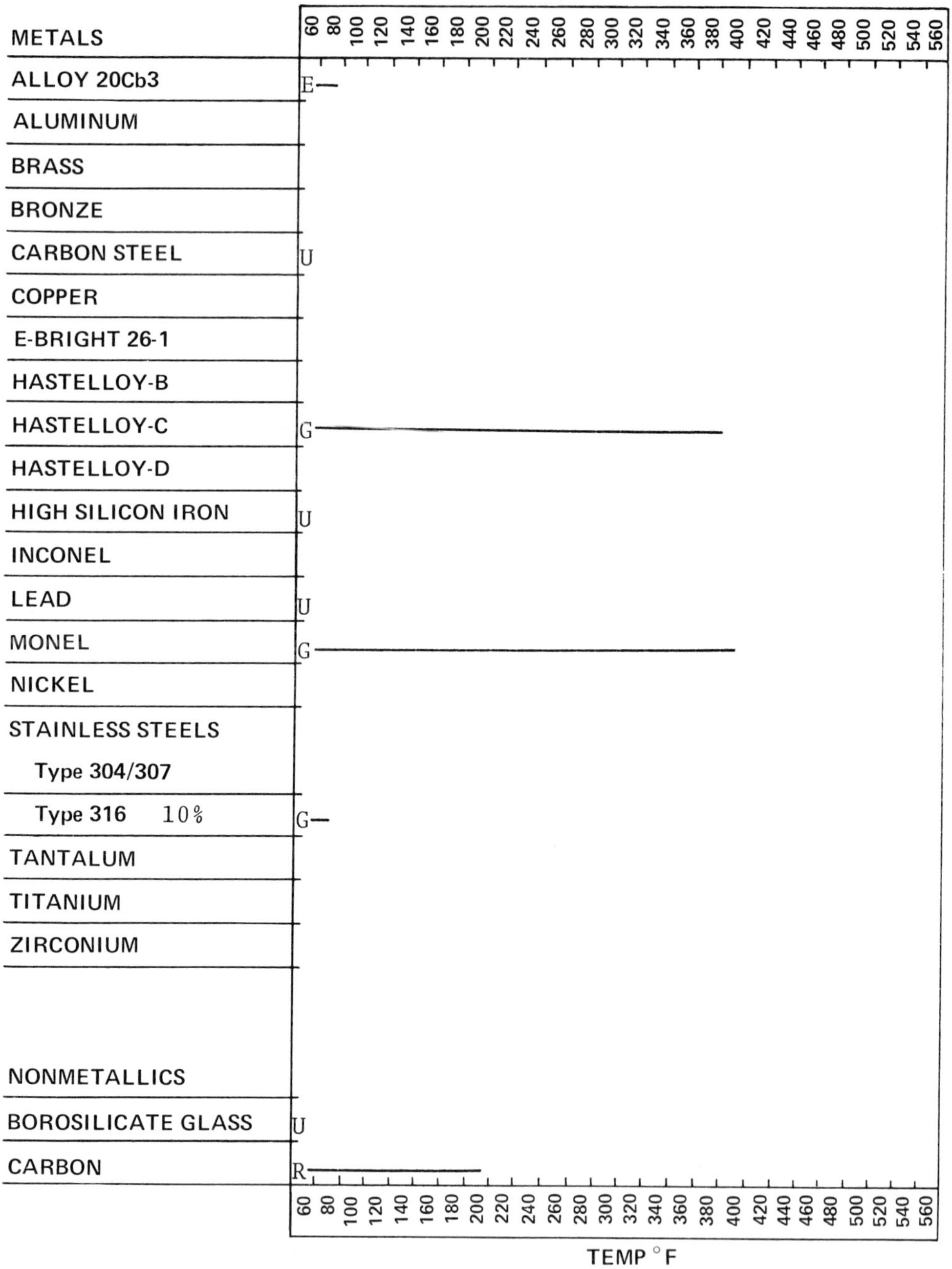

FOR METALS

E = < 2 Mils Penetration/Year; G = < 20 Mils Penetration/Year

S = < 50 Mils Penetration/Year; U = > 50 Mils Penetration/Year

FOR NONMETALLICS

R = Resistant

U = Unsatisfactory

AMMONIUM BIFLUORIDE

TEMP °F (scale 60–460)

PLASTICS	Rating	Temperature range on chart (°F, approx.)
ABS	R	60–140
ACRYLICS		
ASBESTOS REINF., EPOXY		
ASBESTOS REINF., PHENOLIC		
CHLORINATED POLYETHER	R	60–220
CPVC	R–U	R 60–140, U at 150
EPOXY		
FLUOROCARBONS FEP	R	60–400
FLUOROCARBONS TFE	R	60–460
FURFURYL ALCOHOL		
NORYL		
PHENOLIC		
POLYAMIDES–NYLON		
POLYESTERS		
BISPHENOL A-FUMURATE		
HYDROGENATED BISPHENOL-A-BISPHENOL-A		
ISOPHTHALIC		
CHLORINATED POLYESTERS		
POLYSTYRENES		
POLYETHYLENE	R	60–140
POLYPROPYLENE	R	60–210
PVC–TYPE 1	R–U	
PVC–TYPE 2	R–U	
POLYVINYLIDENE CHLORIDE	R	60–150
VINYLIDENE FLUORIDE	R	60–280
VINYL ESTER		

RUBBERS	Rating	Temperature range on chart (°F, approx.)
BUTYL GR-1	U	
FLUORO ELASTOMERS VITON A		
FLUORO ELASTOMERS KEL-F 3700		
HARD RUBBER	R	60–200
HYPALON		
KOROSEAL		
NATURAL RUBBER (GRS)	U	
NEOPRENE GR-M (CR)	U	
NITRILE BUNA N (NBR) 10%	R	60
NORDEL	R	60–300
POLYBUTADIENE		
SILICONE		
POLYURETHANE		

TEMP °F

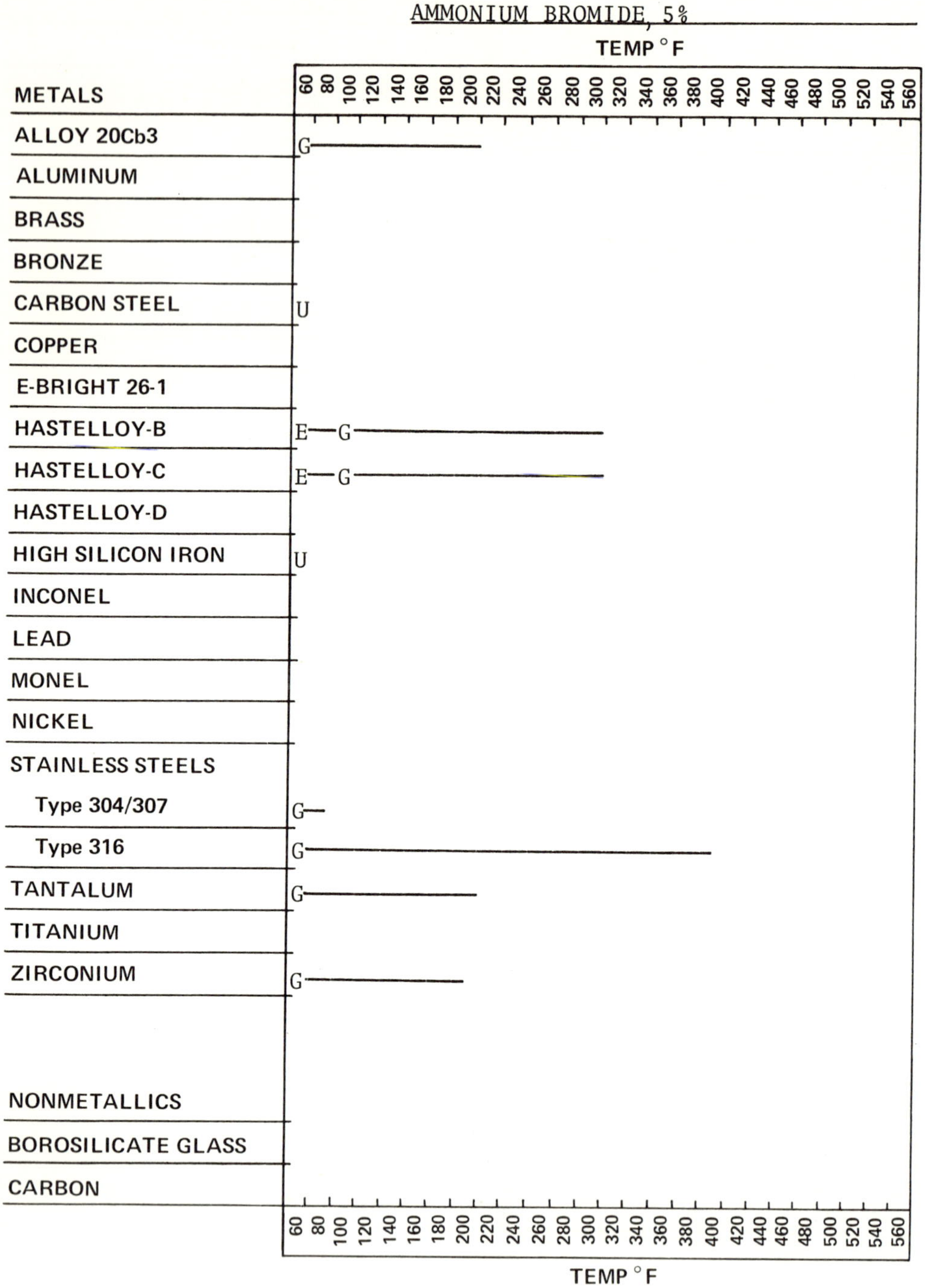

FOR METALS

E = < 2 Mils Penetration/Year; G = < 20 Mils Penetration/Year

S = < 50 Mils Penetration/Year; U = > 50 Mils Penetration/Year

FOR NONMETALLICS

R = Resistant

U = Unsatisfactory

AMMONIUM BROMIDE, 5%

TEMP °F (scale 60–460)

PLASTICS	TEMP °F
ABS	
ACRYLICS	
ASBESTOS REINF., EPOXY	
ASBESTOS REINF., PHENOLIC	
CHLORINATED POLYETHER	R (60–250)
CPVC	
EPOXY	
FLUOROCARBONS FEP	R (60–400)
FLUOROCARBONS TFE	R (60–460)
FURFURYL ALCOHOL	R (60–270)
NORYL	
PHENOLIC	
POLYAMIDES–NYLON	
POLYESTERS	
BISPHENOL A-FUMURATE	
HYDROGENATED BISPHENOL-A-BISPHENOL-A	
ISOPHTHALIC	
CHLORINATED POLYESTERS	
POLYSTYRENES	
POLYETHYLENE	
POLYPROPYLENE	
PVC–TYPE 1	R (60–140)
PVC–TYPE 2	
POLYVINYLIDENE CHLORIDE	R (60–180)
VINYLIDENE FLUORIDE	R (60–300)
VINYL ESTER	
RUBBERS	
BUTYL GR-1	
FLUORO ELASTOMERS VITON A	
FLUORO ELASTOMERS KEL-F 3700	
HARD RUBBER	
HYPALON	
KOROSEAL	
NATURAL RUBBER (GRS)	
NEOPRENE GR-M (CR)	
NITRILE BUNA N (NBR)	
NORDEL	R (60–300)
POLYBUTADIENE	
SILICONE	
POLYURETHANE	

TEMP °F

AMMONIUM CARBONATE (SAT)

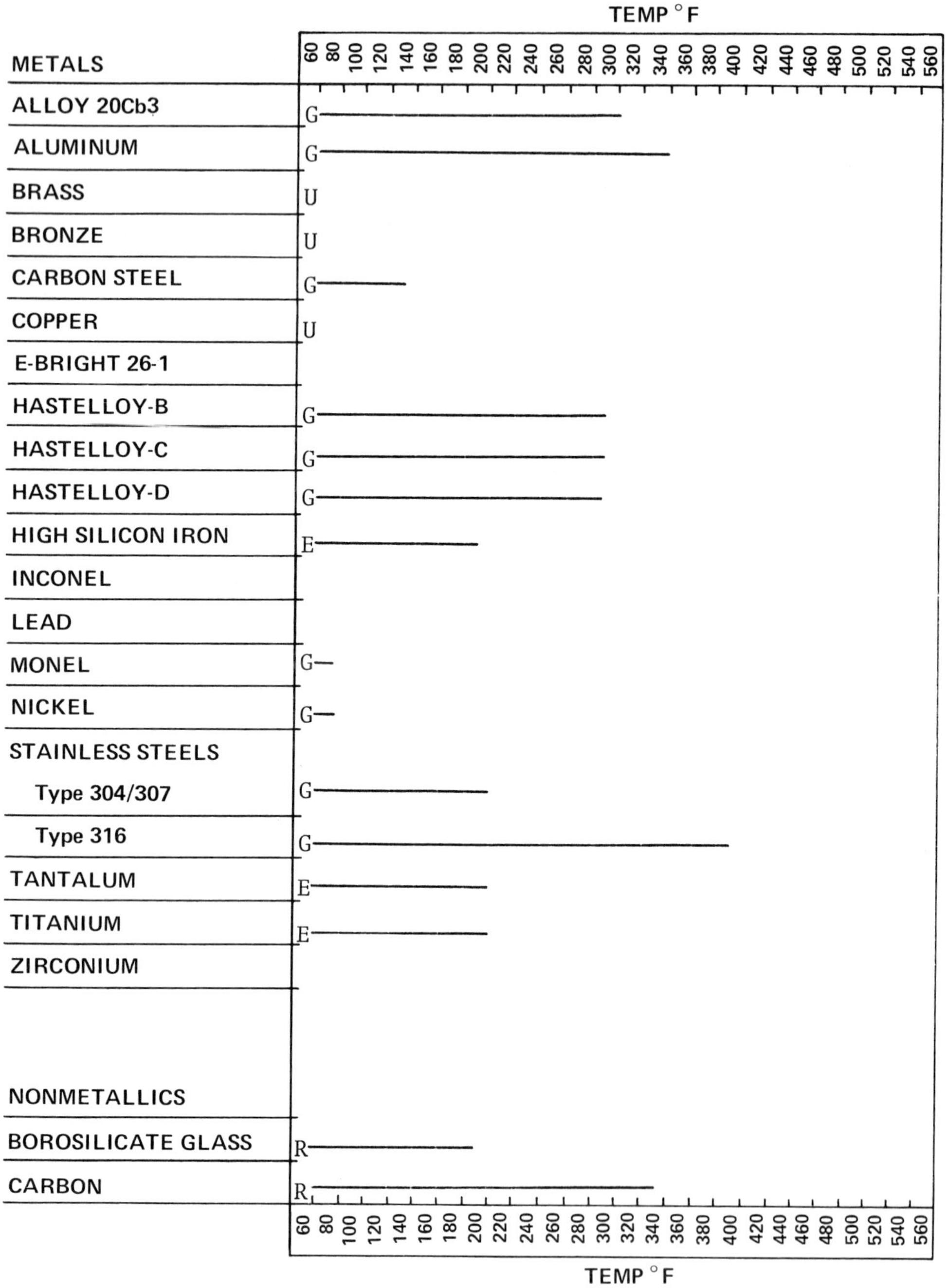

FOR METALS

E = < 2 Mils Penetration/Year; G = < 20 Mils Penetration/Year

S = < 50 Mils Penetration/Year; U = > 50 Mils Penetration/Year

FOR NONMETALLICS

R = Resistant

U = Unsatisfactory

AMMONIUM CARBONATE (SAT)

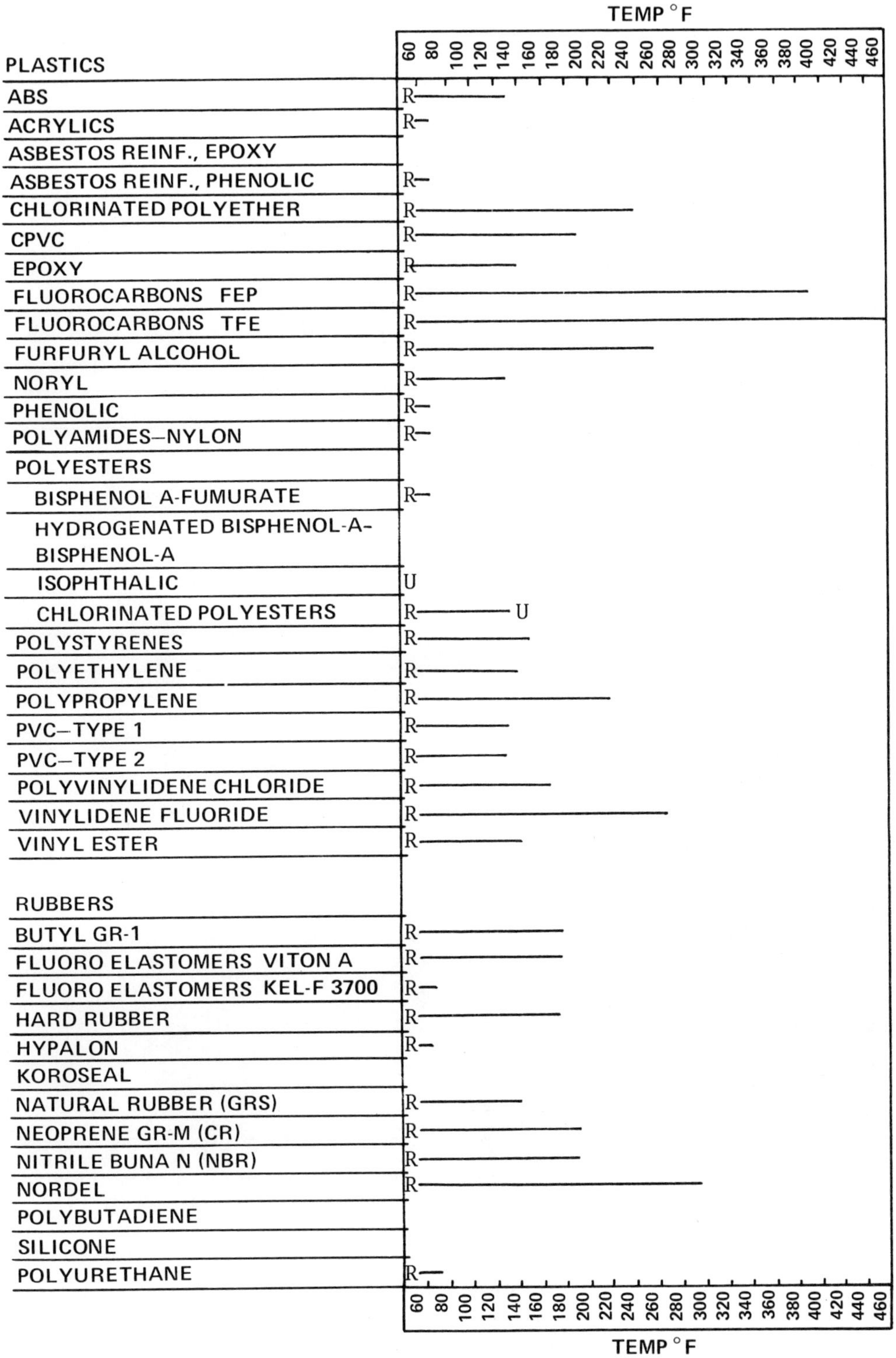

AMMONIUM CHLORIDE (SAT)

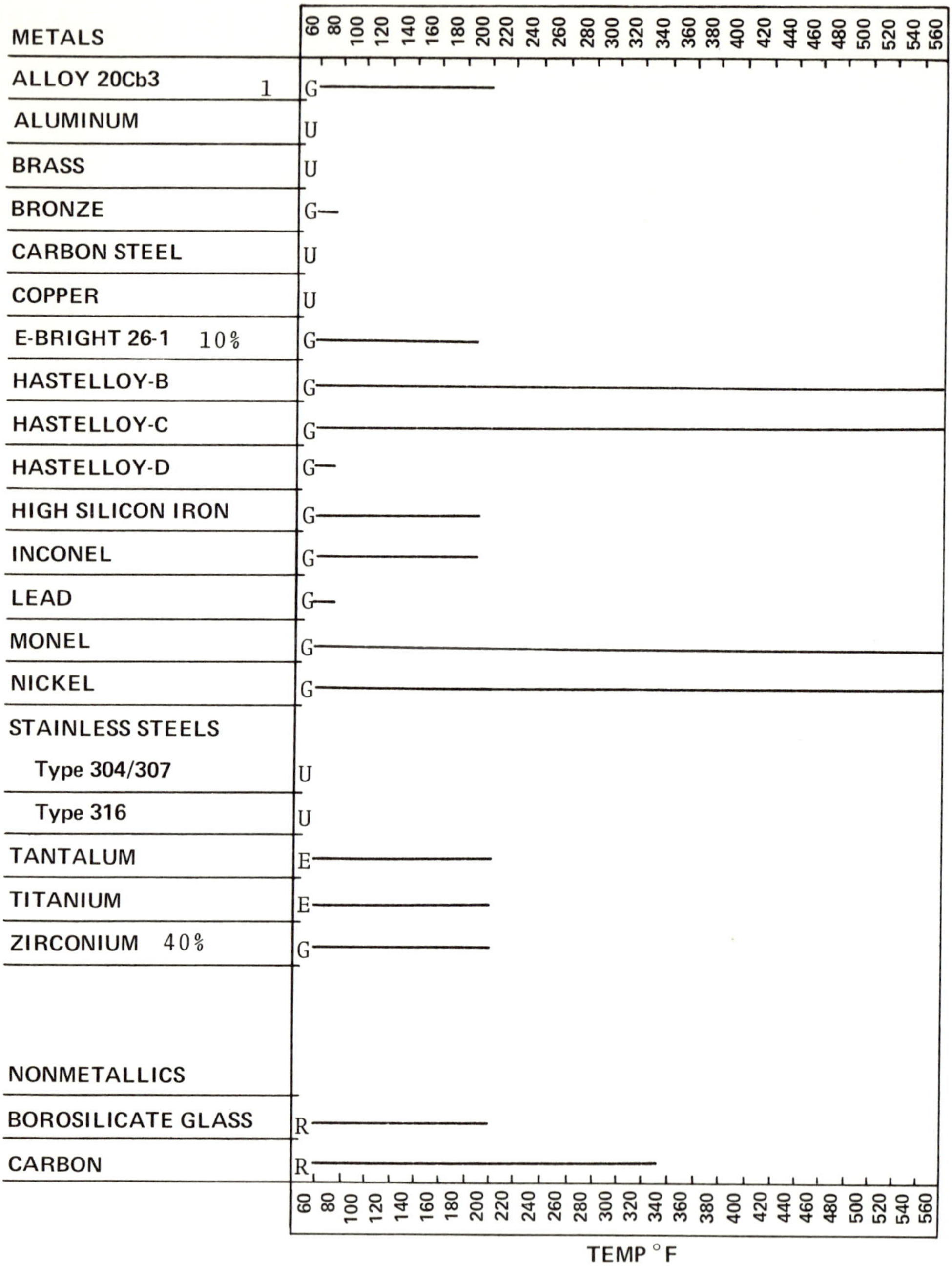

FOR METALS

E = < 2 Mils Penetration/Year; G = < 20 Mils Penetration/Year

S = < 50 Mils Penetration/Year; U = > 50 Mils Penetration/Year

FOR NONMETALLICS

R = Resistant

U = Unsatisfactory

AMMONIUM CHLORIDE (SAT)

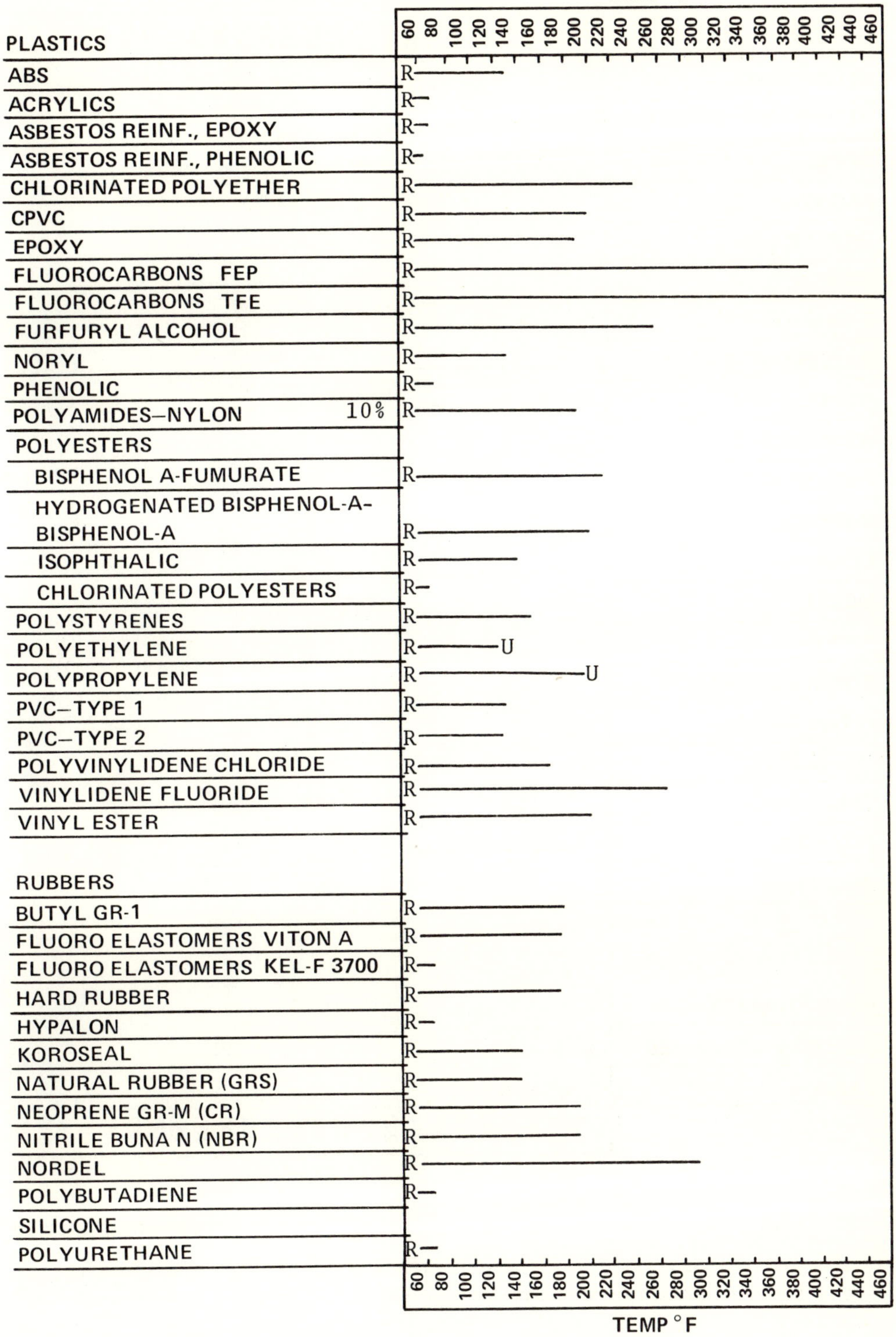

AMMONIUM CITRATE

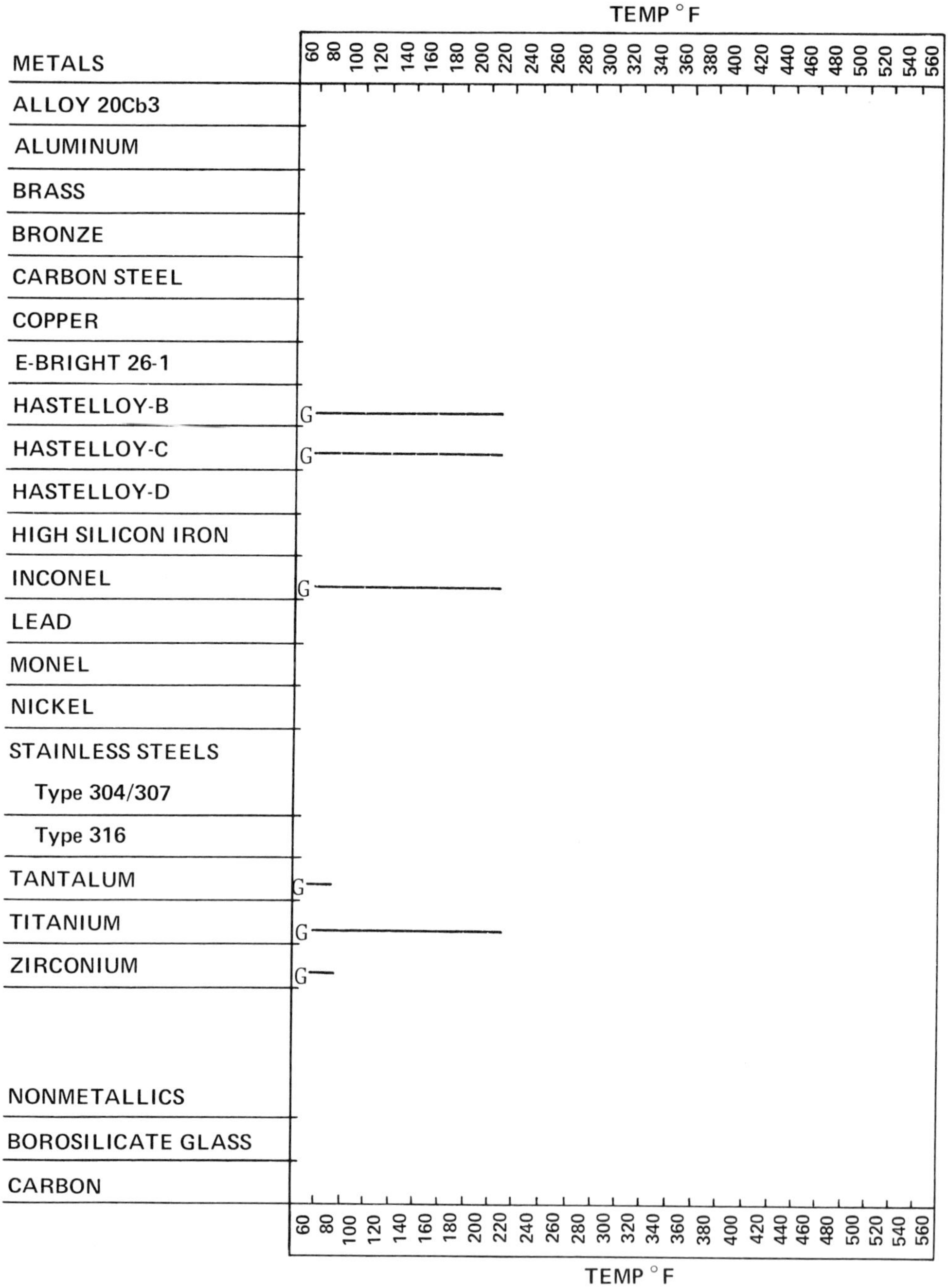

FOR METALS

E = < 2 Mils Penetration/Year; G = < 20 Mils Penetration/Year

S = < 50 Mils Penetration/Year; U = > 50 Mils Penetration/Year

FOR NONMETALLICS

R = Resistant

U = Unsatisfactory

AMMONIUM CITRATE

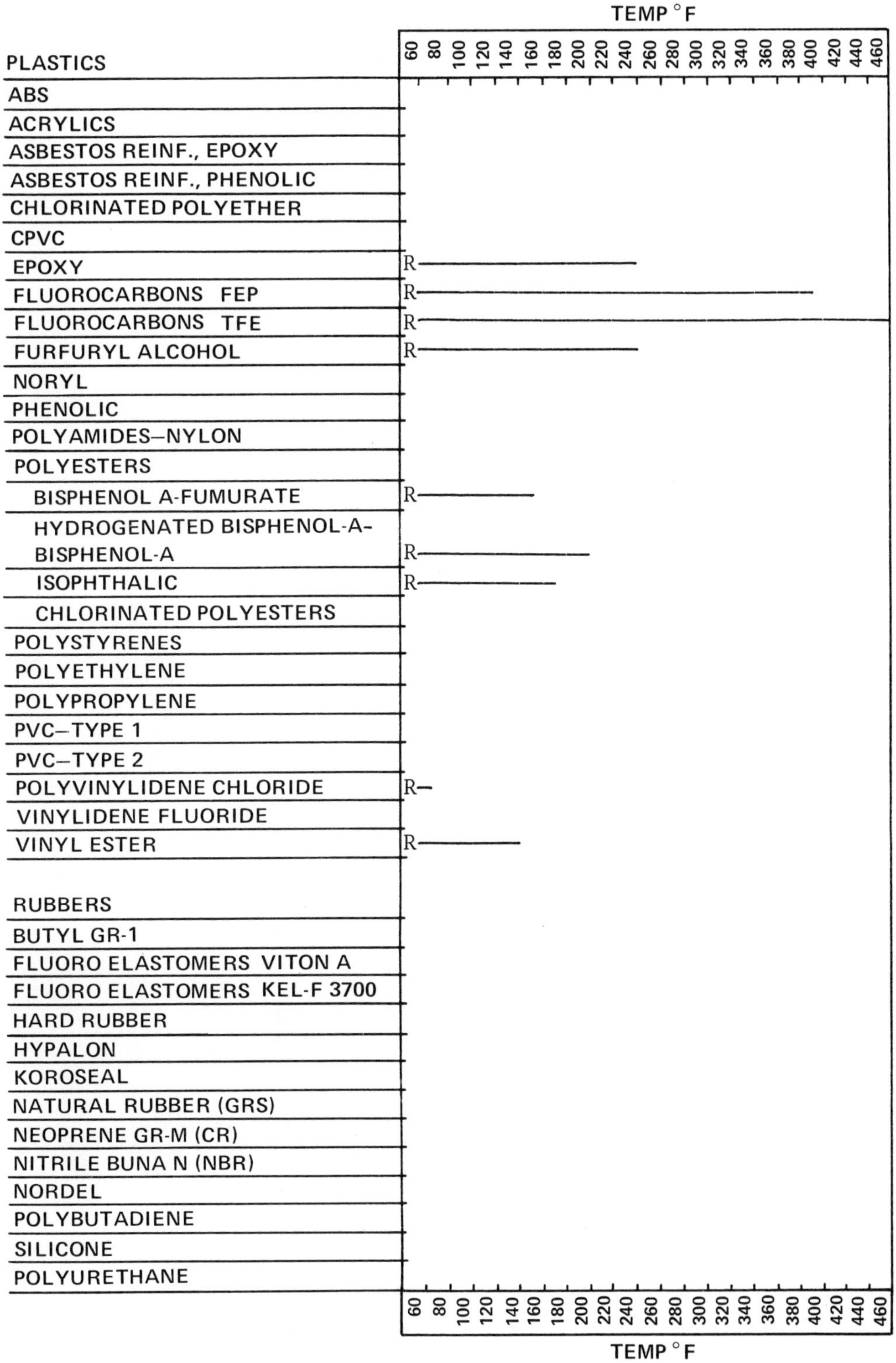

AMMONIUM FLUORIDE, 25%

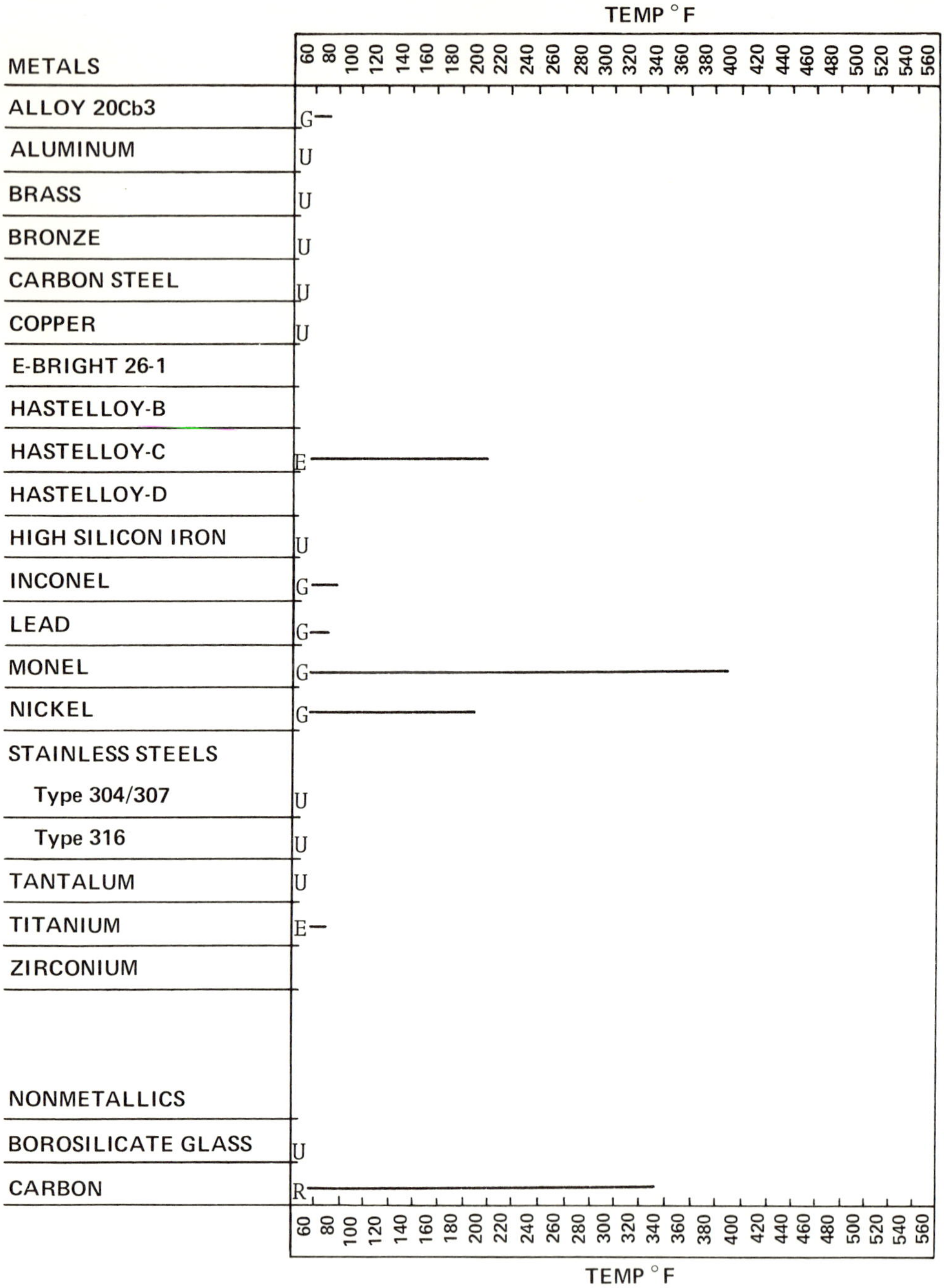

FOR METALS

E = < 2 Mils Penetration/Year; G = < 20 Mils Penetration/Year

S = < 50 Mils Penetration/Year; U = > 50 Mils Penetration/Year

FOR NONMETALLICS

R = Resistant

U = Unsatisfactory

AMMONIUM FLUORIDE, 25%

TEMP °F (scale 60–460)

PLASTICS		Rating (approx. temperature range, °F)
ABS		U (at 60)
ACRYLICS		
ASBESTOS REINF., EPOXY		
ASBESTOS REINF., PHENOLIC		
CHLORINATED POLYETHER		R (60 to ~250)
CPVC		R (60 to ~200)
EPOXY		R (60 to ~150)
FLUOROCARBONS FEP		R (60 to ~400)
FLUOROCARBONS TFE		R (60 to 460)
FURFURYL ALCOHOL		R (60 to ~270)
NORYL		R (60 to ~140)
PHENOLIC		
POLYAMIDES–NYLON		
POLYESTERS		
BISPHENOL A-FUMURATE	10	R (60 to ~120)
HYDROGENATED BISPHENOL-A-BISPHENOL-A		
ISOPHTHALIC	10	R (60 to ~80)
CHLORINATED POLYESTERS	10	R (60 to ~150)
POLYSTYRENES		
POLYETHYLENE		R (60 to ~200), U (~200)
POLYPROPYLENE		R (60 to ~210), U (~220)
PVC–TYPE 1		R (60 to ~140)
PVC–TYPE 2		R (60 to ~75)
POLYVINYLIDENE CHLORIDE		R (60 to ~80)
VINYLIDENE FLUORIDE		R (60 to ~280)
VINYL ESTER	10	R (60 to ~150)

RUBBERS	Rating (approx. temperature range, °F)
BUTYL GR-1	R (60 to ~150)
FLUORO ELASTOMERS VITON A	
FLUORO ELASTOMERS KEL-F 3700	
HARD RUBBER	R (60 to ~210)
HYPALON	
KOROSEAL	
NATURAL RUBBER (GRS)	R (60 to ~80)
NEOPRENE GR-M (CR)	R (60 to ~200)
NITRILE BUNA N (NBR)	R (60 to ~125)
NORDEL	R (60 to ~300)
POLYBUTADIENE	
SILICONE	
POLYURETHANE	

TEMP °F

AMMONIUM FLUORIDE, 10%

TEMP °F: 60, 80, 100, 120, 140, 160, 180, 200, 220, 240, 260, 280, 300, 320, 340, 360, 380, 400, 420, 440, 460, 480, 500, 520, 540, 560

METALS	
ALLOY 20Cb3	E— U
ALUMINUM	U
BRASS	U
BRONZE	U
CARBON STEEL	U
COPPER	U
E-BRIGHT 26-1	
HASTELLOY-B	G
HASTELLOY-C	E
HASTELLOY-D	
HIGH SILICON IRON	U
INCONEL	G—
LEAD	G—
MONEL	G
NICKEL	G
STAINLESS STEELS	
Type 304/307	U
Type 316	G—U
TANTALUM	U
TITANIUM	G—
ZIRCONIUM	

NONMETALLICS	
BOROSILICATE GLASS	U
CARBON	R

TEMP °F: 60, 80, 100, 120, 140, 160, 180, 200, 220, 240, 260, 280, 300, 320, 340, 360, 380, 400, 420, 440, 460, 480, 500, 520, 540, 560

FOR METALS

E = < 2 Mils Penetration/Year; G = < 20 Mils Penetration/Year

S = < 50 Mils Penetration/Year; U = > 50 Mils Penetration/Year

FOR NONMETALLICS

R = Resistant

U = Unsatisfactory

AMMONIUM FLUORIDE, 10%

TEMP °F (60–460)

PLASTICS		
ABS		
ACRYLICS		
ASBESTOS REINF., EPOXY		
ASBESTOS REINF., PHENOLIC		
CHLORINATED POLYETHER		R —
CPVC		R —
EPOXY		
FLUOROCARBONS FEP		R —
FLUOROCARBONS TFE		R —
FURFURYL ALCOHOL		R —
NORYL		R —
PHENOLIC		
POLYAMIDES—NYLON		
POLYESTERS		
BISPHENOL A-FUMURATE	10	R —
HYDROGENATED BISPHENOL-A-BISPHENOL-A		
ISOPHTHALIC	10	R —
CHLORINATED POLYESTERS	10	R —
POLYSTYRENES		
POLYETHYLENE		R — U
POLYPROPYLENE		R — U
PVC—TYPE 1		R —
PVC—TYPE 2		R —
POLYVINYLIDENE CHLORIDE		R —
VINYLIDENE FLUORIDE		R —
VINYL ESTER		
RUBBERS		
BUTYL GR-1		R —
FLUORO ELASTOMERS VITON A		
FLUORO ELASTOMERS KEL-F 3700		
HARD RUBBER		R —
HYPALON		
KOROSEAL		
NATURAL RUBBER (GRS)		R —
NEOPRENE GR-M (CR)		R —
NITRILE BUNA N (NBR)		R —
NORDEL		R —
POLYBUTADIENE		
SILICONE		
POLYURETHANE		

TEMP °F (60–460)

AMMONIUM HYDROXIDE (SAT)

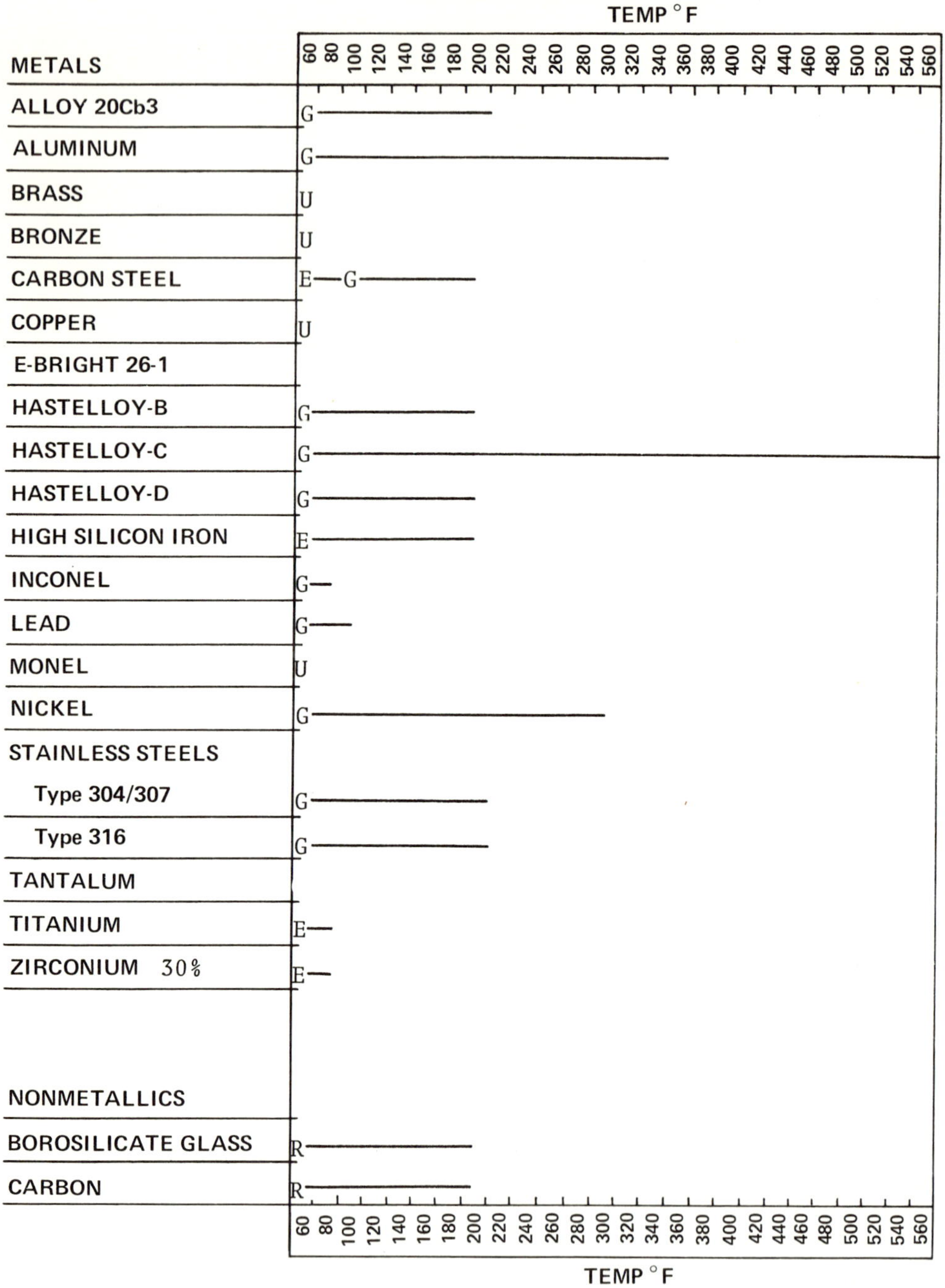

FOR METALS

E = < 2 Mils Penetration/Year; G = < 20 Mils Penetration/Year

S = < 50 Mils Penetration/Year; U = > 50 Mils Penetration/Year

FOR NONMETALLICS

R = Resistant

U = Unsatisfactory

AMMONIUM HYDROXIDE (SAT)

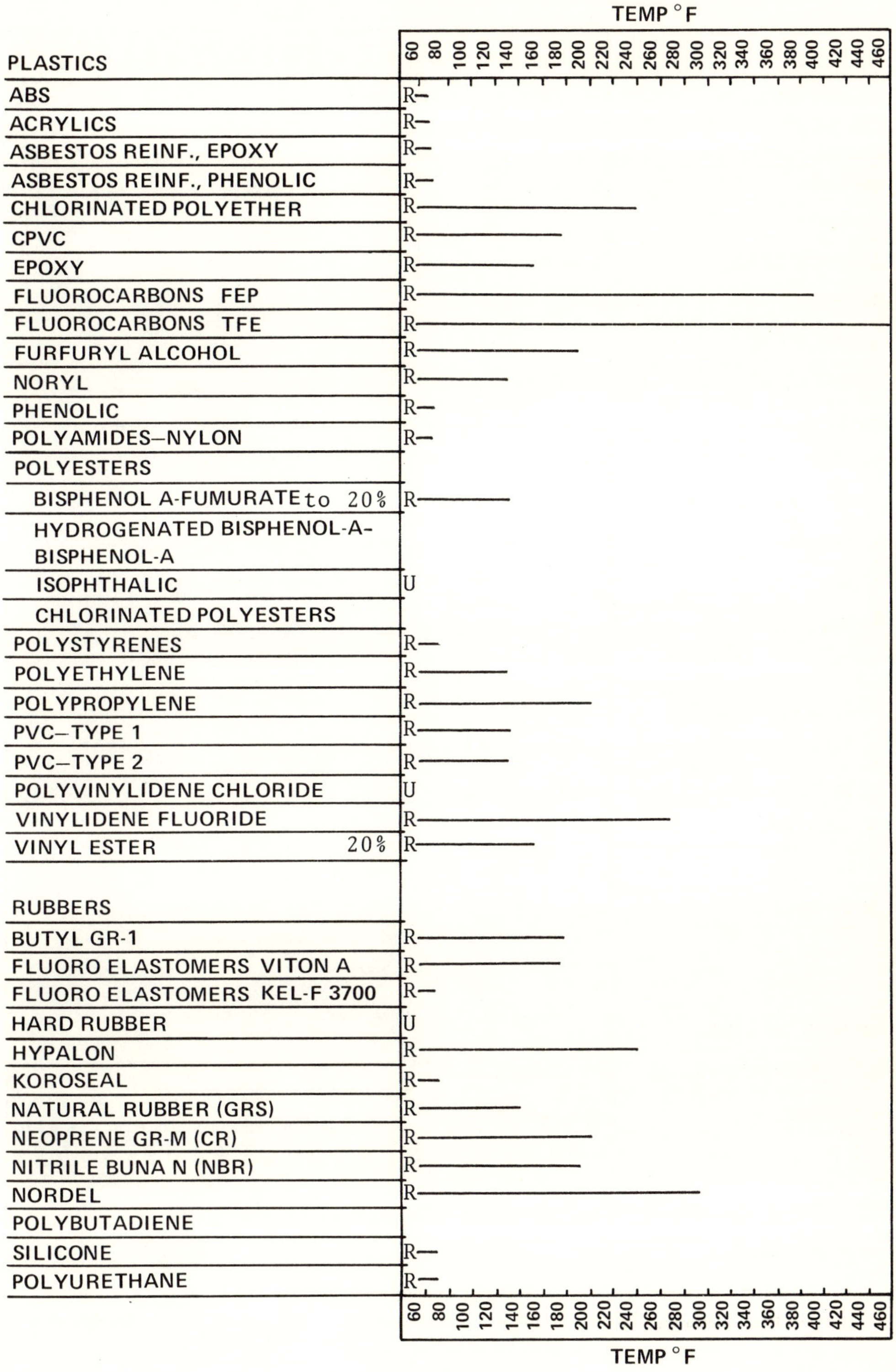

AMMONIUM HYDROXIDE, 25%

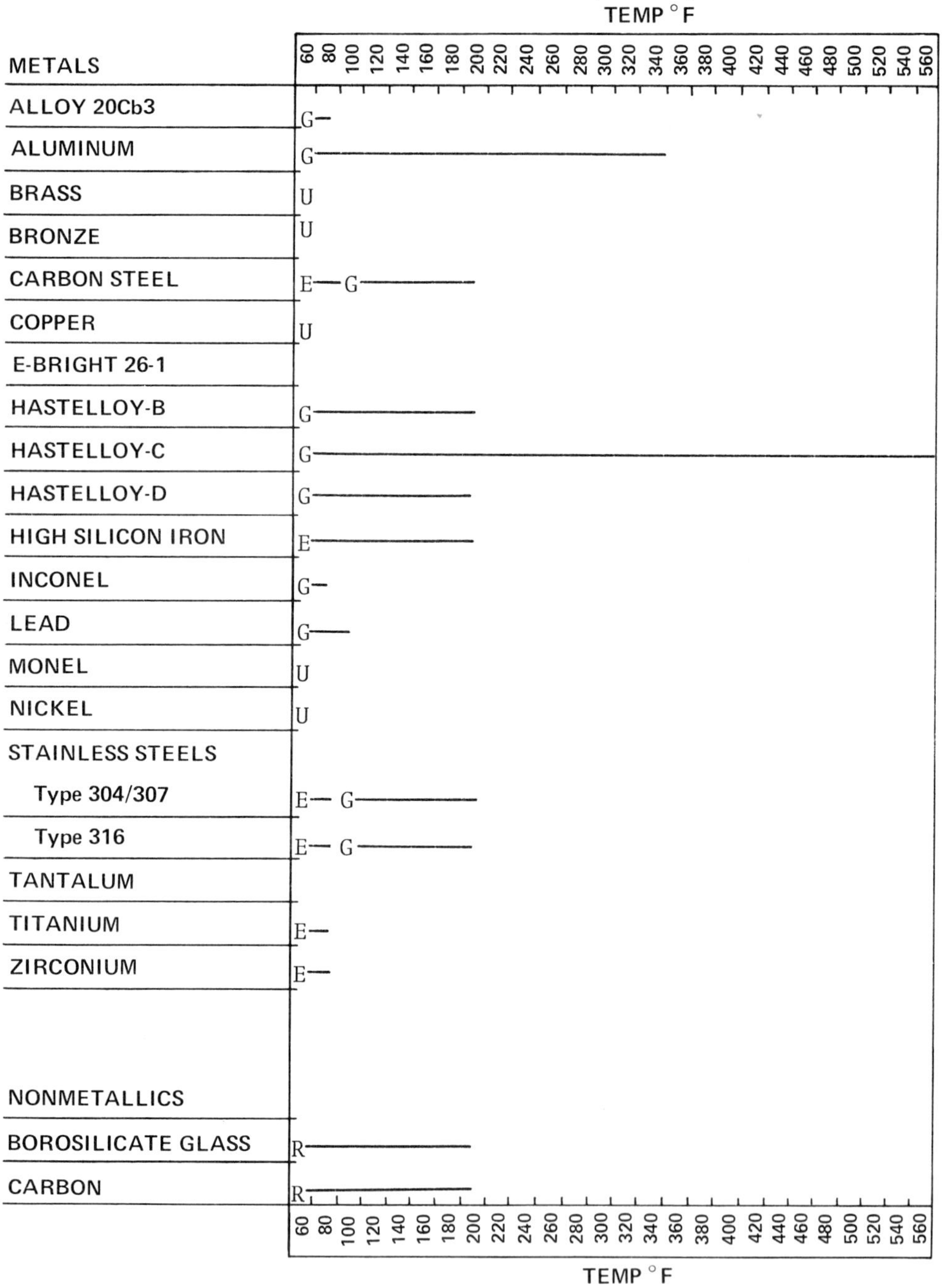

FOR METALS

E = < 2 Mils Penetration/Year; G = < 20 Mils Penetration/Year

S = < 50 Mils Penetration/Year; U = > 50 Mils Penetration/Year

FOR NONMETALLICS

R = Resistant

U = Unsatisfactory

AMMONIUM HYDROXIDE, 25%

TEMP °F (60–460)

PLASTICS	
ABS	R—
ACRYLICS	R—
ASBESTOS REINF., EPOXY	R—
ASBESTOS REINF., PHENOLIC	R–U
CHLORINATED POLYETHER	R—
CPVC	R—
EPOXY	R—U
FLUOROCARBONS FEP	R—
FLUOROCARBONS TFE	R—
FURFURYL ALCOHOL	R—
NORYL	R—
PHENOLIC	R—
POLYAMIDES–NYLON	R—
POLYESTERS	
BISPHENOL A-FUMURATE	R—
HYDROGENATED BISPHENOL-A-BISPHENOL-A	
ISOPHTHALIC	U
CHLORINATED POLYESTERS	R—
POLYSTYRENES	R—
POLYETHYLENE	R—U
POLYPROPYLENE	R—
PVC–TYPE 1	R—
PVC–TYPE 2	R—
POLYVINYLIDENE CHLORIDE	U
VINYLIDENE FLUORIDE	R—
VINYL ESTER	R—

RUBBERS	
BUTYL GR-1	R—
FLUORO ELASTOMERS VITON A	R—
FLUORO ELASTOMERS KEL-F 3700	R—
HARD RUBBER	U
HYPALON	R—
KOROSEAL	R—
NATURAL RUBBER (GRS)	U
NEOPRENE GR-M (CR)	R—
NITRILE BUNA N (NBR)	R—
NORDEL	R—
POLYBUTADIENE	
SILICONE	R—
POLYURETHANE	R—

TEMP °F (60–460)

AMMONIUM HYDROXIDE, 10%

TEMP °F: 60, 80, 100, 120, 140, 160, 180, 200, 220, 240, 260, 280, 300, 320, 340, 360, 380, 400, 420, 440, 460, 480, 500, 520, 540, 560

METALS	
ALLOY 20Cb3	G—
ALUMINUM	G—
BRASS	U
BRONZE	U
CARBON STEEL	E— G—
COPPER	U
E-BRIGHT 26-1	
HASTELLOY-B	G—
HASTELLOY-C	E— G—
HASTELLOY-D	G—
HIGH SILICON IRON	E—
INCONEL	G—
LEAD	G—
MONEL	U
NICKEL	U
STAINLESS STEELS	
Type 304/307	E— G—
Type 316	E— G—
TANTALUM	
TITANIUM	
ZIRCONIUM	E—

NONMETALLICS	
BOROSILICATE GLASS	R—
CARBON	R—

TEMP °F

FOR METALS

E = < 2 Mils Penetration/Year; G = < 20 Mils Penetration/Year

S = < 50 Mils Penetration/Year; U = > 50 Mils Penetration/Year

FOR NONMETALLICS

R = Resistant

U = Unsatisfactory

AMMONIUM HYDROXIDE, 10%

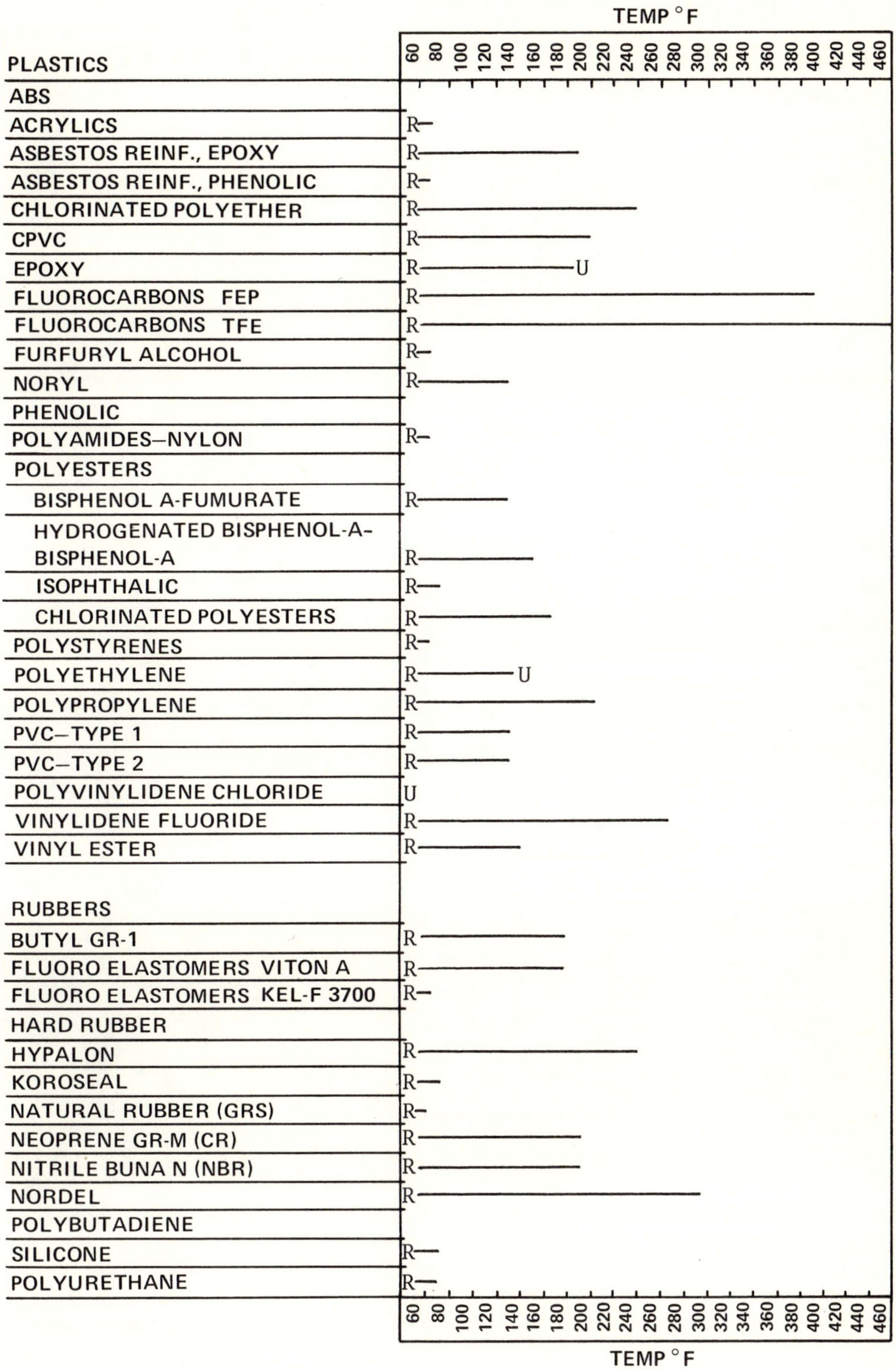

AMMONIUM METAPHOSPHATE

TEMP °F (60, 80, 100, 120, 140, 160, 180, 200, 220, 240, 260, 280, 300, 320, 340, 360, 380, 400, 420, 440, 460, 480, 500, 520, 540, 560)

METALS	
ALLOY 20Cb3	G—
ALUMINUM	G—
BRASS	U
BRONZE	U
CARBON STEEL	G—
COPPER	U
E-BRIGHT 26-1	
HASTELLOY-B	G—
HASTELLOY-C	G—
HASTELLOY-D	
HIGH SILICON IRON	G—
INCONEL	G—
LEAD	
MONEL	G—
NICKEL	G—
STAINLESS STEELS	
Type 304/307	G—
Type 316	G——
TANTALUM	G——————————
TITANIUM	
ZIRCONIUM	

NONMETALLICS	
BOROSILICATE GLASS	R——————————
CARBON	R————————————————————

TEMP °F (60, 80, 100, 120, 140, 160, 180, 200, 220, 240, 260, 280, 300, 320, 340, 360, 380, 400, 420, 440, 460, 480, 500, 520, 540, 560)

FOR METALS

E = < 2 Mils Penetration/Year; G = < 20 Mils Penetration/Year
S = < 50 Mils Penetration/Year; U = > 50 Mils Penetration/Year

FOR NONMETALLICS

R = Resistant
U = Unsatisfactory

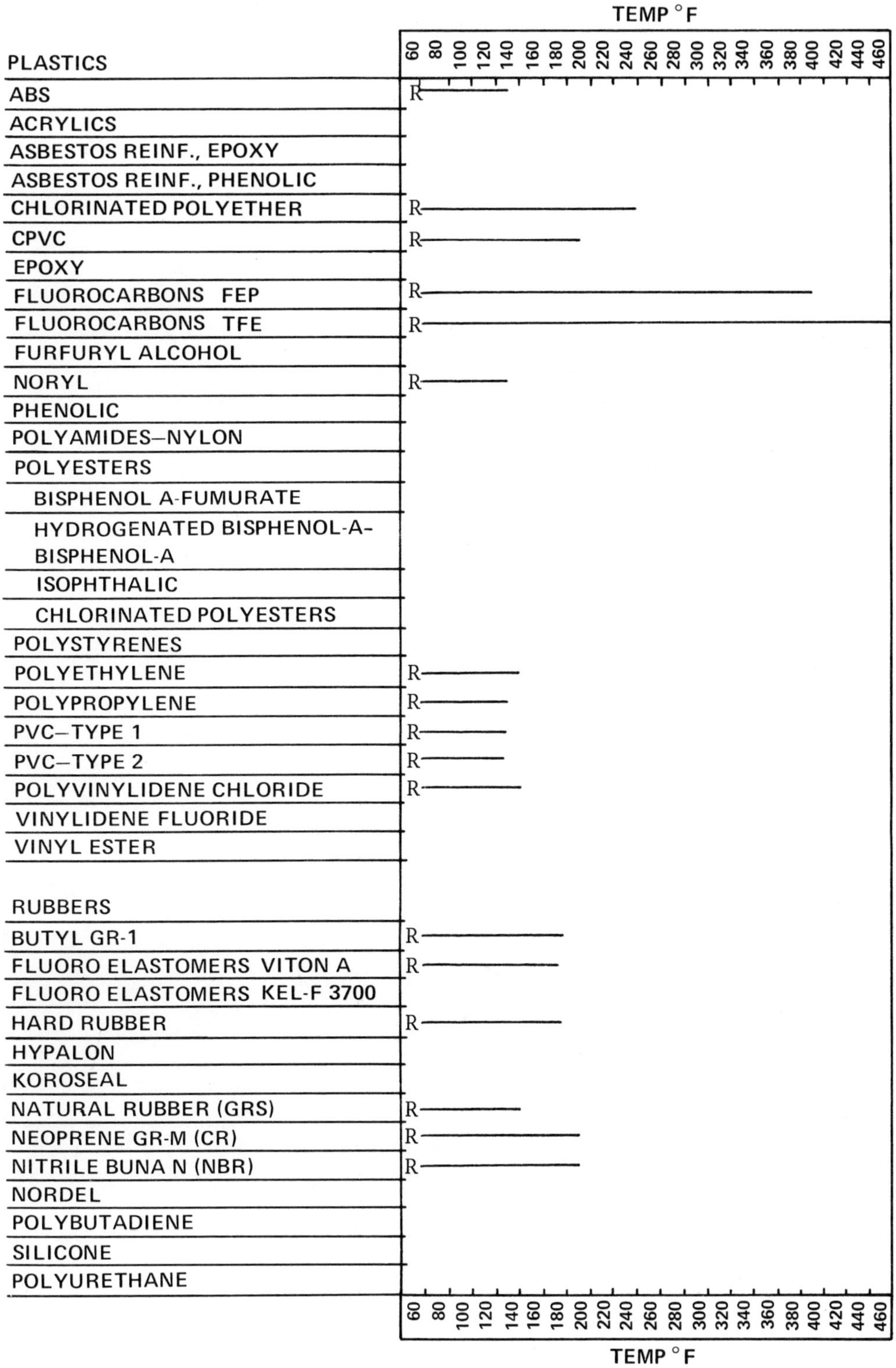
TEMP °F
60 80 100 120 140 160 180 200 220 240 260 280 300 320 340 360 380 400 420 440 460
PLASTICS
ABS R
ACRYLICS
ASBESTOS REINF., EPOXY
ASBESTOS REINF., PHENOLIC
CHLORINATED POLYETHER R
CPVC R
EPOXY
FLUOROCARBONS FEP R
FLUOROCARBONS TFE R
FURFURYL ALCOHOL
NORYL R
PHENOLIC
POLYAMIDES–NYLON
POLYESTERS
BISPHENOL A-FUMURATE
HYDROGENATED BISPHENOL-A-BISPHENOL-A
ISOPHTHALIC
CHLORINATED POLYESTERS
POLYSTYRENES
POLYETHYLENE R
POLYPROPYLENE R
PVC–TYPE 1 R
PVC–TYPE 2 R
POLYVINYLIDENE CHLORIDE R
VINYLIDENE FLUORIDE
VINYL ESTER
RUBBERS
BUTYL GR-1 R
FLUORO ELASTOMERS VITON A R
FLUORO ELASTOMERS KEL-F 3700
HARD RUBBER R
HYPALON
KOROSEAL
NATURAL RUBBER (GRS) R
NEOPRENE GR-M (CR) R
NITRILE BUNA N (NBR) R
NORDEL
POLYBUTADIENE
SILICONE
POLYURETHANE
60 80 100 120 140 160 180 200 220 240 260 280 300 320 340 360 380 400 420 440 460
TEMP °F

AMMONIUM NITRATE

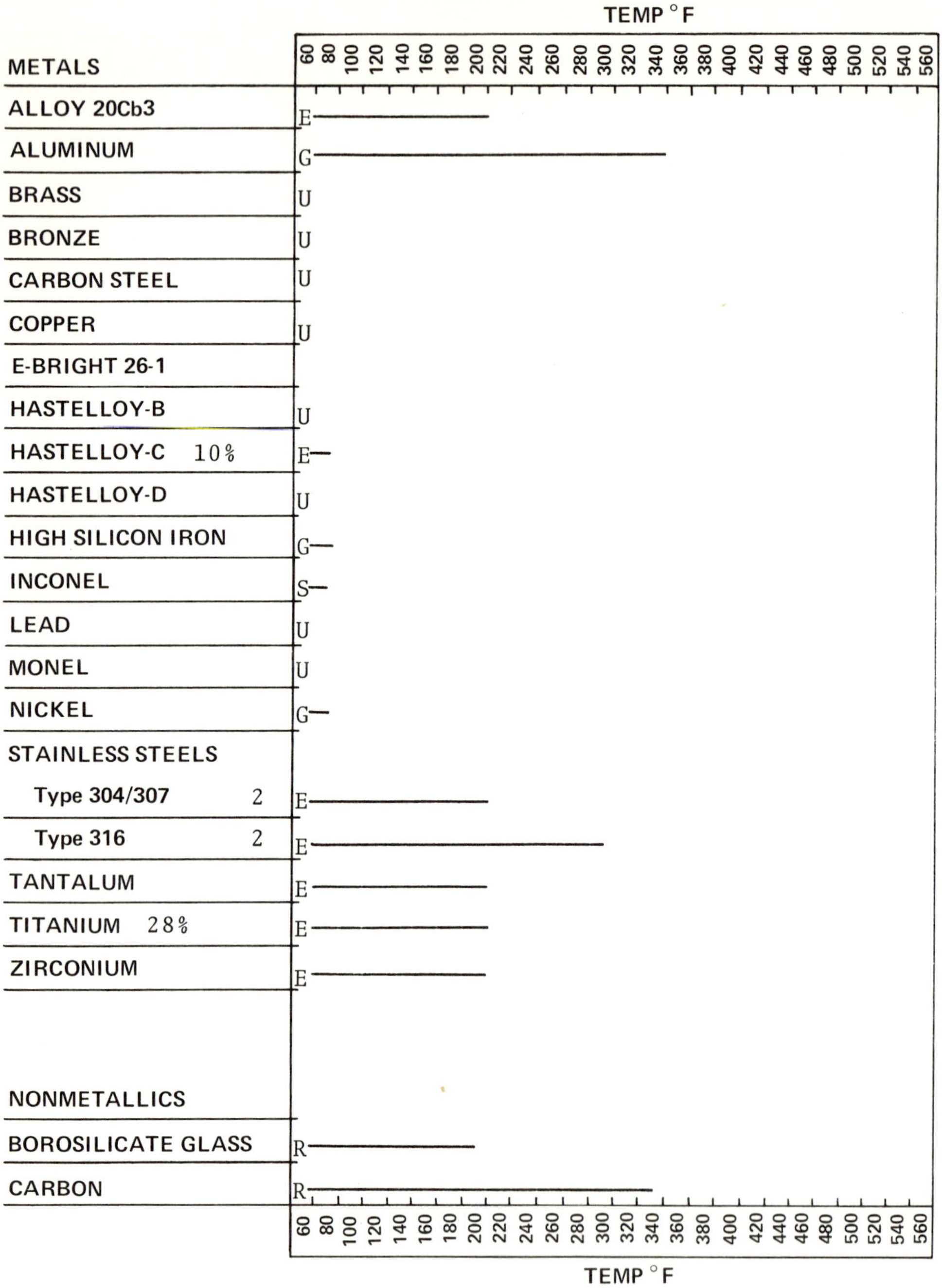

FOR METALS

E = < 2 Mils Penetration/Year; G = < 20 Mils Penetration/Year

S = < 50 Mils Penetration/Year; U = > 50 Mils Penetration/Year

FOR NONMETALLICS

R = Resistant

U = Unsatisfactory

AMMONIUM NITRATE

PLASTICS		TEMP °F
ABS		R — 140
ACRYLICS		R — 70
ASBESTOS REINF., EPOXY		R — 250
ASBESTOS REINF., PHENOLIC		R — 300
CHLORINATED POLYETHER		R — 250
CPVC		R — 210
EPOXY	25%	R — 250
FLUOROCARBONS FEP		R — 400
FLUOROCARBONS TFE		R — 460
FURFURYL ALCOHOL		R — 270
NORYL		R — 140
PHENOLIC		R — 200
POLYAMIDES—NYLON	10%	R — 220
POLYESTERS		
BISPHENOL A-FUMURATE		R — 220
HYDROGENATED BISPHENOL-A-BISPHENOL-A		R — 210
ISOPHTHALIC		R — 80
CHLORINATED POLYESTERS		R — 200
POLYSTYRENES		R — 80
POLYETHYLENE		R — 140 U
POLYPROPYLENE		R — 200 U
PVC—TYPE 1		R — 140
PVC—TYPE 2		R — 140
POLYVINYLIDENE CHLORIDE		R — 130 U
VINYLIDENE FLUORIDE		R — 280
VINYL ESTER		R — 250

RUBBERS	TEMP °F
BUTYL GR-1	R — 180
FLUORO ELASTOMERS VITON A	U
FLUORO ELASTOMERS KEL-F 3700	R — 80
HARD RUBBER	R — 150
HYPALON	R — 80
KOROSEAL	R — 150
NATURAL RUBBER (GRS)	R — 150
NEOPRENE GR-M (CR)	R — 200
NITRILE BUNA N (NBR)	R — 180
NORDEL	R — 300
POLYBUTADIENE	R — 80
SILICONE	
POLYURETHANE	

AMMONIUM OXALATE 30%-20%

TEMP °F (60–560)

METALS	Rating
ALLOY 20Cb3	
ALUMINUM	
BRASS	
BRONZE	
CARBON STEEL	U
COPPER	
E-BRIGHT 26-1	
HASTELLOY-B	G (line from 60 to approx. 200)
HASTELLOY-C	G (line from 60 to approx. 200)
HASTELLOY-D	
HIGH SILICON IRON	
INCONEL	
LEAD	
MONEL	
NICKEL	
STAINLESS STEELS Type 304/307	
Type 316	G (line from 60 to approx. 400)
TANTALUM	E (short line at 60)
TITANIUM	
ZIRCONIUM	

NONMETALLICS	
BOROSILICATE GLASS	
CARBON	

TEMP °F

FOR METALS

E = < 2 Mils Penetration/Year; G = < 20 Mils Penetration/Year

S = < 50 Mils Penetration/Year; U = > 50 Mils Penetration/Year

FOR NONMETALLICS

R = Resistant

U = Unsatisfactory

AMMONIUM OXALATE 30%-20%

TEMP °F: 60 80 100 120 140 160 180 200 220 240 260 280 300 320 340 360 380 400 420 440 460

PLASTICS	
ABS	
ACRYLICS	
ASBESTOS REINF., EPOXY	
ASBESTOS REINF., PHENOLIC	
CHLORINATED POLYETHER	R
CPVC	
EPOXY	
FLUOROCARBONS FEP	R
FLUOROCARBONS TFE	R
FURFURYL ALCOHOL	
NORYL	
PHENOLIC	
POLYAMIDES–NYLON	
POLYESTERS	
BISPHENOL A-FUMURATE	
HYDROGENATED BISPHENOL-A-BISPHENOL-A	
ISOPHTHALIC	
CHLORINATED POLYESTERS	
POLYSTYRENES	
POLYETHYLENE	
POLYPROPYLENE	R
PVC–TYPE 1	R
PVC–TYPE 2	
POLYVINYLIDENE CHLORIDE	R
VINYLIDENE FLUORIDE	
VINYL ESTER	
RUBBERS	
BUTYL GR-1	
FLUORO ELASTOMERS VITON A	
FLUORO ELASTOMERS KEL-F 3700	
HARD RUBBER	
HYPALON	R
KOROSEAL	
NATURAL RUBBER (GRS)	
NEOPRENE GR-M (CR)	
NITRILE BUNA N (NBR)	
NORDEL	R
POLYBUTADIENE	
SILICONE	
POLYURETHANE	

TEMP °F: 60 80 100 120 140 160 180 200 220 240 260 280 300 320 340 360 380 400 420 440 460

AMMONIUM OXALATE, 10%

TEMP °F

METALS	Rating (TEMP °F, 60–560)
ALLOY 20Cb3	
ALUMINUM	
BRASS	
BRONZE	
CARBON STEEL	U (60)
COPPER	
E-BRIGHT 26-1	
HASTELLOY-B	E (60–120) — G (120–200)
HASTELLOY-C	E (60–120) — G (120–200)
HASTELLOY-D	
HIGH SILICON IRON	
INCONEL	
LEAD	
MONEL	
NICKEL	
STAINLESS STEELS	
Type 304/307	G (60)
Type 316	G (60–400)
TANTALUM	E (60)
TITANIUM	
ZIRCONIUM	
NONMETALLICS	
BOROSILICATE GLASS	
CARBON	

TEMP °F

FOR METALS

E = < 2 Mils Penetration/Year; G = < 20 Mils Penetration/Year

S = < 50 Mils Penetration/Year; U = > 50 Mils Penetration/Year

FOR NONMETALLICS

R = Resistant

U = Unsatisfactory

AMMONIUM OXALATE, 10%

TEMP °F

PLASTICS	Rating	Temp °F range (approx., from bar)
ABS		
ACRYLICS		
ASBESTOS REINF., EPOXY		
ASBESTOS REINF., PHENOLIC		
CHLORINATED POLYETHER	R	60–240
CPVC		
EPOXY		
FLUOROCARBONS FEP	R	60–400
FLUOROCARBONS TFE	R	60–460
FURFURYL ALCOHOL		
NORYL		
PHENOLIC		
POLYAMIDES–NYLON		
POLYESTERS		
BISPHENOL A-FUMURATE		
HYDROGENATED BISPHENOL-A-BISPHENOL-A		
ISOPHTHALIC		
CHLORINATED POLYESTERS		
POLYSTYRENES		
POLYETHYLENE		
POLYPROPYLENE	R	60–220
PVC–TYPE 1	R	60–140
PVC–TYPE 2		
POLYVINYLIDENE CHLORIDE	R	60–180
VINYLIDENE FLUORIDE		
VINYL ESTER		

RUBBERS	Rating	Temp °F range (approx., from bar)
BUTYL GR-1		
FLUORO ELASTOMERS VITON A		
FLUORO ELASTOMERS KEL-F 3700		
HARD RUBBER		
HYPALON	R	60–240
KOROSEAL		
NATURAL RUBBER (GRS)		
NEOPRENE GR-M (CR)		
NITRILE BUNA N (NBR)		
NORDEL	R	60–300
POLYBUTADIENE		
SILICONE		
POLYURETHANE		

Temperature scale: 60 80 100 120 140 160 180 200 220 240 260 280 300 320 340 360 380 400 420 440 460

TEMP °F

AMMONIUM PERSULFATE

TEMP °F (60–560)

METALS	Concentration	Rating
ALLOY 20Cb3		G
ALUMINUM		G
BRASS		
BRONZE		U
CARBON STEEL		U
COPPER		
E-BRIGHT 26-1		
HASTELLOY-B		U
HASTELLOY-C	10%	E
HASTELLOY-D		
HIGH SILICON IRON		G
INCONEL		
LEAD		G
MONEL		U
NICKEL		U
STAINLESS STEELS		
Type 304/307	10%	U
Type 316	10%	G
TANTALUM	5%	E
TITANIUM		G
ZIRCONIUM		

NONMETALLICS	Rating
BOROSILICATE GLASS	R
CARBON	R

TEMP °F (60, 80, 100, 120, 140, 160, 180, 200, 220, 240, 260, 280, 300, 320, 340, 360, 380, 400, 420, 440, 460, 480, 500, 520, 540, 560)

FOR METALS

E = < 2 Mils Penetration/Year; G = < 20 Mils Penetration/Year

S = < 50 Mils Penetration/Year; U = > 50 Mils Penetration/Year

FOR NONMETALLICS

R = Resistant

U = Unsatisfactory

AMMONIUM PERSULFATE

TEMP °F

PLASTICS		60 80 100 120 140 160 180 200 220 240 260 280 300 320 340 360 380 400 420 440 460
ABS		R
ACRYLICS		R
ASBESTOS REINF., EPOXY		
ASBESTOS REINF., PHENOLIC		
CHLORINATED POLYETHER		R
CPVC		R
EPOXY		R
FLUOROCARBONS FEP		R
FLUOROCARBONS TFE		R
FURFURYL ALCOHOL	sat	R
NORYL		R
PHENOLIC		
POLYAMIDES–NYLON		
POLYESTERS		
BISPHENOL A-FUMURATE	sat	R
HYDROGENATED BISPHENOL-A-BISPHENOL-A		R
ISOPHTHALIC		R
CHLORINATED POLYESTERS		R
POLYSTYRENES		
POLYETHYLENE		R
POLYPROPYLENE		R
PVC–TYPE 1		R
PVC–TYPE 2		R
POLYVINYLIDENE CHLORIDE		R
VINYLIDENE FLUORIDE		R
VINYL ESTER	sat	R
RUBBERS		
BUTYL GR-1		R
FLUORO ELASTOMERS VITON A		R
FLUORO ELASTOMERS KEL-F 3700		
HARD RUBBER		R
HYPALON		R
KOROSEAL		
NATURAL RUBBER (GRS)		R
NEOPRENE GR-M (CR)		R
NITRILE BUNA N (NBR)		R
NORDEL		R
POLYBUTADIENE		
SILICONE		
POLYURETHANE		U

60 80 100 120 140 160 180 200 220 240 260 280 300 320 340 360 380 400 420 440 460

TEMP °F

AMMONIUM PHOSPHATE

TEMP °F (scale 60–560 in steps of 20)

METALS	Concentration	Rating	Range (°F)
ALLOY 20Cb3		G	60–200
ALUMINUM		U	
BRASS		U	
BRONZE		U	
CARBON STEEL		U	
COPPER		U	
E-BRIGHT 26-1			
HASTELLOY-B			
HASTELLOY-C	10%	G	60–200
HASTELLOY-D			
HIGH SILICON IRON		G	60–80
INCONEL	10%	G	60–200
LEAD		G	60–80
MONEL	30%	G	60–200
NICKEL	30%	G	60–200
STAINLESS STEELS			
Type 304/307	40%	E	60–120
Type 316	40%	G	60–120
TANTALUM	10%	E	60–80
TITANIUM	10%	E	60–200
ZIRCONIUM	10%	E	60–200

NONMETALLICS	Rating	Range (°F)
BOROSILICATE GLASS	R	60–80
CARBON	R	60–340

TEMP °F

FOR METALS

E = < 2 Mils Penetration/Year; G = < 20 Mils Penetration/Year

S = < 50 Mils Penetration/Year; U = > 50 Mils Penetration/Year

FOR NONMETALLICS

R = Resistant

U = Unsatisfactory

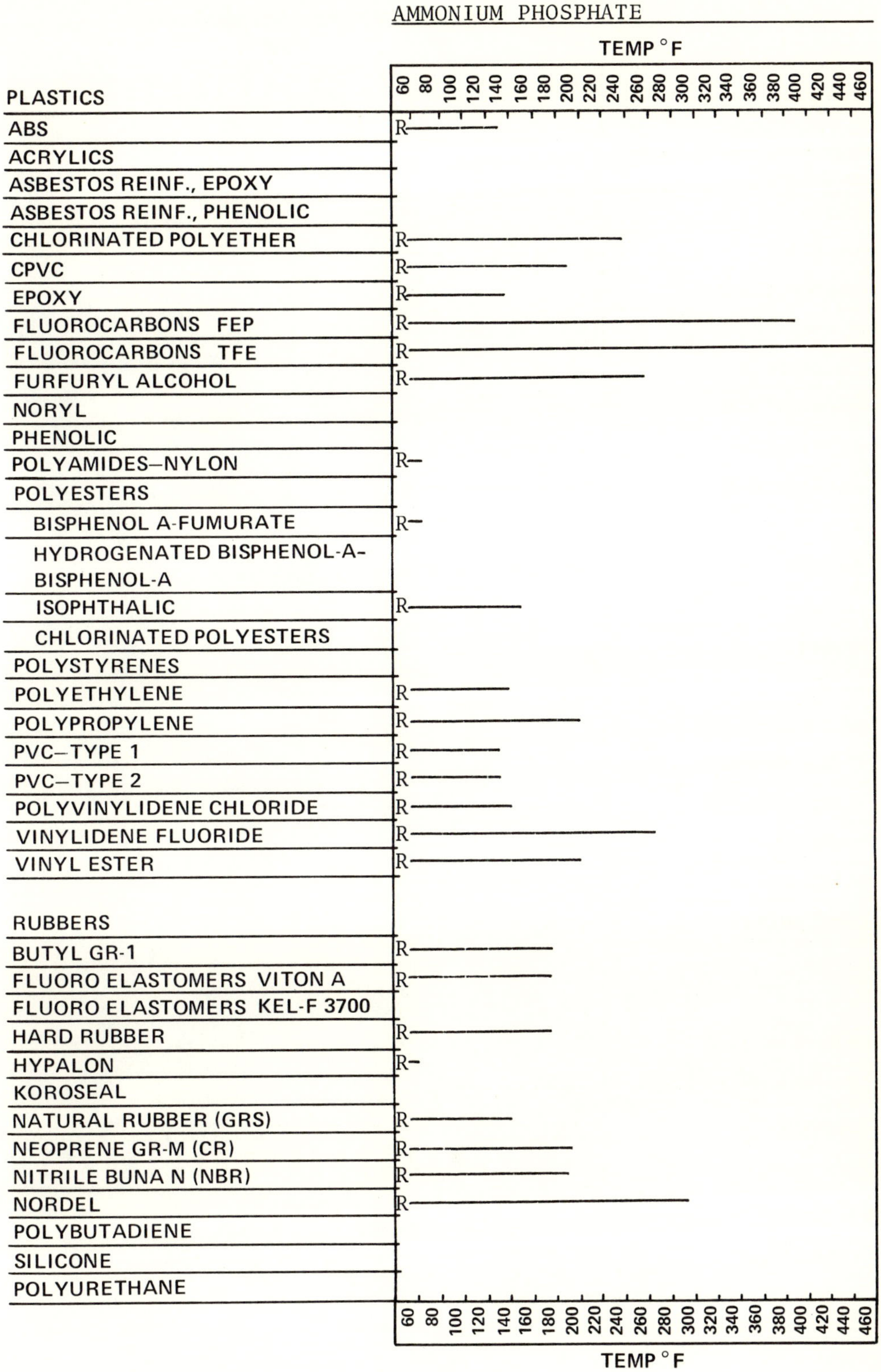
AMMONIUM PHOSPHATE
TEMP °F
60 80 100 120 140 160 180 200 220 240 260 280 300 320 340 360 380 400 420 440 460
PLASTICS
ABS R
ACRYLICS
ASBESTOS REINF., EPOXY
ASBESTOS REINF., PHENOLIC
CHLORINATED POLYETHER R
CPVC R
EPOXY R
FLUOROCARBONS FEP R
FLUOROCARBONS TFE R
FURFURYL ALCOHOL R
NORYL
PHENOLIC
POLYAMIDES—NYLON R
POLYESTERS
BISPHENOL A-FUMURATE R
HYDROGENATED BISPHENOL-A-BISPHENOL-A
ISOPHTHALIC R
CHLORINATED POLYESTERS
POLYSTYRENES
POLYETHYLENE R
POLYPROPYLENE R
PVC—TYPE 1 R
PVC—TYPE 2 R
POLYVINYLIDENE CHLORIDE R
VINYLIDENE FLUORIDE R
VINYL ESTER R
RUBBERS
BUTYL GR-1 R
FLUORO ELASTOMERS VITON A R
FLUORO ELASTOMERS KEL-F 3700
HARD RUBBER R
HYPALON R
KOROSEAL
NATURAL RUBBER (GRS) R
NEOPRENE GR-M (CR) R
NITRILE BUNA N (NBR) R
NORDEL R
POLYBUTADIENE
SILICONE
POLYURETHANE
60 80 100 120 140 160 180 200 220 240 260 280 300 320 340 360 380 400 420 440 460
TEMP °F

AMMONIUM SULFATE, 40%-10%

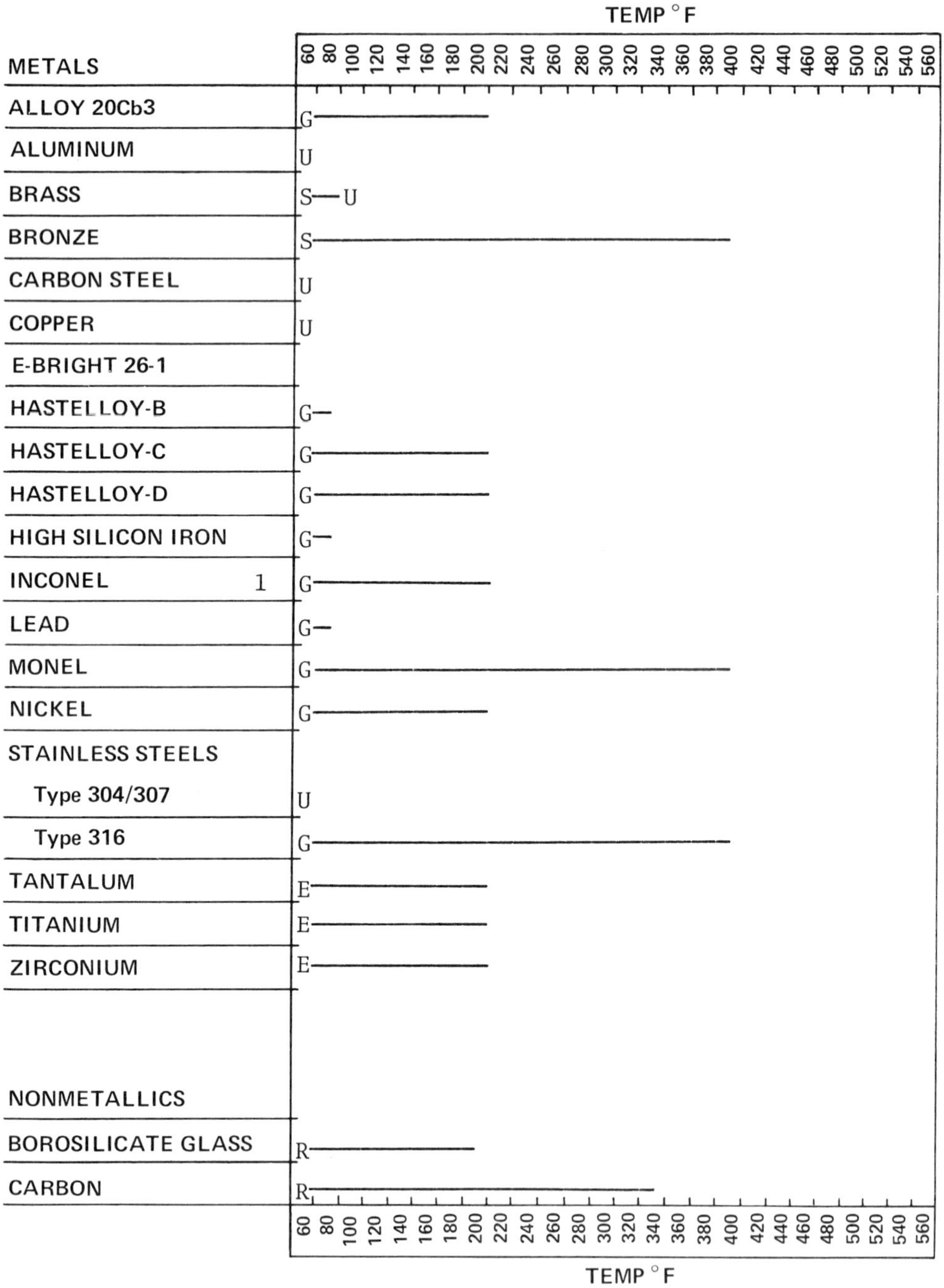

FOR METALS

E = < 2 Mils Penetration/Year; G = < 20 Mils Penetration/Year

S = < 50 Mils Penetration/Year; U = > 50 Mils Penetration/Year

FOR NONMETALLICS

R = Resistant

U = Unsatisfactory

AMMONIUM SULFATE, 40%-10%

TEMP °F: 60 80 100 120 140 160 180 200 220 240 260 280 300 320 340 360 380 400 420 440 460

PLASTICS		
ABS		R
ACRYLICS		R
ASBESTOS REINF., EPOXY		R
ASBESTOS REINF., PHENOLIC		R
CHLORINATED POLYETHER		R
CPVC		R
EPOXY		R
FLUOROCARBONS FEP		R
FLUOROCARBONS TFE		R
FURFURYL ALCOHOL	sat	R
NORYL		R
PHENOLIC		R
POLYAMIDES—NYLON	10%	R
POLYESTERS		
BISPHENOL A-FUMURATE	sat	R
HYDROGENATED BISPHENOL-A-BISPHENOL-A		
ISOPHTHALIC	10%	R
CHLORINATED POLYESTERS		R
POLYSTYRENES		
POLYETHYLENE		R
POLYPROPYLENE		R
PVC—TYPE 1		R
PVC—TYPE 2		R
POLYVINYLIDENE CHLORIDE		R
VINYLIDENE FLUORIDE		R
VINYL ESTER	sat	R

RUBBERS	
BUTYL GR-1	R
FLUORO ELASTOMERS VITON A	U
FLUORO ELASTOMERS KEL-F 3700	R
HARD RUBBER	R
HYPALON	R
KOROSEAL	R
NATURAL RUBBER (GRS)	R
NEOPRENE GR-M (CR)	R
NITRILE BUNA N (NBR)	R
NORDEL	R
POLYBUTADIENE	R
SILICONE	
POLYURETHANE	

TEMP °F: 60 80 100 120 140 160 180 200 220 240 260 280 300 320 340 360 380 400 420 440 460

AMMONIUM SULFIDE

TEMP °F

METALS	60 80 100 120 140 160 180 200 220 240 260 280 300 320 340 360 380 400 420 440 460 480 500 520 540 560
ALLOY 20Cb3	G————
ALUMINUM	E——G——U
BRASS	U
BRONZE	U
CARBON STEEL	
COPPER	U
E-BRIGHT 26-1	
HASTELLOY-B	
HASTELLOY-C	
HASTELLOY-D	
HIGH SILICON IRON	
INCONEL	
LEAD	
MONEL	
NICKEL	
STAINLESS STEELS	
Type 304/307	G————
Type 316	G————————
TANTALUM	G—
TITANIUM	
ZIRCONIUM	
NONMETALLICS	
BOROSILICATE GLASS	
CARBON	R——————
	60 80 100 120 140 160 180 200 220 240 260 280 300 320 340 360 380 400 420 440 460 480 500 520 540 560

TEMP °F

FOR METALS

E = < 2 Mils Penetration/Year; G = < 20 Mils Penetration/Year

S = < 50 Mils Penetration/Year; U = > 50 Mils Penetration/Year

FOR NONMETALLICS

R = Resistant

U = Unsatisfactory

TEMP °F (60–460)

PLASTICS	
ABS	R
ACRYLICS	
ASBESTOS REINF., EPOXY	
ASBESTOS REINF., PHENOLIC	
CHLORINATED POLYETHER	R
CPVC	R
EPOXY	
FLUOROCARBONS FEP	R
FLUOROCARBONS TFE	R
FURFURYL ALCOHOL	R
NORYL	
PHENOLIC	
POLYAMIDES—NYLON	
POLYESTERS	
BISPHENOL A-FUMURATE	R
HYDROGENATED BISPHENOL-A-BISPHENOL-A	
ISOPHTHALIC	U
CHLORINATED POLYESTERS 25%	R
POLYSTYRENES	
POLYETHYLENE	R
POLYPROPYLENE	R
PVC—TYPE 1	R
PVC—TYPE 2	R
POLYVINYLIDENE CHLORIDE	R
VINYLIDENE FLUORIDE	R
VINYL ESTER	

RUBBERS	
BUTYL GR-1	
FLUORO ELASTOMERS VITON A	U
FLUORO ELASTOMERS KEL-F 3700	
HARD RUBBER	R
HYPALON	
KOROSEAL	
NATURAL RUBBER (GRS)	
NEOPRENE GR-M (CR)	
NITRILE BUNA N (NBR)	R
NORDEL	R
POLYBUTADIENE	
SILICONE	
POLYURETHANE	

TEMP °F (60–460)

AMMONIUM SULFITE

TEMP °F (60–560)

METALS	60 – 560 °F
ALLOY 20Cb3	
ALUMINUM	
BRASS	
BRONZE	
CARBON STEEL	
COPPER	
E-BRIGHT 26-1	
HASTELLOY-B	
HASTELLOY-C	E——
HASTELLOY-D	
HIGH SILICON IRON	
INCONEL	G— U
LEAD	
MONEL	G—U
NICKEL	U
STAINLESS STEELS	
Type 304/307	G————————
Type 316	G————————
TANTALUM	E————————
TITANIUM	
ZIRCONIUM	
NONMETALLICS	
BOROSILICATE GLASS	
CARBON	

TEMP °F: 60, 80, 100, 120, 140, 160, 180, 200, 220, 240, 260, 280, 300, 320, 340, 360, 380, 400, 420, 440, 460, 480, 500, 520, 540, 560

FOR METALS

E = < 2 Mils Penetration/Year; G = < 20 Mils Penetration/Year

S = < 50 Mils Penetration/Year; U = > 50 Mils Penetration/Year

FOR NONMETALLICS

R = Resistant

U = Unsatisfactory

AMMONIUM SULFITE

PLASTICS	TEMP °F (60–460)
ABS	
ACRYLICS	
ASBESTOS REINF., EPOXY	
ASBESTOS REINF., PHENOLIC	
CHLORINATED POLYETHER	
CPVC	
EPOXY	
FLUOROCARBONS FEP	R (60–400)
FLUOROCARBONS TFE	R (60–460)
FURFURYL ALCOHOL	
NORYL	
PHENOLIC	
POLYAMIDES–NYLON	
POLYESTERS	
BISPHENOL A-FUMURATE	R (60–80)
HYDROGENATED BISPHENOL-A-BISPHENOL-A	
ISOPHTHALIC	U
CHLORINATED POLYESTERS	
POLYSTYRENES	
POLYETHYLENE	
POLYPROPYLENE	
PVC–TYPE 1	
PVC–TYPE 2	
POLYVINYLIDENE CHLORIDE	
VINYLIDENE FLUORIDE	
VINYL ESTER	

RUBBERS	TEMP °F (60–460)
BUTYL GR-1	
FLUORO ELASTOMERS VITON A	
FLUORO ELASTOMERS KEL-F 3700	
HARD RUBBER	
HYPALON	
KOROSEAL	
NATURAL RUBBER (GRS)	
NEOPRENE GR-M (CR)	
NITRILE BUNA N (NBR)	
NORDEL	
POLYBUTADIENE	
SILICONE	
POLYURETHANE	

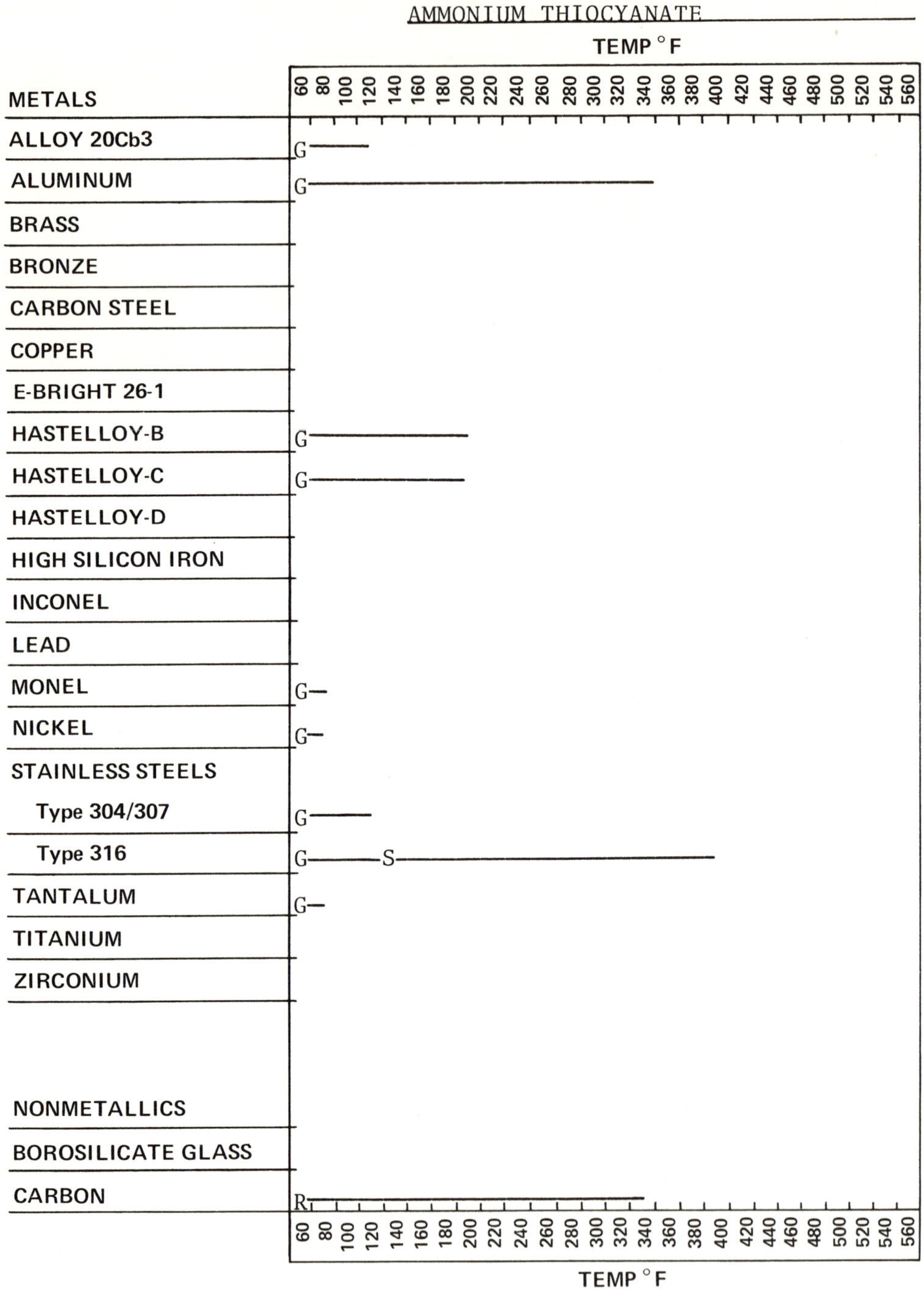

FOR METALS

E = < 2 Mils Penetration/Year; G = < 20 Mils Penetration/Year

S = < 50 Mils Penetration/Year; U = > 50 Mils Penetration/Year

FOR NONMETALLICS

R = Resistant

U = Unsatisfactory

AMMONIUM THIOCYANATE

PLASTICS	TEMP °F
ABS	R
ACRYLICS	
ASBESTOS REINF., EPOXY	
ASBESTOS REINF., PHENOLIC	
CHLORINATED POLYETHER	R
CPVC	R
EPOXY	R
FLUOROCARBONS FEP	R
FLUOROCARBONS TFE	R
FURFURYL ALCOHOL	R
NORYL	
PHENOLIC	
POLYAMIDES—NYLON	
POLYESTERS	
BISPHENOL A-FUMURATE 50%	R
HYDROGENATED BISPHENOL-A-BISPHENOL-A	
ISOPHTHALIC	R
CHLORINATED POLYESTERS 20%	R
POLYSTYRENES	
POLYETHYLENE	R
POLYPROPYLENE	R
PVC—TYPE 1	R
PVC—TYPE 2	R
POLYVINYLIDENE CHLORIDE	R
VINYLIDENE FLUORIDE	R
VINYL ESTER	R
RUBBERS	
BUTYL GR-1	
FLUORO ELASTOMERS VITON A	R
FLUORO ELASTOMERS KEL-F 3700	
HARD RUBBER	
HYPALON	
KOROSEAL	
NATURAL RUBBER (GRS)	
NEOPRENE GR-M (CR)	
NITRILE BUNA N (NBR)	R
NORDEL	R
POLYBUTADIENE	
SILICONE	
POLYURETHANE	

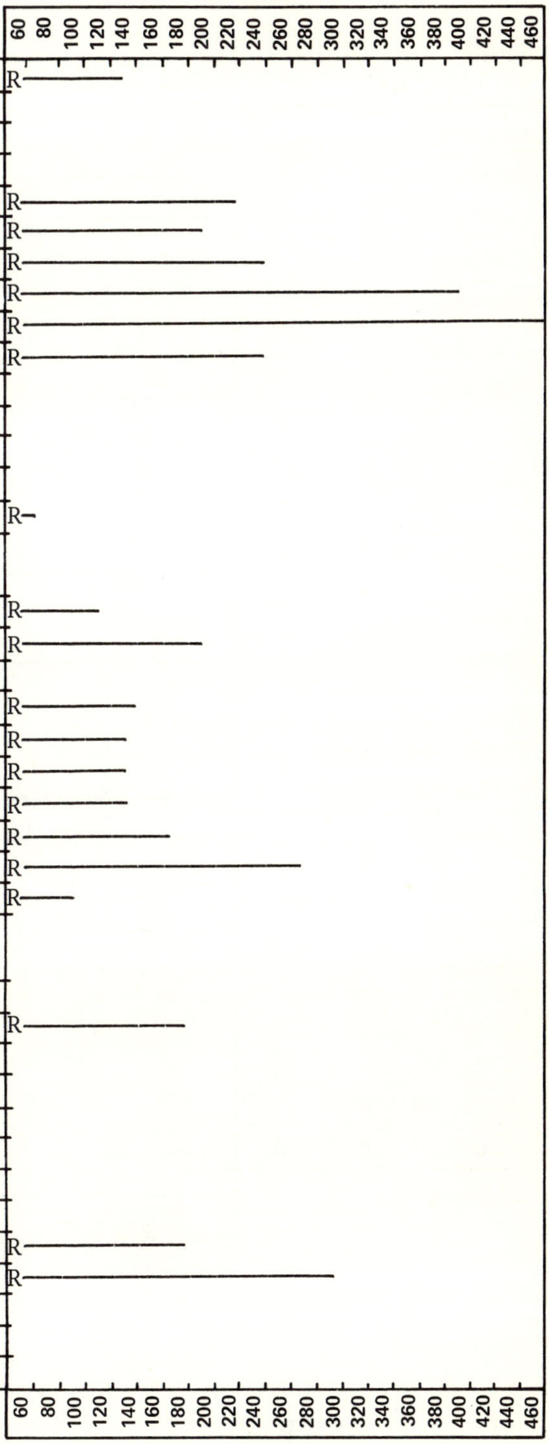

AMYL ACETATE

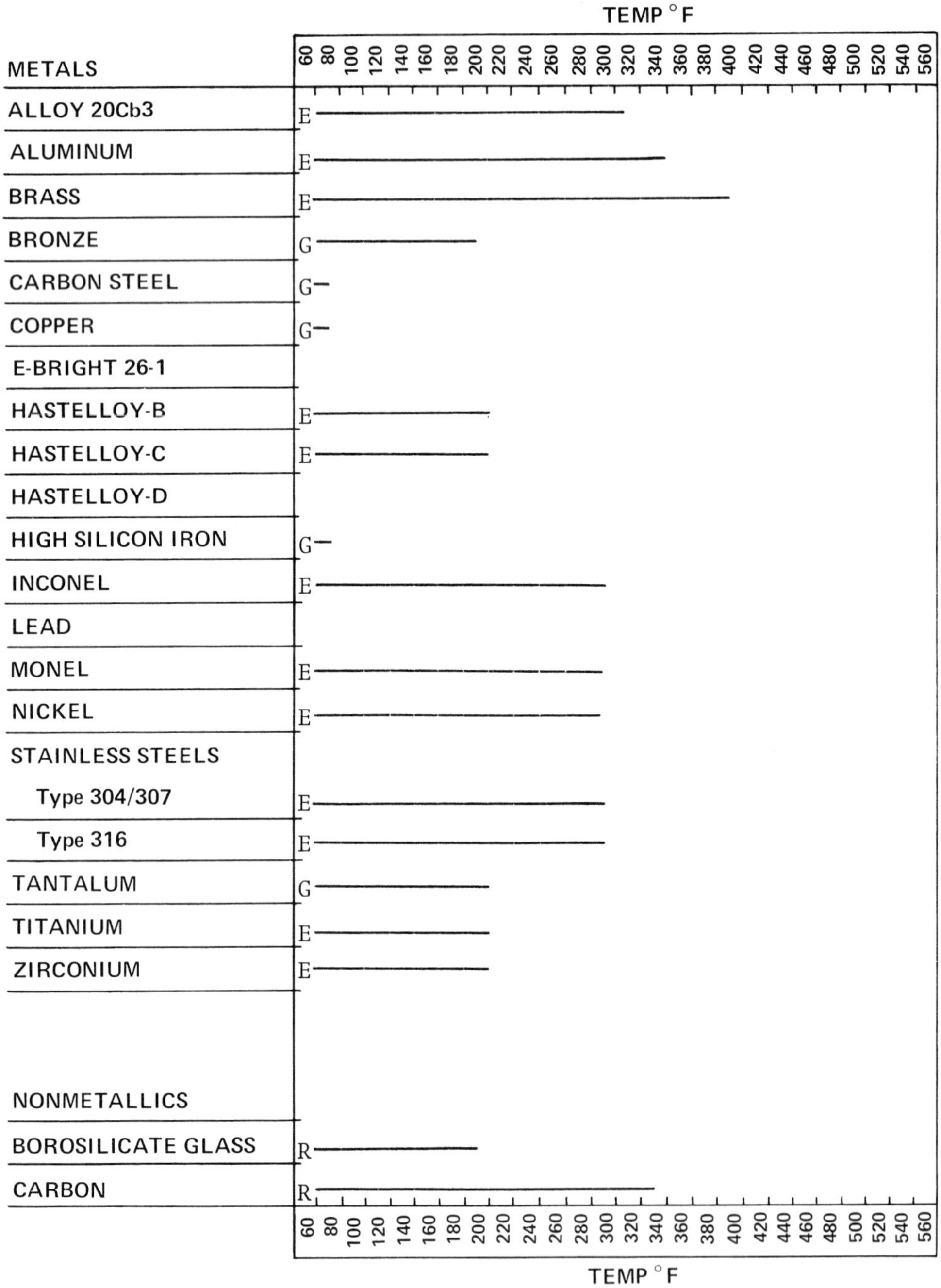

FOR METALS

E = < 2 Mils Penetration/Year; G = < 20 Mils Penetration/Year

S = < 50 Mils Penetration/Year; U = > 50 Mils Penetration/Year

FOR NONMETALLICS

R = Resistant

U = Unsatisfactory

AMYL ACETATE

TEMP °F (60 80 100 120 140 160 180 200 220 240 260 280 300 320 340 360 380 400 420 440 460)

PLASTICS	
ABS	U
ACRYLICS	U
ASBESTOS REINF., EPOXY	
ASBESTOS REINF., PHENOLIC	
CHLORINATED POLYETHER	R————
CPVC	U
EPOXY	R–
FLUOROCARBONS FEP	R————————————
FLUOROCARBONS TFE	R——————————————
FURFURYL ALCOHOL	R———————
NORYL	
PHENOLIC	
POLYAMIDES–NYLON	R—————
POLYESTERS	
BISPHENOL A-FUMURATE	R–
HYDROGENATED BISPHENOL-A-BISPHENOL-A	U
ISOPHTHALIC	U
CHLORINATED POLYESTERS	R————— U
POLYSTYRENES	
POLYETHYLENE	U
POLYPROPYLENE	U
PVC–TYPE 1	U
PVC–TYPE 2	U
POLYVINYLIDENE CHLORIDE	R——
VINYLIDENE FLUORIDE	R————— U
VINYL ESTER	R——

RUBBERS	
BUTYL GR-1	U
FLUORO ELASTOMERS VITON A	U
FLUORO ELASTOMERS KEL-F 3700	
HARD RUBBER	
HYPALON	U
KOROSEAL	U
NATURAL RUBBER (GRS)	U
NEOPRENE GR-M (CR)	U
NITRILE BUNA N (NBR)	U
NORDEL	U
POLYBUTADIENE	
SILICONE	
POLYURETHANE	U

TEMP °F (60 80 100 120 140 160 180 200 220 240 260 280 300 320 340 360 380 400 420 440 460)

AMYL ALCOHOL

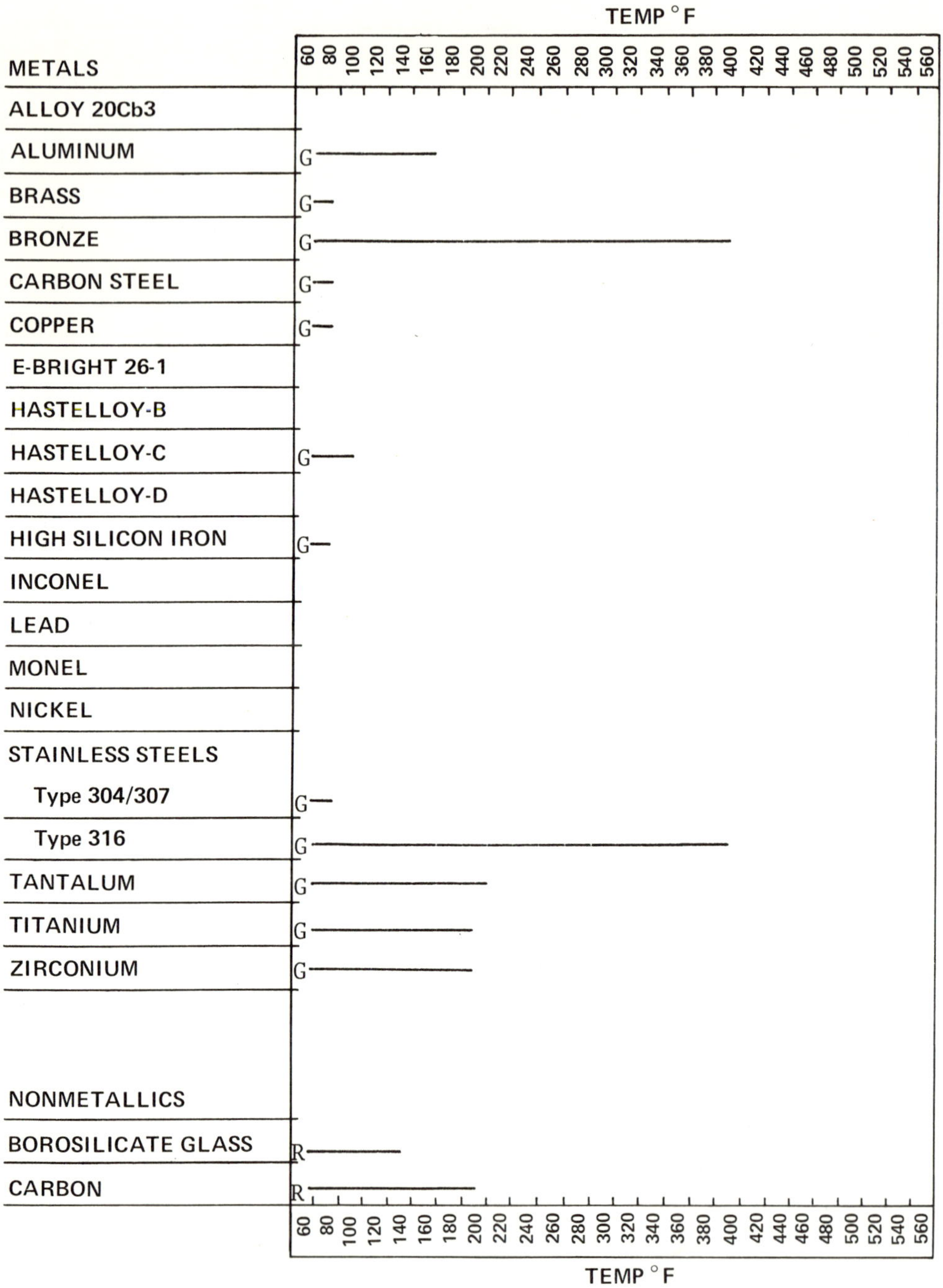

FOR METALS

E = < 2 Mils Penetration/Year; G = < 20 Mils Penetration/Year

S = < 50 Mils Penetration/Year; U = > 50 Mils Penetration/Year

FOR NONMETALLICS

R = Resistant

U = Unsatisfactory

AMYL ALCOHOL

TEMP °F (60–460)

PLASTICS	Rating
ABS	R
ACRYLICS	R
ASBESTOS REINF., EPOXY	
ASBESTOS REINF., PHENOLIC	
CHLORINATED POLYETHER	R
CPVC	R — U
EPOXY	R
FLUOROCARBONS FEP	R
FLUOROCARBONS TFE	R
FURFURYL ALCOHOL	R
NORYL	
PHENOLIC	
POLYAMIDES–NYLON	R
POLYESTERS	
BISPHENOL A-FUMURATE	R
HYDROGENATED BISPHENOL-A-BISPHENOL-A	R
ISOPHTHALIC	R
CHLORINATED POLYESTERS	R
POLYSTYRENES	R
POLYETHYLENE	R
POLYPROPYLENE	R
PVC–TYPE 1	R
PVC–TYPE 2	U
POLYVINYLIDENE CHLORIDE	R
VINYLIDENE FLUORIDE	R
VINYL ESTER	R

RUBBERS	Rating
BUTYL GR-1	R
FLUORO ELASTOMERS VITON A	R
FLUORO ELASTOMERS KEL-F 3700	
HARD RUBBER	R
HYPALON	R
KOROSEAL	R
NATURAL RUBBER (GRS)	R
NEOPRENE GR-M (CR)	R
NITRILE BUNA N (NBR)	R
NORDEL	R
POLYBUTADIENE	
SILICONE	
POLYURETHANE	U

TEMP °F (60–460)

AMYL CHLORIDE

TEMP °F: 60, 80, 100, 120, 140, 160, 180, 200, 220, 240, 260, 280, 300, 320, 340, 360, 380, 400, 420, 440, 460, 480, 500, 520, 540, 560

METALS	
ALLOY 20Cb3	G
ALUMINUM	U
BRASS	G
BRONZE	G
CARBON STEEL	U
COPPER	G
E-BRIGHT 26-1	
HASTELLOY-B	E
HASTELLOY-C	E
HASTELLOY-D	
HIGH SILICON IRON	G
INCONEL	U
LEAD	U
MONEL	G
NICKEL	G
STAINLESS STEELS	
Type 304/307	G
Type 316	G
TANTALUM	G
TITANIUM	
ZIRCONIUM	G

NONMETALLICS	
BOROSILICATE GLASS	R
CARBON	

TEMP °F: 60, 80, 100, 120, 140, 160, 180, 200, 220, 240, 260, 280, 300, 320, 340, 360, 380, 400, 420, 440, 460, 480, 500, 520, 540, 560

FOR METALS

E = < 2 Mils Penetration/Year; G = < 20 Mils Penetration/Year

S = < 50 Mils Penetration/Year; U = > 50 Mils Penetration/Year

FOR NONMETALLICS

R = Resistant

U = Unsatisfactory

AMYL CHLORIDE

TEMP °F: 60 80 100 120 140 160 180 200 220 240 260 280 300 320 340 360 380 400 420 440 460

PLASTICS	
ABS	U
ACRYLICS	
ASBESTOS REINF., EPOXY	
ASBESTOS REINF., PHENOLIC	
CHLORINATED POLYETHER	R
CPVC	U
EPOXY	R
FLUOROCARBONS FEP	R
FLUOROCARBONS TFE	R
FURFURYL ALCOHOL	U
NORYL	
PHENOLIC	
POLYAMIDES–NYLON	
POLYESTERS	
BISPHENOL A-FUMURATE	U
HYDROGENATED BISPHENOL-A-BISPHENOL-A	R
ISOPHTHALIC	U
CHLORINATED POLYESTERS	U
POLYSTYRENES	
POLYETHYLENE	U
POLYPROPYLENE	U
PVC–TYPE 1	U
PVC–TYPE 2	U
POLYVINYLIDENE CHLORIDE	R
VINYLIDENE FLUORIDE	R
VINYL ESTER	R

RUBBERS	
BUTYL GR-1	
FLUORO ELASTOMERS VITON A	R
FLUORO ELASTOMERS KEL-F 3700	
HARD RUBBER	
HYPALON	U
KOROSEAL	
NATURAL RUBBER (GRS)	U
NEOPRENE GR-M (CR)	U
NITRILE BUNA N (NBR)	U
NORDEL	
POLYBUTADIENE	
SILICONE	
POLYURETHANE	

TEMP °F: 60 80 100 120 140 160 180 200 220 240 260 280 300 320 340 360 380 400 420 440 460

ANILINE

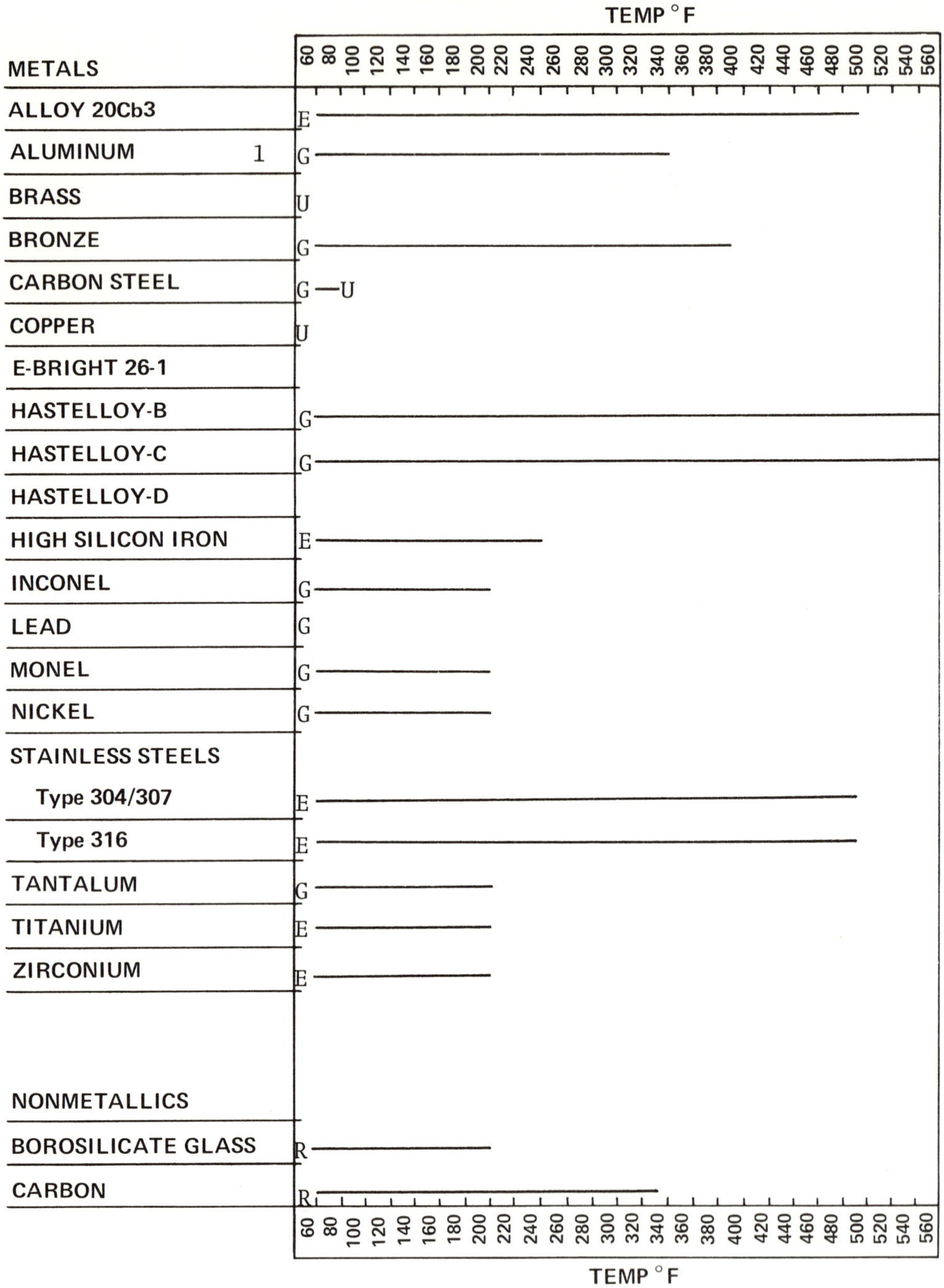

FOR METALS

E = < 2 Mils Penetration/Year; G = < 20 Mils Penetration/Year

S = < 50 Mils Penetration/Year; U = > 50 Mils Penetration/Year

FOR NONMETALLICS

R = Resistant

U = Unsatisfactory

ANILINE

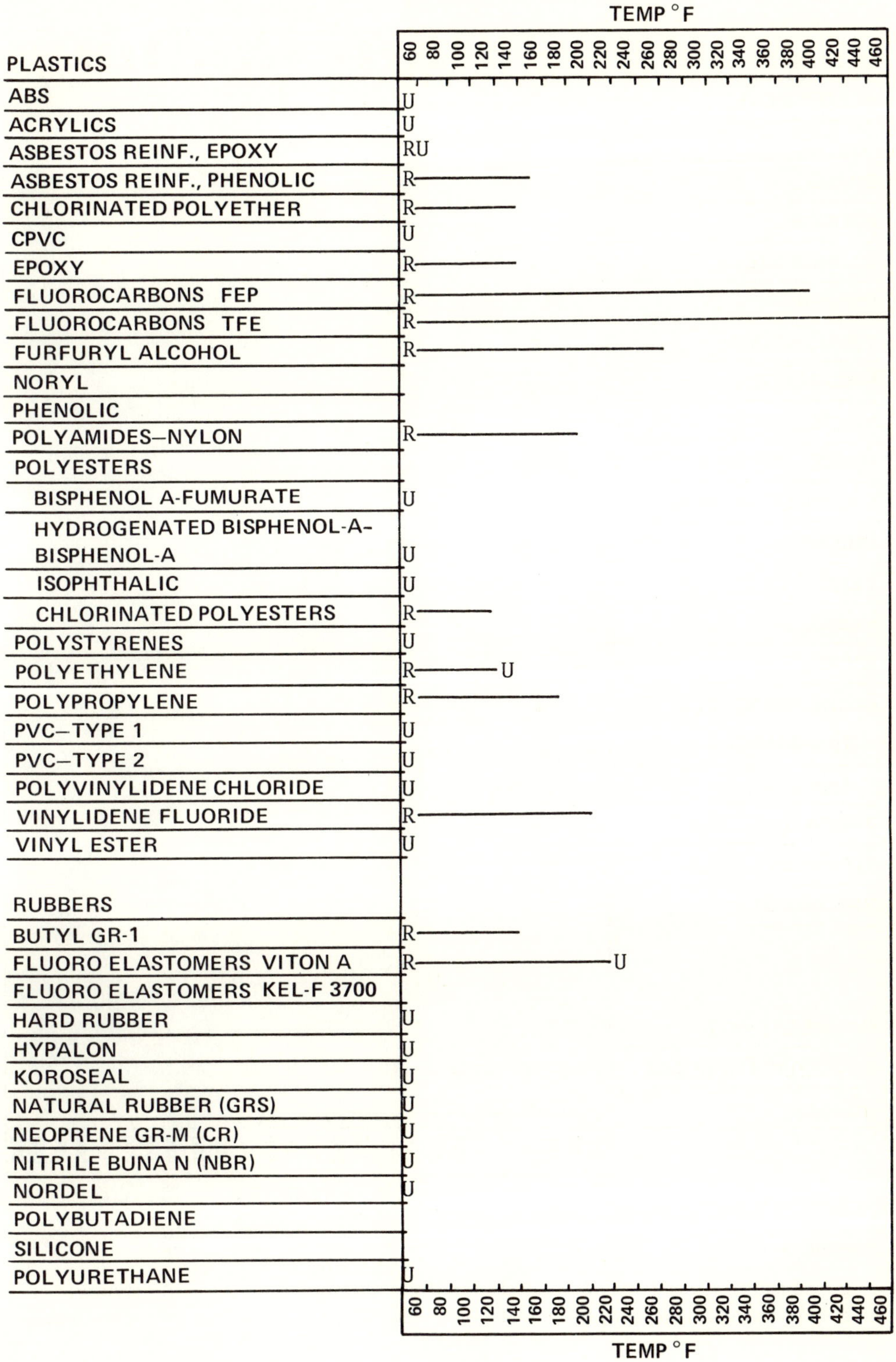
TEMP °F
60 80 100 120 140 160 180 200 220 240 260 280 300 320 340 360 380 400 420 440 460
PLASTICS
ABS U
ACRYLICS U
ASBESTOS REINF., EPOXY RU
ASBESTOS REINF., PHENOLIC R
CHLORINATED POLYETHER R
CPVC U
EPOXY R
FLUOROCARBONS FEP R
FLUOROCARBONS TFE R
FURFURYL ALCOHOL R
NORYL
PHENOLIC
POLYAMIDES—NYLON R
POLYESTERS
BISPHENOL A-FUMURATE U
HYDROGENATED BISPHENOL-A-BISPHENOL-A U
ISOPHTHALIC U
CHLORINATED POLYESTERS R
POLYSTYRENES U
POLYETHYLENE R U
POLYPROPYLENE R
PVC—TYPE 1 U
PVC—TYPE 2 U
POLYVINYLIDENE CHLORIDE U
VINYLIDENE FLUORIDE R
VINYL ESTER U
RUBBERS
BUTYL GR-1 R
FLUORO ELASTOMERS VITON A R U
FLUORO ELASTOMERS KEL-F 3700
HARD RUBBER U
HYPALON U
KOROSEAL U
NATURAL RUBBER (GRS) U
NEOPRENE GR-M (CR) U
NITRILE BUNA N (NBR) U
NORDEL U
POLYBUTADIENE
SILICONE
POLYURETHANE U
60 80 100 120 140 160 180 200 220 240 260 280 300 320 340 360 380 400 420 440 460
TEMP °F

ANILINE HYDROCHLORIDE

TEMP °F (60–560)

METALS		Rating / Temp °F
ALLOY 20Cb3		
ALUMINUM		U
BRASS		
BRONZE		
CARBON STEEL		U
COPPER		
E-BRIGHT 26-1		
HASTELLOY-B		U
HASTELLOY-C		U
HASTELLOY-D		
HIGH SILICON IRON		G (60–80)
INCONEL		U
LEAD		U
MONEL		G (60–80)
NICKEL		G (60–80)
STAINLESS STEELS		
Type 304/307		U
Type 316		U
TANTALUM	20%	G (60–200)
TITANIUM	20%	E (60–200)
ZIRCONIUM		G (60–200)

NONMETALLICS	Rating / Temp °F
BOROSILICATE GLASS	R (60–200)
CARBON	R (60–340)

TEMP °F

FOR METALS

E = < 2 Mils Penetration/Year; G = < 20 Mils Penetration/Year

S = < 50 Mils Penetration/Year; U = > 50 Mils Penetration/Year

FOR NONMETALLICS

R = Resistant

U = Unsatisfactory

ANILINE HYDROCHLORIDE

TEMP °F: 60 80 100 120 140 160 180 200 220 240 260 280 300 320 340 360 380 400 420 440 460

PLASTICS		Rating
ABS		U
ACRYLICS		
ASBESTOS REINF., EPOXY		
ASBESTOS REINF., PHENOLIC		
CHLORINATED POLYETHER	10%	R
CPVC		U
EPOXY		U
FLUOROCARBONS FEP		R
FLUOROCARBONS TFE		R
FURFURYL ALCOHOL		R
NORYL		
PHENOLIC		
POLYAMIDES–NYLON		
POLYESTERS		
BISPHENOL A-FUMURATE		R
HYDROGENATED BISPHENOL-A-BISPHENOL-A		R
ISOPHTHALIC		R
CHLORINATED POLYESTERS		
POLYSTYRENES		
POLYETHYLENE		U
POLYPROPYLENE		R—U
PVC–TYPE 1		U
PVC–TYPE 2		U
POLYVINYLIDENE CHLORIDE		R
VINYLIDENE FLUORIDE		R
VINYL ESTER		

RUBBERS	Rating
BUTYL GR-1	U
FLUORO ELASTOMERS VITON A	R
FLUORO ELASTOMERS KEL-F 3700	
HARD RUBBER	U
HYPALON	
KOROSEAL	
NATURAL RUBBER (GRS)	R
NEOPRENE GR-M (CR)	
NITRILE BUNA N (NBR)	U
NORDEL	
POLYBUTADIENE	
SILICONE	
POLYURETHANE	

TEMP °F: 60 80 100 120 140 160 180 200 220 240 260 280 300 320 340 360 380 400 420 440 460

ANTHRAQUINONE

METALS	TEMP °F
ALLOY 20Cb3	G
ALUMINUM	G
BRASS	G
BRONZE	G
CARBON STEEL	
COPPER	G
E-BRIGHT 26-1	
HASTELLOY-B	G
HASTELLOY-C	G
HASTELLOY-D	
HIGH SILICON IRON	G
INCONEL	G
LEAD	
MONEL	G
NICKEL	G
STAINLESS STEELS	
Type 304/307	G
Type 316	
TANTALUM	G
TITANIUM	
ZIRCONIUM	G
NONMETALLICS	
BOROSILICATE GLASS	
CARBON	R

TEMP °F: 60 80 100 120 140 160 180 200 220 240 260 280 300 320 340 360 380 400 420 440 460 480 500 520 540 560

FOR METALS

E = < 2 Mils Penetration/Year; G = < 20 Mils Penetration/Year

S = < 50 Mils Penetration/Year; U = > 50 Mils Penetration/Year

FOR NONMETALLICS

R = Resistant

U = Unsatisfactory

ANTHRAQUINONE

TEMP °F (60–460)

PLASTICS	Rating
ABS	R
ACRYLICS	
ASBESTOS REINF., EPOXY	
ASBESTOS REINF., PHENOLIC	
CHLORINATED POLYETHER	R
CPVC	R
EPOXY	
FLUOROCARBONS FEP	R
FLUOROCARBONS TFE	R
FURFURYL ALCOHOL	
NORYL	
PHENOLIC	
POLYAMIDES–NYLON	
POLYESTERS	
BISPHENOL A-FUMURATE	
HYDROGENATED BISPHENOL-A-BISPHENOL-A	
ISOPHTHALIC	
CHLORINATED POLYESTERS	
POLYSTYRENES	
POLYETHYLENE	R
POLYPROPYLENE	U
PVC–TYPE 1	R
PVC–TYPE 2	R
POLYVINYLIDENE CHLORIDE	
VINYLIDENE FLUORIDE	R
VINYL ESTER	

RUBBERS	Rating
BUTYL GR-1	
FLUORO ELASTOMERS VITON A	R
FLUORO ELASTOMERS KEL-F 3700	
HARD RUBBER	
HYPALON	
KOROSEAL	
NATURAL RUBBER (GRS)	
NEOPRENE GR-M (CR)	
NITRILE BUNA N (NBR)	
NORDEL	
POLYBUTADIENE	
SILICONE	
POLYURETHANE	

TEMP °F (60–460)

ANTHRAQUINONE SULFONIC ACID

TEMP °F: 60, 80, 100, 120, 140, 160, 180, 200, 220, 240, 260, 280, 300, 320, 340, 360, 380, 400, 420, 440, 460, 480, 500, 520, 540, 560

METALS	
ALLOY 20Cb3	
ALUMINUM	
BRASS	
BRONZE	
CARBON STEEL	
COPPER	G—
E-BRIGHT 26-1	
HASTELLOY-B	G—
HASTELLOY-C	G—
HASTELLOY-D	
HIGH SILICON IRON	G—
INCONEL	
LEAD	
MONEL	G—
NICKEL	G—
STAINLESS STEELS	
Type 304/307	
Type 316	
TANTALUM	G———————
TITANIUM	
ZIRCONIUM	G——————

NONMETALLICS	
BOROSILICATE GLASS	
CARBON	

TEMP °F

FOR METALS

E = < 2 Mils Penetration/Year; G = < 20 Mils Penetration/Year

S = < 50 Mils Penetration/Year; U = > 50 Mils Penetration/Year

FOR NONMETALLICS

R = Resistant

U = Unsatisfactory

PLASTICS	TEMP °F (60–460)
ABS	R—140
ACRYLICS	
ASBESTOS REINF., EPOXY	
ASBESTOS REINF., PHENOLIC	
CHLORINATED POLYETHER	R—180
CPVC	R—200
EPOXY	
FLUOROCARBONS FEP	R—400
FLUOROCARBONS TFE	R—460
FURFURYL ALCOHOL	
NORYL	
PHENOLIC	
POLYAMIDES—NYLON	
POLYESTERS	
BISPHENOL A-FUMURATE	R—210
HYDROGENATED BISPHENOL-A-BISPHENOL-A	
ISOPHTHALIC	
CHLORINATED POLYESTERS	
POLYSTYRENES	
POLYETHYLENE	R—150
POLYPROPYLENE	U
PVC—TYPE 1	R—140
PVC—TYPE 2	R—140
POLYVINYLIDENE CHLORIDE	R—80
VINYLIDENE FLUORIDE	R—280
VINYL ESTER	

RUBBERS	TEMP °F (60–460)
BUTYL GR-1	
FLUORO ELASTOMERS VITON A	R—190
FLUORO ELASTOMERS KEL-F 3700	
HARD RUBBER	
HYPALON	
KOROSEAL	
NATURAL RUBBER (GRS)	
NEOPRENE GR-M (CR)	
NITRILE BUNA N (NBR)	
NORDEL	
POLYBUTADIENE	
SILICONE	
POLYURETHANE	

ANTIMONY TRICHLORIDE

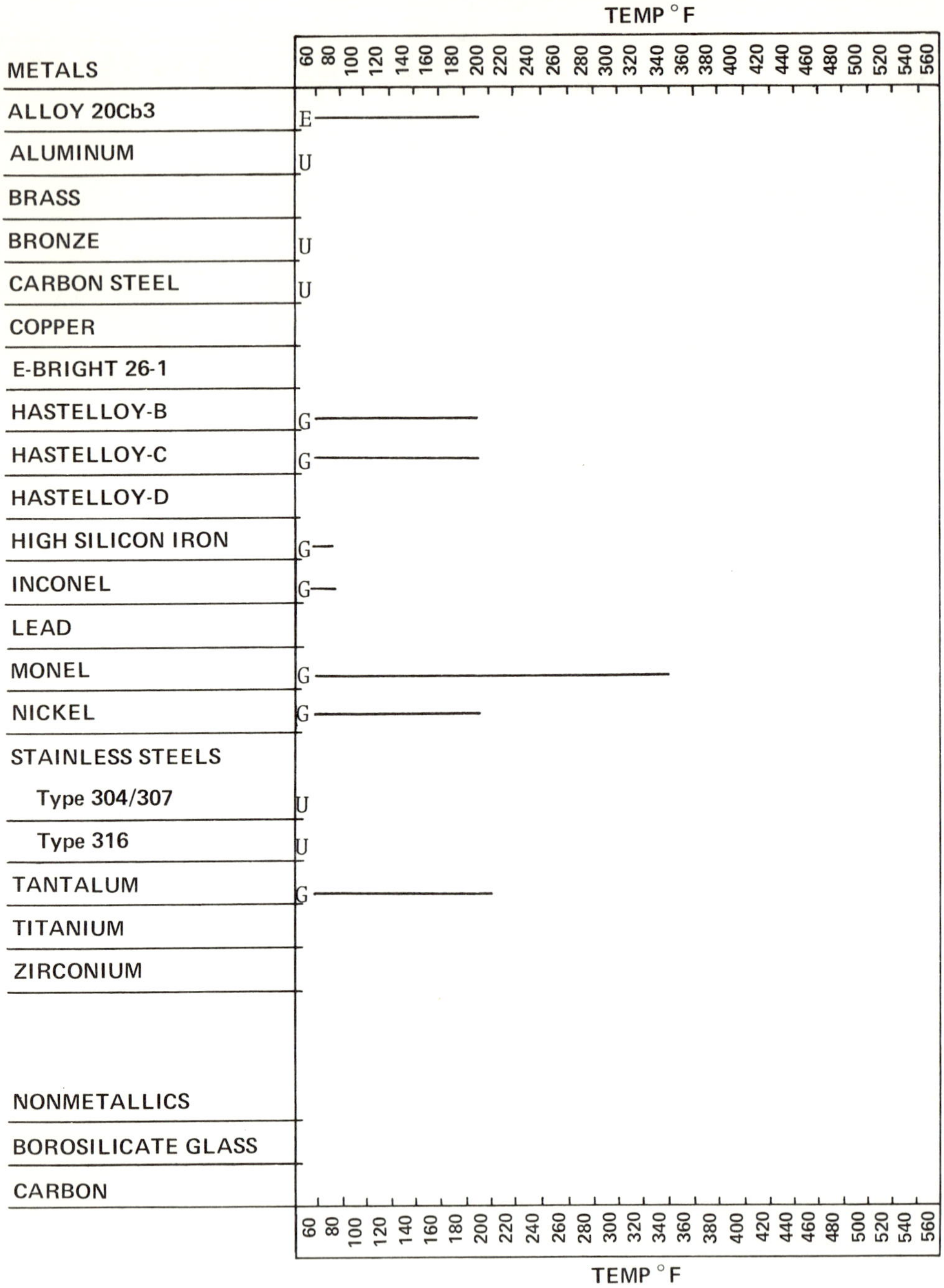

FOR METALS

E = < 2 Mils Penetration/Year; G = < 20 Mils Penetration/Year

S = < 50 Mils Penetration/Year; U = > 50 Mils Penetration/Year

FOR NONMETALLICS

R = Resistant

U = Unsatisfactory

ANTIMONY TRICHLORIDE

PLASTICS	Rating	Max. TEMP °F (end of bar)
ABS	R	140
ACRYLICS		
ASBESTOS REINF., EPOXY		
ASBESTOS REINF., PHENOLIC		
CHLORINATED POLYETHER	R	220
CPVC	R	200
EPOXY	R	180
FLUOROCARBONS FEP	R	400
FLUOROCARBONS TFE	R	460
FURFURYL ALCOHOL	R	250
NORYL		
PHENOLIC		
POLYAMIDES–NYLON		
POLYESTERS		
BISPHENOL A-FUMURATE	R	220
HYDROGENATED BISPHENOL-A-BISPHENOL-A		
ISOPHTHALIC	R	160
CHLORINATED POLYESTERS 50%	R	200
POLYSTYRENES		
POLYETHYLENE	R	150
POLYPROPYLENE	R	180
PVC–TYPE 1	R	140
PVC–TYPE 2	R	140
POLYVINYLIDENE CHLORIDE	R	150
VINYLIDENE FLUORIDE	R	75
VINYL ESTER	R	160

RUBBERS	Rating	Max. TEMP °F (end of bar)
BUTYL GR-1		
FLUORO ELASTOMERS VITON A	R	185
FLUORO ELASTOMERS KEL-F 3700		
HARD RUBBER		
HYPALON	U	
KOROSEAL		
NATURAL RUBBER (GRS)		
NEOPRENE GR-M (CR)	U	
NITRILE BUNA N (NBR)		
NORDEL	R	300
POLYBUTADIENE		
SILICONE		
POLYURETHANE		

TEMP °F: 60 80 100 120 140 160 180 200 220 240 260 280 300 320 340 360 380 400 420 440 460

AQUA REGIA

TEMP °F (scale 60–560 °F)

METALS	Rating
ALLOY 20Cb3	U (60)
ALUMINUM	U (60)
BRASS	U (60)
BRONZE	U (60)
CARBON STEEL	U (60)
COPPER	U (60)
E-BRIGHT 26-1	
HASTELLOY-B	U (60)
HASTELLOY-C	U (60)
HASTELLOY-D	
HIGH SILICON IRON	U (60)
INCONEL	U (60)
LEAD	
MONEL	U (60)
NICKEL	U (60)
STAINLESS STEELS	
Type 304/307	U (60)
Type 316	U (60)
TANTALUM	E (60–180)
TITANIUM	E (60) — S (100–180)
ZIRCONIUM	U (60)

NONMETALLICS	Rating
BOROSILICATE GLASS	R (60–200)
CARBON	

TEMP °F

FOR METALS

E = < 2 Mils Penetration/Year; G = < 20 Mils Penetration/Year

S = < 50 Mils Penetration/Year; U = > 50 Mils Penetration/Year

FOR NONMETALLICS

R = Resistant

U = Unsatisfactory

AQUA REGIA

TEMP °F (60, 80, 100, 120, 140, 160, 180, 200, 220, 240, 260, 280, 300, 320, 340, 360, 380, 400, 420, 440, 460)

PLASTICS	
ABS	U
ACRYLICS	
ASBESTOS REINF., EPOXY	
ASBESTOS REINF., PHENOLIC	
CHLORINATED POLYETHER	R—
CPVC	R—U
EPOXY	U
FLUOROCARBONS FEP	R————————————
FLUOROCARBONS TFE	R——————————————
FURFURYL ALCOHOL	U
NORYL	
PHENOLIC	
POLYAMIDES—NYLON	
POLYESTERS	
BISPHENOL A-FUMURATE	U
HYDROGENATED BISPHENOL-A-BISPHENOL-A	
ISOPHTHALIC	U
CHLORINATED POLYESTERS	
POLYSTYRENES	
POLYETHYLENE	U
POLYPROPYLENE	U
PVC—TYPE 1	RU
PVC—TYPE 2	U
POLYVINYLIDENE CHLORIDE	R—
VINYLIDENE FLUORIDE	R–
VINYL ESTER	U

RUBBERS	
BUTYL GR-1	U
FLUORO ELASTOMERS VITON A	R——————
FLUORO ELASTOMERS KEL-F 3700	
HARD RUBBER	U
HYPALON	
KOROSEAL	U
NATURAL RUBBER (GRS)	U
NEOPRENE GR-M (CR)	U
NITRILE BUNA N (NBR)	U
NORDEL	
POLYBUTADIENE	
SILICONE	
POLYURETHANE	U

TEMP °F (60, 80, 100, 120, 140, 160, 180, 200, 220, 240, 260, 280, 300, 320, 340, 360, 380, 400, 420, 440, 460)

ARSENIC ACID

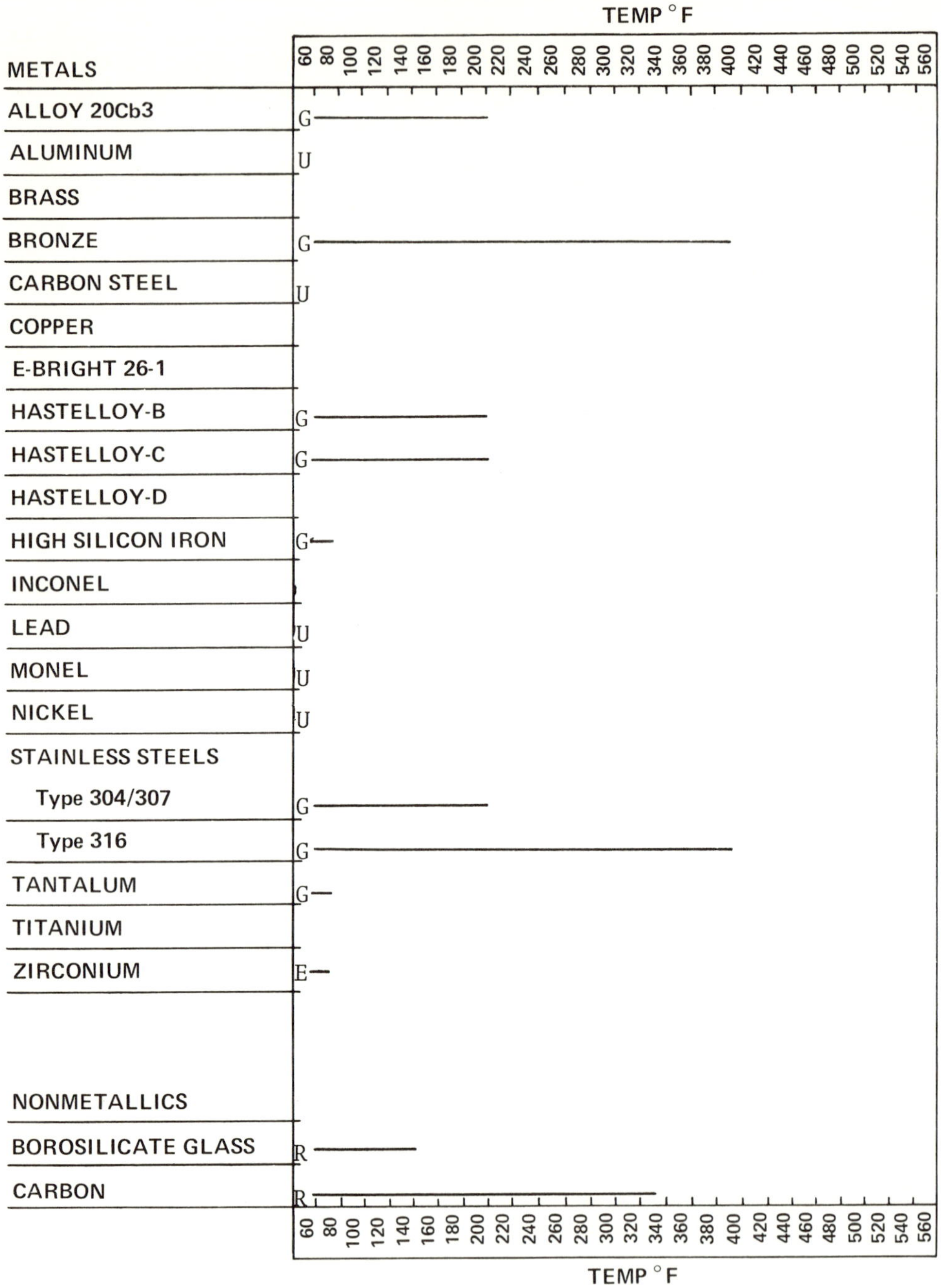

FOR METALS

E = < 2 Mils Penetration/Year; G = < 20 Mils Penetration/Year

S = < 50 Mils Penetration/Year; U = > 50 Mils Penetration/Year

FOR NONMETALLICS

R = Resistant

U = Unsatisfactory

ARSENIC ACID

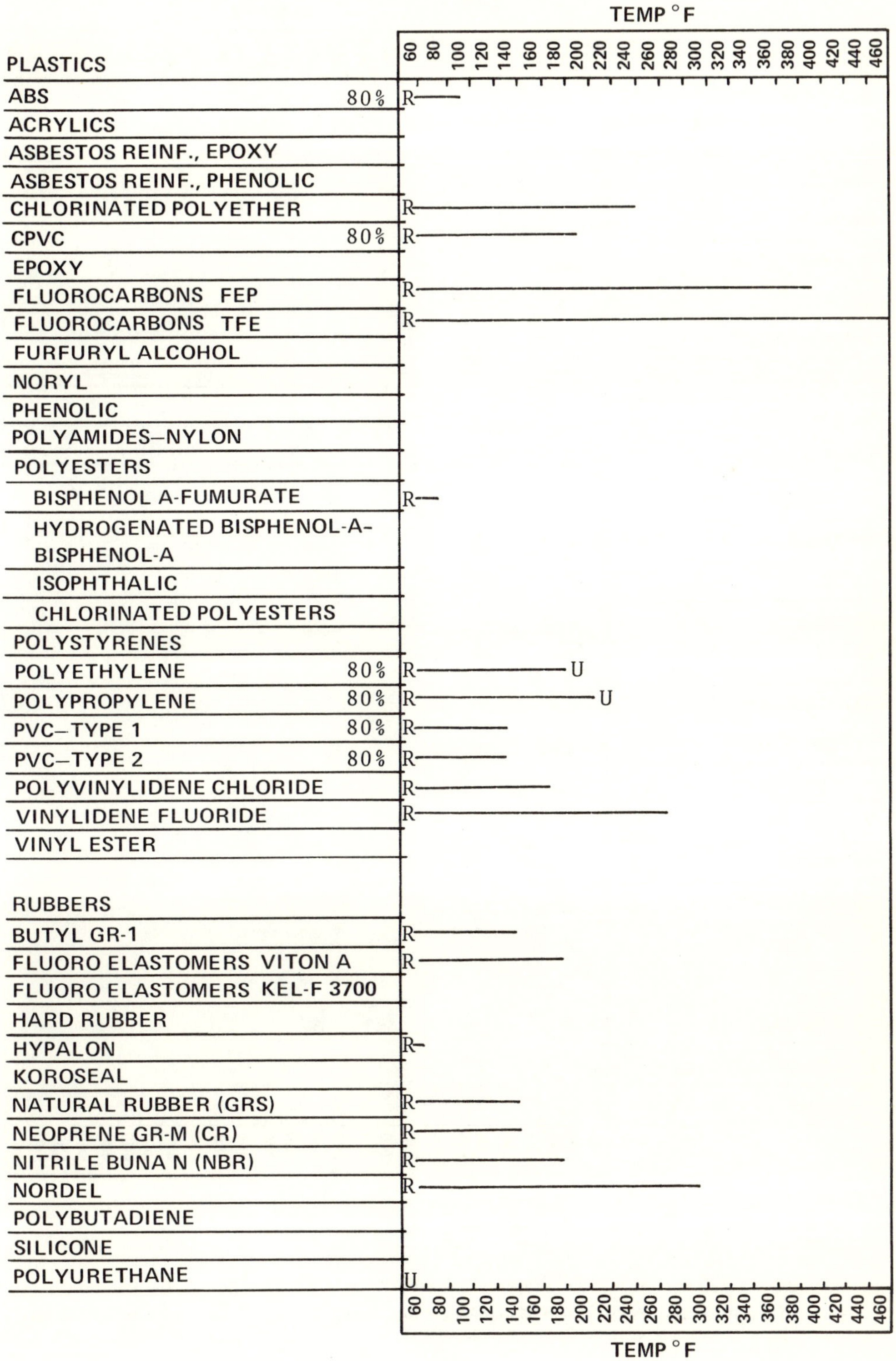

BARIUM CARBONATE (SAT)

TEMP °F

METALS	60 – 560 °F
ALLOY 20Cb3	G
ALUMINUM	U
BRASS	G
BRONZE	G
CARBON STEEL	G
COPPER	
E-BRIGHT 26-1	
HASTELLOY-B	G
HASTELLOY-C	G
HASTELLOY-D	G
HIGH SILICON IRON	G
INCONEL	G
LEAD	
MONEL	G
NICKEL	G
STAINLESS STEELS	
Type 304/307	G
Type 316	G
TANTALUM	G
TITANIUM	E
ZIRCONIUM	E
NONMETALLICS	
BOROSILICATE GLASS	
CARBON	

TEMP °F: 60 80 100 120 140 160 180 200 220 240 260 280 300 320 340 360 380 400 420 440 460 480 500 520 540 560

FOR METALS

E = < 2 Mils Penetration/Year; G = < 20 Mils Penetration/Year

S = < 50 Mils Penetration/Year; U = > 50 Mils Penetration/Year

FOR NONMETALLICS

R = Resistant

U = Unsatisfactory

Please see page 1 for notes on calibration.

BARIUM CARBONATE (SAT)

TEMP °F: 60 80 100 120 140 160 180 200 220 240 260 280 300 320 340 360 380 400 420 440 460

PLASTICS	
ABS	R
ACRYLICS	R
ASBESTOS REINF., EPOXY	
ASBESTOS REINF., PHENOLIC	
CHLORINATED POLYETHER	R
CPVC	R
EPOXY	R
FLUOROCARBONS FEP	R
FLUOROCARBONS TFE	R
FURFURYL ALCOHOL	R
NORYL	R
PHENOLIC	
POLYAMIDES—NYLON	
POLYESTERS	
BISPHENOL A-FUMURATE sat	R
HYDROGENATED BISPHENOL-A-BISPHENOL-A	R
ISOPHTHALIC	R
CHLORINATED POLYESTERS	R
POLYSTYRENES	
POLYETHYLENE	R
POLYPROPYLENE	R
PVC—TYPE 1	R
PVC—TYPE 2	R
POLYVINYLIDENE CHLORIDE	R
VINYLIDENE FLUORIDE	R
VINYL ESTER sat	R

RUBBERS	
BUTYL GR-1	
FLUORO ELASTOMERS VITON A	R
FLUORO ELASTOMERS KEL-F 3700	
HARD RUBBER	
HYPALON	R
KOROSEAL	
NATURAL RUBBER (GRS)	R
NEOPRENE GR-M (CR)	R
NITRILE BUNA N (NBR)	
NORDEL	R
POLYBUTADIENE	
SILICONE	
POLYURETHANE	

TEMP °F: 60 80 100 120 140 160 180 200 220 240 260 280 300 320 340 360 380 400 420 440 460

BARIUM CHLORIDE

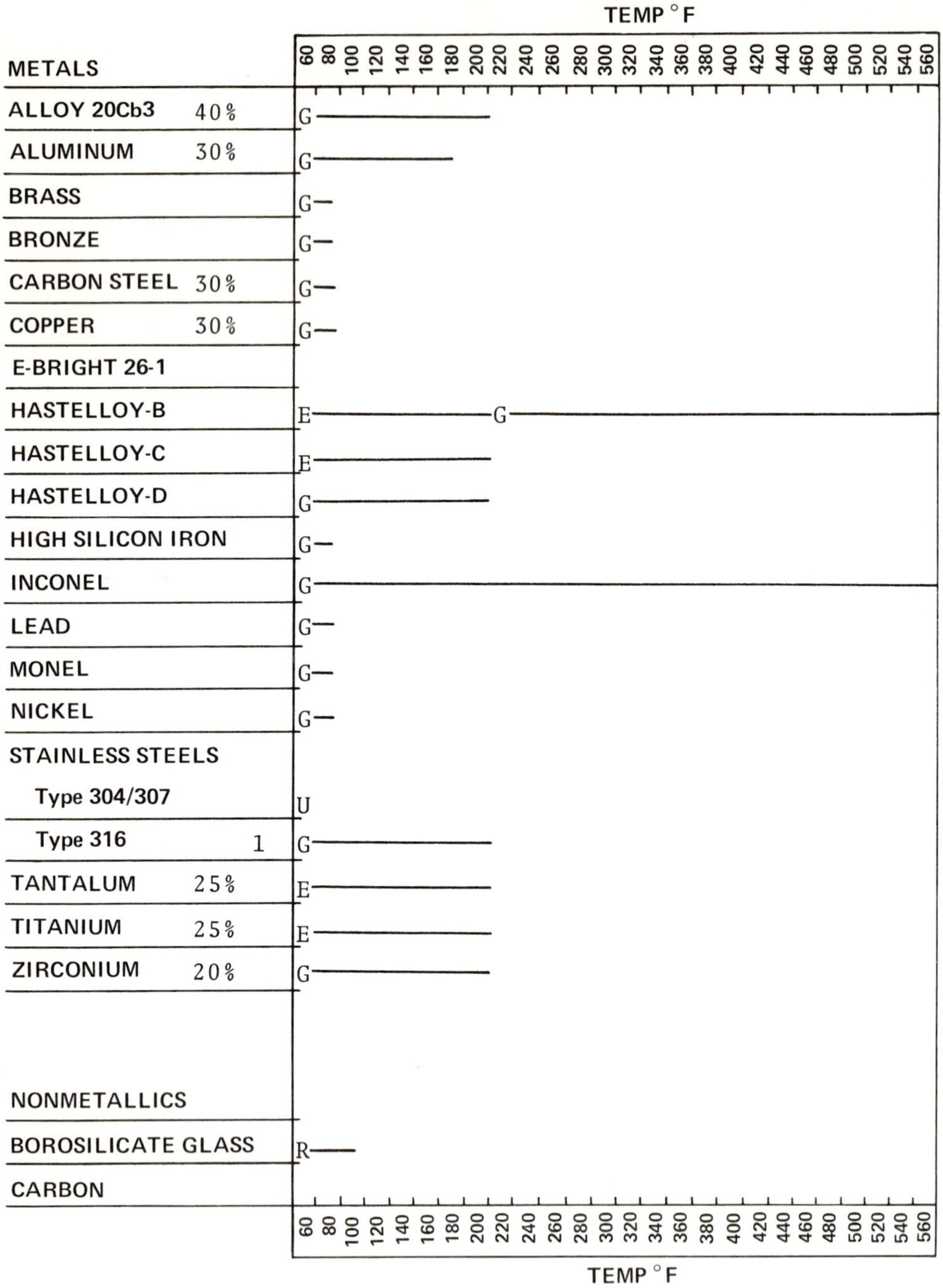

FOR METALS

E = < 2 Mils Penetration/Year; G = < 20 Mils Penetration/Year

S = < 50 Mils Penetration/Year; U = > 50 Mils Penetration/Year

FOR NONMETALLICS

R = Resistant

U = Unsatisfactory

TEMP °F

PLASTICS	Concentration	Rating	Temp. °F (approx. limit of bar)
ABS		R	140
ACRYLICS		R	80
ASBESTOS REINF., EPOXY			
ASBESTOS REINF., PHENOLIC			
CHLORINATED POLYETHER		R	250
CPVC		R	180
EPOXY		R	250
FLUOROCARBONS FEP		R	400
FLUOROCARBONS TFE		R	460
FURFURYL ALCOHOL		R	270
NORYL	25%	R	140
PHENOLIC			
POLYAMIDES–NYLON	10%	R	200
POLYESTERS			
BISPHENOL A-FUMURATE	sat	R	220
HYDROGENATED BISPHENOL-A-BISPHENOL-A		R	210
ISOPHTHALIC	sat	R	150
CHLORINATED POLYESTERS		R	250
POLYSTYRENES			
POLYETHYLENE		R	150
POLYPROPYLENE		R	210
PVC–TYPE 1		R	140
PVC–TYPE 2		R	140
POLYVINYLIDENE CHLORIDE		R	180
VINYLIDENE FLUORIDE		R	280
VINYL ESTER	sat	R	210

RUBBERS	Concentration	Rating	Temp. °F (approx. limit of bar)
BUTYL GR-1		R	180
FLUORO ELASTOMERS VITON A		R	180
FLUORO ELASTOMERS KEL-F 3700			
HARD RUBBER		R	180
HYPALON		R	250
KOROSEAL		R	150
NATURAL RUBBER (GRS)		R	150
NEOPRENE GR-M (CR)		R	200
NITRILE BUNA N (NBR)		R	200
NORDEL		R	300
POLYBUTADIENE			
SILICONE			
POLYURETHANE		R	80

TEMP °F

BARIUM HYDROXIDE

TEMP °F (scale 60–560)

METALS	Concentration	Rating	Approx. upper limit of line (°F)
ALLOY 20Cb3	50%	E	~200
ALUMINUM		U	
BRASS		G	~80
BRONZE		U	
CARBON STEEL		G	~100
COPPER		G	~80
E-BRIGHT 26-1			
HASTELLOY-B		G	~270
HASTELLOY-C		G	~270
HASTELLOY-D		G	~270
HIGH SILICON IRON			
INCONEL		G	~80
LEAD		U	
MONEL		E	~80
NICKEL			
STAINLESS STEELS			
Type 304/307		G	~200
Type 316		G	~400
TANTALUM		G	~80
TITANIUM		E	~220
ZIRCONIUM	50%	E	~200

NONMETALLICS	Rating	Approx. upper limit of line (°F)
BOROSILICATE GLASS	R	~180
CARBON		

FOR METALS

E = < 2 Mils Penetration/Year; G = < 20 Mils Penetration/Year

S = < 50 Mils Penetration/Year; U = > 50 Mils Penetration/Year

FOR NONMETALLICS

R = Resistant

U = Unsatisfactory

BARIUM HYDROXIDE

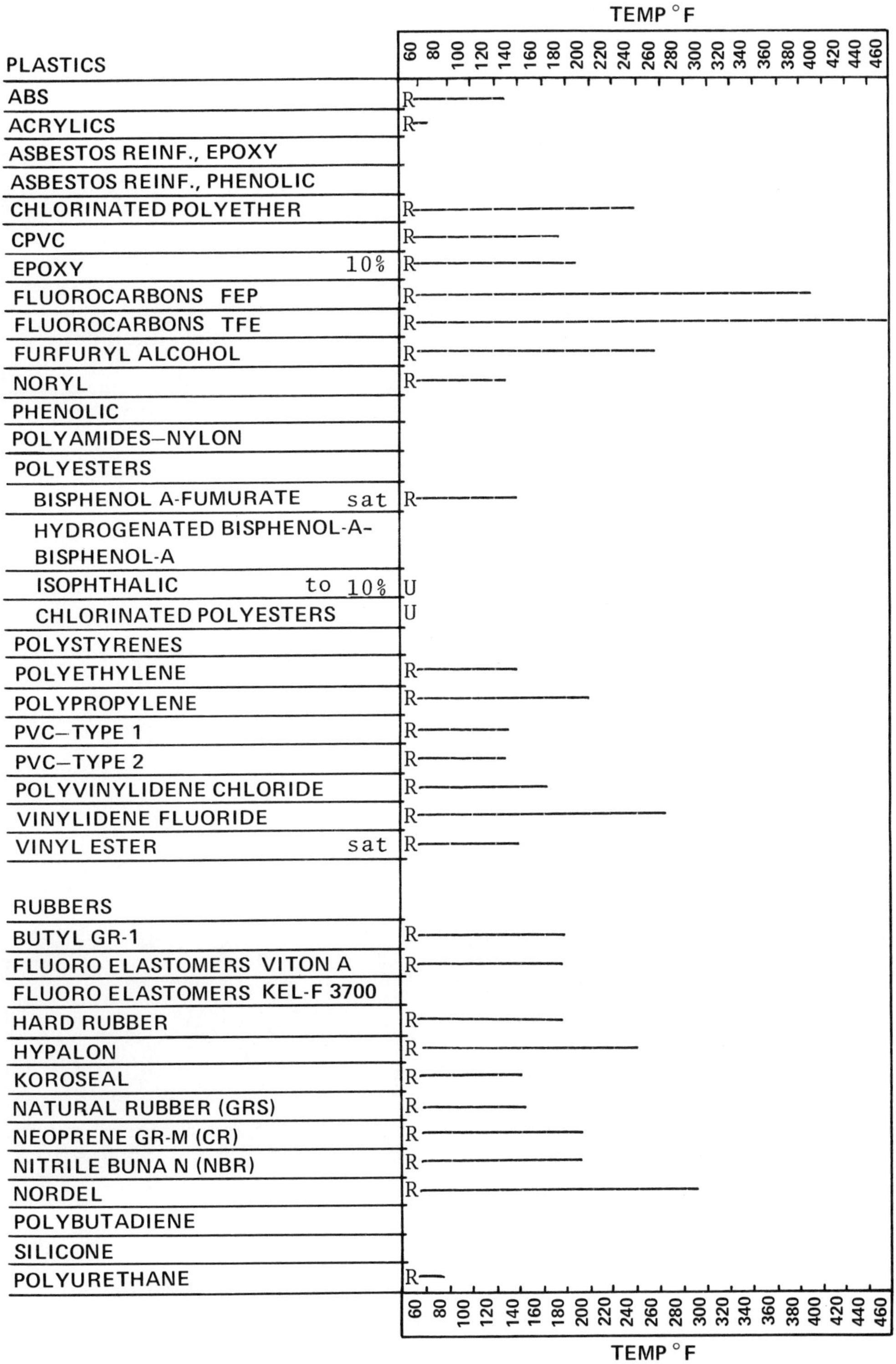

BARIUM NITRATE

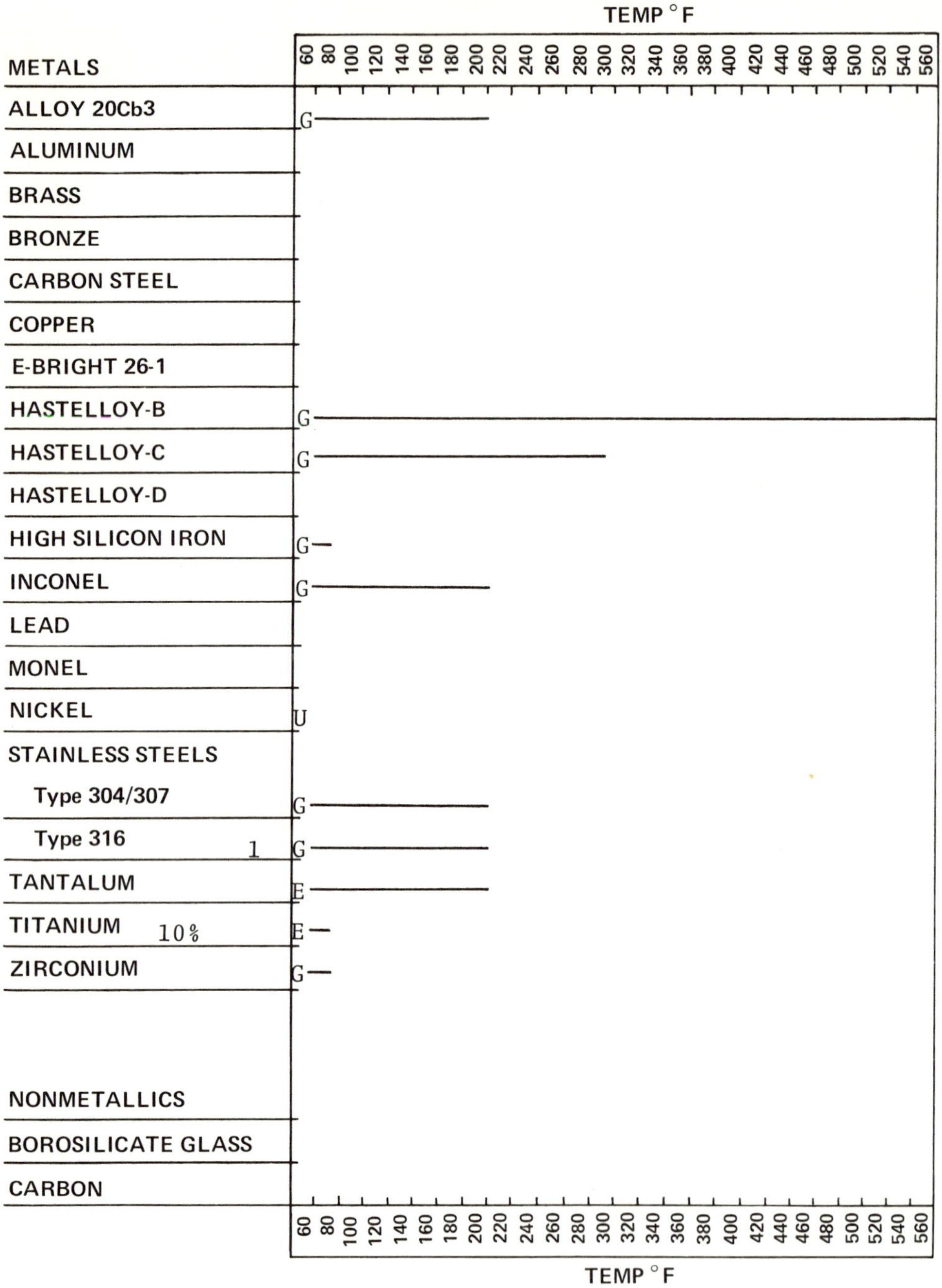

FOR METALS

E = < 2 Mils Penetration/Year; G = < 20 Mils Penetration/Year

S = < 50 Mils Penetration/Year; U = > 50 Mils Penetration/Year

FOR NONMETALLICS

R = Resistant

U = Unsatisfactory

BARIUM NITRATE

TEMP °F (60–460)

PLASTICS	
ABS	
ACRYLICS	
ASBESTOS REINF., EPOXY	
ASBESTOS REINF., PHENOLIC	
CHLORINATED POLYETHER	
CPVC	
EPOXY	
FLUOROCARBONS FEP	R
FLUOROCARBONS TFE	R
FURFURYL ALCOHOL	
NORYL	
PHENOLIC	
POLYAMIDES—NYLON	
POLYESTERS	
BISPHENOL A-FUMURATE	
HYDROGENATED BISPHENOL-A-BISPHENOL-A	
ISOPHTHALIC	
CHLORINATED POLYESTERS	
POLYSTYRENES	
POLYETHYLENE	
POLYPROPYLENE	
PVC—TYPE 1	R
PVC—TYPE 2	
POLYVINYLIDENE CHLORIDE	
VINYLIDENE FLUORIDE	
VINYL ESTER	

RUBBERS	
BUTYL GR-1	
FLUORO ELASTOMERS VITON A	R
FLUORO ELASTOMERS KEL-F 3700	
HARD RUBBER	
HYPALON	
KOROSEAL	R
NATURAL RUBBER (GRS)	
NEOPRENE GR-M (CR)	R
NITRILE BUNA N (NBR)	R
NORDEL	
POLYBUTADIENE	
SILICONE	
POLYURETHANE	

TEMP °F (60–460)

BARIUM SULFATE

TEMP °F

METALS	60–560 °F
ALLOY 20Cb3	G
ALUMINUM	G
BRASS	G
BRONZE	G
CARBON STEEL	S
COPPER	G
E-BRIGHT 26-1	
HASTELLOY-B	G
HASTELLOY-C	
HASTELLOY-D	
HIGH SILICON IRON	G
INCONEL	G
LEAD	
MONEL	G
NICKEL	G
STAINLESS STEELS	
Type 304/307	G
Type 316	G
TANTALUM	G
TITANIUM	E
ZIRCONIUM	E
NONMETALLICS	
BOROSILICATE GLASS	
CARBON	

TEMP °F

FOR METALS

E = < 2 Mils Penetration/Year; G = < 20 Mils Penetration/Year

S = < 50 Mils Penetration/Year; U = > 50 Mils Penetration/Year

FOR NONMETALLICS

R = Resistant

U = Unsatisfactory

BARIUM SULFATE

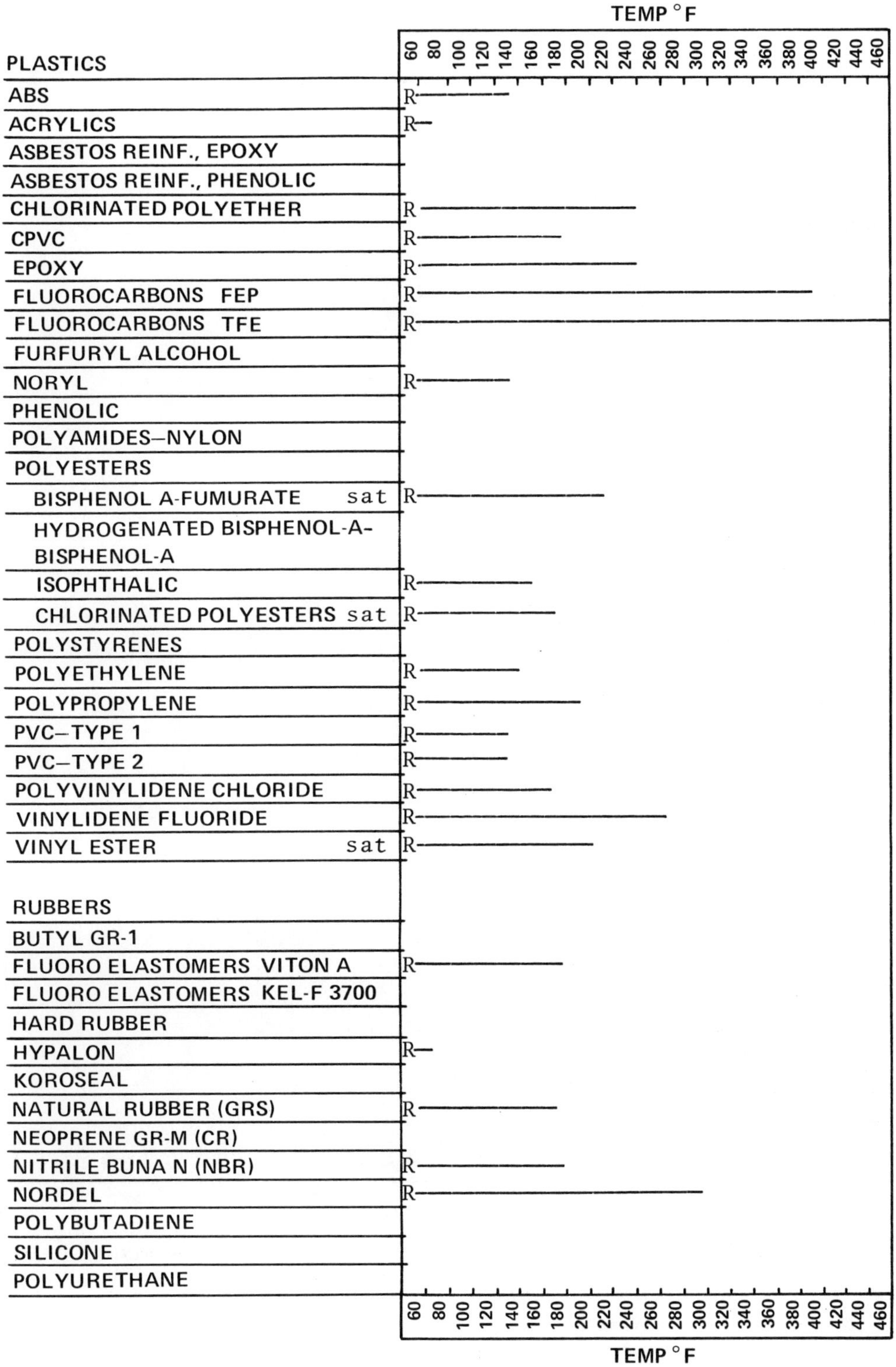

BARIUM SULFIDE

METALS	Rating	TEMP °F (range of line)
ALLOY 20Cb3	G	60–200
ALUMINUM	U	
BRASS	U	
BRONZE	U	
CARBON STEEL	G	60–100
COPPER	U	
E-BRIGHT 26-1		
HASTELLOY-B		
HASTELLOY-C		
HASTELLOY-D		
HIGH SILICON IRON	G	60–80
INCONEL		
LEAD	U	
MONEL	S	60–100
NICKEL	E	60–100
STAINLESS STEELS		
Type 304/307	G	60–210
Type 316	G	60–210
TANTALUM	G	60–80
TITANIUM	G	60–80
ZIRCONIUM	G	60–80

NONMETALLICS	Rating	TEMP °F (range of line)
BOROSILICATE GLASS	R	60–100
CARBON		

TEMP °F scale: 60 80 100 120 140 160 180 200 220 240 260 280 300 320 340 360 380 400 420 440 460 480 500 520 540 560

FOR METALS

E = < 2 Mils Penetration/Year; G = < 20 Mils Penetration/Year

S = < 50 Mils Penetration/Year; U = > 50 Mils Penetration/Year

FOR NONMETALLICS

R = Resistant

U = Unsatisfactory

BARIUM SULFIDE

TEMP °F (60–460)

PLASTICS		
ABS		R
ACRYLICS		
ASBESTOS REINF., EPOXY		
ASBESTOS REINF., PHENOLIC		
CHLORINATED POLYETHER		R
CPVC		R
EPOXY		R
FLUOROCARBONS FEP		R
FLUOROCARBONS TFE		R
FURFURYL ALCOHOL	sat	R
NORYL		R
PHENOLIC		
POLYAMIDES–NYLON		
POLYESTERS		
BISPHENOL A-FUMURATE		R
HYDROGENATED BISPHENOL-A-BISPHENOL-A		
ISOPHTHALIC		R
CHLORINATED POLYESTERS		U
POLYSTYRENES		
POLYETHYLENE		R
POLYPROPYLENE		R
PVC–TYPE 1		R
PVC–TYPE 2		R
POLYVINYLIDENE CHLORIDE		R
VINYLIDENE FLUORIDE		R
VINYL ESTER	sat	R
RUBBERS		
BUTYL GR-1		R
FLUORO ELASTOMERS VITON A		R
FLUORO ELASTOMERS KEL-F 3700		
HARD RUBBER		R
HYPALON		U
KOROSEAL		R
NATURAL RUBBER (GRS)		R
NEOPRENE GR-M (CR)		R
NITRILE BUNA N (NBR)		R
NORDEL		
POLYBUTADIENE		
SILICONE		
POLYURETHANE		R

TEMP °F

BEER

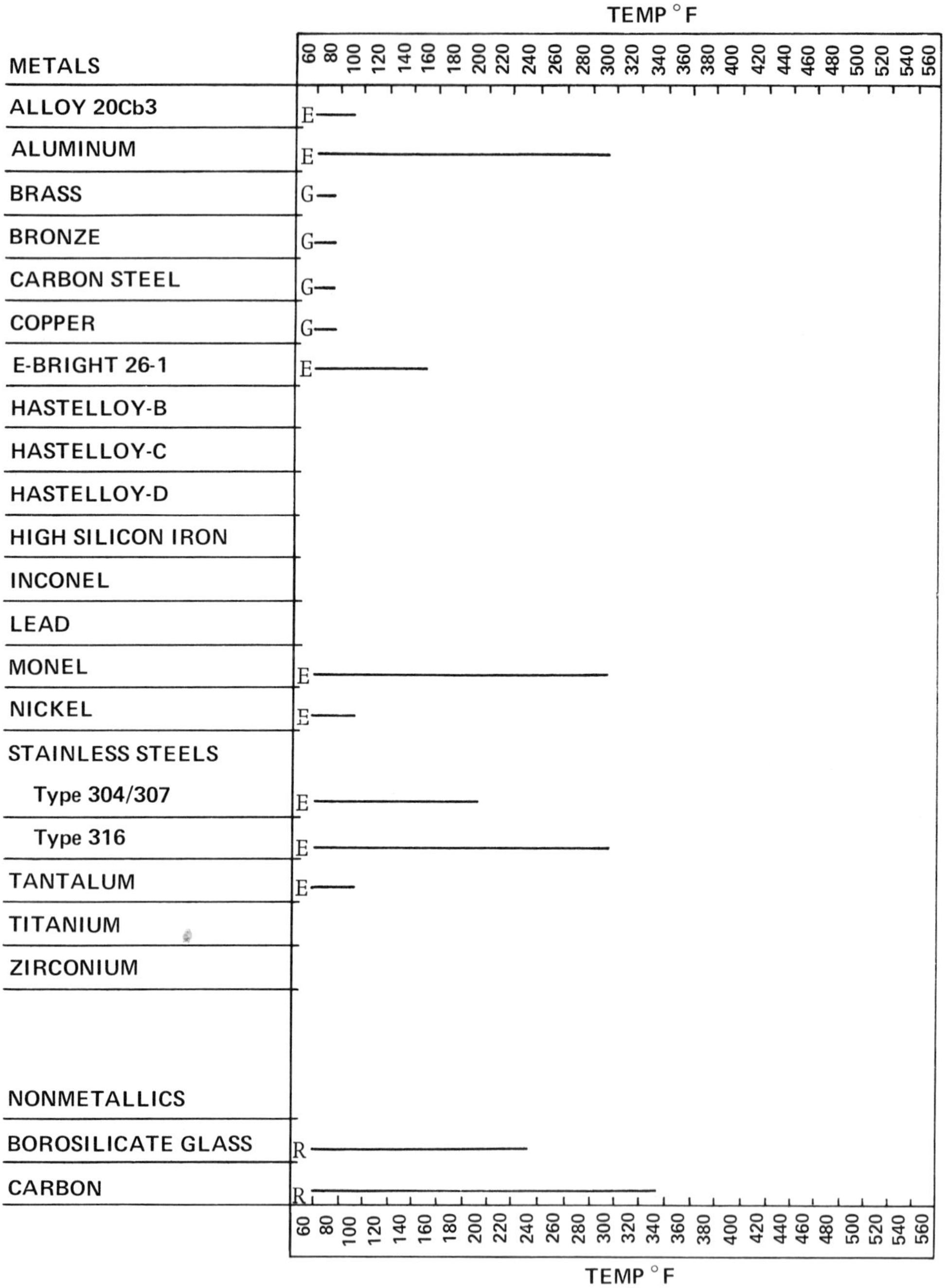

FOR METALS

E = < 2 Mils Penetration/Year; G = < 20 Mils Penetration/Year

S = < 50 Mils Penetration/Year; U = > 50 Mils Penetration/Year

FOR NONMETALLICS

R = Resistant

U = Unsatisfactory

BEER

TEMP °F (60–460)

PLASTICS	
ABS	R
ACRYLICS	U
ASBESTOS REINF., EPOXY	
ASBESTOS REINF., PHENOLIC	
CHLORINATED POLYETHER	R
CPVC	R
EPOXY	R
FLUOROCARBONS FEP	R
FLUOROCARBONS TFE	R
FURFURYL ALCOHOL	
NORYL	
PHENOLIC	
POLYAMIDES—NYLON	R
POLYESTERS	
BISPHENOL A-FUMURATE	R
HYDROGENATED BISPHENOL-A-BISPHENOL-A	
ISOPHTHALIC	
CHLORINATED POLYESTERS	
POLYSTYRENES	
POLYETHYLENE	R
POLYPROPYLENE	R
PVC—TYPE 1	R
PVC—TYPE 2	R
POLYVINYLIDENE CHLORIDE	R
VINYLIDENE FLUORIDE	R
VINYL ESTER	R
RUBBERS	
BUTYL GR-1	R
FLUORO ELASTOMERS VITON A	R
FLUORO ELASTOMERS KEL-F 3700	
HARD RUBBER	R
HYPALON	R
KOROSEAL	
NATURAL RUBBER (GRS)	R
NEOPRENE GR-M (CR)	R
NITRILE BUNA N (NBR)	R
NORDEL	
POLYBUTADIENE	
SILICONE	
POLYURETHANE	U

TEMP °F

BEET SUGAR LIQUORS

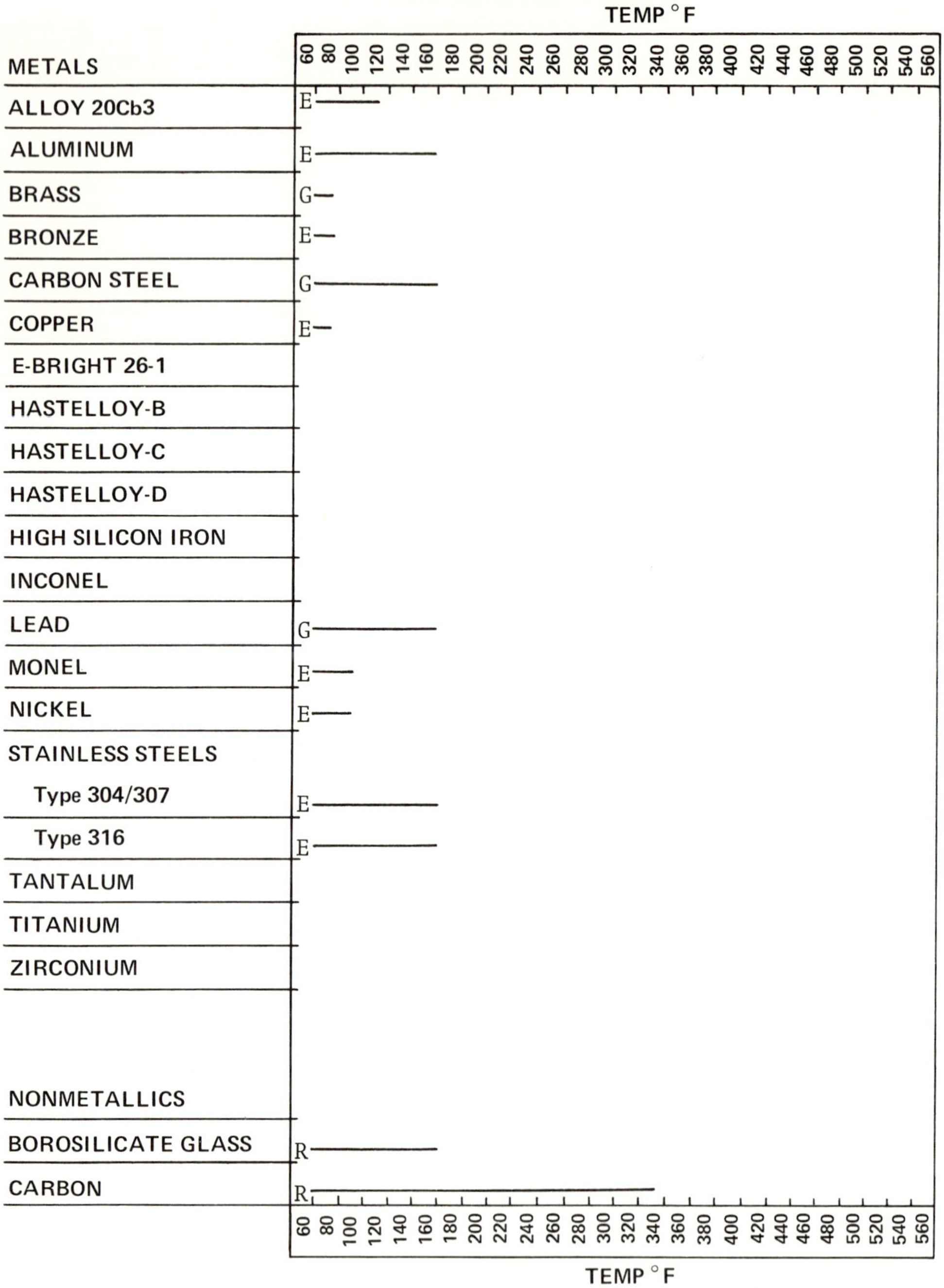

FOR METALS

E = < 2 Mils Penetration/Year; G = < 20 Mils Penetration/Year

S = < 50 Mils Penetration/Year; U = > 50 Mils Penetration/Year

FOR NONMETALLICS

R = Resistant

U = Unsatisfactory

BEET SUGAR LIQUORS

TEMP °F (60–460)

PLASTICS	
ABS	R
ACRYLICS	
ASBESTOS REINF., EPOXY	
ASBESTOS REINF., PHENOLIC	
CHLORINATED POLYETHER	R
CPVC	R
EPOXY	
FLUOROCARBONS FEP	R
FLUOROCARBONS TFE	R
FURFURYL ALCOHOL	
NORYL	
PHENOLIC	
POLYAMIDES—NYLON	
POLYESTERS	
BISPHENOL A-FUMURATE	R
HYDROGENATED BISPHENOL-A-BISPHENOL-A	
ISOPHTHALIC	
CHLORINATED POLYESTERS	
POLYSTYRENES	
POLYETHYLENE	R
POLYPROPYLENE	R
PVC—TYPE 1	R
PVC—TYPE 2	R
POLYVINYLIDENE CHLORIDE	R
VINYLIDENE FLUORIDE	
VINYL ESTER	R

RUBBERS	
BUTYL GR-1	R
FLUORO ELASTOMERS VITON A	R
FLUORO ELASTOMERS KEL-F 3700	
HARD RUBBER	R
HYPALON	R
KOROSEAL	
NATURAL RUBBER (GRS)	R
NEOPRENE GR-M (CR)	R
NITRILE BUNA N (NBR)	R
NORDEL	
POLYBUTADIENE	
SILICONE	
POLYURETHANE	U

TEMP °F (60–460)

BENZALDEHYDE

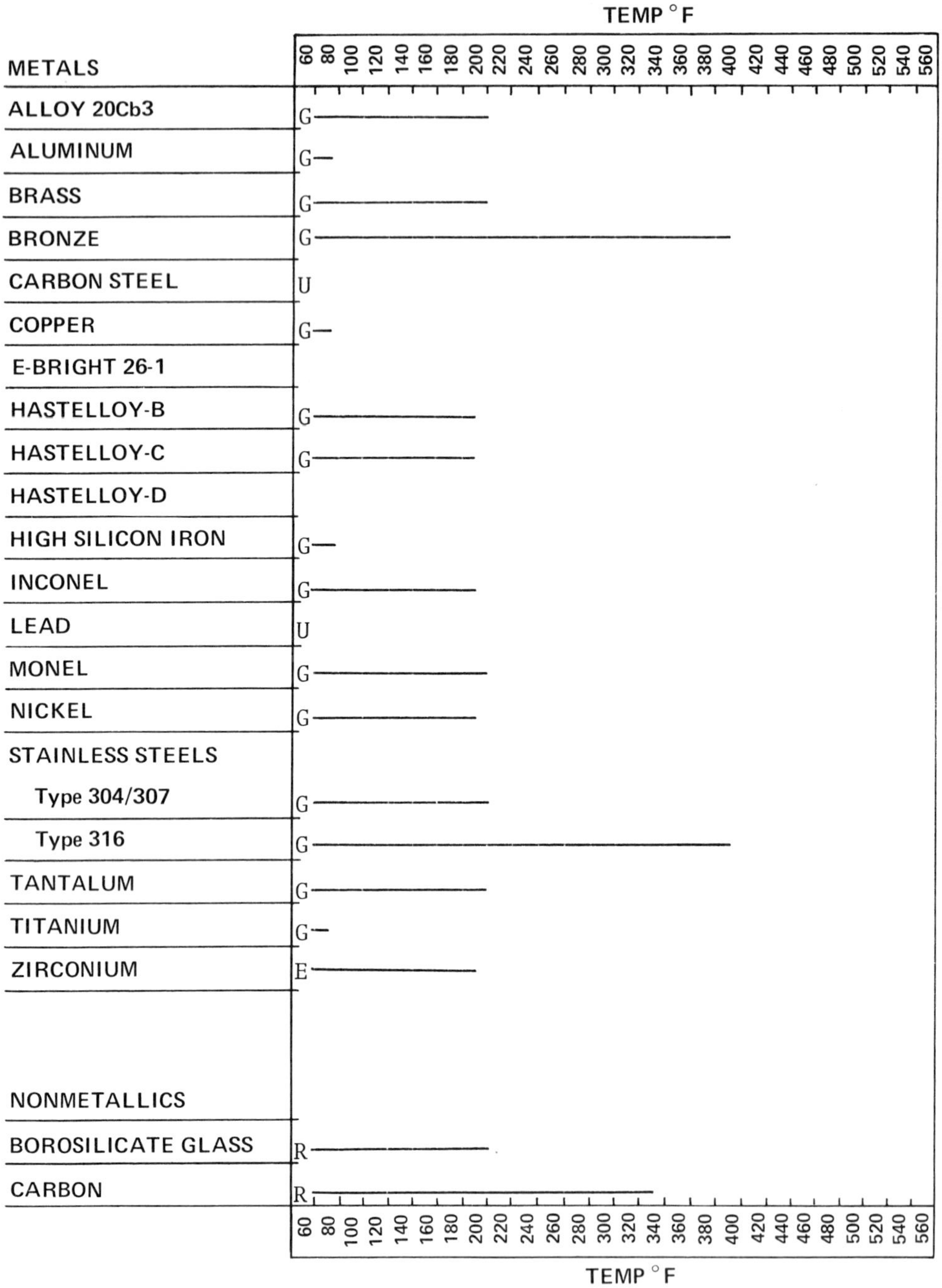

FOR METALS

E = < 2 Mils Penetration/Year; G = < 20 Mils Penetration/Year

S = < 50 Mils Penetration/Year; U = > 50 Mils Penetration/Year

FOR NONMETALLICS

R = Resistant

U = Unsatisfactory

BENZALDEHYDE

TEMP °F (60, 80, 100, 120, 140, 160, 180, 200, 220, 240, 260, 280, 300, 320, 340, 360, 380, 400, 420, 440, 460)

PLASTICS	
ABS	U
ACRYLICS	
ASBESTOS REINF., EPOXY	RU
ASBESTOS REINF., PHENOLIC	RU
CHLORINATED POLYETHER	R—
CPVC	U
EPOXY	U
FLUOROCARBONS FEP	R—
FLUOROCARBONS TFE	R—
FURFURYL ALCOHOL	R—
NORYL	
PHENOLIC	
POLYAMIDES–NYLON	R—
POLYESTERS	
BISPHENOL A-FUMURATE	U
HYDROGENATED BISPHENOL-A-BISPHENOL-A	U
ISOPHTHALIC	U
CHLORINATED POLYESTERS	U
POLYSTYRENES	
POLYETHYLENE	U
POLYPROPYLENE	R—
PVC–TYPE 1	U
PVC–TYPE 2	U
POLYVINYLIDENE CHLORIDE	U
VINYLIDENE FLUORIDE	R—
VINYL ESTER	U

RUBBERS	
BUTYL GR-1	R—
FLUORO ELASTOMERS VITON A	U
FLUORO ELASTOMERS KEL-F 3700	
HARD RUBBER	U
HYPALON	U
KOROSEAL	
NATURAL RUBBER (GRS)	U
NEOPRENE GR-M (CR)	U
NITRILE BUNA N (NBR)	U
NORDEL	
POLYBUTADIENE	
SILICONE	
POLYURETHANE	U

TEMP °F

BENZENE, BENZOL

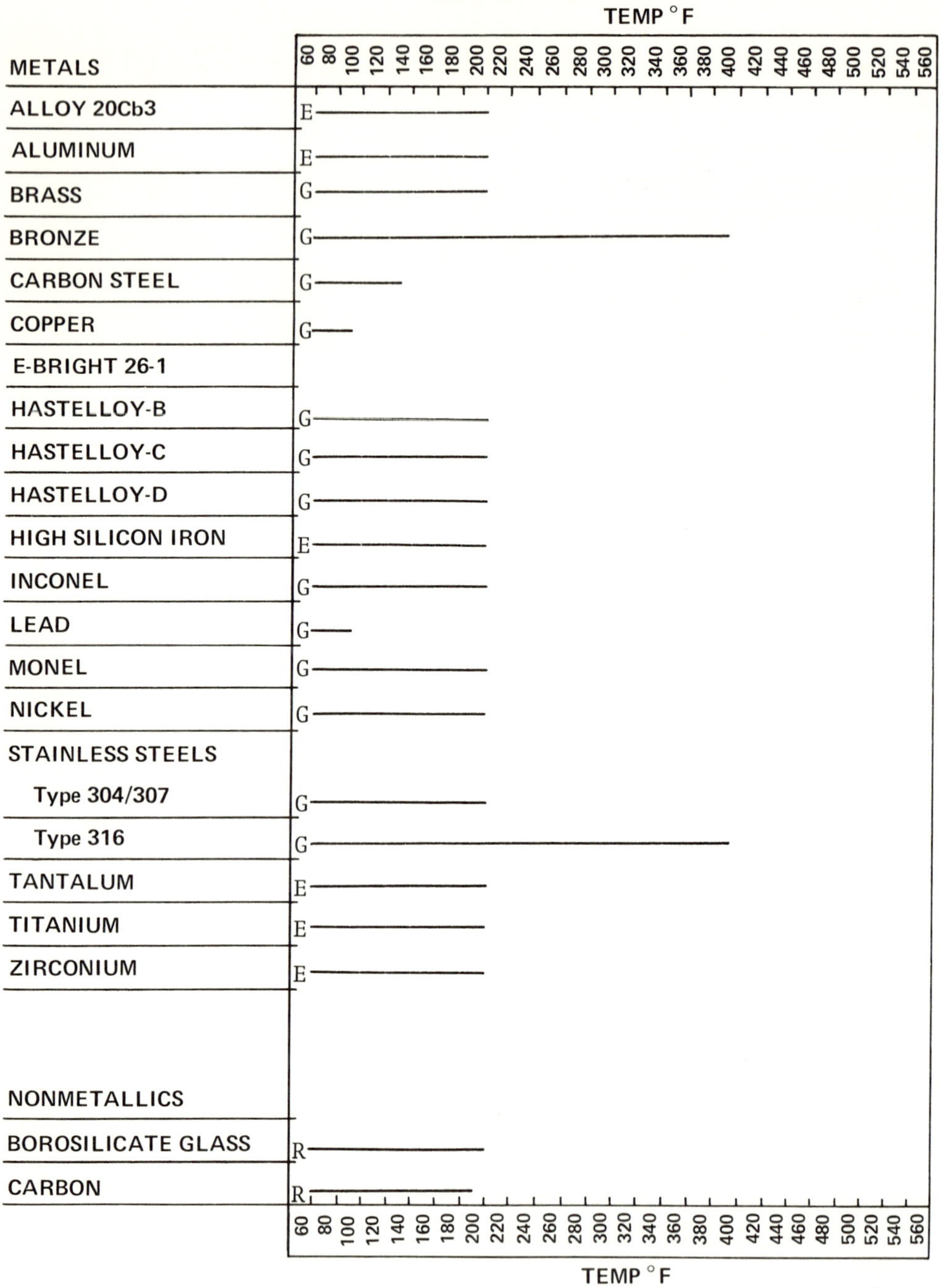

FOR METALS

E = < 2 Mils Penetration/Year; G = < 20 Mils Penetration/Year

S = < 50 Mils Penetration/Year; U = > 50 Mils Penetration/Year

FOR NONMETALLICS

R = Resistant

U = Unsatisfactory

BENZENE, BENZOL

TEMP °F (60–460)

PLASTICS	
ABS	U
ACRYLICS	U
ASBESTOS REINF., EPOXY	RU
ASBESTOS REINF., PHENOLIC	R——
CHLORINATED POLYETHER	R——U
CPVC	U
EPOXY	R——
FLUOROCARBONS FEP	R——
FLUOROCARBONS TFE	R——
FURFURYL ALCOHOL	R——
NORYL	
PHENOLIC	R—
POLYAMIDES–NYLON	R—
POLYESTERS	
BISPHENOL A-FUMURATE	U
HYDROGENATED BISPHENOL-A-BISPHENOL-A	U
ISOPHTHALIC	U
CHLORINATED POLYESTERS	R—
POLYSTYRENES	U
POLYETHYLENE	U
POLYPROPYLENE	R——
PVC–TYPE 1	U
PVC–TYPE 2	U
POLYVINYLIDENE CHLORIDE	R—
VINYLIDENE FLUORIDE	R——U
VINYL ESTER	U
RUBBERS	
BUTYL GR-1	U
FLUORO ELASTOMERS VITON A	R——
FLUORO ELASTOMERS KEL-F 3700	R—
HARD RUBBER	U
HYPALON	U
KOROSEAL	U
NATURAL RUBBER (GRS)	U
NEOPRENE GR-M (CR)	U
NITRILE BUNA N (NBR)	U
NORDEL	U
POLYBUTADIENE	
SILICONE	
POLYURETHANE	U

TEMP °F (60–460)

BENZENE SULFONIC ACID, 10%

TEMP °F (scale: 60, 80, 100, 120, 140, 160, 180, 200, 220, 240, 260, 280, 300, 320, 340, 360, 380, 400, 420, 440, 460, 480, 500, 520, 540, 560)

METALS	Rating	Temperature range (°F, as charted)
ALLOY 20Cb3	G	60–210
ALUMINUM	U	60
BRASS	G	60–70
BRONZE	G	60–70
CARBON STEEL	U	60
COPPER		
E-BRIGHT 26-1		
HASTELLOY-B	G	60–210
HASTELLOY-C	G	60–210
HASTELLOY-D	G	60–210
HIGH SILICON IRON	G	60–70
INCONEL	G	60–210
LEAD	G	60–70
MONEL	G	60–210
NICKEL	G → U	G 60–180; U at ~190
STAINLESS STEELS		
Type 304/307	G	60–210
Type 316	G	60–210
TANTALUM	G	60–210
TITANIUM		
ZIRCONIUM	E	60–210

NONMETALLICS	Rating	Temperature range (°F, as charted)
BOROSILICATE GLASS	R	60–200
CARBON	R	60–340

FOR METALS

E = < 2 Mils Penetration/Year; G = < 20 Mils Penetration/Year

S = < 50 Mils Penetration/Year; U = > 50 Mils Penetration/Year

FOR NONMETALLICS

R = Resistant

U = Unsatisfactory

BENZENE SULFONIC ACID, 10%

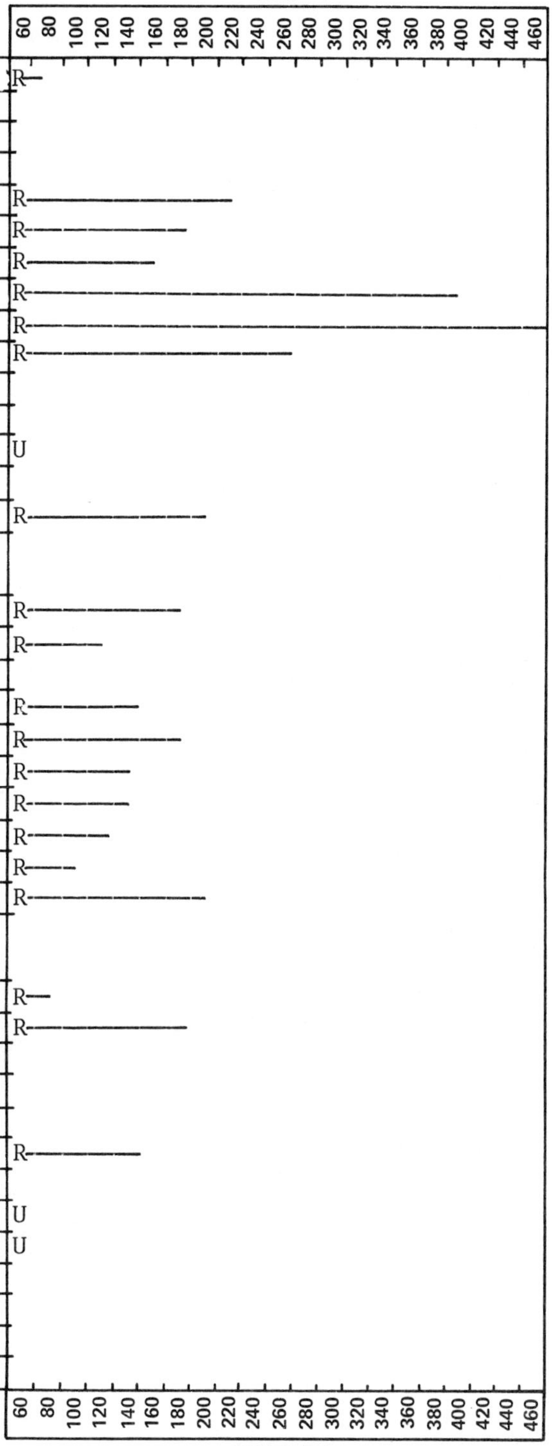

PLASTICS	Conc.	Rating
ABS	10%	R
ACRYLICS		
ASBESTOS REINF., EPOXY		
ASBESTOS REINF., PHENOLIC		
CHLORINATED POLYETHER		R
CPVC		R
EPOXY		R
FLUOROCARBONS FEP		R
FLUOROCARBONS TFE		R
FURFURYL ALCOHOL	sat	R
NORYL		
PHENOLIC		
POLYAMIDES—NYLON		U
POLYESTERS		
BISPHENOL A-FUMURATE		R
HYDROGENATED BISPHENOL-A-BISPHENOL-A		
ISOPHTHALIC	30%	R
CHLORINATED POLYESTERS	all	R
POLYSTYRENES		
POLYETHYLENE		R
POLYPROPYLENE		R
PVC—TYPE 1		R
PVC—TYPE 2		R
POLYVINYLIDENE CHLORIDE		R
VINYLIDENE FLUORIDE		R
VINYL ESTER		R

RUBBERS	Conc.	Rating
BUTYL GR-1		R
FLUORO ELASTOMERS VITON A		R
FLUORO ELASTOMERS KEL-F 3700		
HARD RUBBER		
HYPALON		
KOROSEAL		R
NATURAL RUBBER (GRS)		
NEOPRENE GR-M (CR)		U
NITRILE BUNA N (NBR)		U
NORDEL		
POLYBUTADIENE		
SILICONE		
POLYURETHANE		

BENZOIC ACID

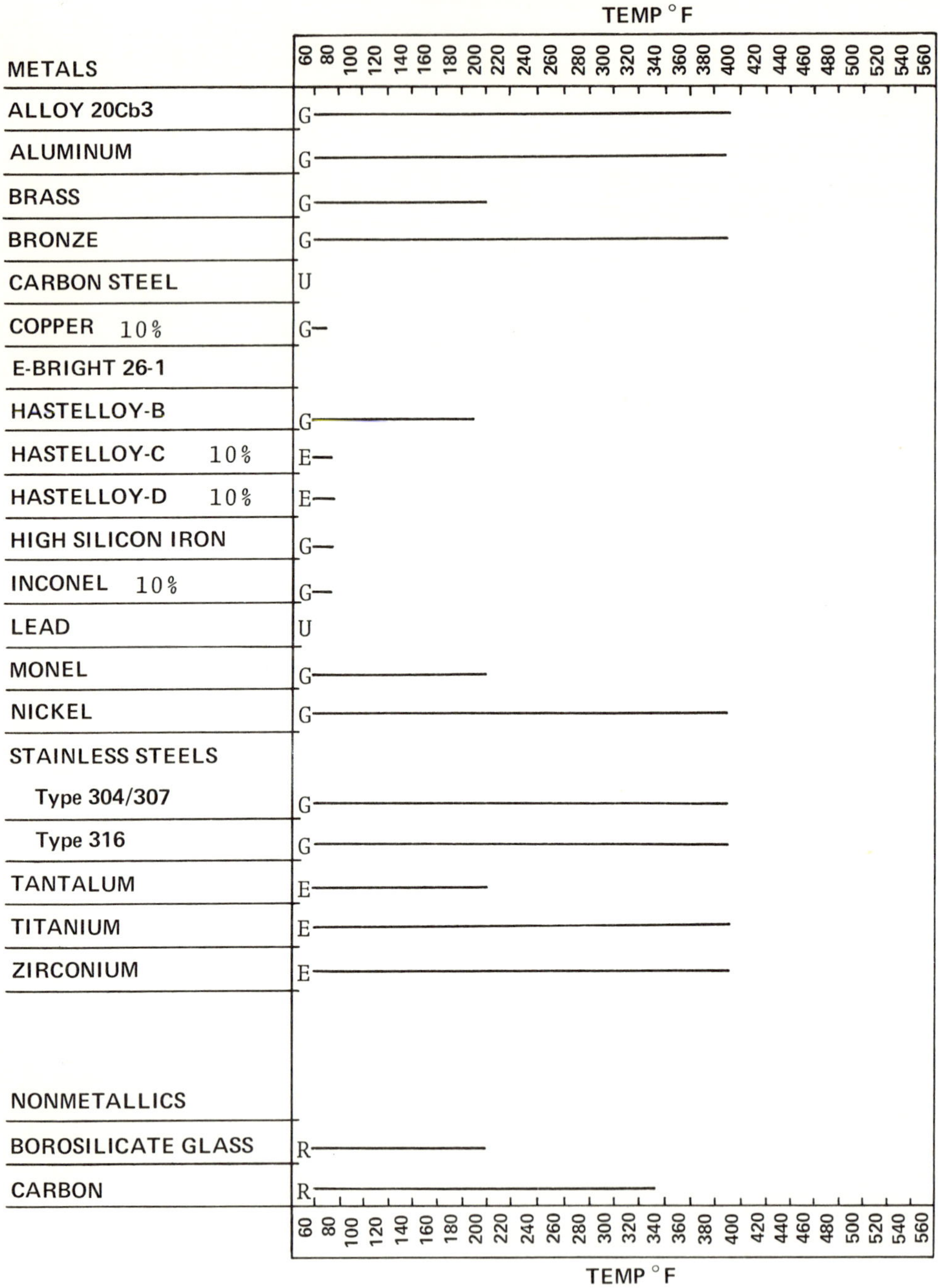

FOR METALS

E = < 2 Mils Penetration/Year; G = < 20 Mils Penetration/Year

S = < 50 Mils Penetration/Year; U = > 50 Mils Penetration/Year

FOR NONMETALLICS

R = Resistant

U = Unsatisfactory

TEMP °F (scale 60–460)

PLASTICS	Conc.	Rating / range (TEMP °F)
ABS		R — to 140
ACRYLICS		U
ASBESTOS REINF., EPOXY		
ASBESTOS REINF., PHENOLIC		
CHLORINATED POLYETHER		R — to 250
CPVC		R — to 200
EPOXY		R — to 200
FLUOROCARBONS FEP		R — to 400
FLUOROCARBONS TFE		R — to 460
FURFURYL ALCOHOL	sat	R — to 270
NORYL		
PHENOLIC		
POLYAMIDES–NYLON		R — to 80
POLYESTERS		
BISPHENOL A-FUMURATE	sat	R — to 180
HYDROGENATED BISPHENOL-A-BISPHENOL-A		R — to 210
ISOPHTHALIC	sat	R — to 180
CHLORINATED POLYESTERS		R — to 250
POLYSTYRENES		
POLYETHYLENE		R — to 190, U at 200
POLYPROPYLENE		R — to 190, U at 200
PVC–TYPE 1		R — to 140
PVC–TYPE 2		R — to 140
POLYVINYLIDENE CHLORIDE		R — to 125
VINYLIDENE FLUORIDE		R — to 250
VINYL ESTER	sat	R — to 180

RUBBERS	Rating / range (TEMP °F)
BUTYL GR-1	
FLUORO ELASTOMERS VITON A	R — to 185
FLUORO ELASTOMERS KEL-F 3700	
HARD RUBBER	R — to 150
HYPALON	U
KOROSEAL	R — to 150
NATURAL RUBBER (GRS)	R — to 150
NEOPRENE GR-M (CR)	R — to 150
NITRILE BUNA N (NBR)	U
NORDEL	R — to 300
POLYBUTADIENE	
SILICONE	
POLYURETHANE	U

BENZYL ALCOHOL

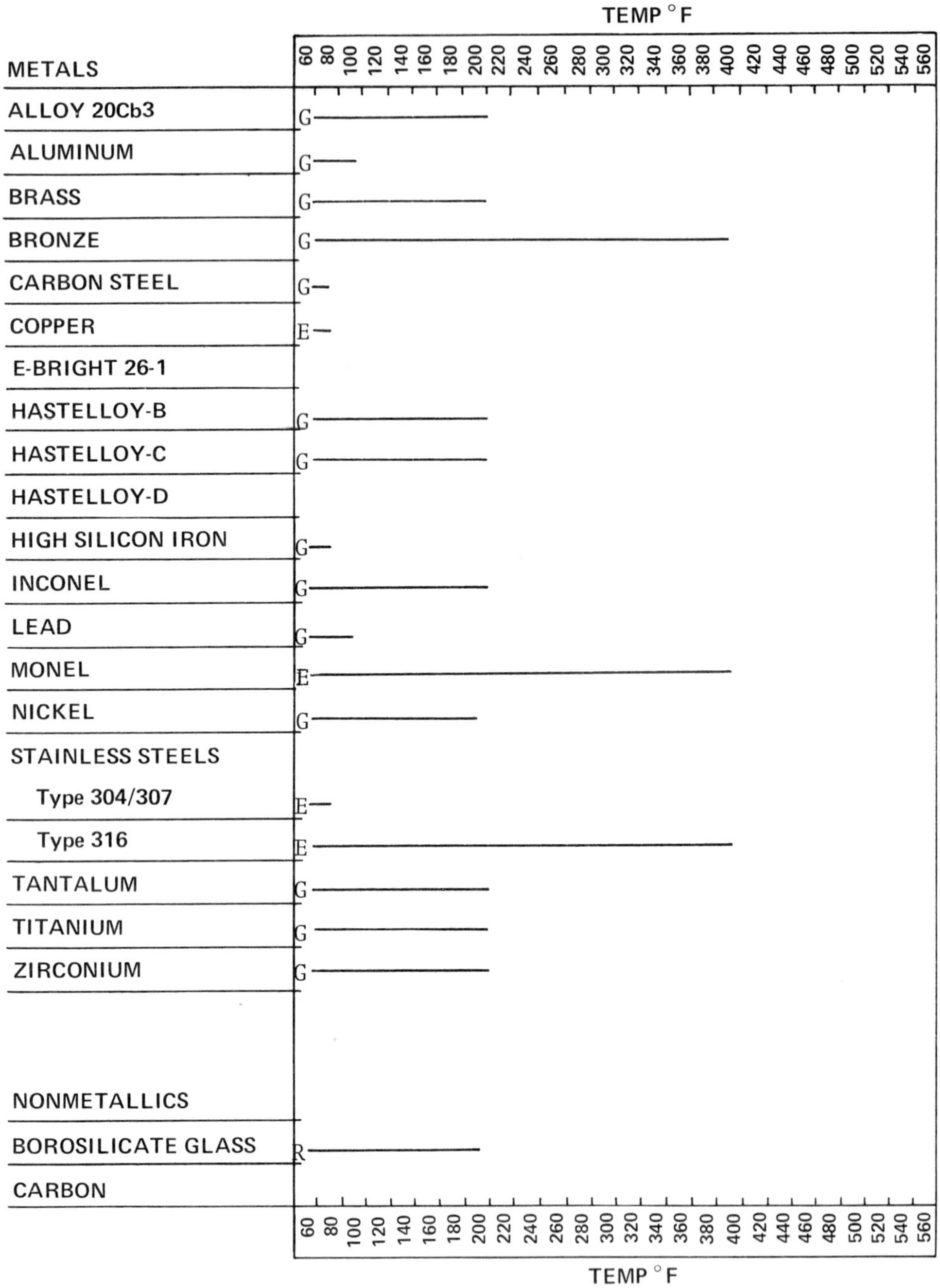

FOR METALS

E = < 2 Mils Penetration/Year; G = < 20 Mils Penetration/Year

S = < 50 Mils Penetration/Year; U = > 50 Mils Penetration/Year

FOR NONMETALLICS

R = Resistant

U = Unsatisfactory

BENZYL ALCOHOL

TEMP °F (60–460)

PLASTICS	Rating
ABS	U
ACRYLICS	
ASBESTOS REINF., EPOXY	
ASBESTOS REINF., PHENOLIC	
CHLORINATED POLYETHER	
CPVC	
EPOXY	U
FLUOROCARBONS FEP	R (60 to ~400)
FLUOROCARBONS TFE	R (60 to 460)
FURFURYL ALCOHOL	R (60 to ~270)
NORYL	
PHENOLIC	
POLYAMIDES–NYLON	R (60 to ~75)
POLYESTERS	
BISPHENOL A-FUMURATE	U
HYDROGENATED BISPHENOL-A-BISPHENOL-A	
ISOPHTHALIC	U
CHLORINATED POLYESTERS	
POLYSTYRENES	
POLYETHYLENE	U
POLYPROPYLENE	R (60 to ~140), U (~160)
PVC–TYPE 1	
PVC–TYPE 2	
POLYVINYLIDENE CHLORIDE	
VINYLIDENE FLUORIDE	R (60 to ~280)
VINYL ESTER	R (60 to ~90)

RUBBERS	Rating
BUTYL GR-1	R (60 to ~180)
FLUORO ELASTOMERS VITON A	R (60 to ~350)
FLUORO ELASTOMERS KEL-F 3700	
HARD RUBBER	
HYPALON	U
KOROSEAL	
NATURAL RUBBER (GRS)	U
NEOPRENE GR-M (CR)	U
NITRILE BUNA N (NBR)	U
NORDEL	
POLYBUTADIENE	
SILICONE	
POLYURETHANE	U

BENZYL CHLORIDE

METALS	TEMP °F
ALLOY 20Cb3	G
ALUMINUM	U
BRASS	U
BRONZE	U
CARBON STEEL	U
COPPER	U
E-BRIGHT 26-1	
HASTELLOY-B	G
HASTELLOY-C	
HASTELLOY-D	
HIGH SILICON IRON	G
INCONEL	G
LEAD	G
MONEL	G
NICKEL	G
STAINLESS STEELS	
Type 304/307	G
Type 316	G
TANTALUM	G
TITANIUM	
ZIRCONIUM	

NONMETALLICS	TEMP °F
BOROSILICATE GLASS	R
CARBON	

TEMP °F: 60 80 100 120 140 160 180 200 220 240 260 280 300 320 340 360 380 400 420 440 460 480 500 520 540 560

FOR METALS

E = < 2 Mils Penetration/Year; G = < 20 Mils Penetration/Year

S = < 50 Mils Penetration/Year; U = > 50 Mils Penetration/Year

FOR NONMETALLICS

R = Resistant

U = Unsatisfactory

BENZYL CHLORIDE

TEMP °F: 60 80 100 120 140 160 180 200 220 240 260 280 300 320 340 360 380 400 420 440 460

PLASTICS	
ABS	U
ACRYLICS	
ASBESTOS REINF., EPOXY	
ASBESTOS REINF., PHENOLIC	
CHLORINATED POLYETHER	R
CPVC	
EPOXY	
FLUOROCARBONS FEP	R
FLUOROCARBONS TFE	R
FURFURYL ALCOHOL	R
NORYL	
PHENOLIC	
POLYAMIDES–NYLON	R
POLYESTERS	
BISPHENOL A-FUMURATE	U
HYDROGENATED BISPHENOL-A-BISPHENOL-A	U
ISOPHTHALIC	U
CHLORINATED POLYESTERS	U
POLYSTYRENES	
POLYETHYLENE	
POLYPROPYLENE	R
PVC–TYPE 1	
PVC–TYPE 2	
POLYVINYLIDENE CHLORIDE	
VINYLIDENE FLUORIDE	R
VINYL ESTER	R

RUBBERS	
BUTYL GR-1	U
FLUORO ELASTOMERS VITON A	
FLUORO ELASTOMERS KEL-F 3700	
HARD RUBBER	
HYPALON	
KOROSEAL	
NATURAL RUBBER (GRS)	U
NEOPRENE GR-M (CR)	U
NITRILE BUNA N (NBR)	U
NORDEL	
POLYBUTADIENE	
SILICONE	
POLYURETHANE	

TEMP °F: 60 80 100 120 140 160 180 200 220 240 260 280 300 320 340 360 380 400 420 440 460

BISMUTH CARBONATE

TEMP °F

METALS	60	80	100	120	140	160	180	200	220	240	260	280	300	320	340	360	380	400	420	440	460	480	500	520	540	560
ALLOY 20Cb3																										
ALUMINUM																										
BRASS																										
BRONZE																										
CARBON STEEL																										
COPPER																										
E-BRIGHT 26-1																										
HASTELLOY-B																										
HASTELLOY-C																										
HASTELLOY-D																										
HIGH SILICON IRON																										
INCONEL																										
LEAD																										
MONEL	G—																									
NICKEL																										
STAINLESS STEELS																										
Type 304/307																										
Type 316																										
TANTALUM	G—																									
TITANIUM																										
ZIRCONIUM																										
NONMETALLICS																										
BOROSILICATE GLASS																										
CARBON																										

60 80 100 120 140 160 180 200 220 240 260 280 300 320 340 360 380 400 420 440 460 480 500 520 540 560

TEMP °F

FOR METALS

E = < 2 Mils Penetration/Year; G = < 20 Mils Penetration/Year

S = < 50 Mils Penetration/Year; U = > 50 Mils Penetration/Year

FOR NONMETALLICS

R = Resistant

U = Unsatisfactory

BISMUTH CARBONATE

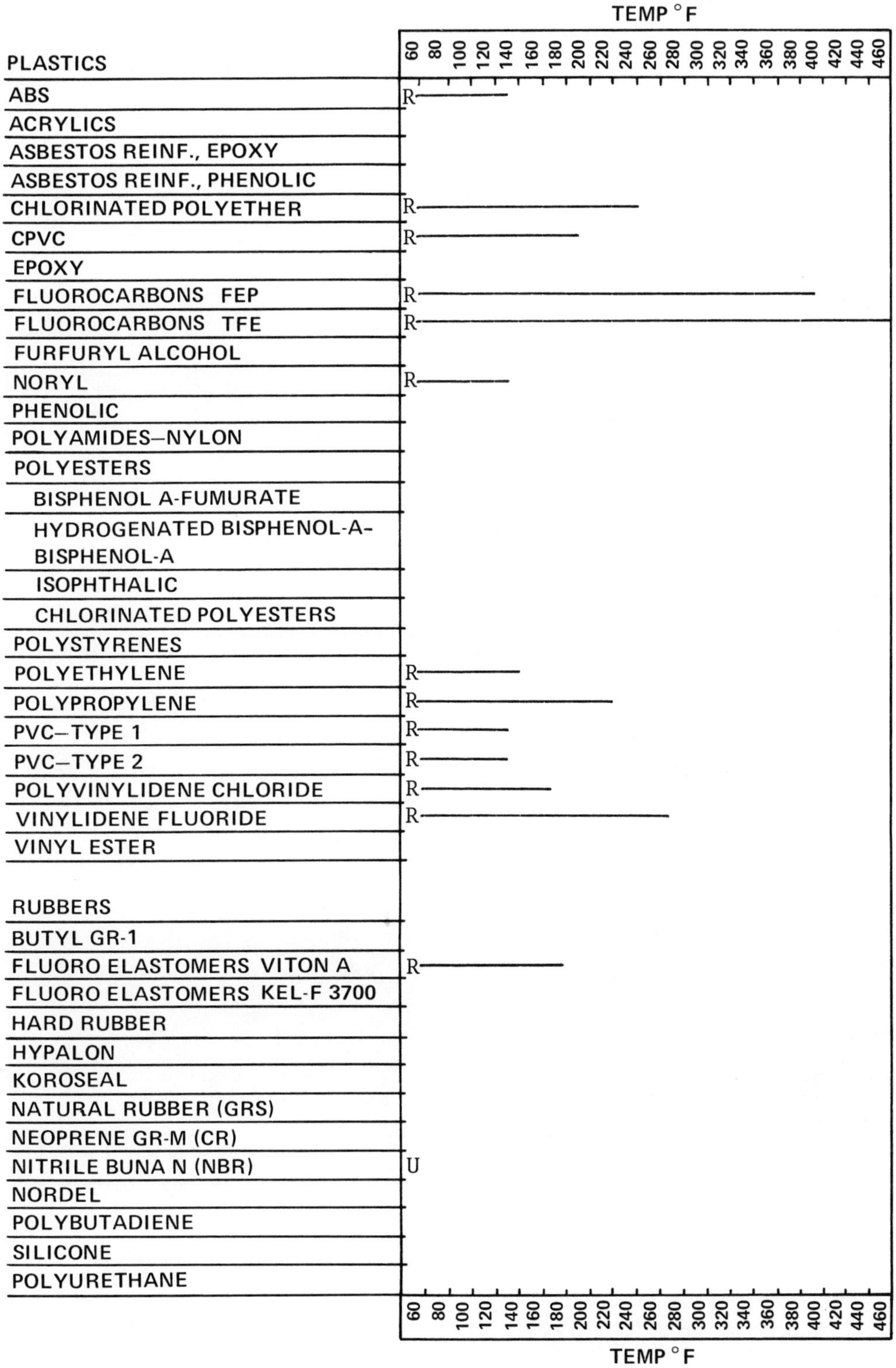

BLACK LIQUOR

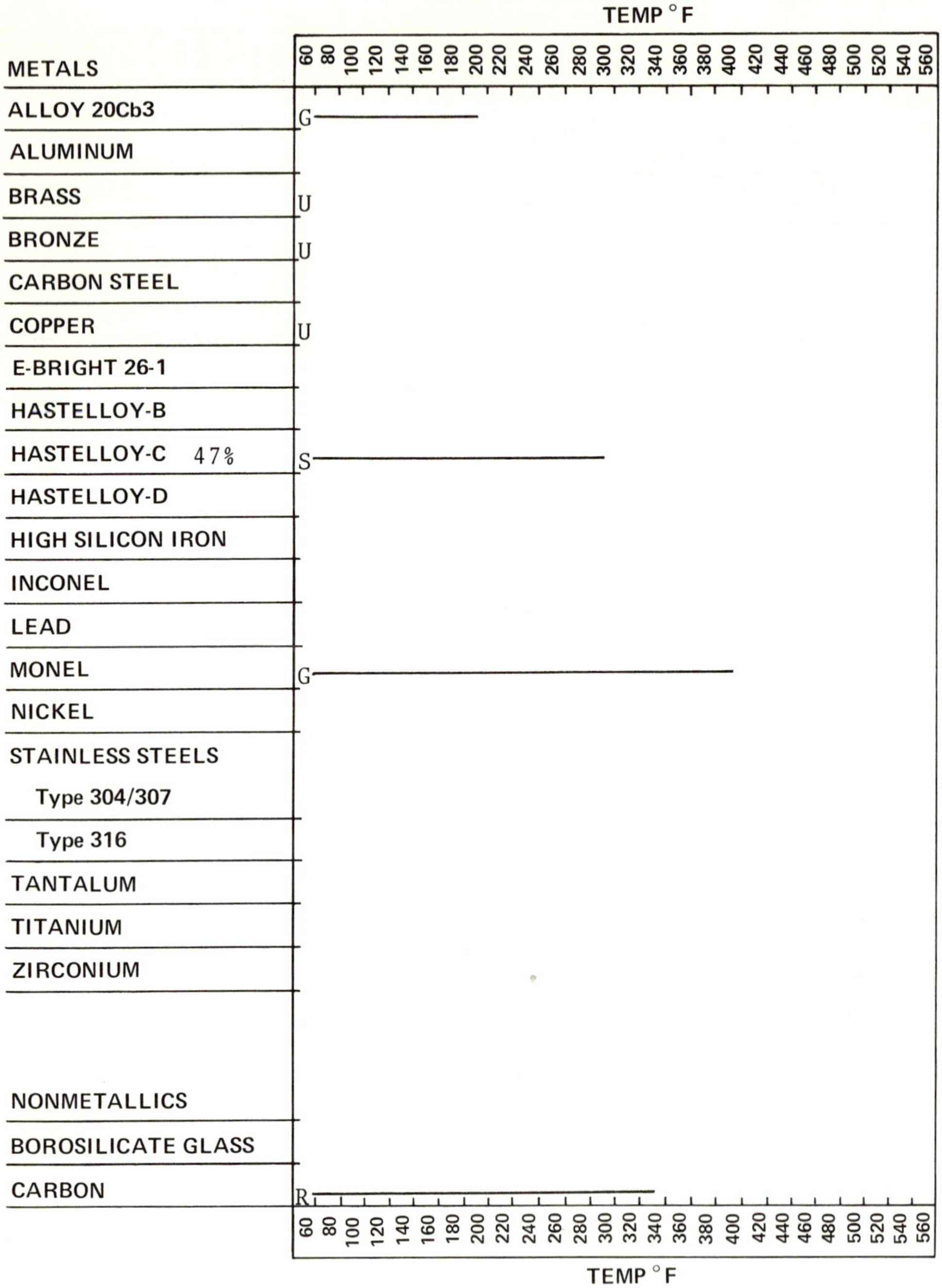

FOR METALS

E = < 2 Mils Penetration/Year; G = < 20 Mils Penetration/Year

S = < 50 Mils Penetration/Year; U = > 50 Mils Penetration/Year

FOR NONMETALLICS

R = Resistant

U = Unsatisfactory

BLACK LIQUOR

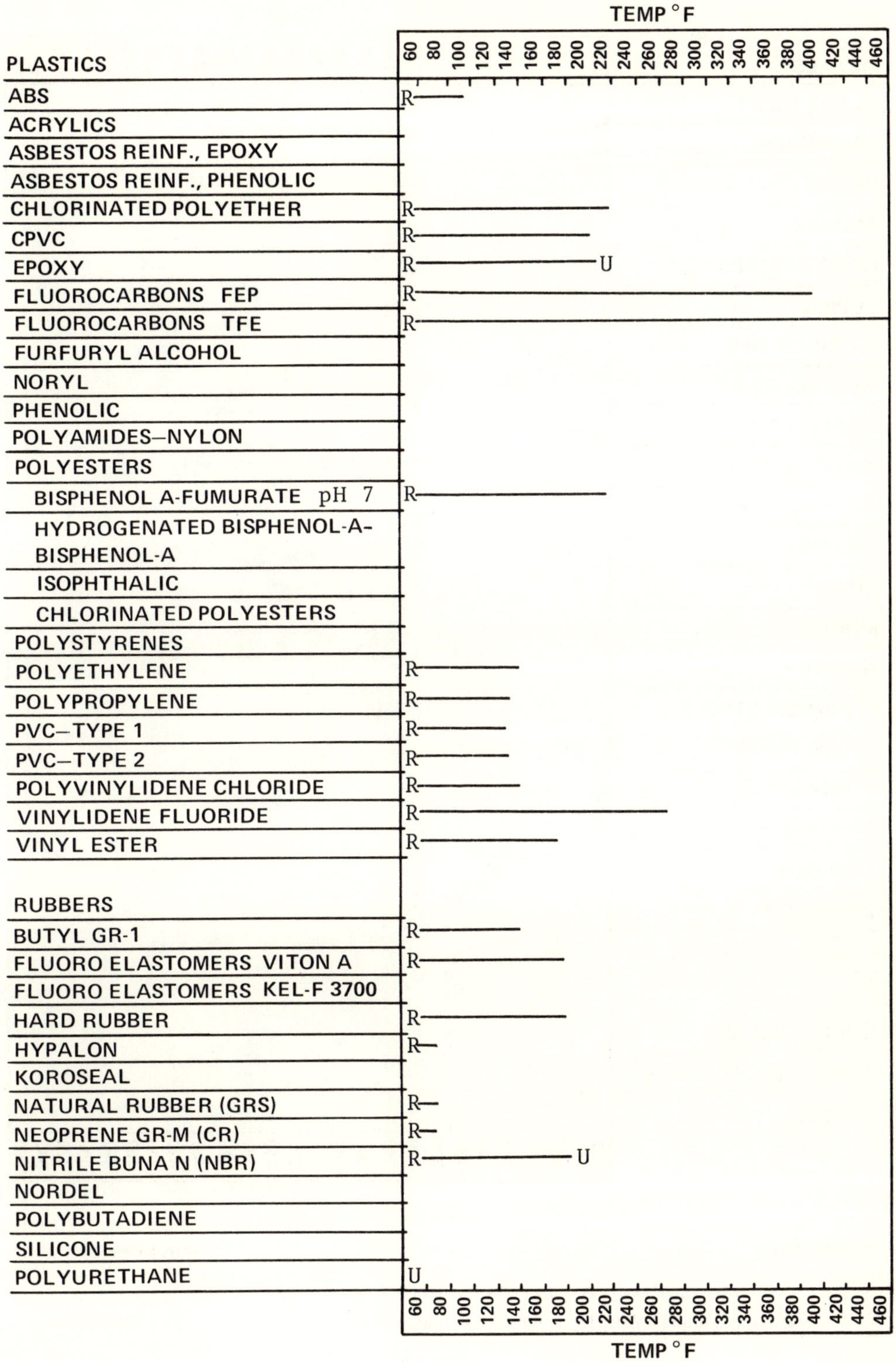

BLEACH, 12.5% ACTIVE CHLORINE

TEMP °F: 60, 80, 100, 120, 140, 160, 180, 200, 220, 240, 260, 280, 300, 320, 340, 360, 380, 400, 420, 440, 460, 480, 500, 520, 540, 560

METALS	60 °F
ALLOY 20Cb3	
ALUMINUM	U
BRASS	
BRONZE	
CARBON STEEL	U
COPPER	
E-BRIGHT 26-1	
HASTELLOY-B	
HASTELLOY-C	E———
HASTELLOY-D	
HIGH SILICON IRON	
INCONEL	
LEAD	
MONEL	
NICKEL	
STAINLESS STEELS Type 304/307	
Type 316	U
TANTALUM	
TITANIUM	
ZIRCONIUM	

NONMETALLICS	60 °F
BOROSILICATE GLASS	
CARBON	

TEMP °F

FOR METALS

E = < 2 Mils Penetration/Year; G = < 20 Mils Penetration/Year

S = < 50 Mils Penetration/Year; U = > 50 Mils Penetration/Year

FOR NONMETALLICS

R = Resistant

U = Unsatisfactory

BLEACH, 12.5% ACTIVE CHLORINE

TEMP °F

PLASTICS	60–460 °F
ABS	U
ACRYLICS	R
ASBESTOS REINF., EPOXY	
ASBESTOS REINF., PHENOLIC	
CHLORINATED POLYETHER	R
CPVC	R
EPOXY	
FLUOROCARBONS FEP	R
FLUOROCARBONS TFE	R
FURFURYL ALCOHOL	
NORYL	
PHENOLIC	
POLYAMIDES—NYLON	
POLYESTERS	
BISPHENOL A-FUMURATE	
HYDROGENATED BISPHENOL-A-BISPHENOL-A	
ISOPHTHALIC	U
CHLORINATED POLYESTERS	
POLYSTYRENES	
POLYETHYLENE	R
POLYPROPYLENE	R
PVC—TYPE 1	R
PVC—TYPE 2	R
POLYVINYLIDENE CHLORIDE	
VINYLIDENE FLUORIDE	
VINYL ESTER	
RUBBERS	
BUTYL GR-1	R
FLUORO ELASTOMERS VITON A	R
FLUORO ELASTOMERS KEL-F 3700	
HARD RUBBER	R
HYPALON	R
KOROSEAL	
NATURAL RUBBER (GRS)	R
NEOPRENE GR-M (CR)	R
NITRILE BUNA N (NBR)	U
NORDEL	
POLYBUTADIENE	
SILICONE	
POLYURETHANE	U

TEMP °F

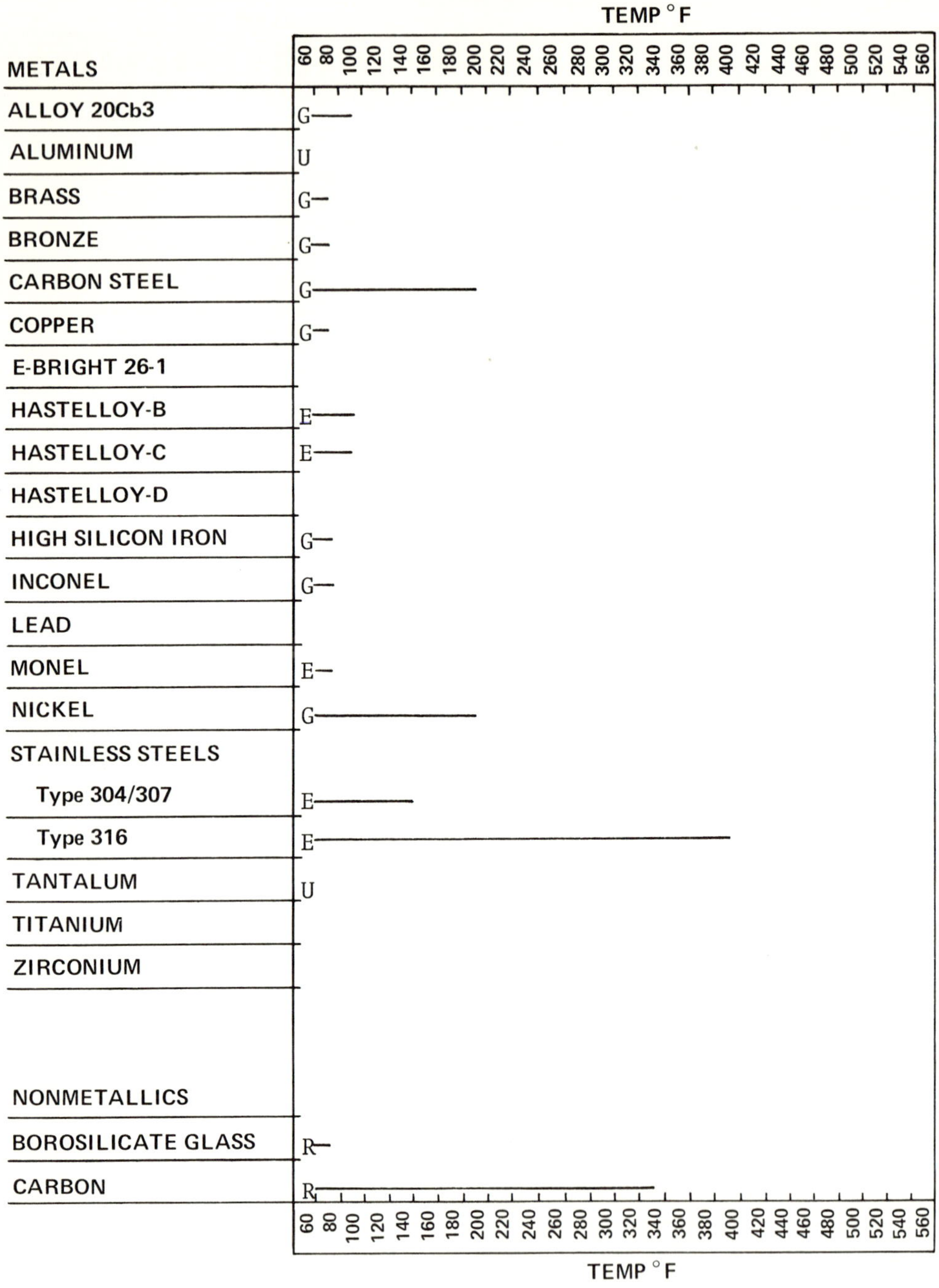

FOR METALS

E = < 2 Mils Penetration/Year; G = < 20 Mils Penetration/Year

S = < 50 Mils Penetration/Year; U = > 50 Mils Penetration/Year

FOR NONMETALLICS

R = Resistant

U = Unsatisfactory

PLASTICS	TEMP °F (60–460)
ABS	R
ACRYLICS	R
ASBESTOS REINF., EPOXY	
ASBESTOS REINF., PHENOLIC	
CHLORINATED POLYETHER	R
CPVC	R
EPOXY	R
FLUOROCARBONS FEP	R
FLUOROCARBONS TFE	R
FURFURYL ALCOHOL	
NORYL	R
PHENOLIC	
POLYAMIDES—NYLON	R
POLYESTERS	
BISPHENOL A-FUMURATE	R
HYDROGENATED BISPHENOL-A-BISPHENOL-A	
ISOPHTHALIC sat	R
CHLORINATED POLYESTERS	
POLYSTYRENES	
POLYETHYLENE	R
POLYPROPYLENE	R
PVC—TYPE 1	R
PVC—TYPE 2	R
POLYVINYLIDENE CHLORIDE	
VINYLIDENE FLUORIDE	R
VINYL ESTER	R

RUBBERS	TEMP °F (60–460)
BUTYL GR-1	R
FLUORO ELASTOMERS VITON A	R
FLUORO ELASTOMERS KEL-F 3700	
HARD RUBBER	R
HYPALON	R
KOROSEAL	
NATURAL RUBBER (GRS)	R
NEOPRENE GR-M (CR)	R
NITRILE BUNA N (NBR)	R
NORDEL	R
POLYBUTADIENE	
SILICONE	
POLYURETHANE	R

BORIC ACID

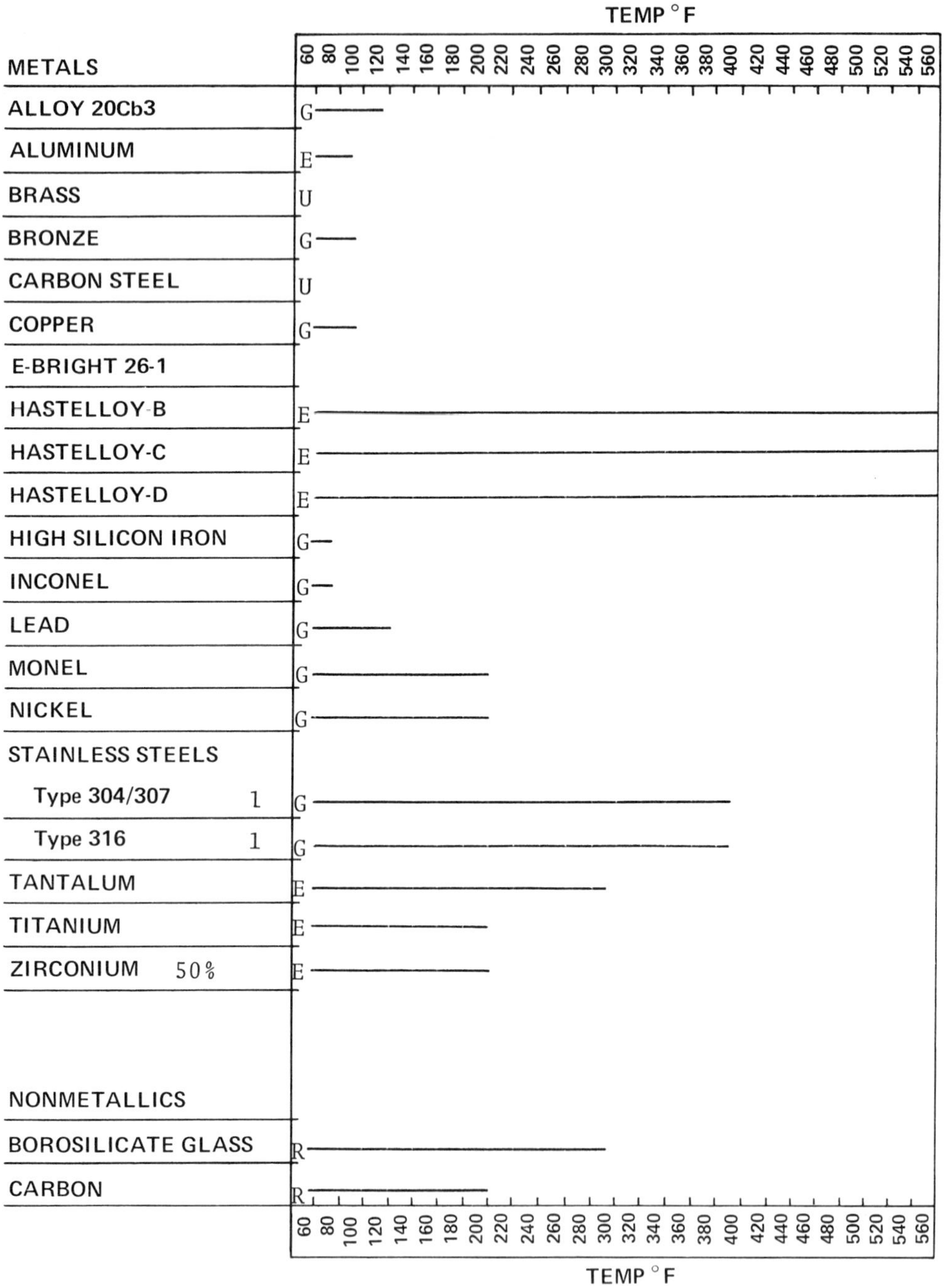

FOR METALS

E = < 2 Mils Penetration/Year; G = < 20 Mils Penetration/Year

S = < 50 Mils Penetration/Year; U = > 50 Mils Penetration/Year

FOR NONMETALLICS

R = Resistant

U = Unsatisfactory

BORIC ACID

PLASTICS		Rating
ABS		R
ACRYLICS		R
ASBESTOS REINF., EPOXY		
ASBESTOS REINF., PHENOLIC		
CHLORINATED POLYETHER		R
CPVC		R
EPOXY	4%	R
FLUOROCARBONS FEP		R
FLUOROCARBONS TFE		R
FURFURYL ALCOHOL	sat	R
NORYL		R
PHENOLIC		
POLYAMIDES–NYLON		R
POLYESTERS		
BISPHENOL A-FUMURATE	sat	R
HYDROGENATED BISPHENOL-A-BISPHENOL-A		R
ISOPHTHALIC	sat	R
CHLORINATED POLYESTERS	sat	R
POLYSTYRENES		
POLYETHYLENE		R
POLYPROPYLENE		R
PVC–TYPE 1		R
PVC–TYPE 2		R
POLYVINYLIDENE CHLORIDE		
VINYLIDENE FLUORIDE		R
VINYL ESTER	sat	R

RUBBERS	Rating
BUTYL GR-1	R
FLUORO ELASTOMERS VITON A	R
FLUORO ELASTOMERS KEL-F 3700	
HARD RUBBER	R
HYPALON	R
KOROSEAL	R
NATURAL RUBBER (GRS)	R
NEOPRENE GR-M (CR)	R
NITRILE BUNA N (NBR)	R
NORDEL	R
POLYBUTADIENE	
SILICONE	
POLYURETHANE	R

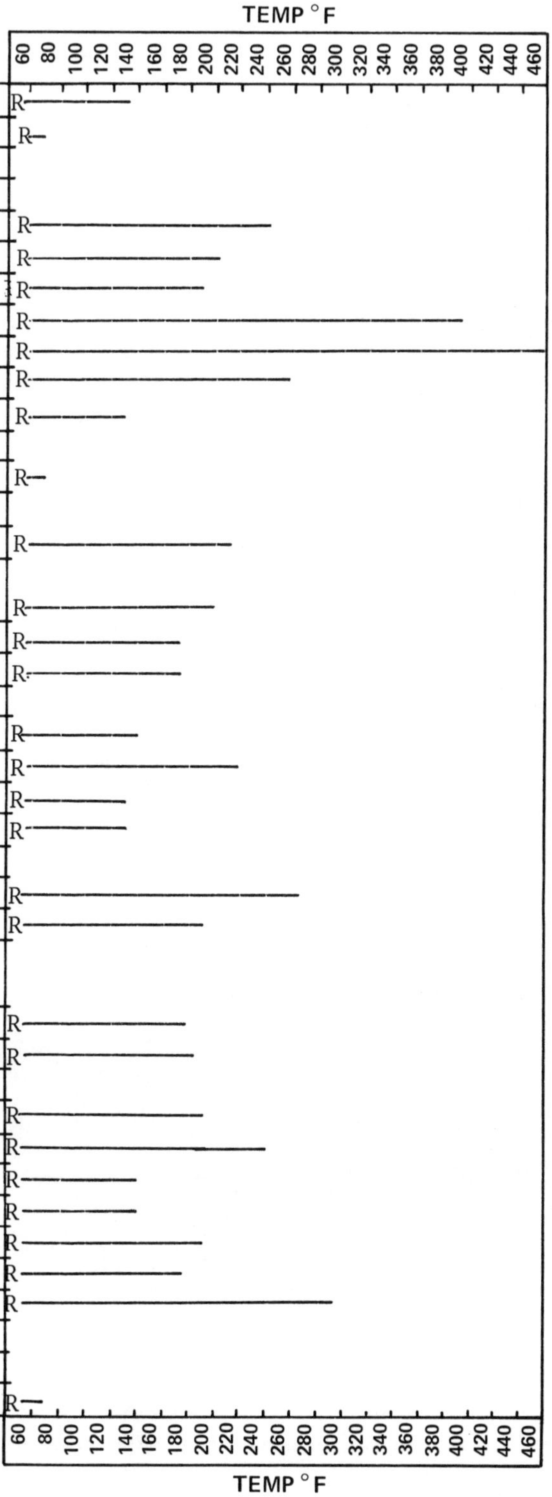

BRINE, ACID

TEMP °F: 60, 80, 100, 120, 140, 160, 180, 200, 220, 240, 260, 280, 300, 320, 340, 360, 380, 400, 420, 440, 460, 480, 500, 520, 540, 560

METALS	Rating
ALLOY 20Cb3	
ALUMINUM	
BRASS	G
BRONZE	G
CARBON STEEL	
COPPER	G
E-BRIGHT 26-1	
HASTELLOY-B	
HASTELLOY-C	E
HASTELLOY-D	
HIGH SILICON IRON	G
INCONEL	
LEAD	
MONEL	
NICKEL	
STAINLESS STEELS	
Type 304/307	
Type 316	
TANTALUM	
TITANIUM	
ZIRCONIUM	

NONMETALLICS	Rating
BOROSILICATE GLASS	R
CARBON	R

TEMP °F

FOR METALS

E = < 2 Mils Penetration/Year; G = < 20 Mils Penetration/Year

S = < 50 Mils Penetration/Year; U = > 50 Mils Penetration/Year

FOR NONMETALLICS

R = Resistant

U = Unsatisfactory

TEMP °F (60–460)

PLASTICS		
ABS		R——
ACRYLICS		
ASBESTOS REINF., EPOXY		
ASBESTOS REINF., PHENOLIC		
CHLORINATED POLYETHER		R——
CPVC		
EPOXY		R——U
FLUOROCARBONS FEP		R——
FLUOROCARBONS TFE		R——
FURFURYL ALCOHOL		
NORYL		
PHENOLIC		
POLYAMIDES—NYLON		
POLYESTERS		
BISPHENOL A-FUMURATE		R——
HYDROGENATED BISPHENOL-A-BISPHENOL-A		R——
ISOPHTHALIC	sat	R——
CHLORINATED POLYESTERS		
POLYSTYRENES		
POLYETHYLENE		R——
POLYPROPYLENE		R——
PVC—TYPE 1		R——
PVC—TYPE 2		R——
POLYVINYLIDENE CHLORIDE		R——
VINYLIDENE FLUORIDE		R——
VINYL ESTER	sat	R——

RUBBERS	
BUTYL GR-1	
FLUORO ELASTOMERS VITON A	R——
FLUORO ELASTOMERS KEL-F 3700	
HARD RUBBER	
HYPALON	R——
KOROSEAL	
NATURAL RUBBER (GRS)	
NEOPRENE GR-M (CR)	R——
NITRILE BUNA N (NBR)	R——
NORDEL	
POLYBUTADIENE	
SILICONE	
POLYURETHANE	

TEMP °F (60–460)

BROMIC ACID

TEMP °F

METALS	60	80	100	120	140	160	180	200	220	240	260	280	300	320	340	360	380	400	420	440	460	480	500	520	540	560
ALLOY 20Cb3																										
ALUMINUM																										
BRASS																										
BRONZE																										
CARBON STEEL																										
COPPER																										
E-BRIGHT 26-1																										
HASTELLOY-B																										
HASTELLOY-C																										
HASTELLOY-D																										
HIGH SILICON IRON																										
INCONEL																										
LEAD																										
MONEL	U																									
NICKEL	U																									
STAINLESS STEELS Type 304/307																										
Type 316																										
TANTALUM	G—																									
TITANIUM																										
ZIRCONIUM																										
NONMETALLICS																										
BOROSILICATE GLASS																										
CARBON																										

TEMP °F

FOR METALS

E = < 2 Mils Penetration/Year; G = < 20 Mils Penetration/Year

S = < 50 Mils Penetration/Year; U = > 50 Mils Penetration/Year

FOR NONMETALLICS

R = Resistant

U = Unsatisfactory

BROMIC ACID

TEMP °F: 60 80 100 120 140 160 180 200 220 240 260 280 300 320 340 360 380 400 420 440 460

PLASTICS	
ABS	U
ACRYLICS	
ASBESTOS REINF., EPOXY	
ASBESTOS REINF., PHENOLIC	
CHLORINATED POLYETHER	
CPVC	R
EPOXY	R
FLUOROCARBONS FEP	R
FLUOROCARBONS TFE	R
FURFURYL ALCOHOL	
NORYL	
PHENOLIC	
POLYAMIDES—NYLON	
POLYESTERS	
BISPHENOL A-FUMURATE	
HYDROGENATED BISPHENOL-A-BISPHENOL-A	
ISOPHTHALIC	
CHLORINATED POLYESTERS	
POLYSTYRENES	
POLYETHYLENE	R
POLYPROPYLENE	R
PVC—TYPE 1	R
PVC—TYPE 2	R
POLYVINYLIDENE CHLORIDE	
VINYLIDENE FLUORIDE	R
VINYL ESTER	

RUBBERS	
BUTYL GR-1	
FLUORO ELASTOMERS VITON A	R
FLUORO ELASTOMERS KEL-F 3700	
HARD RUBBER	
HYPALON	R
KOROSEAL	
NATURAL RUBBER (GRS)	
NEOPRENE GR-M (CR)	R
NITRILE BUNA N (NBR)	
NORDEL	
POLYBUTADIENE	
SILICONE	
POLYURETHANE	

TEMP °F: 60 80 100 120 140 160 180 200 220 240 260 280 300 320 340 360 380 400 420 440 460

BROMINE, LIQUID

METALS	TEMP °F (60–560)
ALLOY 20Cb3	
ALUMINUM	G
BRASS	
BRONZE	
CARBON STEEL	
COPPER	
E-BRIGHT 26-1	
HASTELLOY-B	
HASTELLOY-C	
HASTELLOY-D	
HIGH SILICON IRON	U
INCONEL	
LEAD	
MONEL	
NICKEL	
STAINLESS STEELS	
Type 304/307	U
Type 316	U
TANTALUM	E
TITANIUM	U
ZIRCONIUM	
NONMETALLICS	
BOROSILICATE GLASS	R
CARBON	U

FOR METALS

E = < 2 Mils Penetration/Year; G = < 20 Mils Penetration/Year

S = < 50 Mils Penetration/Year; U = > 50 Mils Penetration/Year

FOR NONMETALLICS

R = Resistant

U = Unsatisfactory

BROMINE, LIQUID

TEMP °F (scale 60–460: 60, 80, 100, 120, 140, 160, 180, 200, 220, 240, 260, 280, 300, 320, 340, 360, 380, 400, 420, 440, 460)

PLASTICS	Rating	Range (°F)
ABS	U	
ACRYLICS	U	
ASBESTOS REINF., EPOXY		
ASBESTOS REINF., PHENOLIC		
CHLORINATED POLYETHER	U	
CPVC	U	
EPOXY	U	
FLUOROCARBONS FEP	R	60 to ~400
FLUOROCARBONS TFE	R	60 to 460
FURFURYL ALCOHOL	RU	
NORYL		
PHENOLIC		
POLYAMIDES–NYLON	U	
POLYESTERS		
BISPHENOL A-FUMURATE	U	
HYDROGENATED BISPHENOL-A-BISPHENOL-A	U	
ISOPHTHALIC	U	
CHLORINATED POLYESTERS	U	
POLYSTYRENES	U	
POLYETHYLENE	U	
POLYPROPYLENE	U	
PVC–TYPE 1	U	
PVC–TYPE 2	U	
POLYVINYLIDENE CHLORIDE	U	
VINYLIDENE FLUORIDE	R	60 to ~150
VINYL ESTER	U	

RUBBERS	Rating	Range (°F)
BUTYL GR-1		
FLUORO ELASTOMERS VITON A	R	60 to ~360
FLUORO ELASTOMERS KEL-F 3700		
HARD RUBBER		
HYPALON	U	
KOROSEAL		
NATURAL RUBBER (GRS)		
NEOPRENE GR-M (CR)	U	
NITRILE BUNA N (NBR)	U	
NORDEL		
POLYBUTADIENE		
SILICONE		
POLYURETHANE	U	

BROMINE WATER (SAT)

TEMP °F (60–560)

METALS	Rating
ALLOY 20Cb3	
ALUMINUM	
BRASS	
BRONZE	
CARBON STEEL	
COPPER	
E-BRIGHT 26-1	
HASTELLOY B	
HASTELLOY-C	
HASTELLOY-D	
HIGH SILICON IRON	U
INCONEL	
LEAD	
MONEL	
NICKEL	
STAINLESS STEELS	
Type 304/307	U
Type 316	U
TANTALUM	E—
TITANIUM	E—
ZIRCONIUM	U

NONMETALLICS	Rating
BOROSILICATE GLASS	R—
CARBON	R—

TEMP °F

FOR METALS

E = < 2 Mils Penetration/Year; G = < 20 Mils Penetration/Year

S = < 50 Mils Penetration/Year; U = > 50 Mils Penetration/Year

FOR NONMETALLICS

R = Resistant

U = Unsatisfactory

BROMINE WATER (SAT)

PLASTICS	Rating	TEMP °F (line extends to)
ABS	U	
ACRYLICS		
ASBESTOS REINF., EPOXY		
ASBESTOS REINF., PHENOLIC		
CHLORINATED POLYETHER	U	
CPVC	U	
EPOXY	R	100
FLUOROCARBONS FEP	R	400
FLUOROCARBONS TFE	R	460
FURFURYL ALCOHOL	R	250
NORYL		
PHENOLIC		
POLYAMIDES—NYLON	R	200
POLYESTERS		
BISPHENOL A-FUMURATE	R	180
HYDROGENATED BISPHENOL-A-BISPHENOL-A		
ISOPHTHALIC		
CHLORINATED POLYESTERS		
POLYSTYRENES		
POLYETHYLENE	U	
POLYPROPYLENE	U	
PVC—TYPE 1	R	140
PVC—TYPE 2	U	
POLYVINYLIDENE CHLORIDE	U	
VINYLIDENE FLUORIDE	R	210
VINYL ESTER		

RUBBERS	Rating	TEMP °F
BUTYL GR-1		
FLUORO ELASTOMERS VITON A	U	
FLUORO ELASTOMERS KEL-F 3700		
HARD RUBBER		
HYPALON	U	
KOROSEAL	U	
NATURAL RUBBER (GRS)		
NEOPRENE GR-M (CR)	U	
NITRILE BUNA N (NBR)	U	
NORDEL		
POLYBUTADIENE		
SILICONE		
POLYURETHANE	U	

TEMP °F: 60 80 100 120 140 160 180 200 220 240 260 280 300 320 340 360 380 400 420 440 460

BROMINE WATER

METALS	TEMP °F (60 80 100 120 140 160 180 200 220 240 260 280 300 320 340 360 380 400 420 440 460 480 500 520 540 560)
ALLOY 20Cb3	
ALUMINUM	
BRASS	
BRONZE	
CARBON STEEL	
COPPER	
E-BRIGHT 26-1 1%	E—
HASTELLOY-B	
HASTELLOY-C	
HASTELLOY-D	
HIGH SILICON IRON	U
INCONEL	U
LEAD	
MONEL	U
NICKEL	U
STAINLESS STEELS	
Type 304/307	U
Type 316	U
TANTALUM	E—
TITANIUM	E—
ZIRCONIUM	G—————
NONMETALLICS	
BOROSILICATE GLASS	
CARBON	

FOR METALS

E = < 2 Mils Penetration/Year; G = < 20 Mils Penetration/Year

S = < 50 Mils Penetration/Year; U = > 50 Mils Penetration/Year

FOR NONMETALLICS

R = Resistant

U = Unsatisfactory

BROMINE WATER

PLASTICS		TEMP °F (60–460)
ABS		U
ACRYLICS		
ASBESTOS REINF., EPOXY		
ASBESTOS REINF., PHENOLIC		
CHLORINATED POLYETHER		U
CPVC		U
EPOXY	4%	R (60–200) U (220)
FLUOROCARBONS FEP		R (60–400)
FLUOROCARBONS TFE		R (60–460)
FURFURYL ALCOHOL		
NORYL		
PHENOLIC		
POLYAMIDES–NYLON		
POLYESTERS		
BISPHENOL A-FUMURATE		
HYDROGENATED BISPHENOL-A-BISPHENOL-A		
ISOPHTHALIC		
CHLORINATED POLYESTERS		
POLYSTYRENES		
POLYETHYLENE		U
POLYPROPYLENE		U
PVC–TYPE 1		R (60–140)
PVC–TYPE 2		U
POLYVINYLIDENE CHLORIDE		
VINYLIDENE FLUORIDE		R (60–220)
VINYL ESTER		

RUBBERS	TEMP °F (60–460)
BUTYL GR-1	
FLUORO ELASTOMERS VITON A	R (60–200)
FLUORO ELASTOMERS KEL-F 3700	
HARD RUBBER	
HYPALON	
KOROSEAL	U
NATURAL RUBBER (GRS)	
NEOPRENE GR-M (CR)	U
NITRILE BUNA N (NBR)	U
NORDEL	
POLYBUTADIENE	
SILICONE	
POLYURETHANE	

BUTADIENE (BUTYLENE)

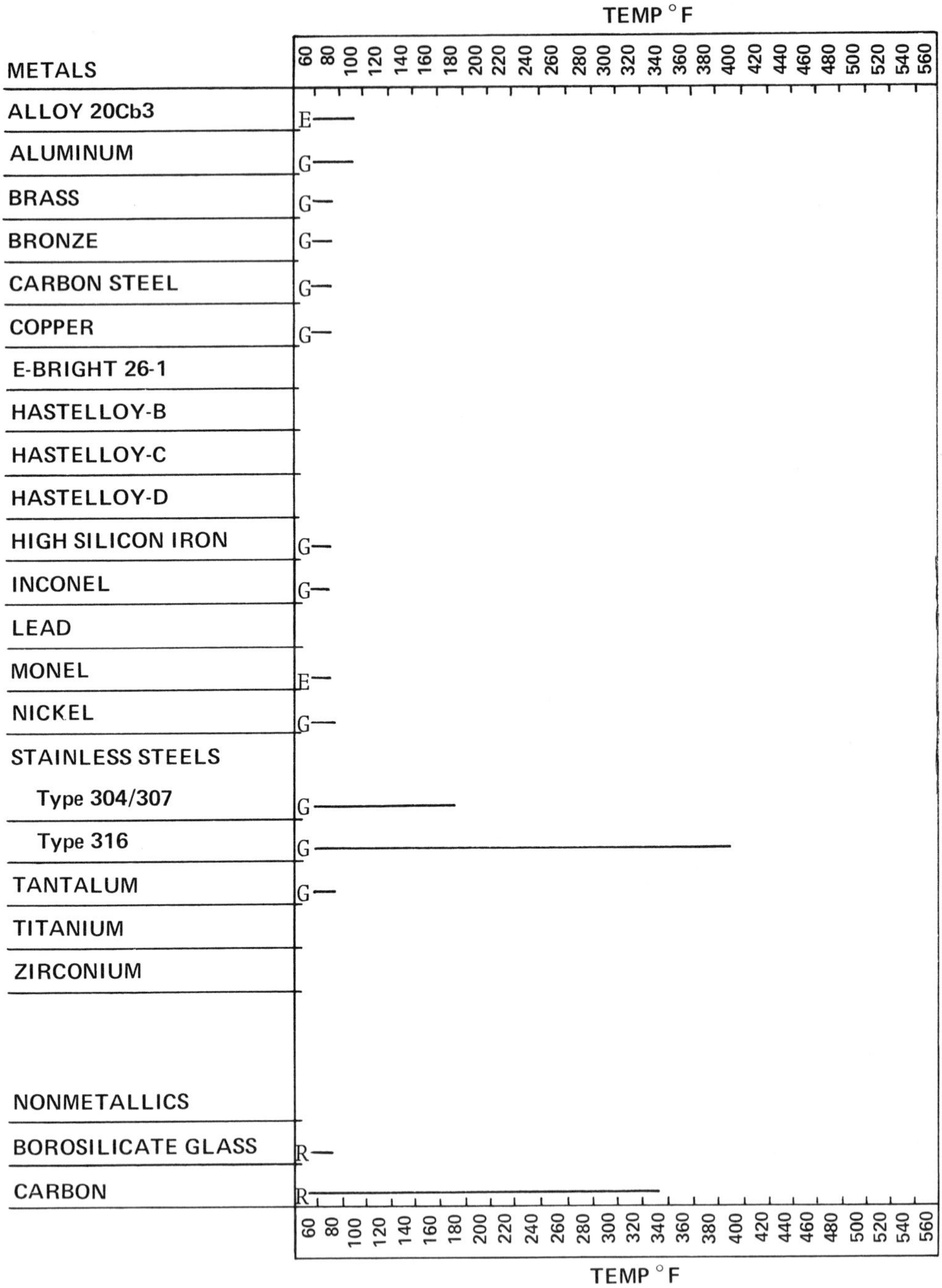

FOR METALS

E = < 2 Mils Penetration/Year; G = < 20 Mils Penetration/Year

S = < 50 Mils Penetration/Year; U = > 50 Mils Penetration/Year

FOR NONMETALLICS

R = Resistant

U = Unsatisfactory

BUTADIENE (BUTYLENE)

TEMP °F (60–460)

PLASTICS	Rating
ABS	U
ACRYLICS	
ASBESTOS REINF., EPOXY	
ASBESTOS REINF., PHENOLIC	
CHLORINATED POLYETHER	R
CPVC	R
EPOXY	R
FLUOROCARBONS FEP	R
FLUOROCARBONS TFE	R
FURFURYL ALCOHOL	
NORYL	
PHENOLIC	
POLYAMIDES–NYLON	
POLYESTERS	
BISPHENOL A-FUMURATE	
HYDROGENATED BISPHENOL-A-BISPHENOL-A	
ISOPHTHALIC	
CHLORINATED POLYESTERS	
POLYSTYRENES	
POLYETHYLENE	U
POLYPROPYLENE	U
PVC–TYPE 1	R
PVC–TYPE 2	RU
POLYVINYLIDENE CHLORIDE	U
VINYLIDENE FLUORIDE	R
VINYL ESTER	

RUBBERS	Rating
BUTYL GR-1	
FLUORO ELASTOMERS VITON A	R
FLUORO ELASTOMERS KEL-F 3700	
HARD RUBBER	
HYPALON	U
KOROSEAL	
NATURAL RUBBER (GRS)	U
NEOPRENE GR-M (CR)	U
NITRILE BUNA N (NBR)	R
NORDEL	
POLYBUTADIENE	
SILICONE	
POLYURETHANE	U

TEMP °F (60–460)

BUTANE

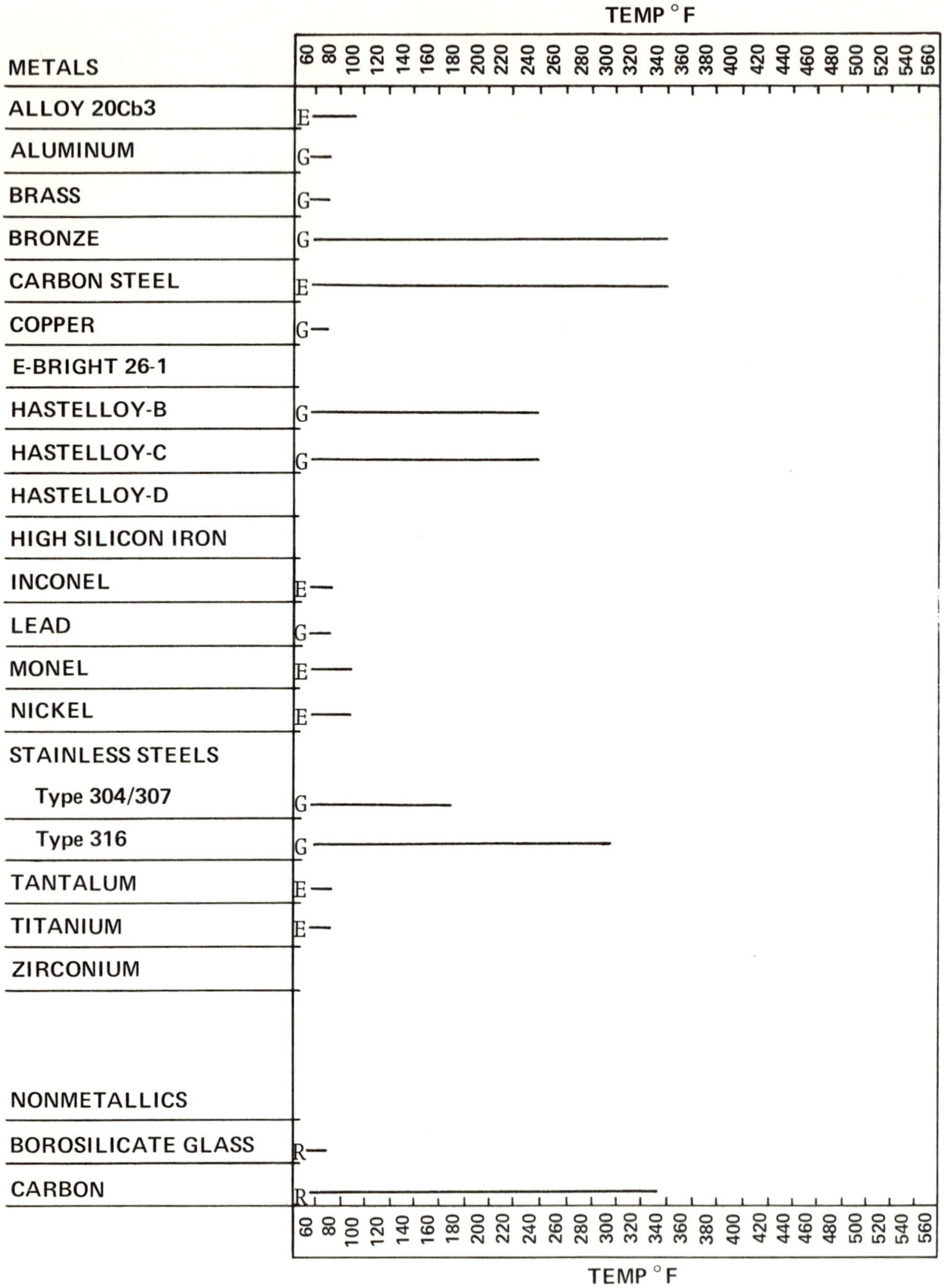

FOR METALS

E = < 2 Mils Penetration/Year; G = < 20 Mils Penetration/Year

S = < 50 Mils Penetration/Year; U = > 50 Mils Penetration/Year

FOR NONMETALLICS

R = Resistant

U = Unsatisfactory

BUTANE

TEMP °F: 60 80 100 120 140 160 180 200 220 240 260 280 300 320 340 360 380 400 420 440 460

PLASTICS	
ABS	R
ACRYLICS	
ASBESTOS REINF., EPOXY	
ASBESTOS REINF., PHENOLIC	
CHLORINATED POLYETHER	R
CPVC	R
EPOXY	R
FLUOROCARBONS FEP	R
FLUOROCARBONS TFE	R
FURFURYL ALCOHOL	
NORYL	
PHENOLIC	
POLYAMIDES–NYLON	
POLYESTERS	
BISPHENOL A-FUMURATE	
HYDROGENATED BISPHENOL-A-BISPHENOL-A	
ISOPHTHALIC	
CHLORINATED POLYESTERS	
POLYSTYRENES	
POLYETHYLENE	U
POLYPROPYLENE	R
PVC–TYPE 1	R
PVC–TYPE 2	R
POLYVINYLIDENE CHLORIDE	R
VINYLIDENE FLUORIDE	R
VINYL ESTER	

RUBBERS	
BUTYL GR-1	U
FLUORO ELASTOMERS VITON A	R
FLUORO ELASTOMERS KEL-F 3700	
HARD RUBBER	U
HYPALON	R
KOROSEAL	
NATURAL RUBBER (GRS)	U
NEOPRENE GR-M (CR)	R
NITRILE BUNA N (NBR)	U
NORDEL	R
POLYBUTADIENE	
SILICONE	
POLYURETHANE	R

TEMP °F: 60 80 100 120 140 160 180 200 220 240 260 280 300 320 340 360 380 400 420 440 460

BUTYL ACETATE

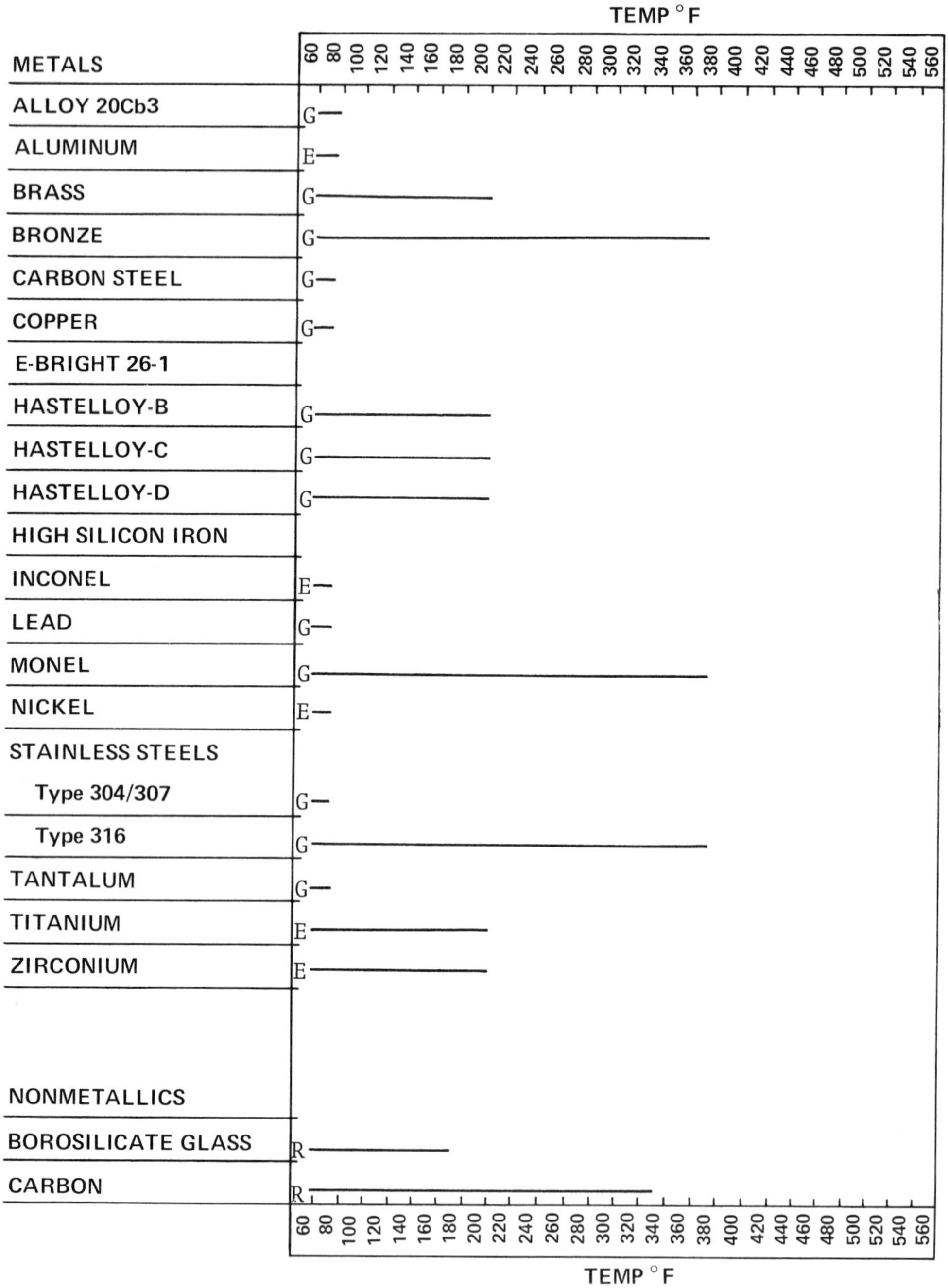

FOR METALS

E = < 2 Mils Penetration/Year; G = < 20 Mils Penetration/Year

S = < 50 Mils Penetration/Year; U = > 50 Mils Penetration/Year

FOR NONMETALLICS

R = Resistant

U = Unsatisfactory

BUTYL ACETATE

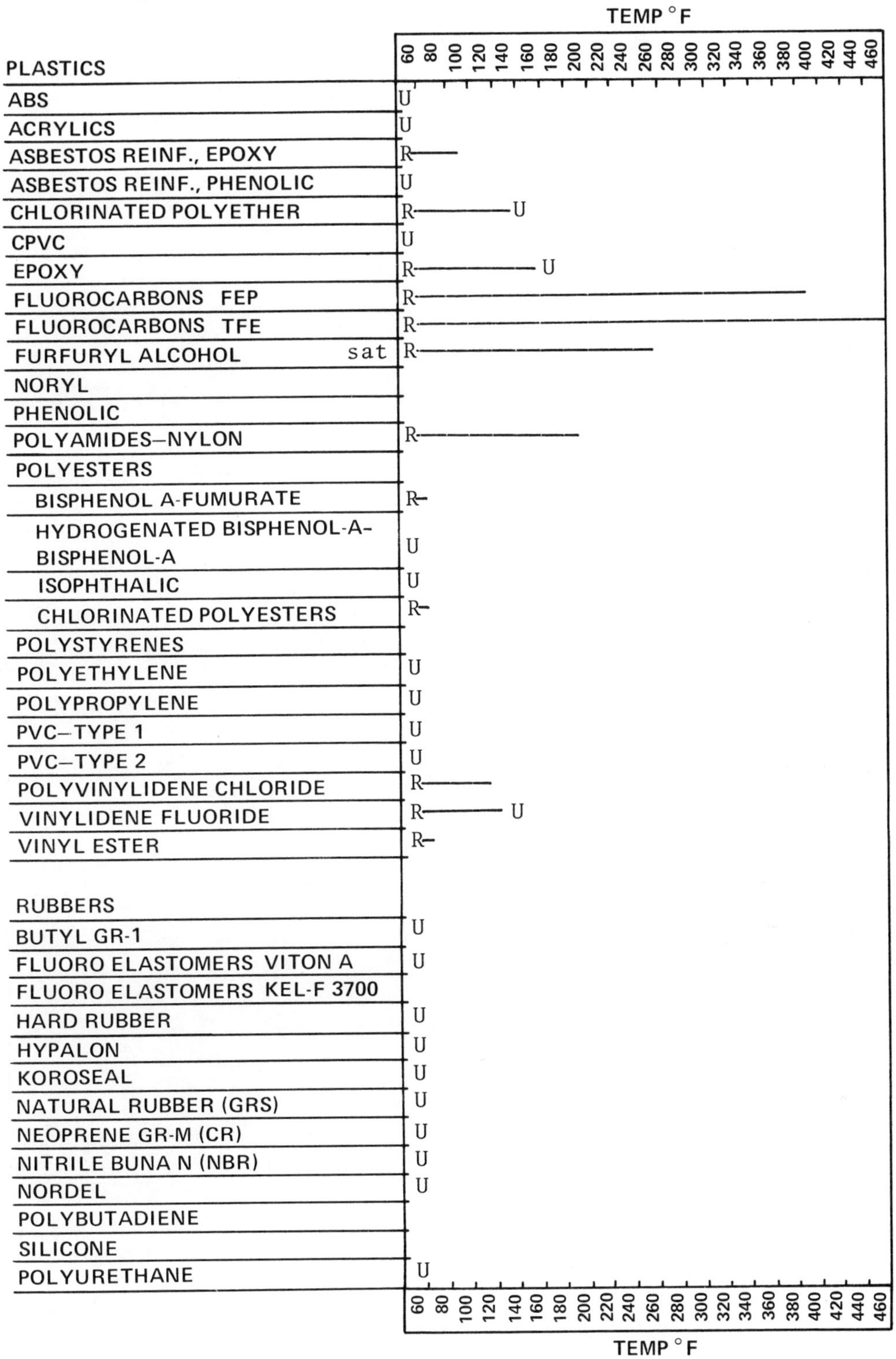

BUTYL ALCOHOL

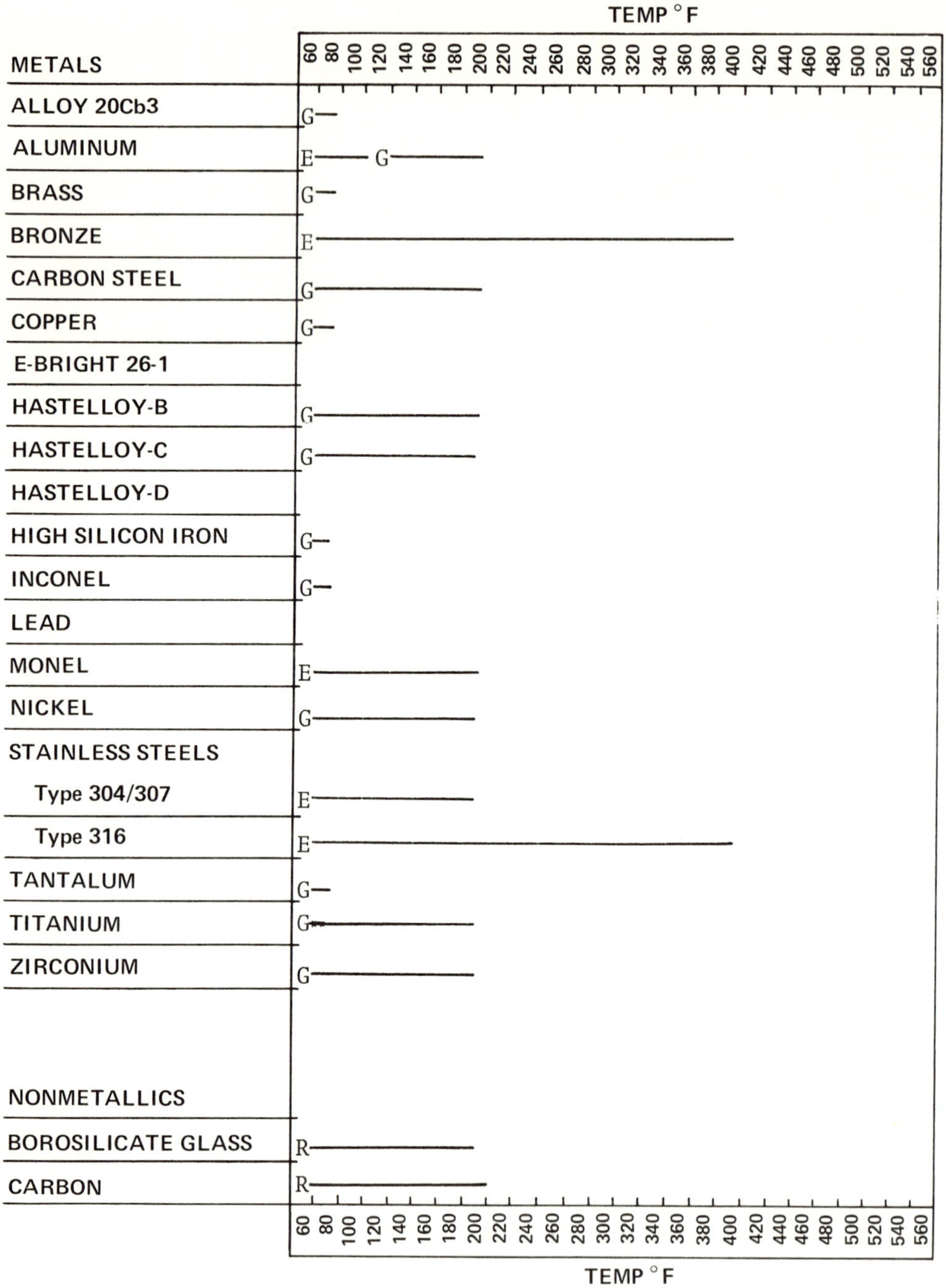

FOR METALS

E = < 2 Mils Penetration/Year; G = < 20 Mils Penetration/Year

S = < 50 Mils Penetration/Year; U = > 50 Mils Penetration/Year

FOR NONMETALLICS

R = Resistant

U = Unsatisfactory

BUTYL ALCOHOL

TEMP °F: 60, 80, 100, 120, 140, 160, 180, 200, 220, 240, 260, 280, 300, 320, 340, 360, 380, 400, 420, 440, 460

PLASTICS	
ABS	U
ACRYLICS	R—
ASBESTOS REINF., EPOXY	
ASBESTOS REINF., PHENOLIC	
CHLORINATED POLYETHER	R—
CPVC	R— U
EPOXY	R— U
FLUOROCARBONS FEP	R—
FLUOROCARBONS TFE	R—
FURFURYL ALCOHOL	R—
NORYL	
PHENOLIC	
POLYAMIDES—NYLON	U
POLYESTERS	
BISPHENOL A-FUMURATE	R—
HYDROGENATED BISPHENOL-A-BISPHENOL-A	
ISOPHTHALIC	R—
CHLORINATED POLYESTERS	
POLYSTYRENES	
POLYETHYLENE	R—
POLYPROPYLENE	R—
PVC—TYPE 1	RU
PVC—TYPE 2	U
POLYVINYLIDENE CHLORIDE	R—
VINYLIDENE FLUORIDE	R—
VINYL ESTER	R—

RUBBERS	
BUTYL GR-1	R—
FLUORO ELASTOMERS VITON A	R—
FLUORO ELASTOMERS KEL-F 3700	
HARD RUBBER	R—
HYPALON	R—
KOROSEAL	R—
NATURAL RUBBER (GRS)	R—
NEOPRENE GR-M (CR)	U
NITRILE BUNA N (NBR)	U
NORDEL	R—
POLYBUTADIENE	
SILICONE	
POLYURETHANE	

TEMP °F: 60, 80, 100, 120, 140, 160, 180, 200, 220, 240, 260, 280, 300, 320, 340, 360, 380, 400, 420, 440, 460

BUTYL ALCOHOL, PRIMARY

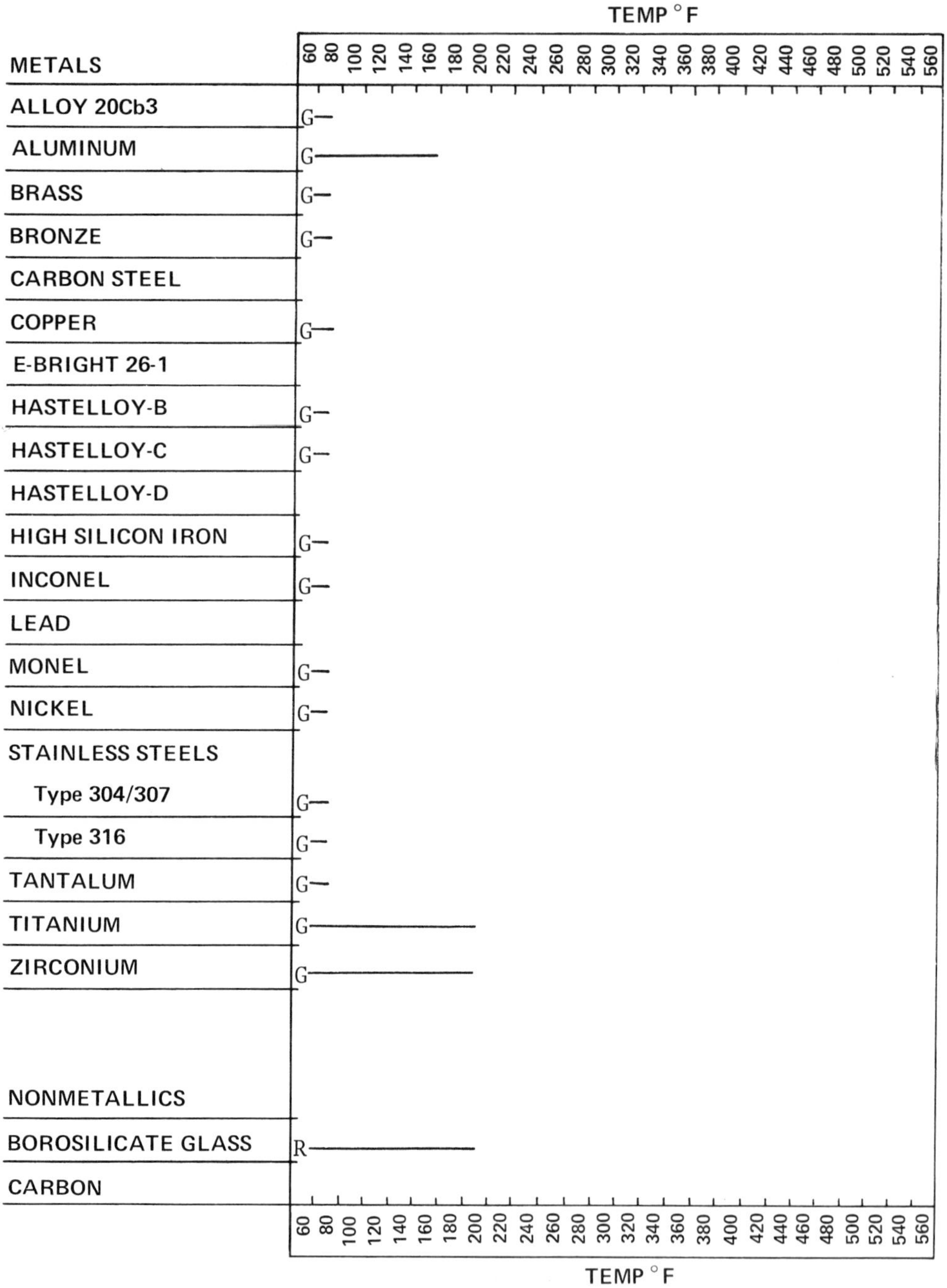

FOR METALS

E = < 2 Mils Penetration/Year; G = < 20 Mils Penetration/Year

S = < 50 Mils Penetration/Year; U = > 50 Mils Penetration/Year

FOR NONMETALLICS

R = Resistant

U = Unsatisfactory

PLASTICS	TEMP °F
ABS	U
ACRYLICS	
ASBESTOS REINF., EPOXY	
ASBESTOS REINF., PHENOLIC	
CHLORINATED POLYETHER	R (60–220)
CPVC	R (60–140) U
EPOXY	R (60–140) U
FLUOROCARBONS FEP	R (60–400)
FLUOROCARBONS TFE	R (60–460)
FURFURYL ALCOHOL	
NORYL	
PHENOLIC	
POLYAMIDES–NYLON	
POLYESTERS	
BISPHENOL A-FUMURATE	
HYDROGENATED BISPHENOL-A-BISPHENOL-A	
ISOPHTHALIC	
CHLORINATED POLYESTERS	R (60–100)
POLYSTYRENES	
POLYETHYLENE	R (60–150)
POLYPROPYLENE	R (60–150)
PVC–TYPE 1	R (60–140)
PVC–TYPE 2	U
POLYVINYLIDENE CHLORIDE	R (60–80)
VINYLIDENE FLUORIDE	R (60–280)
VINYL ESTER	

RUBBERS	TEMP °F
BUTYL GR-1	
FLUORO ELASTOMERS VITON A	R (60–190)
FLUORO ELASTOMERS KEL-F 3700	
HARD RUBBER	
HYPALON	R (60–80)
KOROSEAL	R (60–150)
NATURAL RUBBER (GRS)	
NEOPRENE GR-M (CR)	R (60–80)
NITRILE BUNA N (NBR)	R (60–190)
NORDEL	
POLYBUTADIENE	
SILICONE	
POLYURETHANE	

BUTYL ALCOHOL, SECONDARY

TEMP °F

METALS	Rating
ALLOY 20Cb3	G
ALUMINUM	G
BRASS	G
BRONZE	G
CARBON STEEL	
COPPER	G
E-BRIGHT 26-1	
HASTELLOY-B	G
HASTELLOY-C	G
HASTELLOY-D	
HIGH SILICON IRON	G
INCONEL	G
LEAD	
MONEL	G
NICKEL	G
STAINLESS STEELS	
Type 304/307	G
Type 316	G
TANTALUM	G
TITANIUM	G
ZIRCONIUM	G

NONMETALLICS	Rating
BOROSILICATE GLASS	R
CARBON	

TEMP °F: 60 80 100 120 140 160 180 200 220 240 260 280 300 320 340 360 380 400 420 440 460 480 500 520 540 560

FOR METALS

E = < 2 Mils Penetration/Year; G = < 20 Mils Penetration/Year

S = < 50 Mils Penetration/Year; U = > 50 Mils Penetration/Year

FOR NONMETALLICS

R = Resistant

U = Unsatisfactory

BUTYL ALCOHOL, SECONDARY

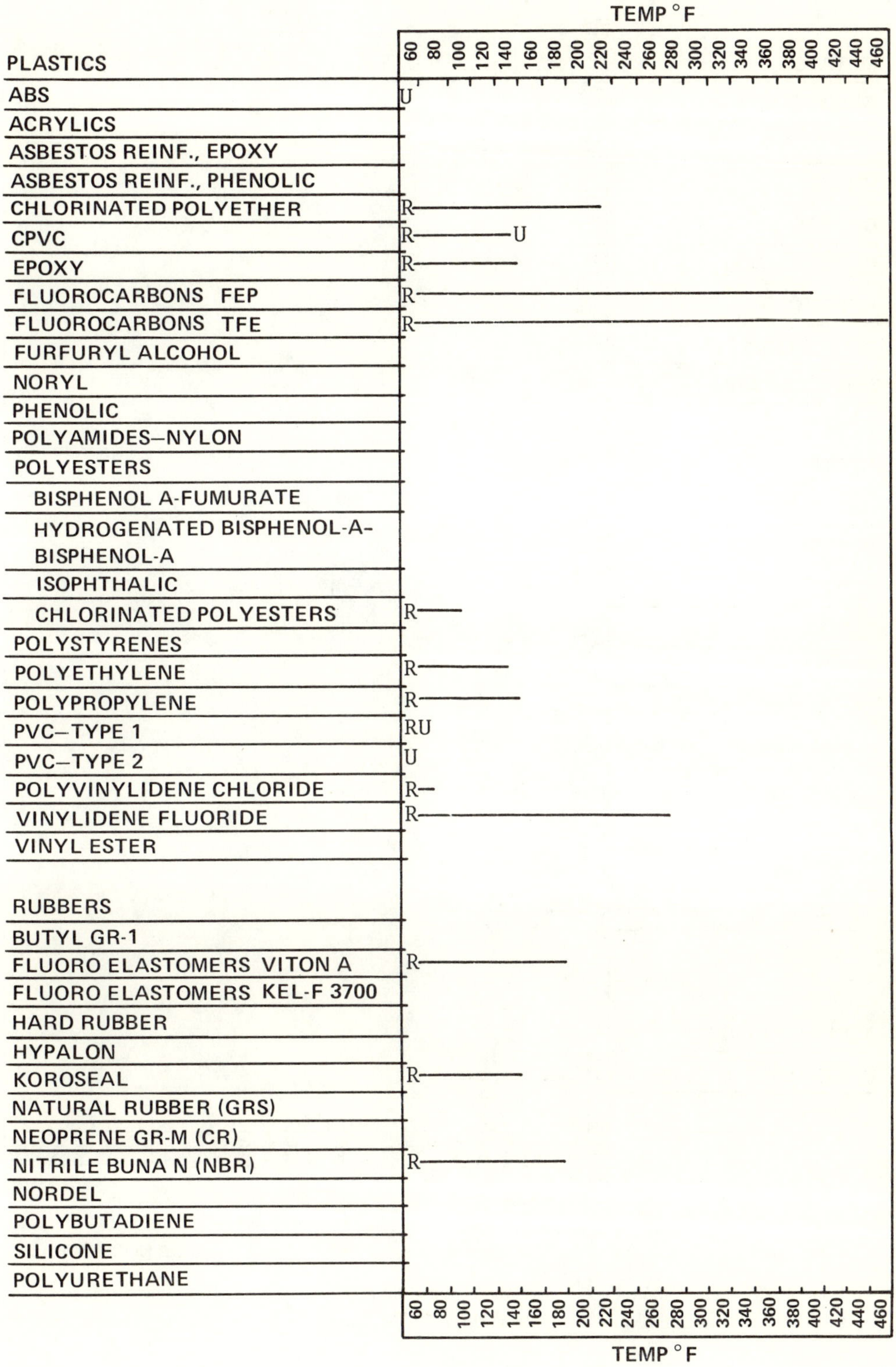

BUTYL ALCOHOL, TERTIARY

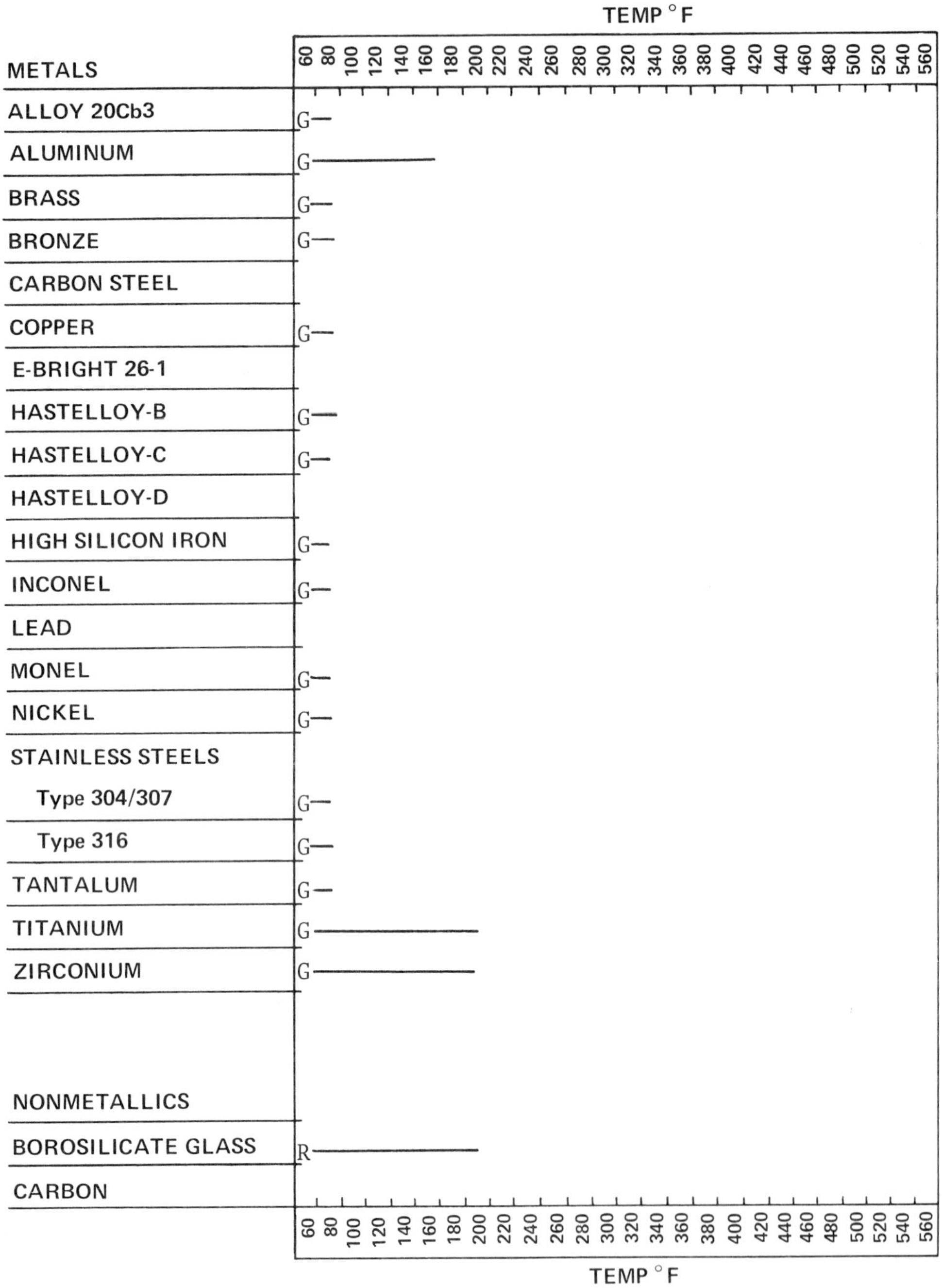

FOR METALS

E = < 2 Mils Penetration/Year; G = < 20 Mils Penetration/Year

S = < 50 Mils Penetration/Year; U = > 50 Mils Penetration/Year

FOR NONMETALLICS

R = Resistant

U = Unsatisfactory

BUTYL ALCOHOL, TERTIARY

PLASTICS	TEMP °F
ABS	U
ACRYLICS	
ASBESTOS REINF., EPOXY	
ASBESTOS REINF., PHENOLIC	
CHLORINATED POLYETHER	R
CPVC	R
EPOXY	R
FLUOROCARBONS FEP	R
FLUOROCARBONS TFE	R
FURFURYL ALCOHOL	
NORYL	
PHENOLIC	
POLYAMIDES–NYLON	
POLYESTERS	
BISPHENOL A-FUMURATE	
HYDROGENATED BISPHENOL-A-BISPHENOL-A	
ISOPHTHALIC	
CHLORINATED POLYESTERS	R
POLYSTYRENES	
POLYETHYLENE	R
POLYPROPYLENE	R
PVC–TYPE 1	R
PVC–TYPE 2	R–U
POLYVINYLIDENE CHLORIDE	R
VINYLIDENE FLUORIDE	R
VINYL ESTER	
RUBBERS	
BUTYL GR-1	
FLUORO ELASTOMERS VITON A	
FLUORO ELASTOMERS KEL-F 3700	
HARD RUBBER	
HYPALON	
KOROSEAL	
NATURAL RUBBER (GRS)	
NEOPRENE GR-M (CR)	
NITRILE BUNA N (NBR)	
NORDEL	
POLYBUTADIENE	
SILICONE	
POLYURETHANE	

n-BUTYLAMINE

METALS	TEMP °F (60–560)
ALLOY 20Cb3	
ALUMINUM	
BRASS	
BRONZE	G (60–400)
CARBON STEEL	
COPPER	
E-BRIGHT 26-1	
HASTELLOY-B	
HASTELLOY-C	
HASTELLOY-D	
HIGH SILICON IRON	
INCONEL	
LEAD	
MONEL	
NICKEL	
STAINLESS STEELS Type 304/307	
Type 316	G (60–400)
TANTALUM	
TITANIUM	
ZIRCONIUM	
NONMETALLICS	
BOROSILICATE GLASS	
CARBON	

FOR METALS

E = < 2 Mils Penetration/Year; G = < 20 Mils Penetration/Year

S = < 50 Mils Penetration/Year; U = > 50 Mils Penetration/Year

FOR NONMETALLICS

R = Resistant

U = Unsatisfactory

n-BUTYLAMINE

TEMP °F (60–460)

PLASTICS	60 80 100 120 140 160 180 200 220 240 260 280 300 320 340 360 380 400 420 440 460
ABS	
ACRYLICS	
ASBESTOS REINF., EPOXY	
ASBESTOS REINF., PHENOLIC	
CHLORINATED POLYETHER	
CPVC	
EPOXY	
FLUOROCARBONS FEP	R (60–400)
FLUOROCARBONS TFE	R (60–460)
FURFURYL ALCOHOL	
NORYL	
PHENOLIC	
POLYAMIDES–NYLON	R (60–200)
POLYESTERS	
BISPHENOL A-FUMURATE	U
HYDROGENATED BISPHENOL-A-BISPHENOL-A	
ISOPHTHALIC	
CHLORINATED POLYESTERS	
POLYSTYRENES	
POLYETHYLENE	
POLYPROPYLENE	
PVC–TYPE 1	
PVC–TYPE 2	
POLYVINYLIDENE CHLORIDE	
VINYLIDENE FLUORIDE	RU
VINYL ESTER	
RUBBERS	
BUTYL GR-1	
FLUORO ELASTOMERS VITON A	U
FLUORO ELASTOMERS KEL-F 3700	
HARD RUBBER	
HYPALON	
KOROSEAL	
NATURAL RUBBER (GRS)	
NEOPRENE GR-M (CR)	
NITRILE BUNA N (NBR)	
NORDEL	
POLYBUTADIENE	
SILICONE	
POLYURETHANE	

TEMP °F

BUTYL BROMIDE

TEMP °F (60–560)

METALS	Rating	Temp °F
ALLOY 20Cb3		
ALUMINUM		
BRASS		
BRONZE		
CARBON STEEL	U	60
COPPER		
E-BRIGHT 26-1		
HASTELLOY-B		
HASTELLOY-C		
HASTELLOY-D		
HIGH SILICON IRON		
INCONEL		
LEAD		
MONEL		
NICKEL		
STAINLESS STEELS Type 304/307		
Type 316		
TANTALUM		
TITANIUM		
ZIRCONIUM		
NONMETALLICS		
BOROSILICATE GLASS	R	60–280 (line)
CARBON		

FOR METALS

E = < 2 Mils Penetration/Year; G = < 20 Mils Penetration/Year

S = < 50 Mils Penetration/Year; U = > 50 Mils Penetration/Year

FOR NONMETALLICS

R = Resistant

U = Unsatisfactory

BUTYL BROMIDE

TEMP °F (60–460)

PLASTICS	TEMP °F
ABS	
ACRYLICS	
ASBESTOS REINF., EPOXY	
ASBESTOS REINF., PHENOLIC	
CHLORINATED POLYETHER	R (60–180)
CPVC	
EPOXY	
FLUOROCARBONS FEP	R (60–400)
FLUOROCARBONS TFE	R (60–460)
FURFURYL ALCOHOL	
NORYL	
PHENOLIC	
POLYAMIDES—NYLON	
POLYESTERS	
BISPHENOL A-FUMURATE	
HYDROGENATED BISPHENOL-A-BISPHENOL-A	
ISOPHTHALIC	
CHLORINATED POLYESTERS	
POLYSTYRENES	
POLYETHYLENE	
POLYPROPYLENE	
PVC—TYPE 1	
PVC—TYPE 2	
POLYVINYLIDENE CHLORIDE	
VINYLIDENE FLUORIDE	R (60–280)
VINYL ESTER	

RUBBERS	TEMP °F
BUTYL GR-1	
FLUORO ELASTOMERS VITON A	
FLUORO ELASTOMERS KEL-F 3700	
HARD RUBBER	
HYPALON	
KOROSEAL	
NATURAL RUBBER (GRS)	
NEOPRENE GR-M (CR)	
NITRILE BUNA N (NBR)	
NORDEL	
POLYBUTADIENE	
SILICONE	
POLYURETHANE	

TEMP °F (60–460)

BUTYL CHLORIDE

TEMP °F: 60, 80, 100, 120, 140, 160, 180, 200, 220, 240, 260, 280, 300, 320, 340, 360, 380, 400, 420, 440, 460, 480, 500, 520, 540, 560

METALS	
ALLOY 20Cb3	
ALUMINUM	G—
BRASS	G—
BRONZE	G—
CARBON STEEL	G—
COPPER	G—
E-BRIGHT 26-1	
HASTELLOY-B	G————————————
HASTELLOY-C	G—
HASTELLOY-D	
HIGH SILICON IRON	
INCONEL	G—
LEAD	
MONEL	G—————————————————
NICKEL	G—
STAINLESS STEELS	
Type 304/307	G—
Type 316	G—
TANTALUM	G———————
TITANIUM	G—
ZIRCONIUM	G—

NONMETALLICS	
BOROSILICATE GLASS	
CARBON	

TEMP °F: 60, 80, 100, 120, 140, 160, 180, 200, 220, 240, 260, 280, 300, 320, 340, 360, 380, 400, 420, 440, 460, 480, 500, 520, 540, 560

FOR METALS

E = < 2 Mils Penetration/Year; G = < 20 Mils Penetration/Year

S = < 50 Mils Penetration/Year; U = > 50 Mils Penetration/Year

FOR NONMETALLICS

R = Resistant

U = Unsatisfactory

BUTYL CHLORIDE

TEMP °F: 60 80 100 120 140 160 180 200 220 240 260 280 300 320 340 360 380 400 420 440 460

PLASTICS	
ABS	
ACRYLICS	
ASBESTOS REINF., EPOXY	
ASBESTOS REINF., PHENOLIC	
CHLORINATED POLYETHER	R
CPVC	
EPOXY	
FLUOROCARBONS FEP	R
FLUOROCARBONS TFE	R
FURFURYL ALCOHOL	
NORYL	
PHENOLIC	
POLYAMIDES—NYLON	
POLYESTERS	
BISPHENOL A-FUMURATE	
HYDROGENATED BISPHENOL-A-BISPHENOL-A	
ISOPHTHALIC	
CHLORINATED POLYESTERS	
POLYSTYRENES	
POLYETHYLENE	
POLYPROPYLENE	
PVC—TYPE 1	
PVC—TYPE 2	
POLYVINYLIDENE CHLORIDE	U
VINYLIDENE FLUORIDE	R
VINYL ESTER	

RUBBERS	
BUTYL GR-1	
FLUORO ELASTOMERS VITON A	
FLUORO ELASTOMERS KEL-F 3700	
HARD RUBBER	
HYPALON	
KOROSEAL	
NATURAL RUBBER (GRS)	
NEOPRENE GR-M (CR)	
NITRILE BUNA N (NBR)	
NORDEL	
POLYBUTADIENE	
SILICONE	
POLYURETHANE	

TEMP °F: 60 80 100 120 140 160 180 200 220 240 260 280 300 320 340 360 380 400 420 440 460

BUTYLENE (BUTADIENE)

TEMP °F

METALS	Rating	Temperature range (°F)
ALLOY 20Cb3	E	60–100
ALUMINUM	G	60–100
BRASS	G	60–80
BRONZE	G	60–80
CARBON STEEL	G	60–80
COPPER	G	60–80
E-BRIGHT 26-1		
HASTELLOY-B		
HASTELLOY-C		
HASTELLOY-D		
HIGH SILICON IRON	G	60–80
INCONEL	G	60–80
LEAD		
MONEL	E	60–80
NICKEL	G	60–80
STAINLESS STEELS		
Type 304/307	G	60–180
Type 316	G	60–400
TANTALUM	G	60–80
TITANIUM		
ZIRCONIUM		

NONMETALLICS	Rating	Temperature range (°F)
BOROSILICATE GLASS		
CARBON	R	60–340

TEMP °F

FOR METALS

E = < 2 Mils Penetration/Year; G = < 20 Mils Penetration/Year

S = < 50 Mils Penetration/Year; U = > 50 Mils Penetration/Year

FOR NONMETALLICS

R = Resistant

U = Unsatisfactory

BUTYLENE (BUTADIENE)

TEMP °F: 60, 80, 100, 120, 140, 160, 180, 200, 220, 240, 260, 280, 300, 320, 340, 360, 380, 400, 420, 440, 460

PLASTICS	
ABS	U
ACRYLICS	
ASBESTOS REINF., EPOXY	
ASBESTOS REINF., PHENOLIC	
CHLORINATED POLYETHER	R
CPVC	R
EPOXY	R
FLUOROCARBONS FEP	R
FLUOROCARBONS TFE	R
FURFURYL ALCOHOL	
NORYL	
PHENOLIC	
POLYAMIDES–NYLON	
POLYESTERS	
BISPHENOL A-FUMURATE	
HYDROGENATED BISPHENOL-A-BISPHENOL-A	
ISOPHTHALIC	
CHLORINATED POLYESTERS	
POLYSTYRENES	
POLYETHYLENE	U
POLYPROPYLENE	U
PVC–TYPE 1	R
PVC–TYPE 2	RU
POLYVINYLIDENE CHLORIDE	U
VINYLIDENE FLUORIDE	R
VINYL ESTER	

RUBBERS	
BUTYL GR-1	
FLUORO ELASTOMERS VITON A	R
FLUORO ELASTOMERS KEL-F 3700	
HARD RUBBER	
HYPALON	U
KOROSEAL	
NATURAL RUBBER (GRS)	U
NEOPRENE GR-M (CR)	U
NITRILE BUNA N (NBR)	RU
NORDEL	
POLYBUTADIENE	
SILICONE	
POLYURETHANE	U

TEMP °F: 60, 80, 100, 120, 140, 160, 180, 200, 220, 240, 260, 280, 300, 320, 340, 360, 380, 400, 420, 440, 460

n-BUTYL MERCAPTAN

METALS	Rating	TEMP °F
ALLOY 20Cb3	G	60–200
ALUMINUM	G	60–200
BRASS	U	60
BRONZE	U	60
CARBON STEEL		
COPPER	U	60
E-BRIGHT 26-1		
HASTELLOY-B	G	60–300
HASTELLOY-C	G	60–300
HASTELLOY-D		
HIGH SILICON IRON		
INCONEL	G	60–200
LEAD		
MONEL		
NICKEL	G	60–200
STAINLESS STEELS		
Type 304/307	G	60–200
Type 316	G	60–200
TANTALUM		
TITANIUM		
ZIRCONIUM		
NONMETALLICS		
BOROSILICATE GLASS		
CARBON		

FOR METALS

E = < 2 Mils Penetration/Year; G = < 20 Mils Penetration/Year

S = < 50 Mils Penetration/Year; U = > 50 Mils Penetration/Year

FOR NONMETALLICS

R = Resistant

U = Unsatisfactory

n-BUTYL MERCAPTAN

TEMP °F (60–460)

PLASTICS	Rating (TEMP °F)
ABS	
ACRYLICS	
ASBESTOS REINF., EPOXY	
ASBESTOS REINF., PHENOLIC	
CHLORINATED POLYETHER	
CPVC	
EPOXY	
FLUOROCARBONS FEP	R 60–400
FLUOROCARBONS TFE	R 60–460
FURFURYL ALCOHOL	
NORYL	
PHENOLIC	
POLYAMIDES—NYLON	
POLYESTERS	
BISPHENOL A-FUMURATE	
HYDROGENATED BISPHENOL-A-BISPHENOL-A	
ISOPHTHALIC	
CHLORINATED POLYESTERS	
POLYSTYRENES	
POLYETHYLENE	
POLYPROPYLENE	
PVC—TYPE 1	
PVC—TYPE 2	
POLYVINYLIDENE CHLORIDE	U 60
VINYLIDENE FLUORIDE	R 60–280
VINYL ESTER	

RUBBERS	TEMP °F
BUTYL GR-1	
FLUORO ELASTOMERS VITON A	
FLUORO ELASTOMERS KEL-F 3700	
HARD RUBBER	
HYPALON	
KOROSEAL	
NATURAL RUBBER (GRS)	
NEOPRENE GR-M (CR)	
NITRILE BUNA N (NBR)	
NORDEL	
POLYBUTADIENE	
SILICONE	
POLYURETHANE	

TEMP °F

BUTYL PHENOL

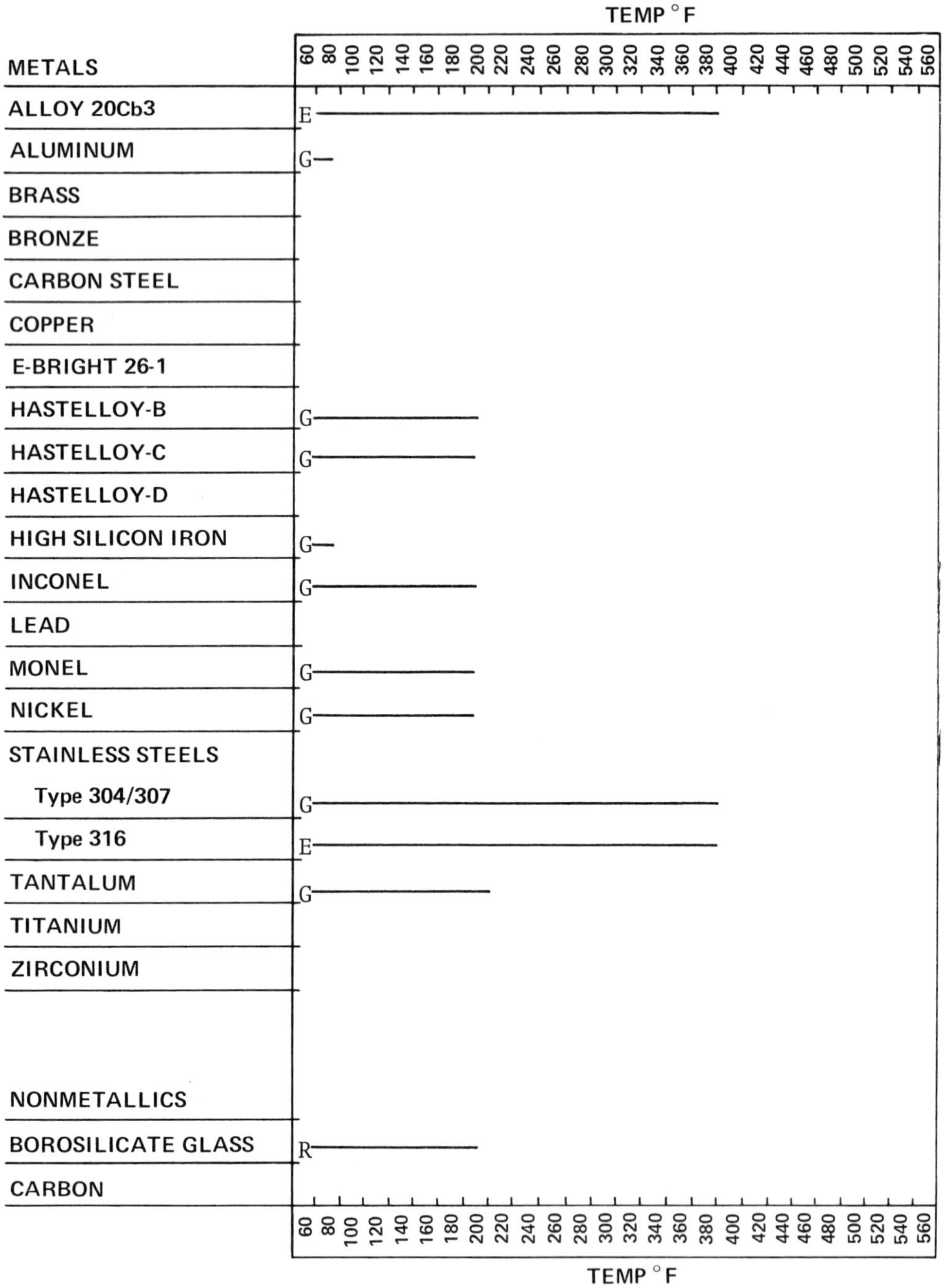

FOR METALS

E = < 2 Mils Penetration/Year; G = < 20 Mils Penetration/Year

S = < 50 Mils Penetration/Year; U = > 50 Mils Penetration/Year

FOR NONMETALLICS

R = Resistant

U = Unsatisfactory

BUTYL PHENOL

TEMP °F (60–460)

PLASTICS	TEMP °F
ABS	U
ACRYLICS	
ASBESTOS REINF., EPOXY	
ASBESTOS REINF., PHENOLIC	
CHLORINATED POLYETHER	R——— (60 to 180)
CPVC	U
EPOXY	
FLUOROCARBONS FEP	R——— (60 to 400)
FLUOROCARBONS TFE	R——— (60 to 460+)
FURFURYL ALCOHOL	
NORYL	
PHENOLIC	
POLYAMIDES–NYLON	
POLYESTERS	
BISPHENOL A-FUMURATE	
HYDROGENATED BISPHENOL-A-BISPHENOL-A	
ISOPHTHALIC	
CHLORINATED POLYESTERS	
POLYSTYRENES	
POLYETHYLENE	U
POLYPROPYLENE	U
PVC–TYPE 1	RU
PVC–TYPE 2	U
POLYVINYLIDENE CHLORIDE	R— (60 to 80)
VINYLIDENE FLUORIDE	R——— (60 to 220)
VINYL ESTER	

RUBBERS	TEMP °F
BUTYL GR-1	
FLUORO ELASTOMERS VITON A	
FLUORO ELASTOMERS KEL-F 3700	
HARD RUBBER	
HYPALON	
KOROSEAL	
NATURAL RUBBER (GRS)	
NEOPRENE GR-M (CR)	
NITRILE BUNA N (NBR)	
NORDEL	
POLYBUTADIENE	
SILICONE	
POLYURETHANE	

TEMP °F (60–460)

BUTYL PHTHALATE

TEMP °F

METALS	60 – 560 °F
ALLOY 20Cb3	G
ALUMINUM	U
BRASS	G
BRONZE	G
CARBON STEEL	
COPPER	G
E-BRIGHT 26-1	
HASTELLOY-B	G
HASTELLOY-C	G
HASTELLOY-D	
HIGH SILICON IRON	G
INCONEL	G
LEAD	
MONEL	G
NICKEL	G
STAINLESS STEELS	
Type 304/307	G
Type 316	G
TANTALUM	G
TITANIUM	G
ZIRCONIUM	G
NONMETALLICS	
BOROSILICATE GLASS	
CARBON	

TEMP °F: 60, 80, 100, 120, 140, 160, 180, 200, 220, 240, 260, 280, 300, 320, 340, 360, 380, 400, 420, 440, 460, 480, 500, 520, 540, 560

FOR METALS

E = < 2 Mils Penetration/Year; G = < 20 Mils Penetration/Year

S = < 50 Mils Penetration/Year; U = > 50 Mils Penetration/Year

FOR NONMETALLICS

R = Resistant

U = Unsatisfactory

TEMP °F (60, 80, 100, 120, 140, 160, 180, 200, 220, 240, 260, 280, 300, 320, 340, 360, 380, 400, 420, 440, 460)

PLASTICS	
ABS	
ACRYLICS	
ASBESTOS REINF., EPOXY	
ASBESTOS REINF., PHENOLIC	
CHLORINATED POLYETHER	
CPVC	
EPOXY	
FLUOROCARBONS FEP	R
FLUOROCARBONS TFE	R
FURFURYL ALCOHOL	
NORYL	
PHENOLIC	
POLYAMIDES–NYLON	R
POLYESTERS	
BISPHENOL A-FUMURATE	
HYDROGENATED BISPHENOL-A-BISPHENOL-A	
ISOPHTHALIC	
CHLORINATED POLYESTERS	R
POLYSTYRENES	
POLYETHYLENE	
POLYPROPYLENE	R
PVC–TYPE 1	
PVC–TYPE 2	
POLYVINYLIDENE CHLORIDE	R
VINYLIDENE FLUORIDE	R
VINYL ESTER	

RUBBERS	
BUTYL GR-1	
FLUORO ELASTOMERS VITON A	RU
FLUORO ELASTOMERS KEL-F 3700	
HARD RUBBER	
HYPALON	
KOROSEAL	
NATURAL RUBBER (GRS)	
NEOPRENE GR-M (CR)	
NITRILE BUNA N (NBR)	U
NORDEL	
POLYBUTADIENE	
SILICONE	
POLYURETHANE	

TEMP °F

BUTYNE DIOL

TEMP °F

METALS	60	80	100	120	140	160	180	200	220	240	260	280	300	320	340	360	380	400	420	440	460	480	500	520	540	560
ALLOY 20Cb3																										
ALUMINUM																										
BRASS																										
BRONZE																										
CARBON STEEL																										
COPPER																										
E-BRIGHT 26-1																										
HASTELLOY-B																										
HASTELLOY-C																										
HASTELLOY-D																										
HIGH SILICON IRON																										
INCONEL																										
LEAD																										
MONEL																										
NICKEL																										
STAINLESS STEELS																										
Type 304/307																										
Type 316																										
TANTALUM																										
TITANIUM																										
ZIRCONIUM																										
NONMETALLICS																										
BOROSILICATE GLASS																										
CARBON																										
	60	80	100	120	140	160	180	200	220	240	260	280	300	320	340	360	380	400	420	440	460	480	500	520	540	560

TEMP °F

FOR METALS

E = < 2 Mils Penetration/Year; G = < 20 Mils Penetration/Year

S = < 50 Mils Penetration/Year; U = > 50 Mils Penetration/Year

FOR NONMETALLICS

R = Resistant

U = Unsatisfactory

BUTYNE DIOL

TEMP °F (60–460)

PLASTICS	
ABS	U
ACRYLICS	
ASBESTOS REINF., EPOXY	
ASBESTOS REINF., PHENOLIC	
CHLORINATED POLYETHER	R
CPVC	R
EPOXY	
FLUOROCARBONS FEP	R
FLUOROCARBONS TFE	R
FURFURYL ALCOHOL	
NORYL	
PHENOLIC	
POLYAMIDES—NYLON	
POLYESTERS	
BISPHENOL A-FUMURATE	
HYDROGENATED BISPHENOL-A-BISPHENOL-A	
ISOPHTHALIC	
CHLORINATED POLYESTERS	
POLYSTYRENES	
POLYETHYLENE	R
POLYPROPYLENE	R
PVC—TYPE 1	R—U
PVC—TYPE 2	U
POLYVINYLIDENE CHLORIDE	
VINYLIDENE FLUORIDE	
VINYL ESTER	

RUBBERS	
BUTYL GR-1	
FLUORO ELASTOMERS VITON A	
FLUORO ELASTOMERS KEL-F 3700	
HARD RUBBER	
HYPALON	
KOROSEAL	
NATURAL RUBBER (GRS)	
NEOPRENE GR-M (CR)	
NITRILE BUNA N (NBR)	
NORDEL	
POLYBUTADIENE	
SILICONE	
POLYURETHANE	

TEMP °F (60–460)

BUTYRIC ACID

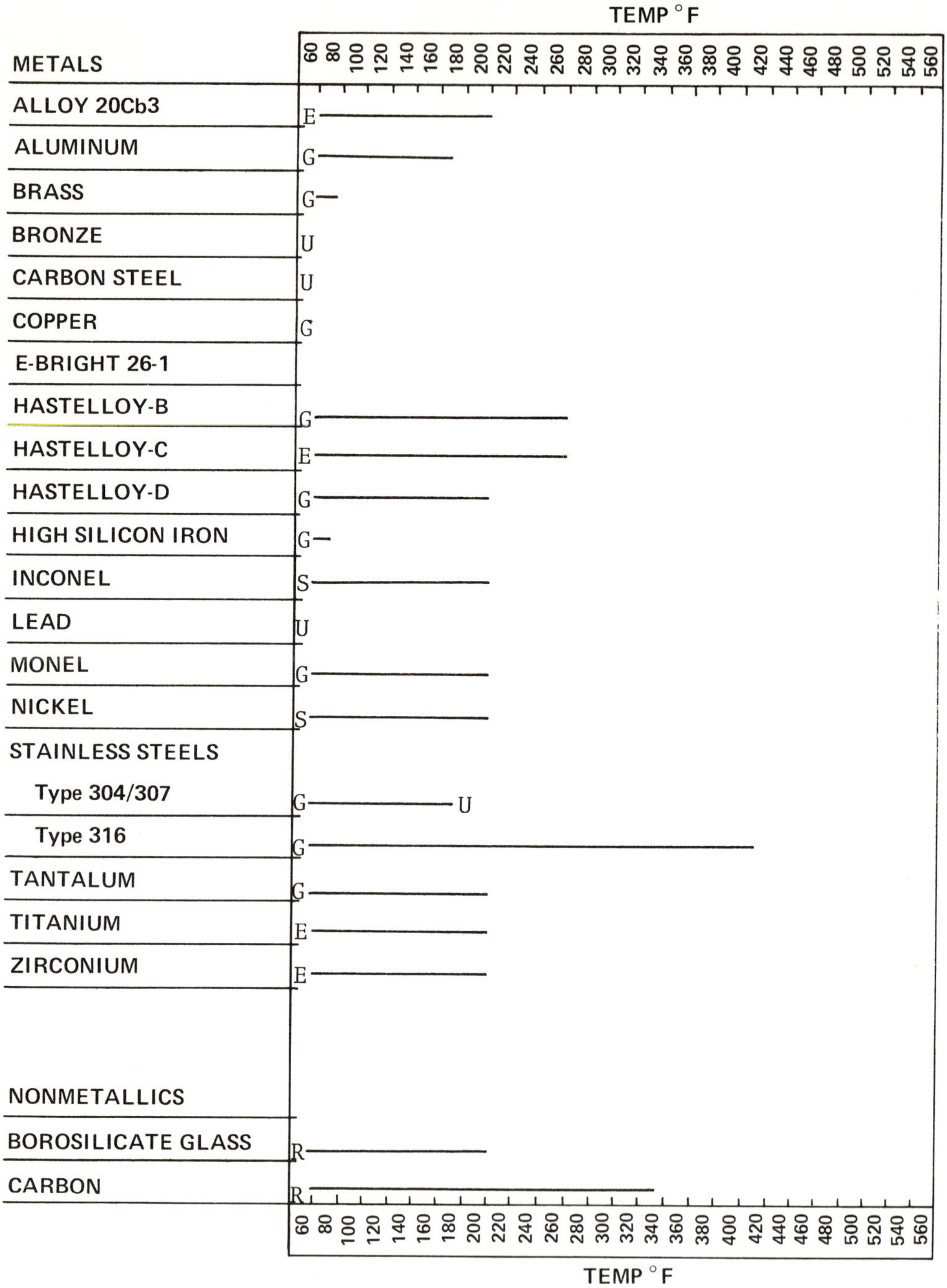

FOR METALS

E = < 2 Mils Penetration/Year; G = < 20 Mils Penetration/Year

S = < 50 Mils Penetration/Year; U = > 50 Mils Penetration/Year

FOR NONMETALLICS

R = Resistant

U = Unsatisfactory

BUTYRIC ACID

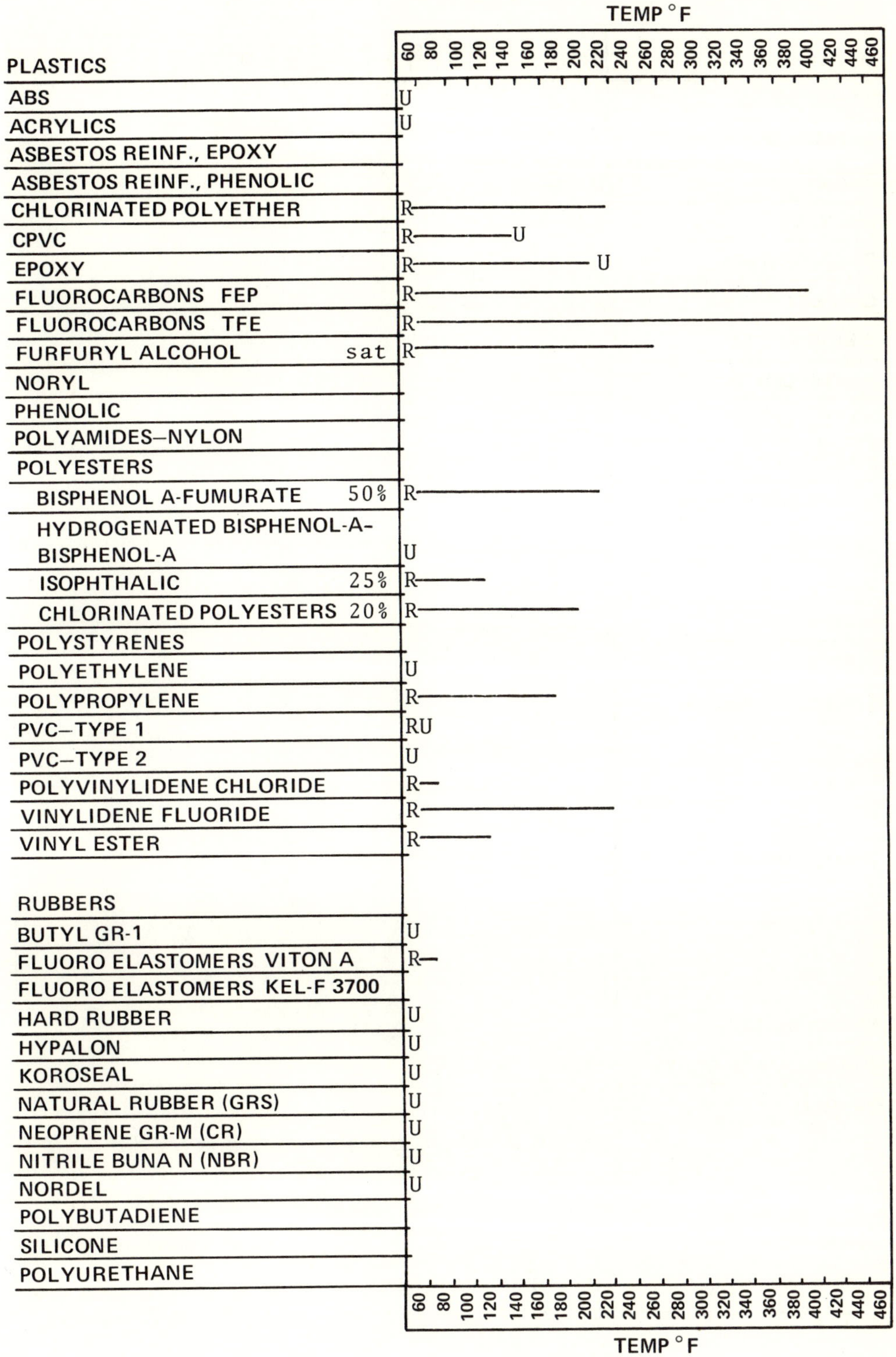

CALCIUM BICARBONATE

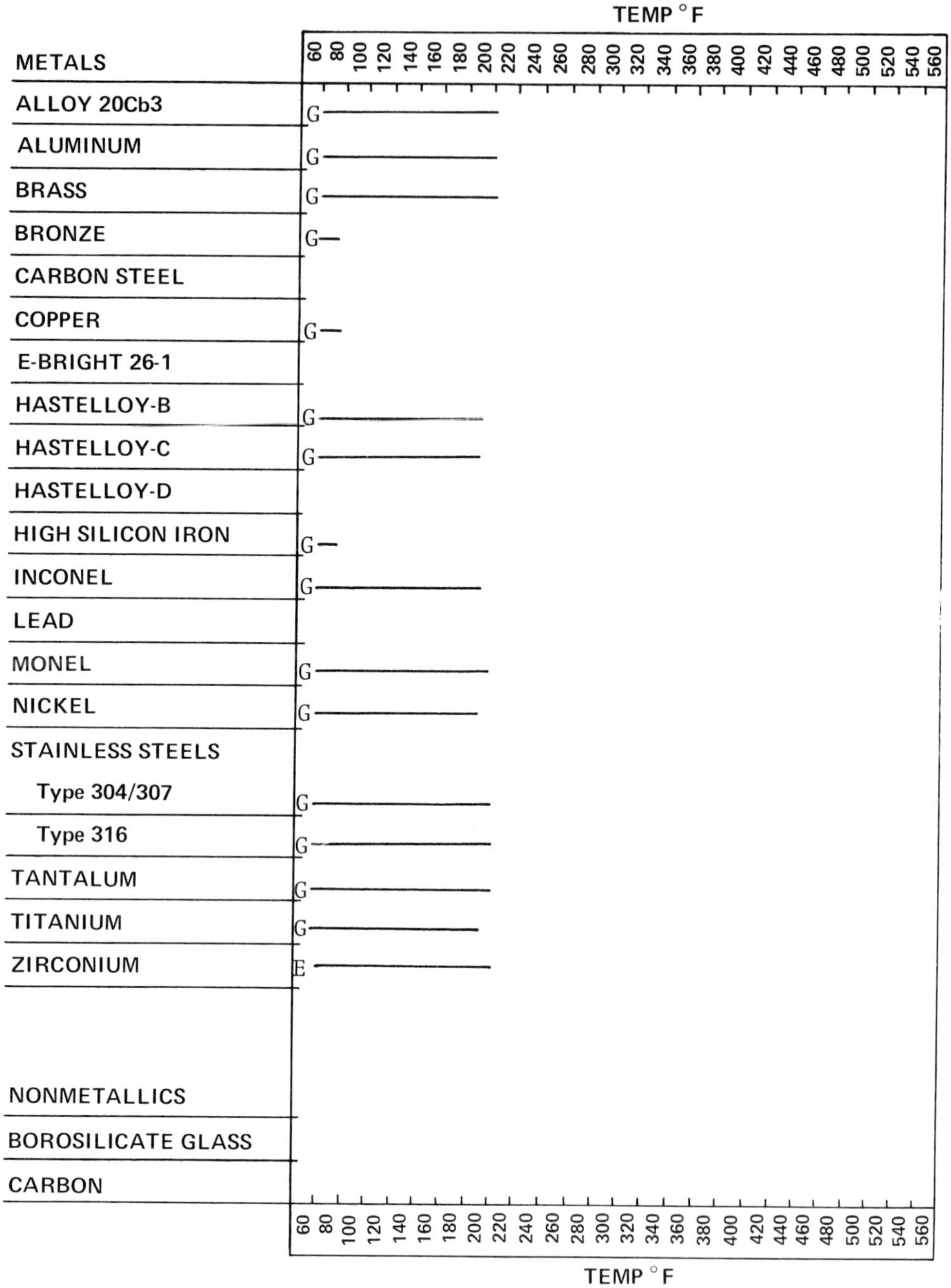

FOR METALS

E = < 2 Mils Penetration/Year; G = < 20 Mils Penetration/Year
S = < 50 Mils Penetration/Year; U = > 50 Mils Penetration/Year

FOR NONMETALLICS

R = Resistant
U = Unsatisfactory

Please see page 1 for notes on calibration.

CALCIUM BICARBONATE

TEMP °F (scale: 60, 80, 100, 120, 140, 160, 180, 200, 220, 240, 260, 280, 300, 320, 340, 360, 380, 400, 420, 440, 460)

PLASTICS	Rating (TEMP °F)
ABS	
ACRYLICS	
ASBESTOS REINF., EPOXY	
ASBESTOS REINF., PHENOLIC	
CHLORINATED POLYETHER	
CPVC	
EPOXY	
FLUOROCARBONS FEP	R (60–400)
FLUOROCARBONS TFE	R (60–460)
FURFURYL ALCOHOL	
NORYL	
PHENOLIC	
POLYAMIDES—NYLON	
POLYESTERS	
BISPHENOL A-FUMURATE	
HYDROGENATED BISPHENOL-A-BISPHENOL-A	
ISOPHTHALIC	
CHLORINATED POLYESTERS	
POLYSTYRENES	
POLYETHYLENE	
POLYPROPYLENE	
PVC—TYPE 1	
PVC—TYPE 2	
POLYVINYLIDENE CHLORIDE	R (60–80)
VINYLIDENE FLUORIDE	
VINYL ESTER	

RUBBERS	TEMP °F
BUTYL GR-1	
FLUORO ELASTOMERS VITON A	
FLUORO ELASTOMERS KEL-F 3700	
HARD RUBBER	
HYPALON	
KOROSEAL	
NATURAL RUBBER (GRS)	
NEOPRENE GR-M (CR)	
NITRILE BUNA N (NBR)	
NORDEL	
POLYBUTADIENE	
SILICONE	
POLYURETHANE	

CALCIUM BISULFATE

TEMP °F (60–560)

METALS	Rating (from 60 °F)
ALLOY 20Cb3	
ALUMINUM	U
BRASS	
BRONZE	U
CARBON STEEL	U
COPPER	U
E-BRIGHT 26-1	
HASTELLOY-B	
HASTELLOY-C	
HASTELLOY-D	
HIGH SILICON IRON	
INCONEL	
LEAD	
MONEL	U
NICKEL	U
STAINLESS STEELS	
Type 304/307	U
Type 316	G (60–200 °F)
TANTALUM	
TITANIUM	
ZIRCONIUM	

NONMETALLICS	Rating
BOROSILICATE GLASS	
CARBON	

TEMP °F

FOR METALS

E = < 2 Mils Penetration/Year; G = < 20 Mils Penetration/Year

S = < 50 Mils Penetration/Year; U = > 50 Mils Penetration/Year

FOR NONMETALLICS

R = Resistant

U = Unsatisfactory

CALCIUM BISULFATE

PLASTICS	TEMP °F (60–460)
ABS	
ACRYLICS	
ASBESTOS REINF., EPOXY	
ASBESTOS REINF., PHENOLIC	
CHLORINATED POLYETHER	
CPVC	
EPOXY	
FLUOROCARBONS FEP	R 60–400
FLUOROCARBONS TFE	R 60–460
FURFURYL ALCOHOL	
NORYL	
PHENOLIC	
POLYAMIDES—NYLON	
POLYESTERS	
BISPHENOL A-FUMURATE	
HYDROGENATED BISPHENOL-A-BISPHENOL-A	
ISOPHTHALIC	
CHLORINATED POLYESTERS	
POLYSTYRENES	
POLYETHYLENE	
POLYPROPYLENE	R 60–220
PVC—TYPE 1	
PVC—TYPE 2	
POLYVINYLIDENE CHLORIDE	
VINYLIDENE FLUORIDE	R 60–280
VINYL ESTER	

RUBBERS	TEMP °F (60–460)
BUTYL GR-1	
FLUORO ELASTOMERS VITON A	
FLUORO ELASTOMERS KEL-F 3700	
HARD RUBBER	
HYPALON	
KOROSEAL	
NATURAL RUBBER (GRS)	
NEOPRENE GR-M (CR)	
NITRILE BUNA N (NBR)	
NORDEL	
POLYBUTADIENE	
SILICONE	
POLYURETHANE	

CALCIUM BISULFIDE

TEMP °F (60–560)

METALS	TEMP °F
ALLOY 20Cb3	
ALUMINUM	
BRASS	
BRONZE	
CARBON STEEL	
COPPER	
E-BRIGHT 26-1	
HASTELLOY-B	
HASTELLOY-C	
HASTELLOY-D	
HIGH SILICON IRON	
INCONEL	
LEAD	
MONEL	
NICKEL	
STAINLESS STEELS	
Type 304/307	
Type 316	
TANTALUM	
TITANIUM	
ZIRCONIUM	

NONMETALLICS	Rating (TEMP °F)
BOROSILICATE GLASS	
CARBON	R (60–340)

FOR METALS

E = < 2 Mils Penetration/Year; G = < 20 Mils Penetration/Year

S = < 50 Mils Penetration/Year; U = > 50 Mils Penetration/Year

FOR NONMETALLICS

R = Resistant

U = Unsatisfactory

CALCIUM BISULFIDE

TEMP °F (60 80 100 120 140 160 180 200 220 240 260 280 300 320 340 360 380 400 420 440 460)

PLASTICS	
ABS	
ACRYLICS	
ASBESTOS REINF., EPOXY	
ASBESTOS REINF., PHENOLIC	
CHLORINATED POLYETHER	R
CPVC	R
EPOXY	
FLUOROCARBONS FEP	R
FLUOROCARBONS TFE	R
FURFURYL ALCOHOL	
NORYL	
PHENOLIC	
POLYAMIDES—NYLON	R
POLYESTERS	
BISPHENOL A-FUMURATE	R
HYDROGENATED BISPHENOL-A-BISPHENOL-A	
ISOPHTHALIC sat	R
CHLORINATED POLYESTERS	
POLYSTYRENES	
POLYETHYLENE	R
POLYPROPYLENE	R
PVC—TYPE 1	R
PVC—TYPE 2	R
POLYVINYLIDENE CHLORIDE	
VINYLIDENE FLUORIDE	R
VINYL ESTER	
RUBBERS	
BUTYL GR-1	
FLUORO ELASTOMERS VITON A	R
FLUORO ELASTOMERS KEL-F 3700	
HARD RUBBER	
HYPALON	
KOROSEAL	
NATURAL RUBBER (GRS)	
NEOPRENE GR-M (CR)	
NITRILE BUNA N (NBR)	R
NORDEL	
POLYBUTADIENE	
SILICONE	
POLYURETHANE	

TEMP °F (60 80 100 120 140 160 180 200 220 240 260 280 300 320 340 360 380 400 420 440 460)

CALCIUM BISULFITE

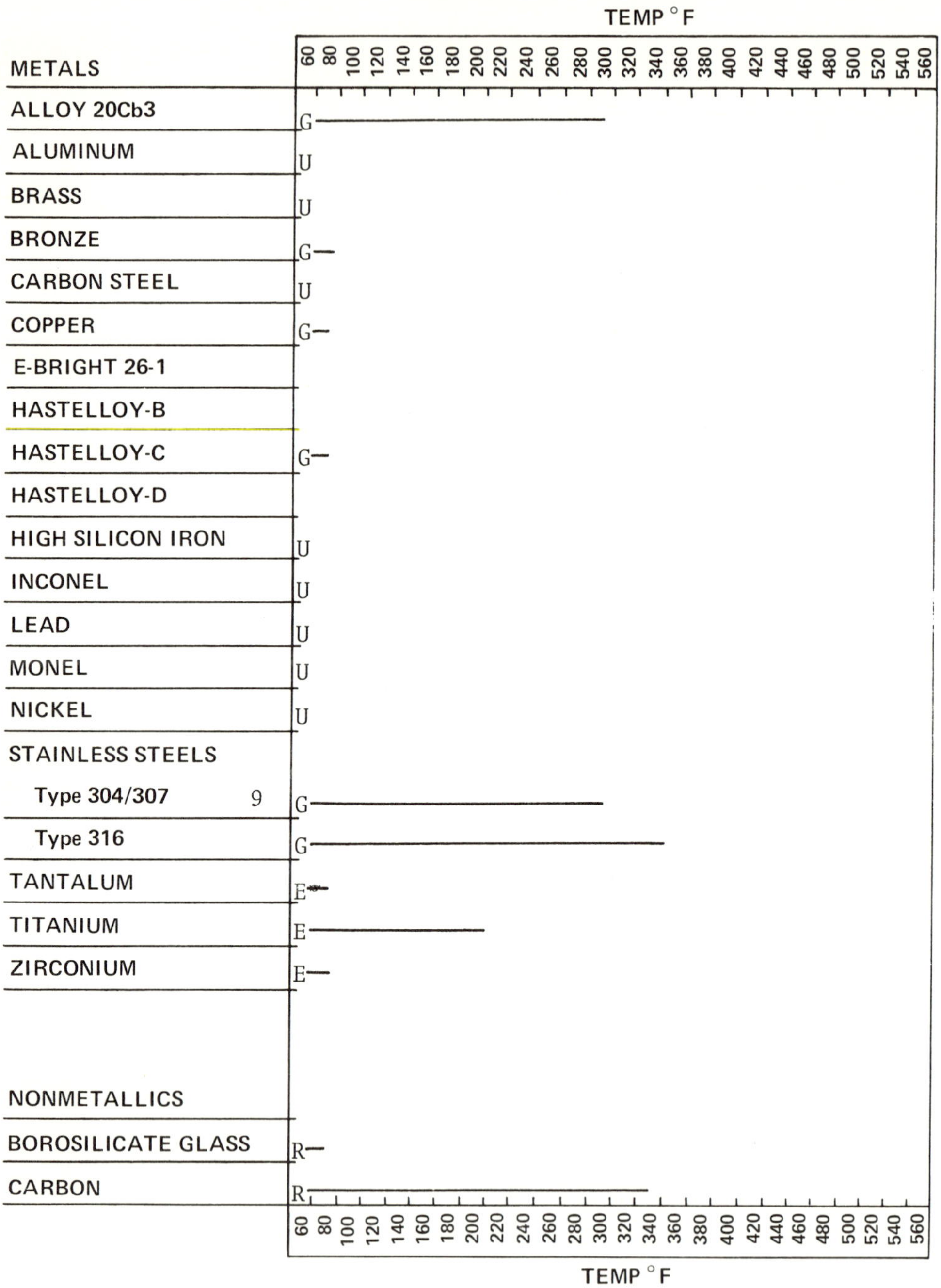

FOR METALS

E = < 2 Mils Penetration/Year; G = < 20 Mils Penetration/Year
S = < 50 Mils Penetration/Year; U = > 50 Mils Penetration/Year

FOR NONMETALLICS

R = Resistant
U = Unsatisfactory

CALCIUM BISULFITE

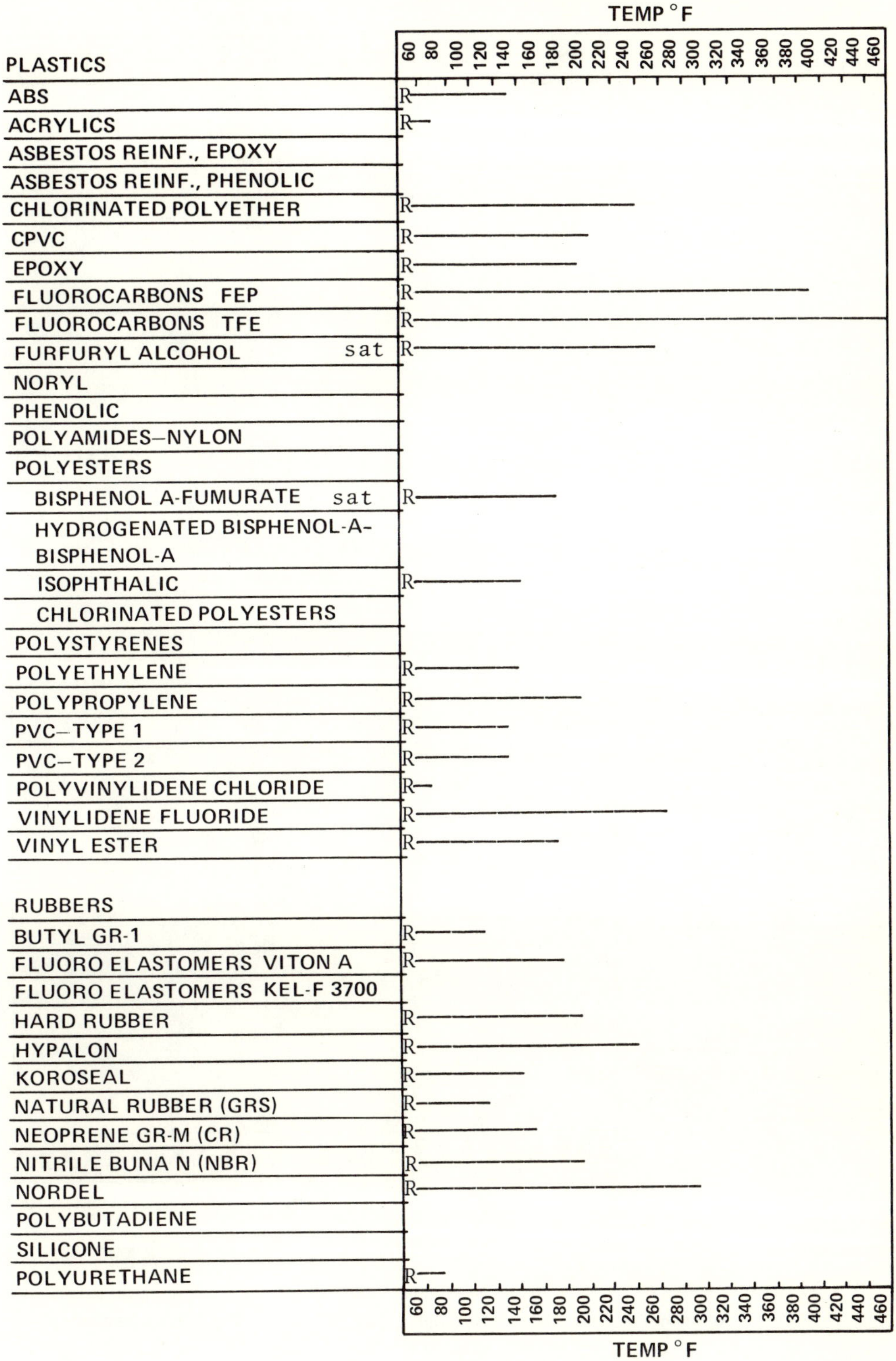

CALCIUM BROMIDE, 38%

TEMP °F: 60, 80, 100, 120, 140, 160, 180, 200, 220, 240, 260, 280, 300, 320, 340, 360, 380, 400, 420, 440, 460, 480, 500, 520, 540, 560

METALS	
ALLOY 20Cb3	G
ALUMINUM	
BRASS	G
BRONZE	G
CARBON STEEL	
COPPER	G
E-BRIGHT 26-1	
HASTELLOY-B	E
HASTELLOY-C	G
HASTELLOY-D	
HIGH SILICON IRON	G
INCONEL	G
LEAD	
MONEL	G
NICKEL	G
STAINLESS STEELS	
Type 304/307 1,3	E
Type 316	
TANTALUM	G
TITANIUM	G
ZIRCONIUM	G

NONMETALLICS	
BOROSILICATE GLASS	
CARBON	

TEMP °F: 60, 80, 100, 120, 140, 160, 180, 200, 220, 240, 260, 280, 300, 320, 340, 360, 380, 400, 420, 440, 460, 480, 500, 520, 540, 560

FOR METALS

E = < 2 Mils Penetration/Year; G = < 20 Mils Penetration/Year

S = < 50 Mils Penetration/Year; U = > 50 Mils Penetration/Year

FOR NONMETALLICS

R = Resistant

U = Unsatisfactory

CALCIUM BROMIDE, 38%

TEMP °F: 60 80 100 120 140 160 180 200 220 240 260 280 300 320 340 360 380 400 420 440 460

PLASTICS	TEMP °F
ABS	
ACRYLICS	
ASBESTOS REINF., EPOXY	
ASBESTOS REINF., PHENOLIC	
CHLORINATED POLYETHER	
CPVC	
EPOXY	
FLUOROCARBONS FEP	R (60–400)
FLUOROCARBONS TFE	R (60–460)
FURFURYL ALCOHOL	
NORYL	
PHENOLIC	
POLYAMIDES—NYLON	
POLYESTERS	
BISPHENOL A-FUMURATE	
HYDROGENATED BISPHENOL-A-BISPHENOL-A	
ISOPHTHALIC	
CHLORINATED POLYESTERS	
POLYSTYRENES	
POLYETHYLENE	
POLYPROPYLENE	
PVC—TYPE 1	
PVC—TYPE 2	
POLYVINYLIDENE CHLORIDE	R (60–80)
VINYLIDENE FLUORIDE	
VINYL ESTER	

RUBBERS	TEMP °F
BUTYL GR-1	
FLUORO ELASTOMERS VITON A	
FLUORO ELASTOMERS KEL-F 3700	
HARD RUBBER	
HYPALON	
KOROSEAL	
NATURAL RUBBER (GRS)	
NEOPRENE GR-M (CR)	
NITRILE BUNA N (NBR)	
NORDEL	
POLYBUTADIENE	
SILICONE	
POLYURETHANE	

TEMP °F: 60 80 100 120 140 160 180 200 220 240 260 280 300 320 340 360 380 400 420 440 460

CALCIUM CARBONATE

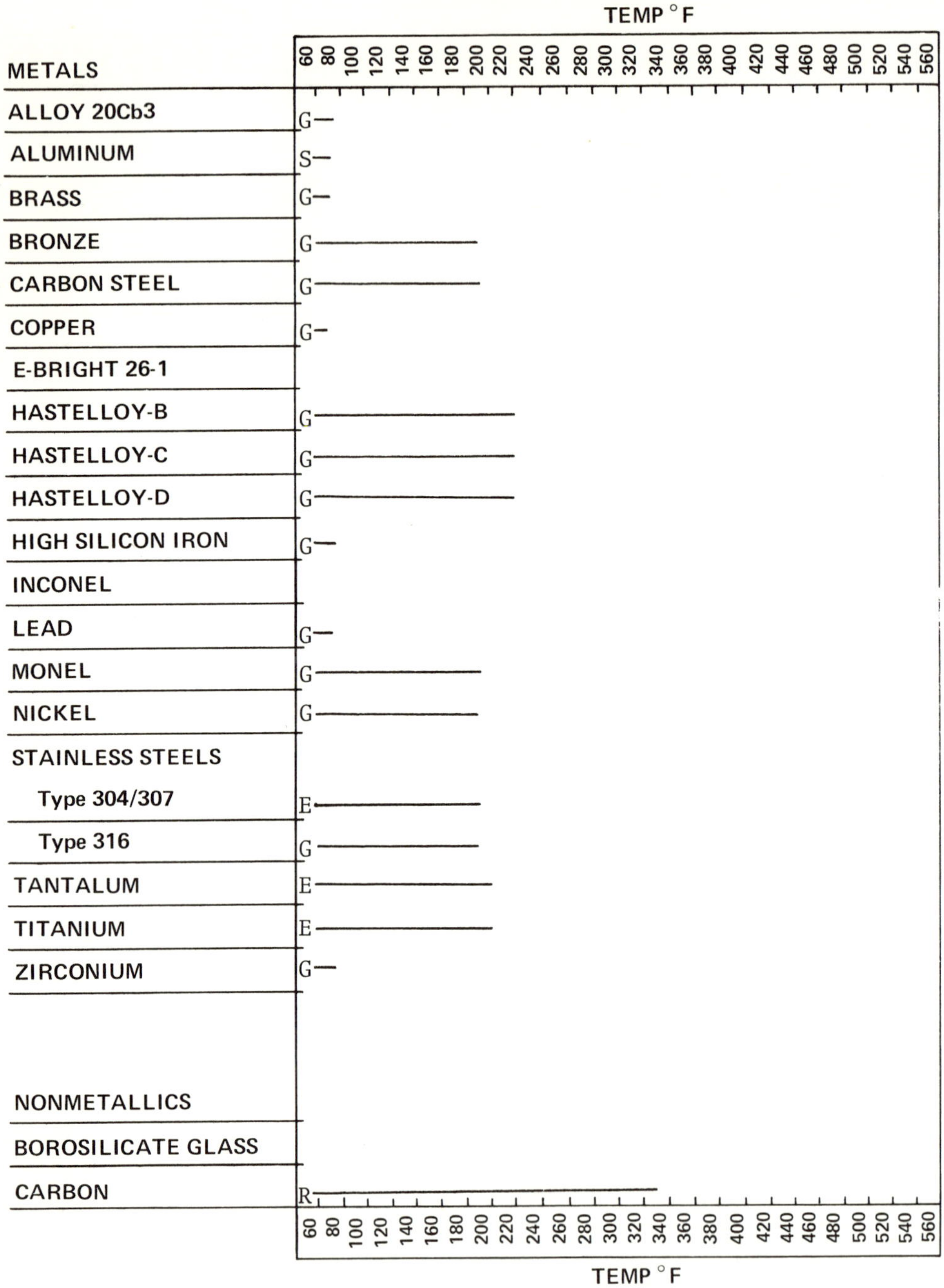

FOR METALS

E = < 2 Mils Penetration/Year; G = < 20 Mils Penetration/Year

S = < 50 Mils Penetration/Year; U = > 50 Mils Penetration/Year

FOR NONMETALLICS

R = Resistant

U = Unsatisfactory

CALCIUM CARBONATE

TEMP °F (60–460)

PLASTICS	Rating	Bar extends to approx. (°F)
ABS	R	100
ACRYLICS	R	80
ASBESTOS REINF., EPOXY		
ASBESTOS REINF., PHENOLIC		
CHLORINATED POLYETHER	R	250
CPVC	R	210
EPOXY	R	300
FLUOROCARBONS FEP	R	400
FLUOROCARBONS TFE	R	460
FURFURYL ALCOHOL		
NORYL	R	140
PHENOLIC		
POLYAMIDES–NYLON	R	80
POLYESTERS		
BISPHENOL A-FUMURATE	R	210
HYDROGENATED BISPHENOL-A-BISPHENOL-A		
ISOPHTHALIC	R	160
CHLORINATED POLYESTERS		
POLYSTYRENES		
POLYETHYLENE	R	150
POLYPROPYLENE	R	210
PVC–TYPE 1	R	140
PVC–TYPE 2	R	140
POLYVINYLIDENE CHLORIDE	R	180
VINYLIDENE FLUORIDE	R	280
VINYL ESTER	R	180

RUBBERS	Rating	Bar extends to approx. (°F)
BUTYL GR-1	R	150
FLUORO ELASTOMERS VITON A	R	190
FLUORO ELASTOMERS KEL-F 3700	R	80
HARD RUBBER		
HYPALON	R	80
KOROSEAL		
NATURAL RUBBER (GRS)	R	190
NEOPRENE GR-M (CR)	U	
NITRILE BUNA N (NBR)	R	190
NORDEL	R	300
POLYBUTADIENE		
SILICONE		
POLYURETHANE		

TEMP °F

CALCIUM CHLORATE

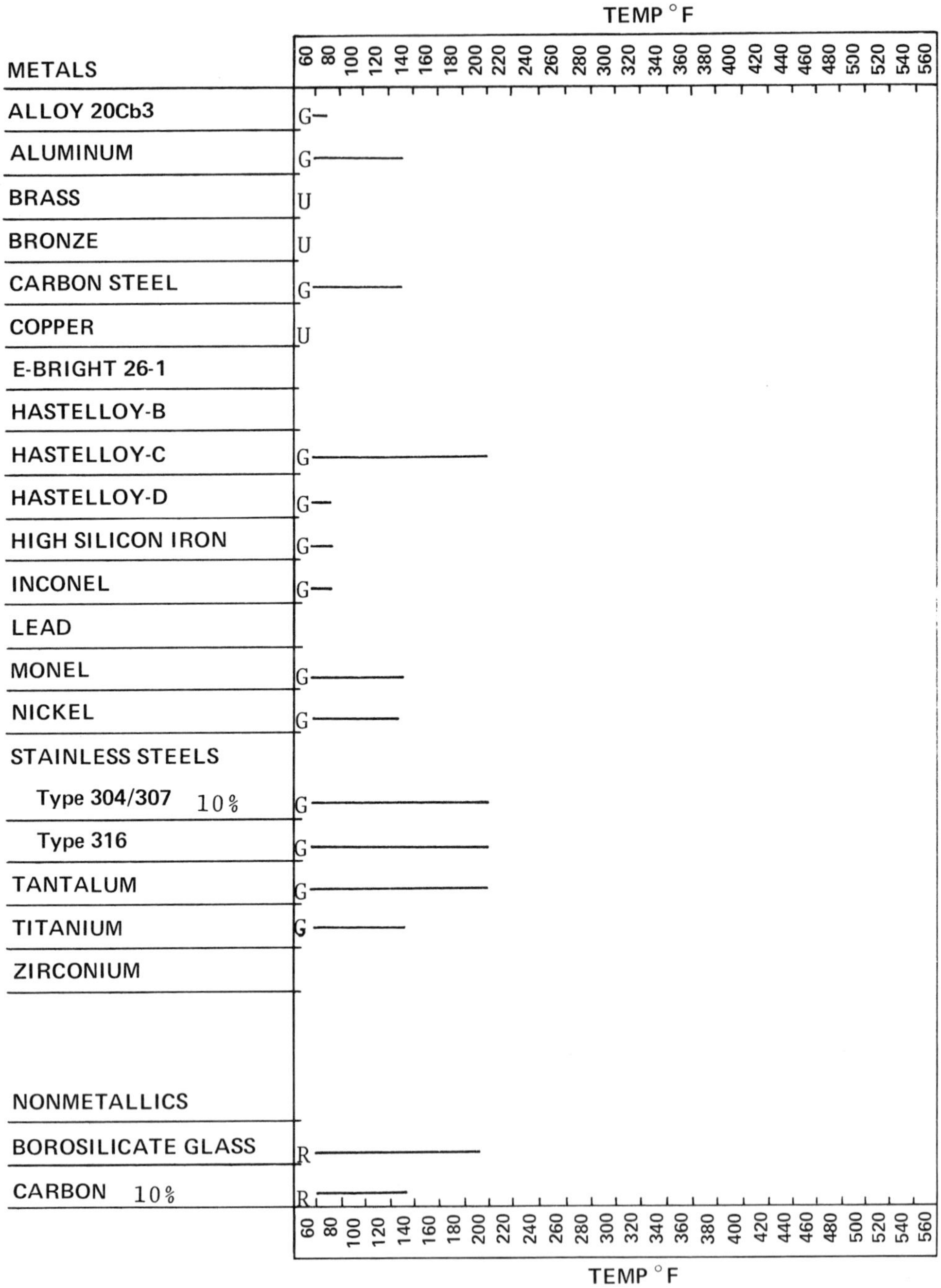

FOR METALS

E = < 2 Mils Penetration/Year; G = < 20 Mils Penetration/Year

S = < 50 Mils Penetration/Year; U = > 50 Mils Penetration/Year

FOR NONMETALLICS

R = Resistant

U = Unsatisfactory

CALCIUM CHLORATE

PLASTICS		TEMP °F
ABS		R to 140
ACRYLICS		R to 80
ASBESTOS REINF., EPOXY		R to 70
ASBESTOS REINF., PHENOLIC		
CHLORINATED POLYETHER		R to 140, U at 150
CPVC		R to 180
EPOXY		R to 200
FLUOROCARBONS FEP		R to 400
FLUOROCARBONS TFE		R to 460
FURFURYL ALCOHOL		R to 250
NORYL		
PHENOLIC		
POLYAMIDES–NYLON		
POLYESTERS		
BISPHENOL A-FUMURATE	sat	R to 200
HYDROGENATED BISPHENOL-A-BISPHENOL-A		R to 210
ISOPHTHALIC	sat	R to 160
CHLORINATED POLYESTERS		R to 250
POLYSTYRENES		
POLYETHYLENE		R to 150
POLYPROPYLENE		R to 225
PVC–TYPE 1		R to 140
PVC–TYPE 2		R to 140
POLYVINYLIDENE CHLORIDE		R to 150
VINYLIDENE FLUORIDE		R to 280
VINYL ESTER	sat	R to 250

RUBBERS	TEMP °F
BUTYL GR-1	R to 190
FLUORO ELASTOMERS VITON A	R to 190
FLUORO ELASTOMERS KEL-F 3700	R to 80
HARD RUBBER	
HYPALON	R to 80
KOROSEAL	
NATURAL RUBBER (GRS)	R to 150
NEOPRENE GR-M (CR)	R to 200
NITRILE BUNA N (NBR)	R to 200
NORDEL	
POLYBUTADIENE	
SILICONE	
POLYURETHANE	

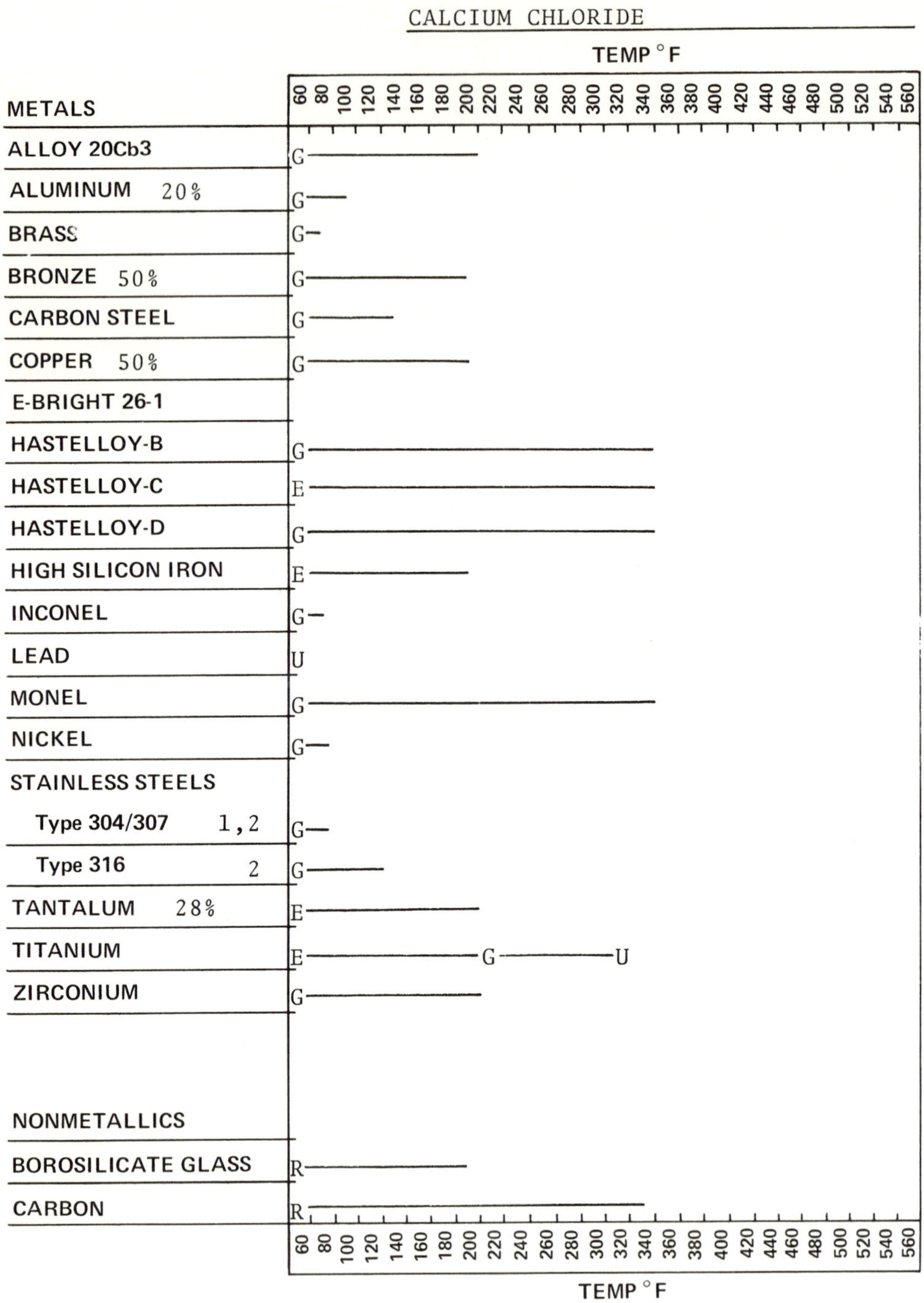

FOR METALS

E = < 2 Mils Penetration/Year; G = < 20 Mils Penetration/Year

S = < 50 Mils Penetration/Year; U = > 50 Mils Penetration/Year

FOR NONMETALLICS

R = Resistant

U = Unsatisfactory

CALCIUM CHLORIDE

TEMP °F

PLASTICS	Concentration	Rating	Bar extends to (°F, approx.)
ABS		R	140
ACRYLICS		R	80
ASBESTOS REINF., EPOXY		R	250
ASBESTOS REINF., PHENOLIC		R	300
CHLORINATED POLYETHER		R	250
CPVC		R	180
EPOXY	37.5%	R — U	U at 200
FLUOROCARBONS FEP		R	400
FLUOROCARBONS TFE		R	460+
FURFURYL ALCOHOL	sat	R	260
NORYL		R	140
PHENOLIC		R	80
POLYAMIDES–NYLON	10%	R	200
POLYESTERS			
BISPHENOL A-FUMURATE	sat	R	220
HYDROGENATED BISPHENOL-A-BISPHENOL-A		R	210
ISOPHTHALIC	sat	R	180
CHLORINATED POLYESTERS		R	250
POLYSTYRENES		R	160
POLYETHYLENE		R — U	U at 140
POLYPROPYLENE		R	220
PVC–TYPE 1		R	140
PVC–TYPE 2		R	140
POLYVINYLIDENE CHLORIDE		R	180
VINYLIDENE FLUORIDE		R	280
VINYL ESTER	sat	R	180

RUBBERS	Concentration	Rating	Bar extends to (°F, approx.)
BUTYL GR-1		R	180
FLUORO ELASTOMERS VITON A		R	180
FLUORO ELASTOMERS KEL-F 3700		R	80
HARD RUBBER		R	180
HYPALON		R	80
KOROSEAL		R	150
NATURAL RUBBER (GRS)		R	150
NEOPRENE GR-M (CR)		R	150
NITRILE BUNA N (NBR)		R	180
NORDEL		R	300
POLYBUTADIENE		R	80
SILICONE			
POLYURETHANE		R	80

TEMP °F

CALCIUM FLUORIDE

TEMP °F: 60 80 100 120 140 160 180 200 220 240 260 280 300 320 340 360 380 400 420 440 460 480 500 520 540 560

METALS	
ALLOY 20Cb3	G
ALUMINUM	G
BRASS	G
BRONZE	G
CARBON STEEL	
COPPER	G
E-BRIGHT 26-1	
HASTELLOY-B	G
HASTELLOY-C	G
HASTELLOY-D	
HIGH SILICON IRON	
INCONEL	G
LEAD	
MONEL	
NICKEL	G
STAINLESS STEELS	
Type 304/307	G
Type 316	G
TANTALUM	U
TITANIUM	
ZIRCONIUM	U

NONMETALLICS	
BOROSILICATE GLASS	U
CARBON	

TEMP °F: 60 80 100 120 140 160 180 200 220 240 260 280 300 320 340 360 380 400 420 440 460 480 500 520 540 560

FOR METALS

E = < 2 Mils Penetration/Year; G = < 20 Mils Penetration/Year

S = < 50 Mils Penetration/Year; U = > 50 Mils Penetration/Year

FOR NONMETALLICS

R = Resistant

U = Unsatisfactory

CALCIUM FLUORIDE

TEMP °F: 60 80 100 120 140 160 180 200 220 240 260 280 300 320 340 360 380 400 420 440 460

PLASTICS	CALCIUM FLUORIDE
ABS	
ACRYLICS	
ASBESTOS REINF., EPOXY	
ASBESTOS REINF., PHENOLIC	
CHLORINATED POLYETHER	
CPVC	
EPOXY	
FLUOROCARBONS FEP	R
FLUOROCARBONS TFE	R
FURFURYL ALCOHOL	
NORYL	
PHENOLIC	
POLYAMIDES—NYLON	
POLYESTERS	
BISPHENOL A-FUMURATE	
HYDROGENATED BISPHENOL-A-BISPHENOL-A	
ISOPHTHALIC	
CHLORINATED POLYESTERS	
POLYSTYRENES	
POLYETHYLENE	
POLYPROPYLENE	
PVC—TYPE 1	
PVC—TYPE 2	
POLYVINYLIDENE CHLORIDE	
VINYLIDENE FLUORIDE	
VINYL ESTER	

RUBBERS	CALCIUM FLUORIDE
BUTYL GR-1	
FLUORO ELASTOMERS VITON A	
FLUORO ELASTOMERS KEL-F 3700	
HARD RUBBER	
HYPALON	
KOROSEAL	
NATURAL RUBBER (GRS)	
NEOPRENE GR-M (CR)	
NITRILE BUNA N (NBR)	
NORDEL	
POLYBUTADIENE	
SILICONE	
POLYURETHANE	

TEMP °F: 60 80 100 120 140 160 180 200 220 240 260 280 300 320 340 360 380 400 420 440 460

CALCIUM HYDROXIDE

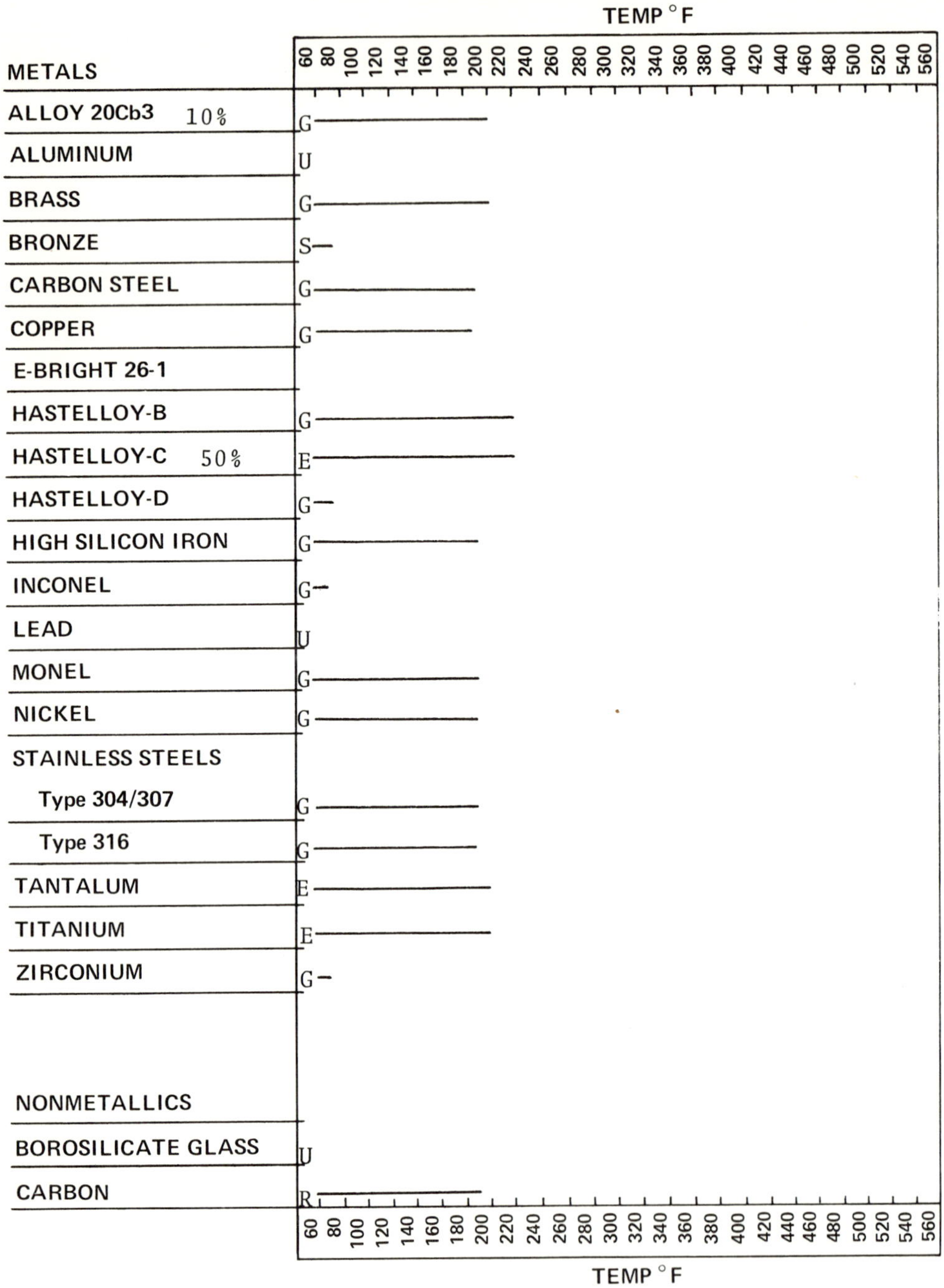

FOR METALS

E = < 2 Mils Penetration/Year; G = < 20 Mils Penetration/Year

S = < 50 Mils Penetration/Year; U = > 50 Mils Penetration/Year

FOR NONMETALLICS

R = Resistant

U = Unsatisfactory

CALCIUM HYDROXIDE

TEMP °F (60–460)

PLASTICS		
ABS		R
ACRYLICS		R
ASBESTOS REINF., EPOXY		R
ASBESTOS REINF., PHENOLIC		U
CHLORINATED POLYETHER		R
CPVC		R
EPOXY		R
FLUOROCARBONS FEP		R
FLUOROCARBONS TFE		R
FURFURYL ALCOHOL	sat	R
NORYL		
PHENOLIC		
POLYAMIDES–NYLON		R
POLYESTERS		
BISPHENOL A-FUMURATE	sat	R
HYDROGENATED BISPHENOL-A-BISPHENOL-A		
ISOPHTHALIC		R
CHLORINATED POLYESTERS		
POLYSTYRENES		
POLYETHYLENE		R
POLYPROPYLENE		R
PVC–TYPE 1		R
PVC–TYPE 2		R
POLYVINYLIDENE CHLORIDE		R
VINYLIDENE FLUORIDE		R
VINYL ESTER	25%	R

RUBBERS		
BUTYL GR-1		R
FLUORO ELASTOMERS VITON A		R
FLUORO ELASTOMERS KEL-F 3700		R
HARD RUBBER		R
HYPALON		R
KOROSEAL		R
NATURAL RUBBER (GRS)		R
NEOPRENE GR-M (CR)		R
NITRILE BUNA N (NBR)		R
NORDEL		R
POLYBUTADIENE		
SILICONE		
POLYURETHANE		R

TEMP °F (60–460)

CALCIUM HYPOCHLORITE (SAT)

TEMP °F (60–560)

METALS	Rating
ALLOY 20Cb3	
ALUMINUM	U
BRASS	U
BRONZE	U
CARBON STEEL	U
COPPER	U
E-BRIGHT 26-1	
HASTELLOY-B	U
HASTELLOY-C 50%	E — G —
HASTELLOY-D	U
HIGH SILICON IRON	G —
INCONEL	U
LEAD	U
MONEL	U
NICKEL	U
STAINLESS STEELS	
Type 304/307	U
Type 316	G —
TANTALUM	G —
TITANIUM	E —
ZIRCONIUM	S —

NONMETALLICS	Rating
BOROSILICATE GLASS	R —
CARBON	R —

TEMP °F

FOR METALS

E = < 2 Mils Penetration/Year; G = < 20 Mils Penetration/Year

S = < 50 Mils Penetration/Year; U = > 50 Mils Penetration/Year

FOR NONMETALLICS

R = Resistant

U = Unsatisfactory

CALCIUM HYPOCHLORITE (SAT)

TEMP °F: 60 80 100 120 140 160 180 200 220 240 260 280 300 320 340 360 380 400 420 440 460

PLASTICS		
ABS		R—
ACRYLICS		R—
ASBESTOS REINF., EPOXY		R—
ASBESTOS REINF., PHENOLIC		U
CHLORINATED POLYETHER		R—
CPVC		R—
EPOXY	70%	R—U
FLUOROCARBONS FEP		R—
FLUOROCARBONS TFE		R—
FURFURYL ALCOHOL	to 20%	R—
NORYL	20%	R—
PHENOLIC		U
POLYAMIDES—NYLON		U
POLYESTERS		
BISPHENOL A-FUMURATE	20%	R—
HYDROGENATED BISPHENOL-A-BISPHENOL-A	10%	R—
ISOPHTHALIC	10%	R—
CHLORINATED POLYESTERS	20%	R—
POLYSTYRENES		R—
POLYETHYLENE		R—
POLYPROPYLENE		R—
PVC—TYPE 1		R—
PVC—TYPE 2		R—
POLYVINYLIDENE CHLORIDE		R—
VINYLIDENE FLUORIDE		R—
VINYL ESTER	sat	R—

RUBBERS	
BUTYL GR-1	R—
FLUORO ELASTOMERS VITON A	R—
FLUORO ELASTOMERS KEL-F 3700	R—
HARD RUBBER	R—
HYPALON	R—
KOROSEAL	
NATURAL RUBBER (GRS)	R—
NEOPRENE GR-M (CR)	U
NITRILE BUNA N (NBR)	RU
NORDEL	R—
POLYBUTADIENE	R—
SILICONE	
POLYURETHANE	U

TEMP °F: 60 80 100 120 140 160 180 200 220 240 260 280 300 320 340 360 380 400 420 440 460

CALCIUM NITRATE

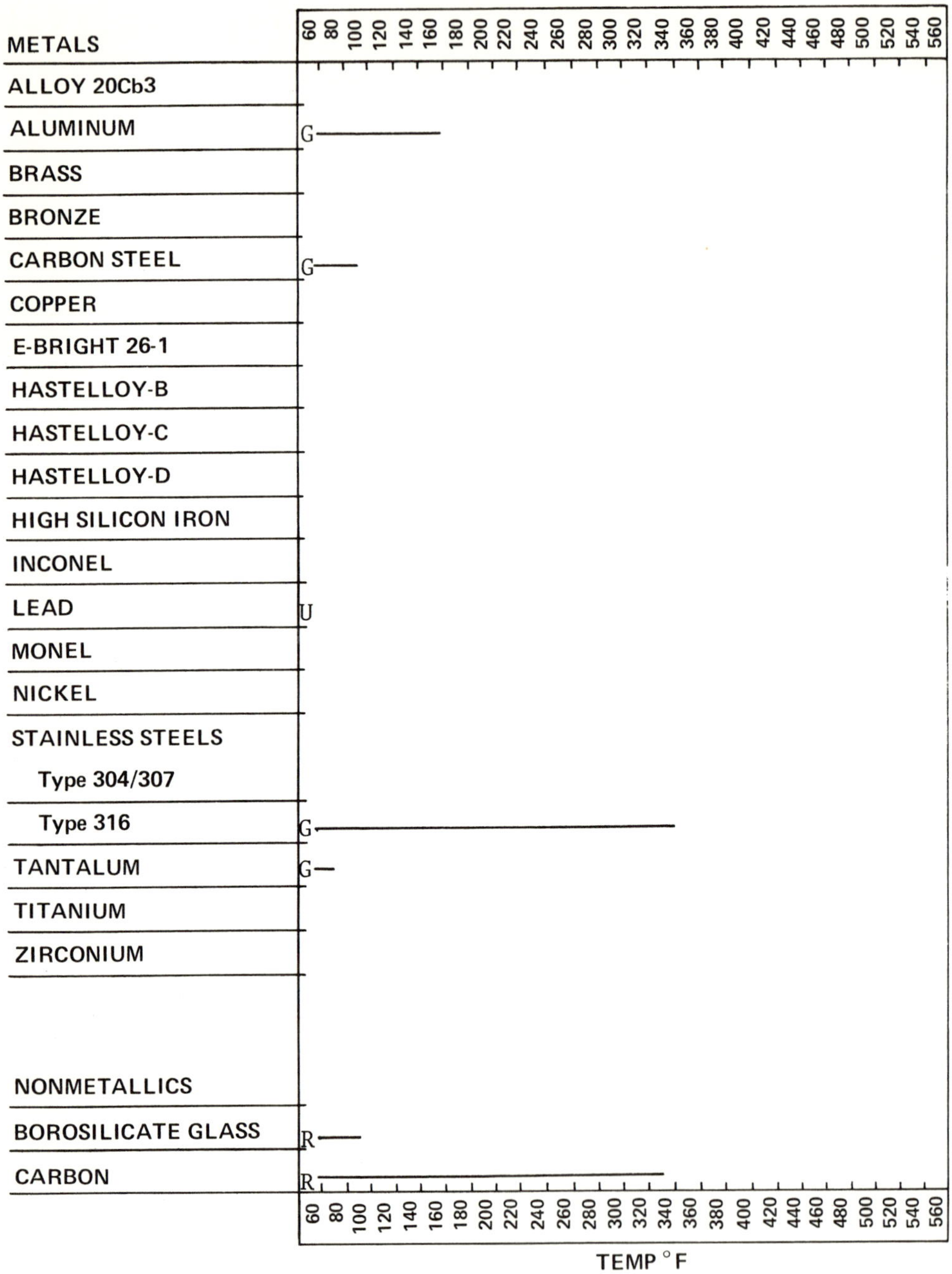

FOR METALS

E = < 2 Mils Penetration/Year; G = < 20 Mils Penetration/Year

S = < 50 Mils Penetration/Year; U = > 50 Mils Penetration/Year

FOR NONMETALLICS

R = Resistant

U = Unsatisfactory

CALCIUM NITRATE

TEMP °F (60–460)

PLASTICS		
ABS		R
ACRYLICS		
ASBESTOS REINF., EPOXY		
ASBESTOS REINF., PHENOLIC		
CHLORINATED POLYETHER		R
CPVC		R
EPOXY		R
FLUOROCARBONS FEP		R
FLUOROCARBONS TFE		R
FURFURYL ALCOHOL	sat	R
NORYL		R
PHENOLIC		
POLYAMIDES–NYLON	10%	R
POLYESTERS		
BISPHENOL A-FUMURATE	sat	R
HYDROGENATED BISPHENOL-A-BISPHENOL-A		
ISOPHTHALIC	sat	R
CHLORINATED POLYESTERS		
POLYSTYRENES		
POLYETHYLENE		R
POLYPROPYLENE		R
PVC–TYPE 1		R
PVC–TYPE 2		R
POLYVINYLIDENE CHLORIDE		R
VINYLIDENE FLUORIDE		R
VINYL ESTER		R

RUBBERS	
BUTYL GR-1	R
FLUORO ELASTOMERS VITON A	R
FLUORO ELASTOMERS KEL-F 3700	R
HARD RUBBER	R
HYPALON	R
KOROSEAL	R
NATURAL RUBBER (GRS)	R
NEOPRENE GR-M (CR)	R
NITRILE BUNA N (NBR)	R
NORDEL	R
POLYBUTADIENE	
SILICONE	
POLYURETHANE	R

TEMP °F

CALCIUM OXIDE

TEMP °F

METALS	60	80	100	120	140	160	180	200	220	240	260	280	300	320	340	360	380	400	420	440	460	480	500	520	540	560
ALLOY 20Cb3	G	—																								
ALUMINUM	G	—																								
BRASS																										
BRONZE																										
CARBON STEEL																										
COPPER																										
E-BRIGHT 26-1																										
HASTELLOY-B																										
HASTELLOY-C	G	—																								
HASTELLOY-D																										
HIGH SILICON IRON																										
INCONEL																										
LEAD																										
MONEL	G	—																								
NICKEL	G	—																								
STAINLESS STEELS																										
Type 304/307	G	—																								
Type 316	G	—																								
TANTALUM																										
TITANIUM																										
ZIRCONIUM																										
NONMETALLICS																										
BOROSILICATE GLASS																										
CARBON	R	—	—	—	—	—	—	—	—	—	—	—	—	—	—											

TEMP °F

FOR METALS

E = < 2 Mils Penetration/Year; G = < 20 Mils Penetration/Year

S = < 50 Mils Penetration/Year; U = > 50 Mils Penetration/Year

FOR NONMETALLICS

R = Resistant

U = Unsatisfactory

CALCIUM OXIDE

PLASTICS		TEMP °F (60–460)
ABS	sat	R to ~140
ACRYLICS		
ASBESTOS REINF., EPOXY		
ASBESTOS REINF., PHENOLIC		
CHLORINATED POLYETHER		R to ~250
CPVC		R to ~185
EPOXY		
FLUOROCARBONS FEP		R to ~400
FLUOROCARBONS TFE		R to 460+
FURFURYL ALCOHOL		
NORYL		
PHENOLIC		
POLYAMIDES–NYLON		
POLYESTERS		
BISPHENOL A-FUMURATE		
HYDROGENATED BISPHENOL-A-BISPHENOL-A		
ISOPHTHALIC		
CHLORINATED POLYESTERS		
POLYSTYRENES		
POLYETHYLENE		R to ~160
POLYPROPYLENE		R to ~225
PVC–TYPE 1		R to ~140
PVC–TYPE 2		R to ~140
POLYVINYLIDENE CHLORIDE		R to ~180
VINYLIDENE FLUORIDE		R to ~250
VINYL ESTER		

RUBBERS	TEMP °F (60–460)
BUTYL GR-1	
FLUORO ELASTOMERS VITON A	
FLUORO ELASTOMERS KEL-F 3700	
HARD RUBBER	
HYPALON	
KOROSEAL	
NATURAL RUBBER (GRS)	
NEOPRENE GR-M (CR)	
NITRILE BUNA N (NBR)	
NORDEL	
POLYBUTADIENE	
SILICONE	
POLYURETHANE	

CALCIUM SULFATE

TEMP °F: 60, 80, 100, 120, 140, 160, 180, 200, 220, 240, 260, 280, 300, 320, 340, 360, 380, 400, 420, 440, 460, 480, 500, 520, 540, 560

METALS		
ALLOY 20Cb3		G
ALUMINUM	1	G
BRASS		G
BRONZE		G
CARBON STEEL		U
COPPER		G
E-BRIGHT 26-1		
HASTELLOY-B	10%	G
HASTELLOY-C	10%	G
HASTELLOY-D	10%	G
HIGH SILICON IRON		G
INCONEL	1	G
LEAD		
MONEL		G
NICKEL		G
STAINLESS STEELS		
Type 304/307		G
Type 316		G
TANTALUM		G
TITANIUM		E
ZIRCONIUM		E

NONMETALLICS	
BOROSILICATE GLASS	
CARBON	R

TEMP °F: 60, 80, 100, 120, 140, 160, 180, 200, 220, 240, 260, 280, 300, 320, 340, 360, 380, 400, 420, 440, 460, 480, 500, 520, 540, 560

FOR METALS

E = < 2 Mils Penetration/Year; G = < 20 Mils Penetration/Year

S = < 50 Mils Penetration/Year; U = > 50 Mils Penetration/Year

FOR NONMETALLICS

R = Resistant

U = Unsatisfactory

CALCIUM SULFATE

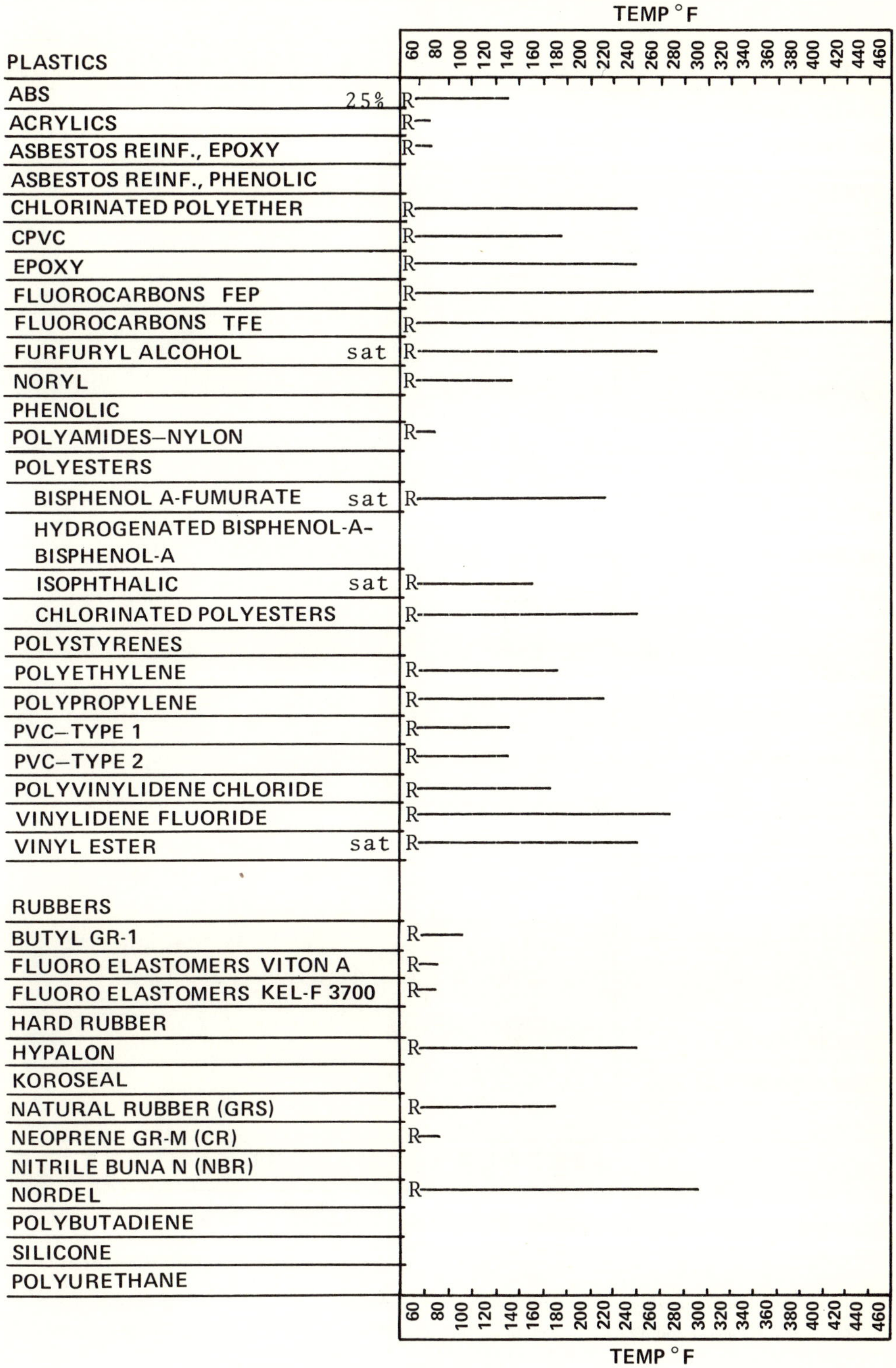

CANE SUGAR LIQUORS

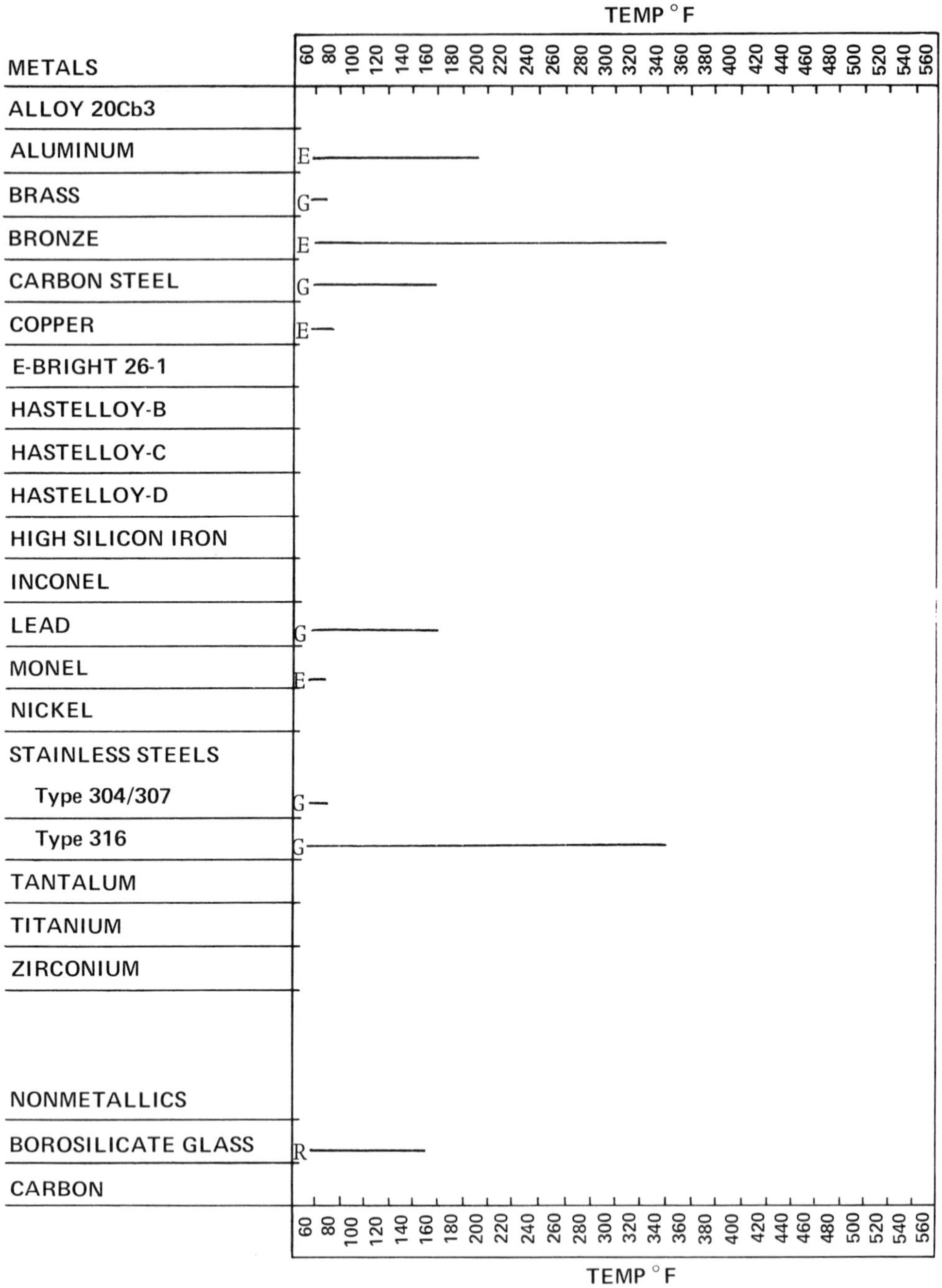

FOR METALS

E = < 2 Mils Penetration/Year; G = < 20 Mils Penetration/Year

S = < 50 Mils Penetration/Year; U = > 50 Mils Penetration/Year

FOR NONMETALLICS

R = Resistant

U = Unsatisfactory

CANE SUGAR LIQUORS

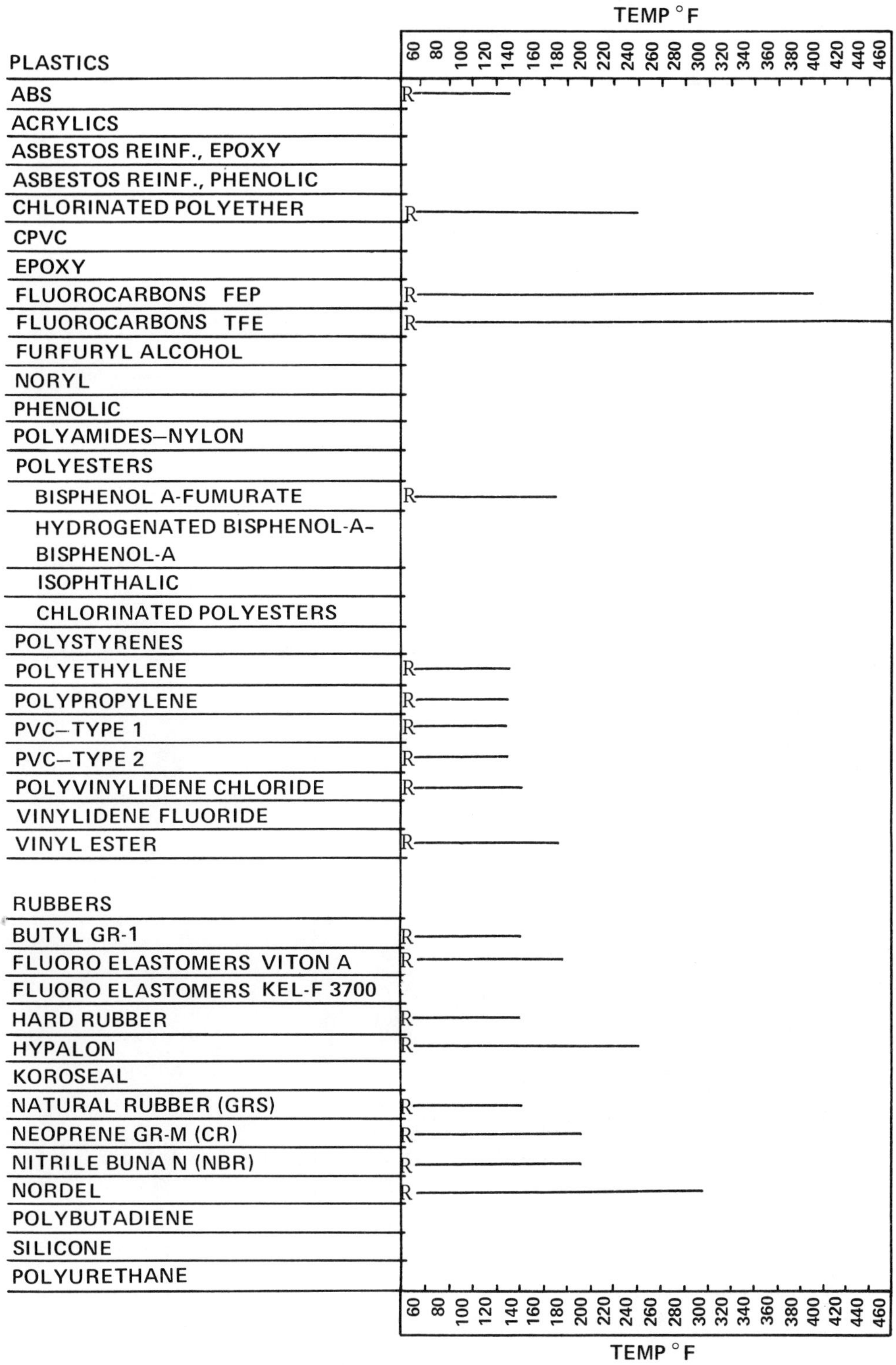

CAPRYLIC ACID

TEMP °F: 60 80 100 120 140 160 180 200 220 240 260 280 300 320 340 360 380 400 420 440 460 480 500 520 540 560

METALS		Rating
ALLOY 20Cb3		G
ALUMINUM		G
BRASS		U
BRONZE		U
CARBON STEEL		
COPPER		U
E-BRIGHT 26-1		
HASTELLOY-B		G
HASTELLOY-C		G
HASTELLOY-D		
HIGH SILICON IRON		G
INCONEL		G
LEAD		
MONEL		G
NICKEL	1	G
STAINLESS STEELS		
Type 304/307	1	G
Type 316		G
TANTALUM		G
TITANIUM		G
ZIRCONIUM		G

NONMETALLICS	Rating
BOROSILICATE GLASS	
CARBON	R

TEMP °F: 60 80 100 120 140 160 180 200 220 240 260 280 300 320 340 360 380 400 420 440 460 480 500 520 540 560

FOR METALS

E = < 2 Mils Penetration/Year; G = < 20 Mils Penetration/Year

S = < 50 Mils Penetration/Year; U = > 50 Mils Penetration/Year

FOR NONMETALLICS

R = Resistant

U = Unsatisfactory

CAPRYLIC ACID

TEMP °F (60–460)

PLASTICS	Rating
ABS	
ACRYLICS	
ASBESTOS REINF., EPOXY	
ASBESTOS REINF., PHENOLIC	
CHLORINATED POLYETHER	R
CPVC	
EPOXY	U
FLUOROCARBONS FEP	R
FLUOROCARBONS TFE	R
FURFURYL ALCOHOL	R
NORYL	
PHENOLIC	
POLYAMIDES–NYLON	
POLYESTERS	
BISPHENOL A-FUMURATE sat	R
HYDROGENATED BISPHENOL-A-BISPHENOL-A	
ISOPHTHALIC sat	R
CHLORINATED POLYESTERS	R
POLYSTYRENES	
POLYETHYLENE	
POLYPROPYLENE	
PVC–TYPE 1	
PVC–TYPE 2	
POLYVINYLIDENE CHLORIDE	R
VINYLIDENE FLUORIDE	R
VINYL ESTER	R

RUBBERS	Rating
BUTYL GR-1	
FLUORO ELASTOMERS VITON A	
FLUORO ELASTOMERS KEL-F 3700	
HARD RUBBER	
HYPALON	
KOROSEAL	
NATURAL RUBBER (GRS)	
NEOPRENE GR-M (CR)	
NITRILE BUNA N (NBR)	
NORDEL	
POLYBUTADIENE	
SILICONE	
POLYURETHANE	

TEMP °F

CARBON BISULFIDE (DISULFIDE)

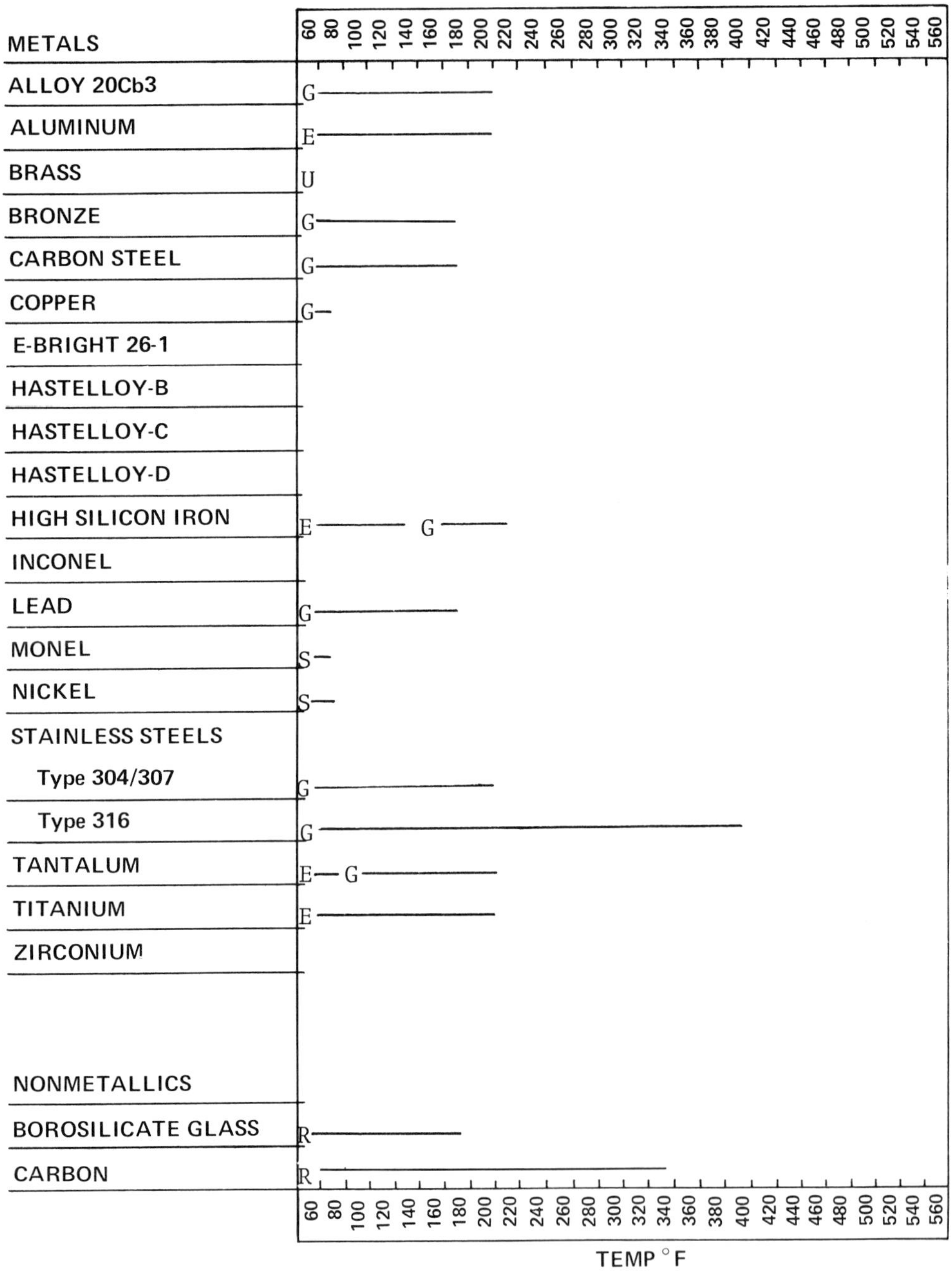

FOR METALS

E = < 2 Mils Penetration/Year; G = < 20 Mils Penetration/Year

S = < 50 Mils Penetration/Year; U = > 50 Mils Penetration/Year

FOR NONMETALLICS

R = Resistant

U = Unsatisfactory

CARBON BISULFIDE (DISULFIDE)

TEMP °F (scale 60–460)

PLASTICS	60 – 460 °F
ABS	U
ACRYLICS	R—
ASBESTOS REINF., EPOXY	
ASBESTOS REINF., PHENOLIC	
CHLORINATED POLYETHER	U
CPVC	U
EPOXY	R—
FLUOROCARBONS FEP	R—
FLUOROCARBONS TFE	R—
FURFURYL ALCOHOL	R—
NORYL	
PHENOLIC	
POLYAMIDES—NYLON	R—
POLYESTERS	
BISPHENOL A-FUMURATE	U
HYDROGENATED BISPHENOL-A-BISPHENOL-A	
ISOPHTHALIC	U
CHLORINATED POLYESTERS	U
POLYSTYRENES	U
POLYETHYLENE	U
POLYPROPYLENE	RU
PVC—TYPE 1	U
PVC—TYPE 2	U
POLYVINYLIDENE CHLORIDE	R—
VINYLIDENE FLUORIDE	R—
VINYL ESTER	U
RUBBERS	
BUTYL GR-1	U
FLUORO ELASTOMERS VITON A	R—
FLUORO ELASTOMERS KEL-F 3700	
HARD RUBBER	U
HYPALON	U
KOROSEAL	U
NATURAL RUBBER (GRS)	U
NEOPRENE GR-M (CR)	U
NITRILE BUNA N (NBR)	U
NORDEL	
POLYBUTADIENE	
SILICONE	
POLYURETHANE	

TEMP °F

CARBON DIOXIDE, DRY

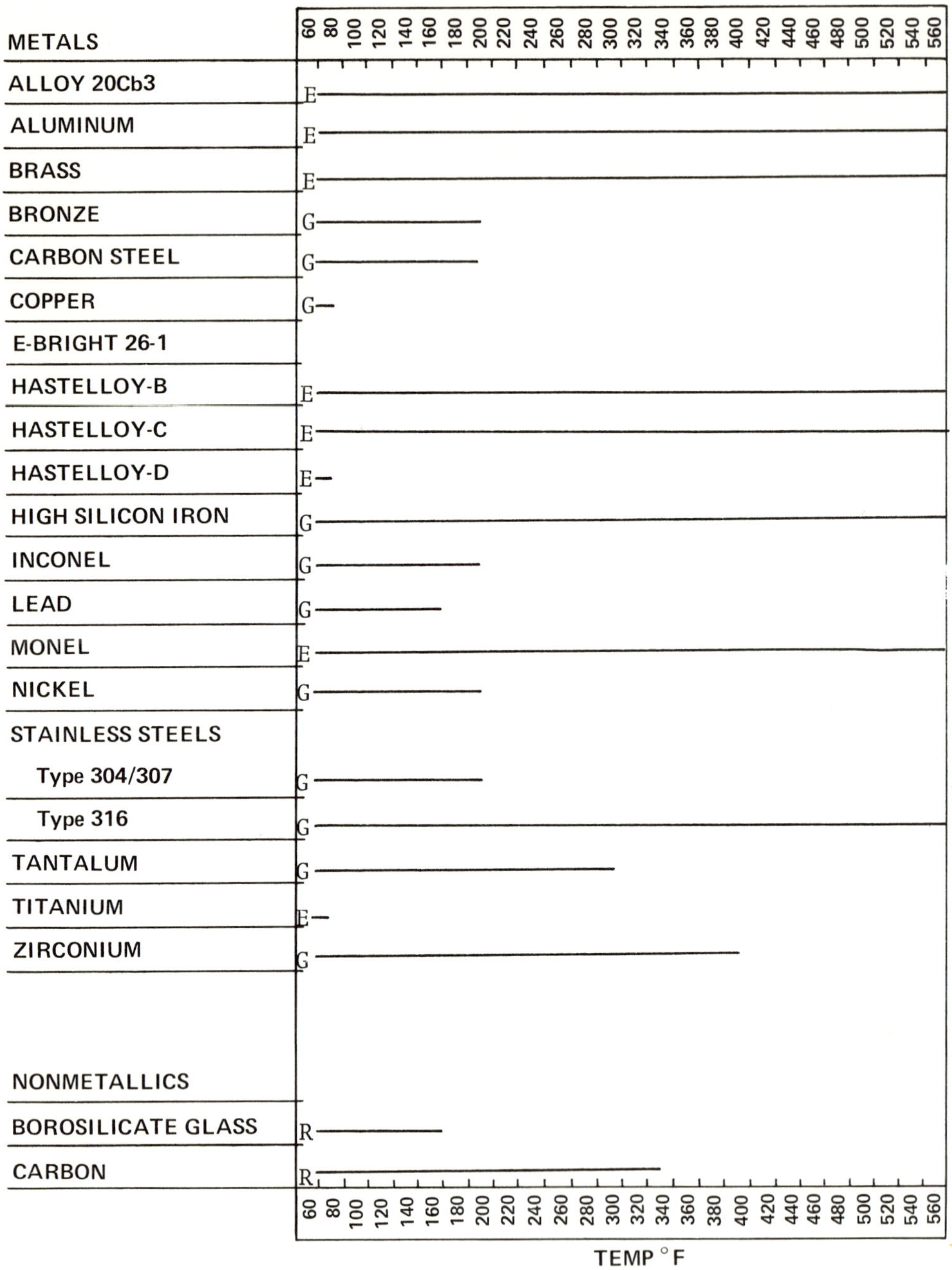

FOR METALS

E = < 2 Mils Penetration/Year; G = < 20 Mils Penetration/Year

S = < 50 Mils Penetration/Year; U = > 50 Mils Penetration/Year

FOR NONMETALLICS

R = Resistant

U = Unsatisfactory

CARBON DIOXIDE, DRY

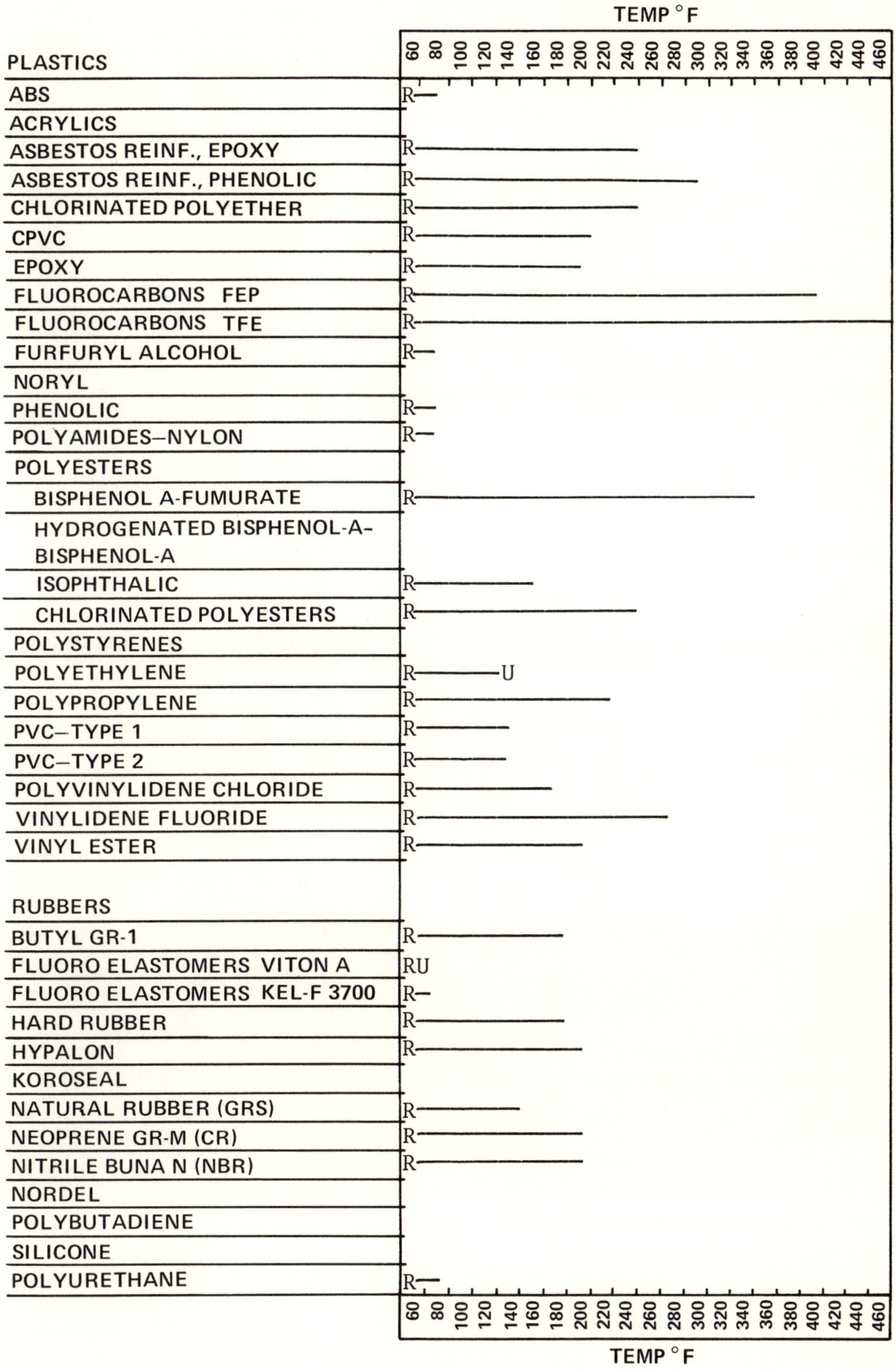

CARBON DIOXIDE, WET

TEMP °F (60–560)

METALS		Rating
ALLOY 20Cb3		G
ALUMINUM		E
BRASS		U
BRONZE	4	G
CARBON STEEL		S
COPPER	4	G
E-BRIGHT 26-1		
HASTELLOY-B		E
HASTELLOY-C		G
HASTELLOY-D		G
HIGH SILICON IRON		G
INCONEL		G
LEAD		G
MONEL	4	G
NICKEL		G
STAINLESS STEELS		
Type 304/307		G
Type 316		G
TANTALUM		E
TITANIUM		E
ZIRCONIUM		

NONMETALLICS	Rating
BOROSILICATE GLASS	R
CARBON	R

TEMP °F

FOR METALS

E = < 2 Mils Penetration/Year; G = < 20 Mils Penetration/Year

S = < 50 Mils Penetration/Year; U = > 50 Mils Penetration/Year

FOR NONMETALLICS

R = Resistant

U = Unsatisfactory

CARBON DIOXIDE, WET

TEMP °F (scale 60–460)

PLASTICS	Rating	Approx. upper temp (°F)
ABS	R	140
ACRYLICS	R	80
ASBESTOS REINF., EPOXY	R	250
ASBESTOS REINF., PHENOLIC	R	300
CHLORINATED POLYETHER	R	250
CPVC	R	180
EPOXY		
FLUOROCARBONS FEP	R	400
FLUOROCARBONS TFE	R	460
FURFURYL ALCOHOL	R	80
NORYL	R	140
PHENOLIC	R	80
POLYAMIDES—NYLON	R	80
POLYESTERS		
BISPHENOL A-FUMURATE		
HYDROGENATED BISPHENOL-A-BISPHENOL-A	R	210
ISOPHTHALIC	R	160
CHLORINATED POLYESTERS		
POLYSTYRENES	R	80
POLYETHYLENE	R	140
POLYPROPYLENE	R	150
PVC—TYPE 1	R	140
PVC—TYPE 2	R	140
POLYVINYLIDENE CHLORIDE	R	80
VINYLIDENE FLUORIDE	R	280
VINYL ESTER	R	215

RUBBERS	Rating	Approx. upper temp (°F)
BUTYL GR-1	R	185
FLUORO ELASTOMERS VITON A	RU	
FLUORO ELASTOMERS KEL-F 3700	R	70
HARD RUBBER	R	185
HYPALON	R	200
KOROSEAL		
NATURAL RUBBER (GRS)	R	150
NEOPRENE GR-M (CR)	R	200
NITRILE BUNA N (NBR)	R	200
NORDEL		
POLYBUTADIENE	R	80
SILICONE		
POLYURETHANE	R	80

TEMP °F: 60 80 100 120 140 160 180 200 220 240 260 280 300 320 340 360 380 400 420 440 460 480 500 520 540 560

METALS	Rating
ALLOY 20Cb3	G
ALUMINUM	E
BRASS	U
BRONZE	G
CARBON STEEL	G
COPPER	G
E-BRIGHT 26-1	
HASTELLOY-B	
HASTELLOY-C	
HASTELLOY-D	
HIGH SILICON IRON	E, G
INCONEL	
LEAD	G
MONEL	S
NICKEL	S
STAINLESS STEELS	
Type 304/307	G
Type 316	G
TANTALUM	E, G
TITANIUM	E
ZIRCONIUM	

NONMETALLICS	Rating
BOROSILICATE GLASS	R
CARBON	R

TEMP °F: 60 80 100 120 140 160 180 200 220 240 260 280 300 320 340 360 380 400 420 440 460 480 500 520 540 560

FOR METALS

E = < 2 Mils Penetration/Year; G = < 20 Mils Penetration/Year

S = < 50 Mils Penetration/Year; U = > 50 Mils Penetration/Year

FOR NONMETALLICS

R = Resistant

U = Unsatisfactory

CARBON DISULFIDE (BISULFIDE)

TEMP °F (60–460)

PLASTICS	60 80 100 120 140 160 180 200 220 240 260 280 300 320 340 360 380 400 420 440 460
ABS	U
ACRYLICS	R—
ASBESTOS REINF., EPOXY	
ASBESTOS REINF., PHENOLIC	
CHLORINATED POLYETHER	U
CPVC	U
EPOXY	R——
FLUOROCARBONS FEP	R————————————————
FLUOROCARBONS TFE	R——————————————————
FURFURYL ALCOHOL	R——————————
NORYL	
PHENOLIC	
POLYAMIDES—NYLON	R—
POLYESTERS	
BISPHENOL A-FUMURATE	U
HYDROGENATED BISPHENOL-A-BISPHENOL-A	U
ISOPHTHALIC	U
CHLORINATED POLYESTERS	U
POLYSTYRENES	U
POLYETHYLENE	U
POLYPROPYLENE	RU
PVC—TYPE 1	U
PVC—TYPE 2	U
POLYVINYLIDENE CHLORIDE	R—
VINYLIDENE FLUORIDE	R—
VINYL ESTER	U
RUBBERS	
BUTYL GR-1	U
FLUORO ELASTOMERS VITON A	R———————
FLUORO ELASTOMERS KEL-F 3700	
HARD RUBBER	U
HYPALON	U
KOROSEAL	U
NATURAL RUBBER (GRS)	U
NEOPRENE GR-M (CR)	U
NITRILE BUNA N (NBR)	U
NORDEL	
POLYBUTADIENE	
SILICONE	
POLYURETHANE	

TEMP °F (60–460)

CARBON MONOXIDE

TEMP °F: 60, 80, 100, 120, 140, 160, 180, 200, 220, 240, 260, 280, 300, 320, 340, 360, 380, 400, 420, 440, 460, 480, 500, 520, 540, 560

METALS	Rating
ALLOY 20Cb3	E
ALUMINUM	E
BRASS	E
BRONZE	E
CARBON STEEL	G
COPPER	G
E-BRIGHT 26-1	
HASTELLOY-B	E
HASTELLOY-C	E
HASTELLOY-D	E
HIGH SILICON IRON	
INCONEL	E
LEAD	G
MONEL	E
NICKEL	E
STAINLESS STEELS	
Type 304/307	E
Type 316	E
TANTALUM	G
TITANIUM	
ZIRCONIUM	

NONMETALLICS	Rating
BOROSILICATE GLASS	R
CARBON	R

TEMP °F: 60, 80, 100, 120, 140, 160, 180, 200, 220, 240, 260, 280, 300, 320, 340, 360, 380, 400, 420, 440, 460, 480, 500, 520, 540, 560

FOR METALS

E = $<$ 2 Mils Penetration/Year; G = $<$ 20 Mils Penetration/Year

S = $<$ 50 Mils Penetration/Year; U = $>$ 50 Mils Penetration/Year

FOR NONMETALLICS

R = Resistant

U = Unsatisfactory

CARBON MONOXIDE

TEMP °F (60–460)

PLASTICS	
ABS	R
ACRYLICS	
ASBESTOS REINF., EPOXY	
ASBESTOS REINF., PHENOLIC	
CHLORINATED POLYETHER	R
CPVC	R
EPOXY	
FLUOROCARBONS FEP	R
FLUOROCARBONS TFE	R
FURFURYL ALCOHOL	
NORYL	
PHENOLIC	
POLYAMIDES—NYLON	R
POLYESTERS	
BISPHENOL A-FUMURATE	R
HYDROGENATED BISPHENOL-A-BISPHENOL-A	
ISOPHTHALIC	R
CHLORINATED POLYESTERS	
POLYSTYRENES	
POLYETHYLENE	R
POLYPROPYLENE	R
PVC—TYPE 1	R
PVC—TYPE 2	R
POLYVINYLIDENE CHLORIDE	R
VINYLIDENE FLUORIDE	R
VINYL ESTER	R
RUBBERS	
BUTYL GR-1	R
FLUORO ELASTOMERS VITON A	R
FLUORO ELASTOMERS KEL-F 3700	
HARD RUBBER	U
HYPALON	R
KOROSEAL	
NATURAL RUBBER (GRS)	U
NEOPRENE GR-M (CR)	R
NITRILE BUNA N (NBR)	R
NORDEL	R
POLYBUTADIENE	
SILICONE	
POLYURETHANE	R

TEMP °F

CARBON TETRACHLORIDE

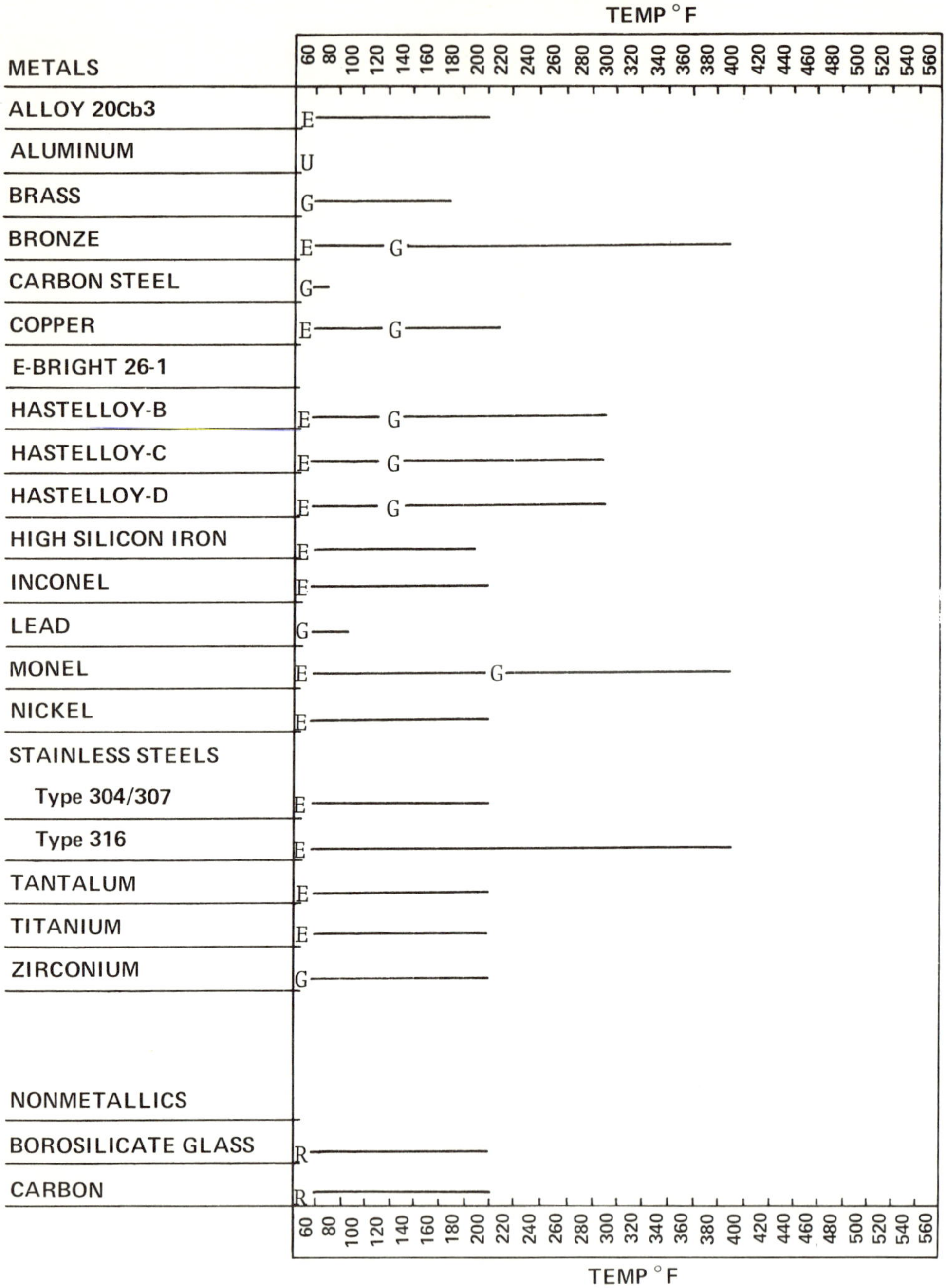

FOR METALS

E = < 2 Mils Penetration/Year; G = < 20 Mils Penetration/Year

S = < 50 Mils Penetration/Year; U = > 50 Mils Penetration/Year

FOR NONMETALLICS

R = Resistant

U = Unsatisfactory

CARBON TETRACHLORIDE

TEMP °F (60–460)

PLASTICS	Rating	Line extends to (°F, approx.)
ABS	U	
ACRYLICS	U	
ASBESTOS REINF., EPOXY	U	
ASBESTOS REINF., PHENOLIC	R	300
CHLORINATED POLYETHER	R	200
CPVC	RU	
EPOXY	R ... U	180 (U at 180)
FLUOROCARBONS FEP	R	400
FLUOROCARBONS TFE	R	460+
FURFURYL ALCOHOL	R	270
NORYL		
PHENOLIC		
POLYAMIDES–NYLON	R	200
POLYESTERS		
BISPHENOL A-FUMURATE	R	80
HYDROGENATED BISPHENOL-A-BISPHENOL-A	U	
ISOPHTHALIC	U	
CHLORINATED POLYESTERS	R	125
POLYSTYRENES	U	
POLYETHYLENE	U	
POLYPROPYLENE	U	
PVC–TYPE 1	RU	
PVC–TYPE 2	U	
POLYVINYLIDENE CHLORIDE	R	150
VINYLIDENE FLUORIDE	R	280
VINYL ESTER	R	180

RUBBERS	Rating	Line extends to (°F, approx.)
BUTYL GR-1	U	
FLUORO ELASTOMERS VITON A	R	350
FLUORO ELASTOMERS KEL-F 3700		
HARD RUBBER	U	
HYPALON	U	
KOROSEAL	U	
NATURAL RUBBER (GRS)	U	
NEOPRENE GR-M (CR)	U	
NITRILE BUNA N (NBR)	RU	
NORDEL	U	
POLYBUTADIENE		
SILICONE	U	
POLYURETHANE	U	

TEMP °F (60–460)

CARBONIC ACID

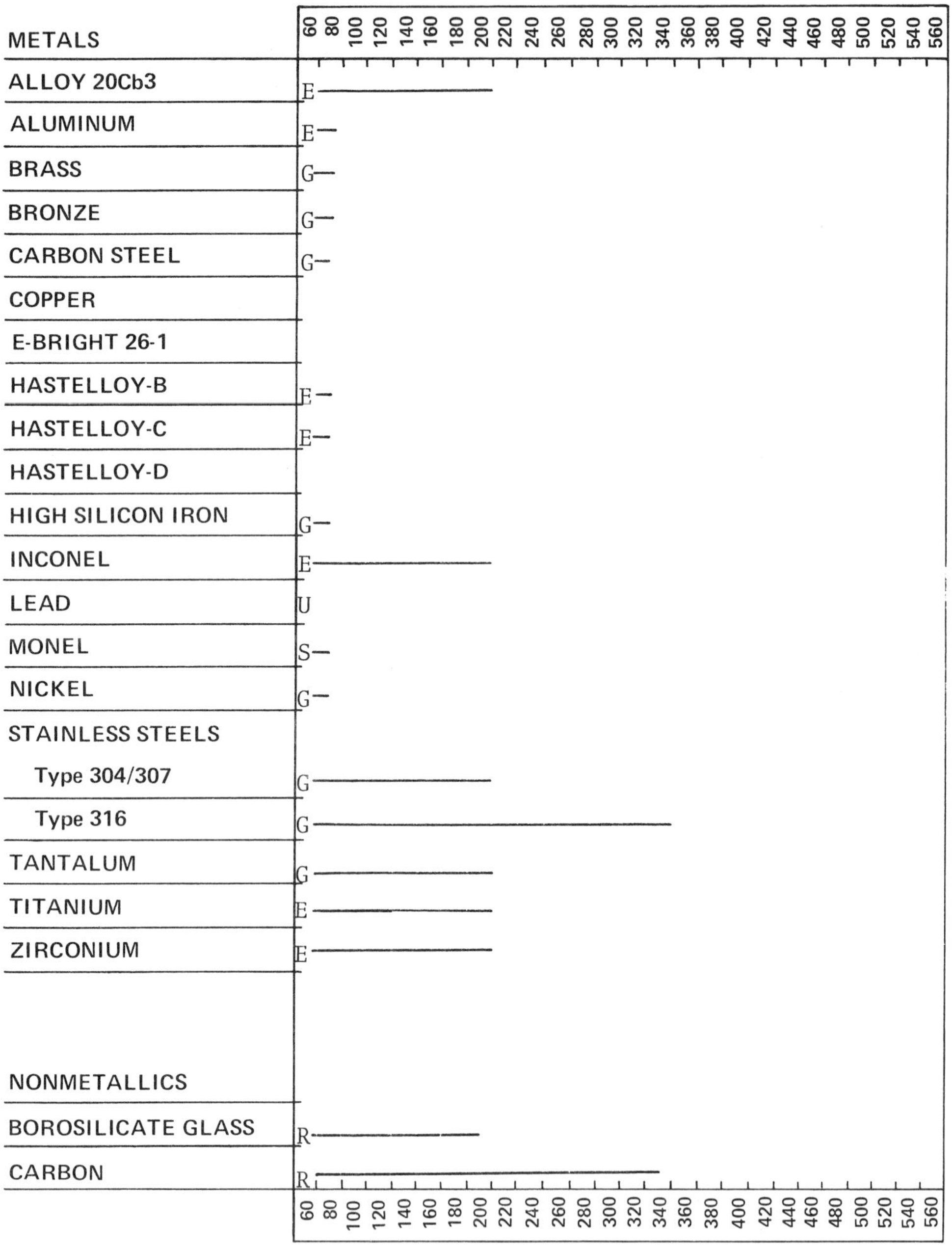

FOR METALS

E = < 2 Mils Penetration/Year; G = < 20 Mils Penetration/Year
S = < 50 Mils Penetration/Year; U = > 50 Mils Penetration/Year

FOR NONMETALLICS

R = Resistant
U = Unsatisfactory

CARBONIC ACID

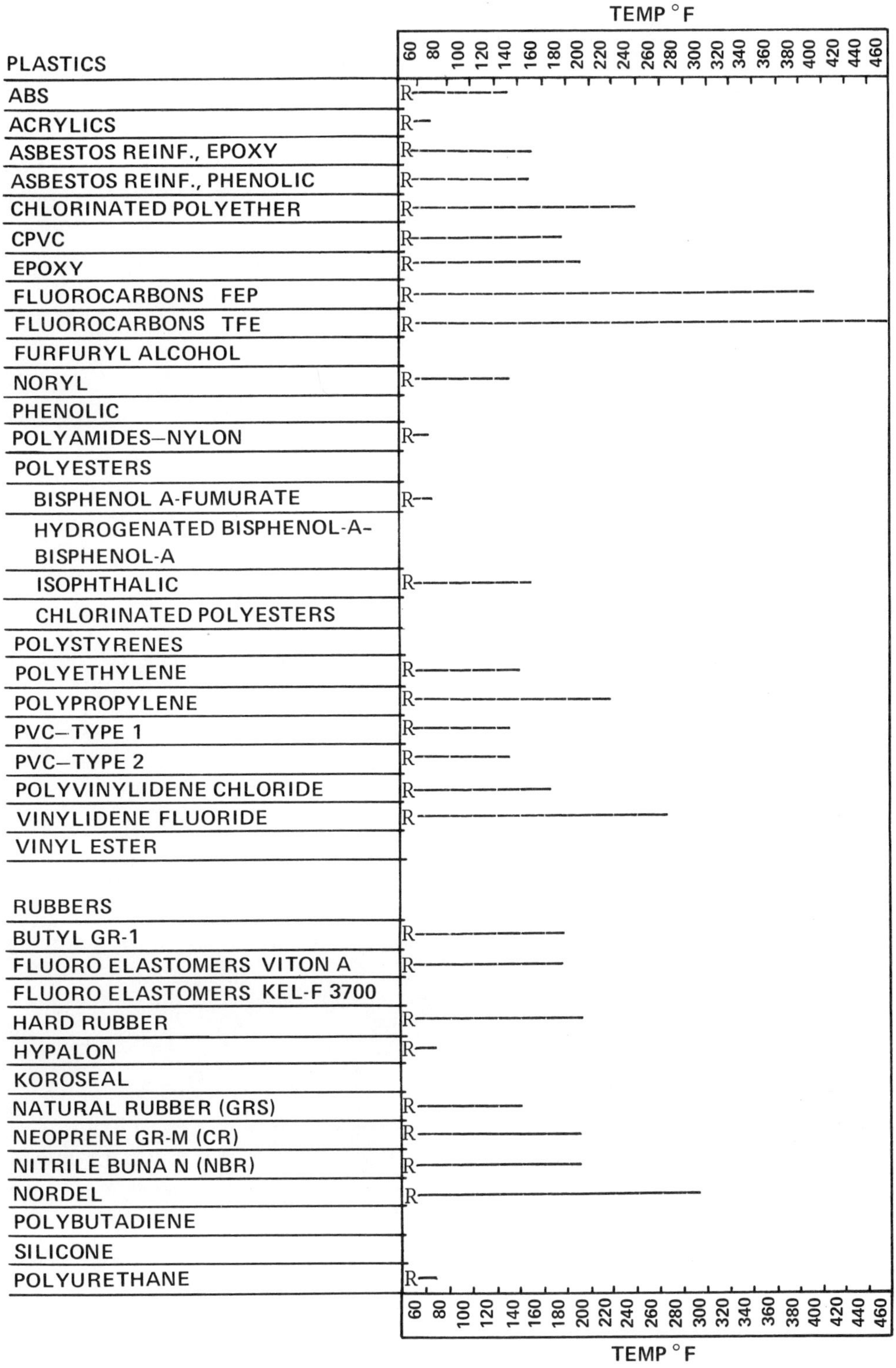

CASTOR OIL

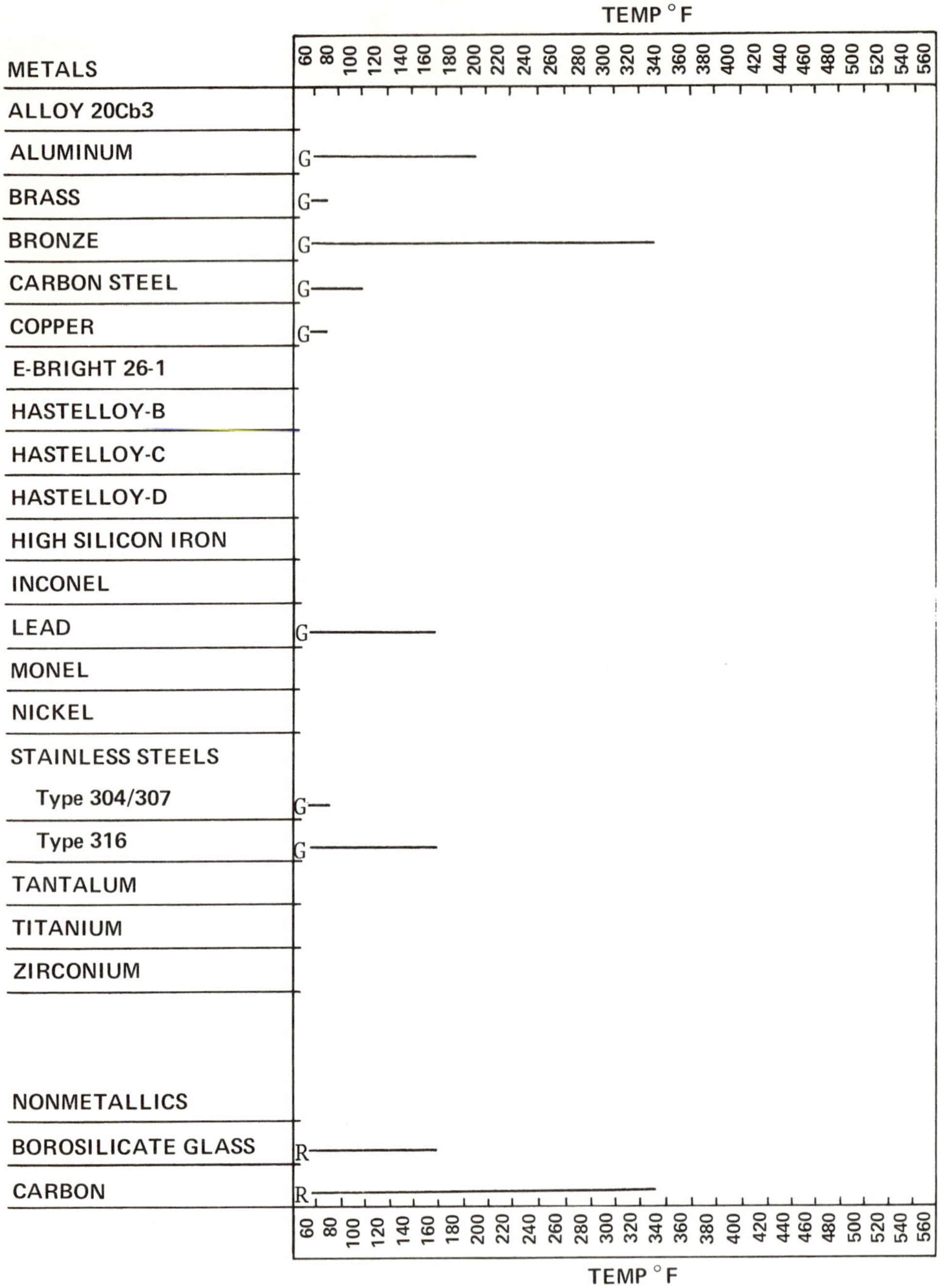

FOR METALS

E = < 2 Mils Penetration/Year; G = < 20 Mils Penetration/Year

S = < 50 Mils Penetration/Year; U = > 50 Mils Penetration/Year

FOR NONMETALLICS

R = Resistant

U = Unsatisfactory

CASTOR OIL

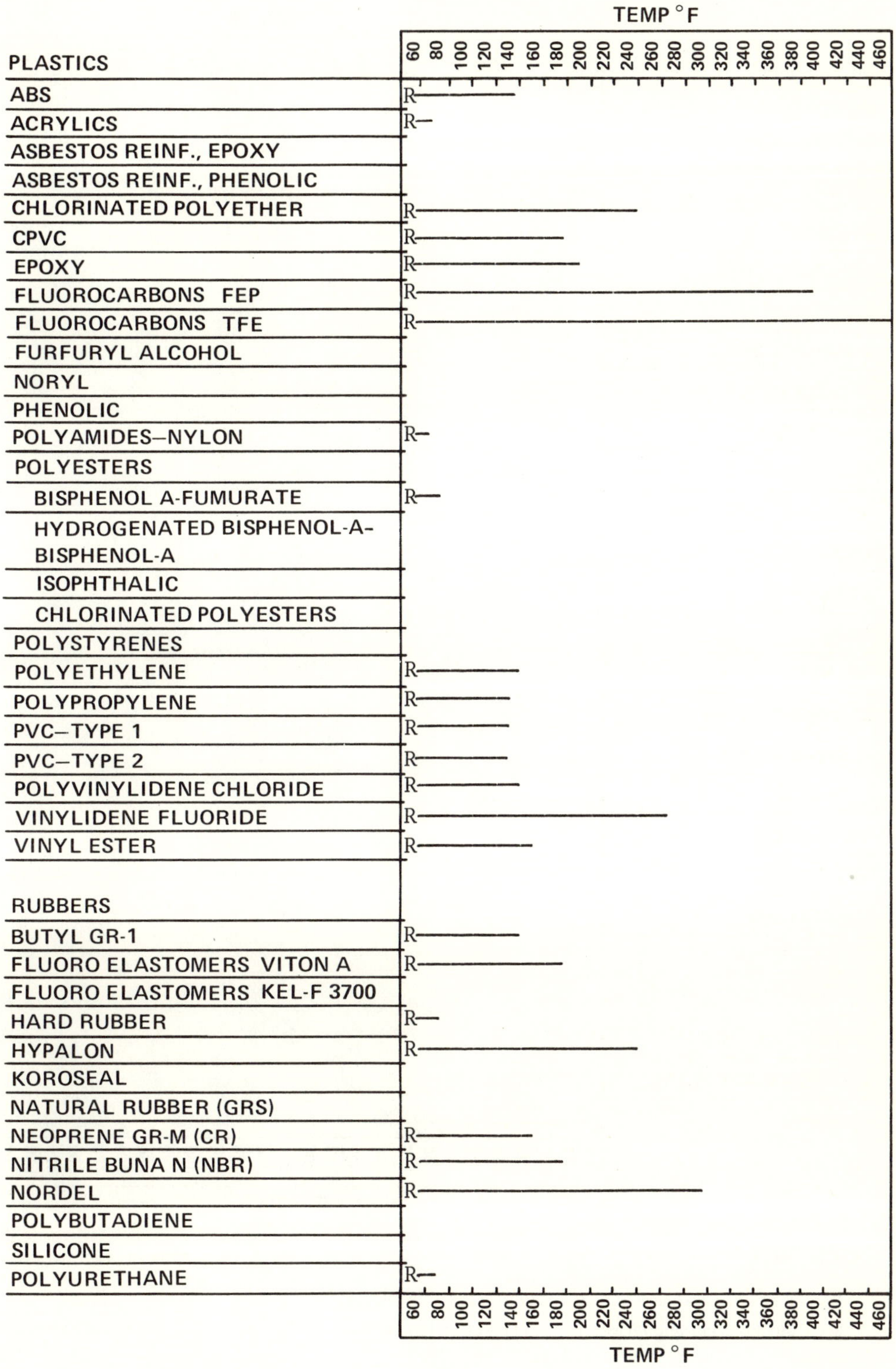

TEMP °F
60 80 100 120 140 160 180 200 220 240 260 280 300 320 340 360 380 400 420 440 460
PLASTICS
ABS R
ACRYLICS R
ASBESTOS REINF., EPOXY
ASBESTOS REINF., PHENOLIC
CHLORINATED POLYETHER R
CPVC R
EPOXY R
FLUOROCARBONS FEP R
FLUOROCARBONS TFE R
FURFURYL ALCOHOL
NORYL
PHENOLIC
POLYAMIDES–NYLON R
POLYESTERS
BISPHENOL A-FUMURATE R
HYDROGENATED BISPHENOL-A-BISPHENOL-A
ISOPHTHALIC
CHLORINATED POLYESTERS
POLYSTYRENES
POLYETHYLENE R
POLYPROPYLENE R
PVC–TYPE 1 R
PVC–TYPE 2 R
POLYVINYLIDENE CHLORIDE R
VINYLIDENE FLUORIDE R
VINYL ESTER R
RUBBERS
BUTYL GR-1 R
FLUORO ELASTOMERS VITON A R
FLUORO ELASTOMERS KEL-F 3700
HARD RUBBER R
HYPALON R
KOROSEAL
NATURAL RUBBER (GRS)
NEOPRENE GR-M (CR) R
NITRILE BUNA N (NBR) R
NORDEL R
POLYBUTADIENE
SILICONE
POLYURETHANE R
60 80 100 120 140 160 180 200 220 240 260 280 300 320 340 360 380 400 420 440 460
TEMP °F

TEMP °F

60 80 100 120 140 160 180 200 220 240 260 280 300 320 340 360 380 400 420 440 460 480 500 520 540 560

METALS	
ALLOY 20Cb3	
ALUMINUM	U
BRASS	
BRONZE	
CARBON STEEL	U
COPPER	
E-BRIGHT 26-1	
HASTELLOY-B	
HASTELLOY-C	
HASTELLOY-D	
HIGH SILICON IRON	
INCONEL	
LEAD	U
MONEL	
NICKEL	
STAINLESS STEELS Type 304/307	
Type 316	G
TANTALUM	
TITANIUM	
ZIRCONIUM	

NONMETALLICS	
BOROSILICATE GLASS	
CARBON	R

60 80 100 120 140 160 180 200 220 240 260 280 300 320 340 360 380 400 420 440 460 480 500 520 540 560

TEMP °F

FOR METALS

E = < 2 Mils Penetration/Year; G = < 20 Mils Penetration/Year

S = < 50 Mils Penetration/Year; U = > 50 Mils Penetration/Year

FOR NONMETALLICS

R = Resistant

U = Unsatisfactory

CAUSTIC POTASH (POTASSIUM HYDROXIDE)

TEMP °F (60–460)

PLASTICS		TEMP °F
ABS		R
ACRYLICS		
ASBESTOS REINF., EPOXY		
ASBESTOS REINF., PHENOLIC		
CHLORINATED POLYETHER		R
CPVC		R
EPOXY		
FLUOROCARBONS FEP		R
FLUOROCARBONS TFE		R
FURFURYL ALCOHOL	sat	R
NORYL		
PHENOLIC		
POLYAMIDES–NYLON		R
POLYESTERS		
BISPHENOL A-FUMURATE		
HYDROGENATED BISPHENOL-A-BISPHENOL-A		
ISOPHTHALIC		
CHLORINATED POLYESTERS		
POLYSTYRENES		
POLYETHYLENE		R
POLYPROPYLENE		R
PVC–TYPE 1		R
PVC–TYPE 2		R
POLYVINYLIDENE CHLORIDE		R
VINYLIDENE FLUORIDE		R
VINYL ESTER		
RUBBERS		
BUTYL GR-1		R
FLUORO ELASTOMERS VITON A		R
FLUORO ELASTOMERS KEL-F 3700		
HARD RUBBER		R
HYPALON		R
KOROSEAL		
NATURAL RUBBER (GRS)		R
NEOPRENE GR-M (CR)		R
NITRILE BUNA N (NBR)		R
NORDEL		
POLYBUTADIENE		
SILICONE		
POLYURETHANE		

TEMP °F (60–460)

CELLOSOLVE

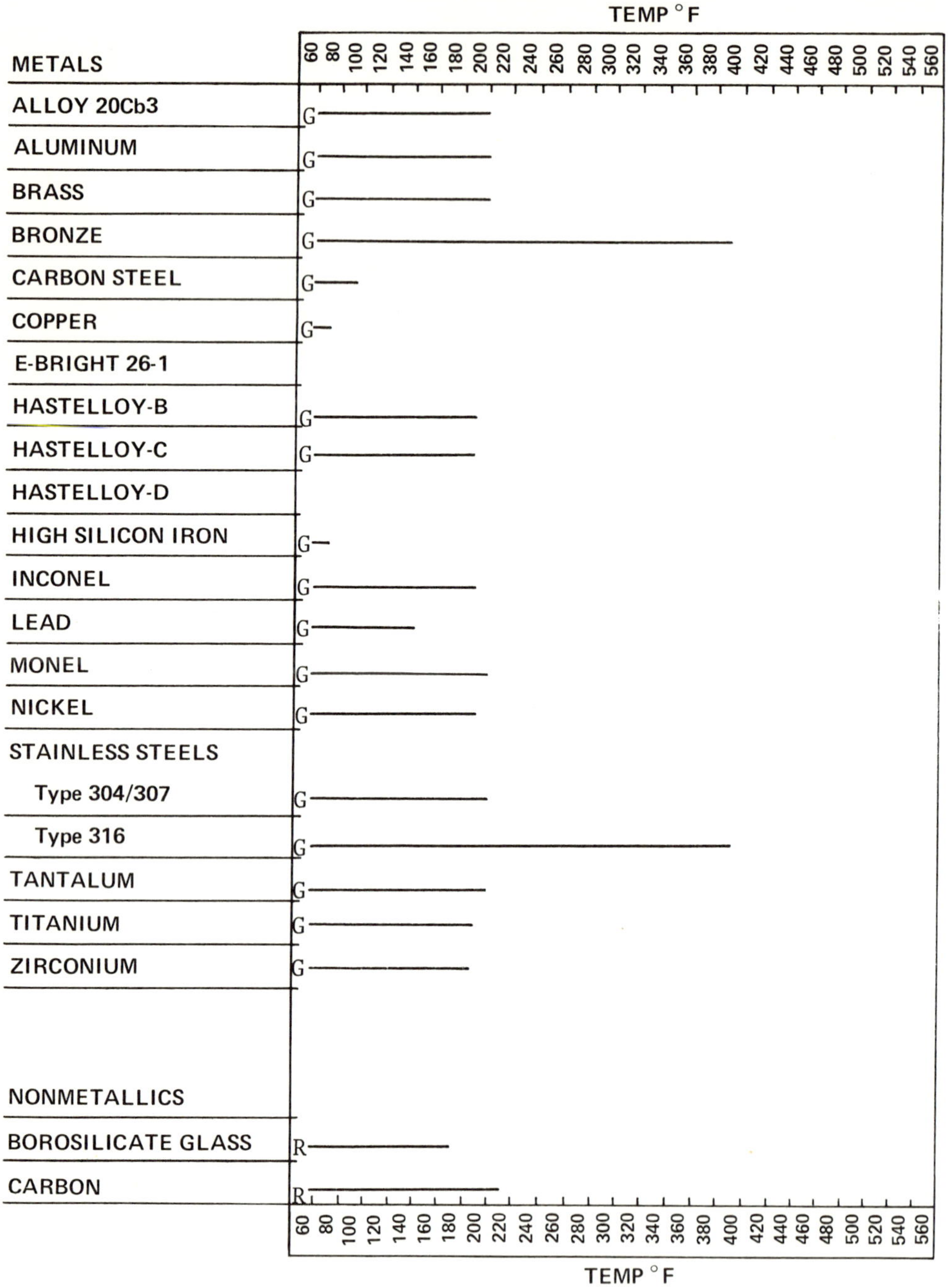

FOR METALS

E = < 2 Mils Penetration/Year; G = < 20 Mils Penetration/Year

S = < 50 Mils Penetration/Year; U = > 50 Mils Penetration/Year

FOR NONMETALLICS

R = Resistant

U = Unsatisfactory

CELLOSOLVE

TEMP °F: 60 80 100 120 140 160 180 200 220 240 260 280 300 320 340 360 380 400 420 440 460

PLASTICS	
ABS	U
ACRYLICS	
ASBESTOS REINF., EPOXY	
ASBESTOS REINF., PHENOLIC	
CHLORINATED POLYETHER	R
CPVC	R
EPOXY	R
FLUOROCARBONS FEP	R
FLUOROCARBONS TFE	R
FURFURYL ALCOHOL	
NORYL	
PHENOLIC	
POLYAMIDES–NYLON	R
POLYESTERS	
BISPHENOL A-FUMURATE	R
HYDROGENATED BISPHENOL-A-BISPHENOL-A	
ISOPHTHALIC	
CHLORINATED POLYESTERS	R
POLYSTYRENES	
POLYETHYLENE	U
POLYPROPYLENE	R
PVC–TYPE 1	RU
PVC–TYPE 2	U
POLYVINYLIDENE CHLORIDE	R
VINYLIDENE FLUORIDE	R
VINYL ESTER	

RUBBERS	
BUTYL GR-1	R
FLUORO ELASTOMERS VITON A	U
FLUORO ELASTOMERS KEL-F 3700	
HARD RUBBER	
HYPALON	
KOROSEAL	
NATURAL RUBBER (GRS)	U
NEOPRENE GR-M (CR)	U
NITRILE BUNA N (NBR)	U
NORDEL	R
POLYBUTADIENE	
SILICONE	
POLYURETHANE	

TEMP °F: 60 80 100 120 140 160 180 200 220 240 260 280 300 320 340 360 380 400 420 440 460

CELLULOSE ACETATE

METALS	Rating	TEMP °F (range shown)
ALLOY 20Cb3	G	60–80
ALUMINUM	E	60–80
BRASS	G	60–80
BRONZE	G	60–400
CARBON STEEL		
COPPER	G	60–80
E-BRIGHT 26-1		
HASTELLOY-B	G	60–80
HASTELLOY-C	G	60–80
HASTELLOY-D	G	60–80
HIGH SILICON IRON	G	60–80
INCONEL	E	60–80
LEAD		
MONEL	E	60–80
NICKEL	E	60–80
STAINLESS STEELS		
Type 304/307	G	60–80
Type 316	G	60–400
TANTALUM	G	60–80
TITANIUM	G	60–80
ZIRCONIUM	G	60–80

NONMETALLICS	Rating	TEMP °F (range shown)
BOROSILICATE GLASS		
CARBON	R	60–340

FOR METALS

E = < 2 Mils Penetration/Year; G = < 20 Mils Penetration/Year

S = < 50 Mils Penetration/Year; U = > 50 Mils Penetration/Year

FOR NONMETALLICS

R = Resistant

U = Unsatisfactory

CELLULOSE ACETATE

TEMP °F (scale 60–460)

PLASTICS	Rating	Bar extends to (°F)
ABS		
ACRYLICS		
ASBESTOS REINF., EPOXY		
ASBESTOS REINF., PHENOLIC		
CHLORINATED POLYETHER		
CPVC		
EPOXY		
FLUOROCARBONS FEP	R	400
FLUOROCARBONS TFE	R	460
FURFURYL ALCOHOL		
NORYL		
PHENOLIC		
POLYAMIDES–NYLON		
POLYESTERS		
BISPHENOL A-FUMURATE		
HYDROGENATED BISPHENOL-A-BISPHENOL-A		
ISOPHTHALIC		
CHLORINATED POLYESTERS		
POLYSTYRENES		
POLYETHYLENE		
POLYPROPYLENE		
PVC–TYPE 1		
PVC–TYPE 2		
POLYVINYLIDENE CHLORIDE	R	80
VINYLIDENE FLUORIDE	U	
VINYL ESTER		

RUBBERS	Rating	Bar extends to (°F)
BUTYL GR-1		
FLUORO ELASTOMERS VITON A		
FLUORO ELASTOMERS KEL-F 3700		
HARD RUBBER		
HYPALON		
KOROSEAL		
NATURAL RUBBER (GRS)		
NEOPRENE GR-M (CR)	R	200
NITRILE BUNA N (NBR)	U	
NORDEL	R	300
POLYBUTADIENE		
SILICONE		
POLYURETHANE		

TEMP °F

CHLORACETIC ACID

TEMP °F (60 80 100 120 140 160 180 200 220 240 260 280 300 320 340 360 380 400 420 440 460 480 500 520 540 560)

METALS	Rating
ALLOY 20Cb3	G—
ALUMINUM	U
BRASS	U
BRONZE	U
CARBON STEEL	U
COPPER	U
E-BRIGHT 26-1	
HASTELLOY-B	E————————————
HASTELLOY-C	E———— G————
HASTELLOY-D	
HIGH SILICON IRON	G—
INCONEL	S—
LEAD	
MONEL	S——————
NICKEL	G— S————
STAINLESS STEELS	
Type 304/307	U
Type 316	U
TANTALUM	E————————————
TITANIUM	E——————
ZIRCONIUM	E——————

NONMETALLICS	Rating
BOROSILICATE GLASS	
CARBON	R———————————

TEMP °F (60 80 100 120 140 160 180 200 220 240 260 280 300 320 340 360 380 400 420 440 460 480 500 520 540 560)

FOR METALS

E = < 2 Mils Penetration/Year; G = < 20 Mils Penetration/Year

S = < 50 Mils Penetration/Year; U = > 50 Mils Penetration/Year

FOR NONMETALLICS

R = Resistant

U = Unsatisfactory

TEMP °F: 60 80 100 120 140 160 180 200 220 240 260 280 300 320 340 360 380 400 420 440 460

PLASTICS	
ABS	U
ACRYLICS	U
ASBESTOS REINF., EPOXY	
ASBESTOS REINF., PHENOLIC	
CHLORINATED POLYETHER	R
CPVC	U
EPOXY	RU
FLUOROCARBONS FEP	R
FLUOROCARBONS TFE	R
FURFURYL ALCOHOL	R
NORYL	
PHENOLIC	
POLYAMIDES—NYLON	U
POLYESTERS	
BISPHENOL A-FUMURATE	R
HYDROGENATED BISPHENOL-A-BISPHENOL-A	
ISOPHTHALIC to 25%	R
CHLORINATED POLYESTERS	
POLYSTYRENES	
POLYETHYLENE	U
POLYPROPYLENE	R
PVC—TYPE 1	R
PVC—TYPE 2	RU
POLYVINYLIDENE CHLORIDE	R
VINYLIDENE FLUORIDE	R
VINYL ESTER 25%	R

RUBBERS	
BUTYL GR-1	R
FLUORO ELASTOMERS VITON A	U
FLUORO ELASTOMERS KEL-F 3700	
HARD RUBBER	U
HYPALON	R
KOROSEAL	U
NATURAL RUBBER (GRS)	U
NEOPRENE GR-M (CR)	R
NITRILE BUNA N (NBR)	U
NORDEL	
POLYBUTADIENE	
SILICONE	
POLYURETHANE	U

TEMP °F: 60 80 100 120 140 160 180 200 220 240 260 280 300 320 340 360 380 400 420 440 460

CHLORACETIC ACID, 50% WATER

TEMP °F (60 – 560)

METALS	Rating
ALLOY 20Cb3	
ALUMINUM	U
BRASS	
BRONZE	
CARBON STEEL	
COPPER	
E-BRIGHT 26-1	
HASTELLOY-B	
HASTELLOY-C	G
HASTELLOY-D	
HIGH SILICON IRON	G
INCONEL	
LEAD	
MONEL	G
NICKEL	
STAINLESS STEELS	
Type 304/307	U
Type 316	U
TANTALUM	E
TITANIUM	E
ZIRCONIUM	E

NONMETALLICS	Rating
BOROSILICATE GLASS	
CARBON	R

TEMP °F

FOR METALS

E = < 2 Mils Penetration/Year; G = < 20 Mils Penetration/Year

S = < 50 Mils Penetration/Year; U = > 50 Mils Penetration/Year

FOR NONMETALLICS

R = Resistant

U = Unsatisfactory

CHLORACETIC ACID, 50% WATER

PLASTICS	TEMP °F (60–460)
ABS	
ACRYLICS	U
ASBESTOS REINF., EPOXY	
ASBESTOS REINF., PHENOLIC	
CHLORINATED POLYETHER	R (60 to ~220)
CPVC	
EPOXY 8%	R (60 to ~140) — U
FLUOROCARBONS FEP	R (60 to ~400)
FLUOROCARBONS TFE	R (60 to 460)
FURFURYL ALCOHOL	
NORYL	
PHENOLIC	
POLYAMIDES—NYLON	U
POLYESTERS	
BISPHENOL A-FUMURATE	R (60 to ~140)
HYDROGENATED BISPHENOL-A-BISPHENOL-A	R (60 to ~80)
ISOPHTHALIC	
CHLORINATED POLYESTERS	
POLYSTYRENES	
POLYETHYLENE	U
POLYPROPYLENE	R (60 to ~70)
PVC—TYPE 1	R (60 to ~140)
PVC—TYPE 2	
POLYVINYLIDENE CHLORIDE	R (60 to ~120)
VINYLIDENE FLUORIDE	R (60 to ~220)
VINYL ESTER	R (60 to ~150)

RUBBERS	TEMP °F (60–460)
BUTYL GR-1	R (60 to ~150)
FLUORO ELASTOMERS VITON A	U
FLUORO ELASTOMERS KEL-F 3700	
HARD RUBBER	U
HYPALON	
KOROSEAL	U
NATURAL RUBBER (GRS)	U
NEOPRENE GR-M (CR)	U
NITRILE BUNA N (NBR)	U
NORDEL	
POLYBUTADIENE	
SILICONE	
POLYURETHANE	

CHLORAL HYDRATE

METALS	TEMP °F (60–560)
ALLOY 20Cb3	
ALUMINUM	
BRASS	
BRONZE	
CARBON STEEL	
COPPER	
E-BRIGHT 26-1	
HASTELLOY-B	
HASTELLOY-C	
HASTELLOY-D	
HIGH SILICON IRON	
INCONEL	
LEAD	
MONEL	
NICKEL	
STAINLESS STEELS Type 304/307	
Type 316	
TANTALUM	
TITANIUM	
ZIRCONIUM	
NONMETALLICS	
BOROSILICATE GLASS	
CARBON	R (60–200)

FOR METALS

E = < 2 Mils Penetration/Year; G = < 20 Mils Penetration/Year

S = < 50 Mils Penetration/Year; U = > 50 Mils Penetration/Year

FOR NONMETALLICS

R = Resistant

U = Unsatisfactory

CHLORAL HYDRATE

TEMP °F (60–460)

PLASTICS	Rating
ABS	R
ACRYLICS	
ASBESTOS REINF., EPOXY	
ASBESTOS REINF., PHENOLIC	
CHLORINATED POLYETHER	R
CPVC	R
EPOXY	
FLUOROCARBONS FEP	R
FLUOROCARBONS TFE	R
FURFURYL ALCOHOL	
NORYL	
PHENOLIC	
POLYAMIDES–NYLON	
POLYESTERS	
BISPHENOL A-FUMURATE	
HYDROGENATED BISPHENOL-A-BISPHENOL-A	
ISOPHTHALIC	
CHLORINATED POLYESTERS	
POLYSTYRENES	
POLYETHYLENE	U
POLYPROPYLENE	U
PVC–TYPE 1	R
PVC–TYPE 2	R
POLYVINYLIDENE CHLORIDE	
VINYLIDENE FLUORIDE	R
VINYL ESTER	

RUBBERS	Rating
BUTYL GR-1	
FLUORO ELASTOMERS VITON A	U
FLUORO ELASTOMERS KEL-F 3700	
HARD RUBBER	
HYPALON	
KOROSEAL	
NATURAL RUBBER (GRS)	
NEOPRENE GR-M (CR)	
NITRILE BUNA N (NBR)	
NORDEL	
POLYBUTADIENE	
SILICONE	
POLYURETHANE	

TEMP °F

CHLORIC ACID, 20%

TEMP °F (60–560)

METALS	
ALLOY 20Cb3	G——
ALUMINUM	
BRASS	U
BRONZE	U
CARBON STEEL	
COPPER	U
E-BRIGHT 26-1	
HASTELLOY-B	U
HASTELLOY-C 3%	E—
HASTELLOY-D	
HIGH SILICON IRON	
INCONEL	U
LEAD	
MONEL	U
NICKEL	U
STAINLESS STEELS	
Type 304/307	U
Type 316	U
TANTALUM	G—
TITANIUM	
ZIRCONIUM	

NONMETALLICS	
BOROSILICATE GLASS	
CARBON	

TEMP °F

FOR METALS

E = < 2 Mils Penetration/Year; G = < 20 Mils Penetration/Year

S = < 50 Mils Penetration/Year; U = > 50 Mils Penetration/Year

FOR NONMETALLICS

R = Resistant

U = Unsatisfactory

CHLORIC ACID, 20%

TEMP °F: 60, 80, 100, 120, 140, 160, 180, 200, 220, 240, 260, 280, 300, 320, 340, 360, 380, 400, 420, 440, 460

PLASTICS	Rating (TEMP °F)
ABS	R 60–140
ACRYLICS	
ASBESTOS REINF., EPOXY	
ASBESTOS REINF., PHENOLIC	
CHLORINATED POLYETHER	
CPVC	R 60–180
EPOXY	
FLUOROCARBONS FEP	R 60–400
FLUOROCARBONS TFE	R 60–460
FURFURYL ALCOHOL	
NORYL	
PHENOLIC	
POLYAMIDES–NYLON	
POLYESTERS	
BISPHENOL A-FUMURATE	
HYDROGENATED BISPHENOL-A-BISPHENOL-A	
ISOPHTHALIC	U 60
CHLORINATED POLYESTERS	
POLYSTYRENES	
POLYETHYLENE	R 60–140, U 160
POLYPROPYLENE	R 60–120, U 140
PVC–TYPE 1	R 60–140
PVC–TYPE 2	R 60–140
POLYVINYLIDENE CHLORIDE	
VINYLIDENE FLUORIDE	
VINYL ESTER	

RUBBERS	TEMP °F
BUTYL GR-1	
FLUORO ELASTOMERS VITON A	
FLUORO ELASTOMERS KEL-F 3700	
HARD RUBBER	
HYPALON	
KOROSEAL	
NATURAL RUBBER (GRS)	
NEOPRENE GR-M (CR)	
NITRILE BUNA N (NBR)	
NORDEL	
POLYBUTADIENE	
SILICONE	
POLYURETHANE	

TEMP °F: 60, 80, 100, 120, 140, 160, 180, 200, 220, 240, 260, 280, 300, 320, 340, 360, 380, 400, 420, 440, 460

CHLORINE DIOXIDE, 15%

TEMP °F (60–560)

METALS	Rating
ALLOY 20Cb3	U
ALUMINUM	G
BRASS	
BRONZE	
CARBON STEEL	U
COPPER	
E-BRIGHT 26-1	
HASTELLOY-B	G
HASTELLOY-C	E
HASTELLOY-D	
HIGH SILICON IRON	
INCONEL	
LEAD	
MONEL	
NICKEL	
STAINLESS STEELS	
Type 304/307	U
Type 316	U
TANTALUM	E
TITANIUM	E
ZIRCONIUM	

NONMETALLICS	Rating
BOROSILICATE GLASS	R
CARBON	R

TEMP °F

FOR METALS

E = < 2 Mils Penetration/Year; G = < 20 Mils Penetration/Year

S = < 50 Mils Penetration/Year; U = > 50 Mils Penetration/Year

FOR NONMETALLICS

R = Resistant

U = Unsatisfactory

CHLORINE DIOXIDE, 15%

TEMP °F: 60 80 100 120 140 160 180 200 220 240 260 280 300 320 340 360 380 400 420 440 460

PLASTICS	
ABS	
ACRYLICS	
ASBESTOS REINF., EPOXY	
ASBESTOS REINF., PHENOLIC	
CHLORINATED POLYETHER	
CPVC	
EPOXY	R——
FLUOROCARBONS FEP	R——
FLUOROCARBONS TFE	R——
FURFURYL ALCOHOL 20%	R——
NORYL	
PHENOLIC	
POLYAMIDES—NYLON	
POLYESTERS	
BISPHENOL A-FUMURATE	R—— U
HYDROGENATED BISPHENOL-A-BISPHENOL-A	R—
ISOPHTHALIC	U
CHLORINATED POLYESTERS	
POLYSTYRENES	
POLYETHYLENE	
POLYPROPYLENE	U
PVC—TYPE 1	
PVC—TYPE 2	
POLYVINYLIDENE CHLORIDE	R——
VINYLIDENE FLUORIDE	R——
VINYL ESTER	R——

RUBBERS	
BUTYL GR-1	U
FLUORO ELASTOMERS VITON A	
FLUORO ELASTOMERS KEL-F 3700	
HARD RUBBER	
HYPALON	
KOROSEAL	
NATURAL RUBBER (GRS)	U
NEOPRENE GR-M (CR)	U
NITRILE BUNA N (NBR)	U
NORDEL	
POLYBUTADIENE	
SILICONE	
POLYURETHANE	

TEMP °F: 60 80 100 120 140 160 180 200 220 240 260 280 300 320 340 360 380 400 420 440 460

CHLORINE GAS, DRY

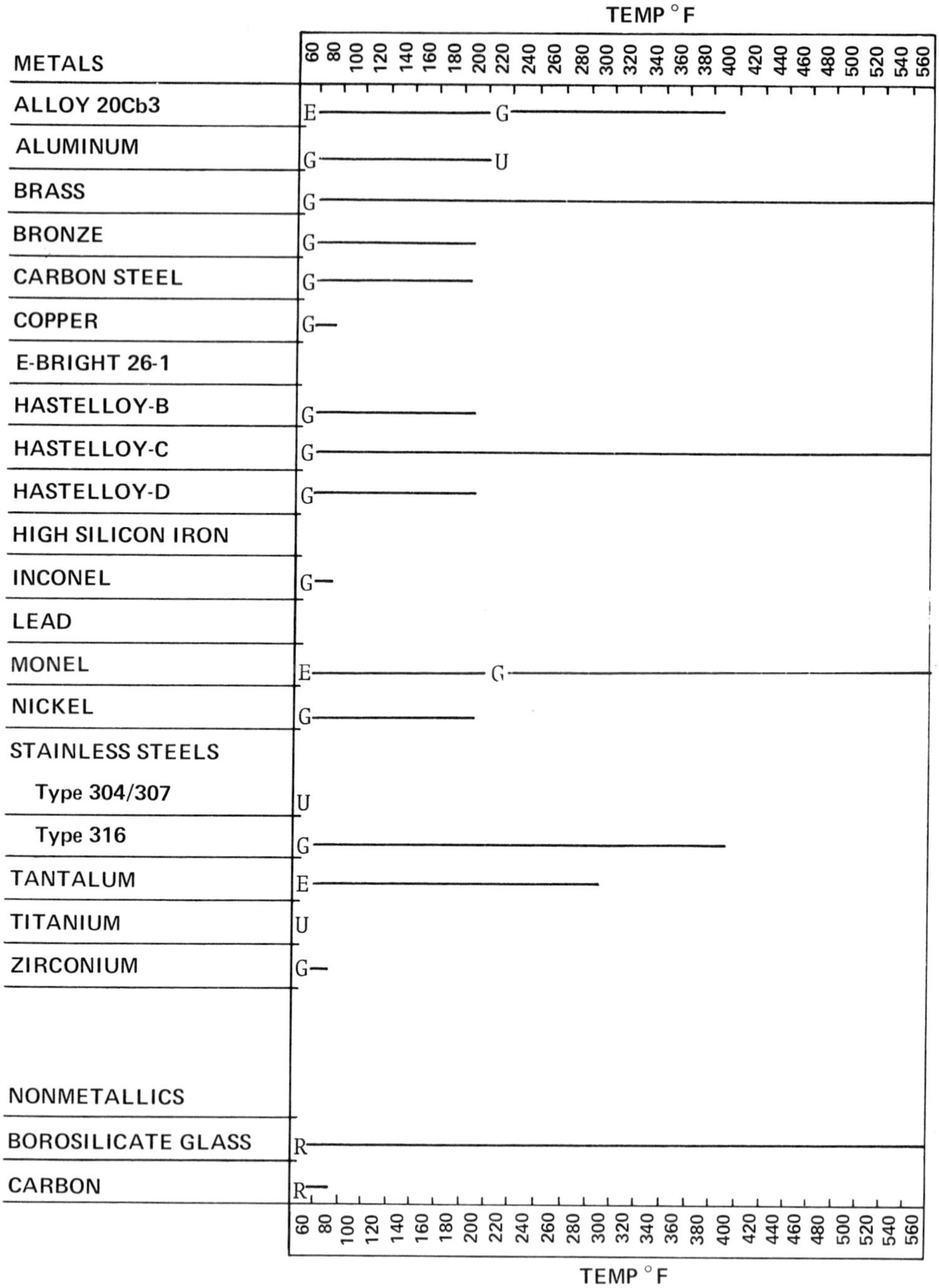

FOR METALS

E = < 2 Mils Penetration/Year; G = < 20 Mils Penetration/Year

S = < 50 Mils Penetration/Year; U = > 50 Mils Penetration/Year

FOR NONMETALLICS

R = Resistant

U = Unsatisfactory

CHLORINE GAS, DRY

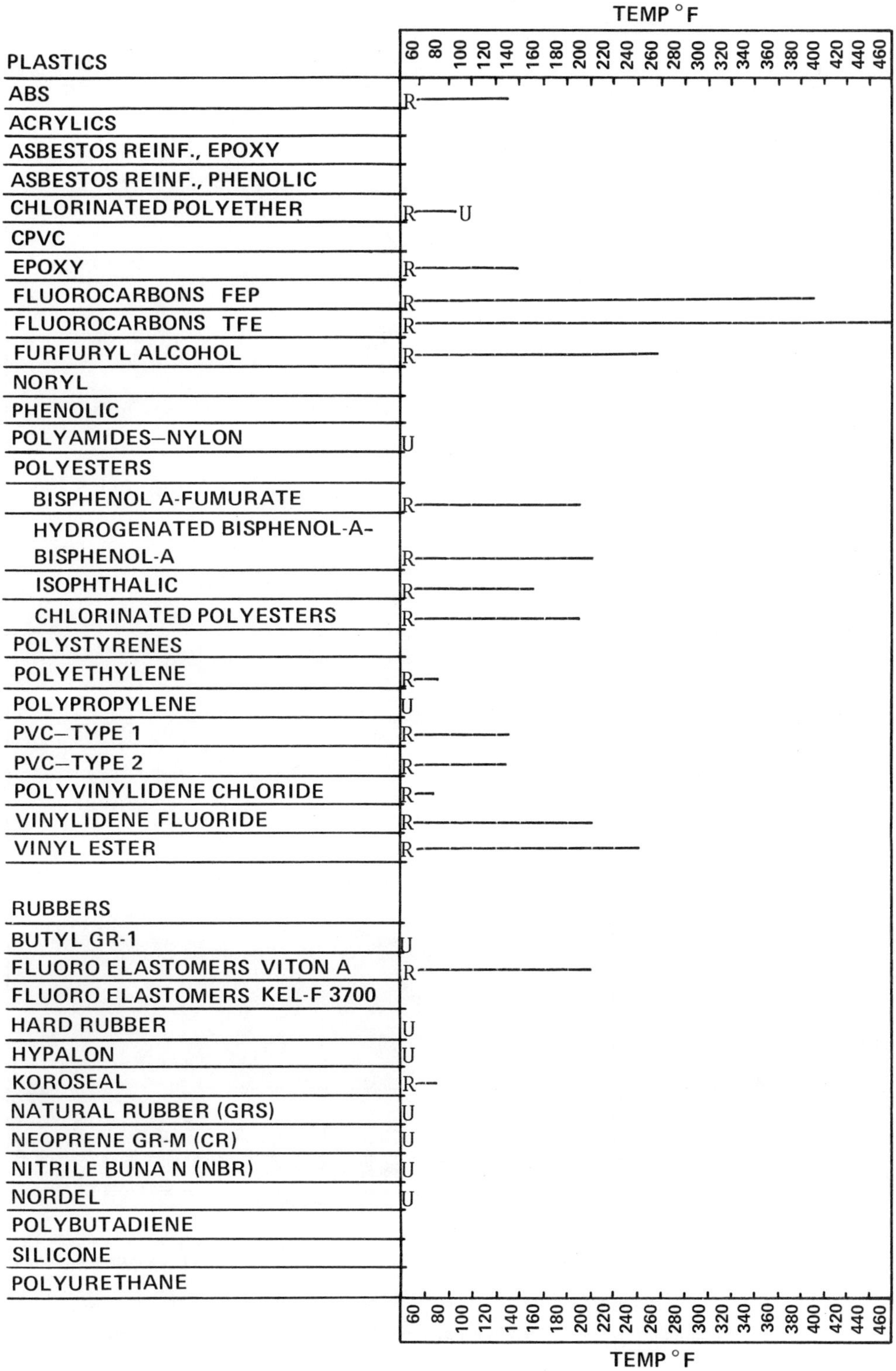

CHLORINE GAS, WET

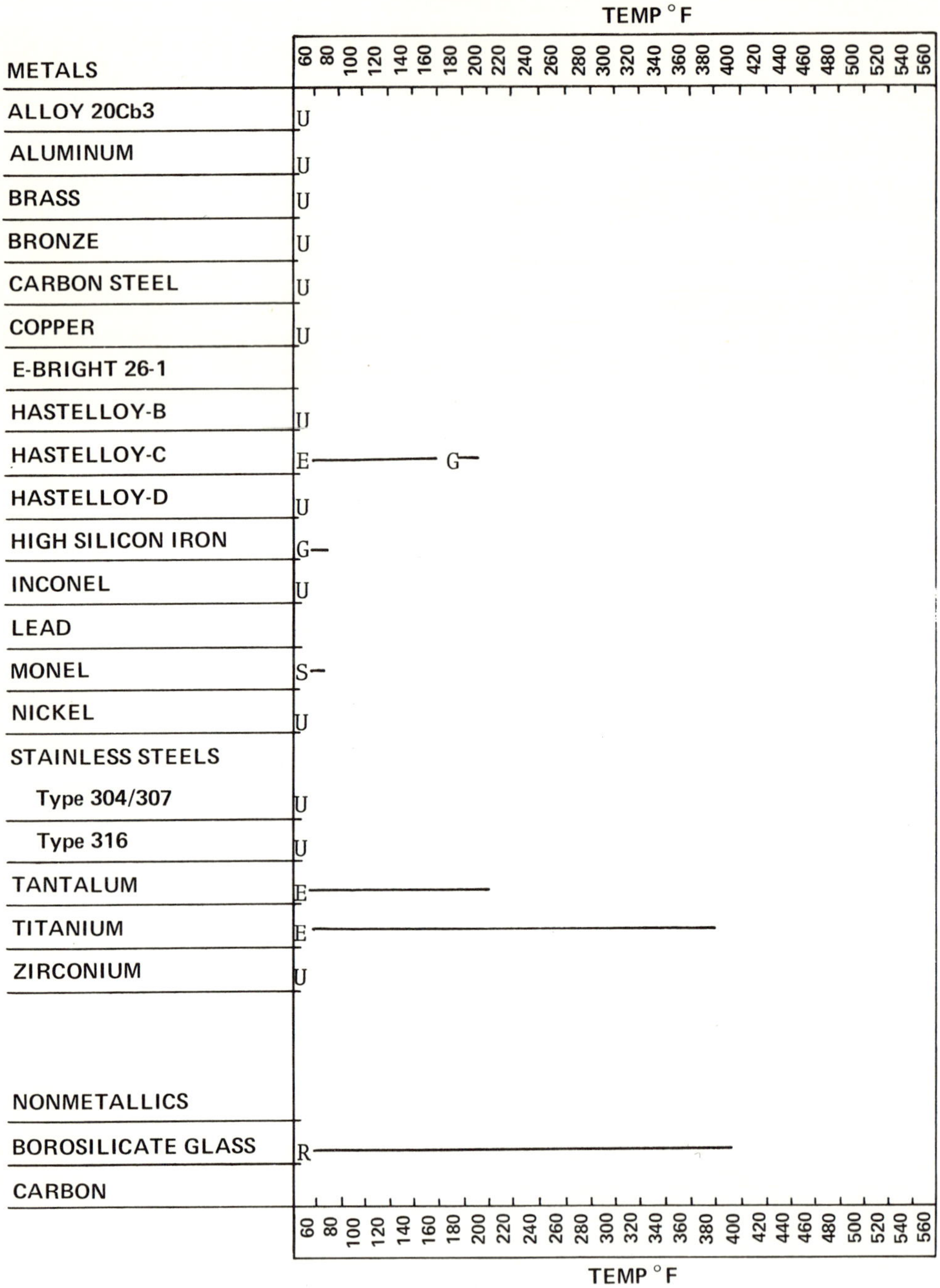

FOR METALS

E = < 2 Mils Penetration/Year; G = < 20 Mils Penetration/Year

S = < 50 Mils Penetration/Year; U = > 50 Mils Penetration/Year

FOR NONMETALLICS

R = Resistant

U = Unsatisfactory

CHLORINE GAS, WET

PLASTICS	TEMP °F (60–460)
ABS	R
ACRYLICS	
ASBESTOS REINF., EPOXY	R
ASBESTOS REINF., PHENOLIC	U
CHLORINATED POLYETHER	R–U
CPVC	
EPOXY	U
FLUOROCARBONS FEP	R
FLUOROCARBONS TFE	R
FURFURYL ALCOHOL	R
NORYL	
PHENOLIC	U
POLYAMIDES–NYLON	U
POLYESTERS	
BISPHENOL A-FUMURATE	R
HYDROGENATED BISPHENOL-A-BISPHENOL-A	R
ISOPHTHALIC	R
CHLORINATED POLYESTERS	R
POLYSTYRENES	R
POLYETHYLENE	R
POLYPROPYLENE	U
PVC–TYPE 1	R–U
PVC–TYPE 2	U
POLYVINYLIDENE CHLORIDE	R
VINYLIDENE FLUORIDE	R
VINYL ESTER	R
RUBBERS	
BUTYL GR-1	
FLUORO ELASTOMERS VITON A	R
FLUORO ELASTOMERS KEL-F 3700	R
HARD RUBBER	R
HYPALON	
KOROSEAL	R
NATURAL RUBBER (GRS)	R
NEOPRENE GR-M (CR)	U
NITRILE BUNA N (NBR)	U
NORDEL	R
POLYBUTADIENE	U
SILICONE	
POLYURETHANE	

TEMP °F: 60 80 100 120 140 160 180 200 220 240 260 280 300 320 340 360 380 400 420 440 460

CHLORINE, LIQUID

TEMP °F: 60 80 100 120 140 160 180 200 220 240 260 280 300 320 340 360 380 400 420 440 460 480 500 520 540 560

METALS	
ALLOY 20Cb3	
ALUMINUM	
BRASS	
BRONZE	
CARBON STEEL	G—
COPPER	
E-BRIGHT 26-1	
HASTELLOY-B	
HASTELLOY-C	E——
HASTELLOY-D	
HIGH SILICON IRON	
INCONEL	
LEAD	G—
MONEL	G——
NICKEL	
STAINLESS STEELS	
Type 304/307 1	G——
Type 316	
TANTALUM	G—
TITANIUM	
ZIRCONIUM	

NONMETALLICS	
BOROSILICATE GLASS	R——
CARBON	

TEMP °F: 60 80 100 120 140 160 180 200 220 240 260 280 300 320 340 360 380 400 420 440 460 480 500 520 540 560

FOR METALS

E = < 2 Mils Penetration/Year; G = < 20 Mils Penetration/Year

S = < 50 Mils Penetration/Year; U = > 50 Mils Penetration/Year

FOR NONMETALLICS

R = Resistant

U = Unsatisfactory

CHLORINE, LIQUID

TEMP °F (60–460)

PLASTICS	Rating	Temp range (°F)
ABS		
ACRYLICS		
ASBESTOS REINF., EPOXY		
ASBESTOS REINF., PHENOLIC		
CHLORINATED POLYETHER	U	60
CPVC	U	60
EPOXY		
FLUOROCARBONS FEP	R	60–400
FLUOROCARBONS TFE	R	60–460
FURFURYL ALCOHOL	U	60
NORYL		
PHENOLIC		
POLYAMIDES–NYLON	U	60
POLYESTERS		
BISPHENOL A-FUMURATE		
HYDROGENATED BISPHENOL-A-BISPHENOL-A		
ISOPHTHALIC		
CHLORINATED POLYESTERS		
POLYSTYRENES		
POLYETHYLENE	U	60
POLYPROPYLENE	U	60
PVC–TYPE 1	U	60
PVC–TYPE 2	U	60
POLYVINYLIDENE CHLORIDE	U	60
VINYLIDENE FLUORIDE	R	60–200
VINYL ESTER		

RUBBERS	Rating	Temp range (°F)
BUTYL GR-1	U	60
FLUORO ELASTOMERS VITON A	R	60–180
FLUORO ELASTOMERS KEL-F 3700		
HARD RUBBER	U	60
HYPALON		
KOROSEAL		
NATURAL RUBBER (GRS)	U	60
NEOPRENE GR-M (CR)	U	60
NITRILE BUNA N (NBR)	U	60
NORDEL		
POLYBUTADIENE		
SILICONE		
POLYURETHANE		

TEMP °F

CHLORINE WATER, (SAT)

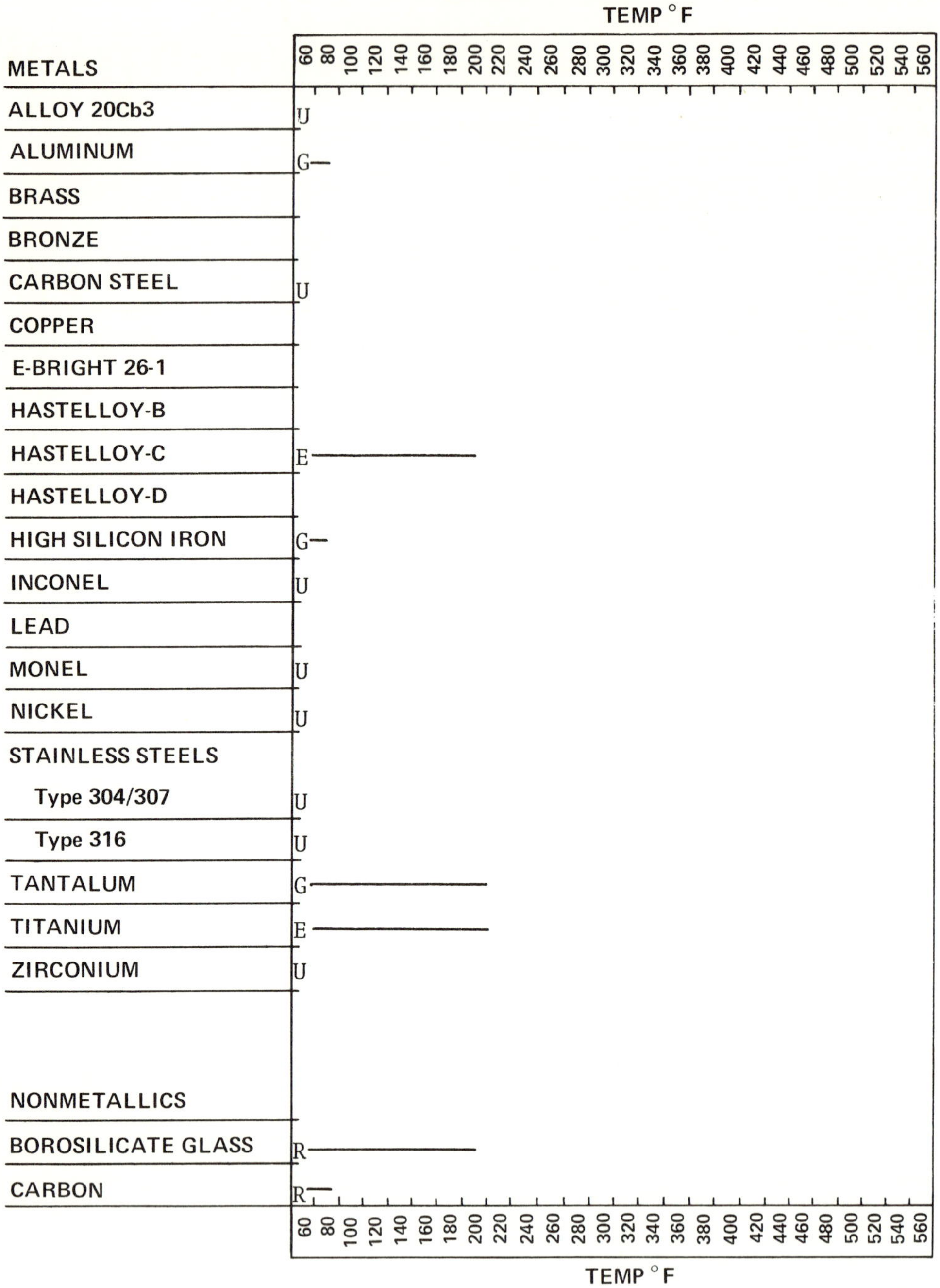

FOR METALS

E = < 2 Mils Penetration/Year; G = < 20 Mils Penetration/Year

S = < 50 Mils Penetration/Year; U = > 50 Mils Penetration/Year

FOR NONMETALLICS

R = Resistant

U = Unsatisfactory

CHLORINE WATER, (SAT)

TEMP °F: 60 80 100 120 140 160 180 200 220 240 260 280 300 320 340 360 380 400 420 440 460

PLASTICS	Rating
ABS	R
ACRYLICS	R
ASBESTOS REINF., EPOXY	RU
ASBESTOS REINF., PHENOLIC	RU
CHLORINATED POLYETHER	R
CPVC	R
EPOXY	U
FLUOROCARBONS FEP	R
FLUOROCARBONS TFE	R
FURFURYL ALCOHOL	R
NORYL	
PHENOLIC	
POLYAMIDES—NYLON	U
POLYESTERS	
BISPHENOL A-FUMURATE	R
HYDROGENATED BISPHENOL-A-BISPHENOL-A	
ISOPHTHALIC	R
CHLORINATED POLYESTERS	R
POLYSTYRENES	
POLYETHYLENE	U
POLYPROPYLENE	R
PVC—TYPE 1	R
PVC—TYPE 2	R
POLYVINYLIDENE CHLORIDE	R
VINYLIDENE FLUORIDE	R
VINYL ESTER	R

RUBBERS	Rating
BUTYL GR-1	U
FLUORO ELASTOMERS VITON A	R
FLUORO ELASTOMERS KEL-F 3700	
HARD RUBBER	R
HYPALON	R
KOROSEAL	R
NATURAL RUBBER (GRS)	R
NEOPRENE GR-M (CR)	U
NITRILE BUNA N (NBR)	U
NORDEL	
POLYBUTADIENE	
SILICONE	
POLYURETHANE	U

TEMP °F

CHLOROBENZENE (PHENYLCHLORIDE)

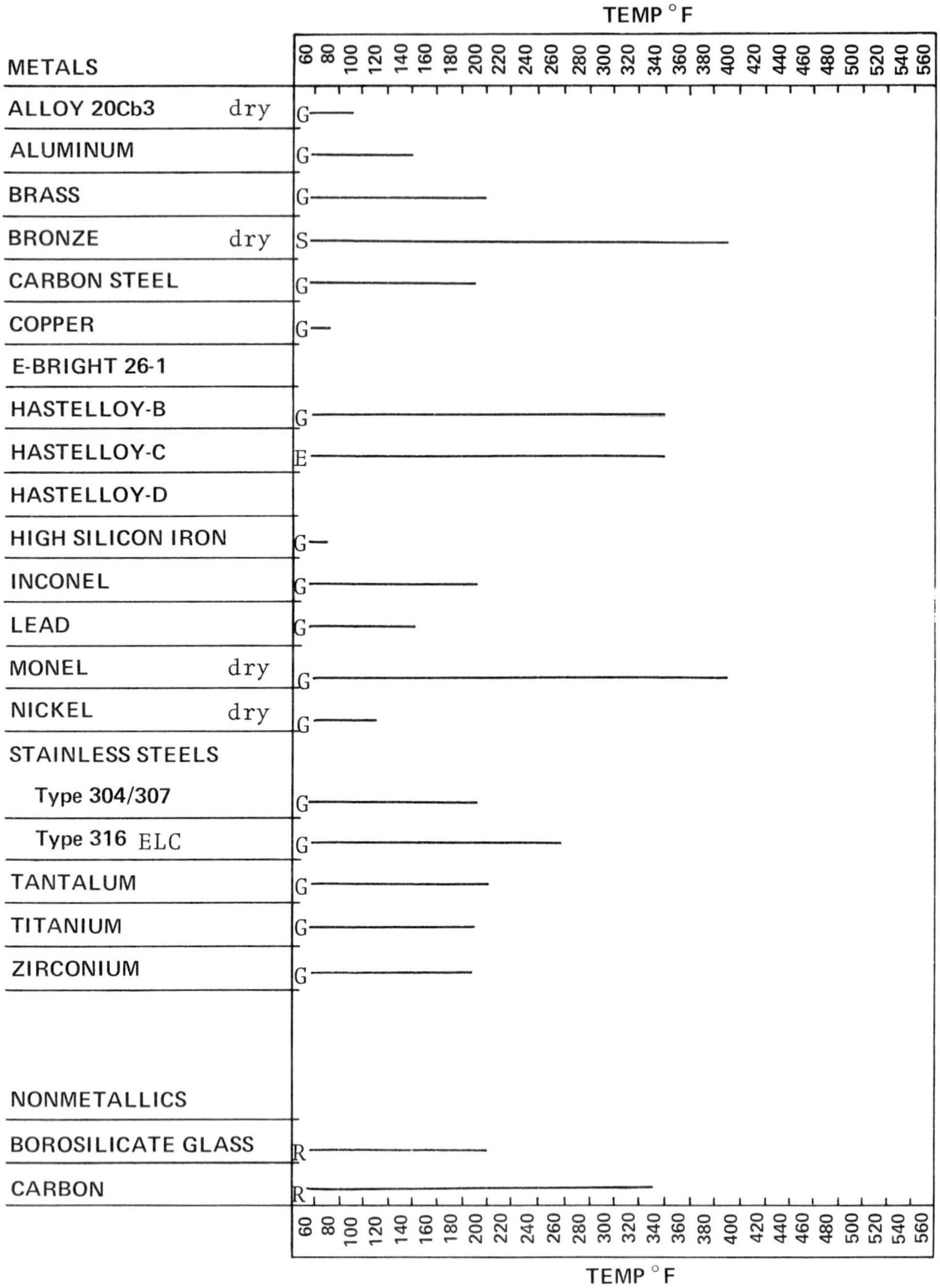

FOR METALS

E = < 2 Mils Penetration/Year; G = < 20 Mils Penetration/Year

S = < 50 Mils Penetration/Year; U = > 50 Mils Penetration/Year

FOR NONMETALLICS

R = Resistant

U = Unsatisfactory

CHLOROBENZENE (PHENYLCHLORIDE)

TEMP °F (60–460)

PLASTICS	TEMP °F
ABS	U
ACRYLICS	U
ASBESTOS REINF., EPOXY	U
ASBESTOS REINF., PHENOLIC	R——
CHLORINATED POLYETHER	R—— U
CPVC	U
EPOXY	R—— U
FLUOROCARBONS FEP	R——
FLUOROCARBONS TFE	R——
FURFURYL ALCOHOL	R——
NORYL	
PHENOLIC	
POLYAMIDES—NYLON	R——
POLYESTERS	
BISPHENOL A-FUMURATE	U
HYDROGENATED BISPHENOL-A-BISPHENOL-A	
ISOPHTHALIC	U
CHLORINATED POLYESTERS	U
POLYSTYRENES	
POLYETHYLENE	U
POLYPROPYLENE	U
PVC—TYPE 1	U
PVC—TYPE 2	U
POLYVINYLIDENE CHLORIDE	R—
VINYLIDENE FLUORIDE	R——
VINYL ESTER	R——
RUBBERS	
BUTYL GR-1	U
FLUORO ELASTOMERS VITON A	R——
FLUORO ELASTOMERS KEL-F 3700	
HARD RUBBER	U
HYPALON	U
KOROSEAL	U
NATURAL RUBBER (GRS)	U
NEOPRENE GR-M (CR)	U
NITRILE BUNA N (NBR)	U
NORDEL	U
POLYBUTADIENE	
SILICONE	
POLYURETHANE	U

TEMP °F

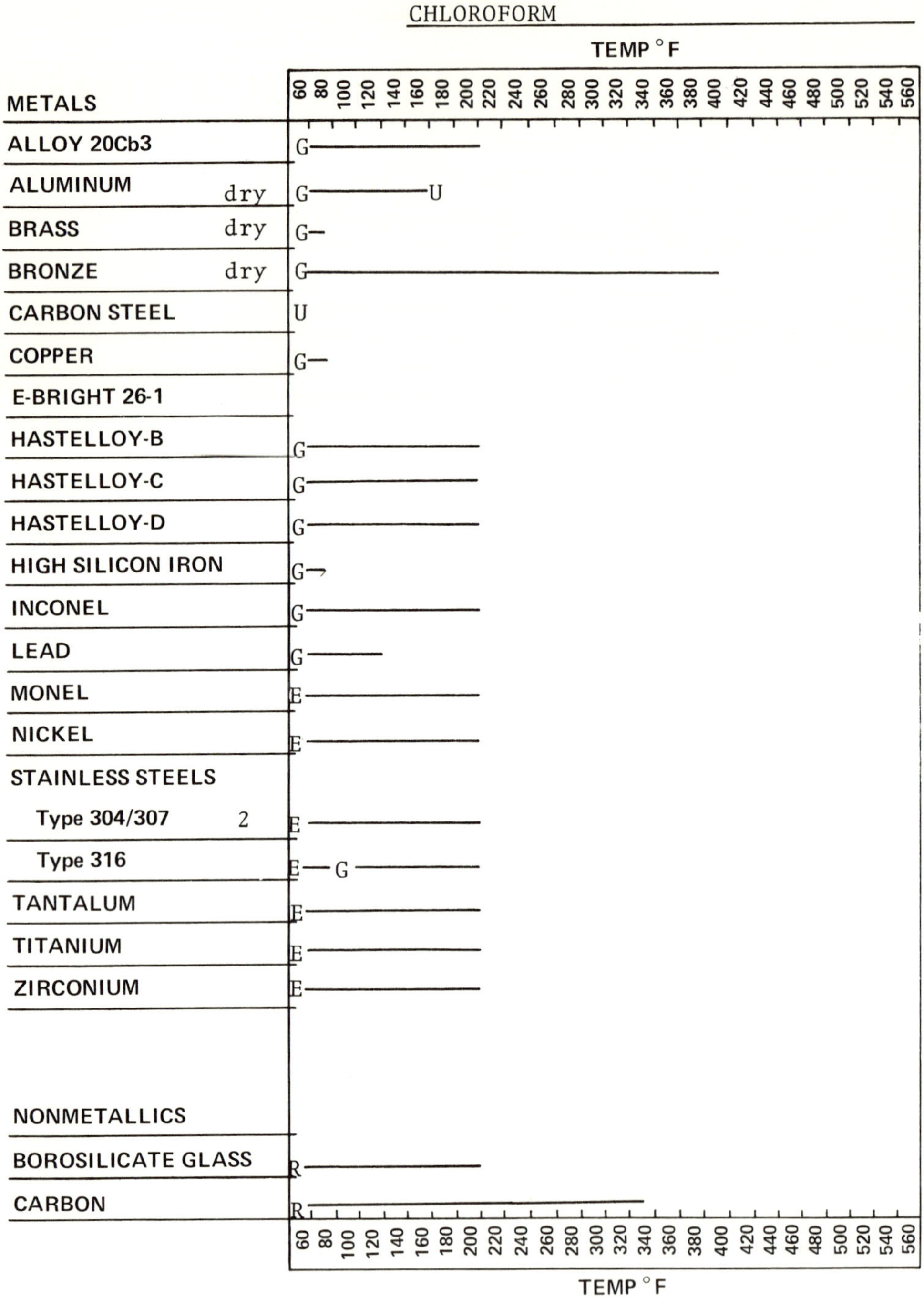

FOR METALS

E = < 2 Mils Penetration/Year; G = < 20 Mils Penetration/Year

S = < 50 Mils Penetration/Year; U = > 50 Mils Penetration/Year

FOR NONMETALLICS

R = Resistant

U = Unsatisfactory

CHLOROFORM

TEMP °F (scale 60–460)

PLASTICS	Rating	Bar extends to (°F, approx.)
ABS	U	
ACRYLICS	U	
ASBESTOS REINF., EPOXY	U	
ASBESTOS REINF., PHENOLIC	R	~300
CHLORINATED POLYETHER	R	~75
CPVC	U	
EPOXY	R — U	R to ~120, then U
FLUOROCARBONS FEP	R	~400
FLUOROCARBONS TFE	R	460+
FURFURYL ALCOHOL	R	~270
NORYL		
PHENOLIC		
POLYAMIDES—NYLON	R	~125
POLYESTERS		
BISPHENOL A-FUMURATE	U	
HYDROGENATED BISPHENOL-A-BISPHENOL-A	U	
ISOPHTHALIC	U	
CHLORINATED POLYESTERS	U	
POLYSTYRENES		
POLYETHYLENE	U	
POLYPROPYLENE	U	
PVC—TYPE 1	U	
PVC—TYPE 2	U	
POLYVINYLIDENE CHLORIDE	U	
VINYLIDENE FLUORIDE	R	~255
VINYL ESTER	U	

RUBBERS	Rating	Bar extends to (°F, approx.)
BUTYL GR-1	U	
FLUORO ELASTOMERS VITON A	R	~190
FLUORO ELASTOMERS KEL-F 3700		
HARD RUBBER	U	
HYPALON	U	
KOROSEAL	U	
NATURAL RUBBER (GRS)	U	
NEOPRENE GR-M (CR)	U	
NITRILE BUNA N (NBR)	U	
NORDEL	U	
POLYBUTADIENE		
SILICONE		
POLYURETHANE	U	

CHLOROSULFONIC ACID, 100%

TEMP °F: 60, 80, 100, 120, 140, 160, 180, 200, 220, 240, 260, 280, 300, 320, 340, 360, 380, 400, 420, 440, 460, 480, 500, 520, 540, 560

METALS		Rating
ALLOY 20Cb3		G
ALUMINUM	dry	G
BRASS		U
BRONZE		
CARBON STEEL		G
COPPER		
E-BRIGHT 26-1		
HASTELLOY-B		E
HASTELLOY-C		E
HASTELLOY-D		
HIGH SILICON IRON		G
INCONEL		G
LEAD		
MONEL		G
NICKEL		G
STAINLESS STEELS		
Type 304/307		U
Type 316		U
TANTALUM		G
TITANIUM		E
ZIRCONIUM		

NONMETALLICS	Rating
BOROSILICATE GLASS	R
CARBON	R

TEMP °F: 60, 80, 100, 120, 140, 160, 180, 200, 220, 240, 260, 280, 300, 320, 340, 360, 380, 400, 420, 440, 460, 480, 500, 520, 540, 560

FOR METALS

E = < 2 Mils Penetration/Year; G = < 20 Mils Penetration/Year

S = < 50 Mils Penetration/Year; U = > 50 Mils Penetration/Year

FOR NONMETALLICS

R = Resistant

U = Unsatisfactory

CHLOROSULFONIC ACID, 100%

TEMP °F (scale: 60, 80, 100, 120, 140, 160, 180, 200, 220, 240, 260, 280, 300, 320, 340, 360, 380, 400, 420, 440, 460)

PLASTICS		Rating
ABS		U
ACRYLICS		
ASBESTOS REINF., EPOXY		
ASBESTOS REINF., PHENOLIC		
CHLORINATED POLYETHER		U
CPVC		U
EPOXY		U
FLUOROCARBONS FEP		R (60–400 °F)
FLUOROCARBONS TFE		R (60–460 °F)
FURFURYL ALCOHOL	sat	R (60–260 °F)
NORYL		
PHENOLIC		
POLYAMIDES–NYLON		
POLYESTERS		
BISPHENOL A-FUMURATE		U
HYDROGENATED BISPHENOL-A-BISPHENOL-A		
ISOPHTHALIC		U
CHLORINATED POLYESTERS		U
POLYSTYRENES		
POLYETHYLENE		U
POLYPROPYLENE		U
PVC–TYPE 1		RU
PVC–TYPE 2		RU
POLYVINYLIDENE CHLORIDE		U
VINYLIDENE FLUORIDE		R (60 °F) —— U (120 °F)
VINYL ESTER		U

RUBBERS	Rating
BUTYL GR-1	U
FLUORO ELASTOMERS VITON A	U
FLUORO ELASTOMERS KEL-F 3700	
HARD RUBBER	U
HYPALON	U
KOROSEAL	
NATURAL RUBBER (GRS)	U
NEOPRENE GR-M (CR)	U
NITRILE BUNA N (NBR)	U
NORDEL	U
POLYBUTADIENE	
SILICONE	
POLYURETHANE	U

TEMP °F

CHLOROX BLEACH SOLUTION, 5.5% CHLORINE

TEMP °F (60–560)

METALS	Rating (from 60°F)
ALLOY 20Cb3	
ALUMINUM	U
BRASS	
BRONZE	
CARBON STEEL	U
COPPER	
E-BRIGHT 26-1	
HASTELLOY-B	
HASTELLOY-C	
HASTELLOY-D	
HIGH SILICON IRON	
INCONEL	
LEAD	
MONEL	
NICKEL	
STAINLESS STEELS Type 304/307	
Type 316	G—
TANTALUM	
TITANIUM	
ZIRCONIUM	

NONMETALLICS	Rating (from 60°F)
BOROSILICATE GLASS	R———— (to ~200)
CARBON	

TEMP °F

FOR METALS

E = < 2 Mils Penetration/Year; G = < 20 Mils Penetration/Year

S = < 50 Mils Penetration/Year; U = > 50 Mils Penetration/Year

FOR NONMETALLICS

R = Resistant

U = Unsatisfactory

CHLOROX BLEACH SOLUTION, 5.5% CHLORINE

TEMP °F (60, 80, 100, 120, 140, 160, 180, 200, 220, 240, 260, 280, 300, 320, 340, 360, 380, 400, 420, 440, 460)

PLASTICS	
ABS	R
ACRYLICS	
ASBESTOS REINF., EPOXY	
ASBESTOS REINF., PHENOLIC	
CHLORINATED POLYETHER	
CPVC	
EPOXY	
FLUOROCARBONS FEP	R
FLUOROCARBONS TFE	R
FURFURYL ALCOHOL	
NORYL	
PHENOLIC	
POLYAMIDES—NYLON	U
POLYESTERS	
BISPHENOL A-FUMURATE	
HYDROGENATED BISPHENOL-A-BISPHENOL-A	
ISOPHTHALIC	
CHLORINATED POLYESTERS	
POLYSTYRENES	
POLYETHYLENE	
POLYPROPYLENE	R
PVC—TYPE 1	
PVC—TYPE 2	
POLYVINYLIDENE CHLORIDE	R
VINYLIDENE FLUORIDE	R
VINYL ESTER	

RUBBERS	
BUTYL GR-1	R
FLUORO ELASTOMERS VITON A	
FLUORO ELASTOMERS KEL-F 3700	
HARD RUBBER	R
HYPALON	
KOROSEAL	
NATURAL RUBBER (GRS)	U
NEOPRENE GR-M (CR)	U
NITRILE BUNA N (NBR)	R
NORDEL	
POLYBUTADIENE	
SILICONE	
POLYURETHANE	

TEMP °F (60, 80, 100, 120, 140, 160, 180, 200, 220, 240, 260, 280, 300, 320, 340, 360, 380, 400, 420, 440, 460)

CHROME ALUM

TEMP °F (scale: 60, 80, 100, 120, 140, 160, 180, 200, 220, 240, 260, 280, 300, 320, 340, 360, 380, 400, 420, 440, 460, 480, 500, 520, 540, 560)

METALS	Rating	Temp °F range
ALLOY 20Cb3	E	60–100
ALUMINUM	U	60
BRASS		
BRONZE		
CARBON STEEL	U	60
COPPER		
E-BRIGHT 26-1		
HASTELLOY-B		
HASTELLOY-C		
HASTELLOY-D		
HIGH SILICON IRON		
INCONEL		
LEAD	G	60–80
MONEL	G	60–100
NICKEL		
STAINLESS STEELS		
Type 304/307	G	60–100
Type 316	G	60–100
TANTALUM		
TITANIUM		
ZIRCONIUM		

NONMETALLICS	Rating	Temp °F range
BOROSILICATE GLASS	R	60–140
CARBON		

FOR METALS

E = < 2 Mils Penetration/Year; G = < 20 Mils Penetration/Year

S = < 50 Mils Penetration/Year; U = > 50 Mils Penetration/Year

FOR NONMETALLICS

R = Resistant

U = Unsatisfactory

CHROME ALUM

TEMP °F (60–460)

PLASTICS	
ABS	
ACRYLICS	
ASBESTOS REINF., EPOXY	
ASBESTOS REINF., PHENOLIC	
CHLORINATED POLYETHER	R—
CPVC	R—
EPOXY	
FLUOROCARBONS FEP	R—
FLUOROCARBONS TFE	R—
FURFURYL ALCOHOL	
NORYL	R—
PHENOLIC	
POLYAMIDES–NYLON	
POLYESTERS	
BISPHENOL A-FUMURATE	
HYDROGENATED BISPHENOL-A-BISPHENOL-A	
ISOPHTHALIC	
CHLORINATED POLYESTERS	
POLYSTYRENES	
POLYETHYLENE	R—U
POLYPROPYLENE	R—U
PVC–TYPE 1	R—
PVC–TYPE 2	R—
POLYVINYLIDENE CHLORIDE	
VINYLIDENE FLUORIDE	
VINYL ESTER	

RUBBERS	
BUTYL GR-1	R—
FLUORO ELASTOMERS VITON A	R—
FLUORO ELASTOMERS KEL-F 3700	R—
HARD RUBBER	R—
HYPALON	R--
KOROSEAL	
NATURAL RUBBER (GRS)	R—
NEOPRENE GR-M (CR)	R—
NITRILE BUNA N (NBR)	R—
NORDEL	
POLYBUTADIENE	R—
SILICONE	
POLYURETHANE	R—

TEMP °F (60–460)

CHROMIC ACID, 50%

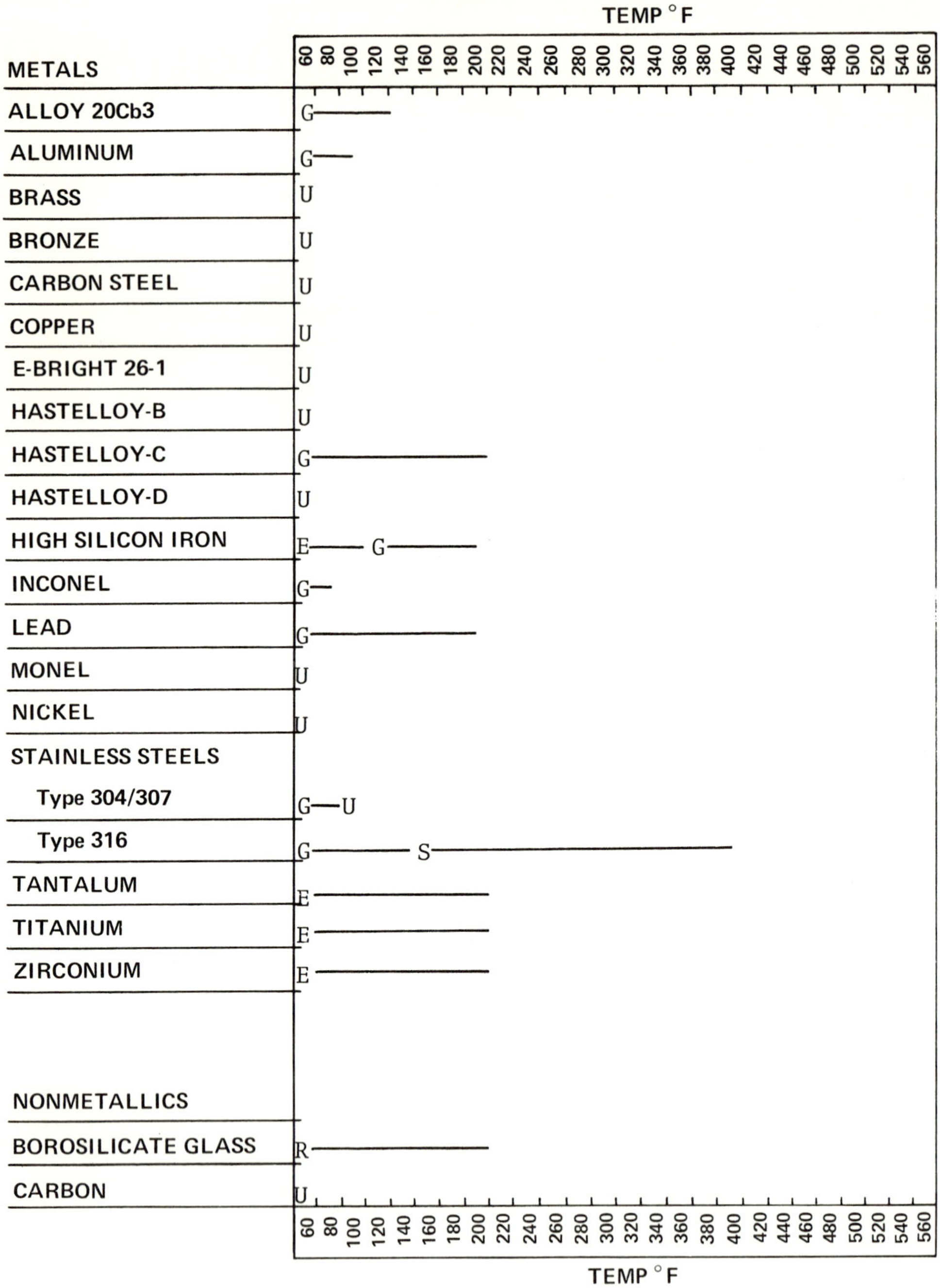

FOR METALS

E = < 2 Mils Penetration/Year; G = < 20 Mils Penetration/Year

S = < 50 Mils Penetration/Year; U = > 50 Mils Penetration/Year

FOR NONMETALLICS

R = Resistant

U = Unsatisfactory

CHROMIC ACID, 50%

TEMP °F: 60 80 100 120 140 160 180 200 220 240 260 280 300 320 340 360 380 400 420 440 460

PLASTICS	
ABS	U
ACRYLICS	U
ASBESTOS REINF., EPOXY	U
ASBESTOS REINF., PHENOLIC	U
CHLORINATED POLYETHER	R
CPVC	R
EPOXY	U
FLUOROCARBONS FEP	R
FLUOROCARBONS TFE	R
FURFURYL ALCOHOL	U
NORYL	
PHENOLIC	U
POLYAMIDES–NYLON	U
POLYESTERS	
BISPHENOL A-FUMURATE	U
HYDROGENATED BISPHENOL-A-BISPHENOL-A	U
ISOPHTHALIC	U
CHLORINATED POLYESTERS	R
POLYSTYRENES	U
POLYETHYLENE	R
POLYPROPYLENE	R
PVC–TYPE 1	U
PVC–TYPE 2	U
POLYVINYLIDENE CHLORIDE	R
VINYLIDENE FLUORIDE	R
VINYL ESTER	

RUBBERS	
BUTYL GR-1	U
FLUORO ELASTOMERS VITON A	R
FLUORO ELASTOMERS KEL-F 3700	R
HARD RUBBER	
HYPALON	R
KOROSEAL	U
NATURAL RUBBER (GRS)	U
NEOPRENE GR-M (CR)	U
NITRILE BUNA N (NBR)	R
NORDEL	
POLYBUTADIENE	U
SILICONE	
POLYURETHANE	U

TEMP °F: 60 80 100 120 140 160 180 200 220 240 260 280 300 320 340 360 380 400 420 440 460

CHROMIC ACID, 40%

TEMP °F: 60, 80, 100, 120, 140, 160, 180, 200, 220, 240, 260, 280, 300, 320, 340, 360, 380, 400, 420, 440, 460, 480, 500, 520, 540, 560

METALS	
ALLOY 20Cb3	G
ALUMINUM	G
BRASS	U
BRONZE	U
CARBON STEEL	U
COPPER	U
E-BRIGHT 26-1	G
HASTELLOY-B	U
HASTELLOY-C	G
HASTELLOY-D	U
HIGH SILICON IRON	E G
INCONEL	G
LEAD	G
MONEL	U
NICKEL	U
STAINLESS STEELS	
Type 304/307 1	E
Type 316	G
TANTALUM	G
TITANIUM	E
ZIRCONIUM	E
NONMETALLICS	
BOROSILICATE GLASS	R
CARBON	U

TEMP °F

FOR METALS

E = < 2 Mils Penetration/Year; G = < 20 Mils Penetration/Year

S = < 50 Mils Penetration/Year; U = > 50 Mils Penetration/Year

FOR NONMETALLICS

R = Resistant

U = Unsatisfactory

CHROMIC ACID, 40%

TEMP °F (60, 80, 100, 120, 140, 160, 180, 200, 220, 240, 260, 280, 300, 320, 340, 360, 380, 400, 420, 440, 460)

PLASTICS	
ABS	U
ACRYLICS	U
ASBESTOS REINF., EPOXY	U
ASBESTOS REINF., PHENOLIC	U
CHLORINATED POLYETHER	R————
CPVC	R————
EPOXY	U
FLUOROCARBONS FEP	R————
FLUOROCARBONS TFE	R————
FURFURYL ALCOHOL	U
NORYL	
PHENOLIC	
POLYAMIDES—NYLON	U
POLYESTERS	
BISPHENOL A-FUMURATE	U
HYDROGENATED BISPHENOL-A-BISPHENOL-A	U
ISOPHTHALIC	U
CHLORINATED POLYESTERS	R———— U
POLYSTYRENES	U
POLYETHYLENE	R————
POLYPROPYLENE	R———— U
PVC—TYPE 1	R————
PVC—TYPE 2	U
POLYVINYLIDENE CHLORIDE	R————
VINYLIDENE FLUORIDE	R————
VINYL ESTER	

RUBBERS	
BUTYL GR-1	U
FLUORO ELASTOMERS VITON A	R————
FLUORO ELASTOMERS KEL-F 3700	R—
HARD RUBBER	
HYPALON	R————
KOROSEAL	U
NATURAL RUBBER (GRS)	U
NEOPRENE GR-M (CR)	U
NITRILE BUNA N (NBR)	R———— U
NORDEL	
POLYBUTADIENE	U
SILICONE	
POLYURETHANE	U

TEMP °F (60, 80, 100, 120, 140, 160, 180, 200, 220, 240, 260, 280, 300, 320, 340, 360, 380, 400, 420, 440, 460)

CHROMIC ACID, 30%

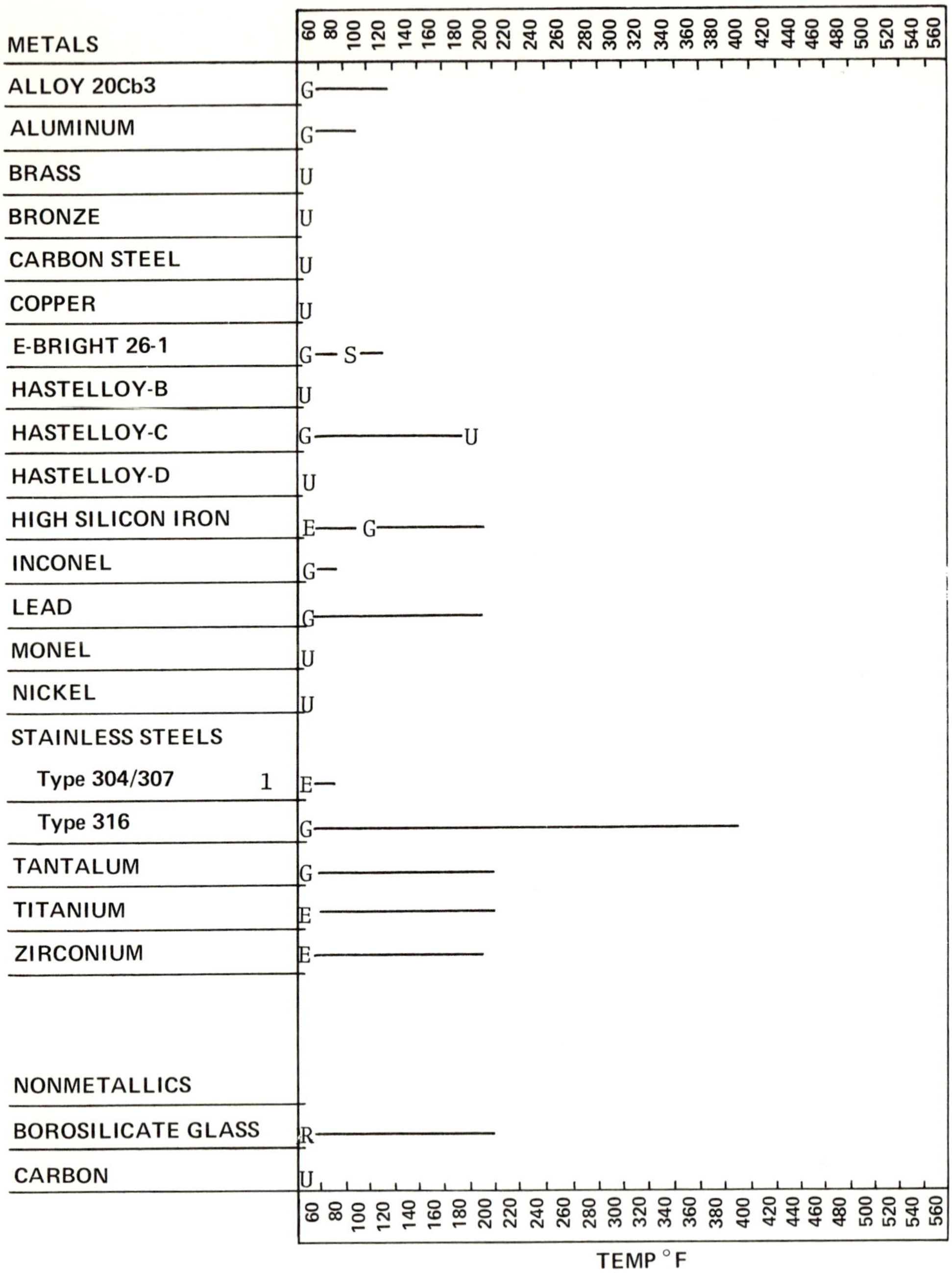

FOR METALS

E = < 2 Mils Penetration/Year; G = < 20 Mils Penetration/Year

S = < 50 Mils Penetration/Year; U = > 50 Mils Penetration/Year

FOR NONMETALLICS

R = Resistant

U = Unsatisfactory

CHROMIC ACID, 30%

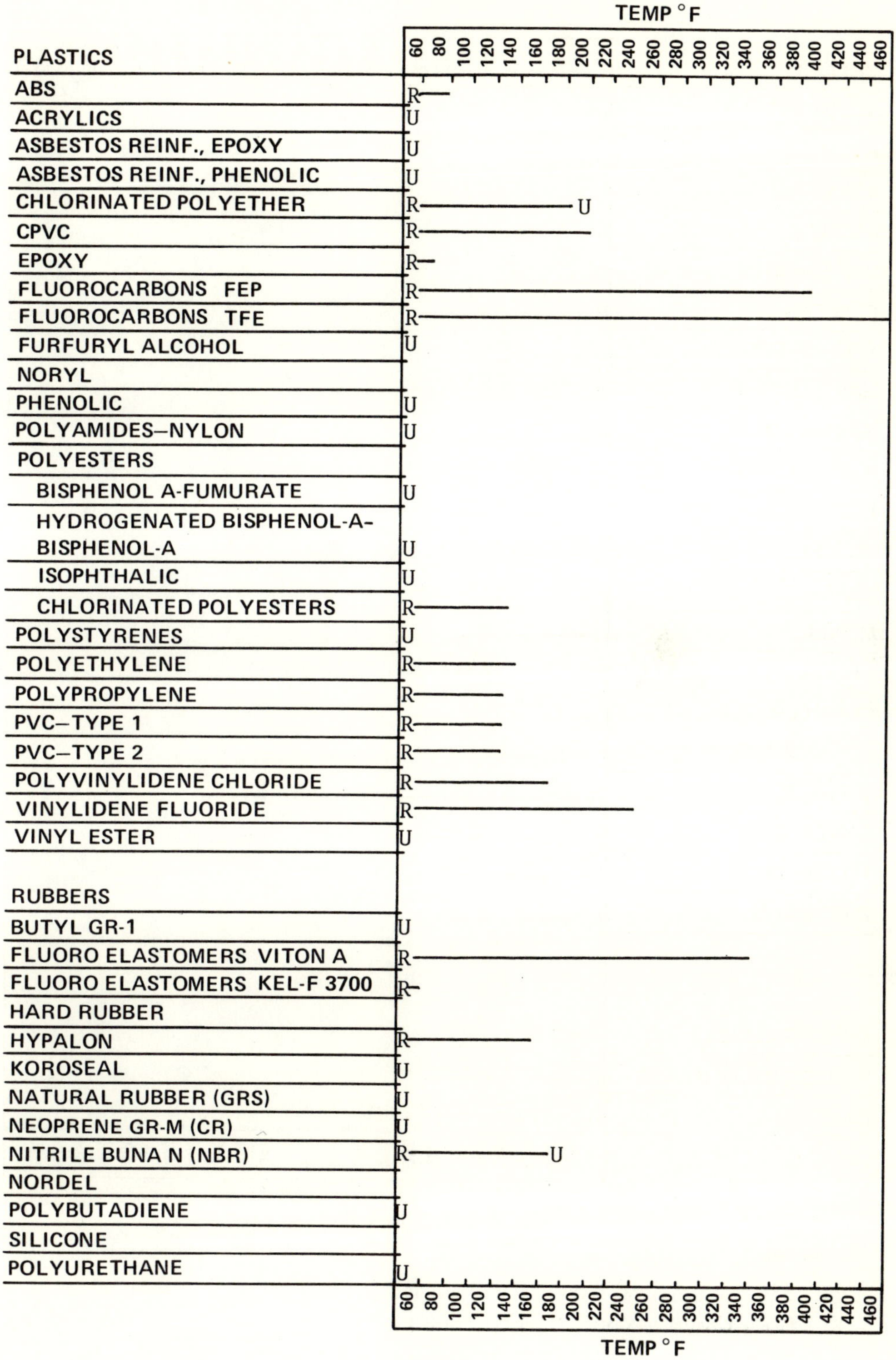

CHROMIC ACID, 10%

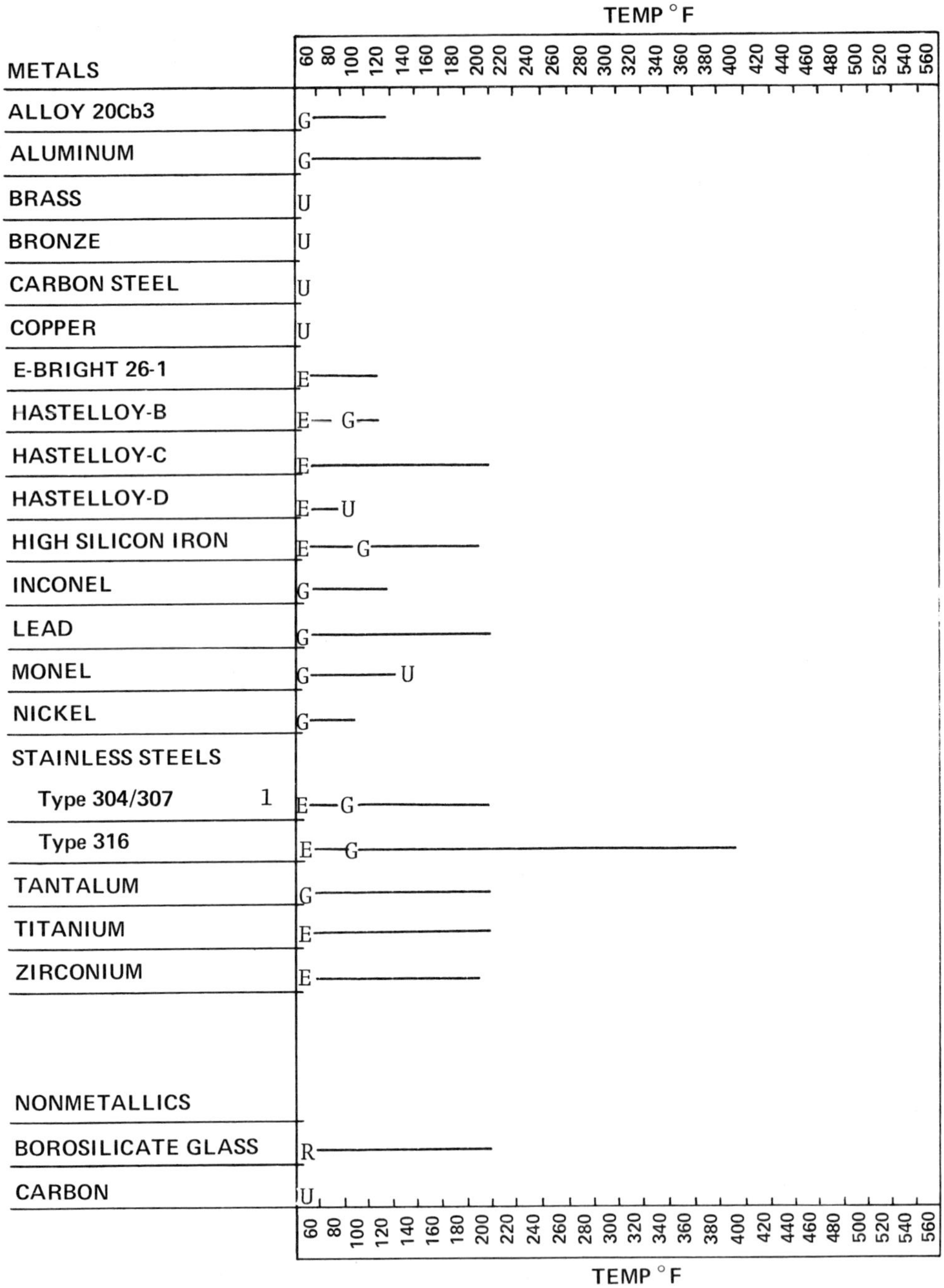

FOR METALS

E = < 2 Mils Penetration/Year; G = < 20 Mils Penetration/Year

S = < 50 Mils Penetration/Year; U = > 50 Mils Penetration/Year

FOR NONMETALLICS

R = Resistant

U = Unsatisfactory

CHROMIC ACID, 10%

TEMP °F (60–460)

PLASTICS	
ABS	R
ACRYLICS	U
ASBESTOS REINF., EPOXY	U
ASBESTOS REINF., PHENOLIC	U
CHLORINATED POLYETHER	R
CPVC	R
EPOXY	R — U
FLUOROCARBONS FEP	R
FLUOROCARBONS TFE	R
FURFURYL ALCOHOL	R
NORYL	
PHENOLIC	U
POLYAMIDES—NYLON	U
POLYESTERS	
BISPHENOL A-FUMURATE	U
HYDROGENATED BISPHENOL-A-BISPHENOL-A	
ISOPHTHALIC	U
CHLORINATED POLYESTERS	R
POLYSTYRENES	U
POLYETHYLENE	R
POLYPROPYLENE	R
PVC—TYPE 1	R
PVC—TYPE 2	R
POLYVINYLIDENE CHLORIDE	R
VINYLIDENE FLUORIDE	R
VINYL ESTER	R

RUBBERS	
BUTYL GR-1	R
FLUORO ELASTOMERS VITON A	R
FLUORO ELASTOMERS KEL-F 3700	R
HARD RUBBER	
HYPALON	R
KOROSEAL	R
NATURAL RUBBER (GRS)	U
NEOPRENE GR-M (CR)	U
NITRILE BUNA N (NBR)	R
NORDEL	
POLYBUTADIENE	U
SILICONE	
POLYURETHANE	U

TEMP °F (60–460)

CHROMYL CHLORIDE

METALS	Rating	TEMP °F
ALLOY 20Cb3	G	60–200
ALUMINUM	G	60–200
BRASS		
BRONZE		
CARBON STEEL		
COPPER		
E-BRIGHT 26-1		
HASTELLOY-B	G	60–200
HASTELLOY-C	G	60–200
HASTELLOY-D		
HIGH SILICON IRON	G	60–200
INCONEL	G	60–200
LEAD		
MONEL	G	60–200
NICKEL	G	60–200
STAINLESS STEELS		
Type 304/307	G	60–200
Type 316	G	60–200
TANTALUM	G	60–200
TITANIUM	G	60
ZIRCONIUM		
NONMETALLICS		
BOROSILICATE GLASS		
CARBON		

FOR METALS

E = < 2 Mils Penetration/Year; G = < 20 Mils Penetration/Year

S = < 50 Mils Penetration/Year; U = > 50 Mils Penetration/Year

FOR NONMETALLICS

R = Resistant

U = Unsatisfactory

TEMP °F (60, 80, 100, 120, 140, 160, 180, 200, 220, 240, 260, 280, 300, 320, 340, 360, 380, 400, 420, 440, 460)

PLASTICS	
ABS	
ACRYLICS	
ASBESTOS REINF., EPOXY	
ASBESTOS REINF., PHENOLIC	
CHLORINATED POLYETHER	
CPVC	
EPOXY	
FLUOROCARBONS FEP	R————————————————
FLUOROCARBONS TFE	R————————————————————
FURFURYL ALCOHOL	
NORYL	
PHENOLIC	
POLYAMIDES–NYLON	
POLYESTERS	
BISPHENOL A-FUMURATE	
HYDROGENATED BISPHENOL-A-BISPHENOL-A	
ISOPHTHALIC	
CHLORINATED POLYESTERS	
POLYSTYRENES	
POLYETHYLENE	
POLYPROPYLENE	R————
PVC–TYPE 1	
PVC–TYPE 2	
POLYVINYLIDENE CHLORIDE	
VINYLIDENE FLUORIDE	R———U
VINYL ESTER	

RUBBERS	
BUTYL GR-1	
FLUORO ELASTOMERS VITON A	
FLUORO ELASTOMERS KEL-F 3700	
HARD RUBBER	
HYPALON	
KOROSEAL	
NATURAL RUBBER (GRS)	
NEOPRENE GR-M (CR)	
NITRILE BUNA N (NBR)	
NORDEL	
POLYBUTADIENE	
SILICONE	
POLYURETHANE	

TEMP °F (60, 80, 100, 120, 140, 160, 180, 200, 220, 240, 260, 280, 300, 320, 340, 360, 380, 400, 420, 440, 460)

CITRIC ACID

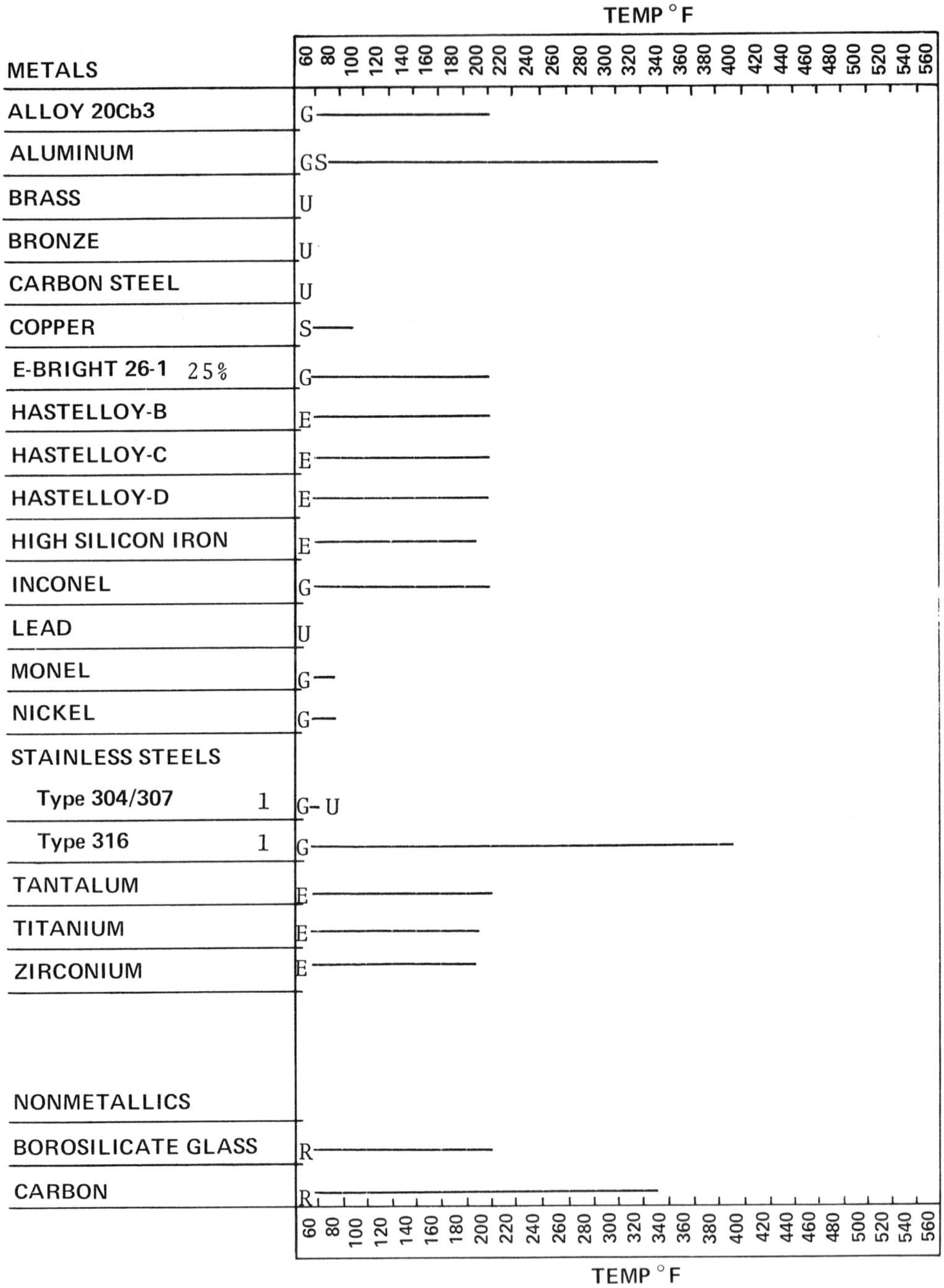

FOR METALS

E = < 2 Mils Penetration/Year; G = < 20 Mils Penetration/Year

S = < 50 Mils Penetration/Year; U = > 50 Mils Penetration/Year

FOR NONMETALLICS

R = Resistant

U = Unsatisfactory

CITRIC ACID

TEMP °F: 60 80 100 120 140 160 180 200 220 240 260 280 300 320 340 360 380 400 420 440 460

PLASTICS		
ABS	25%	R
ACRYLICS		R
ASBESTOS REINF., EPOXY		R
ASBESTOS REINF., PHENOLIC		R
CHLORINATED POLYETHER		R
CPVC		R
EPOXY	32%	R U
FLUOROCARBONS FEP		R
FLUOROCARBONS TFE		R
FURFURYL ALCOHOL		R
NORYL		
PHENOLIC		
POLYAMIDES–NYLON		R
POLYESTERS		
BISPHENOL A-FUMURATE	sat	R
HYDROGENATED BISPHENOL-A-BISPHENOL-A		R
ISOPHTHALIC	sat	R
CHLORINATED POLYESTERS		R
POLYSTYRENES		R
POLYETHYLENE		R
POLYPROPYLENE		R
PVC–TYPE 1		R
PVC–TYPE 2		R
POLYVINYLIDENE CHLORIDE		R
VINYLIDENE FLUORIDE		R
VINYL ESTER	sat	R

RUBBERS		
BUTYL GR-1		R
FLUORO ELASTOMERS VITON A		R
FLUORO ELASTOMERS KEL-F 3700		
HARD RUBBER		
HYPALON		R
KOROSEAL		R
NATURAL RUBBER (GRS)		
NEOPRENE GR-M (CR)		R
NITRILE BUNA N (NBR)		R
NORDEL		R
POLYBUTADIENE		
SILICONE		
POLYURETHANE		

TEMP °F: 60 80 100 120 140 160 180 200 220 240 260 280 300 320 340 360 380 400 420 440 460

CITRIC ACID, 10%

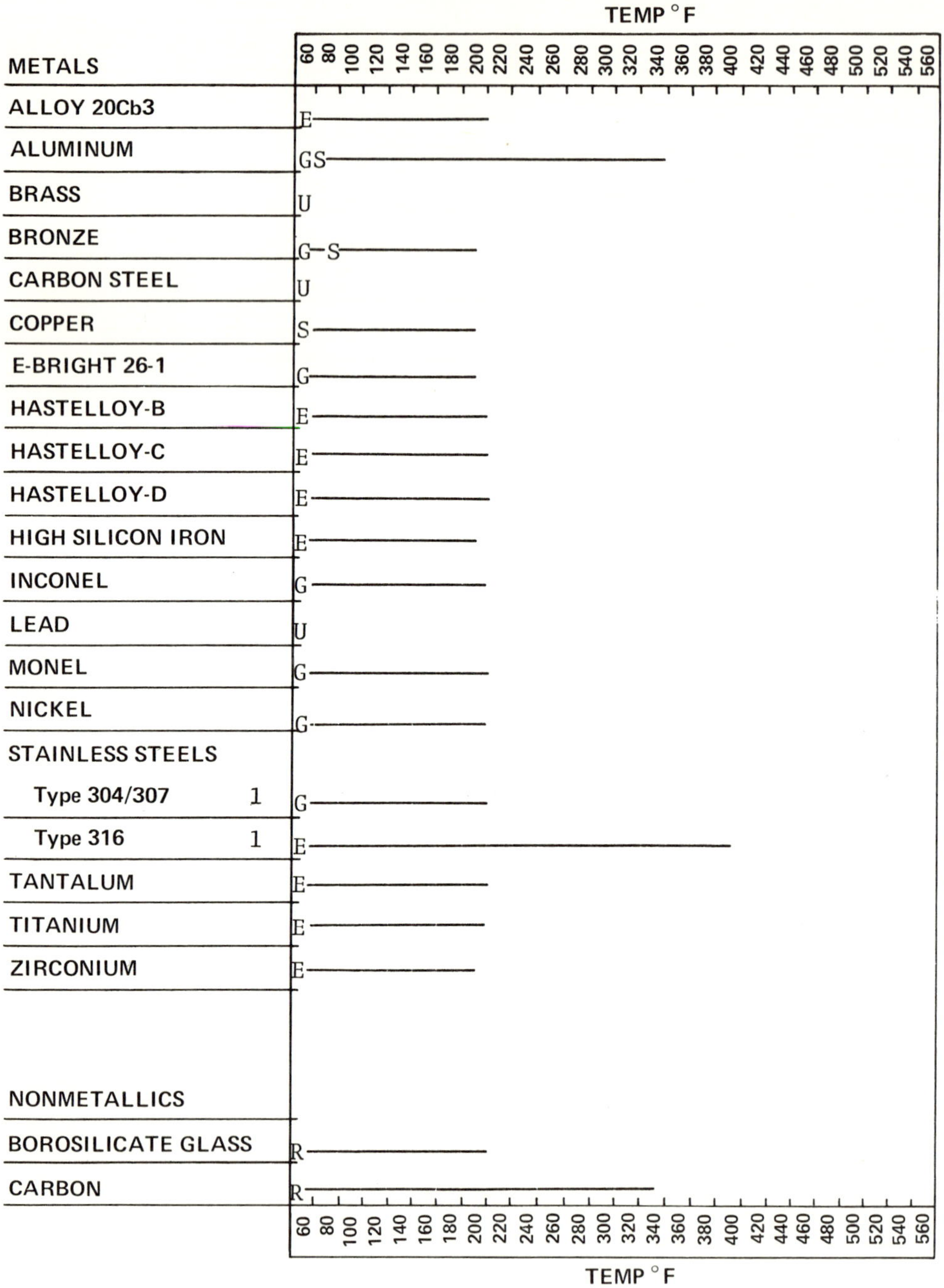

FOR METALS

E = < 2 Mils Penetration/Year; G = < 20 Mils Penetration/Year

S = < 50 Mils Penetration/Year; U = > 50 Mils Penetration/Year

FOR NONMETALLICS

R = Resistant

U = Unsatisfactory

CITRIC ACID, 10%

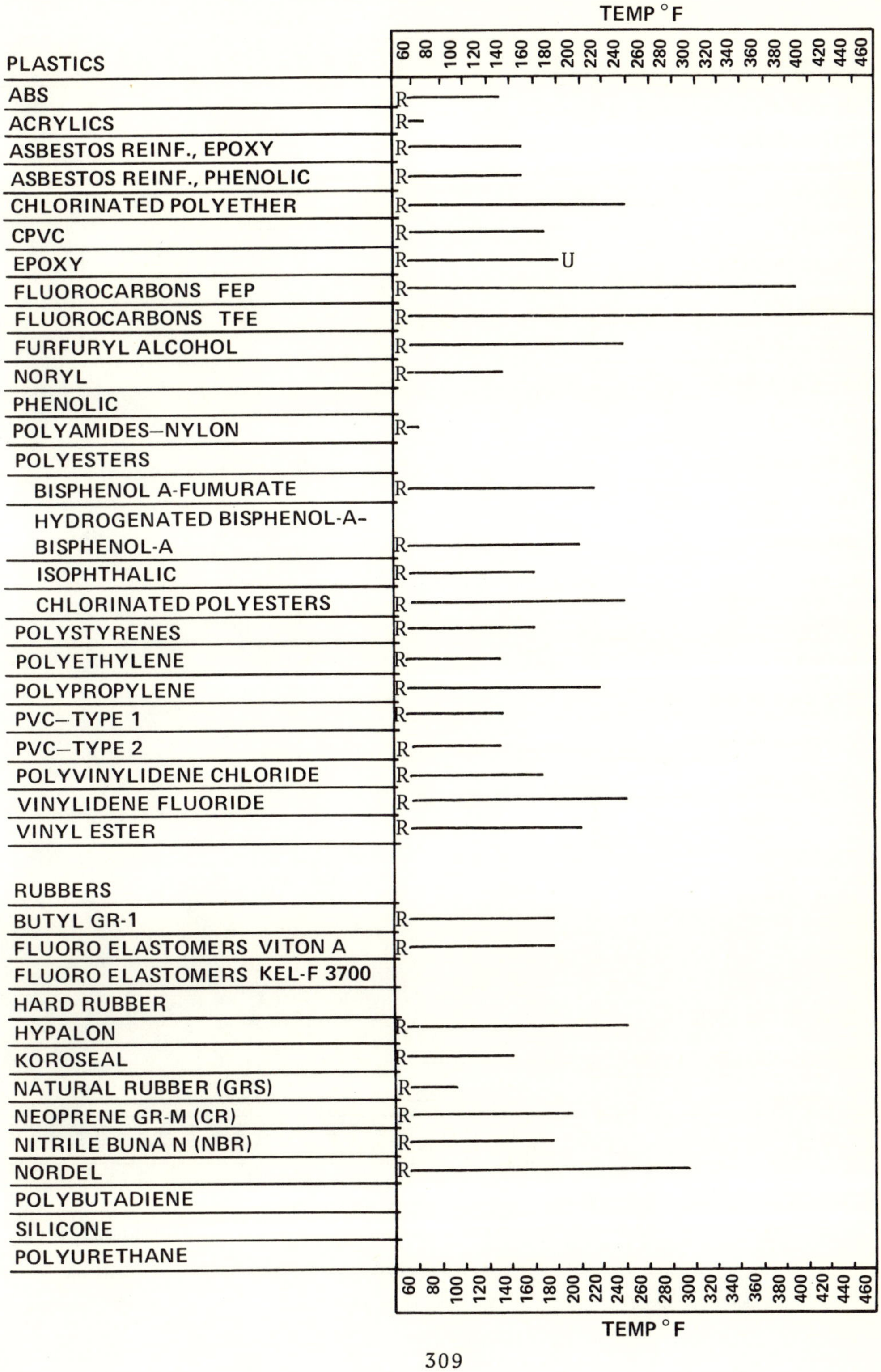

COCONUT OIL

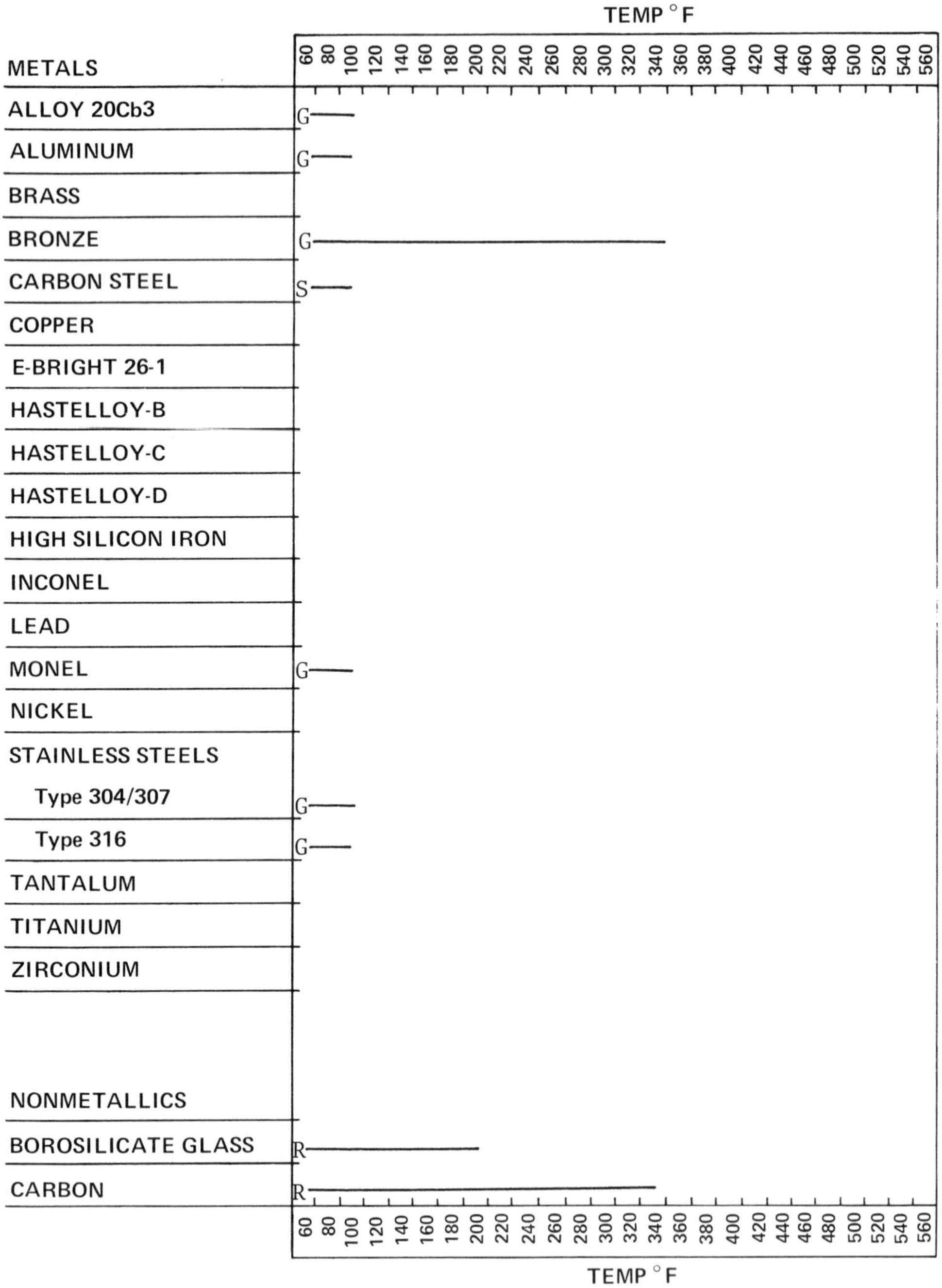

FOR METALS

E = < 2 Mils Penetration/Year; G = < 20 Mils Penetration/Year

S = < 50 Mils Penetration/Year; U = > 50 Mils Penetration/Year

FOR NONMETALLICS

R = Resistant

U = Unsatisfactory

COCONUT OIL

TEMP °F: 60, 80, 100, 120, 140, 160, 180, 200, 220, 240, 260, 280, 300, 320, 340, 360, 380, 400, 420, 440, 460

PLASTICS	
ABS	R
ACRYLICS	
ASBESTOS REINF., EPOXY	
ASBESTOS REINF., PHENOLIC	
CHLORINATED POLYETHER	R
CPVC	R
EPOXY	
FLUOROCARBONS FEP	R
FLUOROCARBONS TFE	R
FURFURYL ALCOHOL	
NORYL	
PHENOLIC	
POLYAMIDES–NYLON	
POLYESTERS	
BISPHENOL A-FUMURATE	R
HYDROGENATED BISPHENOL-A-BISPHENOL-A	
ISOPHTHALIC	R
CHLORINATED POLYESTERS	
POLYSTYRENES	
POLYETHYLENE	R
POLYPROPYLENE	R
PVC–TYPE 1	R
PVC–TYPE 2	R
POLYVINYLIDENE CHLORIDE	R
VINYLIDENE FLUORIDE	R
VINYL ESTER	R

RUBBERS	
BUTYL GR-1	R
FLUORO ELASTOMERS VITON A	R
FLUORO ELASTOMERS KEL-F 3700	
HARD RUBBER	U
HYPALON	
KOROSEAL	
NATURAL RUBBER (GRS)	U
NEOPRENE GR-M (CR)	
NITRILE BUNA N (NBR)	R
NORDEL	
POLYBUTADIENE	
SILICONE	
POLYURETHANE	

TEMP °F: 60, 80, 100, 120, 140, 160, 180, 200, 220, 240, 260, 280, 300, 320, 340, 360, 380, 400, 420, 440, 460

COAL GAS

TEMP °F

METALS	60	80	100	120	140	160	180	200	220	240	260	280	300	320	340	360	380	400	420	440	460	480	500	520	540	560
ALLOY 20Cb3																										
ALUMINUM																										
BRASS																										
BRONZE																										
CARBON STEEL																										
COPPER																										
E-BRIGHT 26-1																										
HASTELLOY-B																										
HASTELLOY-C																										
HASTELLOY-D																										
HIGH SILICON IRON																										
INCONEL																										
LEAD																										
MONEL																										
NICKEL																										
STAINLESS STEELS Type 304/307																										
Type 316																										
TANTALUM																										
TITANIUM																										
ZIRCONIUM																										
NONMETALLICS																										
BOROSILICATE GLASS																										
CARBON																										
	60	80	100	120	140	160	180	200	220	240	260	280	300	320	340	360	380	400	420	440	460	480	500	520	540	560

TEMP °F

FOR METALS

E = < 2 Mils Penetration/Year; G = < 20 Mils Penetration/Year

S = < 50 Mils Penetration/Year; U = > 50 Mils Penetration/Year

FOR NONMETALLICS

R = Resistant

U = Unsatisfactory

COAL GAS

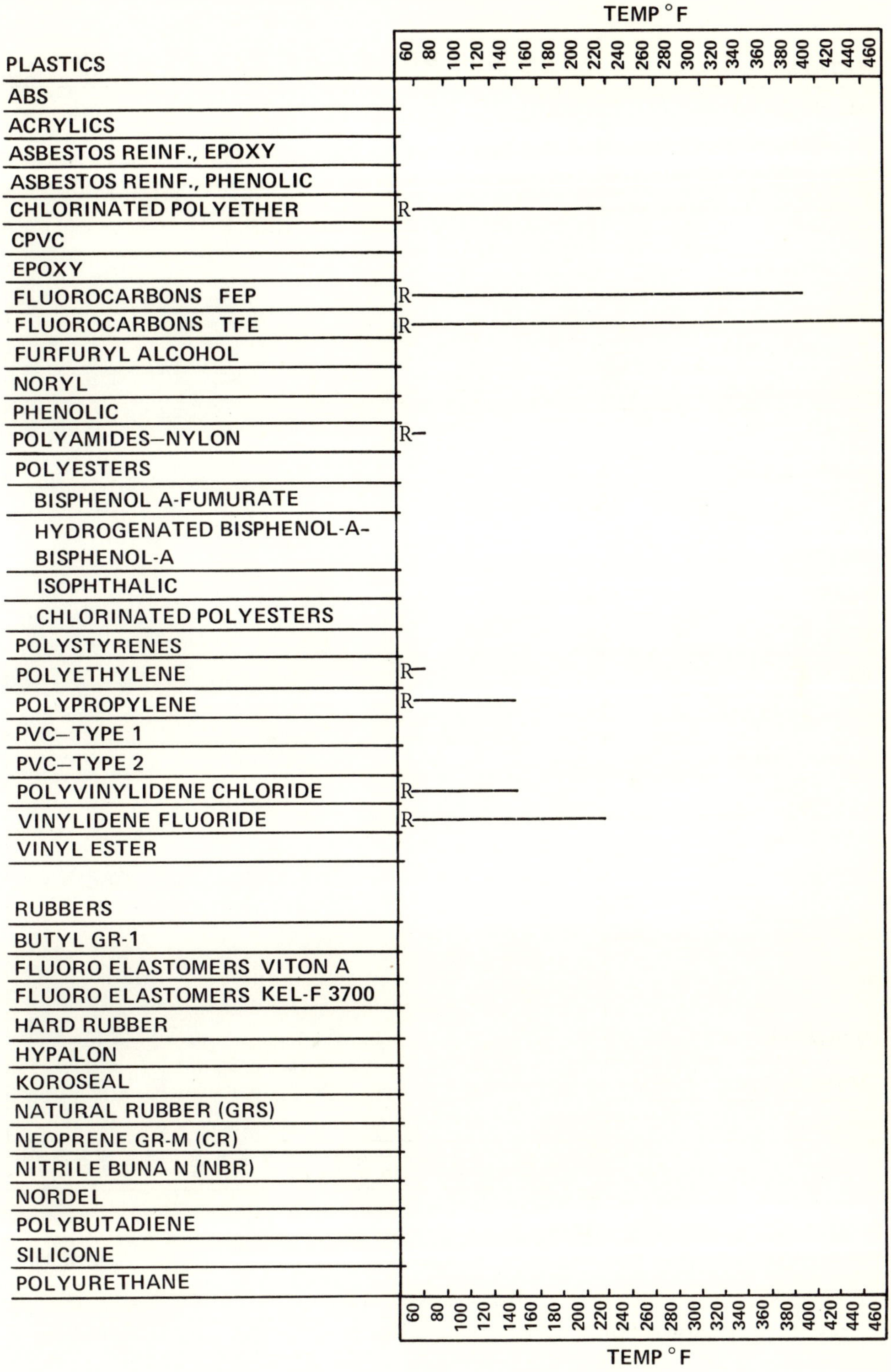
TEMP °F
60 80 100 120 140 160 180 200 220 240 260 280 300 320 340 360 380 400 420 440 460
PLASTICS
ABS
ACRYLICS
ASBESTOS REINF., EPOXY
ASBESTOS REINF., PHENOLIC
CHLORINATED POLYETHER
R
CPVC
EPOXY
FLUOROCARBONS FEP
R
FLUOROCARBONS TFE
R
FURFURYL ALCOHOL
NORYL
PHENOLIC
POLYAMIDES–NYLON
R
POLYESTERS
BISPHENOL A-FUMURATE
HYDROGENATED BISPHENOL-A-BISPHENOL-A
ISOPHTHALIC
CHLORINATED POLYESTERS
POLYSTYRENES
POLYETHYLENE
R
POLYPROPYLENE
R
PVC–TYPE 1
PVC–TYPE 2
POLYVINYLIDENE CHLORIDE
R
VINYLIDENE FLUORIDE
R
VINYL ESTER
RUBBERS
BUTYL GR-1
FLUORO ELASTOMERS VITON A
FLUORO ELASTOMERS KEL-F 3700
HARD RUBBER
HYPALON
KOROSEAL
NATURAL RUBBER (GRS)
NEOPRENE GR-M (CR)
NITRILE BUNA N (NBR)
NORDEL
POLYBUTADIENE
SILICONE
POLYURETHANE
60 80 100 120 140 160 180 200 220 240 260 280 300 320 340 360 380 400 420 440 460
TEMP °F

COKE OVEN GAS

TEMP °F

METALS	Rating (range, °F)
ALLOY 20Cb3	E (60–100)
ALUMINUM	G (60–100)
BRASS	S (60–80)
BRONZE	S (60–340)
CARBON STEEL	G (60–100)
COPPER	S (60–80)
E-BRIGHT 26-1	
HASTELLOY-B	
HASTELLOY-C	
HASTELLOY-D	
HIGH SILICON IRON	
INCONEL	
LEAD	
MONEL	G (60–100)
NICKEL	
STAINLESS STEELS	
Type 304/307	G (60–100)
Type 316	G (60–100)
TANTALUM	
TITANIUM	
ZIRCONIUM	
NONMETALLICS	
BOROSILICATE GLASS	
CARBON	

TEMP °F: 60 80 100 120 140 160 180 200 220 240 260 280 300 320 340 360 380 400 420 440 460 480 500 520 540 560

FOR METALS

E = < 2 Mils Penetration/Year; G = < 20 Mils Penetration/Year

S = < 50 Mils Penetration/Year; U = > 50 Mils Penetration/Year

FOR NONMETALLICS

R = Resistant

U = Unsatisfactory

COKE OVEN GAS

PLASTICS		TEMP °F
ABS		
ACRYLICS		
ASBESTOS REINF., EPOXY		
ASBESTOS REINF., PHENOLIC		
CHLORINATED POLYETHER	R	60–220
CPVC		
EPOXY		
FLUOROCARBONS FEP	R	60–400
FLUOROCARBONS TFE	R	60–460
FURFURYL ALCOHOL		
NORYL		
PHENOLIC		
POLYAMIDES—NYLON		
POLYESTERS		
BISPHENOL A-FUMURATE		
HYDROGENATED BISPHENOL-A-BISPHENOL-A		
ISOPHTHALIC		
CHLORINATED POLYESTERS		
POLYSTYRENES		
POLYETHYLENE		
POLYPROPYLENE	R	60–70
PVC—TYPE 1	R	60–140
PVC—TYPE 2		
POLYVINYLIDENE CHLORIDE		
VINYLIDENE FLUORIDE		
VINYL ESTER		

RUBBERS		TEMP °F
BUTYL GR-1		
FLUORO ELASTOMERS VITON A	R	60–180
FLUORO ELASTOMERS KEL-F 3700		
HARD RUBBER		
HYPALON		
KOROSEAL		
NATURAL RUBBER (GRS)		
NEOPRENE GR-M (CR)	R	60–200
NITRILE BUNA N (NBR)	U	
NORDEL	R	60–300
POLYBUTADIENE		
SILICONE		
POLYURETHANE		

COPPER ACETATE

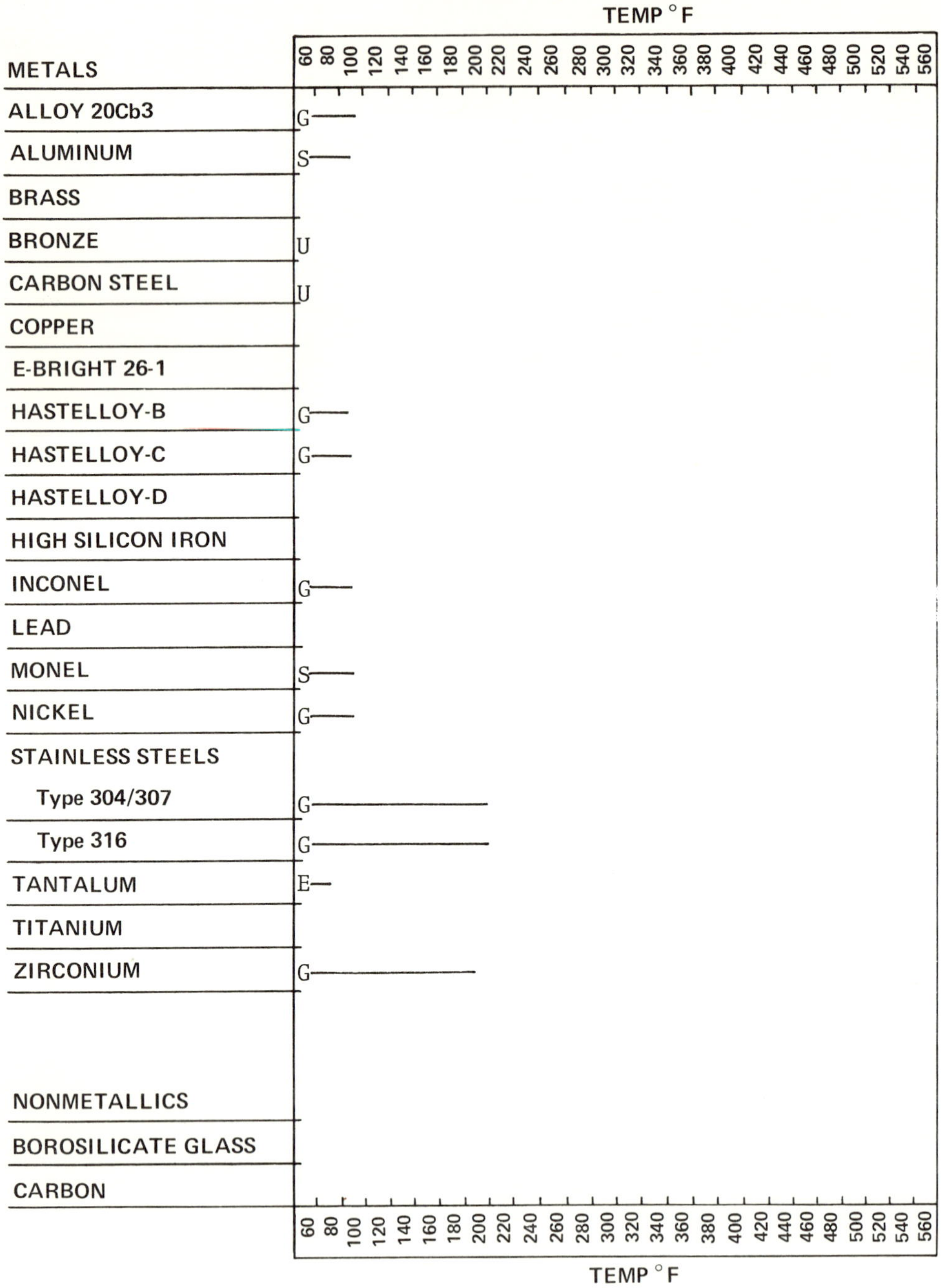

FOR METALS

E = < 2 Mils Penetration/Year; G = < 20 Mils Penetration/Year

S = < 50 Mils Penetration/Year; U = > 50 Mils Penetration/Year

FOR NONMETALLICS

R = Resistant

U = Unsatisfactory

COPPER ACETATE

PLASTICS		TEMP °F (60–460)
ABS		
ACRYLICS		
ASBESTOS REINF., EPOXY		
ASBESTOS REINF., PHENOLIC		
CHLORINATED POLYETHER		
CPVC		
EPOXY		R 60–200
FLUOROCARBONS FEP		R 60–400
FLUOROCARBONS TFE		R 60–460
FURFURYL ALCOHOL	sat	R 60–260
NORYL		
PHENOLIC		
POLYAMIDES–NYLON		
POLYESTERS		
BISPHENOL A-FUMURATE		R 60–180
HYDROGENATED BISPHENOL-A-BISPHENOL-A		R 60–210
ISOPHTHALIC	sat	R 60–160
CHLORINATED POLYESTERS		
POLYSTYRENES		
POLYETHYLENE		
POLYPROPYLENE		
PVC–TYPE 1		
PVC–TYPE 2		
POLYVINYLIDENE CHLORIDE		
VINYLIDENE FLUORIDE		
VINYL ESTER		

RUBBERS	TEMP °F (60–460)
BUTYL GR-1	
FLUORO ELASTOMERS VITON A	
FLUORO ELASTOMERS KEL-F 3700	
HARD RUBBER	
HYPALON	
KOROSEAL	
NATURAL RUBBER (GRS)	
NEOPRENE GR-M (CR)	
NITRILE BUNA N (NBR)	
NORDEL	
POLYBUTADIENE	
SILICONE	
POLYURETHANE	

COPPER CARBONATE

TEMP °F (60–560)

METALS	
ALLOY 20Cb3	
ALUMINUM	U
BRASS	
BRONZE	
CARBON STEEL	
COPPER	
E-BRIGHT 26-1	
HASTELLOY-B	
HASTELLOY-C	
HASTELLOY-D	
HIGH SILICON IRON	
INCONEL	G—
LEAD	
MONEL	U
NICKEL	U
STAINLESS STEELS	
Type 304/307 10%	G—
Type 316 10%	G—
TANTALUM	E—
TITANIUM	E—
ZIRCONIUM	

NONMETALLICS	
BOROSILICATE GLASS	
CARBON	R (60–340)

TEMP °F

FOR METALS

E = < 2 Mils Penetration/Year; G = < 20 Mils Penetration/Year

S = < 50 Mils Penetration/Year; U = > 50 Mils Penetration/Year

FOR NONMETALLICS

R = Resistant

U = Unsatisfactory

COPPER CARBONATE

TEMP °F: 60 80 100 120 140 160 180 200 220 240 260 280 300 320 340 360 380 400 420 440 460

PLASTICS	
ABS	
ACRYLICS	
ASBESTOS REINF., EPOXY	
ASBESTOS REINF., PHENOLIC	
CHLORINATED POLYETHER	R
CPVC	R
EPOXY	
FLUOROCARBONS FEP	R
FLUOROCARBONS TFE	R
FURFURYL ALCOHOL	
NORYL	
PHENOLIC	
POLYAMIDES–NYLON	
POLYESTERS	
BISPHENOL A-FUMURATE	
HYDROGENATED BISPHENOL-A-BISPHENOL-A	
ISOPHTHALIC	
CHLORINATED POLYESTERS	
POLYSTYRENES	
POLYETHYLENE	R
POLYPROPYLENE	R
PVC–TYPE 1	R
PVC–TYPE 2	R
POLYVINYLIDENE CHLORIDE	R
VINYLIDENE FLUORIDE	
VINYL ESTER	
RUBBERS	
BUTYL GR-1	
FLUORO ELASTOMERS VITON A	R
FLUORO ELASTOMERS KEL-F 3700	
HARD RUBBER	
HYPALON	
KOROSEAL	
NATURAL RUBBER (GRS)	
NEOPRENE GR-M (CR)	
NITRILE BUNA N (NBR)	U
NORDEL	
POLYBUTADIENE	
SILICONE	
POLYURETHANE	

TEMP °F: 60 80 100 120 140 160 180 200 220 240 260 280 300 320 340 360 380 400 420 440 460

COPPER CHLORIDE

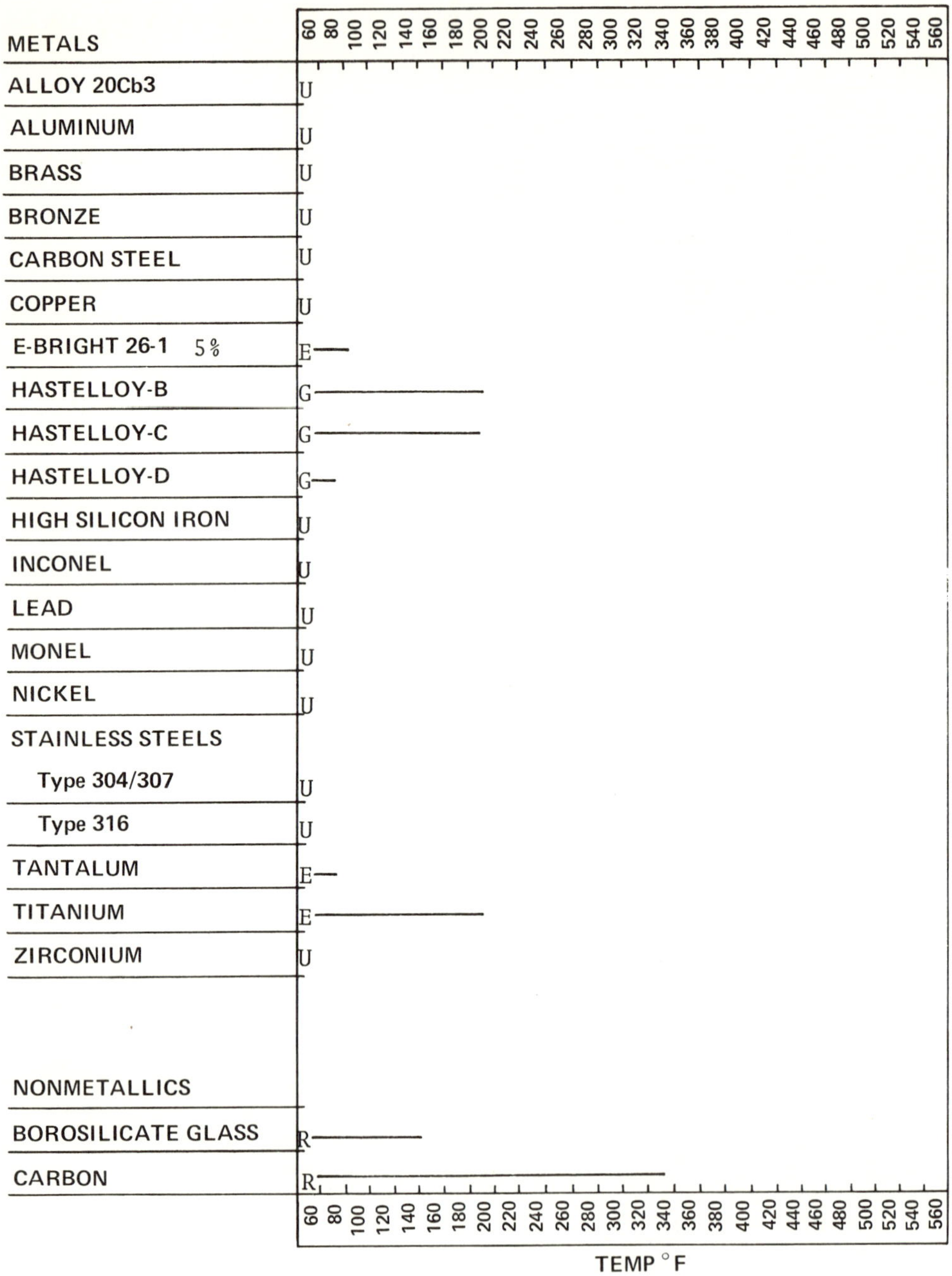

FOR METALS

E = < 2 Mils Penetration/Year; G = < 20 Mils Penetration/Year

S = < 50 Mils Penetration/Year; U = > 50 Mils Penetration/Year

FOR NONMETALLICS

R = Resistant

U = Unsatisfactory

COPPER CHLORIDE

Material	Concentration	Rating
PLASTICS		
ABS		R
ACRYLICS		R
ASBESTOS REINF., EPOXY		R
ASBESTOS REINF., PHENOLIC		
CHLORINATED POLYETHER		R
CPVC		R
EPOXY		R
FLUOROCARBONS FEP		R
FLUOROCARBONS TFE		R
FURFURYL ALCOHOL	sat	R
NORYL		R
PHENOLIC		
POLYAMIDES—NYLON		R
POLYESTERS		
BISPHENOL A-FUMURATE	sat	R
HYDROGENATED BISPHENOL-A-BISPHENOL-A		R
ISOPHTHALIC	satR	R
CHLORINATED POLYESTERS		R
POLYSTYRENES		
POLYETHYLENE		R
POLYPROPYLENE		R
PVC—TYPE 1		R
PVC—TYPE 2		R
POLYVINYLIDENE CHLORIDE		R
VINYLIDENE FLUORIDE		R
VINYL ESTER		R
RUBBERS		
BUTYL GR-1		R
FLUORO ELASTOMERS VITON A		R
FLUORO ELASTOMERS KEL-F 3700		R
HARD RUBBER		R
HYPALON		R
KOROSEAL		R
NATURAL RUBBER (GRS)		R
NEOPRENE GR-M (CR)		R
NITRILE BUNA N (NBR)		R
NORDEL		R
POLYBUTADIENE		R
SILICONE		
POLYURETHANE		R

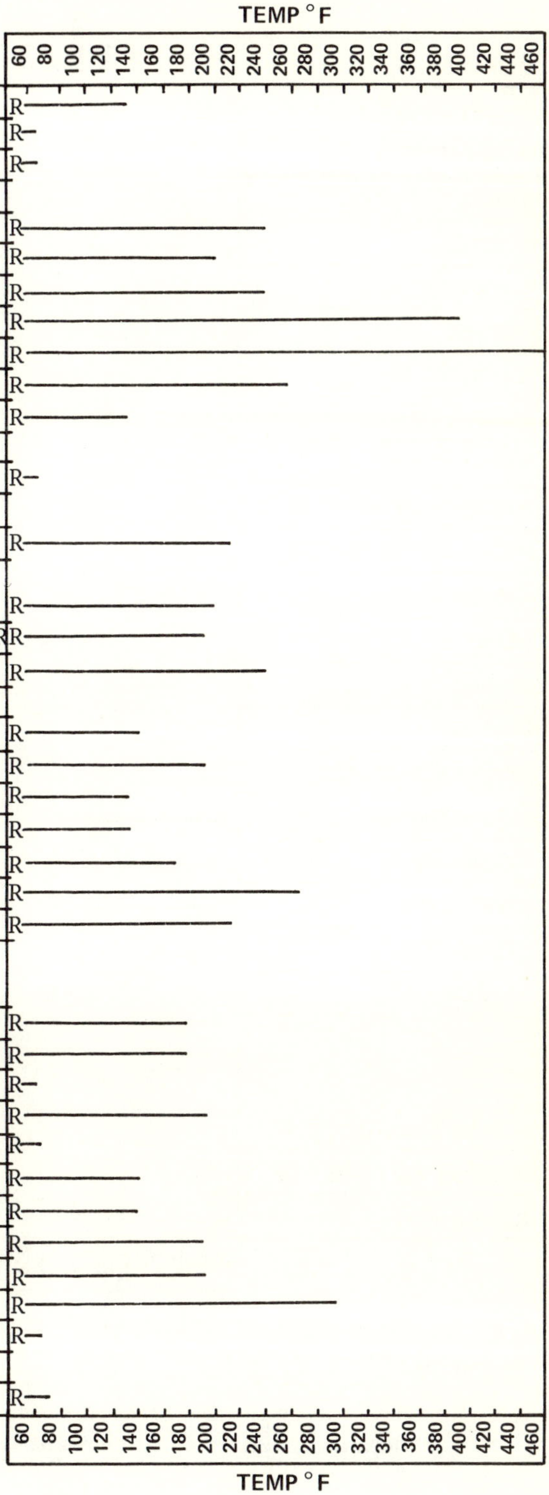

COPPER CYANIDE

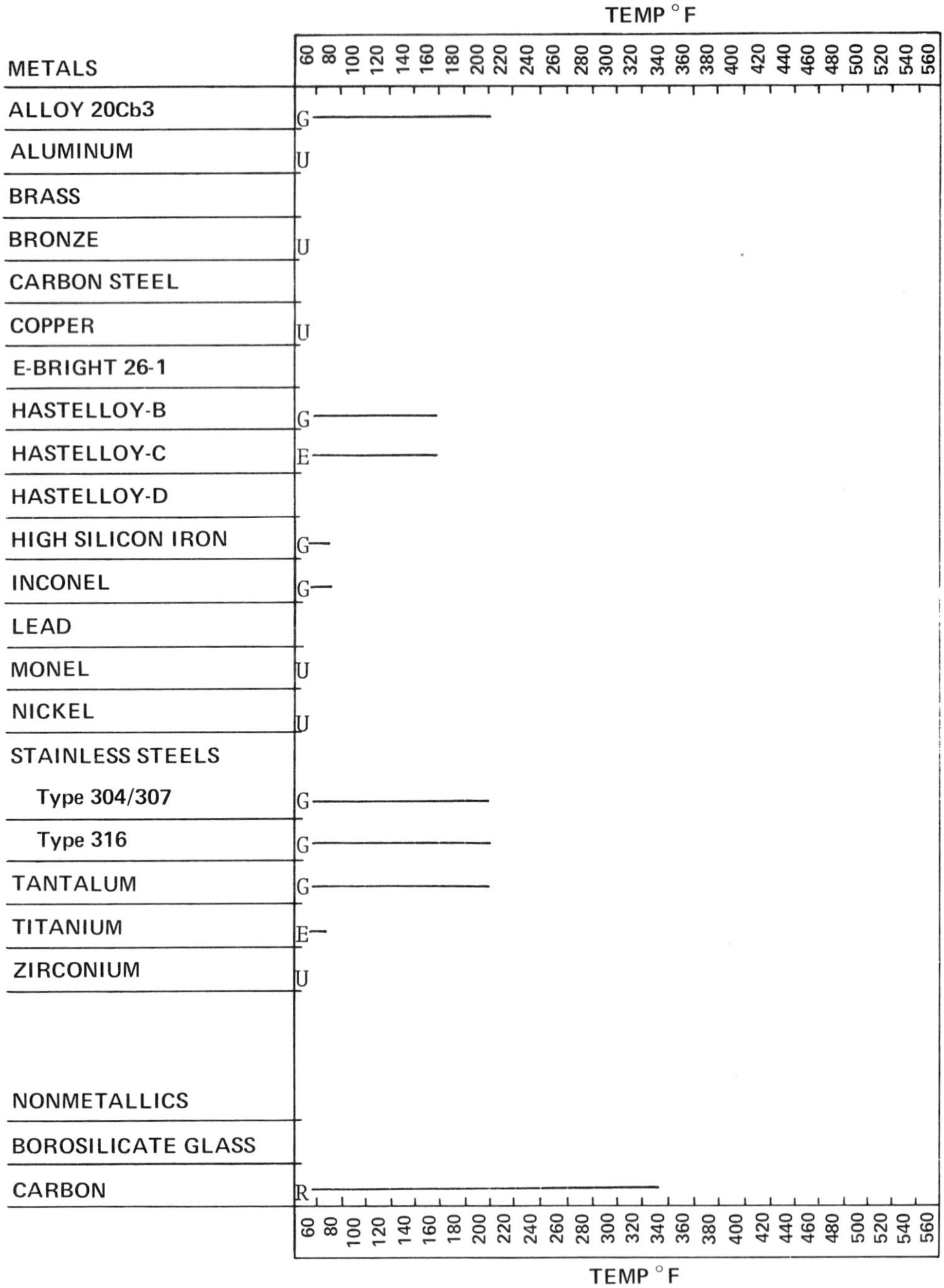

FOR METALS

E = < 2 Mils Penetration/Year; G = < 20 Mils Penetration/Year

S = < 50 Mils Penetration/Year; U = > 50 Mils Penetration/Year

FOR NONMETALLICS

R = Resistant

U = Unsatisfactory

COPPER CYANIDE

TEMP °F: 60 80 100 120 140 160 180 200 220 240 260 280 300 320 340 360 380 400 420 440 460

PLASTICS		
ABS		R
ACRYLICS		
ASBESTOS REINF., EPOXY		
ASBESTOS REINF., PHENOLIC		
CHLORINATED POLYETHER	10%	R
CPVC		R
EPOXY		R—U
FLUOROCARBONS FEP		R
FLUOROCARBONS TFE		R
FURFURYL ALCOHOL		R
NORYL		R
PHENOLIC		
POLYAMIDES–NYLON		
POLYESTERS		
BISPHENOL A-FUMURATE	sat	R
HYDROGENATED BISPHENOL-A-BISPHENOL-A		R
ISOPHTHALIC	sat	R
CHLORINATED POLYESTERS		R
POLYSTYRENES		
POLYETHYLENE		R
POLYPROPYLENE		R
PVC–TYPE 1		R
PVC–TYPE 2		R
POLYVINYLIDENE CHLORIDE		R
VINYLIDENE FLUORIDE		R
VINYL ESTER		R

RUBBERS		
BUTYL GR-1		
FLUORO ELASTOMERS VITON A		R
FLUORO ELASTOMERS KEL-F 3700		
HARD RUBBER		R
HYPALON		R
KOROSEAL		
NATURAL RUBBER (GRS)		
NEOPRENE GR-M (CR)		R
NITRILE BUNA N (NBR)		R
NORDEL		R
POLYBUTADIENE		
SILICONE		
POLYURETHANE		R

TEMP °F: 60 80 100 120 140 160 180 200 220 240 260 280 300 320 340 360 380 400 420 440 460

COPPER FLUORIDE

TEMP °F: 60, 80, 100, 120, 140, 160, 180, 200, 220, 240, 260, 280, 300, 320, 340, 360, 380, 400, 420, 440, 460, 480, 500, 520, 540, 560

METALS	TEMP °F
ALLOY 20Cb3	
ALUMINUM	
BRASS	
BRONZE	
CARBON STEEL	
COPPER	
E-BRIGHT 26-1	
HASTELLOY-B	
HASTELLOY-C	
HASTELLOY-D	
HIGH SILICON IRON	
INCONEL	
LEAD	
MONEL	
NICKEL	
STAINLESS STEELS Type 304/307	
Type 316	
TANTALUM	U (60)
TITANIUM	
ZIRCONIUM	

NONMETALLICS	TEMP °F
BOROSILICATE GLASS	
CARBON	R (60–340)

FOR METALS

E = < 2 Mils Penetration/Year; G = < 20 Mils Penetration/Year

S = < 50 Mils Penetration/Year; U = > 50 Mils Penetration/Year

FOR NONMETALLICS

R = Resistant

U = Unsatisfactory

COPPER FLUORIDE

TEMP °F

60 80 100 120 140 160 180 200 220 240 260 280 300 320 340 360 380 400 420 440 460

PLASTICS		
ABS	2%	R—
ACRYLICS		
ASBESTOS REINF., EPOXY		
ASBESTOS REINF., PHENOLIC		
CHLORINATED POLYETHER		R—
CPVC		R—
EPOXY		R—
FLUOROCARBONS FEP		R—
FLUOROCARBONS TFE		R—
FURFURYL ALCOHOL		
NORYL		
PHENOLIC		
POLYAMIDES—NYLON		
POLYESTERS		
BISPHENOL A-FUMURATE		
HYDROGENATED BISPHENOL-A-BISPHENOL-A		
ISOPHTHALIC		R—U
CHLORINATED POLYESTERS		
POLYSTYRENES		
POLYETHYLENE		R—
POLYPROPYLENE		R—
PVC—TYPE 1		R—
PVC—TYPE 2		R—
POLYVINYLIDENE CHLORIDE		R—
VINYLIDENE FLUORIDE		R—
VINYL ESTER		
RUBBERS		
BUTYL GR-1		
FLUORO ELASTOMERS VITON A		R—
FLUORO ELASTOMERS KEL-F 3700		
HARD RUBBER		
HYPALON		R—
KOROSEAL		
NATURAL RUBBER (GRS)		
NEOPRENE GR-M (CR)		
NITRILE BUNA N (NBR)		R—
NORDEL		
POLYBUTADIENE		
SILICONE		
POLYURETHANE		

60 80 100 120 140 160 180 200 220 240 260 280 300 320 340 360 380 400 420 440 460

TEMP °F

COPPER NITRATE

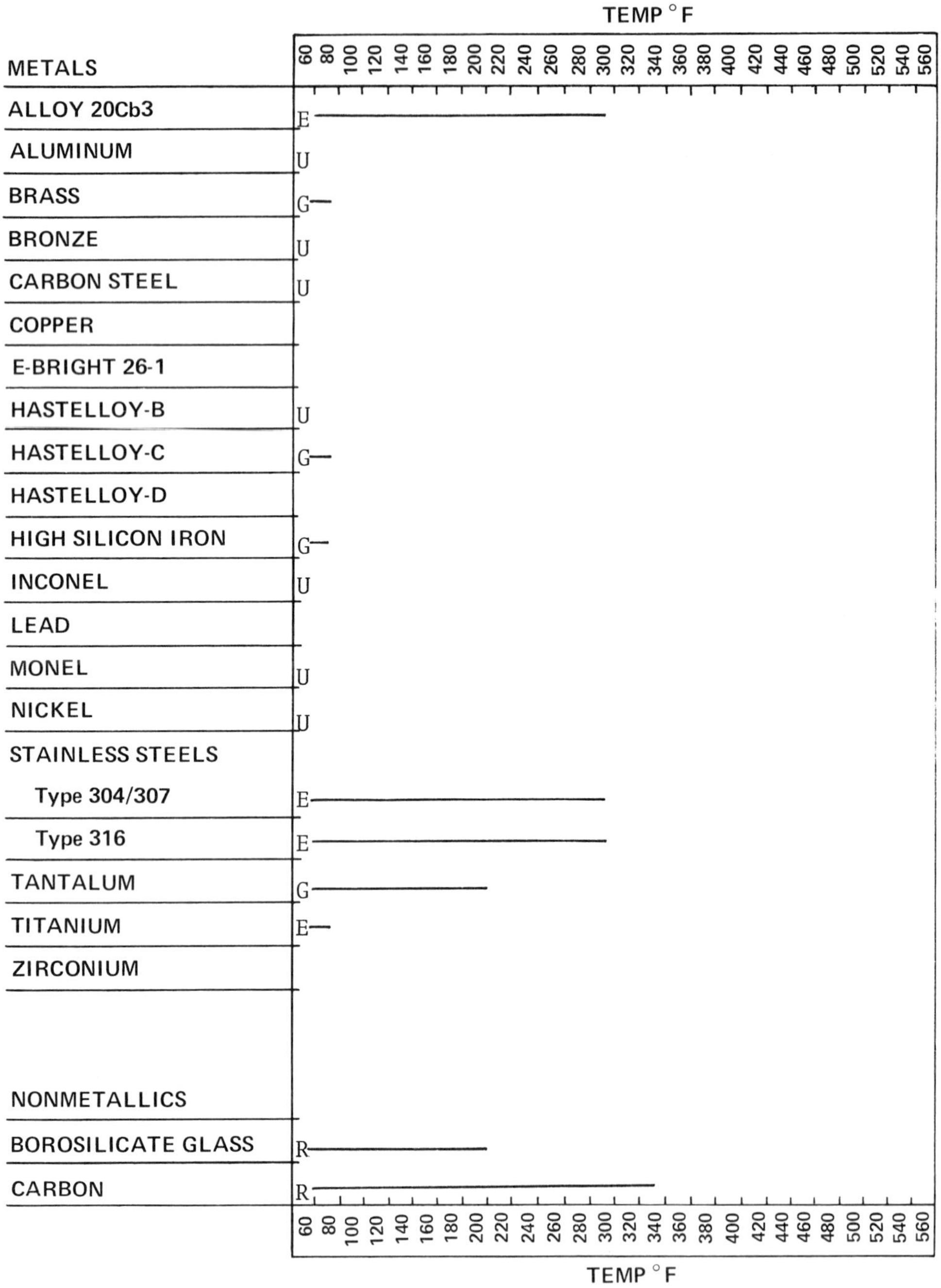

FOR METALS

E = < 2 Mils Penetration/Year; G = < 20 Mils Penetration/Year

S = < 50 Mils Penetration/Year; U = > 50 Mils Penetration/Year

FOR NONMETALLICS

R = Resistant

U = Unsatisfactory

COPPER NITRATE

PLASTICS		TEMP °F
ABS		R to 140
ACRYLICS		
ASBESTOS REINF., EPOXY		
ASBESTOS REINF., PHENOLIC		
CHLORINATED POLYETHER		R to 250
CPVC		R to 185
EPOXY		R to 250
FLUOROCARBONS FEP		R to 400
FLUOROCARBONS TFE		R to 460
FURFURYL ALCOHOL	sat	R to 270
NORYL		R to 140
PHENOLIC		
POLYAMIDES—NYLON		
POLYESTERS		
BISPHENOL A-FUMURATE	sat	R to 220
HYDROGENATED BISPHENOL-A-BISPHENOL-A		R to 210
ISOPHTHALIC	sat	R to 160
CHLORINATED POLYESTERS	sat	R to 140
POLYSTYRENES		
POLYETHYLENE		R to 150
POLYPROPYLENE		R to 200
PVC—TYPE 1		R to 140
PVC—TYPE 2		R to 140
POLYVINYLIDENE CHLORIDE		R to 180
VINYLIDENE FLUORIDE		R to 280
VINYL ESTER		R to 210

RUBBERS	TEMP °F
BUTYL GR-1	R to 185
FLUORO ELASTOMERS VITON A	R to 185
FLUORO ELASTOMERS KEL-F 3700	
HARD RUBBER	R to 185
HYPALON	
KOROSEAL	R to 150
NATURAL RUBBER (GRS)	R to 150
NEOPRENE GR-M (CR)	R to 200
NITRILE BUNA N (NBR)	R to 200
NORDEL	
POLYBUTADIENE	
SILICONE	
POLYURETHANE	

COPPER SULFATE

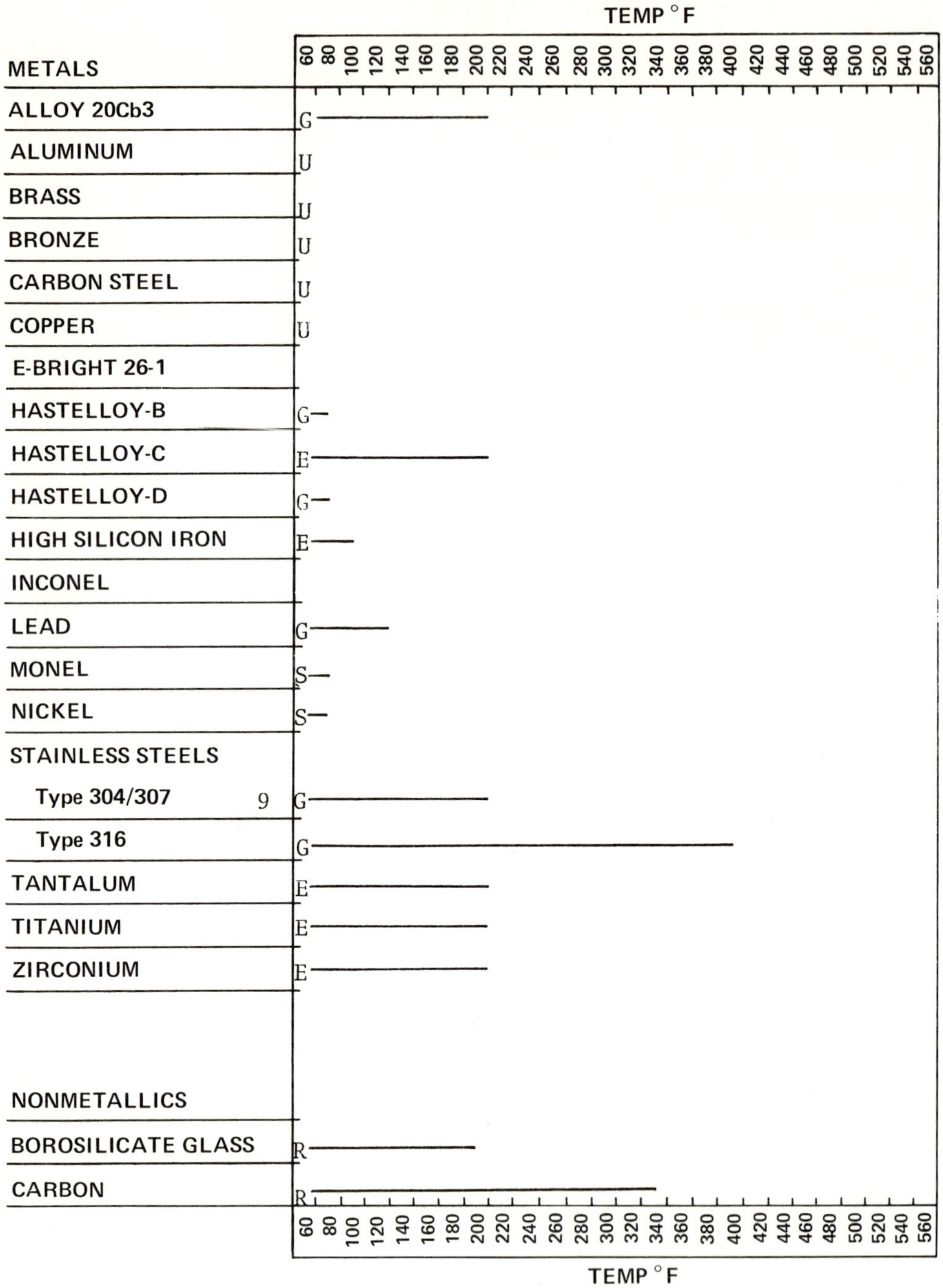

FOR METALS

E = < 2 Mils Penetration/Year; G = < 20 Mils Penetration/Year

S = < 50 Mils Penetration/Year; U = > 50 Mils Penetration/Year

FOR NONMETALLICS

R = Resistant

U = Unsatisfactory

COPPER SULFATE

TEMP °F (60–460)

PLASTICS		
ABS		R
ACRYLICS		U
ASBESTOS REINF., EPOXY		R
ASBESTOS REINF., PHENOLIC		R
CHLORINATED POLYETHER		R
CPVC		R
EPOXY	17%	R U
FLUOROCARBONS FEP		R
FLUOROCARBONS TFE		R
FURFURYL ALCOHOL	sat	R
NORYL		R
PHENOLIC		R
POLYAMIDES–NYLON		U
POLYESTERS		
BISPHENOL A-FUMURATE	sat	R
HYDROGENATED BISPHENOL-A-BISPHENOL-A		R
ISOPHTHALIC	sat	R
CHLORINATED POLYESTERS		R
POLYSTYRENES		R
POLYETHYLENE		R
POLYPROPYLENE		R
PVC–TYPE 1		R
PVC–TYPE 2		R
POLYVINYLIDENE CHLORIDE		R
VINYLIDENE FLUORIDE		R
VINYL ESTER	sat	R
RUBBERS		
BUTYL GR-1		R
FLUORO ELASTOMERS VITON A		R
FLUORO ELASTOMERS KEL-F 3700		R
HARD RUBBER		R
HYPALON		R
KOROSEAL		R
NATURAL RUBBER (GRS)		R
NEOPRENE GR-M (CR)		R
NITRILE BUNA N (NBR)		R
NORDEL		R
POLYBUTADIENE		R
SILICONE		
POLYURETHANE		R

TEMP °F (60–460)

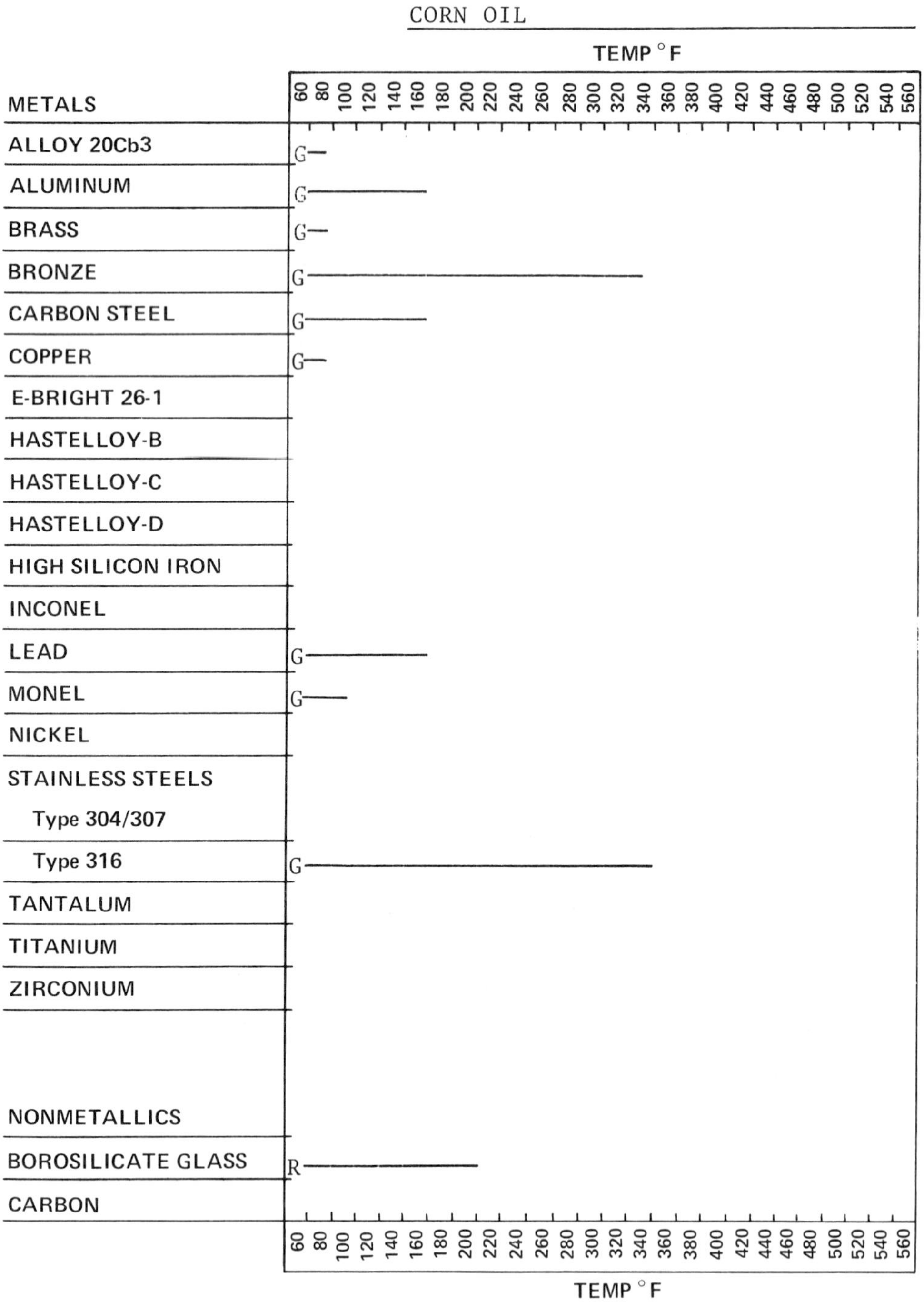

FOR METALS

E = < 2 Mils Penetration/Year; G = < 20 Mils Penetration/Year

S = < 50 Mils Penetration/Year; U = > 50 Mils Penetration/Year

FOR NONMETALLICS

R = Resistant

U = Unsatisfactory

CORN OIL

TEMP °F (60–460)

PLASTICS	60 – 460 °F
ABS	R
ACRYLICS	
ASBESTOS REINF., EPOXY	
ASBESTOS REINF., PHENOLIC	
CHLORINATED POLYETHER	R
CPVC	
EPOXY	
FLUOROCARBONS FEP	R
FLUOROCARBONS TFE	R
FURFURYL ALCOHOL	
NORYL	
PHENOLIC	
POLYAMIDES—NYLON	R
POLYESTERS	
BISPHENOL A-FUMURATE	R
HYDROGENATED BISPHENOL-A-BISPHENOL-A	
ISOPHTHALIC	
CHLORINATED POLYESTERS	
POLYSTYRENES	
POLYETHYLENE	R
POLYPROPYLENE	R
PVC—TYPE 1	R
PVC—TYPE 2	
POLYVINYLIDENE CHLORIDE	R
VINYLIDENE FLUORIDE	R
VINYL ESTER	R
RUBBERS	
BUTYL GR-1	R
FLUORO ELASTOMERS VITON A	R
FLUORO ELASTOMERS KEL-F 3700	
HARD RUBBER	
HYPALON	R
KOROSEAL	
NATURAL RUBBER (GRS)	U
NEOPRENE GR-M (CR)	R
NITRILE BUNA N (NBR)	U
NORDEL	
POLYBUTADIENE	
SILICONE	
POLYURETHANE	

TEMP °F: 60 80 100 120 140 160 180 200 220 240 260 280 300 320 340 360 380 400 420 440 460

COTTONSEED OIL

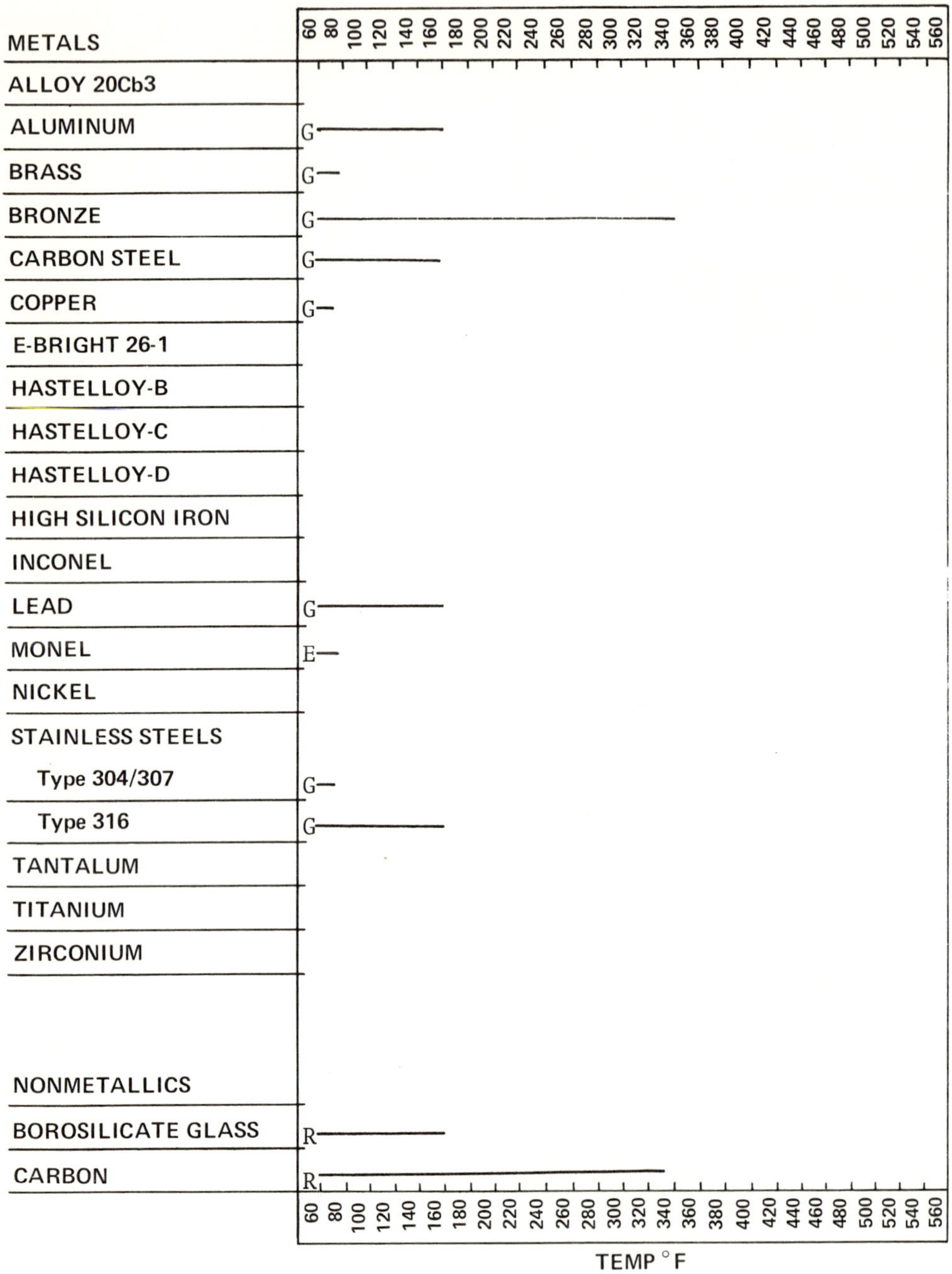

FOR METALS

E = < 2 Mils Penetration/Year; G = < 20 Mils Penetration/Year

S = < 50 Mils Penetration/Year; U = > 50 Mils Penetration/Year

FOR NONMETALLICS

R = Resistant

U = Unsatisfactory

COTTONSEED OIL

PLASTICS	TEMP °F
ABS	R
ACRYLICS	R
ASBESTOS REINF., EPOXY	
ASBESTOS REINF., PHENOLIC	
CHLORINATED POLYETHER	R
CPVC	R
EPOXY	
FLUOROCARBONS FEP	R
FLUOROCARBONS TFE	R
FURFURYL ALCOHOL	
NORYL	
PHENOLIC	
POLYAMIDES—NYLON	R
POLYESTERS	
BISPHENOL A-FUMURATE	R
HYDROGENATED BISPHENOL-A-BISPHENOL-A	
ISOPHTHALIC	R
CHLORINATED POLYESTERS	R
POLYSTYRENES	
POLYETHYLENE	R
POLYPROPYLENE	R
PVC—TYPE 1	R
PVC—TYPE 2	R
POLYVINYLIDENE CHLORIDE	R
VINYLIDENE FLUORIDE	R
VINYL ESTER	R

RUBBERS	TEMP °F
BUTYL GR-1	R
FLUORO ELASTOMERS VITON A	R
FLUORO ELASTOMERS KEL-F 3700	
HARD RUBBER	U
HYPALON	R
KOROSEAL	
NATURAL RUBBER (GRS)	U
NEOPRENE GR-M (CR)	R
NITRILE BUNA N (NBR)	R
NORDEL	R
POLYBUTADIENE	
SILICONE	
POLYURETHANE	R

TEMP °F: 60 80 100 120 140 160 180 200 220 240 260 280 300 320 340 360 380 400 420 440 460

CRESOL

TEMP °F: 60 80 100 120 140 160 180 200 220 240 260 280 300 320 340 360 380 400 420 440 460 480 500 520 540 560

METALS	Rating
ALLOY 20Cb3	
ALUMINUM	G
BRASS	
BRONZE	
CARBON STEEL	G
COPPER	
E-BRIGHT 26-1	
HASTELLOY-B	
HASTELLOY-C	
HASTELLOY-D	
HIGH SILICON IRON	
INCONEL	
LEAD	U
MONEL	
NICKEL	
STAINLESS STEELS	
Type 304/307	E
Type 316	G
TANTALUM	
TITANIUM	
ZIRCONIUM	

NONMETALLICS	Rating
BOROSILICATE GLASS	R
CARBON	

TEMP °F: 60 80 100 120 140 160 180 200 220 240 260 280 300 320 340 360 380 400 420 440 460 480 500 520 540 560

FOR METALS

E = < 2 Mils Penetration/Year; G = < 20 Mils Penetration/Year

S = < 50 Mils Penetration/Year; U = > 50 Mils Penetration/Year

FOR NONMETALLICS

R = Resistant

U = Unsatisfactory

CRESOL

TEMP °F (60, 80, 100, 120, 140, 160, 180, 200, 220, 240, 260, 280, 300, 320, 340, 360, 380, 400, 420, 440, 460)

PLASTICS	
ABS	U
ACRYLICS	U
ASBESTOS REINF., EPOXY	
ASBESTOS REINF., PHENOLIC	
CHLORINATED POLYETHER	R ——————
CPVC	U
EPOXY	
FLUOROCARBONS FEP	R ——————————————————————————————
FLUOROCARBONS TFE	R ——————————————————————————————————————
FURFURYL ALCOHOL	R ————————————————
NORYL	
PHENOLIC	
POLYAMIDES–NYLON	U
POLYESTERS	
BISPHENOL A-FUMURATE	
HYDROGENATED BISPHENOL-A-BISPHENOL-A	
ISOPHTHALIC	
CHLORINATED POLYESTERS	
POLYSTYRENES	
POLYETHYLENE	U
POLYPROPYLENE	U
PVC–TYPE 1	R ————— U
PVC–TYPE 2	U
POLYVINYLIDENE CHLORIDE	
VINYLIDENE FLUORIDE	R ————————————
VINYL ESTER	

RUBBERS	
BUTYL GR-1	U
FLUORO ELASTOMERS VITON A	U
FLUORO ELASTOMERS KEL-F 3700	
HARD RUBBER	
HYPALON	U
KOROSEAL	
NATURAL RUBBER (GRS)	U
NEOPRENE GR-M (CR)	U
NITRILE BUNA N (NBR)	U
NORDEL	
POLYBUTADIENE	
SILICONE	
POLYURETHANE	U

TEMP °F (60, 80, 100, 120, 140, 160, 180, 200, 220, 240, 260, 280, 300, 320, 340, 360, 380, 400, 420, 440, 460)

CRESYLIC ACID

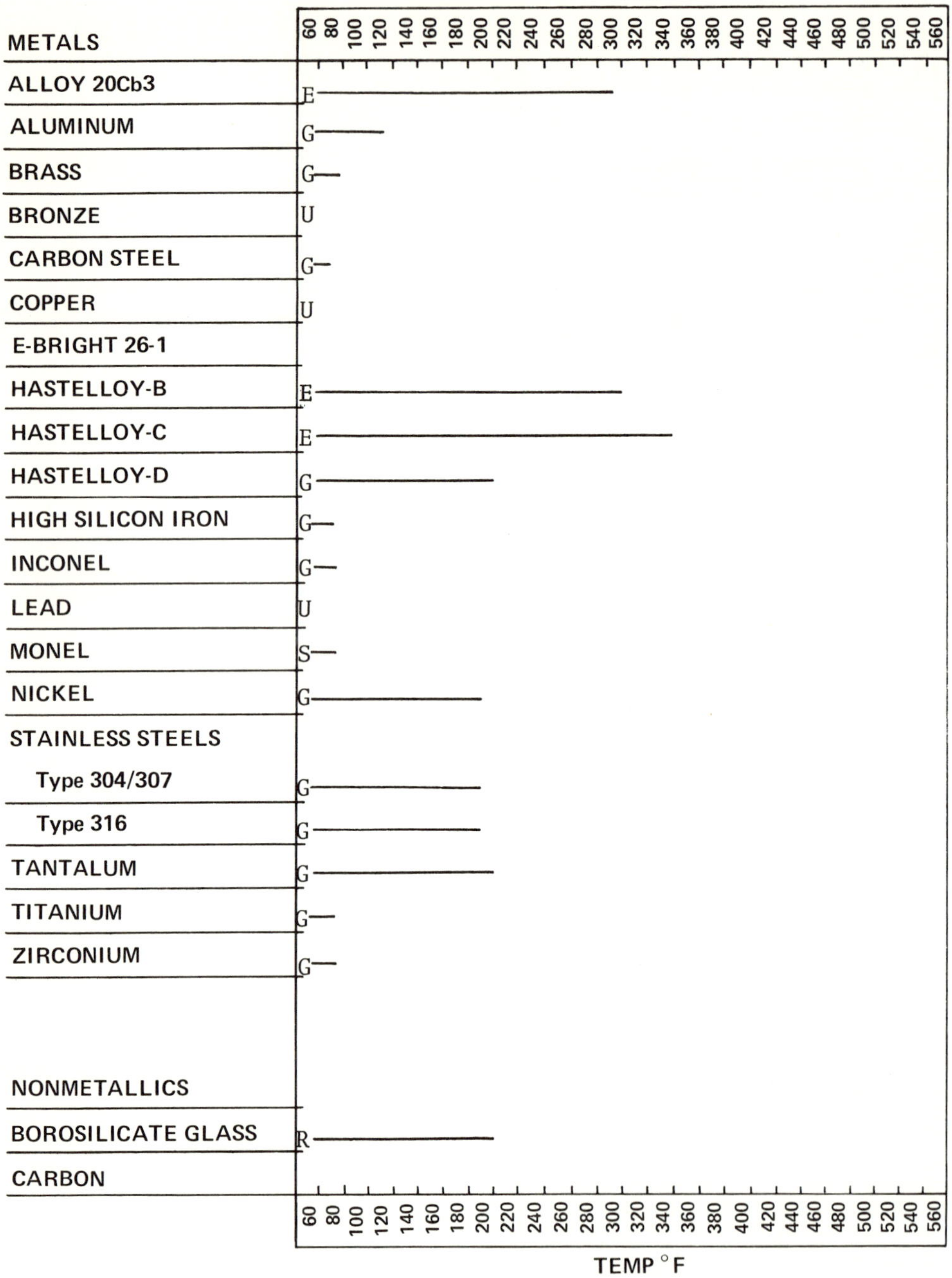

FOR METALS

E = < 2 Mils Penetration/Year; G = < 20 Mils Penetration/Year

S = < 50 Mils Penetration/Year; U = > 50 Mils Penetration/Year

FOR NONMETALLICS

R = Resistant

U = Unsatisfactory

TEMP °F (60–460)

PLASTICS	Rating	Line extends to (°F)
ABS	U	
ACRYLICS		
ASBESTOS REINF., EPOXY		
ASBESTOS REINF., PHENOLIC		
CHLORINATED POLYETHER	R	150
CPVC		
EPOXY	U	
FLUOROCARBONS FEP	R	400
FLUOROCARBONS TFE	R	460
FURFURYL ALCOHOL	R	250
NORYL		
PHENOLIC		
POLYAMIDES—NYLON	U	
POLYESTERS		
BISPHENOL A-FUMURATE sat	U	
HYDROGENATED BISPHENOL-A-BISPHENOL-A		
ISOPHTHALIC	U	
CHLORINATED POLYESTERS		
POLYSTYRENES		
POLYETHYLENE	U	
POLYPROPYLENE	U	
PVC—TYPE 1	U	
PVC—TYPE 2	U	
POLYVINYLIDENE CHLORIDE		
VINYLIDENE FLUORIDE	R	220
VINYL ESTER		

RUBBERS	Rating	Line extends to (°F)
BUTYL GR-1	U	
FLUORO ELASTOMERS VITON A	R	220
FLUORO ELASTOMERS KEL-F 3700		
HARD RUBBER		
HYPALON	U	
KOROSEAL		
NATURAL RUBBER (GRS)	U	
NEOPRENE GR-M (CR)	U	
NITRILE BUNA N (NBR)	U	
NORDEL		
POLYBUTADIENE		
SILICONE		
POLYURETHANE	U	

TEMP °F (60–460)

CRESYLIC ACID, 50%

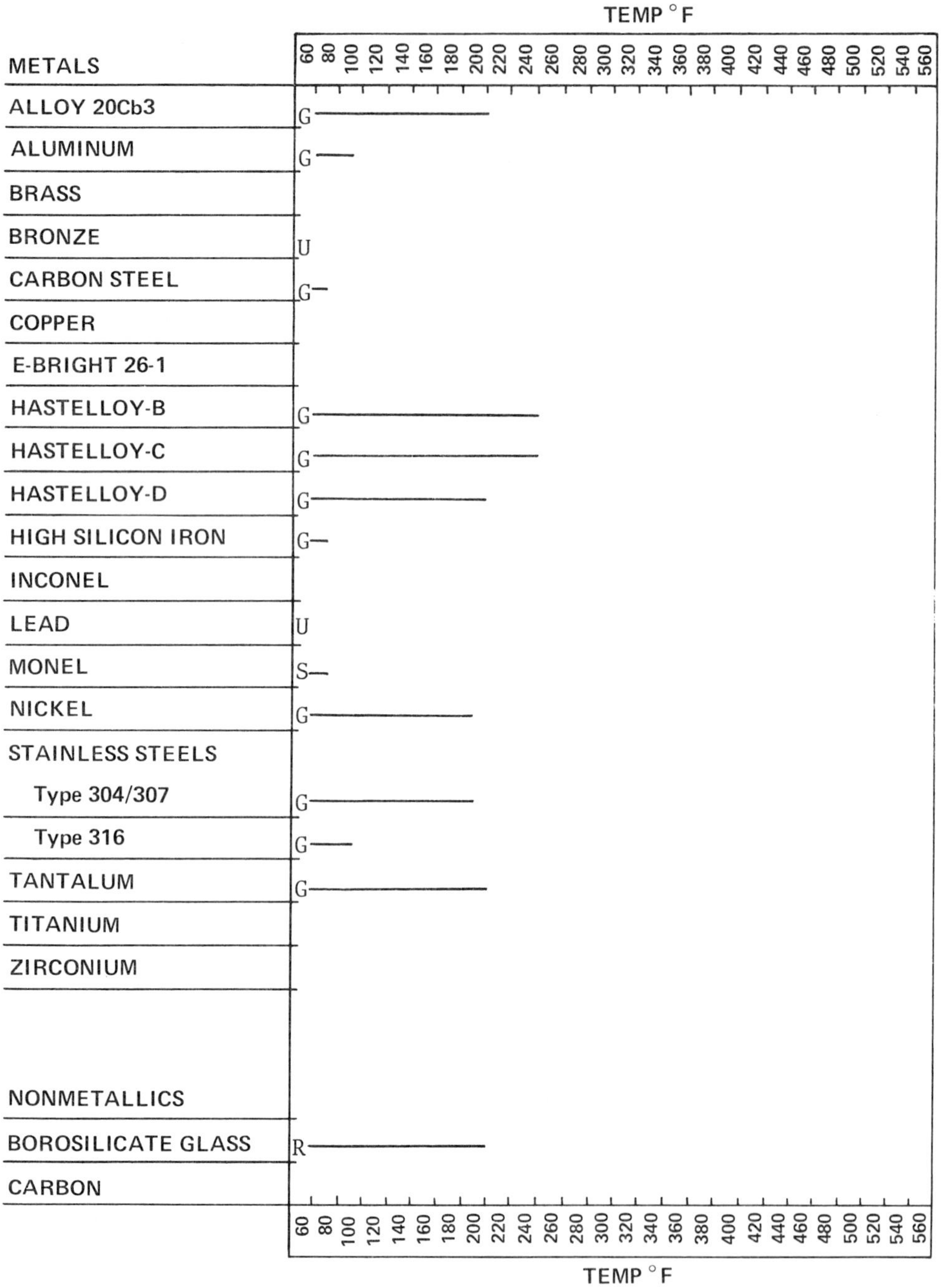

FOR METALS

E = < 2 Mils Penetration/Year; G = < 20 Mils Penetration/Year

S = < 50 Mils Penetration/Year; U = > 50 Mils Penetration/Year

FOR NONMETALLICS

R = Resistant

U = Unsatisfactory

CRESYLIC ACID, 50%

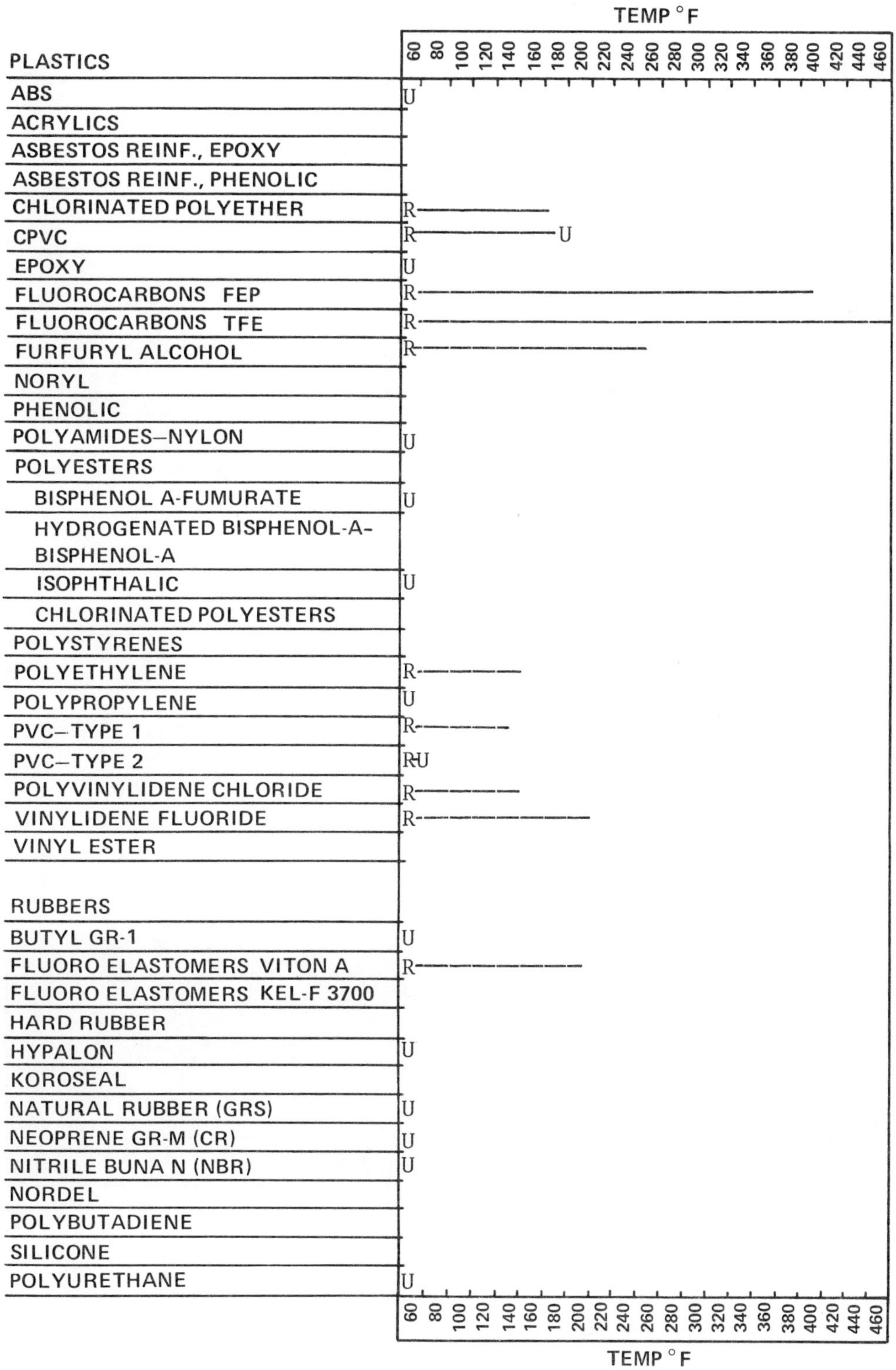

CROTON ALDEHYDE

TEMP °F: 60 80 100 120 140 160 180 200 220 240 260 280 300 320 340 360 380 400 420 440 460 480 500 520 540 560

METALS		
ALLOY 20Cb3		G
ALUMINUM	1	G
BRASS		G
BRONZE		G
CARBON STEEL		
COPPER		G
E-BRIGHT 26-1		
HASTELLOY-B		G
HASTELLOY-C		G
HASTELLOY-D		
HIGH SILICON IRON		G
INCONEL		G
LEAD		
MONEL		G
NICKEL		G
STAINLESS STEELS		
Type 304/307	83%	E
Type 316	83%	E
TANTALUM		G
TITANIUM		
ZIRCONIUM		G

NONMETALLICS	
BOROSILICATE GLASS	
CARBON	R

TEMP °F: 60 80 100 120 140 160 180 200 220 240 260 280 300 320 340 360 380 400 420 440 460 480 500 520 540 560

FOR METALS

E = < 2 Mils Penetration/Year; G = < 20 Mils Penetration/Year

S = < 50 Mils Penetration/Year; U = > 50 Mils Penetration/Year

FOR NONMETALLICS

R = Resistant

U = Unsatisfactory

CROTON ALDEHYDE

TEMP °F (60–460)

PLASTICS	
ABS	U
ACRYLICS	
ASBESTOS REINF., EPOXY	
ASBESTOS REINF., PHENOLIC	
CHLORINATED POLYETHER	U
CPVC	U
EPOXY	
FLUOROCARBONS FEP	R
FLUOROCARBONS TFE	R
FURFURYL ALCOHOL	
NORYL	
PHENOLIC	
POLYAMIDES–NYLON	
POLYESTERS	
BISPHENOL A-FUMURATE	
HYDROGENATED BISPHENOL-A-BISPHENOL-A	
ISOPHTHALIC	
CHLORINATED POLYESTERS	
POLYSTYRENES	
POLYETHYLENE	U
POLYPROPYLENE	U
PVC–TYPE 1	U
PVC–TYPE 2	U
POLYVINYLIDENE CHLORIDE	U
VINYLIDENE FLUORIDE	R
VINYL ESTER	

RUBBERS	
BUTYL GR-1	
FLUORO ELASTOMERS VITON A	
FLUORO ELASTOMERS KEL-F 3700	
HARD RUBBER	
HYPALON	
KOROSEAL	
NATURAL RUBBER (GRS)	
NEOPRENE GR-M (CR)	
NITRILE BUNA N (NBR)	
NORDEL	
POLYBUTADIENE	
SILICONE	
POLYURETHANE	

TEMP °F (60–460)

CRUDE OIL

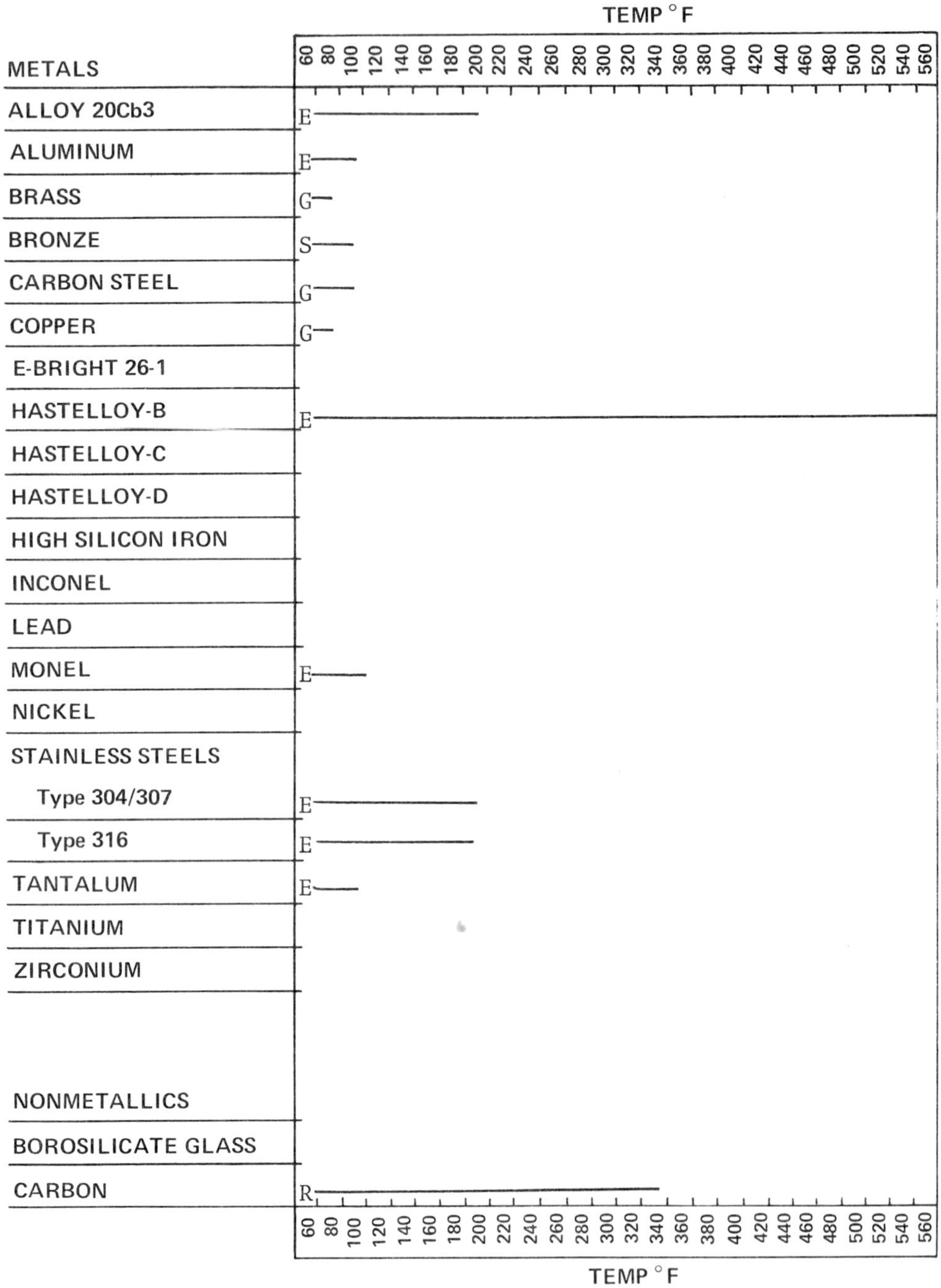

FOR METALS

E = < 2 Mils Penetration/Year; G = < 20 Mils Penetration/Year

S = < 50 Mils Penetration/Year; U = > 50 Mils Penetration/Year

FOR NONMETALLICS

R = Resistant

U = Unsatisfactory

CRUDE OIL

TEMP °F: 60 80 100 120 140 160 180 200 220 240 260 280 300 320 340 360 380 400 420 440 460

PLASTICS	
ABS	R
ACRYLICS	
ASBESTOS REINF., EPOXY	
ASBESTOS REINF., PHENOLIC	
CHLORINATED POLYETHER	R
CPVC	R
EPOXY	R
FLUOROCARBONS FEP	R
FLUOROCARBONS TFE	R
FURFURYL ALCOHOL	
NORYL	
PHENOLIC	
POLYAMIDES–NYLON	R
POLYESTERS	
BISPHENOL A-FUMURATE	R
HYDROGENATED BISPHENOL-A-BISPHENOL-A	
ISOPHTHALIC	R
CHLORINATED POLYESTERS	R
POLYSTYRENES	
POLYETHYLENE	R
POLYPROPYLENE	R
PVC–TYPE 1	R
PVC–TYPE 2	R
POLYVINYLIDENE CHLORIDE	R
VINYLIDENE FLUORIDE	R
VINYL ESTER	R

RUBBERS	
BUTYL GR-1	U
FLUORO ELASTOMERS VITON A	R
FLUORO ELASTOMERS KEL-F 3700	
HARD RUBBER	R
HYPALON	U
KOROSEAL	
NATURAL RUBBER (GRS)	U
NEOPRENE GR-M (CR)	U
NITRILE BUNA N (NBR)	R
NORDEL	
POLYBUTADIENE	
SILICONE	
POLYURETHANE	R

TEMP °F: 60 80 100 120 140 160 180 200 220 240 260 280 300 320 340 360 380 400 420 440 460

CYANOGEN GAS

TEMP °F

METALS	60	80	100	120	140	160	180	200	220	240	260	280	300	320	340	360	380	400	420	440	460	480	500	520	540	560
ALLOY 20Cb3																										
ALUMINUM																										
BRASS																										
BRONZE																										
CARBON STEEL																										
COPPER																										
E-BRIGHT 26-1																										
HASTELLOY-B																										
HASTELLOY-C																										
HASTELLOY-D																										
HIGH SILICON IRON																										
INCONEL	G																									
LEAD																										
MONEL	G																									
NICKEL	G																									
STAINLESS STEELS																										
Type 304/307	G																									
Type 316	G																									
TANTALUM																										
TITANIUM																										
ZIRCONIUM																										
NONMETALLICS																										
BOROSILICATE GLASS																										
CARBON																										

TEMP °F

FOR METALS

E = < 2 Mils Penetration/Year; G = < 20 Mils Penetration/Year

S = < 50 Mils Penetration/Year; U = > 50 Mils Penetration/Year

FOR NONMETALLICS

R = Resistant

U = Unsatisfactory

CYANOGEN GAS

TEMP °F (60, 80, 100, 120, 140, 160, 180, 200, 220, 240, 260, 280, 300, 320, 340, 360, 380, 400, 420, 440, 460)

PLASTICS	TEMP °F
ABS	
ACRYLICS	
ASBESTOS REINF., EPOXY	
ASBESTOS REINF., PHENOLIC	
CHLORINATED POLYETHER	
CPVC	
EPOXY	
FLUOROCARBONS FEP	R (60–400)
FLUOROCARBONS TFE	R (60–460)
FURFURYL ALCOHOL	
NORYL	
PHENOLIC	
POLYAMIDES–NYLON	
POLYESTERS	
BISPHENOL A-FUMURATE	
HYDROGENATED BISPHENOL-A-BISPHENOL-A	
ISOPHTHALIC	
CHLORINATED POLYESTERS	
POLYSTYRENES	
POLYETHYLENE	
POLYPROPYLENE	
PVC–TYPE 1	
PVC–TYPE 2	
POLYVINYLIDENE CHLORIDE	
VINYLIDENE FLUORIDE	
VINYL ESTER	
RUBBERS	
BUTYL GR-1	
FLUORO ELASTOMERS VITON A	
FLUORO ELASTOMERS KEL-F 3700	
HARD RUBBER	R (60–200)
HYPALON	
KOROSEAL	
NATURAL RUBBER (GRS)	
NEOPRENE GR-M (CR)	R (60–200)
NITRILE BUNA N (NBR)	
NORDEL	R (60–300)
POLYBUTADIENE	
SILICONE	
POLYURETHANE	

TEMP °F (60, 80, 100, 120, 140, 160, 180, 200, 220, 240, 260, 280, 300, 320, 340, 360, 380, 400, 420, 440, 460)

CYANOACETIC ACID

TEMP °F: 60 80 100 120 140 160 180 200 220 240 260 280 300 320 340 360 380 400 420 440 460 480 500 520 540 560

METALS	TEMP °F
ALLOY 20Cb3	G—
ALUMINUM	
BRASS	U
BRONZE	U
CARBON STEEL	
COPPER	U
E-BRIGHT 26-1	
HASTELLOY-B	
HASTELLOY-C	
HASTELLOY-D	
HIGH SILICON IRON	
INCONEL	
LEAD	
MONEL	
NICKEL	
STAINLESS STEELS Type 304/307	
Type 316	
TANTALUM	G————————
TITANIUM	G———————
ZIRCONIUM	G———————
NONMETALLICS	
BOROSILICATE GLASS	
CARBON	

TEMP °F: 60 80 100 120 140 160 180 200 220 240 260 280 300 320 340 360 380 400 420 440 460 480 500 520 540 560

FOR METALS

E = < 2 Mils Penetration/Year; G = < 20 Mils Penetration/Year

S = < 50 Mils Penetration/Year; U = > 50 Mils Penetration/Year

FOR NONMETALLICS

R = Resistant

U = Unsatisfactory

TEMP °F

PLASTICS	Resistance (60–460 °F)
ABS	
ACRYLICS	
ASBESTOS REINF., EPOXY	
ASBESTOS REINF., PHENOLIC	
CHLORINATED POLYETHER	
CPVC	
EPOXY	
FLUOROCARBONS FEP	R (60–400)
FLUOROCARBONS TFE	R (60–460)
FURFURYL ALCOHOL	
NORYL	
PHENOLIC	
POLYAMIDES–NYLON	
POLYESTERS	
BISPHENOL A-FUMURATE	
HYDROGENATED BISPHENOL-A-BISPHENOL-A	
ISOPHTHALIC	
CHLORINATED POLYESTERS	
POLYSTYRENES	
POLYETHYLENE	
POLYPROPYLENE	
PVC–TYPE 1	
PVC–TYPE 2	
POLYVINYLIDENE CHLORIDE	R (60)
VINYLIDENE FLUORIDE	
VINYL ESTER	

RUBBERS	
BUTYL GR-1	
FLUORO ELASTOMERS VITON A	
FLUORO ELASTOMERS KEL-F 3700	
HARD RUBBER	
HYPALON	
KOROSEAL	
NATURAL RUBBER (GRS)	
NEOPRENE GR-M (CR)	
NITRILE BUNA N (NBR)	
NORDEL	
POLYBUTADIENE	
SILICONE	
POLYURETHANE	

TEMP °F

CYCLOHEXANE

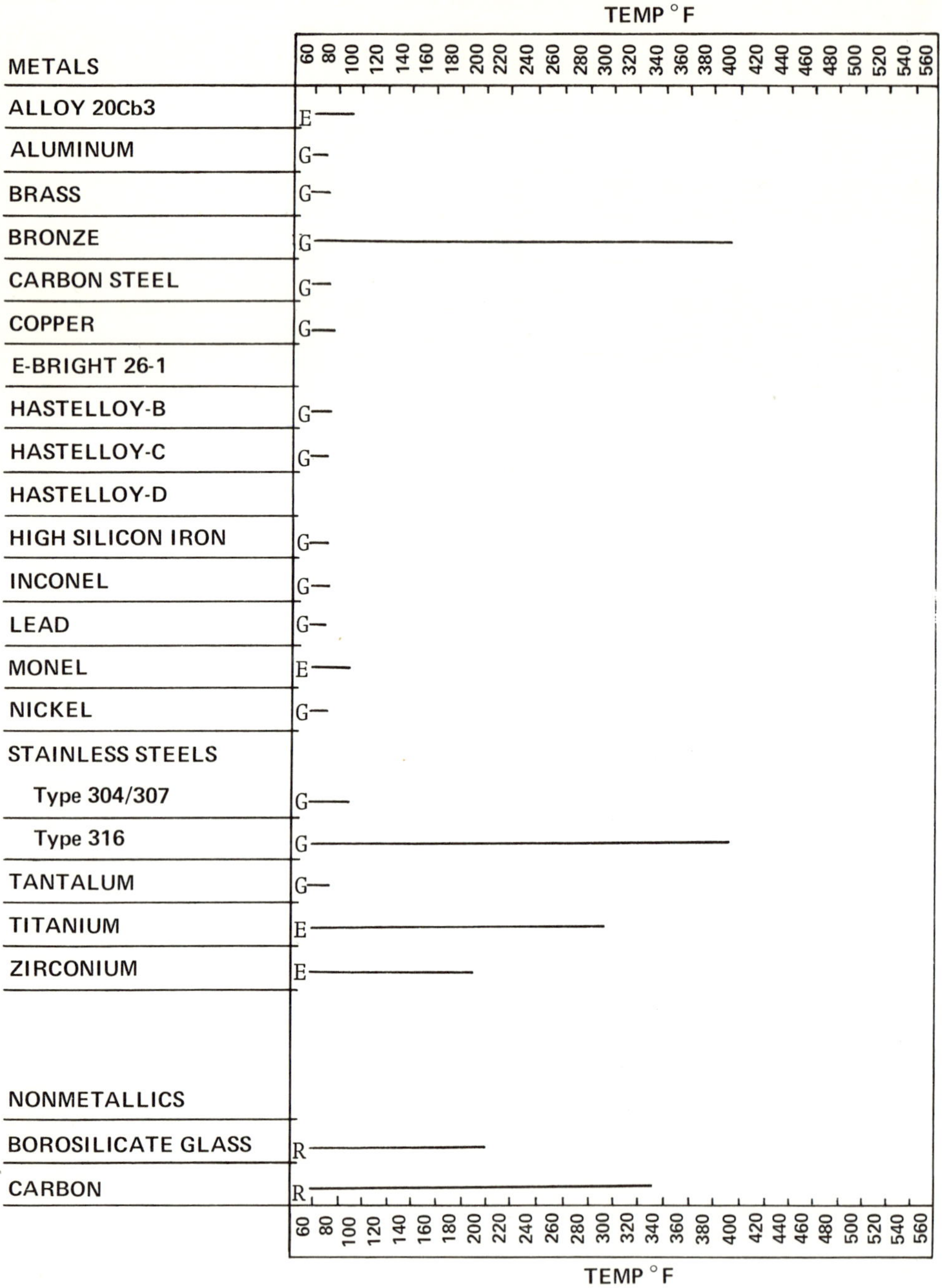

FOR METALS

E = < 2 Mils Penetration/Year; G = < 20 Mils Penetration/Year

S = < 50 Mils Penetration/Year; U = > 50 Mils Penetration/Year

FOR NONMETALLICS

R = Resistant

U = Unsatisfactory

CYCLOHEXANE

TEMP °F (60 80 100 120 140 160 180 200 220 240 260 280 300 320 340 360 380 400 420 440 460)

PLASTICS	
ABS	R—
ACRYLICS	
ASBESTOS REINF., EPOXY	
ASBESTOS REINF., PHENOLIC	
CHLORINATED POLYETHER	R—
CPVC	
EPOXY	
FLUOROCARBONS FEP	R—
FLUOROCARBONS TFE	R—
FURFURYL ALCOHOL	
NORYL	
PHENOLIC	
POLYAMIDES–NYLON	R—
POLYESTERS	
BISPHENOL A-FUMURATE	U
HYDROGENATED BISPHENOL-A-BISPHENOL-A	R—
ISOPHTHALIC	R—
CHLORINATED POLYESTERS	R—
POLYSTYRENES	
POLYETHYLENE	R—
POLYPROPYLENE	U
PVC–TYPE 1	
PVC–TYPE 2	
POLYVINYLIDENE CHLORIDE	R—
VINYLIDENE FLUORIDE	R—
VINYL ESTER	R—

RUBBERS	
BUTYL GR-1	U
FLUORO ELASTOMERS VITON A	R—
FLUORO ELASTOMERS KEL-F 3700	
HARD RUBBER	
HYPALON	U
KOROSEAL	
NATURAL RUBBER (GRS)	U
NEOPRENE GR-M (CR)	U
NITRILE BUNA N (NBR)	R—
NORDEL	U
POLYBUTADIENE	
SILICONE	
POLYURETHANE	R—

TEMP °F (60 80 100 120 140 160 180 200 220 240 260 280 300 320 340 360 380 400 420 440 460)

CYCLOHEXANOL

TEMP °F

METALS	60	80	100	120	140	160	180	200	220	240	260	280	300	320	340	360	380	400	420	440	460	480	500	520	540	560
ALLOY 20Cb3	G—																									
ALUMINUM	U																									
BRASS	G—																									
BRONZE	G—																									
CARBON STEEL	G—																									
COPPER	G—																									
E-BRIGHT 26-1																										
HASTELLOY-B	G—																									
HASTELLOY-C	G—																									
HASTELLOY-D																										
HIGH SILICON IRON	G—																									
INCONEL	G—																									
LEAD																										
MONEL	G—																									
NICKEL	G—																									
STAINLESS STEELS																										
Type 304/307	G—																									
Type 316	G—																									
TANTALUM	G—																									
TITANIUM	G—																									
ZIRCONIUM	G—																									
NONMETALLICS																										
BOROSILICATE GLASS																										
CARBON																										

TEMP °F

FOR METALS

E = < 2 Mils Penetration/Year; G = < 20 Mils Penetration/Year

S = < 50 Mils Penetration/Year; U = > 50 Mils Penetration/Year

FOR NONMETALLICS

R = Resistant

U = Unsatisfactory

CYCLOHEXANOL

TEMP °F (60, 80, 100, 120, 140, 160, 180, 200, 220, 240, 260, 280, 300, 320, 340, 360, 380, 400, 420, 440, 460)

PLASTICS	
ABS	R—
ACRYLICS	
ASBESTOS REINF., EPOXY	
ASBESTOS REINF., PHENOLIC	
CHLORINATED POLYETHER	R————————
CPVC	U
EPOXY	
FLUOROCARBONS FEP	R————————————————
FLUOROCARBONS TFE	R———————————————————
FURFURYL ALCOHOL	
NORYL	
PHENOLIC	
POLYAMIDES–NYLON	R——————
POLYESTERS	
BISPHENOL A-FUMURATE	
HYDROGENATED BISPHENOL-A-BISPHENOL-A	
ISOPHTHALIC	
CHLORINATED POLYESTERS	
POLYSTYRENES	
POLYETHYLENE	R————
POLYPROPYLENE	R————U
PVC–TYPE 1	U
PVC–TYPE 2	U
POLYVINYLIDENE CHLORIDE	R—
VINYLIDENE FLUORIDE	R———————
VINYL ESTER	

RUBBERS	
BUTYL GR-1	
FLUORO ELASTOMERS VITON A	R——————
FLUORO ELASTOMERS KEL-F 3700	
HARD RUBBER	
HYPALON	U
KOROSEAL	
NATURAL RUBBER (GRS)	
NEOPRENE GR-M (CR)	
NITRILE BUNA N (NBR)	RU
NORDEL	
POLYBUTADIENE	
SILICONE	
POLYURETHANE	

TEMP °F (60, 80, 100, 120, 140, 160, 180, 200, 220, 240, 260, 280, 300, 320, 340, 360, 380, 400, 420, 440, 460)

CYCLOHEXANONE

METALS	TEMP °F (60 80 100 120 140 160 180 200 220 240 260 280 300 320 340 360 380 400 420 440 460 480 500 520 540 560)
ALLOY 20Cb3	G—
ALUMINUM	G—
BRASS	G—
BRONZE	G—
CARBON STEEL	U
COPPER	G—
E-BRIGHT 26-1	
HASTELLOY-B	G—
HASTELLOY-C	G—
HASTELLOY-D	
HIGH SILICON IRON	G—
INCONEL	G—
LEAD	G—
MONEL	G—
NICKEL	G—
STAINLESS STEELS	
Type 304/307	G—
Type 316	G—
TANTALUM	G—
TITANIUM	G—
ZIRCONIUM	G—
NONMETALLICS	
BOROSILICATE GLASS	R———————— (to 240)
CARBON	

TEMP °F

FOR METALS

E = < 2 Mils Penetration/Year; G = < 20 Mils Penetration/Year

S = < 50 Mils Penetration/Year; U = > 50 Mils Penetration/Year

FOR NONMETALLICS

R = Resistant

U = Unsatisfactory

CYCLOHEXANONE

PLASTICS	TEMP °F (60–460)
ABS	U
ACRYLICS	
ASBESTOS REINF., EPOXY	
ASBESTOS REINF., PHENOLIC	
CHLORINATED POLYETHER	R— (60–80)
CPVC	U
EPOXY	
FLUOROCARBONS FEP	R———— (60–400)
FLUOROCARBONS TFE	R———— (60–460)
FURFURYL ALCOHOL	R—— (60–100)
NORYL	
PHENOLIC	
POLYAMIDES–NYLON	R———— (60–200)
POLYESTERS	
BISPHENOL A-FUMURATE	U
HYDROGENATED BISPHENOL-A-BISPHENOL-A	U
ISOPHTHALIC	U
CHLORINATED POLYESTERS	
POLYSTYRENES	
POLYETHYLENE	U
POLYPROPYLENE	RU
PVC–TYPE 1	U
PVC–TYPE 2	U
POLYVINYLIDENE CHLORIDE	U
VINYLIDENE FLUORIDE	R———U (60–120)
VINYL ESTER	

RUBBERS	TEMP °F (60–460)
BUTYL GR-1	U
FLUORO ELASTOMERS VITON A	U
FLUORO ELASTOMERS KEL-F 3700	
HARD RUBBER	
HYPALON	
KOROSEAL	
NATURAL RUBBER (GRS)	U
NEOPRENE GR-M (CR)	U
NITRILE BUNA N (NBR)	U
NORDEL	
POLYBUTADIENE	
SILICONE	
POLYURETHANE	

DETERGENTS

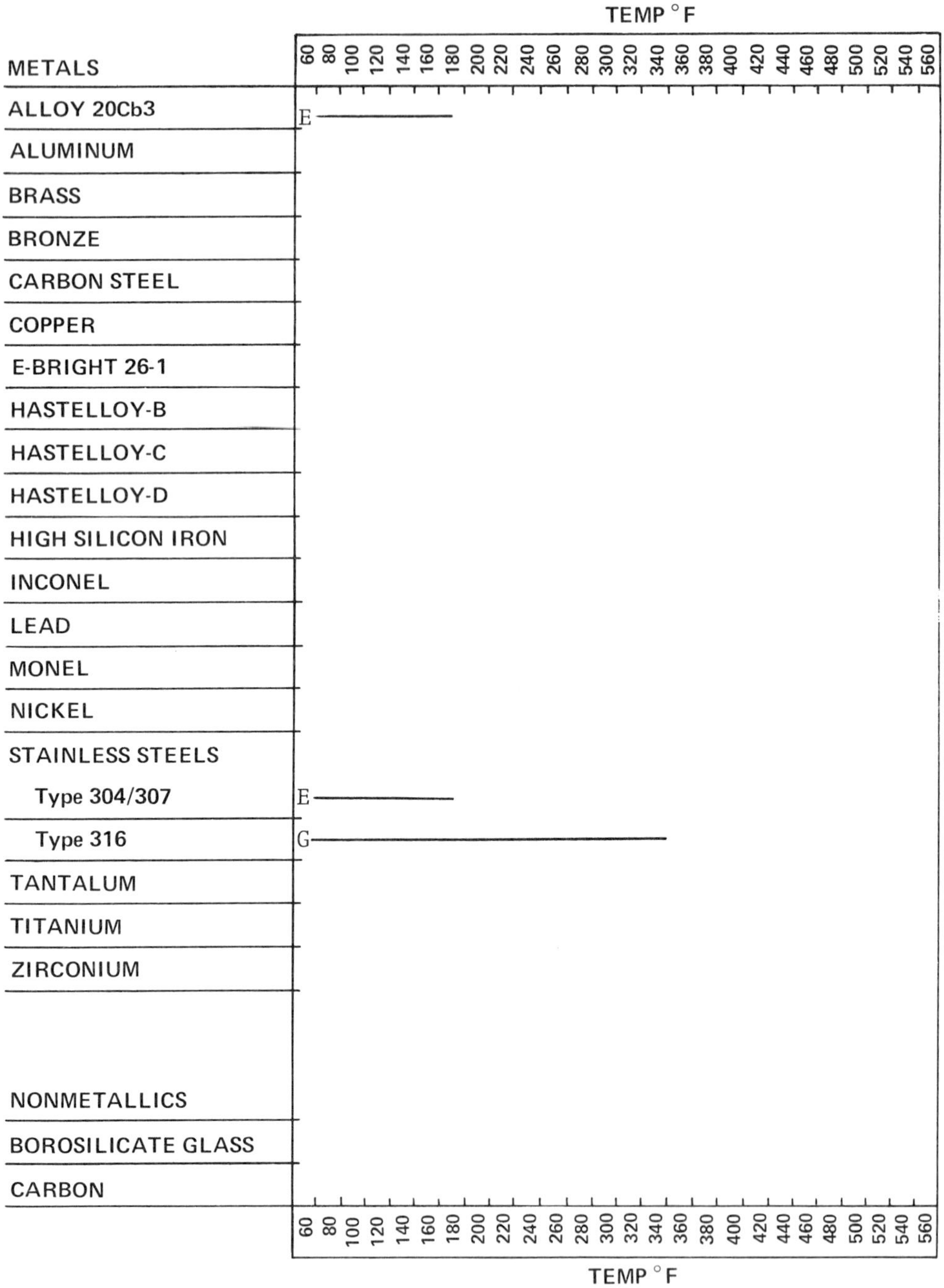

FOR METALS

E = < 2 Mils Penetration/Year; G = < 20 Mils Penetration/Year

S = < 50 Mils Penetration/Year; U = > 50 Mils Penetration/Year

FOR NONMETALLICS

R = Resistant

U = Unsatisfactory

Please see page 1 for notes on calibration.

DETERGENTS

TEMP °F (60 80 100 120 140 160 180 200 220 240 260 280 300 320 340 360 380 400 420 440 460)

PLASTICS	
ABS	
ACRYLICS	
ASBESTOS REINF., EPOXY	
ASBESTOS REINF., PHENOLIC	
CHLORINATED POLYETHER	R
CPVC	R
EPOXY	R
FLUOROCARBONS FEP	R
FLUOROCARBONS TFE	R
FURFURYL ALCOHOL	
NORYL	
PHENOLIC	
POLYAMIDES—NYLON	
POLYESTERS	
BISPHENOL A-FUMURATE	
HYDROGENATED BISPHENOL-A-BISPHENOL-A	
ISOPHTHALIC	
CHLORINATED POLYESTERS	
POLYSTYRENES	
POLYETHYLENE	R
POLYPROPYLENE	R
PVC—TYPE 1	R
PVC—TYPE 2	R
POLYVINYLIDENE CHLORIDE	
VINYLIDENE FLUORIDE	
VINYL ESTER	
RUBBERS	
BUTYL GR-1	
FLUORO ELASTOMERS VITON A	R
FLUORO ELASTOMERS KEL-F 3700	
HARD RUBBER	
HYPALON	
KOROSEAL	
NATURAL RUBBER (GRS)	
NEOPRENE GR-M (CR)	R
NITRILE BUNA N (NBR)	R
NORDEL	R
POLYBUTADIENE	
SILICONE	
POLYURETHANE	

TEMP °F (60 80 100 120 140 160 180 200 220 240 260 280 300 320 340 360 380 400 420 440 460)

DETERGENT SOLUTION (HEAVY DUTY)

TEMP °F: 60 80 100 120 140 160 180 200 220 240 260 280 300 320 340 360 380 400 420 440 460 480 500 520 540 560

METALS	TEMP °F
ALLOY 20Cb3	E
ALUMINUM	
BRASS	
BRONZE	
CARBON STEEL	
COPPER	
E-BRIGHT 26-1	
HASTELLOY-B	
HASTELLOY-C	
HASTELLOY-D	
HIGH SILICON IRON	
INCONEL	
LEAD	
MONEL	
NICKEL	
STAINLESS STEELS	
Type 304/307	E
Type 316	E
TANTALUM	
TITANIUM	
ZIRCONIUM	

NONMETALLICS	TEMP °F
BOROSILICATE GLASS	R
CARBON	

TEMP °F: 60 80 100 120 140 160 180 200 220 240 260 280 300 320 340 360 380 400 420 440 460 480 500 520 540 560

FOR METALS

E = < 2 Mils Penetration/Year; G = < 20 Mils Penetration/Year

S = < 50 Mils Penetration/Year; U = > 50 Mils Penetration/Year

FOR NONMETALLICS

R = Resistant

U = Unsatisfactory

DETERGENT SOLUTION (HEAVY DUTY)

TEMP °F: 60 80 100 120 140 160 180 200 220 240 260 280 300 320 340 360 380 400 420 440 460

PLASTICS	
ABS	
ACRYLICS	
ASBESTOS REINF., EPOXY	
ASBESTOS REINF., PHENOLIC	
CHLORINATED POLYETHER	
CPVC	R
EPOXY	R
FLUOROCARBONS FEP	R
FLUOROCARBONS TFE	R
FURFURYL ALCOHOL	
NORYL	
PHENOLIC	
POLYAMIDES—NYLON	
POLYESTERS	
BISPHENOL A-FUMURATE	
HYDROGENATED BISPHENOL-A-BISPHENOL-A	
ISOPHTHALIC	
CHLORINATED POLYESTERS	
POLYSTYRENES	
POLYETHYLENE	
POLYPROPYLENE	R
PVC—TYPE 1	R
PVC—TYPE 2	R
POLYVINYLIDENE CHLORIDE	
VINYLIDENE FLUORIDE	
VINYL ESTER	

RUBBERS	
BUTYL GR-1	
FLUORO ELASTOMERS VITON A	R
FLUORO ELASTOMERS KEL-F 3700	
HARD RUBBER	
HYPALON	
KOROSEAL	
NATURAL RUBBER (GRS)	
NEOPRENE GR-M (CR)	
NITRILE BUNA N (NBR)	R
NORDEL	
POLYBUTADIENE	
SILICONE	
POLYURETHANE	

TEMP °F: 60 80 100 120 140 160 180 200 220 240 260 280 300 320 340 360 380 400 420 440 460

DEXTRIN

TEMP °F: 60 80 100 120 140 160 180 200 220 240 260 280 300 320 340 360 380 400 420 440 460 480 500 520 540 560

METALS	
ALLOY 20Cb3	
ALUMINUM	G—
BRASS	
BRONZE	
CARBON STEEL	
COPPER	
E-BRIGHT 26-1	
HASTELLOY-B	
HASTELLOY-C	
HASTELLOY-D	
HIGH SILICON IRON	
INCONEL	
LEAD	
MONEL	G—
NICKEL	G—
STAINLESS STEELS	
Type 304/307	
Type 316	G————————————— (60 to 400)
TANTALUM	G—
TITANIUM	
ZIRCONIUM	U

NONMETALLICS	
BOROSILICATE GLASS	
CARBON	

TEMP °F: 60 80 100 120 140 160 180 200 220 240 260 280 300 320 340 360 380 400 420 440 460 480 500 520 540 560

FOR METALS

E = < 2 Mils Penetration/Year; G = < 20 Mils Penetration/Year

S = < 50 Mils Penetration/Year; U = > 50 Mils Penetration/Year

FOR NONMETALLICS

R = Resistant

U = Unsatisfactory

DEXTRIN

TEMP °F (scale 60–460)

PLASTICS	Rating	TEMP °F
ABS	R	60–140
ACRYLICS		
ASBESTOS REINF., EPOXY		
ASBESTOS REINF., PHENOLIC		
CHLORINATED POLYETHER	R	60–250
CPVC	R	60–180
EPOXY		
FLUOROCARBONS FEP	R	60–400
FLUOROCARBONS TFE	R	60–460
FURFURYL ALCOHOL		
NORYL		
PHENOLIC		
POLYAMIDES—NYLON		
POLYESTERS		
BISPHENOL A-FUMURATE		
HYDROGENATED BISPHENOL-A-BISPHENOL-A		
ISOPHTHALIC		
CHLORINATED POLYESTERS		
POLYSTYRENES		
POLYETHYLENE	R	60–150
POLYPROPYLENE	R	60–150
PVC—TYPE 1	R	60–140
PVC—TYPE 2	R	60–140
POLYVINYLIDENE CHLORIDE		
VINYLIDENE FLUORIDE	R	60–230
VINYL ESTER		

RUBBERS	Rating	TEMP °F
BUTYL GR-1		
FLUORO ELASTOMERS VITON A		
FLUORO ELASTOMERS KEL-F 3700		
HARD RUBBER		
HYPALON		
KOROSEAL		
NATURAL RUBBER (GRS)		
NEOPRENE GR-M (CR)	R	60–200
NITRILE BUNA N (NBR)		
NORDEL		
POLYBUTADIENE		
SILICONE		
POLYURETHANE		

DEXTROSE

TEMP °F

60 80 100 120 140 160 180 200 220 240 260 280 300 320 340 360 380 400 420 440 460 480 500 520 540 560

METALS	
ALLOY 20Cb3	
ALUMINUM	G
BRASS	
BRONZE	
CARBON STEEL	
COPPER	
E-BRIGHT 26-1	
HASTELLOY-B	
HASTELLOY-C	G
HASTELLOY-D	
HIGH SILICON IRON	
INCONEL	
LEAD	
MONEL	G
NICKEL	G
STAINLESS STEELS Type 304/307	
Type 316	
TANTALUM	G
TITANIUM	
ZIRCONIUM	

NONMETALLICS	
BOROSILICATE GLASS	R
CARBON	R

60 80 100 120 140 160 180 200 220 240 260 280 300 320 340 360 380 400 420 440 460 480 500 520 540 560

TEMP °F

FOR METALS

E = < 2 Mils Penetration/Year; G = < 20 Mils Penetration/Year

S = < 50 Mils Penetration/Year; U = > 50 Mils Penetration/Year

FOR NONMETALLICS

R = Resistant

U = Unsatisfactory

DEXTROSE

PLASTICS	TEMP °F (60–460)
ABS	R
ACRYLICS	R
ASBESTOS REINF., EPOXY	
ASBESTOS REINF., PHENOLIC	
CHLORINATED POLYETHER	R
CPVC	R
EPOXY	
FLUOROCARBONS FEP	R
FLUOROCARBONS TFE	R
FURFURYL ALCOHOL	
NORYL	
PHENOLIC	
POLYAMIDES–NYLON	
POLYESTERS	
BISPHENOL A-FUMURATE	
HYDROGENATED BISPHENOL-A-BISPHENOL-A	
ISOPHTHALIC	
CHLORINATED POLYESTERS	
POLYSTYRENES	
POLYETHYLENE	R
POLYPROPYLENE	R
PVC–TYPE 1	R
PVC–TYPE 2	R
POLYVINYLIDENE CHLORIDE	R
VINYLIDENE FLUORIDE	R
VINYL ESTER	

RUBBERS	TEMP °F (60–460)
BUTYL GR-1	
FLUORO ELASTOMERS VITON A	R
FLUORO ELASTOMERS KEL-F 3700	
HARD RUBBER	
HYPALON	R
KOROSEAL	
NATURAL RUBBER (GRS)	
NEOPRENE GR-M (CR)	R
NITRILE BUNA N (NBR)	R
NORDEL	
POLYBUTADIENE	
SILICONE	
POLYURETHANE	U

DIACETONE ALCOHOL

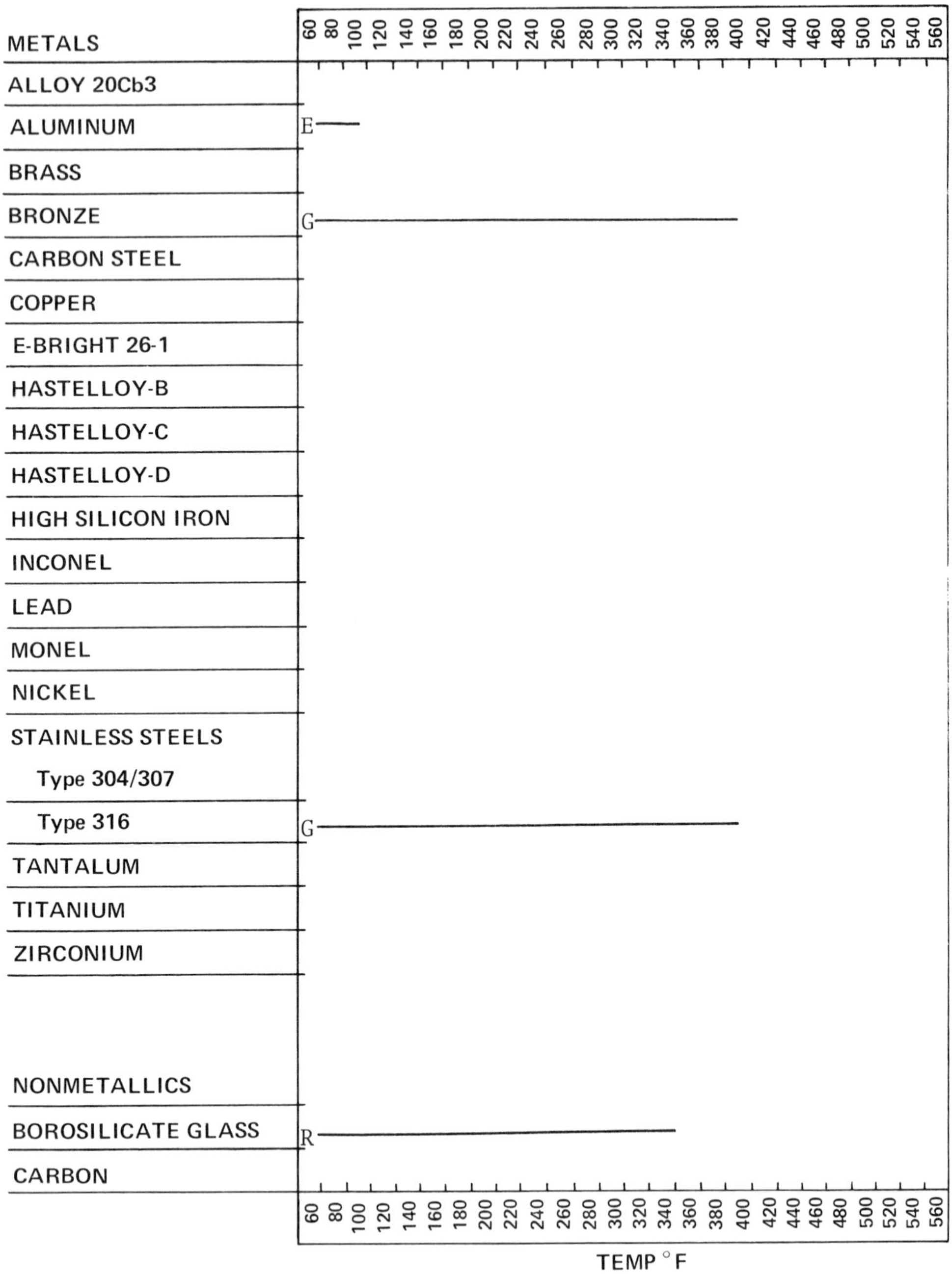

FOR METALS

E = < 2 Mils Penetration/Year; G = < 20 Mils Penetration/Year

S = < 50 Mils Penetration/Year; U = > 50 Mils Penetration/Year

FOR NONMETALLICS

R = Resistant

U = Unsatisfactory

DIACETONE ALCOHOL

TEMP °F (60–460)

PLASTICS	Rating
ABS	
ACRYLICS	
ASBESTOS REINF., EPOXY	
ASBESTOS REINF., PHENOLIC	
CHLORINATED POLYETHER	
CPVC	
EPOXY	R (60–180)
FLUOROCARBONS FEP	R (60–400)
FLUOROCARBONS TFE	R (60–460)
FURFURYL ALCOHOL	
NORYL	
PHENOLIC	
POLYAMIDES–NYLON	R (60)
POLYESTERS	
BISPHENOL A-FUMURATE	
HYDROGENATED BISPHENOL-A-BISPHENOL-A	
ISOPHTHALIC	
CHLORINATED POLYESTERS	
POLYSTYRENES	
POLYETHYLENE	R (60)
POLYPROPYLENE	R (60–200)
PVC–TYPE 1	
PVC–TYPE 2	
POLYVINYLIDENE CHLORIDE	R (60)
VINYLIDENE FLUORIDE	R (60)——U (120)
VINYL ESTER	

RUBBERS	Rating
BUTYL GR-1	
FLUORO ELASTOMERS VITON A	U (60)
FLUORO ELASTOMERS KEL-F 3700	
HARD RUBBER	
HYPALON	
KOROSEAL	
NATURAL RUBBER (GRS)	
NEOPRENE GR-M (CR)	
NITRILE BUNA N (NBR)	U (60)
NORDEL	
POLYBUTADIENE	
SILICONE	
POLYURETHANE	

TEMP °F (60–460)

DIAZO SALTS

TEMP °F: 60 80 100 120 140 160 180 200 220 240 260 280 300 320 340 360 380 400 420 440 460 480 500 520 540 560

METALS	
ALLOY 20Cb3	
ALUMINUM	
BRASS	
BRONZE	
CARBON STEEL	
COPPER	
E-BRIGHT 26-1	
HASTELLOY-B	
HASTELLOY-C	
HASTELLOY-D	
HIGH SILICON IRON	
INCONEL	
LEAD	
MONEL	
NICKEL	
STAINLESS STEELS	
Type 304/307	
Type 316	G
TANTALUM	
TITANIUM	
ZIRCONIUM	

NONMETALLICS	
BOROSILICATE GLASS	R
CARBON	R

TEMP °F: 60 80 100 120 140 160 180 200 220 240 260 280 300 320 340 360 380 400 420 440 460 480 500 520 540 560

FOR METALS

E = < 2 Mils Penetration/Year; G = < 20 Mils Penetration/Year

S = < 50 Mils Penetration/Year; U = > 50 Mils Penetration/Year

FOR NONMETALLICS

R = Resistant

U = Unsatisfactory

PLASTICS	TEMP °F (60–460)
ABS	R
ACRYLICS	
ASBESTOS REINF., EPOXY	
ASBESTOS REINF., PHENOLIC	
CHLORINATED POLYETHER	R
CPVC	R
EPOXY	
FLUOROCARBONS FEP	R
FLUOROCARBONS TFE	R
FURFURYL ALCOHOL	
NORYL	
PHENOLIC	
POLYAMIDES—NYLON	
POLYESTERS	
BISPHENOL A-FUMURATE	
HYDROGENATED BISPHENOL-A-BISPHENOL-A	
ISOPHTHALIC	
CHLORINATED POLYESTERS	
POLYSTYRENES	
POLYETHYLENE	R
POLYPROPYLENE	R
PVC—TYPE 1	R
PVC—TYPE 2	R
POLYVINYLIDENE CHLORIDE	R
VINYLIDENE FLUORIDE	R
VINYL ESTER	

RUBBERS	TEMP °F (60–460)
BUTYL GR-1	
FLUORO ELASTOMERS VITON A	
FLUORO ELASTOMERS KEL-F 3700	
HARD RUBBER	
HYPALON	
KOROSEAL	
NATURAL RUBBER (GRS)	
NEOPRENE GR-M (CR)	
NITRILE BUNA N (NBR)	
NORDEL	
POLYBUTADIENE	
SILICONE	
POLYURETHANE	

DIBUTYL PHTHALATE

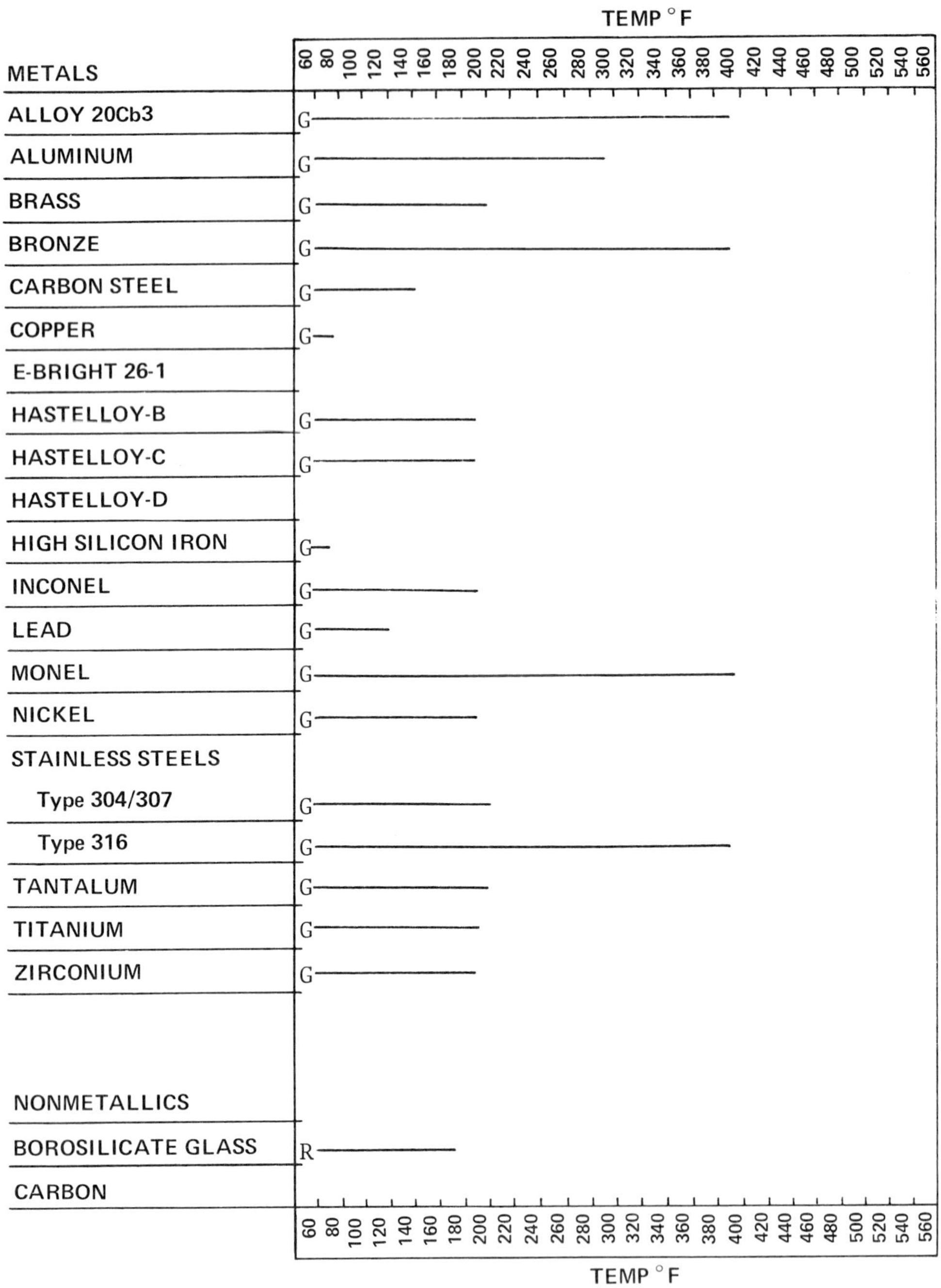

FOR METALS

E = < 2 Mils Penetration/Year; G = < 20 Mils Penetration/Year

S = < 50 Mils Penetration/Year; U = > 50 Mils Penetration/Year

FOR NONMETALLICS

R = Resistant

U = Unsatisfactory

DIBUTYL PHTHALATE

TEMP °F (60–460)

PLASTICS	60 80 100 120 140 160 180 200 220 240 260 280 300 320 340 360 380 400 420 440 460
ABS	U
ACRYLICS	
ASBESTOS REINF., EPOXY	
ASBESTOS REINF., PHENOLIC	
CHLORINATED POLYETHER	
CPVC	U
EPOXY	R—
FLUOROCARBONS FEP	R—
FLUOROCARBONS TFE	R—
FURFURYL ALCOHOL	
NORYL	
PHENOLIC	
POLYAMIDES–NYLON	R—
POLYESTERS	
BISPHENOL A-FUMURATE	R—
HYDROGENATED BISPHENOL-A-BISPHENOL-A	R—
ISOPHTHALIC	R—
CHLORINATED POLYESTERS	R—
POLYSTYRENES	
POLYETHYLENE	U
POLYPROPYLENE	R— U
PVC–TYPE 1	U
PVC–TYPE 2	U
POLYVINYLIDENE CHLORIDE	R—
VINYLIDENE FLUORIDE	U
VINYL ESTER	

RUBBERS	60 80 100 120 140 160 180 200 220 240 260 280 300 320 340 360 380 400 420 440 460
BUTYL GR-1	R—
FLUORO ELASTOMERS VITON A	U
FLUORO ELASTOMERS KEL-F 3700	
HARD RUBBER	
HYPALON	U
KOROSEAL	
NATURAL RUBBER (GRS)	U
NEOPRENE GR-M (CR)	U
NITRILE BUNA N (NBR)	U
NORDEL	R—
POLYBUTADIENE	
SILICONE	
POLYURETHANE	U

TEMP °F

DICHLOROACETIC ACID

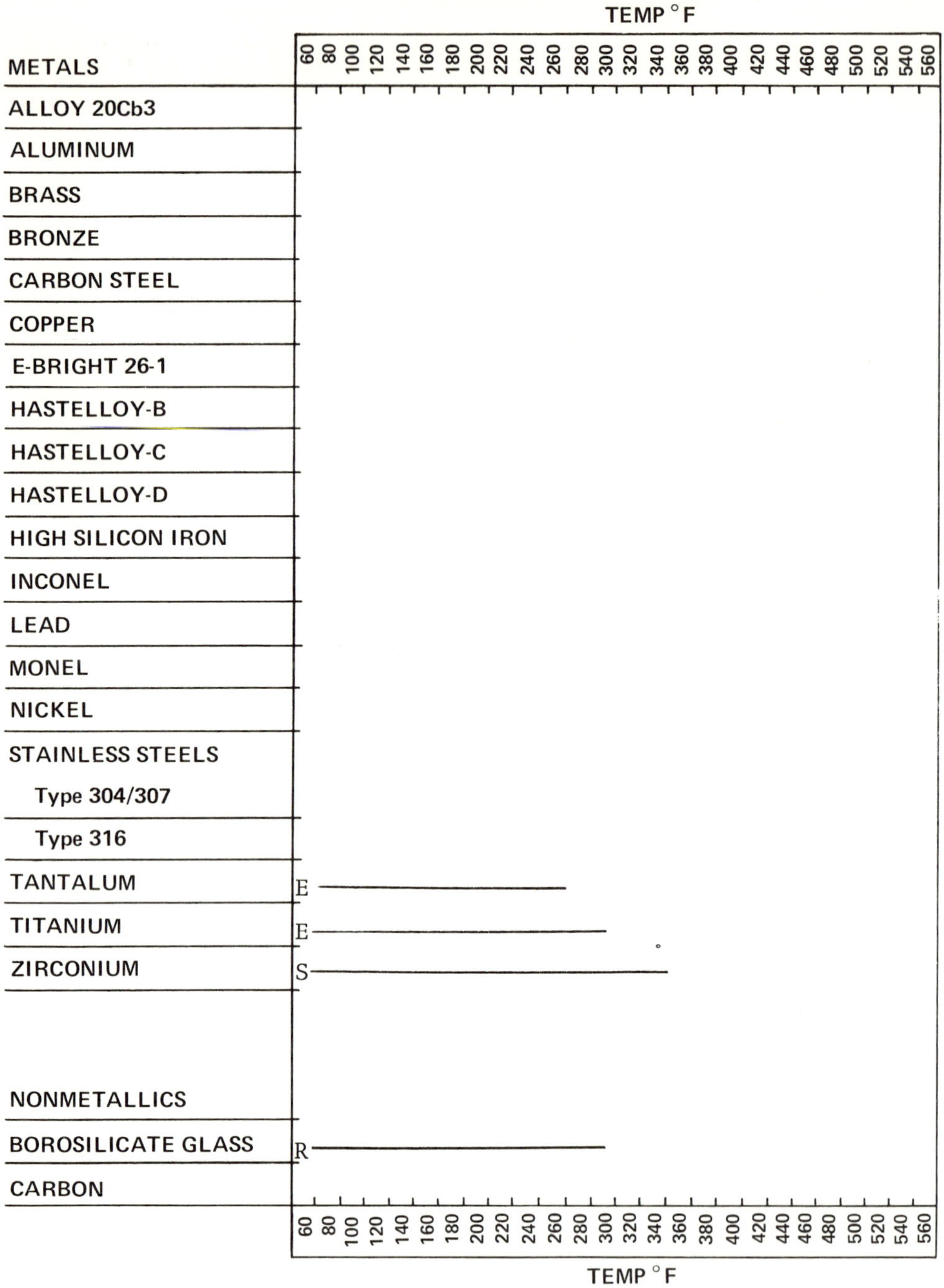

FOR METALS

E = < 2 Mils Penetration/Year; G = < 20 Mils Penetration/Year

S = < 50 Mils Penetration/Year; U = > 50 Mils Penetration/Year

FOR NONMETALLICS

R = Resistant

U = Unsatisfactory

DICHLOROACETIC ACID

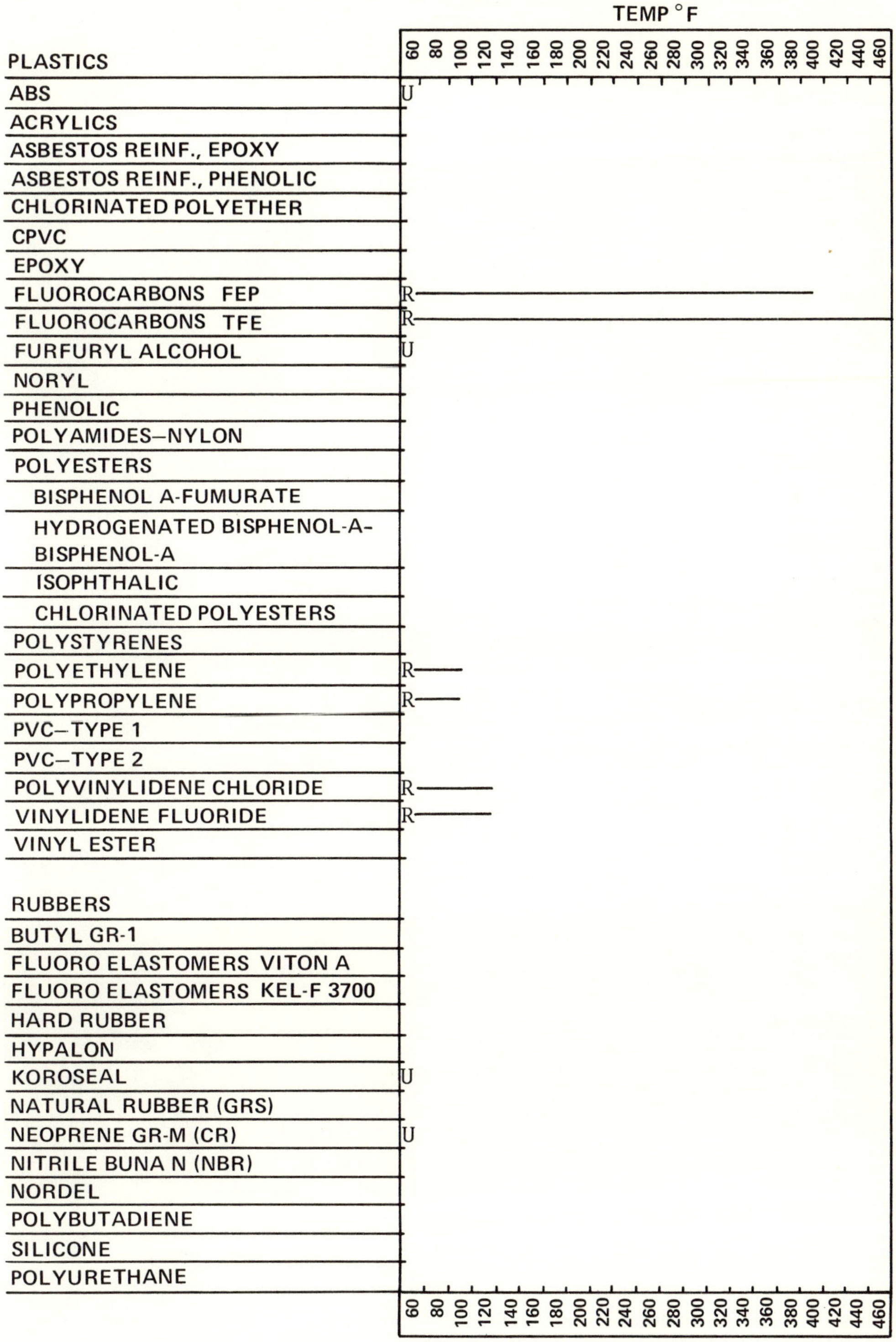

DICHLOROBENZENE

METALS	TEMP °F (60–560)
ALLOY 20Cb3	
ALUMINUM	
BRASS	
BRONZE	
CARBON STEEL	
COPPER	
E-BRIGHT 26-1	
HASTELLOY-B	
HASTELLOY-C	
HASTELLOY-D	
HIGH SILICON IRON	
INCONEL	
LEAD	
MONEL	
NICKEL	
STAINLESS STEELS Type 304/307	
Type 316	
TANTALUM	
TITANIUM	
ZIRCONIUM	
NONMETALLICS	
BOROSILICATE GLASS	R————
CARBON	

FOR METALS

E = < 2 Mils Penetration/Year; G = < 20 Mils Penetration/Year

S = < 50 Mils Penetration/Year; U = > 50 Mils Penetration/Year

FOR NONMETALLICS

R = Resistant

U = Unsatisfactory

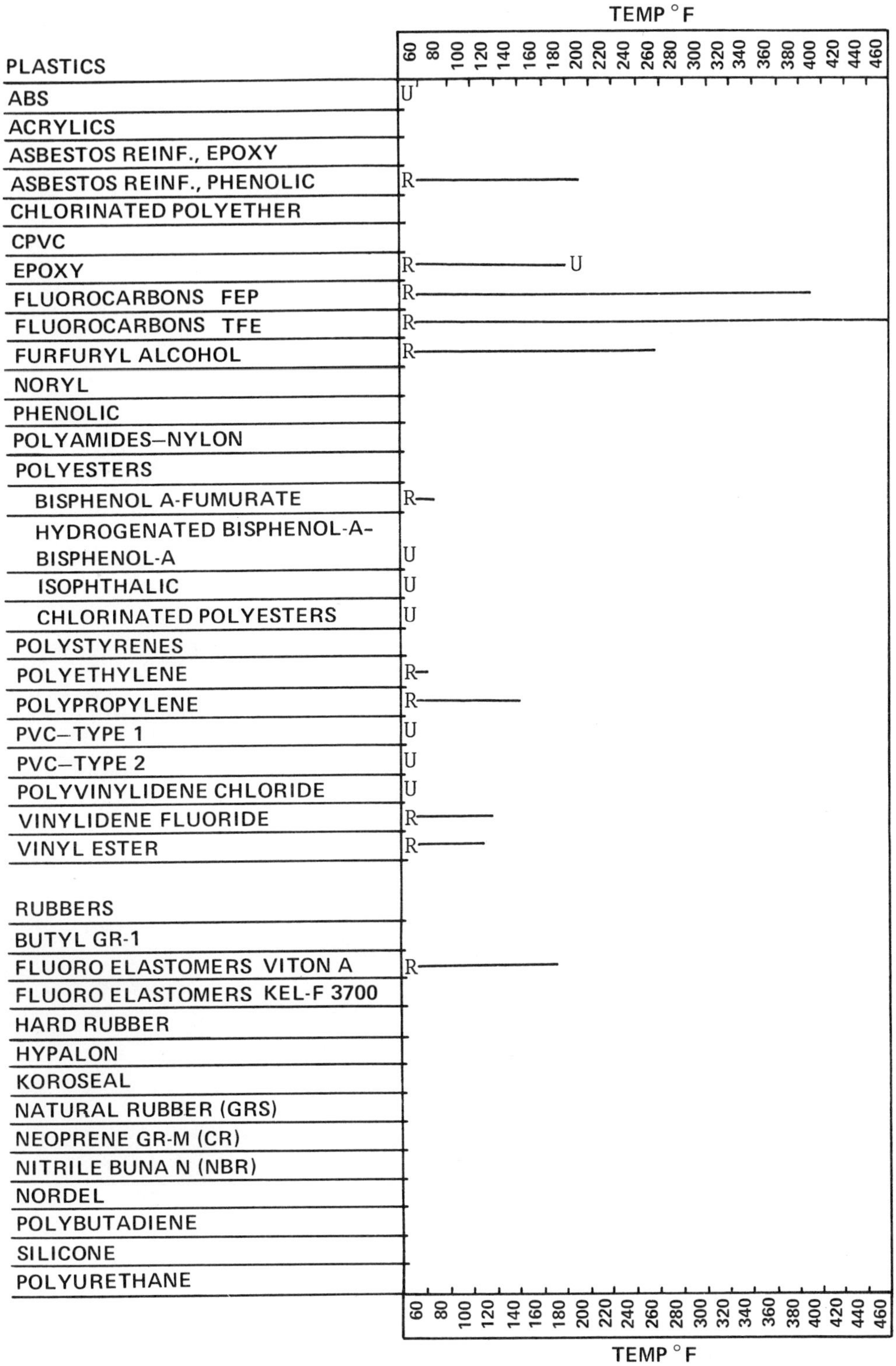
TEMP °F
60 80 100 120 140 160 180 200 220 240 260 280 300 320 340 360 380 400 420 440 460
PLASTICS
ABS U
ACRYLICS
ASBESTOS REINF., EPOXY
ASBESTOS REINF., PHENOLIC R
CHLORINATED POLYETHER
CPVC
EPOXY R U
FLUOROCARBONS FEP R
FLUOROCARBONS TFE R
FURFURYL ALCOHOL R
NORYL
PHENOLIC
POLYAMIDES–NYLON
POLYESTERS
BISPHENOL A-FUMURATE R
HYDROGENATED BISPHENOL-A-BISPHENOL-A U
ISOPHTHALIC U
CHLORINATED POLYESTERS U
POLYSTYRENES
POLYETHYLENE R
POLYPROPYLENE R
PVC–TYPE 1 U
PVC–TYPE 2 U
POLYVINYLIDENE CHLORIDE U
VINYLIDENE FLUORIDE R
VINYL ESTER R
RUBBERS
BUTYL GR-1
FLUORO ELASTOMERS VITON A R
FLUORO ELASTOMERS KEL-F 3700
HARD RUBBER
HYPALON
KOROSEAL
NATURAL RUBBER (GRS)
NEOPRENE GR-M (CR)
NITRILE BUNA N (NBR)
NORDEL
POLYBUTADIENE
SILICONE
POLYURETHANE
60 80 100 120 140 160 180 200 220 240 260 280 300 320 340 360 380 400 420 440 460
TEMP °F

DICHLOROETHANE (ETHYLENEDICHLORIDE)

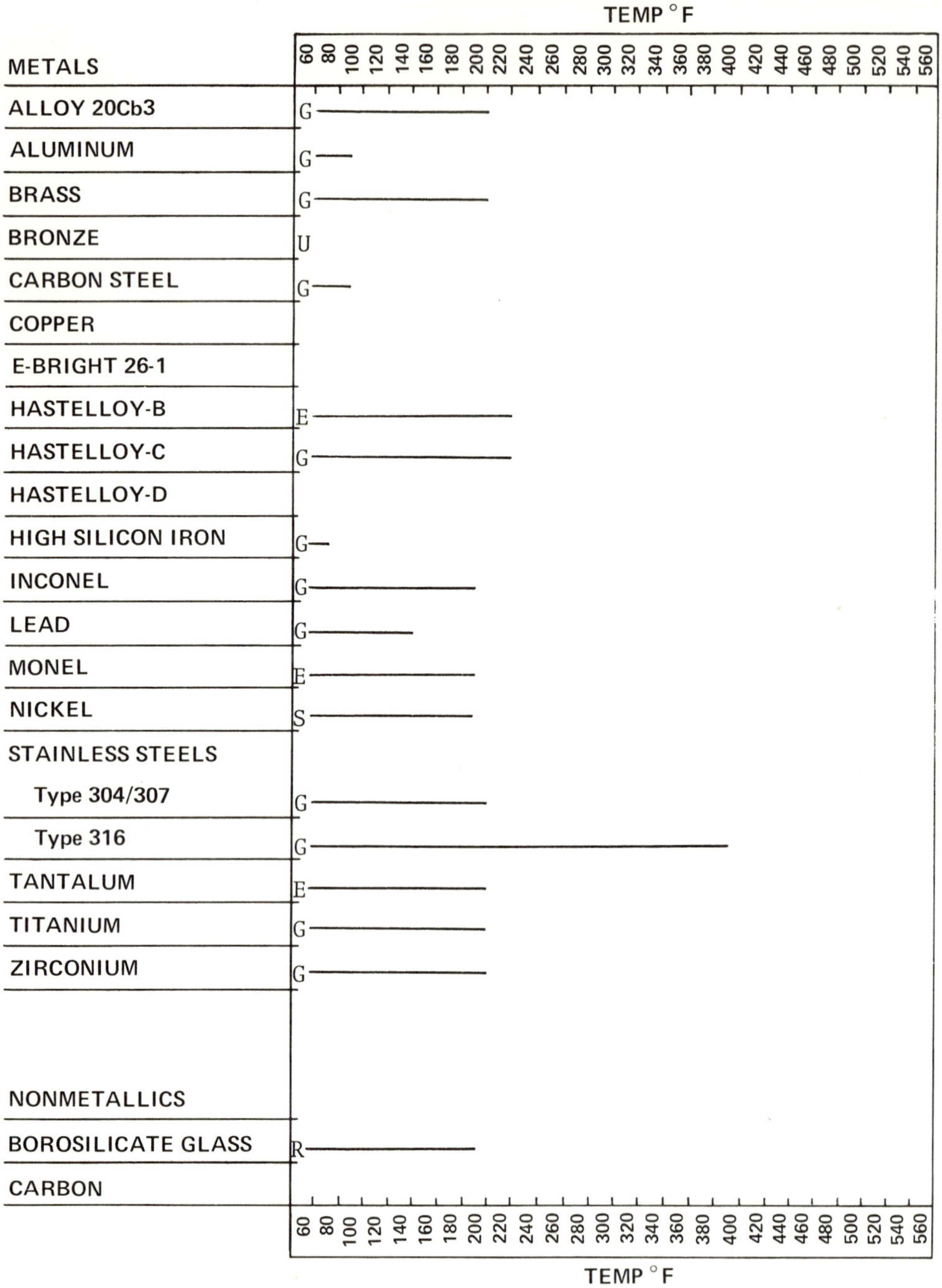

FOR METALS

E = < 2 Mils Penetration/Year; G = < 20 Mils Penetration/Year

S = < 50 Mils Penetration/Year; U = > 50 Mils Penetration/Year

FOR NONMETALLICS

R = Resistant

U = Unsatisfactory

DICHLORETHANE (ETHYLENEDICHLORIDE)

TEMP °F: 60, 80, 100, 120, 140, 160, 180, 200, 220, 240, 260, 280, 300, 320, 340, 360, 380, 400, 420, 440, 460

PLASTICS	
ABS	U
ACRYLICS	U
ASBESTOS REINF., EPOXY	U
ASBESTOS REINF., PHENOLIC	R
CHLORINATED POLYETHER	U
CPVC	U
EPOXY	U
FLUOROCARBONS FEP	R
FLUOROCARBONS TFE	R
FURFURYL ALCOHOL	R
NORYL	
PHENOLIC	
POLYAMIDES–NYLON	R
POLYESTERS	
BISPHENOL A-FUMURATE	U
HYDROGENATED BISPHENOL-A-BISPHENOL-A	U
ISOPHTHALIC	U
CHLORINATED POLYESTERS	U
POLYSTYRENES	
POLYETHYLENE	U
POLYPROPYLENE	R
PVC–TYPE 1	U
PVC–TYPE 2	U
POLYVINYLIDENE CHLORIDE	R
VINYLIDENE FLUORIDE	R
VINYL ESTER	R
RUBBERS	
BUTYL GR-1	U
FLUORO ELASTOMERS VITON A	R
FLUORO ELASTOMERS KEL-F 3700	
HARD RUBBER	
HYPALON	U
KOROSEAL	
NATURAL RUBBER (GRS)	U
NEOPRENE GR-M (CR)	U
NITRILE BUNA N (NBR)	U
NORDEL	
POLYBUTADIENE	
SILICONE	
POLYURETHANE	

TEMP °F: 60, 80, 100, 120, 140, 160, 180, 200, 220, 240, 260, 280, 300, 320, 340, 360, 380, 400, 420, 440, 460

DICHLORETHYLENE

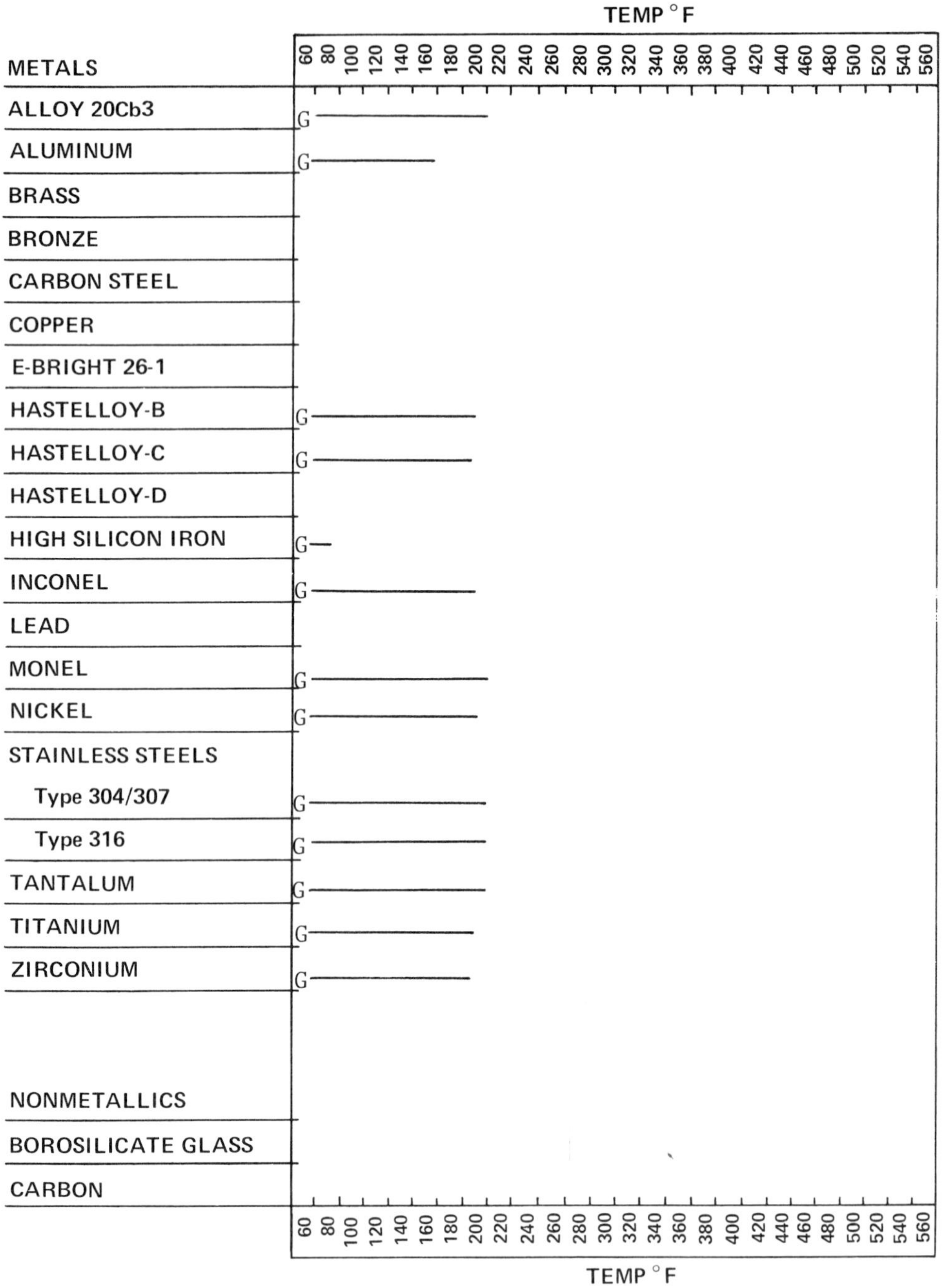

FOR METALS

E = < 2 Mils Penetration/Year; G = < 20 Mils Penetration/Year

S = < 50 Mils Penetration/Year; U = > 50 Mils Penetration/Year

FOR NONMETALLICS

R = Resistant

U = Unsatisfactory

TEMP °F (60, 80, 100, 120, 140, 160, 180, 200, 220, 240, 260, 280, 300, 320, 340, 360, 380, 400, 420, 440, 460)

PLASTICS	
ABS	
ACRYLICS	
ASBESTOS REINF., EPOXY	
ASBESTOS REINF., PHENOLIC	
CHLORINATED POLYETHER	
CPVC	
EPOXY	RU
FLUOROCARBONS FEP	R
FLUOROCARBONS TFE	R
FURFURYL ALCOHOL	
NORYL	
PHENOLIC	
POLYAMIDES–NYLON	
POLYESTERS	
BISPHENOL A-FUMURATE	U
HYDROGENATED BISPHENOL-A-BISPHENOL-A	
ISOPHTHALIC	U
CHLORINATED POLYESTERS	
POLYSTYRENES	
POLYETHYLENE	U
POLYPROPYLENE	R
PVC–TYPE 1	
PVC–TYPE 2	
POLYVINYLIDENE CHLORIDE	
VINYLIDENE FLUORIDE	R
VINYL ESTER	U

RUBBERS	
BUTYL GR-1	
FLUORO ELASTOMERS VITON A	R
FLUORO ELASTOMERS KEL-F 3700	
HARD RUBBER	
HYPALON	
KOROSEAL	
NATURAL RUBBER (GRS)	
NEOPRENE GR-M (CR)	
NITRILE BUNA N (NBR)	U
NORDEL	
POLYBUTADIENE	
SILICONE	
POLYURETHANE	

TEMP °F

DIESEL FUELS

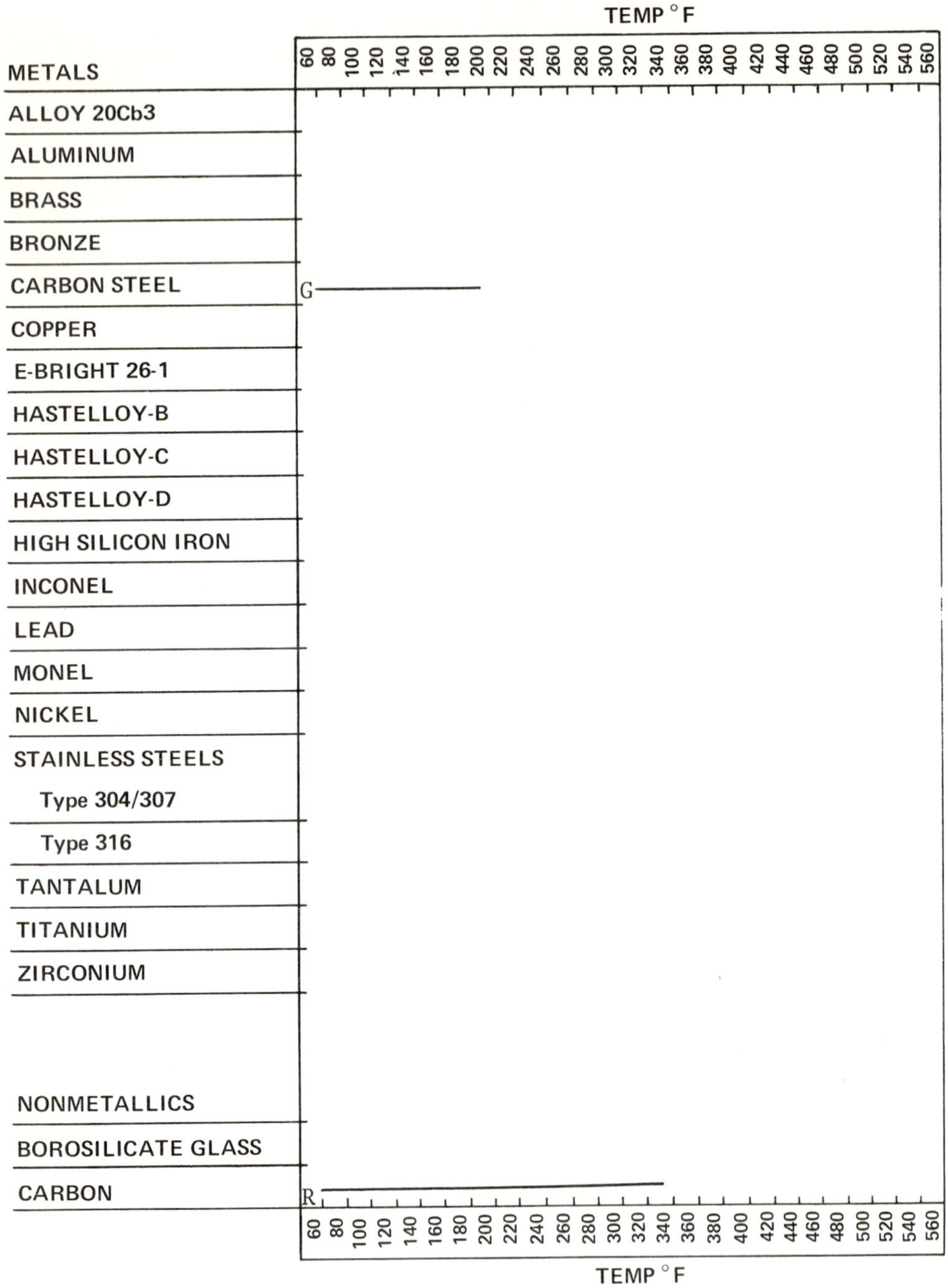

FOR METALS

E = < 2 Mils Penetration/Year; G = < 20 Mils Penetration/Year

S = < 50 Mils Penetration/Year; U = > 50 Mils Penetration/Year

FOR NONMETALLICS

R = Resistant

U = Unsatisfactory

PLASTICS	TEMP °F (60–460)
ABS	
ACRYLICS	
ASBESTOS REINF., EPOXY	
ASBESTOS REINF., PHENOLIC	
CHLORINATED POLYETHER	R
CPVC	
EPOXY	R
FLUOROCARBONS FEP	R
FLUOROCARBONS TFE	R
FURFURYL ALCOHOL	R
NORYL	
PHENOLIC	
POLYAMIDES–NYLON	R
POLYESTERS	
BISPHENOL A-FUMURATE	
HYDROGENATED BISPHENOL-A-BISPHENOL-A	
ISOPHTHALIC	R
CHLORINATED POLYESTERS	R
POLYSTYRENES	
POLYETHYLENE	R
POLYPROPYLENE	R
PVC–TYPE 1	R
PVC–TYPE 2	
POLYVINYLIDENE CHLORIDE	R
VINYLIDENE FLUORIDE	R
VINYL ESTER	R

RUBBERS	TEMP °F (60–460)
BUTYL GR-1	
FLUORO ELASTOMERS VITON A	R
FLUORO ELASTOMERS KEL-F 3700	
HARD RUBBER	
HYPALON	R
KOROSEAL	
NATURAL RUBBER (GRS)	
NEOPRENE GR-M (CR)	R
NITRILE BUNA N (NBR)	R
NORDEL	
POLYBUTADIENE	
SILICONE	
POLYURETHANE	

DIETHYLAMINE

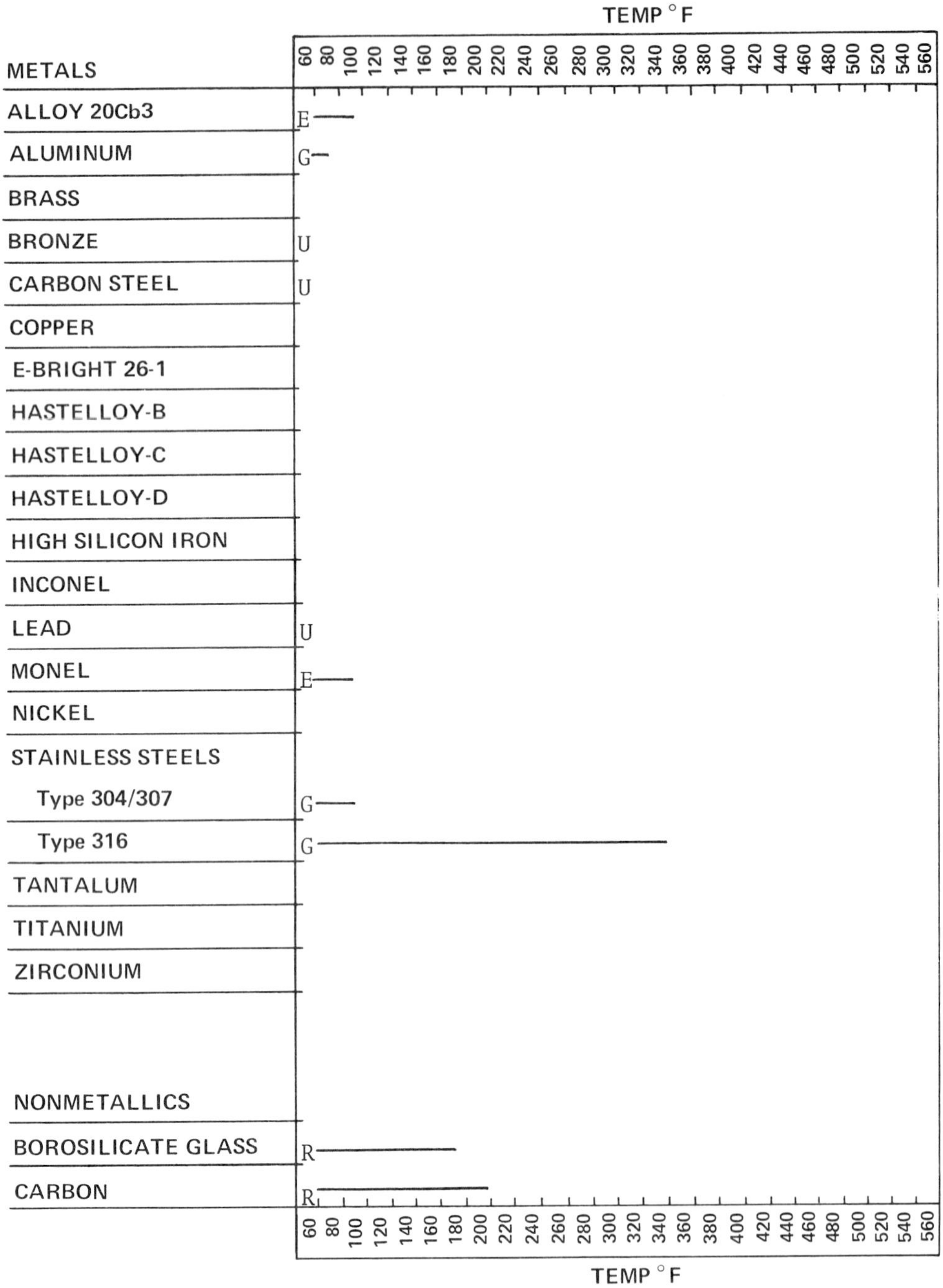

FOR METALS

E = < 2 Mils Penetration/Year; G = < 20 Mils Penetration/Year

S = < 50 Mils Penetration/Year; U = > 50 Mils Penetration/Year

FOR NONMETALLICS

R = Resistant

U = Unsatisfactory

DIETHYLAMINE

TEMP °F

PLASTICS	60–460 °F
ABS	U
ACRYLICS	
ASBESTOS REINF., EPOXY	
ASBESTOS REINF., PHENOLIC	
CHLORINATED POLYETHER	R
CPVC	
EPOXY	
FLUOROCARBONS FEP	R
FLUOROCARBONS TFE	R
FURFURYL ALCOHOL	R
NORYL	
PHENOLIC	
POLYAMIDES—NYLON	
POLYESTERS	
BISPHENOL A-FUMURATE	U
HYDROGENATED BISPHENOL-A-BISPHENOL-A	
ISOPHTHALIC	R
CHLORINATED POLYESTERS	
POLYSTYRENES	
POLYETHYLENE	
POLYPROPYLENE	R
PVC—TYPE 1	U
PVC—TYPE 2	
POLYVINYLIDENE CHLORIDE	U
VINYLIDENE FLUORIDE	R
VINYL ESTER	
RUBBERS	
BUTYL GR-1	R
FLUORO ELASTOMERS VITON A	RU
FLUORO ELASTOMERS KEL-F 3700	
HARD RUBBER	
HYPALON	U
KOROSEAL	
NATURAL RUBBER (GRS)	R
NEOPRENE GR-M (CR)	R
NITRILE BUNA N (NBR)	U
NORDEL	
POLYBUTADIENE	
SILICONE	
POLYURETHANE	

TEMP °F (60, 80, 100, 120, 140, 160, 180, 200, 220, 240, 260, 280, 300, 320, 340, 360, 380, 400, 420, 440, 460)

DIETHYL CELLOSOLVE

TEMP °F: 60 80 100 120 140 160 180 200 220 240 260 280 300 320 340 360 380 400 420 440 460 480 500 520 540 560

METALS	
ALLOY 20Cb3	
ALUMINUM	
BRASS	
BRONZE	
CARBON STEEL	
COPPER	
E-BRIGHT 26-1	
HASTELLOY-B	
HASTELLOY-C	
HASTELLOY-D	
HIGH SILICON IRON	
INCONEL	
LEAD	
MONEL	
NICKEL	
STAINLESS STEELS	
Type 304/307	G
Type 316	
TANTALUM	
TITANIUM	
ZIRCONIUM	

NONMETALLICS	
BOROSILICATE GLASS	
CARBON	R

TEMP °F: 60 80 100 120 140 160 180 200 220 240 260 280 300 320 340 360 380 400 420 440 460 480 500 520 540 560

FOR METALS

E = < 2 Mils Penetration/Year; G = < 20 Mils Penetration/Year

S = < 50 Mils Penetration/Year; U = > 50 Mils Penetration/Year

FOR NONMETALLICS

R = Resistant

U = Unsatisfactory

DIETHYL CELLOSOLVE

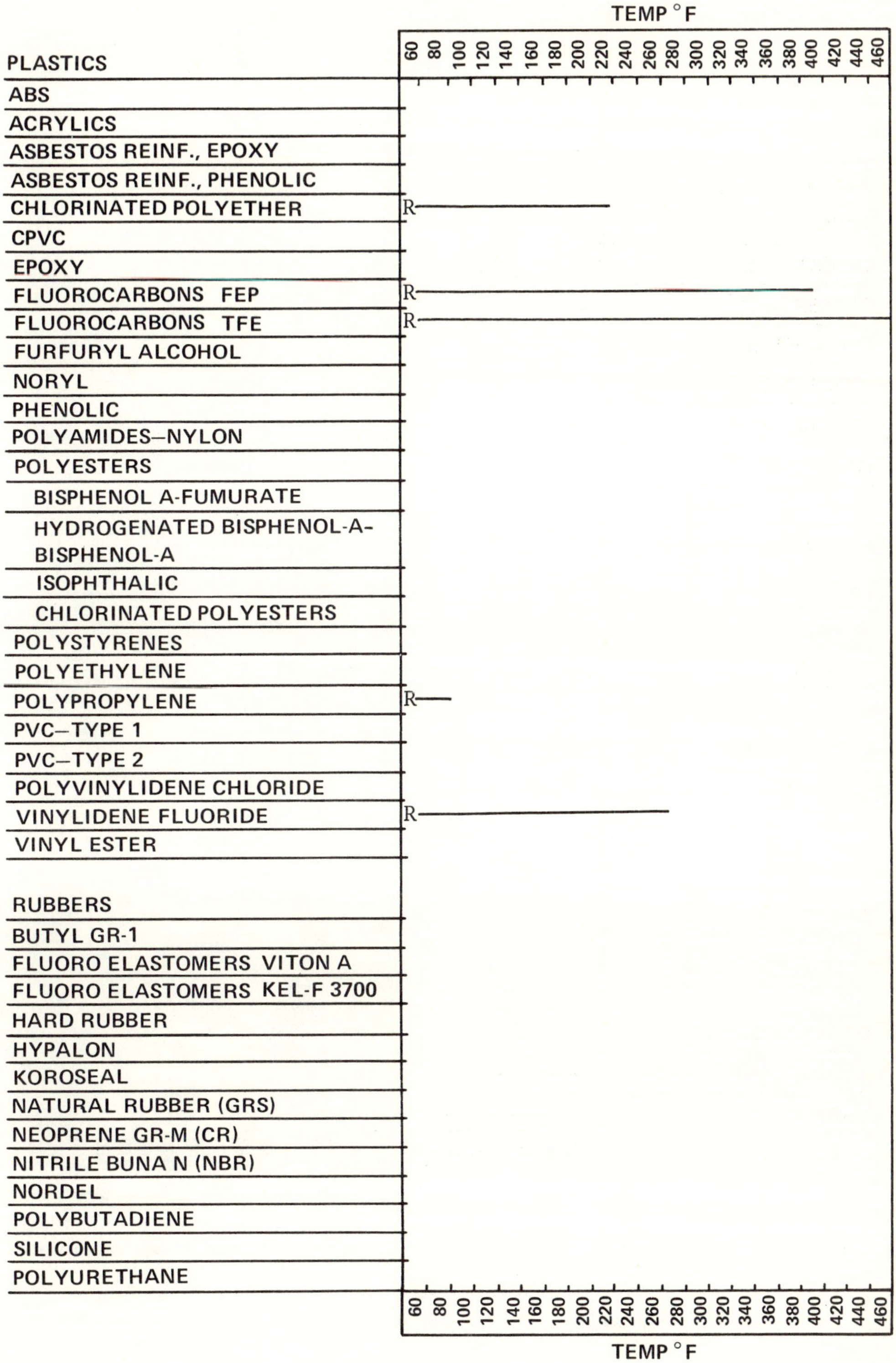

DIETHYL ETHER

METALS	TEMP °F
ALLOY 20Cb3	G
ALUMINUM	G
BRASS	G
BRONZE	G
CARBON STEEL	G
COPPER	G
E-BRIGHT 26-1	
HASTELLOY-B	
HASTELLOY-C	
HASTELLOY-D	
HIGH SILICON IRON	G
INCONEL	G
LEAD	G
MONEL	G
NICKEL	G
STAINLESS STEELS	
Type 304/307	G
Type 316	G
TANTALUM	G
TITANIUM	G
ZIRCONIUM	
NONMETALLICS	
BOROSILICATE GLASS	R
CARBON	R

TEMP °F: 60 80 100 120 140 160 180 200 220 240 260 280 300 320 340 360 380 400 420 440 460 480 500 520 540 560

FOR METALS

E = < 2 Mils Penetration/Year; G = < 20 Mils Penetration/Year

S = < 50 Mils Penetration/Year; U = > 50 Mils Penetration/Year

FOR NONMETALLICS

R = Resistant

U = Unsatisfactory

DIETHYL ETHER

TEMP °F: 60 80 100 120 140 160 180 200 220 240 260 280 300 320 340 360 380 400 420 440 460

PLASTICS	
ABS	U
ACRYLICS	
ASBESTOS REINF., EPOXY	
ASBESTOS REINF., PHENOLIC	
CHLORINATED POLYETHER	R————
CPVC	
EPOXY	U
FLUOROCARBONS FEP	R————
FLUOROCARBONS TFE	R————
FURFURYL ALCOHOL	R————
NORYL	
PHENOLIC	
POLYAMIDES—NYLON	R—
POLYESTERS	
BISPHENOL A-FUMURATE	U
HYDROGENATED BISPHENOL-A-BISPHENOL-A	U
ISOPHTHALIC	U
CHLORINATED POLYESTERS	U
POLYSTYRENES	
POLYETHYLENE	R—
POLYPROPYLENE	R—— U
PVC—TYPE 1	
PVC—TYPE 2	
POLYVINYLIDENE CHLORIDE	U
VINYLIDENE FLUORIDE	R————
VINYL ESTER	U
RUBBERS	
BUTYL GR-1	U
FLUORO ELASTOMERS VITON A	U
FLUORO ELASTOMERS KEL-F 3700	
HARD RUBBER	
HYPALON	
KOROSEAL	U
NATURAL RUBBER (GRS)	U
NEOPRENE GR-M (CR)	U
NITRILE BUNA N (NBR)	U
NORDEL	
POLYBUTADIENE	
SILICONE	
POLYURETHANE	

TEMP °F: 60 80 100 120 140 160 180 200 220 240 260 280 300 320 340 360 380 400 420 440 460

DIETHYLENE TRIAMINE

METALS	60	80	100	120	140	160	180	200	220	240	260	280	300	320	340	360	380	400	420	440	460	480	500	520	540	560
ALLOY 20Cb3																										
ALUMINUM	G—																									
BRASS	U																									
BRONZE	U																									
CARBON STEEL																										
COPPER	U																									
E-BRIGHT 26-1																										
HASTELLOY-B																										
HASTELLOY-C																										
HASTELLOY-D																										
HIGH SILICON IRON																										
INCONEL																										
LEAD																										
MONEL																										
NICKEL																										
STAINLESS STEELS Type 304/307																										
Type 316																										
TANTALUM	G—																									
TITANIUM																										
ZIRCONIUM																										
NONMETALLICS																										
BOROSILICATE GLASS	R	—	—	—	—	—	—	—	—	—	—	—	—													
CARBON																										

TEMP °F

FOR METALS

E = < 2 Mils Penetration/Year; G = < 20 Mils Penetration/Year

S = < 50 Mils Penetration/Year; U = > 50 Mils Penetration/Year

FOR NONMETALLICS

R = Resistant

U = Unsatisfactory

TEMP °F

PLASTICS	60–460 °F
ABS	
ACRYLICS	
ASBESTOS REINF., EPOXY	
ASBESTOS REINF., PHENOLIC	
CHLORINATED POLYETHER	
CPVC	
EPOXY	R (60–80) — U (100)
FLUOROCARBONS FEP	R (60–400)
FLUOROCARBONS TFE	R (60–460)
FURFURYL ALCOHOL	
NORYL	
PHENOLIC	
POLYAMIDES–NYLON	
POLYESTERS	
BISPHENOL A-FUMURATE	
HYDROGENATED BISPHENOL-A-BISPHENOL-A	
ISOPHTHALIC	
CHLORINATED POLYESTERS	
POLYSTYRENES	
POLYETHYLENE	
POLYPROPYLENE	R (60–120)
PVC–TYPE 1	
PVC–TYPE 2	
POLYVINYLIDENE CHLORIDE	
VINYLIDENE FLUORIDE	R (60–140)
VINYL ESTER	

RUBBERS	60–460 °F
BUTYL GR-1	
FLUORO ELASTOMERS VITON A	
FLUORO ELASTOMERS KEL-F 3700	
HARD RUBBER	
HYPALON	
KOROSEAL	
NATURAL RUBBER (GRS)	
NEOPRENE GR-M (CR)	
NITRILE BUNA N (NBR)	
NORDEL	
POLYBUTADIENE	
SILICONE	
POLYURETHANE	

TEMP °F

DIGLYCOLIC ACID

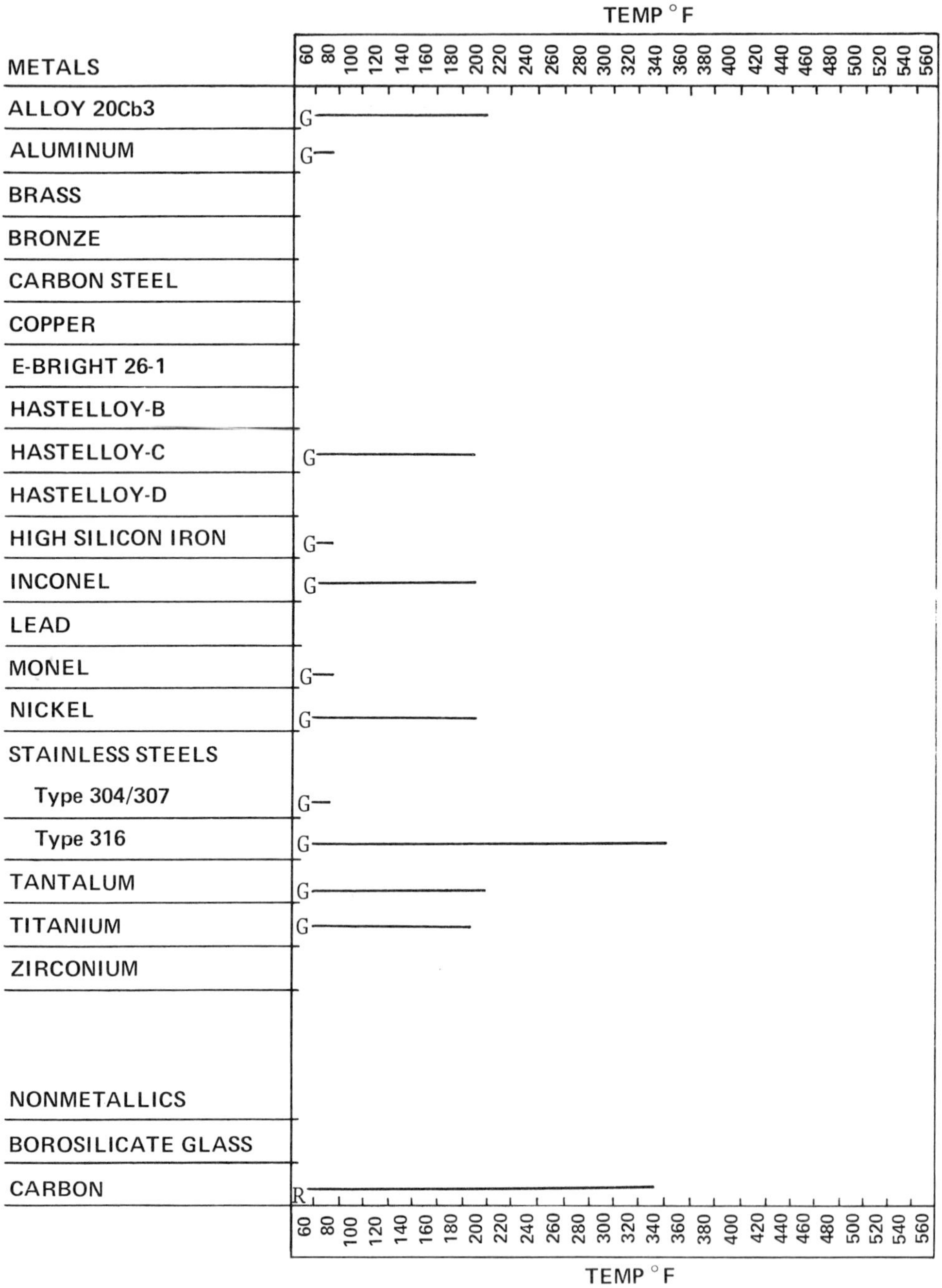

FOR METALS

E = < 2 Mils Penetration/Year; G = < 20 Mils Penetration/Year

S = < 50 Mils Penetration/Year; U = > 50 Mils Penetration/Year

FOR NONMETALLICS

R = Resistant

U = Unsatisfactory

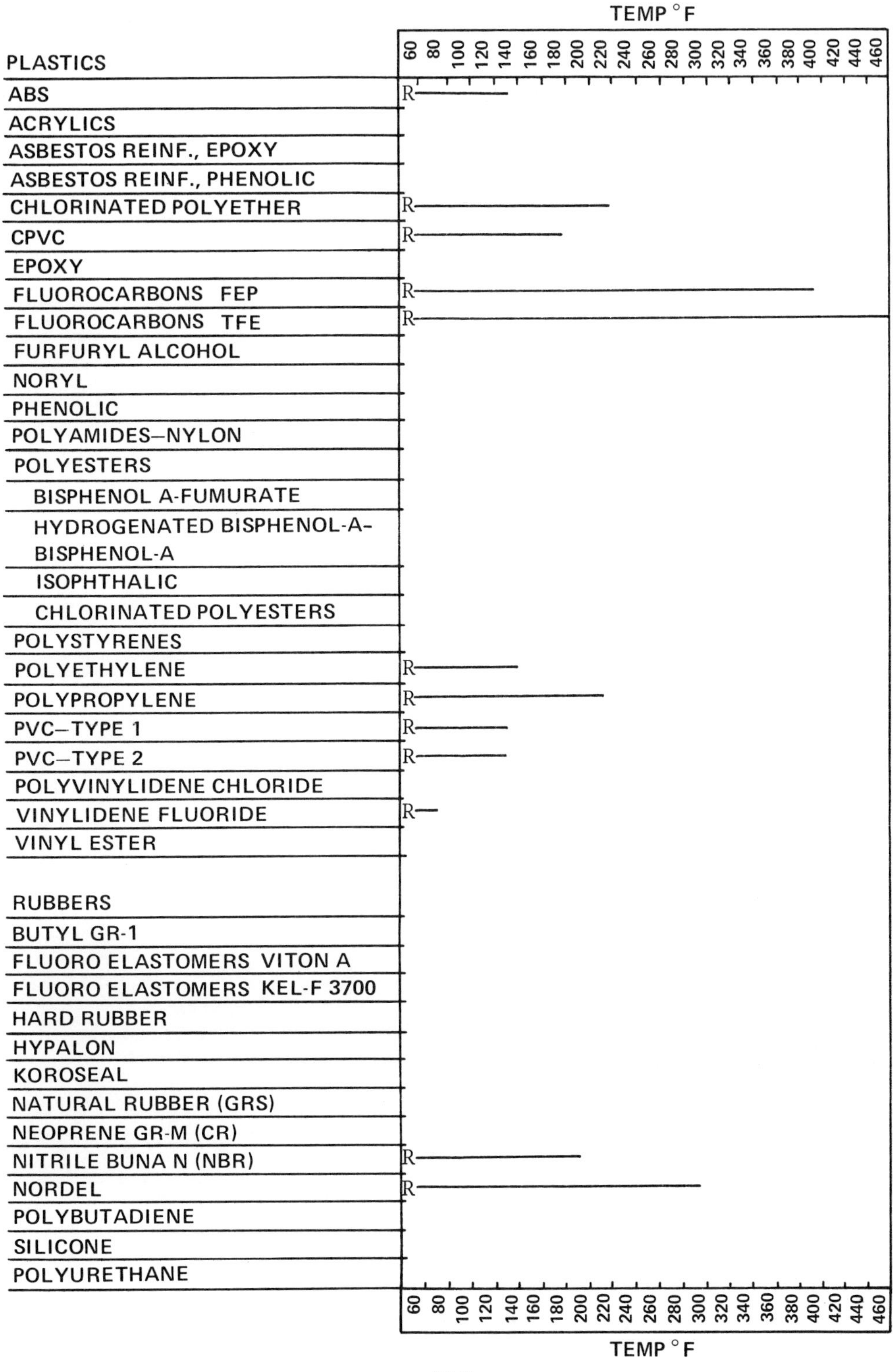
TEMP °F
60 80 100 120 140 160 180 200 220 240 260 280 300 320 340 360 380 400 420 440 460
PLASTICS
ABS R
ACRYLICS
ASBESTOS REINF., EPOXY
ASBESTOS REINF., PHENOLIC
CHLORINATED POLYETHER R
CPVC R
EPOXY
FLUOROCARBONS FEP R
FLUOROCARBONS TFE R
FURFURYL ALCOHOL
NORYL
PHENOLIC
POLYAMIDES–NYLON
POLYESTERS
BISPHENOL A-FUMURATE
HYDROGENATED BISPHENOL-A-BISPHENOL-A
ISOPHTHALIC
CHLORINATED POLYESTERS
POLYSTYRENES
POLYETHYLENE R
POLYPROPYLENE R
PVC–TYPE 1 R
PVC–TYPE 2 R
POLYVINYLIDENE CHLORIDE
VINYLIDENE FLUORIDE R
VINYL ESTER
RUBBERS
BUTYL GR-1
FLUORO ELASTOMERS VITON A
FLUORO ELASTOMERS KEL-F 3700
HARD RUBBER
HYPALON
KOROSEAL
NATURAL RUBBER (GRS)
NEOPRENE GR-M (CR)
NITRILE BUNA N (NBR) R
NORDEL R
POLYBUTADIENE
SILICONE
POLYURETHANE
60 80 100 120 140 160 180 200 220 240 260 280 300 320 340 360 380 400 420 440 460
TEMP °F

DIMETHYLAMINE

TEMP °F

METALS	60	80	100	120	140	160	180	200	220	240	260	280	300	320	340	360	380	400	420	440	460	480	500	520	540	560
ALLOY 20Cb3																										
ALUMINUM																										
BRASS	U																									
BRONZE	U																									
CARBON STEEL																										
COPPER	U																									
E-BRIGHT 26-1																										
HASTELLOY-B																										
HASTELLOY-C																										
HASTELLOY-D																										
HIGH SILICON IRON																										
INCONEL																										
LEAD																										
MONEL																										
NICKEL																										
STAINLESS STEELS Type 304/307																										
Type 316																										
TANTALUM	G—																									
TITANIUM																										
ZIRCONIUM																										
NONMETALLICS																										
BOROSILICATE GLASS																										
CARBON																										

TEMP °F

FOR METALS

E = < 2 Mils Penetration/Year; G = < 20 Mils Penetration/Year

S = < 50 Mils Penetration/Year; U = > 50 Mils Penetration/Year

FOR NONMETALLICS

R = Resistant

U = Unsatisfactory

DIMETHYLAMINE

TEMP °F

PLASTICS	60 – 460
ABS	U
ACRYLICS	
ASBESTOS REINF., EPOXY	
ASBESTOS REINF., PHENOLIC	
CHLORINATED POLYETHER	R
CPVC	U
EPOXY	U
FLUOROCARBONS FEP	R
FLUOROCARBONS TFE	R
FURFURYL ALCOHOL	
NORYL	
PHENOLIC	
POLYAMIDES—NYLON	
POLYESTERS	
BISPHENOL A-FUMURATE	
HYDROGENATED BISPHENOL-A-BISPHENOL-A	
ISOPHTHALIC	
CHLORINATED POLYESTERS	
POLYSTYRENES	
POLYETHYLENE	U
POLYPROPYLENE	R
PVC—TYPE 1	R
PVC—TYPE 2	U
POLYVINYLIDENE CHLORIDE	
VINYLIDENE FLUORIDE	R
VINYL ESTER	R

RUBBERS	60 – 460
BUTYL GR-1	
FLUORO ELASTOMERS VITON A	U
FLUORO ELASTOMERS KEL-F 3700	
HARD RUBBER	
HYPALON	
KOROSEAL	
NATURAL RUBBER (GRS)	
NEOPRENE GR-M (CR)	
NITRILE BUNA N (NBR)	
NORDEL	
POLYBUTADIENE	
SILICONE	
POLYURETHANE	

TEMP °F

DIMETHYLAMINE, AQUEOUS

METALS	TEMP °F (60 80 100 120 140 160 180 200 220 240 260 280 300 320 340 360 380 400 420 440 460 480 500 520 540 560)
ALLOY 20Cb3	
ALUMINUM	
BRASS	
BRONZE	
CARBON STEEL	
COPPER	
E-BRIGHT 26-1	
HASTELLOY-B	
HASTELLOY-C	
HASTELLOY-D	
HIGH SILICON IRON	
INCONEL	
LEAD	
MONEL	
NICKEL	
STAINLESS STEELS Type 304/307	
Type 316	
TANTALUM	
TITANIUM	
ZIRCONIUM	
NONMETALLICS	
BOROSILICATE GLASS	R (60–200)
CARBON	

FOR METALS

E = < 2 Mils Penetration/Year; G = < 20 Mils Penetration/Year

S = < 50 Mils Penetration/Year; U = > 50 Mils Penetration/Year

FOR NONMETALLICS

R = Resistant

U = Unsatisfactory

DIMETHYLAMINE, AQUEOUS

TEMP °F (60–460)

PLASTICS	Rating
ABS	
ACRYLICS	
ASBESTOS REINF., EPOXY	
ASBESTOS REINF., PHENOLIC	
CHLORINATED POLYETHER	R (60 to ~220)
CPVC	
EPOXY	U
FLUOROCARBONS FEP	R (60 to ~400)
FLUOROCARBONS TFE	R (60 to 460)
FURFURYL ALCOHOL	
NORYL	
PHENOLIC	
POLYAMIDES—NYLON	
POLYESTERS	
BISPHENOL A-FUMURATE	
HYDROGENATED BISPHENOL-A-BISPHENOL-A	
ISOPHTHALIC	
CHLORINATED POLYESTERS	U
POLYSTYRENES	
POLYETHYLENE	
POLYPROPYLENE	R (60 to ~80)
PVC—TYPE 1	U
PVC—TYPE 2	U
POLYVINYLIDENE CHLORIDE	
VINYLIDENE FLUORIDE	R (60 to ~150) — U (~160)
VINYL ESTER	

RUBBERS	Rating
BUTYL GR-1	
FLUORO ELASTOMERS VITON A	
FLUORO ELASTOMERS KEL-F 3700	
HARD RUBBER	
HYPALON	
KOROSEAL	
NATURAL RUBBER (GRS)	
NEOPRENE GR-M (CR)	
NITRILE BUNA N (NBR)	
NORDEL	
POLYBUTADIENE	
SILICONE	
POLYURETHANE	

TEMP °F

DIMETHYL ANILINE

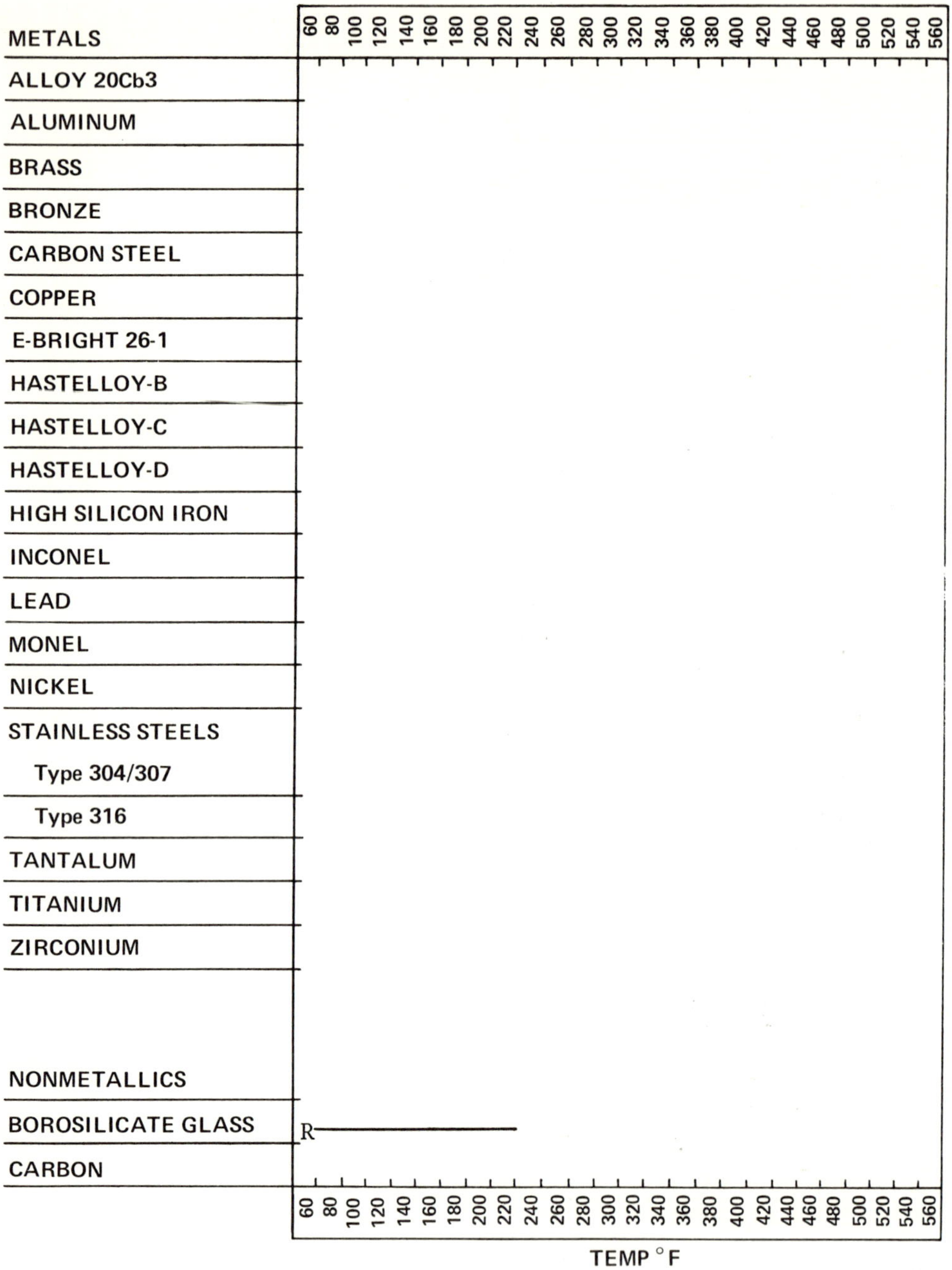

FOR METALS

E = < 2 Mils Penetration/Year; G = < 20 Mils Penetration/Year

S = < 50 Mils Penetration/Year; U = > 50 Mils Penetration/Year

FOR NONMETALLICS

R = Resistant

U = Unsatisfactory

DIMETHYL ANILINE

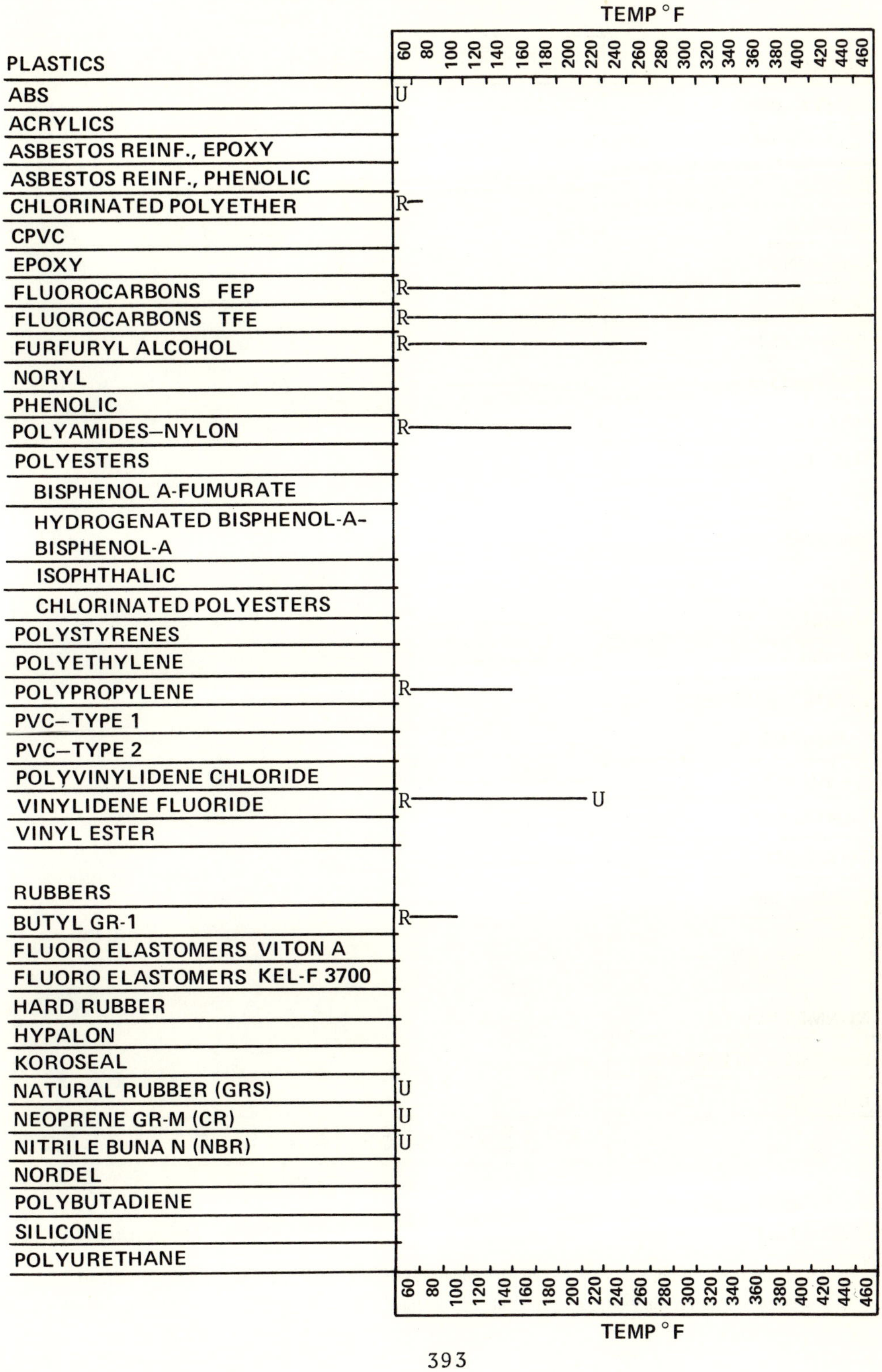

DIMETHYL ETHER

TEMP °F (60–560)

METALS	Rating
ALLOY 20Cb3	G
ALUMINUM	
BRASS	G
BRONZE	G
CARBON STEEL	
COPPER	G
E-BRIGHT 26-1	
HASTELLOY-B	G
HASTELLOY-C	G
HASTELLOY-D	
HIGH SILICON IRON	G
INCONEL	G
LEAD	
MONEL	G
NICKEL	G
STAINLESS STEELS	
Type 304/307	G
Type 316	G
TANTALUM	G
TITANIUM	E
ZIRCONIUM	E

NONMETALLICS	Rating
BOROSILICATE GLASS	
CARBON	

TEMP °F

FOR METALS

E = < 2 Mils Penetration/Year; G = < 20 Mils Penetration/Year

S = < 50 Mils Penetration/Year; U = > 50 Mils Penetration/Year

FOR NONMETALLICS

R = Resistant

U = Unsaturated

DIMETHYL ETHER

TEMP °F: 60 80 100 120 140 160 180 200 220 240 260 280 300 320 340 360 380 400 420 440 460

PLASTICS	DIMETHYL ETHER
ABS	
ACRYLICS	
ASBESTOS REINF., EPOXY	
ASBESTOS REINF., PHENOLIC	
CHLORINATED POLYETHER	
CPVC	
EPOXY	
FLUOROCARBONS FEP	R
FLUOROCARBONS TFE	R
FURFURYL ALCOHOL	
NORYL	
PHENOLIC	
POLYAMIDES—NYLON	
POLYESTERS	
BISPHENOL A-FUMURATE	
HYDROGENATED BISPHENOL-A-BISPHENOL-A	
ISOPHTHALIC	
CHLORINATED POLYESTERS	
POLYSTYRENES	
POLYETHYLENE	
POLYPROPYLENE	
PVC—TYPE 1	
PVC—TYPE 2	
POLYVINYLIDENE CHLORIDE	
VINYLIDENE FLUORIDE	
VINYL ESTER	
RUBBERS	
BUTYL GR-1	
FLUORO ELASTOMERS VITON A	
FLUORO ELASTOMERS KEL-F 3700	
HARD RUBBER	
HYPALON	
KOROSEAL	
NATURAL RUBBER (GRS)	
NEOPRENE GR-M (CR)	
NITRILE BUNA N (NBR)	
NORDEL	
POLYBUTADIENE	
SILICONE	
POLYURETHANE	

TEMP °F: 60 80 100 120 140 160 180 200 220 240 260 280 300 320 340 360 380 400 420 440 460

DIMETHYL FORMAMIDE

TEMP °F (60–560)

METALS	TEMP °F
ALLOY 20Cb3	
ALUMINUM	
BRASS	
BRONZE	
CARBON STEEL	
COPPER	
E-BRIGHT 26-1	
HASTELLOY-B	
HASTELLOY-C	
HASTELLOY-D	
HIGH SILICON IRON	
INCONEL	
LEAD	
MONEL	G (60–400)
NICKEL	
STAINLESS STEELS	
Type 304/307	
Type 316	G (60–400)
TANTALUM	
TITANIUM	
ZIRCONIUM	
NONMETALLICS	
BOROSILICATE GLASS	
CARBON	

FOR METALS

E = < 2 Mils Penetration/Year; G = < 20 Mils Penetration/Year

S = < 50 Mils Penetration/Year; U = > 50 Mils Penetration/Year

FOR NONMETALLICS

R = Resistant

U = Unsatisfactory

DIMETHYL FORMAMIDE

TEMP °F (60–460)

PLASTICS	
ABS	
ACRYLICS	
ASBESTOS REINF., EPOXY	
ASBESTOS REINF., PHENOLIC	
CHLORINATED POLYETHER	R
CPVC	U
EPOXY	U
FLUOROCARBONS FEP	R
FLUOROCARBONS TFE	R
FURFURYL ALCOHOL	U
NORYL	
PHENOLIC	
POLYAMIDES–NYLON	
POLYESTERS	
BISPHENOL A-FUMURATE	U
HYDROGENATED BISPHENOL-A-BISPHENOL-A	U
ISOPHTHALIC	U
CHLORINATED POLYESTERS	U
POLYSTYRENES	
POLYETHYLENE	R
POLYPROPYLENE	R
PVC–TYPE 1	U
PVC–TYPE 2	U
POLYVINYLIDENE CHLORIDE	
VINYLIDENE FLUORIDE	
VINYL ESTER	U
RUBBERS	
BUTYL GR-1	R
FLUORO ELASTOMERS VITON A	U
FLUORO ELASTOMERS KEL-F 3700	
HARD RUBBER	
HYPALON	
KOROSEAL	
NATURAL RUBBER (GRS)	U
NEOPRENE GR-M (CR)	U
NITRILE BUNA N (NBR)	RU
NORDEL	
POLYBUTADIENE	
SILICONE	
POLYURETHANE	

TEMP °F (60–460)

DIMETHYL PHTHALATE

METALS	TEMP °F (60 80 100 120 140 160 180 200 220 240 260 280 300 320 340 360 380 400 420 440 460 480 500 520 540 560)
ALLOY 20Cb3	
ALUMINUM	
BRASS	
BRONZE	
CARBON STEEL	
COPPER	
E-BRIGHT 26-1	
HASTELLOY-B	
HASTELLOY-C	
HASTELLOY-D	
HIGH SILICON IRON	
INCONEL	
LEAD	
MONEL	
NICKEL	
STAINLESS STEELS	
Type 304/307	
Type 316	
TANTALUM	
TITANIUM	
ZIRCONIUM	
NONMETALLICS	
BOROSILICATE GLASS	R————————————
CARBON	

TEMP °F: 60 80 100 120 140 160 180 200 220 240 260 280 300 320 340 360 380 400 420 440 460 480 500 520 540 560

FOR METALS

E = < 2 Mils Penetration/Year; G = < 20 Mils Penetration/Year

S = < 50 Mils Penetration/Year; U = > 50 Mils Penetration/Year

FOR NONMETALLICS

R = Resistant

U = Unsatisfactory

TEMP °F (60–460)

PLASTICS	Rating
ABS	U
ACRYLICS	
ASBESTOS REINF., EPOXY	
ASBESTOS REINF., PHENOLIC	
CHLORINATED POLYETHER	
CPVC	
EPOXY	R to 240
FLUOROCARBONS FEP	R to 400
FLUOROCARBONS TFE	R to 460
FURFURYL ALCOHOL	R to 260
NORYL	
PHENOLIC	
POLYAMIDES–NYLON	
POLYESTERS	
BISPHENOL A-FUMURATE	R to 140
HYDROGENATED BISPHENOL-A-BISPHENOL-A	
ISOPHTHALIC	U
CHLORINATED POLYESTERS	
POLYSTYRENES	
POLYETHYLENE	
POLYPROPYLENE	R to 140
PVC–TYPE 1	
PVC–TYPE 2	
POLYVINYLIDENE CHLORIDE	R to 60
VINYLIDENE FLUORIDE	R to 100, U above
VINYL ESTER	R to 180

RUBBERS	Rating
BUTYL GR-1	
FLUORO ELASTOMERS VITON A	R to 240
FLUORO ELASTOMERS KEL-F 3700	
HARD RUBBER	
HYPALON	
KOROSEAL	
NATURAL RUBBER (GRS)	
NEOPRENE GR-M (CR)	
NITRILE BUNA N (NBR)	
NORDEL	
POLYBUTADIENE	
SILICONE	
POLYURETHANE	

TEMP °F (60–460)

DIOCTYL PHTHALATE

TEMP °F (60–560)

METALS	Rating
ALLOY 20Cb3	
ALUMINUM	
BRASS	
BRONZE	
CARBON STEEL	G—
COPPER	
E-BRIGHT 26-1	
HASTELLOY-B	
HASTELLOY-C	
HASTELLOY-D	
HIGH SILICON IRON	
INCONEL	
LEAD	
MONEL	
NICKEL	
STAINLESS STEELS Type 304/307	
Type 316	G—
TANTALUM	
TITANIUM	
ZIRCONIUM	
NONMETALLICS	
BOROSILICATE GLASS	R————————
CARBON	

TEMP °F: 60 80 100 120 140 160 180 200 220 240 260 280 300 320 340 360 380 400 420 440 460 480 500 520 540 560

FOR METALS

E = < 2 Mils Penetration/Year; G = < 20 Mils Penetration/Year

S = < 50 Mils Penetration/Year; U = > 50 Mils Penetration/Year

FOR NONMETALLICS

R = Resistant

U = Unsatisfactory

TEMP °F: 60, 80, 100, 120, 140, 160, 180, 200, 220, 240, 260, 280, 300, 320, 340, 360, 380, 400, 420, 440, 460

PLASTICS	
ABS	U
ACRYLICS	
ASBESTOS REINF., EPOXY	
ASBESTOS REINF., PHENOLIC	
CHLORINATED POLYETHER	R
CPVC	U
EPOXY	
FLUOROCARBONS FEP	R
FLUOROCARBONS TFE	R
FURFURYL ALCOHOL	
NORYL	
PHENOLIC	
POLYAMIDES—NYLON	R
POLYESTERS	
BISPHENOL A-FUMURATE	R
HYDROGENATED BISPHENOL-A-BISPHENOL-A	
ISOPHTHALIC	R
CHLORINATED POLYESTERS	
POLYSTYRENES	
POLYETHYLENE	U
POLYPROPYLENE	U
PVC—TYPE 1	U
PVC—TYPE 2	U
POLYVINYLIDENE CHLORIDE	R
VINYLIDENE FLUORIDE	R
VINYL ESTER	R
RUBBERS	
BUTYL GR-1	R
FLUORO ELASTOMERS VITON A	RU
FLUORO ELASTOMERS KEL-F 3700	
HARD RUBBER	
HYPALON	U
KOROSEAL	
NATURAL RUBBER (GRS)	U
NEOPRENE GR-M (CR)	U
NITRILE BUNA N (NBR)	U
NORDEL	
POLYBUTADIENE	
SILICONE	
POLYURETHANE	U

TEMP °F: 60, 80, 100, 120, 140, 160, 180, 200, 220, 240, 260, 280, 300, 320, 340, 360, 380, 400, 420, 440, 460

DIOXANE

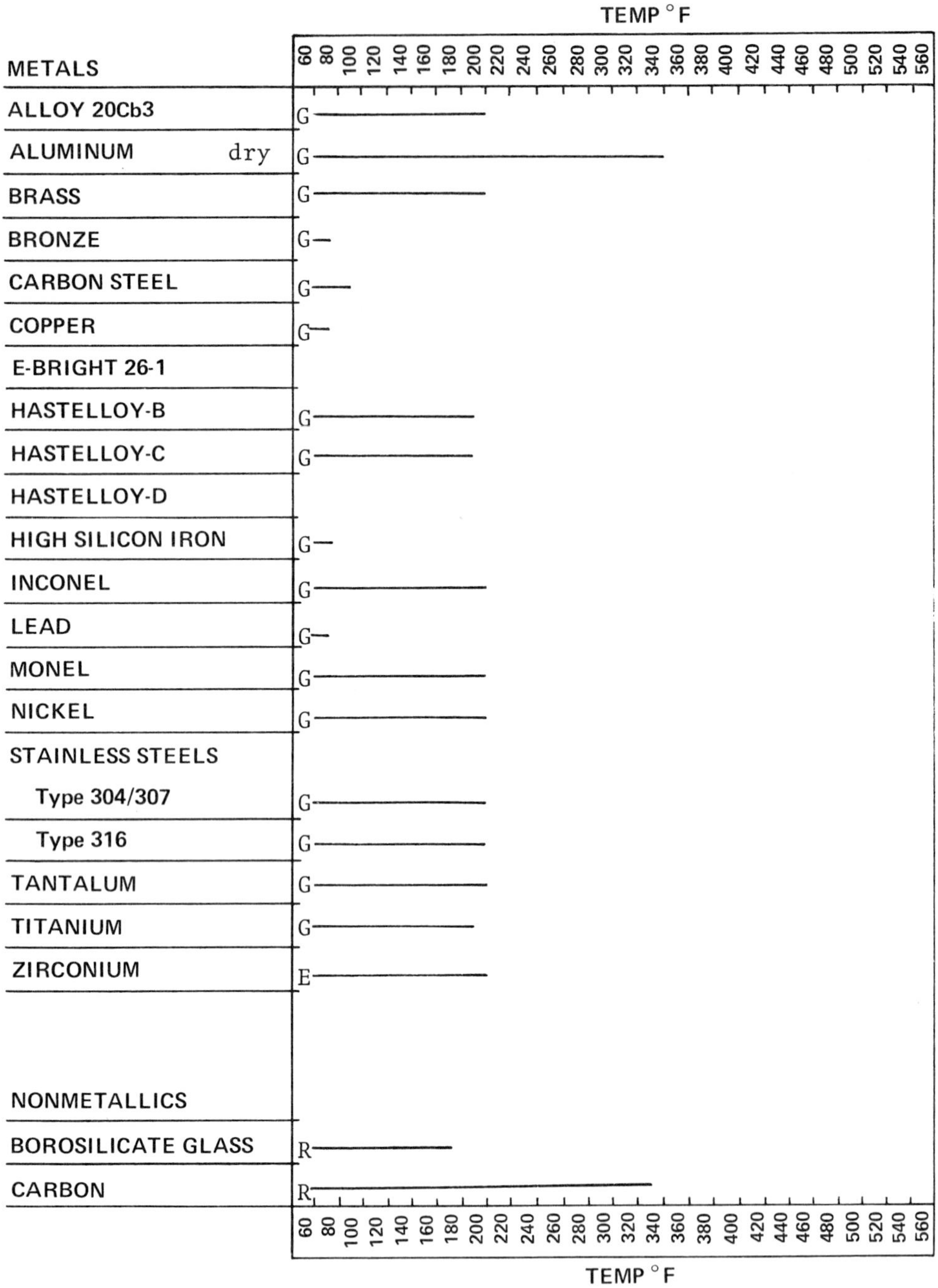

FOR METALS

E = < 2 Mils Penetration/Year; G = < 20 Mils Penetration/Year

S = < 50 Mils Penetration/Year; U = > 50 Mils Penetration/Year

FOR NONMETALLICS

R = Resistant

U = Unsatisfactory

DIOXANE

TEMP °F

60 80 100 120 140 160 180 200 220 240 260 280 300 320 340 360 380 400 420 440 460

PLASTICS	
ABS	U
ACRYLICS	
ASBESTOS REINF., EPOXY	
ASBESTOS REINF., PHENOLIC	
CHLORINATED POLYETHER	R—
CPVC	
EPOXY	R——U
FLUOROCARBONS FEP	R——
FLUOROCARBONS TFE	R——
FURFURYL ALCOHOL	R——
NORYL	
PHENOLIC	
POLYAMIDES—NYLON	
POLYESTERS	
BISPHENOL A-FUMURATE	U
HYDROGENATED BISPHENOL-A-BISPHENOL-A	
ISOPHTHALIC	
CHLORINATED POLYESTERS	
POLYSTYRENES	
POLYETHYLENE	R——
POLYPROPYLENE	R——U
PVC—TYPE 1	
PVC—TYPE 2	
POLYVINYLIDENE CHLORIDE	U
VINYLIDENE FLUORIDE	U
VINYL ESTER	

RUBBERS	
BUTYL GR-1	U
FLUORO ELASTOMERS VITON A	U
FLUORO ELASTOMERS KEL-F 3700	
HARD RUBBER	
HYPALON	
KOROSEAL	
NATURAL RUBBER (GRS)	U
NEOPRENE GR-M (CR)	U
NITRILE BUNA N (NBR)	U
NORDEL	
POLYBUTADIENE	
SILICONE	
POLYURETHANE	

60 80 100 120 140 160 180 200 220 240 260 280 300 320 340 360 380 400 420 440 460

TEMP °F

DIPHENYL (DOWTHERM)

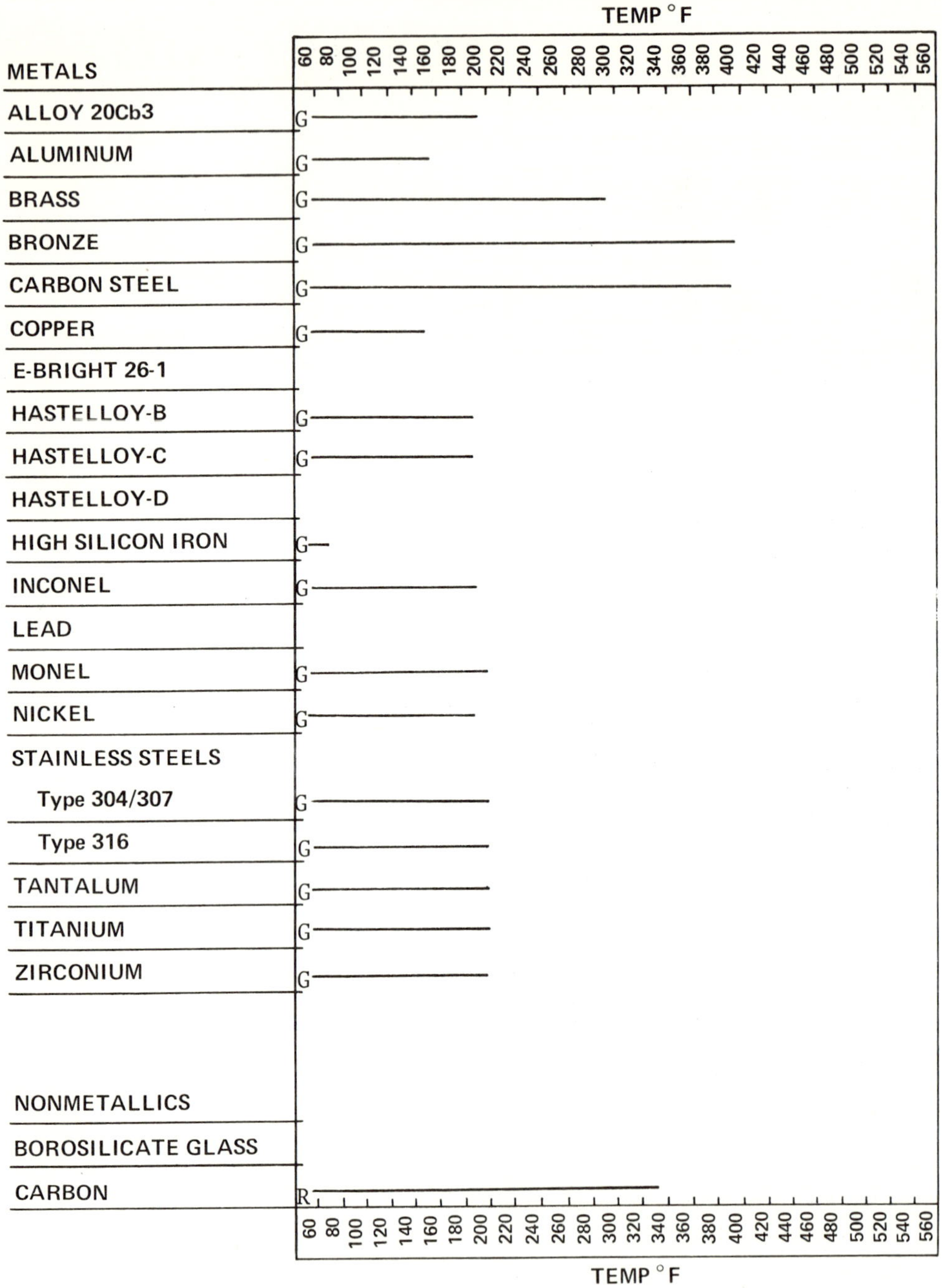

FOR METALS

E = < 2 Mils Penetration/Year; G = < 20 Mils Penetration/Year

S = < 50 Mils Penetration/Year; U = > 50 Mils Penetration/Year

FOR NONMETALLICS

R = Resistant

U = Unsatisfactory

DIPHENYL (DOWTHERM)

PLASTICS	TEMP °F (60–460)
ABS	
ACRYLICS	
ASBESTOS REINF., EPOXY	
ASBESTOS REINF., PHENOLIC	
CHLORINATED POLYETHER	
CPVC	
EPOXY	
FLUOROCARBONS FEP	R (60 to ~400)
FLUOROCARBONS TFE	R (60 to 460)
FURFURYL ALCOHOL	
NORYL	
PHENOLIC	
POLYAMIDES–NYLON	
POLYESTERS	
BISPHENOL A-FUMURATE	
HYDROGENATED BISPHENOL-A-BISPHENOL-A	
ISOPHTHALIC	
CHLORINATED POLYESTERS	
POLYSTYRENES	
POLYETHYLENE	
POLYPROPYLENE	
PVC–TYPE 1	
PVC–TYPE 2	
POLYVINYLIDENE CHLORIDE	
VINYLIDENE FLUORIDE	
VINYL ESTER	
RUBBERS	
BUTYL GR-1	
FLUORO ELASTOMERS VITON A	R (60 to ~350)
FLUORO ELASTOMERS KEL-F 3700	
HARD RUBBER	
HYPALON	
KOROSEAL	
NATURAL RUBBER (GRS)	
NEOPRENE GR-M (CR)	
NITRILE BUNA N (NBR)	
NORDEL	
POLYBUTADIENE	
SILICONE	
POLYURETHANE	

DISODIUM PHOSPHATE

TEMP °F

METALS	60	80	100	120	140	160	180	200	220	240	260	280	300	320	340	360	380	400	420	440	460	480	500	520	540	560
ALLOY 20Cb3																										
ALUMINUM																										
BRASS																										
BRONZE																										
CARBON STEEL																										
COPPER																										
E-BRIGHT 26-1																										
HASTELLOY-B																										
HASTELLOY-C																										
HASTELLOY-D																										
HIGH SILICON IRON																										
INCONEL																										
LEAD																										
MONEL																										
NICKEL																										
STAINLESS STEELS																										
Type 304/307																										
Type 316																										
TANTALUM																										
TITANIUM																										
ZIRCONIUM																										
NONMETALLICS																										
BOROSILICATE GLASS																										
CARBON																										

60 80 100 120 140 160 180 200 220 240 260 280 300 320 340 360 380 400 420 440 460 480 500 520 540 560

TEMP °F

FOR METALS

E = < 2 Mils Penetration/Year; G = < 20 Mils Penetration/Year

S = < 50 Mils Penetration/Year; U = > 50 Mils Penetration/Year

FOR NONMETALLICS

R = Resistant

U = Unsatisfactory

DISODIUM PHOSPHATE

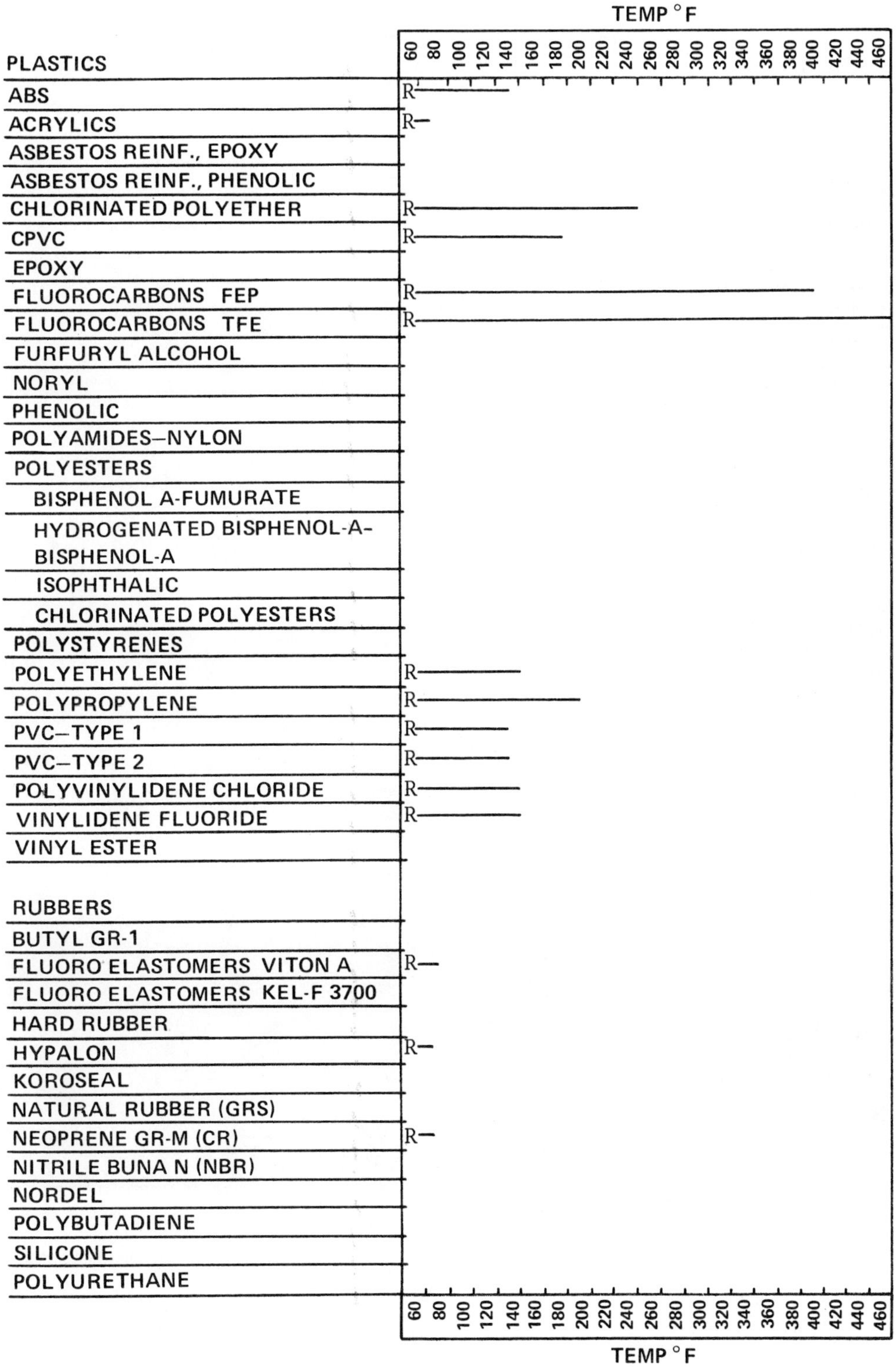

EPICHLORHYDRIN

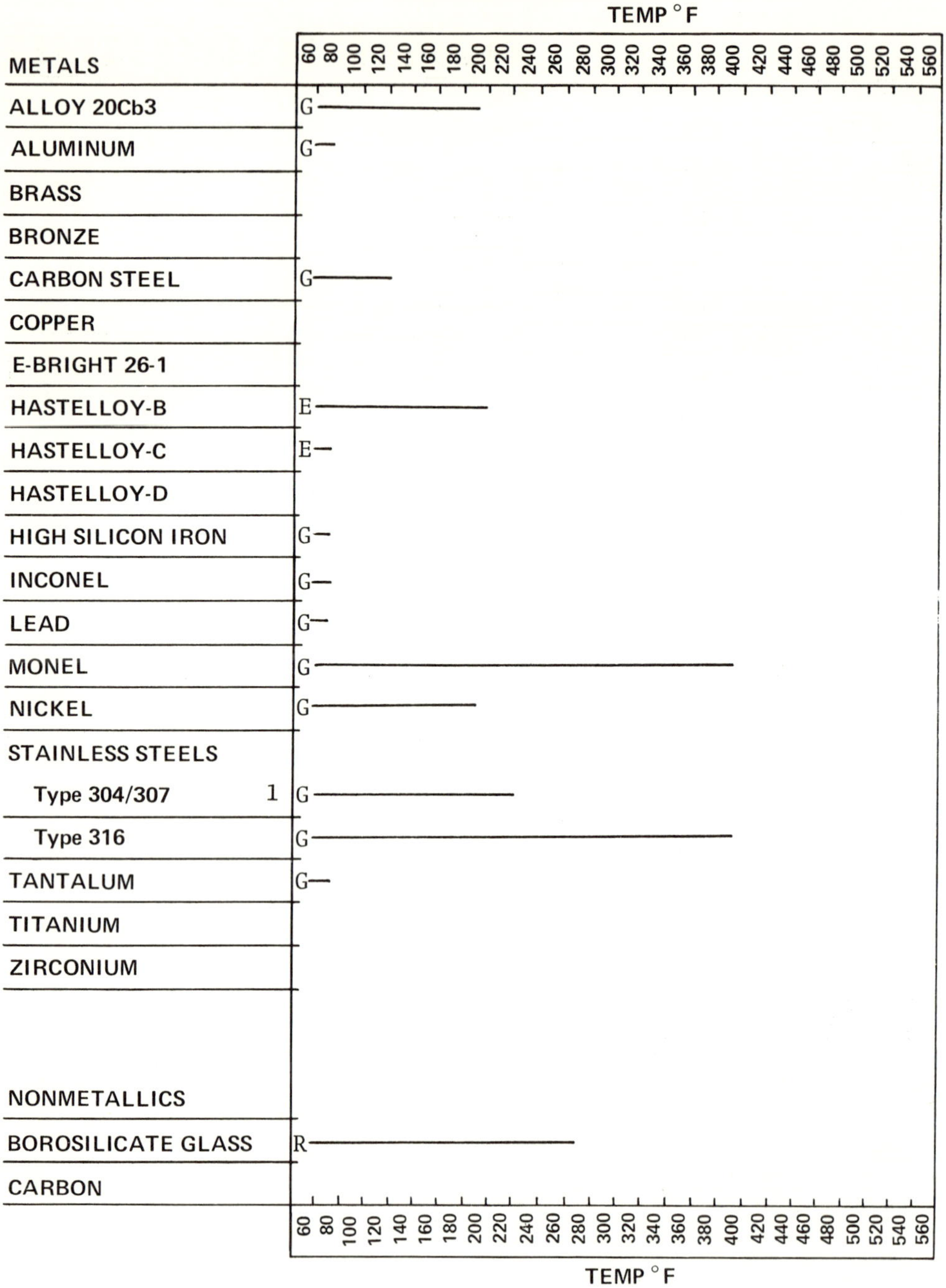

FOR METALS

E = < 2 Mils Penetration/Year; G = < 20 Mils Penetration/Year

S = < 50 Mils Penetration/Year; U = > 50 Mils Penetration/Year

FOR NONMETALLICS

R = Resistant

U = Unsatisfactory

EPICHLORHYDRIN

Please see page 1 for notes on calibration.

TEMP °F

PLASTICS	60–460 °F
ABS	
ACRYLICS	
ASBESTOS REINF., EPOXY	
ASBESTOS REINF., PHENOLIC	
CHLORINATED POLYETHER	
CPVC	
EPOXY	
FLUOROCARBONS FEP	R
FLUOROCARBONS TFE	R
FURFURYL ALCOHOL	
NORYL	
PHENOLIC	
POLYAMIDES—NYLON	R
POLYESTERS	
BISPHENOL A-FUMURATE	
HYDROGENATED BISPHENOL-A-BISPHENOL-A	
ISOPHTHALIC	
CHLORINATED POLYESTERS	
POLYSTYRENES	
POLYETHYLENE	
POLYPROPYLENE	R
PVC—TYPE 1	
PVC—TYPE 2	
POLYVINYLIDENE CHLORIDE	
VINYLIDENE FLUORIDE	R
VINYL ESTER	R

RUBBERS	60–460 °F
BUTYL GR-1	U
FLUORO ELASTOMERS VITON A	U
FLUORO ELASTOMERS KEL-F 3700	
HARD RUBBER	
HYPALON	
KOROSEAL	
NATURAL RUBBER (GRS)	U
NEOPRENE GR-M (CR)	U
NITRILE BUNA N (NBR)	U
NORDEL	
POLYBUTADIENE	
SILICONE	
POLYURETHANE	

TEMP °F: 60 80 100 120 140 160 180 200 220 240 260 280 300 320 340 360 380 400 420 440 460

ESTERS (GENERAL)

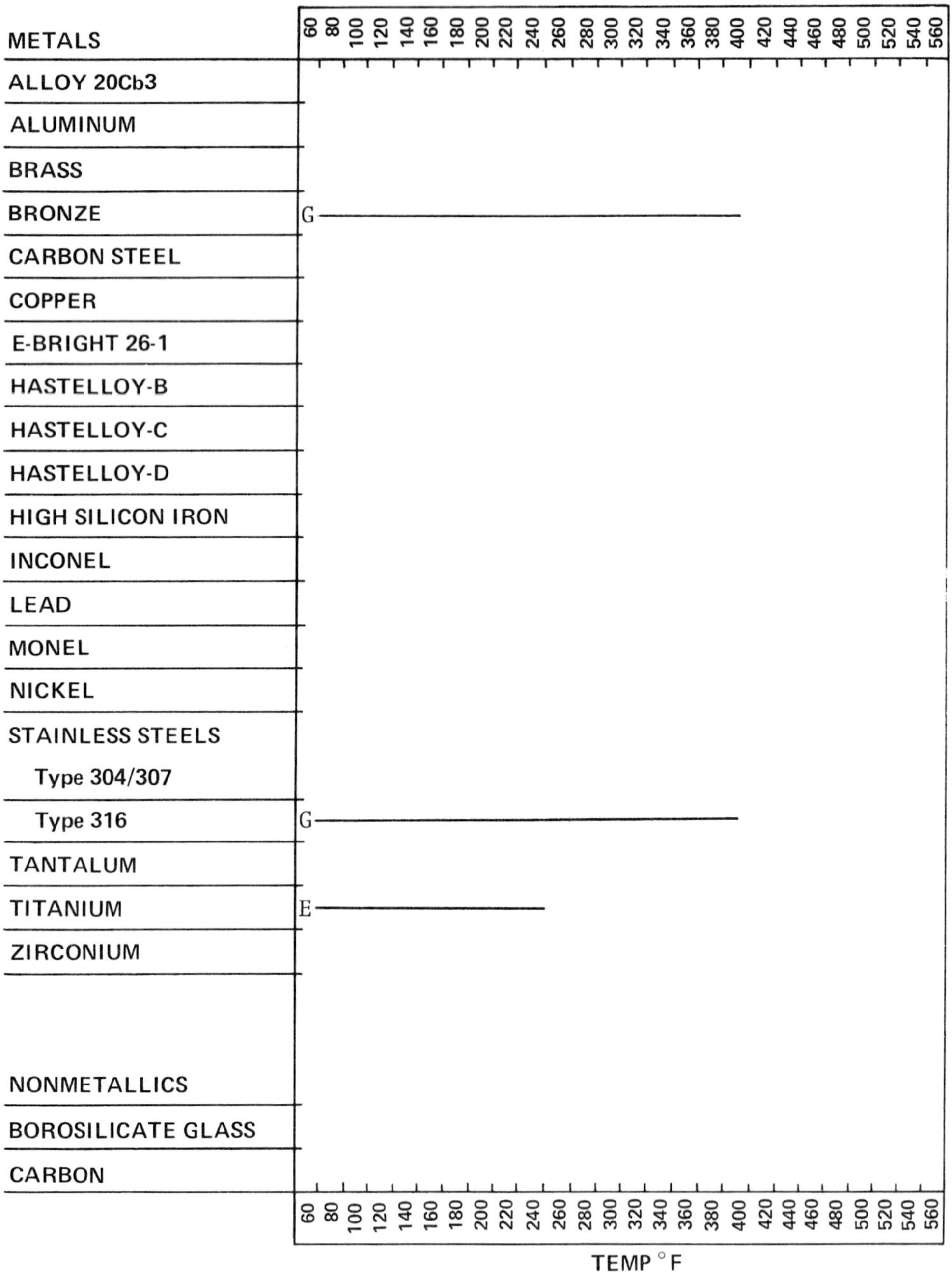

FOR METALS

E = < 2 Mils Penetration/Year; G = < 20 Mils Penetration/Year

S = < 50 Mils Penetration/Year; U = > 50 Mils Penetration/Year

FOR NONMETALLICS

R = Resistant

U = Unsatisfactory

ESTERS (GENERAL)

TEMP °F

PLASTICS	60 80 100 120 140 160 180 200 220 240 260 280 300 320 340 360 380 400 420 440 460
ABS	
ACRYLICS	
ASBESTOS REINF., EPOXY	
ASBESTOS REINF., PHENOLIC	
CHLORINATED POLYETHER	
CPVC	U
EPOXY	R——————
FLUOROCARBONS FEP	R————————————————————
FLUOROCARBONS TFE	R—————————————————————————
FURFURYL ALCOHOL	R———————————
NORYL	
PHENOLIC	
POLYAMIDES–NYLON	
POLYESTERS	
BISPHENOL A-FUMURATE	
HYDROGENATED BISPHENOL-A-BISPHENOL-A	
ISOPHTHALIC	R—————
CHLORINATED POLYESTERS	R———
POLYSTYRENES	
POLYETHYLENE	U
POLYPROPYLENE	
PVC–TYPE 1	U
PVC–TYPE 2	U
POLYVINYLIDENE CHLORIDE	
VINYLIDENE FLUORIDE	R—————— U
VINYL ESTER	R—————

RUBBERS	
BUTYL GR-1	
FLUORO ELASTOMERS VITON A	
FLUORO ELASTOMERS KEL-F 3700	
HARD RUBBER	
HYPALON	
KOROSEAL	
NATURAL RUBBER (GRS)	
NEOPRENE GR-M (CR)	
NITRILE BUNA N (NBR)	
NORDEL	
POLYBUTADIENE	
SILICONE	
POLYURETHANE	

60 80 100 120 140 160 180 200 220 240 260 280 300 320 340 360 380 400 420 440 460

TEMP °F

ETHERS (GENERAL)

TEMP °F

METALS	Rating	Temp °F
ALLOY 20Cb3	G	60–100
ALUMINUM	G	60–80
BRASS	G	60–80
BRONZE	G	60–200
CARBON STEEL	G	60–200
COPPER	G	60–80
E-BRIGHT 26-1		
HASTELLOY-B	G	60–200
HASTELLOY-C	G	60–200
HASTELLOY-D		
HIGH SILICON IRON	G	60–80
INCONEL	G	60–80
LEAD	G	60–80
MONEL	G	60–80
NICKEL	G	60–80
STAINLESS STEELS Type 304/307	E	60–200
Type 316	E	60–200
TANTALUM	G	60–200
TITANIUM	G	60–80
ZIRCONIUM		

NONMETALLICS	Rating	Temp °F
BOROSILICATE GLASS	R	60–140
CARBON	R	60–340

TEMP °F

FOR METALS

E = < 2 Mils Penetration/Year; G = < 20 Mils Penetration/Year

S = < 50 Mils Penetration/Year; U = > 50 Mils Penetration/Year

FOR NONMETALLICS

R = Resistant

U = Unsatisfactory

TEMP °F (60–460)

PLASTICS	Rating
ABS	U
ACRYLICS	U
ASBESTOS REINF., EPOXY	
ASBESTOS REINF., PHENOLIC	
CHLORINATED POLYETHER	R
CPVC	U
EPOXY	
FLUOROCARBONS FEP	R
FLUOROCARBONS TFE	R
FURFURYL ALCOHOL	
NORYL	
PHENOLIC	
POLYAMIDES—NYLON	R
POLYESTERS	
BISPHENOL A-FUMURATE	
HYDROGENATED BISPHENOL-A-BISPHENOL-A	
ISOPHTHALIC	
CHLORINATED POLYESTERS	
POLYSTYRENES	
POLYETHYLENE	U
POLYPROPYLENE	U
PVC—TYPE 1	U
PVC—TYPE 2	U
POLYVINYLIDENE CHLORIDE	
VINYLIDENE FLUORIDE	R
VINYL ESTER	

RUBBERS	Rating
BUTYL GR-1	U
FLUORO ELASTOMERS VITON A	U
FLUORO ELASTOMERS KEL-F 3700	
HARD RUBBER	
HYPALON	U
KOROSEAL	
NATURAL RUBBER (GRS)	U
NEOPRENE GR-M (CR)	U
NITRILE BUNA N (NBR)	R
NORDEL	
POLYBUTADIENE	
SILICONE	
POLYURETHANE	U

TEMP °F

ETHYL ACETATE

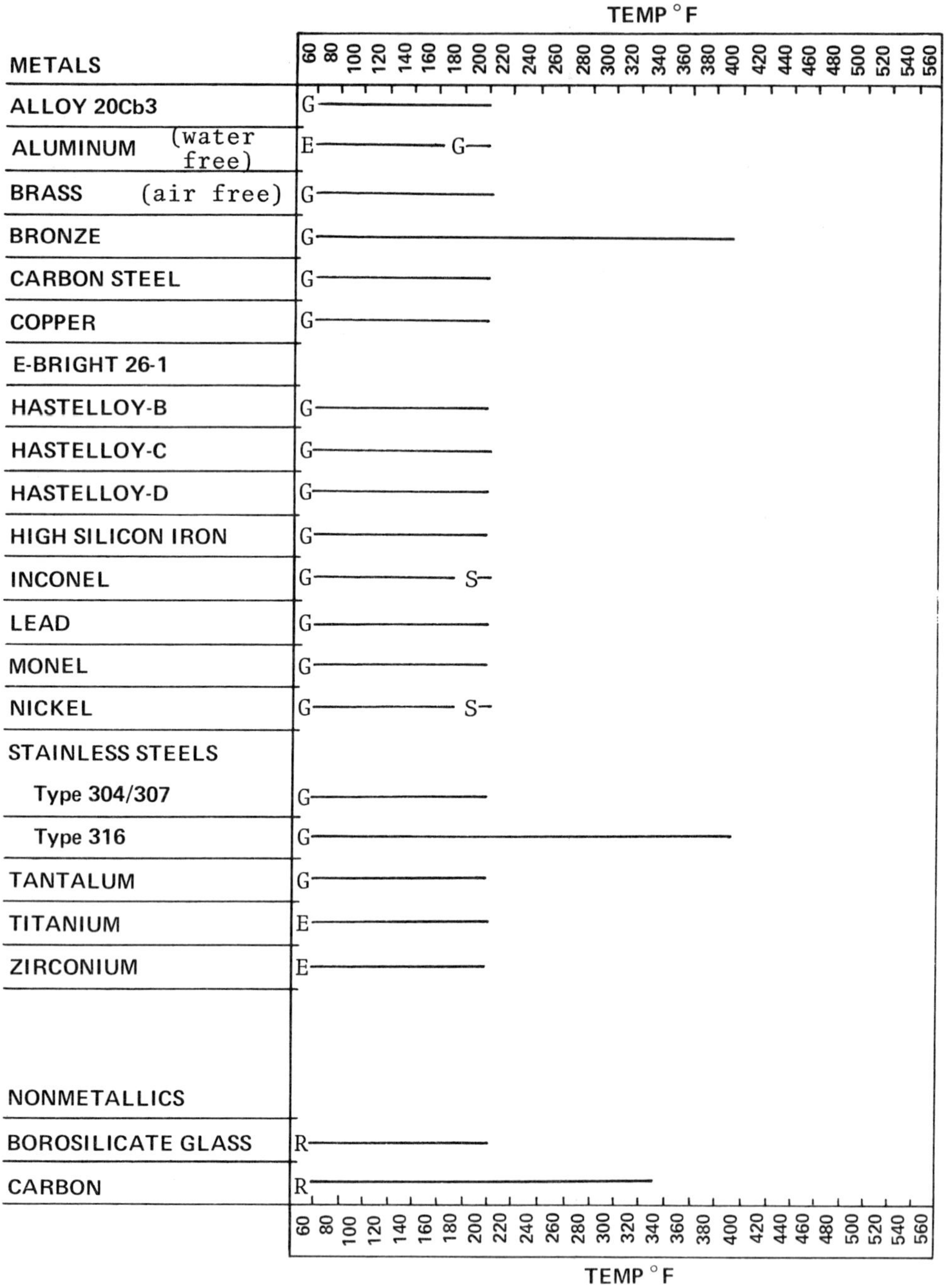

FOR METALS

E = < 2 Mils Penetration/Year; G = < 20 Mils Penetration/Year

S = < 50 Mils Penetration/Year; U = > 50 Mils Penetration/Year

FOR NONMETALLICS

R = Resistant

U = Unsatisfactory

ETHYL ACETATE

TEMP °F (scale 60–460)

PLASTICS		Rating (line extends to, °F)
ABS		U
ACRYLICS		U
ASBESTOS REINF., EPOXY		R (to 160)
ASBESTOS REINF., PHENOLIC		U
CHLORINATED POLYETHER		R (to 80)
CPVC		U
EPOXY		R (to 150)
FLUOROCARBONS FEP		R (to 400)
FLUOROCARBONS TFE		R (to 460)
FURFURYL ALCOHOL	sat	R (to 270)
NORYL		
PHENOLIC		
POLYAMIDES–NYLON		R (to 200)
POLYESTERS		
BISPHENOL A-FUMURATE		U
HYDROGENATED BISPHENOL-A-BISPHENOL-A		U
ISOPHTHALIC		U
CHLORINATED POLYESTERS		U
POLYSTYRENES		
POLYETHYLENE		U
POLYPROPYLENE		R (to 140)
PVC–TYPE 1		U
PVC–TYPE 2		U
POLYVINYLIDENE CHLORIDE		R (to 80)
VINYLIDENE FLUORIDE		R (to 150)
VINYL ESTER		U

RUBBERS	Rating (line extends to, °F)
BUTYL GR-1	R (to 100)
FLUORO ELASTOMERS VITON A	U
FLUORO ELASTOMERS KEL-F 3700	
HARD RUBBER	
HYPALON	U
KOROSEAL	U
NATURAL RUBBER (GRS)	U
NEOPRENE GR-M (CR)	U
NITRILE BUNA N (NBR)	U
NORDEL	U
POLYBUTADIENE	
SILICONE	
POLYURETHANE	U

TEMP °F

ETHYL ACETOACETATE

TEMP °F

METALS	60	80	100	120	140	160	180	200	220	240	260	280	300	320	340	360	380	400	420	440	460	480	500	520	540	560
ALLOY 20Cb3																										
ALUMINUM	G—																									
BRASS																										
BRONZE																										
CARBON STEEL	G—																									
COPPER																										
E-BRIGHT 26-1																										
HASTELLOY-B																										
HASTELLOY-C																										
HASTELLOY-D																										
HIGH SILICON IRON																										
INCONEL																										
LEAD																										
MONEL																										
NICKEL																										
STAINLESS STEELS																										
Type 304/307																										
Type 316	G	—	—	—	—	—																				
TANTALUM	G—																									
TITANIUM																										
ZIRCONIUM																										
NONMETALLICS																										
BOROSILICATE GLASS	R	—	—	—	—	—	—	—	—																	
CARBON																										

TEMP °F

FOR METALS

E = < 2 Mils Penetration/Year; G = < 20 Mils Penetration/Year

S = < 50 Mils Penetration/Year; U = > 50 Mils Penetration/Year

FOR NONMETALLICS

R = Resistant

U = Unsatisfactory

PLASTICS	TEMP °F (60–460)
ABS	U (60)
ACRYLICS	
ASBESTOS REINF., EPOXY	R (60–160)
ASBESTOS REINF., PHENOLIC	R (60–160)
CHLORINATED POLYETHER	R (60–80)
CPVC	R (60–80)
EPOXY	
FLUOROCARBONS FEP	R (60–400)
FLUOROCARBONS TFE	R (60–460)
FURFURYL ALCOHOL	
NORYL	
PHENOLIC	
POLYAMIDES–NYLON	
POLYESTERS	
BISPHENOL A-FUMURATE	
HYDROGENATED BISPHENOL-A-BISPHENOL-A	
ISOPHTHALIC	
CHLORINATED POLYESTERS sat	R (60–80)
POLYSTYRENES	
POLYETHYLENE	
POLYPROPYLENE	
PVC–TYPE 1	
PVC–TYPE 2	
POLYVINYLIDENE CHLORIDE	U (60)
VINYLIDENE FLUORIDE	R (60–180) U (180)
VINYL ESTER	

RUBBERS	TEMP °F (60–460)
BUTYL GR-1	R (60–80)
FLUORO ELASTOMERS VITON A	U (60)
FLUORO ELASTOMERS KEL-F 3700	
HARD RUBBER	
HYPALON	
KOROSEAL	
NATURAL RUBBER (GRS)	R (60–80)
NEOPRENE GR-M (CR)	U (60)
NITRILE BUNA N (NBR)	U (60)
NORDEL	
POLYBUTADIENE	
SILICONE	
POLYURETHANE	

ETHYL ACRYLATE

TEMP °F: 60, 80, 100, 120, 140, 160, 180, 200, 220, 240, 260, 280, 300, 320, 340, 360, 380, 400, 420, 440, 460, 480, 500, 520, 540, 560

METALS	
ALLOY 20Cb3	
ALUMINUM	
BRASS	
BRONZE	
CARBON STEEL	G
COPPER	
E-BRIGHT 26-1	
HASTELLOY-B	
HASTELLOY-C	
HASTELLOY-D	
HIGH SILICON IRON	
INCONEL	
LEAD	
MONEL	
NICKEL	
STAINLESS STEELS	
Type 304/307	G
Type 316	G
TANTALUM	G
TITANIUM	
ZIRCONIUM	

NONMETALLICS	
BOROSILICATE GLASS	R
CARBON	

TEMP °F: 60, 80, 100, 120, 140, 160, 180, 200, 220, 240, 260, 280, 300, 320, 340, 360, 380, 400, 420, 440, 460, 480, 500, 520, 540, 560

FOR METALS

E = < 2 Mils Penetration/Year; G = < 20 Mils Penetration/Year

S = < 50 Mils Penetration/Year; U = > 50 Mils Penetration/Year

FOR NONMETALLICS

R = Resistant

U = Unsatisfactory

ETHYL ACRYLATE

TEMP °F (60, 80, 100, 120, 140, 160, 180, 200, 220, 240, 260, 280, 300, 320, 340, 360, 380, 400, 420, 440, 460)

PLASTICS	TEMP °F
ABS	U
ACRYLICS	
ASBESTOS REINF., EPOXY	
ASBESTOS REINF., PHENOLIC	
CHLORINATED POLYETHER	R——————
CPVC	U
EPOXY	R——U
FLUOROCARBONS FEP	R——————————————
FLUOROCARBONS TFE	R————————————————————
FURFURYL ALCOHOL	
NORYL	
PHENOLIC	
POLYAMIDES—NYLON	
POLYESTERS	
BISPHENOL A-FUMURATE	
HYDROGENATED BISPHENOL-A-BISPHENOL-A	U
ISOPHTHALIC	
CHLORINATED POLYESTERS	
POLYSTYRENES	
POLYETHYLENE	U
POLYPROPYLENE	R——
PVC—TYPE 1	U
PVC—TYPE 2	U
POLYVINYLIDENE CHLORIDE	
VINYLIDENE FLUORIDE	R——————U
VINYL ESTER	U

RUBBERS	TEMP °F
BUTYL GR-1	U
FLUORO ELASTOMERS VITON A	U
FLUORO ELASTOMERS KEL-F 3700	
HARD RUBBER	
HYPALON	U
KOROSEAL	
NATURAL RUBBER (GRS)	U
NEOPRENE GR-M (CR)	U
NITRILE BUNA N (NBR)	U
NORDEL	
POLYBUTADIENE	
SILICONE	
POLYURETHANE	

TEMP °F (60, 80, 100, 120, 140, 160, 180, 200, 220, 240, 260, 280, 300, 320, 340, 360, 380, 400, 420, 440, 460)

ETHYL ALCOHOL

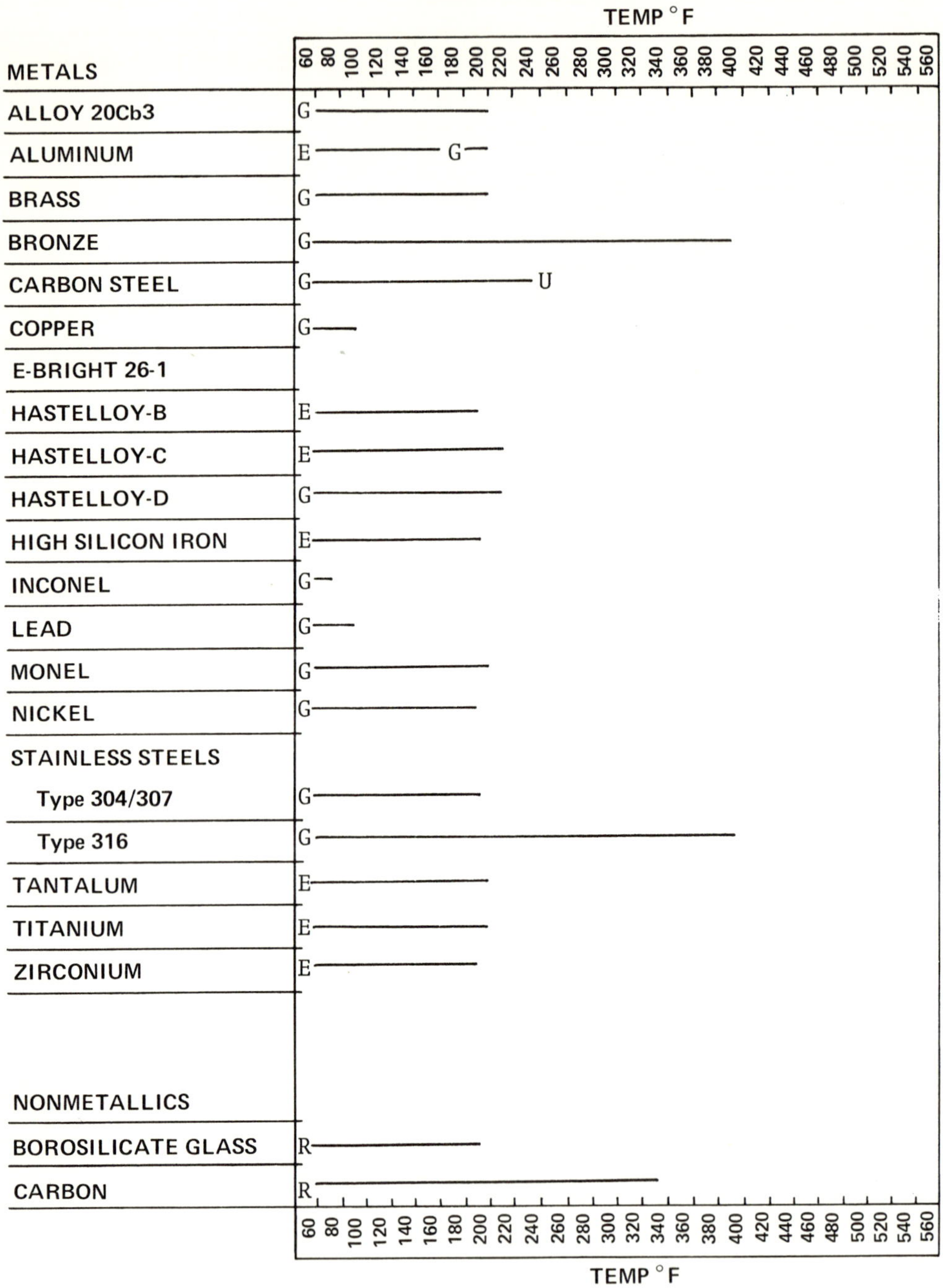

FOR METALS

E = < 2 Mils Penetration/Year; G = < 20 Mils Penetration/Year

S = < 50 Mils Penetration/Year; U = > 50 Mils Penetration/Year

FOR NONMETALLICS

R = Resistant

U = Unsatisfactory

ETHYL ALCOHOL

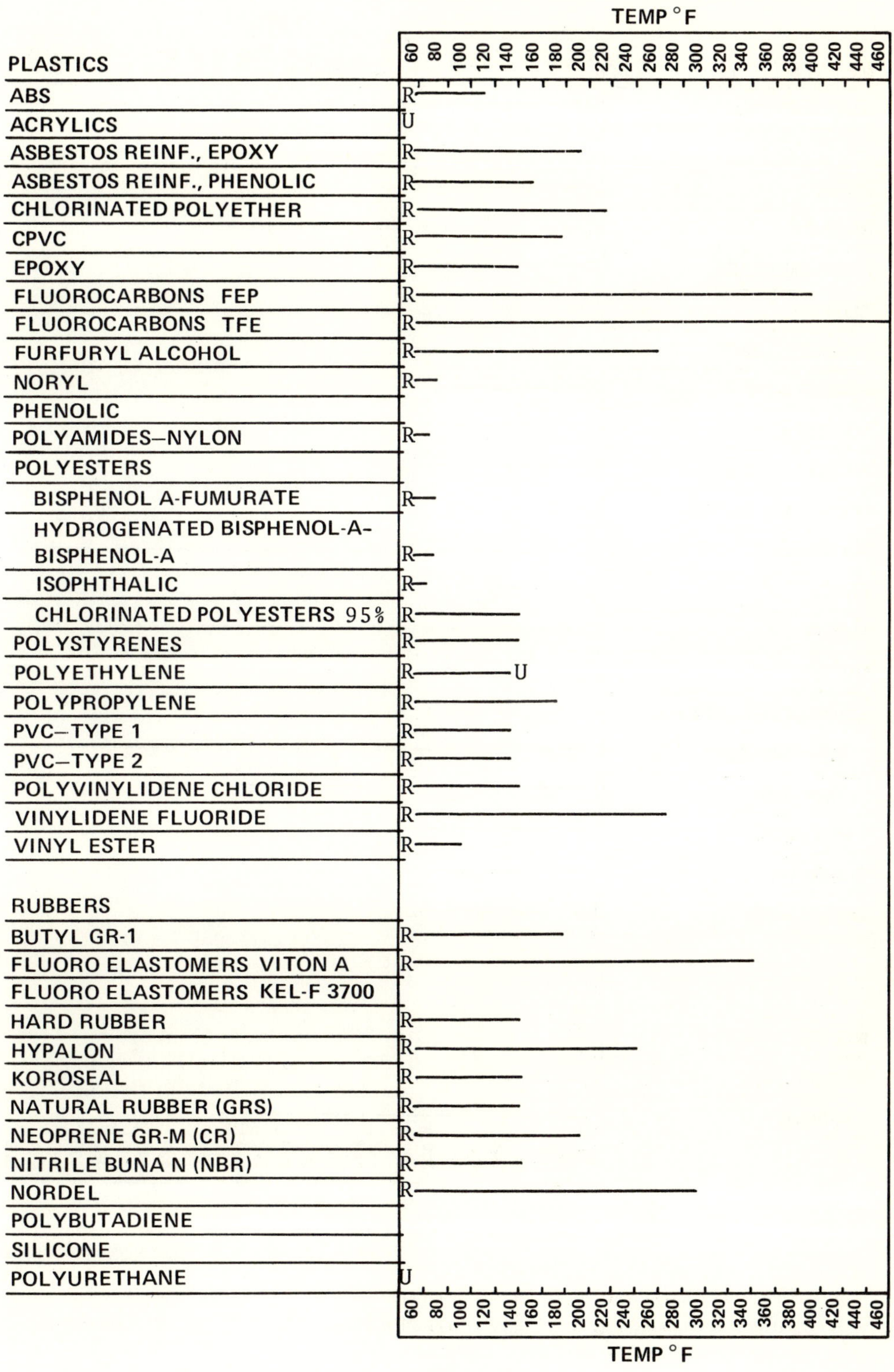

ETHYL CHLORIDE

TEMP °F: 60, 80, 100, 120, 140, 160, 180, 200, 220, 240, 260, 280, 300, 320, 340, 360, 380, 400, 420, 440, 460, 480, 500, 520, 540, 560

METALS		Rating
ALLOY 20Cb3		E
ALUMINUM	(dry)	G
BRASS	(dry)	E
BRONZE		G
CARBON STEEL		G
COPPER		G
E-BRIGHT 26-1		
HASTELLOY-B		G
HASTELLOY-C		G
HASTELLOY-D		G
HIGH SILICON IRON		G
INCONEL	(dry)	E
LEAD		G
MONEL		G
NICKEL	(dry)	E
STAINLESS STEELS		
Type 304/307	2	E
Type 316		E
TANTALUM		E
TITANIUM		E
ZIRCONIUM		E

NONMETALLICS	Rating
BOROSILICATE GLASS	R
CARBON	R

FOR METALS

E = < 2 Mils Penetration/Year; G = < 20 Mils Penetration/Year

S = < 50 Mils Penetration/Year; U = > 50 Mils Penetration/Year

FOR NONMETALLICS

R = Resistant

U = Unsatisfactory

ETHYL CHLORIDE

TEMP °F: 60, 80, 100, 120, 140, 160, 180, 200, 220, 240, 260, 280, 300, 320, 340, 360, 380, 400, 420, 440, 460

PLASTICS	Rating (TEMP °F)
ABS	U
ACRYLICS	U
ASBESTOS REINF., EPOXY	
ASBESTOS REINF., PHENOLIC	
CHLORINATED POLYETHER	R (60–220)
CPVC	U
EPOXY	RU
FLUOROCARBONS FEP	R (60–400)
FLUOROCARBONS TFE	R (60–460)
FURFURYL ALCOHOL	R (60–260)
NORYL	
PHENOLIC	
POLYAMIDES—NYLON	R–
POLYESTERS	
BISPHENOL A-FUMURATE	
HYDROGENATED BISPHENOL-A-BISPHENOL-A	
ISOPHTHALIC	U
CHLORINATED POLYESTERS	R–
POLYSTYRENES	
POLYETHYLENE	U
POLYPROPYLENE	U
PVC—TYPE 1	U
PVC—TYPE 2	U
POLYVINYLIDENE CHLORIDE	U
VINYLIDENE FLUORIDE	R (60–280)
VINYL ESTER	U

RUBBERS	Rating (TEMP °F)
BUTYL GR-1	U
FLUORO ELASTOMERS VITON A	R (60–180)
FLUORO ELASTOMERS KEL-F 3700	
HARD RUBBER	U
HYPALON	U
KOROSEAL	
NATURAL RUBBER (GRS)	U
NEOPRENE GR-M (CR)	U
NITRILE BUNA N (NBR)	R (60–180)
NORDEL	
POLYBUTADIENE	
SILICONE	
POLYURETHANE	U

TEMP °F: 60, 80, 100, 120, 140, 160, 180, 200, 220, 240, 260, 280, 300, 320, 340, 360, 380, 400, 420, 440, 460

ETHYL ETHER

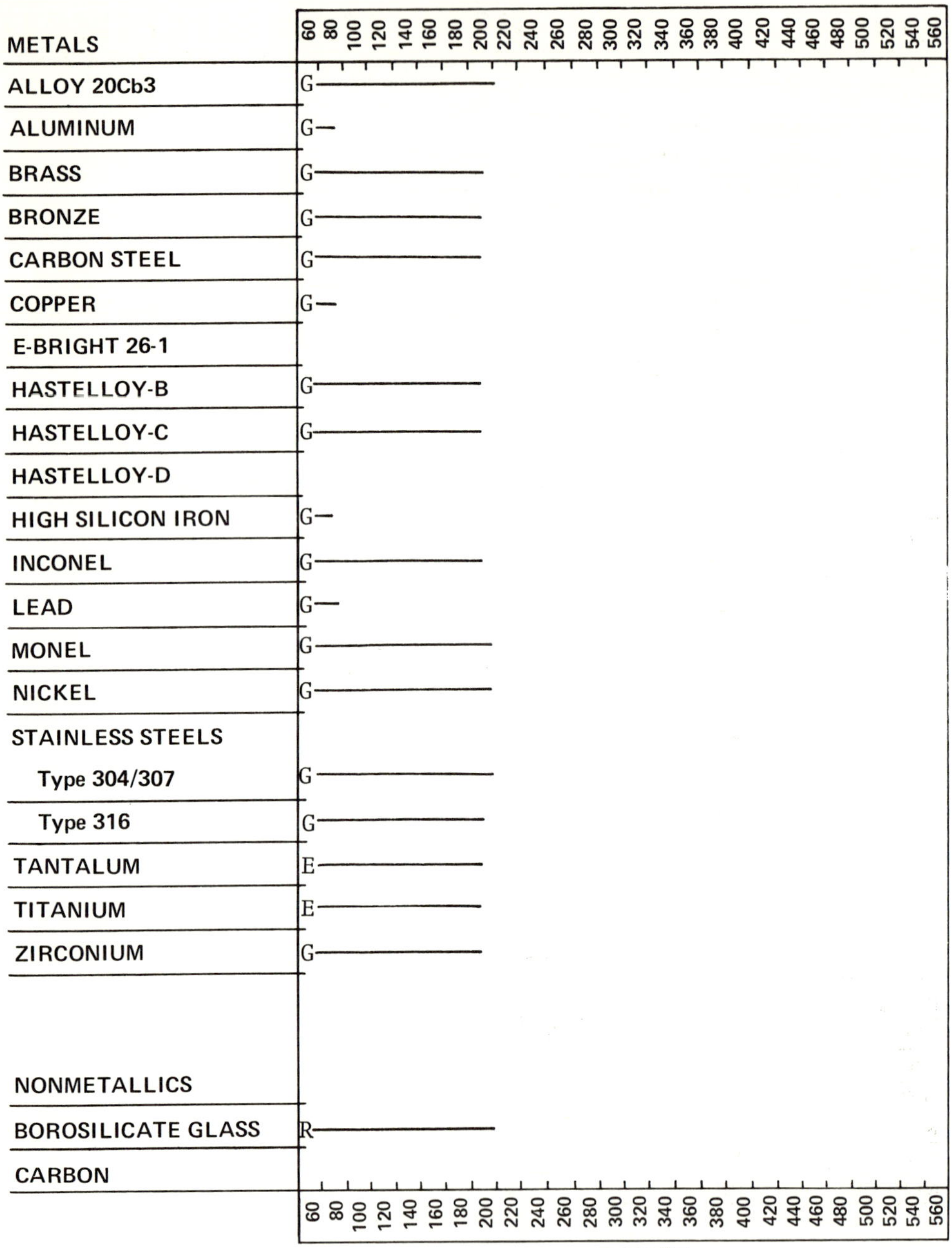

FOR METALS

E = < 2 Mils Penetration/Year; G = < 20 Mils Penetration/Year

S = < 50 Mils Penetration/Year; U = > 50 Mils Penetration/Year

FOR NONMETALLICS

R = Resistant

U = Unsatisfactory

ETHYL ETHER

PLASTICS	TEMP °F (60–460)
ABS	U
ACRYLICS	U
ASBESTOS REINF., EPOXY	
ASBESTOS REINF., PHENOLIC	
CHLORINATED POLYETHER	
CPVC	U
EPOXY	R—U
FLUOROCARBONS FEP	R—
FLUOROCARBONS TFE	R—
FURFURYL ALCOHOL	R—
NORYL	
PHENOLIC	
POLYAMIDES–NYLON	R—
POLYESTERS	
BISPHENOL A-FUMURATE	U
HYDROGENATED BISPHENOL-A-BISPHENOL-A	
ISOPHTHALIC	
CHLORINATED POLYESTERS	
POLYSTYRENES	
POLYETHYLENE	U
POLYPROPYLENE	U
PVC–TYPE 1	U
PVC–TYPE 2	U
POLYVINYLIDENE CHLORIDE	
VINYLIDENE FLUORIDE	R—U
VINYL ESTER	U
RUBBERS	
BUTYL GR-1	U
FLUORO ELASTOMERS VITON A	U
FLUORO ELASTOMERS KEL-F 3700	
HARD RUBBER	
HYPALON	U
KOROSEAL	
NATURAL RUBBER (GRS)	U
NEOPRENE GR-M (CR)	U
NITRILE BUNA N (NBR)	R—
NORDEL	
POLYBUTADIENE	
SILICONE	
POLYURETHANE	U

ETHYLENE BROMIDE

TEMP °F: 60 80 100 120 140 160 180 200 220 240 260 280 300 320 340 360 380 400 420 440 460 480 500 520 540 560

METALS	Rating
ALLOY 20Cb3	E
ALUMINUM (dry)	G
BRASS	
BRONZE	
CARBON STEEL	G
COPPER	
E-BRIGHT 26-1	
HASTELLOY-B	E
HASTELLOY-C	E
HASTELLOY-D	E
HIGH SILICON IRON	G
INCONEL	G
LEAD	G
MONEL	G
NICKEL	G
STAINLESS STEELS	
Type 304/307 (dry)	E
Type 316	E
TANTALUM	G
TITANIUM	G
ZIRCONIUM	G

NONMETALLICS	Rating
BOROSILICATE GLASS	R
CARBON	

TEMP °F: 60 80 100 120 140 160 180 200 220 240 260 280 300 320 340 360 380 400 420 440 460 480 500 520 540 560

FOR METALS

E = < 2 Mils Penetration/Year; G = < 20 Mils Penetration/Year

S = < 50 Mils Penetration/Year; U = > 50 Mils Penetration/Year

FOR NONMETALLICS

R = Resistant

U = Unsatisfactory

ETHYLENE BROMIDE

PLASTICS	TEMP °F (60–460)
ABS	U
ACRYLICS	
ASBESTOS REINF., EPOXY	
ASBESTOS REINF., PHENOLIC	
CHLORINATED POLYETHER	R—
CPVC	U
EPOXY	
FLUOROCARBONS FEP	R———— [line to 400]
FLUOROCARBONS TFE	R———— [line to 460]
FURFURYL ALCOHOL	
NORYL	
PHENOLIC	
POLYAMIDES–NYLON	
POLYESTERS	
BISPHENOL A-FUMURATE	U
HYDROGENATED BISPHENOL-A-BISPHENOL-A	
ISOPHTHALIC	
CHLORINATED POLYESTERS	U
POLYSTYRENES	
POLYETHYLENE	U
POLYPROPYLENE	U
PVC–TYPE 1	U
PVC–TYPE 2	U
POLYVINYLIDENE CHLORIDE	R—
VINYLIDENE FLUORIDE	R———— [line to 280]
VINYL ESTER	
RUBBERS	
BUTYL GR-1	U
FLUORO ELASTOMERS VITON A	R—
FLUORO ELASTOMERS KEL-F 3700	
HARD RUBBER	
HYPALON	U
KOROSEAL	
NATURAL RUBBER (GRS)	U
NEOPRENE GR-M (CR)	U
NITRILE BUNA N (NBR)	U
NORDEL	
POLYBUTADIENE	
SILICONE	
POLYURETHANE	

60 80 100 120 140 160 180 200 220 240 260 280 300 320 340 360 380 400 420 440 460
TEMP °F

ETHYLENE CHLORIDE

TEMP °F (60–560)

METALS		Rating
ALLOY 20Cb3		G
ALUMINUM	(dry)	G
BRASS		
BRONZE		
CARBON STEEL		G
COPPER		
E-BRIGHT 26-1		
HASTELLOY-B		E
HASTELLOY-C		
HASTELLOY-D		
HIGH SILICON IRON		G
INCONEL		G
LEAD		
MONEL		G
NICKEL		G
STAINLESS STEELS		
Type 304/307	(dry)	G
Type 316	(dry)	G
TANTALUM		G
TITANIUM		E
ZIRCONIUM		U

NONMETALLICS	Rating
BOROSILICATE GLASS	R
CARBON	

FOR METALS

E = < 2 Mils Penetration/Year; G = < 20 Mils Penetration/Year

S = < 50 Mils Penetration/Year; U = > 50 Mils Penetration/Year

FOR NONMETALLICS

R = Resistant

U = Unsatisfactory

ETHYLENE CHLORIDE

TEMP °F: 60, 80, 100, 120, 140, 160, 180, 200, 220, 240, 260, 280, 300, 320, 340, 360, 380, 400, 420, 440, 460

PLASTICS	
ABS	U
ACRYLICS	
ASBESTOS REINF., EPOXY	U
ASBESTOS REINF., PHENOLIC	R—
CHLORINATED POLYETHER	R—U
CPVC	
EPOXY	U
FLUOROCARBONS FEP	R—
FLUOROCARBONS TFE	R—
FURFURYL ALCOHOL	R—
NORYL	
PHENOLIC	
POLYAMIDES—NYLON	R—
POLYESTERS	
BISPHENOL A-FUMURATE	U
HYDROGENATED BISPHENOL-A-BISPHENOL-A	
ISOPHTHALIC	U
CHLORINATED POLYESTERS	
POLYSTYRENES	
POLYETHYLENE	U
POLYPROPYLENE	U
PVC—TYPE 1	U
PVC—TYPE 2	U
POLYVINYLIDENE CHLORIDE	R—
VINYLIDENE FLUORIDE	R—
VINYL ESTER	

RUBBERS	
BUTYL GR-1	R—
FLUORO ELASTOMERS VITON A	R—
FLUORO ELASTOMERS KEL-F 3700	
HARD RUBBER	
HYPALON	
KOROSEAL	U
NATURAL RUBBER (GRS)	U
NEOPRENE GR-M (CR)	U
NITRILE BUNA N (NBR)	U
NORDEL	
POLYBUTADIENE	
SILICONE	
POLYURETHANE	

TEMP °F: 60, 80, 100, 120, 140, 160, 180, 200, 220, 240, 260, 280, 300, 320, 340, 360, 380, 400, 420, 440, 460

ETHYLENE CHLOROHYDRIN

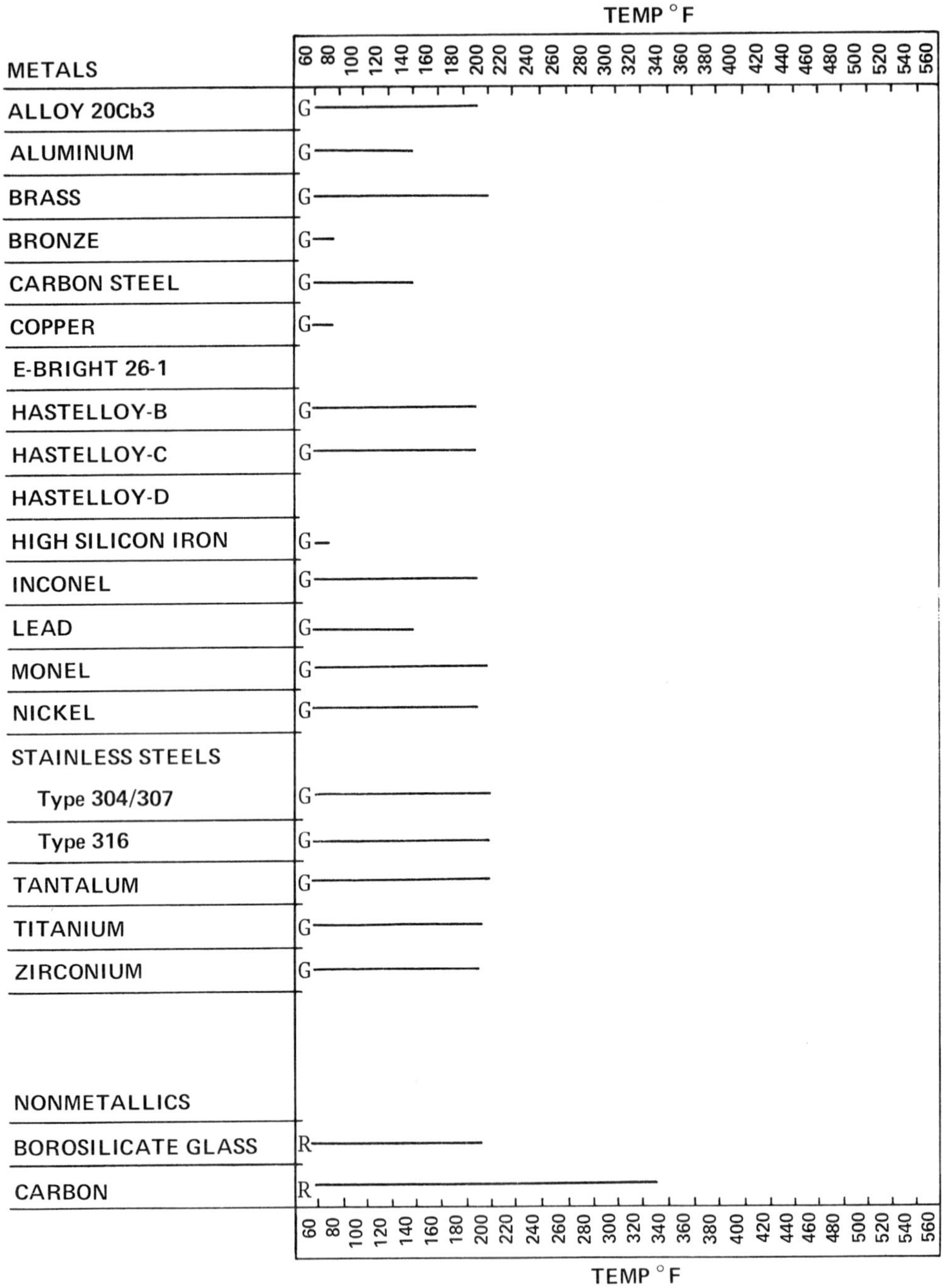

FOR METALS

E = < 2 Mils Penetration/Year; G = < 20 Mils Penetration/Year

S = < 50 Mils Penetration/Year; U = > 50 Mils Penetration/Year

FOR NONMETALLICS

R = Resistant

U = Unsatisfactory

ETHYLENE CHLOROHYDRIN

TEMP °F (60 80 100 120 140 160 180 200 220 240 260 280 300 320 340 360 380 400 420 440 460)

PLASTICS	
ABS	U
ACRYLICS	
ASBESTOS REINF., EPOXY	R—
ASBESTOS REINF., PHENOLIC	R—
CHLORINATED POLYETHER	R—
CPVC	U
EPOXY	U
FLUOROCARBONS FEP	R—
FLUOROCARBONS TFE	R—
FURFURYL ALCOHOL	R—
NORYL	
PHENOLIC	
POLYAMIDES—NYLON	U
POLYESTERS	
BISPHENOL A-FUMURATE	R—
HYDROGENATED BISPHENOL-A-BISPHENOL-A	R—
ISOPHTHALIC	U
CHLORINATED POLYESTERS	R—
POLYSTYRENES	
POLYETHYLENE	U
POLYPROPYLENE	R—
PVC—TYPE 1	U
PVC—TYPE 2	U
POLYVINYLIDENE CHLORIDE	U
VINYLIDENE FLUORIDE	R—U
VINYL ESTER	R—

RUBBERS	
BUTYL GR-1	R—
FLUORO ELASTOMERS VITON A	R—
FLUORO ELASTOMERS KEL-F 3700	
HARD RUBBER	
HYPALON	U
KOROSEAL	
NATURAL RUBBER (GRS)	U
NEOPRENE GR-M (CR)	U
NITRILE BUNA N (NBR)	U
NORDEL	
POLYBUTADIENE	
SILICONE	
POLYURETHANE	U

TEMP °F (60 80 100 120 140 160 180 200 220 240 260 280 300 320 340 360 380 400 420 440 460)

ETHYLENE DIAMINE

TEMP °F

60 80 100 120 140 160 180 200 220 240 260 280 300 320 340 360 380 400 420 440 460 480 500 520 540 560

METALS	
ALLOY 20Cb3	G
ALUMINUM	G
BRASS	U
BRONZE	G
CARBON STEEL	G
COPPER	U
E-BRIGHT 26-1	
HASTELLOY-B	G
HASTELLOY-C	
HASTELLOY-D	
HIGH SILICON IRON	
INCONEL	
LEAD	
MONEL	
NICKEL	
STAINLESS STEELS	
Type 304/307	G
Type 316	G
TANTALUM	G
TITANIUM	
ZIRCONIUM	

NONMETALLICS	
BOROSILICATE GLASS	R
CARBON	R

60 80 100 120 140 160 180 200 220 240 260 280 300 320 340 360 380 400 420 440 460 480 500 520 540 560

TEMP °F

FOR METALS

E = < 2 Mils Penetration/Year; G = < 20 Mils Penetration/Year

S = < 50 Mils Penetration/Year; U = > 50 Mils Penetration/Year

FOR NONMETALLICS

R = Resistant

U = Unsatisfactory

ETHYLENE DIAMINE

TEMP °F: 60 80 100 120 140 160 180 200 220 240 260 280 300 320 340 360 380 400 420 440 460

PLASTICS	
ABS	U
ACRYLICS	
ASBESTOS REINF., EPOXY	
ASBESTOS REINF., PHENOLIC	
CHLORINATED POLYETHER	R—
CPVC	
EPOXY	R—U
FLUOROCARBONS FEP	R—
FLUOROCARBONS TFE	R—
FURFURYL ALCOHOL	R—
NORYL	
PHENOLIC	
POLYAMIDES–NYLON	U
POLYESTERS	
BISPHENOL A-FUMURATE	
HYDROGENATED BISPHENOL-A-BISPHENOL-A	R—
ISOPHTHALIC	U
CHLORINATED POLYESTERS	U
POLYSTYRENES	
POLYETHYLENE	R—
POLYPROPYLENE	R—
PVC–TYPE 1	
PVC–TYPE 2	
POLYVINYLIDENE CHLORIDE	
VINYLIDENE FLUORIDE	R—U
VINYL ESTER	
RUBBERS	
BUTYL GR-1	R—
FLUORO ELASTOMERS VITON A	R—
FLUORO ELASTOMERS KEL-F 3700	
HARD RUBBER	
HYPALON	
KOROSEAL	
NATURAL RUBBER (GRS)	R—
NEOPRENE GR-M (CR)	R—
NITRILE BUNA N (NBR)	R—
NORDEL	
POLYBUTADIENE	
SILICONE	
POLYURETHANE	

TEMP °F: 60 80 100 120 140 160 180 200 220 240 260 280 300 320 340 360 380 400 420 440 460

ETHYLENE DICHLORIDE (DICHLORETHANE)

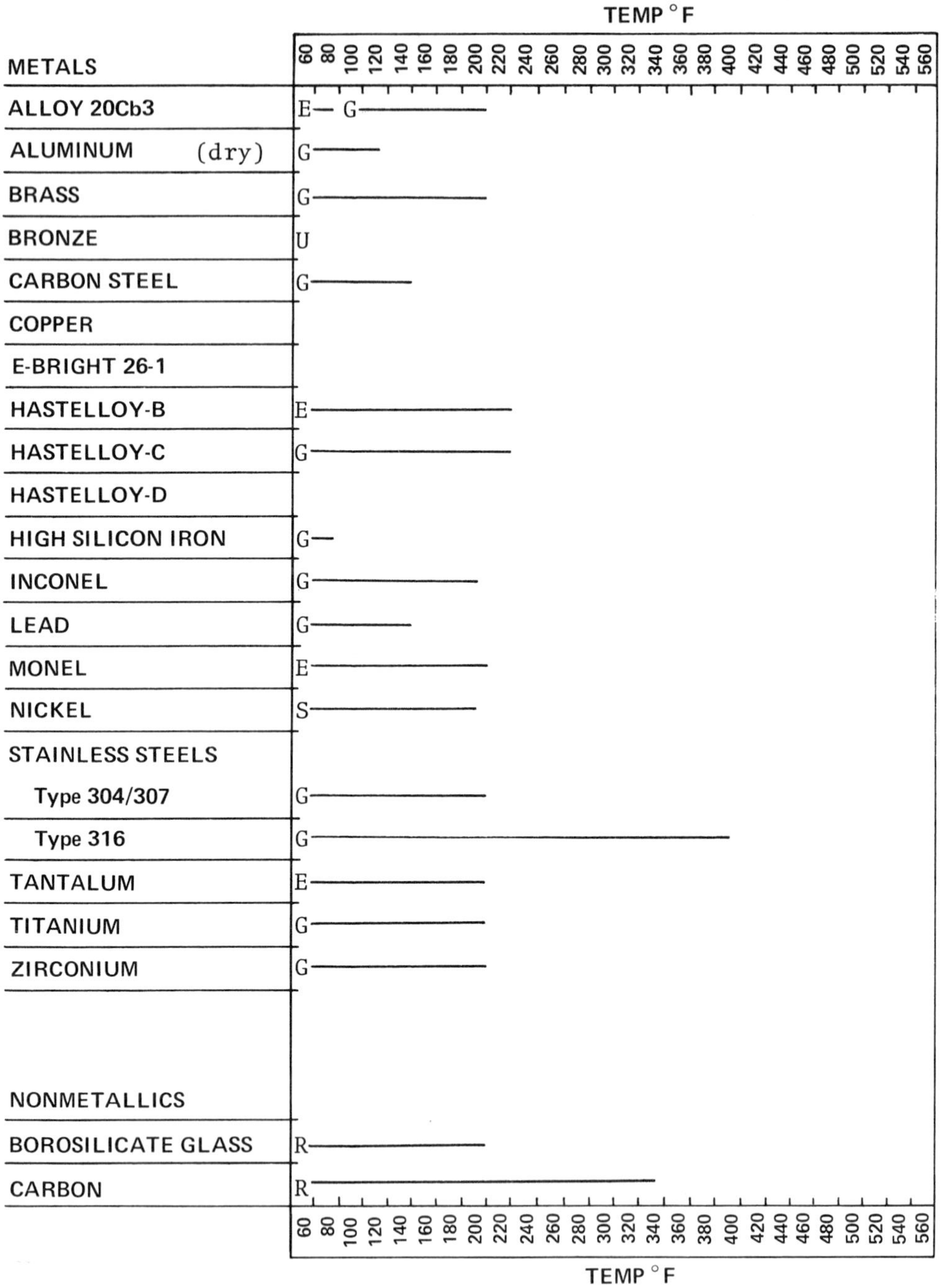

FOR METALS

E = < 2 Mils Penetration/Year; G = < 20 Mils Penetration/Year

S = < 50 Mils Penetration/Year; U = > 50 Mils Penetration/Year

FOR NONMETALLICS

R = Resistant

U = Unsatisfactory

ETHYLENE DICHLORIDE (DICHLORETHANE)

PLASTICS	Rating	TEMP °F (line extends to approx.)
ABS	U	
ACRYLICS	U	
ASBESTOS REINF., EPOXY	U	
ASBESTOS REINF., PHENOLIC	R	300
CHLORINATED POLYETHER	U	
CPVC	U	
EPOXY	U	
FLUOROCARBONS FEP	R	400
FLUOROCARBONS TFE	R	460
FURFURYL ALCOHOL	R	260
NORYL		
PHENOLIC		
POLYAMIDES—NYLON	R	80
POLYESTERS		
BISPHENOL A-FUMURATE	U	
HYDROGENATED BISPHENOL-A-BISPHENOL-A	U	
ISOPHTHALIC	U	
CHLORINATED POLYESTERS	U	
POLYSTYRENES		
POLYETHYLENE	U	
POLYPROPYLENE	R	80
PVC—TYPE 1	U	
PVC—TYPE 2	U	
POLYVINYLIDENE CHLORIDE	R	80
VINYLIDENE FLUORIDE	R	280
VINYL ESTER	R	80
RUBBERS		
BUTYL GR-1	U	
FLUORO ELASTOMERS VITON A	R	200
FLUORO ELASTOMERS KEL-F 3700		
HARD RUBBER	U	
HYPALON	U	
KOROSEAL		
NATURAL RUBBER (GRS)	U	
NEOPRENE GR-M (CR)	U	
NITRILE BUNA N (NBR)	U	
NORDEL		
POLYBUTADIENE		
SILICONE		
POLYURETHANE	U	

TEMP °F: 60 80 100 120 140 160 180 200 220 240 260 280 300 320 340 360 380 400 420 440 460

ETHYLENE GLYCOL

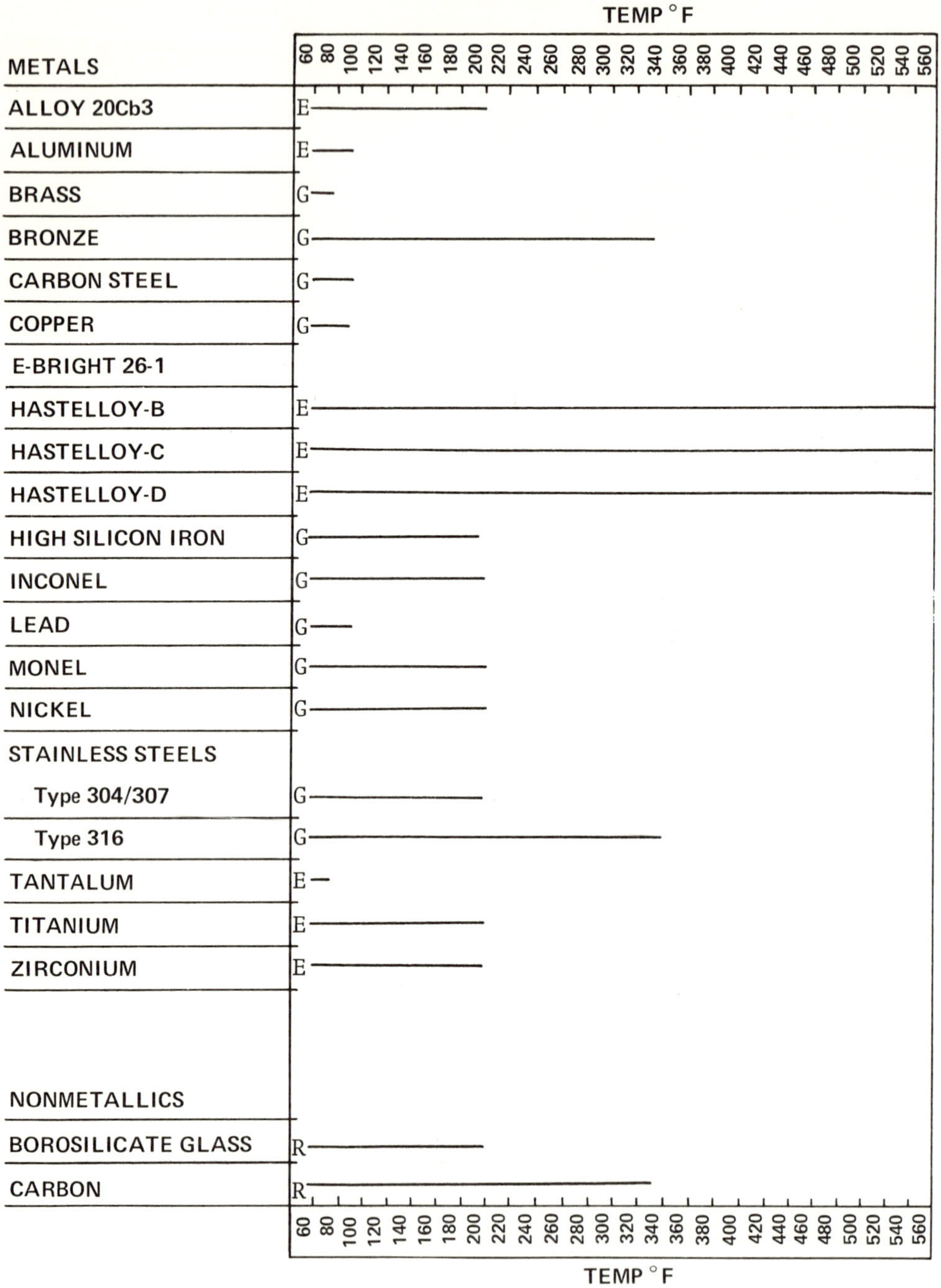

FOR METALS

E = < 2 Mils Penetration/Year; G = < 20 Mils Penetration/Year

S = < 50 Mils Penetration/Year; U = > 50 Mils Penetration/Year

FOR NONMETALLICS

R = Resistant

U = Unsatisfactory

TEMP °F (60–460)

PLASTICS	
ABS	R
ACRYLICS	R
ASBESTOS REINF., EPOXY	R
ASBESTOS REINF., PHENOLIC	R
CHLORINATED POLYETHER	R
CPVC	R
EPOXY	R
FLUOROCARBONS FEP	R
FLUOROCARBONS TFE	R
FURFURYL ALCOHOL	R
NORYL	R
PHENOLIC	
POLYAMIDES–NYLON	R
POLYESTERS	
BISPHENOL A-FUMURATE	R
HYDROGENATED BISPHENOL-A-BISPHENOL-A	R
ISOPHTHALIC 25%	R—U
CHLORINATED POLYESTERS	R
POLYSTYRENES	R
POLYETHYLENE	R
POLYPROPYLENE	R
PVC–TYPE 1	R
PVC–TYPE 2	R
POLYVINYLIDENE CHLORIDE	R
VINYLIDENE FLUORIDE	R
VINYL ESTER	R
RUBBERS	
BUTYL GR-1	R
FLUORO ELASTOMERS VITON A	R
FLUORO ELASTOMERS KEL-F 3700	
HARD RUBBER	R
HYPALON	R
KOROSEAL	R
NATURAL RUBBER (GRS)	R
NEOPRENE GR-M (CR)	R
NITRILE BUNA N (NBR)	R
NORDEL	R
POLYBUTADIENE	
SILICONE	
POLYURETHANE	R

TEMP °F (60–460)

ETHYLENE OXIDE

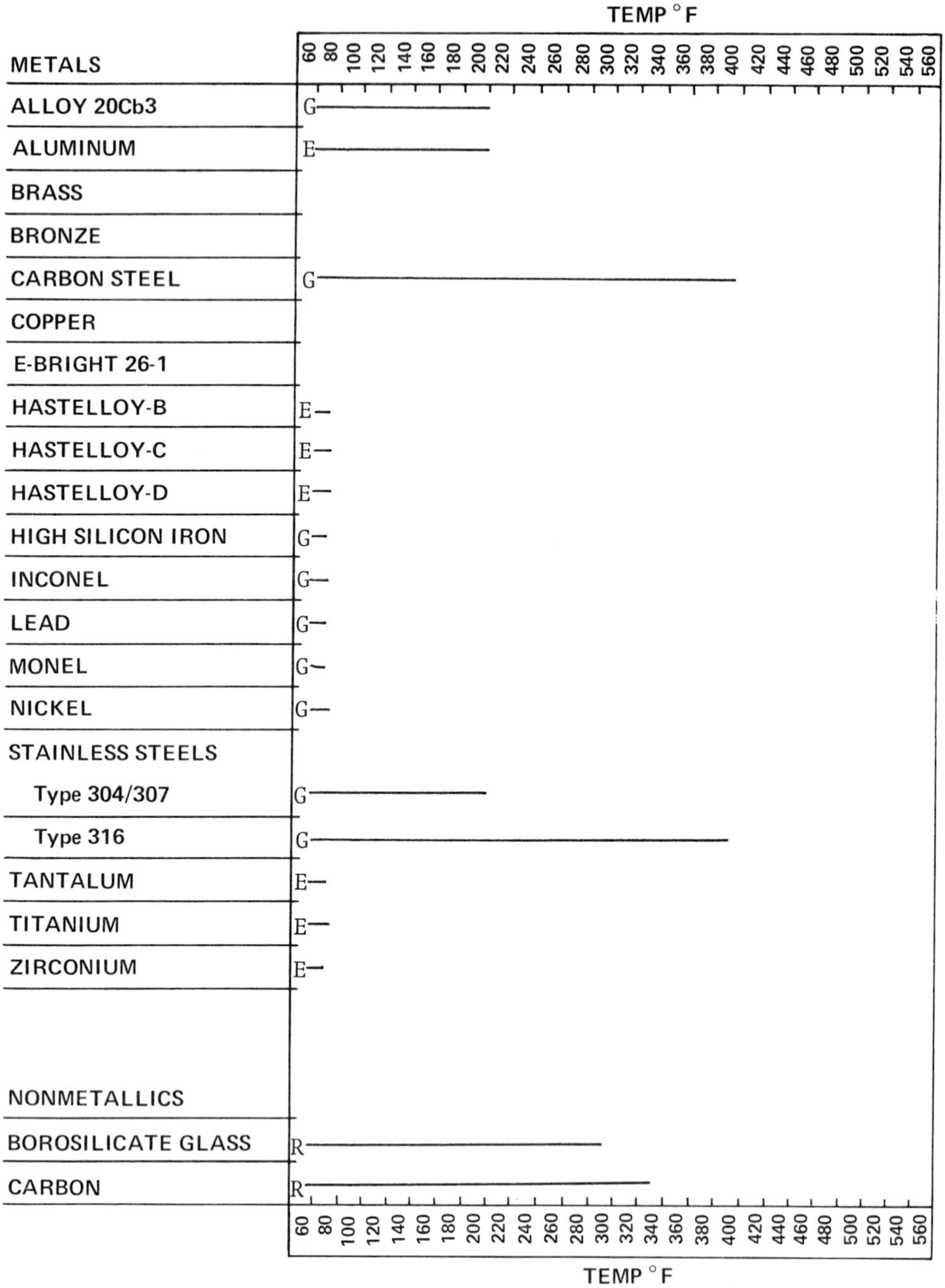

FOR METALS

E = < 2 Mils Penetration/Year; G = < 20 Mils Penetration/Year

S = < 50 Mils Penetration/Year; U = > 50 Mils Penetration/Year

FOR NONMETALLICS

R = Resistant

U = Unsatisfactory

ETHYLENE OXIDE

TEMP °F

PLASTICS	60–460 °F
ABS	U
ACRYLICS	
ASBESTOS REINF., EPOXY	
ASBESTOS REINF., PHENOLIC	
CHLORINATED POLYETHER	U
CPVC	U
EPOXY	U
FLUOROCARBONS FEP	R
FLUOROCARBONS TFE	R
FURFURYL ALCOHOL	R
NORYL	
PHENOLIC	
POLYAMIDES—NYLON	R
POLYESTERS	
BISPHENOL A-FUMURATE	
HYDROGENATED BISPHENOL-A-BISPHENOL-A	
ISOPHTHALIC	
CHLORINATED POLYESTERS	
POLYSTYRENES	
POLYETHYLENE	U
POLYPROPYLENE	U
PVC—TYPE 1	U
PVC—TYPE 2	U
POLYVINYLIDENE CHLORIDE	U
VINYLIDENE FLUORIDE	R
VINYL ESTER	

RUBBERS	60–460 °F
BUTYL GR-1	U
FLUORO ELASTOMERS VITON A	U
FLUORO ELASTOMERS KEL-F 3700	
HARD RUBBER	
HYPALON	U
KOROSEAL	
NATURAL RUBBER (GRS)	U
NEOPRENE GR-M (CR)	U
NITRILE BUNA N (NBR)	U
NORDEL	
POLYBUTADIENE	
SILICONE	
POLYURETHANE	U

TEMP °F

FATTY ACIDS

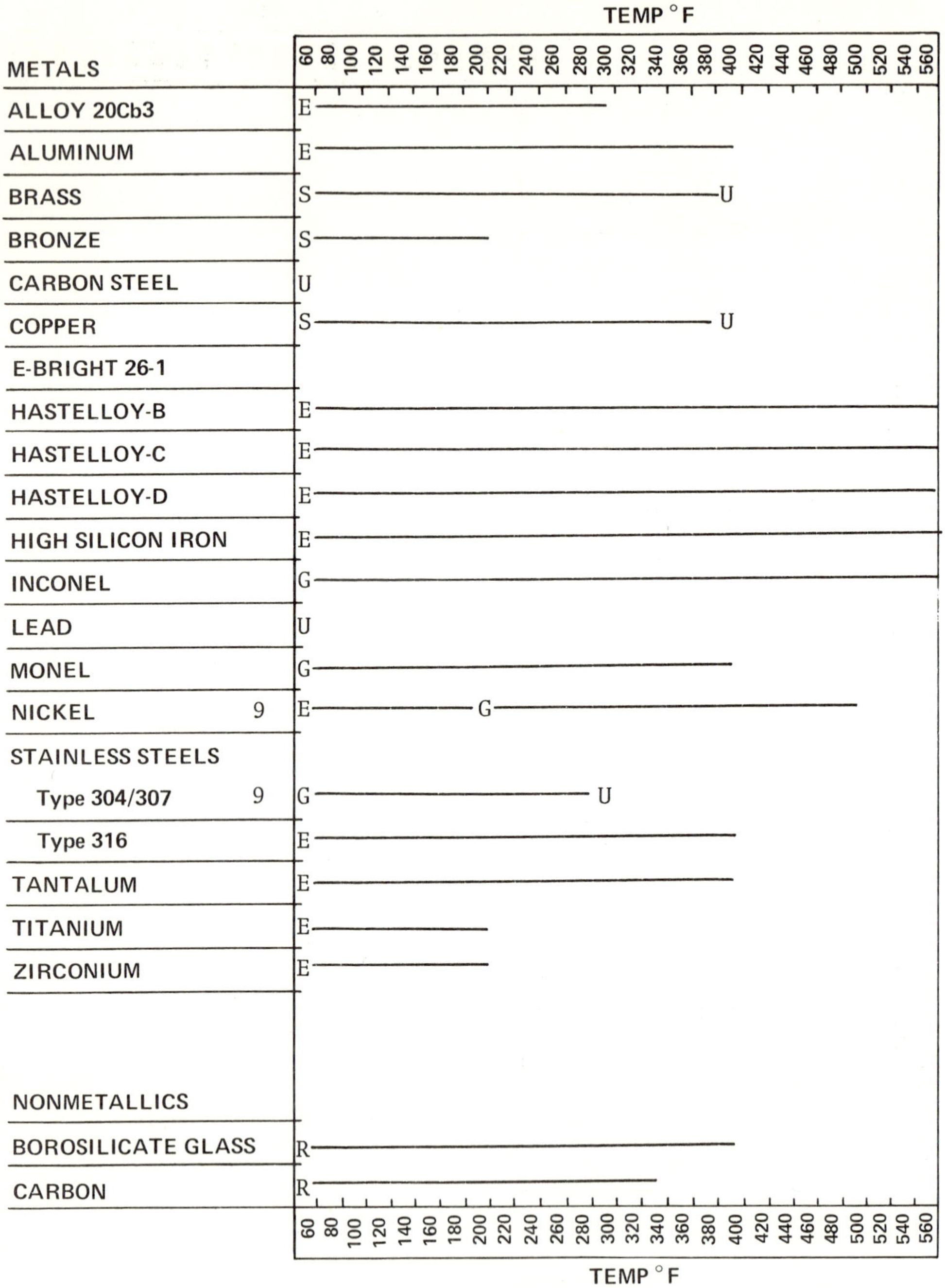

FOR METALS

E = < 2 Mils Penetration/Year; G = < 20 Mils Penetration/Year

S = < 50 Mils Penetration/Year; U = > 50 Mils Penetration/Year

FOR NONMETALLICS

R = Resistant

U = Unsatisfactory

FATTY ACIDS

Please see page 1 for notes on calibration.

TEMP °F

PLASTICS		Rating	Temp °F
ABS		R	140
ACRYLICS		R	80
ASBESTOS REINF., EPOXY		R	80
ASBESTOS REINF., PHENOLIC		R	80
CHLORINATED POLYETHER		R	150
CPVC		R	180
EPOXY		R	200
FLUOROCARBONS FEP		R	400
FLUOROCARBONS TFE		R	460
FURFURYL ALCOHOL	sat	R	270
NORYL			
PHENOLIC		R	80
POLYAMIDES–NYLON		R	80
POLYESTERS			
BISPHENOL A-FUMURATE	sat	R	200
HYDROGENATED BISPHENOL-A-BISPHENOL-A		R	210
ISOPHTHALIC		R	180
CHLORINATED POLYESTERS		R	250
POLYSTYRENES		R	160
POLYETHYLENE		R	150
POLYPROPYLENE		R	140
PVC–TYPE 1		R	140
PVC–TYPE 2		R	140
POLYVINYLIDENE CHLORIDE		R	150
VINYLIDENE FLUORIDE		R	280
VINYL ESTER		R	250

RUBBERS	Rating	Temp °F
BUTYL GR-1	R	150
FLUORO ELASTOMERS VITON A	R	180
FLUORO ELASTOMERS KEL-F 3700	R	80
HARD RUBBER	R	80
HYPALON	R	250
KOROSEAL		
NATURAL RUBBER (GRS)	U	
NEOPRENE GR-M (CR)	U	
NITRILE BUNA N (NBR)	R	180
NORDEL		
POLYBUTADIENE	R	80
SILICONE		
POLYURETHANE		

TEMP °F

FERRIC CHLORIDE

TEMP °F (60–560)

METALS	Rating
ALLOY 20Cb3	U
ALUMINUM	U
BRASS	U
BRONZE	G—
CARBON STEEL	U
COPPER	G—
E-BRIGHT 26-1	E—
HASTELLOY-B	G—U
HASTELLOY-C	G— U
HASTELLOY-D	G— U
HIGH SILICON IRON	U
INCONEL	U
LEAD	U
MONEL	U
NICKEL	U
STAINLESS STEELS	
Type 304/307	U
Type 316	U
TANTALUM	E———————
TITANIUM	E————————————
ZIRCONIUM	U
NONMETALLICS	
BOROSILICATE GLASS	R———————————
CARBON	R——————————————

TEMP °F (60–560)

FOR METALS

E = < 2 Mils Penetration/Year; G = < 20 Mils Penetration/Year

S = < 50 Mils Penetration/Year; U = > 50 Mils Penetration/Year

FOR NONMETALLICS

R = Resistant

U = Unsatisfactory

FERRIC CHLORIDE

TEMP °F (scale 60, 80, 100, 120, 140, 160, 180, 200, 220, 240, 260, 280, 300, 320, 340, 360, 380, 400, 420, 440, 460)

PLASTICS		
ABS		R
ACRYLICS		R
ASBESTOS REINF., EPOXY		R
ASBESTOS REINF., PHENOLIC		R
CHLORINATED POLYETHER		R
CPVC		R
EPOXY		R
FLUOROCARBONS FEP		R
FLUOROCARBONS TFE		R
FURFURYL ALCOHOL	sat	R
NORYL		R
PHENOLIC		
POLYAMIDES—NYLON	10%	R
POLYESTERS		
BISPHENOL A-FUMURATE	sat	R
HYDROGENATED BISPHENOL-A-BISPHENOL-A		R
ISOPHTHALIC	sat	R
CHLORINATED POLYESTERS		R
POLYSTYRENES		R
POLYETHYLENE		R — U
POLYPROPYLENE		R — U
PVC—TYPE 1		R
PVC—TYPE 2		R
POLYVINYLIDENE CHLORIDE		R
VINYLIDENE FLUORIDE		R
VINYL ESTER	sat	R

RUBBERS	
BUTYL GR-1	R
FLUORO ELASTOMERS VITON A	R
FLUORO ELASTOMERS KEL-F 3700	
HARD RUBBER	R
HYPALON	R
KOROSEAL	R
NATURAL RUBBER (GRS)	R
NEOPRENE GR-M (CR)	R
NITRILE BUNA N (NBR)	R
NORDEL	R
POLYBUTADIENE	
SILICONE	
POLYURETHANE	R

TEMP °F

FERRIC CHLORIDE, 50% IN WATER

TEMP °F: 60, 80, 100, 120, 140, 160, 180, 200, 220, 240, 260, 280, 300, 320, 340, 360, 380, 400, 420, 440, 460, 480, 500, 520, 540, 560

METALS	
ALLOY 20Cb3	U
ALUMINUM	U
BRASS	
BRONZE	U
CARBON STEEL	U
COPPER	U
E-BRIGHT 26-1	
HASTELLOY-B	U
HASTELLOY-C	S—
HASTELLOY-D	U
HIGH SILICON IRON	
INCONEL	S—
LEAD	U
MONEL	U
NICKEL	U
STAINLESS STEELS	
Type 304/307	U
Type 316	U
TANTALUM	E
TITANIUM	E
ZIRCONIUM	U

NONMETALLICS	
BOROSILICATE GLASS	R
CARBON	R

TEMP °F

FOR METALS

E = < 2 Mils Penetration/Year; G = < 20 Mils Penetration/Year

S = < 50 Mils Penetration/Year; U = > 50 Mils Penetration/Year

FOR NONMETALLICS

R = Resistant

U = Unsatisfactory

FERRIC CHLORIDE, 50% IN WATER

TEMP °F: 60 80 100 120 140 160 180 200 220 240 260 280 300 320 340 360 380 400 420 440 460

PLASTICS	
ABS	
ACRYLICS	
ASBESTOS REINF., EPOXY	
ASBESTOS REINF., PHENOLIC	
CHLORINATED POLYETHER	R
CPVC	
EPOXY	R
FLUOROCARBONS FEP	R
FLUOROCARBONS TFE	R
FURFURYL ALCOHOL	R
NORYL	
PHENOLIC	
POLYAMIDES—NYLON	U
POLYESTERS	
BISPHENOL A-FUMURATE	R
HYDROGENATED BISPHENOL-A-BISPHENOL-A	R
ISOPHTHALIC	R
CHLORINATED POLYESTERS	
POLYSTYRENES	R
POLYETHYLENE	
POLYPROPYLENE	R — U
PVC—TYPE 1	R
PVC—TYPE 2	
POLYVINYLIDENE CHLORIDE	R
VINYLIDENE FLUORIDE	R
VINYL ESTER	
RUBBERS	
BUTYL GR-1	R
FLUORO ELASTOMERS VITON A	R
FLUORO ELASTOMERS KEL-F 3700	
HARD RUBBER	R
HYPALON	R
KOROSEAL	R
NATURAL RUBBER (GRS)	R
NEOPRENE GR-M (CR)	R
NITRILE BUNA N (NBR)	R
NORDEL	R
POLYBUTADIENE	
SILICONE	
POLYURETHANE	R

TEMP °F: 60 80 100 120 140 160 180 200 220 240 260 280 300 320 340 360 380 400 420 440 460

FERRIC HYDROXIDE

TEMP °F

METALS	Rating (TEMP °F)
ALLOY 20Cb3	
ALUMINUM	
BRASS	
BRONZE	
CARBON STEEL	
COPPER	
E-BRIGHT 26-1	
HASTELLOY-B	
HASTELLOY-C	
HASTELLOY-D	
HIGH SILICON IRON	
INCONEL	G (60)
LEAD	
MONEL	G (60)
NICKEL	G (60)
STAINLESS STEELS	
Type 304/307	E (60)
Type 316	E (60)
TANTALUM	E (60)
TITANIUM	
ZIRCONIUM	
NONMETALLICS	
BOROSILICATE GLASS	R (60–140)
CARBON	

TEMP °F: 60, 80, 100, 120, 140, 160, 180, 200, 220, 240, 260, 280, 300, 320, 340, 360, 380, 400, 420, 440, 460, 480, 500, 520, 540, 560

FOR METALS

E = < 2 Mils Penetration/Year; G = < 20 Mils Penetration/Year

S = < 50 Mils Penetration/Year; U = > 50 Mils Penetration/Year

FOR NONMETALLICS

R = Resistant

U = Unsatisfactory

FERRIC HYDROXIDE

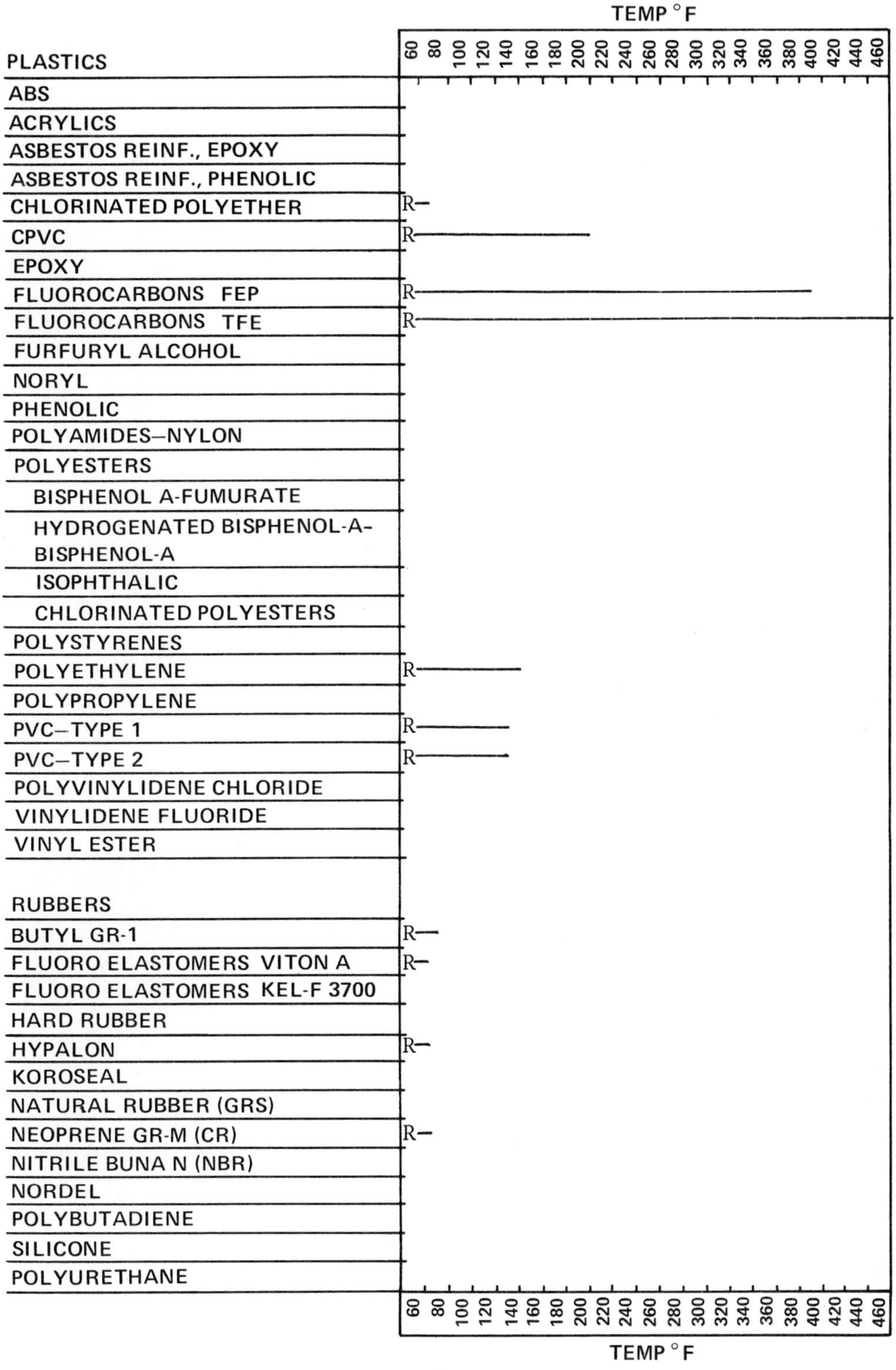

FERRIC NITRATE, 10%-50%

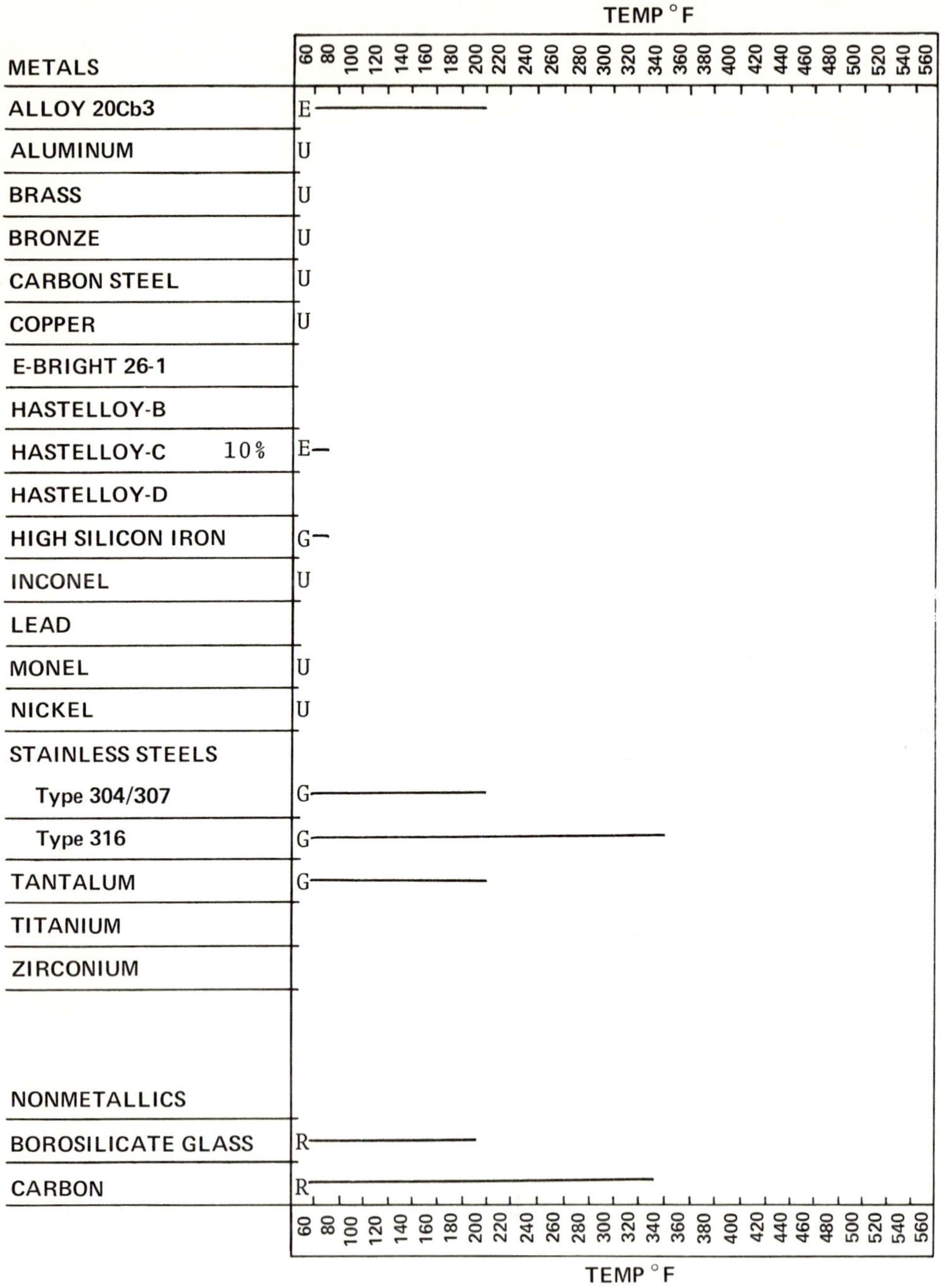

FOR METALS

E = < 2 Mils Penetration/Year; G = < 20 Mils Penetration/Year

S = < 50 Mils Penetration/Year; U = > 50 Mils Penetration/Year

FOR NONMETALLICS

R = Resistant

U = Unsatisfactory

FERRIC NITRATE, 10%-50%

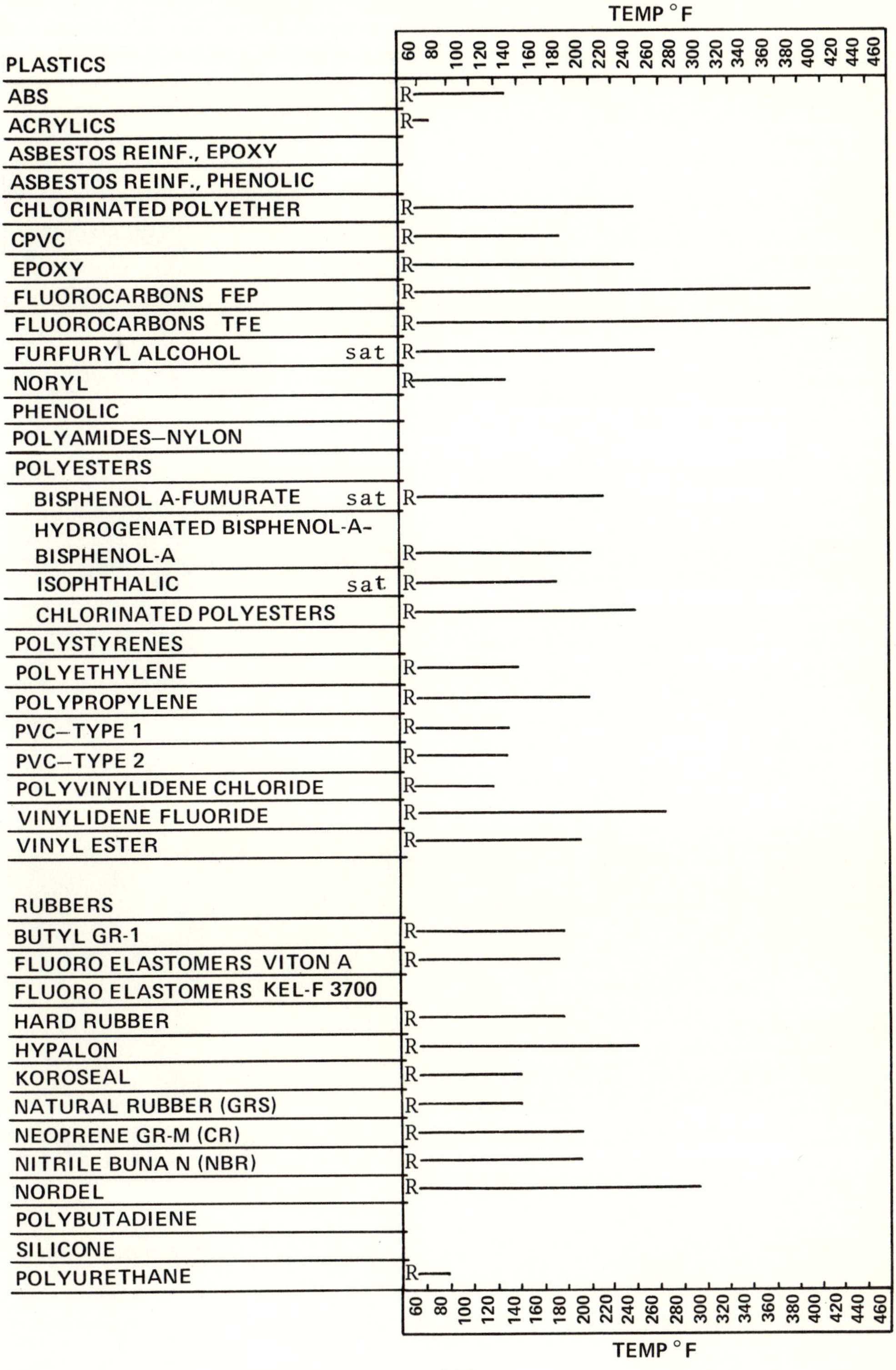

FERRIC SULFATE

TEMP °F: 60 80 100 120 140 160 180 200 220 240 260 280 300 320 340 360 380 400 420 440 460 480 500 520 540 560

METALS		
ALLOY 20Cb3		E
ALUMINUM		U
BRASS		U
BRONZE		U
CARBON STEEL		U
COPPER		S
E-BRIGHT 26-1		
HASTELLOY-B		U
HASTELLOY-C	to 30%	E
HASTELLOY-D	to 30%	E
HIGH SILICON IRON		G
INCONEL	10%	U
LEAD		G
MONEL		G
NICKEL	10%+	U
STAINLESS STEELS		
Type 304/307	9	G
Type 316	10%	E
TANTALUM	sat.	E
TITANIUM	10%	E
ZIRCONIUM	10%	E

NONMETALLICS		
BOROSILICATE GLASS		R
CARBON		R

TEMP °F: 60 80 100 120 140 160 180 200 220 240 260 280 300 320 340 360 380 400 420 440 460 480 500 520 540 560

FOR METALS

E = < 2 Mils Penetration/Year; G = < 20 Mils Penetration/Year

S = < 50 Mils Penetration/Year; U = > 50 Mils Penetration/Year

FOR NONMETALLICS

R = Resistant

U = Unsatisfactory

FERRIC SULFATE

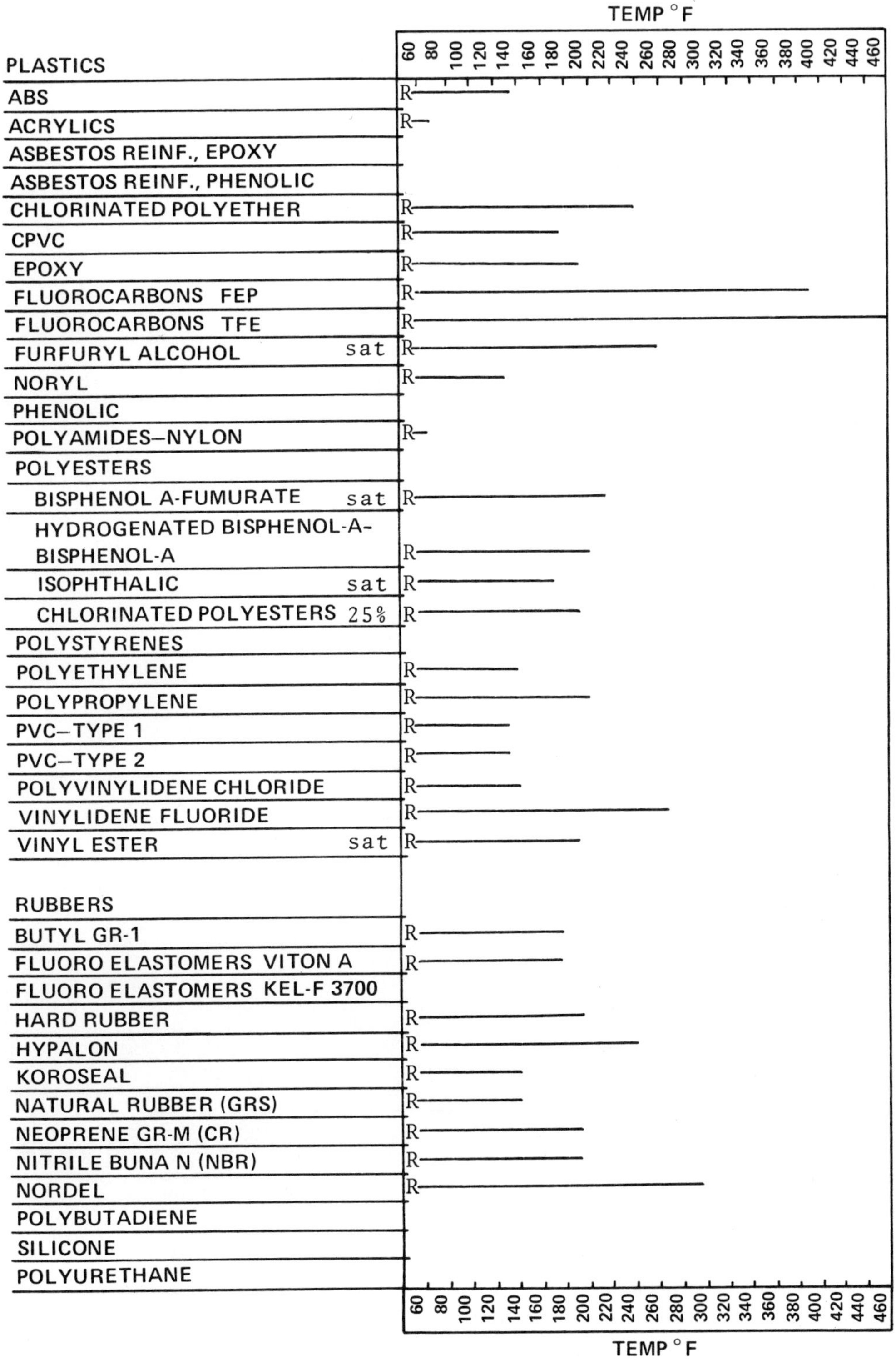

FERROUS CHLORIDE (SAT)

TEMP °F: 60 80 100 120 140 160 180 200 220 240 260 280 300 320 340 360 380 400 420 440 460 480 500 520 540 560

METALS	
ALLOY 20Cb3	S
ALUMINUM	U
BRASS	U
BRONZE	G
CARBON STEEL	U
COPPER	G
E-BRIGHT 26-1	
HASTELLOY-B	G
HASTELLOY-C	G
HASTELLOY-D	G
HIGH SILICON IRON	G
INCONEL	U
LEAD	U
MONEL	U
NICKEL	U
STAINLESS STEELS	
Type 304/307	U
Type 316	U
TANTALUM	E
TITANIUM	E
ZIRCONIUM	E
NONMETALLICS	
BOROSILICATE GLASS	R
CARBON	R

TEMP °F: 60 80 100 120 140 160 180 200 220 240 260 280 300 320 340 360 380 400 420 440 460 480 500 520 540 560

FOR METALS

E = < 2 Mils Penetration/Year; G = < 20 Mils Penetration/Year

S = < 50 Mils Penetration/Year; U = > 50 Mils Penetration/Year

FOR NONMETALLICS

R = Resistant

U = Unsatisfactory

FERROUS CHLORIDE (SAT)

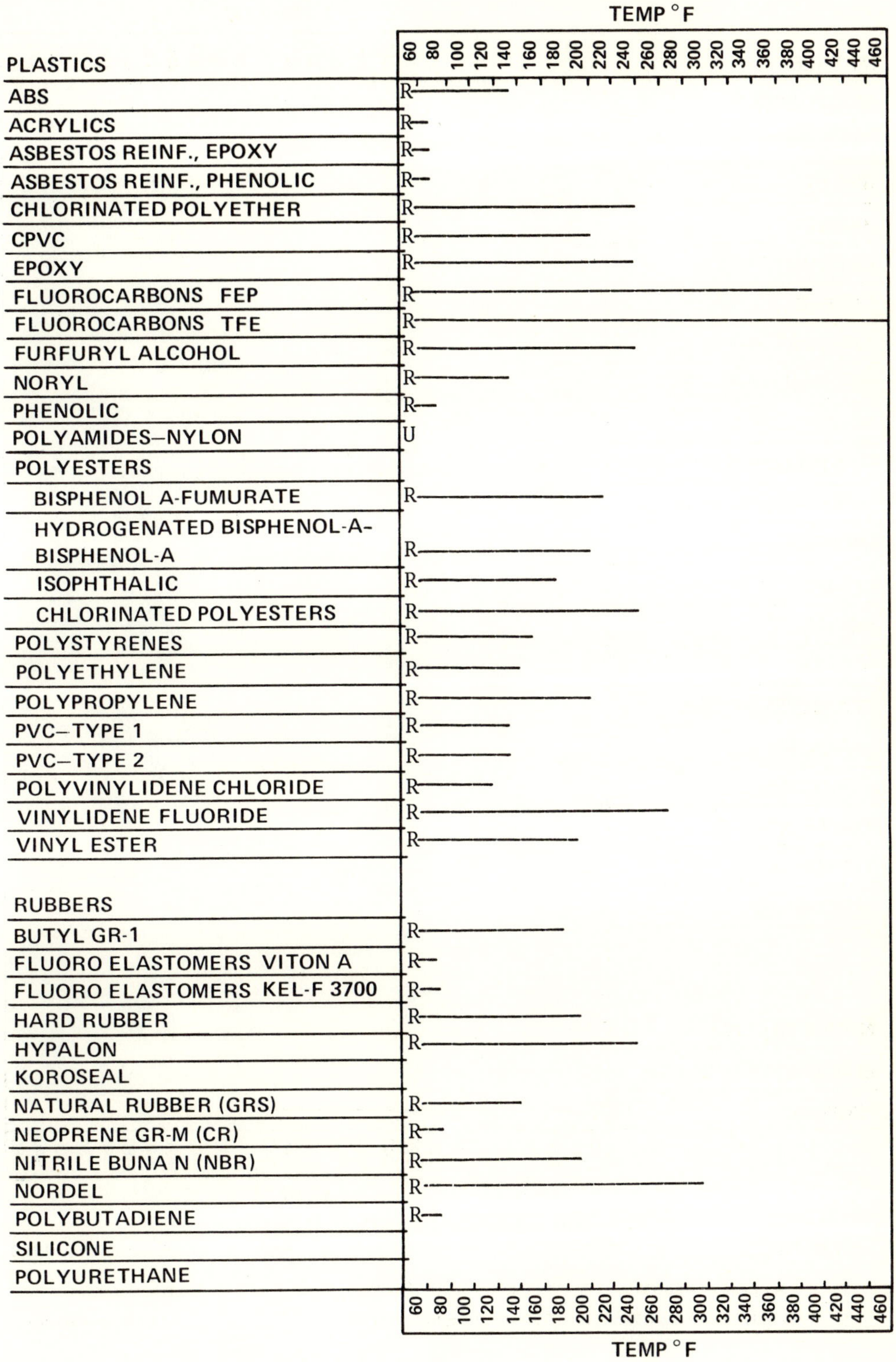

FERROUS NITRATE

TEMP °F (60–560)

METALS	Rating
ALLOY 20Cb3	
ALUMINUM	
BRASS	
BRONZE	
CARBON STEEL	
COPPER	
E-BRIGHT 26-1	
HASTELLOY-B	
HASTELLOY-C	
HASTELLOY-D	
HIGH SILICON IRON	
INCONEL	
LEAD	
MONEL	
NICKEL	
STAINLESS STEELS Type 304/307	
Type 316	
TANTALUM	
TITANIUM	
ZIRCONIUM	

NONMETALLICS	Rating
BOROSILICATE GLASS	
CARBON	R (60–340)

FOR METALS

E = < 2 Mils Penetration/Year; G = < 20 Mils Penetration/Year

S = < 50 Mils Penetration/Year; U = > 50 Mils Penetration/Year

FOR NONMETALLICS

R = Resistant

U = Unsatisfactory

FERROUS NITRATE

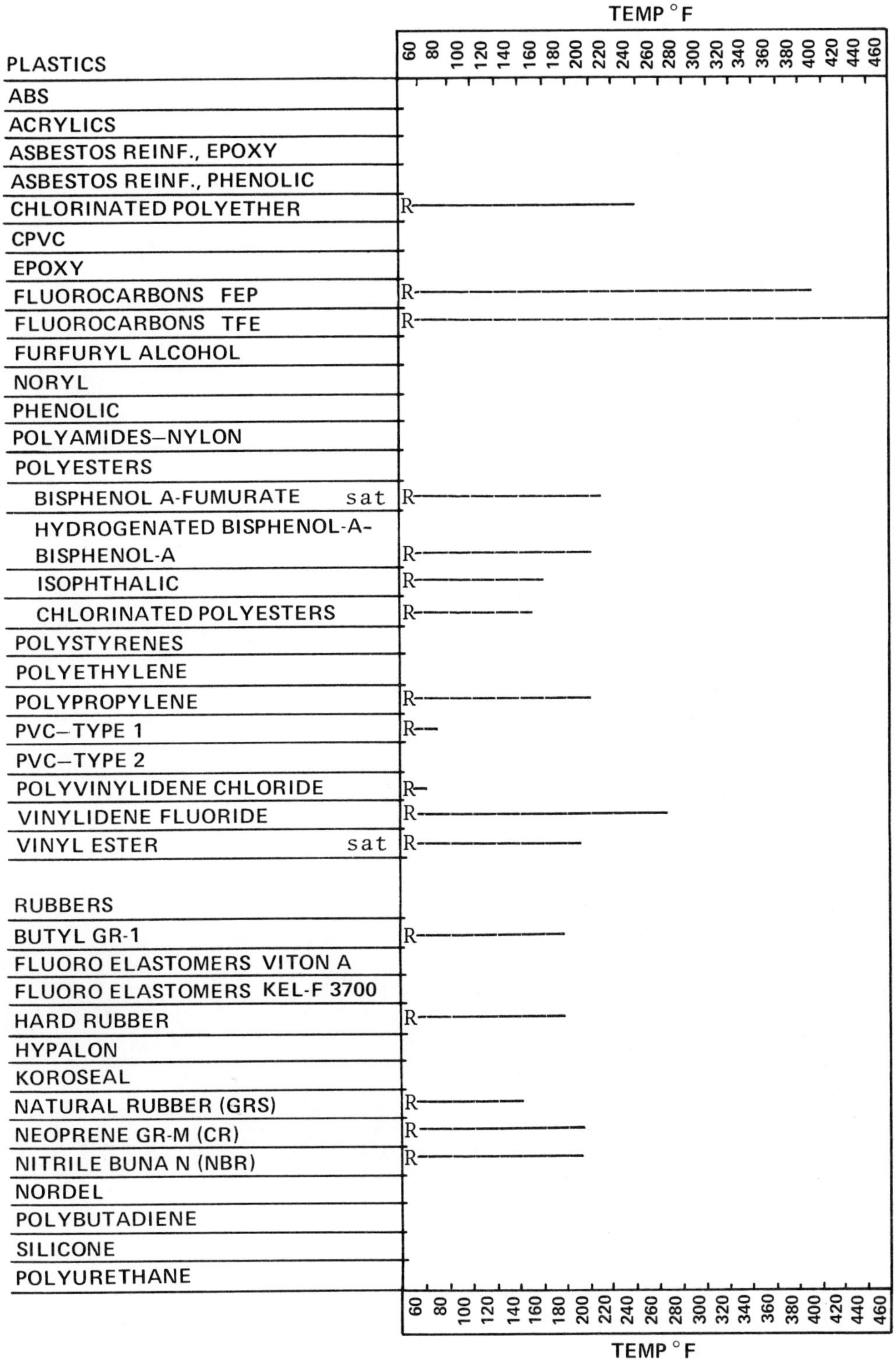

FERROUS SULFATE

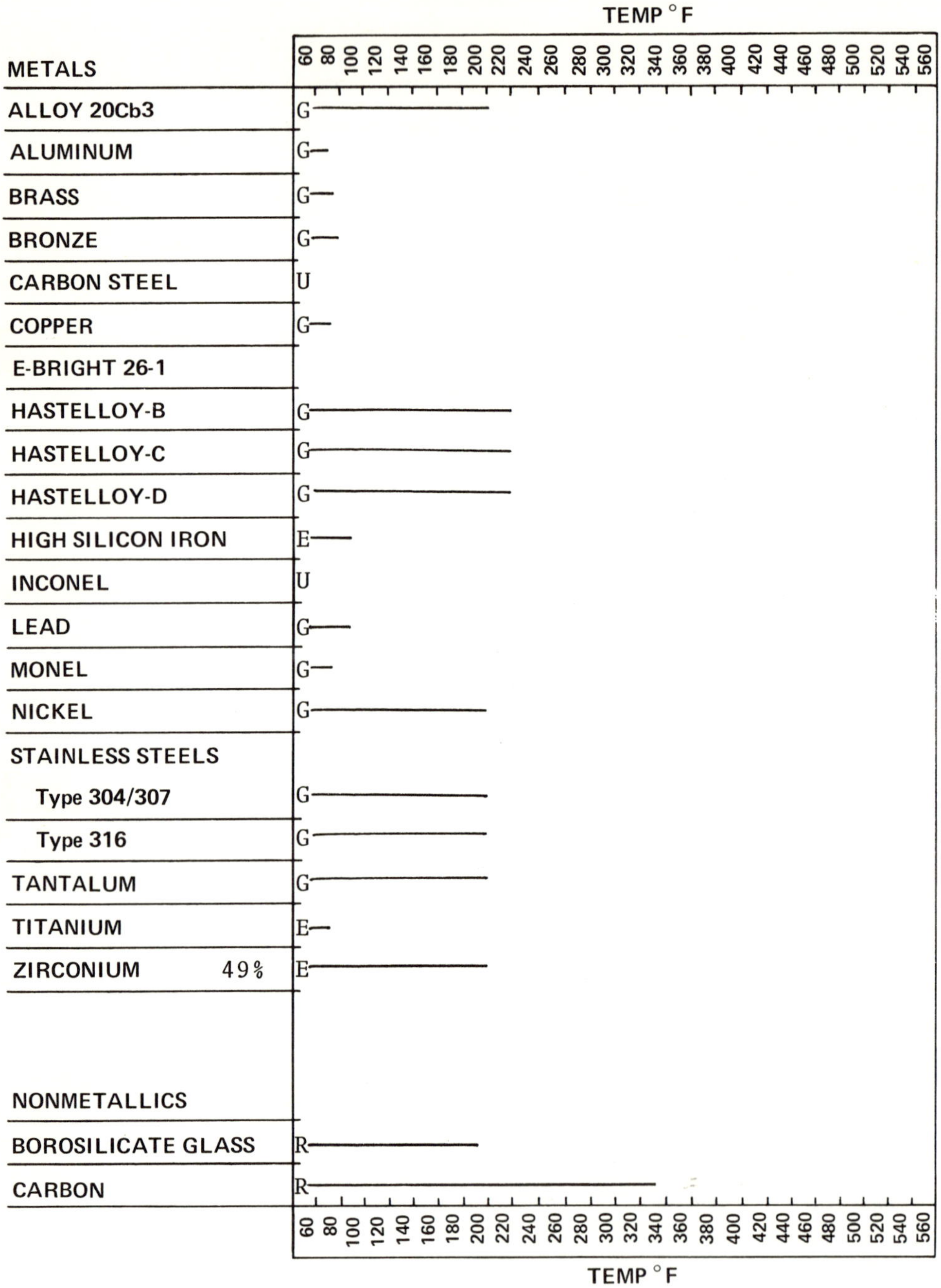

FOR METALS

E = < 2 Mils Penetration/Year; G = < 20 Mils Penetration/Year

S = < 50 Mils Penetration/Year; U = > 50 Mils Penetration/Year

FOR NONMETALLICS

R = Resistant

U = Unsatisfactory

FERROUS SULFATE

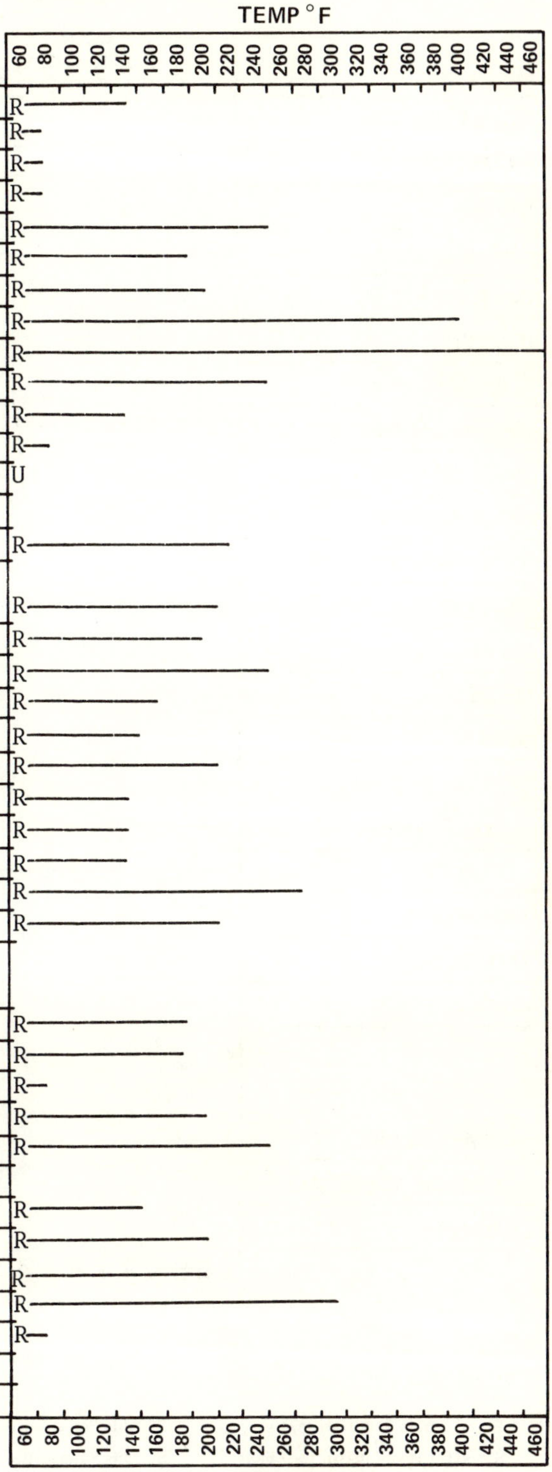

PLASTICS	
ABS	R
ACRYLICS	R
ASBESTOS REINF., EPOXY	R
ASBESTOS REINF., PHENOLIC	R
CHLORINATED POLYETHER	R
CPVC	R
EPOXY	R
FLUOROCARBONS FEP	R
FLUOROCARBONS TFE	R
FURFURYL ALCOHOL	R
NORYL	R
PHENOLIC	R
POLYAMIDES–NYLON	U
POLYESTERS	
BISPHENOL A-FUMURATE sat	R
HYDROGENATED BISPHENOL-A-BISPHENOL-A	R
ISOPHTHALIC sat	R
CHLORINATED POLYESTERS	R
POLYSTYRENES	R
POLYETHYLENE	R
POLYPROPYLENE	R
PVC–TYPE 1	R
PVC–TYPE 2	R
POLYVINYLIDENE CHLORIDE	R
VINYLIDENE FLUORIDE	R
VINYL ESTER	R
RUBBERS	
BUTYL GR-1	R
FLUORO ELASTOMERS VITON A	R
FLUORO ELASTOMERS KEL-F 3700	R
HARD RUBBER	R
HYPALON	R
KOROSEAL	
NATURAL RUBBER (GRS)	R
NEOPRENE GR-M (CR)	R
NITRILE BUNA N (NBR)	R
NORDEL	R
POLYBUTADIENE	R
SILICONE	
POLYURETHANE	

FISH SOLUBLES

TEMP °F

METALS	Rating (TEMP °F, 60–560)
ALLOY 20Cb3	E (60–100)
ALUMINUM	
BRASS	
BRONZE	G (60–340)
CARBON STEEL	
COPPER	
E-BRIGHT 26-1	
HASTELLOY-B	
HASTELLOY-C	
HASTELLOY-D	
HIGH SILICON IRON	
INCONEL	
LEAD	
MONEL	
NICKEL	
STAINLESS STEELS	
Type 304/307	E (60–80)
Type 316	E (60–80); G (100–300)
TANTALUM	
TITANIUM	
ZIRCONIUM	

NONMETALLICS	Rating (TEMP °F, 60–560)
BOROSILICATE GLASS	
CARBON	

TEMP °F

FOR METALS

E = < 2 Mils Penetration/Year; G = < 20 Mils Penetration/Year

S = < 50 Mils Penetration/Year; U = > 50 Mils Penetration/Year

FOR NONMETALLICS

R = Resistant

U = Unsatisfactory

FISH SOLUBLES

TEMP °F (60 80 100 120 140 160 180 200 220 240 260 280 300 320 340 360 380 400 420 440 460)

PLASTICS	
ABS	R
ACRYLICS	
ASBESTOS REINF., EPOXY	
ASBESTOS REINF., PHENOLIC	
CHLORINATED POLYETHER	R
CPVC	R
EPOXY	
FLUOROCARBONS FEP	R
FLUOROCARBONS TFE	R
FURFURYL ALCOHOL	
NORYL	
PHENOLIC	
POLYAMIDES–NYLON	
POLYESTERS	
BISPHENOL A-FUMURATE	
HYDROGENATED BISPHENOL-A-BISPHENOL-A	
ISOPHTHALIC	
CHLORINATED POLYESTERS	
POLYSTYRENES	
POLYETHYLENE	R
POLYPROPYLENE	R
PVC–TYPE 1	R
PVC–TYPE 2	R
POLYVINYLIDENE CHLORIDE	R
VINYLIDENE FLUORIDE	
VINYL ESTER	

RUBBERS	
BUTYL GR-1	
FLUORO ELASTOMERS VITON A	
FLUORO ELASTOMERS KEL-F 3700	
HARD RUBBER	
HYPALON	
KOROSEAL	
NATURAL RUBBER (GRS)	
NEOPRENE GR-M (CR)	
NITRILE BUNA N (NBR)	R
NORDEL	
POLYBUTADIENE	
SILICONE	
POLYURETHANE	

TEMP °F (60 80 100 120 140 160 180 200 220 240 260 280 300 320 340 360 380 400 420 440 460)

FLUORINE GAS, WET

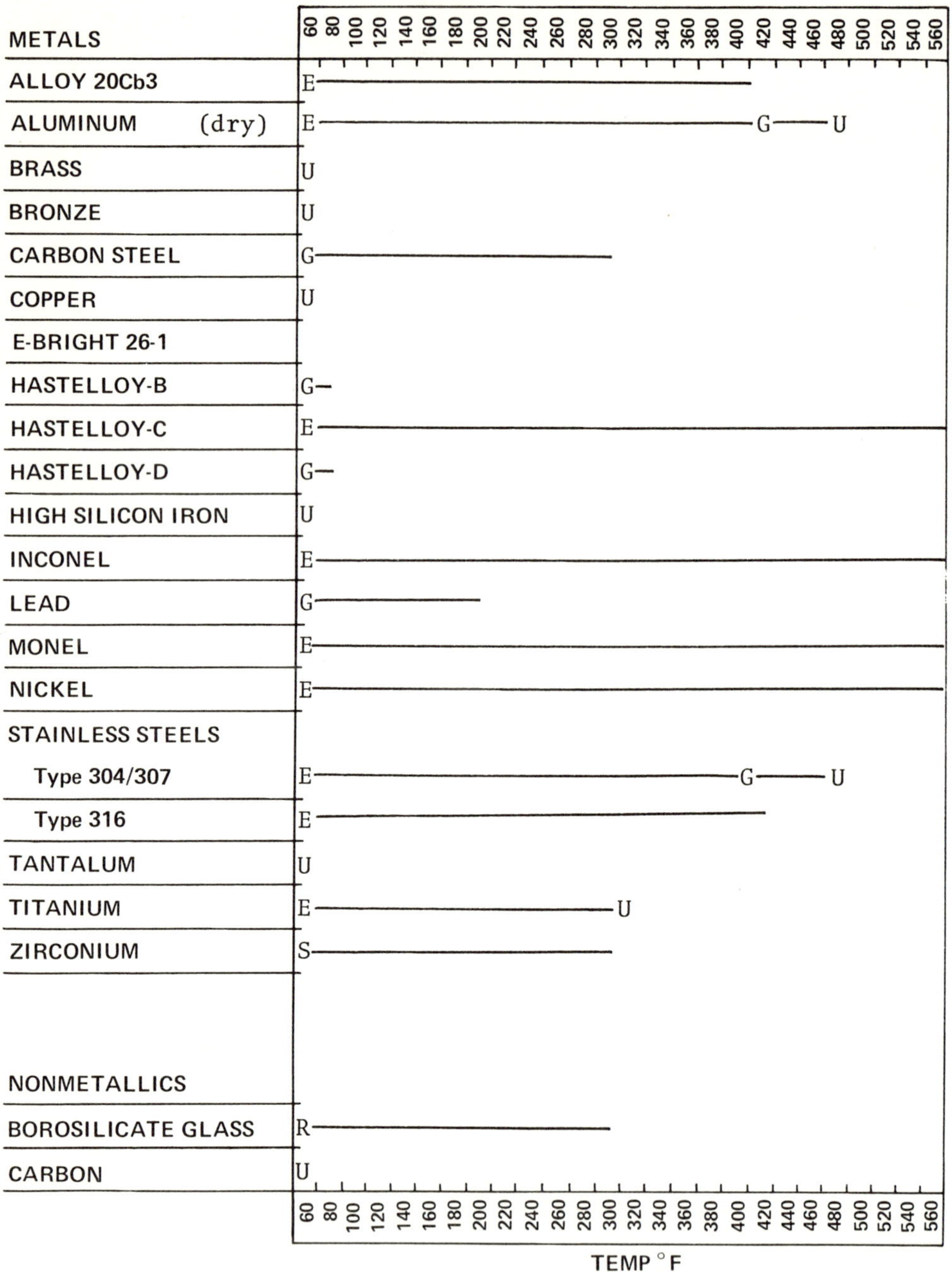

FOR METALS

E = < 2 Mils Penetration/Year; G = < 20 Mils Penetration/Year

S = < 50 Mils Penetration/Year; U = > 50 Mils Penetration/Year

FOR NONMETALLICS

R = Resistant

U = Unsatisfactory

FLUORINE GAS, WET

PLASTICS	TEMP °F (60–460)
ABS	R (60–80)
ACRYLICS	
ASBESTOS REINF., EPOXY	
ASBESTOS REINF., PHENOLIC	
CHLORINATED POLYETHER	U
CPVC	U
EPOXY	R (60–80)
FLUOROCARBONS FEP	R (60–200)
FLUOROCARBONS TFE	R (60–300)
FURFURYL ALCOHOL	
NORYL	
PHENOLIC	
POLYAMIDES–NYLON	
POLYESTERS	
BISPHENOL A-FUMURATE	
HYDROGENATED BISPHENOL-A-BISPHENOL-A	
ISOPHTHALIC	
CHLORINATED POLYESTERS	
POLYSTYRENES	
POLYETHYLENE	RU
POLYPROPYLENE	U
PVC–TYPE 1	RU
PVC–TYPE 2	U
POLYVINYLIDENE CHLORIDE	U
VINYLIDENE FLUORIDE	R (60–80)
VINYL ESTER	
RUBBERS	
BUTYL GR-1	U
FLUORO ELASTOMERS VITON A	U
FLUORO ELASTOMERS KEL-F 3700	
HARD RUBBER	
HYPALON	
KOROSEAL	
NATURAL RUBBER (GRS)	U
NEOPRENE GR-M (CR)	
NITRILE BUNA N (NBR)	U
NORDEL	
POLYBUTADIENE	
SILICONE	
POLYURETHANE	

FLUOBORIC ACID

TEMP °F: 60, 80, 100, 120, 140, 160, 180, 200, 220, 240, 260, 280, 300, 320, 340, 360, 380, 400, 420, 440, 460, 480, 500, 520, 540, 560

METALS		Rating
ALLOY 20Cb3	30%	E—
ALUMINUM		U
BRASS		
BRONZE		
CARBON STEEL		
COPPER		
E-BRIGHT 26-1		
HASTELLOY-B	32%	E——
HASTELLOY-C		E——
HASTELLOY-D		
HIGH SILICON IRON		
INCONEL		
LEAD		
MONEL		G——
NICKEL		
STAINLESS STEELS		
Type 304/307	30%	G—
Type 316	30%	E— G——
TANTALUM		U
TITANIUM		U
ZIRCONIUM		U

NONMETALLICS	
BOROSILICATE GLASS	
CARBON	

TEMP °F: 60, 80, 100, 120, 140, 160, 180, 200, 220, 240, 260, 280, 300, 320, 340, 360, 380, 400, 420, 440, 460, 480, 500, 520, 540, 560

FOR METALS

E = < 2 Mils Penetration/Year; G = < 20 Mils Penetration/Year

S = < 50 Mils Penetration/Year; U = > 50 Mils Penetration/Year

FOR NONMETALLICS

R = Resistant

U = Unsatisfactory

FLUOBORIC ACID

PLASTICS			TEMP °F (60–460)
ABS		R	60–140
ACRYLICS			
ASBESTOS REINF., EPOXY			
ASBESTOS REINF., PHENOLIC			
CHLORINATED POLYETHER		R	60–150
CPVC		R	60–210
EPOXY		U	
FLUOROCARBONS FEP		R	60–400
FLUOROCARBONS TFE		R	60–460
FURFURYL ALCOHOL	sat	R	60–270
NORYL			
PHENOLIC			
POLYAMIDES–NYLON			
POLYESTERS			
BISPHENOL A-FUMURATE	10 sat	R	60–120
HYDROGENATED BISPHENOL-A-BISPHENOL-A			
ISOPHTHALIC	sat 10,	R	60–150
CHLORINATED POLYESTERS			
POLYSTYRENES			
POLYETHYLENE		R	60–150
POLYPROPYLENE		R	60–200
PVC–TYPE 1		R	60–140
PVC–TYPE 2		R	60–140
POLYVINYLIDENE CHLORIDE		R	60–180
VINYLIDENE FLUORIDE		R	60–280
VINYL ESTER	10	R	60–200

RUBBERS			TEMP °F (60–460)
BUTYL GR-1		U	
FLUORO ELASTOMERS VITON A		R	60–80
FLUORO ELASTOMERS KEL-F 3700			
HARD RUBBER			
HYPALON		R	60–80
KOROSEAL			
NATURAL RUBBER (GRS)		R	60–150
NEOPRENE GR-M (CR)		R	60–80
NITRILE BUNA N (NBR)		R	60–150
NORDEL		R	60–300
POLYBUTADIENE			
SILICONE			
POLYURETHANE			

FLUOSILICIC ACID

TEMP °F: 60 80 100 120 140 160 180 200 220 240 260 280 300 320 340 360 380 400 420 440 460 480 500 520 540 560

METALS		
ALLOY 20Cb3	2	G
ALUMINUM		U
BRASS (air free)		G
BRONZE		G
CARBON STEEL		U
COPPER		
E-BRIGHT 26-1		U
HASTELLOY-B	2	G
HASTELLOY-C		G
HASTELLOY-D		G
HIGH SILICON IRON		U
INCONEL		E
LEAD		G
MONEL	2	E
NICKEL		E
STAINLESS STEELS		
Type 304/307		U
Type 316	1	G
TANTALUM		U
TITANIUM		U
ZIRCONIUM		U

NONMETALLICS	
BOROSILICATE GLASS	U
CARBON	

TEMP °F: 60 80 100 120 140 160 180 200 220 240 260 280 300 320 340 360 380 400 420 440 460 480 500 520 540 560

FOR METALS

E = < 2 Mils Penetration/Year; G = < 20 Mils Penetration/Year

S = < 50 Mils Penetration/Year; U = > 50 Mils Penetration/Year

FOR NONMETALLICS

R = Resistant

U = Unsatisfactory

TEMP °F (60–460)

PLASTICS	Concentration	Rating
ABS		R
ACRYLICS		R
ASBESTOS REINF., EPOXY		
ASBESTOS REINF., PHENOLIC		R
CHLORINATED POLYETHER		R
CPVC	25%	R
EPOXY	10%	R
FLUOROCARBONS FEP		R
FLUOROCARBONS TFE		R
FURFURYL ALCOHOL	10	R
NORYL		R
PHENOLIC		
POLYAMIDES–NYLON		
POLYESTERS	10	
BISPHENOL A-FUMURATE	10%	R
HYDROGENATED BISPHENOL-A-BISPHENOL-A	23%	R
ISOPHTHALIC	sat 10	R
CHLORINATED POLYESTERS	25%	R
POLYSTYRENES		
POLYETHYLENE		R–U
POLYPROPYLENE		R
PVC–TYPE 1		R
PVC–TYPE 2		R
POLYVINYLIDENE CHLORIDE		R
VINYLIDENE FLUORIDE		R
VINYL ESTER	30% 10	R

RUBBERS	Concentration	Rating
BUTYL GR-1		R
FLUORO ELASTOMERS VITON A		RU
FLUORO ELASTOMERS KEL-F 3700		
HARD RUBBER		R
HYPALON		R
KOROSEAL		R
NATURAL RUBBER (GRS)		R
NEOPRENE GR-M (CR)		R–U
NITRILE BUNA N (NBR)		R
NORDEL		R
POLYBUTADIENE		
SILICONE		
POLYURETHANE		

TEMP °F (60–460)

FORMALDEHYDE, 50% SOLUTION

TEMP °F: 60 80 100 120 140 160 180 200 220 240 260 280 300 320 340 360 380 400 420 440 460 480 500 520 540 560

METALS	
ALLOY 20Cb3	G
ALUMINUM 1	G
BRASS	G
BRONZE	G
CARBON STEEL	U
COPPER	G
E-BRIGHT 26-1	
HASTELLOY-B	G
HASTELLOY-C	G
HASTELLOY-D	G
HIGH SILICON IRON	G
INCONEL	G
LEAD	
MONEL	G
NICKEL	G
STAINLESS STEELS	
Type 304/307	E G
Type 316	E G
TANTALUM	G
TITANIUM	E
ZIRCONIUM	E

NONMETALLICS	
BOROSILICATE GLASS	R
CARBON	R

TEMP °F: 60 80 100 120 140 160 180 200 220 240 260 280 300 320 340 360 380 400 420 440 460 480 500 520 540 560

FOR METALS

E = < 2 Mils Penetration/Year; G = < 20 Mils Penetration/Year

S = < 50 Mils Penetration/Year; U = > 50 Mils Penetration/Year

FOR NONMETALLICS

R = Resistant

U = Unsatisfactory

FORMALDEHYDE, 50% SOLUTION

TEMP °F (60–460)

PLASTICS	
ABS	R
ACRYLICS	R
ASBESTOS REINF., EPOXY	R
ASBESTOS REINF., PHENOLIC	R
CHLORINATED POLYETHER	R
CPVC	R
EPOXY	R — U
FLUOROCARBONS FEP	R
FLUOROCARBONS TFE	R
FURFURYL ALCOHOL	
NORYL	
PHENOLIC	
POLYAMIDES–NYLON	R
POLYESTERS	
BISPHENOL A-FUMURATE	R
HYDROGENATED BISPHENOL-A-BISPHENOL-A	
ISOPHTHALIC	R — U
CHLORINATED POLYESTERS	
POLYSTYRENES	R — U
POLYETHYLENE	R
POLYPROPYLENE	R
PVC–TYPE 1	R
PVC–TYPE 2	R — U
POLYVINYLIDENE CHLORIDE	R
VINYLIDENE FLUORIDE	
VINYL ESTER	R

RUBBERS	
BUTYL GR-1	
FLUORO ELASTOMERS VITON A	U
FLUORO ELASTOMERS KEL-F 3700	
HARD RUBBER	
HYPALON	U
KOROSEAL	R
NATURAL RUBBER (GRS)	U
NEOPRENE GR-M (CR)	R
NITRILE BUNA N (NBR)	U
NORDEL	R
POLYBUTADIENE	
SILICONE	
POLYURETHANE	

TEMP °F (60–460)

FORMALDEHYDE, 37% SOLUTION

TEMP °F (60–560)

METALS	Rating
ALLOY 20Cb3	E
ALUMINUM 1	G
BRASS	E
BRONZE	G
CARBON STEEL	U
COPPER	G
E-BRIGHT 26-1	
HASTELLOY-B	G
HASTELLOY-C	G
HASTELLOY-D	G
HIGH SILICON IRON	G
INCONEL	G
LEAD	G
MONEL	G
NICKEL	G
STAINLESS STEELS	
Type 304/307	E G
Type 316	E
TANTALUM	G
TITANIUM	E
ZIRCONIUM	E

NONMETALLICS	Rating
BOROSILICATE GLASS	R
CARBON	R

TEMP °F

FOR METALS

E = < 2 Mils Penetration/Year; G = < 20 Mils Penetration/Year

S = < 50 Mils Penetration/Year; U = > 50 Mils Penetration/Year

FOR NONMETALLICS

R = Resistant

U = Unsatisfactory

FORMALDEHYDE, 37% SOLUTION

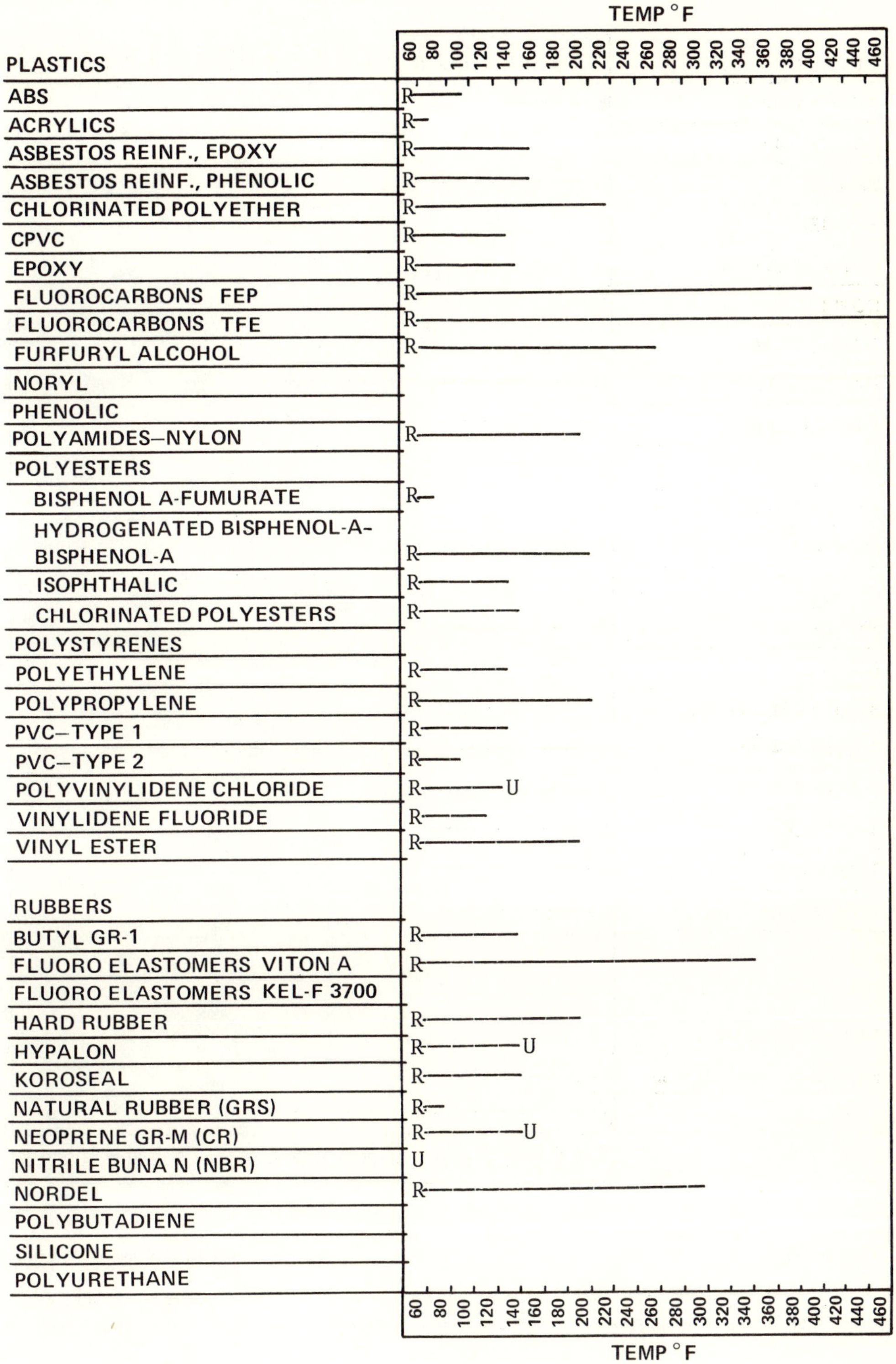

FORMALDEHYDE, 35% SOLUTION

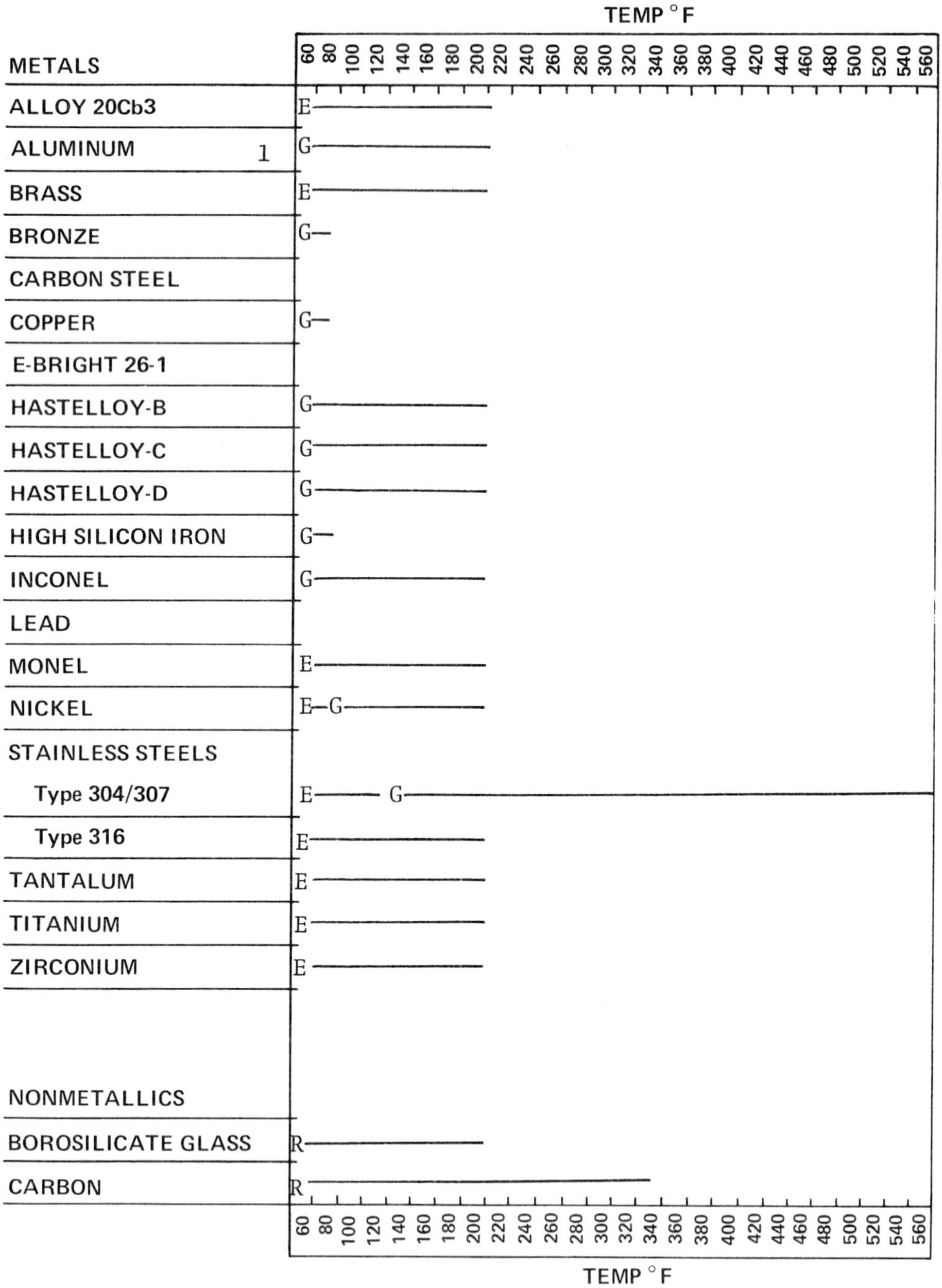

FOR METALS

E = < 2 Mils Penetration/Year; G = < 20 Mils Penetration/Year
S = < 50 Mils Penetration/Year; U = > 50 Mils Penetration/Year

FOR NONMETALLICS

R = Resistant
U = Unsatisfactory

FORMALDEHYDE, 35% SOLUTION

TEMP °F

PLASTICS		Rating	Temperature range (°F)
ABS	30%	R	60–100
ACRYLICS		R	60–80
ASBESTOS REINF., EPOXY		R	60–160
ASBESTOS REINF., PHENOLIC		R	60–160
CHLORINATED POLYETHER		R	60–220
CPVC		R	60–140
EPOXY		R—U	R 60–100, U 120
FLUOROCARBONS FEP		R	60–400
FLUOROCARBONS TFE		R	60–460
FURFURYL ALCOHOL		R	60–160
NORYL			
PHENOLIC			
POLYAMIDES–NYLON		R	60–200
POLYESTERS			
BISPHENOL A-FUMURATE		R	60–80
HYDROGENATED BISPHENOL-A-BISPHENOL-A		R	60–210
ISOPHTHALIC		R	60–140
CHLORINATED POLYESTERS			
POLYSTYRENES			
POLYETHYLENE		R	60–140
POLYPROPYLENE		R	60–210
PVC–TYPE 1		R	60–140
PVC–TYPE 2		R	60–100
POLYVINYLIDENE CHLORIDE			
VINYLIDENE FLUORIDE		R	60–120
VINYL ESTER		R	60–150

RUBBERS	Rating	Temperature range (°F)
BUTYL GR-1		
FLUORO ELASTOMERS VITON A	U	60
FLUORO ELASTOMERS KEL-F 3700		
HARD RUBBER		
HYPALON	R	60–80
KOROSEAL	R	60–150
NATURAL RUBBER (GRS)		
NEOPRENE GR-M (CR)	R	60–80
NITRILE BUNA N (NBR)	U	60
NORDEL	R	60–300
POLYBUTADIENE		
SILICONE		
POLYURETHANE		

TEMP °F

FORMIC ACID (ANHYDROUS)

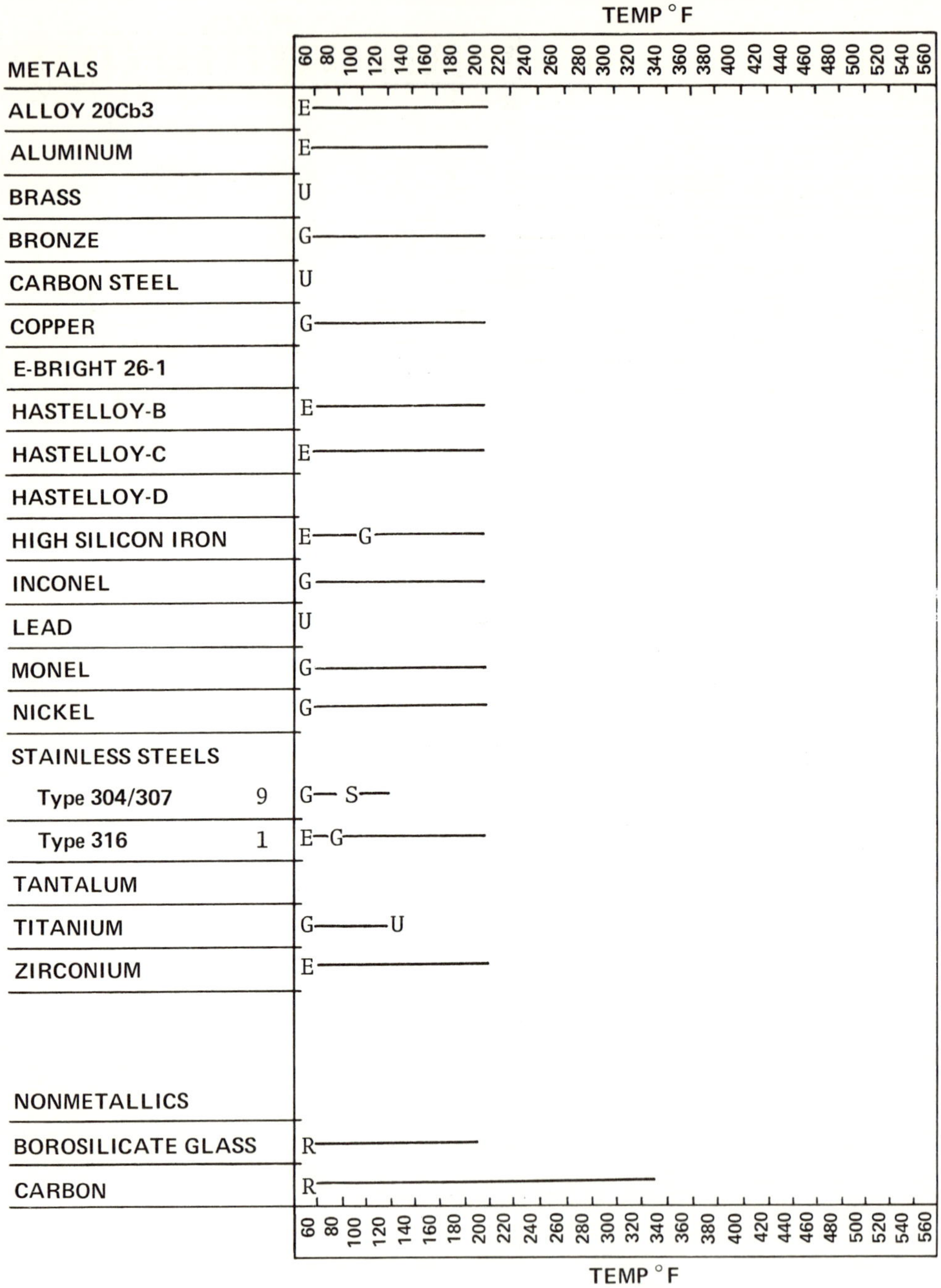

FOR METALS

E = < 2 Mils Penetration/Year; G = < 20 Mils Penetration/Year

S = < 50 Mils Penetration/Year; U = > 50 Mils Penetration/Year

FOR NONMETALLICS

R = Resistant

U = Unsatisfactory

FORMIC ACID (ANHYDROUS)

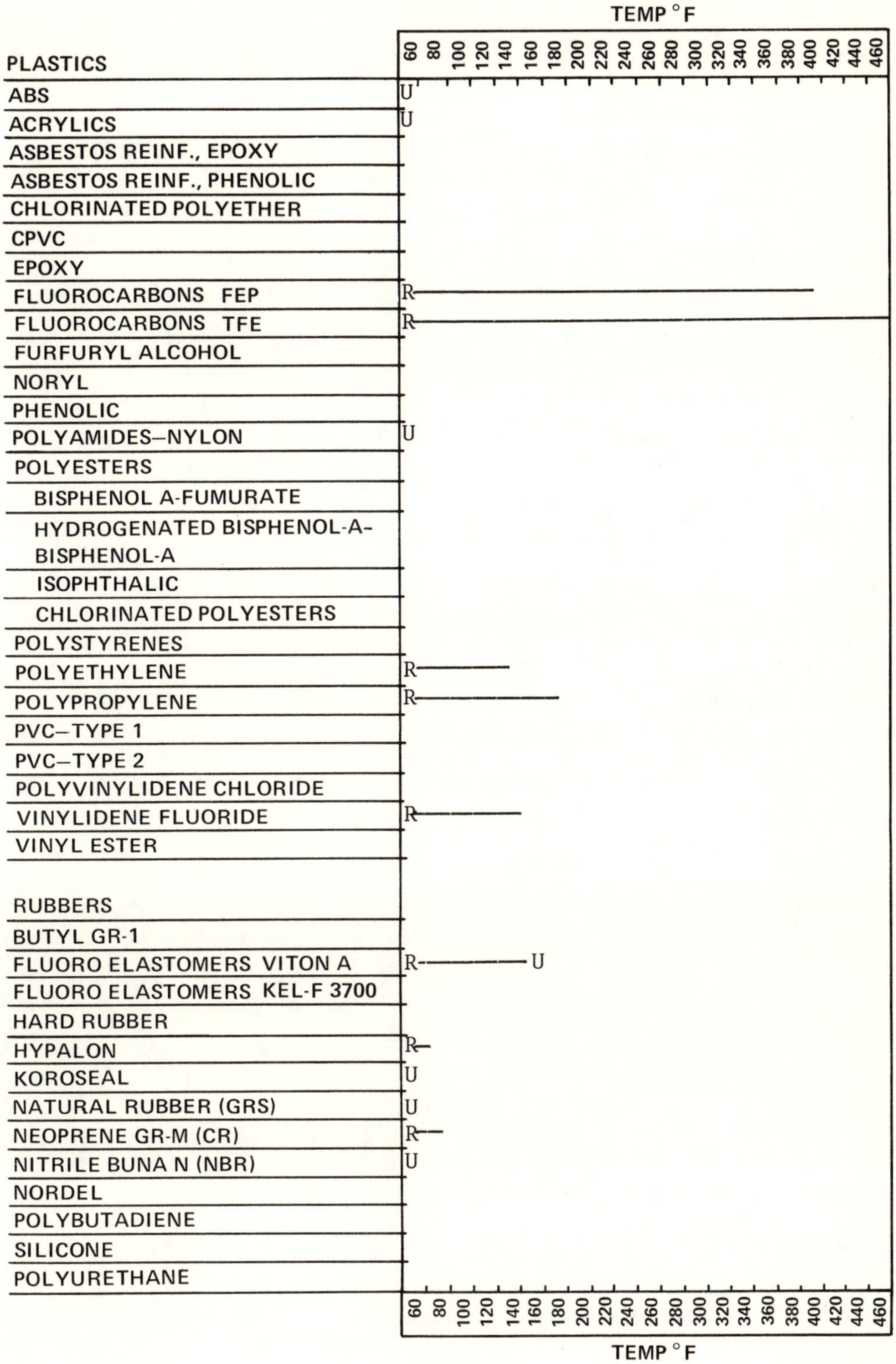

FORMIC ACID, 10%-85%

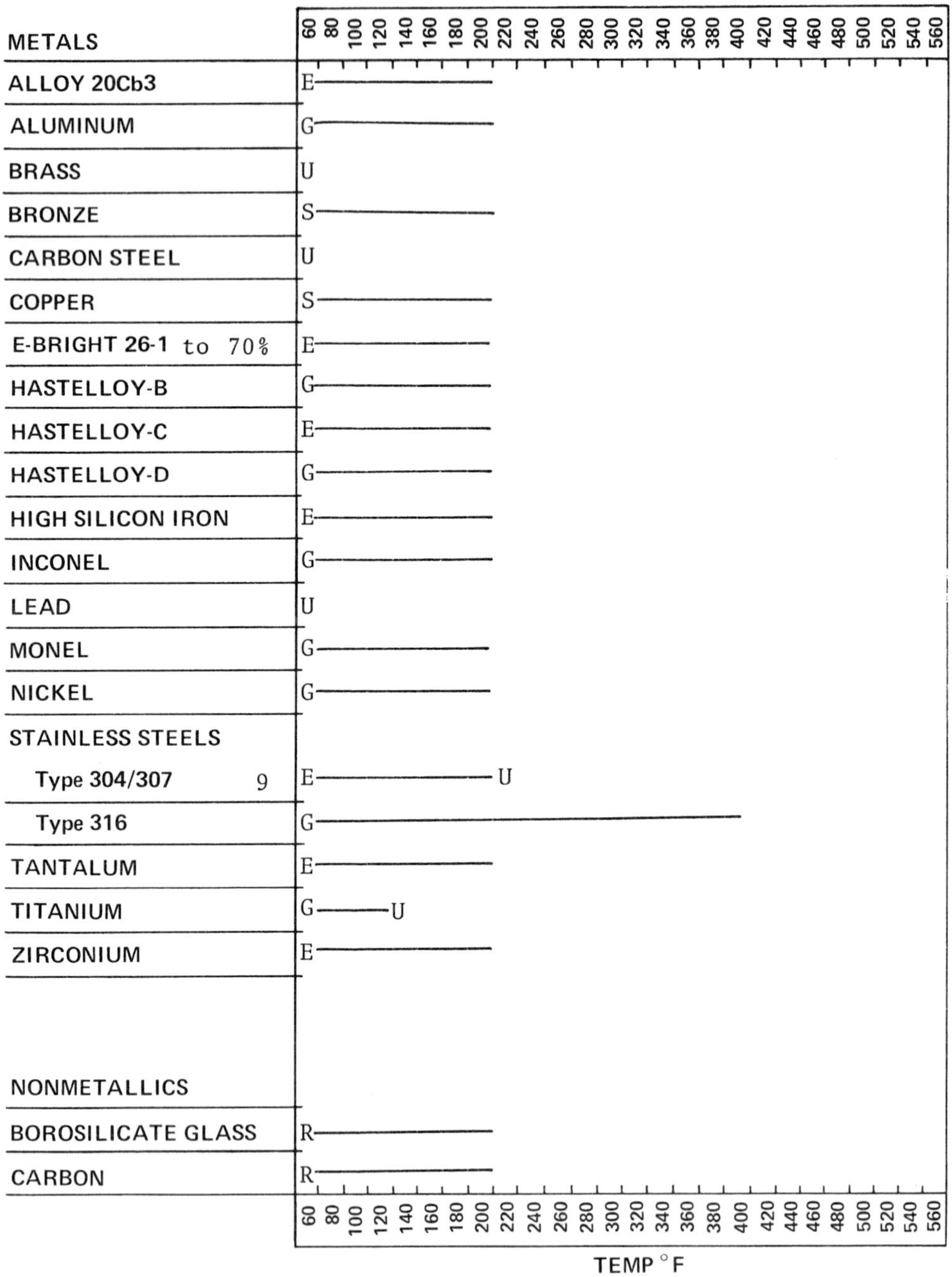

FOR METALS

E = < 2 Mils Penetration/Year; G = < 20 Mils Penetration/Year

S = < 50 Mils Penetration/Year; U = > 50 Mils Penetration/Year

FOR NONMETALLICS

R = Resistant

U = Unsatisfactory

FORMIC ACID, 10%-85%

TEMP °F (scale: 60, 80, 100, 120, 140, 160, 180, 200, 220, 240, 260, 280, 300, 320, 340, 360, 380, 400, 420, 440, 460)

PLASTICS		Rating
ABS		U
ACRYLICS		U
ASBESTOS REINF., EPOXY		R——
ASBESTOS REINF., PHENOLIC		R——
CHLORINATED POLYETHER		R——
CPVC		R—— U
EPOXY	88%	RU
FLUOROCARBONS FEP		R——
FLUOROCARBONS TFE		R——
FURFURYL ALCOHOL	sat	R——
NORYL	10%	R——
PHENOLIC		
POLYAMIDES–NYLON		U
POLYESTERS		
BISPHENOL A-FUMURATE	10%	R——
HYDROGENATED BISPHENOL-A-BISPHENOL-A	25%	R—
ISOPHTHALIC	25%	U
CHLORINATED POLYESTERS	25%	R—— U
POLYSTYRENES		
POLYETHYLENE		R——
POLYPROPYLENE		R——
PVC–TYPE 1		R—
PVC–TYPE 2		R—
POLYVINYLIDENE CHLORIDE		R——
VINYLIDENE FLUORIDE		R——
VINYL ESTER		R—

RUBBERS	Rating
BUTYL GR-1	R——
FLUORO ELASTOMERS VITON A	R——
FLUORO ELASTOMERS KEL-F 3700	
HARD RUBBER	R——
HYPALON	R—
KOROSEAL	U
NATURAL RUBBER (GRS)	U
NEOPRENE GR-M (CR)	R——
NITRILE BUNA N (NBR)	U
NORDEL	R——
POLYBUTADIENE	
SILICONE	
POLYURETHANE	

TEMP °F

FREON F - 11

TEMP °F

METALS	60 – 560 °F
ALLOY 20Cb3	
ALUMINUM	G
BRASS	G
BRONZE	G
CARBON STEEL	U
COPPER	G
E-BRIGHT 26-1	
HASTELLOY-B	
HASTELLOY-C	
HASTELLOY-D	
HIGH SILICON IRON	
INCONEL	
LEAD	G
MONEL	E
NICKEL	
STAINLESS STEELS	
Type 304/307	G
Type 316	G
TANTALUM	
TITANIUM	
ZIRCONIUM	
NONMETALLICS	
BOROSILICATE GLASS	R
CARBON	R

TEMP °F

FOR METALS

E = < 2 Mils Penetration/Year; G = < 20 Mils Penetration/Year

S = < 50 Mils Penetration/Year; U = > 50 Mils Penetration/Year

FOR NONMETALLICS

R = Resistant

U = Unsatisfactory

FREON F - 11

TEMP °F (60–460)

PLASTICS	
ABS	U
ACRYLICS	
ASBESTOS REINF., EPOXY	
ASBESTOS REINF., PHENOLIC	
CHLORINATED POLYETHER	R
CPVC	R
EPOXY	R
FLUOROCARBONS FEP	R
FLUOROCARBONS TFE	R
FURFURYL ALCOHOL	
NORYL	
PHENOLIC	
POLYAMIDES–NYLON	
POLYESTERS	
BISPHENOL A-FUMURATE	R
HYDROGENATED BISPHENOL-A-BISPHENOL-A	
ISOPHTHALIC	
CHLORINATED POLYESTERS	
POLYSTYRENES	
POLYETHYLENE	
POLYPROPYLENE	
PVC–TYPE 1	R
PVC–TYPE 2	U
POLYVINYLIDENE CHLORIDE	
VINYLIDENE FLUORIDE	R
VINYL ESTER	R

RUBBERS	
BUTYL GR-1	U
FLUORO ELASTOMERS VITON A	R
FLUORO ELASTOMERS KEL-F 3700	
HARD RUBBER	
HYPALON	R
KOROSEAL	
NATURAL RUBBER (GRS)	U
NEOPRENE GR-M (CR)	R
NITRILE BUNA N (NBR)	R
NORDEL	
POLYBUTADIENE	
SILICONE	
POLYURETHANE	R

TEMP °F (60–460)

FREON F - 12

TEMP °F: 60, 80, 100, 120, 140, 160, 180, 200, 220, 240, 260, 280, 300, 320, 340, 360, 380, 400, 420, 440, 460, 480, 500, 520, 540, 560

METALS	
ALLOY 20Cb3	
ALUMINUM	G—
BRASS	G—
BRONZE	G—
CARBON STEEL	U
COPPER	G—
E-BRIGHT 26-1	
HASTELLOY-B	
HASTELLOY-C	
HASTELLOY-D	
HIGH SILICON IRON	
INCONEL	G—
LEAD	G—
MONEL	G—
NICKEL	G—
STAINLESS STEELS	
Type 304/307	G—
Type 316	G—
TANTALUM	G—
TITANIUM	G—
ZIRCONIUM	

NONMETALLICS	
BOROSILICATE GLASS	R—
CARBON	R—

TEMP °F

FOR METALS

E = < 2 Mils Penetration/Year; G = < 20 Mils Penetration/Year

S = < 50 Mils Penetration/Year; U = > 50 Mils Penetration/Year

FOR NONMETALLICS

R = Resistant

U = Unsatisfactory

FREON F - 12

TEMP °F (60–460)

PLASTICS	Rating
ABS	R
ACRYLICS	
ASBESTOS REINF., EPOXY	
ASBESTOS REINF., PHENOLIC	
CHLORINATED POLYETHER	R
CPVC	R
EPOXY	R
FLUOROCARBONS FEP	R
FLUOROCARBONS TFE	R
FURFURYL ALCOHOL	
NORYL	
PHENOLIC	
POLYAMIDES—NYLON	
POLYESTERS	
BISPHENOL A-FUMURATE	
HYDROGENATED BISPHENOL-A-BISPHENOL-A	
ISOPHTHALIC	
CHLORINATED POLYESTERS	
POLYSTYRENES	
POLYETHYLENE	R
POLYPROPYLENE	R
PVC—TYPE 1	R
PVC—TYPE 2	R
POLYVINYLIDENE CHLORIDE	
VINYLIDENE FLUORIDE	R
VINYL ESTER	

RUBBERS	Rating
BUTYL GR-1	U
FLUORO ELASTOMERS VITON A	R
FLUORO ELASTOMERS KEL-F 3700	
HARD RUBBER	
HYPALON	R
KOROSEAL	
NATURAL RUBBER (GRS)	U
NEOPRENE GR-M (CR)	R
NITRILE BUNA N (NBR)	R
NORDEL	
POLYBUTADIENE	
SILICONE	
POLYURETHANE	R

TEMP °F

FREON F - 22

TEMP °F: 60, 80, 100, 120, 140, 160, 180, 200, 220, 240, 260, 280, 300, 320, 340, 360, 380, 400, 420, 440, 460, 480, 500, 520, 540, 560

METALS	
ALLOY 20Cb3	
ALUMINUM	G
BRASS	G
BRONZE	G
CARBON STEEL	U
COPPER	G
E-BRIGHT 26-1	
HASTELLOY-B	
HASTELLOY-C	G
HASTELLOY-D	
HIGH SILICON IRON	
INCONEL	G
LEAD	G
MONEL	G
NICKEL	G
STAINLESS STEELS	
Type 304/307	G
Type 316	G
TANTALUM	G
TITANIUM	G
ZIRCONIUM	
NONMETALLICS	
BOROSILICATE GLASS	R
CARBON	R

TEMP °F: 60, 80, 100, 120, 140, 160, 180, 200, 220, 240, 260, 280, 300, 320, 340, 360, 380, 400, 420, 440, 460, 480, 500, 520, 540, 560

FOR METALS

E = < 2 Mils Penetration/Year; G = < 20 Mils Penetration/Year

S = < 50 Mils Penetration/Year; U = > 50 Mils Penetration/Year

FOR NONMETALLICS

R = Resistant

U = Unsatisfactory

FREON F - 22

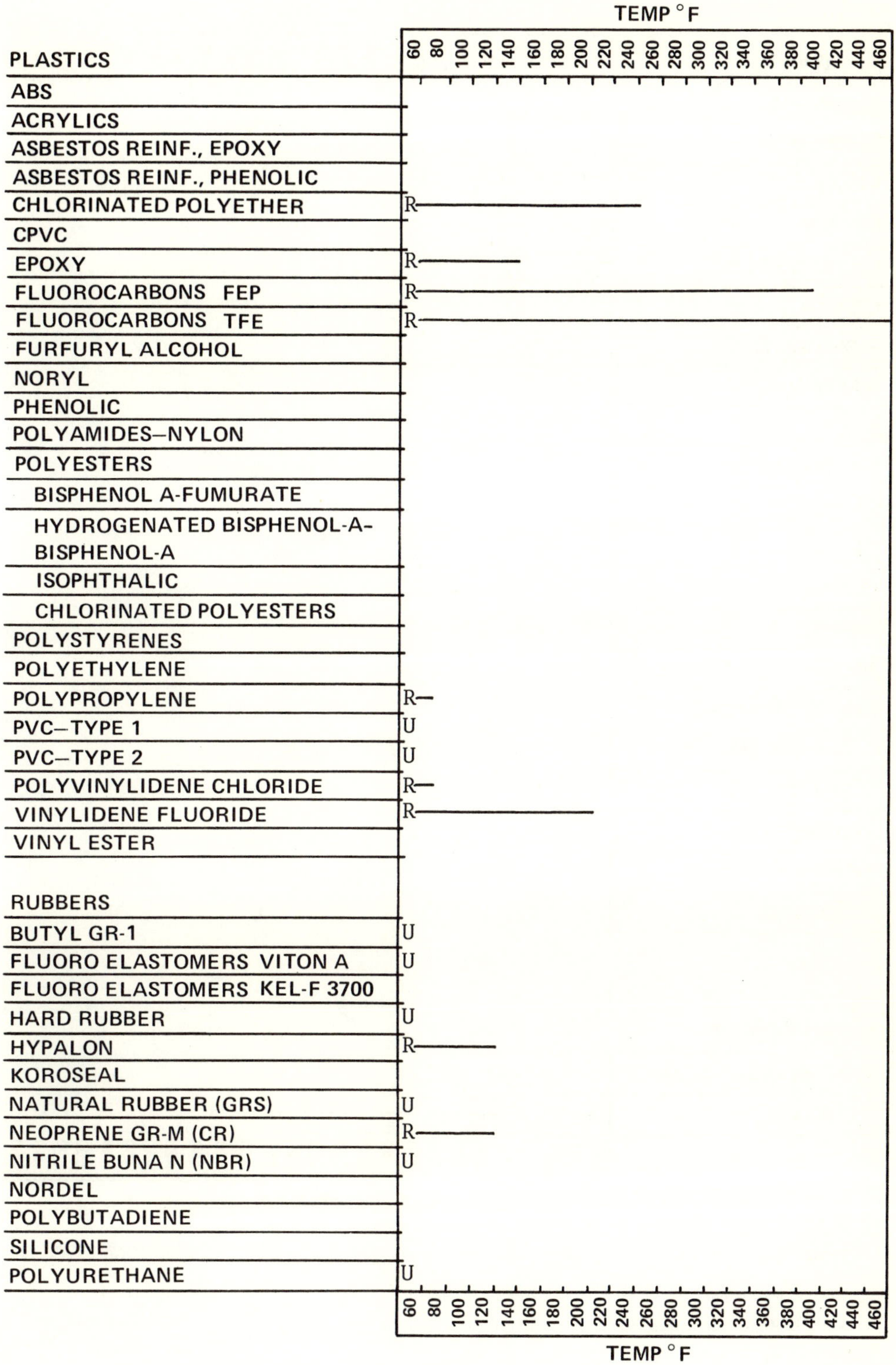
TEMP °F
60 80 100 120 140 160 180 200 220 240 260 280 300 320 340 360 380 400 420 440 460
PLASTICS
ABS
ACRYLICS
ASBESTOS REINF., EPOXY
ASBESTOS REINF., PHENOLIC
CHLORINATED POLYETHER R
CPVC
EPOXY R
FLUOROCARBONS FEP R
FLUOROCARBONS TFE R
FURFURYL ALCOHOL
NORYL
PHENOLIC
POLYAMIDES–NYLON
POLYESTERS
BISPHENOL A-FUMURATE
HYDROGENATED BISPHENOL-A-BISPHENOL-A
ISOPHTHALIC
CHLORINATED POLYESTERS
POLYSTYRENES
POLYETHYLENE
POLYPROPYLENE R
PVC–TYPE 1 U
PVC–TYPE 2 U
POLYVINYLIDENE CHLORIDE R
VINYLIDENE FLUORIDE R
VINYL ESTER
RUBBERS
BUTYL GR-1 U
FLUORO ELASTOMERS VITON A U
FLUORO ELASTOMERS KEL-F 3700
HARD RUBBER U
HYPALON R
KOROSEAL
NATURAL RUBBER (GRS) U
NEOPRENE GR-M (CR) R
NITRILE BUNA N (NBR) U
NORDEL
POLYBUTADIENE
SILICONE
POLYURETHANE U
60 80 100 120 140 160 180 200 220 240 260 280 300 320 340 360 380 400 420 440 460
TEMP °F

FRUCTOSE

TEMP °F: 60, 80, 100, 120, 140, 160, 180, 200, 220, 240, 260, 280, 300, 320, 340, 360, 380, 400, 420, 440, 460, 480, 500, 520, 540, 560

METALS	TEMP °F
ALLOY 20Cb3	
ALUMINUM	
BRASS	
BRONZE	
CARBON STEEL	
COPPER	
E-BRIGHT 26-1	
HASTELLOY-B	
HASTELLOY-C	
HASTELLOY-D	
HIGH SILICON IRON	
INCONEL	
LEAD	
MONEL	
NICKEL	
STAINLESS STEELS Type 304/307	
Type 316	
TANTALUM	
TITANIUM	
ZIRCONIUM	
NONMETALLICS	
BOROSILICATE GLASS	
CARBON	R (60–340)

TEMP °F

FOR METALS

E = < 2 Mils Penetration/Year; G = < 20 Mils Penetration/Year

S = < 50 Mils Penetration/Year; U = > 50 Mils Penetration/Year

FOR NONMETALLICS

R = Resistant

U = Unsatisfactory

PLASTICS	TEMP °F (60–460)
ABS	R (to 140)
ACRYLICS	
ASBESTOS REINF., EPOXY	
ASBESTOS REINF., PHENOLIC	
CHLORINATED POLYETHER	R (to 250)
CPVC	R (to 180)
EPOXY	
FLUOROCARBONS FEP	R (to 400)
FLUOROCARBONS TFE	R (to 460)
FURFURYL ALCOHOL	
NORYL	
PHENOLIC	
POLYAMIDES–NYLON	
POLYESTERS	
BISPHENOL A-FUMURATE	
HYDROGENATED BISPHENOL-A-BISPHENOL-A	
ISOPHTHALIC	
CHLORINATED POLYESTERS	
POLYSTYRENES	
POLYETHYLENE	R (to 150)
POLYPROPYLENE	R (to 220)
PVC–TYPE 1	R (to 140)
PVC–TYPE 2	R (to 140)
POLYVINYLIDENE CHLORIDE	R (to 180)
VINYLIDENE FLUORIDE	R (to 280)
VINYL ESTER	

RUBBERS	TEMP °F (60–460)
BUTYL GR-1	
FLUORO ELASTOMERS VITON A	R (to 180)
FLUORO ELASTOMERS KEL-F 3700	
HARD RUBBER	
HYPALON	
KOROSEAL	
NATURAL RUBBER (GRS)	
NEOPRENE GR-M (CR)	
NITRILE BUNA N (NBR)	R (to 180)
NORDEL	
POLYBUTADIENE	
SILICONE	
POLYURETHANE	

FRUIT JUICES, PULP

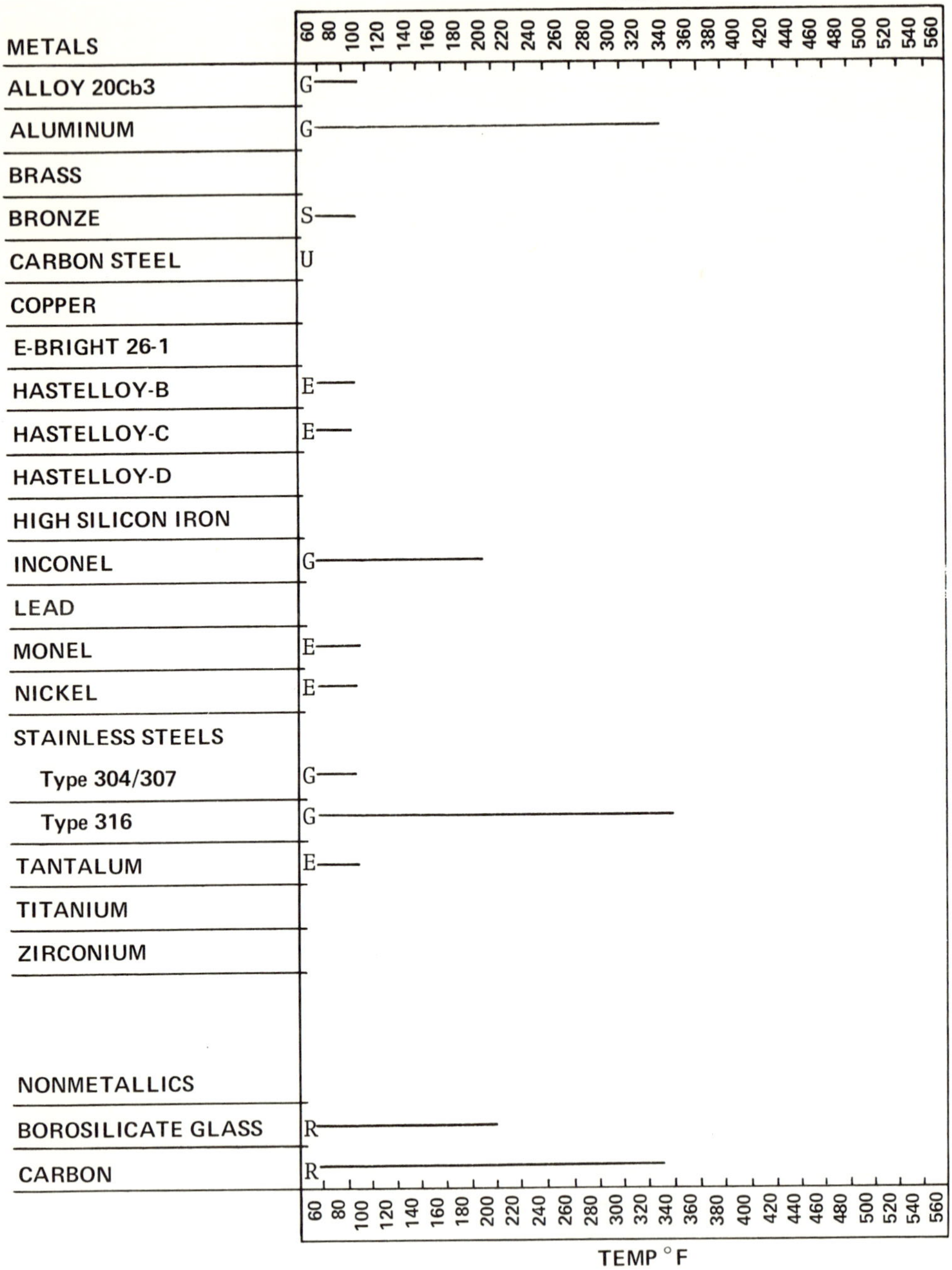

FOR METALS

E = < 2 Mils Penetration/Year; G = < 20 Mils Penetration/Year

S = < 50 Mils Penetration/Year; U = > 50 Mils Penetration/Year

FOR NONMETALLICS

R = Resistant

U = Unsatisfactory

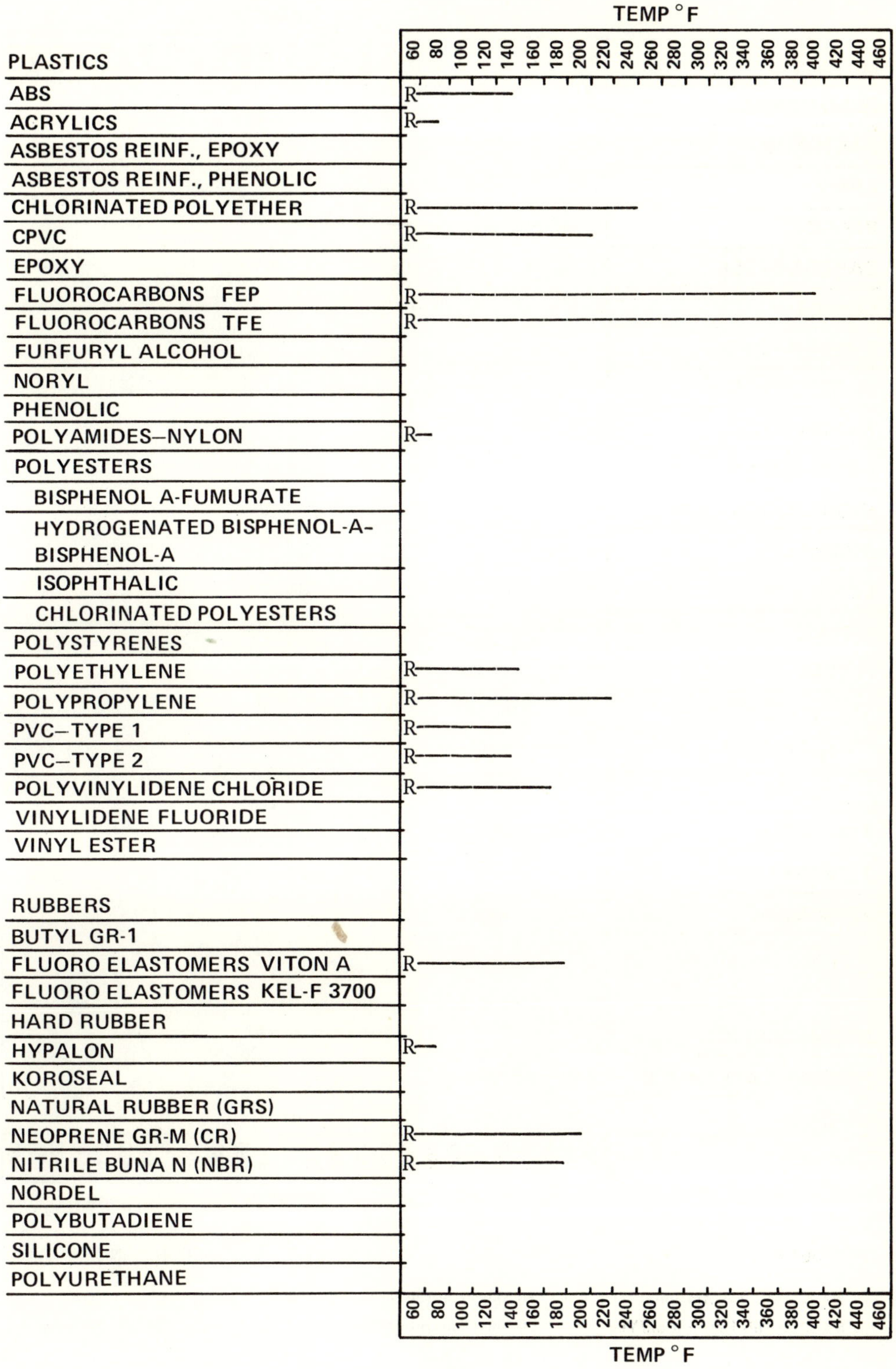
TEMP °F
60 80 100 120 140 160 180 200 220 240 260 280 300 320 340 360 380 400 420 440 460
PLASTICS
ABS R
ACRYLICS R
ASBESTOS REINF., EPOXY
ASBESTOS REINF., PHENOLIC
CHLORINATED POLYETHER R
CPVC R
EPOXY
FLUOROCARBONS FEP R
FLUOROCARBONS TFE R
FURFURYL ALCOHOL
NORYL
PHENOLIC
POLYAMIDES–NYLON R
POLYESTERS
BISPHENOL A-FUMURATE
HYDROGENATED BISPHENOL-A-BISPHENOL-A
ISOPHTHALIC
CHLORINATED POLYESTERS
POLYSTYRENES
POLYETHYLENE R
POLYPROPYLENE R
PVC–TYPE 1 R
PVC–TYPE 2 R
POLYVINYLIDENE CHLORIDE R
VINYLIDENE FLUORIDE
VINYL ESTER
RUBBERS
BUTYL GR-1
FLUORO ELASTOMERS VITON A R
FLUORO ELASTOMERS KEL-F 3700
HARD RUBBER
HYPALON R
KOROSEAL
NATURAL RUBBER (GRS)
NEOPRENE GR-M (CR) R
NITRILE BUNA N (NBR) R
NORDEL
POLYBUTADIENE
SILICONE
POLYURETHANE
60 80 100 120 140 160 180 200 220 240 260 280 300 320 340 360 380 400 420 440 460
TEMP °F

FUEL OIL

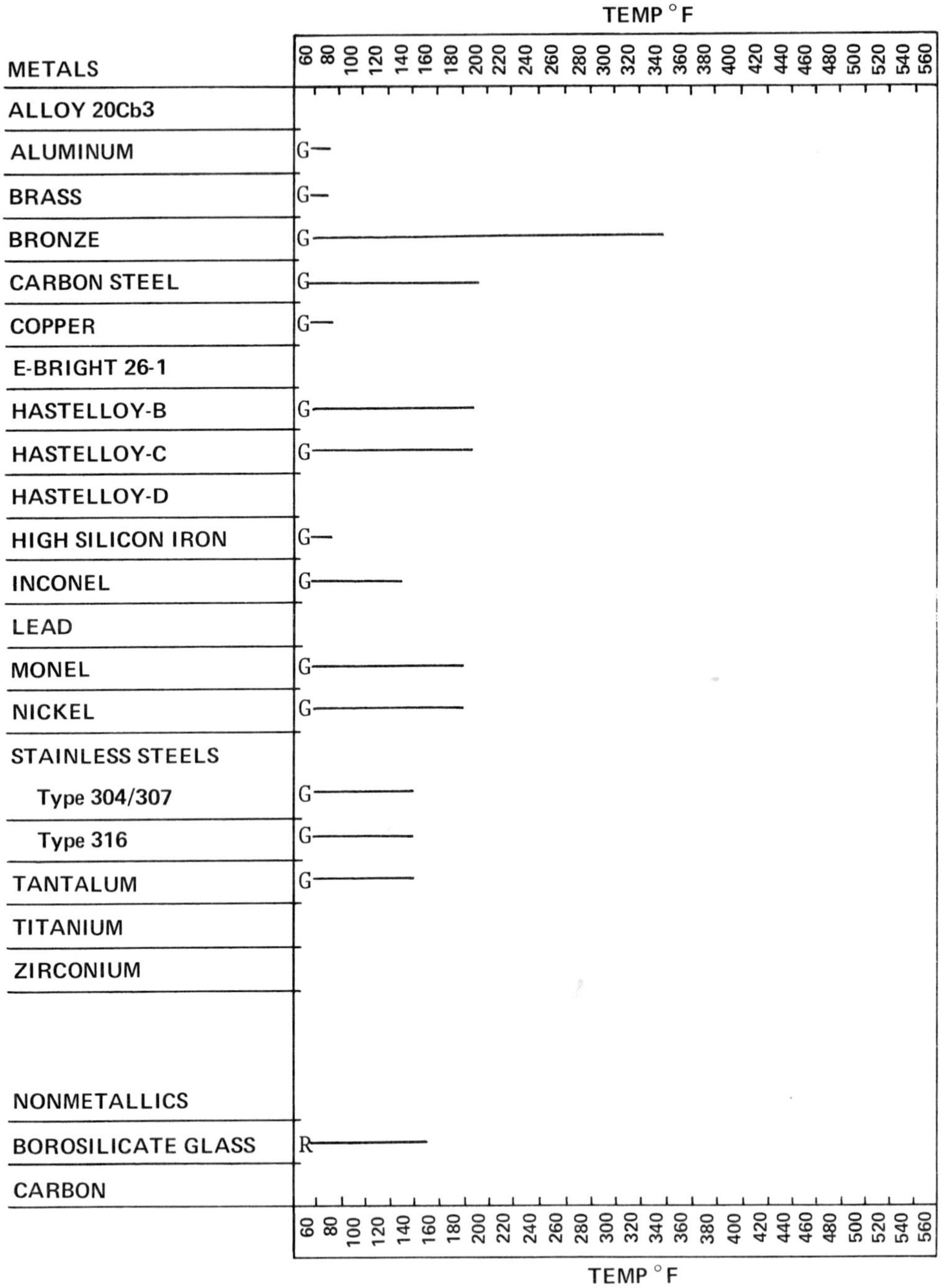

FOR METALS

E = < 2 Mils Penetration/Year; G = < 20 Mils Penetration/Year

S = < 50 Mils Penetration/Year; U = > 50 Mils Penetration/Year

FOR NONMETALLICS

R = Resistant

U = Unsatisfactory

FUEL OIL

TEMP °F: 60 80 100 120 140 160 180 200 220 240 260 280 300 320 340 360 380 400 420 440 460

PLASTICS	
ABS	
ACRYLICS	
ASBESTOS REINF., EPOXY	
ASBESTOS REINF., PHENOLIC	
CHLORINATED POLYETHER	R—
CPVC	
EPOXY	R—
FLUOROCARBONS FEP	R—
FLUOROCARBONS TFE	R—
FURFURYL ALCOHOL	R—
NORYL	
PHENOLIC	
POLYAMIDES–NYLON	R—
POLYESTERS	
BISPHENOL A-FUMURATE	R—
HYDROGENATED BISPHENOL-A-BISPHENOL-A	
ISOPHTHALIC	R—
CHLORINATED POLYESTERS	R—
POLYSTYRENES	
POLYETHYLENE	U
POLYPROPYLENE	R—U
PVC–TYPE 1	R—
PVC–TYPE 2	R—
POLYVINYLIDENE CHLORIDE	R—
VINYLIDENE FLUORIDE	R—
VINYL ESTER	R—
RUBBERS	
BUTYL GR-1	U
FLUORO ELASTOMERS VITON A	R—
FLUORO ELASTOMERS KEL-F 3700	
HARD RUBBER	
HYPALON	R—
KOROSEAL	
NATURAL RUBBER (GRS)	U
NEOPRENE GR-M (CR)	R—
NITRILE BUNA N (NBR)	R—
NORDEL	
POLYBUTADIENE	
SILICONE	
POLYURETHANE	R—

TEMP °F: 60 80 100 120 140 160 180 200 220 240 260 280 300 320 340 360 380 400 420 440 460

FURFURAL

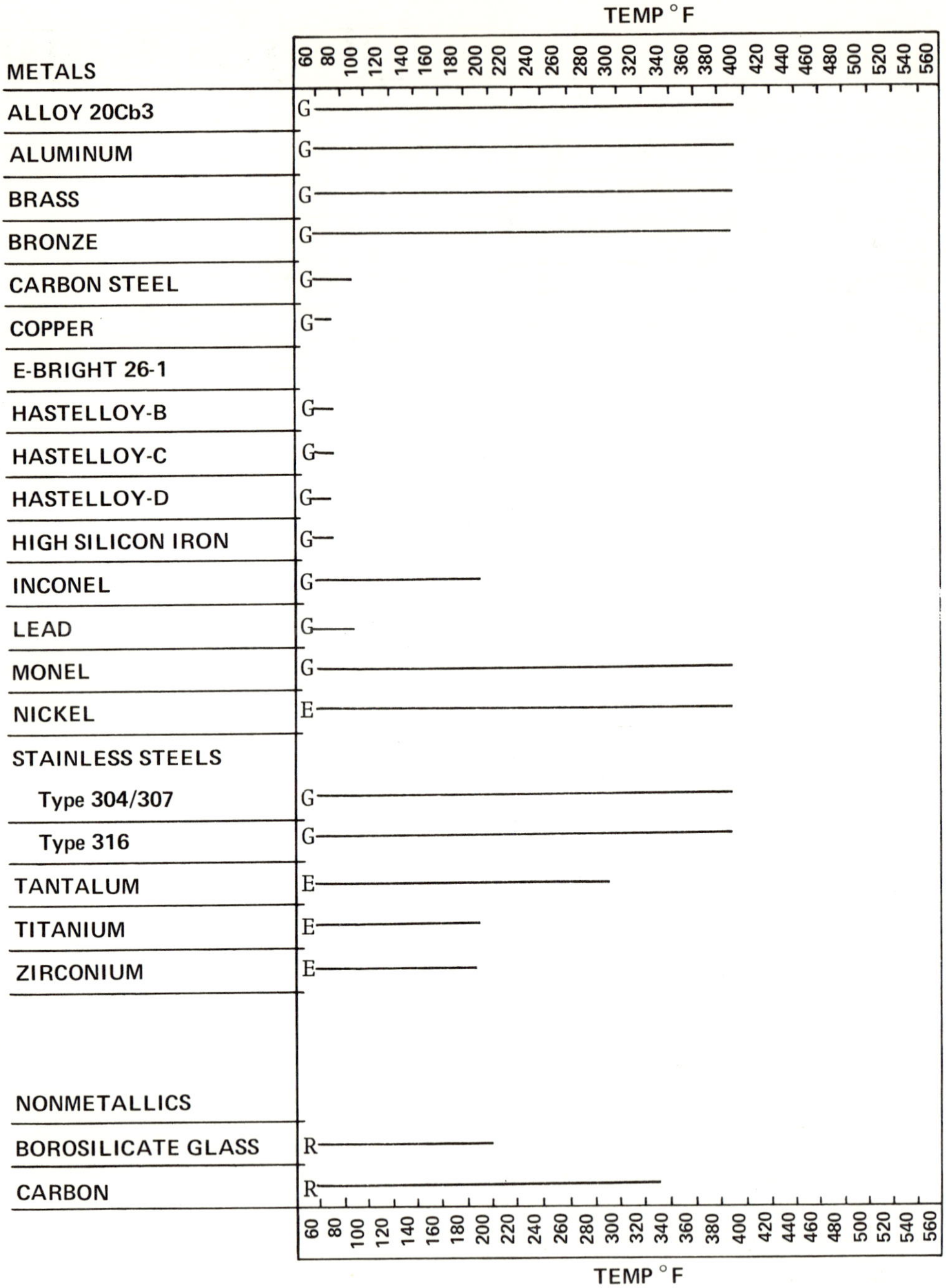

FOR METALS

E = < 2 Mils Penetration/Year; G = < 20 Mils Penetration/Year

S = < 50 Mils Penetration/Year; U = > 50 Mils Penetration/Year

FOR NONMETALLICS

R = Resistant

U = Unsatisfactory

TEMP °F (60–460)

PLASTICS	60 80 100 120 140 160 180 200 220 240 260 280 300 320 340 360 380 400 420 440 460
ABS	U
ACRYLICS	
ASBESTOS REINF., EPOXY	
ASBESTOS REINF., PHENOLIC	
CHLORINATED POLYETHER	R———
CPVC	U
EPOXY	U
FLUOROCARBONS FEP	R———————————
FLUOROCARBONS TFE	R——————————————
FURFURYL ALCOHOL	R———————
NORYL	
PHENOLIC	
POLYAMIDES–NYLON	U
POLYESTERS	
BISPHENOL A-FUMURATE	U
HYDROGENATED BISPHENOL-A-BISPHENOL-A	
ISOPHTHALIC 10%	U
CHLORINATED POLYESTERS	U
POLYSTYRENES	
POLYETHYLENE	U
POLYPROPYLENE	U
PVC–TYPE 1	U
PVC–TYPE 2	U
POLYVINYLIDENE CHLORIDE	R—
VINYLIDENE FLUORIDE	R——U
VINYL ESTER	U
RUBBERS	
BUTYL GR-1	R————
FLUORO ELASTOMERS VITON A	U
FLUORO ELASTOMERS KEL-F 3700	
HARD RUBBER	
HYPALON	R—
KOROSEAL	
NATURAL RUBBER (GRS)	U
NEOPRENE GR-M (CR)	R—————
NITRILE BUNA N (NBR)	U
NORDEL	
POLYBUTADIENE	
SILICONE	
POLYURETHANE	U

TEMP °F (60–460)

GALLIC ACID

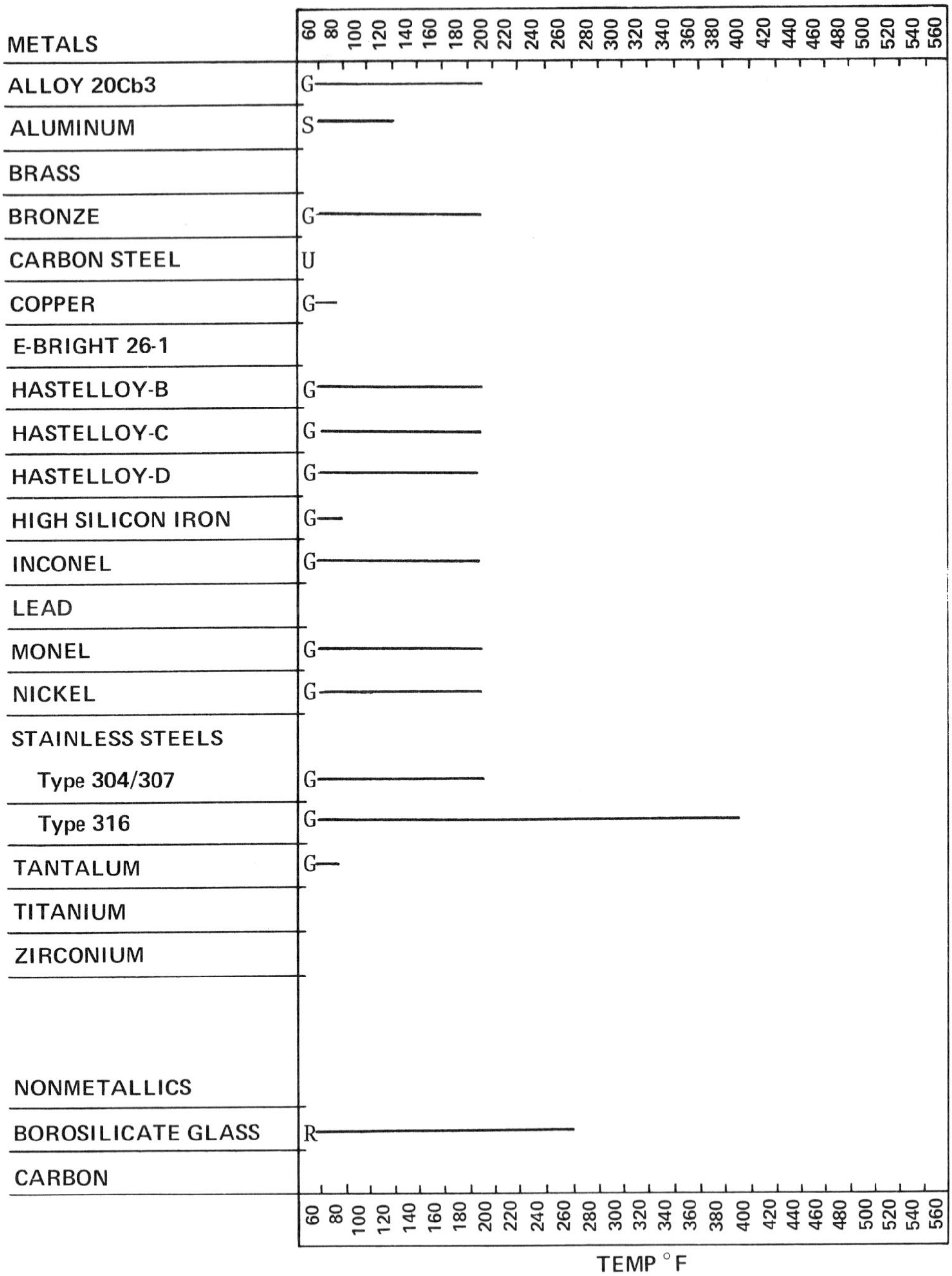

FOR METALS

E = < 2 Mils Penetration/Year; G = < 20 Mils Penetration/Year

S = < 50 Mils Penetration/Year; U = > 50 Mils Penetration/Year

FOR NONMETALLICS

R = Resistant

U = Unsatisfactory

GALLIC ACID

Please see page 1 for notes on calibration.

TEMP °F

PLASTICS		Rating	Temp range (°F)
ABS		R	60–140
ACRYLICS			
ASBESTOS REINF., EPOXY			
ASBESTOS REINF., PHENOLIC			
CHLORINATED POLYETHER		R	60–250
CPVC		R	60–185
EPOXY			
FLUOROCARBONS FEP		R	60–400
FLUOROCARBONS TFE		R	60–460
FURFURYL ALCOHOL	sat	R	60–260
NORYL			
PHENOLIC			
POLYAMIDES—NYLON		R	60–80
POLYESTERS			
BISPHENOL A-FUMURATE			
HYDROGENATED BISPHENOL-A-BISPHENOL-A			
ISOPHTHALIC			
CHLORINATED POLYESTERS		R	60–200
POLYSTYRENES			
POLYETHYLENE		R	60–150
POLYPROPYLENE		R	60–175
PVC—TYPE 1		R	60–140
PVC—TYPE 2		R	60–140
POLYVINYLIDENE CHLORIDE		R	60–150
VINYLIDENE FLUORIDE		R	60–125
VINYL ESTER			

RUBBERS	Rating	Temp range (°F)
BUTYL GR-1	R	60–150
FLUORO ELASTOMERS VITON A	R	60–185
FLUORO ELASTOMERS KEL-F 3700		
HARD RUBBER	R	60–200
HYPALON	U	
KOROSEAL		
NATURAL RUBBER (GRS)	R	60–150
NEOPRENE GR-M (CR)	U	
NITRILE BUNA N (NBR)	RU	
NORDEL	R	60–300
POLYBUTADIENE		
SILICONE		
POLYURETHANE	U	

TEMP °F

GAS, MANUFACTURED

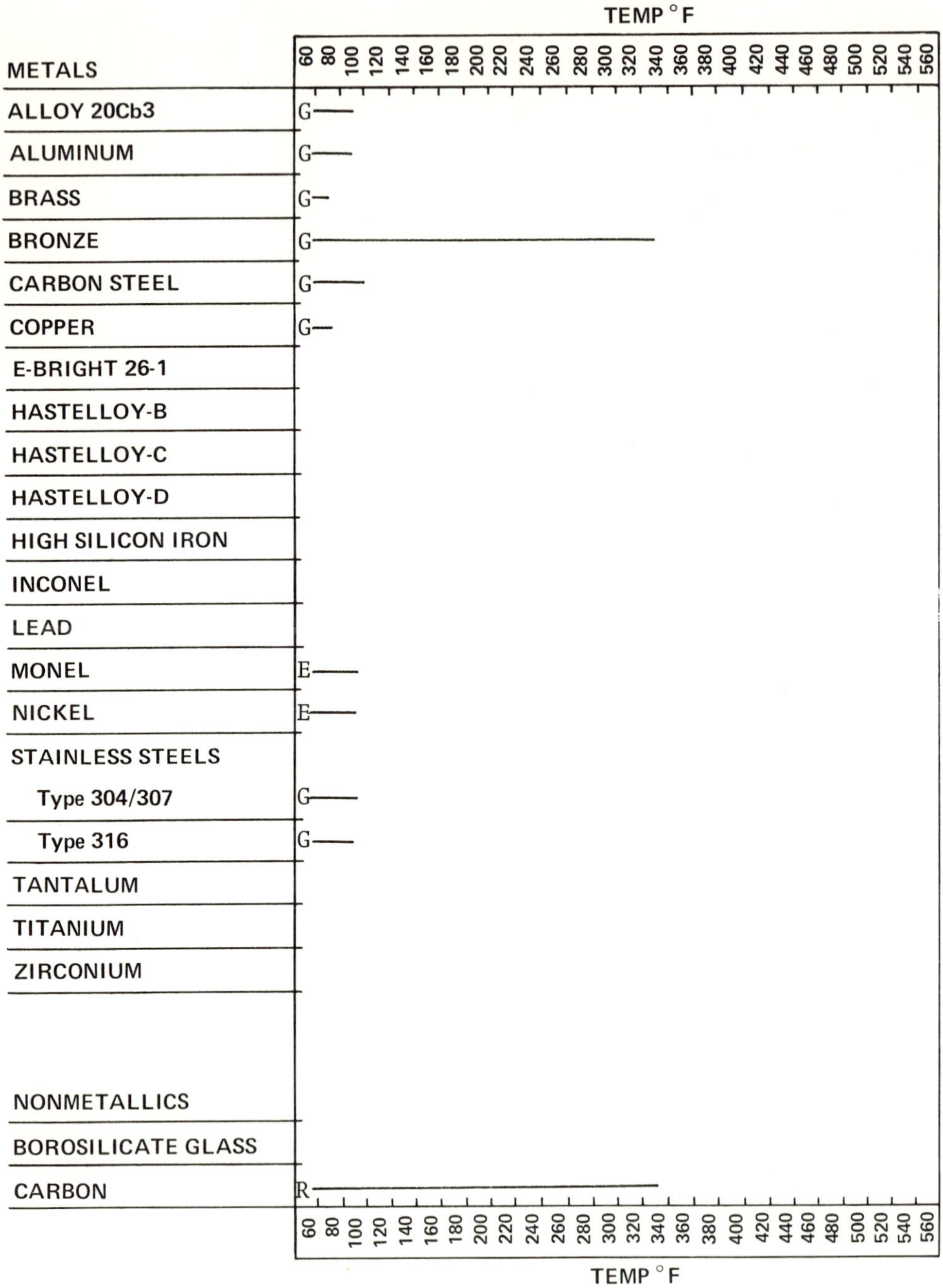

FOR METALS

E = < 2 Mils Penetration/Year; G = < 20 Mils Penetration/Year

S = < 50 Mils Penetration/Year; U = > 50 Mils Penetration/Year

FOR NONMETALLICS

R = Resistant

U = Unsatisfactory

GAS, MANUFACTURED

TEMP °F (60, 80, 100, 120, 140, 160, 180, 200, 220, 240, 260, 280, 300, 320, 340, 360, 380, 400, 420, 440, 460)

PLASTICS	
ABS	U
ACRYLICS	
ASBESTOS REINF., EPOXY	
ASBESTOS REINF., PHENOLIC	
CHLORINATED POLYETHER	R
CPVC	R
EPOXY	
FLUOROCARBONS FEP	R
FLUOROCARBONS TFE	R
FURFURYL ALCOHOL	
NORYL	
PHENOLIC	
POLYAMIDES—NYLON	
POLYESTERS	
BISPHENOL A-FUMURATE	
HYDROGENATED BISPHENOL-A-BISPHENOL-A	
ISOPHTHALIC	
CHLORINATED POLYESTERS	
POLYSTYRENES	
POLYETHYLENE	U
POLYPROPYLENE	R
PVC—TYPE 1	R
PVC—TYPE 2	R
POLYVINYLIDENE CHLORIDE	R
VINYLIDENE FLUORIDE	R
VINYL ESTER	

RUBBERS	
BUTYL GR-1	
FLUORO ELASTOMERS VITON A	
FLUORO ELASTOMERS KEL-F 3700	
HARD RUBBER	
HYPALON	
KOROSEAL	
NATURAL RUBBER (GRS)	
NEOPRENE GR-M (CR)	
NITRILE BUNA N (NBR)	R
NORDEL	
POLYBUTADIENE	
SILICONE	
POLYURETHANE	

TEMP °F (60, 80, 100, 120, 140, 160, 180, 200, 220, 240, 260, 280, 300, 320, 340, 360, 380, 400, 420, 440, 460)

GAS, NATURAL

TEMP °F: 60, 80, 100, 120, 140, 160, 180, 200, 220, 240, 260, 280, 300, 320, 340, 360, 380, 400, 420, 440, 460, 480, 500, 520, 540, 560

METALS	
ALLOY 20Cb3	G
ALUMINUM	G
BRASS	G
BRONZE	G
CARBON STEEL	G
COPPER	G
E-BRIGHT 26-1	
HASTELLOY-B	
HASTELLOY-C	
HASTELLOY-D	
HIGH SILICON IRON	
INCONEL	
LEAD	
MONEL	G
NICKEL	G
STAINLESS STEELS	
Type 304/307	G
Type 316	G
TANTALUM	
TITANIUM	
ZIRCONIUM	

NONMETALLICS	
BOROSILICATE GLASS	
CARBON	R

TEMP °F: 60, 80, 100, 120, 140, 160, 180, 200, 220, 240, 260, 280, 300, 320, 340, 360, 380, 400, 420, 440, 460, 480, 500, 520, 540, 560

FOR METALS

E = < 2 Mils Penetration/Year; G = < 20 Mils Penetration/Year

S = < 50 Mils Penetration/Year; U = > 50 Mils Penetration/Year

FOR NONMETALLICS

R = Resistant

U = Unsatisfactory

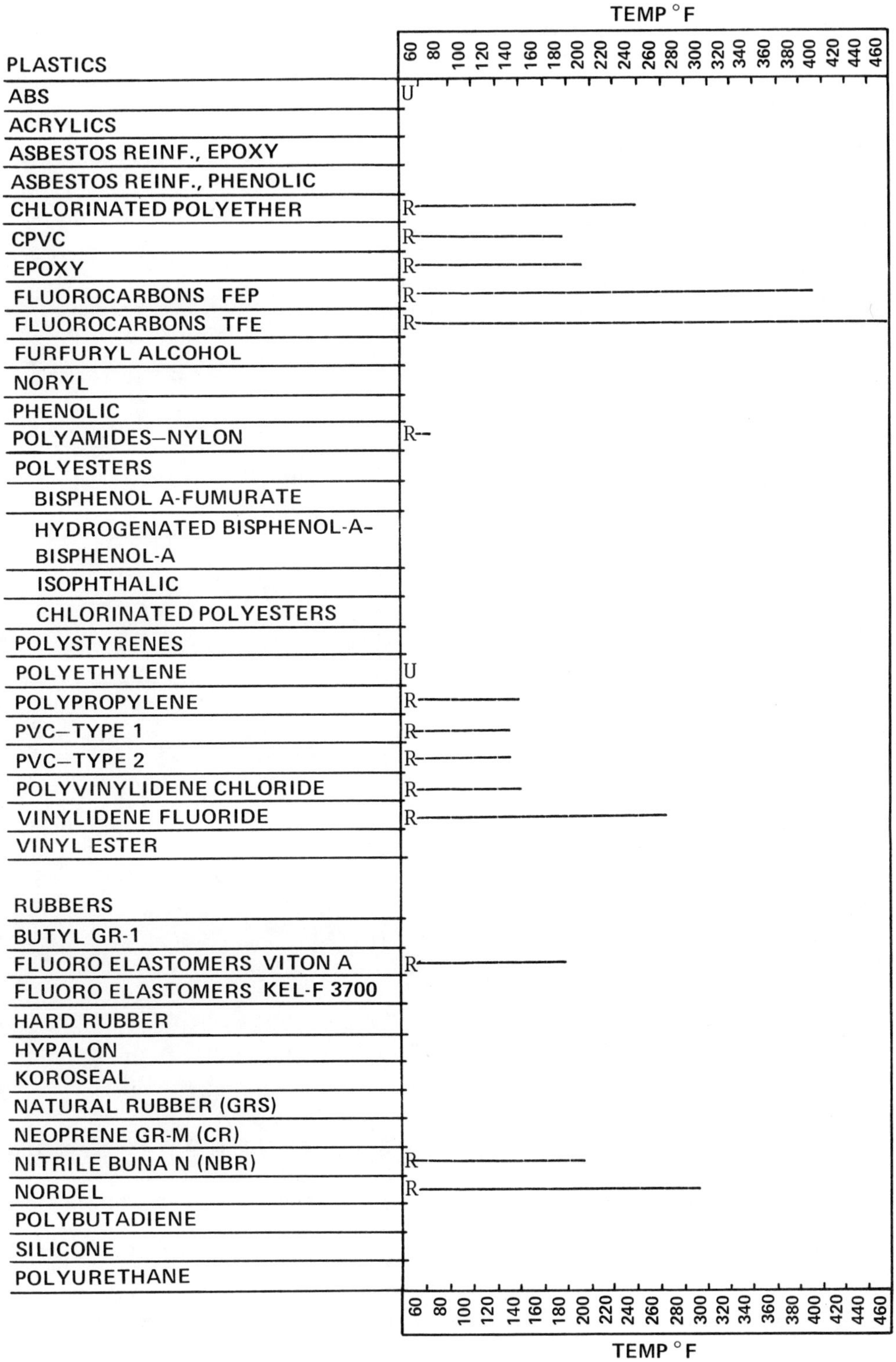
TEMP °F
60 80 100 120 140 160 180 200 220 240 260 280 300 320 340 360 380 400 420 440 460
PLASTICS
ABS U
ACRYLICS
ASBESTOS REINF., EPOXY
ASBESTOS REINF., PHENOLIC
CHLORINATED POLYETHER R
CPVC R
EPOXY R
FLUOROCARBONS FEP R
FLUOROCARBONS TFE R
FURFURYL ALCOHOL
NORYL
PHENOLIC
POLYAMIDES–NYLON R
POLYESTERS
BISPHENOL A-FUMURATE
HYDROGENATED BISPHENOL-A-BISPHENOL-A
ISOPHTHALIC
CHLORINATED POLYESTERS
POLYSTYRENES
POLYETHYLENE U
POLYPROPYLENE R
PVC–TYPE 1 R
PVC–TYPE 2 R
POLYVINYLIDENE CHLORIDE R
VINYLIDENE FLUORIDE R
VINYL ESTER
RUBBERS
BUTYL GR-1
FLUORO ELASTOMERS VITON A R
FLUORO ELASTOMERS KEL-F 3700
HARD RUBBER
HYPALON
KOROSEAL
NATURAL RUBBER (GRS)
NEOPRENE GR-M (CR)
NITRILE BUNA N (NBR) R
NORDEL R
POLYBUTADIENE
SILICONE
POLYURETHANE
60 80 100 120 140 160 180 200 220 240 260 280 300 320 340 360 380 400 420 440 460
TEMP °F

GASOLINE, LEADED

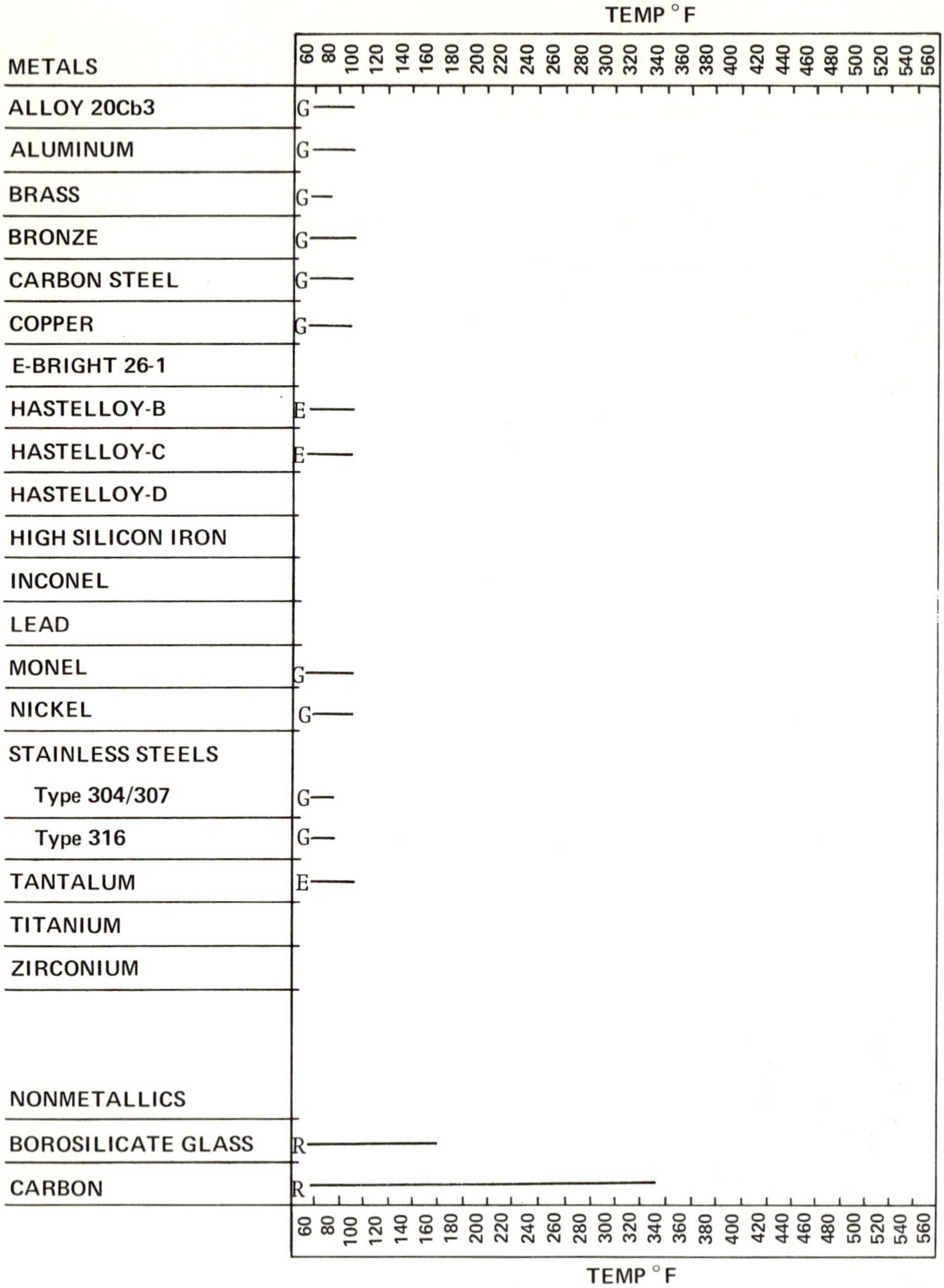

FOR METALS

E = < 2 Mils Penetration/Year; G = < 20 Mils Penetration/Year

S = < 50 Mils Penetration/Year; U = > 50 Mils Penetration/Year

FOR NONMETALLICS

R = Resistant

U = Unsatisfactory

GASOLINE, LEADED

TEMP °F: 60, 80, 100, 120, 140, 160, 180, 200, 220, 240, 260, 280, 300, 320, 340, 360, 380, 400, 420, 440, 460

PLASTICS	
ABS	U
ACRYLICS	
ASBESTOS REINF., EPOXY	R
ASBESTOS REINF., PHENOLIC	R
CHLORINATED POLYETHER	R
CPVC	
EPOXY	R
FLUOROCARBONS FEP	R
FLUOROCARBONS TFE	R
FURFURYL ALCOHOL	R
NORYL	
PHENOLIC	
POLYAMIDES–NYLON	R
POLYESTERS	
BISPHENOL A-FUMURATE	R
HYDROGENATED BISPHENOL-A-BISPHENOL-A	R
ISOPHTHALIC	R
CHLORINATED POLYESTERS	
POLYSTYRENES	
POLYETHYLENE	U
POLYPROPYLENE	U
PVC–TYPE 1	R
PVC–TYPE 2	
POLYVINYLIDENE CHLORIDE	R
VINYLIDENE FLUORIDE	R
VINYL ESTER	R
RUBBERS	
BUTYL GR-1	
FLUORO ELASTOMERS VITON A	R
FLUORO ELASTOMERS KEL-F 3700	
HARD RUBBER	
HYPALON	
KOROSEAL	
NATURAL RUBBER (GRS)	
NEOPRENE GR-M (CR)	
NITRILE BUNA N (NBR)	R
NORDEL	
POLYBUTADIENE	
SILICONE	U
POLYURETHANE	

TEMP °F: 60, 80, 100, 120, 140, 160, 180, 200, 220, 240, 260, 280, 300, 320, 340, 360, 380, 400, 420, 440, 460

GASOLINE, UNLEADED

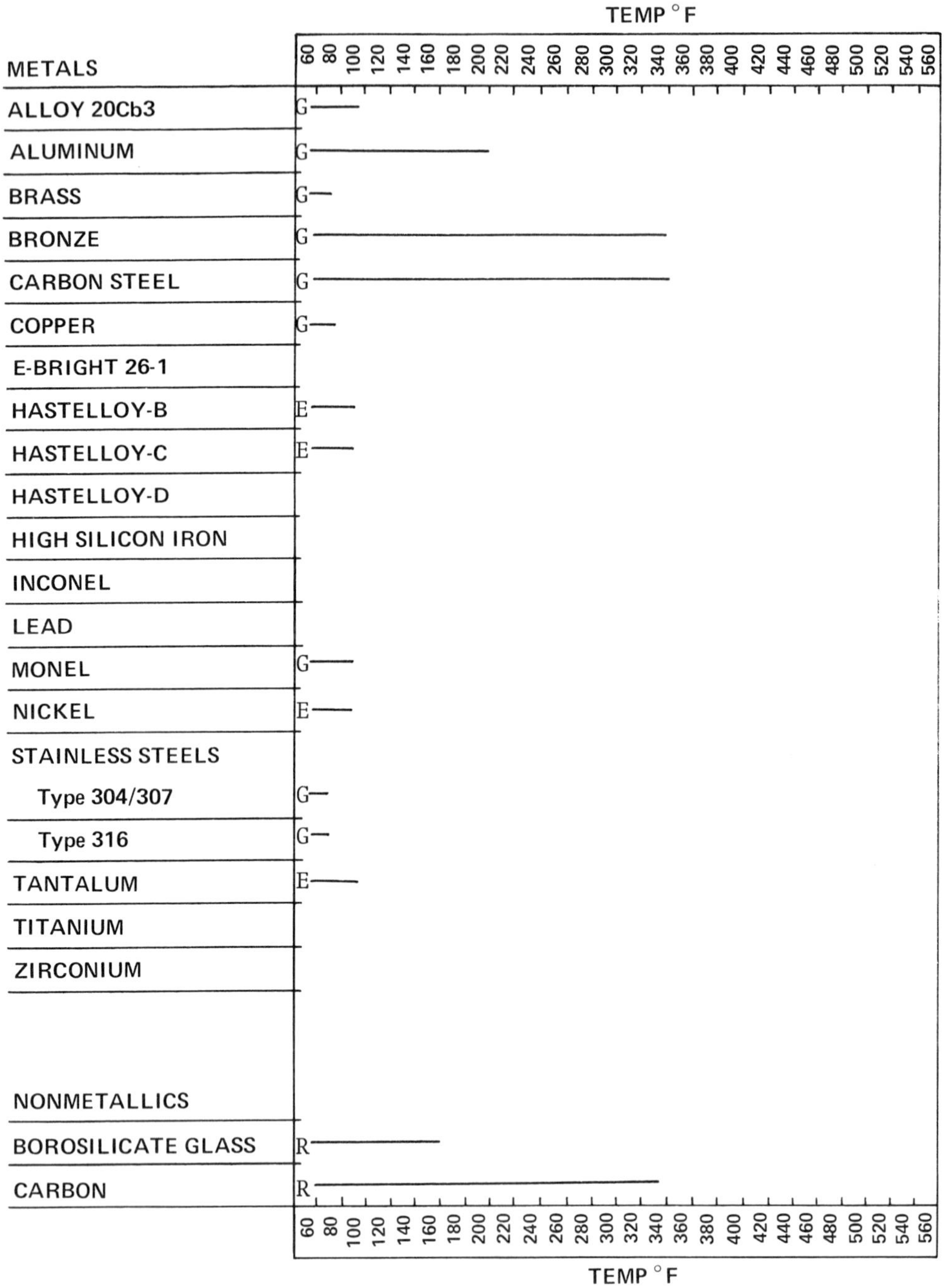

FOR METALS

E = < 2 Mils Penetration/Year; G = < 20 Mils Penetration/Year

S = < 50 Mils Penetration/Year; U = > 50 Mils Penetration/Year

FOR NONMETALLICS

R = Resistant

U = Unsatisfactory

GASOLINE, UNLEADED

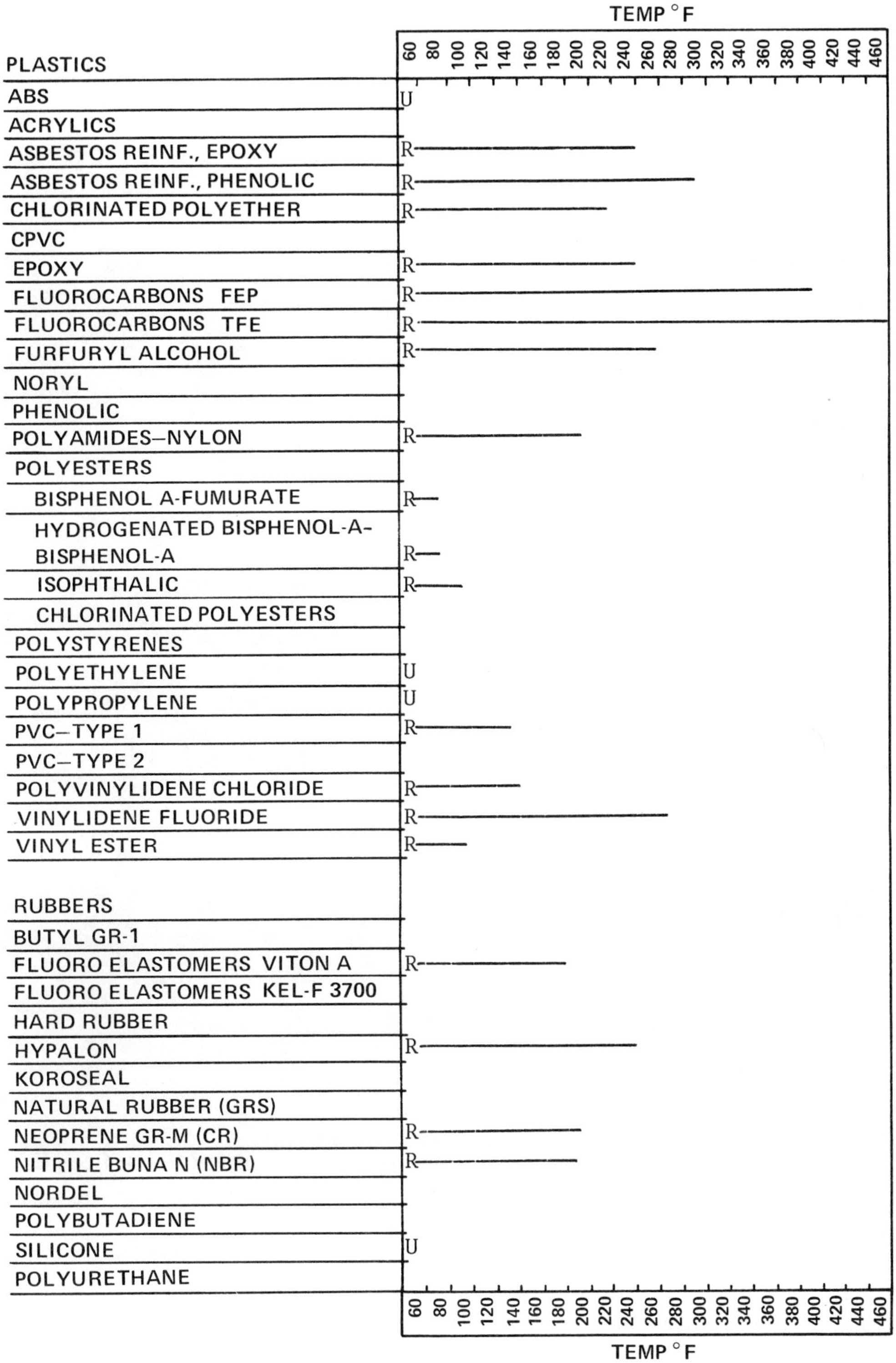

GASOLINE, REFINED

TEMP °F: 60 80 100 120 140 160 180 200 220 240 260 280 300 320 340 360 380 400 420 440 460 480 500 520 540 560

METALS	
ALLOY 20Cb3	G
ALUMINUM	G
BRASS	G
BRONZE	G
CARBON STEEL	G
COPPER	G
E-BRIGHT 26-1	
HASTELLOY-B	G
HASTELLOY-C	G
HASTELLOY-D	
HIGH SILICON IRON	G
INCONEL	
LEAD	
MONEL	G
NICKEL	G
STAINLESS STEELS	
Type 304/307	G
Type 316	G
TANTALUM	G
TITANIUM	
ZIRCONIUM	
NONMETALLICS	
BOROSILICATE GLASS	R
CARBON	R

FOR METALS

E = < 2 Mils Penetration/Year; G = < 20 Mils Penetration/Year

S = < 50 Mils Penetration/Year; U = > 50 Mils Penetration/Year

FOR NONMETALLICS

R = Resistant

U = Unsatisfactory

GASOLINE, REFINED

TEMP °F (60–460)

PLASTICS	TEMP °F
ABS	U
ACRYLICS	R
ASBESTOS REINF., EPOXY	R
ASBESTOS REINF., PHENOLIC	R
CHLORINATED POLYETHER	
CPVC	R
EPOXY	R
FLUOROCARBONS FEP	R
FLUOROCARBONS TFE	R
FURFURYL ALCOHOL	R
NORYL	
PHENOLIC	
POLYAMIDES–NYLON	R
POLYESTERS	
BISPHENOL A-FUMURATE	R
HYDROGENATED BISPHENOL-A-BISPHENOL-A	R
ISOPHTHALIC	R
CHLORINATED POLYESTERS	R
POLYSTYRENES	
POLYETHYLENE	U
POLYPROPYLENE	U
PVC–TYPE 1	R
PVC–TYPE 2	U
POLYVINYLIDENE CHLORIDE	R
VINYLIDENE FLUORIDE	
VINYL ESTER	
RUBBERS	
BUTYL GR-1	
FLUORO ELASTOMERS VITON A	R
FLUORO ELASTOMERS KEL-F 3700	
HARD RUBBER	
HYPALON	U
KOROSEAL	
NATURAL RUBBER (GRS)	U
NEOPRENE GR-M (CR)	
NITRILE BUNA N (NBR)	R
NORDEL	
POLYBUTADIENE	
SILICONE	U
POLYURETHANE	R

TEMP °F (60–460)

GASOLINE, SOUR

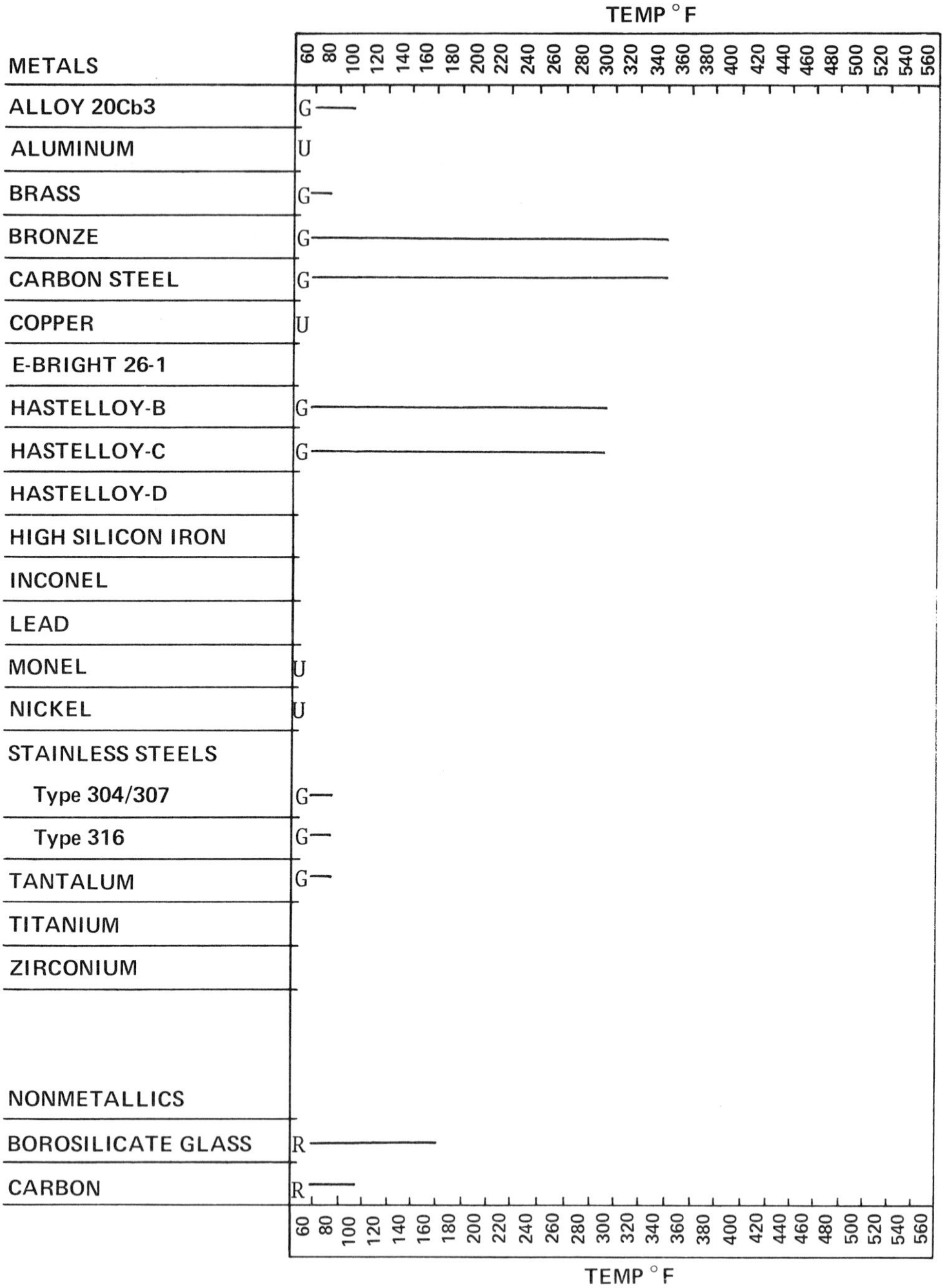

FOR METALS

E = < 2 Mils Penetration/Year; G = < 20 Mils Penetration/Year

S = < 50 Mils Penetration/Year; U = > 50 Mils Penetration/Year

FOR NONMETALLICS

R = Resistant

U = Unsatisfactory

GASOLINE, SOUR

TEMP °F: 60 80 100 120 140 160 180 200 220 240 260 280 300 320 340 360 380 400 420 440 460

PLASTICS	
ABS	U
ACRYLICS	
ASBESTOS REINF., EPOXY	R
ASBESTOS REINF., PHENOLIC	R
CHLORINATED POLYETHER	R
CPVC	R
EPOXY	R
FLUOROCARBONS FEP	R
FLUOROCARBONS TFE	R
FURFURYL ALCOHOL	
NORYL	
PHENOLIC	
POLYAMIDES–NYLON	
POLYESTERS	
BISPHENOL A-FUMURATE	R
HYDROGENATED BISPHENOL-A-BISPHENOL-A	
ISOPHTHALIC	
CHLORINATED POLYESTERS	
POLYSTYRENES	
POLYETHYLENE	U
POLYPROPYLENE	U
PVC–TYPE 1	R
PVC–TYPE 2	R
POLYVINYLIDENE CHLORIDE	
VINYLIDENE FLUORIDE	R
VINYL ESTER	

RUBBERS	
BUTYL GR-1	
FLUORO ELASTOMERS VITON A	R
FLUORO ELASTOMERS KEL-F 3700	
HARD RUBBER	
HYPALON	
KOROSEAL	
NATURAL RUBBER (GRS)	
NEOPRENE GR-M (CR)	
NITRILE BUNA N (NBR)	R
NORDEL	
POLYBUTADIENE	
SILICONE	U
POLYURETHANE	

TEMP °F: 60 80 100 120 140 160 180 200 220 240 260 280 300 320 340 360 380 400 420 440 460

GELATIN

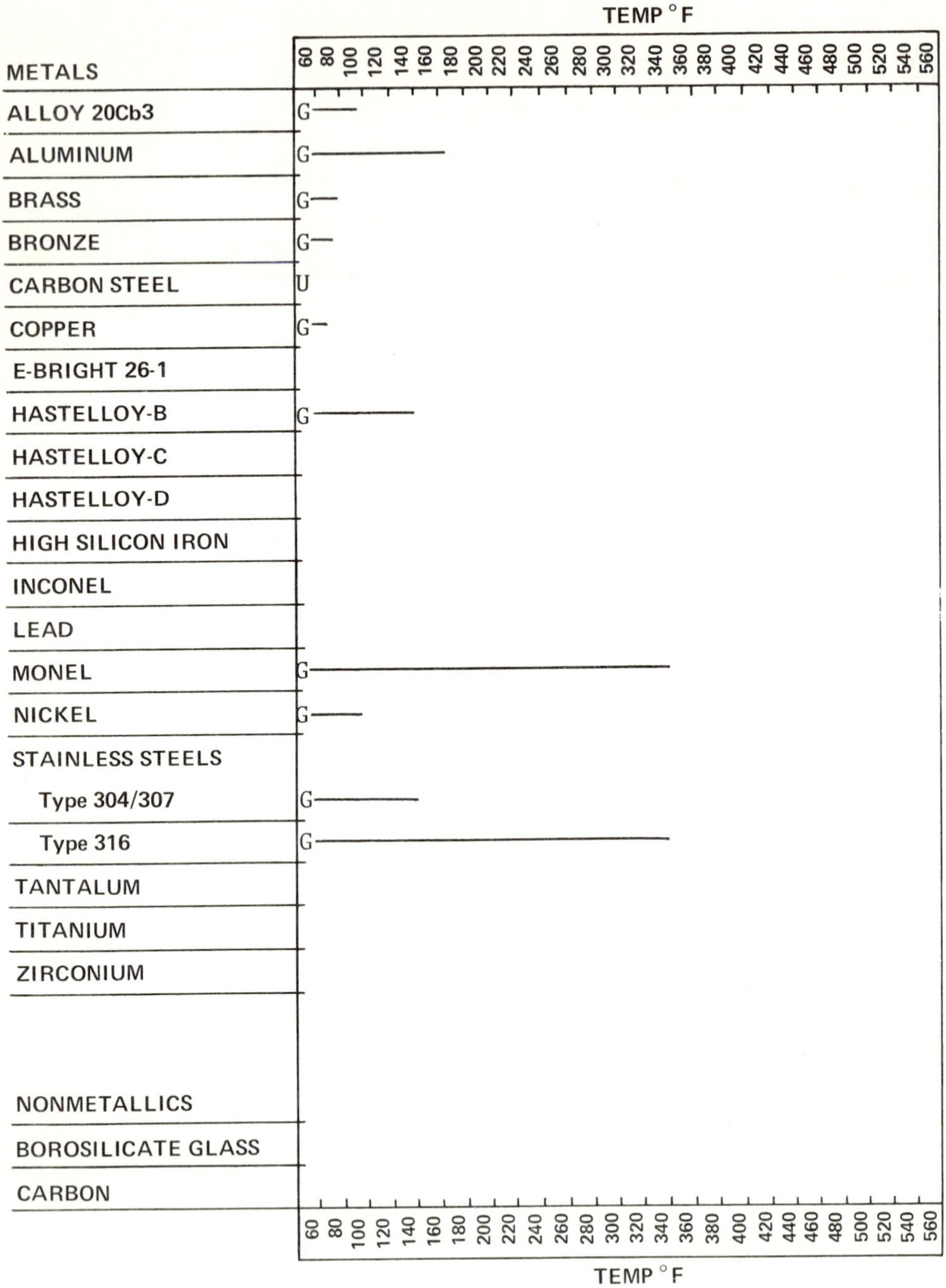

FOR METALS

E = < 2 Mils Penetration/Year; G = < 20 Mils Penetration/Year

S = < 50 Mils Penetration/Year; U = > 50 Mils Penetration/Year

FOR NONMETALLICS

R = Resistant

U = Unsatisfactory

TEMP °F: 60 80 100 120 140 160 180 200 220 240 260 280 300 320 340 360 380 400 420 440 460

PLASTICS	Rating	Line extends to (°F)
ABS	R	140
ACRYLICS		
ASBESTOS REINF., EPOXY		
ASBESTOS REINF., PHENOLIC		
CHLORINATED POLYETHER	R	250
CPVC		
EPOXY		
FLUOROCARBONS FEP	R	400
FLUOROCARBONS TFE	R	460
FURFURYL ALCOHOL		
NORYL		
PHENOLIC		
POLYAMIDES–NYLON	R	80
POLYESTERS		
BISPHENOL A-FUMURATE		
HYDROGENATED BISPHENOL-A-BISPHENOL-A		
ISOPHTHALIC		
CHLORINATED POLYESTERS		
POLYSTYRENES		
POLYETHYLENE	R	140
POLYPROPYLENE	R	180
PVC–TYPE 1	R	140
PVC–TYPE 2	R	140
POLYVINYLIDENE CHLORIDE	R	180
VINYLIDENE FLUORIDE		
VINYL ESTER		

RUBBERS	Rating	Line extends to (°F)
BUTYL GR-1	R	180
FLUORO ELASTOMERS VITON A	R	180
FLUORO ELASTOMERS KEL-F 3700		
HARD RUBBER		
HYPALON	R	80
KOROSEAL		
NATURAL RUBBER (GRS)	R	150
NEOPRENE GR-M (CR)	R	200
NITRILE BUNA N (NBR)	R	180
NORDEL		
POLYBUTADIENE		
SILICONE		
POLYURETHANE		

TEMP °F: 60 80 100 120 140 160 180 200 220 240 260 280 300 320 340 360 380 400 420 440 460

GIN

METALS	TEMP °F (60–560)
ALLOY 20Cb3	
ALUMINUM	
BRASS	
BRONZE	
CARBON STEEL	
COPPER	
E-BRIGHT 26-1	
HASTELLOY-B	
HASTELLOY-C	
HASTELLOY-D	
HIGH SILICON IRON	G (60–80)
INCONEL	
LEAD	
MONEL	
NICKEL	
STAINLESS STEELS	
Type 304/307	G (60–200)
Type 316	G (60–200)
TANTALUM	
TITANIUM	
ZIRCONIUM	
NONMETALLICS	
BOROSILICATE GLASS	R (60–200)
CARBON	R (60–340)

FOR METALS

E = < 2 Mils Penetration/Year; G = < 20 Mils Penetration/Year

S = < 50 Mils Penetration/Year; U = > 50 Mils Penetration/Year

FOR NONMETALLICS

R = Resistant

U = Unsatisfactory

GIN

PLASTICS	TEMP °F (60–460)
ABS	
ACRYLICS	
ASBESTOS REINF., EPOXY	
ASBESTOS REINF., PHENOLIC	
CHLORINATED POLYETHER	R (60–250)
CPVC	
EPOXY	
FLUOROCARBONS FEP	R (60–400)
FLUOROCARBONS TFE	R (60–460)
FURFURYL ALCOHOL	
NORYL	
PHENOLIC	
POLYAMIDES—NYLON	
POLYESTERS	
BISPHENOL A-FUMURATE	
HYDROGENATED BISPHENOL-A-BISPHENOL-A	
ISOPHTHALIC	
CHLORINATED POLYESTERS	
POLYSTYRENES	
POLYETHYLENE	R (60–140)
POLYPROPYLENE	R (60–200)
PVC—TYPE 1	
PVC—TYPE 2	
POLYVINYLIDENE CHLORIDE	R (60–150)
VINYLIDENE FLUORIDE	R (60–220)
VINYL ESTER	

RUBBERS	TEMP °F (60–460)
BUTYL GR-1	
FLUORO ELASTOMERS VITON A	
FLUORO ELASTOMERS KEL-F 3700	
HARD RUBBER	
HYPALON	
KOROSEAL	
NATURAL RUBBER (GRS)	
NEOPRENE GR-M (CR)	
NITRILE BUNA N (NBR)	
NORDEL	
POLYBUTADIENE	
SILICONE	
POLYURETHANE	

GLUCOSE (CORN SYRUP)

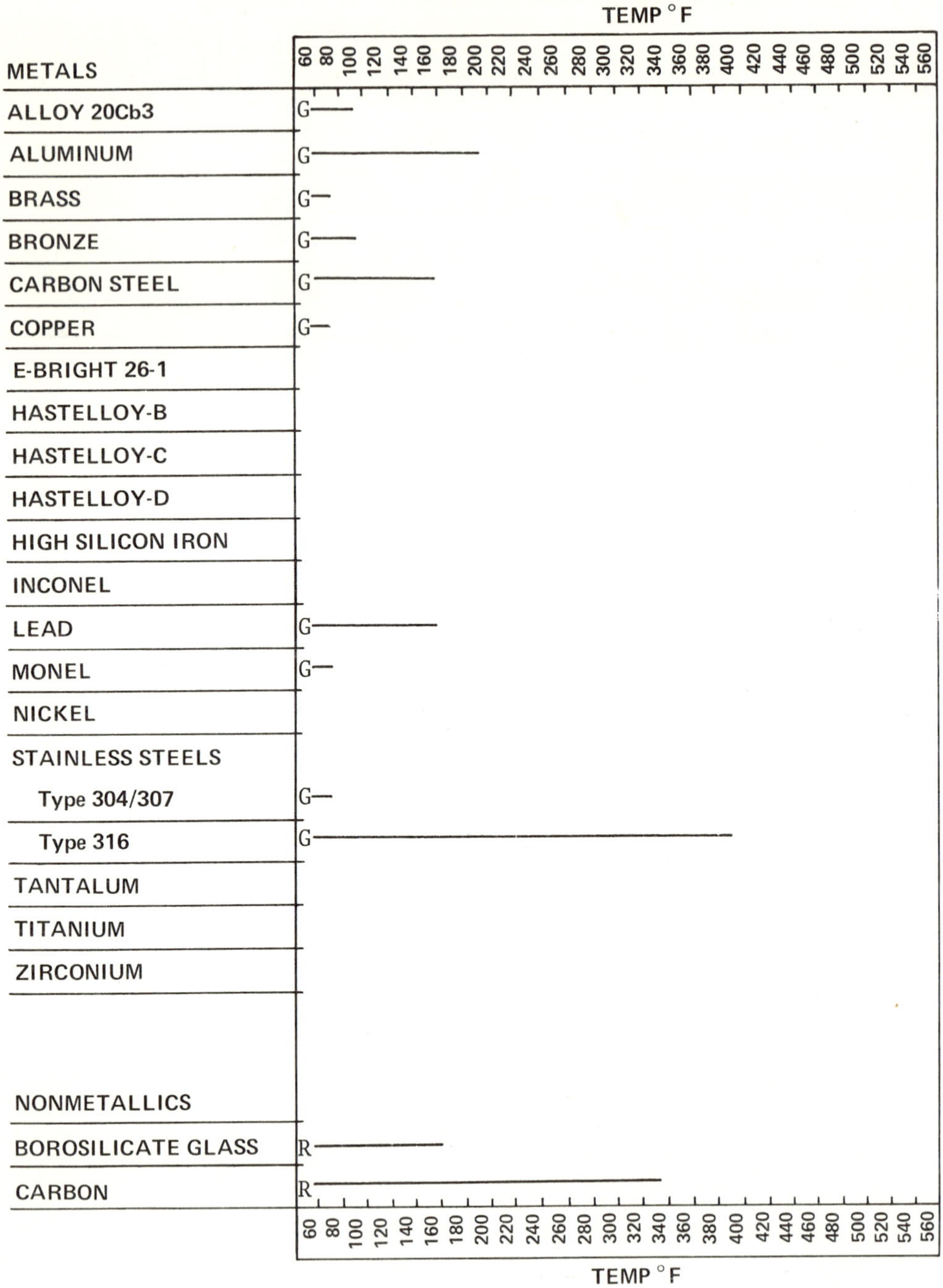

FOR METALS

E = < 2 Mils Penetration/Year; G = < 20 Mils Penetration/Year

S = < 50 Mils Penetration/Year; U = > 50 Mils Penetration/Year

FOR NONMETALLICS

R = Resistant

U = Unsatisfactory

GLUCOSE (CORN SYRUP)

TEMP °F (60, 80, 100, 120, 140, 160, 180, 200, 220, 240, 260, 280, 300, 320, 340, 360, 380, 400, 420, 440, 460)

PLASTICS		
ABS	30%	R
ACRYLICS		R
ASBESTOS REINF., EPOXY		
ASBESTOS REINF., PHENOLIC		
CHLORINATED POLYETHER		R
CPVC		R
EPOXY		R
FLUOROCARBONS FEP		R
FLUOROCARBONS TFE		R
FURFURYL ALCOHOL		
NORYL		
PHENOLIC		
POLYAMIDES—NYLON		R
POLYESTERS		
BISPHENOL A-FUMURATE		R
HYDROGENATED BISPHENOL-A-BISPHENOL-A		
ISOPHTHALIC		R
CHLORINATED POLYESTERS		
POLYSTYRENES		
POLYETHYLENE		R U
POLYPROPYLENE		R
PVC—TYPE 1		R
PVC—TYPE 2		R
POLYVINYLIDENE CHLORIDE		R
VINYLIDENE FLUORIDE		R
VINYL ESTER		R
RUBBERS		
BUTYL GR-1		R
FLUORO ELASTOMERS VITON A		R
FLUORO ELASTOMERS KEL-F 3700		
HARD RUBBER		
HYPALON		R
KOROSEAL		
NATURAL RUBBER (GRS)		R
NEOPRENE GR-M (CR)		R
NITRILE BUNA N (NBR)		R
NORDEL		
POLYBUTADIENE		
SILICONE		
POLYURETHANE		U

TEMP °F (60, 80, 100, 120, 140, 160, 180, 200, 220, 240, 260, 280, 300, 320, 340, 360, 380, 400, 420, 440, 460)

GLUE

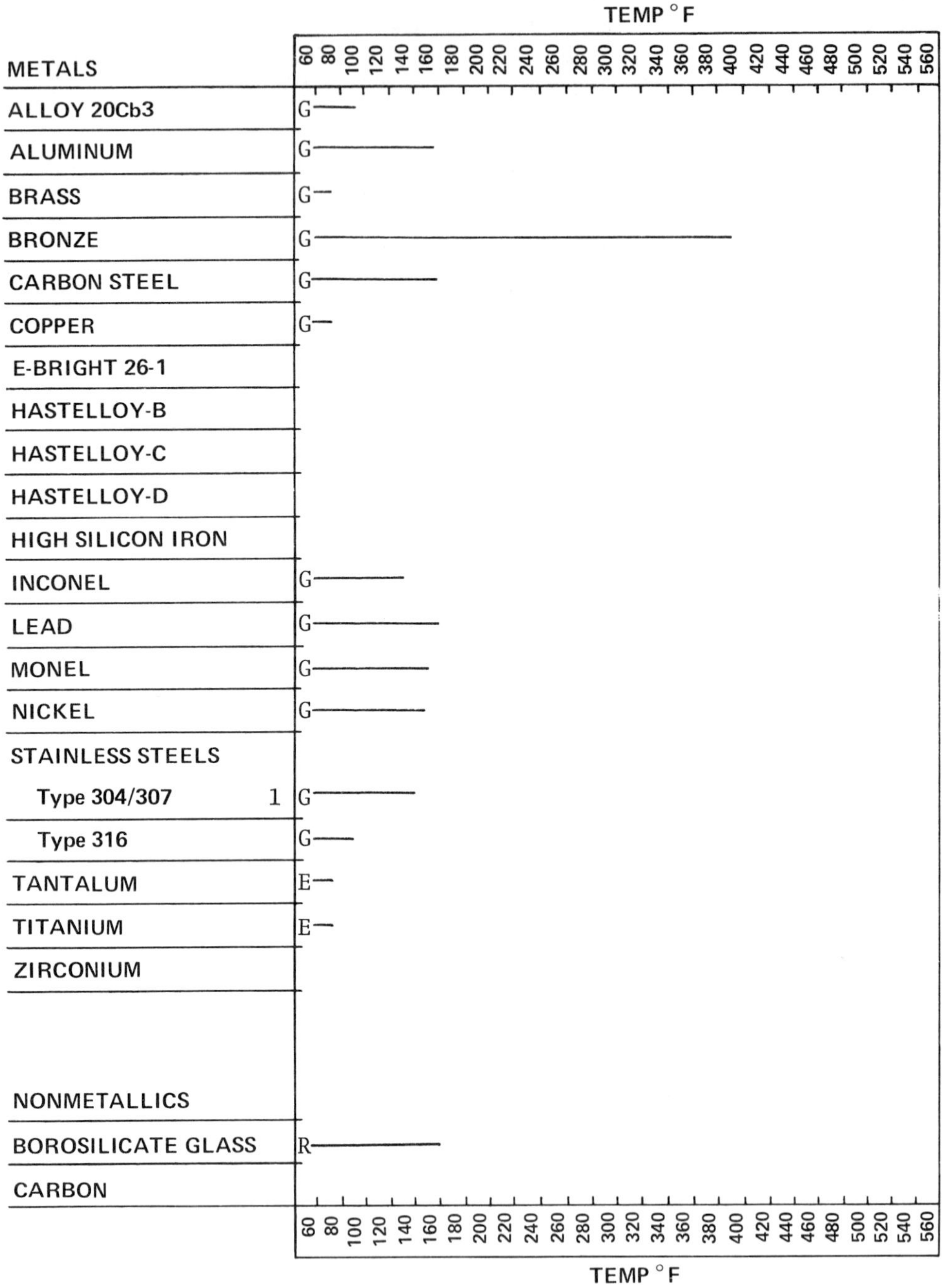

FOR METALS

E = < 2 Mils Penetration/Year; G = < 20 Mils Penetration/Year

S = < 50 Mils Penetration/Year; U = > 50 Mils Penetration/Year

FOR NONMETALLICS

R = Resistant

U = Unsatisfactory

GLUE

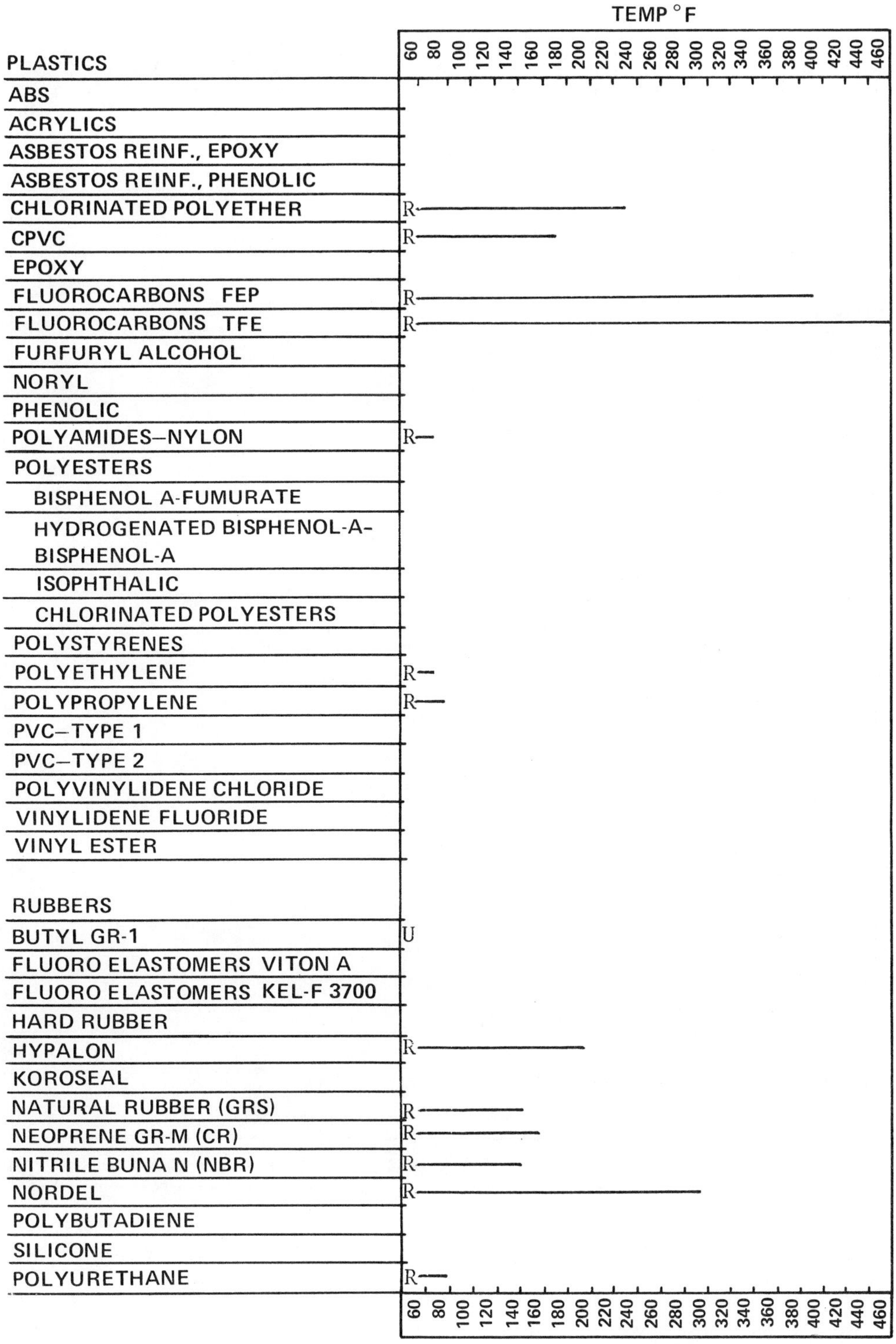
TEMP °F
60 80 100 120 140 160 180 200 220 240 260 280 300 320 340 360 380 400 420 440 460
PLASTICS
ABS
ACRYLICS
ASBESTOS REINF., EPOXY
ASBESTOS REINF., PHENOLIC
CHLORINATED POLYETHER R
CPVC R
EPOXY
FLUOROCARBONS FEP R
FLUOROCARBONS TFE R
FURFURYL ALCOHOL
NORYL
PHENOLIC
POLYAMIDES—NYLON R
POLYESTERS
BISPHENOL A-FUMURATE
HYDROGENATED BISPHENOL-A-BISPHENOL-A
ISOPHTHALIC
CHLORINATED POLYESTERS
POLYSTYRENES
POLYETHYLENE R
POLYPROPYLENE R
PVC—TYPE 1
PVC—TYPE 2
POLYVINYLIDENE CHLORIDE
VINYLIDENE FLUORIDE
VINYL ESTER
RUBBERS
BUTYL GR-1 U
FLUORO ELASTOMERS VITON A
FLUORO ELASTOMERS KEL-F 3700
HARD RUBBER
HYPALON R
KOROSEAL
NATURAL RUBBER (GRS) R
NEOPRENE GR-M (CR) R
NITRILE BUNA N (NBR) R
NORDEL R
POLYBUTADIENE
SILICONE
POLYURETHANE R
60 80 100 120 140 160 180 200 220 240 260 280 300 320 340 360 380 400 420 440 460
TEMP °F

GLYCERINE, GLYCEROL

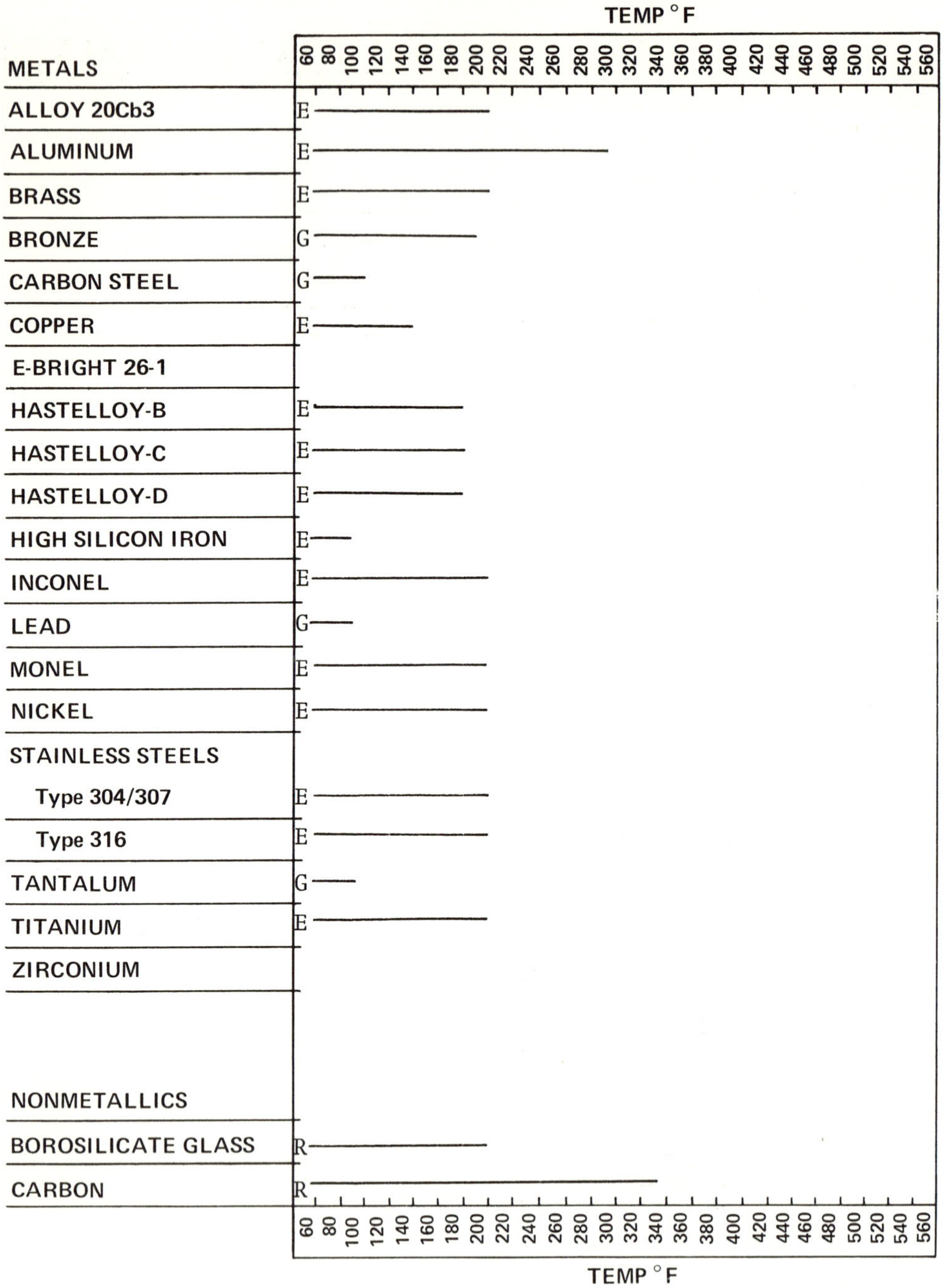

FOR METALS

E = < 2 Mils Penetration/Year; G = < 20 Mils Penetration/Year

S = < 50 Mils Penetration/Year; U = > 50 Mils Penetration/Year

FOR NONMETALLICS

R = Resistant

U = Unsatisfactory

TEMP °F

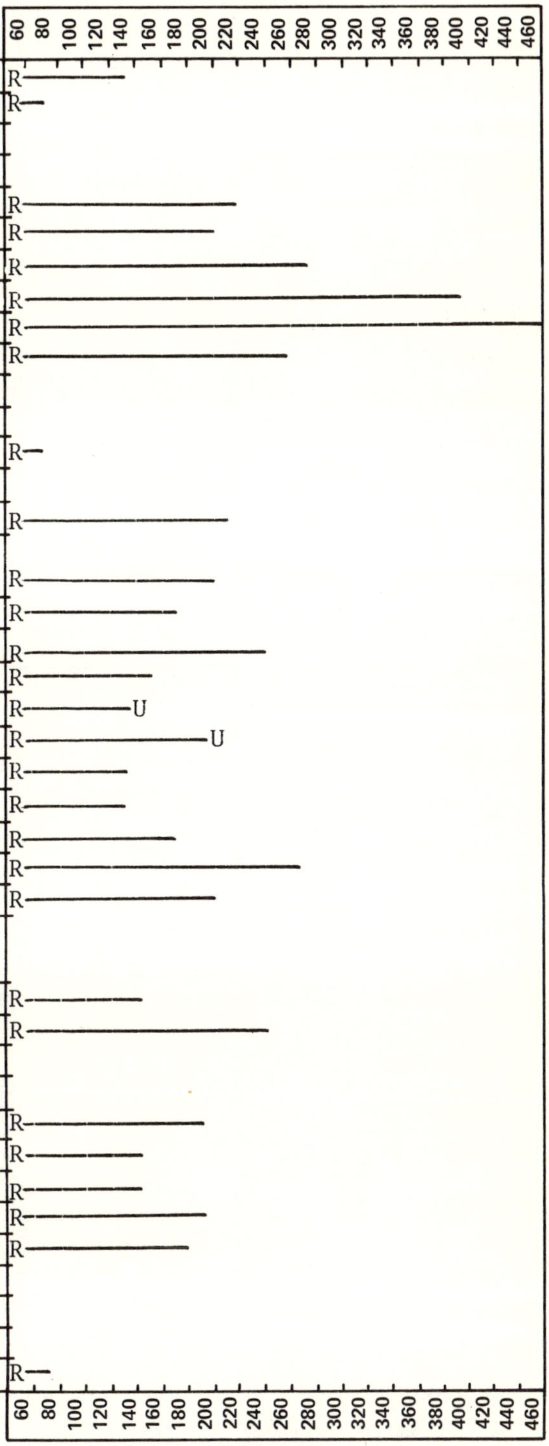

PLASTICS	
ABS	R
ACRYLICS	R
ASBESTOS REINF., EPOXY	
ASBESTOS REINF., PHENOLIC	
CHLORINATED POLYETHER	R
CPVC	R
EPOXY	R
FLUOROCARBONS FEP	R
FLUOROCARBONS TFE	R
FURFURYL ALCOHOL	R
NORYL	
PHENOLIC	
POLYAMIDES–NYLON	R
POLYESTERS	
BISPHENOL A-FUMURATE	R
HYDROGENATED BISPHENOL-A-BISPHENOL-A	R
ISOPHTHALIC	R
CHLORINATED POLYESTERS 75%	R
POLYSTYRENES	R
POLYETHYLENE	R U
POLYPROPYLENE	R U
PVC–TYPE 1	R
PVC–TYPE 2	R
POLYVINYLIDENE CHLORIDE	R
VINYLIDENE FLUORIDE	R
VINYL ESTER	R
RUBBERS	
BUTYL GR-1	R
FLUORO ELASTOMERS VITON A	R
FLUORO ELASTOMERS KEL-F 3700	
HARD RUBBER	
HYPALON	R
KOROSEAL	R
NATURAL RUBBER (GRS)	R
NEOPRENE GR-M (CR)	R
NITRILE BUNA N (NBR)	R
NORDEL	
POLYBUTADIENE	
SILICONE	
POLYURETHANE	R

TEMP °F

GLYCOLIC ACID

TEMP °F

METALS	60 – 560 °F
ALLOY 20Cb3	G
ALUMINUM	G
BRASS	
BRONZE	
CARBON STEEL	
COPPER	
E-BRIGHT 26-1	
HASTELLOY-B	G
HASTELLOY-C	G
HASTELLOY-D	G
HIGH SILICON IRON	
INCONEL	G
LEAD	
MONEL (air free)	G
NICKEL (air free)	G
STAINLESS STEELS	
Type 304/307	G
Type 316	G
TANTALUM	G
TITANIUM	E
ZIRCONIUM	E

NONMETALLICS	60 – 560 °F
BOROSILICATE GLASS	
CARBON	

TEMP °F

FOR METALS

E = < 2 Mils Penetration/Year; G = < 20 Mils Penetration/Year

S = < 50 Mils Penetration/Year; U = > 50 Mils Penetration/Year

FOR NONMETALLICS

R = Resistant

U = Unsatisfactory

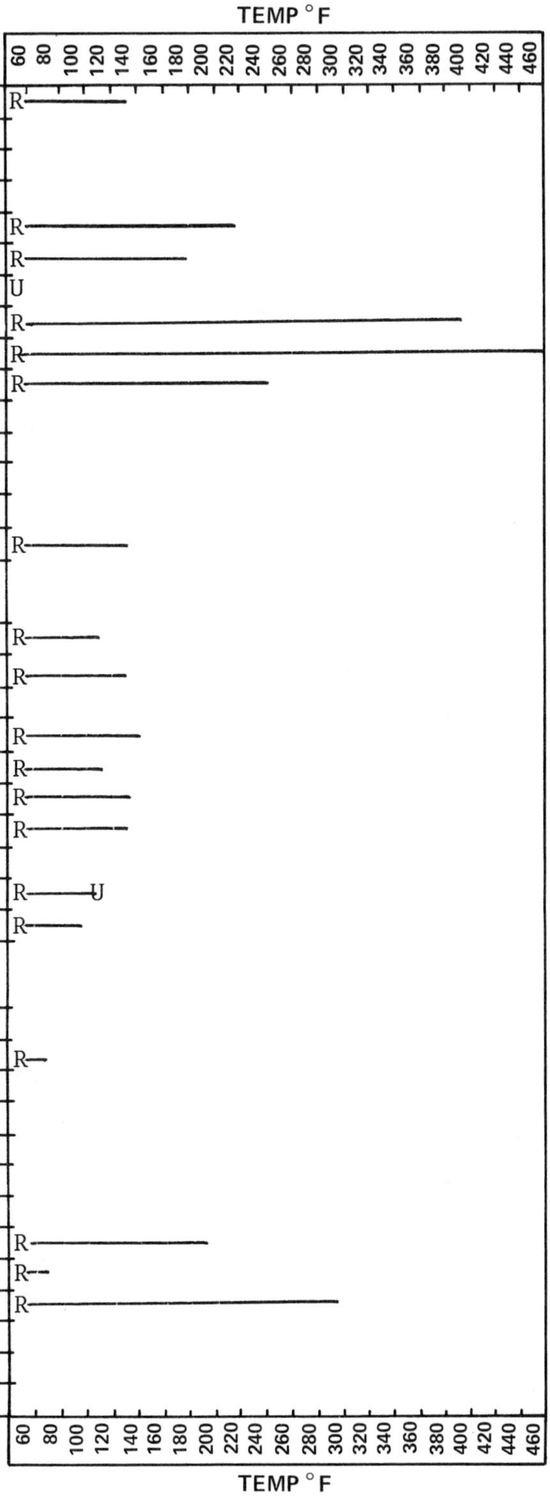

PLASTICS		
ABS	30%	R
ACRYLICS		
ASBESTOS REINF., EPOXY		
ASBESTOS REINF., PHENOLIC		
CHLORINATED POLYETHER		R
CPVC		R
EPOXY		U
FLUOROCARBONS FEP		R
FLUOROCARBONS TFE		R
FURFURYL ALCOHOL		R
NORYL		
PHENOLIC		
POLYAMIDES—NYLON		
POLYESTERS		
BISPHENOL A-FUMURATE		R
HYDROGENATED BISPHENOL-A-BISPHENOL-A		
ISOPHTHALIC		R
CHLORINATED POLYESTERS	35%	R
POLYSTYRENES		
POLYETHYLENE		R
POLYPROPYLENE		R
PVC—TYPE 1		R
PVC—TYPE 2		R
POLYVINYLIDENE CHLORIDE		
VINYLIDENE FLUORIDE		R—U
VINYL ESTER		R

RUBBERS	
BUTYL GR-1	
FLUORO ELASTOMERS VITON A	R
FLUORO ELASTOMERS KEL-F 3700	
HARD RUBBER	
HYPALON	
KOROSEAL	
NATURAL RUBBER (GRS)	
NEOPRENE GR-M (CR)	R
NITRILE BUNA N (NBR)	R
NORDEL	R
POLYBUTADIENE	
SILICONE	
POLYURETHANE	

GLYCOLS

TEMP °F (60–560)

METALS	Rating
ALLOY 20Cb3	G ——
ALUMINUM	G —
BRASS	
BRONZE	G ——
CARBON STEEL	G —
COPPER	
E-BRIGHT 26-1	
HASTELLOY-B	
HASTELLOY-C	
HASTELLOY-D	
HIGH SILICON IRON	
INCONEL	
LEAD	
MONEL	G ——
NICKEL	
STAINLESS STEELS	
Type 304/307	G ——
Type 316	G —
TANTALUM	
TITANIUM	
ZIRCONIUM	

NONMETALLICS	Rating
BOROSILICATE GLASS	
CARBON	

TEMP °F

FOR METALS

E = < 2 Mils Penetration/Year; G = < 20 Mils Penetration/Year

S = < 50 Mils Penetration/Year; U = > 50 Mils Penetration/Year

FOR NONMETALLICS

R = Resistant

U = Unsatisfactory

GLYCOLS

TEMP °F: 60, 80, 100, 120, 140, 160, 180, 200, 220, 240, 260, 280, 300, 320, 340, 360, 380, 400, 420, 440, 460

PLASTICS	Rating
ABS	R
ACRYLICS	
ASBESTOS REINF., EPOXY	
ASBESTOS REINF., PHENOLIC	
CHLORINATED POLYETHER	R — U
CPVC	R
EPOXY	R
FLUOROCARBONS FEP	R
FLUOROCARBONS TFE	R
FURFURYL ALCOHOL	
NORYL	
PHENOLIC	
POLYAMIDES–NYLON	R
POLYESTERS	
BISPHENOL A-FUMURATE	
HYDROGENATED BISPHENOL-A-BISPHENOL-A	
ISOPHTHALIC	
CHLORINATED POLYESTERS	
POLYSTYRENES	
POLYETHYLENE	R
POLYPROPYLENE	R
PVC–TYPE 1	R
PVC–TYPE 2	R
POLYVINYLIDENE CHLORIDE	
VINYLIDENE FLUORIDE	R
VINYL ESTER	R

RUBBERS	Rating
BUTYL GR-1	R
FLUORO ELASTOMERS VITON A	R
FLUORO ELASTOMERS KEL-F 3700	
HARD RUBBER	R
HYPALON	R
KOROSEAL	
NATURAL RUBBER (GRS)	R
NEOPRENE GR-M (CR)	R
NITRILE BUNA N (NBR)	R
NORDEL	
POLYBUTADIENE	
SILICONE	
POLYURETHANE	

TEMP °F: 60, 80, 100, 120, 140, 160, 180, 200, 220, 240, 260, 280, 300, 320, 340, 360, 380, 400, 420, 440, 460

GREEN LIQUOR

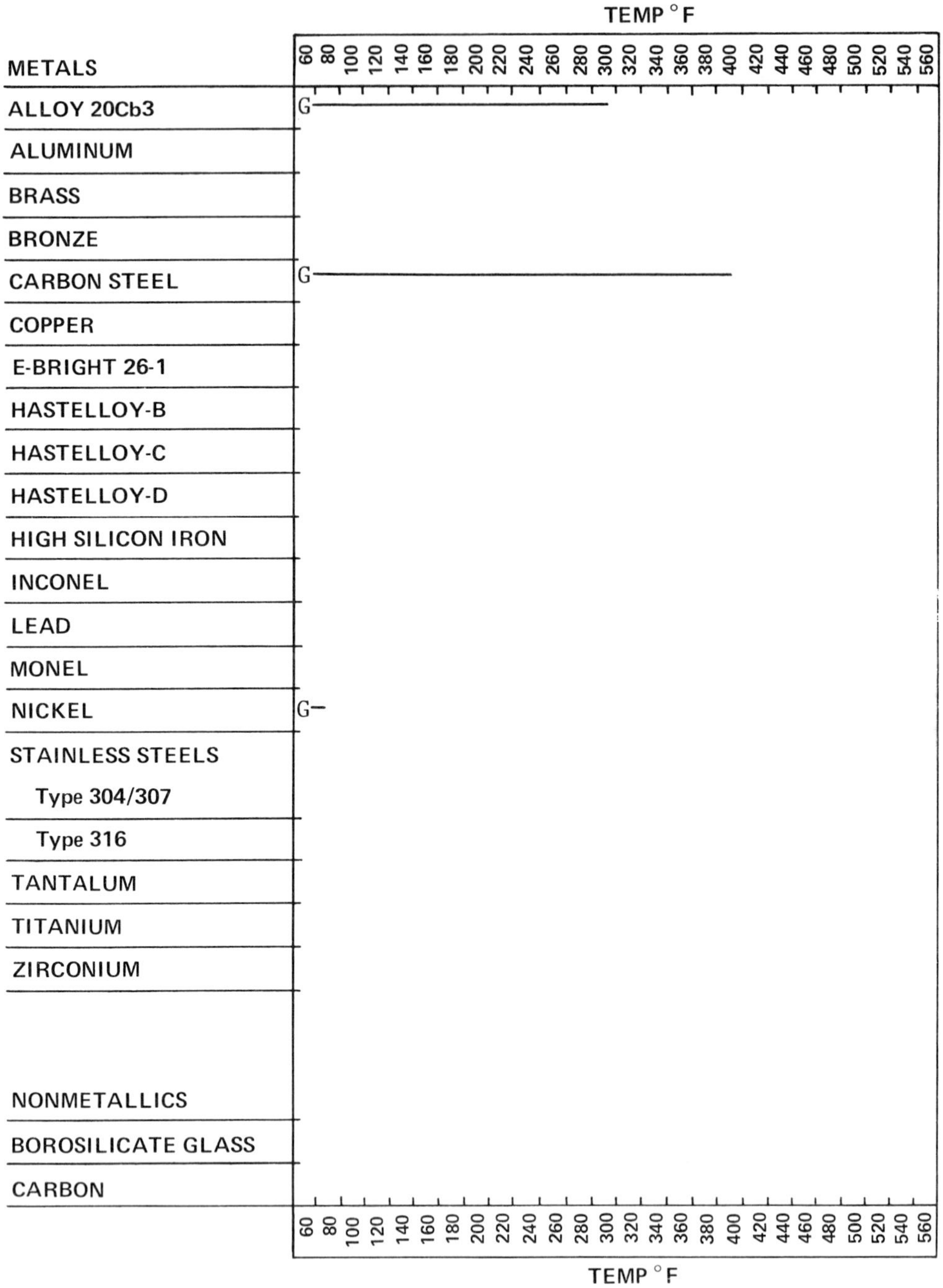

FOR METALS

E = < 2 Mils Penetration/Year; G = < 20 Mils Penetration/Year

S = < 50 Mils Penetration/Year; U = > 50 Mils Penetration/Year

FOR NONMETALLICS

R = Resistant

U = Unsatisfactory

GREEN LIQUOR

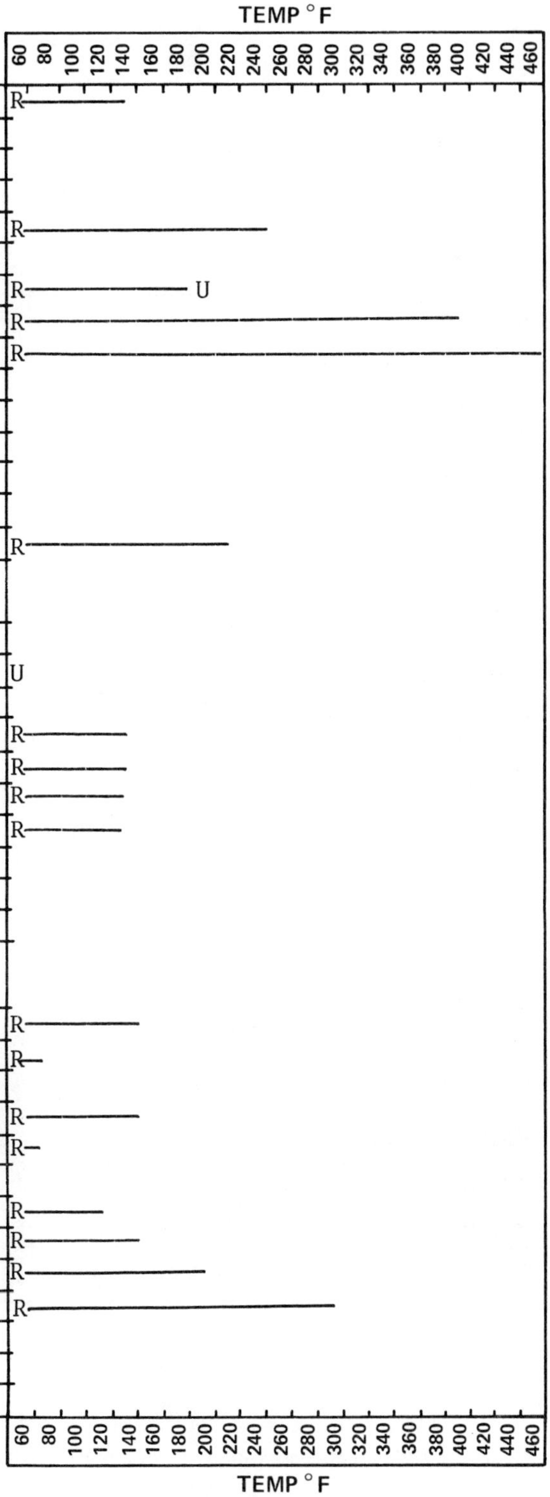

PLASTICS	
ABS	R
ACRYLICS	
ASBESTOS REINF., EPOXY	
ASBESTOS REINF., PHENOLIC	
CHLORINATED POLYETHER	R
CPVC	
EPOXY	R — U
FLUOROCARBONS FEP	R
FLUOROCARBONS TFE	R
FURFURYL ALCOHOL	
NORYL	
PHENOLIC	
POLYAMIDES—NYLON	
POLYESTERS	
BISPHENOL A-FUMURATE	R
HYDROGENATED BISPHENOL-A-BISPHENOL-A	
ISOPHTHALIC	
CHLORINATED POLYESTERS	U
POLYSTYRENES	
POLYETHYLENE	R
POLYPROPYLENE	R
PVC—TYPE 1	R
PVC—TYPE 2	R
POLYVINYLIDENE CHLORIDE	
VINYLIDENE FLUORIDE	
VINYL ESTER	
RUBBERS	
BUTYL GR-1	R
FLUORO ELASTOMERS VITON A	R
FLUORO ELASTOMERS KEL-F 3700	
HARD RUBBER	R
HYPALON	R
KOROSEAL	
NATURAL RUBBER (GRS)	R
NEOPRENE GR-M (CR)	R
NITRILE BUNA N (NBR)	R
NORDEL	R
POLYBUTADIENE	
SILICONE	
POLYURETHANE	

HEPTANE

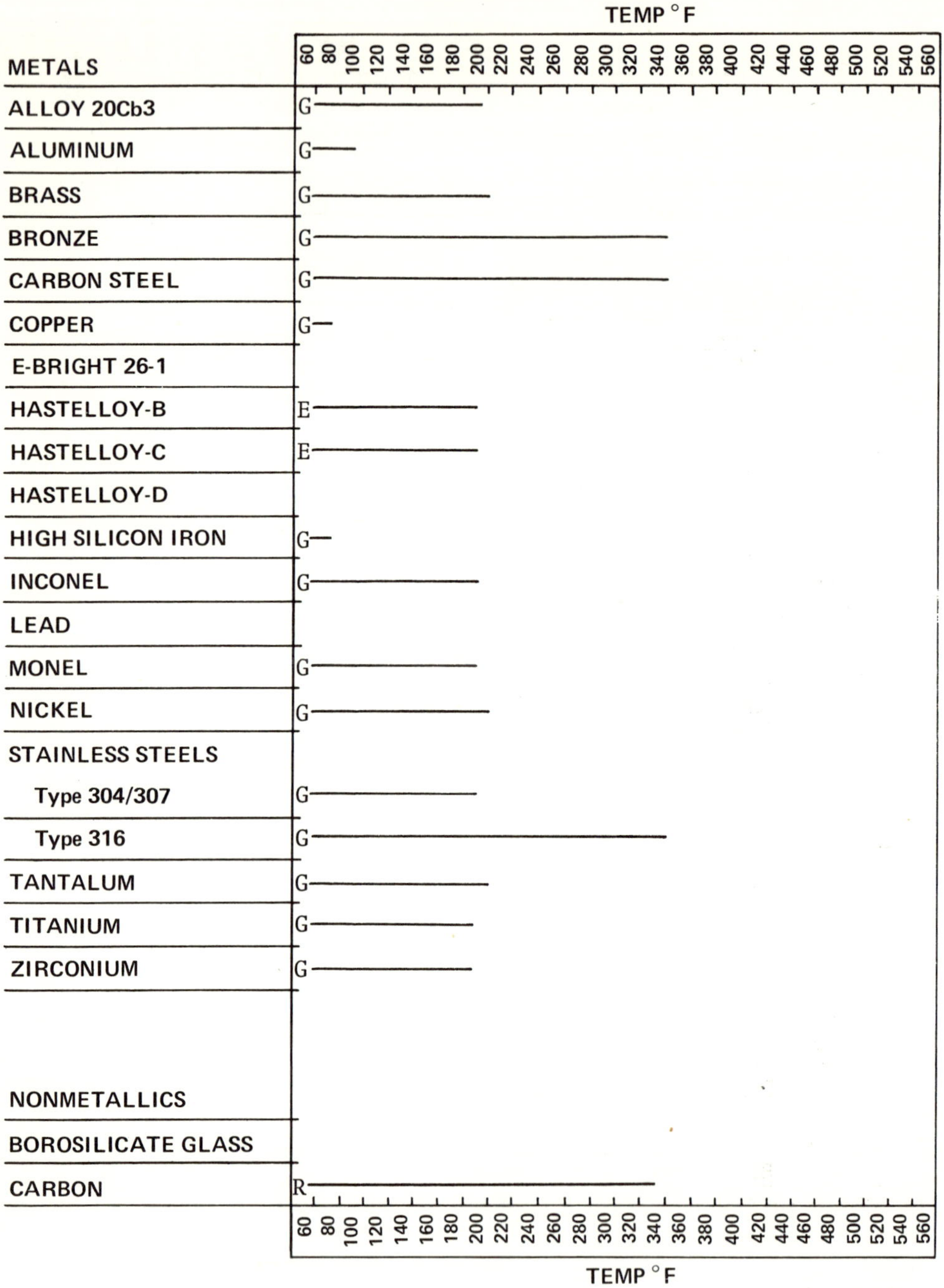

FOR METALS

E = < 2 Mils Penetration/Year; G = < 20 Mils Penetration/Year

S = < 50 Mils Penetration/Year; U = > 50 Mils Penetration/Year

FOR NONMETALLICS

R = Resistant

U = Unsatisfactory

Please see page 1 for notes on calibration.

HEPTANE

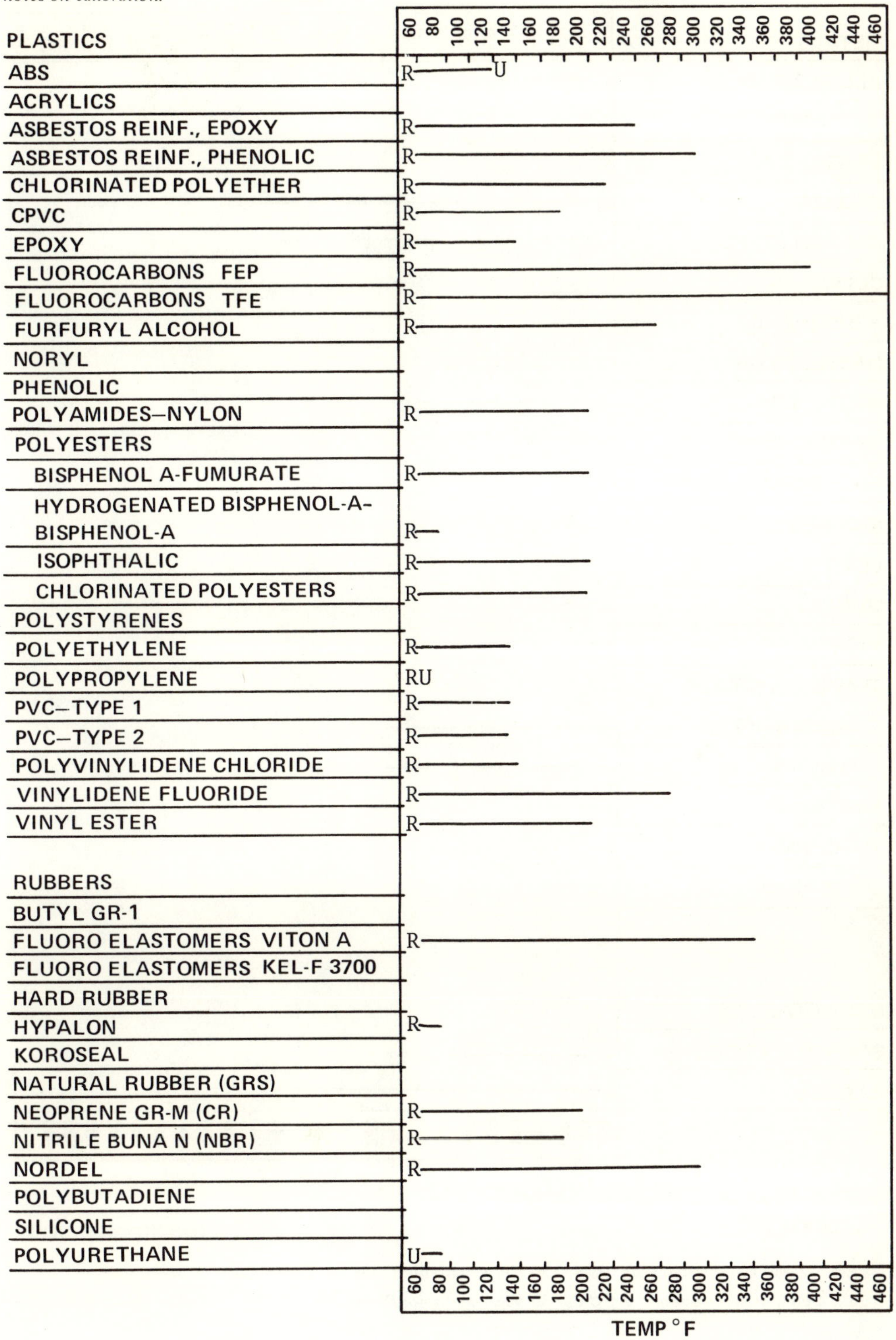

HEXANE

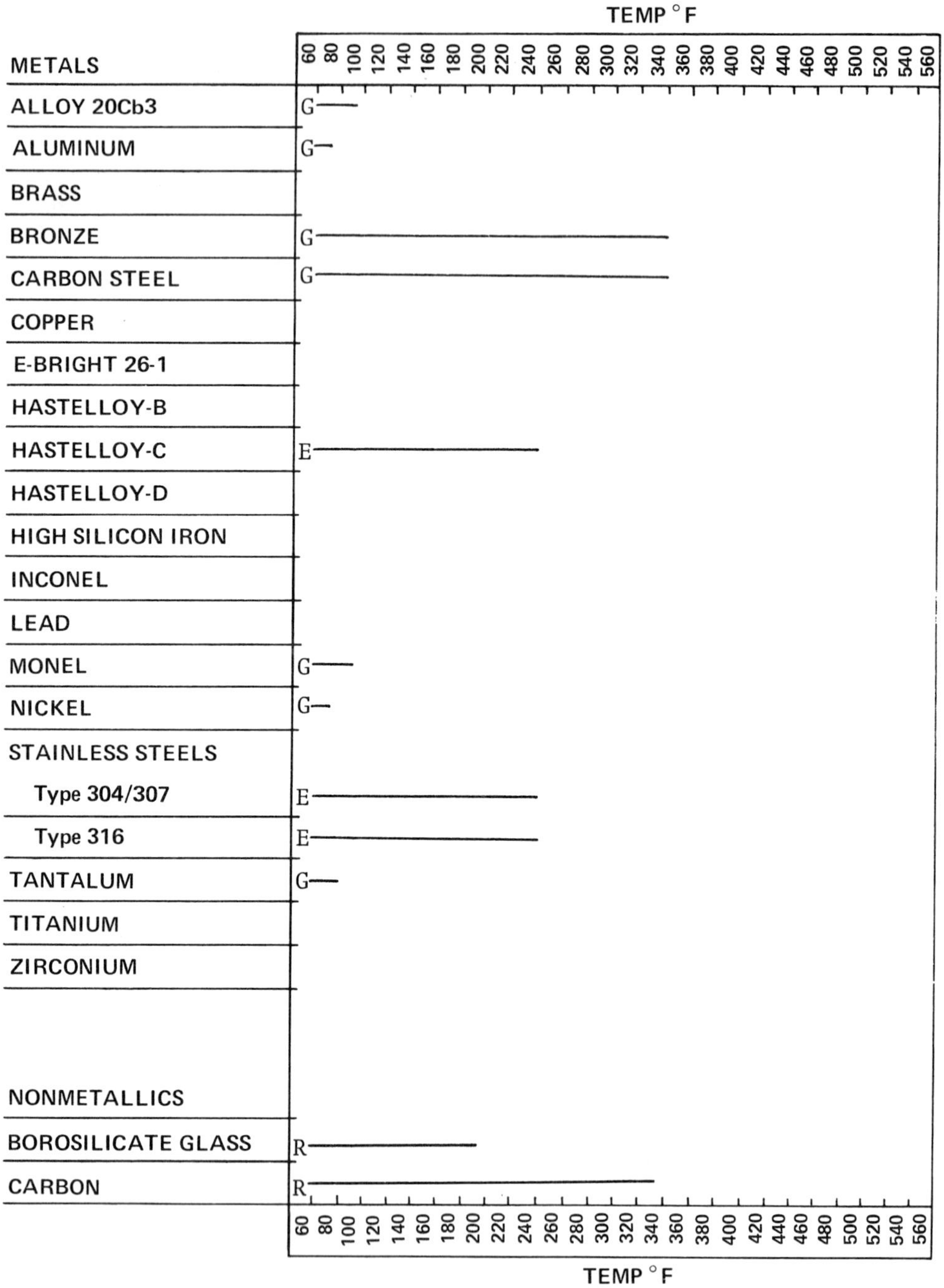

FOR METALS

E = < 2 Mils Penetration/Year; G = < 20 Mils Penetration/Year

S = < 50 Mils Penetration/Year; U = > 50 Mils Penetration/Year

FOR NONMETALLICS

R = Resistant

U = Unsatisfactory

HEXANE

TEMP °F (60–460)

PLASTICS	HEXANE
ABS	U
ACRYLICS	
ASBESTOS REINF., EPOXY	R———
ASBESTOS REINF., PHENOLIC	R———
CHLORINATED POLYETHER	R———
CPVC	R——— U
EPOXY	R———
FLUOROCARBONS FEP	R———
FLUOROCARBONS TFE	R———
FURFURYL ALCOHOL	
NORYL	
PHENOLIC	
POLYAMIDES—NYLON	R—
POLYESTERS	
BISPHENOL A-FUMURATE	R—
HYDROGENATED BISPHENOL-A-BISPHENOL-A	
ISOPHTHALIC	R——
CHLORINATED POLYESTERS	
POLYSTYRENES	
POLYETHYLENE	U
POLYPROPYLENE	R——— U
PVC—TYPE 1	RU
PVC—TYPE 2	RU
POLYVINYLIDENE CHLORIDE	R———
VINYLIDENE FLUORIDE	R———
VINYL ESTER	R———
RUBBERS	
BUTYL GR-1	U
FLUORO ELASTOMERS VITON A	R———
FLUORO ELASTOMERS KEL-F 3700	
HARD RUBBER	
HYPALON	R—
KOROSEAL	
NATURAL RUBBER (GRS)	U
NEOPRENE GR-M (CR)	R—
NITRILE BUNA N (NBR)	R———
NORDEL	U
POLYBUTADIENE	
SILICONE	
POLYURETHANE	R——

TEMP °F (60–460)

HEXANOL, TERTIARY

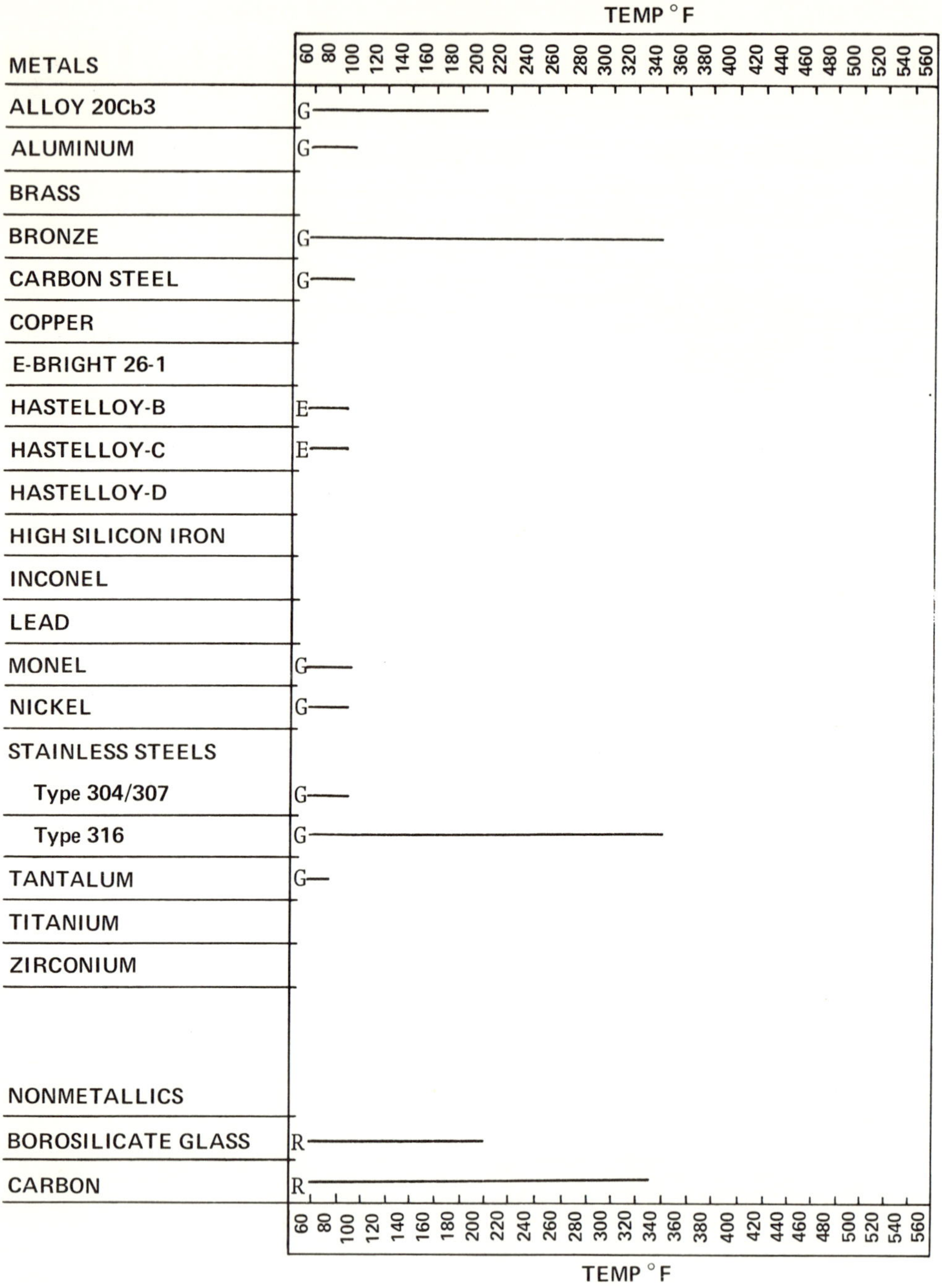

FOR METALS

E = < 2 Mils Penetration/Year; G = < 20 Mils Penetration/Year

S = < 50 Mils Penetration/Year; U = > 50 Mils Penetration/Year

FOR NONMETALLICS

R = Resistant

U = Unsatisfactory

HEXANOL, TERTIARY

TEMP °F (60–460)

PLASTICS	
ABS	U (60)
ACRYLICS	
ASBESTOS REINF., EPOXY	
ASBESTOS REINF., PHENOLIC	R (60 to 200) — U (220)
CHLORINATED POLYETHER	R (60 to 200) — U (220)
CPVC	R (60 to 140) — U (160)
EPOXY	
FLUOROCARBONS FEP	R (60 to 400)
FLUOROCARBONS TFE	R (60 to 460)
FURFURYL ALCOHOL	
NORYL	
PHENOLIC	
POLYAMIDES–NYLON	
POLYESTERS	
BISPHENOL A-FUMURATE	
HYDROGENATED BISPHENOL-A-BISPHENOL-A	
ISOPHTHALIC	
CHLORINATED POLYESTERS	
POLYSTYRENES	
POLYETHYLENE	R (60 to 160)
POLYPROPYLENE	R (60 to 140)
PVC–TYPE 1	R (60 to 140)
PVC–TYPE 2	U (60)
POLYVINYLIDENE CHLORIDE	
VINYLIDENE FLUORIDE	
VINYL ESTER	

RUBBERS	
BUTYL GR-1	U (60)
FLUORO ELASTOMERS VITON A	R (60 to 360)
FLUORO ELASTOMERS KEL-F 3700	
HARD RUBBER	
HYPALON	U (60)
KOROSEAL	
NATURAL RUBBER (GRS)	U (60)
NEOPRENE GR-M (CR)	U (60)
NITRILE BUNA N (NBR)	R (60 to 200)
NORDEL	
POLYBUTADIENE	
SILICONE	
POLYURETHANE	U (60)

TEMP °F

HYDROBROMIC ACID, 50%

TEMP °F (60–560)

METALS	
ALLOY 20Cb3	U
ALUMINUM	U
BRASS	U
BRONZE	U
CARBON STEEL	U
COPPER	U
E-BRIGHT 26-1	
HASTELLOY-B	G
HASTELLOY-C	
HASTELLOY-D	
HIGH SILICON IRON	U
INCONEL	U
LEAD	U
MONEL	U
NICKEL	U
STAINLESS STEELS	
Type 304/307	U
Type 316	U
TANTALUM	E
TITANIUM	E
ZIRCONIUM	U

NONMETALLICS	
BOROSILICATE GLASS	R
CARBON	R

TEMP °F

FOR METALS

E = < 2 Mils Penetration/Year; G = < 20 Mils Penetration/Year

S = < 50 Mils Penetration/Year; U = > 50 Mils Penetration/Year

FOR NONMETALLICS

R = Resistant

U = Unsatisfactory

HYDROBROMIC ACID, 50%

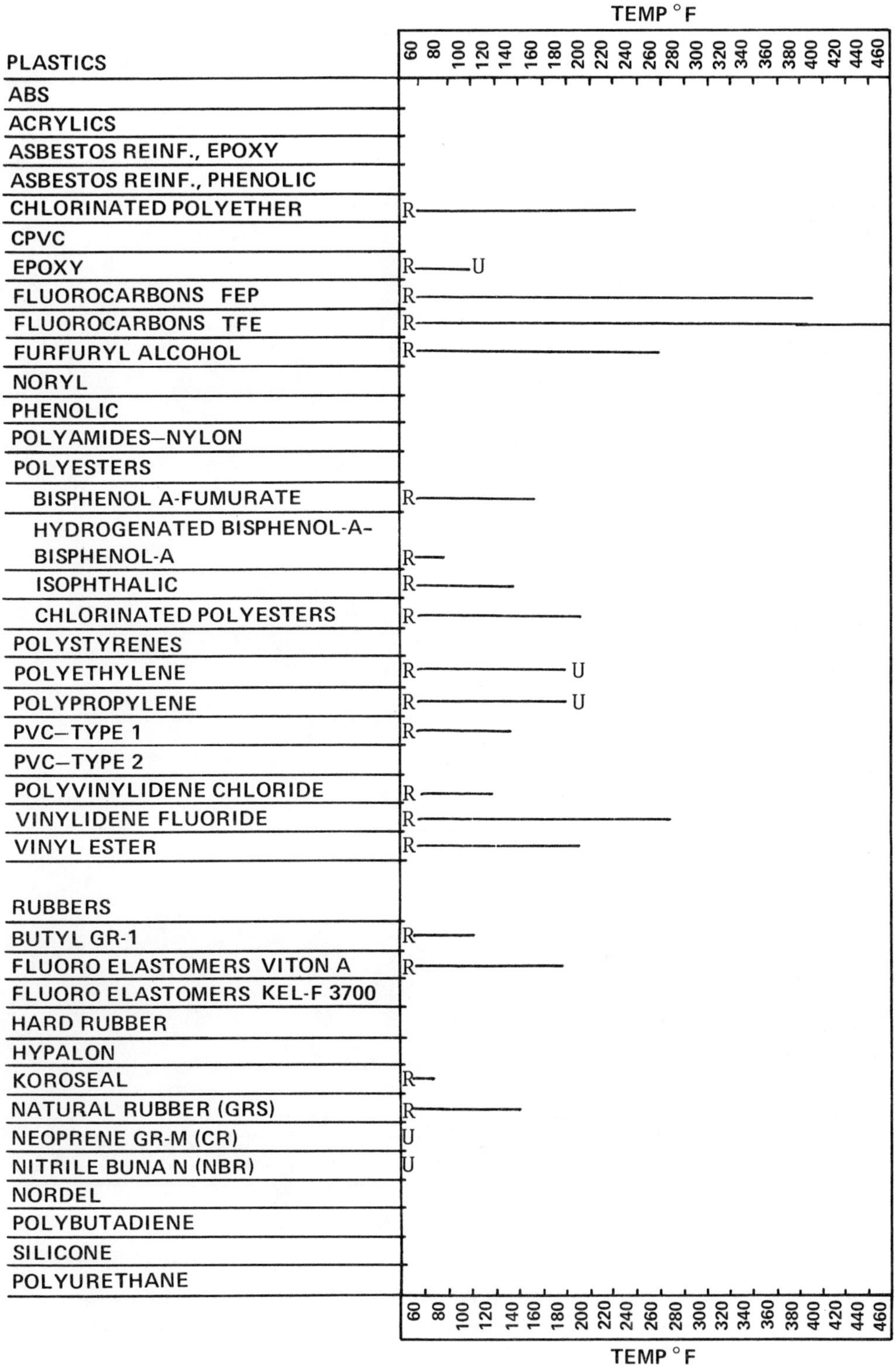

TEMP °F: 60 80 100 120 140 160 180 200 220 240 260 280 300 320 340 360 380 400 420 440 460 480 500 520 540 560

METALS	
ALLOY 20Cb3	U
ALUMINUM	U
BRASS	U
BRONZE	U
CARBON STEEL	U
COPPER	U
E-BRIGHT 26-1	
HASTELLOY-B	G
HASTELLOY-C	
HASTELLOY-D	
HIGH SILICON IRON	
INCONEL	G–U
LEAD	U
MONEL	U
NICKEL	U
STAINLESS STEELS	
Type 304/307	U
Type 316	U
TANTALUM	E
TITANIUM	E
ZIRCONIUM	U

NONMETALLICS	
BOROSILICATE GLASS	R
CARBON	R

TEMP °F: 60 80 100 120 140 160 180 200 220 240 260 280 300 320 340 360 380 400 420 440 460 480 500 520 540 560

FOR METALS

E = < 2 Mils Penetration/Year; G = < 20 Mils Penetration/Year
S = < 50 Mils Penetration/Year; U = > 50 Mils Penetration/Year

FOR NONMETALLICS

R = Resistant
U = Unsatisfactory

HYDROBROMIC ACID, 20%

PLASTICS	TEMP °F
ABS	R
ACRYLICS	
ASBESTOS REINF., EPOXY	
ASBESTOS REINF., PHENOLIC	
CHLORINATED POLYETHER	R
CPVC	R
EPOXY	R
FLUOROCARBONS FEP	R
FLUOROCARBONS TFE	R
FURFURYL ALCOHOL	R
NORYL	R
PHENOLIC	
POLYAMIDES–NYLON	
POLYESTERS	
BISPHENOL A-FUMURATE	R
HYDROGENATED BISPHENOL-A-BISPHENOL-A	R
ISOPHTHALIC	R
CHLORINATED POLYESTERS	
POLYSTYRENES	
POLYETHYLENE	R
POLYPROPYLENE	R
PVC–TYPE 1	R
PVC–TYPE 2	R
POLYVINYLIDENE CHLORIDE	R
VINYLIDENE FLUORIDE	R
VINYL ESTER	R

RUBBERS	TEMP °F
BUTYL GR-1	R
FLUORO ELASTOMERS VITON A	R
FLUORO ELASTOMERS KEL-F 3700	
HARD RUBBER	
HYPALON	R
KOROSEAL	R
NATURAL RUBBER (GRS)	R
NEOPRENE GR-M (CR)	U
NITRILE BUNA N (NBR)	U
NORDEL	
POLYBUTADIENE	
SILICONE	
POLYURETHANE	

HYDROBROMIC ACID DILUTE

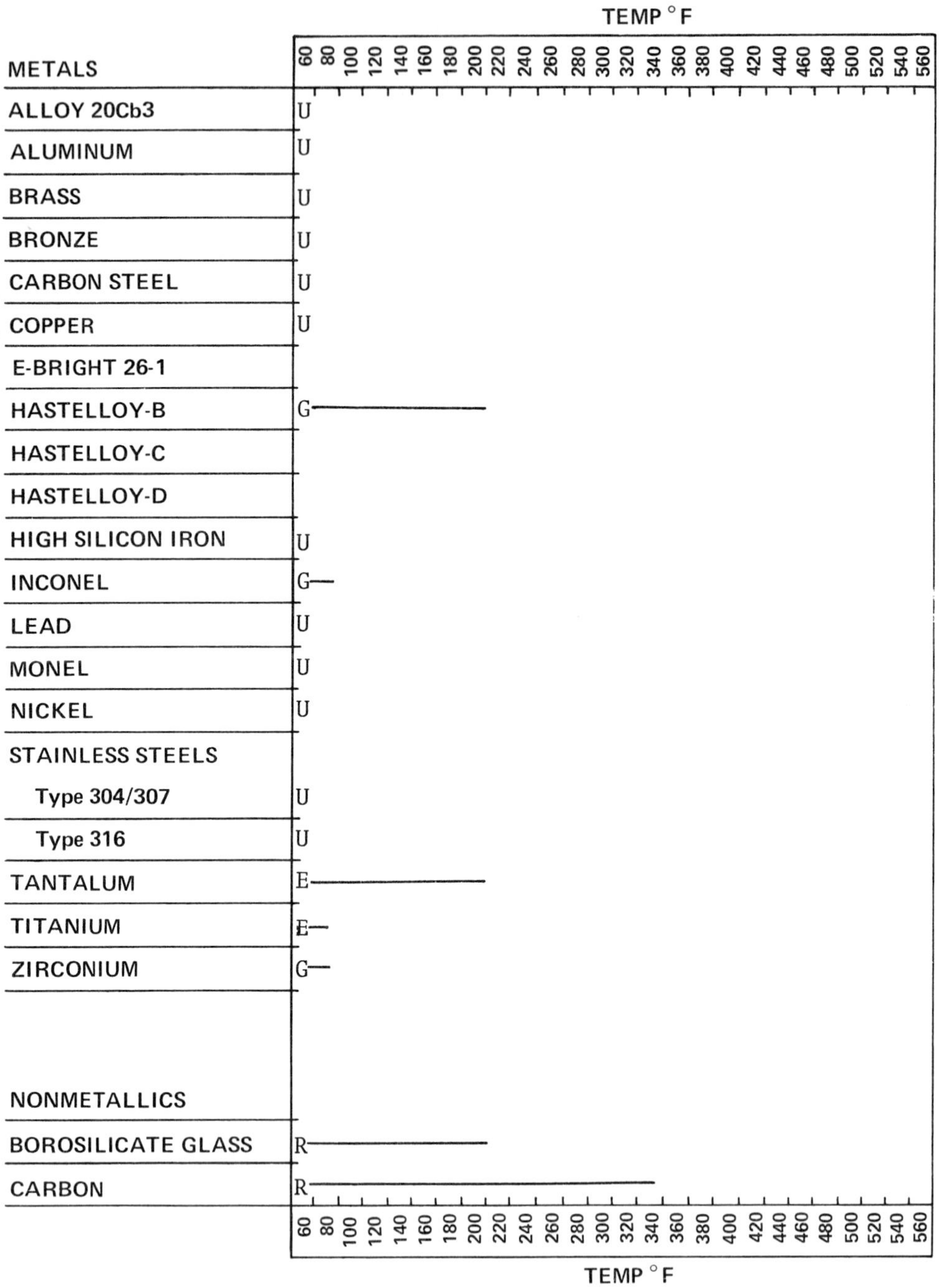

FOR METALS

E = < 2 Mils Penetration/Year; G = < 20 Mils Penetration/Year

S = < 50 Mils Penetration/Year; U = > 50 Mils Penetration/Year

FOR NONMETALLICS

R = Resistant

U = Unsatisfactory

HYDROBROMIC ACID DILUTE

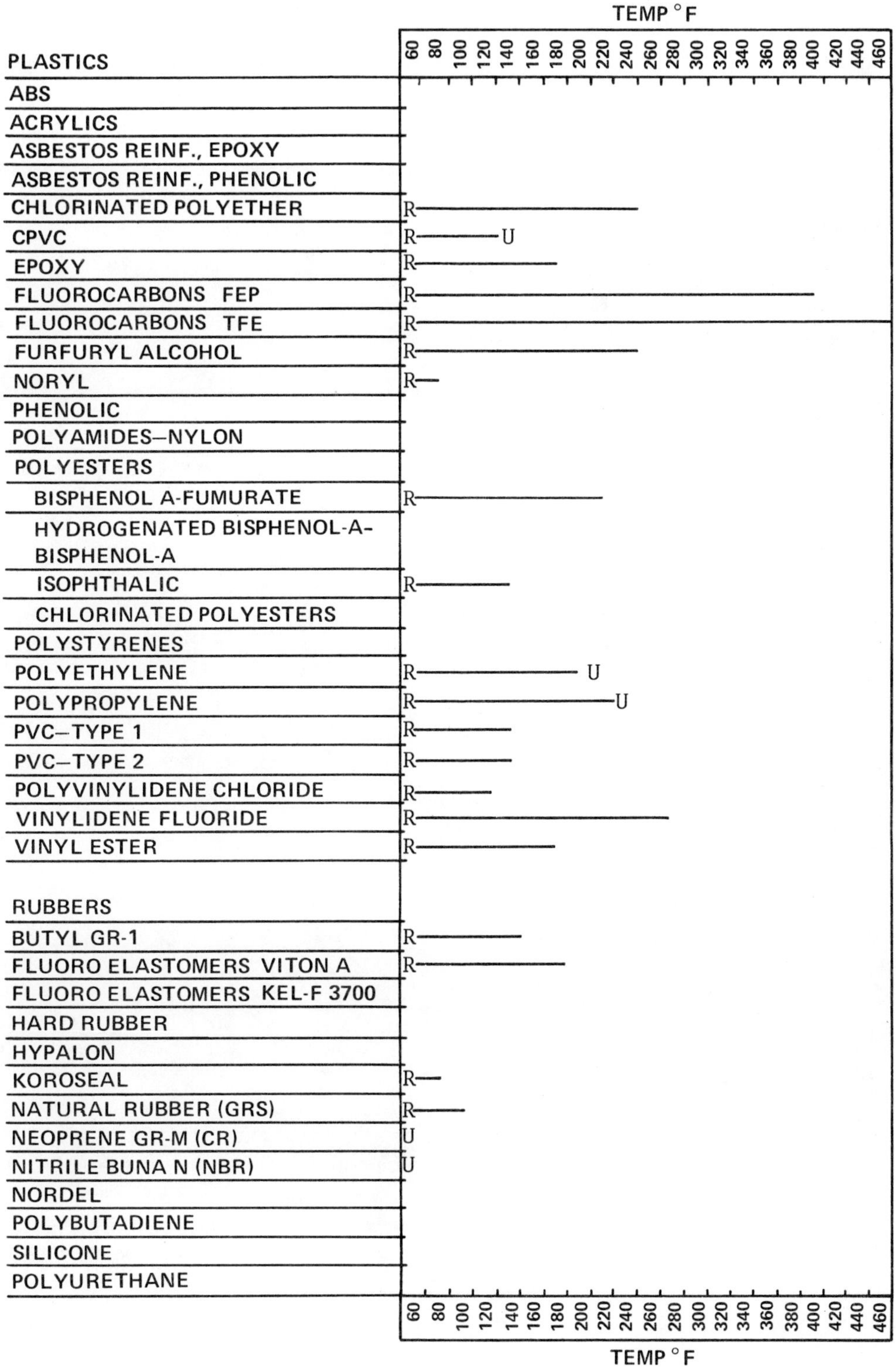

HYDROCHLORIC ACID, 50%

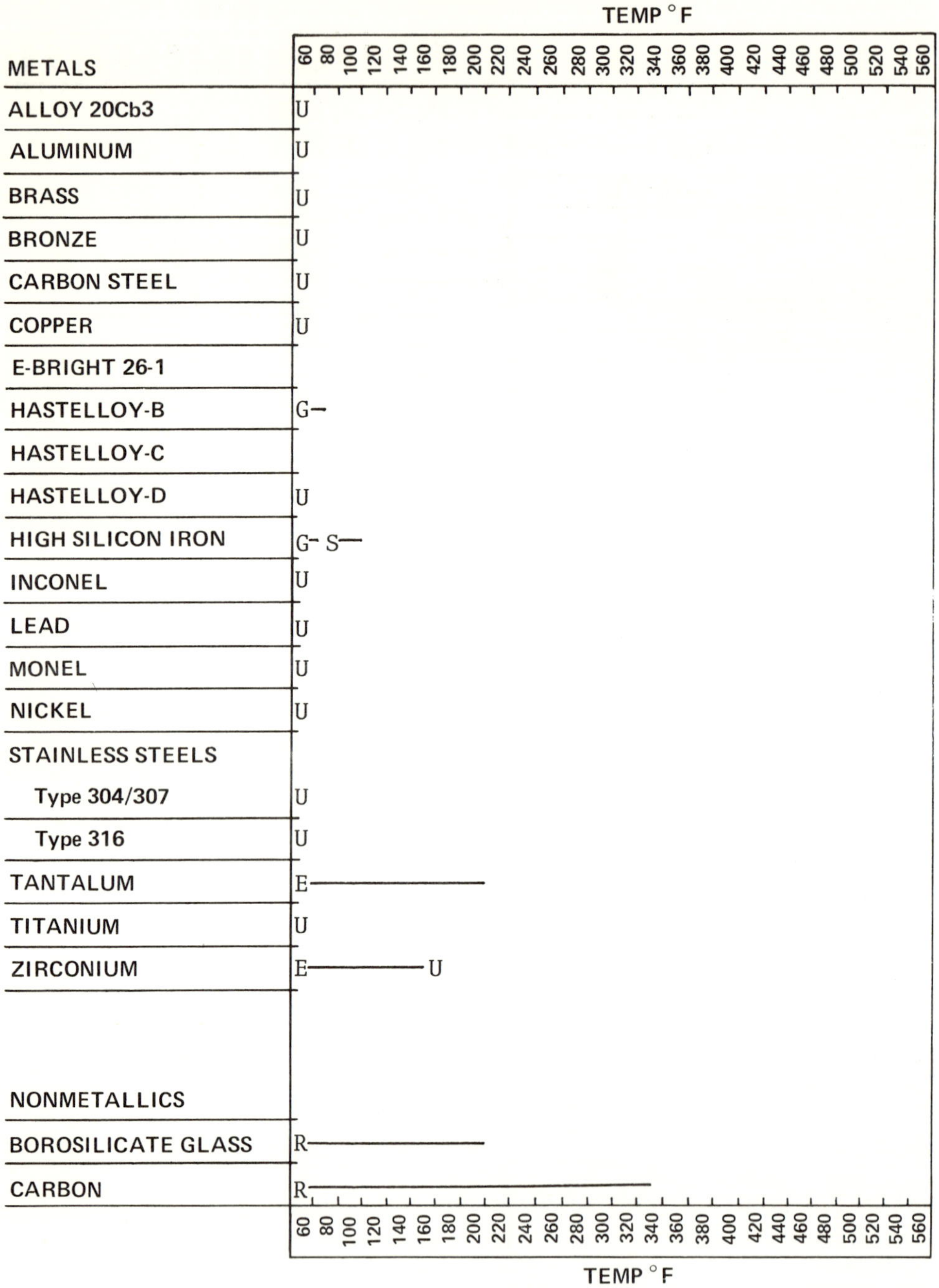

FOR METALS

E = < 2 Mils Penetration/Year; G = < 20 Mils Penetration/Year

S = < 50 Mils Penetration/Year; U = > 50 Mils Penetration/Year

FOR NONMETALLICS

R = Resistant

U = Unsatisfactory

HYDROCHLORIC ACID, 50%

TEMP °F (60–460)

PLASTICS	Rating
ABS	R——U
ACRYLICS	R—
ASBESTOS REINF., EPOXY	R—
ASBESTOS REINF., PHENOLIC	R—
CHLORINATED POLYETHER	R——
CPVC	
EPOXY	R——
FLUOROCARBONS FEP	R——
FLUOROCARBONS TFE	R——
FURFURYL ALCOHOL	R—
NORYL	
PHENOLIC	R—
POLYAMIDES–NYLON	U
POLYESTERS	
BISPHENOL A-FUMURATE	
HYDROGENATED BISPHENOL-A-BISPHENOL-A	
ISOPHTHALIC	
CHLORINATED POLYESTERS	
POLYSTYRENES	U
POLYETHYLENE	R——
POLYPROPYLENE	R——U
PVC–TYPE 1	R——
PVC–TYPE 2	R——
POLYVINYLIDENE CHLORIDE	R——
VINYLIDENE FLUORIDE	R——
VINYL ESTER	

RUBBERS	Rating
BUTYL GR-1	
FLUORO ELASTOMERS VITON A	R—
FLUORO ELASTOMERS KEL-F 3700	R—
HARD RUBBER	R——
HYPALON	R—
KOROSEAL	R—
NATURAL RUBBER (GRS)	R—
NEOPRENE GR-M (CR)	U
NITRILE BUNA N (NBR)	R——
NORDEL	R——
POLYBUTADIENE	R—
SILICONE	U
POLYURETHANE	U

TEMP °F (60–460)

HYDROCHLORIC ACID, 38%

TEMP °F: 60 80 100 120 140 160 180 200 220 240 260 280 300 320 340 360 380 400 420 440 460 480 500 520 540 560

METALS	
ALLOY 20Cb3	U
ALUMINUM	U
BRASS	U
BRONZE	U
CARBON STEEL	U
COPPER	U
E-BRIGHT 26-1	
HASTELLOY-B	E——U
HASTELLOY-C	E—S—U
HASTELLOY-D	S—U
HIGH SILICON IRON	G—S——U
INCONEL	U
LEAD	U
MONEL	U
NICKEL	U
STAINLESS STEELS	
Type 304/307	U
Type 316	U
TANTALUM	E——————
TITANIUM	U
ZIRCONIUM	E———U

NONMETALLICS	
BOROSILICATE GLASS	R——————
CARBON	R——————————

TEMP °F: 60 80 100 120 140 160 180 200 220 240 260 280 300 320 340 360 380 400 420 440 460 480 500 520 540 560

FOR METALS

E = < 2 Mils Penetration/Year; G = < 20 Mils Penetration/Year

S = < 50 Mils Penetration/Year; U = > 50 Mils Penetration/Year

FOR NONMETALLICS

R = Resistant

U = Unsatisfactory

HYDROCHLORIC ACID, 38%

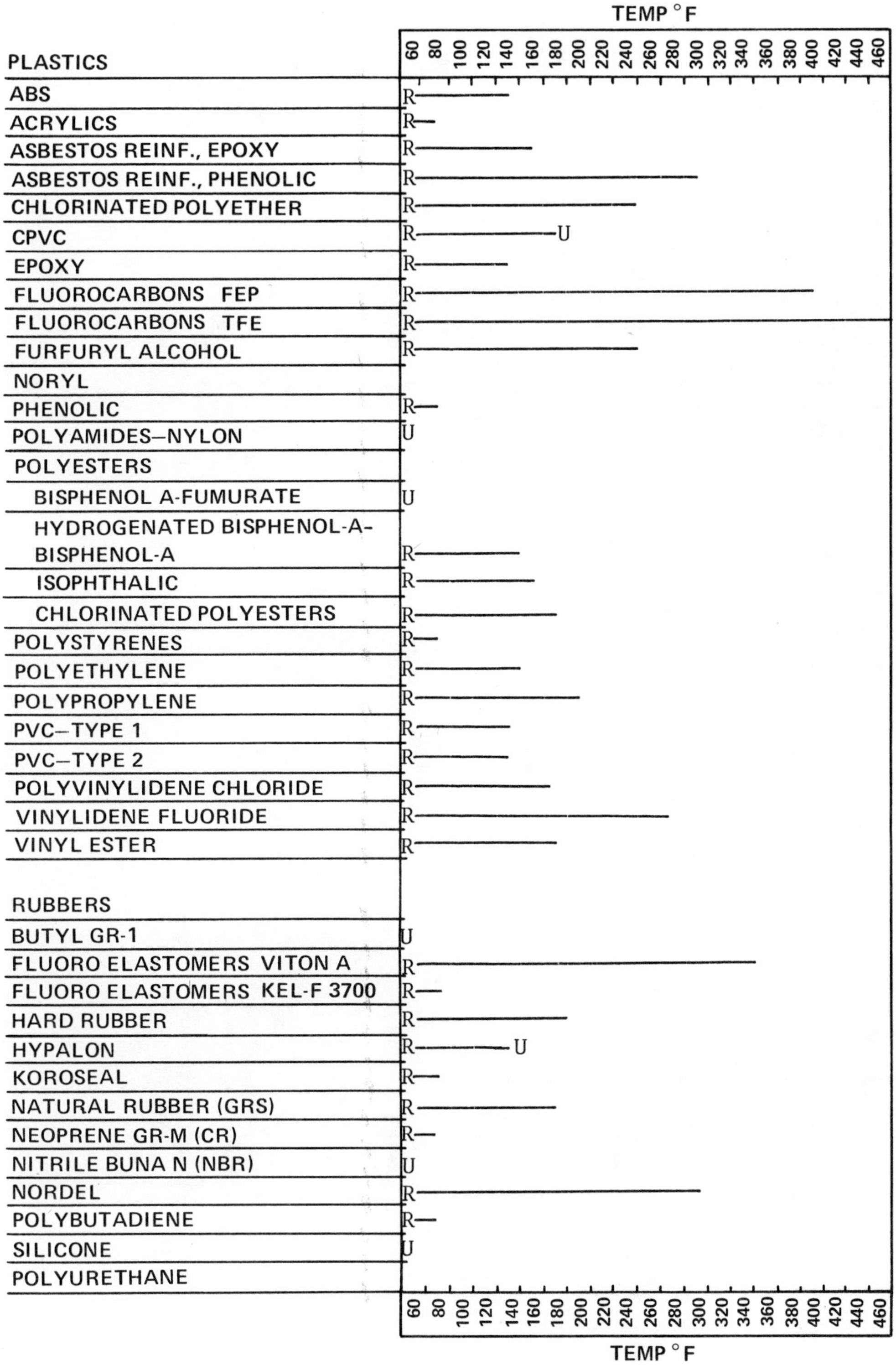

HYDROCHLORIC ACID, 35%

TEMP °F (60, 80, 100, 120, 140, 160, 180, 200, 220, 240, 260, 280, 300, 320, 340, 360, 380, 400, 420, 440, 460, 480, 500, 520, 540, 560)

METALS	
ALLOY 20Cb3	U
ALUMINUM	U
BRASS	U
BRONZE	U
CARBON STEEL	U
COPPER	U
E-BRIGHT 26-1	
HASTELLOY-B	E—
HASTELLOY-C	E—G—S
HASTELLOY-D	U
HIGH SILICON IRON	G—S—U
INCONEL	U
LEAD	U
MONEL	U
NICKEL	U
STAINLESS STEELS Type 304/307	U
Type 316	U
TANTALUM	E—
TITANIUM	U
ZIRCONIUM	E—

NONMETALLICS	
BOROSILICATE GLASS	R—
CARBON	R—

TEMP °F

FOR METALS

E = < 2 Mils Penetration/Year; G = < 20 Mils Penetration/Year

S = < 50 Mils Penetration/Year; U = > 50 Mils Penetration/Year

FOR NONMETALLICS

R = Resistant

U = Unsatisfactory

HYDROCHLORIC ACID, 35%

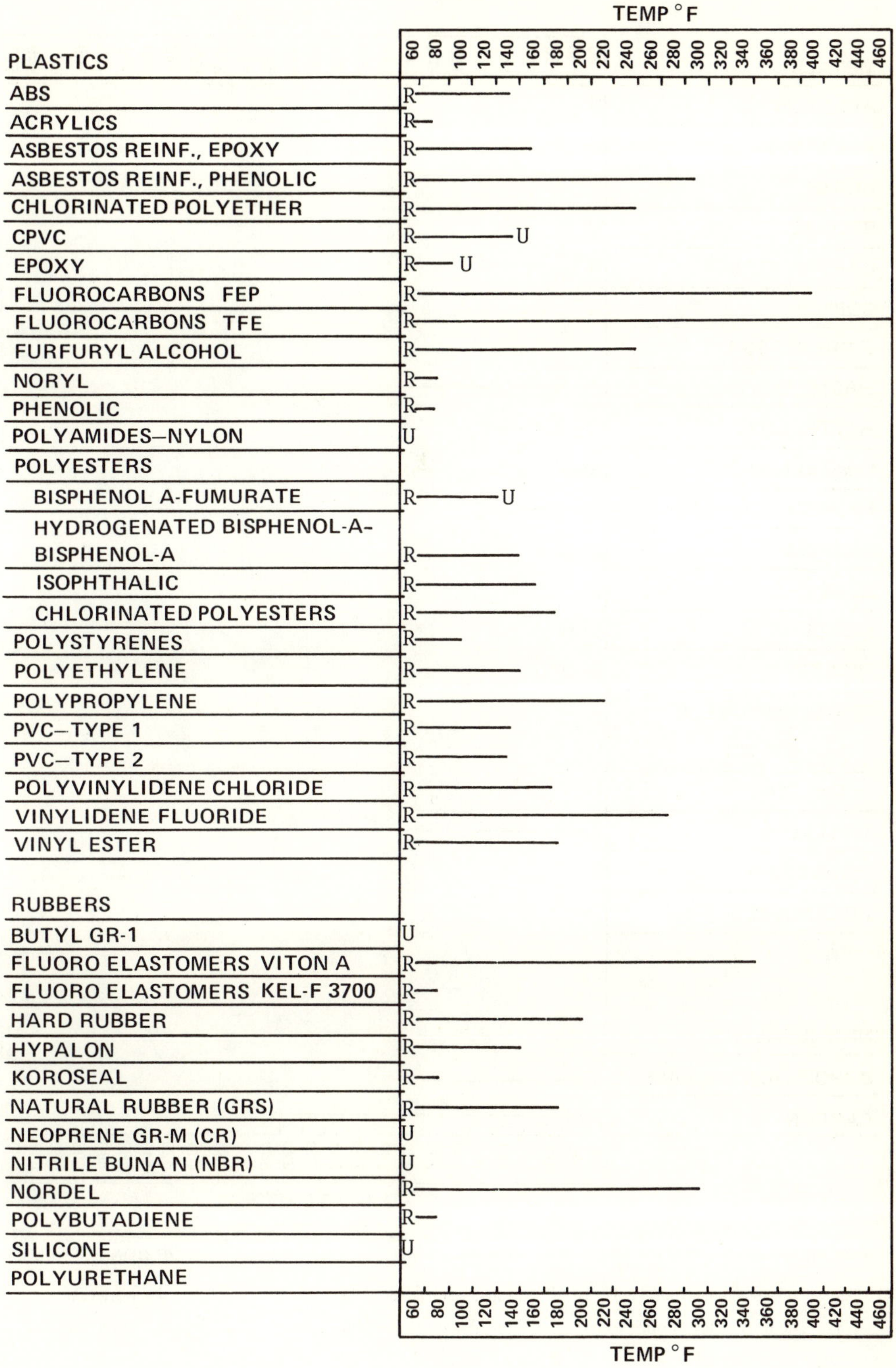

TEMP °F: 60, 80, 100, 120, 140, 160, 180, 200, 220, 240, 260, 280, 300, 320, 340, 360, 380, 400, 420, 440, 460, 480, 500, 520, 540, 560

METALS	
ALLOY 20Cb3	U
ALUMINUM	U
BRASS	U
BRONZE	U
CARBON STEEL	U
COPPER	U
E-BRIGHT 26-1	
HASTELLOY-B	G—— S——
HASTELLOY-C	E—G——U
HASTELLOY-D	S—U
HIGH SILICON IRON	G—S——U
INCONEL	G—U
LEAD	U
MONEL	G- U
NICKEL	G—U
STAINLESS STEELS	
Type 304/307	U
Type 316	U
TANTALUM	E——————
TITANIUM	U
ZIRCONIUM	G——

NONMETALLICS	
BOROSILICATE GLASS	R——————
CARBON	R——————————

TEMP °F: 60, 80, 100, 120, 140, 160, 180, 200, 220, 240, 260, 280, 300, 320, 340, 360, 380, 400, 420, 440, 460, 480, 500, 520, 540, 560

FOR METALS

E = < 2 Mils Penetration/Year; G = < 20 Mils Penetration/Year

S = < 50 Mils Penetration/Year; U = > 50 Mils Penetration/Year

FOR NONMETALLICS

R = Resistant

U = Unsatisfactory

HYDROCHLORIC ACID, 20%

TEMP °F: 60 80 100 120 140 160 180 200 220 240 260 280 300 320 340 360 380 400 420 440 460

PLASTICS	
ABS	R—
ACRYLICS	R—
ASBESTOS REINF., EPOXY	R—
ASBESTOS REINF., PHENOLIC	R—
CHLORINATED POLYETHER	R—
CPVC	R—
EPOXY	R—
FLUOROCARBONS FEP	R—
FLUOROCARBONS TFE	R—
FURFURYL ALCOHOL	R—
NORYL	R—
PHENOLIC	
POLYAMIDES—NYLON	U
POLYESTERS	
BISPHENOL A-FUMURATE	R—U
HYDROGENATED BISPHENOL-A-BISPHENOL-A	R—
ISOPHTHALIC	R—
CHLORINATED POLYESTERS	R—
POLYSTYRENES	R—
POLYETHYLENE	R—
POLYPROPYLENE	R—
PVC—TYPE 1	R—
PVC—TYPE 2	R—
POLYVINYLIDENE CHLORIDE	R—
VINYLIDENE FLUORIDE	R—
VINYL ESTER	R—

RUBBERS	
BUTYL GR-1	U
FLUORO ELASTOMERS VITON A	R—
FLUORO ELASTOMERS KEL-F 3700	R—
HARD RUBBER	R—
HYPALON	R—
KOROSEAL	R—
NATURAL RUBBER (GRS)	R—
NEOPRENE GR-M (CR)	R—
NITRILE BUNA N (NBR)	R—
NORDEL	R—
POLYBUTADIENE	R—
SILICONE	R—
POLYURETHANE	

TEMP °F: 60 80 100 120 140 160 180 200 220 240 260 280 300 320 340 360 380 400 420 440 460

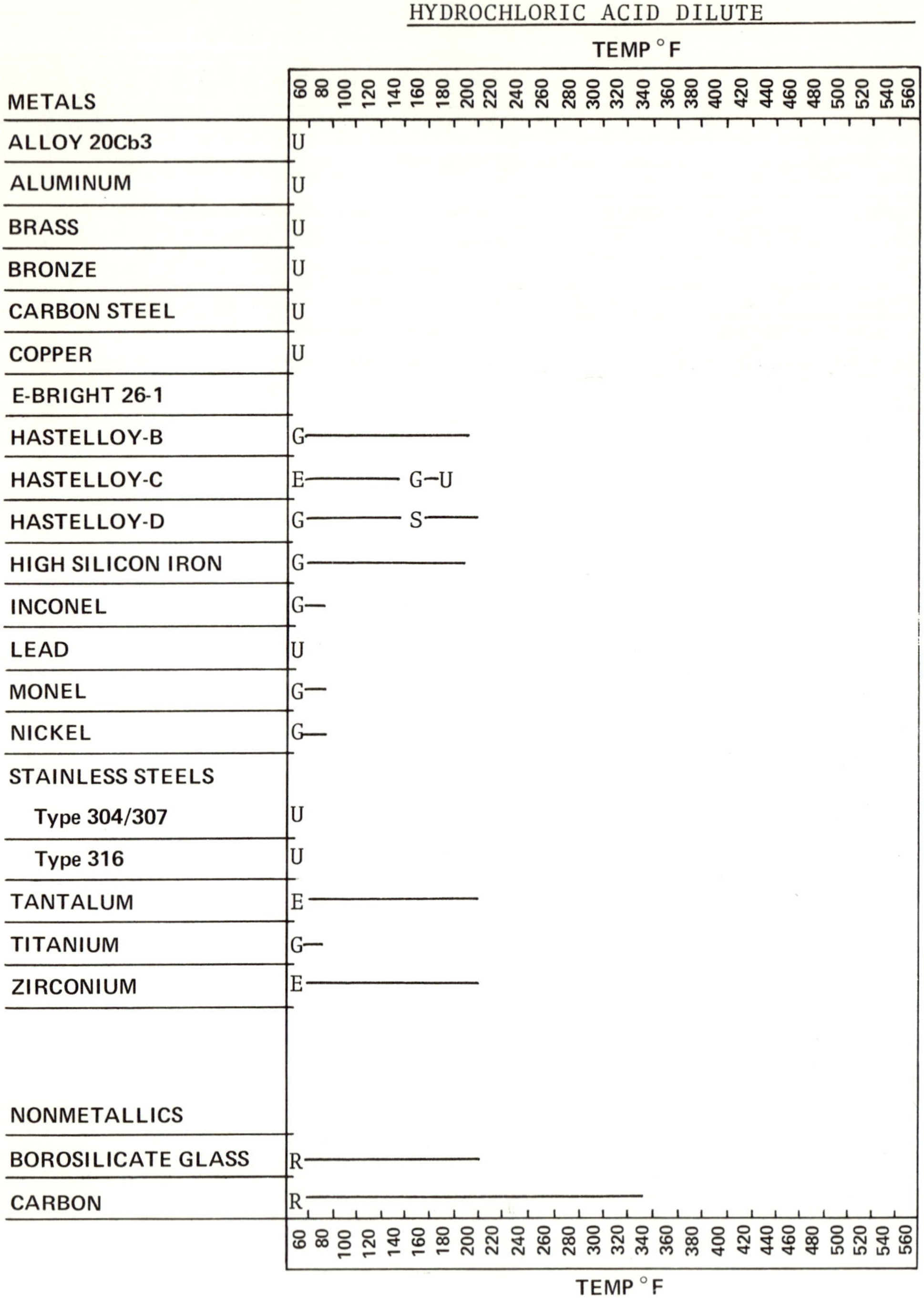

FOR METALS

E = < 2 Mils Penetration/Year; G = < 20 Mils Penetration/Year

S = < 50 Mils Penetration/Year; U = > 50 Mils Penetration/Year

FOR NONMETALLICS

R = Resistant

U = Unsatisfactory

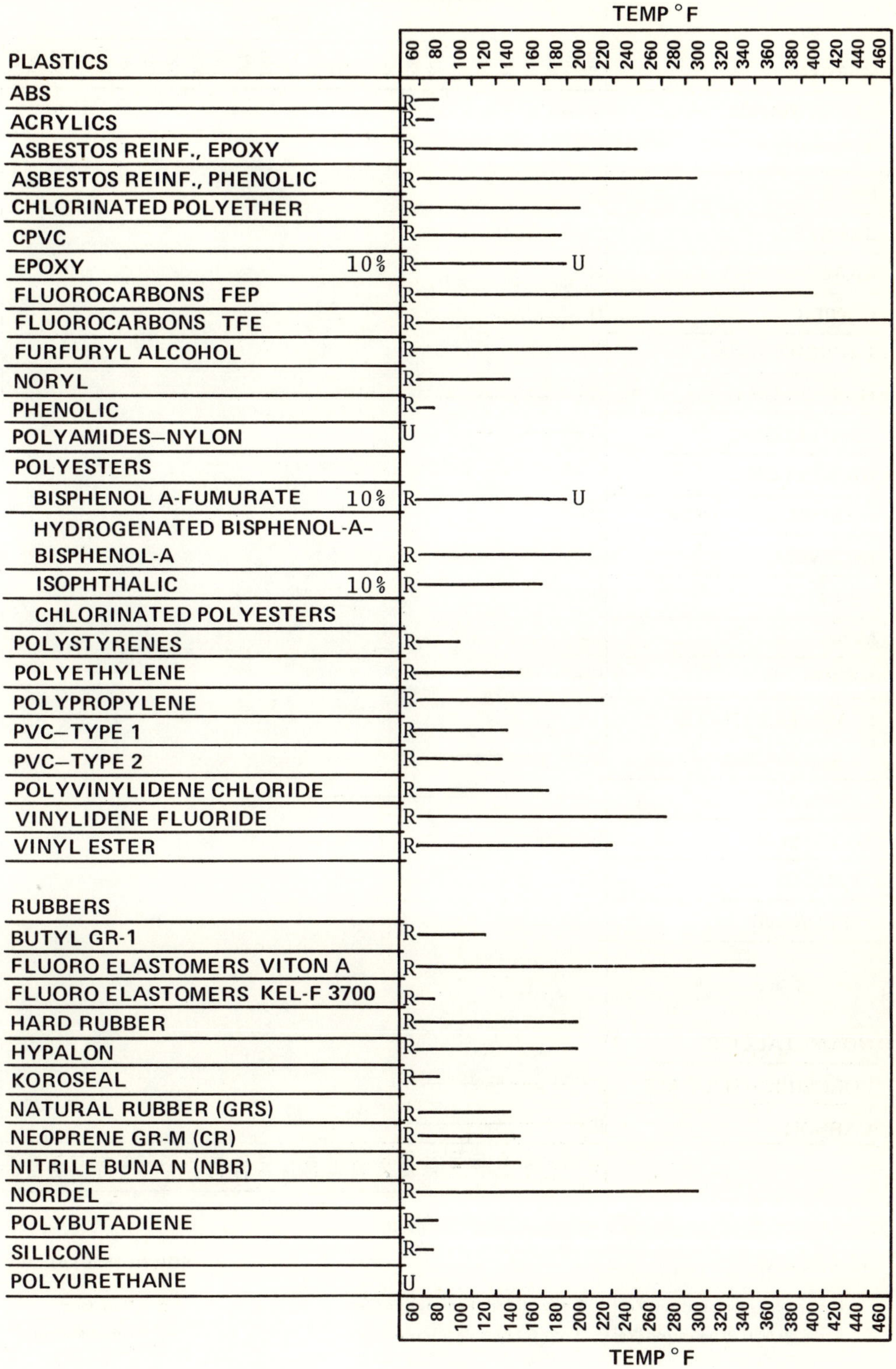
TEMP °F
60 80 100 120 140 160 180 200 220 240 260 280 300 320 340 360 380 400 420 440 460
PLASTICS
ABS R
ACRYLICS R
ASBESTOS REINF., EPOXY R
ASBESTOS REINF., PHENOLIC R
CHLORINATED POLYETHER R
CPVC R
EPOXY 10% R U
FLUOROCARBONS FEP R
FLUOROCARBONS TFE R
FURFURYL ALCOHOL R
NORYL R
PHENOLIC R
POLYAMIDES–NYLON U
POLYESTERS
BISPHENOL A-FUMURATE 10% R U
HYDROGENATED BISPHENOL-A-BISPHENOL-A R
ISOPHTHALIC 10% R
CHLORINATED POLYESTERS
POLYSTYRENES R
POLYETHYLENE R
POLYPROPYLENE R
PVC–TYPE 1 R
PVC–TYPE 2 R
POLYVINYLIDENE CHLORIDE R
VINYLIDENE FLUORIDE R
VINYL ESTER R
RUBBERS
BUTYL GR-1 R
FLUORO ELASTOMERS VITON A R
FLUORO ELASTOMERS KEL-F 3700 R
HARD RUBBER R
HYPALON R
KOROSEAL R
NATURAL RUBBER (GRS) R
NEOPRENE GR-M (CR) R
NITRILE BUNA N (NBR) R
NORDEL R
POLYBUTADIENE R
SILICONE R
POLYURETHANE U
60 80 100 120 140 160 180 200 220 240 260 280 300 320 340 360 380 400 420 440 460
TEMP °F

HYDROCYANIC ACID

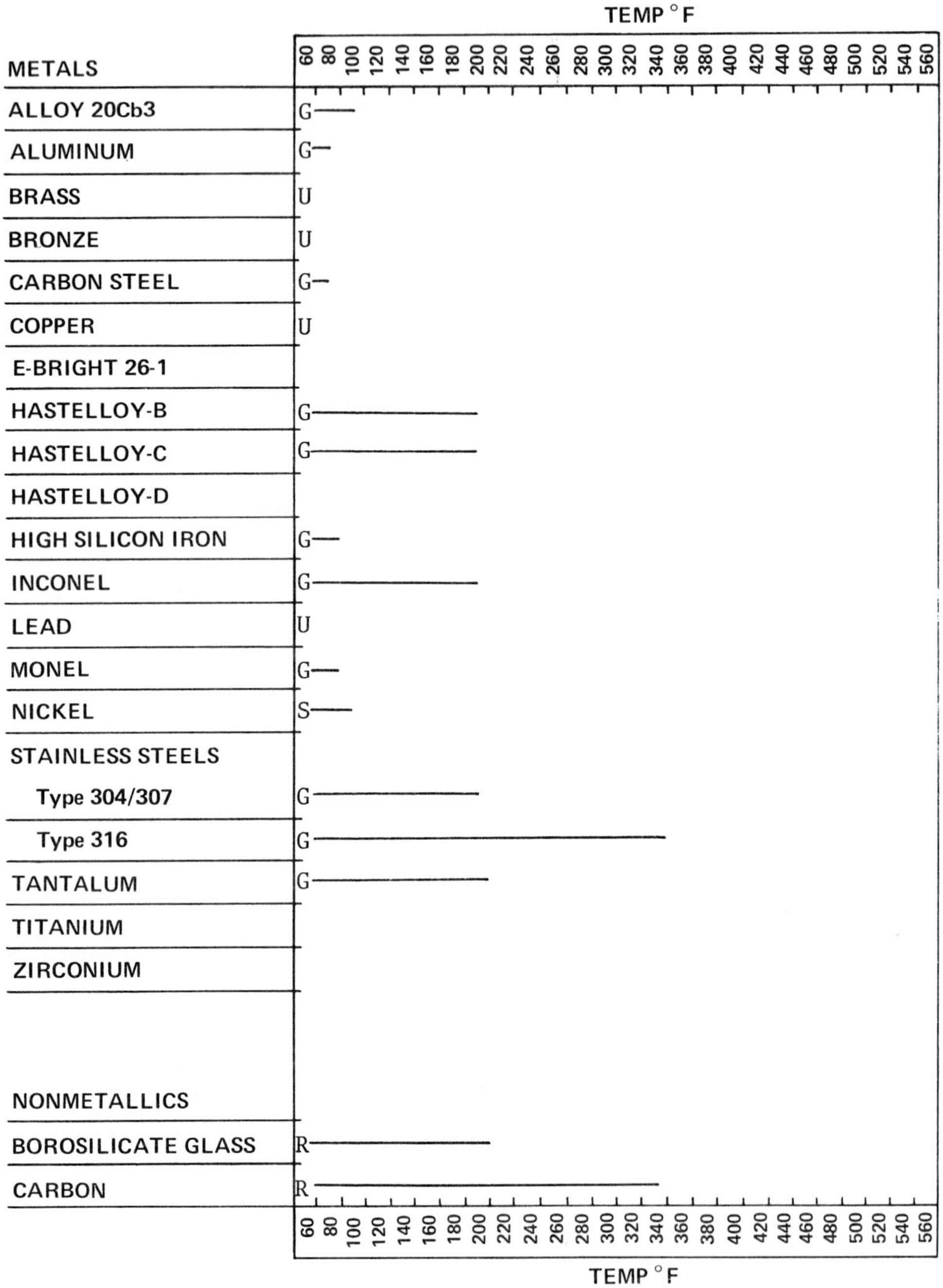

FOR METALS

E = < 2 Mils Penetration/Year; G = < 20 Mils Penetration/Year

S = < 50 Mils Penetration/Year; U = > 50 Mils Penetration/Year

FOR NONMETALLICS

R = Resistant

U = Unsatisfactory

TEMP °F (60–460)

PLASTICS		Rating
ABS		R
ACRYLICS		
ASBESTOS REINF., EPOXY		
ASBESTOS REINF., PHENOLIC		
CHLORINATED POLYETHER		R
CPVC		R
EPOXY		R
FLUOROCARBONS FEP		R
FLUOROCARBONS TFE		R
FURFURYL ALCOHOL	sat	R
NORYL		
PHENOLIC		
POLYAMIDES–NYLON		U
POLYESTERS		
BISPHENOL A-FUMURATE	sat	R
HYDROGENATED BISPHENOL-A-BISPHENOL-A		
ISOPHTHALIC		U
CHLORINATED POLYESTERS		R
POLYSTYRENES		
POLYETHYLENE		R
POLYPROPYLENE		R
PVC–TYPE 1		R
PVC–TYPE 2		R
POLYVINYLIDENE CHLORIDE		R
VINYLIDENE FLUORIDE		R
VINYL ESTER		R

RUBBERS		Rating
BUTYL GR-1		R
FLUORO ELASTOMERS VITON A		R
FLUORO ELASTOMERS KEL-F 3700		
HARD RUBBER		R
HYPALON		R
KOROSEAL		R
NATURAL RUBBER (GRS)		R
NEOPRENE GR-M (CR)		R
NITRILE BUNA N (NBR)	60%	R
NORDEL		R
POLYBUTADIENE		
SILICONE		
POLYURETHANE		

TEMP °F (60–460)

HYDROCYANIC ACID, 10%

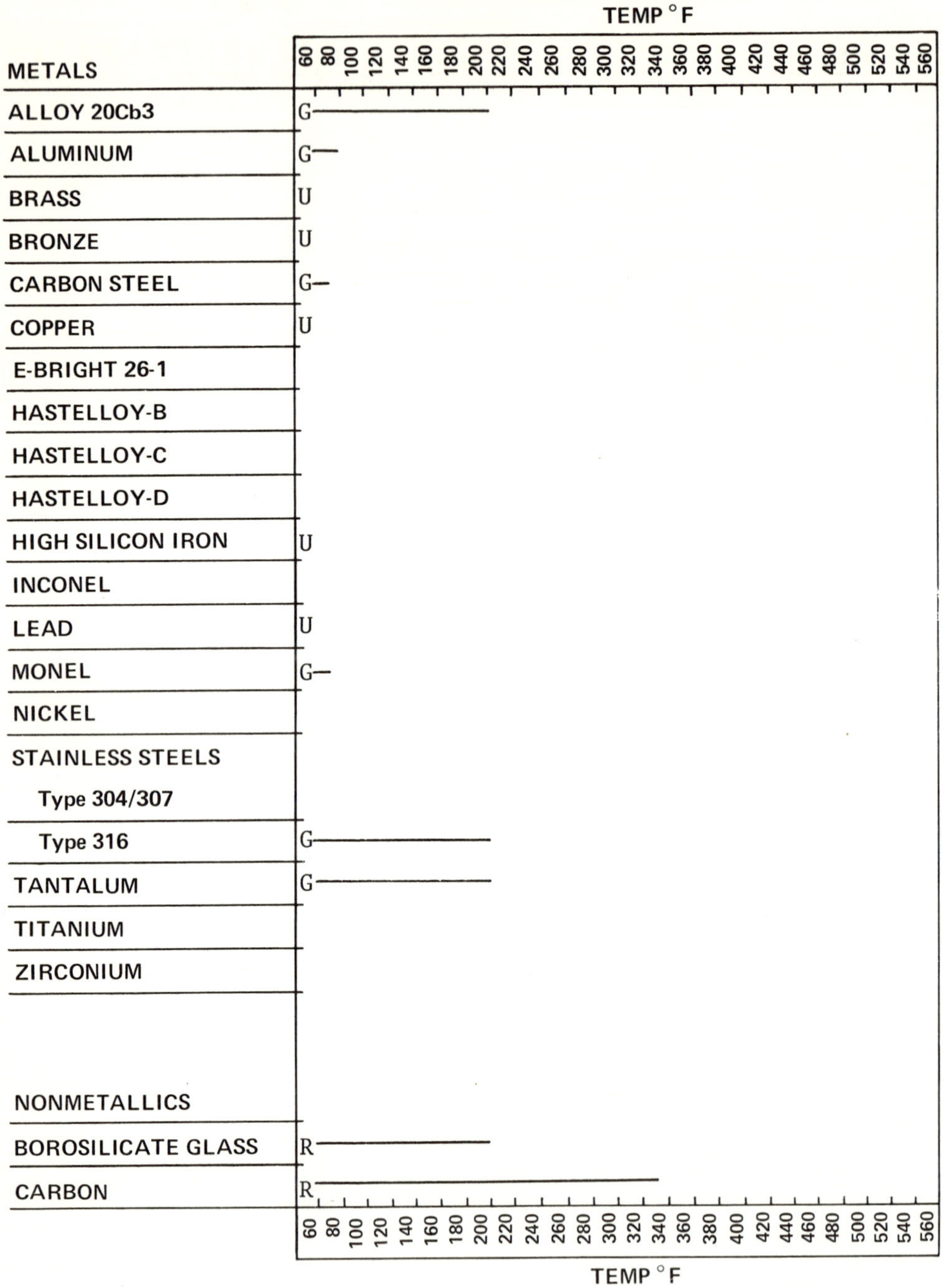

FOR METALS

E = < 2 Mils Penetration/Year; G = < 20 Mils Penetration/Year

S = < 50 Mils Penetration/Year; U = > 50 Mils Penetration/Year

FOR NONMETALLICS

R = Resistant

U = Unsatisfactory

HYDROCYANIC ACID, 10%

TEMP °F (60, 80, 100, 120, 140, 160, 180, 200, 220, 240, 260, 280, 300, 320, 340, 360, 380, 400, 420, 440, 460)

PLASTICS	
ABS	
ACRYLICS	
ASBESTOS REINF., EPOXY	
ASBESTOS REINF., PHENOLIC	
CHLORINATED POLYETHER	R
CPVC	
EPOXY	R
FLUOROCARBONS FEP	R
FLUOROCARBONS TFE	R
FURFURYL ALCOHOL	R
NORYL	
PHENOLIC	
POLYAMIDES–NYLON	U
POLYESTERS	
BISPHENOL A-FUMURATE	R
HYDROGENATED BISPHENOL-A-BISPHENOL-A	
ISOPHTHALIC	R
CHLORINATED POLYESTERS	
POLYSTYRENES	
POLYETHYLENE	R
POLYPROPYLENE	R
PVC–TYPE 1	R
PVC–TYPE 2	R
POLYVINYLIDENE CHLORIDE	R
VINYLIDENE FLUORIDE	R
VINYL ESTER	R

RUBBERS	
BUTYL GR-1	R
FLUORO ELASTOMERS VITON A	R
FLUORO ELASTOMERS KEL-F 3700	
HARD RUBBER	R
HYPALON	R
KOROSEAL	R
NATURAL RUBBER (GRS)	R
NEOPRENE GR-M (CR)	U
NITRILE BUNA N (NBR)	R
NORDEL	R
POLYBUTADIENE	
SILICONE	
POLYURETHANE	

TEMP °F (60, 80, 100, 120, 140, 160, 180, 200, 220, 240, 260, 280, 300, 320, 340, 360, 380, 400, 420, 440, 460)

HYDROFLUORIC ACID, 100%

TEMP °F: 60 80 100 120 140 160 180 200 220 240 260 280 300 320 340 360 380 400 420 440 460 480 500 520 540 560

METALS		TEMP °F
ALLOY 20Cb3		G—
ALUMINUM		U
BRASS		U
BRONZE		G————
CARBON STEEL		G—— S—U
COPPER		G————
E-BRIGHT 26-1		
HASTELLOY-B		G—
HASTELLOY-C		G——————
HASTELLOY-D		G—
HIGH SILICON IRON		U
INCONEL		G———
LEAD		U
MONEL	2	E-G—————
NICKEL	2	G———
STAINLESS STEELS		
Type 304/307		U
Type 316		G—
TANTALUM		U
TITANIUM		U
ZIRCONIUM		U
NONMETALLICS		
BOROSILICATE GLASS		U
CARBON		U

TEMP °F: 60 80 100 120 140 160 180 200 220 240 260 280 300 320 340 360 380 400 420 440 460 480 500 520 540 560

FOR METALS

E = < 2 Mils Penetration/Year; G = < 20 Mils Penetration/Year

S = < 50 Mils Penetration/Year; U = > 50 Mils Penetration/Year

FOR NONMETALLICS

R = Resistant

U = Unsatisfactory

HYDROFLUORIC ACID, 100%

PLASTICS		TEMP °F (60–460)
ABS		U
ACRYLICS		R— (to 80)
ASBESTOS REINF., EPOXY		R— (to 80)
ASBESTOS REINF., PHENOLIC		U
CHLORINATED POLYETHER		U
CPVC		
EPOXY		U
FLUOROCARBONS FEP		R— (to 400)
FLUOROCARBONS TFE		R— (to 460)
FURFURYL ALCOHOL	sat	R— (to 260)
NORYL		U
PHENOLIC		U
POLYAMIDES—NYLON		U
POLYESTERS		
BISPHENOL A-FUMURATE		
HYDROGENATED BISPHENOL-A-BISPHENOL-A		U
ISOPHTHALIC		U
CHLORINATED POLYESTERS		
POLYSTYRENES		R— (to 80)
POLYETHYLENE		R— (to 80)
POLYPROPYLENE		R— (to 200)
PVC—TYPE 1		
PVC—TYPE 2		
POLYVINYLIDENE CHLORIDE		U
VINYLIDENE FLUORIDE		R— (to 210)
VINYL ESTER		

RUBBERS	TEMP °F (60–460)
BUTYL GR-1	U
FLUORO ELASTOMERS VITON A	RU
FLUORO ELASTOMERS KEL-F 3700	
HARD RUBBER	U
HYPALON	R— (to 80)
KOROSEAL	R— (to 80)
NATURAL RUBBER (GRS)	U
NEOPRENE GR-M (CR)	U
NITRILE BUNA N (NBR)	U
NORDEL	
POLYBUTADIENE	
SILICONE	
POLYURETHANE	

HYDROFLUORIC ACID, 70%

TEMP °F (60–560, in steps of 20)

METALS		Rating
ALLOY 20Cb3		U
ALUMINUM		U
BRASS		U
BRONZE		U
CARBON STEEL		U
COPPER		U
E-BRIGHT 26-1		
HASTELLOY-B		G
HASTELLOY-C		G
HASTELLOY-D		
HIGH SILICON IRON		U
INCONEL		U
LEAD		U
MONEL	2	E-G
NICKEL	2	G
STAINLESS STEELS		
Type 304/307		U
Type 316		U
TANTALUM		U
TITANIUM		U
ZIRCONIUM		U

NONMETALLICS	Rating
BOROSILICATE GLASS	U
CARBON	U

TEMP °F

FOR METALS

E = < 2 Mils Penetration/Year; G = < 20 Mils Penetration/Year

S = < 50 Mils Penetration/Year; U = > 50 Mils Penetration/Year

FOR NONMETALLICS

R = Resistant

U = Unsatisfactory

HYDROFLUORIC ACID, 70%

TEMP °F (scale 60–460)

PLASTICS	Rating	Bar extends to (°F)
ABS	U	
ACRYLICS	U	
ASBESTOS REINF., EPOXY		
ASBESTOS REINF., PHENOLIC		
CHLORINATED POLYETHER	U	
CPVC		
EPOXY	U	
FLUOROCARBONS FEP	R	400
FLUOROCARBONS TFE	R	460
FURFURYL ALCOHOL		
NORYL	U	
PHENOLIC		
POLYAMIDES—NYLON	U	
POLYESTERS		
BISPHENOL A-FUMURATE		
HYDROGENATED BISPHENOL-A-BISPHENOL-A	U	
ISOPHTHALIC	U	
CHLORINATED POLYESTERS		
POLYSTYRENES		
POLYETHYLENE	U	
POLYPROPYLENE	R	200
PVC—TYPE 1		
PVC—TYPE 2		
POLYVINYLIDENE CHLORIDE		
VINYLIDENE FLUORIDE	R	210
VINYL ESTER		

RUBBERS	Rating	Bar extends to (°F)
BUTYL GR-1	R	150
FLUORO ELASTOMERS VITON A	R	350
FLUORO ELASTOMERS KEL-F 3700	R	80
HARD RUBBER	U	
HYPALON	R	80
KOROSEAL	R	80
NATURAL RUBBER (GRS)	U	
NEOPRENE GR-M (CR)	R	200
NITRILE BUNA N (NBR)	U	
NORDEL		
POLYBUTADIENE		
SILICONE		
POLYURETHANE		

HYDROFLUORIC ACID, 50%

TEMP °F (60, 80, 100, 120, 140, 160, 180, 200, 220, 240, 260, 280, 300, 320, 340, 360, 380, 400, 420, 440, 460, 480, 500, 520, 540, 560)

METALS		TEMP °F
ALLOY 20Cb3		
ALUMINUM		U
BRASS		U
BRONZE		U
CARBON STEEL		U
COPPER		U
E-BRIGHT 26-1		
HASTELLOY-B		E——
HASTELLOY-C		G————————————
HASTELLOY-D		G——
HIGH SILICON IRON		U
INCONEL		U
LEAD		G——
MONEL	2	E—————— G—————————————————
NICKEL	2	G—S—————U
STAINLESS STEELS		
Type 304/307		U
Type 316		U
TANTALUM		U
TITANIUM		U
ZIRCONIUM		U
NONMETALLICS		
BOROSILICATE GLASS		U
CARBON		R—————————————————————

TEMP °F (60, 80, 100, 120, 140, 160, 180, 200, 220, 240, 260, 280, 300, 320, 340, 360, 380, 400, 420, 440, 460, 480, 500, 520, 540, 560)

FOR METALS

E = < 2 Mils Penetration/Year; G = < 20 Mils Penetration/Year

S = < 50 Mils Penetration/Year; U = > 50 Mils Penetration/Year

FOR NONMETALLICS

R = Resistant

U = Unsatisfactory

TEMP °F (60 80 100 120 140 160 180 200 220 240 260 280 300 320 340 360 380 400 420 440 460)

PLASTICS	
ABS	U
ACRYLICS	
ASBESTOS REINF., EPOXY	
ASBESTOS REINF., PHENOLIC	
CHLORINATED POLYETHER	R
CPVC	U
EPOXY	U
FLUOROCARBONS FEP	R
FLUOROCARBONS TFE	R
FURFURYL ALCOHOL	
NORYL	R
PHENOLIC	
POLYAMIDES–NYLON	U
POLYESTERS	
BISPHENOL A-FUMURATE	
HYDROGENATED BISPHENOL-A-BISPHENOL-A	U
ISOPHTHALIC	U
CHLORINATED POLYESTERS	
POLYSTYRENES	
POLYETHYLENE	R
POLYPROPYLENE	R
PVC–TYPE 1	RU
PVC–TYPE 2	RU
POLYVINYLIDENE CHLORIDE	R
VINYLIDENE FLUORIDE	R
VINYL ESTER	

RUBBERS	
BUTYL GR-1	R
FLUORO ELASTOMERS VITON A	R
FLUORO ELASTOMERS KEL-F 3700	R
HARD RUBBER	U
HYPALON	R
KOROSEAL	R
NATURAL RUBBER (GRS)	R
NEOPRENE GR-M (CR)	R
NITRILE BUNA N (NBR)	U
NORDEL	
POLYBUTADIENE	
SILICONE	
POLYURETHANE	

TEMP °F (60 80 100 120 140 160 180 200 220 240 260 280 300 320 340 360 380 400 420 440 460)

HYDROFLUORIC ACID, 40%

TEMP °F (60–560)

METALS	Note	Rating by temperature (°F)
ALLOY 20Cb3		
ALUMINUM		U (60)
BRASS		U (60)
BRONZE		G (60–80)
CARBON STEEL		U (60)
COPPER		G (60–80)
E-BRIGHT 26-1		
HASTELLOY-B		E (60–120)
HASTELLOY-C		G (60–200)
HASTELLOY-D		E (60–100); G (100–120)
HIGH SILICON IRON		U (60)
INCONEL		U (60)
LEAD		G (60–100); U (100)
MONEL	2	E (60–180); G (180–400)
NICKEL	2	G (60–80); S (80–180); U (180)
STAINLESS STEELS		
Type 304/307		U (60)
Type 316		U (60)
TANTALUM		U (60)
TITANIUM		U (60)
ZIRCONIUM		U (60)

NONMETALLICS	Rating by temperature (°F)
BOROSILICATE GLASS	U (60)
CARBON	R (60–340)

TEMP °F

FOR METALS

E = < 2 Mils Penetration/Year; G = < 20 Mils Penetration/Year

S = < 50 Mils Penetration/Year; U = > 50 Mils Penetration/Year

FOR NONMETALLICS

R = Resistant

U = Unsatisfactory

HYDROFLUORIC ACID, 40%

TEMP °F: 60 80 100 120 140 160 180 200 220 240 260 280 300 320 340 360 380 400 420 440 460

PLASTICS	
ABS	U
ACRYLICS	
ASBESTOS REINF., EPOXY	
ASBESTOS REINF., PHENOLIC	
CHLORINATED POLYETHER	R—
CPVC	R— U
EPOXY	U
FLUOROCARBONS FEP	R—
FLUOROCARBONS TFE	R—
FURFURYL ALCOHOL	
NORYL	R—
PHENOLIC	
POLYAMIDES—NYLON	U
POLYESTERS	
BISPHENOL A-FUMURATE	U
HYDROGENATED BISPHENOL-A-BISPHENOL-A	U
ISOPHTHALIC	U
CHLORINATED POLYESTERS 10	R— U
POLYSTYRENES	R—
POLYETHYLENE	R—
POLYPROPYLENE	R—
PVC—TYPE 1	R— U
PVC—TYPE 2	R— U
POLYVINYLIDENE CHLORIDE	R—
VINYLIDENE FLUORIDE	R—
VINYL ESTER	
RUBBERS	
BUTYL GR-1	R—
FLUORO ELASTOMERS VITON A	R—
FLUORO ELASTOMERS KEL-F 3700	R—
HARD RUBBER	R—
HYPALON	R—
KOROSEAL	R—
NATURAL RUBBER (GRS)	R—
NEOPRENE GR-M (CR)	R—
NITRILE BUNA N (NBR)	U
NORDEL	
POLYBUTADIENE	
SILICONE	
POLYURETHANE	

TEMP °F: 60 80 100 120 140 160 180 200 220 240 260 280 300 320 340 360 380 400 420 440 460

HYDROFLUORIC ACID, 30%

TEMP °F (60–560)

METALS		TEMP °F
ALLOY 20Cb3		G (60–120)
ALUMINUM		U
BRASS		U
BRONZE		G (60–140)
CARBON STEEL		U
COPPER		G (60–140)
E-BRIGHT 26-1		
HASTELLOY-B		G (60–140)
HASTELLOY-C		G (60–200)
HASTELLOY-D		G (60–120)
HIGH SILICON IRON		U
INCONEL		
LEAD		G (60–100)
MONEL	2	E (60–180), G (180–400)
NICKEL	2	G (60), S (80–170), U (170)
STAINLESS STEELS		
Type 304/307		U
Type 316		U
TANTALUM		U
TITANIUM		U
ZIRCONIUM		U

NONMETALLICS	TEMP °F
BOROSILICATE GLASS	U
CARBON	R (60–340)

FOR METALS

E = < 2 Mils Penetration/Year; G = < 20 Mils Penetration/Year

S = < 50 Mils Penetration/Year; U = > 50 Mils Penetration/Year

FOR NONMETALLICS

R = Resistant

U = Unsatisfactory

HYDROFLUORIC ACID, 30%

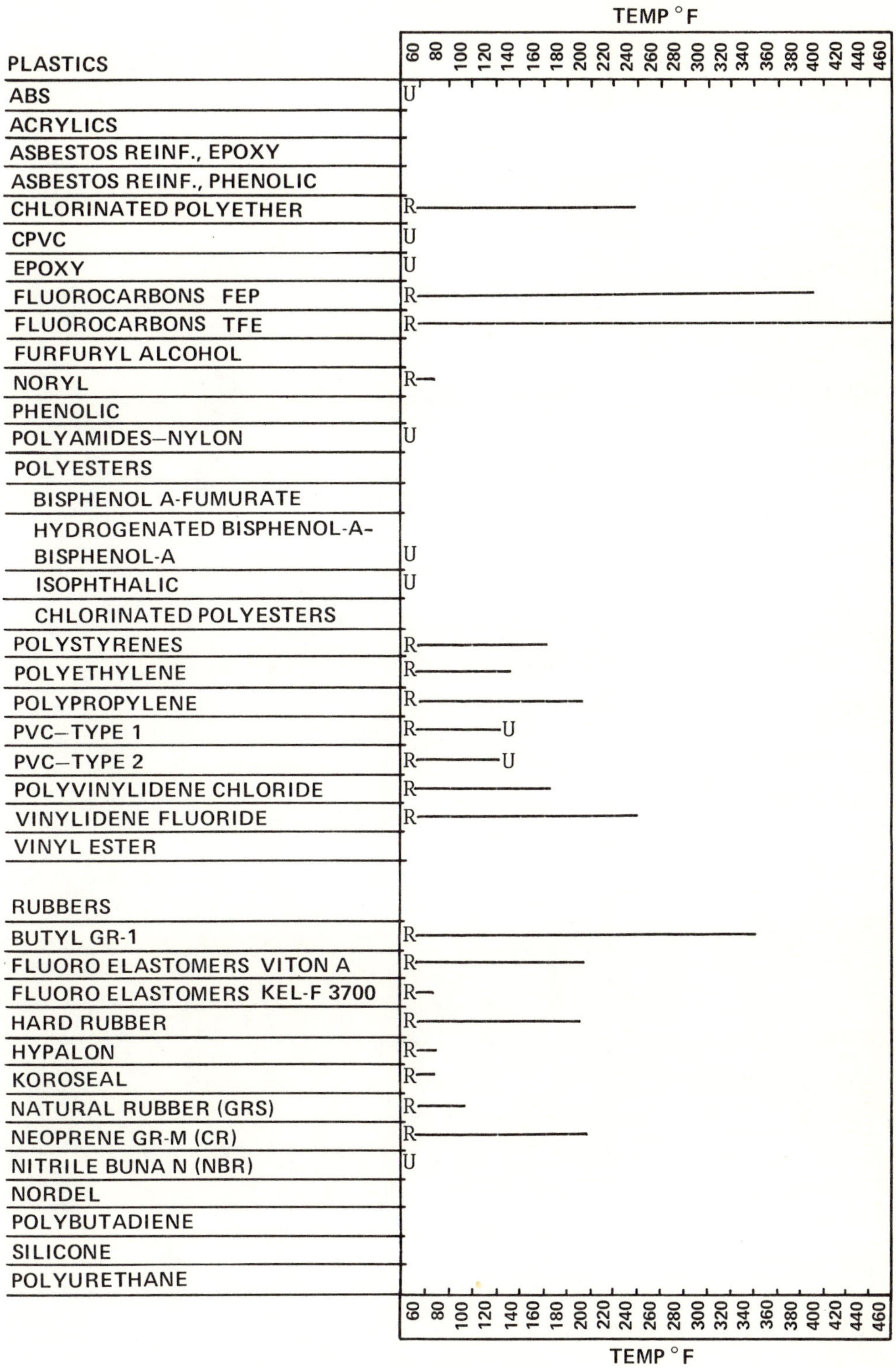

HYDROFLUORIC ACID DILUTE

TEMP °F: 60 80 100 120 140 160 180 200 220 240 260 280 300 320 340 360 380 400 420 440 460 480 500 520 540 560

METALS	
ALLOY 20Cb3	G
ALUMINUM	U
BRASS	U
BRONZE	G
CARBON STEEL	U
COPPER	G
E-BRIGHT 26-1	
HASTELLOY-B	G
HASTELLOY-C	E
HASTELLOY-D	G
HIGH SILICON IRON	U
INCONEL	E-U
LEAD	G
MONEL 2	E
NICKEL	G-S—U
STAINLESS STEELS	
Type 304/307	U
Type 316	U
TANTALUM	U
TITANIUM	U
ZIRCONIUM	U

NONMETALLICS	
BOROSILICATE GLASS	U
CARBON	R

TEMP °F: 60 80 100 120 140 160 180 200 220 240 260 280 300 320 340 360 380 400 420 440 460 480 500 520 540 560

FOR METALS

E = < 2 Mils Penetration/Year; G = < 20 Mils Penetration/Year

S = < 50 Mils Penetration/Year; U = > 50 Mils Penetration/Year

FOR NONMETALLICS

R = Resistant

U = Unsatisfactory

HYDROFLUORIC ACID DILUTE

TEMP °F

PLASTICS			
ABS			U
ACRYLICS			
ASBESTOS REINF., EPOXY			R
ASBESTOS REINF., PHENOLIC			R
CHLORINATED POLYETHER			R
CPVC			R—U
EPOXY		10%	U
FLUOROCARBONS FEP			R
FLUOROCARBONS TFE			R
FURFURYL ALCOHOL			R
NORYL			R
PHENOLIC			
POLYAMIDES—NYLON			U
POLYESTERS			
BISPHENOL A-FUMURATE	20%	10	R
HYDROGENATED BISPHENOL-A-BISPHENOL-A		10	R
ISOPHTHALIC	10%	10	R
CHLORINATED POLYESTERS		20%	R
POLYSTYRENES			R
POLYETHYLENE			R
POLYPROPYLENE			R
PVC—TYPE 1			R—U
PVC—TYPE 2			R
POLYVINYLIDENE CHLORIDE			R
VINYLIDENE FLUORIDE			R
VINYL ESTER	20%	10	R
RUBBERS			
BUTYL GR-1			
FLUORO ELASTOMERS VITON A			R
FLUORO ELASTOMERS KEL-F 3700			R
HARD RUBBER			R
HYPALON			R
KOROSEAL			R
NATURAL RUBBER (GRS)			R
NEOPRENE GR-M (CR)			R
NITRILE BUNA N (NBR)			U
NORDEL			
POLYBUTADIENE			
SILICONE			
POLYURETHANE			

TEMP °F: 60 80 100 120 140 160 180 200 220 240 260 280 300 320 340 360 380 400 420 440 460

(HYDRO) FLUOSILICIC ACID, 50%-10%

TEMP °F (60 80 100 120 140 160 180 200 220 240 260 280 300 320 340 360 380 400 420 440 460 480 500 520 540 560)

METALS		Rating
ALLOY 20Cb3		G——
ALUMINUM		U
BRASS		G—
BRONZE		U
CARBON STEEL		U
COPPER		G—
E-BRIGHT 26-1		
HASTELLOY-B		G—
HASTELLOY-C	35%	G—
HASTELLOY-D		E——
HIGH SILICON IRON		U
INCONEL		G—
LEAD		G—
MONEL		G—
NICKEL		U
STAINLESS STEELS		
Type 304/307		U
Type 316		G————U
TANTALUM		U
TITANIUM		
ZIRCONIUM		
NONMETALLICS		
BOROSILICATE GLASS		U
CARBON		R————————————

TEMP °F (60 80 100 120 140 160 180 200 220 240 260 280 300 320 340 360 380 400 420 440 460 480 500 520 540 560)

FOR METALS

E = < 2 Mils Penetration/Year; G = < 20 Mils Penetration/Year

S = < 50 Mils Penetration/Year; U = > 50 Mils Penetration/Year

FOR NONMETALLICS

R = Resistant

U = Unsatisfactory

(HYDRO) FLUOSILICIC ACID, 50%-10%

TEMP °F: 60 80 100 120 140 160 180 200 220 240 260 280 300 320 340 360 380 400 420 440 460

PLASTICS		
ABS	25%	R
ACRYLICS		
ASBESTOS REINF., EPOXY		
ASBESTOS REINF., PHENOLIC		
CHLORINATED POLYETHER		R
CPVC		U
EPOXY	30%	R
FLUOROCARBONS FEP		R
FLUOROCARBONS TFE		R
FURFURYL ALCOHOL	30%	R
NORYL		
PHENOLIC		
POLYAMIDES—NYLON		
POLYESTERS	10	
BISPHENOL A-FUMURATE	10% 10	R
HYDROGENATED BISPHENOL-A-BISPHENOL-A		
ISOPHTHALIC	30%	U
CHLORINATED POLYESTERS	10%	R
POLYSTYRENES		
POLYETHYLENE		R
POLYPROPYLENE		
PVC—TYPE 1		R
PVC—TYPE 2		R
POLYVINYLIDENE CHLORIDE		R
VINYLIDENE FLUORIDE		R
VINYL ESTER	10 35%	R

RUBBERS	
BUTYL GR-1	R
FLUORO ELASTOMERS VITON A	R
FLUORO ELASTOMERS KEL-F 3700	
HARD RUBBER	R
HYPALON	R
KOROSEAL	R
NATURAL RUBBER (GRS)	R
NEOPRENE GR-M (CR)	R
NITRILE BUNA N (NBR)	R
NORDEL	
POLYBUTADIENE	
SILICONE	
POLYURETHANE	

TEMP °F: 60 80 100 120 140 160 180 200 220 240 260 280 300 320 340 360 380 400 420 440 460

HYDROGEN

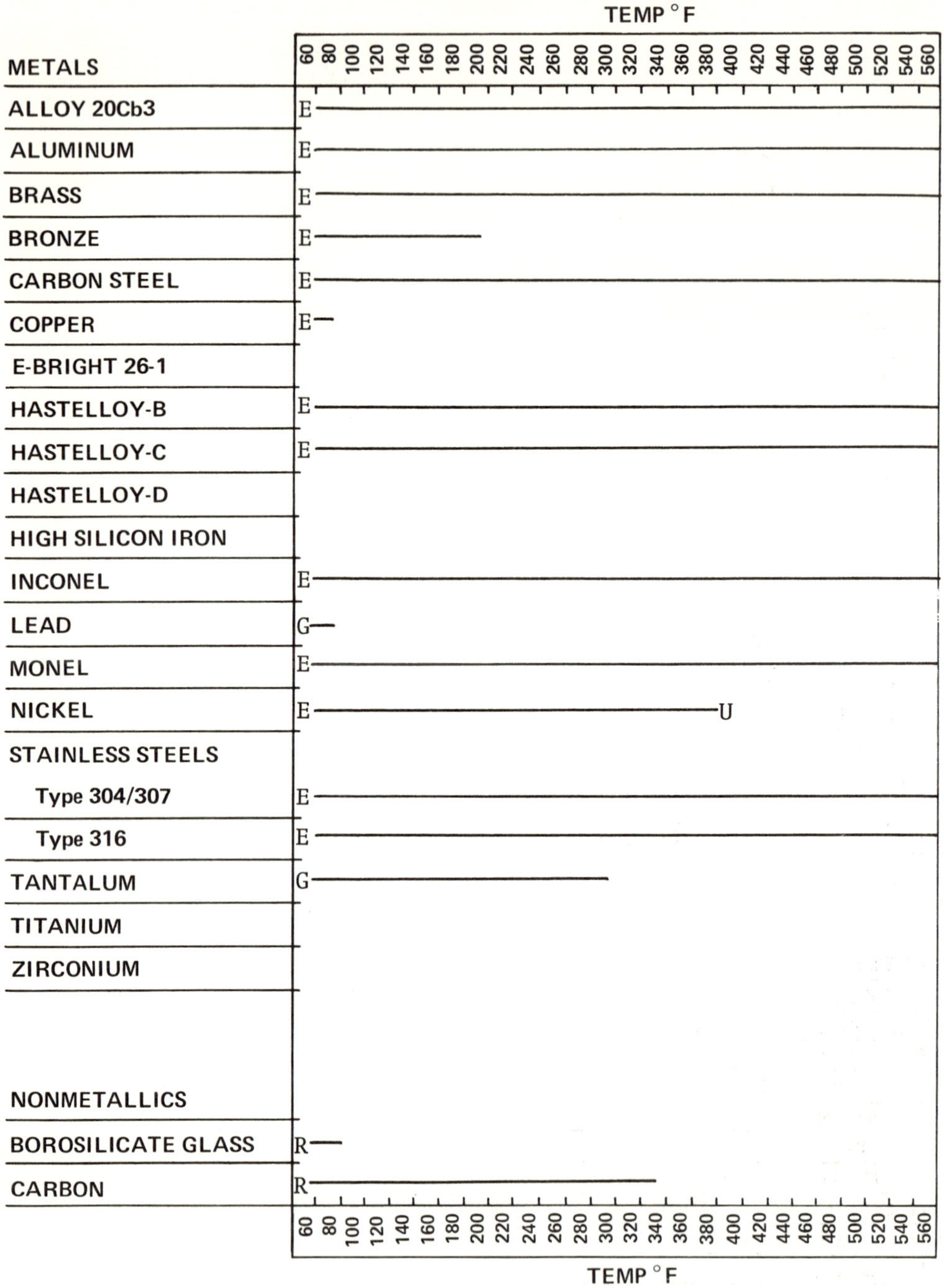

FOR METALS

E = < 2 Mils Penetration/Year; G = < 20 Mils Penetration/Year

S = < 50 Mils Penetration/Year; U = > 50 Mils Penetration/Year

FOR NONMETALLICS

R = Resistant

U = Unsatisfactory

PLASTICS	TEMP °F (60–460)
ABS	R (to ~140)
ACRYLICS	
ASBESTOS REINF., EPOXY	
ASBESTOS REINF., PHENOLIC	
CHLORINATED POLYETHER	R (to ~250)
CPVC	U
EPOXY	R (to ~150)
FLUOROCARBONS FEP	R (to 400)
FLUOROCARBONS TFE	R (to 460 and beyond)
FURFURYL ALCOHOL	
NORYL	
PHENOLIC	
POLYAMIDES–NYLON	R (to ~80)
POLYESTERS	
BISPHENOL A-FUMURATE	
HYDROGENATED BISPHENOL-A-BISPHENOL-A	
ISOPHTHALIC	R (to ~170)
CHLORINATED POLYESTERS	R (to ~250)
POLYSTYRENES	
POLYETHYLENE	R (to ~145)
POLYPROPYLENE	R (to ~210), U (~220)
PVC–TYPE 1	R (to ~150)
PVC–TYPE 2	R (to ~150)
POLYVINYLIDENE CHLORIDE	R (to ~175)
VINYLIDENE FLUORIDE	R (to ~280)
VINYL ESTER	

RUBBERS	TEMP °F (60–460)
BUTYL GR-1	U
FLUORO ELASTOMERS VITON A	R (to ~190)
FLUORO ELASTOMERS KEL-F 3700	
HARD RUBBER	
HYPALON	R (to ~80)
KOROSEAL	
NATURAL RUBBER (GRS)	U
NEOPRENE GR-M (CR)	R (to ~200)
NITRILE BUNA N (NBR)	R (to ~190)
NORDEL	R (to 300)
POLYBUTADIENE	
SILICONE	
POLYURETHANE	R (to ~90)

HYDROGEN CHLORIDE GAS, DRY

TEMP °F (60–560)

METALS		Rating	Line to (°F)
ALLOY 20Cb3		E	560
ALUMINUM		U	
BRASS		G	200
BRONZE			
CARBON STEEL		G	80
COPPER		U	
E-BRIGHT 26-1			
HASTELLOY-B		E	560
HASTELLOY-C		E	560
HASTELLOY-D		U	
HIGH SILICON IRON		G	80
INCONEL		E	560
LEAD			
MONEL		E	400
NICKEL		E	560
STAINLESS STEELS			
Type 304/307	2	E	560
Type 316	2	E	560
TANTALUM		G	300
TITANIUM			
ZIRCONIUM		E	200

NONMETALLICS	Rating	Line to (°F)
BOROSILICATE GLASS	R	200
CARBON	R	340

TEMP °F

FOR METALS

E = < 2 Mils Penetration/Year; G = < 20 Mils Penetration/Year

S = < 50 Mils Penetration/Year; U = > 50 Mils Penetration/Year

FOR NONMETALLICS

R = Resistant

U = Unsatisfactory

HYDROGEN CHLORIDE GAS, DRY

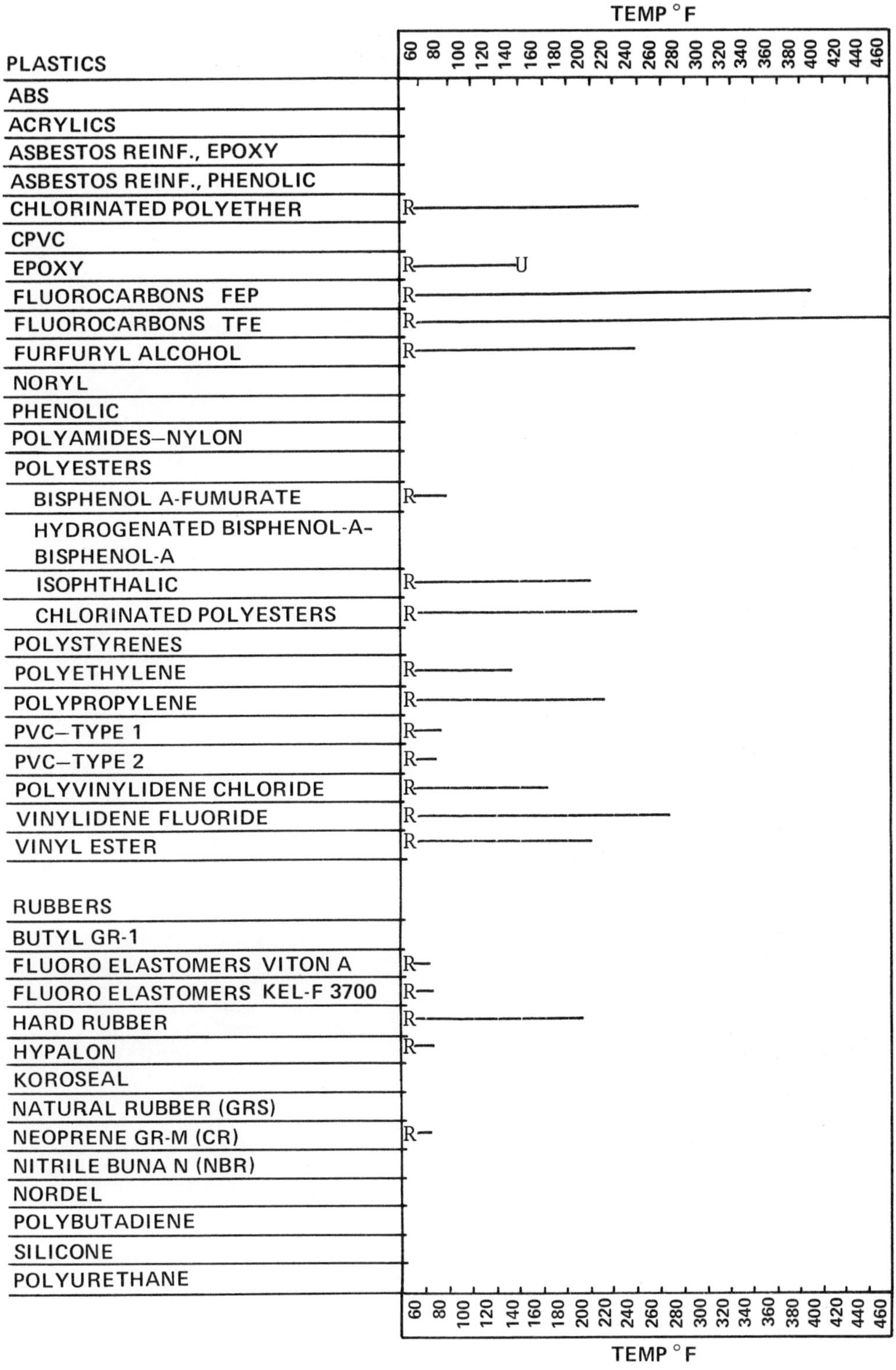

TEMP °F (scale: 60, 80, 100, 120, 140, 160, 180, 200, 220, 240, 260, 280, 300, 320, 340, 360, 380, 400, 420, 440, 460, 480, 500, 520, 540, 560)

METALS	Rating	Temperature range (°F)
ALLOY 20Cb3	G	60
ALUMINUM	E	60
BRASS	G	60
BRONZE		
CARBON STEEL	G	60
COPPER	G	60
E-BRIGHT 26-1		
HASTELLOY-B		
HASTELLOY-C		
HASTELLOY-D		
HIGH SILICON IRON		
INCONEL	G	60
LEAD	G	60
MONEL	G	60
NICKEL	G	60
STAINLESS STEELS		
Type 304/307	G	60
Type 316	E	60
TANTALUM	G	60
TITANIUM		
ZIRCONIUM		

NONMETALLICS	Rating	Temperature range (°F)
BOROSILICATE GLASS	R	60–200
CARBON	R	60–340

TEMP °F

FOR METALS

E = < 2 Mils Penetration/Year; G = < 20 Mils Penetration/Year

S = < 50 Mils Penetration/Year; U = > 50 Mils Penetration/Year

FOR NONMETALLICS

R = Resistant

U = Unsatisfactory

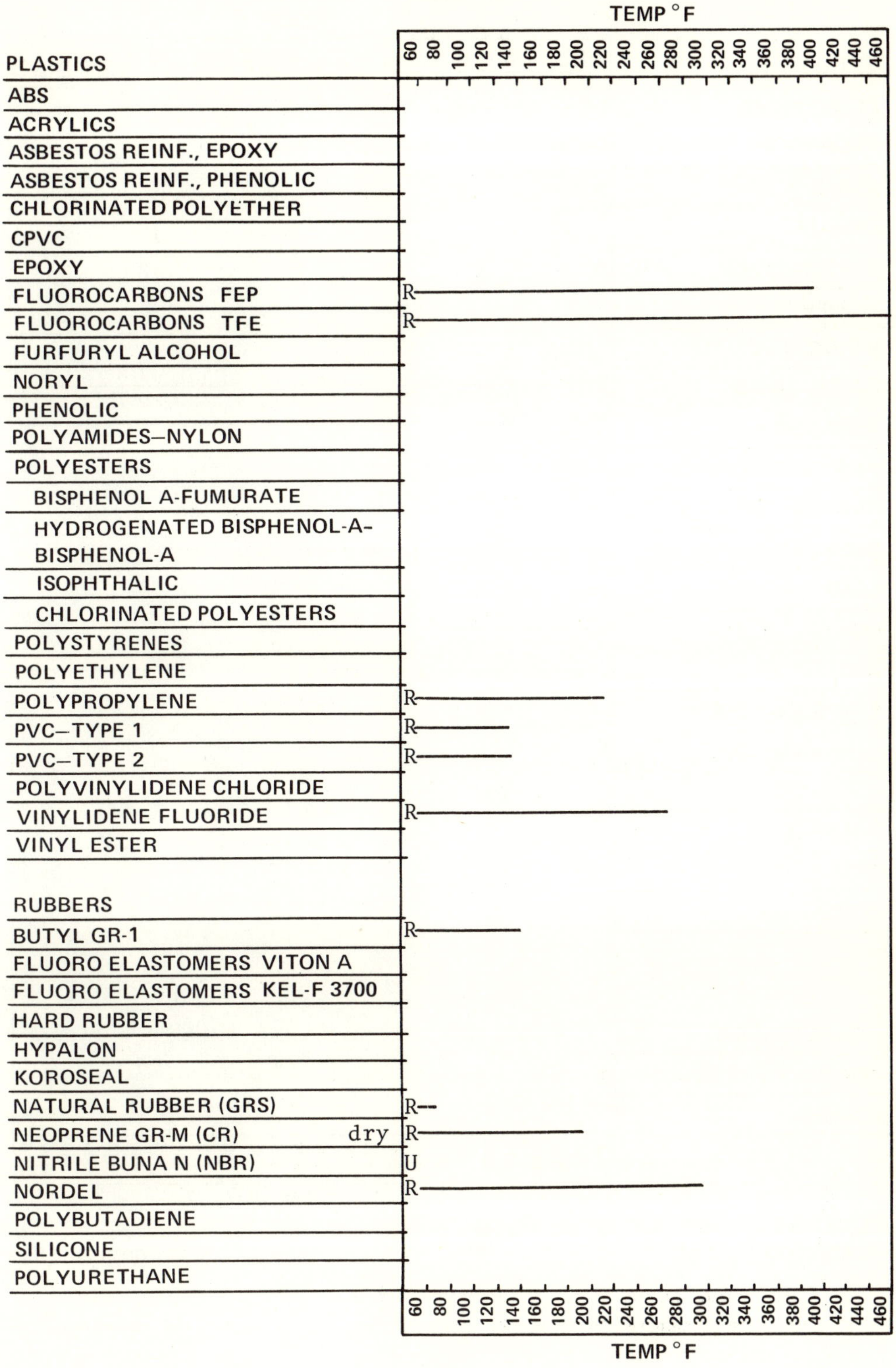
TEMP °F
60 80 100 120 140 160 180 200 220 240 260 280 300 320 340 360 380 400 420 440 460
PLASTICS
ABS
ACRYLICS
ASBESTOS REINF., EPOXY
ASBESTOS REINF., PHENOLIC
CHLORINATED POLYETHER
CPVC
EPOXY
FLUOROCARBONS FEP R
FLUOROCARBONS TFE R
FURFURYL ALCOHOL
NORYL
PHENOLIC
POLYAMIDES–NYLON
POLYESTERS
BISPHENOL A-FUMURATE
HYDROGENATED BISPHENOL-A-BISPHENOL-A
ISOPHTHALIC
CHLORINATED POLYESTERS
POLYSTYRENES
POLYETHYLENE
POLYPROPYLENE R
PVC–TYPE 1 R
PVC–TYPE 2 R
POLYVINYLIDENE CHLORIDE
VINYLIDENE FLUORIDE R
VINYL ESTER
RUBBERS
BUTYL GR-1 R
FLUORO ELASTOMERS VITON A
FLUORO ELASTOMERS KEL-F 3700
HARD RUBBER
HYPALON
KOROSEAL
NATURAL RUBBER (GRS) R
NEOPRENE GR-M (CR) dry R
NITRILE BUNA N (NBR) U
NORDEL R
POLYBUTADIENE
SILICONE
POLYURETHANE
60 80 100 120 140 160 180 200 220 240 260 280 300 320 340 360 380 400 420 440 460
TEMP °F

HYDROGEN FLUORIDE

TEMP °F (60, 80, 100, 120, 140, 160, 180, 200, 220, 240, 260, 280, 300, 320, 340, 360, 380, 400, 420, 440, 460, 480, 500, 520, 540, 560)

METALS	
ALLOY 20Cb3	E——G——
ALUMINUM	G——U
BRASS	G—
BRONZE	
CARBON STEEL 20%	U
COPPER	
E-BRIGHT 26-1	
HASTELLOY-B	E——————————
HASTELLOY-C	G——————————
HASTELLOY-D	
HIGH SILICON IRON	
INCONEL	G——————————
LEAD	G—
MONEL	G——————————
NICKEL	E——G——————————
STAINLESS STEELS	
Type 304/307	E——G——————
Type 316	E——G——————
TANTALUM	
TITANIUM	E—
ZIRCONIUM	E—

NONMETALLICS	
BOROSILICATE GLASS	U
CARBON	

TEMP °F (60, 80, 100, 120, 140, 160, 180, 200, 220, 240, 260, 280, 300, 320, 340, 360, 380, 400, 420, 440, 460, 480, 500, 520, 540, 560)

FOR METALS

E = < 2 Mils Penetration/Year; G = < 20 Mils Penetration/Year

S = < 50 Mils Penetration/Year; U = > 50 Mils Penetration/Year

FOR NONMETALLICS

R = Resistant

U = Unsatisfactory

HYDROGEN FLUORIDE

PLASTICS		Rating	Temp °F (60–460)
ABS			
ACRYLICS			
ASBESTOS REINF., EPOXY			
ASBESTOS REINF., PHENOLIC			
CHLORINATED POLYETHER			
CPVC			
EPOXY		R	60–250
FLUOROCARBONS FEP		R	60–400
FLUOROCARBONS TFE		R	60–460
FURFURYL ALCOHOL		R	60–250
NORYL			
PHENOLIC			
POLYAMIDES–NYLON			
POLYESTERS			
BISPHENOL A-FUMURATE			
HYDROGENATED BISPHENOL-A-BISPHENOL-A			
ISOPHTHALIC	10	R	60–100
CHLORINATED POLYESTERS			
POLYSTYRENES			
POLYETHYLENE			
POLYPROPYLENE		R	60–80
PVC–TYPE 1			
PVC–TYPE 2			
POLYVINYLIDENE CHLORIDE			
VINYLIDENE FLUORIDE			
VINYL ESTER	10	R	60–180

RUBBERS	Rating	Temp °F (60–460)
BUTYL GR-1	U	
FLUORO ELASTOMERS VITON A	R	60–180
FLUORO ELASTOMERS KEL-F 3700		
HARD RUBBER		
HYPALON		
KOROSEAL		
NATURAL RUBBER (GRS)	U	
NEOPRENE GR-M (CR)	U	
NITRILE BUNA N (NBR)	R	60–180
NORDEL		
POLYBUTADIENE		
SILICONE		
POLYURETHANE		

TEMP °F: 60 80 100 120 140 160 180 200 220 240 260 280 300 320 340 360 380 400 420 440 460

HYDROGEN IODIDE

METALS	TEMP °F (60–560)
ALLOY 20Cb3	G
ALUMINUM	U
BRASS	
BRONZE	
CARBON STEEL	
COPPER	
E-BRIGHT 26-1	
HASTELLOY-B	G
HASTELLOY-C	G
HASTELLOY-D	
HIGH SILICON IRON	G
INCONEL	
LEAD	
MONEL	
NICKEL	G
STAINLESS STEELS	
Type 304/307	G
Type 316	G
TANTALUM	E
TITANIUM	
ZIRCONIUM	
NONMETALLICS	
BOROSILICATE GLASS	
CARBON	

Temperature scale (°F): 60, 80, 100, 120, 140, 160, 180, 200, 220, 240, 260, 280, 300, 320, 340, 360, 380, 400, 420, 440, 460, 480, 500, 520, 540, 560

FOR METALS

E = < 2 Mils Penetration/Year; G = < 20 Mils Penetration/Year

S = < 50 Mils Penetration/Year; U = > 50 Mils Penetration/Year

FOR NONMETALLICS

R = Resistant

U = Unsatisfactory

HYDROGEN IODIDE

PLASTICS	TEMP °F (60–460)
ABS	
ACRYLICS	
ASBESTOS REINF., EPOXY	
ASBESTOS REINF., PHENOLIC	
CHLORINATED POLYETHER	
CPVC	
EPOXY	
FLUOROCARBONS FEP	R (60–400)
FLUOROCARBONS TFE	R (60–460)
FURFURYL ALCOHOL	
NORYL	
PHENOLIC	
POLYAMIDES—NYLON	
POLYESTERS	
BISPHENOL A-FUMURATE	
HYDROGENATED BISPHENOL-A-BISPHENOL-A	
ISOPHTHALIC	
CHLORINATED POLYESTERS	
POLYSTYRENES	
POLYETHYLENE	
POLYPROPYLENE	
PVC—TYPE 1	
PVC—TYPE 2	
POLYVINYLIDENE CHLORIDE	
VINYLIDENE FLUORIDE	
VINYL ESTER	
RUBBERS	
BUTYL GR-1	
FLUORO ELASTOMERS VITON A	
FLUORO ELASTOMERS KEL-F 3700	
HARD RUBBER	
HYPALON	
KOROSEAL	
NATURAL RUBBER (GRS)	
NEOPRENE GR-M (CR)	
NITRILE BUNA N (NBR)	
NORDEL	
POLYBUTADIENE	
SILICONE	
POLYURETHANE	

HYDROGEN PEROXIDE, 90%

TEMP °F (60 – 560)

METALS		
ALLOY 20Cb3	7	E—
ALUMINUM	5	E—
BRASS		U
BRONZE		S—
CARBON STEEL		U
COPPER		U
E-BRIGHT 26-1		
HASTELLOY-B		E—
HASTELLOY-C		E—
HASTELLOY-D		
HIGH SILICON IRON		G—
INCONEL	6	G—
LEAD		U
MONEL	6	E—
NICKEL		G—
STAINLESS STEELS		
Type 304/307	8	E—G—S
Type 316	7	E—G—S—
TANTALUM		G—
TITANIUM		G—
ZIRCONIUM		E—

NONMETALLICS	
BOROSILICATE GLASS	R—
CARBON	R—

TEMP °F

FOR METALS

E = < 2 Mils Penetration/Year; G = < 20 Mils Penetration/Year

S = < 50 Mils Penetration/Year; U = > 50 Mils Penetration/Year

FOR NONMETALLICS

R = Resistant

U = Unsatisfactory

HYDROGEN PEROXIDE, 90%

TEMP °F (60 80 100 120 140 160 180 200 220 240 260 280 300 320 340 360 380 400 420 440 460)

PLASTICS	
ABS	U
ACRYLICS	R—
ASBESTOS REINF., EPOXY	
ASBESTOS REINF., PHENOLIC	R—
CHLORINATED POLYETHER	R————
CPVC	
EPOXY	U
FLUOROCARBONS FEP	R————————————
FLUOROCARBONS TFE	R——————————————
FURFURYL ALCOHOL	R—
NORYL	U
PHENOLIC	R—
POLYAMIDES—NYLON	R—
POLYESTERS	
BISPHENOL A-FUMURATE	
HYDROGENATED BISPHENOL-A-BISPHENOL-A	
ISOPHTHALIC	U
CHLORINATED POLYESTERS	
POLYSTYRENES	U
POLYETHYLENE	R——
POLYPROPYLENE	R——— U
PVC—TYPE 1	R————
PVC—TYPE 2	U
POLYVINYLIDENE CHLORIDE	R———
VINYLIDENE FLUORIDE	R———
VINYL ESTER	

RUBBERS	
BUTYL GR-1	
FLUORO ELASTOMERS VITON A	R——————— U
FLUORO ELASTOMERS KEL-F 3700	R—
HARD RUBBER	
HYPALON	U
KOROSEAL	R————
NATURAL RUBBER (GRS)	U
NEOPRENE GR-M (CR)	U
NITRILE BUNA N (NBR)	U
NORDEL	R—————————
POLYBUTADIENE	R—
SILICONE	
POLYURETHANE	

TEMP °F (60 80 100 120 140 160 180 200 220 240 260 280 300 320 340 360 380 400 420 440 460)

HYDROGEN PEROXIDE, 50%

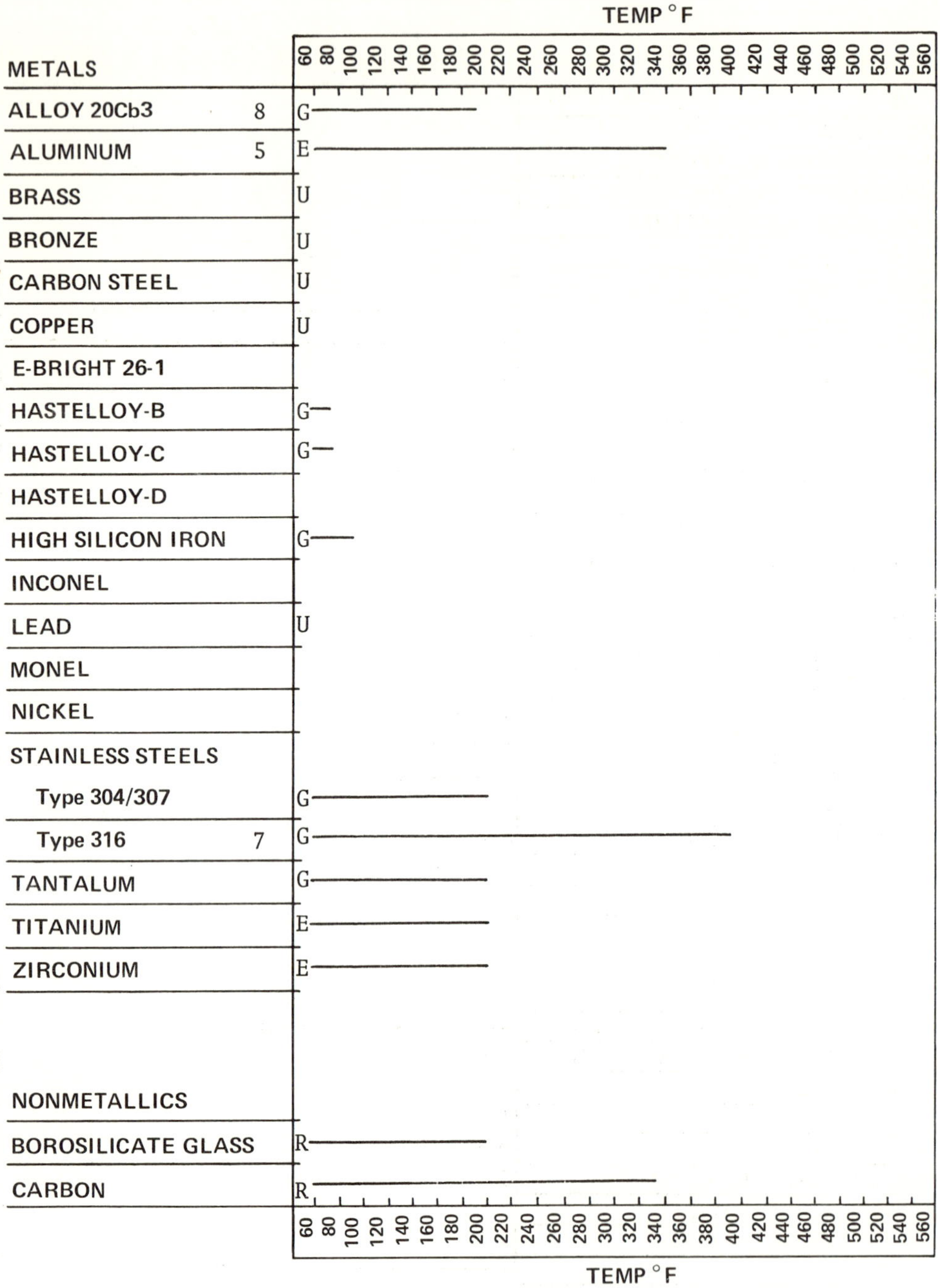

FOR METALS

E = < 2 Mils Penetration/Year; G = < 20 Mils Penetration/Year

S = < 50 Mils Penetration/Year; U = > 50 Mils Penetration/Year

FOR NONMETALLICS

R = Resistant

U = Unsatisfactory

HYDROGEN PEROXIDE, 50%

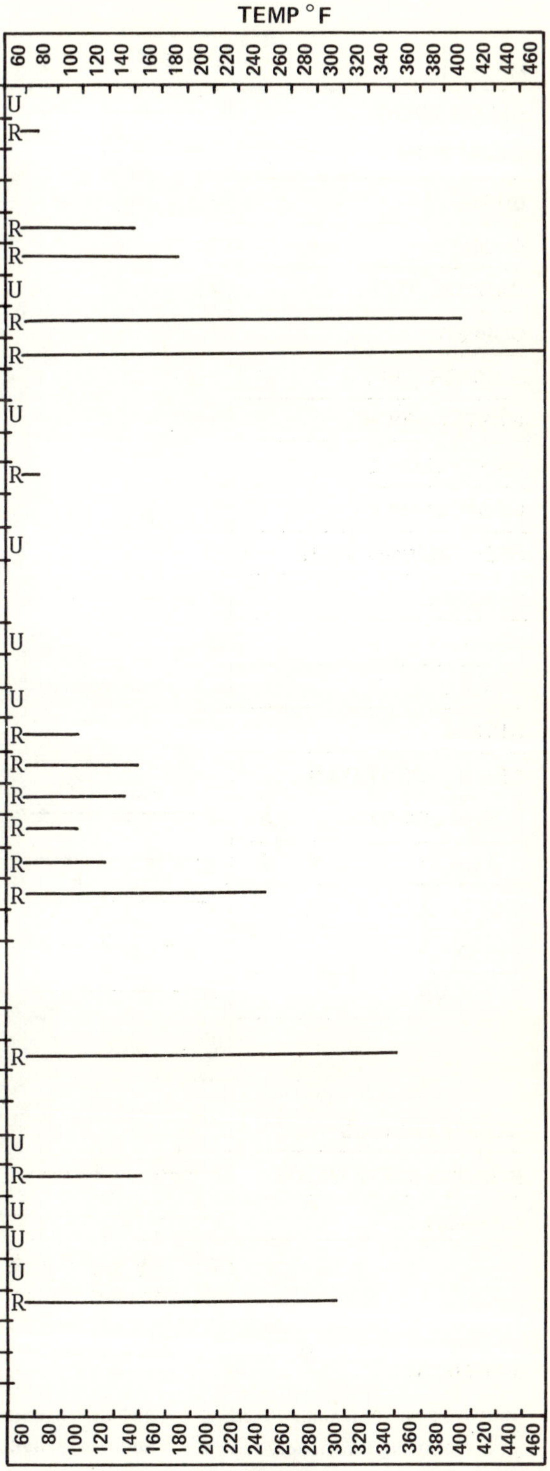

PLASTICS	
ABS	U
ACRYLICS	R
ASBESTOS REINF., EPOXY	
ASBESTOS REINF., PHENOLIC	
CHLORINATED POLYETHER	R
CPVC	R
EPOXY	U
FLUOROCARBONS FEP	R
FLUOROCARBONS TFE	R
FURFURYL ALCOHOL	
NORYL	U
PHENOLIC	
POLYAMIDES–NYLON	R
POLYESTERS	
BISPHENOL A-FUMURATE	U
HYDROGENATED BISPHENOL-A-BISPHENOL-A	
ISOPHTHALIC	U
CHLORINATED POLYESTERS	
POLYSTYRENES	U
POLYETHYLENE	R
POLYPROPYLENE	R
PVC–TYPE 1	R
PVC–TYPE 2	R
POLYVINYLIDENE CHLORIDE	R
VINYLIDENE FLUORIDE	R
VINYL ESTER	
RUBBERS	
BUTYL GR-1	
FLUORO ELASTOMERS VITON A	R
FLUORO ELASTOMERS KEL-F 3700	
HARD RUBBER	
HYPALON	U
KOROSEAL	R
NATURAL RUBBER (GRS)	U
NEOPRENE GR-M (CR)	U
NITRILE BUNA N (NBR)	U
NORDEL	R
POLYBUTADIENE	
SILICONE	
POLYURETHANE	

HYDROGEN PEROXIDE, 30%

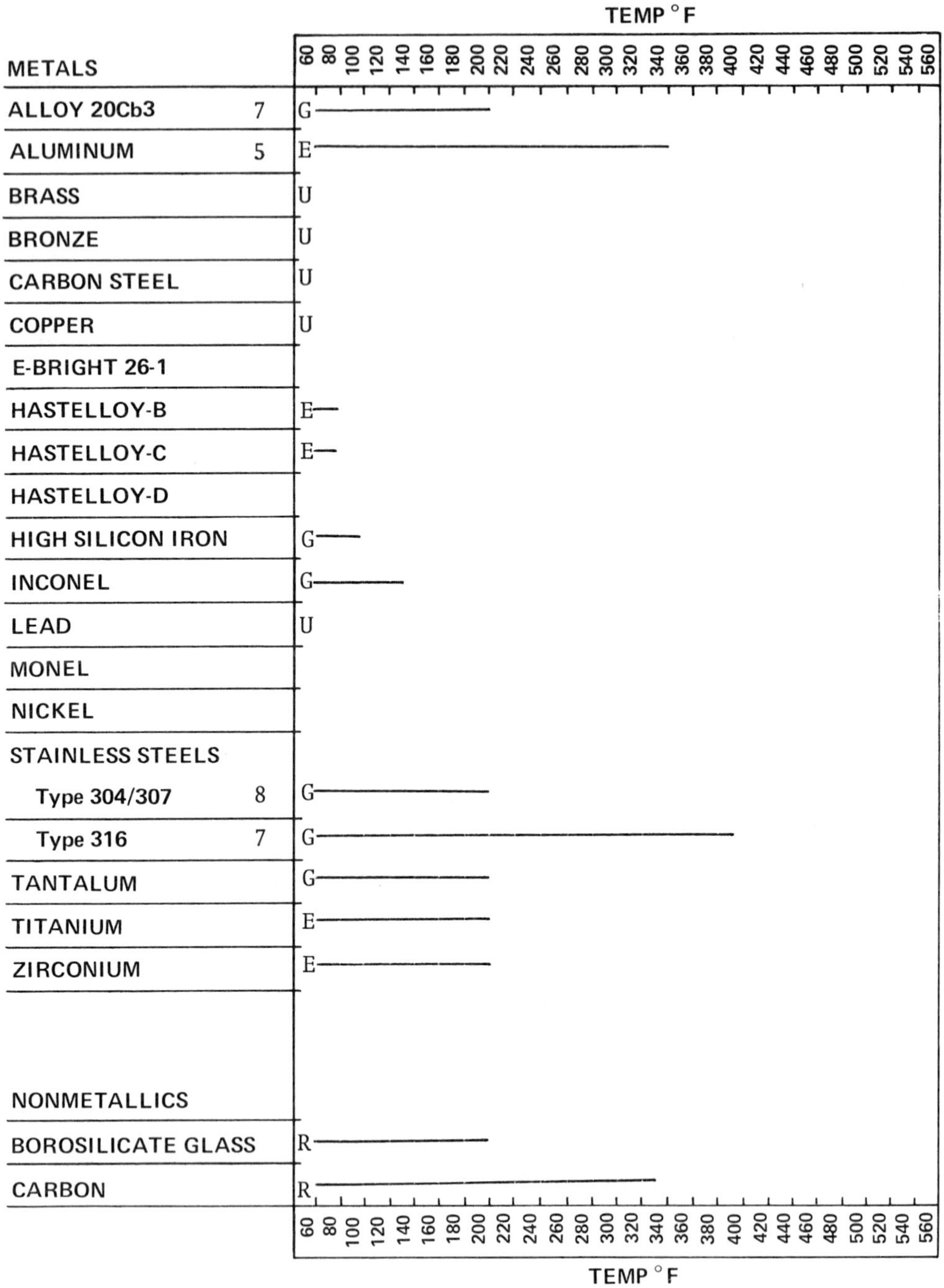

FOR METALS

E = < 2 Mils Penetration/Year; G = < 20 Mils Penetration/Year

S = < 50 Mils Penetration/Year; U = > 50 Mils Penetration/Year

FOR NONMETALLICS

R = Resistant

U = Unsatisfactory

HYDROGEN PEROXIDE, 30%

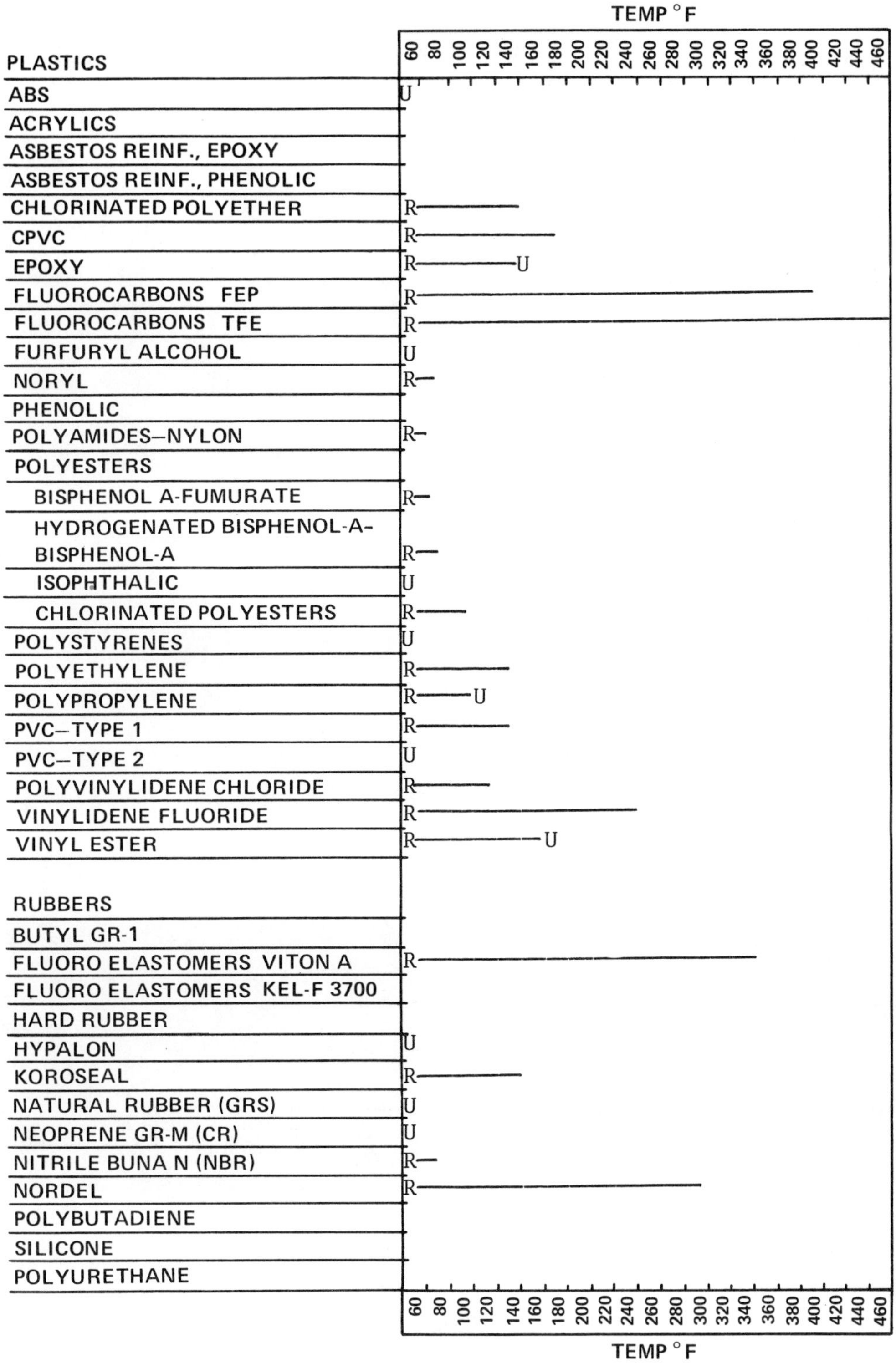

HYDROGEN PEROXIDE DILUTE

TEMP °F: 60 80 100 120 140 160 180 200 220 240 260 280 300 320 340 360 380 400 420 440 460 480 500 520 540 560

METALS		
ALLOY 20Cb3	8	G
ALUMINUM	5	E
BRASS		U
BRONZE		U
CARBON STEEL		U
COPPER		U
E-BRIGHT 26-1		
HASTELLOY-B		E
HASTELLOY-C		E
HASTELLOY-D		
HIGH SILICON IRON		G
INCONEL	6	G
LEAD		U
MONEL	8	G
NICKEL	6	G
STAINLESS STEELS		
Type 304/307	8	G
Type 316	7	G
TANTALUM		G
TITANIUM		E
ZIRCONIUM		E

NONMETALLICS	
BOROSILICATE GLASS	R
CARBON	R

TEMP °F: 60 80 100 120 140 160 180 200 220 240 260 280 300 320 340 360 380 400 420 440 460 480 500 520 540 560

FOR METALS

E = < 2 Mils Penetration/Year; G = < 20 Mils Penetration/Year

S = < 50 Mils Penetration/Year; U = > 50 Mils Penetration/Year

FOR NONMETALLICS

R = Resistant

U = Unsatisfactory

HYDROGEN PEROXIDE DILUTE

TEMP °F (scale 60–460)

PLASTICS	Conc.	Rating	Bar extends to (°F, approx.)
ABS	3%	R	80
ACRYLICS			
ASBESTOS REINF., EPOXY			
ASBESTOS REINF., PHENOLIC			
CHLORINATED POLYETHER		R	150
CPVC			
EPOXY	10%	R	150
FLUOROCARBONS FEP		R	400
FLUOROCARBONS TFE		R	460+
FURFURYL ALCOHOL		U	
NORYL		R	80
PHENOLIC			
POLYAMIDES—NYLON		R	80
POLYESTERS			
BISPHENOL A-FUMURATE		R	150
HYDROGENATED BISPHENOL-A-BISPHENOL-A		R	90
ISOPHTHALIC	5%	R	180
CHLORINATED POLYESTERS	5%	R	210
POLYSTYRENES		R	100
POLYETHYLENE		U	
POLYPROPYLENE		R	125
PVC—TYPE 1		R	145
PVC—TYPE 2			
POLYVINYLIDENE CHLORIDE		R	125
VINYLIDENE FLUORIDE		R	250
VINYL ESTER			

RUBBERS	Conc.	Rating	Bar extends to (°F, approx.)
BUTYL GR-1		U	
FLUORO ELASTOMERS VITON A		R	350
FLUORO ELASTOMERS KEL-F 3700			
HARD RUBBER		U	
HYPALON		R	250
KOROSEAL		R	150
NATURAL RUBBER (GRS)		R	80
NEOPRENE GR-M (CR)		U	
NITRILE BUNA N (NBR)		R	80
NORDEL		R	300
POLYBUTADIENE			
SILICONE		R	80
POLYURETHANE			

HYDROGEN PHOSPHIDE

TEMP °F

METALS	60	80	100	120	140	160	180	200	220	240	260	280	300	320	340	360	380	400	420	440	460	480	500	520	540	560
ALLOY 20Cb3																										
ALUMINUM																										
BRASS																										
BRONZE																										
CARBON STEEL																										
COPPER																										
E-BRIGHT 26-1																										
HASTELLOY-B																										
HASTELLOY-C																										
HASTELLOY-D																										
HIGH SILICON IRON																										
INCONEL																										
LEAD																										
MONEL																										
NICKEL																										
STAINLESS STEELS																										
Type 304/307																										
Type 316																										
TANTALUM																										
TITANIUM																										
ZIRCONIUM																										
NONMETALLICS																										
BOROSILICATE GLASS																										
CARBON																										
	60	80	100	120	140	160	180	200	220	240	260	280	300	320	340	360	380	400	420	440	460	480	500	520	540	560

TEMP °F

FOR METALS

E = < 2 Mils Penetration/Year; G = < 20 Mils Penetration/Year

S = < 50 Mils Penetration/Year; U = > 50 Mils Penetration/Year

FOR NONMETALLICS

R = Resistant

U = Unsatisfactory

TEMP °F (60–460)

PLASTICS	60 … 460
ABS	U
ACRYLICS	
ASBESTOS REINF., EPOXY	
ASBESTOS REINF., PHENOLIC	
CHLORINATED POLYETHER	
CPVC	R———U
EPOXY	
FLUOROCARBONS FEP	R———————————
FLUOROCARBONS TFE	R——————————————
FURFURYL ALCOHOL	
NORYL	
PHENOLIC	
POLYAMIDES—NYLON	
POLYESTERS	
BISPHENOL A-FUMURATE	
HYDROGENATED BISPHENOL-A-BISPHENOL-A	
ISOPHTHALIC	
CHLORINATED POLYESTERS	
POLYSTYRENES	
POLYETHYLENE	R———
POLYPROPYLENE	R———
PVC—TYPE 1	R———
PVC—TYPE 2	R———
POLYVINYLIDENE CHLORIDE	
VINYLIDENE FLUORIDE	R——
VINYL ESTER	

RUBBERS	60 … 460
BUTYL GR-1	
FLUORO ELASTOMERS VITON A	
FLUORO ELASTOMERS KEL-F 3700	
HARD RUBBER	
HYPALON	
KOROSEAL	
NATURAL RUBBER (GRS)	
NEOPRENE GR-M (CR)	
NITRILE BUNA N (NBR)	
NORDEL	
POLYBUTADIENE	
SILICONE	
POLYURETHANE	

TEMP °F (60–460)

HYDROGEN SULFIDE, DRY

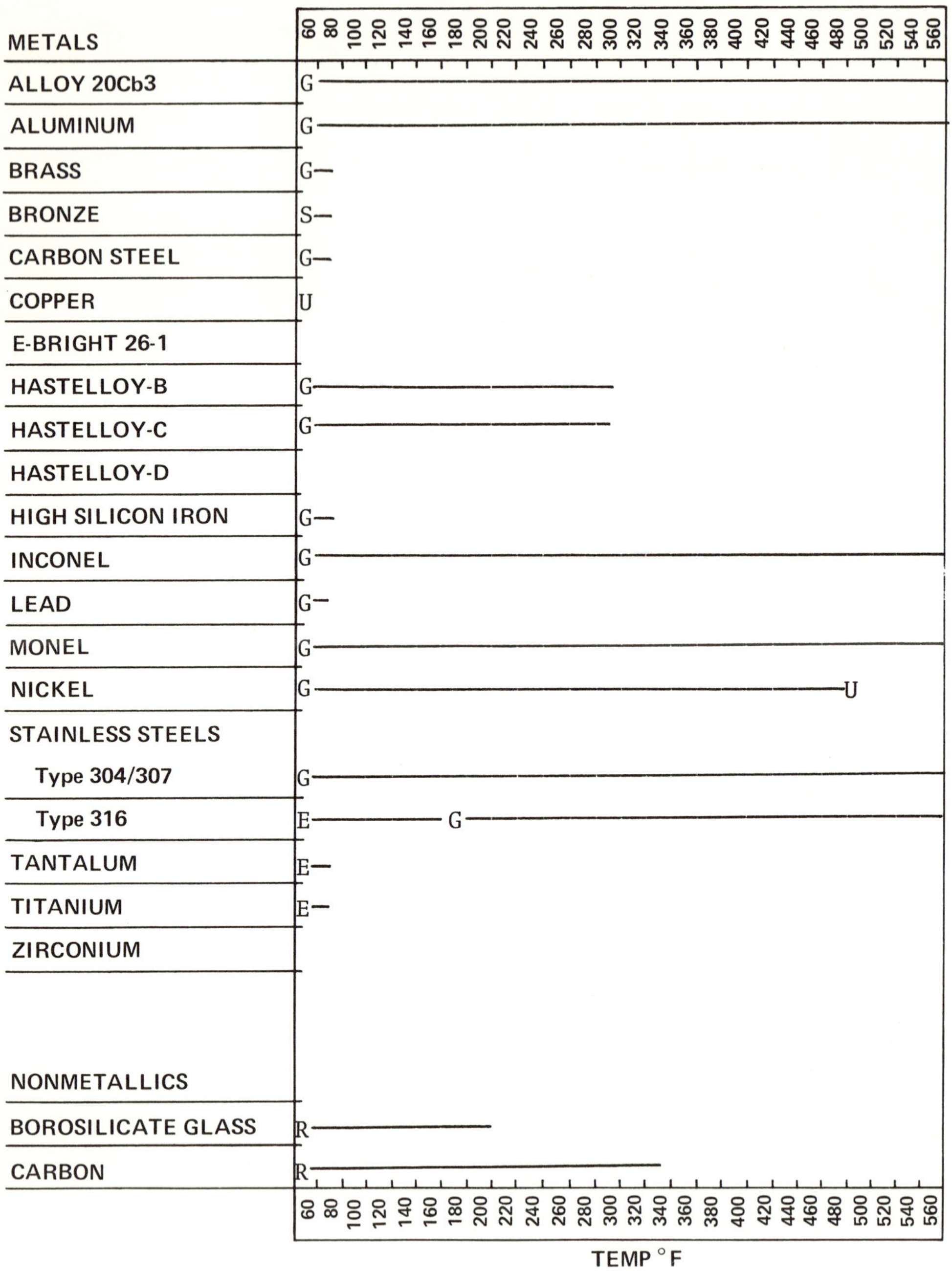

FOR METALS

E = < 2 Mils Penetration/Year; G = < 20 Mils Penetration/Year

S = < 50 Mils Penetration/Year; U = > 50 Mils Penetration/Year

FOR NONMETALLICS

R = Resistant

U = Unsatisfactory

HYDROGEN SULFIDE, DRY

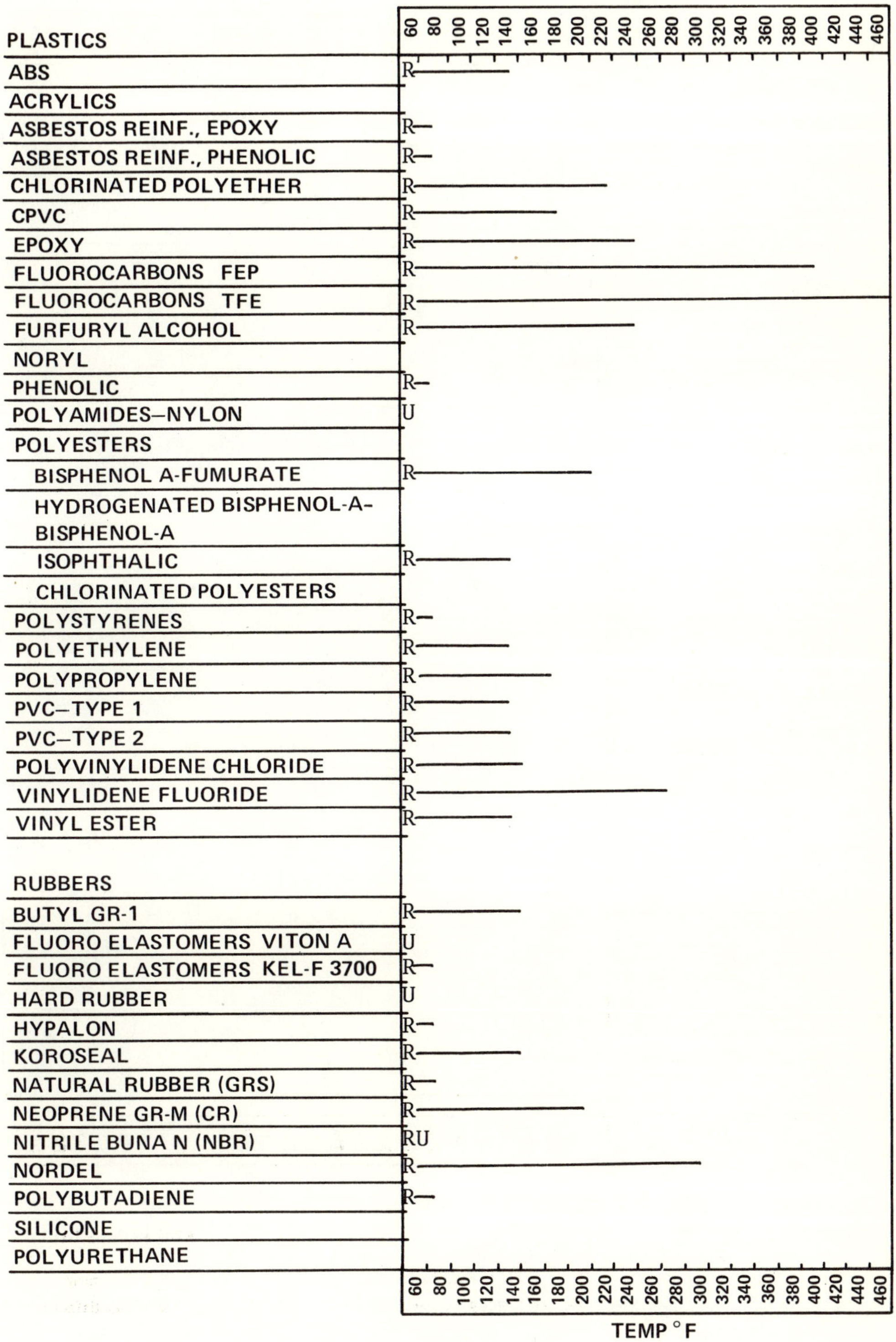

HYDROGEN SULFIDE, WET

TEMP °F: 60 80 100 120 140 160 180 200 220 240 260 280 300 320 340 360 380 400 420 440 460 480 500 520 540 560

METALS	
ALLOY 20Cb3	G—
ALUMINUM	G—
BRASS	U
BRONZE	U
CARBON STEEL	G——
COPPER	U
E-BRIGHT 26-1	
HASTELLOY-B	G——
HASTELLOY-C	E—G——
HASTELLOY-D	
HIGH SILICON IRON	G—
INCONEL	G—
LEAD	U
MONEL	U
NICKEL	U
STAINLESS STEELS	
Type 304/307	U
Type 316	G——
TANTALUM	E—
TITANIUM	E—
ZIRCONIUM	

NONMETALLICS	
BOROSILICATE GLASS	R——
CARBON	R——

TEMP °F: 60 80 100 120 140 160 180 200 220 240 260 280 300 320 340 360 380 400 420 440 460 480 500 520 540 560

FOR METALS

E = < 2 Mils Penetration/Year; G = < 20 Mils Penetration/Year
S = < 50 Mils Penetration/Year; U = > 50 Mils Penetration/Year

FOR NONMETALLICS

R = Resistant
U = Unsatisfactory

HYDROGEN SULFIDE, WET

TEMP °F (60–460)

PLASTICS	Rating
ABS	
ACRYLICS	
ASBESTOS REINF., EPOXY	
ASBESTOS REINF., PHENOLIC	
CHLORINATED POLYETHER	R
CPVC	
EPOXY	R
FLUOROCARBONS FEP	R
FLUOROCARBONS TFE	R
FURFURYL ALCOHOL	R
NORYL	
PHENOLIC	
POLYAMIDES—NYLON	
POLYESTERS	
BISPHENOL A-FUMURATE	R
HYDROGENATED BISPHENOL-A-BISPHENOL-A	
ISOPHTHALIC	R
CHLORINATED POLYESTERS	
POLYSTYRENES	
POLYETHYLENE	R
POLYPROPYLENE	R
PVC—TYPE 1	R
PVC—TYPE 2	R
POLYVINYLIDENE CHLORIDE	R
VINYLIDENE FLUORIDE	R
VINYL ESTER	

RUBBERS	Rating
BUTYL GR-1	R
FLUORO ELASTOMERS VITON A	R
FLUORO ELASTOMERS KEL-F 3700	R
HARD RUBBER	R
HYPALON	R
KOROSEAL	R
NATURAL RUBBER (GRS)	U
NEOPRENE GR-M (CR)	R
NITRILE BUNA N (NBR)	
NORDEL	R
POLYBUTADIENE	
SILICONE	
POLYURETHANE	

TEMP °F

HYDROGEN SULFIDE, AQ. SOL.

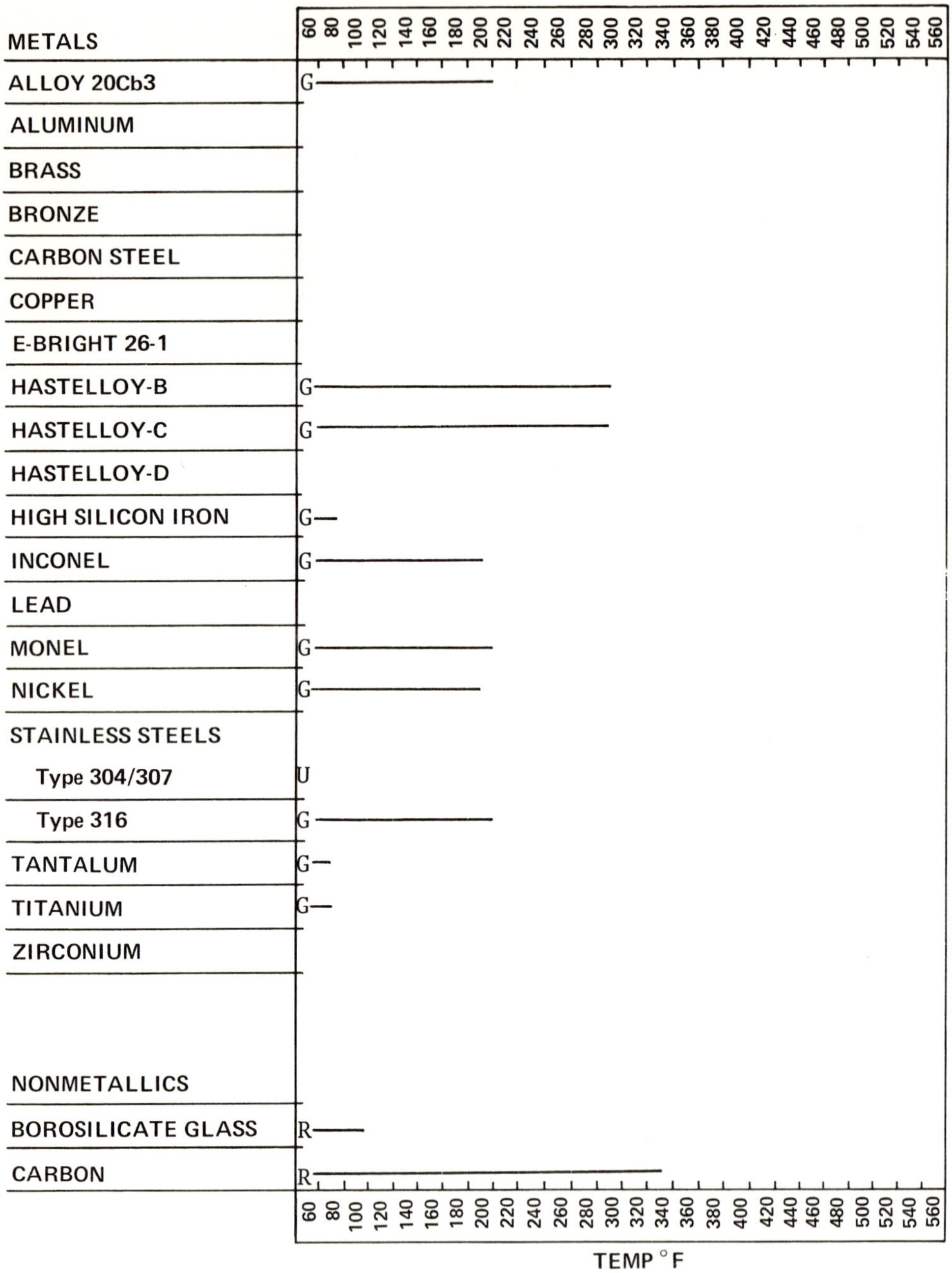

FOR METALS

E = < 2 Mils Penetration/Year; G = < 20 Mils Penetration/Year

S = < 50 Mils Penetration/Year; U = > 50 Mils Penetration/Year

FOR NONMETALLICS

R = Resistant

U = Unsatisfactory

HYDROGEN SULFIDE, AQ. SOL.

TEMP °F

PLASTICS	TEMP °F (60–460)
ABS	R — to 140
ACRYLICS	R — to 80
ASBESTOS REINF., EPOXY	
ASBESTOS REINF., PHENOLIC	
CHLORINATED POLYETHER	R — to 220
CPVC	
EPOXY	R — to 300
FLUOROCARBONS FEP	R — to 400
FLUOROCARBONS TFE	R — to 460
FURFURYL ALCOHOL	R — to 260
NORYL	
PHENOLIC	
POLYAMIDES—NYLON	R — to 80
POLYESTERS	
BISPHENOL A-FUMURATE	
HYDROGENATED BISPHENOL-A-BISPHENOL-A	
ISOPHTHALIC	
CHLORINATED POLYESTERS	R — to 240
POLYSTYRENES	
POLYETHYLENE	R — to 140
POLYPROPYLENE	R — to 180
PVC—TYPE 1	R — to 140
PVC—TYPE 2	R — to 140
POLYVINYLIDENE CHLORIDE	R — to 150
VINYLIDENE FLUORIDE	R — to 220
VINYL ESTER	

RUBBERS	TEMP °F (60–460)
BUTYL GR-1	
FLUORO ELASTOMERS VITON A	U
FLUORO ELASTOMERS KEL-F 3700	
HARD RUBBER	
HYPALON	U
KOROSEAL	
NATURAL RUBBER (GRS)	
NEOPRENE GR-M (CR)	
NITRILE BUNA N (NBR)	U
NORDEL	
POLYBUTADIENE	
SILICONE	
POLYURETHANE	

TEMP °F

HYDROIODIC ACID (49% + 12% IODINE)

TEMP °F: 60, 80, 100, 120, 140, 160, 180, 200, 220, 240, 260, 280, 300, 320, 340, 360, 380, 400, 420, 440, 460, 480, 500, 520, 540, 560

METALS	
ALLOY 20Cb3	
ALUMINUM	U
BRASS	U
BRONZE	
CARBON STEEL	
COPPER	U
E-BRIGHT 26-1	
HASTELLOY-B	G
HASTELLOY-C	G
HASTELLOY-D	
HIGH SILICON IRON	U
INCONEL	U
LEAD	
MONEL	U
NICKEL	U
STAINLESS STEELS	
Type 304/307	U
Type 316	U
TANTALUM	G
TITANIUM	S
ZIRCONIUM	E

NONMETALLICS	
BOROSILICATE GLASS	
CARBON	

TEMP °F

FOR METALS

E = < 2 Mils Penetration/Year; G = < 20 Mils Penetration/Year
S = < 50 Mils Penetration/Year; U = > 50 Mils Penetration/Year

FOR NONMETALLICS

R = Resistant
U = Unsatisfactory

HYDROIODIC ACID (49% + 12% IODINE)

TEMP °F: 60 80 100 120 140 160 180 200 220 240 260 280 300 320 340 360 380 400 420 440 460

PLASTICS	
ABS	
ACRYLICS	
ASBESTOS REINF., EPOXY	
ASBESTOS REINF., PHENOLIC	
CHLORINATED POLYETHER	R
CPVC	
EPOXY	
FLUOROCARBONS FEP	R
FLUOROCARBONS TFE	R
FURFURYL ALCOHOL	
NORYL	
PHENOLIC	
POLYAMIDES—NYLON	
POLYESTERS	
BISPHENOL A-FUMURATE	
HYDROGENATED BISPHENOL-A-BISPHENOL-A	
ISOPHTHALIC 10%	R
CHLORINATED POLYESTERS	
POLYSTYRENES	
POLYETHYLENE	
POLYPROPYLENE	U
PVC—TYPE 1	
PVC—TYPE 2	
POLYVINYLIDENE CHLORIDE	
VINYLIDENE FLUORIDE	R
VINYL ESTER	

RUBBERS	
BUTYL GR-1	
FLUORO ELASTOMERS VITON A	
FLUORO ELASTOMERS KEL-F 3700	
HARD RUBBER	
HYPALON	
KOROSEAL	
NATURAL RUBBER (GRS)	
NEOPRENE GR-M (CR)	
NITRILE BUNA N (NBR)	
NORDEL	R
POLYBUTADIENE	
SILICONE	
POLYURETHANE	

TEMP °F: 60 80 100 120 140 160 180 200 220 240 260 280 300 320 340 360 380 400 420 440 460

HYDROQUINONE

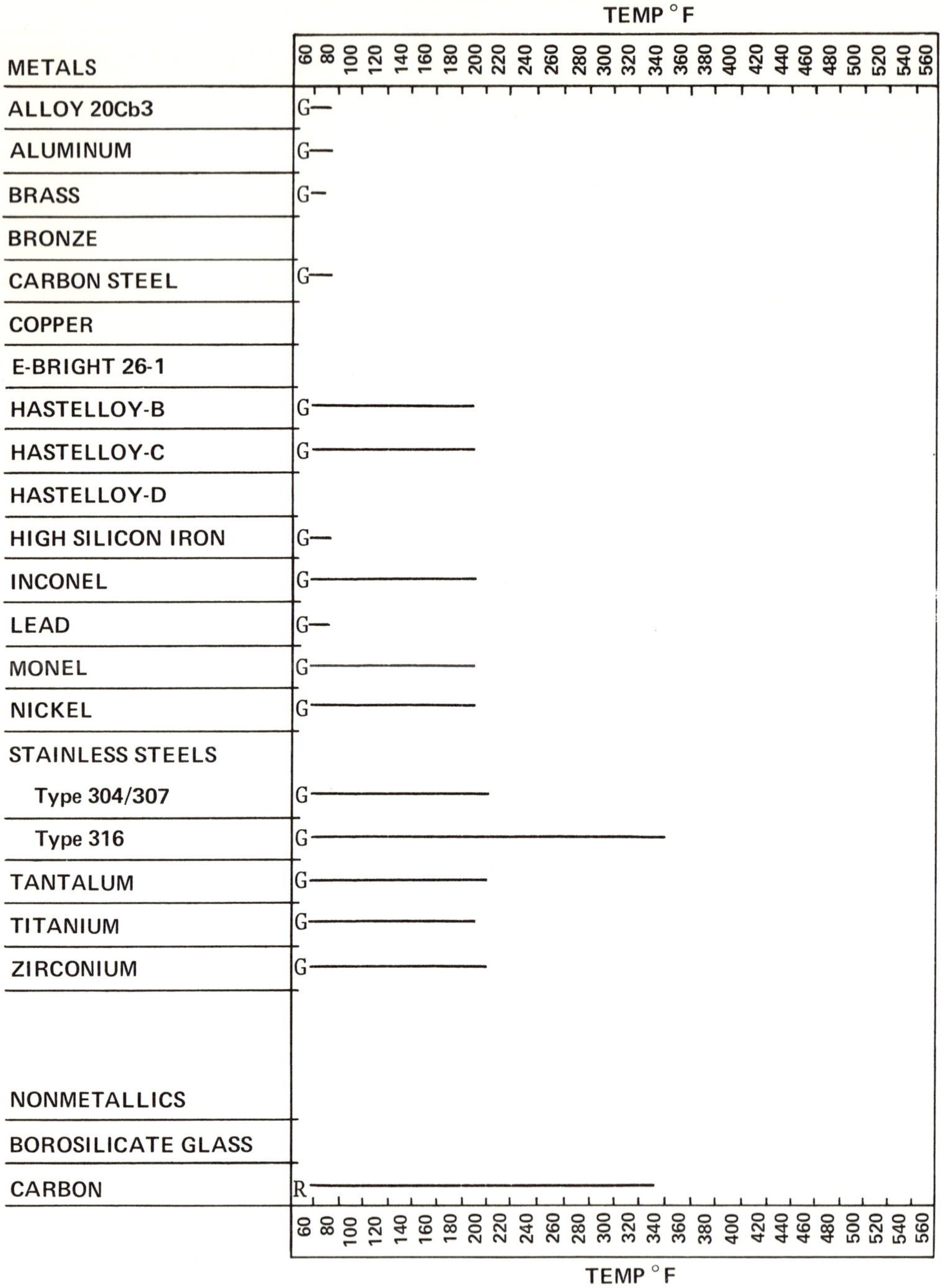

FOR METALS

E = < 2 Mils Penetration/Year; G = < 20 Mils Penetration/Year

S = < 50 Mils Penetration/Year; U = > 50 Mils Penetration/Year

FOR NONMETALLICS

R = Resistant

U = Unsatisfactory

HYDROQUINONE

TEMP °F

PLASTICS	60 80 100 120 140 160 180 200 220 240 260 280 300 320 340 360 380 400 420 440 460
ABS	R
ACRYLICS	
ASBESTOS REINF., EPOXY	
ASBESTOS REINF., PHENOLIC	
CHLORINATED POLYETHER	R
CPVC	R
EPOXY	
FLUOROCARBONS FEP	R
FLUOROCARBONS TFE	R
FURFURYL ALCOHOL	
NORYL	
PHENOLIC	
POLYAMIDES–NYLON	
POLYESTERS	
BISPHENOL A-FUMURATE	
HYDROGENATED BISPHENOL-A-BISPHENOL-A	
ISOPHTHALIC	
CHLORINATED POLYESTERS	
POLYSTYRENES	
POLYETHYLENE	R
POLYPROPYLENE	R
PVC–TYPE 1	R
PVC–TYPE 2	R
POLYVINYLIDENE CHLORIDE	R
VINYLIDENE FLUORIDE	R
VINYL ESTER	
RUBBERS	
BUTYL GR-1	U
FLUORO ELASTOMERS VITON A	R
FLUORO ELASTOMERS KEL-F 3700	
HARD RUBBER	
HYPALON	
KOROSEAL	
NATURAL RUBBER (GRS)	R
NEOPRENE GR-M (CR)	R
NITRILE BUNA N (NBR)	U
NORDEL	R
POLYBUTADIENE	
SILICONE	
POLYURETHANE	

60 80 100 120 140 160 180 200 220 240 260 280 300 320 340 360 380 400 420 440 460

TEMP °F

HYPOCHLOROUS ACID

TEMP °F: 60, 80, 100, 120, 140, 160, 180, 200, 220, 240, 260, 280, 300, 320, 340, 360, 380, 400, 420, 440, 460, 480, 500, 520, 540, 560

METALS	TEMP °F (60–560)
ALLOY 20Cb3	
ALUMINUM	U
BRASS	U
BRONZE	U
CARBON STEEL	U
COPPER	U
E-BRIGHT 26-1	
HASTELLOY-B	G—
HASTELLOY-C	G—
HASTELLOY-D	
HIGH SILICON IRON	
INCONEL	
LEAD	U
MONEL	U
NICKEL	U
STAINLESS STEELS	
Type 304/307	U
Type 316	U
TANTALUM	G————————
TITANIUM	G—
ZIRCONIUM	

NONMETALLICS	TEMP °F (60–560)
BOROSILICATE GLASS	R————
CARBON	

TEMP °F

FOR METALS

E = < 2 Mils Penetration/Year; G = < 20 Mils Penetration/Year

S = < 50 Mils Penetration/Year; U = > 50 Mils Penetration/Year

FOR NONMETALLICS

R = Resistant

U = Unsatisfactory

HYPOCHLOROUS ACID

TEMP °F

PLASTICS		
ABS		R
ACRYLICS		
ASBESTOS REINF., EPOXY		
ASBESTOS REINF., PHENOLIC		
CHLORINATED POLYETHER		R
CPVC		R
EPOXY		R
FLUOROCARBONS FEP		R
FLUOROCARBONS TFE		R
FURFURYL ALCOHOL	10%	U
NORYL		
PHENOLIC		
POLYAMIDES–NYLON		
POLYESTERS		
BISPHENOL A-FUMURATE	20%	R
HYDROGENATED BISPHENOL-A-BISPHENOL-A	50%	R
ISOPHTHALIC		R
CHLORINATED POLYESTERS	10%	R
POLYSTYRENES		
POLYETHYLENE		R
POLYPROPYLENE		R
PVC–TYPE 1		R
PVC–TYPE 2		R
POLYVINYLIDENE CHLORIDE		R
VINYLIDENE FLUORIDE		R
VINYL ESTER	50%	R

RUBBERS	
BUTYL GR-1	U
FLUORO ELASTOMERS VITON A	R
FLUORO ELASTOMERS KEL-F 3700	
HARD RUBBER	R
HYPALON	U
KOROSEAL	
NATURAL RUBBER (GRS)	R
NEOPRENE GR-M (CR)	U
NITRILE BUNA N (NBR)	U
NORDEL	R
POLYBUTADIENE	
SILICONE	
POLYURETHANE	

TEMP °F: 60 80 100 120 140 160 180 200 220 240 260 280 300 320 340 360 380 400 420 440 460

IODINE

TEMP °F: 60 80 100 120 140 160 180 200 220 240 260 280 300 320 340 360 380 400 420 440 460 480 500 520 540 560

METALS		Rating
ALLOY 20Cb3		G—
ALUMINUM		E—
BRASS		U
BRONZE		U
CARBON STEEL	20%	U
COPPER	20%	U
E-BRIGHT 26-1		
HASTELLOY-B		U
HASTELLOY-C	1	E————————————————— (to 500)
HASTELLOY-D		
HIGH SILICON IRON		U
INCONEL		U
LEAD		
MONEL		E—
NICKEL		E—
STAINLESS STEELS		
Type 304/307		U
Type 316		U
TANTALUM	20%	G——
TITANIUM		E—
ZIRCONIUM		

NONMETALLICS	Rating
BOROSILICATE GLASS	
CARBON	U

TEMP °F

FOR METALS

E = < 2 Mils Penetration/Year; G = < 20 Mils Penetration/Year

S = < 50 Mils Penetration/Year; U = > 50 Mils Penetration/Year

FOR NONMETALLICS

R = Resistant

U = Unsatisfactory

Please see page 1 for notes on calibration.

IODINE

TEMP °F: 60 80 100 120 140 160 180 200 220 240 260 280 300 320 340 360 380 400 420 440 460

PLASTICS	
ABS	U
ACRYLICS	
ASBESTOS REINF., EPOXY	
ASBESTOS REINF., PHENOLIC	
CHLORINATED POLYETHER	R
CPVC	U
EPOXY	U
FLUOROCARBONS FEP	R
FLUOROCARBONS TFE	R
FURFURYL ALCOHOL	
NORYL	
PHENOLIC	
POLYAMIDES–NYLON	
POLYESTERS	
BISPHENOL A-FUMURATE	
HYDROGENATED BISPHENOL-A-BISPHENOL-A	
ISOPHTHALIC	
CHLORINATED POLYESTERS	R
POLYSTYRENES	
POLYETHYLENE	U
POLYPROPYLENE	U
PVC–TYPE 1	U
PVC–TYPE 2	U
POLYVINYLIDENE CHLORIDE	
VINYLIDENE FLUORIDE	R
VINYL ESTER	R

RUBBERS	
BUTYL GR-1	
FLUORO ELASTOMERS VITON A	R
FLUORO ELASTOMERS KEL-F 3700	
HARD RUBBER	
HYPALON	U
KOROSEAL	
NATURAL RUBBER (GRS)	U
NEOPRENE GR-M (CR)	
NITRILE BUNA N (NBR)	RU
NORDEL	
POLYBUTADIENE	
SILICONE	
POLYURETHANE	

TEMP °F: 60 80 100 120 140 160 180 200 220 240 260 280 300 320 340 360 380 400 420 440 460

IODINE SOLUTION

TEMP °F: 60, 80, 100, 120, 140, 160, 180, 200, 220, 240, 260, 280, 300, 320, 340, 360, 380, 400, 420, 440, 460, 480, 500, 520, 540, 560

METALS	
ALLOY 20Cb3	G—
ALUMINUM	
BRASS	
BRONZE	
CARBON STEEL	
COPPER	
E-BRIGHT 26-1	
HASTELLOY-B	
HASTELLOY-C	
HASTELLOY-D	
HIGH SILICON IRON	
INCONEL	
LEAD	
MONEL	U
NICKEL	
STAINLESS STEELS	
Type 304/307	U
Type 316	U
TANTALUM	
TITANIUM	G—
ZIRCONIUM	

NONMETALLICS	
BOROSILICATE GLASS	R————
CARBON	

TEMP °F: 60, 80, 100, 120, 140, 160, 180, 200, 220, 240, 260, 280, 300, 320, 340, 360, 380, 400, 420, 440, 460, 480, 500, 520, 540, 560

FOR METALS

E = < 2 Mils Penetration/Year; G = < 20 Mils Penetration/Year
S = < 50 Mils Penetration/Year; U = > 50 Mils Penetration/Year

FOR NONMETALLICS

R = Resistant
U = Unsatisfactory

IODINE SOLUTION

PLASTICS		TEMP °F (60–460)
ABS		
ACRYLICS		
ASBESTOS REINF., EPOXY		
ASBESTOS REINF., PHENOLIC		
CHLORINATED POLYETHER		
CPVC		
EPOXY	sat	R———U
FLUOROCARBONS FEP		R———
FLUOROCARBONS TFE		R———
FURFURYL ALCOHOL		
NORYL		
PHENOLIC		
POLYAMIDES—NYLON		
POLYESTERS		
BISPHENOL A-FUMURATE		
HYDROGENATED BISPHENOL-A-BISPHENOL-A		
ISOPHTHALIC		
CHLORINATED POLYESTERS		
POLYSTYRENES		
POLYETHYLENE		U
POLYPROPYLENE		U
PVC—TYPE 1		
PVC—TYPE 2		
POLYVINYLIDENE CHLORIDE		
VINYLIDENE FLUORIDE		R———
VINYL ESTER		

RUBBERS	TEMP °F (60–460)
BUTYL GR-1	
FLUORO ELASTOMERS VITON A	
FLUORO ELASTOMERS KEL-F 3700	
HARD RUBBER	
HYPALON	
KOROSEAL	
NATURAL RUBBER (GRS)	
NEOPRENE GR-M (CR)	
NITRILE BUNA N (NBR)	
NORDEL	
POLYBUTADIENE	
SILICONE	
POLYURETHANE	

IODINE SOLUTION, 10%

TEMP °F

METALS	60	80	100	120	140	160	180	200	220	240	260	280	300	320	340	360	380	400	420	440	460	480	500	520	540	560
ALLOY 20Cb3																										
ALUMINUM																										
BRASS																										
BRONZE																										
CARBON STEEL																										
COPPER																										
E-BRIGHT 26-1																										
HASTELLOY-B																										
HASTELLOY-C																										
HASTELLOY-D																										
HIGH SILICON IRON																										
INCONEL																										
LEAD																										
MONEL																										
NICKEL																										
STAINLESS STEELS																										
Type 304/307	U																									
Type 316	U																									
TANTALUM																										
TITANIUM																										
ZIRCONIUM																										
NONMETALLICS																										
BOROSILICATE GLASS																										
CARBON																										

TEMP °F

FOR METALS

E = < 2 Mils Penetration/Year; G = < 20 Mils Penetration/Year

S = < 50 Mils Penetration/Year; U = > 50 Mils Penetration/Year

FOR NONMETALLICS

R = Resistant

U = Unsatisfactory

IODINE SOLUTION, 10%

TEMP °F (60–460)

PLASTICS	Rating (60 → 460 °F)
ABS	
ACRYLICS	
ASBESTOS REINF., EPOXY	
ASBESTOS REINF., PHENOLIC	
CHLORINATED POLYETHER	R———U
CPVC	
EPOXY	
FLUOROCARBONS FEP	R——— (to 400)
FLUOROCARBONS TFE	R——— (to 460)
FURFURYL ALCOHOL	
NORYL	
PHENOLIC	
POLYAMIDES–NYLON	
POLYESTERS	
BISPHENOL A-FUMURATE	
HYDROGENATED BISPHENOL-A-BISPHENOL-A	
ISOPHTHALIC	
CHLORINATED POLYESTERS	
POLYSTYRENES	
POLYETHYLENE	U
POLYPROPYLENE	U
PVC–TYPE 1	R—
PVC–TYPE 2	
POLYVINYLIDENE CHLORIDE	
VINYLIDENE FLUORIDE	R———
VINYL ESTER	

RUBBERS	Rating (60 → 460 °F)
BUTYL GR-1	
FLUORO ELASTOMERS VITON A	R———
FLUORO ELASTOMERS KEL-F 3700	
HARD RUBBER	
HYPALON	
KOROSEAL	
NATURAL RUBBER (GRS)	
NEOPRENE GR-M (CR)	
NITRILE BUNA N (NBR)	RU
NORDEL	
POLYBUTADIENE	
SILICONE	
POLYURETHANE	

TEMP °F

IODINE IN ALCOHOL

TEMP °F: 60 80 100 120 140 160 180 200 220 240 260 280 300 320 340 360 380 400 420 440 460 480 500 520 540 560

METALS	
ALLOY 20Cb3	
ALUMINUM	
BRASS	
BRONZE	
CARBON STEEL	
COPPER	
E-BRIGHT 26-1	
HASTELLOY-B	
HASTELLOY-C	G—
HASTELLOY-D	
HIGH SILICON IRON	
INCONEL	
LEAD	
MONEL	
NICKEL	
STAINLESS STEELS	
Type 304/307	
Type 316	G————————
TANTALUM	
TITANIUM 1	G—
ZIRCONIUM	
NONMETALLICS	
BOROSILICATE GLASS	
CARBON	

TEMP °F: 60 80 100 120 140 160 180 200 220 240 260 280 300 320 340 360 380 400 420 440 460 480 500 520 540 560

FOR METALS

E = < 2 Mils Penetration/Year; G = < 20 Mils Penetration/Year

S = < 50 Mils Penetration/Year; U = > 50 Mils Penetration/Year

FOR NONMETALLICS

R = Resistant

U = Unsatisfactory

IODINE IN ALCOHOL

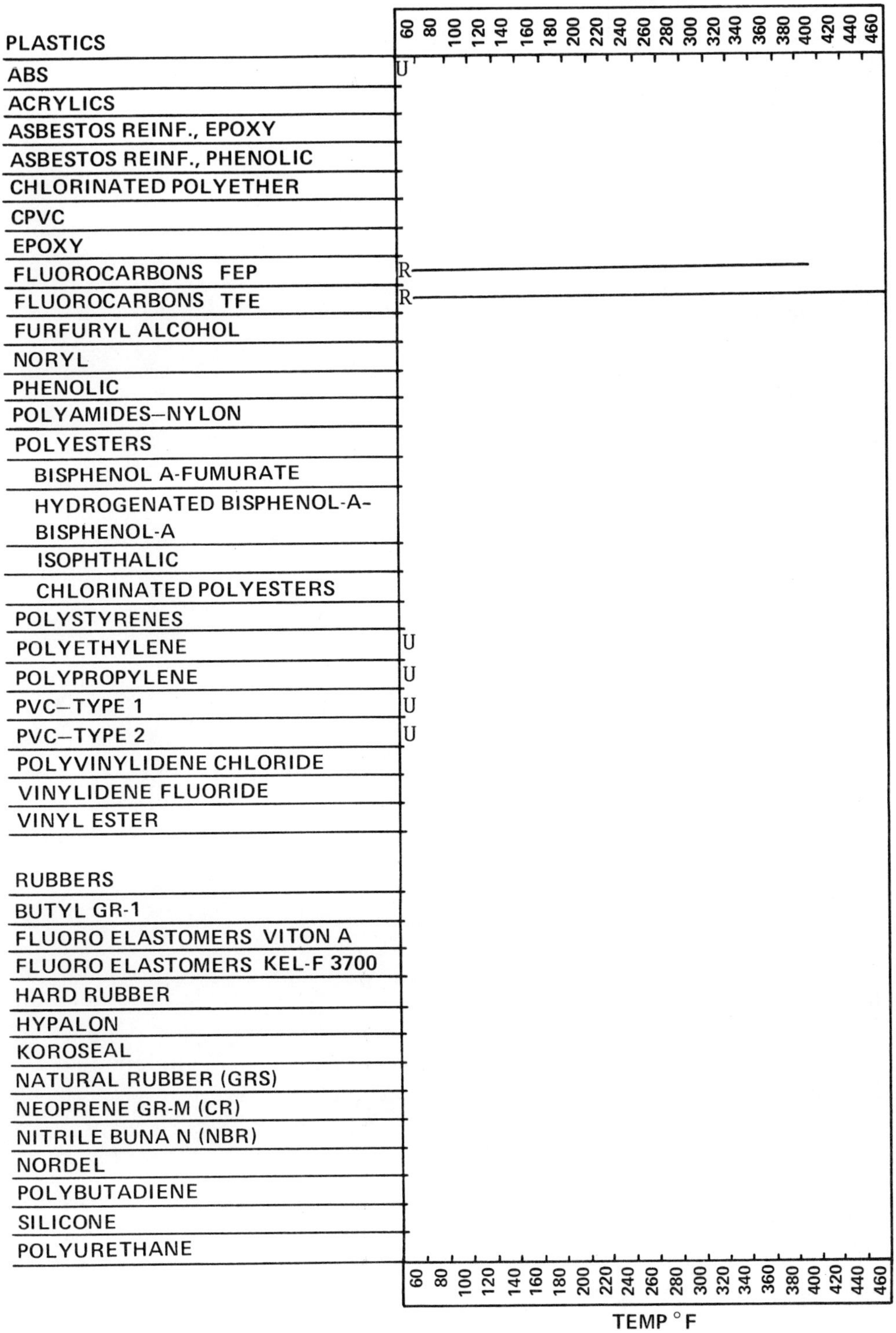

IODOFORM

TEMP °F

METALS		Rating
ALLOY 20Cb3		G
ALUMINUM		G
BRASS		
BRONZE		G
CARBON STEEL		U
COPPER		G
E-BRIGHT 26-1		
HASTELLOY-B		U
HASTELLOY-C		U
HASTELLOY-D		
HIGH SILICON IRON		G
INCONEL		G
LEAD		
MONEL		S
NICKEL		G
STAINLESS STEELS		
Type 304/307	1	E
Type 316		E
TANTALUM		G
TITANIUM		G
ZIRCONIUM		G
NONMETALLICS		
BOROSILICATE GLASS		
CARBON		

TEMP °F: 60 80 100 120 140 160 180 200 220 240 260 280 300 320 340 360 380 400 420 440 460 480 500 520 540 560

FOR METALS

E = < 2 Mils Penetration/Year; G = < 20 Mils Penetration/Year

S = < 50 Mils Penetration/Year; U = > 50 Mils Penetration/Year

FOR NONMETALLICS

R = Resistant

U = Unsatisfactory

IODOFORM

TEMP °F (60, 80, 100, 120, 140, 160, 180, 200, 220, 240, 260, 280, 300, 320, 340, 360, 380, 400, 420, 440, 460)

PLASTICS	
ABS	
ACRYLICS	
ASBESTOS REINF., EPOXY	
ASBESTOS REINF., PHENOLIC	
CHLORINATED POLYETHER	R
CPVC	
EPOXY	
FLUOROCARBONS FEP	R
FLUOROCARBONS TFE	R
FURFURYL ALCOHOL	
NORYL	
PHENOLIC	
POLYAMIDES—NYLON	
POLYESTERS	
BISPHENOL A-FUMURATE	
HYDROGENATED BISPHENOL-A-BISPHENOL-A	
ISOPHTHALIC	
CHLORINATED POLYESTERS	
POLYSTYRENES	
POLYETHYLENE	
POLYPROPYLENE	R
PVC—TYPE 1	
PVC—TYPE 2	
POLYVINYLIDENE CHLORIDE	R
VINYLIDENE FLUORIDE	R
VINYL ESTER	
RUBBERS	
BUTYL GR-1	
FLUORO ELASTOMERS VITON A	R
FLUORO ELASTOMERS KEL-F 3700	
HARD RUBBER	
HYPALON	
KOROSEAL	
NATURAL RUBBER (GRS)	
NEOPRENE GR-M (CR)	
NITRILE BUNA N (NBR)	
NORDEL	
POLYBUTADIENE	
SILICONE	
POLYURETHANE	

TEMP °F (60, 80, 100, 120, 140, 160, 180, 200, 220, 240, 260, 280, 300, 320, 340, 360, 380, 400, 420, 440, 460)

ISOPROPYL ALCOHOL

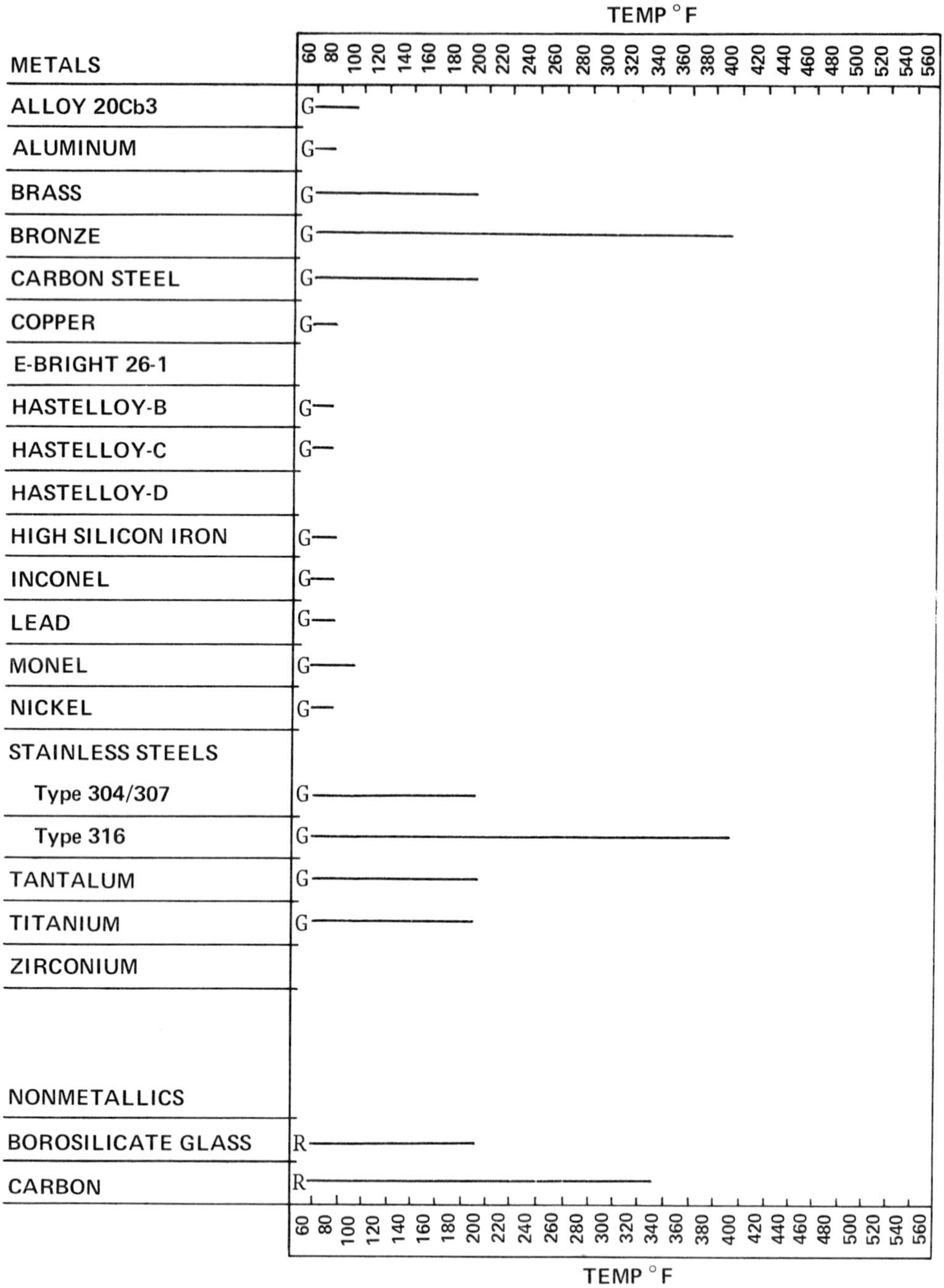

FOR METALS

E = < 2 Mils Penetration/Year; G = < 20 Mils Penetration/Year

S = < 50 Mils Penetration/Year; U = > 50 Mils Penetration/Year

FOR NONMETALLICS

R = Resistant

U = Unsatisfactory

ISOPROPYL ALCOHOL

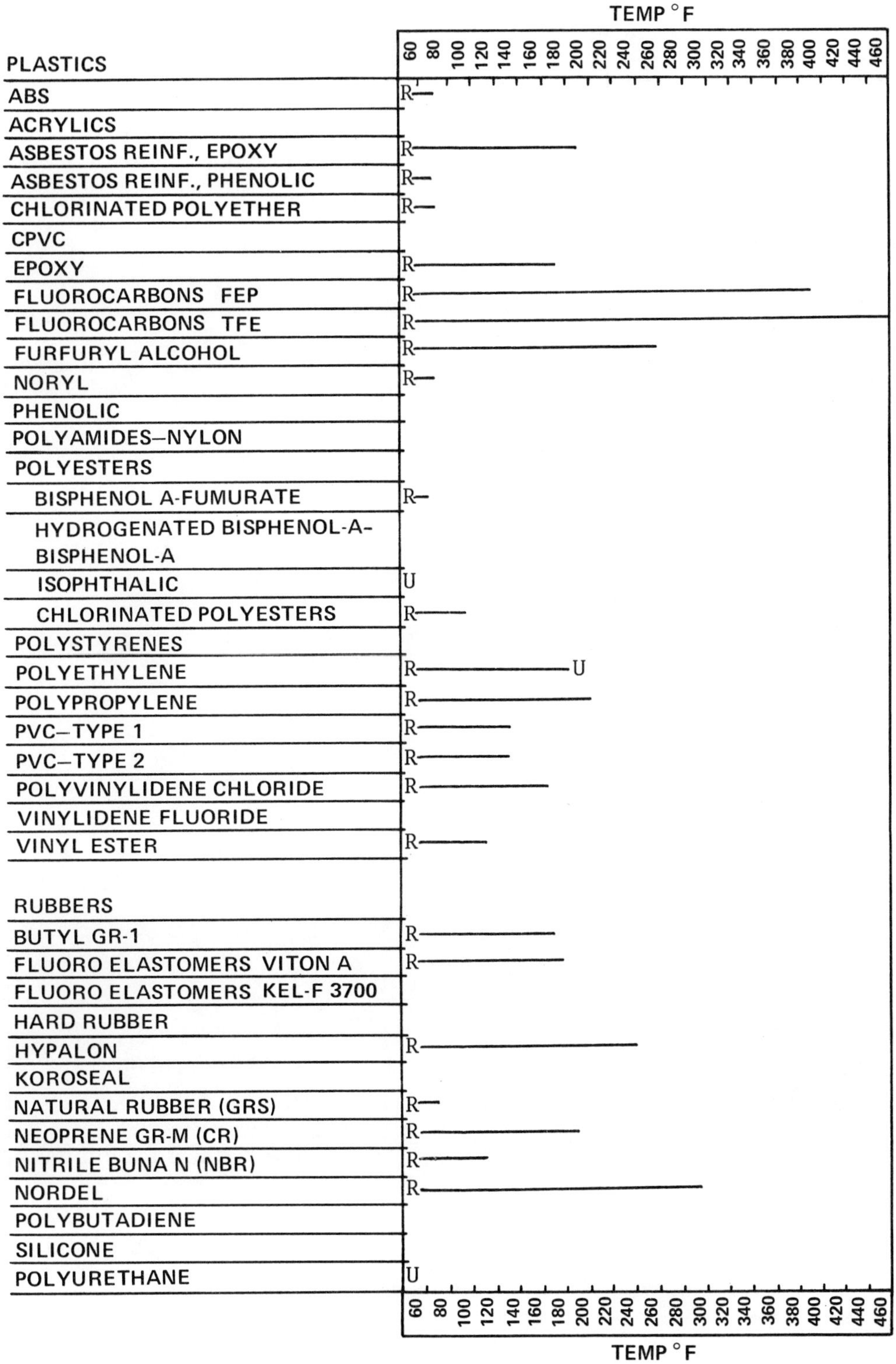

ISOPROPYL ETHER

TEMP °F (60–560)

METALS	TEMP °F
ALLOY 20Cb3	
ALUMINUM	
BRASS	
BRONZE	G (60–200)
CARBON STEEL	
COPPER	
E-BRIGHT 26-1	
HASTELLOY-B	
HASTELLOY-C	
HASTELLOY-D	
HIGH SILICON IRON	
INCONEL	
LEAD	
MONEL	G (60–200)
NICKEL	
STAINLESS STEELS Type 304/307	
Type 316	G (60–200)
TANTALUM	
TITANIUM	
ZIRCONIUM	

NONMETALLICS	TEMP °F
BOROSILICATE GLASS	
CARBON	R (60–340)

FOR METALS

E = < 2 Mils Penetration/Year; G = < 20 Mils Penetration/Year

S = < 50 Mils Penetration/Year; U = > 50 Mils Penetration/Year

FOR NONMETALLICS

R = Resistant

U = Unsatisfactory

ISOPROPYL ETHER

PLASTICS	TEMP °F (60 80 100 120 140 160 180 200 220 240 260 280 300 320 340 360 380 400 420 440 460)
ABS	
ACRYLICS	
ASBESTOS REINF., EPOXY	
ASBESTOS REINF., PHENOLIC	
CHLORINATED POLYETHER	
CPVC	
EPOXY	
FLUOROCARBONS FEP	R——————————— (to 400)
FLUOROCARBONS TFE	R——————————————— (to 460)
FURFURYL ALCOHOL	
NORYL	
PHENOLIC	
POLYAMIDES—NYLON	
POLYESTERS	
BISPHENOL A-FUMURATE	
HYDROGENATED BISPHENOL-A-BISPHENOL-A	
ISOPHTHALIC	
CHLORINATED POLYESTERS	
POLYSTYRENES	
POLYETHYLENE	U
POLYPROPYLENE	U
PVC—TYPE 1	
PVC—TYPE 2	
POLYVINYLIDENE CHLORIDE	
VINYLIDENE FLUORIDE	
VINYL ESTER	
RUBBERS	
BUTYL GR-1	U
FLUORO ELASTOMERS VITON A	U
FLUORO ELASTOMERS KEL-F 3700	
HARD RUBBER	
HYPALON	U
KOROSEAL	U
NATURAL RUBBER (GRS)	R—
NEOPRENE GR-M (CR)	U
NITRILE BUNA N (NBR)	RU
NORDEL	
POLYBUTADIENE	
SILICONE	
POLYURETHANE	R—

TEMP °F

JET FUEL JP-4

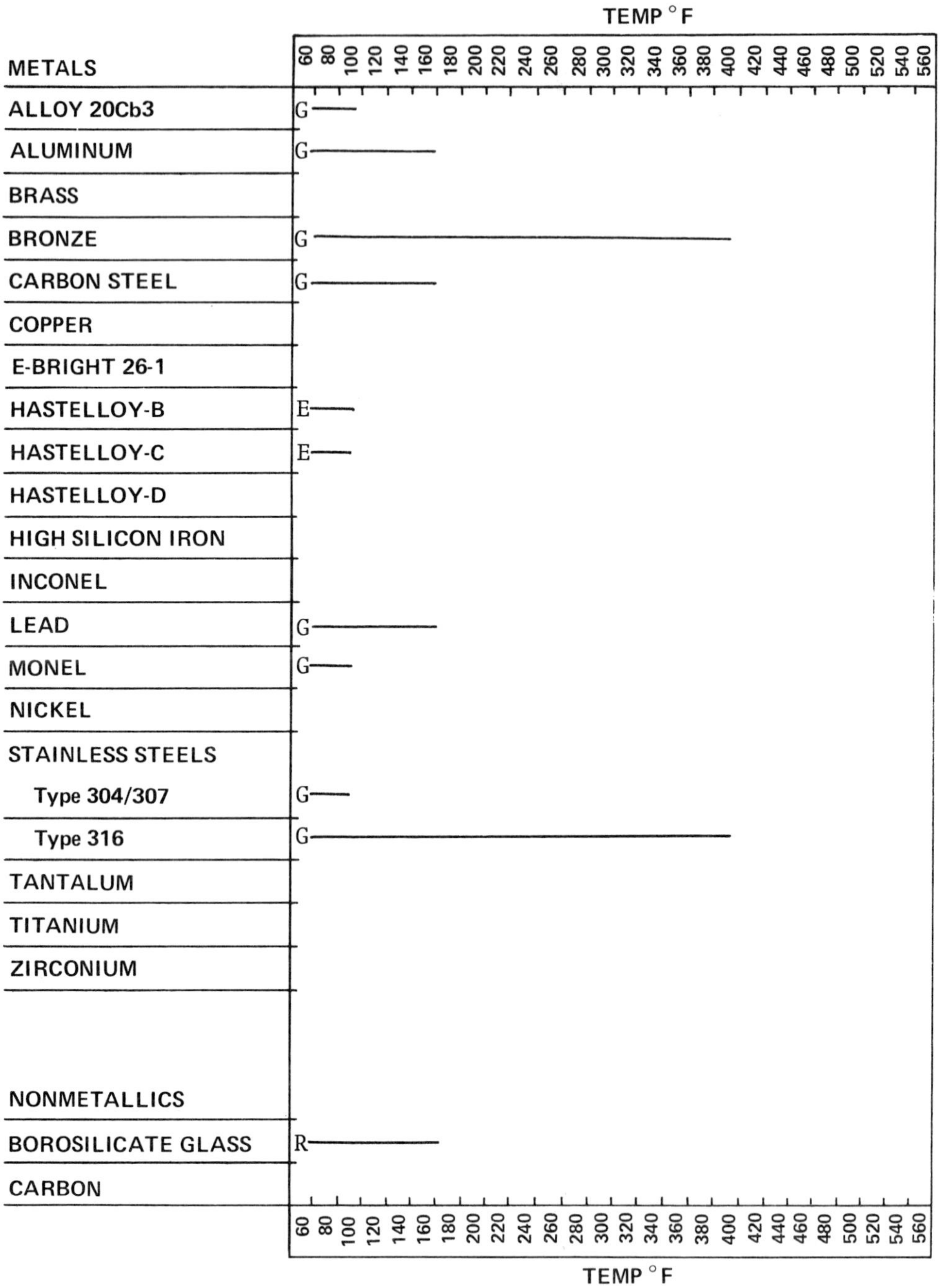

FOR METALS

E = < 2 Mils Penetration/Year; G = < 20 Mils Penetration/Year

S = < 50 Mils Penetration/Year; U = > 50 Mils Penetration/Year

FOR NONMETALLICS

R = Resistant

U = Unsatisfactory

Please see page 1 for notes on calibration.

JET FUEL JP-4

TEMP °F: 60, 80, 100, 120, 140, 160, 180, 200, 220, 240, 260, 280, 300, 320, 340, 360, 380, 400, 420, 440, 460

PLASTICS	
ABS	
ACRYLICS	
ASBESTOS REINF., EPOXY	
ASBESTOS REINF., PHENOLIC	
CHLORINATED POLYETHER	R—
CPVC	R—U
EPOXY	R—
FLUOROCARBONS FEP	R—
FLUOROCARBONS TFE	R—
FURFURYL ALCOHOL	
NORYL	
PHENOLIC	
POLYAMIDES—NYLON	
POLYESTERS	
BISPHENOL A-FUMURATE	R—
HYDROGENATED BISPHENOL-A-BISPHENOL-A	
ISOPHTHALIC	R—
CHLORINATED POLYESTERS	
POLYSTYRENES	
POLYETHYLENE	R—
POLYPROPYLENE	RU
PVC—TYPE 1	R—
PVC—TYPE 2	R—
POLYVINYLIDENE CHLORIDE	R—
VINYLIDENE FLUORIDE	R—
VINYL ESTER	R—

RUBBERS	
BUTYL GR-1	U
FLUORO ELASTOMERS VITON A	R—
FLUORO ELASTOMERS KEL-F 3700	
HARD RUBBER	
HYPALON	U
KOROSEAL	
NATURAL RUBBER (GRS)	U
NEOPRENE GR-M (CR)	U
NITRILE BUNA N (NBR)	R—
NORDEL	U
POLYBUTADIENE	
SILICONE	
POLYURETHANE	U

TEMP °F: 60, 80, 100, 120, 140, 160, 180, 200, 220, 240, 260, 280, 300, 320, 340, 360, 380, 400, 420, 440, 460

JET FUEL JP-5

TEMP °F: 60 80 100 120 140 160 180 200 220 240 260 280 300 320 340 360 380 400 420 440 460 480 500 520 540 560

METALS	
ALLOY 20Cb3	G
ALUMINUM	G
BRASS	
BRONZE	G
CARBON STEEL	G
COPPER	
E-BRIGHT 26-1	
HASTELLOY-B	E
HASTELLOY-C	E
HASTELLOY-D	
HIGH SILICON IRON	
INCONEL	
LEAD	
MONEL	G
NICKEL	
STAINLESS STEELS	
Type 304/307	G
Type 316	G
TANTALUM	
TITANIUM	
ZIRCONIUM	

NONMETALLICS	
BOROSILICATE GLASS	R
CARBON	

TEMP °F: 60 80 100 120 140 160 180 200 220 240 260 280 300 320 340 360 380 400 420 440 460 480 500 520 540 560

FOR METALS

E = < 2 Mils Penetration/Year; G = < 20 Mils Penetration/Year

S = < 50 Mils Penetration/Year; U = > 50 Mils Penetration/Year

FOR NONMETALLICS

R = Resistant

U = Unsatisfactory

JET FUEL JP-5

TEMP °F (60, 80, 100, 120, 140, 160, 180, 200, 220, 240, 260, 280, 300, 320, 340, 360, 380, 400, 420, 440, 460)

PLASTICS	
ABS	
ACRYLICS	
ASBESTOS REINF., EPOXY	
ASBESTOS REINF., PHENOLIC	
CHLORINATED POLYETHER	R—
CPVC	R—U
EPOXY	R—
FLUOROCARBONS FEP	R—
FLUOROCARBONS TFE	R—
FURFURYL ALCOHOL	
NORYL	
PHENOLIC	
POLYAMIDES—NYLON	
POLYESTERS	
BISPHENOL A-FUMURATE	R—
HYDROGENATED BISPHENOL-A-BISPHENOL-A	
ISOPHTHALIC	
CHLORINATED POLYESTERS	
POLYSTYRENES	
POLYETHYLENE	R—
POLYPROPYLENE	RU
PVC—TYPE 1	R—
PVC—TYPE 2	R—
POLYVINYLIDENE CHLORIDE	R—
VINYLIDENE FLUORIDE	R—
VINYL ESTER	

RUBBERS	
BUTYL GR-1	U
FLUORO ELASTOMERS VITON A	R—
FLUORO ELASTOMERS KEL-F 3700	
HARD RUBBER	
HYPALON	U
KOROSEAL	
NATURAL RUBBER (GRS)	U
NEOPRENE GR-M (CR)	U
NITRILE BUNA N (NBR)	R—
NORDEL	
POLYBUTADIENE	
SILICONE	
POLYURETHANE	U

TEMP °F (60, 80, 100, 120, 140, 160, 180, 200, 220, 240, 260, 280, 300, 320, 340, 360, 380, 400, 420, 440, 460)

KEROSENE

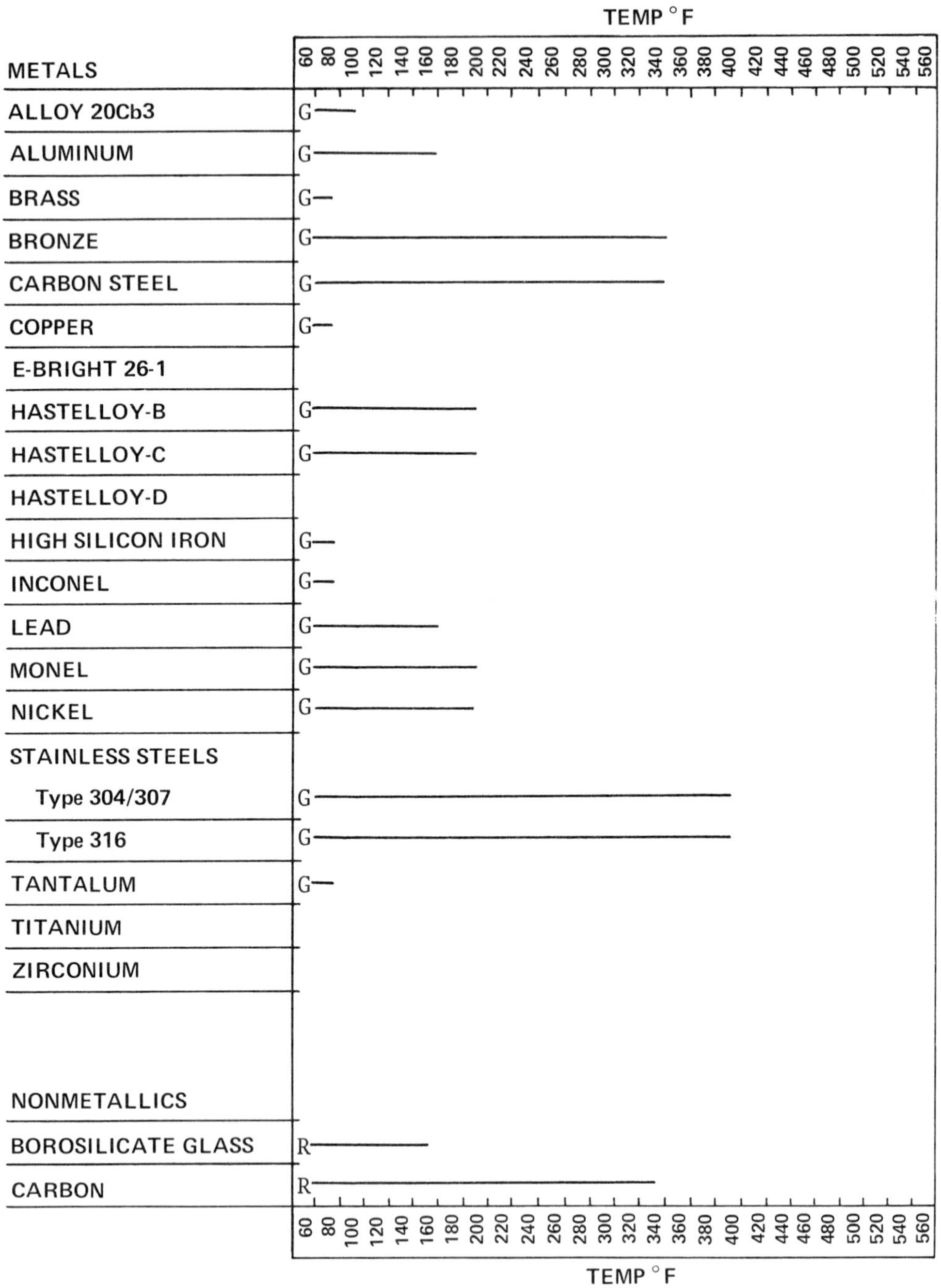

FOR METALS

E = < 2 Mils Penetration/Year; G = < 20 Mils Penetration/Year

S = < 50 Mils Penetration/Year; U = > 50 Mils Penetration/Year

FOR NONMETALLICS

R = Resistant

U = Unsatisfactory

Please see page 1 for notes on calibration.

KEROSENE

TEMP °F (60–460)

Material	Rating
PLASTICS	
ABS	R
ACRYLICS	R
ASBESTOS REINF., EPOXY	R
ASBESTOS REINF., PHENOLIC	R
CHLORINATED POLYETHER	R
CPVC	R
EPOXY	R
FLUOROCARBONS FEP	R
FLUOROCARBONS TFE	R
FURFURYL ALCOHOL	R
NORYL	
PHENOLIC	
POLYAMIDES–NYLON	R
POLYESTERS	
BISPHENOL A-FUMURATE	R
HYDROGENATED BISPHENOL-A-BISPHENOL-A	R
ISOPHTHALIC	R
CHLORINATED POLYESTERS	R
POLYSTYRENES	
POLYETHYLENE	R
POLYPROPYLENE	R — U
PVC–TYPE 1	R
PVC–TYPE 2	R
POLYVINYLIDENE CHLORIDE	R
VINYLIDENE FLUORIDE	R
VINYL ESTER	R
RUBBERS	
BUTYL GR-1	U
FLUORO ELASTOMERS VITON A	R
FLUORO ELASTOMERS KEL-F 3700	
HARD RUBBER	
HYPALON	U
KOROSEAL	
NATURAL RUBBER (GRS)	U
NEOPRENE GR-M (CR)	R
NITRILE BUNA N (NBR)	R
NORDEL	
POLYBUTADIENE	
SILICONE	
POLYURETHANE	U

TEMP °F

KETONES (GENERAL)

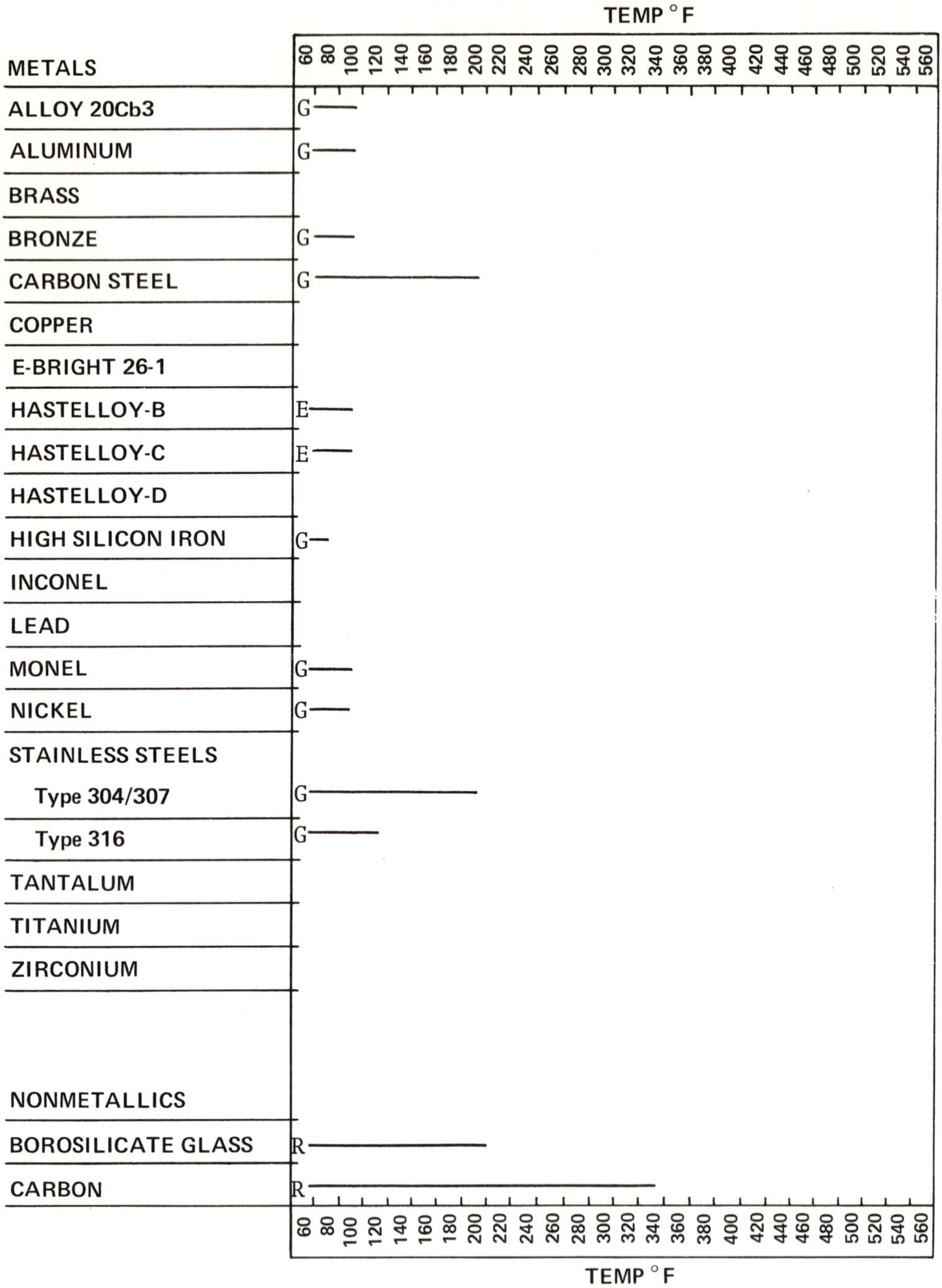

FOR METALS

E = < 2 Mils Penetration/Year; G = < 20 Mils Penetration/Year

S = < 50 Mils Penetration/Year; U = > 50 Mils Penetration/Year

FOR NONMETALLICS

R = Resistant

U = Unsatisfactory

KETONES (GENERAL)

TEMP °F

PLASTICS	60	80	100	120	140	160	180	200	220	240	260	280	300	320	340	360	380	400	420	440	460
ABS																					
ACRYLICS																					
ASBESTOS REINF., EPOXY																					
ASBESTOS REINF., PHENOLIC																					
CHLORINATED POLYETHER																					
CPVC	U																				
EPOXY																					
FLUOROCARBONS FEP	R	—	—	—	—	—	—	—	—	—	—	—	—	—	—	—	—	—			
FLUOROCARBONS TFE	R	—	—	—	—	—	—	—	—	—	—	—	—	—	—	—	—	—	—	—	—
FURFURYL ALCOHOL																					
NORYL																					
PHENOLIC																					
POLYAMIDES–NYLON	R	—																			
POLYESTERS																					
BISPHENOL A-FUMURATE																					
HYDROGENATED BISPHENOL-A-BISPHENOL-A																					
ISOPHTHALIC																					
CHLORINATED POLYESTERS																					
POLYSTYRENES																					
POLYETHYLENE	U																				
POLYPROPYLENE	R	—	—	U																	
PVC–TYPE 1	U																				
PVC–TYPE 2	U																				
POLYVINYLIDENE CHLORIDE	R	—																			
VINYLIDENE FLUORIDE	R	—	—	U																	
VINYL ESTER																					
RUBBERS																					
BUTYL GR-1																					
FLUORO ELASTOMERS VITON A																					
FLUORO ELASTOMERS KEL-F 3700																					
HARD RUBBER																					
HYPALON																					
KOROSEAL																					
NATURAL RUBBER (GRS)																					
NEOPRENE GR-M (CR)	U																				
NITRILE BUNA N (NBR)																					
NORDEL																					
POLYBUTADIENE																					
SILICONE																					
POLYURETHANE																					

60 80 100 120 140 160 180 200 220 240 260 280 300 320 340 360 380 400 420 440 460

TEMP °F

KRAFT LIQUOR

TEMP °F

METALS	60	80	100	120	140	160	180	200	220	240	260	280	300	320	340	360	380	400	420	440	460	480	500	520	540	560
ALLOY 20Cb3																										
ALUMINUM																										
BRASS																										
BRONZE																										
CARBON STEEL																										
COPPER																										
E-BRIGHT 26-1																										
HASTELLOY-B																										
HASTELLOY-C																										
HASTELLOY-D																										
HIGH SILICON IRON																										
INCONEL																										
LEAD																										
MONEL																										
NICKEL																										
STAINLESS STEELS																										
Type 304/307																										
Type 316																										
TANTALUM																										
TITANIUM																										
ZIRCONIUM																										
NONMETALLICS																										
BOROSILICATE GLASS																										
CARBON																										

TEMP °F

FOR METALS

E = < 2 Mils Penetration/Year; G = < 20 Mils Penetration/Year

S = < 50 Mils Penetration/Year; U = > 50 Mils Penetration/Year

FOR NONMETALLICS

R = Resistant

U = Unsatisfactory

KRAFT LIQUOR

TEMP °F (60–460)

PLASTICS	Rating	Line extends to (°F)
ABS	R	140
ACRYLICS		
ASBESTOS REINF., EPOXY		
ASBESTOS REINF., PHENOLIC		
CHLORINATED POLYETHER		
CPVC	R	180
EPOXY		
FLUOROCARBONS FEP	R	400
FLUOROCARBONS TFE	R	460
FURFURYL ALCOHOL		
NORYL		
PHENOLIC		
POLYAMIDES–NYLON		
POLYESTERS		
BISPHENOL A-FUMURATE		
HYDROGENATED BISPHENOL-A-BISPHENOL-A	R	220
ISOPHTHALIC	R	160
CHLORINATED POLYESTERS		
POLYSTYRENES		
POLYETHYLENE	R	150
POLYPROPYLENE	R	140
PVC–TYPE 1	R	140
PVC–TYPE 2	R	140
POLYVINYLIDENE CHLORIDE		
VINYLIDENE FLUORIDE		
VINYL ESTER		

RUBBERS	Rating	Line extends to (°F)
BUTYL GR-1		
FLUORO ELASTOMERS VITON A	R	80
FLUORO ELASTOMERS KEL-F 3700		
HARD RUBBER		
HYPALON	R	80
KOROSEAL		
NATURAL RUBBER (GRS)		
NEOPRENE GR-M (CR)	U	
NITRILE BUNA N (NBR)		
NORDEL		
POLYBUTADIENE		
SILICONE		
POLYURETHANE		

TEMP °F (60–460)

LACTIC ACID

METALS	Notes	Rating	Temperature range (°F)
ALLOY 20Cb3 air free		G	60–300
ALUMINUM	9	G	60–80
BRASS		G	60–80
BRONZE		U	60
CARBON STEEL		U	60
COPPER		G	60–80
E-BRIGHT 26-1			
HASTELLOY-B		G	60–240
HASTELLOY-C		G	60–200
HASTELLOY-D		G	60–240
HIGH SILICON IRON		G	60–80
INCONEL			
LEAD			
MONEL		U	60
NICKEL		U	60
STAINLESS STEELS			
Type 304/307	1,9	G–U	G 60, U 80
Type 316	1,9	G	60–300
TANTALUM		E	60–200
TITANIUM		E	60–300
ZIRCONIUM		E	60–300

NONMETALLICS	Notes	Rating	Temperature range (°F)
BOROSILICATE GLASS		R	60–200
CARBON		R	60–340

TEMP °F

FOR METALS

E = < 2 Mils Penetration/Year; G = < 20 Mils Penetration/Year
S = < 50 Mils Penetration/Year; U = > 50 Mils Penetration/Year

FOR NONMETALLICS

R = Resistant
U = Unsatisfactory

LACTIC ACID

Please see page 1 for notes on calibration.

TEMP °F

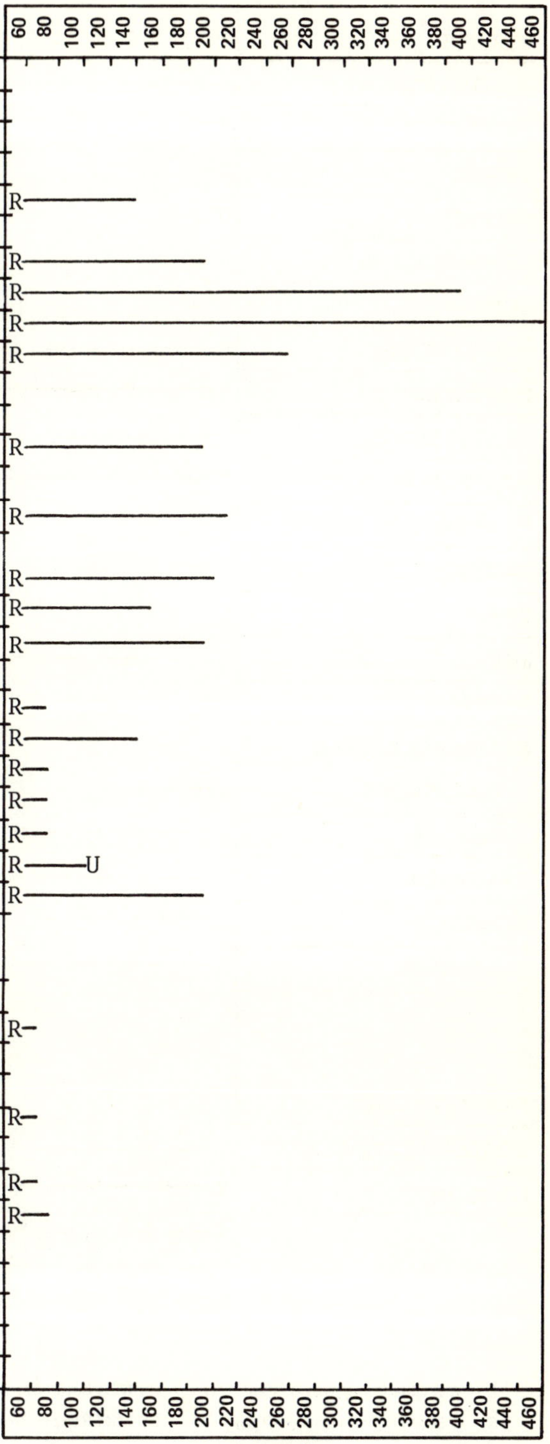

PLASTICS		
ABS		
ACRYLICS		
ASBESTOS REINF., EPOXY		
ASBESTOS REINF., PHENOLIC		
CHLORINATED POLYETHER		R
CPVC		
EPOXY		R
FLUOROCARBONS FEP		R
FLUOROCARBONS TFE		R
FURFURYL ALCOHOL	sat	R
NORYL		
PHENOLIC		
POLYAMIDES–NYLON		R
POLYESTERS		
BISPHENOL A-FUMURATE		R
HYDROGENATED BISPHENOL-A-BISPHENOL-A		R
ISOPHTHALIC	sat	R
CHLORINATED POLYESTERS		R
POLYSTYRENES		
POLYETHYLENE		R
POLYPROPYLENE		R
PVC–TYPE 1		R
PVC–TYPE 2		R
POLYVINYLIDENE CHLORIDE		R
VINYLIDENE FLUORIDE		R—U
VINYL ESTER	sat	R

RUBBERS	
BUTYL GR-1	
FLUORO ELASTOMERS VITON A	R
FLUORO ELASTOMERS KEL-F 3700	
HARD RUBBER	
HYPALON	R
KOROSEAL	
NATURAL RUBBER (GRS)	R
NEOPRENE GR-M (CR)	R
NITRILE BUNA N (NBR)	
NORDEL	
POLYBUTADIENE	
SILICONE	
POLYURETHANE	

LACTIC ACID, 80%

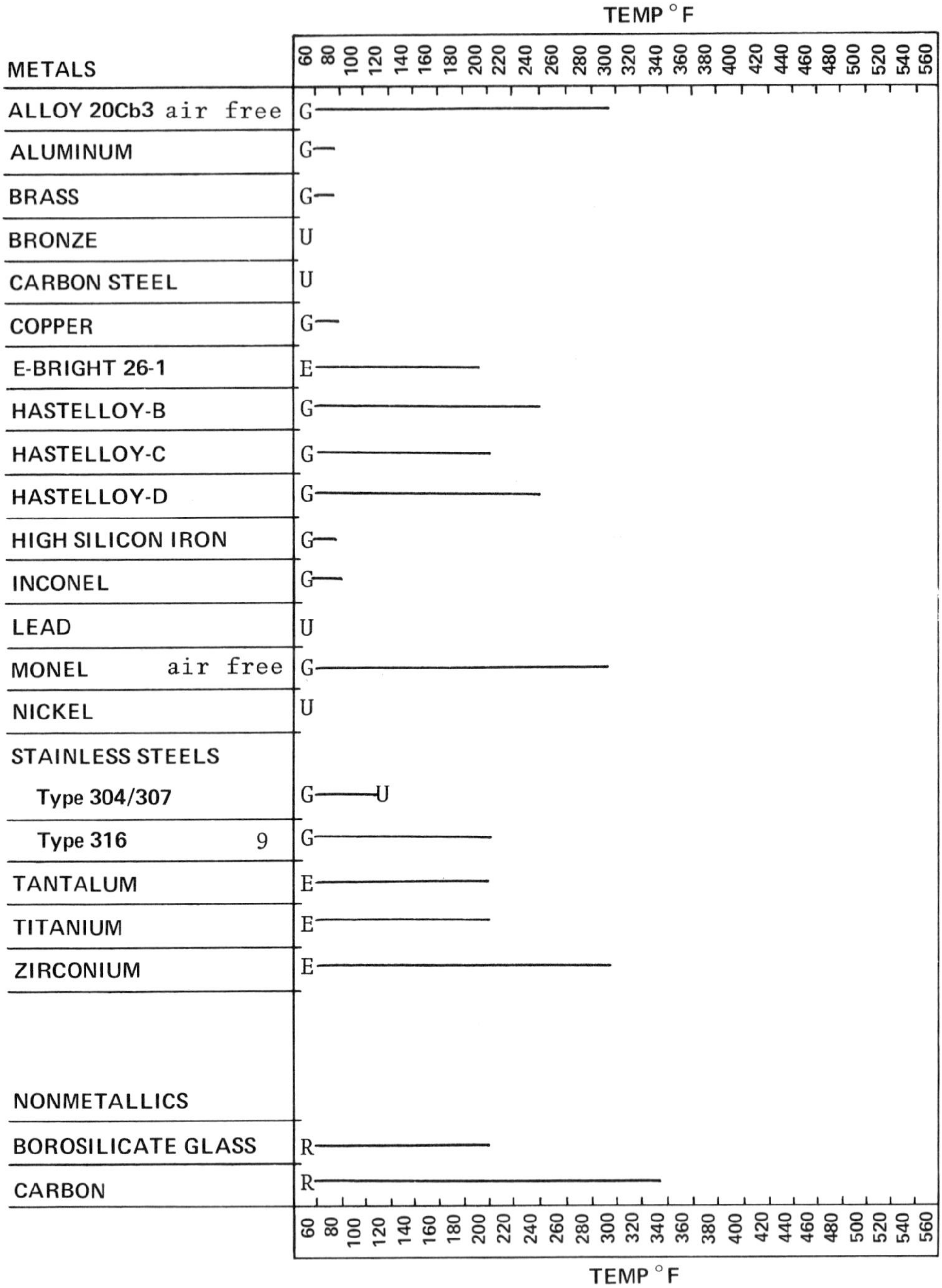

FOR METALS

E = < 2 Mils Penetration/Year; G = < 20 Mils Penetration/Year

S = < 50 Mils Penetration/Year; U = > 50 Mils Penetration/Year

FOR NONMETALLICS

R = Resistant

U = Unsatisfactory

LACTIC ACID, 80%

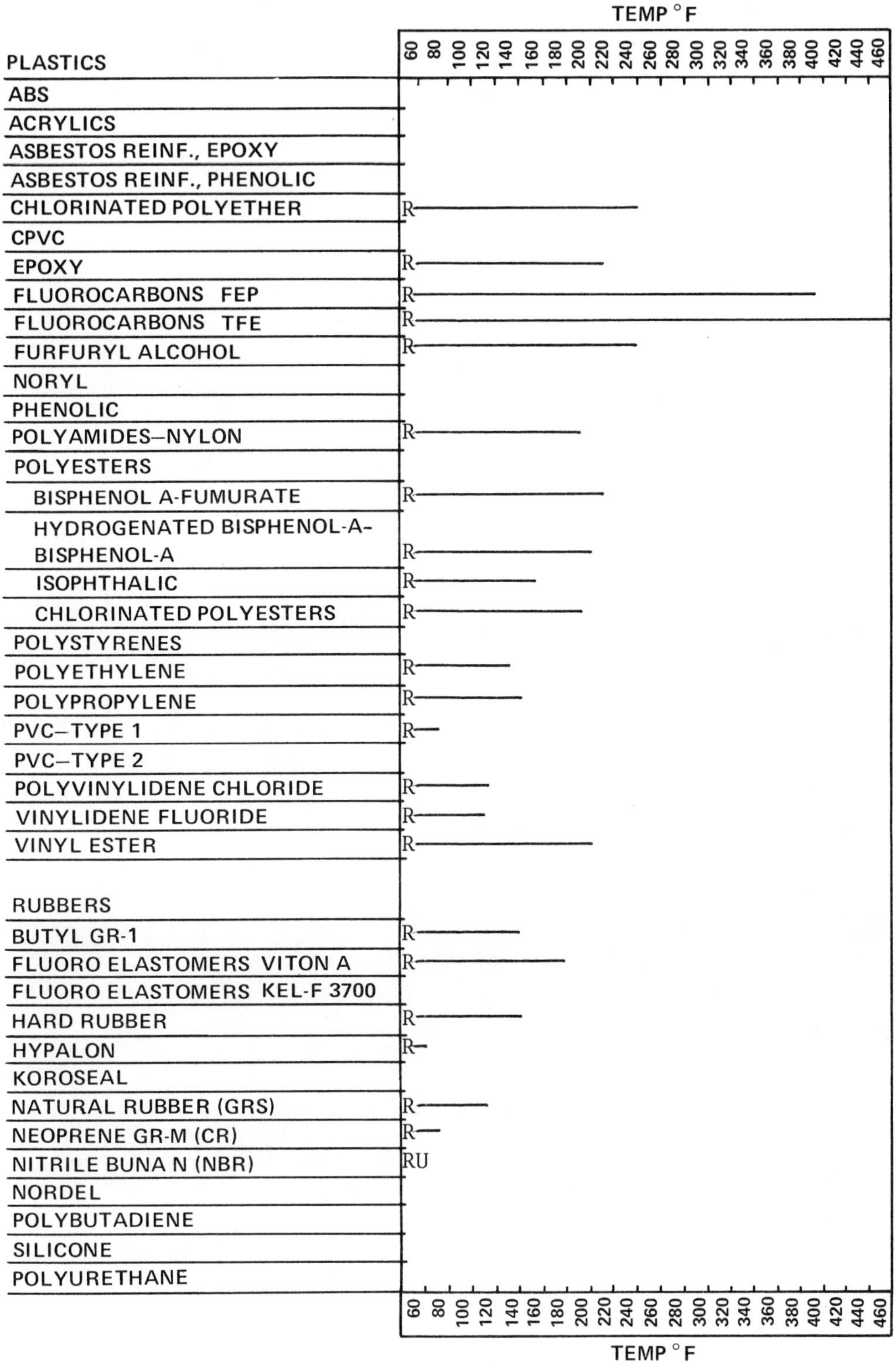

LACTIC ACID, 25%

TEMP °F: 60 80 100 120 140 160 180 200 220 240 260 280 300 320 340 360 380 400 420 440 460 480 500 520 540 560

METALS		
ALLOY 20Cb3	9	E——G——
ALUMINUM		G–U
BRASS		G—
BRONZE		G—
CARBON STEEL		U
COPPER		G—
E-BRIGHT 26-1		
HASTELLOY-B		G——
HASTELLOY-C		E——G——
HASTELLOY-D		G——
HIGH SILICON IRON		G—
INCONEL		G——
LEAD		
MONEL		U
NICKEL		U
STAINLESS STEELS		
Type 304/307	1,9	G——
Type 316	9	E——G——
TANTALUM		E——
TITANIUM		E——
ZIRCONIUM		E——

NONMETALLICS	
BOROSILICATE GLASS	R——
CARBON	R——

TEMP °F: 60 80 100 120 140 160 180 200 220 240 260 280 300 320 340 360 380 400 420 440 460 480 500 520 540 560

FOR METALS

E = < 2 Mils Penetration/Year; G = < 20 Mils Penetration/Year

S = < 50 Mils Penetration/Year; U = > 50 Mils Penetration/Year

FOR NONMETALLICS

R = Resistant

U = Unsatisfactory

LACTIC ACID, 25%

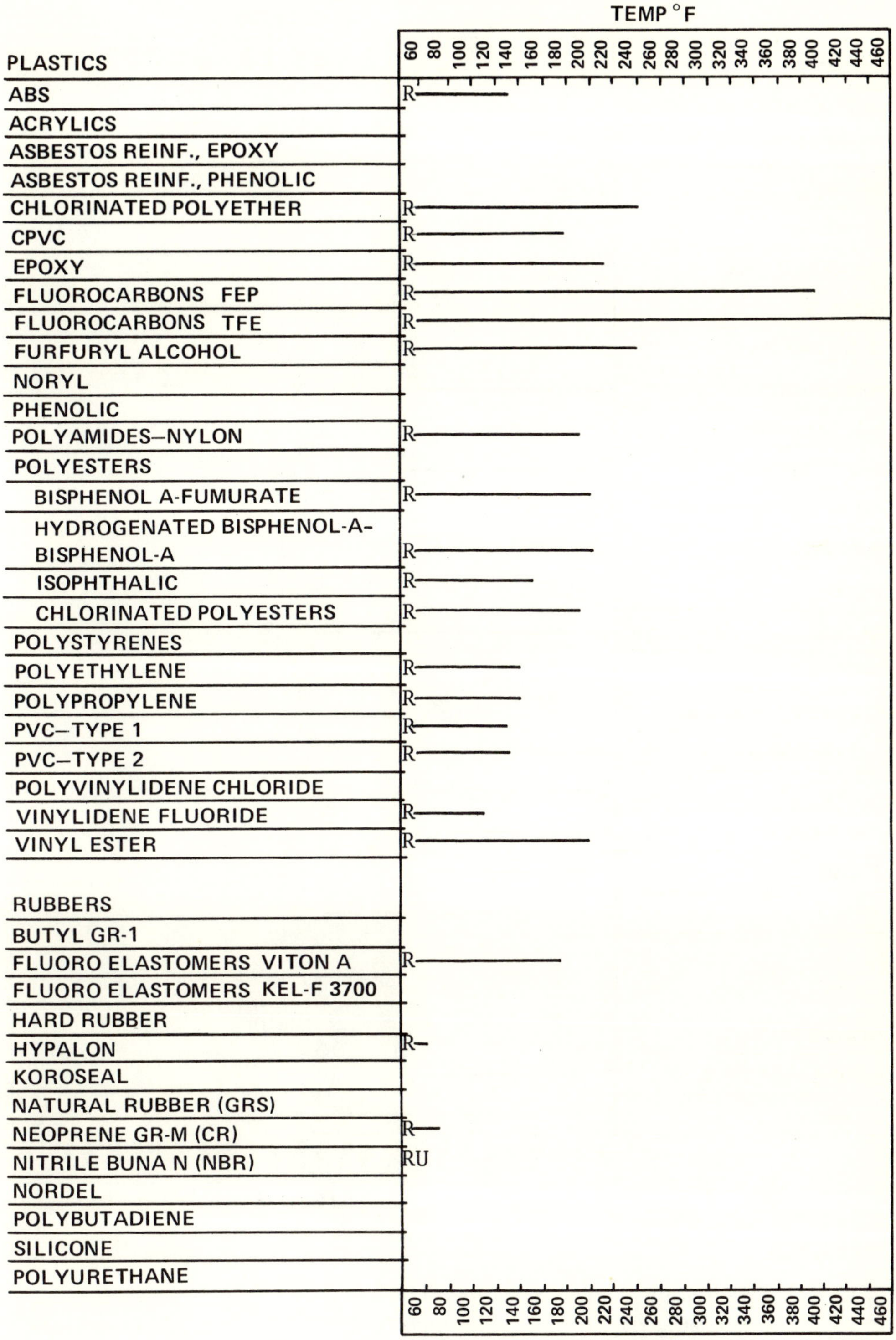

LACTIC ACID, 10%

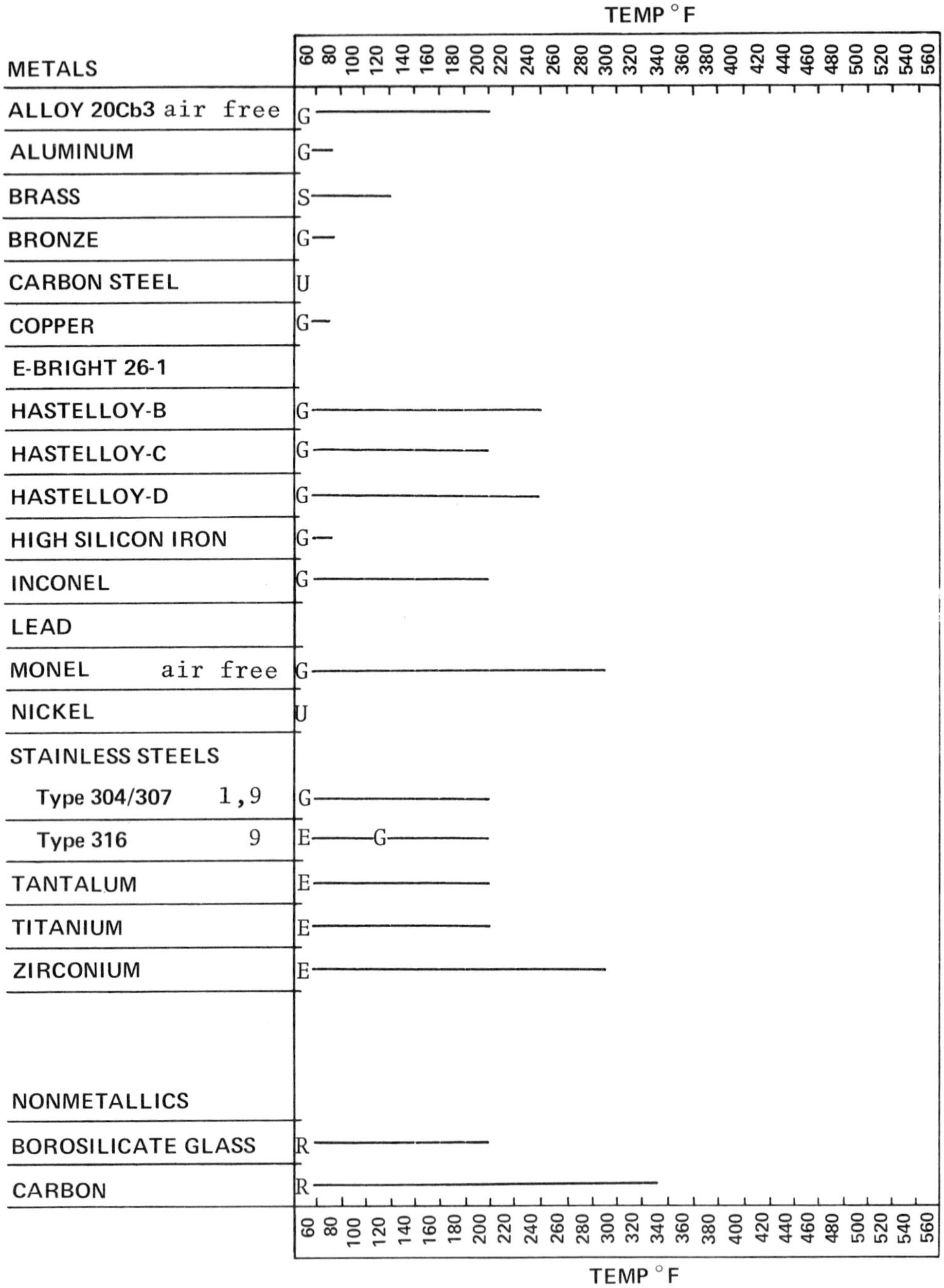

FOR METALS

E = < 2 Mils Penetration/Year; G = < 20 Mils Penetration/Year

S = < 50 Mils Penetration/Year; U = > 50 Mils Penetration/Year

FOR NONMETALLICS

R = Resistant

U = Unsatisfactory

LACTIC ACID, 10%

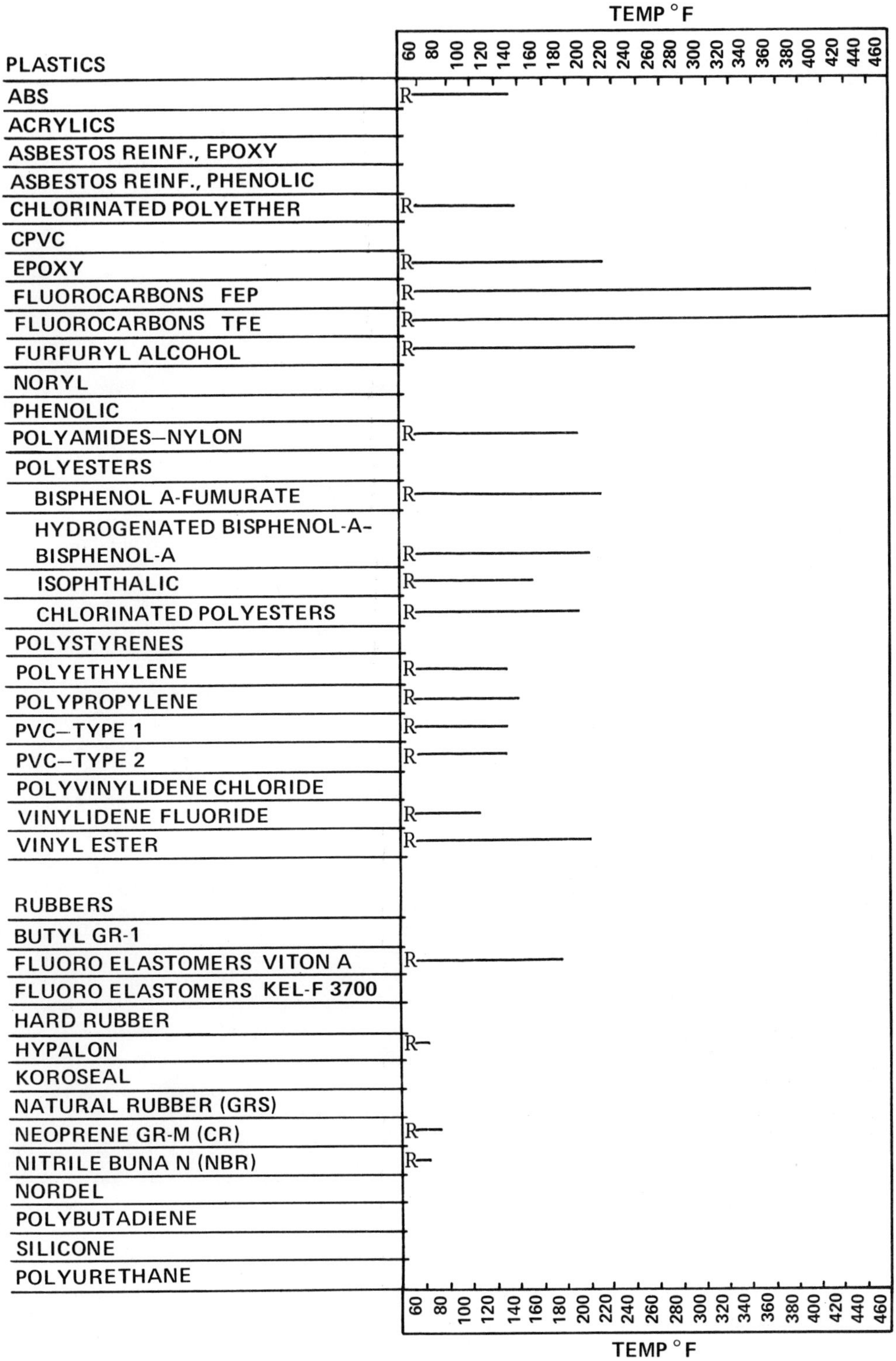

LACTIC ACID, 5%

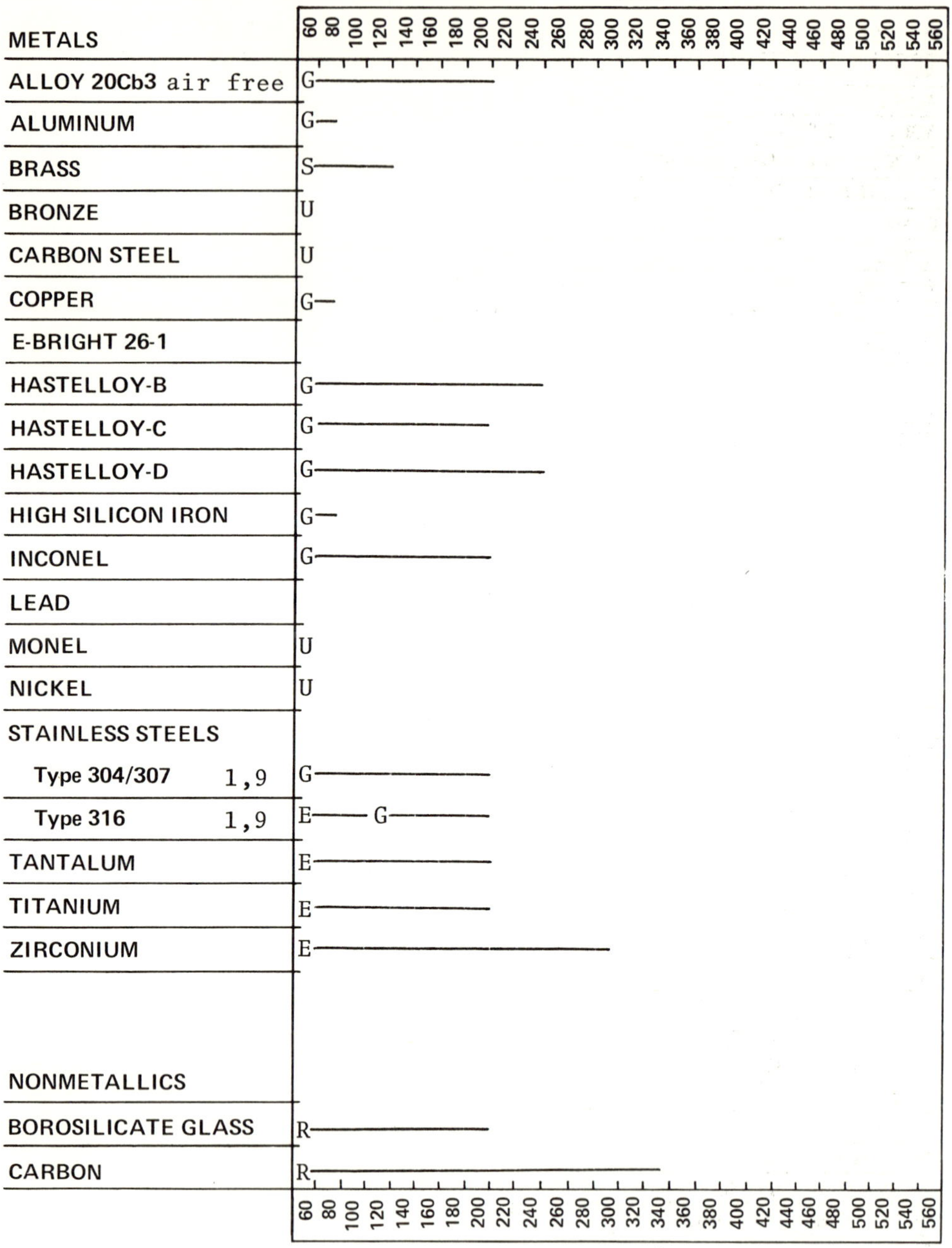

FOR METALS

E = < 2 Mils Penetration/Year; G = < 20 Mils Penetration/Year

S = < 50 Mils Penetration/Year; U = > 50 Mils Penetration/Year

FOR NONMETALLICS

R = Resistant

U = Unsatisfactory

LACTIC ACID, 5%

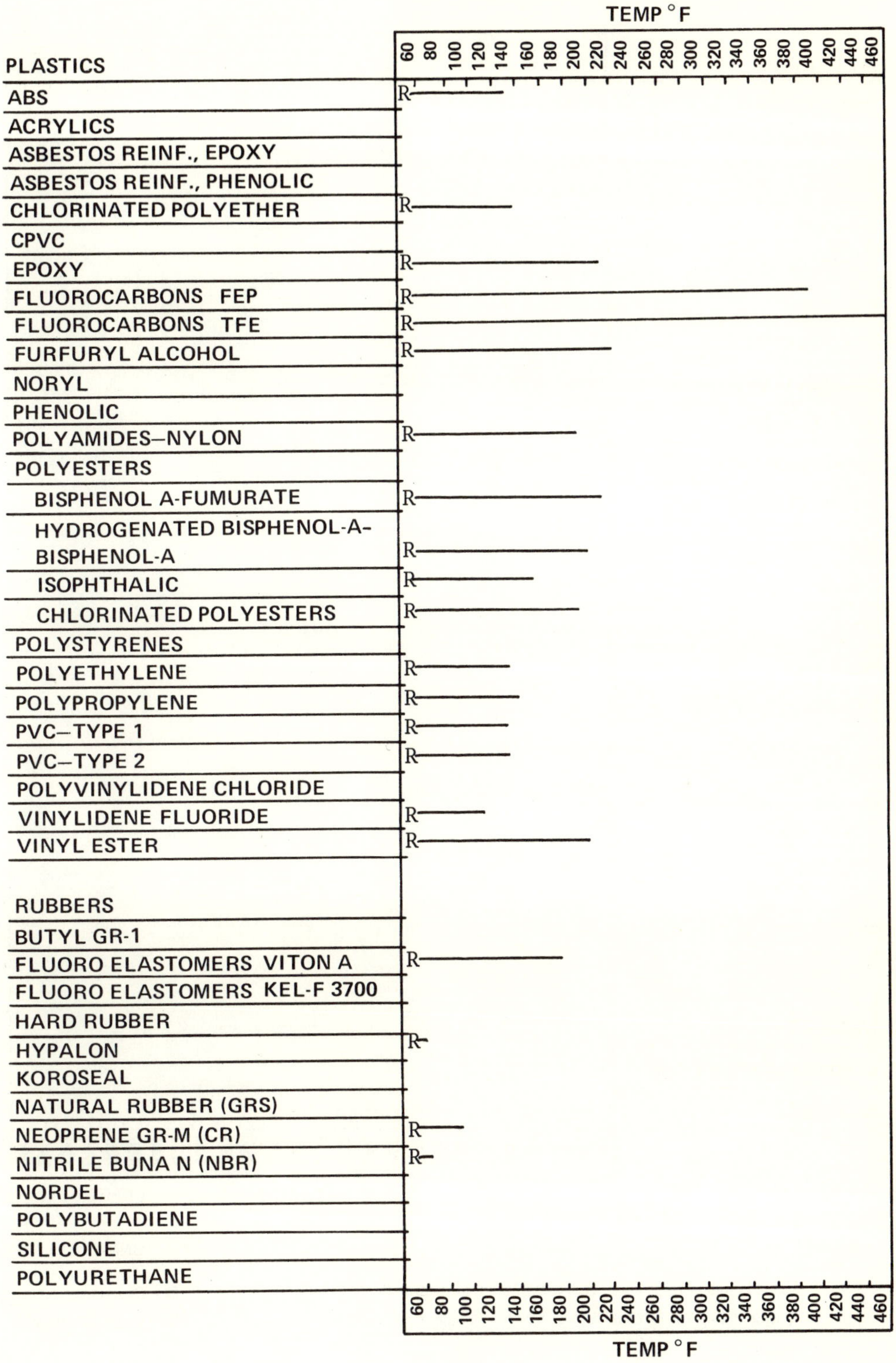

LARD OIL

TEMP °F: 60 80 100 120 140 160 180 200 220 240 260 280 300 320 340 360 380 400 420 440 460 480 500 520 540 560

METALS	
ALLOY 20Cb3	G
ALUMINUM	G
BRASS	
BRONZE	G
CARBON STEEL	S
COPPER	
E-BRIGHT 26-1	
HASTELLOY-B	E
HASTELLOY-C	E
HASTELLOY-D	
HIGH SILICON IRON	
INCONEL	
LEAD	
MONEL	G
NICKEL	
STAINLESS STEELS	
Type 304/307	G
Type 316	G
TANTALUM	E
TITANIUM	
ZIRCONIUM	

NONMETALLICS	
BOROSILICATE GLASS	
CARBON	R

TEMP °F: 60 80 100 120 140 160 180 200 220 240 260 280 300 320 340 360 380 400 420 440 460 480 500 520 540 560

FOR METALS

E = < 2 Mils Penetration/Year; G = < 20 Mils Penetration/Year

S = < 50 Mils Penetration/Year; U = > 50 Mils Penetration/Year

FOR NONMETALLICS

R = Resistant

U = Unsatisfactory

LARD OIL

TEMP °F (60–460)

PLASTICS	
ABS	R
ACRYLICS	
ASBESTOS REINF., EPOXY	
ASBESTOS REINF., PHENOLIC	
CHLORINATED POLYETHER	R
CPVC	R
EPOXY	
FLUOROCARBONS FEP	R
FLUOROCARBONS TFE	R
FURFURYL ALCOHOL	
NORYL	
PHENOLIC	
POLYAMIDES—NYLON	R
POLYESTERS	
BISPHENOL A-FUMURATE	
HYDROGENATED BISPHENOL-A-BISPHENOL-A	
ISOPHTHALIC	
CHLORINATED POLYESTERS	
POLYSTYRENES	
POLYETHYLENE	R
POLYPROPYLENE	R—U
PVC—TYPE 1	R
PVC—TYPE 2	R
POLYVINYLIDENE CHLORIDE	R
VINYLIDENE FLUORIDE	R
VINYL ESTER	

RUBBERS	
BUTYL GR-1	R
FLUORO ELASTOMERS VITON A	R
FLUORO ELASTOMERS KEL-F 3700	
HARD RUBBER	
HYPALON	R
KOROSEAL	
NATURAL RUBBER (GRS)	U
NEOPRENE GR-M (CR)	U
NITRILE BUNA N (NBR)	R
NORDEL	
POLYBUTADIENE	
SILICONE	
POLYURETHANE	R

TEMP °F (60–460)

LAURIC ACID

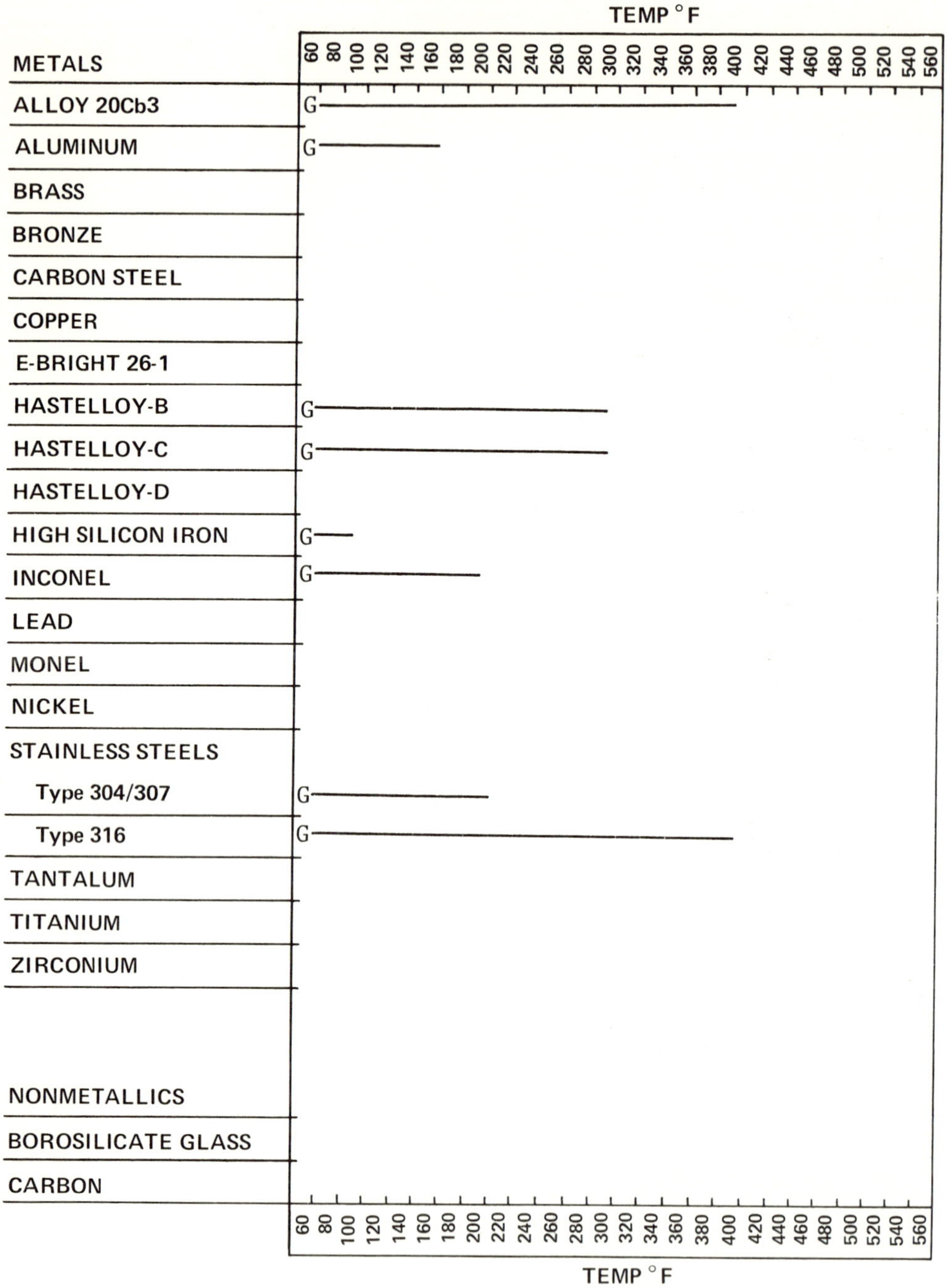

FOR METALS

E = < 2 Mils Penetration/Year; G = < 20 Mils Penetration/Year

S = < 50 Mils Penetration/Year; U = > 50 Mils Penetration/Year

FOR NONMETALLICS

R = Resistant

U = Unsatisfactory

LAURIC ACID

TEMP °F (60–460)

PLASTICS	
ABS	R
ACRYLICS	
ASBESTOS REINF., EPOXY	
ASBESTOS REINF., PHENOLIC	
CHLORINATED POLYETHER	R
CPVC	R
EPOXY	R
FLUOROCARBONS FEP	R
FLUOROCARBONS TFE	R
FURFURYL ALCOHOL	R
NORYL	
PHENOLIC	
POLYAMIDES–NYLON	
POLYESTERS	
BISPHENOL A-FUMURATE	R
HYDROGENATED BISPHENOL-A-BISPHENOL-A	
ISOPHTHALIC sat	R
CHLORINATED POLYESTERS	
POLYSTYRENES	
POLYETHYLENE	R
POLYPROPYLENE	R
PVC–TYPE 1	R
PVC–TYPE 2	R
POLYVINYLIDENE CHLORIDE	R
VINYLIDENE FLUORIDE	R
VINYL ESTER	
RUBBERS	
BUTYL GR-1	
FLUORO ELASTOMERS VITON A	
FLUORO ELASTOMERS KEL-F 3700	
HARD RUBBER	
HYPALON	
KOROSEAL	
NATURAL RUBBER (GRS)	
NEOPRENE GR-M (CR)	
NITRILE BUNA N (NBR)	
NORDEL	
POLYBUTADIENE	
SILICONE	
POLYURETHANE	

TEMP °F (60 80 100 120 140 160 180 200 220 240 260 280 300 320 340 360 380 400 420 440 460)

LAURYL CHLORIDE

TEMP °F

METALS	60	80	100	120	140	160	180	200	220	240	260	280	300	320	340	360	380	400	420	440	460	480	500	520	540	560
ALLOY 20Cb3																										
ALUMINUM																										
BRASS																										
BRONZE																										
CARBON STEEL																										
COPPER																										
E-BRIGHT 26-1																										
HASTELLOY-B																										
HASTELLOY-C																										
HASTELLOY-D																										
HIGH SILICON IRON																										
INCONEL																										
LEAD																										
MONEL																										
NICKEL																										
STAINLESS STEELS Type 304/307																										
Type 316																										
TANTALUM																										
TITANIUM																										
ZIRCONIUM																										
NONMETALLICS																										
BOROSILICATE GLASS																										
CARBON																										
	60	80	100	120	140	160	180	200	220	240	260	280	300	320	340	360	380	400	420	440	460	480	500	520	540	560

TEMP °F

FOR METALS

E = < 2 Mils Penetration/Year; G = < 20 Mils Penetration/Year

S = < 50 Mils Penetration/Year; U = > 50 Mils Penetration/Year

FOR NONMETALLICS

R = Resistant

U = Unsatisfactory

LAURYL CHLORIDE

TEMP °F (60–460)

PLASTICS	Rating
ABS	R
ACRYLICS	
ASBESTOS REINF., EPOXY	
ASBESTOS REINF., PHENOLIC	
CHLORINATED POLYETHER	
CPVC	R
EPOXY	
FLUOROCARBONS FEP	R
FLUOROCARBONS TFE	R
FURFURYL ALCOHOL	R
NORYL	
PHENOLIC	
POLYAMIDES—NYLON	
POLYESTERS	
BISPHENOL A-FUMURATE	
HYDROGENATED BISPHENOL-A-BISPHENOL-A	
ISOPHTHALIC	
CHLORINATED POLYESTERS	R
POLYSTYRENES	
POLYETHYLENE	U
POLYPROPYLENE	U
PVC—TYPE 1	R
PVC—TYPE 2	R
POLYVINYLIDENE CHLORIDE	R
VINYLIDENE FLUORIDE	R
VINYL ESTER	R

RUBBERS	Rating
BUTYL GR-1	
FLUORO ELASTOMERS VITON A	
FLUORO ELASTOMERS KEL-F 3700	
HARD RUBBER	
HYPALON	
KOROSEAL	
NATURAL RUBBER (GRS)	
NEOPRENE GR-M (CR)	
NITRILE BUNA N (NBR)	
NORDEL	
POLYBUTADIENE	
SILICONE	
POLYURETHANE	

TEMP °F (60–460)

LEAD ACETATE

TEMP °F: 60, 80, 100, 120, 140, 160, 180, 200, 220, 240, 260, 280, 300, 320, 340, 360, 380, 400, 420, 440, 460, 480, 500, 520, 540, 560

METALS	
ALLOY 20Cb3	G
ALUMINUM	U
BRASS	
BRONZE	U
CARBON STEEL	U
COPPER	
E-BRIGHT 26-1	
HASTELLOY-B	G
HASTELLOY-C	G
HASTELLOY-D	G
HIGH SILICON IRON	G
INCONEL	G
LEAD	U
MONEL	G
NICKEL	G
STAINLESS STEELS	
Type 304/307	G
Type 316	G
TANTALUM	G
TITANIUM	E
ZIRCONIUM	

NONMETALLICS	
BOROSILICATE GLASS	R
CARBON	R

TEMP °F

FOR METALS

E = < 2 Mils Penetration/Year; G = < 20 Mils Penetration/Year

S = < 50 Mils Penetration/Year; U = > 50 Mils Penetration/Year

FOR NONMETALLICS

R = Resistant

U = Unsatisfactory

LEAD ACETATE

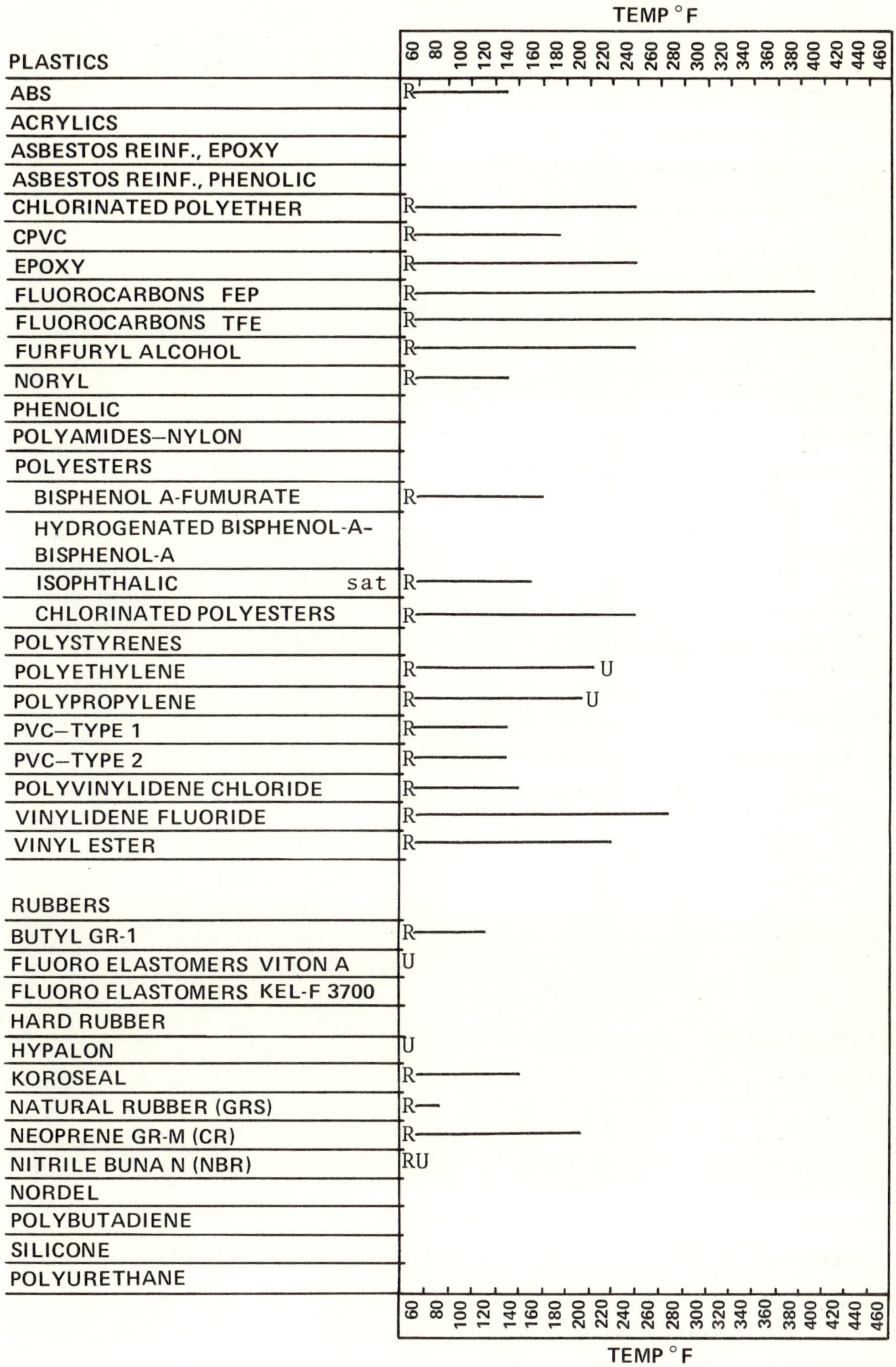

LEAD CHLORIDE

TEMP °F

METALS	Rating
ALLOY 20Cb3	G——
ALUMINUM	U
BRASS	
BRONZE	
CARBON STEEL	
COPPER	
E-BRIGHT 26-1	
HASTELLOY-B	G——
HASTELLOY-C	G——
HASTELLOY-D	G——
HIGH SILICON IRON	
INCONEL	G–
LEAD	
MONEL	
NICKEL	
STAINLESS STEELS	
Type 304/307	G——
Type 316	G——
TANTALUM	G——
TITANIUM	
ZIRCONIUM	
NONMETALLICS	
BOROSILICATE GLASS	
CARBON	

Temperature scale: 60 80 100 120 140 160 180 200 220 240 260 280 300 320 340 360 380 400 420 440 460 480 500 520 540 560

TEMP °F

FOR METALS

E = < 2 Mils Penetration/Year; G = < 20 Mils Penetration/Year

S = < 50 Mils Penetration/Year; U = > 50 Mils Penetration/Year

FOR NONMETALLICS

R = Resistant

U = Unsatisfactory

LEAD CHLORIDE

TEMP °F: 60 80 100 120 140 160 180 200 220 240 260 280 300 320 340 360 380 400 420 440 460

PLASTICS		
ABS		
ACRYLICS		
ASBESTOS REINF., EPOXY		
ASBESTOS REINF., PHENOLIC		
CHLORINATED POLYETHER		
CPVC		R
EPOXY		R
FLUOROCARBONS FEP		R
FLUOROCARBONS TFE		R
FURFURYL ALCOHOL	sat	R
NORYL		
PHENOLIC		
POLYAMIDES—NYLON		
POLYESTERS		
BISPHENOL A-FUMURATE		R
HYDROGENATED BISPHENOL-A-BISPHENOL-A		
ISOPHTHALIC		R
CHLORINATED POLYESTERS		
POLYSTYRENES		
POLYETHYLENE		R
POLYPROPYLENE		
PVC—TYPE 1		R
PVC—TYPE 2		R
POLYVINYLIDENE CHLORIDE		
VINYLIDENE FLUORIDE		
VINYL ESTER		
RUBBERS		
BUTYL GR-1		
FLUORO ELASTOMERS VITON A		R
FLUORO ELASTOMERS KEL-F 3700		
HARD RUBBER		
HYPALON		R
KOROSEAL		
NATURAL RUBBER (GRS)		
NEOPRENE GR-M (CR)		R
NITRILE BUNA N (NBR)		
NORDEL		
POLYBUTADIENE		
SILICONE		
POLYURETHANE		

60 80 100 120 140 160 180 200 220 240 260 280 300 320 340 360 380 400 420 440 460

TEMP °F

LEAD SULFATE

METALS	TEMP °F
ALLOY 20Cb3	G
ALUMINUM	U
BRASS	
BRONZE	
CARBON STEEL	U
COPPER	
E-BRIGHT 26-1	
HASTELLOY-B	G
HASTELLOY-C	G
HASTELLOY-D	G
HIGH SILICON IRON	G
INCONEL	G
LEAD	G
MONEL	G
NICKEL	G
STAINLESS STEELS	
Type 304/307	G
Type 316	G
TANTALUM	G
TITANIUM	
ZIRCONIUM	
NONMETALLICS	
BOROSILICATE GLASS	R
CARBON	

TEMP °F: 60, 80, 100, 120, 140, 160, 180, 200, 220, 240, 260, 280, 300, 320, 340, 360, 380, 400, 420, 440, 460, 480, 500, 520, 540, 560

FOR METALS

E = < 2 Mils Penetration/Year; G = < 20 Mils Penetration/Year

S = < 50 Mils Penetration/Year; U = > 50 Mils Penetration/Year

FOR NONMETALLICS

R = Resistant

U = Unsatisfactory

LEAD SULFATE

TEMP °F

PLASTICS	Rating	Temp range (°F, 60–460 scale)
ABS		
ACRYLICS		
ASBESTOS REINF., EPOXY		
ASBESTOS REINF., PHENOLIC		
CHLORINATED POLYETHER		
CPVC	R	60–180
EPOXY		
FLUOROCARBONS FEP	R	60–400
FLUOROCARBONS TFE	R	60–460
FURFURYL ALCOHOL		
NORYL		
PHENOLIC		
POLYAMIDES–NYLON	R	60–80
POLYESTERS		
BISPHENOL A-FUMURATE		
HYDROGENATED BISPHENOL-A-BISPHENOL-A		
ISOPHTHALIC		
CHLORINATED POLYESTERS		
POLYSTYRENES		
POLYETHYLENE	R	60–140
POLYPROPYLENE		
PVC–TYPE 1	R	60–140
PVC–TYPE 2	R	60–140
POLYVINYLIDENE CHLORIDE	R	60–80
VINYLIDENE FLUORIDE		
VINYL ESTER		

RUBBERS	Rating	Temp range (°F, 60–460 scale)
BUTYL GR-1	R	60–180
FLUORO ELASTOMERS VITON A	R	60–80
FLUORO ELASTOMERS KEL-F 3700		
HARD RUBBER		
HYPALON	R	60–240
KOROSEAL		
NATURAL RUBBER (GRS)	R	60–180
NEOPRENE GR-M (CR)	R	60–180
NITRILE BUNA N (NBR)	R	60–120
NORDEL		
POLYBUTADIENE		
SILICONE		
POLYURETHANE		

TEMP °F

LIME SULFUR

TEMP °F (60–560)

METALS	Rating
ALLOY 20Cb3	
ALUMINUM	U
BRASS	U
BRONZE	S—
CARBON STEEL	
COPPER	S—
E-BRIGHT 26-1	
HASTELLOY-B	
HASTELLOY-C	
HASTELLOY-D	
HIGH SILICON IRON	
INCONEL	
LEAD	
MONEL	G—
NICKEL	
STAINLESS STEELS	
Type 304/307	G—
Type 316	G————————————
TANTALUM	
TITANIUM	
ZIRCONIUM	

NONMETALLICS	Rating
BOROSILICATE GLASS	
CARBON	

TEMP °F

FOR METALS

E = < 2 Mils Penetration/Year; G = < 20 Mils Penetration/Year

S = < 50 Mils Penetration/Year; U = > 50 Mils Penetration/Year

FOR NONMETALLICS

R = Resistant

U = Unsatisfactory

LIME SULFUR

TEMP °F

PLASTICS	60–460 °F
ABS	R
ACRYLICS	
ASBESTOS REINF., EPOXY	
ASBESTOS REINF., PHENOLIC	
CHLORINATED POLYETHER	
CPVC	
EPOXY	
FLUOROCARBONS FEP	R
FLUOROCARBONS TFE	R
FURFURYL ALCOHOL	
NORYL	
PHENOLIC	
POLYAMIDES—NYLON	
POLYESTERS	
BISPHENOL A-FUMURATE	
HYDROGENATED BISPHENOL-A-BISPHENOL-A	
ISOPHTHALIC	
CHLORINATED POLYESTERS	
POLYSTYRENES	
POLYETHYLENE	R
POLYPROPYLENE	R
PVC—TYPE 1	R
PVC—TYPE 2	R
POLYVINYLIDENE CHLORIDE	R
VINYLIDENE FLUORIDE	
VINYL ESTER	

RUBBERS	60–460 °F
BUTYL GR-1	
FLUORO ELASTOMERS VITON A	R
FLUORO ELASTOMERS KEL-F 3700	
HARD RUBBER	
HYPALON	
KOROSEAL	
NATURAL RUBBER (GRS)	
NEOPRENE GR-M (CR)	
NITRILE BUNA N (NBR)	U
NORDEL	R
POLYBUTADIENE	
SILICONE	
POLYURETHANE	

TEMP °F

LINOLEIC ACID

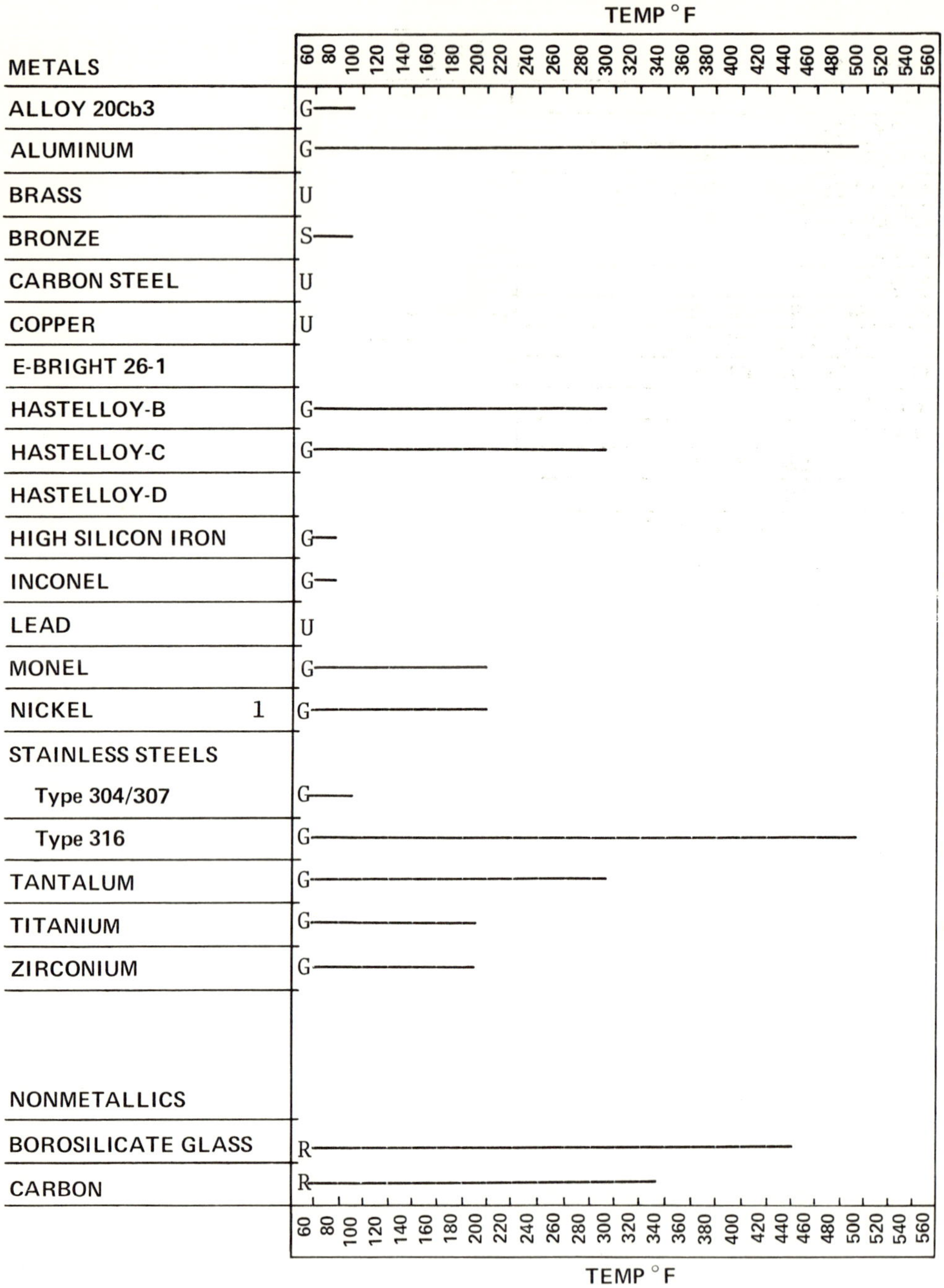

FOR METALS

E = < 2 Mils Penetration/Year; G = < 20 Mils Penetration/Year

S = < 50 Mils Penetration/Year; U = > 50 Mils Penetration/Year

FOR NONMETALLICS

R = Resistant

U = Unsatisfactory

PLASTICS	TEMP °F
ABS	R (60–140)
ACRYLICS	
ASBESTOS REINF., EPOXY	
ASBESTOS REINF., PHENOLIC	
CHLORINATED POLYETHER	R (60–250)
CPVC	R (60–150)
EPOXY	R (60–160)
FLUOROCARBONS FEP	R (60–400)
FLUOROCARBONS TFE	R (60–460)
FURFURYL ALCOHOL	R (60–250)
NORYL	
PHENOLIC	
POLYAMIDES–NYLON	
POLYESTERS	
BISPHENOL A-FUMURATE	
HYDROGENATED BISPHENOL-A-BISPHENOL-A	
ISOPHTHALIC	R (60–160)
CHLORINATED POLYESTERS	
POLYSTYRENES	
POLYETHYLENE	U
POLYPROPYLENE	R (60–140)
PVC–TYPE 1	R (60–140)
PVC–TYPE 2	R (60–140)
POLYVINYLIDENE CHLORIDE	R (60–150)
VINYLIDENE FLUORIDE	R (60–250)
VINYL ESTER	

RUBBERS	TEMP °F
BUTYL GR-1	U
FLUORO ELASTOMERS VITON A	R (60–140) U
FLUORO ELASTOMERS KEL-F 3700	
HARD RUBBER	
HYPALON	U
KOROSEAL	
NATURAL RUBBER (GRS)	U
NEOPRENE GR-M (CR)	U
NITRILE BUNA N (NBR)	R (60–140) U
NORDEL	
POLYBUTADIENE	
SILICONE	
POLYURETHANE	

LINSEED OIL

TEMP °F

METALS	Rating
ALLOY 20Cb3	E
ALUMINUM	G
BRASS	G
BRONZE	G
CARBON STEEL	G
COPPER	G
E-BRIGHT 26-1	
HASTELLOY-B	G
HASTELLOY-C	G
HASTELLOY-D	
HIGH SILICON IRON	
INCONEL	G
LEAD	
MONEL	G
NICKEL	G
STAINLESS STEELS	
Type 304/307	G
Type 316	G
TANTALUM	G
TITANIUM	E
ZIRCONIUM	

NONMETALLICS	Rating
BOROSILICATE GLASS	R
CARBON	R

TEMP °F: 60 80 100 120 140 160 180 200 220 240 260 280 300 320 340 360 380 400 420 440 460 480 500 520 540 560

FOR METALS

E = < 2 Mils Penetration/Year; G = < 20 Mils Penetration/Year

S = < 50 Mils Penetration/Year; U = > 50 Mils Penetration/Year

FOR NONMETALLICS

R = Resistant

U = Unsatisfactory

LINSEED OIL

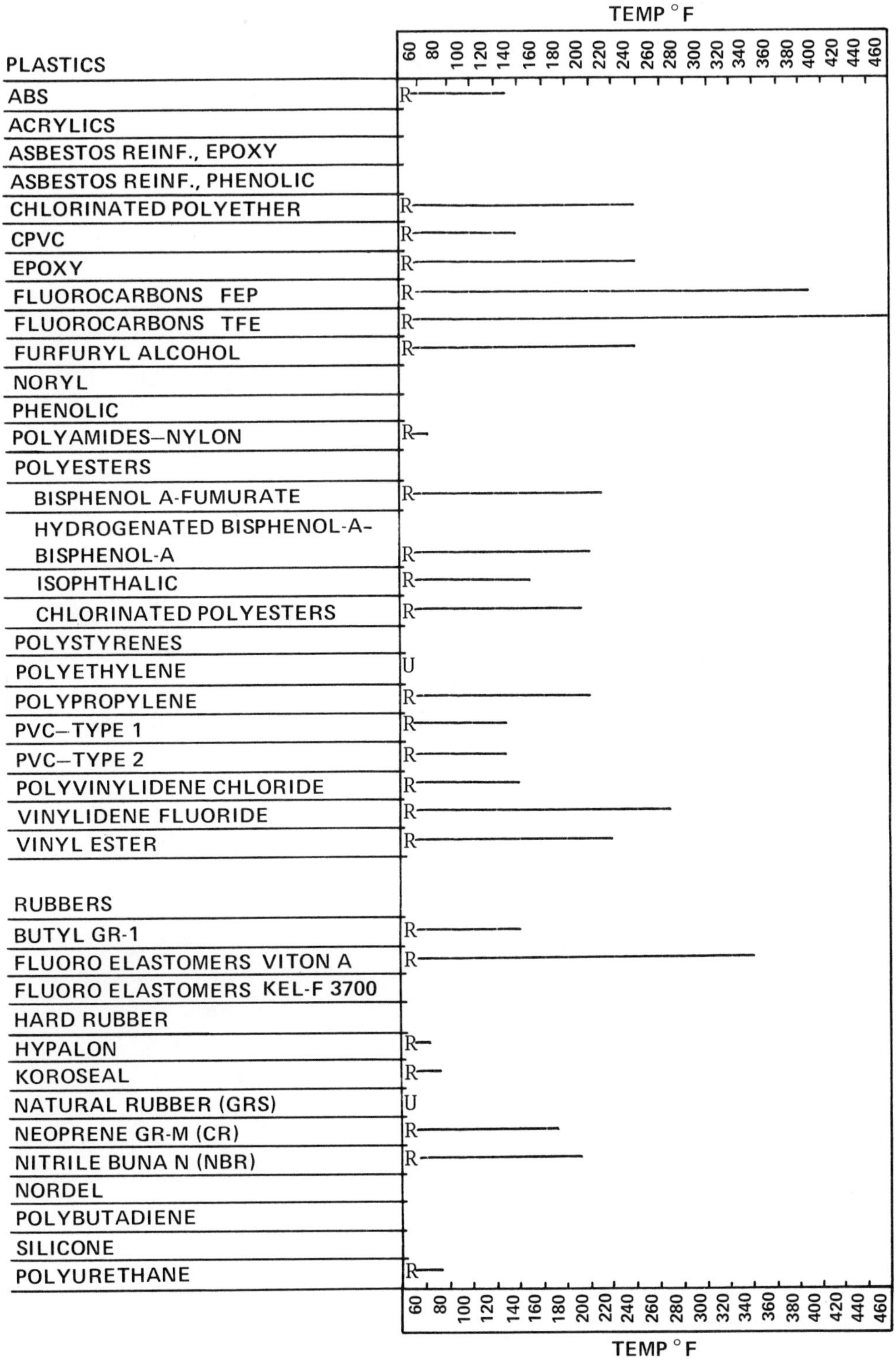
TEMP °F
60 80 100 120 140 160 180 200 220 240 260 280 300 320 340 360 380 400 420 440 460
PLASTICS
ABS R
ACRYLICS
ASBESTOS REINF., EPOXY
ASBESTOS REINF., PHENOLIC
CHLORINATED POLYETHER R
CPVC R
EPOXY R
FLUOROCARBONS FEP R
FLUOROCARBONS TFE R
FURFURYL ALCOHOL R
NORYL
PHENOLIC
POLYAMIDES–NYLON R
POLYESTERS
BISPHENOL A-FUMURATE R
HYDROGENATED BISPHENOL-A-BISPHENOL-A R
ISOPHTHALIC R
CHLORINATED POLYESTERS R
POLYSTYRENES
POLYETHYLENE U
POLYPROPYLENE R
PVC–TYPE 1 R
PVC–TYPE 2 R
POLYVINYLIDENE CHLORIDE R
VINYLIDENE FLUORIDE R
VINYL ESTER R
RUBBERS
BUTYL GR-1 R
FLUORO ELASTOMERS VITON A R
FLUORO ELASTOMERS KEL-F 3700
HARD RUBBER
HYPALON R
KOROSEAL R
NATURAL RUBBER (GRS) U
NEOPRENE GR-M (CR) R
NITRILE BUNA N (NBR) R
NORDEL
POLYBUTADIENE
SILICONE
POLYURETHANE R
60 80 100 120 140 160 180 200 220 240 260 280 300 320 340 360 380 400 420 440 460
TEMP °F

LIQUEURS

METALS	TEMP °F (60–560)
ALLOY 20Cb3	
ALUMINUM	
BRASS	
BRONZE	
CARBON STEEL	
COPPER	
E-BRIGHT 26-1	
HASTELLOY-B	
HASTELLOY-C	
HASTELLOY-D	
HIGH SILICON IRON	
INCONEL	
LEAD	
MONEL	
NICKEL	
STAINLESS STEELS Type 304/307	
Type 316	
TANTALUM	
TITANIUM	
ZIRCONIUM	
NONMETALLICS	
BOROSILICATE GLASS	R (60 to about 200)
CARBON	

FOR METALS

E = < 2 Mils Penetration/Year; G = < 20 Mils Penetration/Year

S = < 50 Mils Penetration/Year; U = > 50 Mils Penetration/Year

FOR NONMETALLICS

R = Resistant

U = Unsatisfactory

LIQUEURS

TEMP °F: 60 80 100 120 140 160 180 200 220 240 260 280 300 320 340 360 380 400 420 440 460

PLASTICS	
ABS	R
ACRYLICS	
ASBESTOS REINF., EPOXY	
ASBESTOS REINF., PHENOLIC	
CHLORINATED POLYETHER	
CPVC	R
EPOXY	
FLUOROCARBONS FEP	R
FLUOROCARBONS TFE	R
FURFURYL ALCOHOL	
NORYL	
PHENOLIC	
POLYAMIDES—NYLON	
POLYESTERS	
BISPHENOL A-FUMURATE	
HYDROGENATED BISPHENOL-A-BISPHENOL-A	
ISOPHTHALIC	
CHLORINATED POLYESTERS	
POLYSTYRENES	
POLYETHYLENE	R
POLYPROPYLENE	R
PVC—TYPE 1	R
PVC—TYPE 2	R
POLYVINYLIDENE CHLORIDE	
VINYLIDENE FLUORIDE	
VINYL ESTER	

RUBBERS	
BUTYL GR-1	
FLUORO ELASTOMERS VITON A	
FLUORO ELASTOMERS KEL-F 3700	
HARD RUBBER	
HYPALON	
KOROSEAL	
NATURAL RUBBER (GRS)	
NEOPRENE GR-M (CR)	
NITRILE BUNA N (NBR)	
NORDEL	
POLYBUTADIENE	
SILICONE	
POLYURETHANE	

TEMP °F: 60 80 100 120 140 160 180 200 220 240 260 280 300 320 340 360 380 400 420 440 460

LITHIUM BROMIDE

METALS	TEMP °F (60–560)
ALLOY 20Cb3	G (60–220)
ALUMINUM	
BRASS	
BRONZE	
CARBON STEEL	
COPPER	
E-BRIGHT 26-1	
HASTELLOY-B	
HASTELLOY-C	
HASTELLOY-D	
HIGH SILICON IRON	
INCONEL	
LEAD	
MONEL	G (60–340)
NICKEL	G (60–140)
STAINLESS STEELS	
Type 304/307	
Type 316	G (60–340)
TANTALUM	
TITANIUM	
ZIRCONIUM	
NONMETALLICS	
BOROSILICATE GLASS	
CARBON	

FOR METALS

E = < 2 Mils Penetration/Year; G = < 20 Mils Penetration/Year

S = < 50 Mils Penetration/Year; U = > 50 Mils Penetration/Year

FOR NONMETALLICS

R = Resistant

U = Unsatisfactory

LITHIUM BROMIDE

TEMP °F

PLASTICS	60 – 460 °F
ABS	
ACRYLICS	
ASBESTOS REINF., EPOXY	
ASBESTOS REINF., PHENOLIC	
CHLORINATED POLYETHER	
CPVC	R
EPOXY	
FLUOROCARBONS FEP	R
FLUOROCARBONS TFE	R
FURFURYL ALCOHOL	
NORYL	
PHENOLIC	
POLYAMIDES–NYLON	
POLYESTERS	
BISPHENOL A-FUMURATE	
HYDROGENATED BISPHENOL-A-BISPHENOL-A	
ISOPHTHALIC	
CHLORINATED POLYESTERS	
POLYSTYRENES	
POLYETHYLENE	R
POLYPROPYLENE	
PVC–TYPE 1	R
PVC–TYPE 2	R
POLYVINYLIDENE CHLORIDE	R
VINYLIDENE FLUORIDE	R
VINYL ESTER	R

RUBBERS	60 – 460 °F
BUTYL GR-1	
FLUORO ELASTOMERS VITON A	R
FLUORO ELASTOMERS KEL-F 3700	
HARD RUBBER	R
HYPALON	R
KOROSEAL	
NATURAL RUBBER (GRS)	
NEOPRENE GR-M (CR)	R
NITRILE BUNA N (NBR)	R
NORDEL	R
POLYBUTADIENE	
SILICONE	
POLYURETHANE	

TEMP °F

LUBRICATING OIL

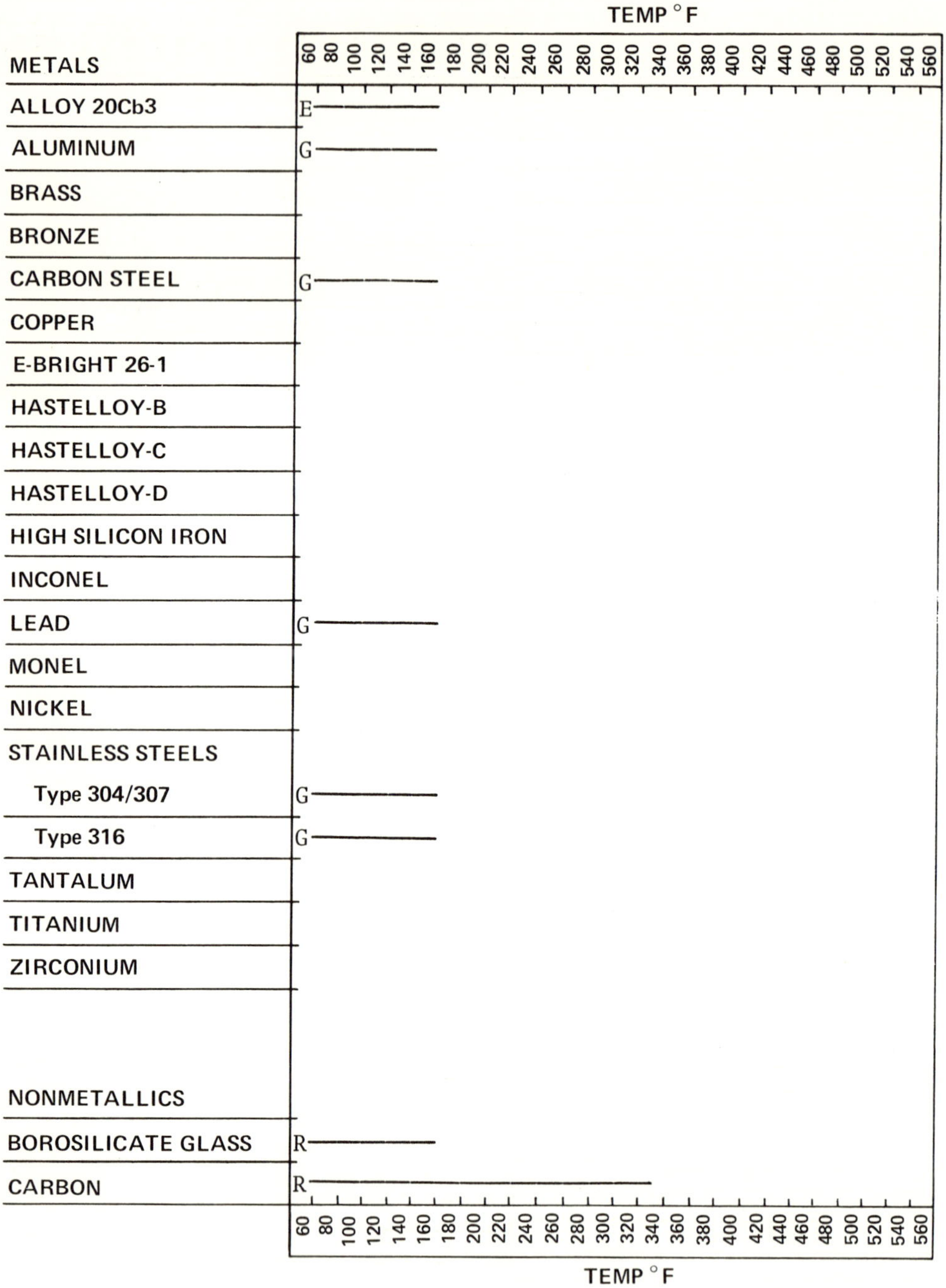

FOR METALS

E = < 2 Mils Penetration/Year; G = < 20 Mils Penetration/Year

S = < 50 Mils Penetration/Year; U = > 50 Mils Penetration/Year

FOR NONMETALLICS

R = Resistant

U = Unsatisfactory

LUBRICATING OIL

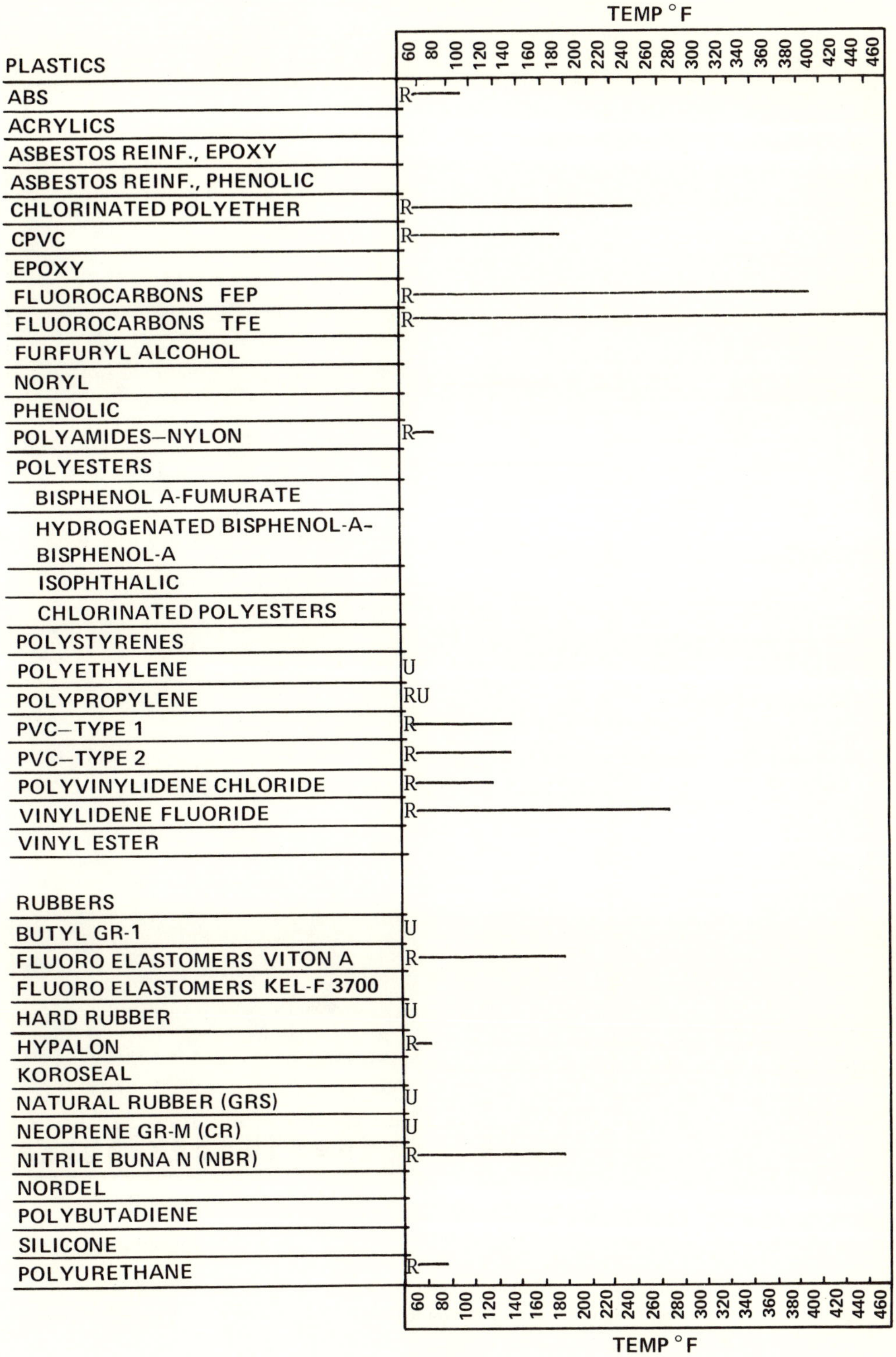

MACHINE OIL

METALS	TEMP °F
ALLOY 20Cb3	
ALUMINUM	
BRASS	
BRONZE	
CARBON STEEL	G (60–200)
COPPER	
E-BRIGHT 26-1	
HASTELLOY-B	G (60–200)
HASTELLOY-C	G (60–200)
HASTELLOY-D	G (60–200)
HIGH SILICON IRON	
INCONEL	
LEAD	
MONEL	
NICKEL	
STAINLESS STEELS	
Type 304/307	G (60–200)
Type 316	G (60–200)
TANTALUM	
TITANIUM	
ZIRCONIUM	
NONMETALLICS	
BOROSILICATE GLASS	R (60–200)
CARBON	

FOR METALS

E = $<$ 2 Mils Penetration/Year; G = $<$ 20 Mils Penetration/Year
S = $<$ 50 Mils Penetration/Year; U = $>$ 50 Mils Penetration/Year

FOR NONMETALLICS

R = Resistant
U = Unsatisfactory

MACHINE OIL

Please see page 1 for notes on calibration.

TEMP °F (scale: 60, 80, 100, 120, 140, 160, 180, 200, 220, 240, 260, 280, 300, 320, 340, 360, 380, 400, 420, 440, 460)

PLASTICS	MACHINE OIL
ABS	
ACRYLICS	
ASBESTOS REINF., EPOXY	
ASBESTOS REINF., PHENOLIC	
CHLORINATED POLYETHER	R (60–250)
CPVC	R (60–190)
EPOXY	
FLUOROCARBONS FEP	R (60–400)
FLUOROCARBONS TFE	R (60–460)
FURFURYL ALCOHOL	
NORYL	
PHENOLIC	
POLYAMIDES–NYLON	
POLYESTERS	
BISPHENOL A-FUMURATE	
HYDROGENATED BISPHENOL-A-BISPHENOL-A	
ISOPHTHALIC	
CHLORINATED POLYESTERS	
POLYSTYRENES	
POLYETHYLENE	U (60)
POLYPROPYLENE	R (60) —— U (120)
PVC–TYPE 1	R (60–140)
PVC–TYPE 2	R (60–140)
POLYVINYLIDENE CHLORIDE	
VINYLIDENE FLUORIDE	
VINYL ESTER	

RUBBERS	MACHINE OIL
BUTYL GR-1	
FLUORO ELASTOMERS VITON A	
FLUORO ELASTOMERS KEL-F 3700	
HARD RUBBER	
HYPALON	
KOROSEAL	
NATURAL RUBBER (GRS)	
NEOPRENE GR-M (CR)	
NITRILE BUNA N (NBR)	
NORDEL	
POLYBUTADIENE	
SILICONE	
POLYURETHANE	

TEMP °F

MAGNESIUM CARBONATE

TEMP °F (60–560)

METALS		Rating (temperature range)
ALLOY 20Cb3		G — (60)
ALUMINUM		
BRASS		
BRONZE		
CARBON STEEL		
COPPER		
E-BRIGHT 26-1		
HASTELLOY-B	10%	G (60–200)
HASTELLOY-C	10%	G (60–200)
HASTELLOY-D	10%	G (60–200)
HIGH SILICON IRON		G — (60)
INCONEL		
LEAD		
MONEL		G — (60)
NICKEL		
STAINLESS STEELS		
Type 304/307		G (60–200)
Type 316		G (60–200)
TANTALUM		G (60–200)
TITANIUM		
ZIRCONIUM		

NONMETALLICS	Rating
BOROSILICATE GLASS	
CARBON	

TEMP °F

FOR METALS

E = < 2 Mils Penetration/Year; G = < 20 Mils Penetration/Year
S = < 50 Mils Penetration/Year; U = > 50 Mils Penetration/Year

FOR NONMETALLICS

R = Resistant
U = Unsatisfactory

MAGNESIUM CARBONATE

TEMP °F (60–460)

PLASTICS	
ABS	R
ACRYLICS	R
ASBESTOS REINF., EPOXY	
ASBESTOS REINF., PHENOLIC	
CHLORINATED POLYETHER	R
CPVC	R
EPOXY	R
FLUOROCARBONS FEP	R
FLUOROCARBONS TFE	R
FURFURYL ALCOHOL	R
NORYL	
PHENOLIC	
POLYAMIDES–NYLON	
POLYESTERS	
BISPHENOL A-FUMURATE sat	R
HYDROGENATED BISPHENOL-A-BISPHENOL-A	
ISOPHTHALIC sat	R
CHLORINATED POLYESTERS	R
POLYSTYRENES	
POLYETHYLENE	R
POLYPROPYLENE	R
PVC–TYPE 1	R
PVC–TYPE 2	R
POLYVINYLIDENE CHLORIDE	R
VINYLIDENE FLUORIDE	R
VINYL ESTER	R

RUBBERS	
BUTYL GR-1	
FLUORO ELASTOMERS VITON A	R
FLUORO ELASTOMERS KEL-F 3700	
HARD RUBBER	R
HYPALON	R
KOROSEAL	
NATURAL RUBBER (GRS)	
NEOPRENE GR-M (CR)	R
NITRILE BUNA N (NBR)	R
NORDEL	R
POLYBUTADIENE	
SILICONE	
POLYURETHANE	

TEMP °F (60–460)

MAGNESIUM CHLORIDE

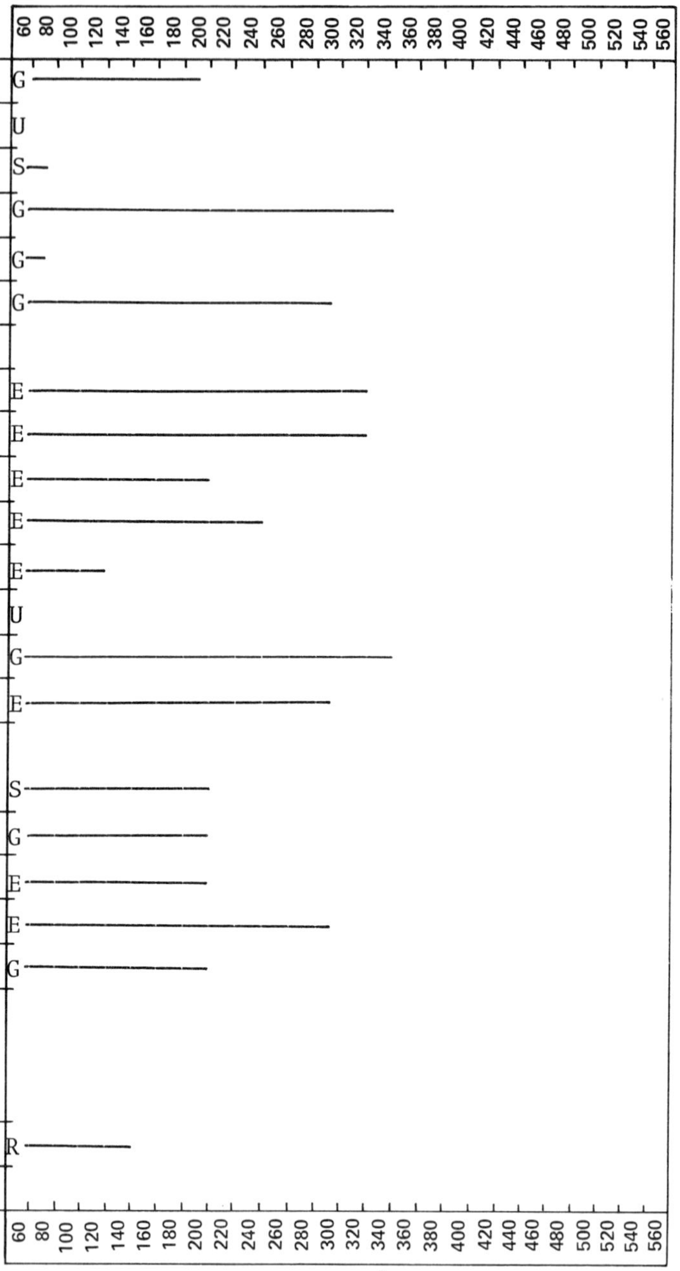

METALS		Rating
ALLOY 20Cb3	50%	G
ALUMINUM		U
BRASS		S
BRONZE		G
CARBON STEEL	30%	G
COPPER		G
E-BRIGHT 26-1		
HASTELLOY-B		E
HASTELLOY-C		E
HASTELLOY-D		E
HIGH SILICON IRON	30%	E
INCONEL	50%	E
LEAD		U
MONEL	50%	G
NICKEL	50%	E
STAINLESS STEELS		
Type 304/307 1,2	50%	S
Type 316 1,2	50%	G
TANTALUM	to 40%	E
TITANIUM	50%	E
ZIRCONIUM	5 to 40%	G

NONMETALLICS		Rating
BOROSILICATE GLASS		R
CARBON		

FOR METALS

E = < 2 Mils Penetration/Year; G = < 20 Mils Penetration/Year

S = < 50 Mils Penetration/Year; U = > 50 Mils Penetration/Year

FOR NONMETALLICS

R = Resistant

U = Unsatisfactory

MAGNESIUM CHLORIDE

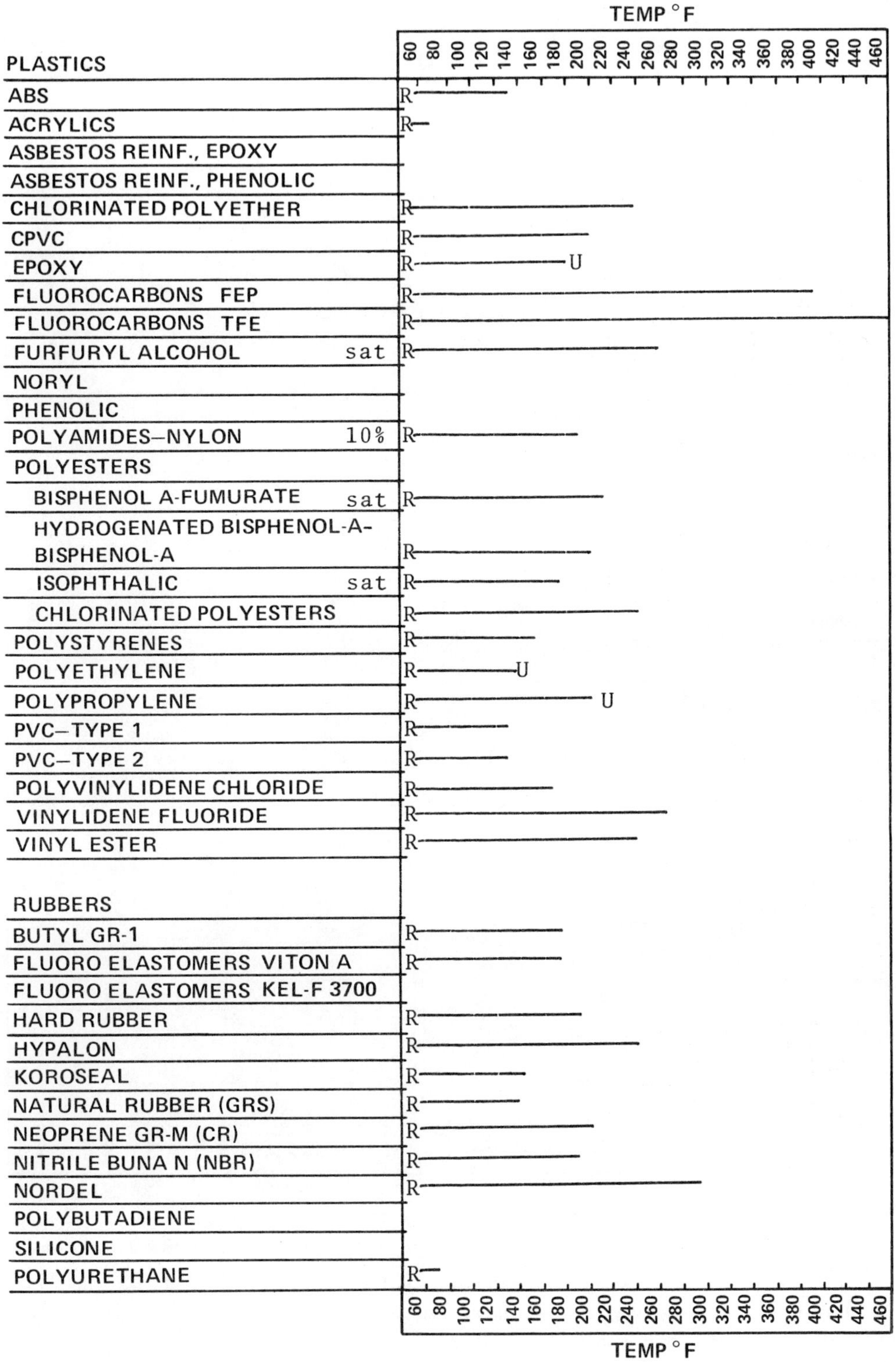

MAGNESIUM CITRATE

TEMP °F

METALS	60	80	100	120	140	160	180	200	220	240	260	280	300	320	340	360	380	400	420	440	460	480	500	520	540	560
ALLOY 20Cb3																										
ALUMINUM																										
BRASS																										
BRONZE																										
CARBON STEEL																										
COPPER																										
E-BRIGHT 26-1																										
HASTELLOY-B																										
HASTELLOY-C																										
HASTELLOY-D																										
HIGH SILICON IRON																										
INCONEL																										
LEAD																										
MONEL																										
NICKEL																										
STAINLESS STEELS																										
Type 304/307																										
Type 316																										
TANTALUM																										
TITANIUM																										
ZIRCONIUM																										
NONMETALLICS																										
BOROSILICATE GLASS																										
CARBON																										

TEMP °F

FOR METALS

E = < 2 Mils Penetration/Year; G = < 20 Mils Penetration/Year

S = < 50 Mils Penetration/Year; U = > 50 Mils Penetration/Year

FOR NONMETALLICS

R = Resistant

U = Unsatisfactory

MAGNESIUM CITRATE

PLASTICS	TEMP °F (60–460)
ABS	
ACRYLICS	
ASBESTOS REINF., EPOXY	
ASBESTOS REINF., PHENOLIC	
CHLORINATED POLYETHER	
CPVC	R 60–180
EPOXY	
FLUOROCARBONS FEP	R 60–400
FLUOROCARBONS TFE	R 60–460
FURFURYL ALCOHOL	
NORYL	
PHENOLIC	
POLYAMIDES–NYLON	
POLYESTERS	
BISPHENOL A-FUMURATE	
HYDROGENATED BISPHENOL-A-BISPHENOL-A	
ISOPHTHALIC	
CHLORINATED POLYESTERS	
POLYSTYRENES	
POLYETHYLENE	R 60–140
POLYPROPYLENE	
PVC–TYPE 1	R 60–140
PVC–TYPE 2	R 60–140
POLYVINYLIDENE CHLORIDE	
VINYLIDENE FLUORIDE	
VINYL ESTER	

RUBBERS	TEMP °F (60–460)
BUTYL GR-1	
FLUORO ELASTOMERS VITON A	
FLUORO ELASTOMERS KEL-F 3700	
HARD RUBBER	
HYPALON	
KOROSEAL	
NATURAL RUBBER (GRS)	
NEOPRENE GR-M (CR)	
NITRILE BUNA N (NBR)	
NORDEL	
POLYBUTADIENE	
SILICONE	
POLYURETHANE	

MAGNESIUM HYDROXIDE

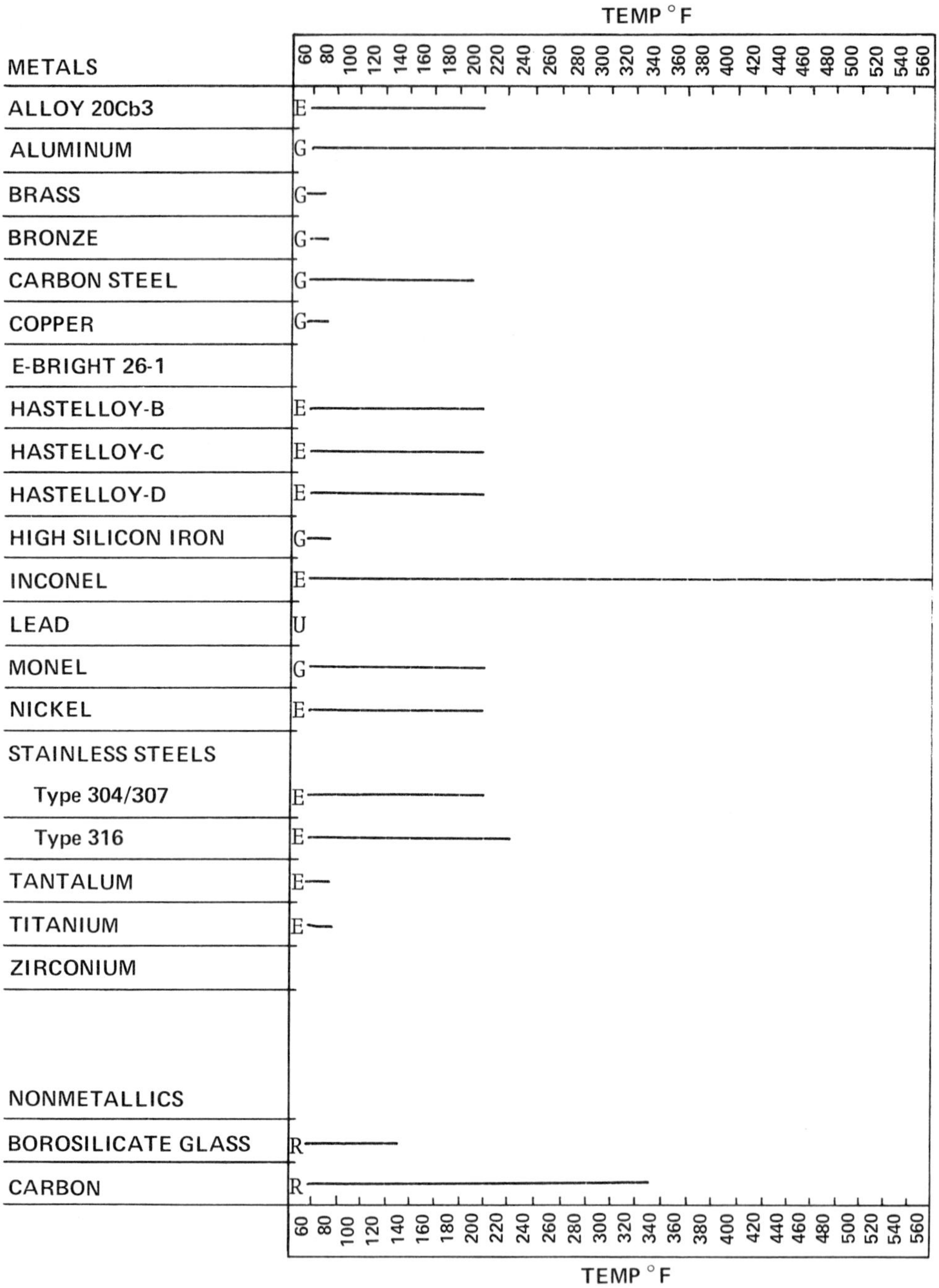

FOR METALS

E = < 2 Mils Penetration/Year; G = < 20 Mils Penetration/Year

S = < 50 Mils Penetration/Year; U = > 50 Mils Penetration/Year

FOR NONMETALLICS

R = Resistant

U = Unsatisfactory

MAGNESIUM HYDROXIDE

TEMP °F (60–460)

PLASTICS	Rating
ABS	R
ACRYLICS	R
ASBESTOS REINF., EPOXY	
ASBESTOS REINF., PHENOLIC	
CHLORINATED POLYETHER	R
CPVC	R
EPOXY	R U
FLUOROCARBONS FEP	R
FLUOROCARBONS TFE	R
FURFURYL ALCOHOL	
NORYL	R
PHENOLIC	
POLYAMIDES–NYLON	R
POLYESTERS	
BISPHENOL A-FUMURATE	R
HYDROGENATED BISPHENOL-A-BISPHENOL-A	
ISOPHTHALIC	U
CHLORINATED POLYESTERS	
POLYSTYRENES	
POLYETHYLENE	R
POLYPROPYLENE	R
PVC–TYPE 1	R
PVC–TYPE 2	R
POLYVINYLIDENE CHLORIDE	R
VINYLIDENE FLUORIDE	R
VINYL ESTER	R

RUBBERS	Rating
BUTYL GR-1	R
FLUORO ELASTOMERS VITON A	R
FLUORO ELASTOMERS KEL-F 3700	
HARD RUBBER	R
HYPALON	R
KOROSEAL	R
NATURAL RUBBER (GRS)	R
NEOPRENE GR-M (CR)	R
NITRILE BUNA N (NBR)	R
NORDEL	R
POLYBUTADIENE	
SILICONE	
POLYURETHANE	R

TEMP °F: 60 80 100 120 140 160 180 200 220 240 260 280 300 320 340 360 380 400 420 440 460

MAGNESIUM NITRATE

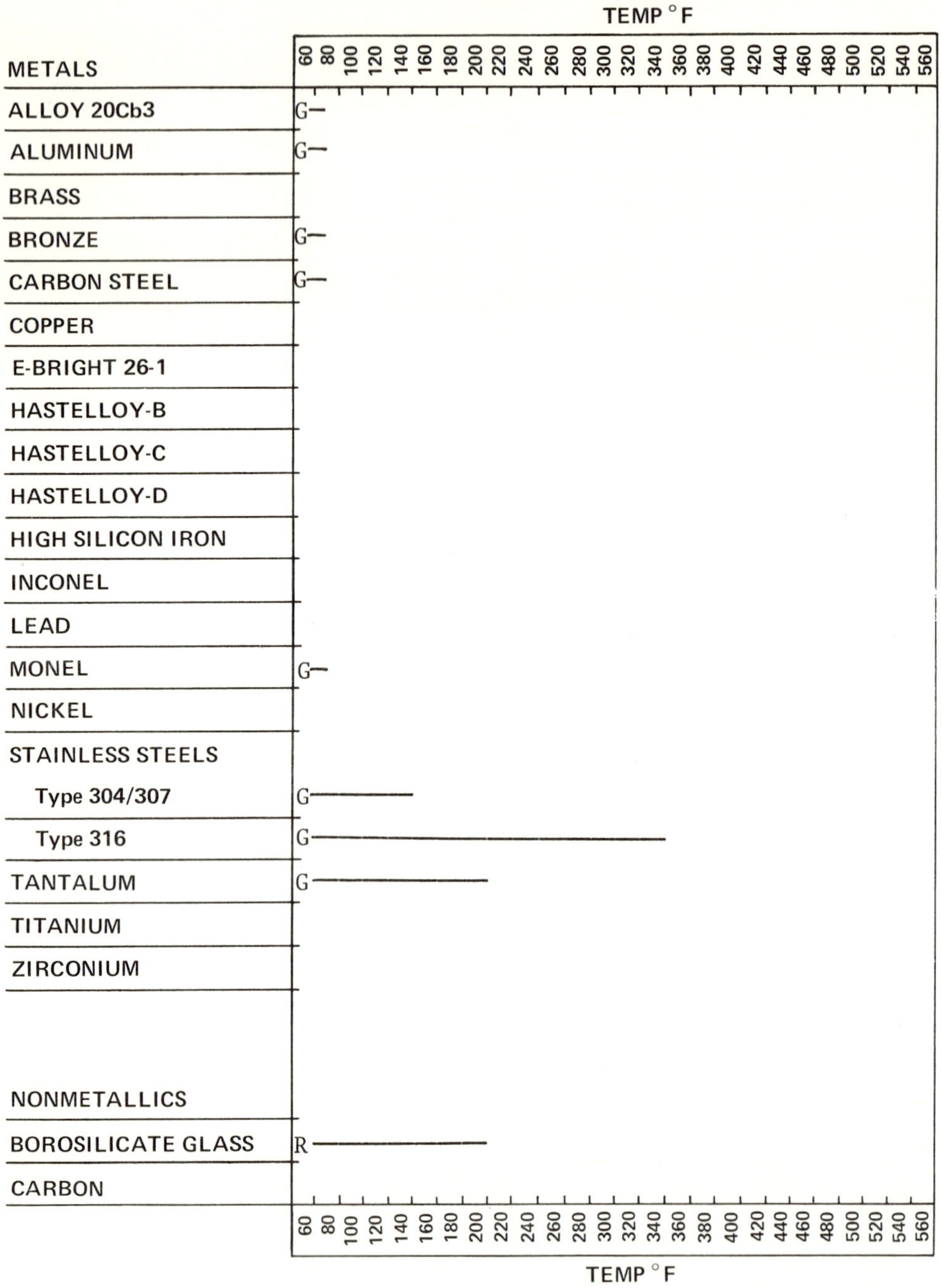

FOR METALS

E = < 2 Mils Penetration/Year; G = < 20 Mils Penetration/Year

S = < 50 Mils Penetration/Year; U = > 50 Mils Penetration/Year

FOR NONMETALLICS

R = Resistant

U = Unsatisfactory

MAGNESIUM NITRATE

TEMP °F: 60, 80, 100, 120, 140, 160, 180, 200, 220, 240, 260, 280, 300, 320, 340, 360, 380, 400, 420, 440, 460

PLASTICS	
ABS	R
ACRYLICS	R
ASBESTOS REINF., EPOXY	
ASBESTOS REINF., PHENOLIC	
CHLORINATED POLYETHER	R
CPVC	R
EPOXY	R
FLUOROCARBONS FEP	R
FLUOROCARBONS TFE	R
FURFURYL ALCOHOL	
NORYL	
PHENOLIC	
POLYAMIDES–NYLON	
POLYESTERS	
BISPHENOL A-FUMURATE	
HYDROGENATED BISPHENOL-A-BISPHENOL-A	R
ISOPHTHALIC sat	R
CHLORINATED POLYESTERS	
POLYSTYRENES	
POLYETHYLENE	R
POLYPROPYLENE	R
PVC–TYPE 1	R
PVC–TYPE 2	R
POLYVINYLIDENE CHLORIDE	R
VINYLIDENE FLUORIDE	R
VINYL ESTER	R

RUBBERS	
BUTYL GR-1	R
FLUORO ELASTOMERS VITON A	R
FLUORO ELASTOMERS KEL-F 3700	
HARD RUBBER	R
HYPALON	R
KOROSEAL	R
NATURAL RUBBER (GRS)	R
NEOPRENE GR-M (CR)	R
NITRILE BUNA N (NBR)	R
NORDEL	R
POLYBUTADIENE	
SILICONE	
POLYURETHANE	

TEMP °F: 60, 80, 100, 120, 140, 160, 180, 200, 220, 240, 260, 280, 300, 320, 340, 360, 380, 400, 420, 440, 460

MAGNESIUM SULFATE

TEMP °F: 60, 80, 100, 120, 140, 160, 180, 200, 220, 240, 260, 280, 300, 320, 340, 360, 380, 400, 420, 440, 460, 480, 500, 520, 540, 560

METALS		
ALLOY 20Cb3		G
ALUMINUM		G
BRASS	30%	G
BRONZE		G
CARBON STEEL	30%	G
COPPER	30%	G
E-BRIGHT 26-1		
HASTELLOY-B	to 50%	E
HASTELLOY-C	to 50%	G
HASTELLOY-D	to 50%	E
HIGH SILICON IRON		G
INCONEL	to 60%	E
LEAD		G
MONEL	to 60%	E
NICKEL	to 60%	E
STAINLESS STEELS		
Type 304/307		G
Type 316		G
TANTALUM	30%	G
TITANIUM		G
ZIRCONIUM	to 50%	E

NONMETALLICS	
BOROSILICATE GLASS	R
CARBON	R

TEMP °F: 60, 80, 100, 120, 140, 160, 180, 200, 220, 240, 260, 280, 300, 320, 340, 360, 380, 400, 420, 440, 460, 480, 500, 520, 540, 560

FOR METALS

E = < 2 Mils Penetration/Year; G = < 20 Mils Penetration/Year

S = < 50 Mils Penetration/Year; U = > 50 Mils Penetration/Year

FOR NONMETALLICS

R = Resistant

U = Unsatisfactory

MAGNESIUM SULFATE

PLASTICS		
ABS		R
ACRYLICS		R
ASBESTOS REINF., EPOXY		
ASBESTOS REINF., PHENOLIC		
CHLORINATED POLYETHER		R
CPVC		R
EPOXY		R
FLUOROCARBONS FEP		R
FLUOROCARBONS TFE		R
FURFURYL ALCOHOL	sat	R
NORYL		R
PHENOLIC		
POLYAMIDES—NYLON		R
POLYESTERS		
BISPHENOL A-FUMURATE	sat	R
HYDROGENATED BISPHENOL-A-BISPHENOL-A		R
ISOPHTHALIC	sat	R
CHLORINATED POLYESTERS	all	R
POLYSTYRENES		R
POLYETHYLENE		R — U
POLYPROPYLENE		R — U
PVC—TYPE 1		R
PVC—TYPE 2		R
POLYVINYLIDENE CHLORIDE		R
VINYLIDENE FLUORIDE		R
VINYL ESTER		R

RUBBERS	
BUTYL GR-1	R
FLUORO ELASTOMERS VITON A	R
FLUORO ELASTOMERS KEL-F 3700	
HARD RUBBER	R
HYPALON	R
KOROSEAL	R
NATURAL RUBBER (GRS)	R
NEOPRENE GR-M (CR)	R
NITRILE BUNA N (NBR)	R
NORDEL	R
POLYBUTADIENE	
SILICONE	
POLYURETHANE	

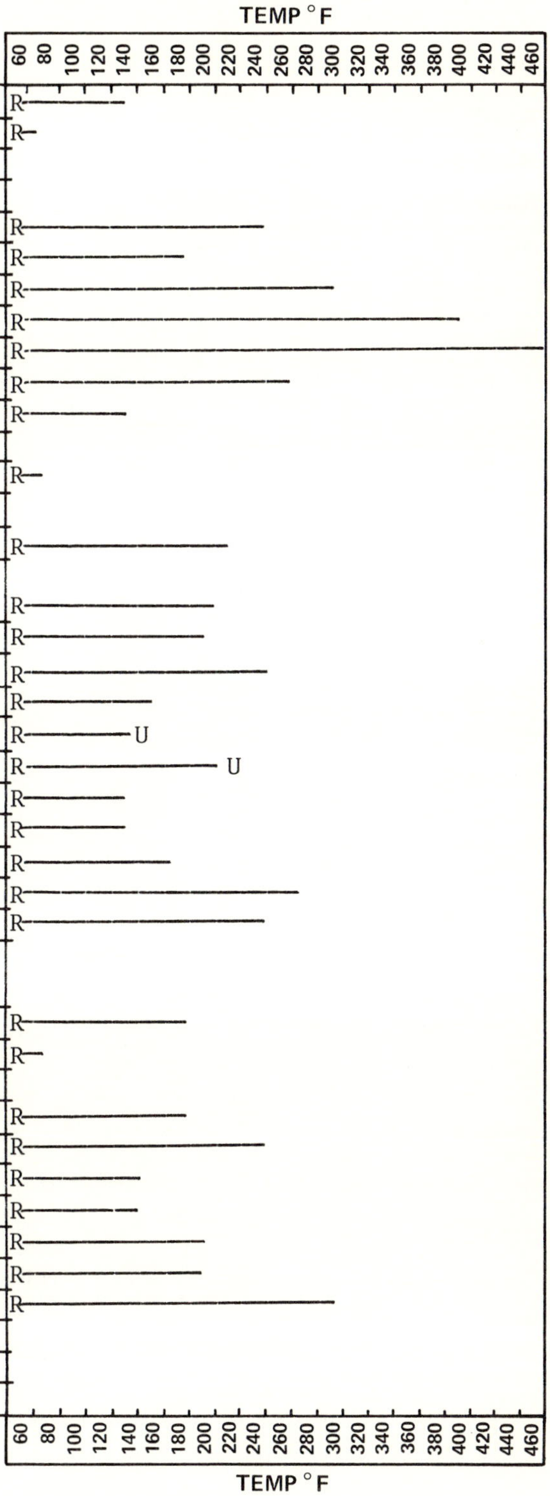

MALEIC ACID

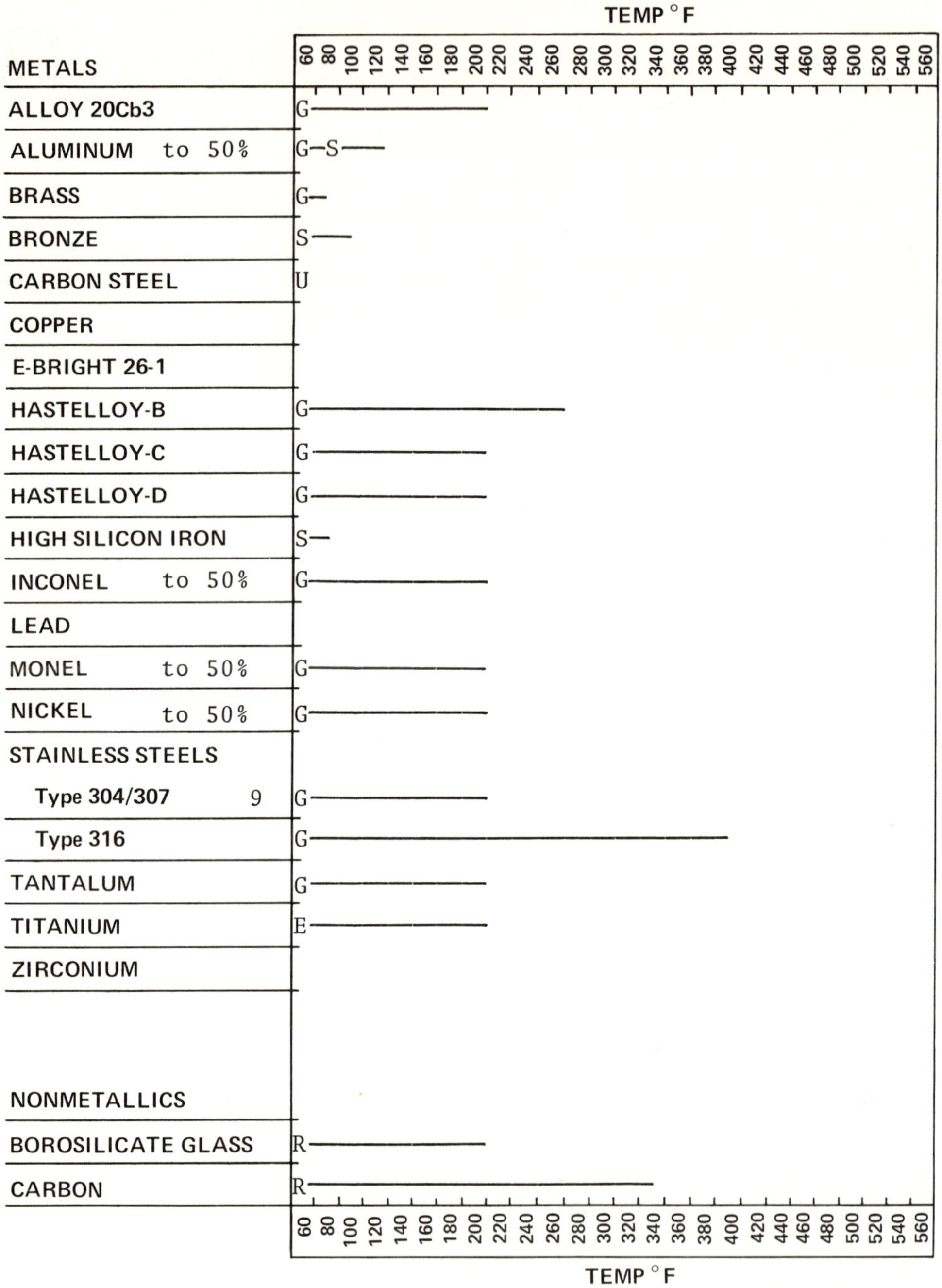

FOR METALS

E = < 2 Mils Penetration/Year; G = < 20 Mils Penetration/Year

S = < 50 Mils Penetration/Year; U = > 50 Mils Penetration/Year

FOR NONMETALLICS

R = Resistant

U = Unsatisfactory

MALEIC ACID

TEMP °F (scale 60–460)

PLASTICS	Concentration	Rating
ABS		R
ACRYLICS		
ASBESTOS REINF., EPOXY		R
ASBESTOS REINF., PHENOLIC		R
CHLORINATED POLYETHER	10%	R
CPVC		R
EPOXY	35%	R
FLUOROCARBONS FEP		R
FLUOROCARBONS TFE		R
FURFURYL ALCOHOL	sat	R
NORYL		
PHENOLIC		
POLYAMIDES—NYLON		U
POLYESTERS		
BISPHENOL A-FUMURATE		R
HYDROGENATED BISPHENOL-A-BISPHENOL-A		R
ISOPHTHALIC	sat	R
CHLORINATED POLYESTERS	all	R
POLYSTYRENES		
POLYETHYLENE		R—U
POLYPROPYLENE		R
PVC—TYPE 1		R
PVC—TYPE 2		R
POLYVINYLIDENE CHLORIDE		R
VINYLIDENE FLUORIDE		R
VINYL ESTER	sat	R

RUBBERS	Concentration	Rating
BUTYL GR-1		U
FLUORO ELASTOMERS VITON A		R
FLUORO ELASTOMERS KEL-F 3700		
HARD RUBBER		R
HYPALON		U
KOROSEAL		R
NATURAL RUBBER (GRS)		R
NEOPRENE GR-M (CR)		U
NITRILE BUNA N (NBR)		U
NORDEL		R
POLYBUTADIENE		
SILICONE		
POLYURETHANE		

TEMP °F

MALIC ACID

TEMP °F: 60 80 100 120 140 160 180 200 220 240 260 280 300 320 340 360 380 400 420 440 460 480 500 520 540 560

METALS		
ALLOY 20Cb3	50%	G
ALUMINUM		G
BRASS		
BRONZE		U
CARBON STEEL		U
COPPER		
E-BRIGHT 26-1		
HASTELLOY-B		G
HASTELLOY-C		G
HASTELLOY-D		G
HIGH SILICON IRON		G
INCONEL		G
LEAD		G
MONEL		G
NICKEL		G
STAINLESS STEELS		
Type 304/307	50%	E
Type 316	all	E
TANTALUM		G
TITANIUM		E
ZIRCONIUM		E

NONMETALLICS	
BOROSILICATE GLASS	R
CARBON	

TEMP °F: 60 80 100 120 140 160 180 200 220 240 260 280 300 320 340 360 380 400 420 440 460 480 500 520 540 560

FOR METALS

E = < 2 Mils Penetration/Year; G = < 20 Mils Penetration/Year

S = < 50 Mils Penetration/Year; U = > 50 Mils Penetration/Year

FOR NONMETALLICS

R = Resistant

U = Unsatisfactory

MALIC ACID

TEMP °F (scale 60–460)

PLASTICS	Rating (bar end, approx. °F)
ABS	R — to ~140
ACRYLICS	
ASBESTOS REINF., EPOXY	
ASBESTOS REINF., PHENOLIC	
CHLORINATED POLYETHER	R — to ~250
CPVC	R — to ~190
EPOXY	
FLUOROCARBONS FEP	R — to ~400
FLUOROCARBONS TFE	R — to 460
FURFURYL ALCOHOL	
NORYL	
PHENOLIC	
POLYAMIDES–NYLON	
POLYESTERS	
BISPHENOL A-FUMURATE	
HYDROGENATED BISPHENOL-A-BISPHENOL-A	
ISOPHTHALIC	
CHLORINATED POLYESTERS 10%	R — to ~100
POLYSTYRENES	
POLYETHYLENE	R — to ~150
POLYPROPYLENE	R — to ~130, U at ~140
PVC–TYPE 1	R — to ~140
PVC–TYPE 2	R — to ~140
POLYVINYLIDENE CHLORIDE	R — to ~80
VINYLIDENE FLUORIDE	R — to ~250
VINYL ESTER	

RUBBERS	Rating (bar end, approx. °F)
BUTYL GR-1	U
FLUORO ELASTOMERS VITON A	R — to ~190
FLUORO ELASTOMERS KEL-F 3700	
HARD RUBBER	R — to ~150
HYPALON	U
KOROSEAL	
NATURAL RUBBER (GRS)	R — to ~80
NEOPRENE GR-M (CR)	U
NITRILE BUNA N (NBR)	R — to ~190
NORDEL	
POLYBUTADIENE	
SILICONE	
POLYURETHANE	

TEMP °F

MERCURIC CHLORIDE

TEMP °F: 60 80 100 120 140 160 180 200 220 240 260 280 300 320 340 360 380 400 420 440 460 480 500 520 540 560

METALS		
ALLOY 20Cb3		G
ALUMINUM		U
BRASS		U
BRONZE		U
CARBON STEEL		U
COPPER		U
E-BRIGHT 26-1		
HASTELLOY-B		U
HASTELLOY-C	10%	G
HASTELLOY-D		
HIGH SILICON IRON		G
INCONEL		
LEAD		U
MONEL		U
NICKEL		S
STAINLESS STEELS		
Type 304/307		U
Type 316	2	G
TANTALUM		E
TITANIUM		G
ZIRCONIUM		E

NONMETALLICS	
BOROSILICATE GLASS	R
CARBON	

TEMP °F: 60 80 100 120 140 160 180 200 220 240 260 280 300 320 340 360 380 400 420 440 460 480 500 520 540 560

FOR METALS

E = < 2 Mils Penetration/Year; G = < 20 Mils Penetration/Year

S = < 50 Mils Penetration/Year; U = > 50 Mils Penetration/Year

FOR NONMETALLICS

R = Resistant

U = Unsatisfactory

MERCURIC CHLORIDE

TEMP °F (scale 60–460)

PLASTICS		
ABS		U
ACRYLICS		
ASBESTOS REINF., EPOXY		
ASBESTOS REINF., PHENOLIC		
CHLORINATED POLYETHER		R
CPVC		R
EPOXY		R
FLUOROCARBONS FEP		R
FLUOROCARBONS TFE		R
FURFURYL ALCOHOL	sat	R
NORYL		R
PHENOLIC		
POLYAMIDES–NYLON		R
POLYESTERS		
BISPHENOL A-FUMURATE	sat	R
HYDROGENATED BISPHENOL-A-BISPHENOL-A		R
ISOPHTHALIC	sat	R
CHLORINATED POLYESTERS		R
POLYSTYRENES		
POLYETHYLENE		R
POLYPROPYLENE		R
PVC–TYPE 1		R
PVC–TYPE 2		R
POLYVINYLIDENE CHLORIDE		R
VINYLIDENE FLUORIDE		R
VINYL ESTER		R

RUBBERS	
BUTYL GR-1	R
FLUORO ELASTOMERS VITON A	R
FLUORO ELASTOMERS KEL-F 3700	
HARD RUBBER	R
HYPALON	R
KOROSEAL	
NATURAL RUBBER (GRS)	R
NEOPRENE GR-M (CR)	U
NITRILE BUNA N (NBR)	R
NORDEL	R
POLYBUTADIENE	
SILICONE	
POLYURETHANE	

TEMP °F

MERCURIC CYANIDE

TEMP °F: 60 80 100 120 140 160 180 200 220 240 260 280 300 320 340 360 380 400 420 440 460 480 500 520 540 560

METALS	Rating
ALLOY 20Cb3	G——
ALUMINUM	U
BRASS	
BRONZE	U
CARBON STEEL	U
COPPER	
E-BRIGHT 26-1	
HASTELLOY-B	G——
HASTELLOY-C	G——
HASTELLOY-D	
HIGH SILICON IRON	G—
INCONEL	
LEAD	
MONEL	G——
NICKEL	G——
STAINLESS STEELS	
Type 304/307	G——
Type 316	G—
TANTALUM	G————————
TITANIUM	E—
ZIRCONIUM	

NONMETALLICS	Rating
BOROSILICATE GLASS	R—————
CARBON	

TEMP °F: 60 80 100 120 140 160 180 200 220 240 260 280 300 320 340 360 380 400 420 440 460 480 500 520 540 560

FOR METALS

E = < 2 Mils Penetration/Year; G = < 20 Mils Penetration/Year

S = < 50 Mils Penetration/Year; U = > 50 Mils Penetration/Year

FOR NONMETALLICS

R = Resistant

U = Unsatisfactory

MERCURIC CYANIDE

TEMP °F (60–460)

PLASTICS	
ABS	U
ACRYLICS	
ASBESTOS REINF., EPOXY	
ASBESTOS REINF., PHENOLIC	
CHLORINATED POLYETHER	R
CPVC	R
EPOXY	
FLUOROCARBONS FEP	R
FLUOROCARBONS TFE	R
FURFURYL ALCOHOL	
NORYL	R
PHENOLIC	
POLYAMIDES—NYLON	
POLYESTERS	
BISPHENOL A-FUMURATE	
HYDROGENATED BISPHENOL-A-BISPHENOL-A	
ISOPHTHALIC	
CHLORINATED POLYESTERS	
POLYSTYRENES	
POLYETHYLENE	R
POLYPROPYLENE	R
PVC—TYPE 1	R
PVC—TYPE 2	R
POLYVINYLIDENE CHLORIDE	R
VINYLIDENE FLUORIDE	R
VINYL ESTER	

RUBBERS	
BUTYL GR-1	R
FLUORO ELASTOMERS VITON A	R
FLUORO ELASTOMERS KEL-F 3700	
HARD RUBBER	
HYPALON	R
KOROSEAL	
NATURAL RUBBER (GRS)	R
NEOPRENE GR-M (CR)	U
NITRILE BUNA N (NBR)	R
NORDEL	
POLYBUTADIENE	
SILICONE	
POLYURETHANE	

TEMP °F (60–460)

MERCURIC NITRATE

TEMP °F: 60 80 100 120 140 160 180 200 220 240 260 280 300 320 340 360 380 400 420 440 460 480 500 520 540 560

METALS		
ALLOY 20Cb3	10%	G
ALUMINUM		U
BRASS	10%	U
BRONZE		
CARBON STEEL		
COPPER		
E-BRIGHT 26-1		
HASTELLOY-B		
HASTELLOY-C		
HASTELLOY-D		
HIGH SILICON IRON		G
INCONEL		
LEAD		
MONEL		
NICKEL		
STAINLESS STEELS Type 304/307		
Type 316	10%	G
TANTALUM		G
TITANIUM		
ZIRCONIUM		

NONMETALLICS		
BOROSILICATE GLASS		
CARBON		

TEMP °F: 60 80 100 120 140 160 180 200 220 240 260 280 300 320 340 360 380 400 420 440 460 480 500 520 540 560

FOR METALS

E = < 2 Mils Penetration/Year; G = < 20 Mils Penetration/Year

S = < 50 Mils Penetration/Year; U = > 50 Mils Penetration/Year

FOR NONMETALLICS

R = Resistant

U = Unsatisfactory

MERCURIC NITRATE

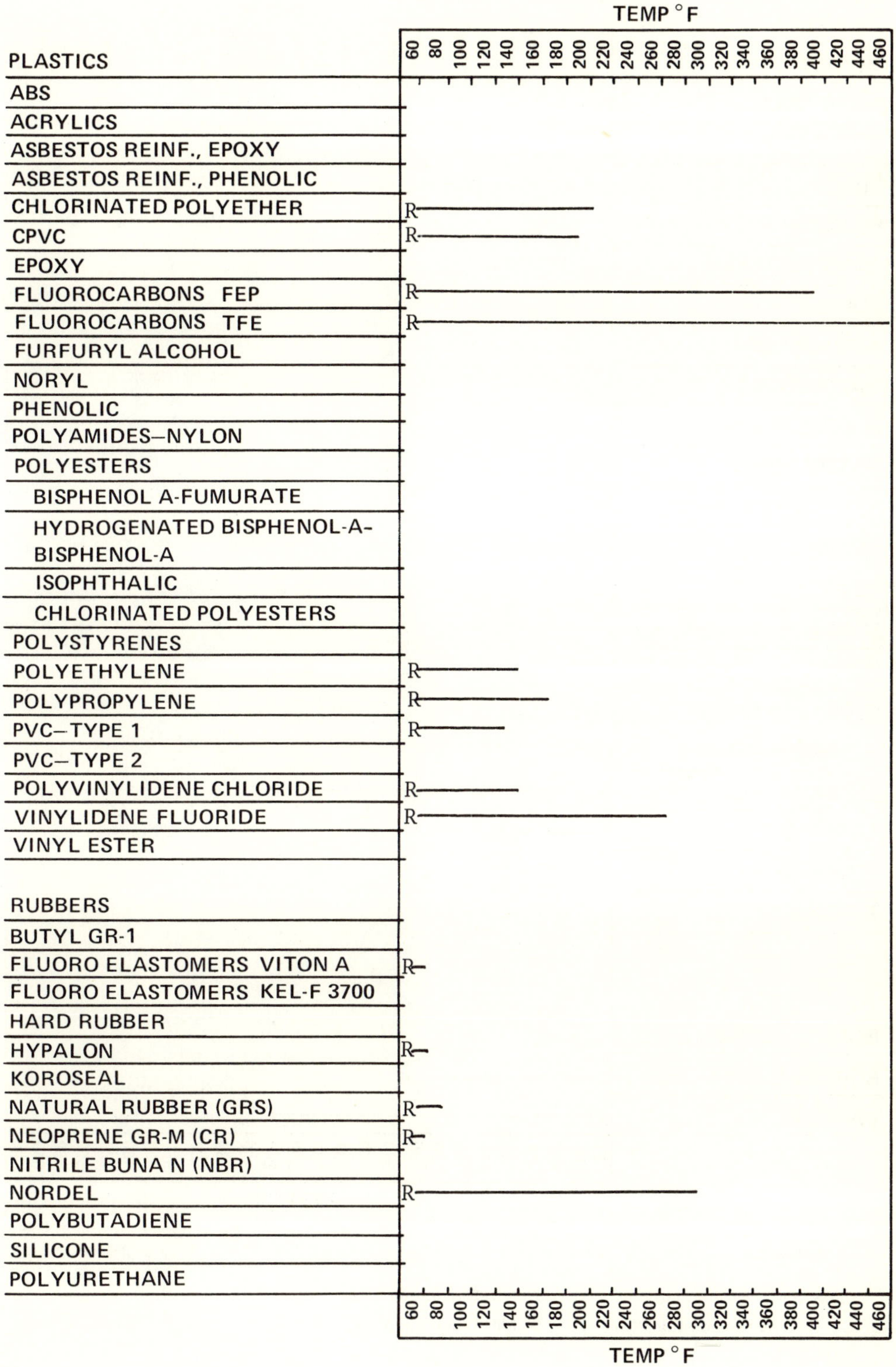

MERCUROUS NITRATE

TEMP °F: 60 80 100 120 140 160 180 200 220 240 260 280 300 320 340 360 380 400 420 440 460 480 500 520 540 560

METALS	
ALLOY 20Cb3	G—
ALUMINUM	U
BRASS	U
BRONZE	
CARBON STEEL	
COPPER	U
E-BRIGHT 26-1	
HASTELLOY-B	
HASTELLOY-C	G—
HASTELLOY-D	
HIGH SILICON IRON	G—
INCONEL	
LEAD	U
MONEL	
NICKEL	
STAINLESS STEELS	
Type 304/307 2	G—
Type 316	G————————————
TANTALUM	G——————
TITANIUM	
ZIRCONIUM	

NONMETALLICS	
BOROSILICATE GLASS	R——————
CARBON	R———————————

TEMP °F: 60 80 100 120 140 160 180 200 220 240 260 280 300 320 340 360 380 400 420 440 460 480 500 520 540 560

FOR METALS

E = < 2 Mils Penetration/Year; G = < 20 Mils Penetration/Year

S = < 50 Mils Penetration/Year; U = > 50 Mils Penetration/Year

FOR NONMETALLICS

R = Resistant

U = Unsatisfactory

TEMP °F (60–460)

PLASTICS	
ABS	U
ACRYLICS	
ASBESTOS REINF., EPOXY	
ASBESTOS REINF., PHENOLIC	
CHLORINATED POLYETHER	R
CPVC	R
EPOXY	
FLUOROCARBONS FEP	R
FLUOROCARBONS TFE	R
FURFURYL ALCOHOL	
NORYL	
PHENOLIC	
POLYAMIDES–NYLON	
POLYESTERS	
BISPHENOL A-FUMURATE	
HYDROGENATED BISPHENOL-A-BISPHENOL-A	
ISOPHTHALIC	
CHLORINATED POLYESTERS	
POLYSTYRENES	
POLYETHYLENE	R
POLYPROPYLENE	R
PVC–TYPE 1	R
PVC–TYPE 2	R
POLYVINYLIDENE CHLORIDE	R
VINYLIDENE FLUORIDE	
VINYL ESTER	

RUBBERS	
BUTYL GR-1	R
FLUORO ELASTOMERS VITON A	
FLUORO ELASTOMERS KEL-F 3700	
HARD RUBBER	
HYPALON	R
KOROSEAL	
NATURAL RUBBER (GRS)	R
NEOPRENE GR-M (CR)	R
NITRILE BUNA N (NBR)	R
NORDEL	R
POLYBUTADIENE	
SILICONE	
POLYURETHANE	

TEMP °F (60–460)

MERCURY

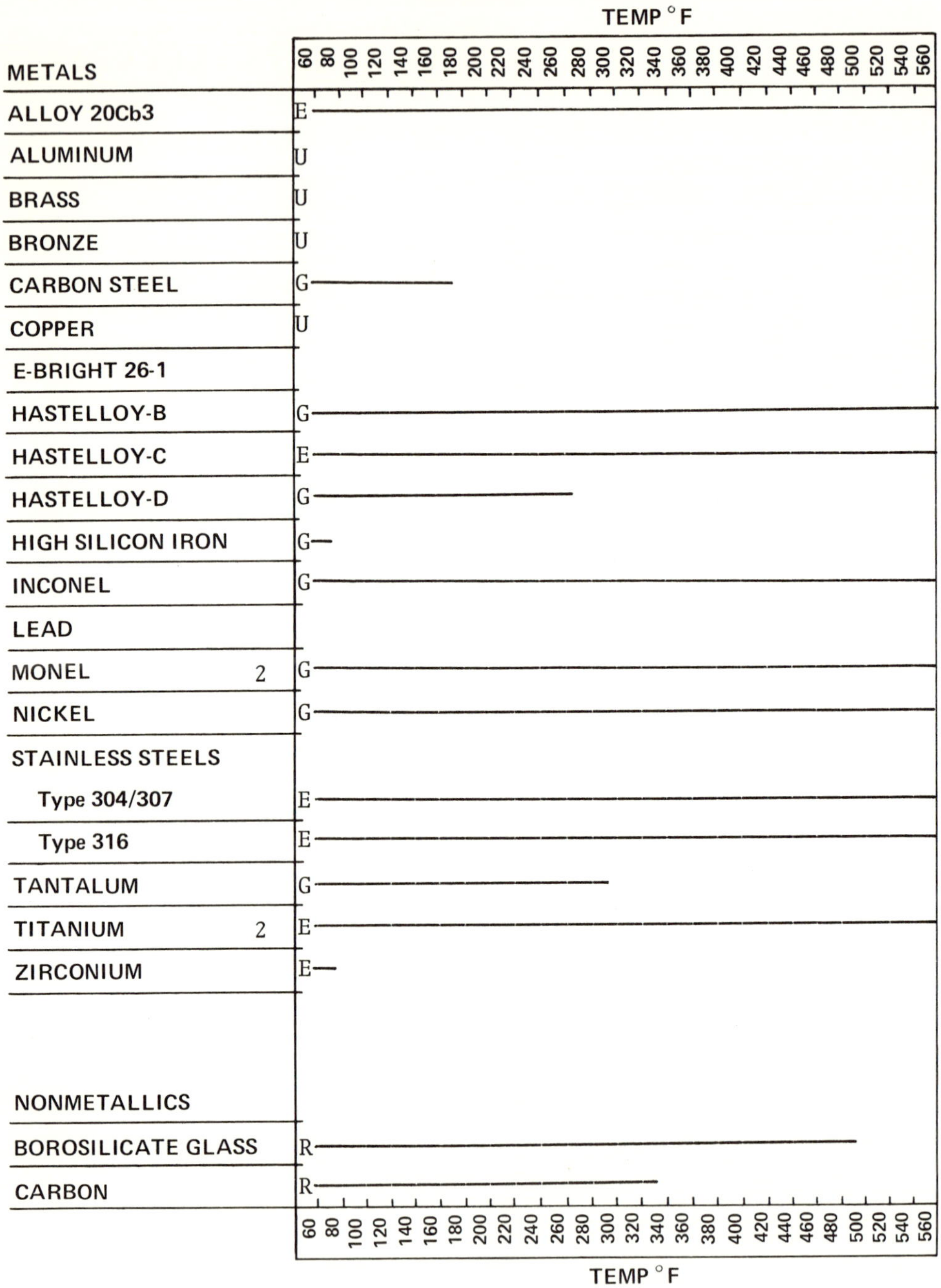

FOR METALS

E = < 2 Mils Penetration/Year; G = < 20 Mils Penetration/Year

S = < 50 Mils Penetration/Year; U = > 50 Mils Penetration/Year

FOR NONMETALLICS

R = Resistant

U = Unsatisfactory

MERCURY

TEMP °F

PLASTICS	Rating
ABS	U
ACRYLICS	
ASBESTOS REINF., EPOXY	
ASBESTOS REINF., PHENOLIC	
CHLORINATED POLYETHER	R
CPVC	R
EPOXY	R
FLUOROCARBONS FEP	R
FLUOROCARBONS TFE	R
FURFURYL ALCOHOL	R
NORYL	
PHENOLIC	
POLYAMIDES–NYLON	R
POLYESTERS	
BISPHENOL A-FUMURATE	R
HYDROGENATED BISPHENOL-A-BISPHENOL-A	
ISOPHTHALIC	R
CHLORINATED POLYESTERS	R
POLYSTYRENES	
POLYETHYLENE	R
POLYPROPYLENE	R
PVC–TYPE 1	R
PVC–TYPE 2	R
POLYVINYLIDENE CHLORIDE	R
VINYLIDENE FLUORIDE	R
VINYL ESTER	R

RUBBERS	Rating
BUTYL GR-1	
FLUORO ELASTOMERS VITON A	R
FLUORO ELASTOMERS KEL-F 3700	
HARD RUBBER	
HYPALON	R
KOROSEAL	
NATURAL RUBBER (GRS)	
NEOPRENE GR-M (CR)	R
NITRILE BUNA N (NBR)	R
NORDEL	
POLYBUTADIENE	
SILICONE	
POLYURETHANE	R

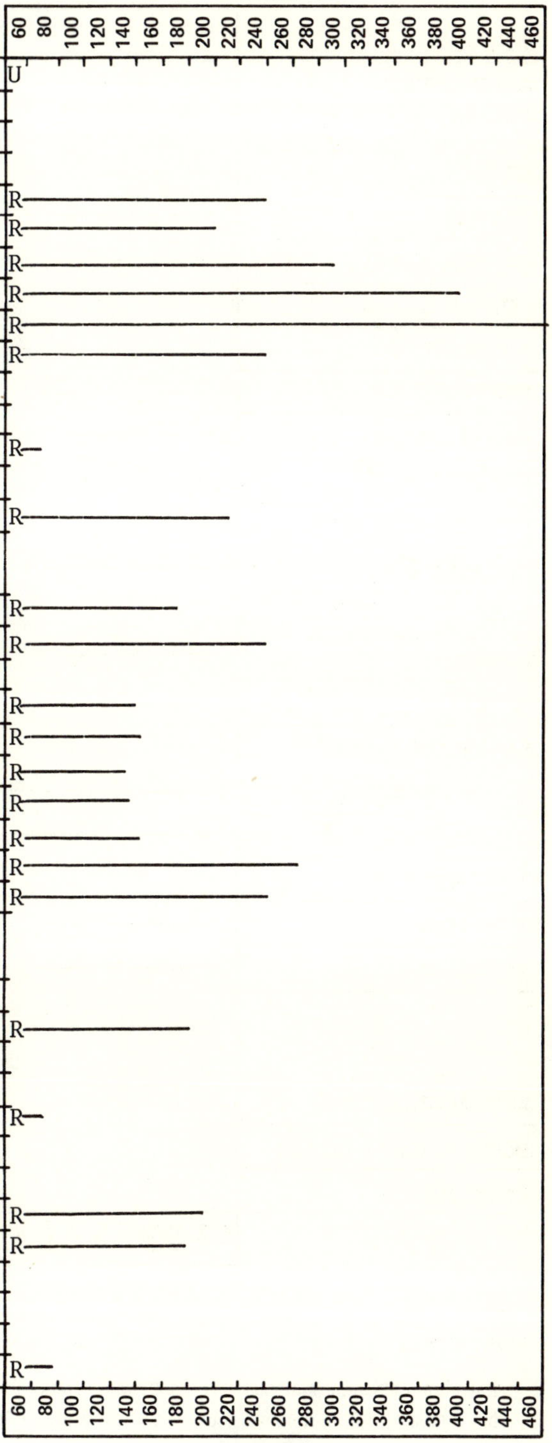

METHANE

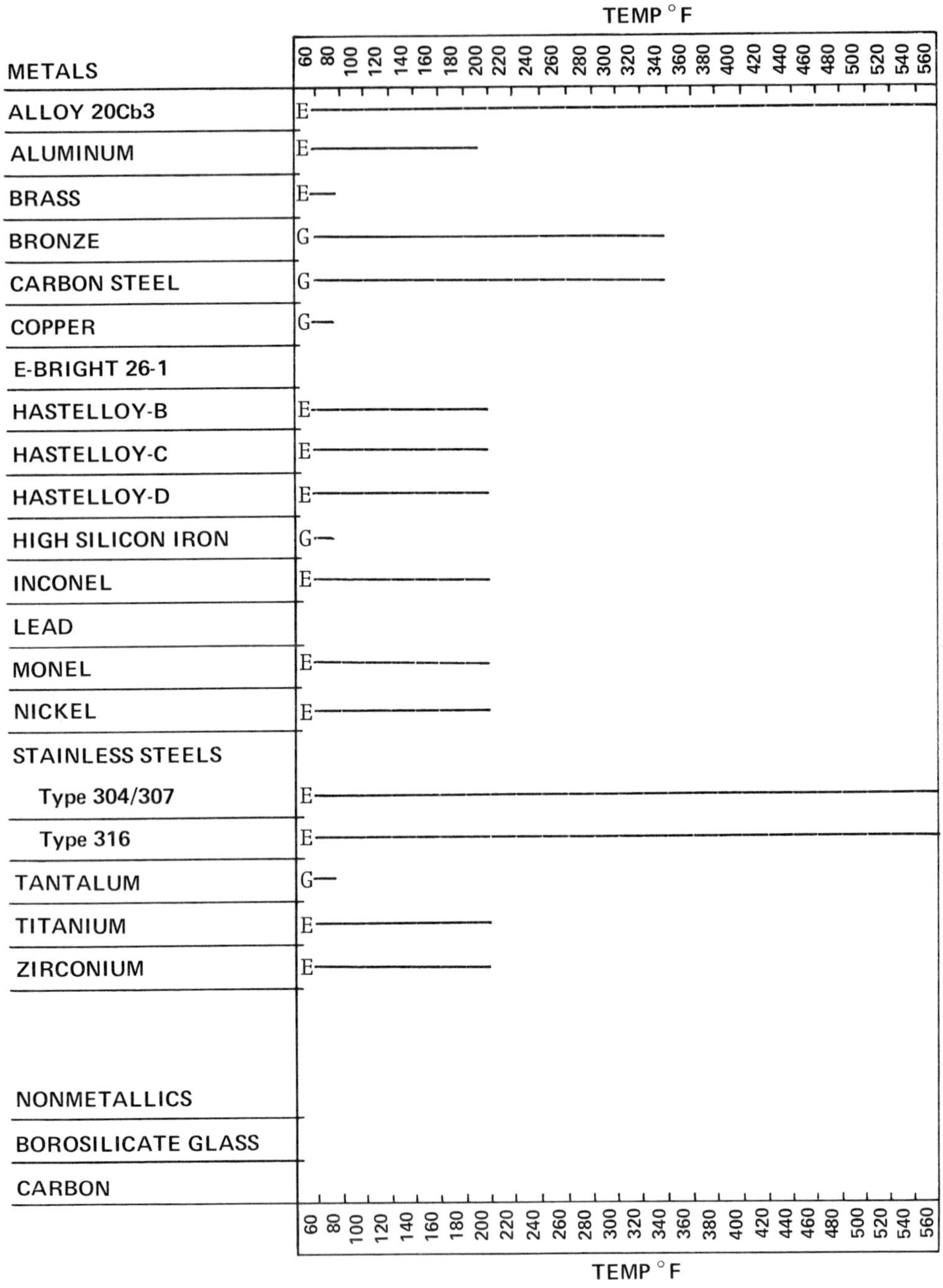

FOR METALS

E = < 2 Mils Penetration/Year; G = < 20 Mils Penetration/Year

S = < 50 Mils Penetration/Year; U = > 50 Mils Penetration/Year

FOR NONMETALLICS

R = Resistant

U = Unsatisfactory

METHANE

PLASTICS	TEMP °F (60–460)
ABS	
ACRYLICS	
ASBESTOS REINF., EPOXY	
ASBESTOS REINF., PHENOLIC	
CHLORINATED POLYETHER	R—
CPVC	R—
EPOXY	
FLUOROCARBONS FEP	R—
FLUOROCARBONS TFE	R—
FURFURYL ALCOHOL	
NORYL	
PHENOLIC	
POLYAMIDES–NYLON	R—
POLYESTERS	
BISPHENOL A-FUMURATE	
HYDROGENATED BISPHENOL-A-BISPHENOL-A	
ISOPHTHALIC	
CHLORINATED POLYESTERS	
POLYSTYRENES	
POLYETHYLENE	U
POLYPROPYLENE	R— U
PVC–TYPE 1	R—
PVC–TYPE 2	R—
POLYVINYLIDENE CHLORIDE	R—
VINYLIDENE FLUORIDE	R—
VINYL ESTER	

RUBBERS	TEMP °F (60–460)
BUTYL GR-1	
FLUORO ELASTOMERS VITON A	R—
FLUORO ELASTOMERS KEL-F 3700	
HARD RUBBER	
HYPALON	R—
KOROSEAL	
NATURAL RUBBER (GRS)	
NEOPRENE GR-M (CR)	R—
NITRILE BUNA N (NBR)	R—
NORDEL	R—
POLYBUTADIENE	
SILICONE	
POLYURETHANE	

METHANE SULPHONIC ACID, 50%

TEMP °F

METALS	60	80	100	120	140	160	180	200	220	240	260	280	300	320	340	360	380	400	420	440	460	480	500	520	540	560
ALLOY 20Cb3																										
ALUMINUM																										
BRASS																										
BRONZE																										
CARBON STEEL																										
COPPER																										
E-BRIGHT 26-1																										
HASTELLOY-B																										
HASTELLOY-C																										
HASTELLOY-D																										
HIGH SILICON IRON																										
INCONEL																										
LEAD																										
MONEL																										
NICKEL																										
STAINLESS STEELS Type 304/307																										
Type 316																										
TANTALUM																										
TITANIUM																										
ZIRCONIUM																										
NONMETALLICS																										
BOROSILICATE GLASS																										
CARBON																										

60 80 100 120 140 160 180 200 220 240 260 280 300 320 340 360 380 400 420 440 460 480 500 520 540 560

TEMP °F

FOR METALS

E = < 2 Mils Penetration/Year; G = < 20 Mils Penetration/Year

S = < 50 Mils Penetration/Year; U = > 50 Mils Penetration/Year

FOR NONMETALLICS

R = Resistant

U = Unsatisfactory

METHANE SULPHONIC ACID, 50%

TEMP °F (60–460)

PLASTICS	
ABS	
ACRYLICS	
ASBESTOS REINF., EPOXY	
ASBESTOS REINF., PHENOLIC	
CHLORINATED POLYETHER	R
CPVC	
EPOXY	
FLUOROCARBONS FEP	R
FLUOROCARBONS TFE	R
FURFURYL ALCOHOL	
NORYL	
PHENOLIC	
POLYAMIDES–NYLON	
POLYESTERS	
BISPHENOL A-FUMURATE	
HYDROGENATED BISPHENOL-A-BISPHENOL-A	
ISOPHTHALIC	
CHLORINATED POLYESTERS	
POLYSTYRENES	
POLYETHYLENE	
POLYPROPYLENE	R
PVC–TYPE 1	
PVC–TYPE 2	
POLYVINYLIDENE CHLORIDE	R
VINYLIDENE FLUORIDE	R
VINYL ESTER	

RUBBERS	
BUTYL GR-1	
FLUORO ELASTOMERS VITON A	
FLUORO ELASTOMERS KEL-F 3700	
HARD RUBBER	
HYPALON	
KOROSEAL	
NATURAL RUBBER (GRS)	
NEOPRENE GR-M (CR)	
NITRILE BUNA N (NBR)	
NORDEL	
POLYBUTADIENE	
SILICONE	
POLYURETHANE	

TEMP °F (60–460)

METHYL ALCOHOL

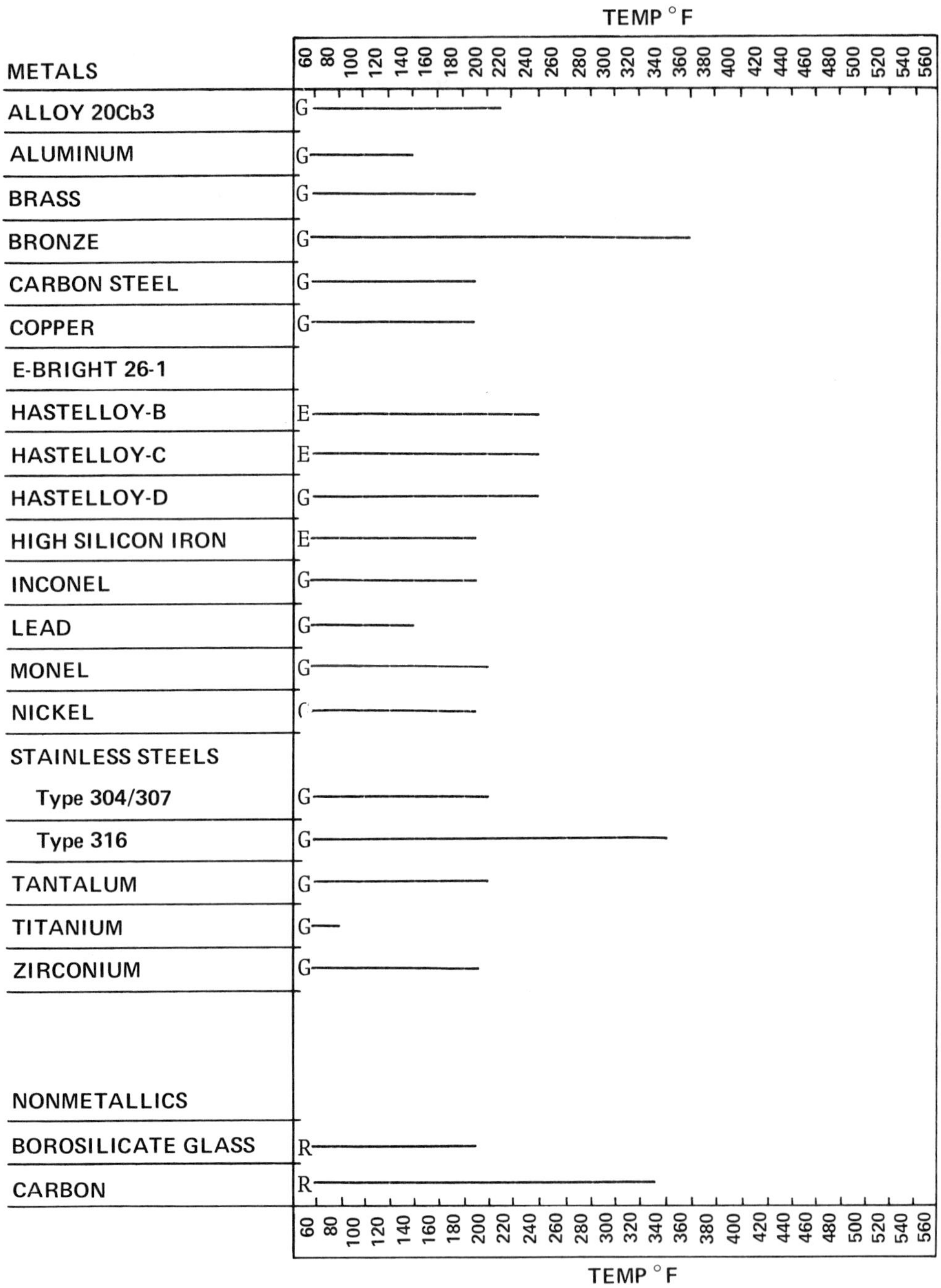

FOR METALS

E = < 2 Mils Penetration/Year; G = < 20 Mils Penetration/Year

S = < 50 Mils Penetration/Year; U = > 50 Mils Penetration/Year

FOR NONMETALLICS

R = Resistant

U = Unsatisfactory

METHYL ALCOHOL

TEMP °F: 60, 80, 100, 120, 140, 160, 180, 200, 220, 240, 260, 280, 300, 320, 340, 360, 380, 400, 420, 440, 460

PLASTICS	
ABS	U
ACRYLICS	
ASBESTOS REINF., EPOXY	R
ASBESTOS REINF., PHENOLIC	RU
CHLORINATED POLYETHER	R
CPVC	R
EPOXY	R— U
FLUOROCARBONS FEP	R
FLUOROCARBONS TFE	R
FURFURYL ALCOHOL	R
NORYL	
PHENOLIC	
POLYAMIDES–NYLON	R
POLYESTERS	
BISPHENOL A-FUMURATE 95%	R
HYDROGENATED BISPHENOL-A-BISPHENOL-A	
ISOPHTHALIC 96%	U
CHLORINATED POLYESTERS	R
POLYSTYRENES	R
POLYETHYLENE	R— U
POLYPROPYLENE	R— U
PVC–TYPE 1	R
PVC–TYPE 2	R
POLYVINYLIDENE CHLORIDE	R
VINYLIDENE FLUORIDE	R
VINYL ESTER	R

RUBBERS	
BUTYL GR-1	R
FLUORO ELASTOMERS VITON A	U
FLUORO ELASTOMERS KEL-F 3700	
HARD RUBBER	U
HYPALON	R
KOROSEAL	R
NATURAL RUBBER (GRS)	R
NEOPRENE GR-M (CR)	R
NITRILE BUNA N (NBR)	R
NORDEL	R
POLYBUTADIENE	
SILICONE	
POLYURETHANE	U

TEMP °F: 60, 80, 100, 120, 140, 160, 180, 200, 220, 240, 260, 280, 300, 320, 340, 360, 380, 400, 420, 440, 460

METHYL BROMIDE

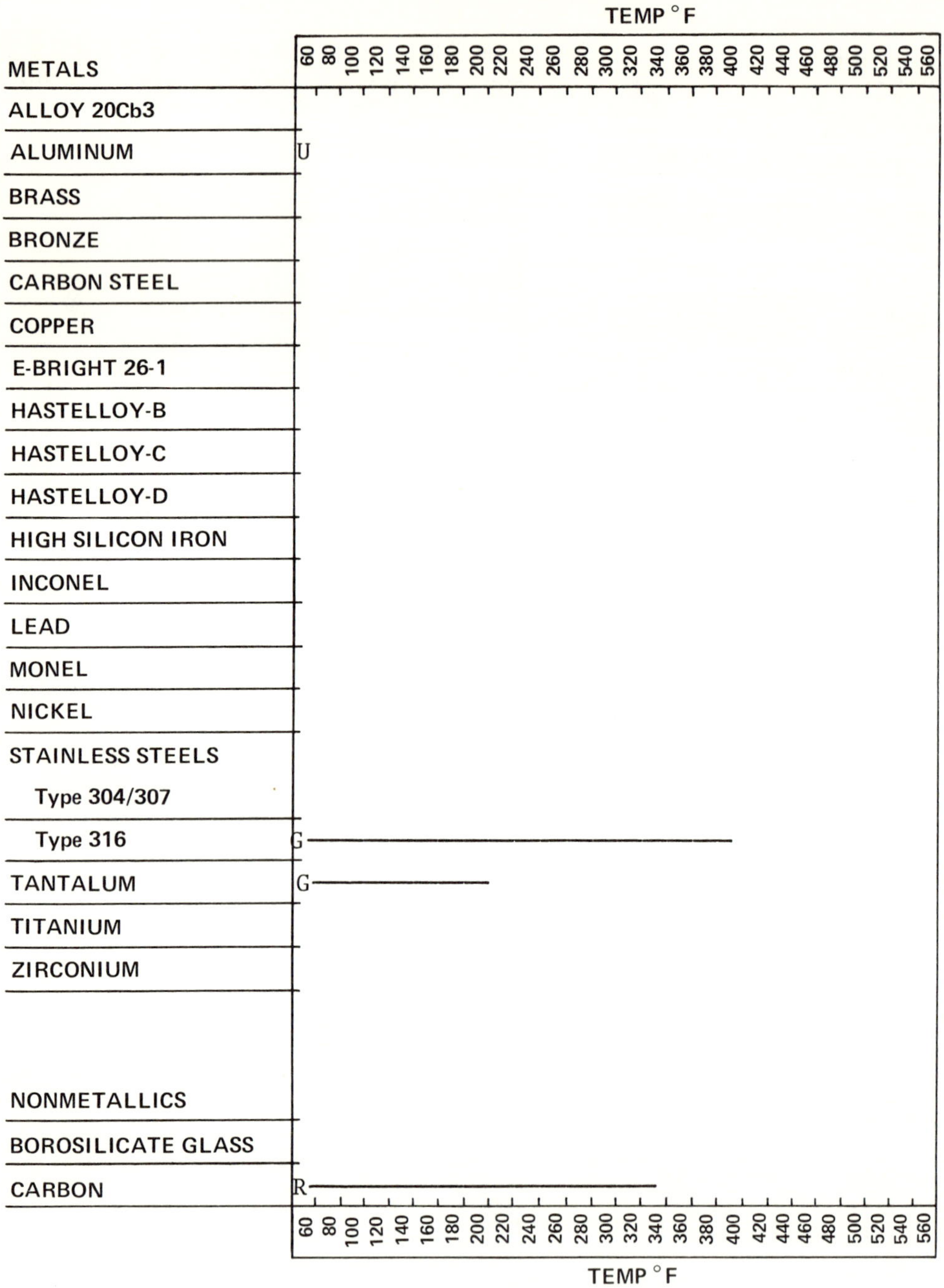

FOR METALS

E = < 2 Mils Penetration/Year; G = < 20 Mils Penetration/Year

S = < 50 Mils Penetration/Year; U = > 50 Mils Penetration/Year

FOR NONMETALLICS

R = Resistant

U = Unsatisfactory

METHYL BROMIDE

PLASTICS	TEMP °F
ABS	
ACRYLICS	
ASBESTOS REINF., EPOXY	
ASBESTOS REINF., PHENOLIC	
CHLORINATED POLYETHER	R (to 240)
CPVC	
EPOXY	
FLUOROCARBONS FEP	R (to 400)
FLUOROCARBONS TFE	R (to 460)
FURFURYL ALCOHOL	
NORYL	
PHENOLIC	
POLYAMIDES—NYLON	U
POLYESTERS	
BISPHENOL A-FUMURATE	
HYDROGENATED BISPHENOL-A-BISPHENOL-A	
ISOPHTHALIC	
CHLORINATED POLYESTERS	
POLYSTYRENES	
POLYETHYLENE	U
POLYPROPYLENE	U
PVC—TYPE 1	
PVC—TYPE 2	
POLYVINYLIDENE CHLORIDE	R (to 120)
VINYLIDENE FLUORIDE	R (to 280)
VINYL ESTER	

RUBBERS	TEMP °F
BUTYL GR-1	
FLUORO ELASTOMERS VITON A	R (to 180)
FLUORO ELASTOMERS KEL-F 3700	
HARD RUBBER	
HYPALON	
KOROSEAL	
NATURAL RUBBER (GRS)	
NEOPRENE GR-M (CR)	
NITRILE BUNA N (NBR)	RU
NORDEL	
POLYBUTADIENE	
SILICONE	
POLYURETHANE	

Temperature scale (°F): 60, 80, 100, 120, 140, 160, 180, 200, 220, 240, 260, 280, 300, 320, 340, 360, 380, 400, 420, 440, 460

METHYL CELLOSOLVE

TEMP °F: 60 80 100 120 140 160 180 200 220 240 260 280 300 320 340 360 380 400 420 440 460 480 500 520 540 560

METALS	
ALLOY 20Cb3	G
ALUMINUM	G
BRASS	
BRONZE	G
CARBON STEEL	S
COPPER	
E-BRIGHT 26-1	
HASTELLOY-B	
HASTELLOY-C	
HASTELLOY-D	
HIGH SILICON IRON	
INCONEL	
LEAD	
MONEL	G
NICKEL	
STAINLESS STEELS	
Type 304/307	G
Type 316	G
TANTALUM	G
TITANIUM	
ZIRCONIUM	

NONMETALLICS	
BOROSILICATE GLASS	
CARBON	R

TEMP °F: 60 80 100 120 140 160 180 200 220 240 260 280 300 320 340 360 380 400 420 440 460 480 500 520 540 560

FOR METALS

E = < 2 Mils Penetration/Year; G = < 20 Mils Penetration/Year

S = < 50 Mils Penetration/Year; U = > 50 Mils Penetration/Year

FOR NONMETALLICS

R = Resistant

U = Unsatisfactory

METHYL CELLOSOLVE

TEMP °F

PLASTICS	60 80 100 120 140 160 180 200 220 240 260 280 300 320 340 360 380 400 420 440 460
ABS	
ACRYLICS	
ASBESTOS REINF., EPOXY	
ASBESTOS REINF., PHENOLIC	
CHLORINATED POLYETHER	R
CPVC	
EPOXY	
FLUOROCARBONS FEP	R
FLUOROCARBONS TFE	R
FURFURYL ALCOHOL	
NORYL	
PHENOLIC	
POLYAMIDES–NYLON	
POLYESTERS	
BISPHENOL A-FUMURATE	
HYDROGENATED BISPHENOL-A-BISPHENOL-A	
ISOPHTHALIC	
CHLORINATED POLYESTERS	
POLYSTYRENES	
POLYETHYLENE	
POLYPROPYLENE	R
PVC–TYPE 1	R
PVC–TYPE 2	
POLYVINYLIDENE CHLORIDE	U
VINYLIDENE FLUORIDE	R
VINYL ESTER	
RUBBERS	
BUTYL GR-1	
FLUORO ELASTOMERS VITON A	U
FLUORO ELASTOMERS KEL-F 3700	
HARD RUBBER	
HYPALON	
KOROSEAL	
NATURAL RUBBER (GRS)	
NEOPRENE GR-M (CR)	R
NITRILE BUNA N (NBR)	U
NORDEL	
POLYBUTADIENE	
SILICONE	
POLYURETHANE	

60 80 100 120 140 160 180 200 220 240 260 280 300 320 340 360 380 400 420 440 460

TEMP °F

METHYL CHLORIDE

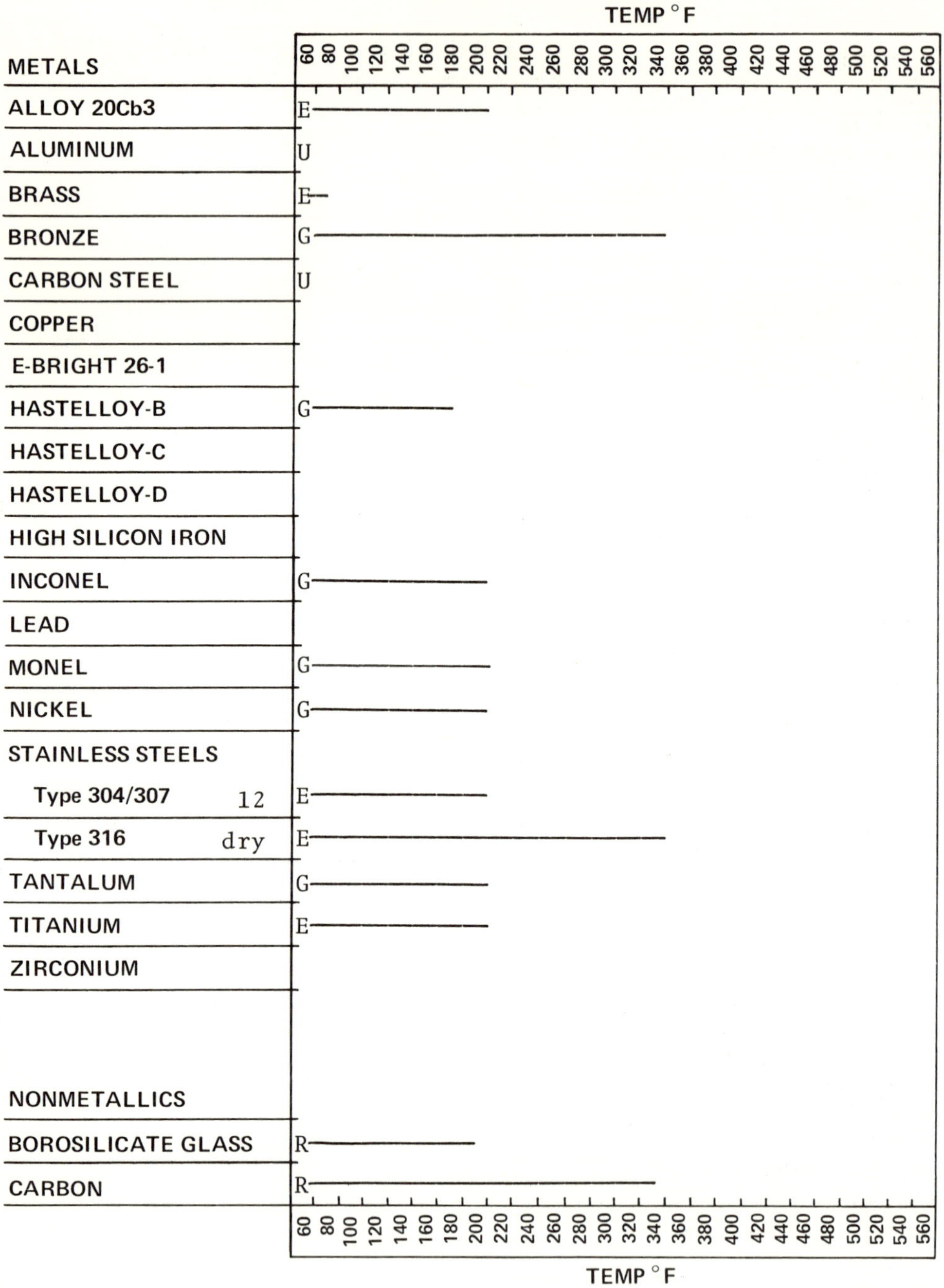

FOR METALS

E = < 2 Mils Penetration/Year; G = < 20 Mils Penetration/Year

S = < 50 Mils Penetration/Year; U = > 50 Mils Penetration/Year

FOR NONMETALLICS

R = Resistant

U = Unsatisfactory

METHYL CHLORIDE

TEMP °F: 60 80 100 120 140 160 180 200 220 240 260 280 300 320 340 360 380 400 420 440 460

PLASTICS	
ABS	U
ACRYLICS	
ASBESTOS REINF., EPOXY	U
ASBESTOS REINF., PHENOLIC	R
CHLORINATED POLYETHER	R
CPVC	U
EPOXY	U
FLUOROCARBONS FEP	R
FLUOROCARBONS TFE	R
FURFURYL ALCOHOL	
NORYL	
PHENOLIC	
POLYAMIDES–NYLON	R
POLYESTERS	
BISPHENOL A-FUMURATE	
HYDROGENATED BISPHENOL-A-BISPHENOL-A	
ISOPHTHALIC	
CHLORINATED POLYESTERS	R
POLYSTYRENES	
POLYETHYLENE	U
POLYPROPYLENE	U
PVC–TYPE 1	U
PVC–TYPE 2	U
POLYVINYLIDENE CHLORIDE	R
VINYLIDENE FLUORIDE	R
VINYL ESTER	

RUBBERS	
BUTYL GR-1	R
FLUORO ELASTOMERS VITON A	R
FLUORO ELASTOMERS KEL-F 3700	
HARD RUBBER	
HYPALON	U
KOROSEAL	
NATURAL RUBBER (GRS)	U
NEOPRENE GR-M (CR)	U
NITRILE BUNA N (NBR)	U
NORDEL	
POLYBUTADIENE	
SILICONE	
POLYURETHANE	U

TEMP °F: 60 80 100 120 140 160 180 200 220 240 260 280 300 320 340 360 380 400 420 440 460

METHYL CHLOROFORM

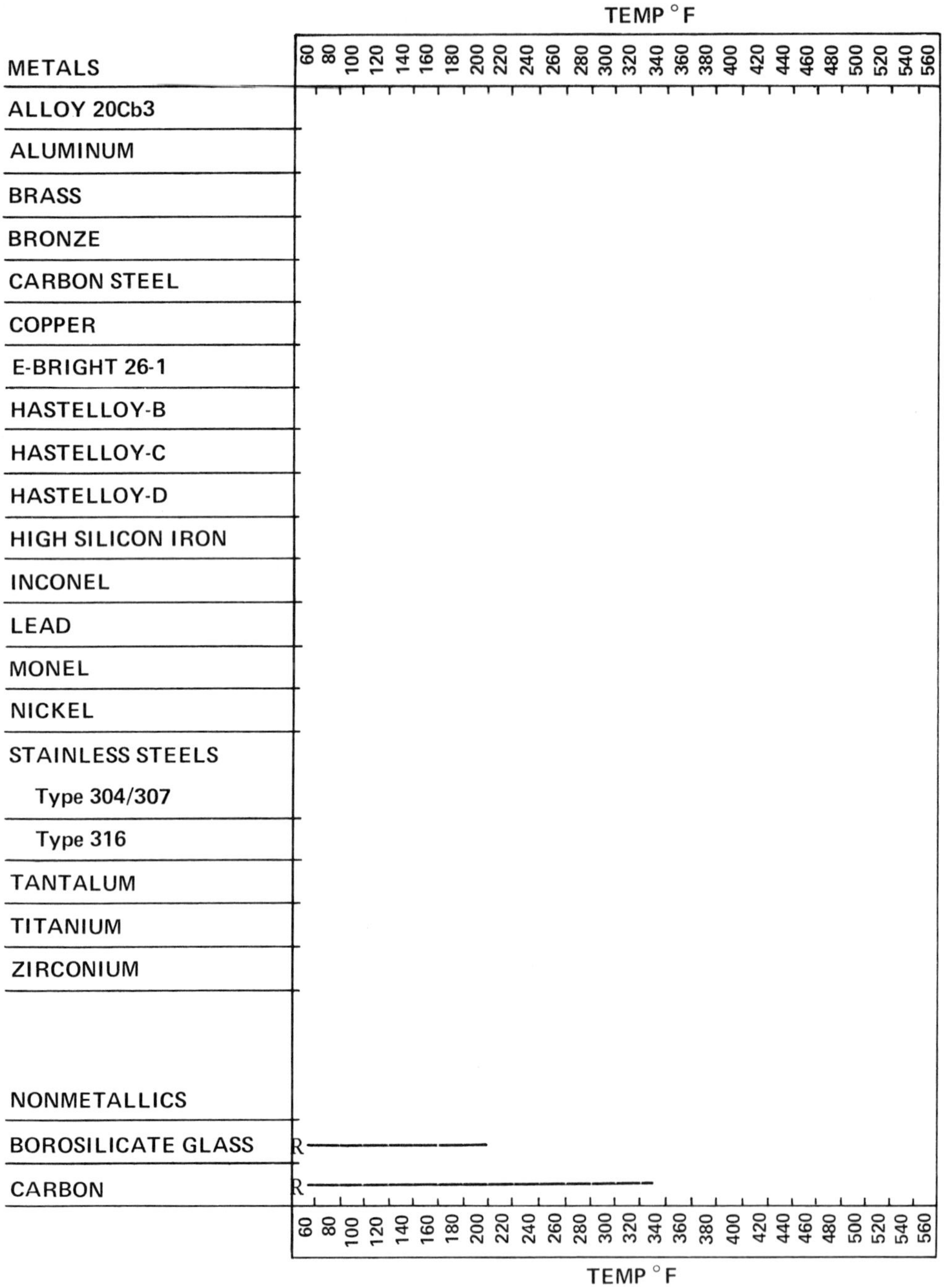

FOR METALS

E = < 2 Mils Penetration/Year; G = < 20 Mils Penetration/Year

S = < 50 Mils Penetration/Year; U = > 50 Mils Penetration/Year

FOR NONMETALLICS

R = Resistant

U = Unsatisfactory

METHYL CHLOROFORM

TEMP °F (60–460)

PLASTICS	60 → 460 °F
ABS	
ACRYLICS	
ASBESTOS REINF., EPOXY	
ASBESTOS REINF., PHENOLIC	
CHLORINATED POLYETHER	R–U
CPVC	
EPOXY	
FLUOROCARBONS FEP	R————————————————
FLUOROCARBONS TFE	R————————————————————
FURFURYL ALCOHOL	
NORYL	
PHENOLIC	
POLYAMIDES–NYLON	
POLYESTERS	
BISPHENOL A-FUMURATE	
HYDROGENATED BISPHENOL-A-BISPHENOL-A	
ISOPHTHALIC	
CHLORINATED POLYESTERS	R—
POLYSTYRENES	
POLYETHYLENE	
POLYPROPYLENE	
PVC–TYPE 1	
PVC–TYPE 2	
POLYVINYLIDENE CHLORIDE	R—
VINYLIDENE FLUORIDE	R———
VINYL ESTER	
RUBBERS	
BUTYL GR-1	
FLUORO ELASTOMERS VITON A	R—
FLUORO ELASTOMERS KEL-F 3700	
HARD RUBBER	
HYPALON	
KOROSEAL	
NATURAL RUBBER (GRS)	
NEOPRENE GR-M (CR)	
NITRILE BUNA N (NBR)	
NORDEL	
POLYBUTADIENE	
SILICONE	
POLYURETHANE	

TEMP °F (60–460)

METHYL ETHYL KETONE

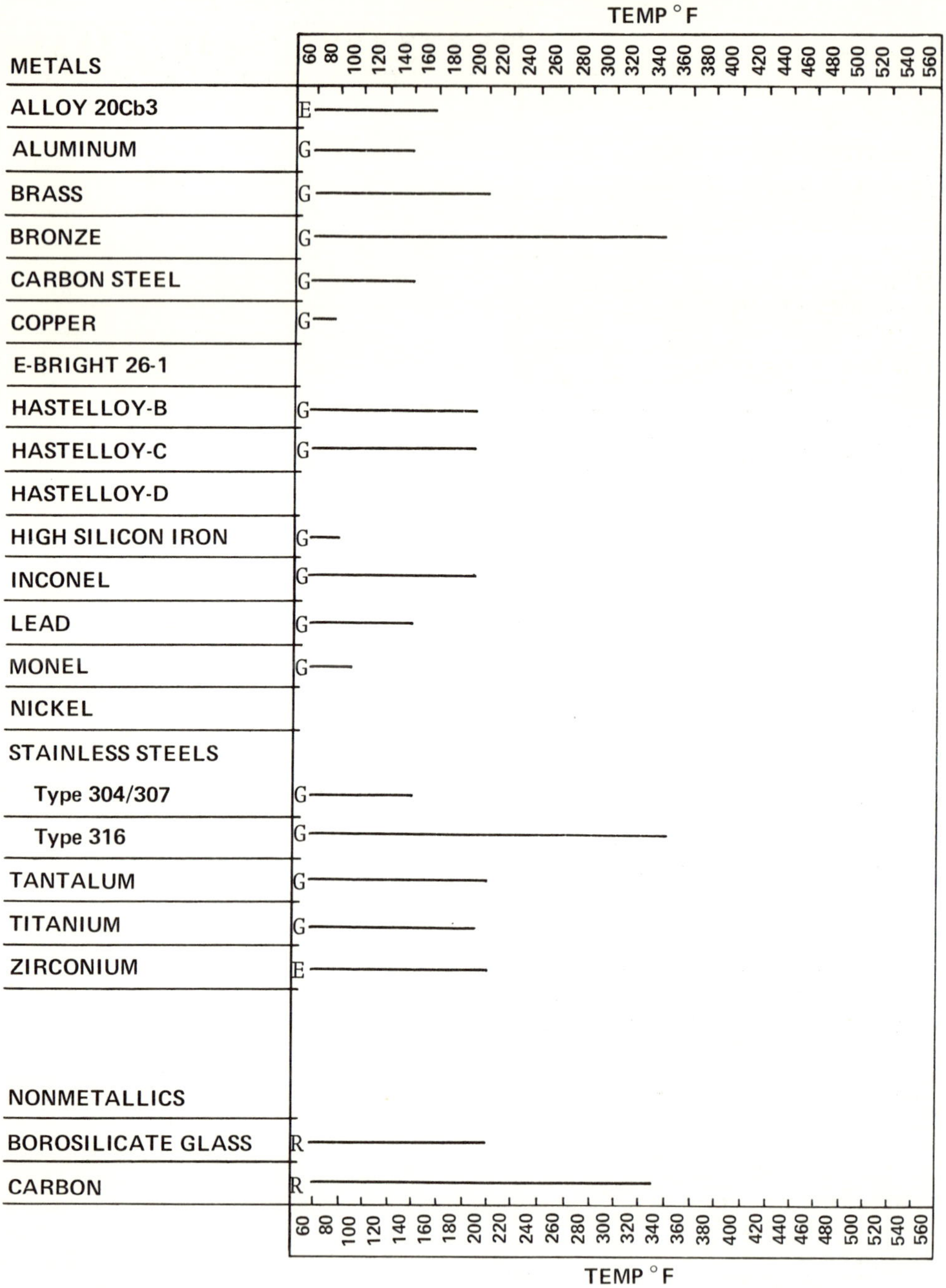

FOR METALS

E = < 2 Mils Penetration/Year; G = < 20 Mils Penetration/Year

S = < 50 Mils Penetration/Year; U = > 50 Mils Penetration/Year

FOR NONMETALLICS

R = Resistant

U = Unsatisfactory

TEMP °F (60–460)

PLASTICS	60 → 460
ABS	U
ACRYLICS	
ASBESTOS REINF., EPOXY	RU
ASBESTOS REINF., PHENOLIC	R————
CHLORINATED POLYETHER	R—
CPVC	U
EPOXY	R— U
FLUOROCARBONS FEP	R————————————
FLUOROCARBONS TFE	R——————————————
FURFURYL ALCOHOL	R———— U
NORYL	
PHENOLIC	
POLYAMIDES–NYLON	R—————
POLYESTERS	
BISPHENOL A-FUMURATE	U
HYDROGENATED BISPHENOL-A-BISPHENOL-A	U
ISOPHTHALIC	U
CHLORINATED POLYESTERS	U
POLYSTYRENES	
POLYETHYLENE	U
POLYPROPYLENE	R——— U
PVC–TYPE 1	U
PVC–TYPE 2	U
POLYVINYLIDENE CHLORIDE	U
VINYLIDENE FLUORIDE	U
VINYL ESTER	U
RUBBERS	
BUTYL GR-1	R——
FLUORO ELASTOMERS VITON A	U
FLUORO ELASTOMERS KEL-F 3700	
HARD RUBBER	U
HYPALON	U
KOROSEAL	U
NATURAL RUBBER (GRS)	U
NEOPRENE GR-M (CR)	U
NITRILE BUNA N (NBR)	U
NORDEL	R—
POLYBUTADIENE	
SILICONE	
POLYURETHANE	U

TEMP °F (60, 80, 100, 120, 140, 160, 180, 200, 220, 240, 260, 280, 300, 320, 340, 360, 380, 400, 420, 440, 460)

METHYL ISOBUTYL CARBINOL

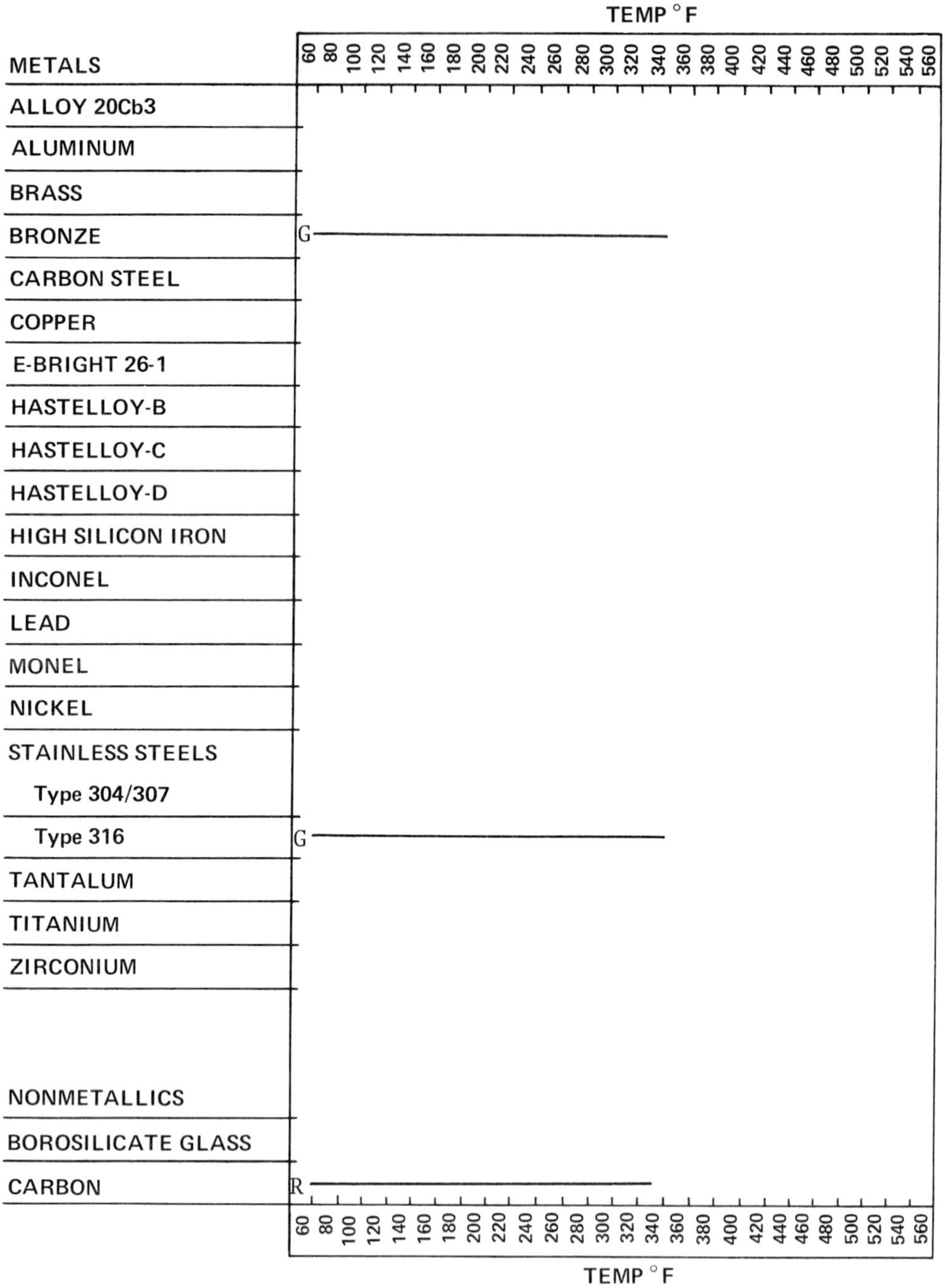

FOR METALS

E = < 2 Mils Penetration/Year; G = < 20 Mils Penetration/Year

S = < 50 Mils Penetration/Year; U = > 50 Mils Penetration/Year

FOR NONMETALLICS

R = Resistant

U = Unsatisfactory

METHYL ISOBUTYL CARBINOL

PLASTICS	TEMP °F (60–460)
ABS	
ACRYLICS	
ASBESTOS REINF., EPOXY	
ASBESTOS REINF., PHENOLIC	
CHLORINATED POLYETHER	
CPVC	U
EPOXY	
FLUOROCARBONS FEP	R (60–400)
FLUOROCARBONS TFE	R (60–460)
FURFURYL ALCOHOL	
NORYL	
PHENOLIC	
POLYAMIDES—NYLON	
POLYESTERS	
BISPHENOL A-FUMURATE	
HYDROGENATED BISPHENOL-A-BISPHENOL-A	
ISOPHTHALIC	
CHLORINATED POLYESTERS	
POLYSTYRENES	
POLYETHYLENE	U
POLYPROPYLENE	R (60–120)
PVC—TYPE 1	U
PVC—TYPE 2	U
POLYVINYLIDENE CHLORIDE	
VINYLIDENE FLUORIDE	
VINYL ESTER	

RUBBERS	TEMP °F (60–460)
BUTYL GR-1	
FLUORO ELASTOMERS VITON A	
FLUORO ELASTOMERS KEL-F 3700	
HARD RUBBER	
HYPALON	
KOROSEAL	
NATURAL RUBBER (GRS)	
NEOPRENE GR-M (CR)	
NITRILE BUNA N (NBR)	
NORDEL	
POLYBUTADIENE	
SILICONE	
POLYURETHANE	

METHYL ISOBUTYL KETONE

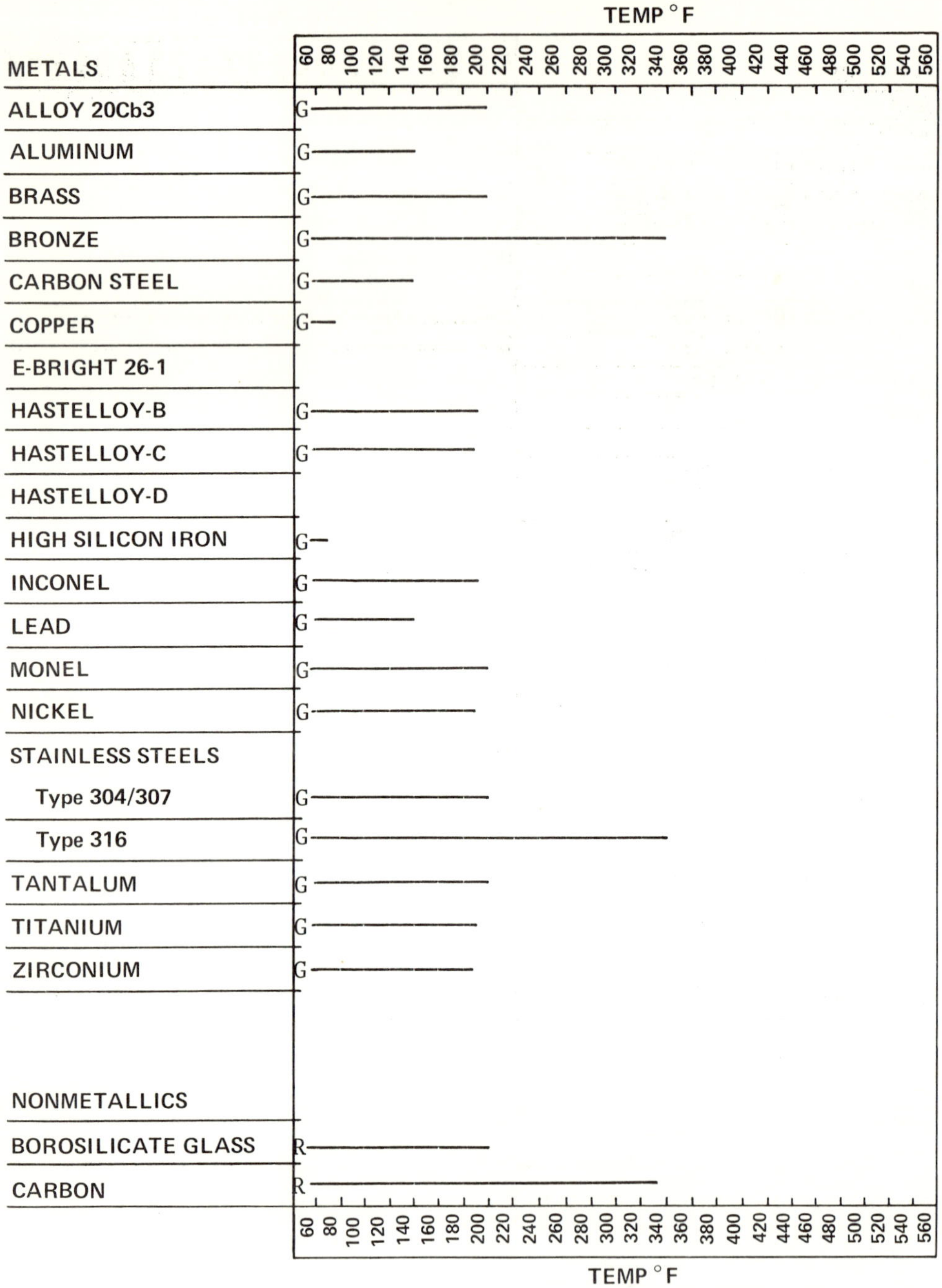

FOR METALS

E = < 2 Mils Penetration/Year; G = < 20 Mils Penetration/Year

S = < 50 Mils Penetration/Year; U = > 50 Mils Penetration/Year

FOR NONMETALLICS

R = Resistant

U = Unsatisfactory

METHYL ISOBUTYL KETONE

TEMP °F (60 80 100 120 140 160 180 200 220 240 260 280 300 320 340 360 380 400 420 440 460)

PLASTICS	Rating (TEMP °F)
ABS	U
ACRYLICS	
ASBESTOS REINF., EPOXY	RU
ASBESTOS REINF., PHENOLIC	R to 160
CHLORINATED POLYETHER	R to 80
CPVC	U
EPOXY	R to 140, U
FLUOROCARBONS FEP	R to 400
FLUOROCARBONS TFE	R to 460
FURFURYL ALCOHOL	R to 250
NORYL	
PHENOLIC	
POLYAMIDES–NYLON	R to 80
POLYESTERS	
BISPHENOL A-FUMURATE	U
HYDROGENATED BISPHENOL-A-BISPHENOL-A	U
ISOPHTHALIC	U
CHLORINATED POLYESTERS	R to 80
POLYSTYRENES	
POLYETHYLENE	U
POLYPROPYLENE	R to 80
PVC–TYPE 1	U
PVC–TYPE 2	U
POLYVINYLIDENE CHLORIDE	R to 80
VINYLIDENE FLUORIDE	R to 120, U
VINYL ESTER	U

RUBBERS	Rating (TEMP °F)
BUTYL GR-1	R to 80
FLUORO ELASTOMERS VITON A	U
FLUORO ELASTOMERS KEL-F 3700	
HARD RUBBER	
HYPALON	U
KOROSEAL	
NATURAL RUBBER (GRS)	U
NEOPRENE GR-M (CR)	U
NITRILE BUNA N (NBR)	U
NORDEL	
POLYBUTADIENE	
SILICONE	
POLYURETHANE	

METHYL METHACRYLATE

TEMP °F (60–560)

METALS	
ALLOY 20Cb3	G
ALUMINUM	G
BRASS	
BRONZE	
CARBON STEEL	U
COPPER	
E-BRIGHT 26-1	
HASTELLOY-B	
HASTELLOY-C	
HASTELLOY-D	
HIGH SILICON IRON	
INCONEL	
LEAD	
MONEL	
NICKEL	
STAINLESS STEELS	
Type 304/307	G
Type 316	G
TANTALUM	G
TITANIUM	
ZIRCONIUM	

NONMETALLICS	
BOROSILICATE GLASS	R
CARBON	

TEMP °F: 60 80 100 120 140 160 180 200 220 240 260 280 300 320 340 360 380 400 420 440 460 480 500 520 540 560

FOR METALS

E = < 2 Mils Penetration/Year; G = < 20 Mils Penetration/Year

S = < 50 Mils Penetration/Year; U = > 50 Mils Penetration/Year

FOR NONMETALLICS

R = Resistant

U = Unsatisfactory

METHYL METHACRYLATE

TEMP °F: 60, 80, 100, 120, 140, 160, 180, 200, 220, 240, 260, 280, 300, 320, 340, 360, 380, 400, 420, 440, 460

PLASTICS	
ABS	
ACRYLICS	
ASBESTOS REINF., EPOXY	
ASBESTOS REINF., PHENOLIC	
CHLORINATED POLYETHER	R
CPVC	
EPOXY	U
FLUOROCARBONS FEP	R
FLUOROCARBONS TFE	R
FURFURYL ALCOHOL	R
NORYL	
PHENOLIC	
POLYAMIDES–NYLON	
POLYESTERS	
BISPHENOL A-FUMURATE	U
HYDROGENATED BISPHENOL-A-BISPHENOL-A	U
ISOPHTHALIC	U
CHLORINATED POLYESTERS	
POLYSTYRENES	
POLYETHYLENE	
POLYPROPYLENE	R
PVC–TYPE 1	R
PVC–TYPE 2	U
POLYVINYLIDENE CHLORIDE	R
VINYLIDENE FLUORIDE	R
VINYL ESTER	

RUBBERS	
BUTYL GR-1	U
FLUORO ELASTOMERS VITON A	U
FLUORO ELASTOMERS KEL-F 3700	
HARD RUBBER	
HYPALON	
KOROSEAL	
NATURAL RUBBER (GRS)	U
NEOPRENE GR-M (CR)	U
NITRILE BUNA N (NBR)	U
NORDEL	
POLYBUTADIENE	
SILICONE	
POLYURETHANE	

TEMP °F: 60, 80, 100, 120, 140, 160, 180, 200, 220, 240, 260, 280, 300, 320, 340, 360, 380, 400, 420, 440, 460

METHYL SALICYLATE

TEMP °F: 60 80 100 120 140 160 180 200 220 240 260 280 300 320 340 360 380 400 420 440 460 480 500 520 540 560

METALS	Rating (TEMP °F)
ALLOY 20Cb3	
ALUMINUM	G — (60–80)
BRASS	
BRONZE	
CARBON STEEL	G — (60–80)
COPPER	G — (60–80)
E-BRIGHT 26-1	
HASTELLOY-B	
HASTELLOY-C	
HASTELLOY-D	
HIGH SILICON IRON	
INCONEL	
LEAD	
MONEL	G — (60–80)
NICKEL	
STAINLESS STEELS	
Type 304/307	
Type 316	
TANTALUM	G — (60–80)
TITANIUM	
ZIRCONIUM	

NONMETALLICS	Rating (TEMP °F)
BOROSILICATE GLASS	R — (60–100)
CARBON	R — (60–340)

TEMP °F

FOR METALS

E = < 2 Mils Penetration/Year; G = < 20 Mils Penetration/Year

S = < 50 Mils Penetration/Year; U = > 50 Mils Penetration/Year

FOR NONMETALLICS

R = Resistant

U = Unsatisfactory

METHYL SALICYLATE

TEMP °F

PLASTICS	60 – 460 °F
ABS	
ACRYLICS	
ASBESTOS REINF., EPOXY	
ASBESTOS REINF., PHENOLIC	
CHLORINATED POLYETHER	R (60–140) U
CPVC	
EPOXY	
FLUOROCARBONS FEP	R (60–400)
FLUOROCARBONS TFE	R (60–460)
FURFURYL ALCOHOL	
NORYL	
PHENOLIC	
POLYAMIDES–NYLON	
POLYESTERS	
BISPHENOL A-FUMURATE	
HYDROGENATED BISPHENOL-A-BISPHENOL-A	
ISOPHTHALIC	
CHLORINATED POLYESTERS	
POLYSTYRENES	
POLYETHYLENE	
POLYPROPYLENE	R (60–125)
PVC–TYPE 1	R (60–73)
PVC–TYPE 2	
POLYVINYLIDENE CHLORIDE	R (60–125)
VINYLIDENE FLUORIDE	R (60–150)
VINYL ESTER	

RUBBERS	60 – 460 °F
BUTYL GR-1	R (60–80)
FLUORO ELASTOMERS VITON A	
FLUORO ELASTOMERS KEL-F 3700	
HARD RUBBER	
HYPALON	
KOROSEAL	
NATURAL RUBBER (GRS)	U (60)
NEOPRENE GR-M (CR)	
NITRILE BUNA N (NBR)	U (60)
NORDEL	
POLYBUTADIENE	
SILICONE	
POLYURETHANE	

TEMP °F

METHYL SULFATE

TEMP °F

METALS	60	80	100	120	140	160	180	200	220	240	260	280	300	320	340	360	380	400	420	440	460	480	500	520	540	560
ALLOY 20Cb3																										
ALUMINUM																										
BRASS																										
BRONZE																										
CARBON STEEL																										
COPPER																										
E-BRIGHT 26-1																										
HASTELLOY-B																										
HASTELLOY-C																										
HASTELLOY-D																										
HIGH SILICON IRON																										
INCONEL																										
LEAD																										
MONEL																										
NICKEL																										
STAINLESS STEELS Type 304/307																										
Type 316																										
TANTALUM																										
TITANIUM																										
ZIRCONIUM																										
NONMETALLICS																										
BOROSILICATE GLASS																										
CARBON																										

60 80 100 120 140 160 180 200 220 240 260 280 300 320 340 360 380 400 420 440 460 480 500 520 540 560

TEMP °F

FOR METALS

E = < 2 Mils Penetration/Year; G = < 20 Mils Penetration/Year

S = < 50 Mils Penetration/Year; U = > 50 Mils Penetration/Year

FOR NONMETALLICS

R = Resistant

U = Unsatisfactory

METHYL SULFATE

PLASTICS	TEMP °F
ABS	R 60–100
ACRYLICS	
ASBESTOS REINF., EPOXY	
ASBESTOS REINF., PHENOLIC	
CHLORINATED POLYETHER	U
CPVC	R 60–140, U at 140
EPOXY	
FLUOROCARBONS FEP	R 60–400
FLUOROCARBONS TFE	R 60–460
FURFURYL ALCOHOL	R 60–260
NORYL	
PHENOLIC	
POLYAMIDES–NYLON	R 60–80
POLYESTERS	
BISPHENOL A-FUMURATE	
HYDROGENATED BISPHENOL-A-BISPHENOL-A	
ISOPHTHALIC	
CHLORINATED POLYESTERS	
POLYSTYRENES	
POLYETHYLENE	R 60–140
POLYPROPYLENE	U
PVC–TYPE 1	RU
PVC–TYPE 2	RU
POLYVINYLIDENE CHLORIDE	
VINYLIDENE FLUORIDE	
VINYL ESTER	

RUBBERS	TEMP °F
BUTYL GR-1	
FLUORO ELASTOMERS VITON A	
FLUORO ELASTOMERS KEL-F 3700	
HARD RUBBER	
HYPALON	
KOROSEAL	U
NATURAL RUBBER (GRS)	
NEOPRENE GR-M (CR)	U
NITRILE BUNA N (NBR)	
NORDEL	
POLYBUTADIENE	
SILICONE	
POLYURETHANE	

TEMP °F: 60 80 100 120 140 160 180 200 220 240 260 280 300 320 340 360 380 400 420 440 460

METHYL SULFURIC ACID

METALS	TEMP °F (60–560)
ALLOY 20Cb3	G—
ALUMINUM	
BRASS	U
BRONZE	U
CARBON STEEL	
COPPER	U
E-BRIGHT 26-1	
HASTELLOY-B	
HASTELLOY-C	
HASTELLOY-D	
HIGH SILICON IRON	G—
INCONEL	U
LEAD	
MONEL	U
NICKEL	U
STAINLESS STEELS	
Type 304/307	
Type 316	G—
TANTALUM	E ————————
TITANIUM	
ZIRCONIUM	
NONMETALLICS	
BOROSILICATE GLASS	
CARBON	

TEMP °F: 60 80 100 120 140 160 180 200 220 240 260 280 300 320 340 360 380 400 420 440 460 480 500 520 540 560

FOR METALS

E = < 2 Mils Penetration/Year; G = < 20 Mils Penetration/Year

S = < 50 Mils Penetration/Year; U = > 50 Mils Penetration/Year

FOR NONMETALLICS

R = Resistant

U = Unsatisfactory

METHYL SULFURIC ACID

TEMP °F: 60 80 100 120 140 160 180 200 220 240 260 280 300 320 340 360 380 400 420 440 460

PLASTICS	
ABS	U
ACRYLICS	
ASBESTOS REINF., EPOXY	
ASBESTOS REINF., PHENOLIC	
CHLORINATED POLYETHER	R—
CPVC	R————U
EPOXY	
FLUOROCARBONS FEP	R————————————————
FLUOROCARBONS TFE	R——————————————————
FURFURYL ALCOHOL	
NORYL	
PHENOLIC	
POLYAMIDES–NYLON	
POLYESTERS	
BISPHENOL A-FUMURATE	
HYDROGENATED BISPHENOL-A-BISPHENOL-A	
ISOPHTHALIC	
CHLORINATED POLYESTERS	
POLYSTYRENES	
POLYETHYLENE	R—————
POLYPROPYLENE	R—————
PVC–TYPE 1	R————
PVC–TYPE 2	R————
POLYVINYLIDENE CHLORIDE	R—
VINYLIDENE FLUORIDE	R———
VINYL ESTER	

RUBBERS	
BUTYL GR-1	
FLUORO ELASTOMERS VITON A	
FLUORO ELASTOMERS KEL-F 3700	
HARD RUBBER	
HYPALON	
KOROSEAL	
NATURAL RUBBER (GRS)	
NEOPRENE GR-M (CR)	
NITRILE BUNA N (NBR)	
NORDEL	
POLYBUTADIENE	
SILICONE	
POLYURETHANE	

TEMP °F: 60 80 100 120 140 160 180 200 220 240 260 280 300 320 340 360 380 400 420 440 460

METHYLENE CHLORIDE

METALS		Rating
ALLOY 20Cb3		E—G
ALUMINUM	copper free	E—G
BRASS		G
BRONZE		G
CARBON STEEL		G
COPPER		G
E-BRIGHT 26-1		S
HASTELLOY-B	2	E
HASTELLOY-C	2	E
HASTELLOY-D		
HIGH SILICON IRON		
INCONEL		G
LEAD		G
MONEL	acid free	E—G
NICKEL	2	G
STAINLESS STEELS		
Type 304/307	2	G
Type 316	1,2	G
TANTALUM	40%	G
TITANIUM		E
ZIRCONIUM		

NONMETALLICS	Rating
BOROSILICATE GLASS	R
CARBON	R

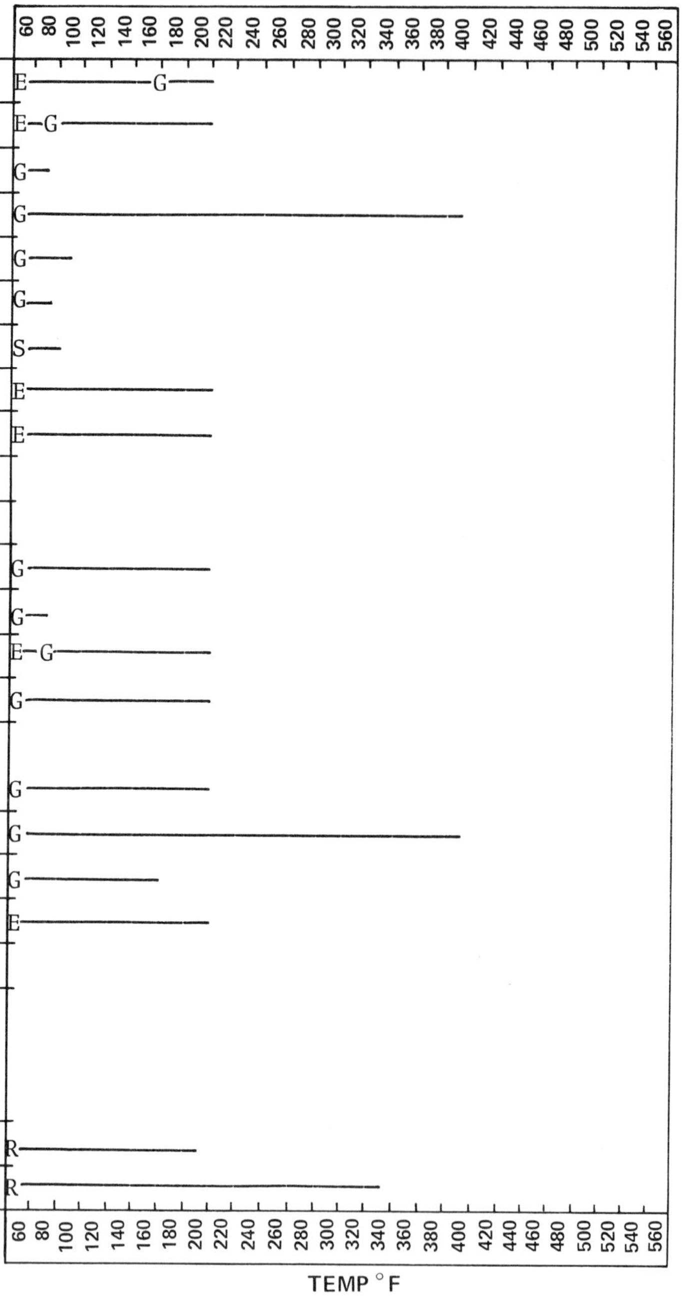

FOR METALS

E = < 2 Mils Penetration/Year; G = < 20 Mils Penetration/Year

S = < 50 Mils Penetration/Year; U = > 50 Mils Penetration/Year

FOR NONMETALLICS

R = Resistant

U = Unsatisfactory

TEMP °F (60 80 100 120 140 160 180 200 220 240 260 280 300 320 340 360 380 400 420 440 460)

PLASTICS	TEMP °F
ABS	U
ACRYLICS	
ASBESTOS REINF., EPOXY	
ASBESTOS REINF., PHENOLIC	
CHLORINATED POLYETHER	U
CPVC	U
EPOXY	R– U
FLUOROCARBONS FEP	R———— (60–400)
FLUOROCARBONS TFE	R———— (60–460)
FURFURYL ALCOHOL	R———— (60–~270)
NORYL	
PHENOLIC	
POLYAMIDES–NYLON	
POLYESTERS	
BISPHENOL A-FUMURATE	U
HYDROGENATED BISPHENOL-A-BISPHENOL-A	U
ISOPHTHALIC	U
CHLORINATED POLYESTERS	U
POLYSTYRENES	
POLYETHYLENE	U
POLYPROPYLENE	RU
PVC–TYPE 1	U
PVC–TYPE 2	U
POLYVINYLIDENE CHLORIDE	U
VINYLIDENE FLUORIDE	R—— (60–120)
VINYL ESTER	U

RUBBERS	TEMP °F
BUTYL GR-1	U
FLUORO ELASTOMERS VITON A	RU
FLUORO ELASTOMERS KEL-F 3700	
HARD RUBBER	
HYPALON	U
KOROSEAL	
NATURAL RUBBER (GRS)	U
NEOPRENE GR-M (CR)	U
NITRILE BUNA N (NBR)	U
NORDEL	U
POLYBUTADIENE	
SILICONE	
POLYURETHANE	U

TEMP °F (60 80 100 120 140 160 180 200 220 240 260 280 300 320 340 360 380 400 420 440 460)

MILK

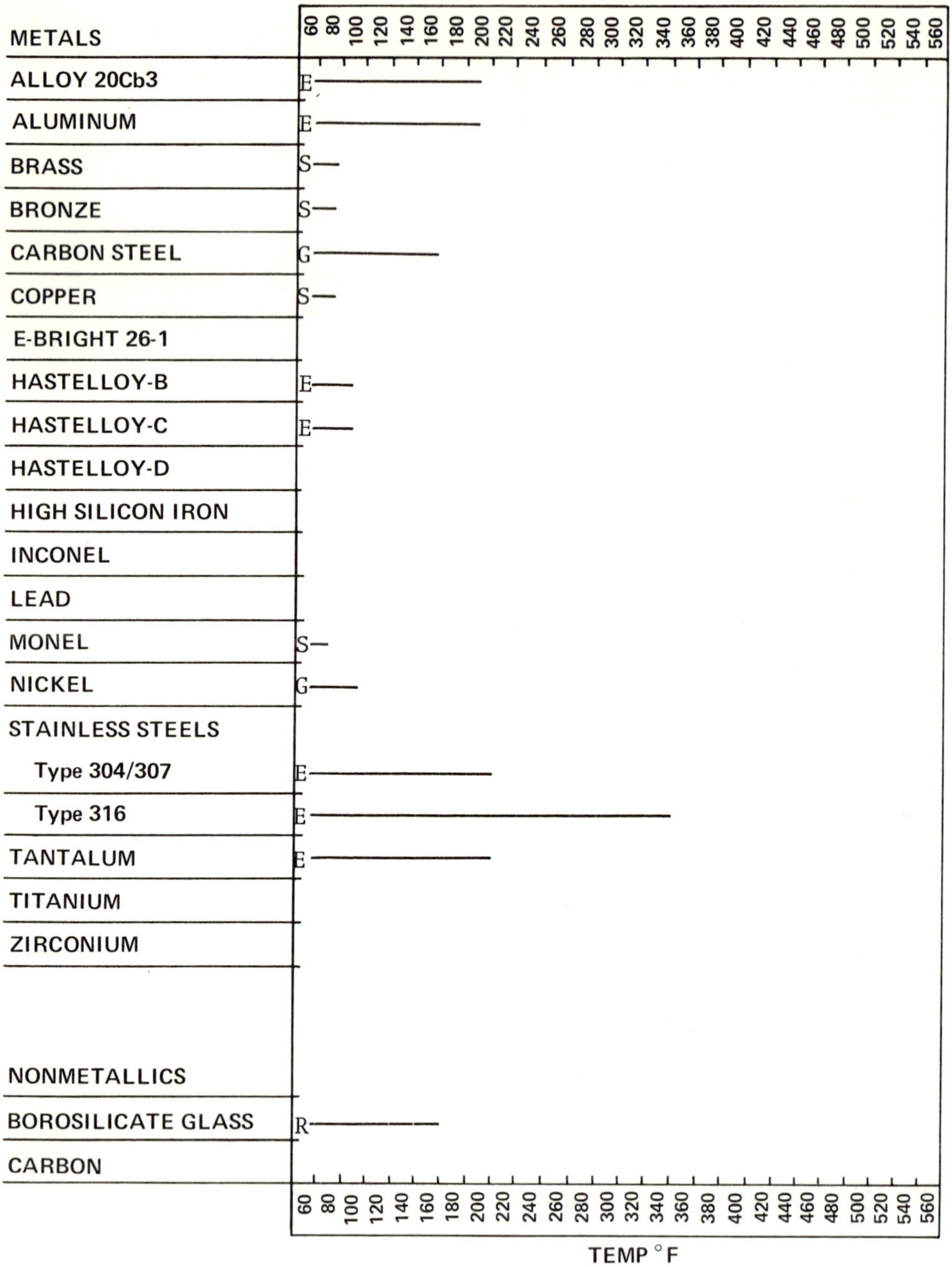

FOR METALS

E = < 2 Mils Penetration/Year; G = < 20 Mils Penetration/Year

S = < 50 Mils Penetration/Year; U = > 50 Mils Penetration/Year

FOR NONMETALLICS

R = Resistant

U = Unsatisfactory

MILK

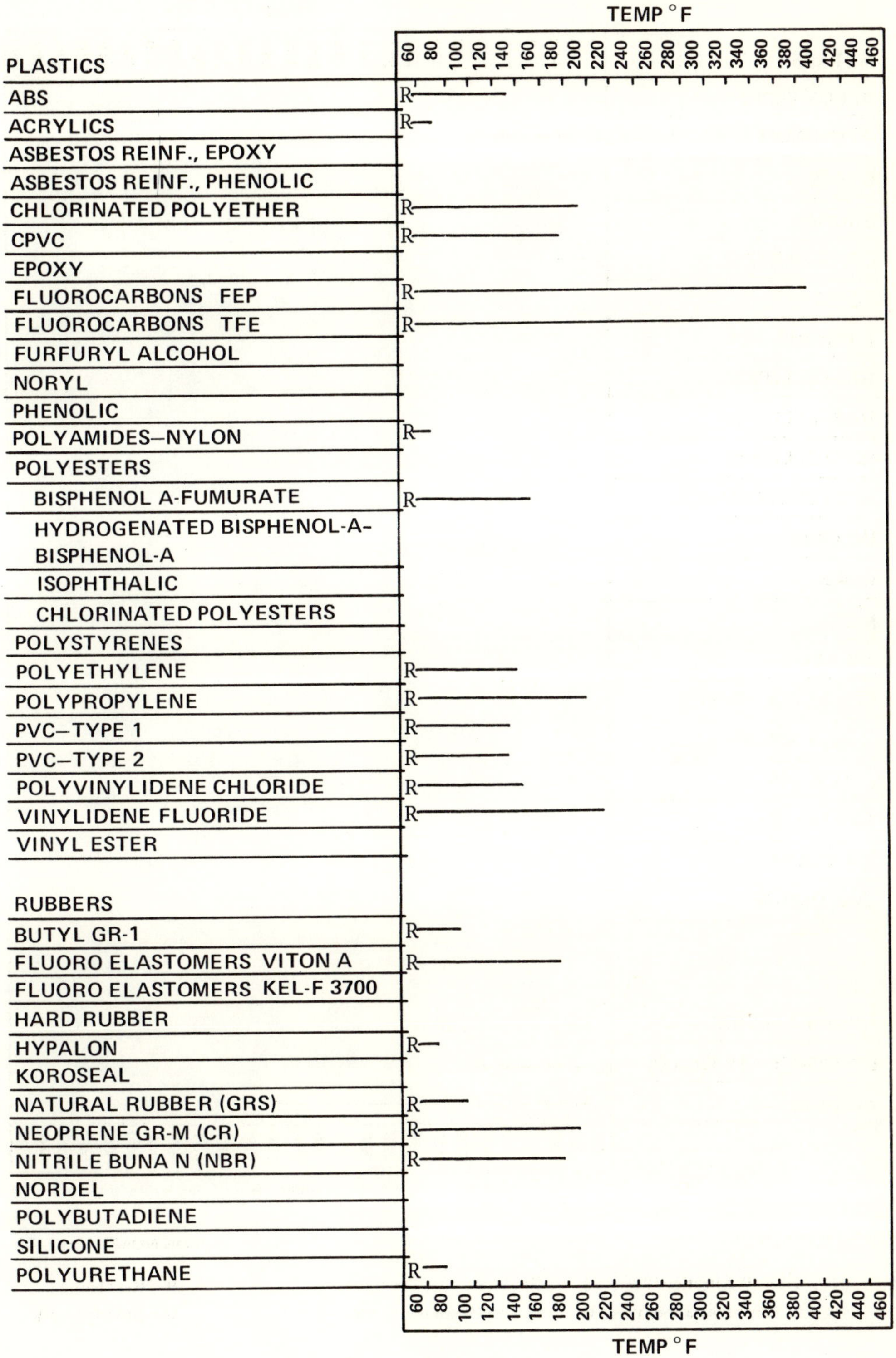
TEMP °F
60 80 100 120 140 160 180 200 220 240 260 280 300 320 340 360 380 400 420 440 460
PLASTICS
ABS R
ACRYLICS R
ASBESTOS REINF., EPOXY
ASBESTOS REINF., PHENOLIC
CHLORINATED POLYETHER R
CPVC R
EPOXY
FLUOROCARBONS FEP R
FLUOROCARBONS TFE R
FURFURYL ALCOHOL
NORYL
PHENOLIC
POLYAMIDES–NYLON R
POLYESTERS
BISPHENOL A-FUMURATE R
HYDROGENATED BISPHENOL-A-BISPHENOL-A
ISOPHTHALIC
CHLORINATED POLYESTERS
POLYSTYRENES
POLYETHYLENE R
POLYPROPYLENE R
PVC–TYPE 1 R
PVC–TYPE 2 R
POLYVINYLIDENE CHLORIDE R
VINYLIDENE FLUORIDE R
VINYL ESTER
RUBBERS
BUTYL GR-1 R
FLUORO ELASTOMERS VITON A R
FLUORO ELASTOMERS KEL-F 3700
HARD RUBBER
HYPALON R
KOROSEAL
NATURAL RUBBER (GRS) R
NEOPRENE GR-M (CR) R
NITRILE BUNA N (NBR) R
NORDEL
POLYBUTADIENE
SILICONE
POLYURETHANE R
60 80 100 120 140 160 180 200 220 240 260 280 300 320 340 360 380 400 420 440 460
TEMP °F

MINERAL OIL

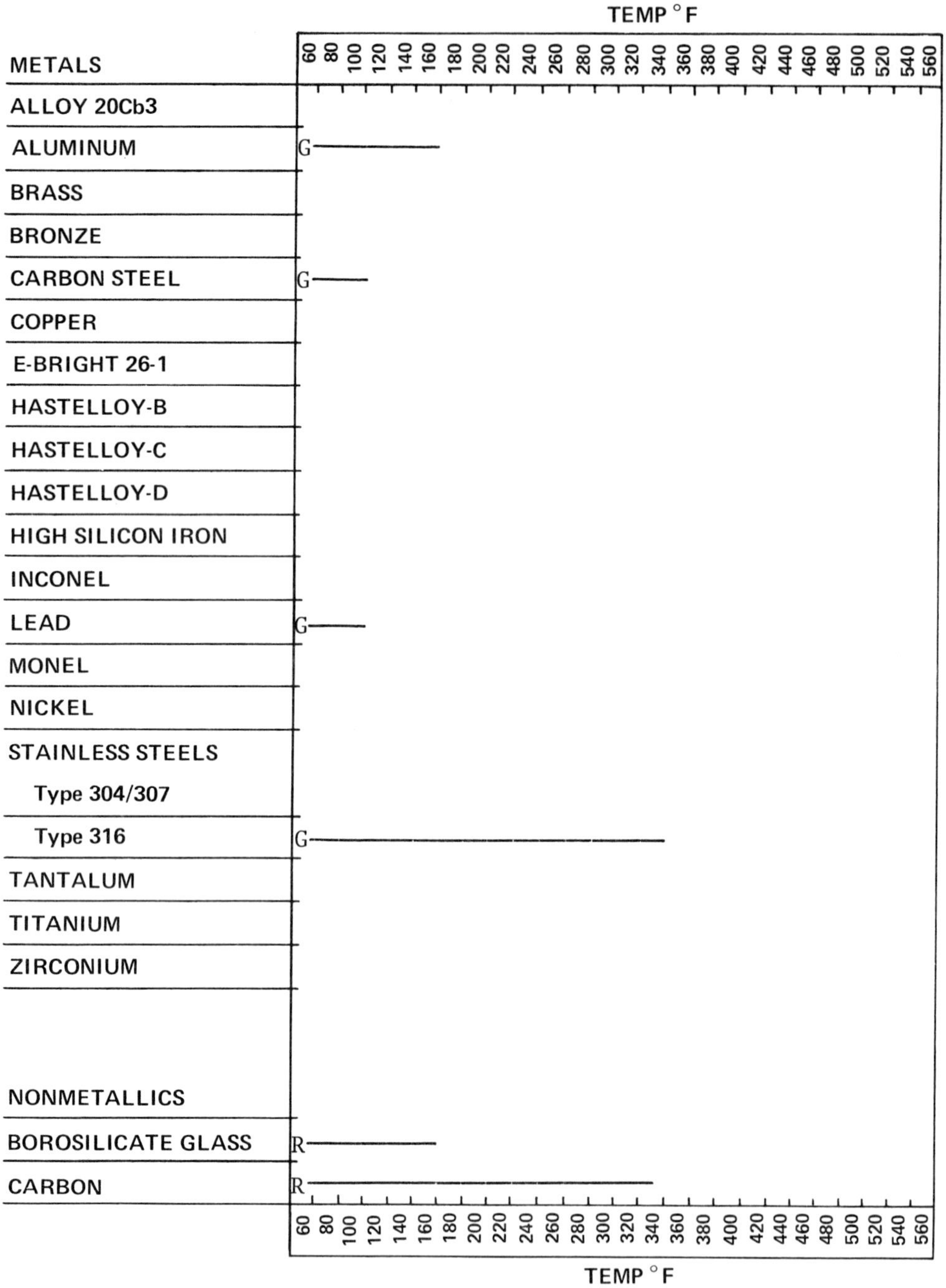

FOR METALS

E = < 2 Mils Penetration/Year; G = < 20 Mils Penetration/Year

S = < 50 Mils Penetration/Year; U = > 50 Mils Penetration/Year

FOR NONMETALLICS

R = Resistant

U = Unsatisfactory

MINERAL OIL

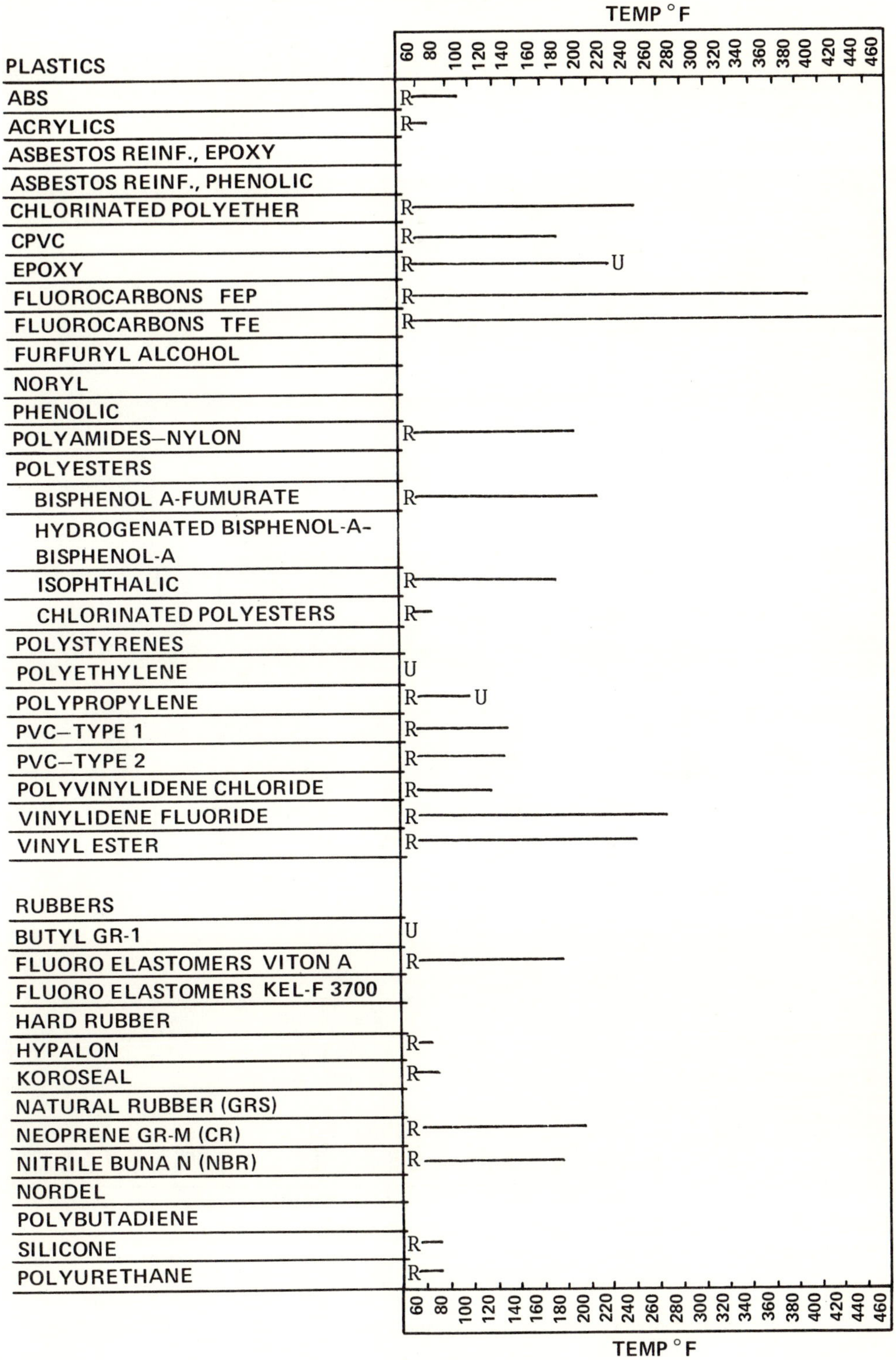

TEMP °F
60 80 100 120 140 160 180 200 220 240 260 280 300 320 340 360 380 400 420 440 460
PLASTICS
ABS
ACRYLICS
ASBESTOS REINF., EPOXY
ASBESTOS REINF., PHENOLIC
CHLORINATED POLYETHER
CPVC
EPOXY
FLUOROCARBONS FEP
FLUOROCARBONS TFE
FURFURYL ALCOHOL
NORYL
PHENOLIC
POLYAMIDES–NYLON
POLYESTERS
BISPHENOL A-FUMURATE
HYDROGENATED BISPHENOL-A-BISPHENOL-A
ISOPHTHALIC
CHLORINATED POLYESTERS
POLYSTYRENES
POLYETHYLENE
POLYPROPYLENE
PVC–TYPE 1
PVC–TYPE 2
POLYVINYLIDENE CHLORIDE
VINYLIDENE FLUORIDE
VINYL ESTER
RUBBERS
BUTYL GR-1
FLUORO ELASTOMERS VITON A
FLUORO ELASTOMERS KEL-F 3700
HARD RUBBER
HYPALON
KOROSEAL
NATURAL RUBBER (GRS)
NEOPRENE GR-M (CR)
NITRILE BUNA N (NBR)
NORDEL
POLYBUTADIENE
SILICONE
POLYURETHANE
R
U

MIXED ACIDS

TEMP °F: 60, 80, 100, 120, 140, 160, 180, 200, 220, 240, 260, 280, 300, 320, 340, 360, 380, 400, 420, 440, 460, 480, 500, 520, 540, 560

METALS	Rating
ALLOY 20Cb3	G
ALUMINUM	U
BRASS	U
BRONZE	U
CARBON STEEL	U
COPPER	
E-BRIGHT 26-1	
HASTELLOY-B	
HASTELLOY-C	
HASTELLOY-D	
HIGH SILICON IRON	G
INCONEL	G
LEAD	
MONEL	U
NICKEL	U
STAINLESS STEELS	
Type 304/307	E
Type 316	E
TANTALUM	G
TITANIUM	
ZIRCONIUM	
NONMETALLICS	
BOROSILICATE GLASS	
CARBON	

TEMP °F

FOR METALS

E = < 2 Mils Penetration/Year; G = < 20 Mils Penetration/Year

S = < 50 Mils Penetration/Year; U = > 50 Mils Penetration/Year

FOR NONMETALLICS

R = Resistant

U = Unsatisfactory

MIXED ACIDS

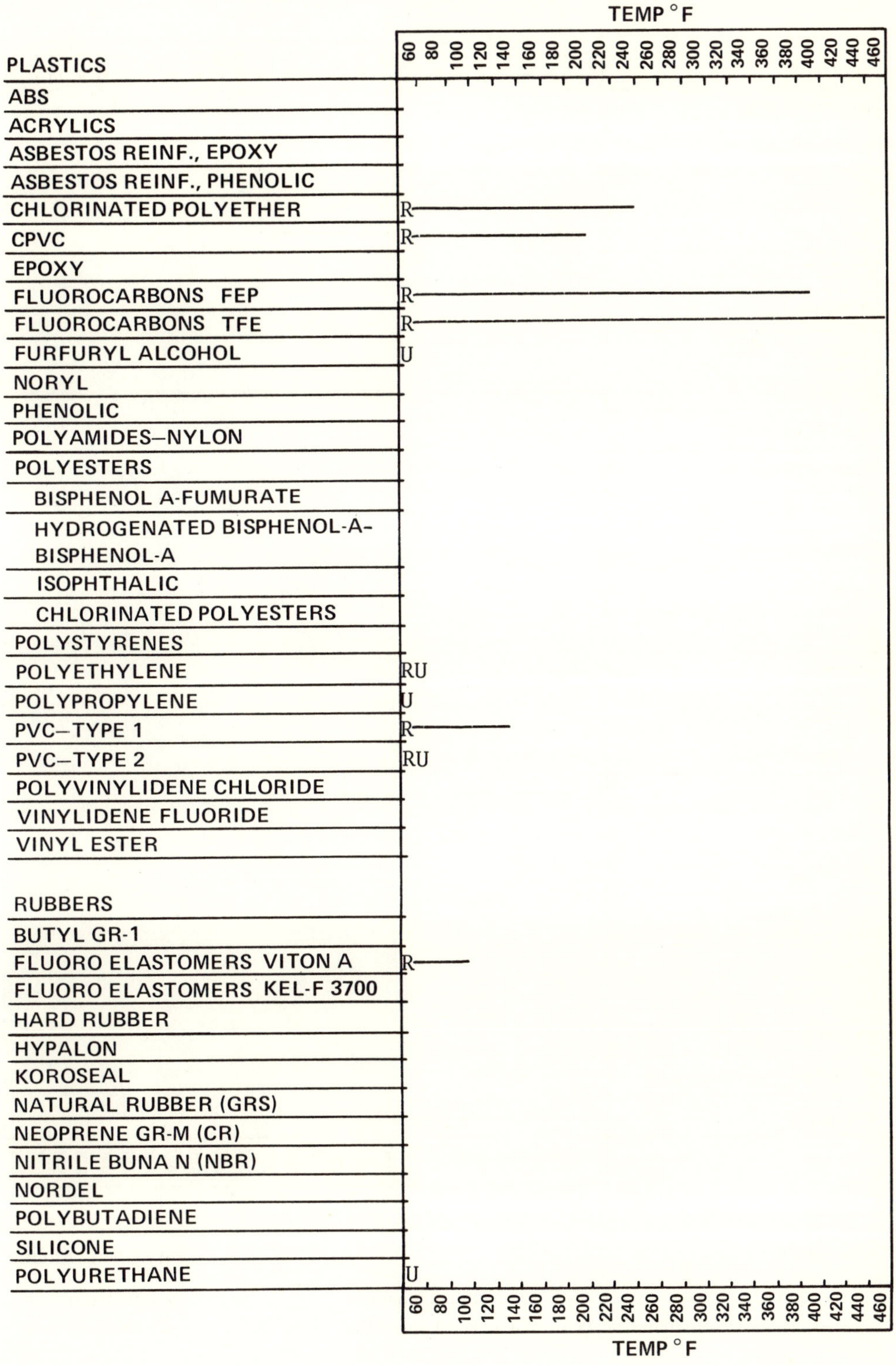
TEMP °F
PLASTICS
60 80 100 120 140 160 180 200 220 240 260 280 300 320 340 360 380 400 420 440 460
ABS
ACRYLICS
ASBESTOS REINF., EPOXY
ASBESTOS REINF., PHENOLIC
CHLORINATED POLYETHER R
CPVC R
EPOXY
FLUOROCARBONS FEP R
FLUOROCARBONS TFE R
FURFURYL ALCOHOL U
NORYL
PHENOLIC
POLYAMIDES–NYLON
POLYESTERS
BISPHENOL A-FUMURATE
HYDROGENATED BISPHENOL-A-BISPHENOL-A
ISOPHTHALIC
CHLORINATED POLYESTERS
POLYSTYRENES
POLYETHYLENE RU
POLYPROPYLENE U
PVC–TYPE 1 R
PVC–TYPE 2 RU
POLYVINYLIDENE CHLORIDE
VINYLIDENE FLUORIDE
VINYL ESTER
RUBBERS
BUTYL GR-1
FLUORO ELASTOMERS VITON A R
FLUORO ELASTOMERS KEL-F 3700
HARD RUBBER
HYPALON
KOROSEAL
NATURAL RUBBER (GRS)
NEOPRENE GR-M (CR)
NITRILE BUNA N (NBR)
NORDEL
POLYBUTADIENE
SILICONE
POLYURETHANE U
60 80 100 120 140 160 180 200 220 240 260 280 300 320 340 360 380 400 420 440 460
TEMP °F

MOLASSES

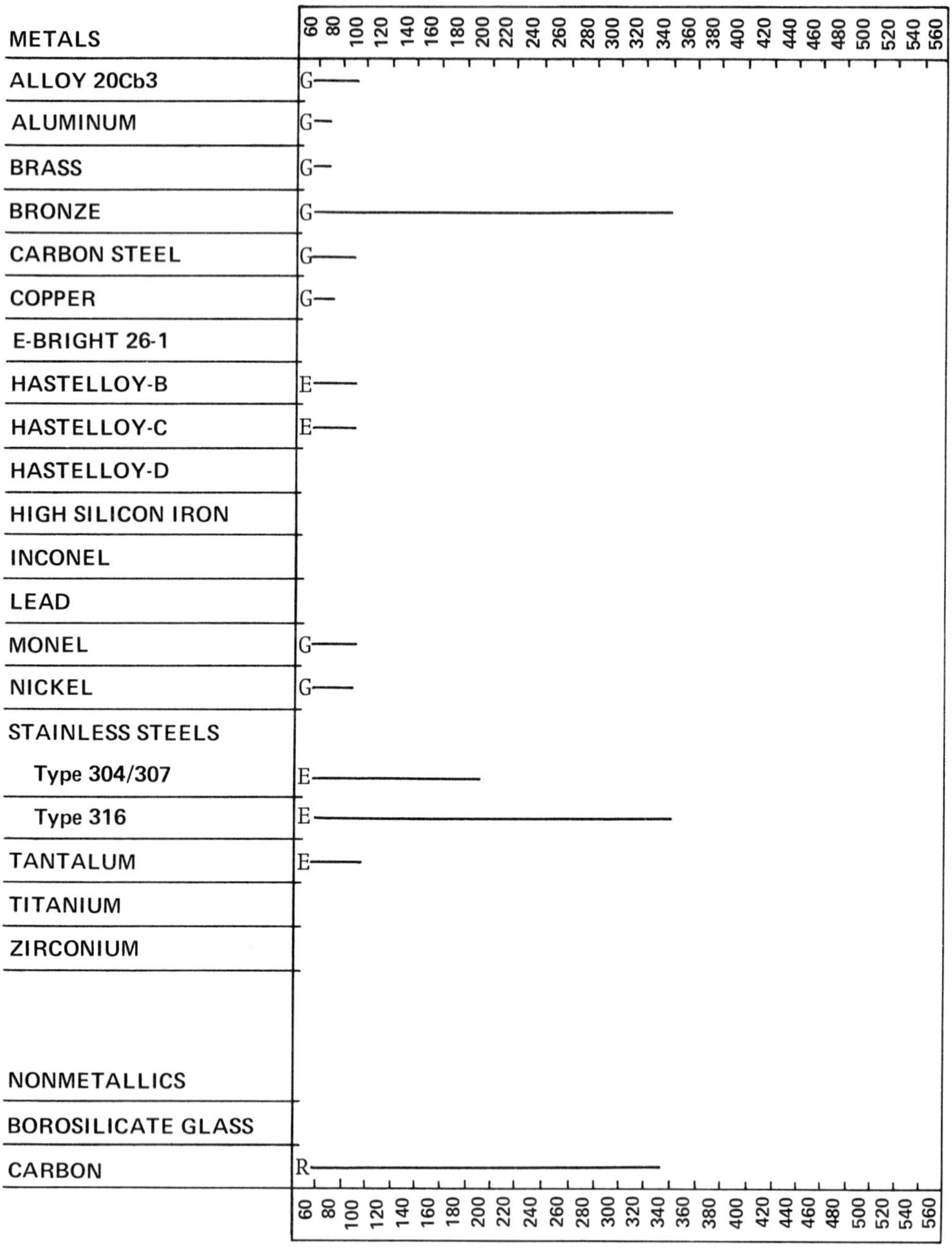

FOR METALS

E = < 2 Mils Penetration/Year; G = < 20 Mils Penetration/Year
S = < 50 Mils Penetration/Year; U = > 50 Mils Penetration/Year

FOR NONMETALLICS

R = Resistant
U = Unsatisfactory

MOLASSES

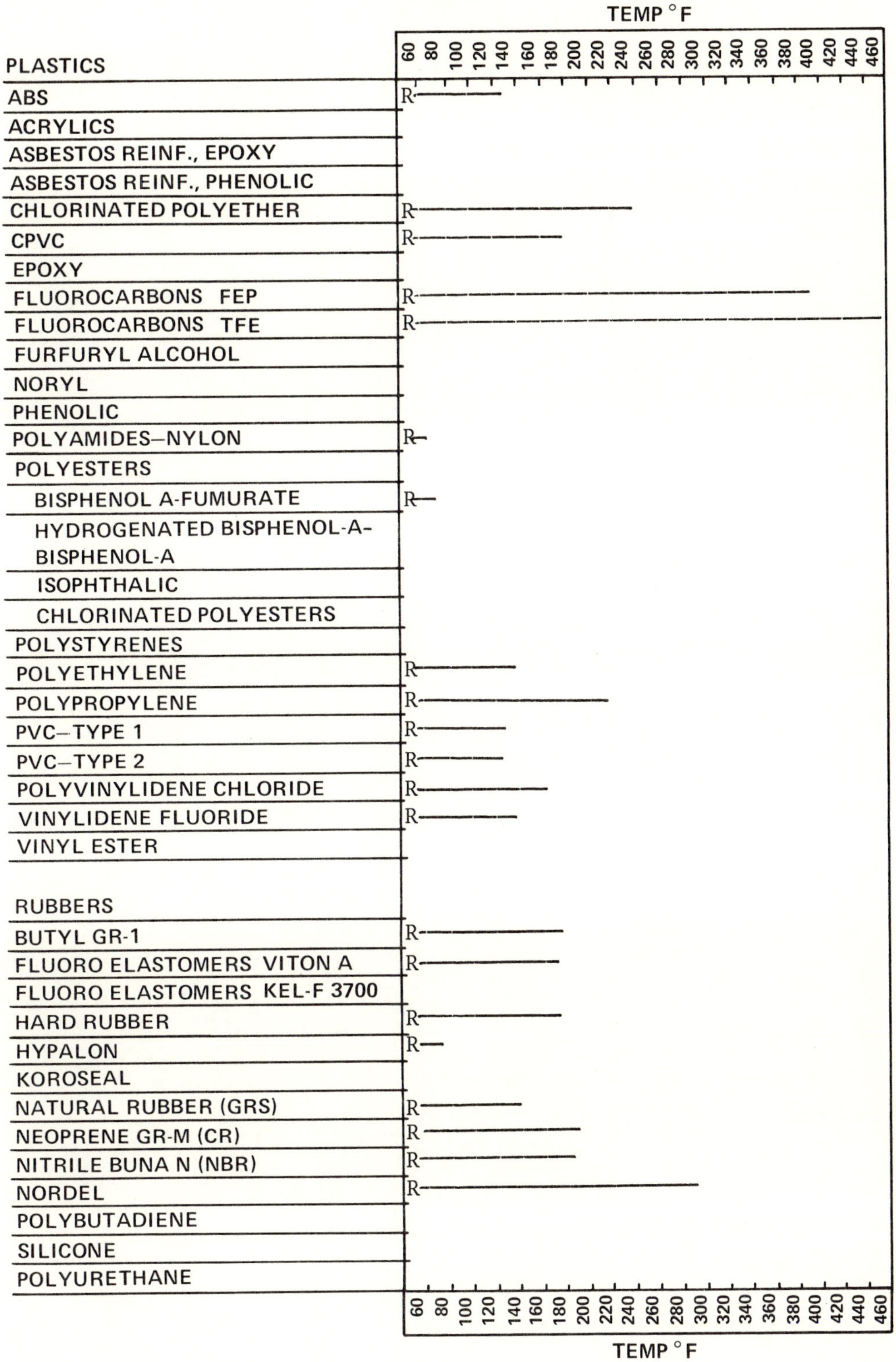
TEMP °F
60 80 100 120 140 160 180 200 220 240 260 280 300 320 340 360 380 400 420 440 460
PLASTICS
ABS R
ACRYLICS
ASBESTOS REINF., EPOXY
ASBESTOS REINF., PHENOLIC
CHLORINATED POLYETHER R
CPVC R
EPOXY
FLUOROCARBONS FEP R
FLUOROCARBONS TFE R
FURFURYL ALCOHOL
NORYL
PHENOLIC
POLYAMIDES—NYLON R
POLYESTERS
BISPHENOL A-FUMURATE R
HYDROGENATED BISPHENOL-A-BISPHENOL-A
ISOPHTHALIC
CHLORINATED POLYESTERS
POLYSTYRENES
POLYETHYLENE R
POLYPROPYLENE R
PVC—TYPE 1 R
PVC—TYPE 2 R
POLYVINYLIDENE CHLORIDE R
VINYLIDENE FLUORIDE R
VINYL ESTER
RUBBERS
BUTYL GR-1 R
FLUORO ELASTOMERS VITON A R
FLUORO ELASTOMERS KEL-F 3700
HARD RUBBER R
HYPALON R
KOROSEAL
NATURAL RUBBER (GRS) R
NEOPRENE GR-M (CR) R
NITRILE BUNA N (NBR) R
NORDEL R
POLYBUTADIENE
SILICONE
POLYURETHANE
60 80 100 120 140 160 180 200 220 240 260 280 300 320 340 360 380 400 420 440 460
TEMP °F

MONOCHLOROBENZENE

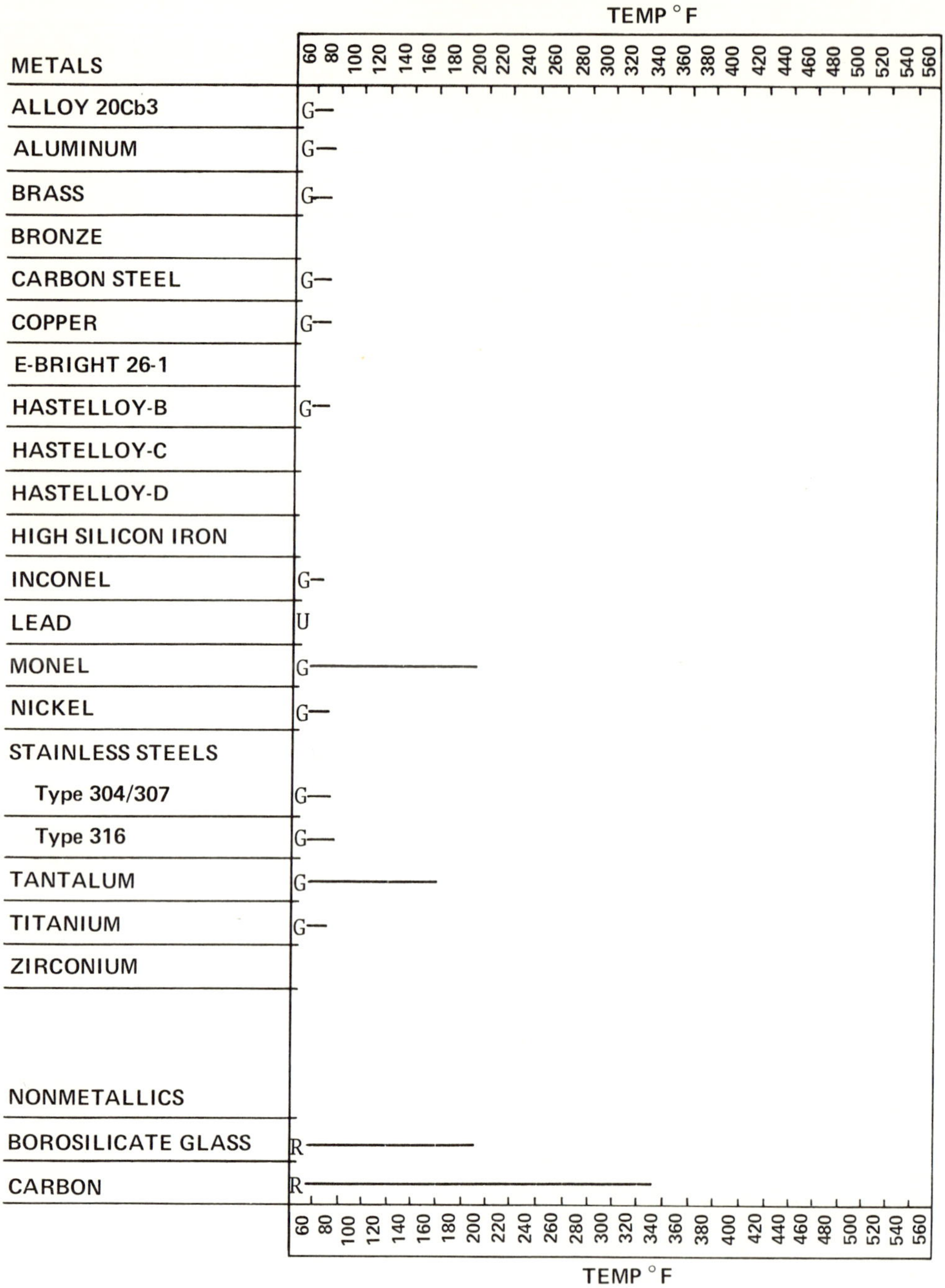

FOR METALS

E = < 2 Mils Penetration/Year; G = < 20 Mils Penetration/Year

S = < 50 Mils Penetration/Year; U = > 50 Mils Penetration/Year

FOR NONMETALLICS

R = Resistant

U = Unsatisfactory

MONOCHLOROBENZENE

TEMP °F: 60 80 100 120 140 160 180 200 220 240 260 280 300 320 340 360 380 400 420 440 460

PLASTICS	
ABS	
ACRYLICS	
ASBESTOS REINF., EPOXY	
ASBESTOS REINF., PHENOLIC	
CHLORINATED POLYETHER	
CPVC	
EPOXY	R—U
FLUOROCARBONS FEP	R—
FLUOROCARBONS TFE	R—
FURFURYL ALCOHOL	R—
NORYL	
PHENOLIC	
POLYAMIDES–NYLON	
POLYESTERS	
BISPHENOL A-FUMURATE	U
HYDROGENATED BISPHENOL-A-BISPHENOL-A	U
ISOPHTHALIC	U
CHLORINATED POLYESTERS	U
POLYSTYRENES	
POLYETHYLENE	U
POLYPROPYLENE	
PVC–TYPE 1	
PVC–TYPE 2	
POLYVINYLIDENE CHLORIDE	U
VINYLIDENE FLUORIDE	R—
VINYL ESTER	R—U
RUBBERS	
BUTYL GR-1	U
FLUORO ELASTOMERS VITON A	R—
FLUORO ELASTOMERS KEL-F 3700	
HARD RUBBER	
HYPALON	
KOROSEAL	
NATURAL RUBBER (GRS)	U
NEOPRENE GR-M (CR)	U
NITRILE BUNA N (NBR)	U
NORDEL	
POLYBUTADIENE	
SILICONE	
POLYURETHANE	

TEMP °F: 60 80 100 120 140 160 180 200 220 240 260 280 300 320 340 360 380 400 420 440 460

MONOETHANOLAMINE

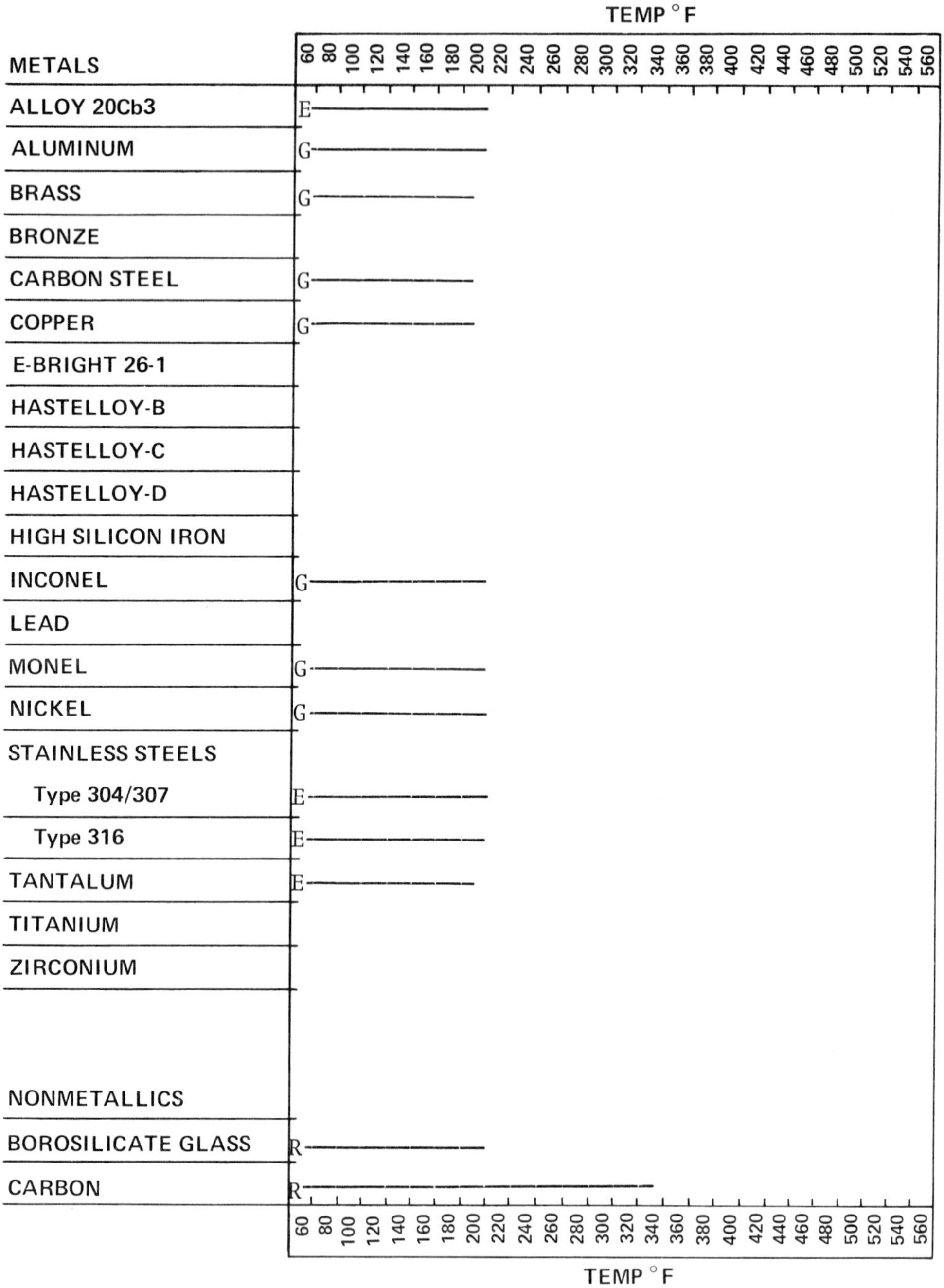

FOR METALS

E = < 2 Mils Penetration/Year; G = < 20 Mils Penetration/Year

S = < 50 Mils Penetration/Year; U = > 50 Mils Penetration/Year

FOR NONMETALLICS

R = Resistant

U = Unsatisfactory

MONOETHANOLAMINE

TEMP °F (60–460)

PLASTICS	
ABS	
ACRYLICS	
ASBESTOS REINF., EPOXY	
ASBESTOS REINF., PHENOLIC	
CHLORINATED POLYETHER	R
CPVC	
EPOXY	
FLUOROCARBONS FEP	R
FLUOROCARBONS TFE	R
FURFURYL ALCOHOL	
NORYL	
PHENOLIC	
POLYAMIDES–NYLON	
POLYESTERS	
BISPHENOL A-FUMURATE	R
HYDROGENATED BISPHENOL-A-BISPHENOL-A	
ISOPHTHALIC	
CHLORINATED POLYESTERS	R
POLYSTYRENES	
POLYETHYLENE	
POLYPROPYLENE	R
PVC–TYPE 1	
PVC–TYPE 2	
POLYVINYLIDENE CHLORIDE	
VINYLIDENE FLUORIDE	U
VINYL ESTER	

RUBBERS	
BUTYL GR-1	R
FLUORO ELASTOMERS VITON A	R
FLUORO ELASTOMERS KEL-F 3700	
HARD RUBBER	R
HYPALON	
KOROSEAL	
NATURAL RUBBER (GRS)	R
NEOPRENE GR-M (CR)	R
NITRILE BUNA N (NBR)	R
NORDEL	
POLYBUTADIENE	
SILICONE	
POLYURETHANE	

TEMP °F

MORPHOLINE

TEMP °F (60, 80, 100, 120, 140, 160, 180, 200, 220, 240, 260, 280, 300, 320, 340, 360, 380, 400, 420, 440, 460, 480, 500, 520, 540, 560)

METALS	Rating / TEMP °F
ALLOY 20Cb3	
ALUMINUM	
BRASS	
BRONZE	G (60–340)
CARBON STEEL	
COPPER	
E-BRIGHT 26-1	
HASTELLOY-B	
HASTELLOY-C	
HASTELLOY-D	
HIGH SILICON IRON	
INCONEL	
LEAD	
MONEL	
NICKEL	
STAINLESS STEELS	
Type 304/307	
Type 316	G (60–340)
TANTALUM	
TITANIUM	
ZIRCONIUM	
NONMETALLICS	
BOROSILICATE GLASS	
CARBON	G (60–200)

TEMP °F

FOR METALS

E = < 2 Mils Penetration/Year; G = < 20 Mils Penetration/Year

S = < 50 Mils Penetration/Year; U = > 50 Mils Penetration/Year

FOR NONMETALLICS

R = Resistant

U = Unsatisfactory

MORPHOLINE

PLASTICS	TEMP °F (60–460)
ABS	
ACRYLICS	
ASBESTOS REINF., EPOXY	
ASBESTOS REINF., PHENOLIC	
CHLORINATED POLYETHER	R (60–80)
CPVC	
EPOXY	
FLUOROCARBONS FEP	R (60–400)
FLUOROCARBONS TFE	R (60–460)
FURFURYL ALCOHOL	
NORYL	
PHENOLIC	
POLYAMIDES–NYLON	R (60–200)
POLYESTERS	
BISPHENOL A-FUMURATE	
HYDROGENATED BISPHENOL-A-BISPHENOL-A	
ISOPHTHALIC	
CHLORINATED POLYESTERS	
POLYSTYRENES	
POLYETHYLENE	R (60–140)
POLYPROPYLENE	R (60–150)
PVC–TYPE 1	
PVC–TYPE 2	
POLYVINYLIDENE CHLORIDE	
VINYLIDENE FLUORIDE	RU
VINYL ESTER	R (60–80)

RUBBERS	TEMP °F (60–460)
BUTYL GR-1	
FLUORO ELASTOMERS VITON A	
FLUORO ELASTOMERS KEL-F 3700	
HARD RUBBER	
HYPALON	
KOROSEAL	
NATURAL RUBBER (GRS)	
NEOPRENE GR-M (CR)	
NITRILE BUNA N (NBR)	
NORDEL	
POLYBUTADIENE	
SILICONE	
POLYURETHANE	

MOTOR OIL

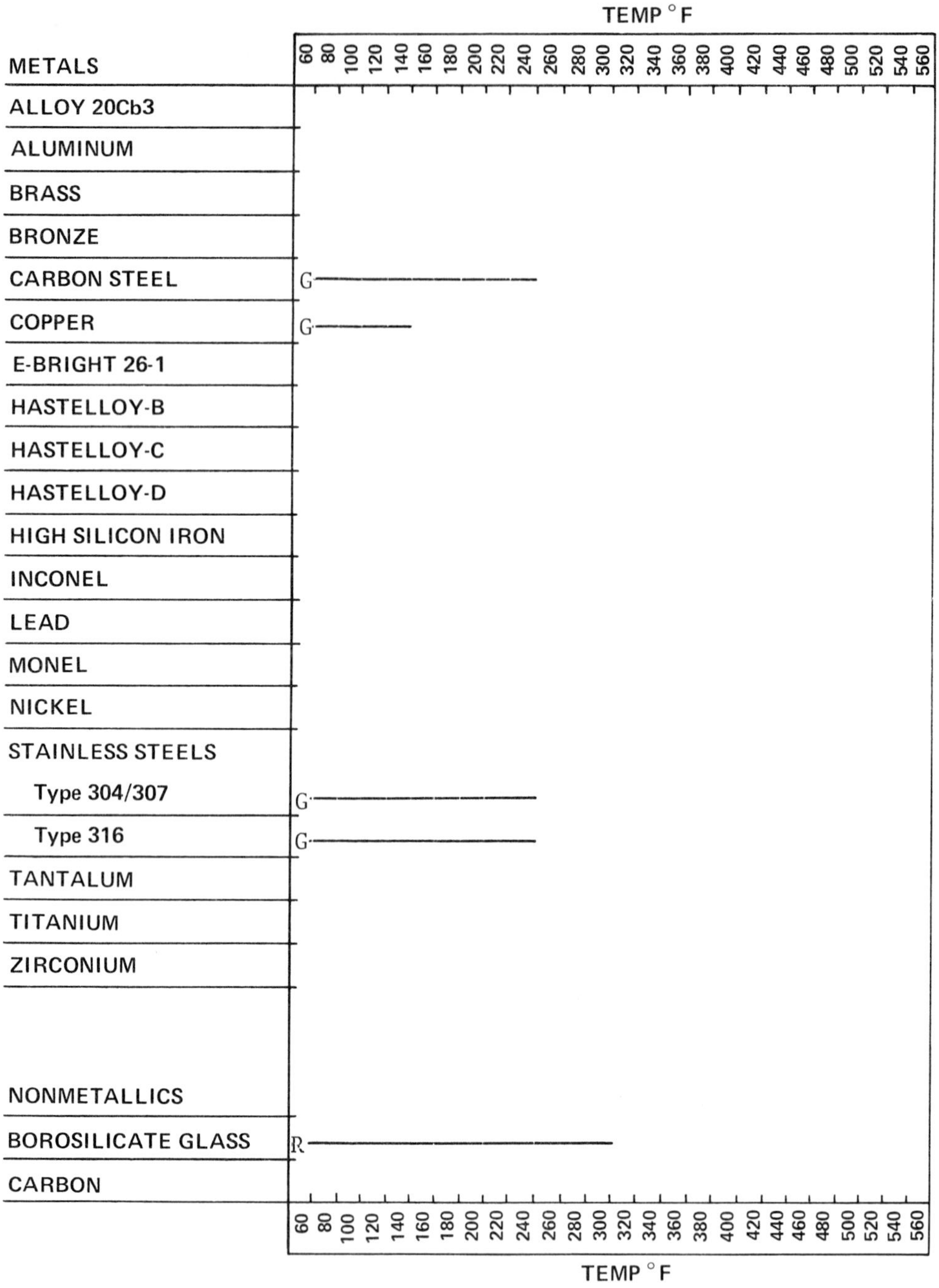

FOR METALS

E = < 2 Mils Penetration/Year; G = < 20 Mils Penetration/Year

S = < 50 Mils Penetration/Year; U = > 50 Mils Penetration/Year

FOR NONMETALLICS

R = Resistant

U = Unsatisfactory

MOTOR OIL

TEMP °F (60–460)

PLASTICS	
ABS	R
ACRYLICS	
ASBESTOS REINF., EPOXY	
ASBESTOS REINF., PHENOLIC	
CHLORINATED POLYETHER	R
CPVC	R
EPOXY	
FLUOROCARBONS FEP	R
FLUOROCARBONS TFE	R
FURFURYL ALCOHOL	
NORYL	
PHENOLIC	
POLYAMIDES–NYLON	R
POLYESTERS	
BISPHENOL A-FUMURATE	
HYDROGENATED BISPHENOL-A-BISPHENOL-A	
ISOPHTHALIC	
CHLORINATED POLYESTERS	
POLYSTYRENES	
POLYETHYLENE	U
POLYPROPYLENE	U
PVC–TYPE 1	R
PVC–TYPE 2	R
POLYVINYLIDENE CHLORIDE	
VINYLIDENE FLUORIDE	
VINYL ESTER	R

RUBBERS	
BUTYL GR-1	
FLUORO ELASTOMERS VITON A	R
FLUORO ELASTOMERS KEL-F 3700	
HARD RUBBER	
HYPALON	
KOROSEAL	
NATURAL RUBBER (GRS)	
NEOPRENE GR-M (CR)	
NITRILE BUNA N (NBR)	R
NORDEL	
POLYBUTADIENE	
SILICONE	
POLYURETHANE	

TEMP °F (60–460)

NAPHTHA

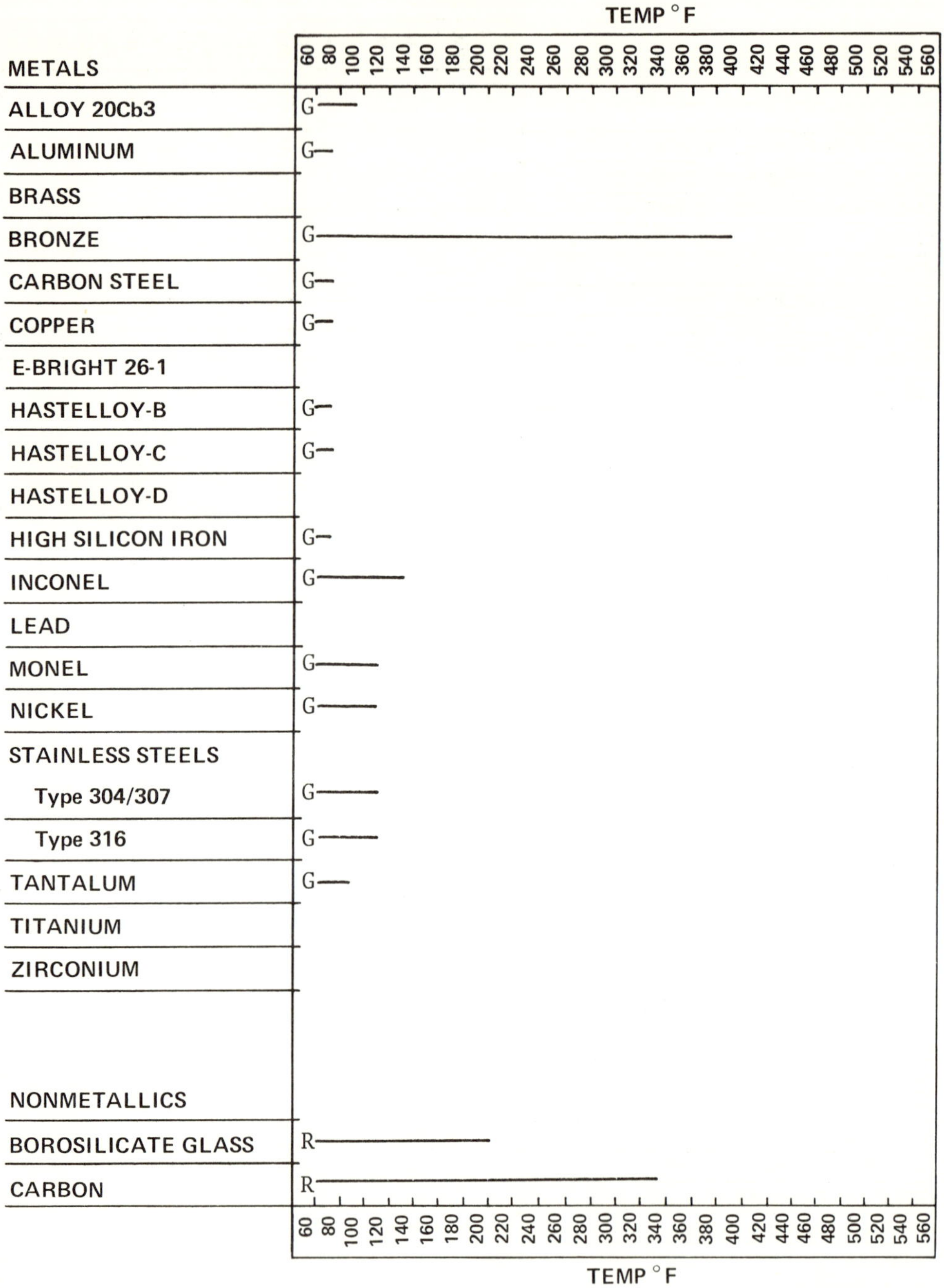

FOR METALS

E = < 2 Mils Penetration/Year; G = < 20 Mils Penetration/Year

S = < 50 Mils Penetration/Year; U = > 50 Mils Penetration/Year

FOR NONMETALLICS

R = Resistant

U = Unsatisfactory

Please see page 1 for notes on calibration.

NAPTHA

PLASTICS	
ABS	R
ACRYLICS	R
ASBESTOS REINF., EPOXY	
ASBESTOS REINF., PHENOLIC	
CHLORINATED POLYETHER	R
CPVC	R—U
EPOXY	R
FLUOROCARBONS FEP	R
FLUOROCARBONS TFE	R
FURFURYL ALCOHOL	R
NORYL	
PHENOLIC	
POLYAMIDES–NYLON	R
POLYESTERS	
BISPHENOL A-FUMURATE	R
HYDROGENATED BISPHENOL-A-BISPHENOL-A	R
ISOPHTHALIC	R
CHLORINATED POLYESTERS	R
POLYSTYRENES	
POLYETHYLENE	RU
POLYPROPYLENE	R—U
PVC–TYPE 1	R
PVC–TYPE 2	R
POLYVINYLIDENE CHLORIDE	R
VINYLIDENE FLUORIDE	R
VINYL ESTER	R

RUBBERS	
BUTYL GR-1	U
FLUORO ELASTOMERS VITON A	R
FLUORO ELASTOMERS KEL-F 3700	
HARD RUBBER	U
HYPALON	U
KOROSEAL	
NATURAL RUBBER (GRS)	U
NEOPRENE GR-M (CR)	U
NITRILE BUNA N (NBR)	RU
NORDEL	U
POLYBUTADIENE	
SILICONE	
POLYURETHANE	R

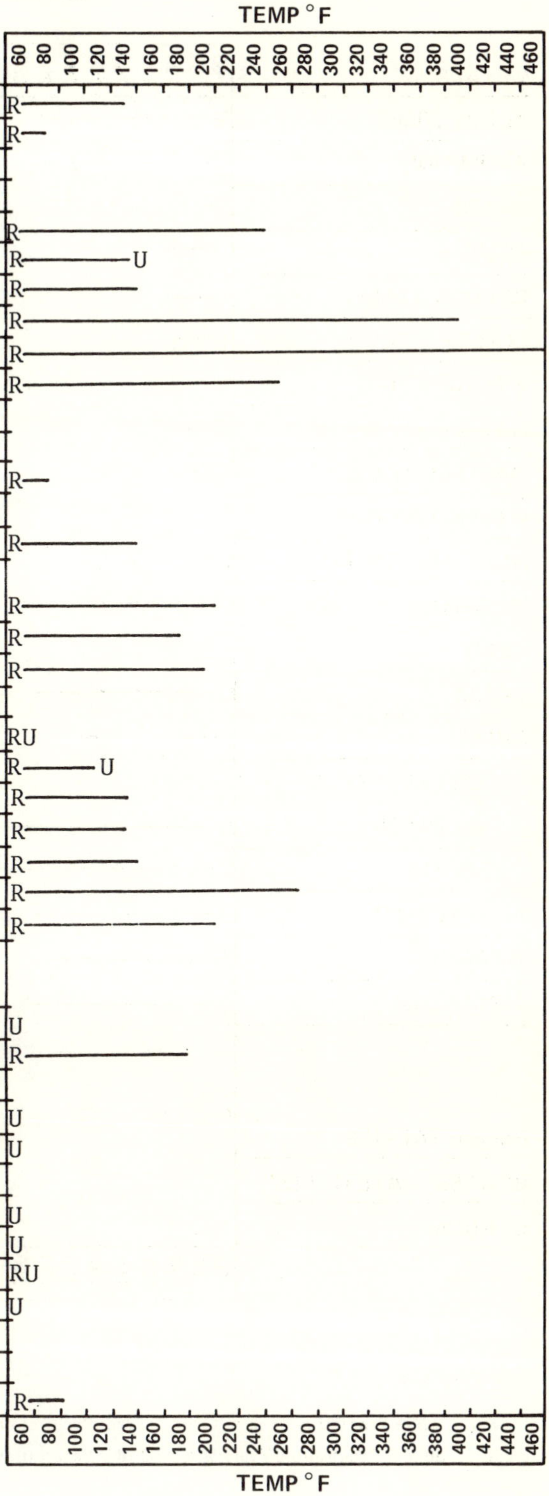

NAPHTHALENE

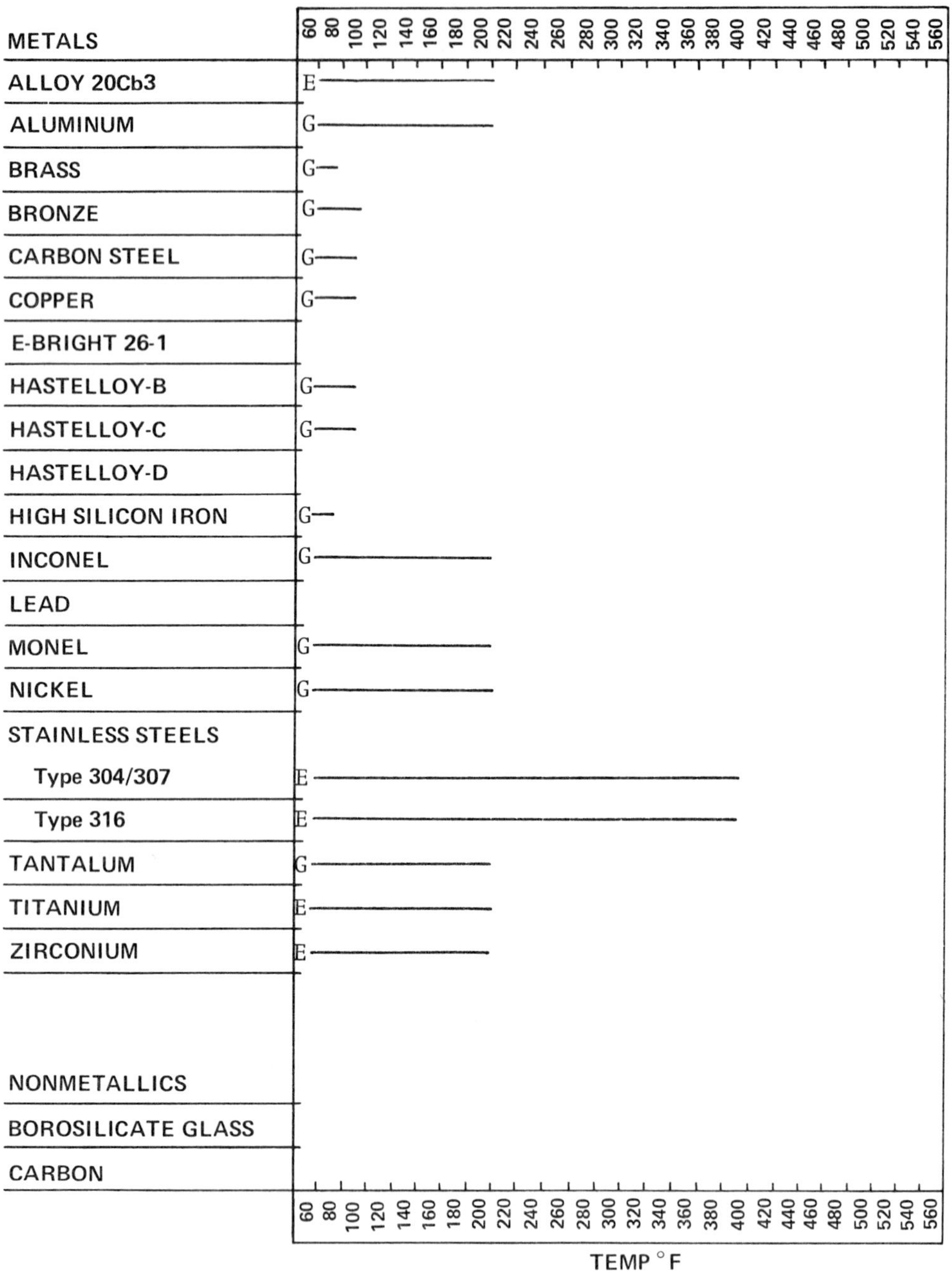

FOR METALS

E = < 2 Mils Penetration/Year; G = < 20 Mils Penetration/Year

S = < 50 Mils Penetration/Year; U = > 50 Mils Penetration/Year

FOR NONMETALLICS

R = Resistant

U = Unsatisfactory

NAPHTHALENE

TEMP °F (60–460)

PLASTICS	
ABS	U
ACRYLICS	
ASBESTOS REINF., EPOXY	RU
ASBESTOS REINF., PHENOLIC	R
CHLORINATED POLYETHER	R
CPVC	U
EPOXY	R
FLUOROCARBONS FEP	R
FLUOROCARBONS TFE	R
FURFURYL ALCOHOL	R
NORYL	
PHENOLIC	
POLYAMIDES–NYLON	R
POLYESTERS	
BISPHENOL A-FUMURATE	R
HYDROGENATED BISPHENOL-A-BISPHENOL-A	R
ISOPHTHALIC	R
CHLORINATED POLYESTERS	R
POLYSTYRENES	
POLYETHYLENE	R
POLYPROPYLENE	R
PVC–TYPE 1	U
PVC–TYPE 2	U
POLYVINYLIDENE CHLORIDE	
VINYLIDENE FLUORIDE	R
VINYL ESTER	R

RUBBERS	
BUTYL GR-1	
FLUORO ELASTOMERS VITON A	R
FLUORO ELASTOMERS KEL-F 3700	
HARD RUBBER	
HYPALON	U
KOROSEAL	
NATURAL RUBBER (GRS)	
NEOPRENE GR-M (CR)	U
NITRILE BUNA N (NBR)	U
NORDEL	
POLYBUTADIENE	
SILICONE	
POLYURETHANE	R

TEMP °F

NICKEL CHLORIDE

TEMP °F: 60 80 100 120 140 160 180 200 220 240 260 280 300 320 340 360 380 400 420 440 460 480 500 520 540 560

METALS		Rating
ALLOY 20Cb3		G—
ALUMINUM		U
BRASS		U
BRONZE		S—
CARBON STEEL		U
COPPER		S—
E-BRIGHT 26-1		
HASTELLOY-B	to 30%	E—
HASTELLOY-C	80%	E—
HASTELLOY-D		
HIGH SILICON IRON		G—
INCONEL		E— G—
LEAD		G—
MONEL		G—
NICKEL		G—
STAINLESS STEELS		
Type 304/307	1,2	G—
Type 316	1,2	G—
TANTALUM		G—
TITANIUM		E—
ZIRCONIUM	5-20%	E—

NONMETALLICS	Rating
BOROSILICATE GLASS	R—
CARBON	R—

60 80 100 120 140 160 180 200 220 240 260 280 300 320 340 360 380 400 420 440 460 480 500 520 540 560

TEMP °F

FOR METALS

E = < 2 Mils Penetration/Year; G = < 20 Mils Penetration/Year

S = < 50 Mils Penetration/Year; U = > 50 Mils Penetration/Year

FOR NONMETALLICS

R = Resistant

U = Unsatisfactory

NICKEL CHLORIDE

TEMP °F

PLASTICS	Concentration	Rating	Max. temp. °F
ABS		R	140
ACRYLICS		R	80
ASBESTOS REINF., EPOXY			
ASBESTOS REINF., PHENOLIC			
CHLORINATED POLYETHER		R	250
CPVC		R	180
EPOXY		R	250
FLUOROCARBONS FEP		R	400
FLUOROCARBONS TFE		R	460
FURFURYL ALCOHOL	sat	R	260
NORYL		R	140
PHENOLIC			
POLYAMIDES—NYLON		U	
POLYESTERS			
BISPHENOL A-FUMURATE	sat	R	220
HYDROGENATED BISPHENOL-A-BISPHENOL-A		R	210
ISOPHTHALIC	sat	R	180
CHLORINATED POLYESTERS	all	R	250
POLYSTYRENES			
POLYETHYLENE		R	150
POLYPROPYLENE		R	220
PVC—TYPE 1		R	140
PVC—TYPE 2		R	140
POLYVINYLIDENE CHLORIDE		R	180
VINYLIDENE FLUORIDE		R	280
VINYL ESTER		R	210

RUBBERS	Rating	Max. temp. °F
BUTYL GR-1	R	150
FLUORO ELASTOMERS VITON A	R	190
FLUORO ELASTOMERS KEL-F 3700		
HARD RUBBER	R	190
HYPALON	R	250
KOROSEAL	R	150
NATURAL RUBBER (GRS)	R	150
NEOPRENE GR-M (CR)	R	200
NITRILE BUNA N (NBR)	R	200
NORDEL	R	300
POLYBUTADIENE		
SILICONE		
POLYURETHANE	R	90

TEMP °F

NICKEL NITRATE

TEMP °F (60–560)

METALS		Rating	Temperature range (°F)
ALLOY 20Cb3		E	60–400
ALUMINUM		U	
BRASS		U	
BRONZE		U	
CARBON STEEL		G	60–80
COPPER			
E-BRIGHT 26-1			
HASTELLOY-B		G	60–560
HASTELLOY-C		G	60–560
HASTELLOY-D			
HIGH SILICON IRON		E	60–200
INCONEL	dry	G	60–560
LEAD		G	60–200
MONEL	dry	G	60–560
NICKEL	dry	G	60–560
STAINLESS STEELS			
Type 304/307		E	60–400
Type 316		E	60–400
TANTALUM		G	60–210
TITANIUM		E	60–80
ZIRCONIUM			

NONMETALLICS	Rating	Temperature range (°F)
BOROSILICATE GLASS	R	60–200
CARBON		

TEMP °F

FOR METALS

E = < 2 Mils Penetration/Year; G = < 20 Mils Penetration/Year

S = < 50 Mils Penetration/Year; U = > 50 Mils Penetration/Year

FOR NONMETALLICS

R = Resistant

U = Unsatisfactory

NICKEL NITRATE

PLASTICS		TEMP °F (60–460)
ABS		R
ACRYLICS		
ASBESTOS REINF., EPOXY		
ASBESTOS REINF., PHENOLIC		
CHLORINATED POLYETHER		R
CPVC		R
EPOXY		R
FLUOROCARBONS FEP		R
FLUOROCARBONS TFE		R
FURFURYL ALCOHOL	sat	R
NORYL		R
PHENOLIC		
POLYAMIDES–NYLON		
POLYESTERS		
BISPHENOL A-FUMURATE	sat	R
HYDROGENATED BISPHENOL-A-BISPHENOL-A		R
ISOPHTHALIC	sat	R
CHLORINATED POLYESTERS	all	R
POLYSTYRENES		R
POLYETHYLENE		R
POLYPROPYLENE		R
PVC–TYPE 1		R
PVC–TYPE 2		R
POLYVINYLIDENE CHLORIDE		R
VINYLIDENE FLUORIDE		R
VINYL ESTER		R

RUBBERS	TEMP °F (60–460)
BUTYL GR-1	R
FLUORO ELASTOMERS VITON A	R
FLUORO ELASTOMERS KEL-F 3700	
HARD RUBBER	R
HYPALON	R
KOROSEAL	R
NATURAL RUBBER (GRS)	R
NEOPRENE GR-M (CR)	R
NITRILE BUNA N (NBR)	R
NORDEL	
POLYBUTADIENE	
SILICONE	
POLYURETHANE	

NICKEL SALT

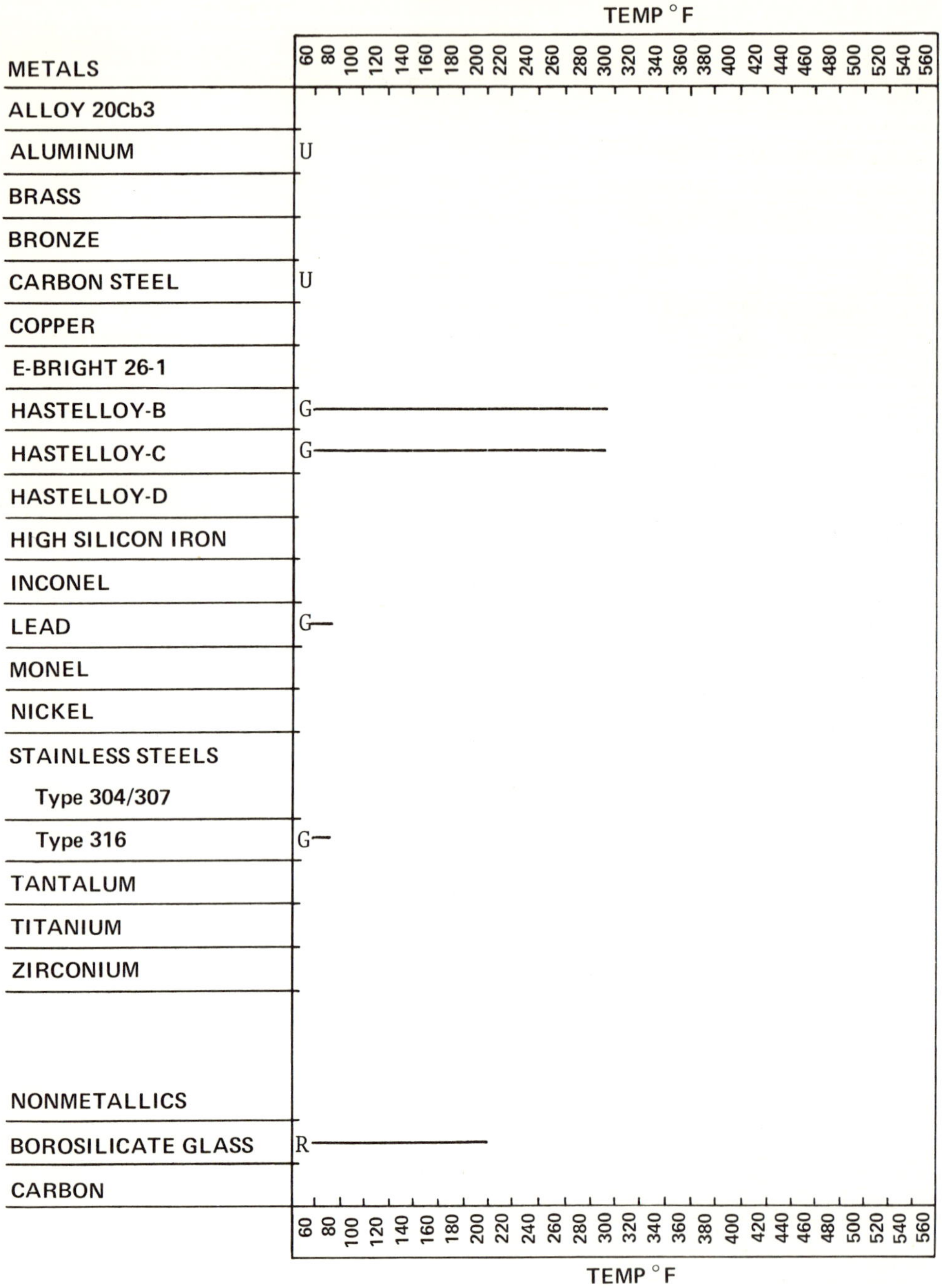

FOR METALS

E = < 2 Mils Penetration/Year; G = < 20 Mils Penetration/Year

S = < 50 Mils Penetration/Year; U = > 50 Mils Penetration/Year

FOR NONMETALLICS

R = Resistant

U = Unsatisfactory

NICKEL SALT

TEMP °F

60 80 100 120 140 160 180 200 220 240 260 280 300 320 340 360 380 400 420 440 460

PLASTICS	
ABS	
ACRYLICS	
ASBESTOS REINF., EPOXY	
ASBESTOS REINF., PHENOLIC	
CHLORINATED POLYETHER	
CPVC	
EPOXY	
FLUOROCARBONS FEP	R
FLUOROCARBONS TFE	R
FURFURYL ALCOHOL	
NORYL	
PHENOLIC	
POLYAMIDES–NYLON	
POLYESTERS	
BISPHENOL A-FUMURATE	
HYDROGENATED BISPHENOL-A-BISPHENOL-A	
ISOPHTHALIC	
CHLORINATED POLYESTERS	
POLYSTYRENES	
POLYETHYLENE	R
POLYPROPYLENE	R
PVC–TYPE 1	
PVC–TYPE 2	
POLYVINYLIDENE CHLORIDE	
VINYLIDENE FLUORIDE	
VINYL ESTER	

RUBBERS	
BUTYL GR-1	R
FLUORO ELASTOMERS VITON A	R
FLUORO ELASTOMERS KEL-F 3700	
HARD RUBBER	R
HYPALON	R
KOROSEAL	
NATURAL RUBBER (GRS)	R
NEOPRENE GR-M (CR)	R
NITRILE BUNA N (NBR)	R
NORDEL	
POLYBUTADIENE	
SILICONE	
POLYURETHANE	

60 80 100 120 140 160 180 200 220 240 260 280 300 320 340 360 380 400 420 440 460

TEMP °F

NICKEL SULFATE

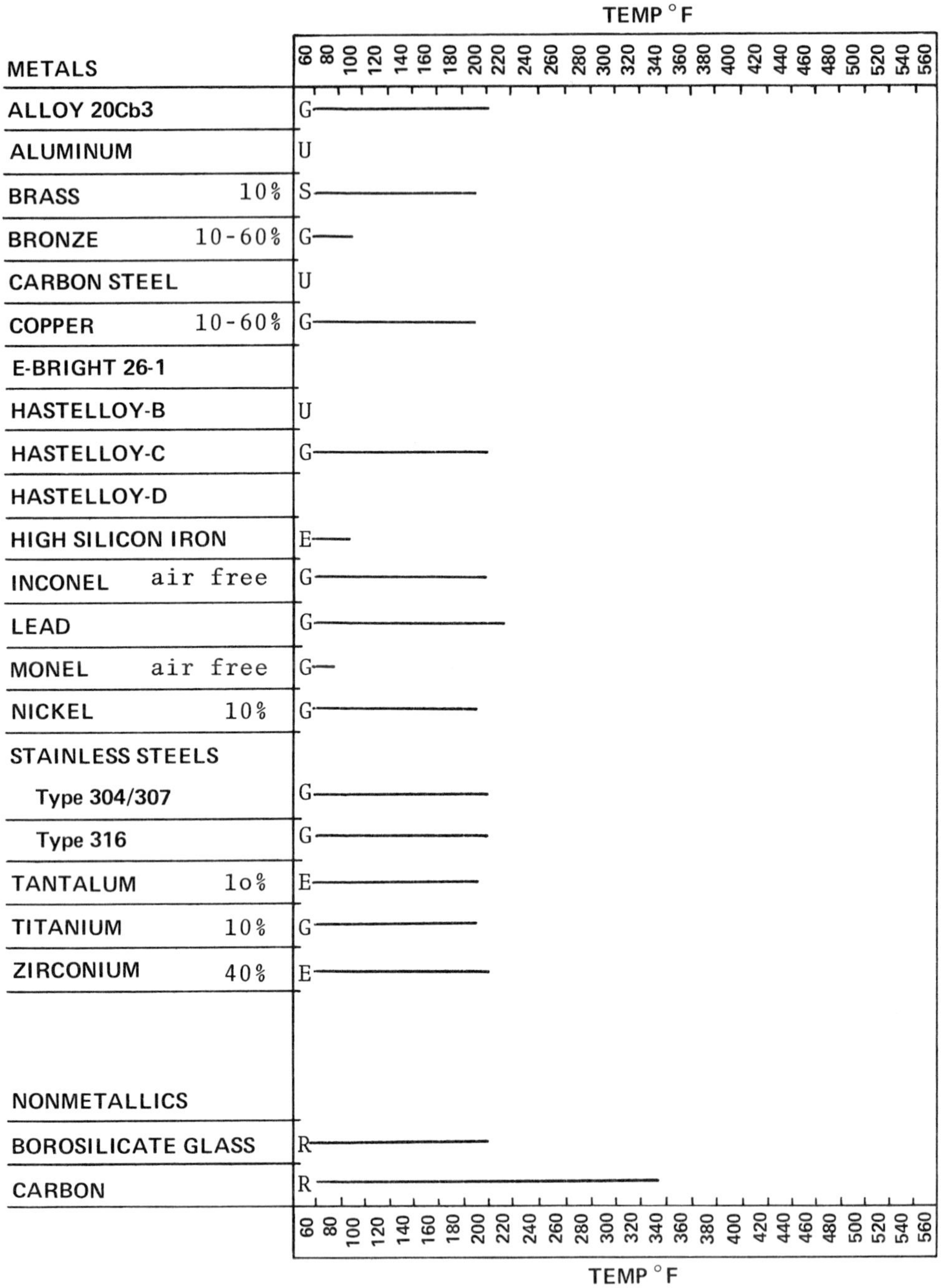

FOR METALS

E = < 2 Mils Penetration/Year; G = < 20 Mils Penetration/Year

S = < 50 Mils Penetration/Year; U = > 50 Mils Penetration/Year

FOR NONMETALLICS

R = Resistant

U = Unsatisfactory

NICKEL SULFATE

PLASTICS	Conc.	TEMP °F (60–460)
ABS		R to 140
ACRYLICS		R to 80
ASBESTOS REINF., EPOXY		
ASBESTOS REINF., PHENOLIC		
CHLORINATED POLYETHER		R to 240
CPVC		R to 180
EPOXY	36%	R to 130, U at 140
FLUOROCARBONS FEP		R to 400
FLUOROCARBONS TFE		R to 460
FURFURYL ALCOHOL	sat	R to 260
NORYL		R to 140
PHENOLIC		
POLYAMIDES–NYLON		
POLYESTERS		
BISPHENOL A-FUMURATE	sat	R to 220
HYDROGENATED BISPHENOL-A-BISPHENOL-A		R to 200
ISOPHTHALIC	sat	R to 180
CHLORINATED POLYESTERS	all	R to 240
POLYSTYRENES		R to 150
POLYETHYLENE		R to 140, U
POLYPROPYLENE		R to 170, U
PVC–TYPE 1		R to 140
PVC–TYPE 2		R to 140
POLYVINYLIDENE CHLORIDE		R to 180
VINYLIDENE FLUORIDE		R to 280
VINYL ESTER		R to 200

RUBBERS	Conc.	TEMP °F (60–460)
BUTYL GR-1		R to 180
FLUORO ELASTOMERS VITON A		R to 180
FLUORO ELASTOMERS KEL-F 3700		
HARD RUBBER		R to 180
HYPALON		R to 240
KOROSEAL		R to 150
NATURAL RUBBER (GRS)		R to 150
NEOPRENE GR-M (CR)		R to 200
NITRILE BUNA N (NBR)		R to 200
NORDEL		R to 300
POLYBUTADIENE		
SILICONE		
POLYURETHANE		R to 90

NICOTINE

TEMP °F

METALS	Rating	TEMP °F (60–560)
ALLOY 20Cb3		
ALUMINUM		
BRASS		
BRONZE		
CARBON STEEL		
COPPER		
E-BRIGHT 26-1		
HASTELLOY-B		
HASTELLOY-C		
HASTELLOY-D		
HIGH SILICON IRON		
INCONEL		
LEAD		
MONEL		
NICKEL		
STAINLESS STEELS		
Type 304/307	G	60–200
Type 316	G	60–200
TANTALUM	G	60–80
TITANIUM		
ZIRCONIUM		

NONMETALLICS	Rating	TEMP °F (60–560)
BOROSILICATE GLASS	R	60–200
CARBON		

TEMP °F

FOR METALS

E = < 2 Mils Penetration/Year; G = < 20 Mils Penetration/Year

S = < 50 Mils Penetration/Year; U = > 50 Mils Penetration/Year

FOR NONMETALLICS

R = Resistant

U = Unsatisfactory

NICOTINE

PLASTICS		TEMP °F (60–460)
ABS		R 60–140
ACRYLICS		
ASBESTOS REINF., EPOXY		
ASBESTOS REINF., PHENOLIC		
CHLORINATED POLYETHER		
CPVC		R 60–180
EPOXY		
FLUOROCARBONS FEP		R 60–400
FLUOROCARBONS TFE		R 60–460
FURFURYL ALCOHOL		
NORYL		
PHENOLIC		
POLYAMIDES–NYLON		R 60–80
POLYESTERS		
BISPHENOL A-FUMURATE		
HYDROGENATED BISPHENOL-A-BISPHENOL-A		
ISOPHTHALIC		
CHLORINATED POLYESTERS		
POLYSTYRENES		
POLYETHYLENE	11	R 60–150
POLYPROPYLENE		R 60–140
PVC–TYPE 1		R 60–140
PVC–TYPE 2		R 60–140
POLYVINYLIDENE CHLORIDE		
VINYLIDENE FLUORIDE		R 60–150
VINYL ESTER		

RUBBERS	TEMP °F (60–460)
BUTYL GR-1	
FLUORO ELASTOMERS VITON A	
FLUORO ELASTOMERS KEL-F 3700	
HARD RUBBER	
HYPALON	
KOROSEAL	
NATURAL RUBBER (GRS)	
NEOPRENE GR-M (CR)	
NITRILE BUNA N (NBR)	
NORDEL	
POLYBUTADIENE	
SILICONE	
POLYURETHANE	

NICOTINE ACID

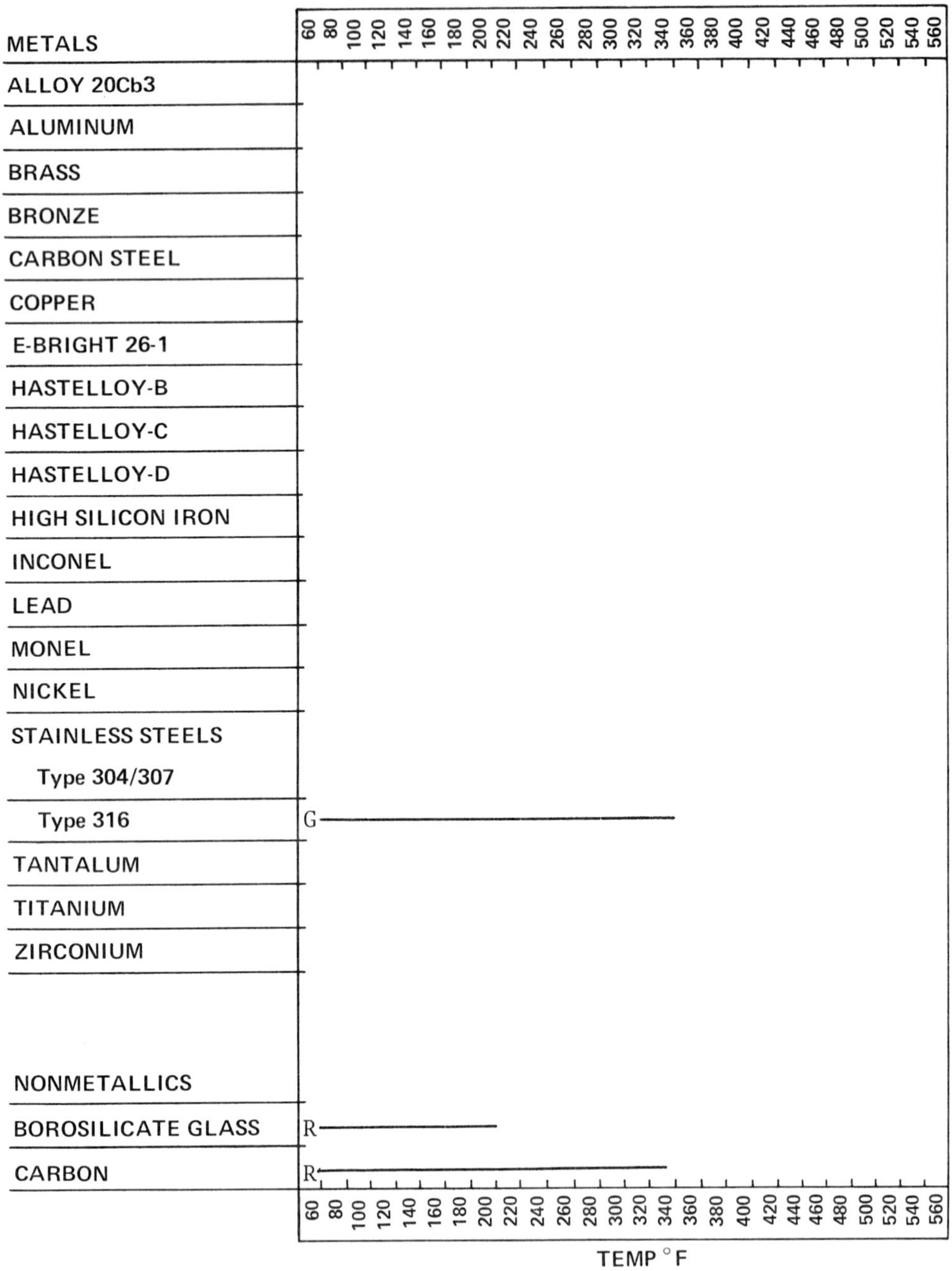

FOR METALS

E = < 2 Mils Penetration/Year; G = < 20 Mils Penetration/Year

S = < 50 Mils Penetration/Year; U = > 50 Mils Penetration/Year

FOR NONMETALLICS

R = Resistant

U = Unsatisfactory

NICOTINE ACID

TEMP °F (60–460)

PLASTICS		
ABS		U
ACRYLICS		
ASBESTOS REINF., EPOXY		
ASBESTOS REINF., PHENOLIC		
CHLORINATED POLYETHER		R
CPVC		R
EPOXY		
FLUOROCARBONS FEP		R
FLUOROCARBONS TFE		R
FURFURYL ALCOHOL		
NORYL		
PHENOLIC		
POLYAMIDES–NYLON		
POLYESTERS		
BISPHENOL A-FUMURATE		R
HYDROGENATED BISPHENOL-A-BISPHENOL-A		
ISOPHTHALIC		
CHLORINATED POLYESTERS		
POLYSTYRENES		
POLYETHYLENE	11	R
POLYPROPYLENE		R
PVC–TYPE 1		R
PVC–TYPE 2		R
POLYVINYLIDENE CHLORIDE		R
VINYLIDENE FLUORIDE		R
VINYL ESTER		

RUBBERS	
BUTYL GR-1	
FLUORO ELASTOMERS VITON A	
FLUORO ELASTOMERS KEL-F 3700	
HARD RUBBER	R
HYPALON	
KOROSEAL	
NATURAL RUBBER (GRS)	
NEOPRENE GR-M (CR)	R
NITRILE BUNA N (NBR)	
NORDEL	R
POLYBUTADIENE	
SILICONE	
POLYURETHANE	

TEMP °F (60–460)

NITRIC ACID, 100% (ANHYDROUS)

TEMP °F: 60, 80, 100, 120, 140, 160, 180, 200, 220, 240, 260, 280, 300, 320, 340, 360, 380, 400, 420, 440, 460, 480, 500, 520, 540, 560

METALS	
ALLOY 20Cb3	E–U
ALUMINUM	E—
BRASS	U
BRONZE	U
CARBON STEEL	U
COPPER	U
E-BRIGHT 26-1	
HASTELLOY-B	U
HASTELLOY-C	G–
HASTELLOY-D	U
HIGH SILICON IRON	E————
INCONEL	
LEAD	U
MONEL	U
NICKEL	U
STAINLESS STEELS	
Type 304/307	E–U
Type 316 9	E——U
TANTALUM	E——————
TITANIUM	G—————
ZIRCONIUM	E—

NONMETALLICS	
BOROSILICATE GLASS	R——————
CARBON	U

TEMP °F: 6C, 80, 100, 120, 140, 160, 180, 200, 220, 240, 260, 280, 300, 320, 340, 360, 380, 400, 420, 440, 460, 480, 500, 520, 540, 560

FOR METALS

E = < 2 Mils Penetration/Year; G = < 20 Mils Penetration/Year

S = < 50 Mils Penetration/Year; U = > 50 Mils Penetration/Year

FOR NONMETALLICS

R = Resistant

U = Unsatisfactory

NITRIC ACID, 100% (ANHYDROUS)

TEMP °F: 60 80 100 120 140 160 180 200 220 240 260 280 300 320 340 360 380 400 420 440 460

PLASTICS	
ABS	U
ACRYLICS	U
ASBESTOS REINF., EPOXY	U
ASBESTOS REINF., PHENOLIC	R
CHLORINATED POLYETHER	U
CPVC	U
EPOXY	U
FLUOROCARBONS FEP	R
FLUOROCARBONS TFE	R
FURFURYL ALCOHOL	U
NORYL	U
PHENOLIC	R
POLYAMIDES–NYLON	R
POLYESTERS	
BISPHENOL A-FUMURATE	
HYDROGENATED BISPHENOL-A-BISPHENOL-A	
ISOPHTHALIC	U
CHLORINATED POLYESTERS	
POLYSTYRENES	U
POLYETHYLENE	U
POLYPROPYLENE	U
PVC–TYPE 1	U
PVC–TYPE 2	U
POLYVINYLIDENE CHLORIDE	U
VINYLIDENE FLUORIDE	R
VINYL ESTER	

RUBBERS	
BUTYL GR-1	U
FLUORO ELASTOMERS VITON A	R
FLUORO ELASTOMERS KEL-F 3700	R
HARD RUBBER	
HYPALON	U
KOROSEAL	U
NATURAL RUBBER (GRS)	U
NEOPRENE GR-M (CR)	U
NITRILE BUNA N (NBR)	U
NORDEL	
POLYBUTADIENE	U
SILICONE	U
POLYURETHANE	

TEMP °F: 60 80 100 120 140 160 180 200 220 240 260 280 300 320 340 360 380 400 420 440 460

NITRIC ACID, 70%

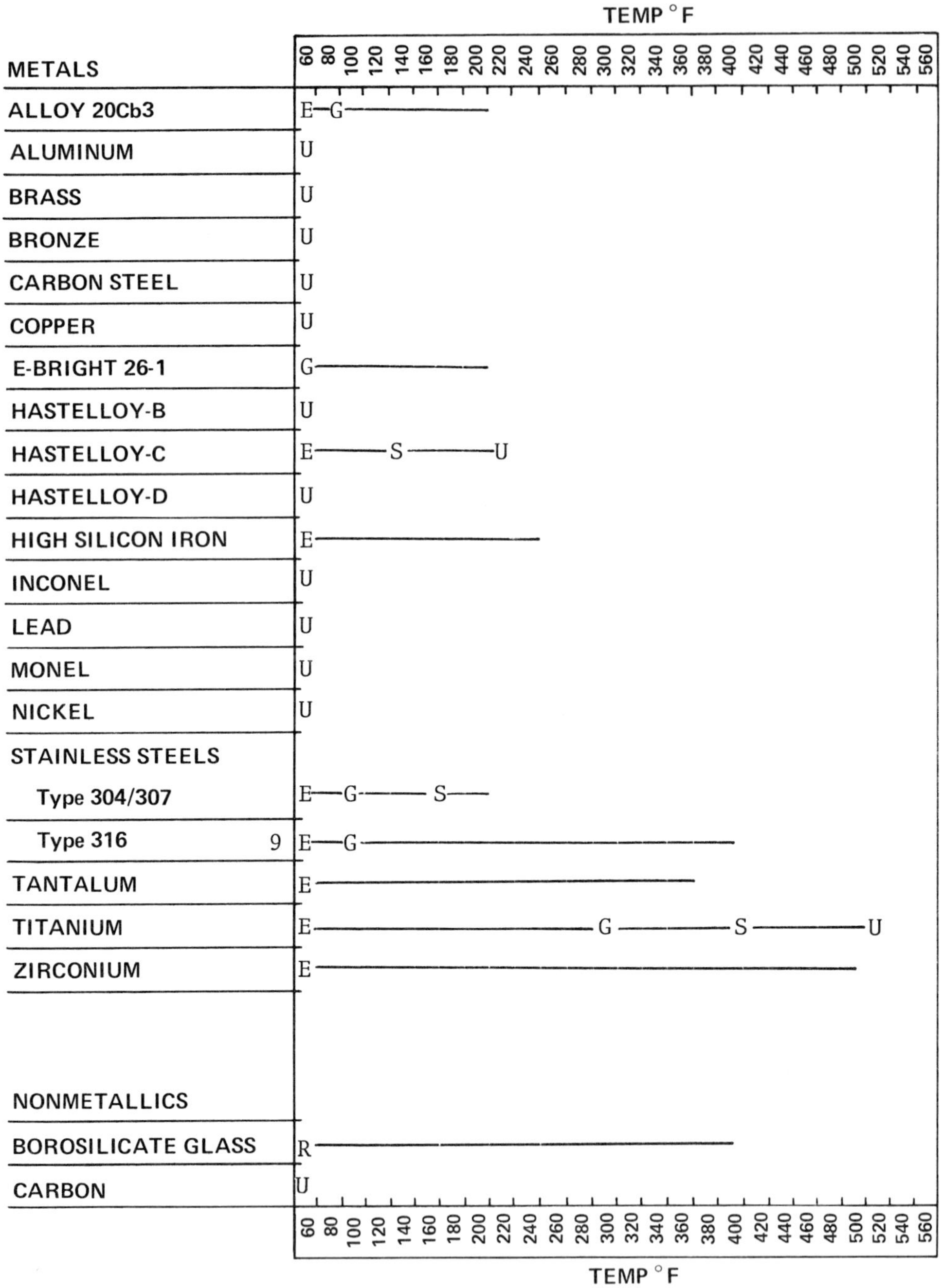

FOR METALS

E = < 2 Mils Penetration/Year; G = < 20 Mils Penetration/Year

S = < 50 Mils Penetration/Year; U = > 50 Mils Penetration/Year

FOR NONMETALLICS

R = Resistant

U = Unsatisfactory

NITRIC ACID, 70%

PLASTICS	TEMP °F (60–460)
ABS	U
ACRYLICS	U
ASBESTOS REINF., EPOXY	
ASBESTOS REINF., PHENOLIC	
CHLORINATED POLYETHER	R—U
CPVC	R (to 180 °F)
EPOXY	U
FLUOROCARBONS FEP	R (to 400 °F)
FLUOROCARBONS TFE	R (to 460 °F)
FURFURYL ALCOHOL	
NORYL	U
PHENOLIC	
POLYAMIDES—NYLON	R (to 80 °F)
POLYESTERS	
BISPHENOL A-FUMURATE	
HYDROGENATED BISPHENOL-A-BISPHENOL-A	
ISOPHTHALIC	U
CHLORINATED POLYESTERS	R (to 80 °F)
POLYSTYRENES	U
POLYETHYLENE	R (to 140 °F)
POLYPROPYLENE	U
PVC—TYPE 1	R (to 140 °F)
PVC—TYPE 2	R (to 140 °F)
POLYVINYLIDENE CHLORIDE	U
VINYLIDENE FLUORIDE	R (to 120 °F)
VINYL ESTER	
RUBBERS	
BUTYL GR-1	R (to 80 °F)
FLUORO ELASTOMERS VITON A	R (to 180 °F)
FLUORO ELASTOMERS KEL-F 3700	R (to 80 °F)
HARD RUBBER	
HYPALON	U
KOROSEAL	U
NATURAL RUBBER (GRS)	U
NEOPRENE GR-M (CR)	U
NITRILE BUNA N (NBR)	U
NORDEL	U
POLYBUTADIENE	U
SILICONE	U
POLYURETHANE	U

NITRIC ACID, 50%

TEMP °F

METALS	60 80 100 120 140 160 180 200 220 240 260 280 300 320 340 360 380 400 420 440 460 480 500 520 540 560
ALLOY 20Cb3	E——G——U
ALUMINUM	U
BRASS	U
BRONZE	U
CARBON STEEL	U
COPPER	U
E-BRIGHT 26-1	G——
HASTELLOY-B	U
HASTELLOY-C	E——G——S——U
HASTELLOY-D	U
HIGH SILICON IRON	E——
INCONEL	G–U
LEAD	U
MONEL	U
NICKEL	U
STAINLESS STEELS	
Type 304/307	E–G——
Type 316 9	E—G——
TANTALUM	E——
TITANIUM	E——
ZIRCONIUM	E——
NONMETALLICS	
BOROSILICATE GLASS	R——
CARBON	U
	60 80 100 120 140 160 180 200 220 240 260 280 300 320 340 360 380 400 420 440 460 480 500 520 540 560

TEMP °F

FOR METALS

E = < 2 Mils Penetration/Year; G = < 20 Mils Penetration/Year

S = < 50 Mils Penetration/Year; U = > 50 Mils Penetration/Year

FOR NONMETALLICS

R = Resistant

U = Unsatisfactory

NITRIC ACID, 50%

PLASTICS	TEMP °F (60–460)
ABS	U
ACRYLICS	U
ASBESTOS REINF., EPOXY	
ASBESTOS REINF., PHENOLIC	
CHLORINATED POLYETHER	R–U
CPVC	R— (60 to 180)
EPOXY	U
FLUOROCARBONS FEP	R— (60 to 400)
FLUOROCARBONS TFE	R— (60 to 460)
FURFURYL ALCOHOL	
NORYL	R–U
PHENOLIC	
POLYAMIDES–NYLON	R— (60 to 80)
POLYESTERS	
BISPHENOL A-FUMURATE	R— (60 to 80)
HYDROGENATED BISPHENOL-A-BISPHENOL-A	
ISOPHTHALIC	U
CHLORINATED POLYESTERS	R— (60 to 140)
POLYSTYRENES	U
POLYETHYLENE	R— (60 to 100)
POLYPROPYLENE	R— (60 to 150)
PVC–TYPE 1	R— (60 to 140)
PVC–TYPE 2	R— (60 to 140)
POLYVINYLIDENE CHLORIDE	R— (60 to 125)
VINYLIDENE FLUORIDE	R— (60 to 175)
VINYL ESTER	U

RUBBERS	TEMP °F (60–460)
BUTYL GR-1	
FLUORO ELASTOMERS VITON A	R— (60 to 185)
FLUORO ELASTOMERS KEL-F 3700	R— (60 to 80)
HARD RUBBER	
HYPALON	R— (60 to 80)
KOROSEAL	U
NATURAL RUBBER (GRS)	U
NEOPRENE GR-M (CR)	U
NITRILE BUNA N (NBR)	U
NORDEL	
POLYBUTADIENE	
SILICONE	U
POLYURETHANE	U

NITRIC ACID, 40%

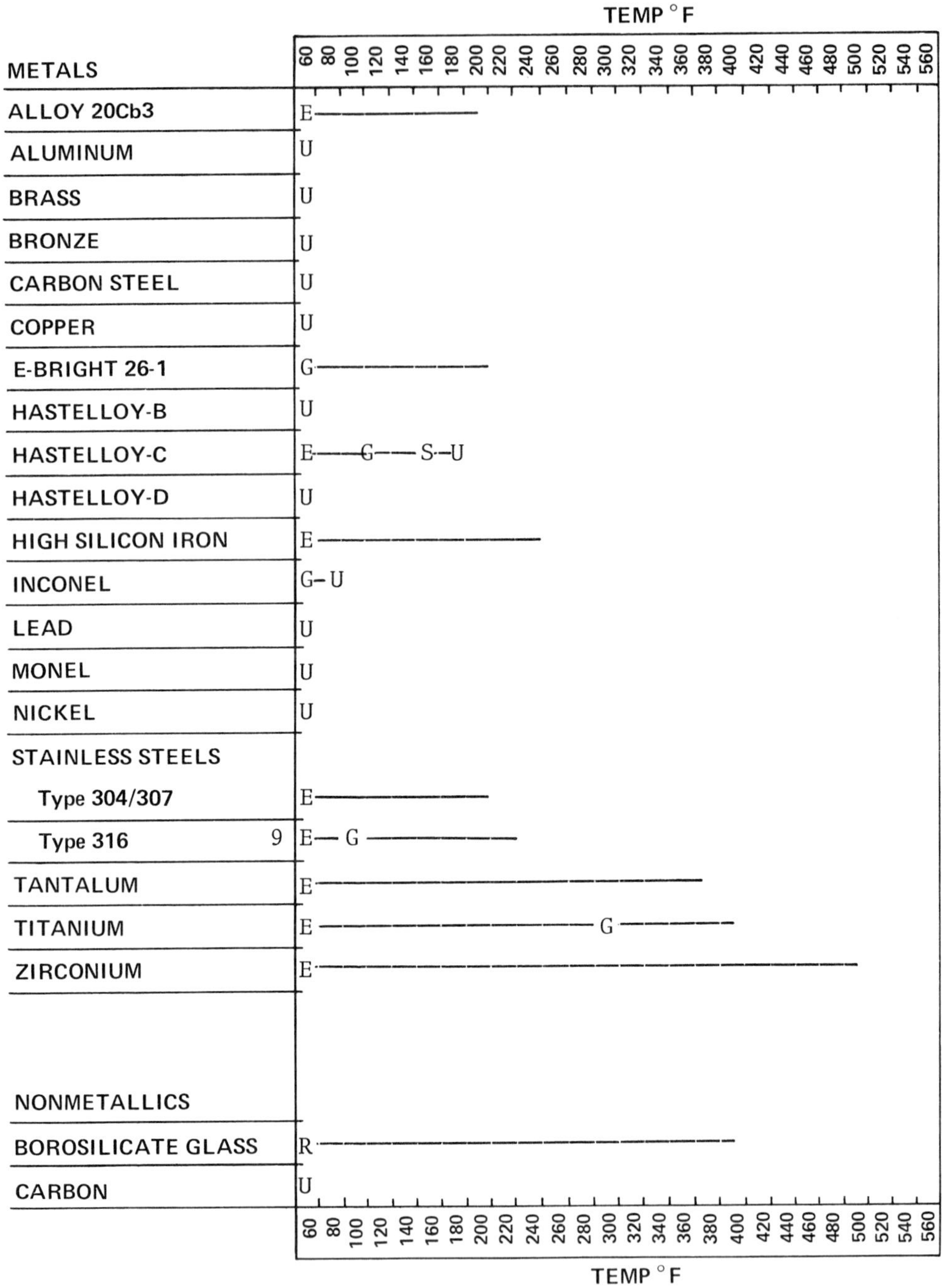

FOR METALS

E = < 2 Mils Penetration/Year; G = < 20 Mils Penetration/Year

S = < 50 Mils Penetration/Year; U = > 50 Mils Penetration/Year

FOR NONMETALLICS

R = Resistant

U = Unsatisfactory

NITRIC ACID, 40%

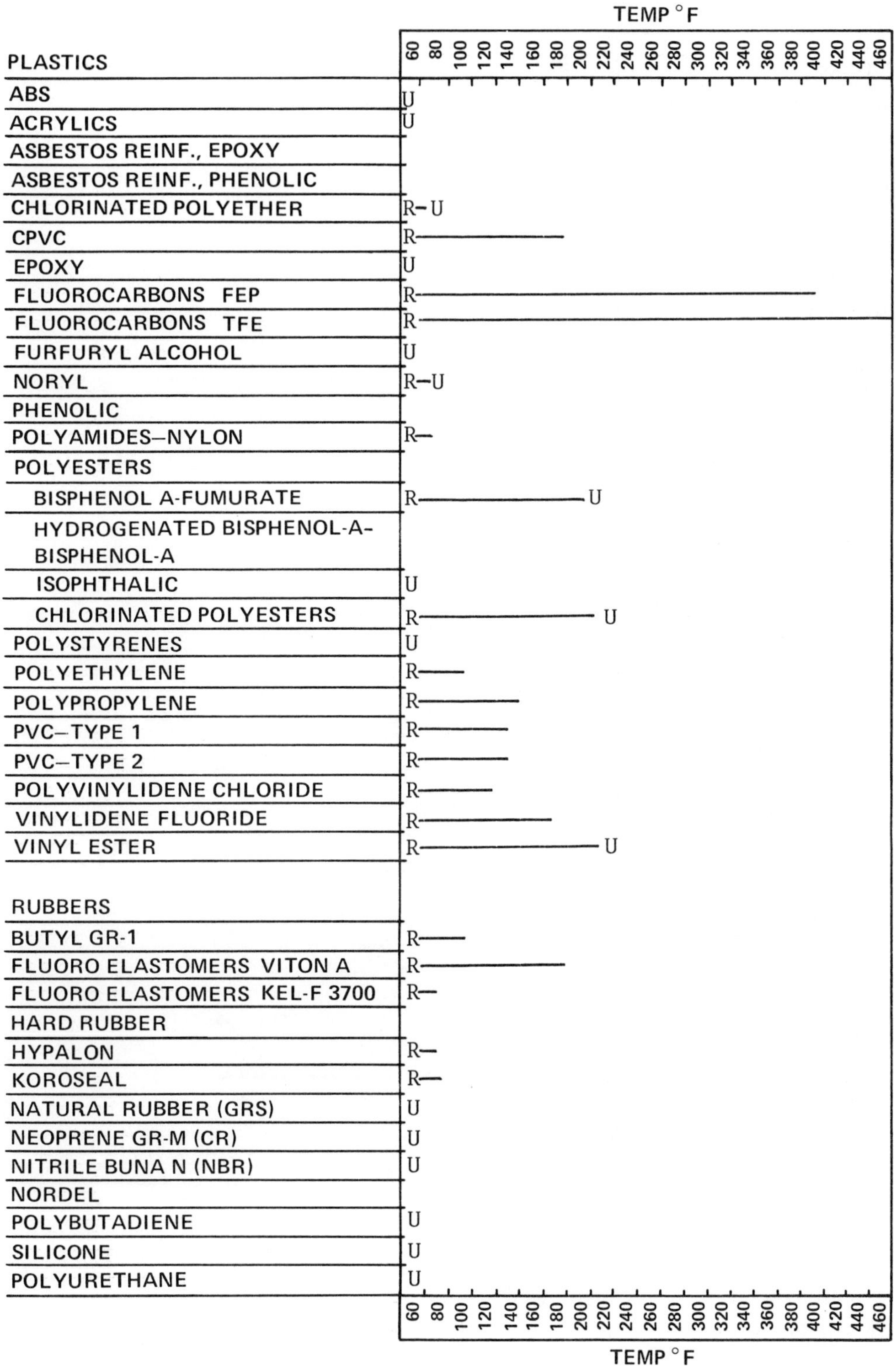

NITRIC ACID, 30%

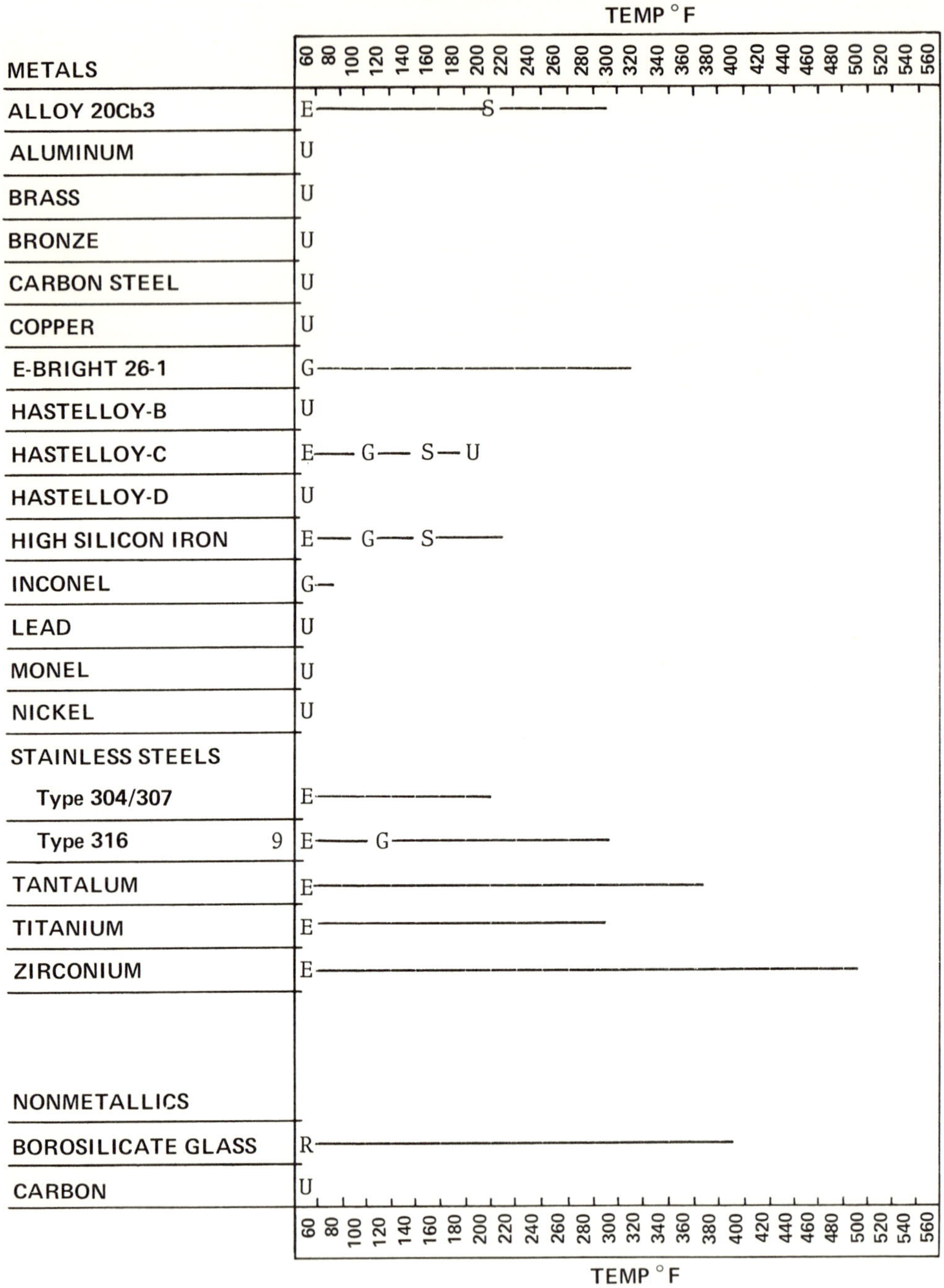

FOR METALS

E = < 2 Mils Penetration/Year; G = < 20 Mils Penetration/Year

S = < 50 Mils Penetration/Year; U = > 50 Mils Penetration/Year

FOR NONMETALLICS

R = Resistant

U = Unsatisfactory

NITRIC ACID, 30%

TEMP °F: 60, 80, 100, 120, 140, 160, 180, 200, 220, 240, 260, 280, 300, 320, 340, 360, 380, 400, 420, 440, 460

PLASTICS	
ABS	U
ACRYLICS	U
ASBESTOS REINF., EPOXY	
ASBESTOS REINF., PHENOLIC	
CHLORINATED POLYETHER	R—U
CPVC	R——
EPOXY	U
FLUOROCARBONS FEP	R——
FLUOROCARBONS TFE	R——
FURFURYL ALCOHOL	
NORYL	R—U
PHENOLIC	
POLYAMIDES—NYLON	R—
POLYESTERS	
BISPHENOL A-FUMURATE	R—
HYDROGENATED BISPHENOL-A-BISPHENOL-A	
ISOPHTHALIC	U
CHLORINATED POLYESTERS	R——U
POLYSTYRENES	U
POLYETHYLENE	R——
POLYPROPYLENE	R——
PVC—TYPE 1	R——
PVC—TYPE 2	R——
POLYVINYLIDENE CHLORIDE	R——
VINYLIDENE FLUORIDE	R——
VINYL ESTER	
RUBBERS	
BUTYL GR-1	R——
FLUORO ELASTOMERS VITON A	R——
FLUORO ELASTOMERS KEL-F 3700	R—
HARD RUBBER	
HYPALON	R——U
KOROSEAL	R—
NATURAL RUBBER (GRS)	U
NEOPRENE GR-M (CR)	U
NITRILE BUNA N (NBR)	U
NORDEL	R—
POLYBUTADIENE	R—
SILICONE	U
POLYURETHANE	U

TEMP °F: 60, 80, 100, 120, 140, 160, 180, 200, 220, 240, 260, 280, 300, 320, 340, 360, 380, 400, 420, 440, 460

NITRIC ACID, 20%

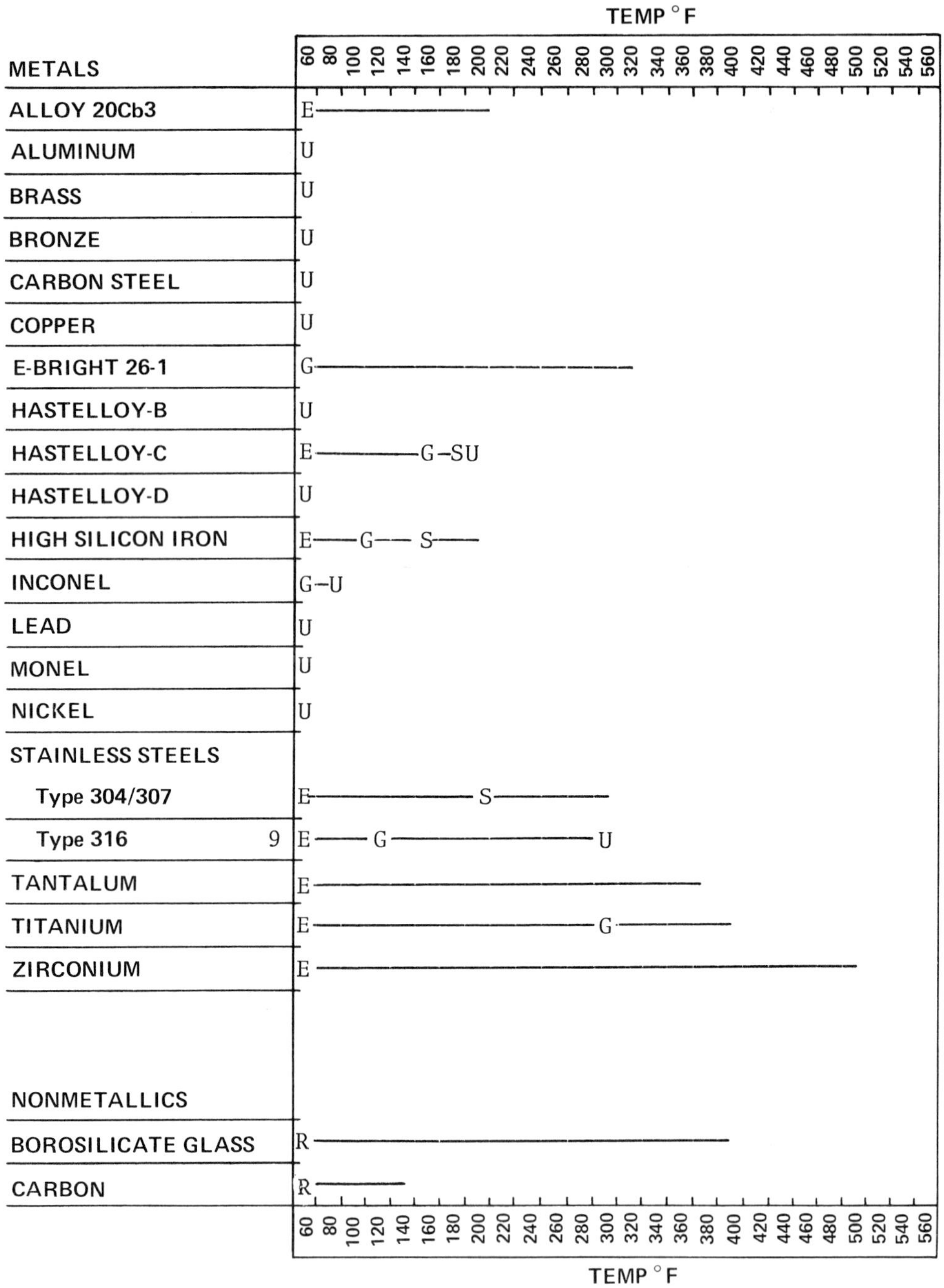

FOR METALS

E = < 2 Mils Penetration/Year; G = < 20 Mils Penetration/Year

S = < 50 Mils Penetration/Year; U = > 50 Mils Penetration/Year

FOR NONMETALLICS

R = Resistant

U = Unsatisfactory

NITRIC ACID, 20%

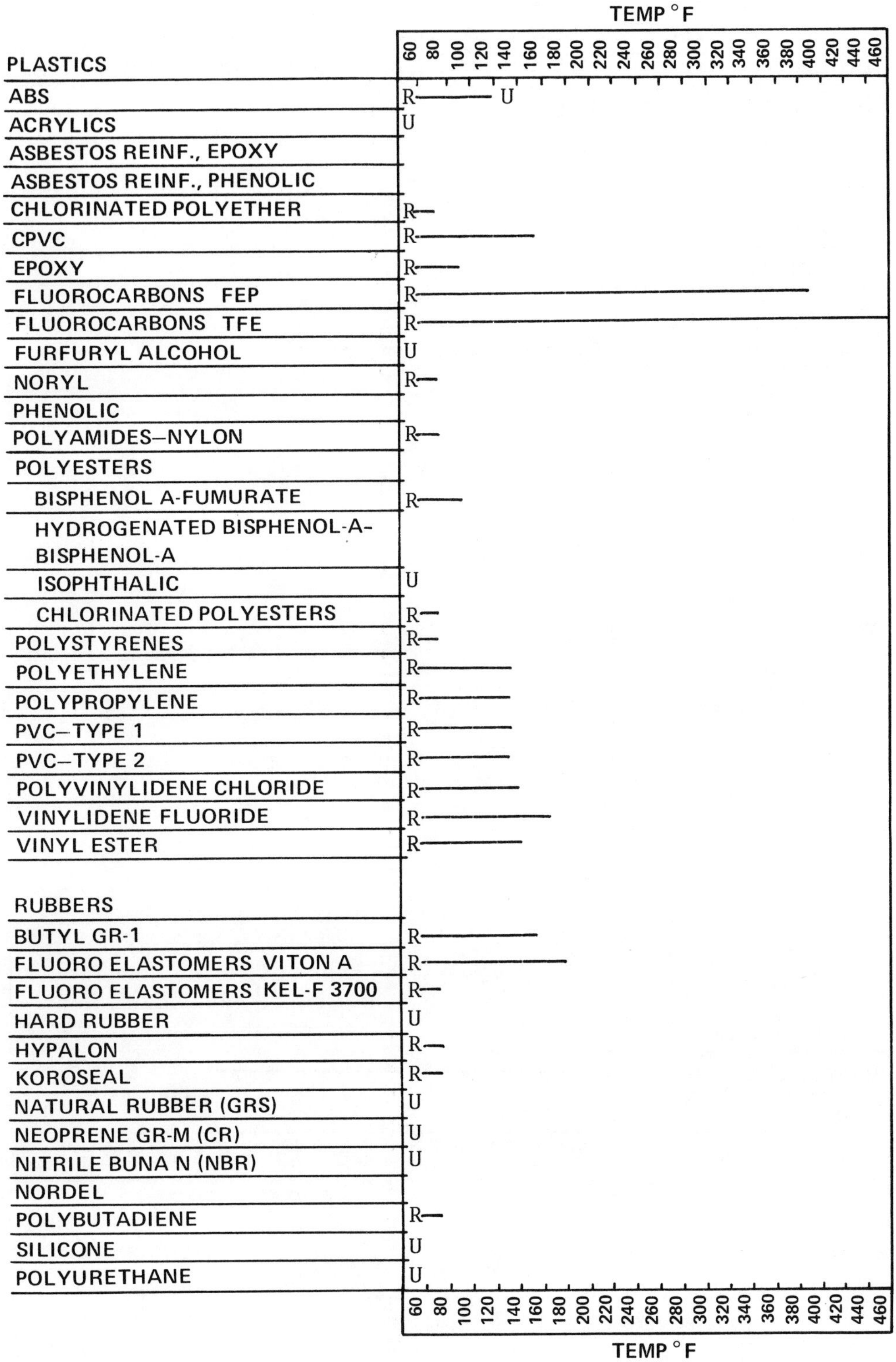

NITRIC ACID, 10%

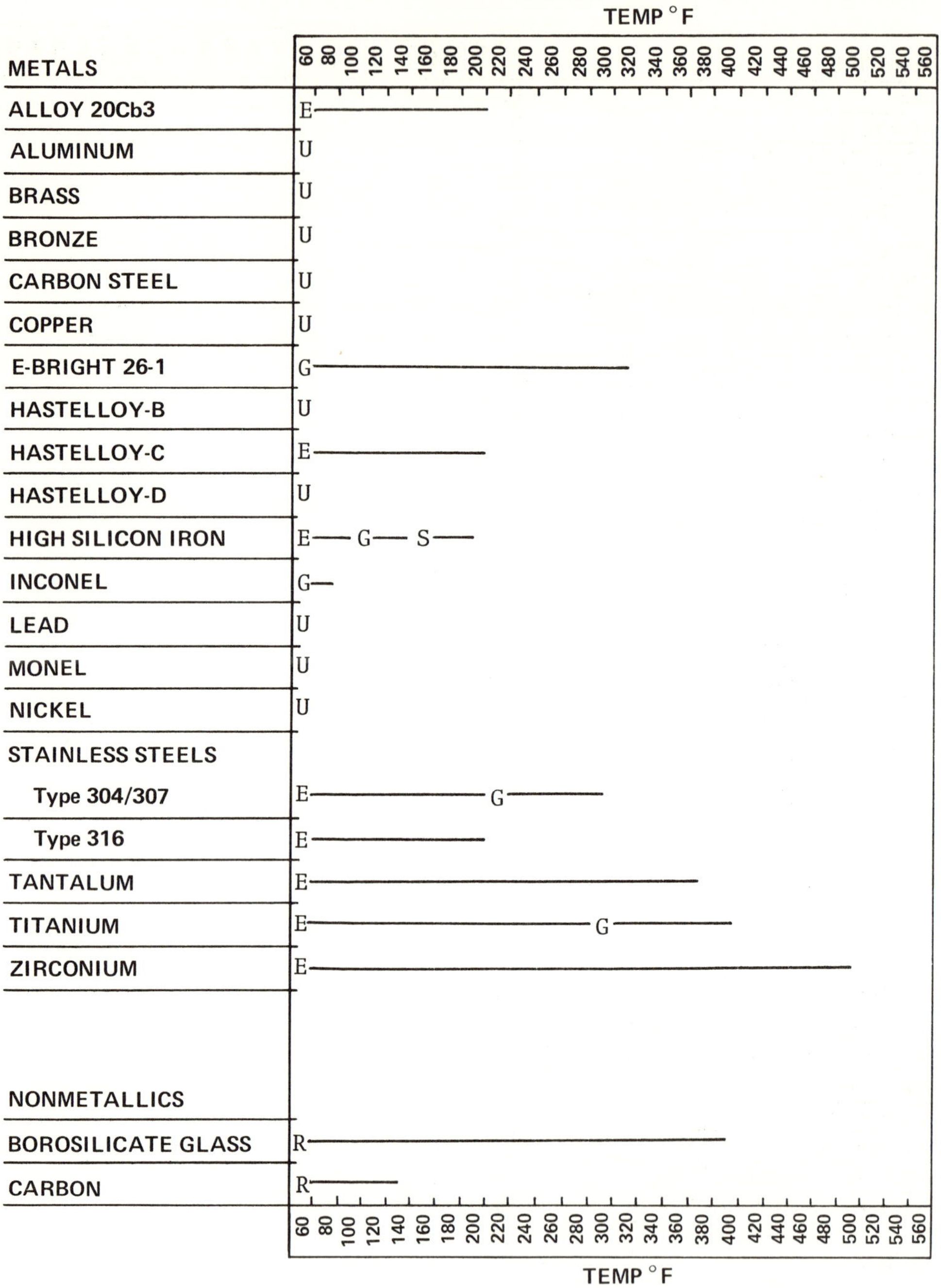

FOR METALS

E = < 2 Mils Penetration/Year; G = < 20 Mils Penetration/Year

S = < 50 Mils Penetration/Year; U = > 50 Mils Penetration/Year

FOR NONMETALLICS

R = Resistant

U = Unsatisfactory

NITRIC ACID, 10%

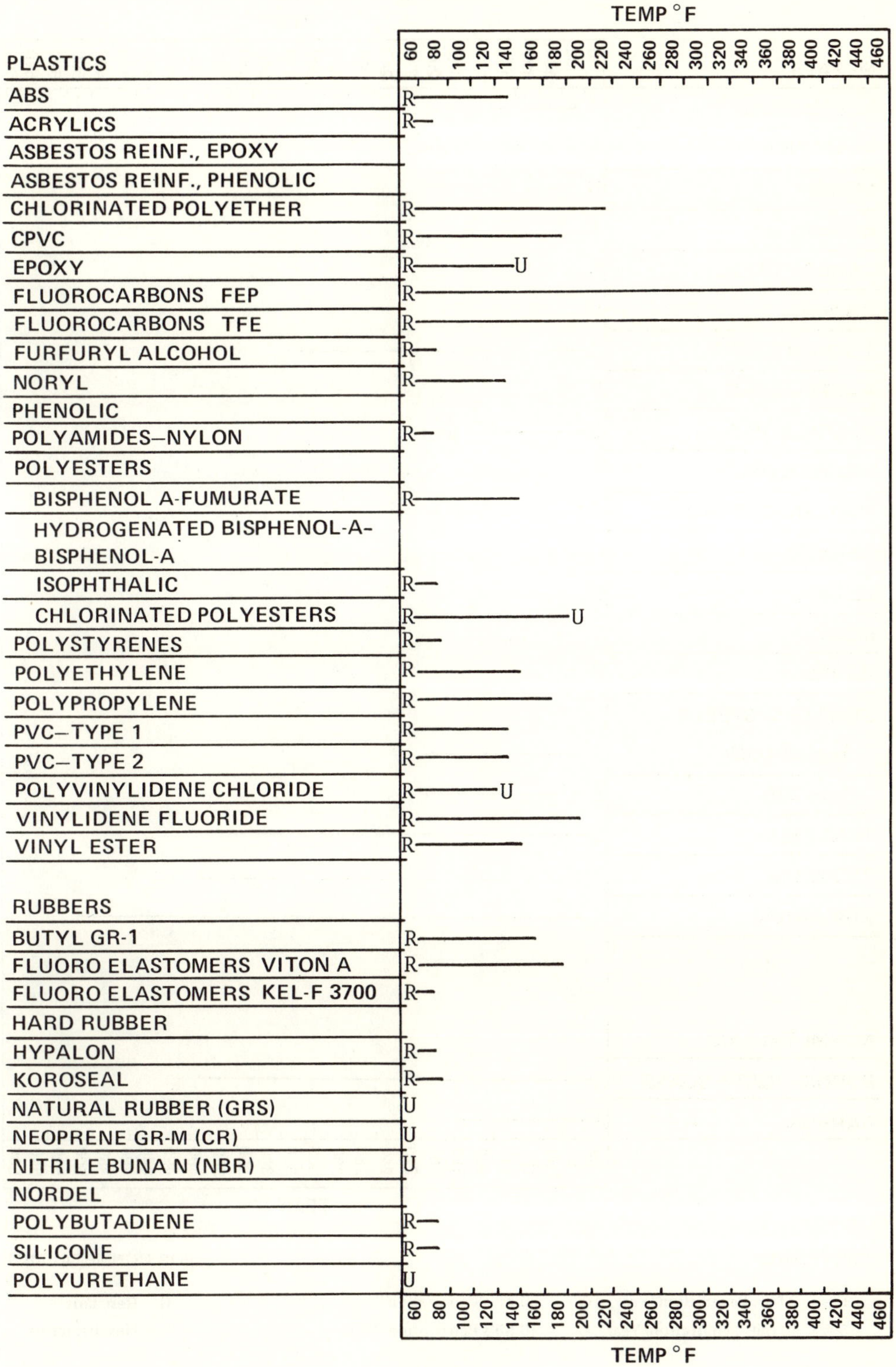

NITRIC ACID, 5%

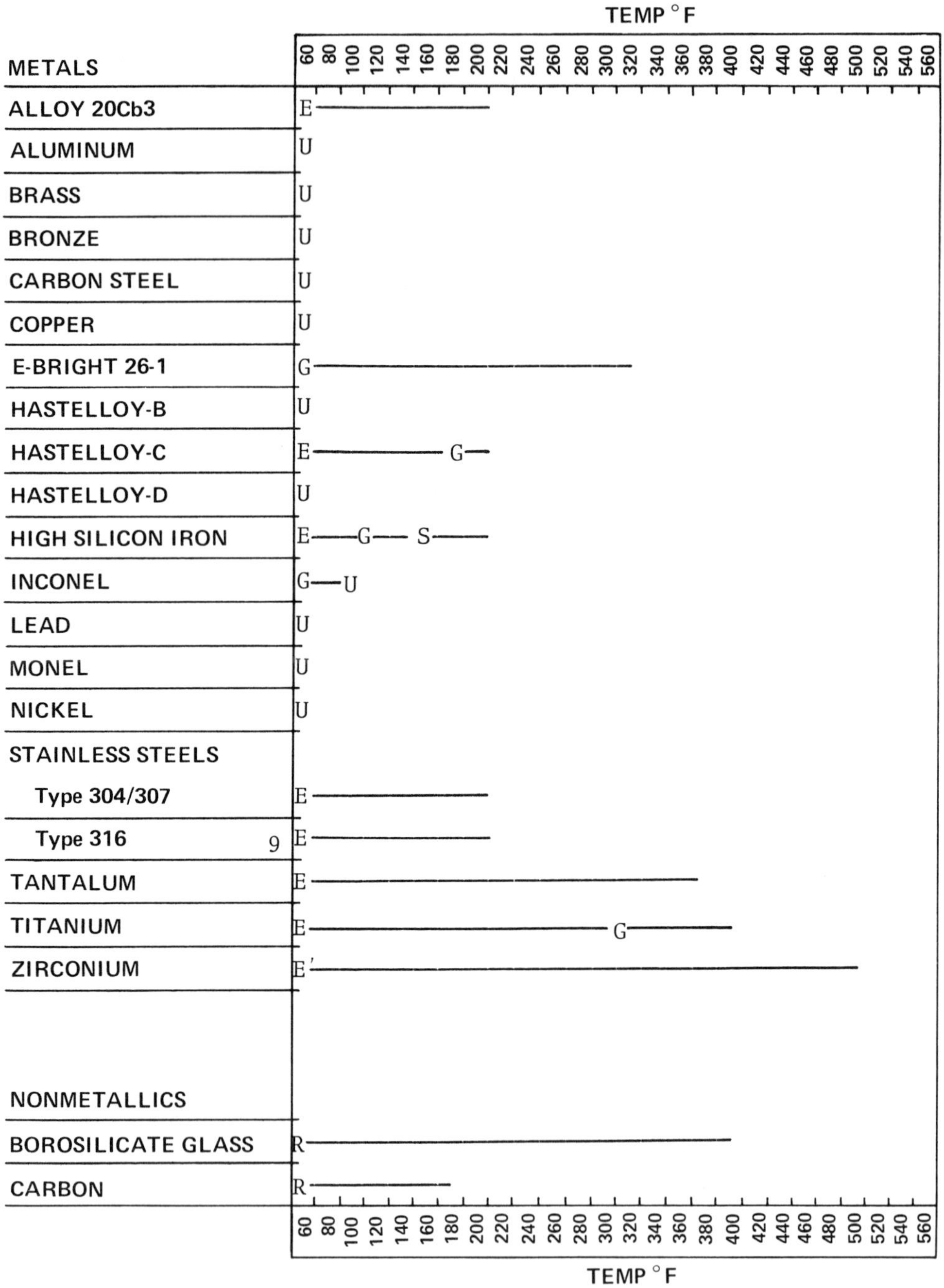

FOR METALS

E = < 2 Mils Penetration/Year; G = < 20 Mils Penetration/Year

S = < 50 Mils Penetration/Year; U = > 50 Mils Penetration/Year

FOR NONMETALLICS

R = Resistant

U = Unsatisfactory

NITRIC ACID, 5%

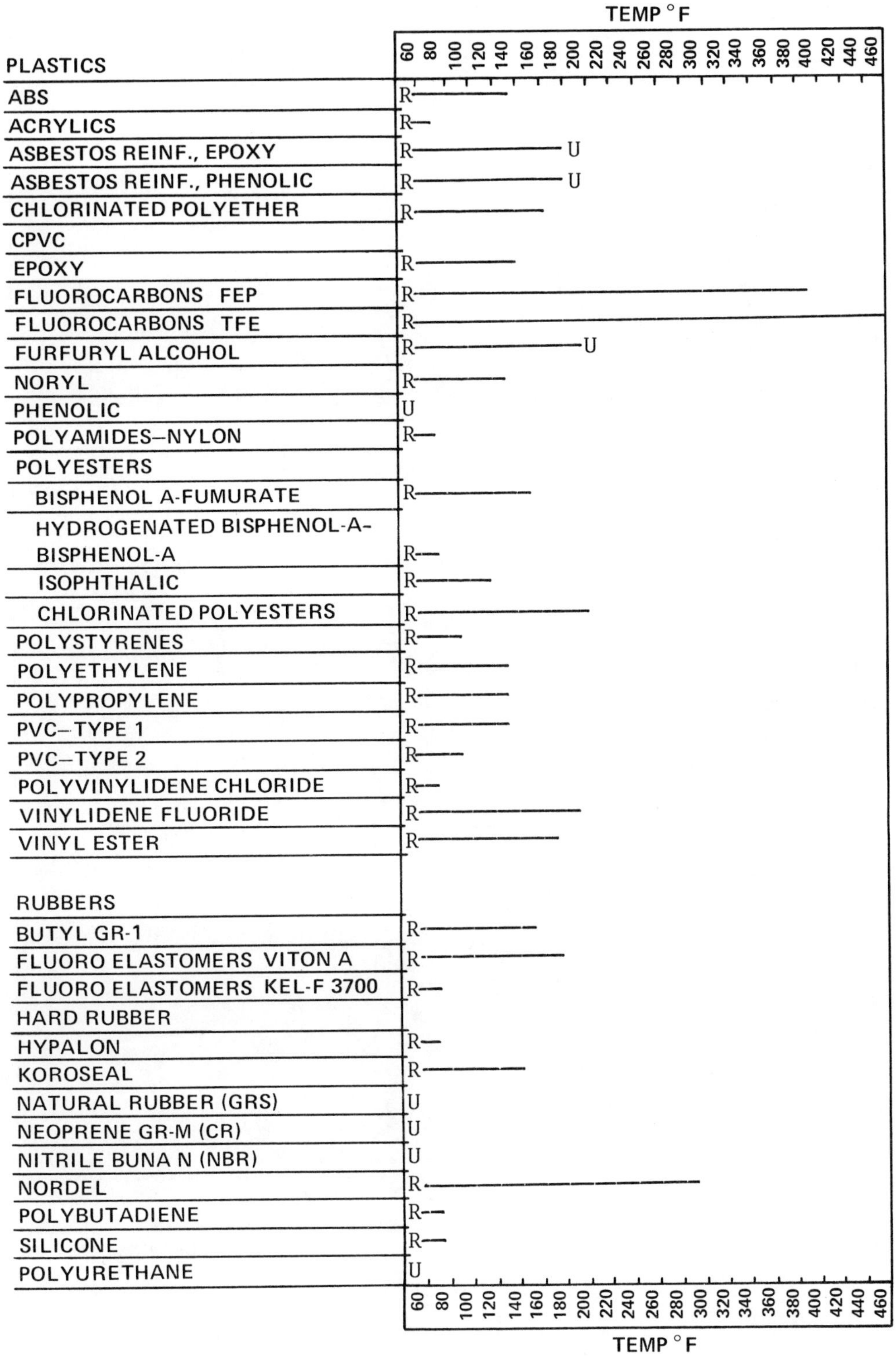

NITRIC ACID FUMING (RED)

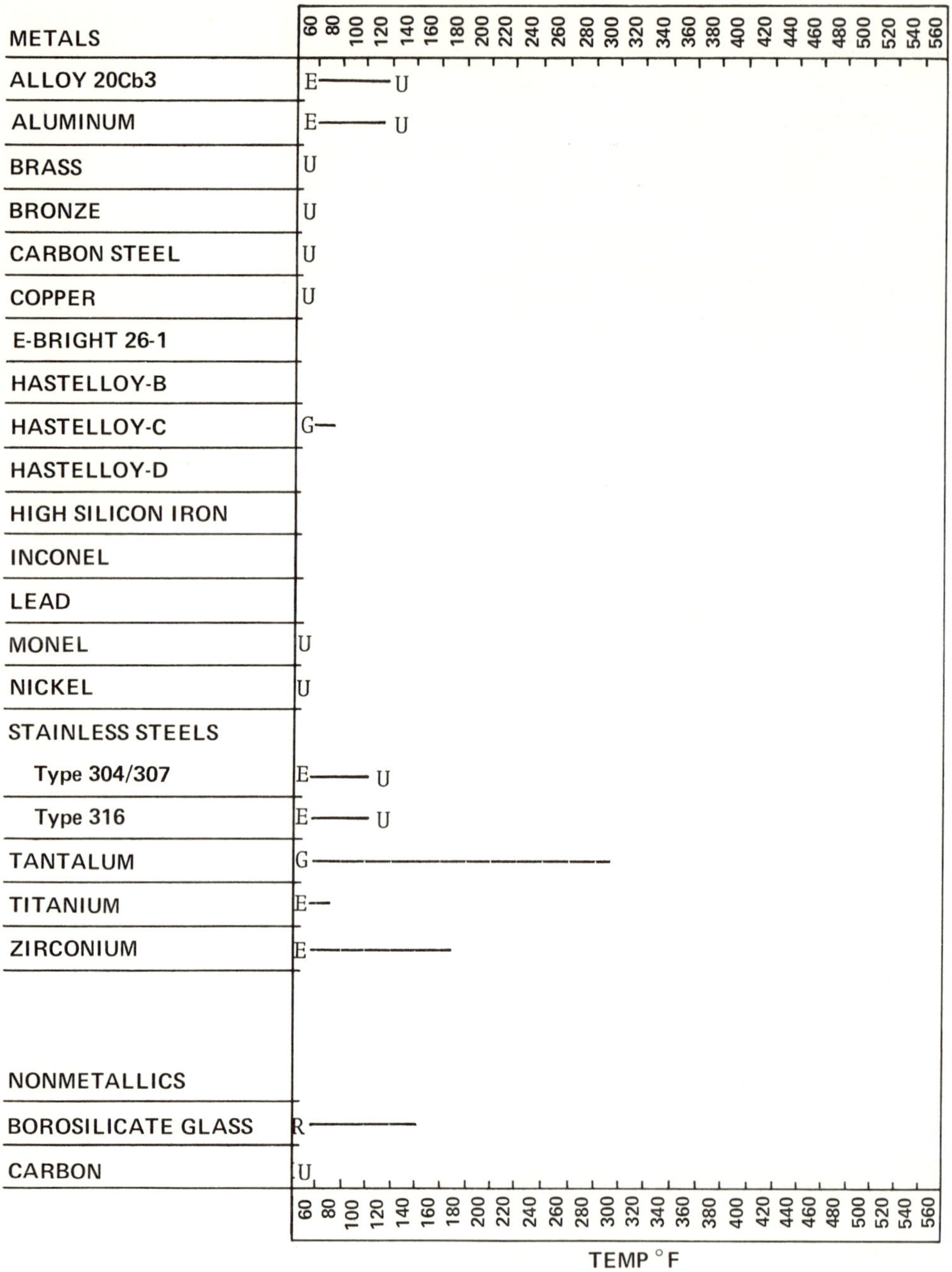

FOR METALS

E = < 2 Mils Penetration/Year; G = < 20 Mils Penetration/Year

S = < 50 Mils Penetration/Year; U = > 50 Mils Penetration/Year

FOR NONMETALLICS

R = Resistant

U = Unsatisfactory

NITRIC ACID FUMING (RED)

TEMP °F (60–460)

PLASTICS	Rating	Temp range (°F)
ABS		
ACRYLICS		
ASBESTOS REINF., EPOXY		
ASBESTOS REINF., PHENOLIC		
CHLORINATED POLYETHER	U	
CPVC		
EPOXY	U	
FLUOROCARBONS FEP	R	60–400
FLUOROCARBONS TFE	R	60–460
FURFURYL ALCOHOL	U	
NORYL	U	
PHENOLIC		
POLYAMIDES–NYLON		
POLYESTERS		
BISPHENOL A-FUMURATE		
HYDROGENATED BISPHENOL-A-BISPHENOL-A		
ISOPHTHALIC	R	60–~150
CHLORINATED POLYESTERS	U	
POLYSTYRENES		
POLYETHYLENE	U	
POLYPROPYLENE	U	
PVC–TYPE 1		
PVC–TYPE 2		
POLYVINYLIDENE CHLORIDE		
VINYLIDENE FLUORIDE	R	60–~120
VINYL ESTER		

RUBBERS	Rating	Temp range (°F)
BUTYL GR-1	U	
FLUORO ELASTOMERS VITON A	U	
FLUORO ELASTOMERS KEL-F 3700		
HARD RUBBER		
HYPALON	U	
KOROSEAL		
NATURAL RUBBER (GRS)	U	
NEOPRENE GR-M (CR)	U	
NITRILE BUNA N (NBR)	U	
NORDEL		
POLYBUTADIENE		
SILICONE	U	
POLYURETHANE	U	

TEMP °F

NITRIC ACID-SULFURIC ACID 50/50

TEMP °F: 60, 80, 100, 120, 140, 160, 180, 200, 220, 240, 260, 280, 300, 320, 340, 360, 380, 400, 420, 440, 460, 480, 500, 520, 540, 560

METALS	Rating
ALLOY 20Cb3	E
ALUMINUM	U
BRASS	
BRONZE	
CARBON STEEL	
COPPER	
E-BRIGHT 26-1	
HASTELLOY-B	
HASTELLOY-C	
HASTELLOY-D	
HIGH SILICON IRON	G
INCONEL	U
LEAD	
MONEL	U
NICKEL	U
STAINLESS STEELS	
Type 304/307	E
Type 316	E
TANTALUM	G
TITANIUM	
ZIRCONIUM	G

NONMETALLICS	Rating
BOROSILICATE GLASS	R
CARBON	U

FOR METALS

E = < 2 Mils Penetration/Year; G = < 20 Mils Penetration/Year

S = < 50 Mils Penetration/Year; U = > 50 Mils Penetration/Year

FOR NONMETALLICS

R = Resistant

U = Unsatisfactory

NITRIC ACID-SULFURIC ACID 50/50

TEMP °F: 60 80 100 120 140 160 180 200 220 240 260 280 300 320 340 360 380 400 420 440 460

PLASTICS	TEMP °F
ABS	U
ACRYLICS	
ASBESTOS REINF., EPOXY	
ASBESTOS REINF., PHENOLIC	
CHLORINATED POLYETHER	
CPVC	
EPOXY	U
FLUOROCARBONS FEP	R
FLUOROCARBONS TFE	R
FURFURYL ALCOHOL	U
NORYL	
PHENOLIC	
POLYAMIDES–NYLON	
POLYESTERS	
BISPHENOL A-FUMURATE	
HYDROGENATED BISPHENOL-A-BISPHENOL-A	
ISOPHTHALIC	U
CHLORINATED POLYESTERS	
POLYSTYRENES	
POLYETHYLENE	
POLYPROPYLENE	U
PVC–TYPE 1	
PVC–TYPE 2	
POLYVINYLIDENE CHLORIDE	
VINYLIDENE FLUORIDE	R
VINYL ESTER	
RUBBERS	
BUTYL GR-1	
FLUORO ELASTOMERS VITON A	
FLUORO ELASTOMERS KEL-F 3700	
HARD RUBBER	
HYPALON	
KOROSEAL	
NATURAL RUBBER (GRS)	
NEOPRENE GR-M (CR)	
NITRILE BUNA N (NBR)	
NORDEL	
POLYBUTADIENE	
SILICONE	U
POLYURETHANE	

TEMP °F: 60 80 100 120 140 160 180 200 220 240 260 280 300 320 340 360 380 400 420 440 460

NITROBENZENE

METALS		Rating
ALLOY 20Cb3		E
ALUMINUM		E–G
BRASS		G
BRONZE		G
CARBON STEEL		G
COPPER		G
E-BRIGHT 26-1		
HASTELLOY-B	85%	U
HASTELLOY-C	85%	G
HASTELLOY-D		
HIGH SILICON IRON		G
INCONEL		G
LEAD		G
MONEL		G
NICKEL	acid free	G
STAINLESS STEELS		
Type 304/307		G
Type 316		G
TANTALUM		G
TITANIUM		E
ZIRCONIUM		
NONMETALLICS		
BOROSILICATE GLASS		R
CARBON		R

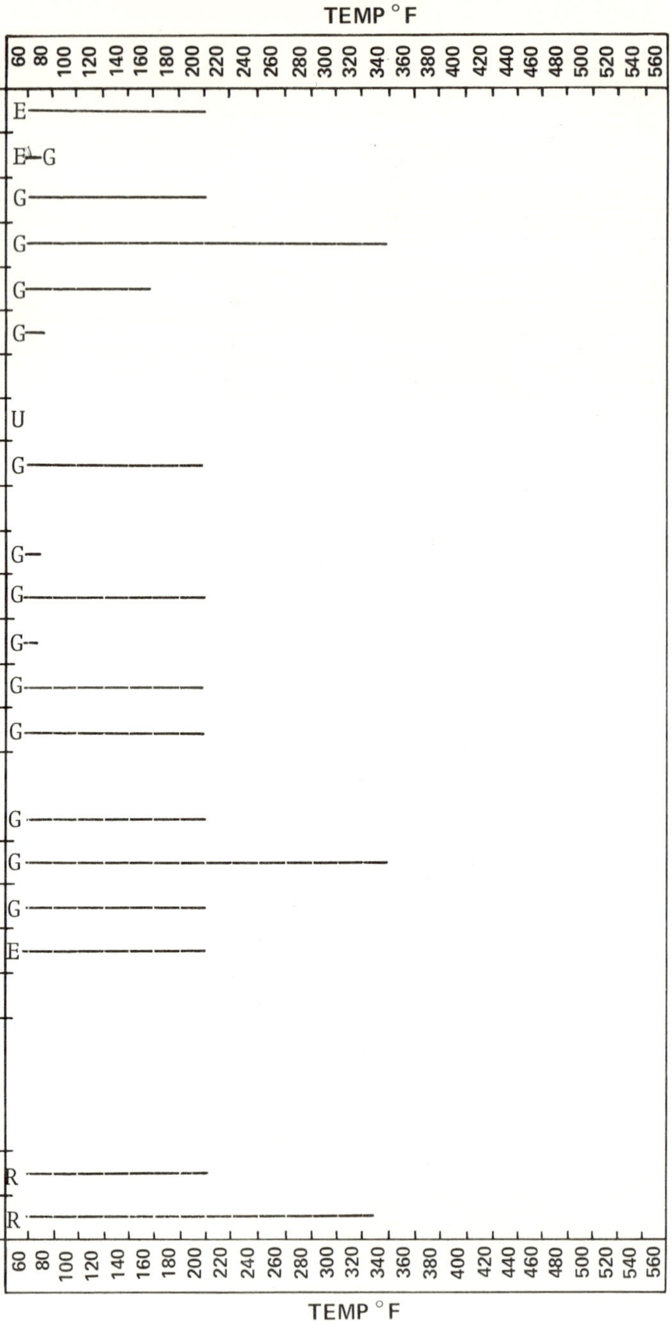

FOR METALS

E = < 2 Mils Penetration/Year; G = < 20 Mils Penetration/Year

S = < 50 Mils Penetration/Year; U = > 50 Mils Penetration/Year

FOR NONMETALLICS

R = Resistant

U = Unsatisfactory

TEMP °F (60, 80, 100, 120, 140, 160, 180, 200, 220, 240, 260, 280, 300, 320, 340, 360, 380, 400, 420, 440, 460)

PLASTICS	TEMP °F
ABS	U
ACRYLICS	
ASBESTOS REINF., EPOXY	
ASBESTOS REINF., PHENOLIC	
CHLORINATED POLYETHER	R—
CPVC	U
EPOXY	U
FLUOROCARBONS FEP	R—————————
FLUOROCARBONS TFE	R——————————
FURFURYL ALCOHOL	R——————
NORYL	
PHENOLIC	
POLYAMIDES—NYLON	R————
POLYESTERS	
BISPHENOL A-FUMURATE	U
HYDROGENATED BISPHENOL-A-BISPHENOL-A	U
ISOPHTHALIC	U
CHLORINATED POLYESTERS	
POLYSTYRENES	
POLYETHYLENE	U
POLYPROPYLENE	R—U
PVC—TYPE 1	U
PVC—TYPE 2	U
POLYVINYLIDENE CHLORIDE	
VINYLIDENE FLUORIDE	R———
VINYL ESTER	R——

RUBBERS	TEMP °F
BUTYL GR-1	R—
FLUORO ELASTOMERS VITON A	RU
FLUORO ELASTOMERS KEL-F 3700	
HARD RUBBER	
HYPALON	U
KOROSEAL	U
NATURAL RUBBER (GRS)	U
NEOPRENE GR-M (CR)	U
NITRILE BUNA N (NBR)	U
NORDEL	R—
POLYBUTADIENE	
SILICONE	
POLYURETHANE	U

TEMP °F (60, 80, 100, 120, 140, 160, 180, 200, 220, 240, 260, 280, 300, 320, 340, 360, 380, 400, 420, 440, 460)

NITROGEN DIOXIDE

TEMP °F

METALS	60	80	100	120	140	160	180	200	220	240	260	280	300	320	340	360	380	400	420	440	460	480	500	520	540	560
ALLOY 20Cb3																										
ALUMINUM																										
BRASS																										
BRONZE																										
CARBON STEEL																										
COPPER																										
E-BRIGHT 26-1																										
HASTELLOY-B																										
HASTELLOY-C																										
HASTELLOY-D																										
HIGH SILICON IRON																										
INCONEL																										
LEAD																										
MONEL																										
NICKEL																										
STAINLESS STEELS Type 304/307																										
Type 316																										
TANTALUM																										
TITANIUM																										
ZIRCONIUM																										
NONMETALLICS																										
BOROSILICATE GLASS																										
CARBON																										
	60	80	100	120	140	160	180	200	220	240	260	280	300	320	340	360	380	400	420	440	460	480	500	520	540	560

TEMP °F

FOR METALS

E = < 2 Mils Penetration/Year; G = < 20 Mils Penetration/Year

S = < 50 Mils Penetration/Year; U = > 50 Mils Penetration/Year

FOR NONMETALLICS

R = Resistant

U = Unsatisfactory

NITROGEN DIOXIDE

TEMP °F: 60 80 100 120 140 160 180 200 220 240 260 280 300 320 340 360 380 400 420 440 460

PLASTICS	
ABS	
ACRYLICS	
ASBESTOS REINF., EPOXY	
ASBESTOS REINF., PHENOLIC	
CHLORINATED POLYETHER	
CPVC	
EPOXY	
FLUOROCARBONS FEP	R
FLUOROCARBONS TFE	R
FURFURYL ALCOHOL	
NORYL	
PHENOLIC	
POLYAMIDES–NYLON	
POLYESTERS	
BISPHENOL A-FUMURATE	
HYDROGENATED BISPHENOL-A-BISPHENOL-A	
ISOPHTHALIC	
CHLORINATED POLYESTERS	
POLYSTYRENES	
POLYETHYLENE	
POLYPROPYLENE	R
PVC–TYPE 1	
PVC–TYPE 2	
POLYVINYLIDENE CHLORIDE	R
VINYLIDENE FLUORIDE	R
VINYL ESTER	

RUBBERS	
BUTYL GR-1	
FLUORO ELASTOMERS VITON A	U
FLUORO ELASTOMERS KEL-F 3700	
HARD RUBBER	
HYPALON	
KOROSEAL	
NATURAL RUBBER (GRS)	
NEOPRENE GR-M (CR)	
NITRILE BUNA N (NBR)	
NORDEL	
POLYBUTADIENE	
SILICONE	
POLYURETHANE	

TEMP °F: 60 80 100 120 140 160 180 200 220 240 260 280 300 320 340 360 380 400 420 440 460

NITROMETHANE

TEMP °F: 60, 80, 100, 120, 140, 160, 180, 200, 220, 240, 260, 280, 300, 320, 340, 360, 380, 400, 420, 440, 460, 480, 500, 520, 540, 560

METALS	60
ALLOY 20Cb3	G—
ALUMINUM	G—
BRASS	
BRONZE	
CARBON STEEL	G—
COPPER	
E-BRIGHT 26-1	
HASTELLOY-B	
HASTELLOY-C	
HASTELLOY-D	
HIGH SILICON IRON	
INCONEL	
LEAD	
MONEL	
NICKEL	
STAINLESS STEELS	
Type 304/307	G—
Type 316	G—
TANTALUM	G—
TITANIUM	
ZIRCONIUM	

NONMETALLICS	60
BOROSILICATE GLASS	R—
CARBON	

TEMP °F

FOR METALS

E = < 2 Mils Penetration/Year; G = < 20 Mils Penetration/Year

S = < 50 Mils Penetration/Year; U = > 50 Mils Penetration/Year

FOR NONMETALLICS

R = Resistant

U = Unsatisfactory

TEMP °F (60, 80, 100, 120, 140, 160, 180, 200, 220, 240, 260, 280, 300, 320, 340, 360, 380, 400, 420, 440, 460)

PLASTICS	
ABS	U
ACRYLICS	
ASBESTOS REINF., EPOXY	
ASBESTOS REINF., PHENOLIC	
CHLORINATED POLYETHER	R—
CPVC	
EPOXY	
FLUOROCARBONS FEP	R—
FLUOROCARBONS TFE	R—
FURFURYL ALCOHOL	
NORYL	
PHENOLIC	
POLYAMIDES—NYLON	R—
POLYESTERS	
BISPHENOL A-FUMURATE	
HYDROGENATED BISPHENOL-A-BISPHENOL-A	
ISOPHTHALIC	
CHLORINATED POLYESTERS	
POLYSTYRENES	
POLYETHYLENE	
POLYPROPYLENE	R—
PVC—TYPE 1	
PVC—TYPE 2	
POLYVINYLIDENE CHLORIDE	R—
VINYLIDENE FLUORIDE	R—
VINYL ESTER	

RUBBERS	
BUTYL GR-1	R—
FLUORO ELASTOMERS VITON A	
FLUORO ELASTOMERS KEL-F 3700	
HARD RUBBER	
HYPALON	
KOROSEAL	
NATURAL RUBBER (GRS)	R—
NEOPRENE GR-M (CR)	U
NITRILE BUNA N (NBR)	U
NORDEL	
POLYBUTADIENE	
SILICONE	
POLYURETHANE	

TEMP °F (60, 80, 100, 120, 140, 160, 180, 200, 220, 240, 260, 280, 300, 320, 340, 360, 380, 400, 420, 440, 460)

NITROUS ACID

TEMP °F (60–560)

METALS		Rating
ALLOY 20Cb3		G—
ALUMINUM	10%	U
BRASS		G—
BRONZE		G————————
CARBON STEEL		U
COPPER		G—
E-BRIGHT 26-1		
HASTELLOY-B		U
HASTELLOY-C		U
HASTELLOY-D		
HIGH SILICON IRON		G—
INCONEL	10%	U
LEAD		
MONEL		U
NICKEL		U
STAINLESS STEELS		
Type 304/307		G—
Type 316		G—
TANTALUM		G—————————
TITANIUM		
ZIRCONIUM		
NONMETALLICS		
BOROSILICATE GLASS		
CARBON		

TEMP °F (60–560)

FOR METALS

E = < 2 Mils Penetration/Year; G = < 20 Mils Penetration/Year
S = < 50 Mils Penetration/Year; U = > 50 Mils Penetration/Year

FOR NONMETALLICS

R = Resistant
U = Unsatisfactory

NITROUS ACID

PLASTICS	Rating	TEMP °F (range of line)
ABS		
ACRYLICS		
ASBESTOS REINF., EPOXY		
ASBESTOS REINF., PHENOLIC		
CHLORINATED POLYETHER		
CPVC		
EPOXY	U	
FLUOROCARBONS FEP	R	60–400
FLUOROCARBONS TFE	R	60–460
FURFURYL ALCOHOL	U	
NORYL		
PHENOLIC		
POLYAMIDES–NYLON		
POLYESTERS		
BISPHENOL A-FUMURATE		
HYDROGENATED BISPHENOL-A-BISPHENOL-A		
ISOPHTHALIC	R	60–100
CHLORINATED POLYESTERS		
POLYSTYRENES		
POLYETHYLENE		
POLYPROPYLENE	U	
PVC–TYPE 1		
PVC–TYPE 2		
POLYVINYLIDENE CHLORIDE		
VINYLIDENE FLUORIDE	R	60–200
VINYL ESTER		

RUBBERS	Rating	TEMP °F (range of line)
BUTYL GR-1		
FLUORO ELASTOMERS VITON A		
FLUORO ELASTOMERS KEL-F 3700	R	60
HARD RUBBER		
HYPALON		
KOROSEAL		
NATURAL RUBBER (GRS)	U	
NEOPRENE GR-M (CR)	U	
NITRILE BUNA N (NBR)		
NORDEL		
POLYBUTADIENE	R	60
SILICONE		
POLYURETHANE		

TEMP °F: 60 80 100 120 140 160 180 200 220 240 260 280 300 320 340 360 380 400 420 440 460

NITROUS ACID, 10%

METALS	TEMP °F (60–560)
ALLOY 20Cb3	G
ALUMINUM	U
BRASS	
BRONZE	
CARBON STEEL	
COPPER	
E-BRIGHT 26-1	
HASTELLOY-B	
HASTELLOY-C	
HASTELLOY-D	
HIGH SILICON IRON	G
INCONEL	
LEAD	
MONEL	U
NICKEL	U
STAINLESS STEELS	
Type 304/307	G
Type 316	G
TANTALUM	
TITANIUM	
ZIRCONIUM	
NONMETALLICS	
BOROSILICATE GLASS	
CARBON	

FOR METALS

E = < 2 Mils Penetration/Year; G = < 20 Mils Penetration/Year

S = < 50 Mils Penetration/Year; U = > 50 Mils Penetration/Year

FOR NONMETALLICS

R = Resistant

U = Unsatisfactory

NITROUS ACID, 10%

TEMP °F (60–460)

PLASTICS	Rating
ABS	
ACRYLICS	
ASBESTOS REINF., EPOXY	
ASBESTOS REINF., PHENOLIC	
CHLORINATED POLYETHER	R
CPVC	
EPOXY	U
FLUOROCARBONS FEP	R
FLUOROCARBONS TFE	R
FURFURYL ALCOHOL	U
NORYL	
PHENOLIC	
POLYAMIDES–NYLON	
POLYESTERS	
BISPHENOL A-FUMURATE	
HYDROGENATED BISPHENOL-A-BISPHENOL-A	
ISOPHTHALIC	R
CHLORINATED POLYESTERS	R
POLYSTYRENES	
POLYETHYLENE	
POLYPROPYLENE	R
PVC–TYPE 1	R
PVC–TYPE 2	
POLYVINYLIDENE CHLORIDE	R
VINYLIDENE FLUORIDE	R
VINYL ESTER	
RUBBERS	
BUTYL GR-1	
FLUORO ELASTOMERS VITON A	
FLUORO ELASTOMERS KEL-F 3700	
HARD RUBBER	
HYPALON	
KOROSEAL	
NATURAL RUBBER (GRS)	
NEOPRENE GR-M (CR)	
NITRILE BUNA N (NBR)	
NORDEL	
POLYBUTADIENE	
SILICONE	
POLYURETHANE	

TEMP °F: 60 80 100 120 140 160 180 200 220 240 260 280 300 320 340 360 380 400 420 440 460

NITROUS ACID, 5%

TEMP °F

METALS	Rating (60–560 °F)
ALLOY 20Cb3	G (60–120)
ALUMINUM	U (60)
BRASS	
BRONZE	
CARBON STEEL	
COPPER	
E-BRIGHT 26-1	
HASTELLOY-B	
HASTELLOY-C	
HASTELLOY-D	
HIGH SILICON IRON	G (60–80)
INCONEL	G (60–80)
LEAD	
MONEL	U (60)
NICKEL	U (60)
STAINLESS STEELS	
Type 304/307	G (60–100)
Type 316	G (60–140)
TANTALUM	G (60–200)
TITANIUM	
ZIRCONIUM	
NONMETALLICS	
BOROSILICATE GLASS	
CARBON	

TEMP °F: 60 80 100 120 140 160 180 200 220 240 260 280 300 320 340 360 380 400 420 440 460 480 500 520 540 560

FOR METALS

E = < 2 Mils Penetration/Year; G = < 20 Mils Penetration/Year

S = < 50 Mils Penetration/Year; U = > 50 Mils Penetration/Year

FOR NONMETALLICS

R = Resistant

U = Unsatisfactory

NITROUS ACID, 5%

PLASTICS	TEMP °F (60–460)
ABS	
ACRYLICS	
ASBESTOS REINF., EPOXY	
ASBESTOS REINF., PHENOLIC	
CHLORINATED POLYETHER	
CPVC	
EPOXY	U
FLUOROCARBONS FEP	R 60–400
FLUOROCARBONS TFE	R 60–460
FURFURYL ALCOHOL	U
NORYL	
PHENOLIC	
POLYAMIDES–NYLON	
POLYESTERS	
BISPHENOL A-FUMURATE	
HYDROGENATED BISPHENOL-A-BISPHENOL-A	
ISOPHTHALIC	R 60–150
CHLORINATED POLYESTERS	
POLYSTYRENES	
POLYETHYLENE	
POLYPROPYLENE	
PVC–TYPE 1	
PVC–TYPE 2	
POLYVINYLIDENE CHLORIDE	
VINYLIDENE FLUORIDE	R 60–210
VINYL ESTER	
RUBBERS	
BUTYL GR-1	
FLUORO ELASTOMERS VITON A	
FLUORO ELASTOMERS KEL-F 3700	
HARD RUBBER	
HYPALON	
KOROSEAL	
NATURAL RUBBER (GRS)	
NEOPRENE GR-M (CR)	
NITRILE BUNA N (NBR)	
NORDEL	
POLYBUTADIENE	
SILICONE	
POLYURETHANE	

TEMP °F: 60 80 100 120 140 160 180 200 220 240 260 280 300 320 340 360 380 400 420 440 460

NITROUS OXIDE

TEMP °F: 60, 80, 100, 120, 140, 160, 180, 200, 220, 240, 260, 280, 300, 320, 340, 360, 380, 400, 420, 440, 460, 480, 500, 520, 540, 560

METALS	
ALLOY 20Cb3	G
ALUMINUM	G
BRASS	G
BRONZE	U
CARBON STEEL	G
COPPER	
E-BRIGHT 26-1	
HASTELLOY-B	
HASTELLOY-C	G
HASTELLOY-D	
HIGH SILICON IRON	
INCONEL	U
LEAD	
MONEL	U
NICKEL	U
STAINLESS STEELS	
Type 304/307	G
Type 316	G
TANTALUM	G
TITANIUM	
ZIRCONIUM	

NONMETALLICS	
BOROSILICATE GLASS	
CARBON	

TEMP °F

FOR METALS

E = < 2 Mils Penetration/Year; G = < 20 Mils Penetration/Year

S = < 50 Mils Penetration/Year; U = > 50 Mils Penetration/Year

FOR NONMETALLICS

R = Resistant

U = Unsatisfactory

PLASTICS	TEMP °F (60 80 100 120 140 160 180 200 220 240 260 280 300 320 340 360 380 400 420 440 460)
ABS	R———
ACRYLICS	
ASBESTOS REINF., EPOXY	
ASBESTOS REINF., PHENOLIC	
CHLORINATED POLYETHER	R——————— U
CPVC	R———U
EPOXY	
FLUOROCARBONS FEP	R————————————————
FLUOROCARBONS TFE	R————————————————————
FURFURYL ALCOHOL	
NORYL	
PHENOLIC	
POLYAMIDES–NYLON	
POLYESTERS	
BISPHENOL A-FUMURATE	
HYDROGENATED BISPHENOL-A-BISPHENOL-A	
ISOPHTHALIC	
CHLORINATED POLYESTERS	
POLYSTYRENES	
POLYETHYLENE	U
POLYPROPYLENE	U
PVC–TYPE 1	R———
PVC–TYPE 2	R———
POLYVINYLIDENE CHLORIDE	R—
VINYLIDENE FLUORIDE	U
VINYL ESTER	

RUBBERS	
BUTYL GR-1	
FLUORO ELASTOMERS VITON A	R—
FLUORO ELASTOMERS KEL-F 3700	
HARD RUBBER	
HYPALON	
KOROSEAL	
NATURAL RUBBER (GRS)	
NEOPRENE GR-M (CR)	
NITRILE BUNA N (NBR)	U
NORDEL	
POLYBUTADIENE	
SILICONE	
POLYURETHANE	

TEMP °F: 60 80 100 120 140 160 180 200 220 240 260 280 300 320 340 360 380 400 420 440 460

OCENOL

TEMP °F

METALS	60	80	100	120	140	160	180	200	220	240	260	280	300	320	340	360	380	400	420	440	460	480	500	520	540	560
ALLOY 20Cb3																										
ALUMINUM																										
BRASS																										
BRONZE																										
CARBON STEEL																										
COPPER																										
E-BRIGHT 26-1																										
HASTELLOY-B																										
HASTELLOY-C																										
HASTELLOY-D																										
HIGH SILICON IRON																										
INCONEL																										
LEAD																										
MONEL																										
NICKEL																										
STAINLESS STEELS Type 304/307																										
Type 316																										
TANTALUM																										
TITANIUM																										
ZIRCONIUM																										
NONMETALLICS																										
BOROSILICATE GLASS																										
CARBON																										

60 80 100 120 140 160 180 200 220 240 260 280 300 320 340 360 380 400 420 440 460 480 500 520 540 560

TEMP °F

FOR METALS

E = < 2 Mils Penetration/Year; G = < 20 Mils Penetration/Year

S = < 50 Mils Penetration/Year; U = > 50 Mils Penetration/Year

FOR NONMETALLICS

R = Resistant

U = Unsatisfactory

Please see page 1 for notes on calibration.

OCENOL

TEMP °F: 60 80 100 120 140 160 180 200 220 240 260 280 300 320 340 360 380 400 420 440 460

PLASTICS	
ABS	R
ACRYLICS	
ASBESTOS REINF., EPOXY	
ASBESTOS REINF., PHENOLIC	
CHLORINATED POLYETHER	
CPVC	R
EPOXY	
FLUOROCARBONS FEP	R
FLUOROCARBONS TFE	R
FURFURYL ALCOHOL	
NORYL	
PHENOLIC	
POLYAMIDES—NYLON	
POLYESTERS	
BISPHENOL A-FUMURATE	
HYDROGENATED BISPHENOL-A-BISPHENOL-A	
ISOPHTHALIC	
CHLORINATED POLYESTERS	
POLYSTYRENES	
POLYETHYLENE	
POLYPROPYLENE	R
PVC—TYPE 1	R
PVC—TYPE 2	R
POLYVINYLIDENE CHLORIDE	
VINYLIDENE FLUORIDE	
VINYL ESTER	

RUBBERS	
BUTYL GR-1	
FLUORO ELASTOMERS VITON A	
FLUORO ELASTOMERS KEL-F 3700	
HARD RUBBER	
HYPALON	
KOROSEAL	
NATURAL RUBBER (GRS)	
NEOPRENE GR-M (CR)	
NITRILE BUNA N (NBR)	
NORDEL	
POLYBUTADIENE	
SILICONE	
POLYURETHANE	

TEMP °F: 60 80 100 120 140 160 180 200 220 240 260 280 300 320 340 360 380 400 420 440 460

OCTANE

METALS	TEMP °F 60	80	100	120	140	160	180	200	220	240	260	280	300	320	340	360	380	400	420	440	460	480	500	520	540	560
ALLOY 20Cb3																										
ALUMINUM																										
BRASS																										
BRONZE																										
CARBON STEEL																										
COPPER																										
E-BRIGHT 26-1																										
HASTELLOY-B																										
HASTELLOY-C																										
HASTELLOY-D																										
HIGH SILICON IRON																										
INCONEL																										
LEAD																										
MONEL																										
NICKEL																										
STAINLESS STEELS Type 304/307																										
Type 316																										
TANTALUM																										
TITANIUM																										
ZIRCONIUM																										
NONMETALLICS																										
BOROSILICATE GLASS																										
CARBON																										
	60	80	100	120	140	160	180	200	220	240	260	280	300	320	340	360	380	400	420	440	460	480	500	520	540	560

TEMP °F

FOR METALS

E = < 2 Mils Penetration/Year; G = < 20 Mils Penetration/Year
S = < 50 Mils Penetration/Year; U = > 50 Mils Penetration/Year

FOR NONMETALLICS

R = Resistant
U = Unsatisfactory

OCTANE

PLASTICS	TEMP °F (60–460)
ABS	
ACRYLICS	
ASBESTOS REINF., EPOXY	
ASBESTOS REINF., PHENOLIC	
CHLORINATED POLYETHER	R
CPVC	
EPOXY	
FLUOROCARBONS FEP	R
FLUOROCARBONS TFE	R
FURFURYL ALCOHOL	
NORYL	
PHENOLIC	
POLYAMIDES—NYLON	
POLYESTERS	
BISPHENOL A-FUMURATE	
HYDROGENATED BISPHENOL-A-BISPHENOL-A	
ISOPHTHALIC	
CHLORINATED POLYESTERS	
POLYSTYRENES	
POLYETHYLENE	
POLYPROPYLENE	R
PVC—TYPE 1	
PVC—TYPE 2	
POLYVINYLIDENE CHLORIDE	R
VINYLIDENE FLUORIDE	R
VINYL ESTER	
RUBBERS	
BUTYL GR-1	
FLUORO ELASTOMERS VITON A	
FLUORO ELASTOMERS KEL-F 3700	
HARD RUBBER	
HYPALON	
KOROSEAL	
NATURAL RUBBER (GRS)	
NEOPRENE GR-M (CR)	
NITRILE BUNA N (NBR)	
NORDEL	
POLYBUTADIENE	
SILICONE	
POLYURETHANE	

OIL AND FATS

TEMP °F (60–560)

METALS	TEMP °F
ALLOY 20Cb3	
ALUMINUM	G— (60)
BRASS	
BRONZE	
CARBON STEEL	
COPPER	
E-BRIGHT 26-1	
HASTELLOY-B	
HASTELLOY-C	
HASTELLOY-D	
HIGH SILICON IRON	
INCONEL	
LEAD	
MONEL	
NICKEL	
STAINLESS STEELS	
Type 304/307	G— (60)
Type 316	G— (60)
TANTALUM	
TITANIUM	
ZIRCONIUM	

NONMETALLICS	TEMP °F
BOROSILICATE GLASS	R—————— (60–200)
CARBON	

TEMP °F

FOR METALS

E = < 2 Mils Penetration/Year; G = < 20 Mils Penetration/Year

S = < 50 Mils Penetration/Year; U = > 50 Mils Penetration/Year

FOR NONMETALLICS

R = Resistant

U = Unsatisfactory

OILS AND FATS

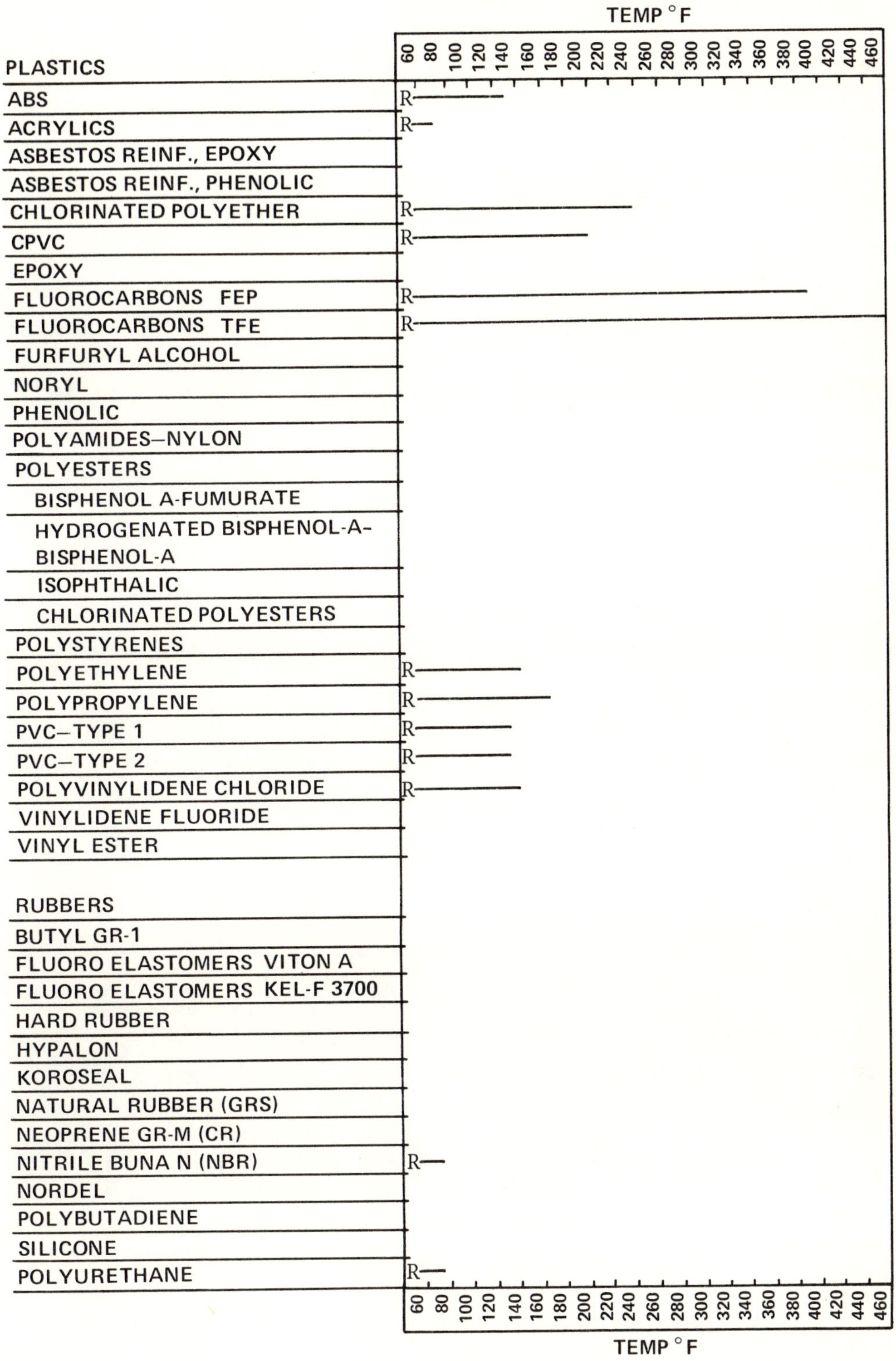

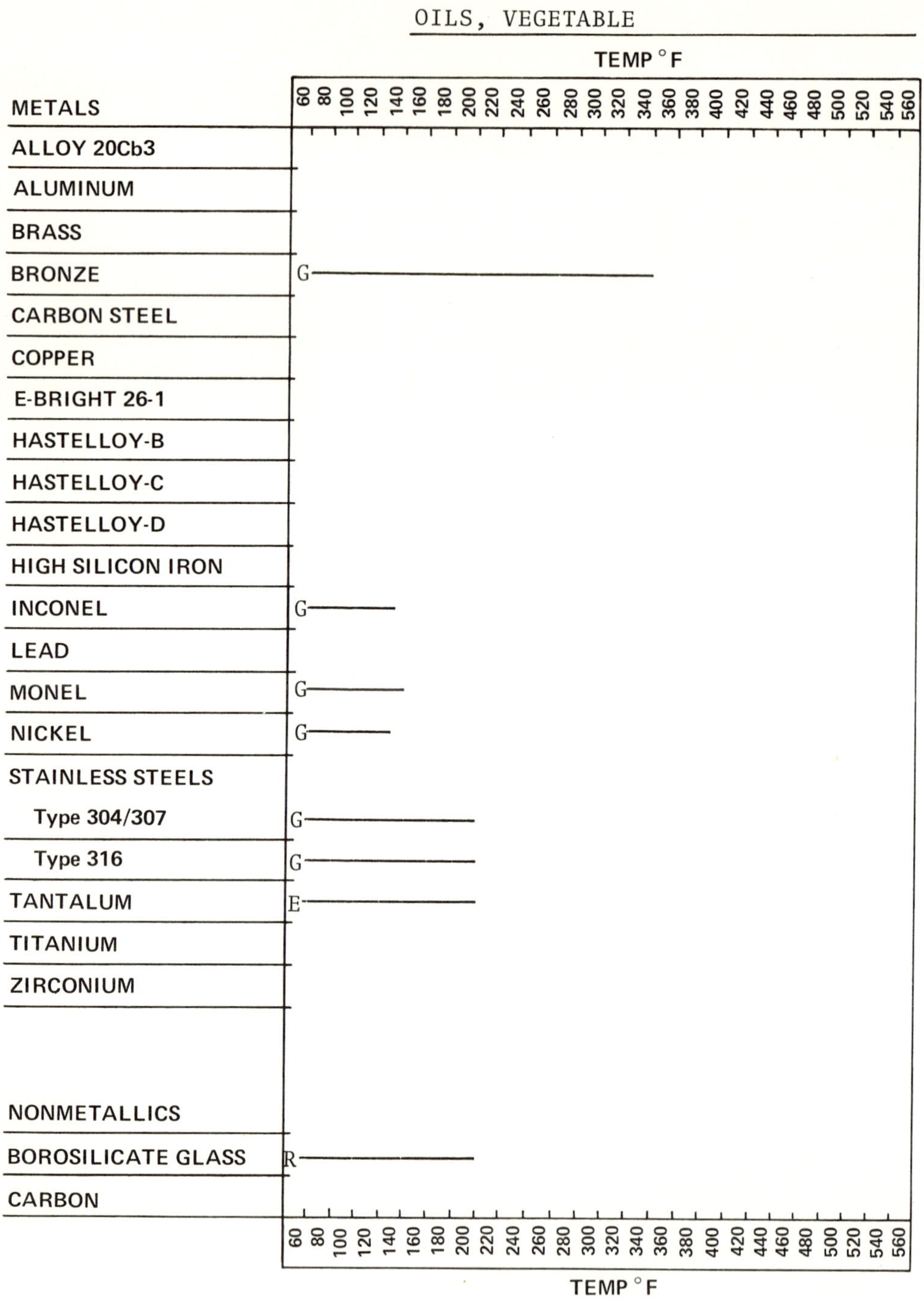

FOR METALS

E = < 2 Mils Penetration/Year; G = < 20 Mils Penetration/Year

S = < 50 Mils Penetration/Year; U = > 50 Mils Penetration/Year

FOR NONMETALLICS

R = Resistant

U = Unsatisfactory

OILS, VEGETABLE

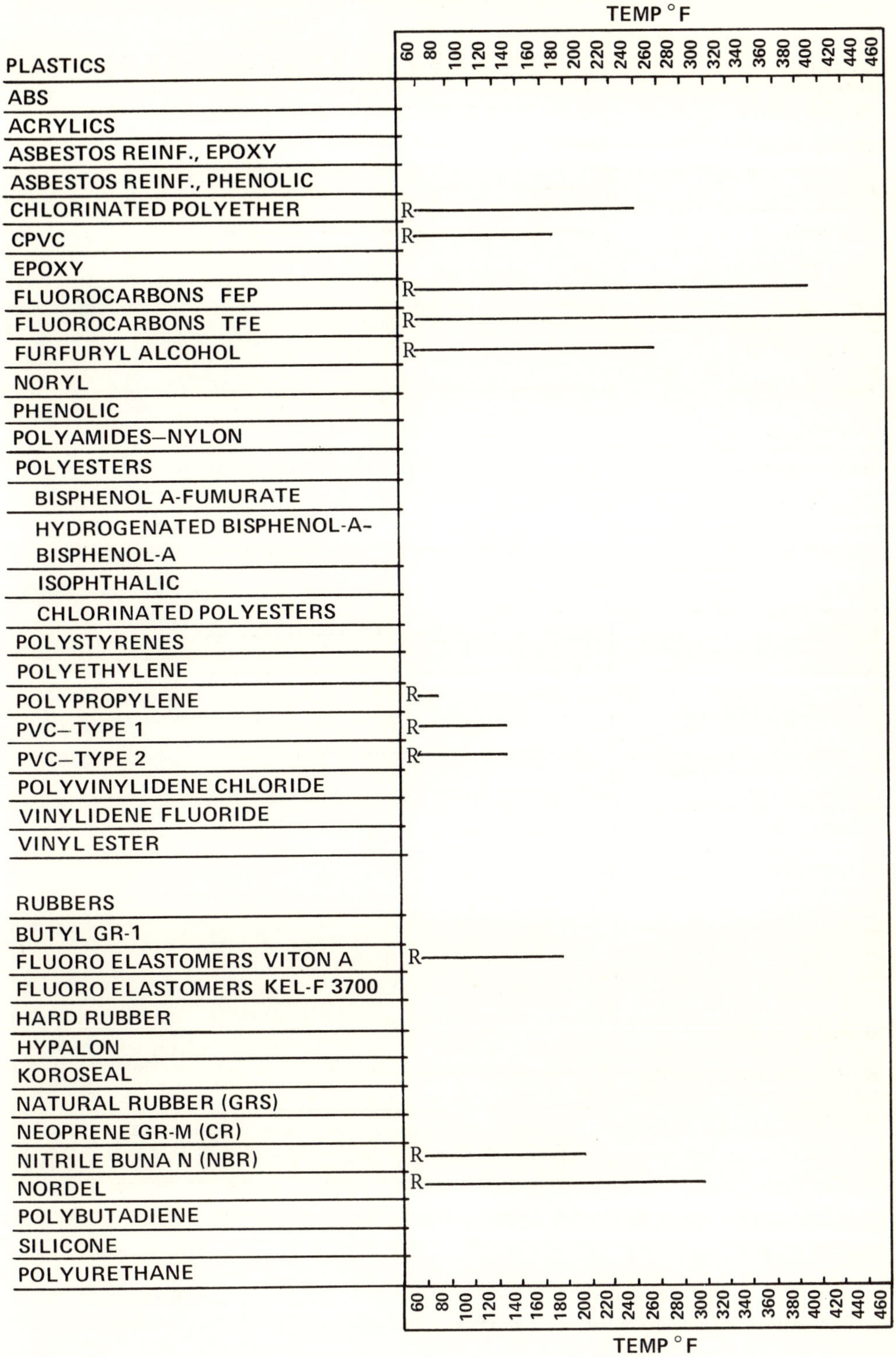

OLEIC ACID

TEMP °F: 60 80 100 120 140 160 180 200 220 240 260 280 300 320 340 360 380 400 420 440 460 480 500 520 540 560

METALS	
ALLOY 20Cb3	G
ALUMINUM	G
BRASS	S
BRONZE	G
CARBON STEEL	G
COPPER	G
E-BRIGHT 26-1	
HASTELLOY-B	G
HASTELLOY-C	G
HASTELLOY-D	G
HIGH SILICON IRON	G
INCONEL	E
LEAD	U
MONEL	E–G
NICKEL	E–G
STAINLESS STEELS	
Type 304/307 1	E — U
Type 316	E — S
TANTALUM	G
TITANIUM	E
ZIRCONIUM	E

NONMETALLICS	
BOROSILICATE GLASS	R
CARBON	R

TEMP °F: 60 80 100 120 140 160 180 200 220 240 260 280 300 320 340 360 380 400 420 440 460 480 500 520 540 560

FOR METALS

E = < 2 Mils Penetration/Year; G = < 20 Mils Penetration/Year

S = < 50 Mils Penetration/Year; U = > 50 Mils Penetration/Year

FOR NONMETALLICS

R = Resistant

U = Unsatisfactory

OLEIC ACID

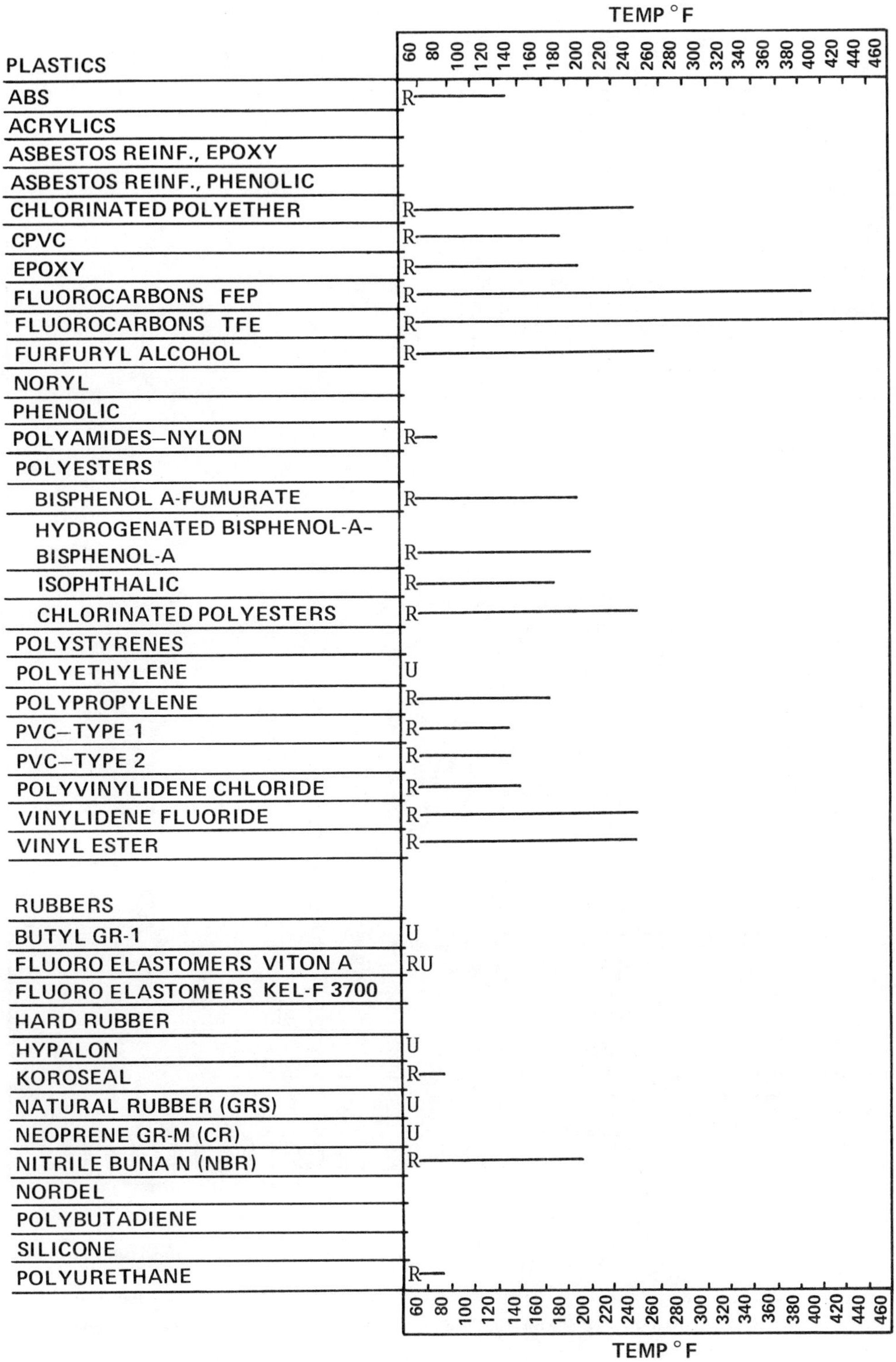
TEMP °F
60 80 100 120 140 160 180 200 220 240 260 280 300 320 340 360 380 400 420 440 460
PLASTICS
ABS R
ACRYLICS
ASBESTOS REINF., EPOXY
ASBESTOS REINF., PHENOLIC
CHLORINATED POLYETHER R
CPVC R
EPOXY R
FLUOROCARBONS FEP R
FLUOROCARBONS TFE R
FURFURYL ALCOHOL R
NORYL
PHENOLIC
POLYAMIDES—NYLON R
POLYESTERS
BISPHENOL A-FUMURATE R
HYDROGENATED BISPHENOL-A-BISPHENOL-A R
ISOPHTHALIC R
CHLORINATED POLYESTERS R
POLYSTYRENES
POLYETHYLENE U
POLYPROPYLENE R
PVC—TYPE 1 R
PVC—TYPE 2 R
POLYVINYLIDENE CHLORIDE R
VINYLIDENE FLUORIDE R
VINYL ESTER R
RUBBERS
BUTYL GR-1 U
FLUORO ELASTOMERS VITON A RU
FLUORO ELASTOMERS KEL-F 3700
HARD RUBBER
HYPALON U
KOROSEAL R
NATURAL RUBBER (GRS) U
NEOPRENE GR-M (CR) U
NITRILE BUNA N (NBR) R
NORDEL
POLYBUTADIENE
SILICONE
POLYURETHANE R
60 80 100 120 140 160 180 200 220 240 260 280 300 320 340 360 380 400 420 440 460
TEMP °F

OLEUM

TEMP °F (60–560, in 20° steps)

METALS	Rating
ALLOY 20Cb3	G
ALUMINUM	G
BRASS	U
BRONZE	U
CARBON STEEL	G
COPPER	
E-BRIGHT 26-1	
HASTELLOY-B to 25%	E
HASTELLOY-C to 40%	E
HASTELLOY-D	
HIGH SILICON IRON	U
INCONEL	
LEAD	G
MONEL	U
NICKEL	
STAINLESS STEELS	
Type 304/307	G
Type 316	G
TANTALUM	
TITANIUM	
ZIRCONIUM	

NONMETALLICS	Rating
BOROSILICATE GLASS	R
CARBON	

TEMP °F

FOR METALS

E = < 2 Mils Penetration/Year; G = < 20 Mils Penetration/Year

S = < 50 Mils Penetration/Year; U = > 50 Mils Penetration/Year

FOR NONMETALLICS

R = Resistant

U = Unsatisfactory

OLEUM

PLASTICS	TEMP °F (60–460)
ABS	U
ACRYLICS	U
ASBESTOS REINF., EPOXY	
ASBESTOS REINF., PHENOLIC	
CHLORINATED POLYETHER	
CPVC	U
EPOXY	U
FLUOROCARBONS FEP	R (60–400)
FLUOROCARBONS TFE	R (60–460)
FURFURYL ALCOHOL	R (60–200) U
NORYL	
PHENOLIC	
POLYAMIDES–NYLON	
POLYESTERS	
BISPHENOL A-FUMURATE	U
HYDROGENATED BISPHENOL-A-BISPHENOL-A	U
ISOPHTHALIC	U
CHLORINATED POLYESTERS all	U
POLYSTYRENES	
POLYETHYLENE	U
POLYPROPYLENE	U
PVC–TYPE 1	U
PVC–TYPE 2	U
POLYVINYLIDENE CHLORIDE	U
VINYLIDENE FLUORIDE	U
VINYL ESTER	U

RUBBERS	TEMP °F (60–460)
BUTYL GR-1	U
FLUORO ELASTOMERS VITON A	R (60–180)
FLUORO ELASTOMERS KEL-F 3700	
HARD RUBBER	
HYPALON	U
KOROSEAL	
NATURAL RUBBER (GRS)	U
NEOPRENE GR-M (CR)	U
NITRILE BUNA N (NBR)	U
NORDEL	
POLYBUTADIENE	
SILICONE	
POLYURETHANE	U

OXALIC ACID (SAT)

TEMP °F: 60 80 100 120 140 160 180 200 220 240 260 280 300 320 340 360 380 400 420 440 460 480 500 520 540 560

METALS		Rating
ALLOY 20Cb3		G
ALUMINUM		G-S
BRASS	air free	S
BRONZE		G
CARBON STEEL		U
COPPER		G
E-BRIGHT 26-1		
HASTELLOY-B		G
HASTELLOY-C		G
HASTELLOY-D		G
HIGH SILICON IRON		G
INCONEL		G
LEAD		U
MONEL		G
NICKEL		G
STAINLESS STEELS		
Type 304/307		U
Type 316		U
TANTALUM		E
TITANIUM		G
ZIRCONIUM		E

NONMETALLICS	Rating
BOROSILICATE GLASS	R
CARBON	R

TEMP °F: 60 80 100 120 140 160 180 200 220 240 260 280 300 320 340 360 380 400 420 440 460 480 500 520 540 560

FOR METALS

E = < 2 Mils Penetration/Year; G = < 20 Mils Penetration/Year

S = < 50 Mils Penetration/Year; U = > 50 Mils Penetration/Year

FOR NONMETALLICS

R = Resistant

U = Unsatisfactory

OXALIC ACID (SAT)

TEMP °F: 60 80 100 120 140 160 180 200 220 240 260 280 300 320 340 360 380 400 420 440 460

PLASTICS		
ABS		R
ACRYLICS		R
ASBESTOS REINF., EPOXY		R
ASBESTOS REINF., PHENOLIC		R
CHLORINATED POLYETHER		R
CPVC		R
EPOXY		R
FLUOROCARBONS FEP		R
FLUOROCARBONS TFE		R
FURFURYL ALCOHOL		R
NORYL		
PHENOLIC		
POLYAMIDES–NYLON		R
POLYESTERS		
BISPHENOL A-FUMURATE	sat	R
HYDROGENATED BISPHENOL-A-BISPHENOL-A		R
ISOPHTHALIC		R
CHLORINATED POLYESTERS	all	R
POLYSTYRENES		R
POLYETHYLENE	11	R U
POLYPROPYLENE		R U
PVC–TYPE 1		R
PVC–TYPE 2		R
POLYVINYLIDENE CHLORIDE		R
VINYLIDENE FLUORIDE		R U
VINYL ESTER		R

RUBBERS	
BUTYL GR-1	
FLUORO ELASTOMERS VITON A	R
FLUORO ELASTOMERS KEL-F 3700	
HARD RUBBER	R
HYPALON	U
KOROSEAL	R
NATURAL RUBBER (GRS)	R
NEOPRENE GR-M (CR)	U
NITRILE BUNA N (NBR)	RU
NORDEL	R
POLYBUTADIENE	
SILICONE	
POLYURETHANE	

TEMP °F: 60 80 100 120 140 160 180 200 220 240 260 280 300 320 340 360 380 400 420 440 460

OXALIC ACID, 50%

TEMP °F: 60 80 100 120 140 160 180 200 220 240 260 280 300 320 340 360 380 400 420 440 460 480 500 520 540 560

METALS	
ALLOY 20Cb3	G
ALUMINUM	G-S U
BRASS	S
BRONZE	G
CARBON STEEL	U
COPPER	G
E-BRIGHT 26-1	
HASTELLOY-B	G
HASTELLOY-C	G
HASTELLOY-D	G
HIGH SILICON IRON	G
INCONEL	E G
LEAD	
MONEL	G
NICKEL	G
STAINLESS STEELS	
Type 304/307	U
Type 316	G
TANTALUM	E
TITANIUM 25%	U
ZIRCONIUM	E

NONMETALLICS	
BOROSILICATE GLASS	R
CARBON	R

TEMP °F

FOR METALS

E = < 2 Mils Penetration/Year; G = < 20 Mils Penetration/Year

S = < 50 Mils Penetration/Year; U = > 50 Mils Penetration/Year

FOR NONMETALLICS

R = Resistant

U = Unsatisfactory

OXALIC ACID, 50%

TEMP °F (60–460)

PLASTICS	Rating
ABS	R
ACRYLICS	R
ASBESTOS REINF., EPOXY	R
ASBESTOS REINF., PHENOLIC	R
CHLORINATED POLYETHER	R
CPVC	R
EPOXY	R
FLUOROCARBONS FEP	R
FLUOROCARBONS TFE	R
FURFURYL ALCOHOL	
NORYL	
PHENOLIC	
POLYAMIDES—NYLON	R
POLYESTERS	
BISPHENOL A-FUMURATE	R
HYDROGENATED BISPHENOL-A-BISPHENOL-A	R
ISOPHTHALIC	R
CHLORINATED POLYESTERS	
POLYSTYRENES	R
POLYETHYLENE	R
POLYPROPYLENE	R
PVC—TYPE 1	R
PVC—TYPE 2	R
POLYVINYLIDENE CHLORIDE	R
VINYLIDENE FLUORIDE	R
VINYL ESTER	R

RUBBERS	Rating
BUTYL GR-1	
FLUORO ELASTOMERS VITON A	R
FLUORO ELASTOMERS KEL-F 3700	
HARD RUBBER	R
HYPALON	R
KOROSEAL	R
NATURAL RUBBER (GRS)	
NEOPRENE GR-M (CR)	U
NITRILE BUNA N (NBR)	U
NORDEL	R
POLYBUTADIENE	
SILICONE	
POLYURETHANE	

OXALIC ACID, 10%

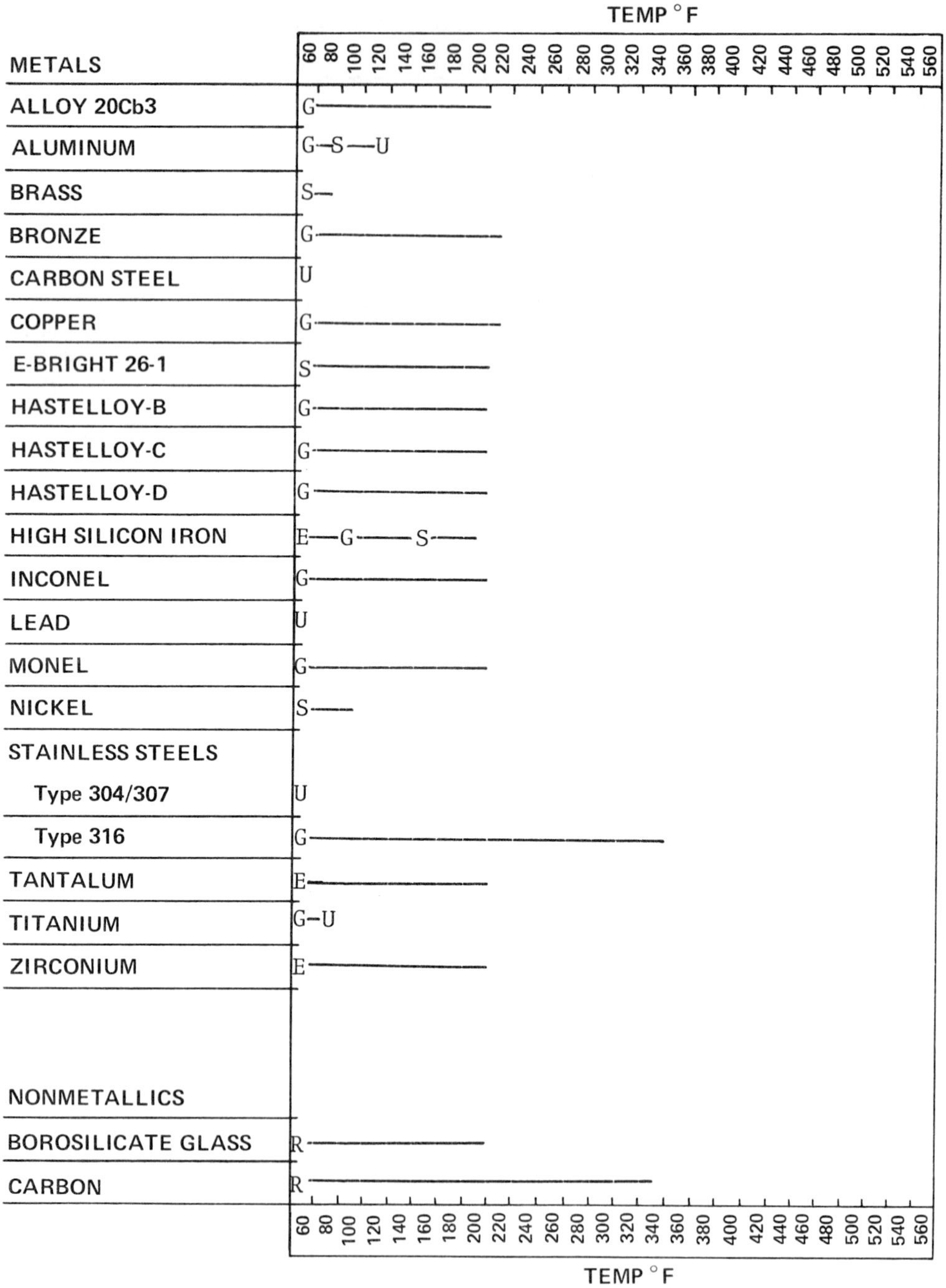

FOR METALS

E = < 2 Mils Penetration/Year; G = < 20 Mils Penetration/Year

S = < 50 Mils Penetration/Year; U = > 50 Mils Penetration/Year

FOR NONMETALLICS

R = Resistant

U = Unsatisfactory

OXALIC ACID, 10%

TEMP °F (60–460)

PLASTICS	Rating
ABS	R
ACRYLICS	R
ASBESTOS REINF., EPOXY	R
ASBESTOS REINF., PHENOLIC	R
CHLORINATED POLYETHER	R
CPVC	R
EPOXY	R
FLUOROCARBONS FEP	R
FLUOROCARBONS TFE	R
FURFURYL ALCOHOL	
NORYL	
PHENOLIC	
POLYAMIDES—NYLON	R
POLYESTERS	
BISPHENOL A-FUMURATE	R
HYDROGENATED BISPHENOL-A-BISPHENOL-A	R
ISOPHTHALIC	R
CHLORINATED POLYESTERS	
POLYSTYRENES	R
POLYETHYLENE	R
POLYPROPYLENE	R
PVC—TYPE 1	R
PVC—TYPE 2	R
POLYVINYLIDENE CHLORIDE	R
VINYLIDENE FLUORIDE	R
VINYL ESTER	R
RUBBERS	
BUTYL GR-1	R
FLUORO ELASTOMERS VITON A	R
FLUORO ELASTOMERS KEL-F 3700	
HARD RUBBER	R
HYPALON	R
KOROSEAL	R
NATURAL RUBBER (GRS)	R
NEOPRENE GR-M (CR)	R
NITRILE BUNA N (NBR)	U
NORDEL	R
POLYBUTADIENE	
SILICONE	
POLYURETHANE	

TEMP °F (60–460)

OXALIC ACID, 5%

TEMP °F (60, 80, 100, 120, 140, 160, 180, 200, 220, 240, 260, 280, 300, 320, 340, 360, 380, 400, 420, 440, 460, 480, 500, 520, 540, 560)

METALS	
ALLOY 20Cb3	G——
ALUMINUM	G–S——U
BRASS	S–
BRONZE	G——
CARBON STEEL	U
COPPER	G——
E-BRIGHT 26-1	
HASTELLOY-B	G——
HASTELLOY-C	G——
HASTELLOY-D	G——
HIGH SILICON IRON	E—G—— S——
INCONEL	G——
LEAD	U
MONEL	G——
NICKEL	S—–
STAINLESS STEELS	
Type 304/307	U
Type 316	G——
TANTALUM	G——
TITANIUM	G–U
ZIRCONIUM	E——

NONMETALLICS	
BOROSILICATE GLASS	R——
CARBON	R——

TEMP °F (60, 80, 100, 120, 140, 160, 180, 200, 220, 240, 260, 280, 300, 320, 340, 360, 380, 400, 420, 440, 460, 480, 500, 520, 540, 560)

FOR METALS

E = < 2 Mils Penetration/Year; G = < 20 Mils Penetration/Year

S = < 50 Mils Penetration/Year; U = > 50 Mils Penetration/Year

FOR NONMETALLICS

R = Resistant

U = Unsatisfactory

OXALIC ACID, 5%

PLASTICS	Rating	Bar extends to (TEMP °F, approx.)
ABS	R	140
ACRYLICS	R	80
ASBESTOS REINF., EPOXY	R	160
ASBESTOS REINF., PHENOLIC	R	160
CHLORINATED POLYETHER	R	150
CPVC	R	140
EPOXY	R	240
FLUOROCARBONS FEP	R	400
FLUOROCARBONS TFE	R	460
FURFURYL ALCOHOL		
NORYL		
PHENOLIC		
POLYAMIDES–NYLON	R	80
POLYESTERS		
BISPHENOL A-FUMURATE	R	220
HYDROGENATED BISPHENOL-A-BISPHENOL-A	R	220
ISOPHTHALIC	R	170
CHLORINATED POLYESTERS		
POLYSTYRENES	R	170
POLYETHYLENE	R	140
POLYPROPYLENE	R	150
PVC–TYPE 1	R	140
PVC–TYPE 2	R	140
POLYVINYLIDENE CHLORIDE	R	180
VINYLIDENE FLUORIDE	R	150
VINYL ESTER	R	210

RUBBERS	Rating	Bar extends to (TEMP °F, approx.)
BUTYL GR-1	R	180
FLUORO ELASTOMERS VITON A	R	180
FLUORO ELASTOMERS KEL-F 3700		
HARD RUBBER	R	180
HYPALON	R	250
KOROSEAL	R	100
NATURAL RUBBER (GRS)	R	150
NEOPRENE GR-M (CR)	R	200
NITRILE BUNA N (NBR)	U	
NORDEL	R	300
POLYBUTADIENE		
SILICONE		
POLYURETHANE		

TEMP °F: 60 80 100 120 140 160 180 200 220 240 260 280 300 320 340 360 380 400 420 440 460

OXYGEN

TEMP °F: 60, 80, 100, 120, 140, 160, 180, 200, 220, 240, 260, 280, 300, 320, 340, 360, 380, 400, 420, 440, 460, 480, 500, 520, 540, 560

METALS	
ALLOY 20Cb3	G——
ALUMINUM	G–
BRASS	G–
BRONZE	G————————————————
CARBON STEEL	G——
COPPER	G–
E-BRIGHT 26-1	
HASTELLOY-B	
HASTELLOY-C	
HASTELLOY-D	
HIGH SILICON IRON	
INCONEL	
LEAD	
MONEL	G–
NICKEL	
STAINLESS STEELS	
Type 304/307	G–
Type 316	G————————————————
TANTALUM	E——————————————————————————U
TITANIUM	
ZIRCONIUM	

NONMETALLICS	
BOROSILICATE GLASS	
CARBON	R————————————————

TEMP °F: 60, 80, 100, 120, 140, 160, 180, 200, 220, 240, 260, 280, 300, 320, 340, 360, 380, 400, 420, 440, 460, 480, 500, 520, 540, 560

FOR METALS

E = < 2 Mils Penetration/Year; G = < 20 Mils Penetration/Year

S = < 50 Mils Penetration/Year; U = > 50 Mils Penetration/Year

FOR NONMETALLICS

R = Resistant

U = Unsatisfactory

OXYGEN

TEMP °F (60, 80, 100, 120, 140, 160, 180, 200, 220, 240, 260, 280, 300, 320, 340, 360, 380, 400, 420, 440, 460)

PLASTICS	OXYGEN
ABS	R——
ACRYLICS	
ASBESTOS REINF., EPOXY	
ASBESTOS REINF., PHENOLIC	
CHLORINATED POLYETHER	R——
CPVC	R——
EPOXY	
FLUOROCARBONS FEP	R——
FLUOROCARBONS TFE	R——
FURFURYL ALCOHOL	
NORYL	
PHENOLIC	
POLYAMIDES–NYLON	
POLYESTERS	
BISPHENOL A-FUMURATE	
HYDROGENATED BISPHENOL-A-BISPHENOL-A	
ISOPHTHALIC	
CHLORINATED POLYESTERS	
POLYSTYRENES	
POLYETHYLENE	RU
POLYPROPYLENE	R——
PVC–TYPE 1	R——
PVC–TYPE 2	R——
POLYVINYLIDENE CHLORIDE	R——
VINYLIDENE FLUORIDE	R——
VINYL ESTER	

RUBBERS	OXYGEN
BUTYL GR-1	R——
FLUORO ELASTOMERS VITON A	R——
FLUORO ELASTOMERS KEL-F 3700	
HARD RUBBER	
HYPALON	
KOROSEAL	
NATURAL RUBBER (GRS)	U
NEOPRENE GR-M (CR)	R——
NITRILE BUNA N (NBR)	R——
NORDEL	R——
POLYBUTADIENE	
SILICONE	
POLYURETHANE	

TEMP °F (60, 80, 100, 120, 140, 160, 180, 200, 220, 240, 260, 280, 300, 320, 340, 360, 380, 400, 420, 440, 460)

OXYGEN GAS

TEMP °F

METALS	60	80	100	120	140	160	180	200	220	240	260	280	300	320	340	360	380	400	420	440	460	480	500	520	540	560
ALLOY 20Cb3																										
ALUMINUM																										
BRASS																										
BRONZE																										
CARBON STEEL																										
COPPER																										
E-BRIGHT 26-1																										
HASTELLOY-B																										
HASTELLOY-C																										
HASTELLOY-D																										
HIGH SILICON IRON																										
INCONEL																										
LEAD																										
MONEL																										
NICKEL																										
STAINLESS STEELS Type 304/307																										
Type 316																										
TANTALUM																										
TITANIUM																										
ZIRCONIUM																										
NONMETALLICS																										
BOROSILICATE GLASS																										
CARBON																										

60 80 100 120 140 160 180 200 220 240 260 280 300 320 340 360 380 400 420 440 460 480 500 520 540 560

TEMP °F

FOR METALS

E = < 2 Mils Penetration/Year; G = < 20 Mils Penetration/Year
S = < 50 Mils Penetration/Year; U = > 50 Mils Penetration/Year

FOR NONMETALLICS

R = Resistant
U = Unsatisfactory

OXYGEN GAS

TEMP °F (60, 80, 100, 120, 140, 160, 180, 200, 220, 240, 260, 280, 300, 320, 340, 360, 380, 400, 420, 440, 460)

PLASTICS	
ABS	
ACRYLICS	
ASBESTOS REINF., EPOXY	
ASBESTOS REINF., PHENOLIC	
CHLORINATED POLYETHER	R
CPVC	R
EPOXY	
FLUOROCARBONS FEP	R
FLUOROCARBONS TFE	R
FURFURYL ALCOHOL	
NORYL	
PHENOLIC	
POLYAMIDES–NYLON	
POLYESTERS	
BISPHENOL A-FUMURATE	
HYDROGENATED BISPHENOL-A-BISPHENOL-A	
ISOPHTHALIC	
CHLORINATED POLYESTERS	
POLYSTYRENES	
POLYETHYLENE	
POLYPROPYLENE	R
PVC–TYPE 1	R
PVC–TYPE 2	
POLYVINYLIDENE CHLORIDE	
VINYLIDENE FLUORIDE	R
VINYL ESTER	

RUBBERS	
BUTYL GR-1	
FLUORO ELASTOMERS VITON A	R
FLUORO ELASTOMERS KEL-F 3700	
HARD RUBBER	
HYPALON	
KOROSEAL	
NATURAL RUBBER (GRS)	
NEOPRENE GR-M (CR)	
NITRILE BUNA N (NBR)	RU
NORDEL	
POLYBUTADIENE	
SILICONE	
POLYURETHANE	

TEMP °F (60, 80, 100, 120, 140, 160, 180, 200, 220, 240, 260, 280, 300, 320, 340, 360, 380, 400, 420, 440, 460)

OZONE

TEMP °F: 60 80 100 120 140 160 180 200 220 240 260 280 300 320 340 360 380 400 420 440 460 480 500 520 540 560

METALS	
ALLOY 20Cb3	G
ALUMINUM	G
BRASS	
BRONZE	G
CARBON STEEL	S
COPPER	
E-BRIGHT 26-1	
HASTELLOY-B	
HASTELLOY-C	
HASTELLOY-D	
HIGH SILICON IRON	
INCONEL	
LEAD	
MONEL	G
NICKEL	
STAINLESS STEELS	
Type 304/307	G
Type 316	G
TANTALUM	
TITANIUM	
ZIRCONIUM	

NONMETALLICS	
BOROSILICATE GLASS	
CARBON	

TEMP °F: 60 80 100 120 140 160 180 200 220 240 260 280 300 320 340 360 380 400 420 440 460 480 500 520 540 560

FOR METALS

E = < 2 Mils Penetration/Year; G = < 20 Mils Penetration/Year

S = < 50 Mils Penetration/Year; U = > 50 Mils Penetration/Year

FOR NONMETALLICS

R = Resistant

U = Unsatisfactory

OZONE

PLASTICS	TEMP °F (60–460)
ABS	U
ACRYLICS	
ASBESTOS REINF., EPOXY	
ASBESTOS REINF., PHENOLIC	
CHLORINATED POLYETHER	R
CPVC	R
EPOXY	
FLUOROCARBONS FEP	R
FLUOROCARBONS TFE	R
FURFURYL ALCOHOL	
NORYL	
PHENOLIC	
POLYAMIDES–NYLON	U
POLYESTERS	
BISPHENOL A-FUMURATE	
HYDROGENATED BISPHENOL-A-BISPHENOL-A	
ISOPHTHALIC	
CHLORINATED POLYESTERS	
POLYSTYRENES	
POLYETHYLENE	U
POLYPROPYLENE	U
PVC–TYPE 1	R
PVC–TYPE 2	R
POLYVINYLIDENE CHLORIDE	R
VINYLIDENE FLUORIDE	R
VINYL ESTER	R
RUBBERS	
BUTYL GR-1	R
FLUORO ELASTOMERS VITON A	R
FLUORO ELASTOMERS KEL-F 3700	
HARD RUBBER	
HYPALON	
KOROSEAL	
NATURAL RUBBER (GRS)	U
NEOPRENE GR-M (CR)	R
NITRILE BUNA N (NBR)	U
NORDEL	
POLYBUTADIENE	
SILICONE	
POLYURETHANE	

PALMITIC ACID

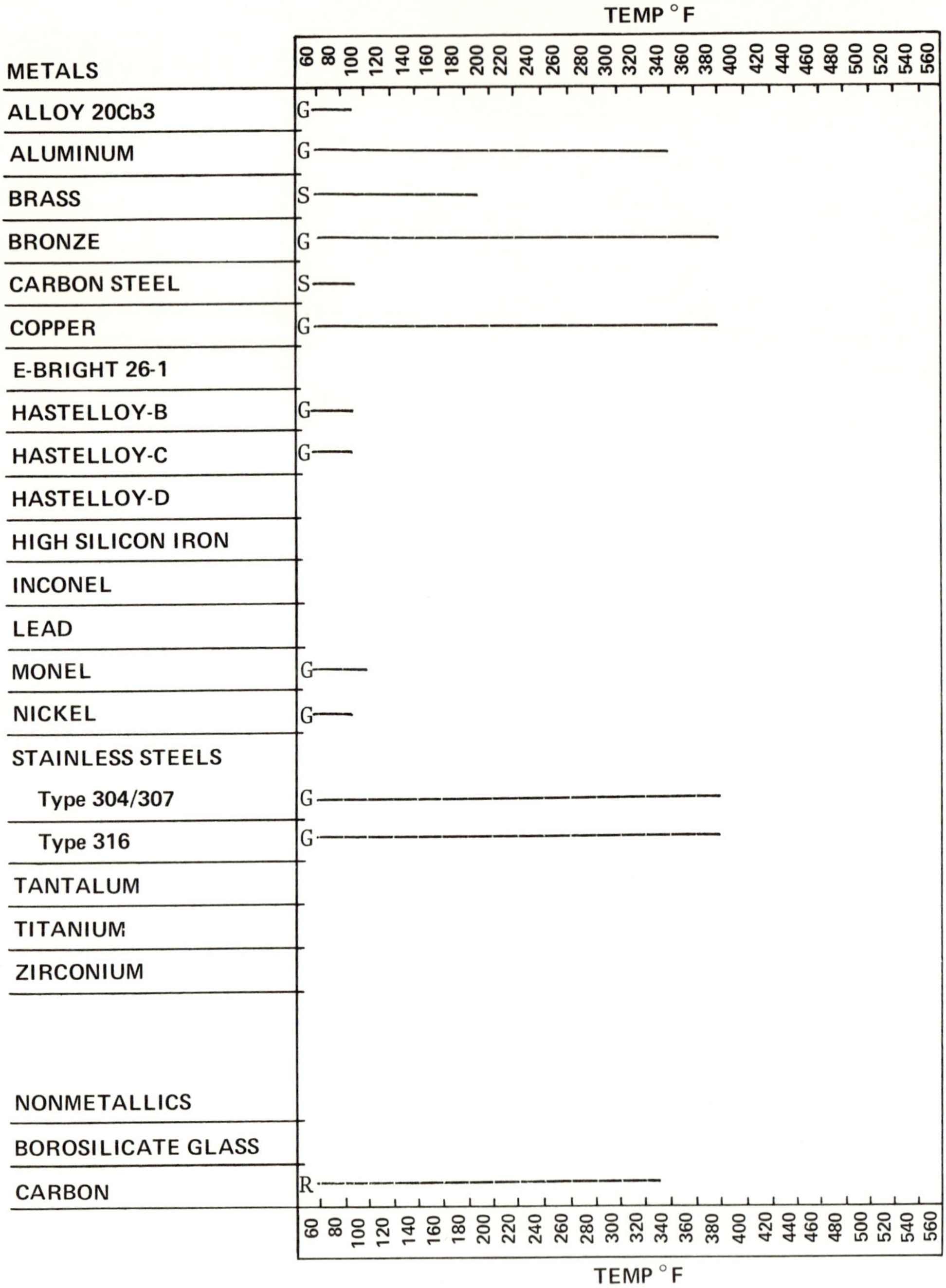

FOR METALS

E = < 2 Mils Penetration/Year; G = < 20 Mils Penetration/Year

S = < 50 Mils Penetration/Year; U = > 50 Mils Penetration/Year

FOR NONMETALLICS

R = Resistant

U = Unsatisfactory

Please see page 1 for notes on calibration.

PALMITIC ACID

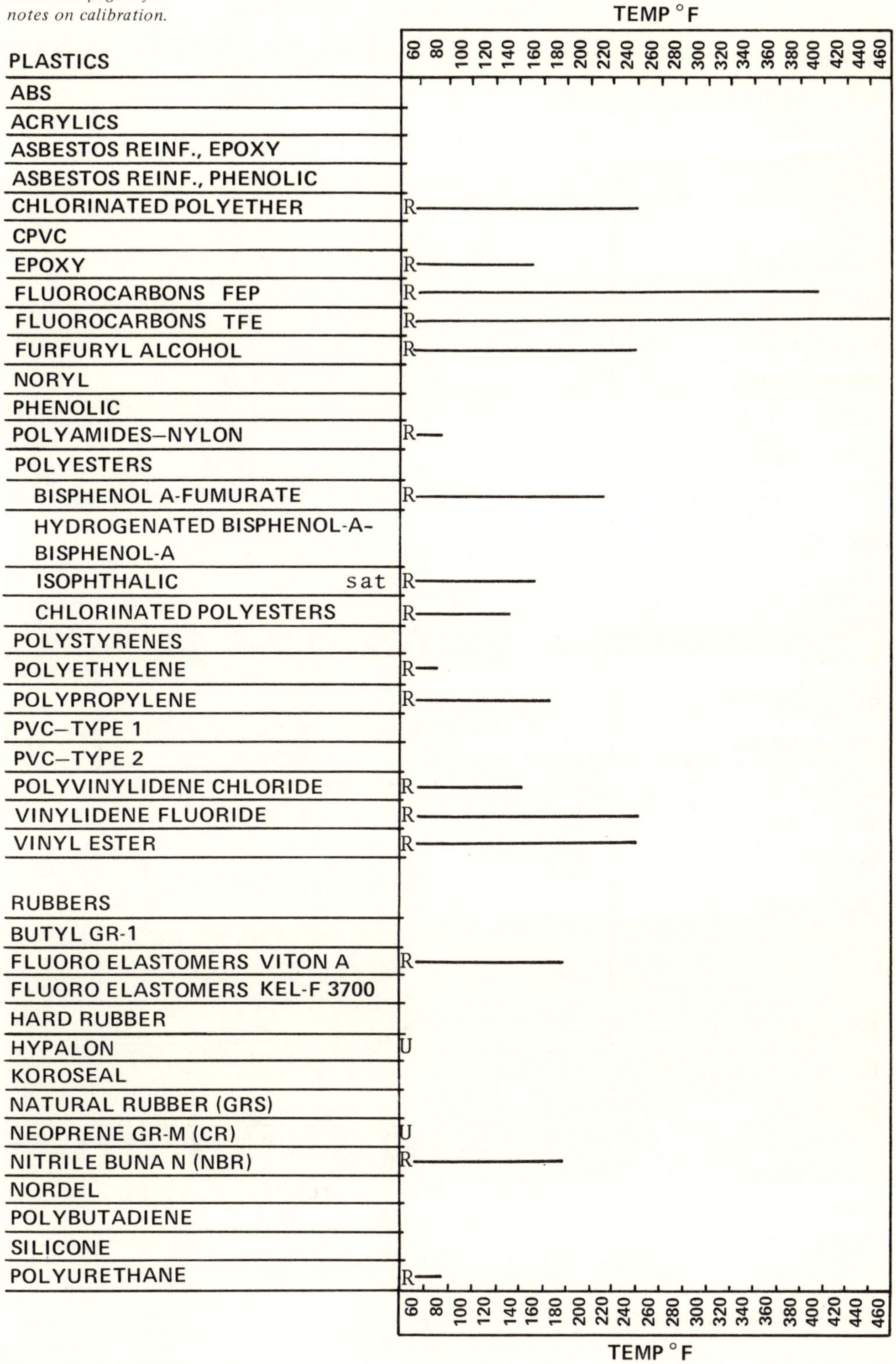

PALMITIC ACID, 70%

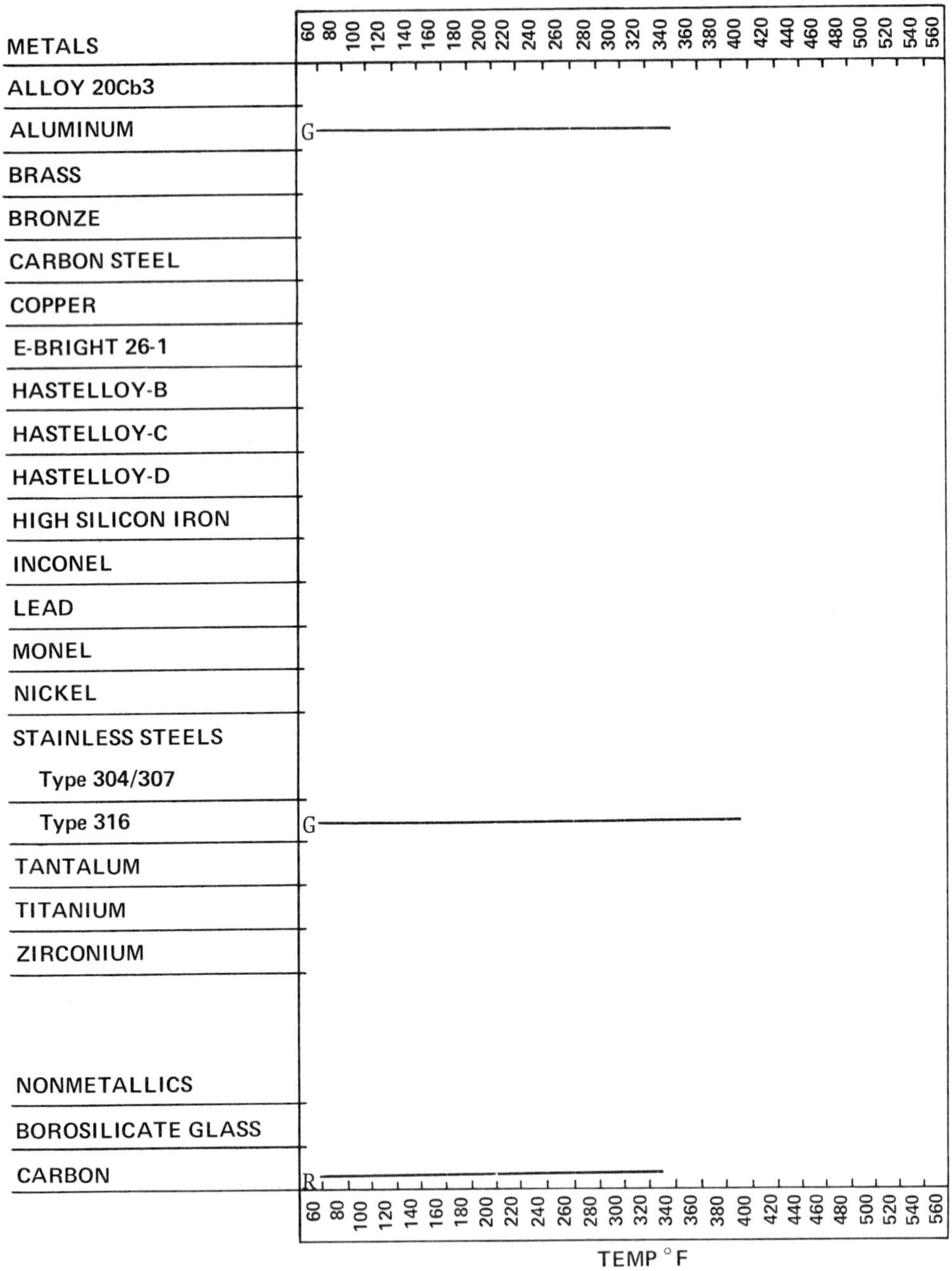

FOR METALS

E = < 2 Mils Penetration/Year; G = < 20 Mils Penetration/Year

S = < 50 Mils Penetration/Year; U = > 50 Mils Penetration/Year

FOR NONMETALLICS

R = Resistant

U = Unsatisfactory

PALMITIC ACID, 70%

TEMP °F (60–460)

PLASTICS	
ABS	U
ACRYLICS	
ASBESTOS REINF., EPOXY	
ASBESTOS REINF., PHENOLIC	
CHLORINATED POLYETHER	R
CPVC	U
EPOXY	R
FLUOROCARBONS FEP	R
FLUOROCARBONS TFE	R
FURFURYL ALCOHOL	R
NORYL	
PHENOLIC	
POLYAMIDES–NYLON	R
POLYESTERS	
BISPHENOL A-FUMURATE	R
HYDROGENATED BISPHENOL-A-BISPHENOL-A	
ISOPHTHALIC	R
CHLORINATED POLYESTERS	
POLYSTYRENES	
POLYETHYLENE	U
POLYPROPYLENE	R
PVC–TYPE 1	RU
PVC–TYPE 2	U
POLYVINYLIDENE CHLORIDE	
VINYLIDENE FLUORIDE	R
VINYL ESTER	R

RUBBERS	
BUTYL GR-1	
FLUORO ELASTOMERS VITON A	R
FLUORO ELASTOMERS KEL-F 3700	
HARD RUBBER	
HYPALON	U
KOROSEAL	
NATURAL RUBBER (GRS)	
NEOPRENE GR-M (CR)	U
NITRILE BUNA N (NBR)	R
NORDEL	
POLYBUTADIENE	
SILICONE	
POLYURETHANE	R

TEMP °F (60–460)

PALMITIC ACID, 10%

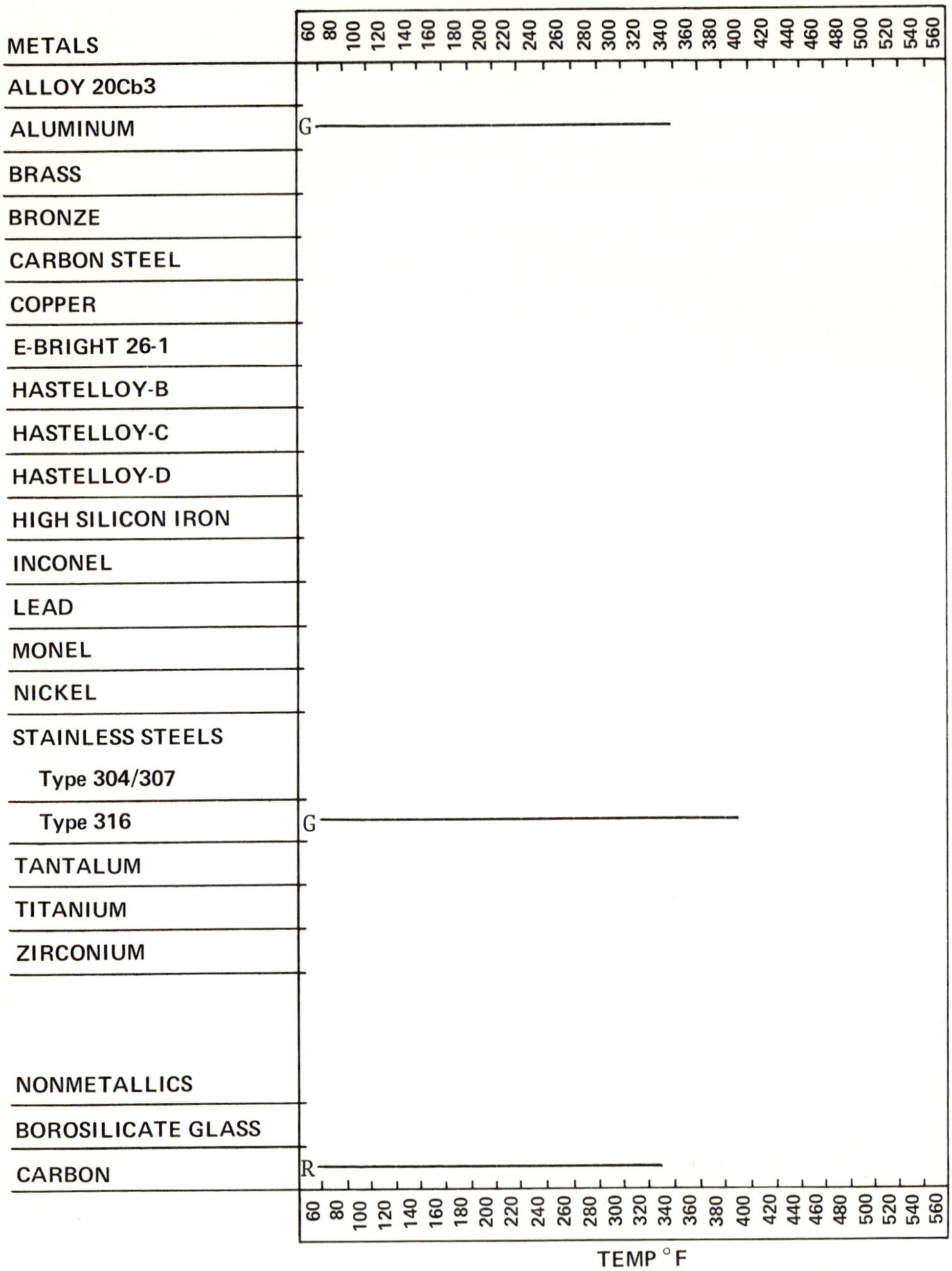

FOR METALS

E = < 2 Mils Penetration/Year; G = < 20 Mils Penetration/Year

S = < 50 Mils Penetration/Year; U = > 50 Mils Penetration/Year

FOR NONMETALLICS

R = Resistant

U = Unsatisfactory

PALMITIC ACID, 10%

PLASTICS	
ABS	R
ACRYLICS	
ASBESTOS REINF., EPOXY	
ASBESTOS REINF., PHENOLIC	
CHLORINATED POLYETHER	R
CPVC	RU
EPOXY	R
FLUOROCARBONS FEP	R
FLUOROCARBONS TFE	R
FURFURYL ALCOHOL	R
NORYL	
PHENOLIC	
POLYAMIDES—NYLON	R
POLYESTERS	
BISPHENOL A-FUMURATE	R
HYDROGENATED BISPHENOL-A-BISPHENOL-A	
ISOPHTHALIC	R
CHLORINATED POLYESTERS	
POLYSTYRENES	
POLYETHYLENE	U
POLYPROPYLENE	R
PVC—TYPE 1	R
PVC—TYPE 2	R
POLYVINYLIDENE CHLORIDE	
VINYLIDENE FLUORIDE	R
VINYL ESTER	R

RUBBERS	
BUTYL GR-1	
FLUORO ELASTOMERS VITON A	R
FLUORO ELASTOMERS KEL-F 3700	
HARD RUBBER	
HYPALON	U
KOROSEAL	
NATURAL RUBBER (GRS)	
NEOPRENE GR-M (CR)	U
NITRILE BUNA N (NBR)	R
NORDEL	
POLYBUTADIENE	
SILICONE	
POLYURETHANE	R

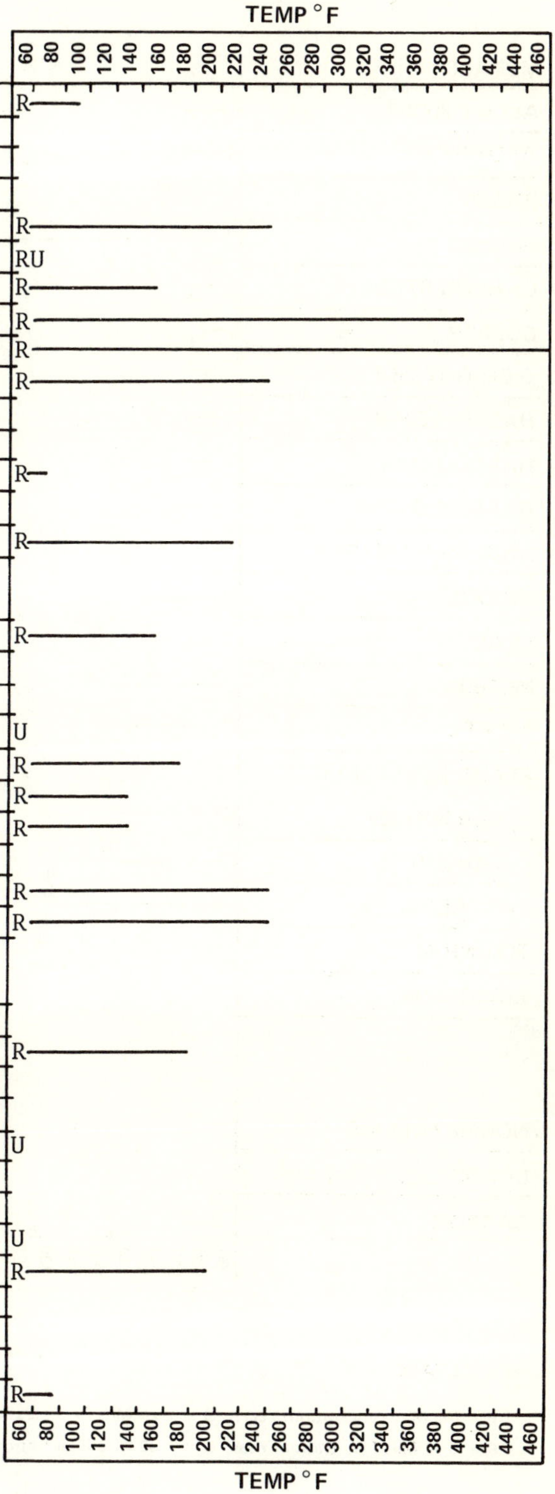

PARAFFIN

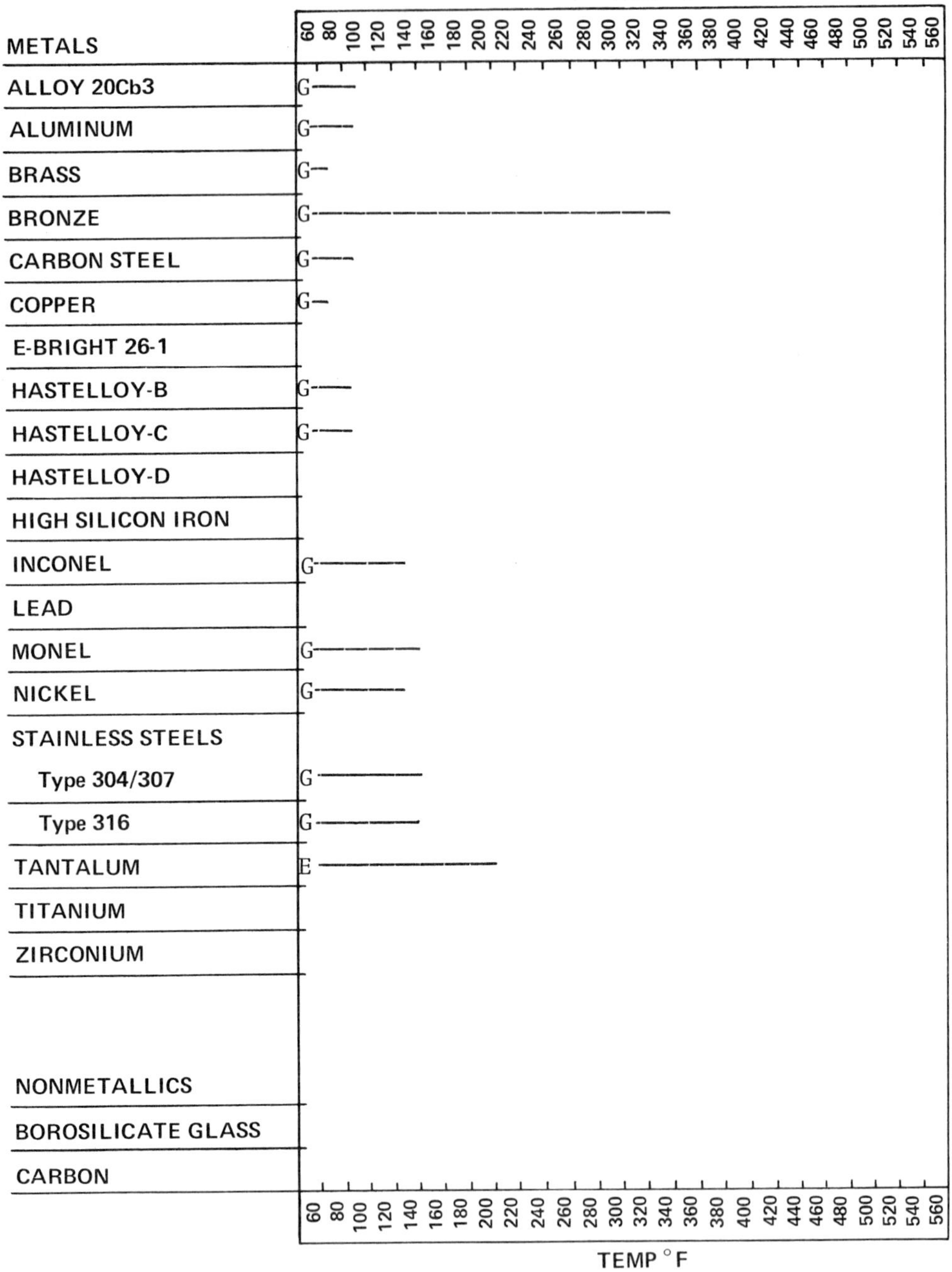

FOR METALS

E = < 2 Mils Penetration/Year; G = < 20 Mils Penetration/Year

S = < 50 Mils Penetration/Year; U = > 50 Mils Penetration/Year

FOR NONMETALLICS

R = Resistant

U = Unsatisfactory

PARAFFIN

PLASTICS	Rating	TEMP °F (to)
ABS	R	80
ACRYLICS		
ASBESTOS REINF., EPOXY		
ASBESTOS REINF., PHENOLIC		
CHLORINATED POLYETHER		
CPVC	R	180
EPOXY		
FLUOROCARBONS FEP	R	400
FLUOROCARBONS TFE	R	460
FURFURYL ALCOHOL	R	220
NORYL		
PHENOLIC		
POLYAMIDES–NYLON		
POLYESTERS		
BISPHENOL A-FUMURATE	R	80
HYDROGENATED BISPHENOL-A-BISPHENOL-A		
ISOPHTHALIC		
CHLORINATED POLYESTERS		
POLYSTYRENES		
POLYETHYLENE	U	
POLYPROPYLENE	R	140
PVC–TYPE 1	R	140
PVC–TYPE 2	R	140
POLYVINYLIDENE CHLORIDE		
VINYLIDENE FLUORIDE		
VINYL ESTER		

RUBBERS	Rating	TEMP °F (to)
BUTYL GR-1		
FLUORO ELASTOMERS VITON A	R	180
FLUORO ELASTOMERS KEL-F 3700		
HARD RUBBER		
HYPALON	R	80
KOROSEAL		
NATURAL RUBBER (GRS)		
NEOPRENE GR-M (CR)	R	80
NITRILE BUNA N (NBR)	R	200
NORDEL		
POLYBUTADIENE		
SILICONE		
POLYURETHANE		

TEMP °F: 60 80 100 120 140 160 180 200 220 240 260 280 300 320 340 360 380 400 420 440 460

PERACETIC ACID, 40%

TEMP °F

METALS	60	80	100	120	140	160	180	200	220	240	260	280	300	320	340	360	380	400	420	440	460	480	500	520	540	560
ALLOY 20Cb3																										
ALUMINUM	G–																									
BRASS																										
BRONZE																										
CARBON STEEL																										
COPPER																										
E-BRIGHT 26-1																										
HASTELLOY-B																										
HASTELLOY-C																										
HASTELLOY-D																										
HIGH SILICON IRON																										
INCONEL																										
LEAD																										
MONEL																										
NICKEL																										
STAINLESS STEELS Type 304/307																										
Type 316																										
TANTALUM	G–																									
TITANIUM																										
ZIRCONIUM																										
NONMETALLICS																										
BOROSILICATE GLASS																										
CARBON																										

TEMP °F

FOR METALS

E = < 2 Mils Penetration/Year; G = < 20 Mils Penetration/Year

S = < 50 Mils Penetration/Year; U = > 50 Mils Penetration/Year

FOR NONMETALLICS

R = Resistant

U = Unsatisfactory

PERACETIC ACID, 40%

TEMP °F

PLASTICS	60	80	100	120	140	160	180	200	220	240	260	280	300	320	340	360	380	400	420	440	460
ABS	U																				
ACRYLICS																					
ASBESTOS REINF., EPOXY																					
ASBESTOS REINF., PHENOLIC																					
CHLORINATED POLYETHER																					
CPVC	U																				
EPOXY																					
FLUOROCARBONS FEP	R	—	—	—	—	—	—	—	—	—	—	—	—	—	—	—	—	—			
FLUOROCARBONS TFE	R	—	—	—	—	—	—	—	—	—	—	—	—	—	—	—	—	—	—	—	—
FURFURYL ALCOHOL																					
NORYL																					
PHENOLIC																					
POLYAMIDES–NYLON																					
POLYESTERS																					
BISPHENOL A-FUMURATE																					
HYDROGENATED BISPHENOL-A-BISPHENOL-A																					
ISOPHTHALIC																					
CHLORINATED POLYESTERS																					
POLYSTYRENES																					
POLYETHYLENE																					
POLYPROPYLENE	U																				
PVC–TYPE 1	R	—	—	—	U																
PVC–TYPE 2	U																				
POLYVINYLIDENE CHLORIDE																					
VINYLIDENE FLUORIDE																					
VINYL ESTER																					
RUBBERS																					
BUTYL GR-1																					
FLUORO ELASTOMERS VITON A																					
FLUORO ELASTOMERS KEL-F 3700																					
HARD RUBBER																					
HYPALON																					
KOROSEAL																					
NATURAL RUBBER (GRS)																					
NEOPRENE GR-M (CR)																					
NITRILE BUNA N (NBR)																					
NORDEL																					
POLYBUTADIENE																					
SILICONE																					
POLYURETHANE																					

TEMP °F

PERCHLORETHYLENE

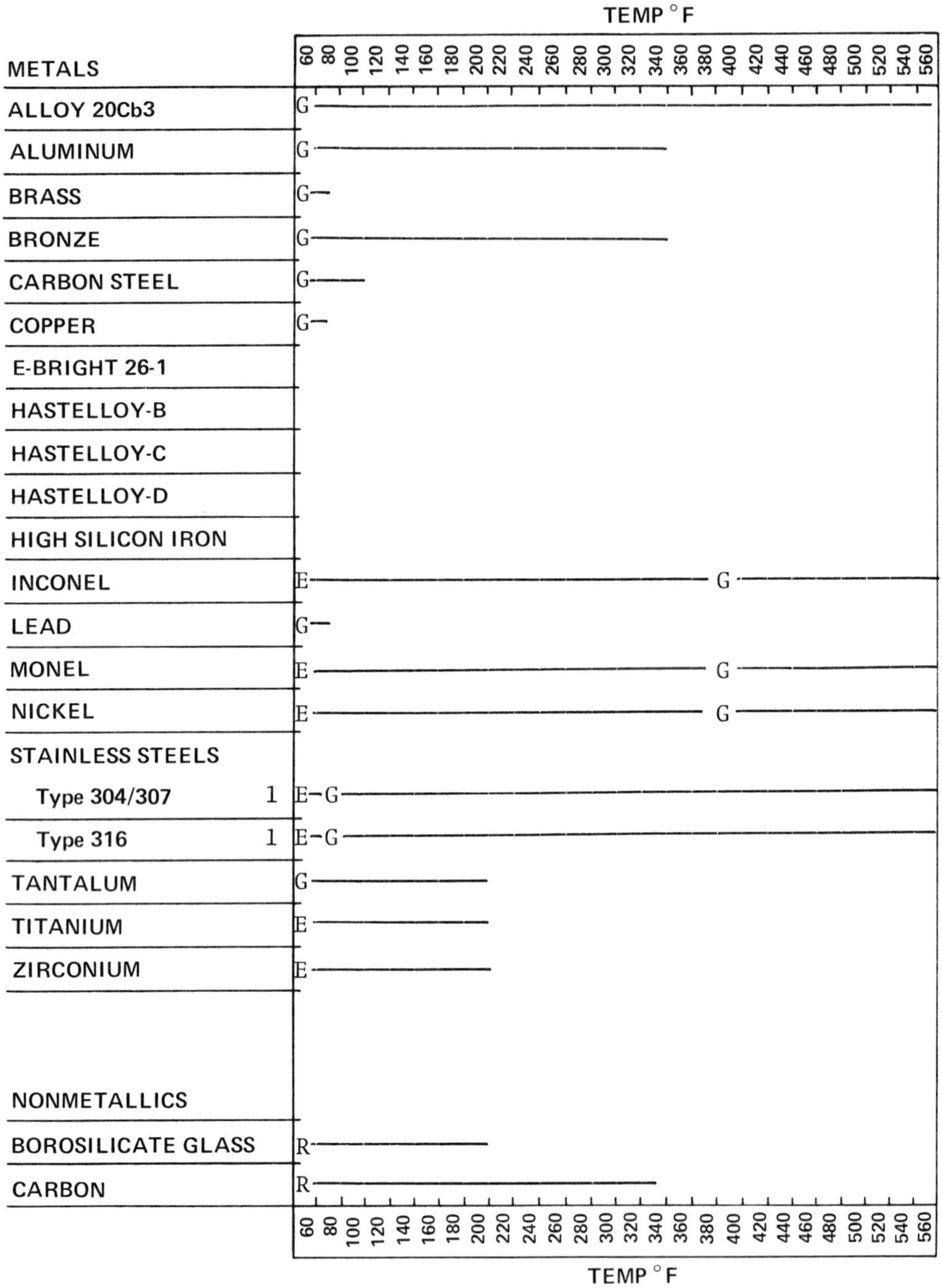

FOR METALS

E = < 2 Mils Penetration/Year; G = < 20 Mils Penetration/Year

S = < 50 Mils Penetration/Year; U = > 50 Mils Penetration/Year

FOR NONMETALLICS

R = Resistant

U = Unsatisfactory

PERCHLORETHYLENE

TEMP °F (60, 80, 100, 120, 140, 160, 180, 200, 220, 240, 260, 280, 300, 320, 340, 360, 380, 400, 420, 440, 460)

PLASTICS	PERCHLORETHYLENE
ABS	U
ACRYLICS	
ASBESTOS REINF., EPOXY	
ASBESTOS REINF., PHENOLIC	
CHLORINATED POLYETHER	
CPVC	
EPOXY	R——
FLUOROCARBONS FEP	R——
FLUOROCARBONS TFE	R——
FURFURYL ALCOHOL	R——
NORYL	
PHENOLIC	
POLYAMIDES—NYLON	U
POLYESTERS	
BISPHENOL A-FUMURATE	R—
HYDROGENATED BISPHENOL-A-BISPHENOL-A	R–U
ISOPHTHALIC	RU
CHLORINATED POLYESTERS	R——U
POLYSTYRENES	
POLYETHYLENE	U
POLYPROPYLENE	U
PVC—TYPE 1	
PVC—TYPE 2	
POLYVINYLIDENE CHLORIDE	U
VINYLIDENE FLUORIDE	R——
VINYL ESTER	R——
RUBBERS	
BUTYL GR-1	U
FLUORO ELASTOMERS VITON A	R——
FLUORO ELASTOMERS KEL-F 3700	
HARD RUBBER	
HYPALON	U
KOROSEAL	
NATURAL RUBBER (GRS)	U
NEOPRENE GR-M (CR)	U
NITRILE BUNA N (NBR)	U
NORDEL	U
POLYBUTADIENE	
SILICONE	
POLYURETHANE	U

TEMP °F (60, 80, 100, 120, 140, 160, 180, 200, 220, 240, 260, 280, 300, 320, 340, 360, 380, 400, 420, 440, 460)

PERCHLORIC ACID, 70%

TEMP °F (60 – 560)

METALS	Rating
ALLOY 20Cb3	
ALUMINUM	U
BRASS	U
BRONZE	G–
CARBON STEEL	U
COPPER	
E-BRIGHT 26-1	
HASTELLOY-B	
HASTELLOY-C	
HASTELLOY-D	
HIGH SILICON IRON	G–
INCONEL	
LEAD	
MONEL	U
NICKEL	
STAINLESS STEELS	
Type 304/307	U
Type 316	U
TANTALUM	E ——
TITANIUM	U
ZIRCONIUM	

NONMETALLICS	Rating
BOROSILICATE GLASS	R ——
CARBON	R ——

TEMP °F

FOR METALS

E = < 2 Mils Penetration/Year; G = < 20 Mils Penetration/Year

S = < 50 Mils Penetration/Year; U = > 50 Mils Penetration/Year

FOR NONMETALLICS

R = Resistant

U = Unsatisfactory

PERCHLORIC ACID, 70%

PLASTICS	Rating	TEMP °F (60–460)
ABS	U	
ACRYLICS		
ASBESTOS REINF., EPOXY		
ASBESTOS REINF., PHENOLIC		
CHLORINATED POLYETHER	U	
CPVC	R	60–180
EPOXY	R–U	
FLUOROCARBONS FEP	R	60–400
FLUOROCARBONS TFE	R	60–460
FURFURYL ALCOHOL	R	60–260
NORYL		
PHENOLIC		
POLYAMIDES—NYLON	U	
POLYESTERS		
BISPHENOL A-FUMURATE		
HYDROGENATED BISPHENOL-A-BISPHENOL-A	U	
ISOPHTHALIC	U	
CHLORINATED POLYESTERS	R	60–80
POLYSTYRENES		
POLYETHYLENE	U	
POLYPROPYLENE	U	
PVC—TYPE 1	RU	
PVC—TYPE 2	RU	
POLYVINYLIDENE CHLORIDE	R	60–120
VINYLIDENE FLUORIDE	R	60–120
VINYL ESTER		

RUBBERS	Rating	TEMP °F (60–460)
BUTYL GR-1		
FLUORO ELASTOMERS VITON A	R	60–180
FLUORO ELASTOMERS KEL-F 3700		
HARD RUBBER		
HYPALON	R	60–80
KOROSEAL	U	
NATURAL RUBBER (GRS)		
NEOPRENE GR-M (CR)	U	
NITRILE BUNA N (NBR)	U	
NORDEL		
POLYBUTADIENE		
SILICONE		
POLYURETHANE	U	

PERCHLORIC ACID, 10%

METALS	Rating	Temp. range (°F, read from chart)
ALLOY 20Cb3		
ALUMINUM	U	60
BRASS	U	60
BRONZE		
CARBON STEEL	U	60
COPPER		
E-BRIGHT 26-1		
HASTELLOY-B		
HASTELLOY-C		
HASTELLOY-D		
HIGH SILICON IRON	G–	60
INCONEL		
LEAD		
MONEL	U	60
NICKEL	U	60
STAINLESS STEELS		
Type 304/307	U	60
Type 316	U	60
TANTALUM	E	60–300
TITANIUM	U	60
ZIRCONIUM		

NONMETALLICS	Rating	Temp. range (°F, read from chart)
BOROSILICATE GLASS	R	60–200
CARBON	R	60–340

TEMP °F: 60 80 100 120 140 160 180 200 220 240 260 280 300 320 340 360 380 400 420 440 460 480 500 520 540 560

FOR METALS

E = < 2 Mils Penetration/Year; G = < 20 Mils Penetration/Year

S = < 50 Mils Penetration/Year; U = > 50 Mils Penetration/Year

FOR NONMETALLICS

R = Resistant

U = Unsatisfactory

PERCHLORIC ACID, 10%

TEMP °F (60–460)

PLASTICS	Rating
ABS	U
ACRYLICS	
ASBESTOS REINF., EPOXY	
ASBESTOS REINF., PHENOLIC	
CHLORINATED POLYETHER	R
CPVC	R
EPOXY	R
FLUOROCARBONS FEP	R
FLUOROCARBONS TFE	R
FURFURYL ALCOHOL	
NORYL	
PHENOLIC	
POLYAMIDES–NYLON	U
POLYESTERS	
BISPHENOL A-FUMURATE	
HYDROGENATED BISPHENOL-A-BISPHENOL-A	U
ISOPHTHALIC	U
CHLORINATED POLYESTERS	R
POLYSTYRENES	
POLYETHYLENE	RU
POLYPROPYLENE	R
PVC–TYPE 1	R
PVC–TYPE 2	RU
POLYVINYLIDENE CHLORIDE	R
VINYLIDENE FLUORIDE	R
VINYL ESTER	R
RUBBERS	
BUTYL GR-1	R
FLUORO ELASTOMERS VITON A	R
FLUORO ELASTOMERS KEL-F 3700	
HARD RUBBER	
HYPALON	R
KOROSEAL	U
NATURAL RUBBER (GRS)	R
NEOPRENE GR-M (CR)	U
NITRILE BUNA N (NBR)	U
NORDEL	
POLYBUTADIENE	
SILICONE	
POLYURETHANE	U

TEMP °F

PETROLATUM

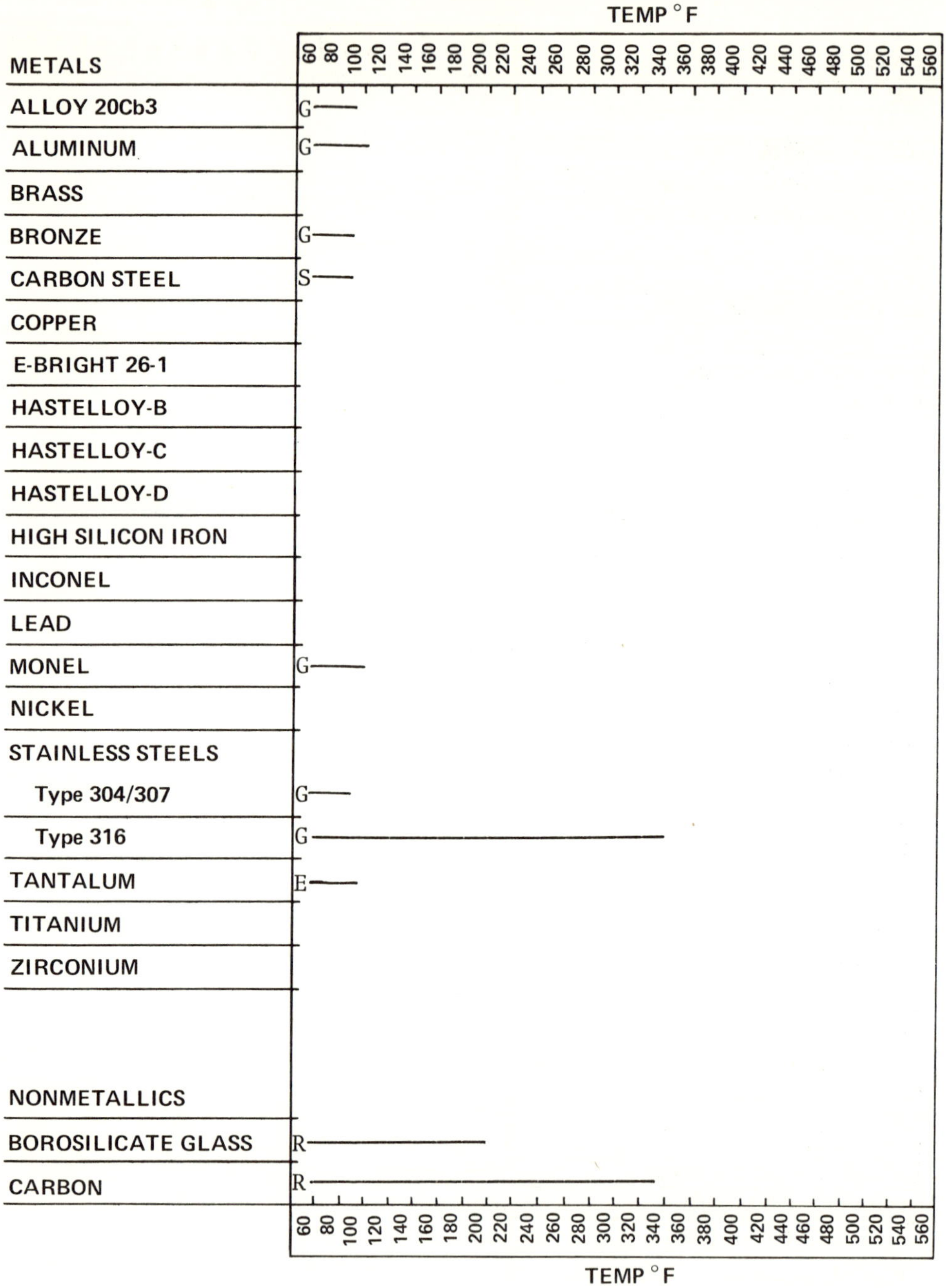

FOR METALS

E = < 2 Mils Penetration/Year; G = < 20 Mils Penetration/Year

S = < 50 Mils Penetration/Year; U = > 50 Mils Penetration/Year

FOR NONMETALLICS

R = Resistant

U = Unsatisfactory

PETROLATUM

TEMP °F (60, 80, 100, 120, 140, 160, 180, 200, 220, 240, 260, 280, 300, 320, 340, 360, 380, 400, 420, 440, 460)

PLASTICS	TEMP °F
ABS	R 60–140
ACRYLICS	
ASBESTOS REINF., EPOXY	
ASBESTOS REINF., PHENOLIC	
CHLORINATED POLYETHER	R 60–250
CPVC	
EPOXY	
FLUOROCARBONS FEP	R 60–400
FLUOROCARBONS TFE	R 60–460+
FURFURYL ALCOHOL	
NORYL	
PHENOLIC	
POLYAMIDES—NYLON	
POLYESTERS	
BISPHENOL A-FUMURATE	
HYDROGENATED BISPHENOL-A-BISPHENOL-A	
ISOPHTHALIC	
CHLORINATED POLYESTERS	
POLYSTYRENES	
POLYETHYLENE	
POLYPROPYLENE	R 60–180
PVC—TYPE 1	
PVC—TYPE 2	
POLYVINYLIDENE CHLORIDE	R 60–130
VINYLIDENE FLUORIDE	R 60–280
VINYL ESTER	
RUBBERS	
BUTYL GR-1	
FLUORO ELASTOMERS VITON A	
FLUORO ELASTOMERS KEL-F 3700	
HARD RUBBER	
HYPALON	
KOROSEAL	
NATURAL RUBBER (GRS)	
NEOPRENE GR-M (CR)	R 60–200
NITRILE BUNA N (NBR)	
NORDEL	
POLYBUTADIENE	
SILICONE	
POLYURETHANE	

TEMP °F (60, 80, 100, 120, 140, 160, 180, 200, 220, 240, 260, 280, 300, 320, 340, 360, 380, 400, 420, 440, 460)

PETROLEUM OILS, REFINED

TEMP °F: 60 80 100 120 140 160 180 200 220 240 260 280 300 320 340 360 380 400 420 440 460 480 500 520 540 560

METALS	Rating (TEMP °F)
ALLOY 20Cb3	
ALUMINUM	G— (from 60)
BRASS	G— (from 60)
BRONZE	G— (from 60)
CARBON STEEL	
COPPER	G— (from 60)
E-BRIGHT 26-1	
HASTELLOY-B	
HASTELLOY-C	
HASTELLOY-D	
HIGH SILICON IRON	
INCONEL	
LEAD	
MONEL	G— (from 60)
NICKEL	
STAINLESS STEELS	
Type 304/307	G— (from 60)
Type 316	G— (from 60)
TANTALUM	
TITANIUM	
ZIRCONIUM	

NONMETALLICS	Rating (TEMP °F)
BOROSILICATE GLASS	
CARBON	R (60 to 340)

TEMP °F

FOR METALS

E = < 2 Mils Penetration/Year; G = < 20 Mils Penetration/Year

S = < 50 Mils Penetration/Year; U = > 50 Mils Penetration/Year

FOR NONMETALLICS

R = Resistant

U = Unsatisfactory

PETROLEUM OILS, REFINED

TEMP °F (60–460)

PLASTICS	Rating
ABS	
ACRYLICS	
ASBESTOS REINF., EPOXY	
ASBESTOS REINF., PHENOLIC	
CHLORINATED POLYETHER	
CPVC	
EPOXY	
FLUOROCARBONS FEP	R
FLUOROCARBONS TFE	R
FURFURYL ALCOHOL	
NORYL	
PHENOLIC	
POLYAMIDES—NYLON	R
POLYESTERS	
BISPHENOL A-FUMURATE	
HYDROGENATED BISPHENOL-A-BISPHENOL-A	
ISOPHTHALIC	
CHLORINATED POLYESTERS	
POLYSTYRENES	
POLYETHYLENE	U
POLYPROPYLENE	R
PVC—TYPE 1	
PVC—TYPE 2	
POLYVINYLIDENE CHLORIDE	
VINYLIDENE FLUORIDE	
VINYL ESTER	

RUBBERS	Rating
BUTYL GR-1	
FLUORO ELASTOMERS VITON A	R
FLUORO ELASTOMERS KEL-F 3700	
HARD RUBBER	
HYPALON	
KOROSEAL	
NATURAL RUBBER (GRS)	
NEOPRENE GR-M (CR)	R
NITRILE BUNA N (NBR)	R
NORDEL	
POLYBUTADIENE	
SILICONE	
POLYURETHANE	

TEMP °F (60–460)

PETROLEUM OILS, SOUR

TEMP °F

METALS	60 °F
ALLOY 20Cb3	
ALUMINUM	U
BRASS	S—
BRONZE	U
CARBON STEEL	
COPPER	U
E-BRIGHT 26-1	
HASTELLOY-B	
HASTELLOY-C	
HASTELLOY-D	
HIGH SILICON IRON	
INCONEL	
LEAD	
MONEL	U
NICKEL	
STAINLESS STEELS	
Type 304/307	G—
Type 316	G—
TANTALUM	
TITANIUM	
ZIRCONIUM	
NONMETALLICS	
BOROSILICATE GLASS	
CARBON	

TEMP °F: 60, 80, 100, 120, 140, 160, 180, 200, 220, 240, 260, 280, 300, 320, 340, 360, 380, 400, 420, 440, 460, 480, 500, 520, 540, 560

FOR METALS

E = < 2 Mils Penetration/Year; G = < 20 Mils Penetration/Year

S = < 50 Mils Penetration/Year; U = > 50 Mils Penetration/Year

FOR NONMETALLICS

R = Resistant

U = Unsatisfactory

PETROLEUM OILS, SOUR

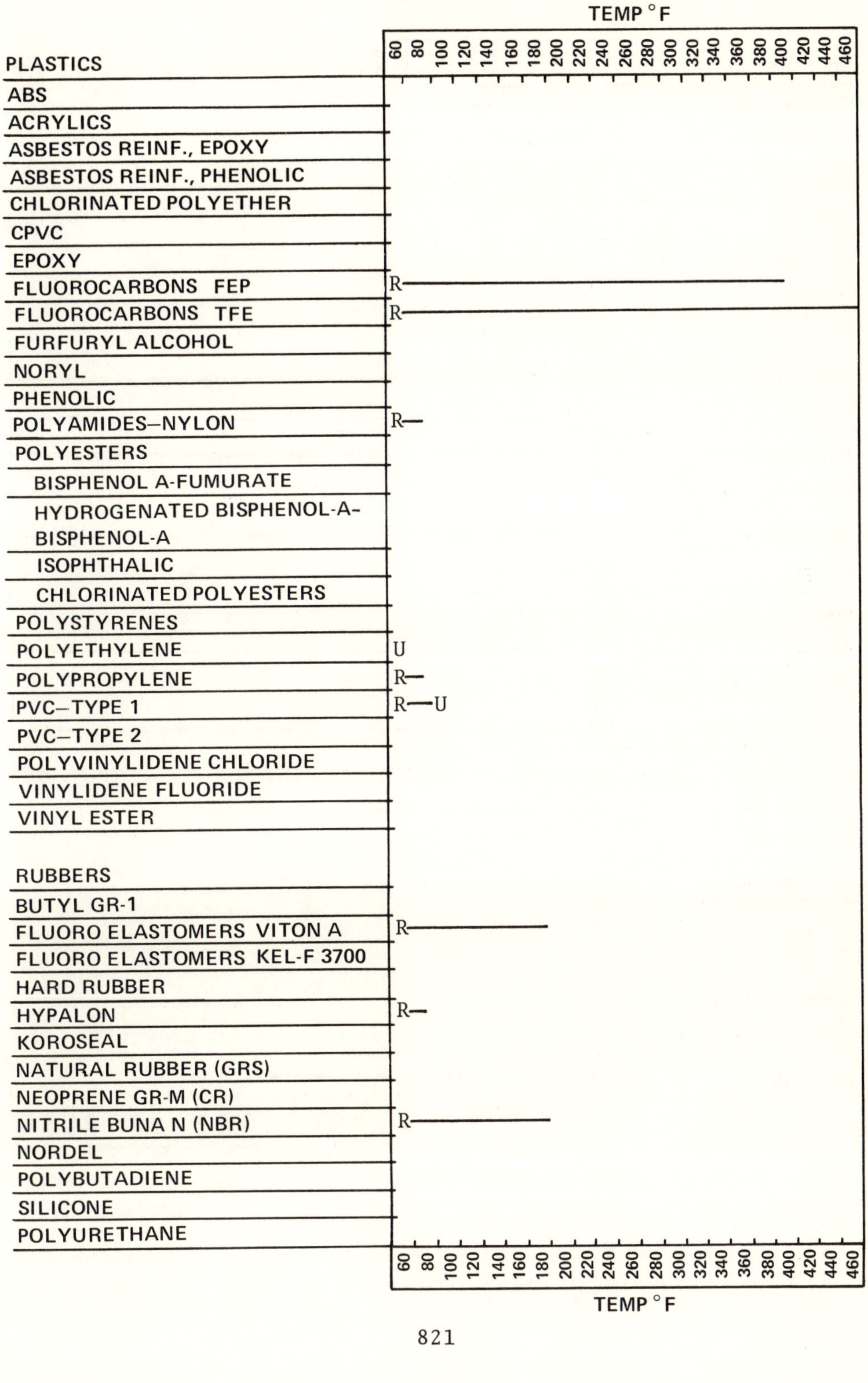

PHENOL (CARBOLIC ACID)

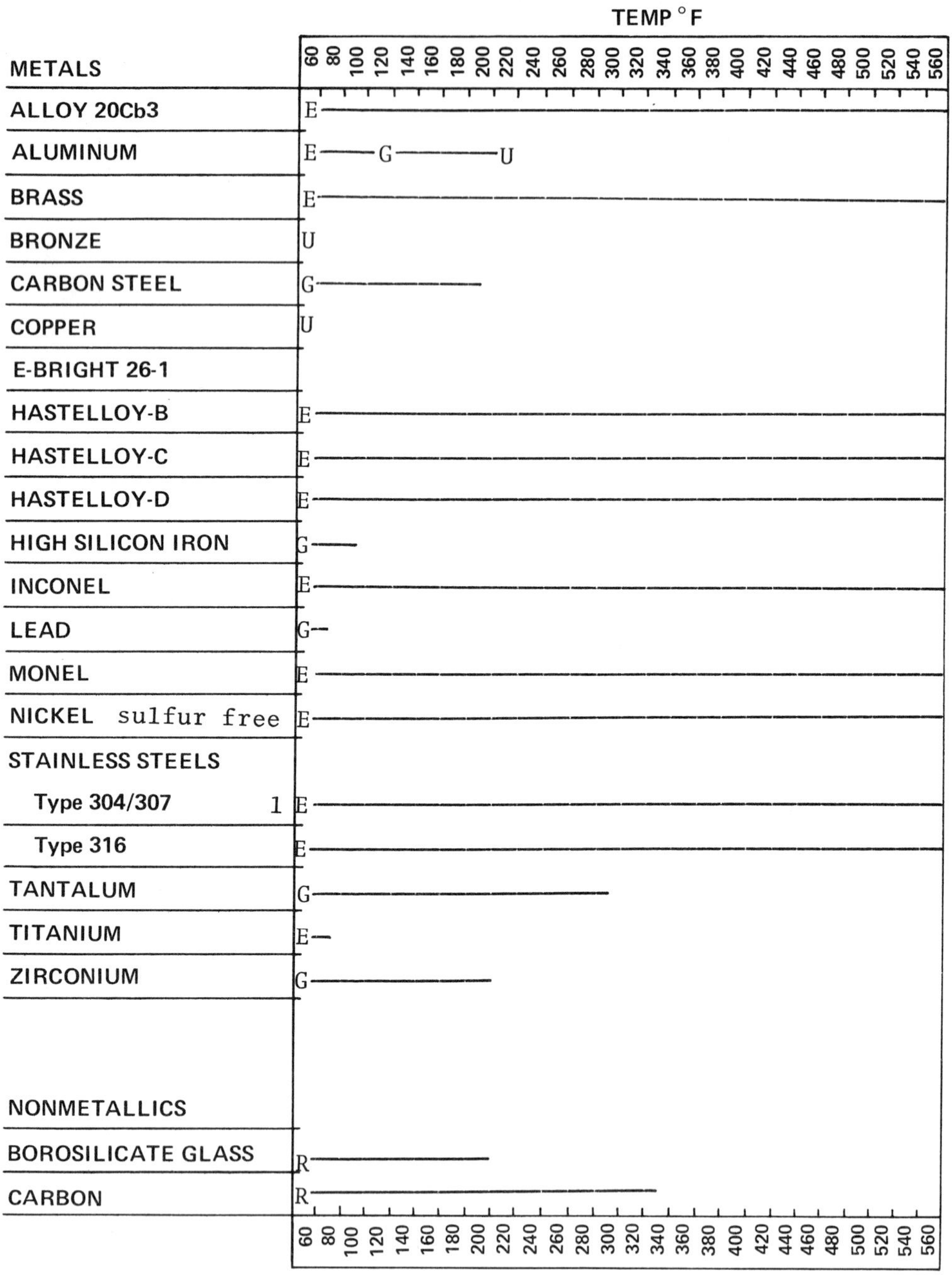

FOR METALS

E = < 2 Mils Penetration/Year; G = < 20 Mils Penetration/Year

S = < 50 Mils Penetration/Year; U = > 50 Mils Penetration/Year

FOR NONMETALLICS

R = Resistant

U = Unsatisfactory

PHENOL (CARBOLIC ACID)

TEMP °F: 60 80 100 120 140 160 180 200 220 240 260 280 300 320 340 360 380 400 420 440 460

PLASTICS		
ABS		U
ACRYLICS		R—
ASBESTOS REINF., EPOXY		
ASBESTOS REINF., PHENOLIC		U
CHLORINATED POLYETHER		R—
CPVC		R———— U
EPOXY		U
FLUOROCARBONS FEP		R————————————————
FLUOROCARBONS TFE		R———————————————————
FURFURYL ALCOHOL	85%	R——————— U
NORYL		
PHENOLIC		U
POLYAMIDES—NYLON		U
POLYESTERS		
BISPHENOL A-FUMURATE	5%	U
HYDROGENATED BISPHENOL-A-BISPHENOL-A		U
ISOPHTHALIC		U
CHLORINATED POLYESTERS	5%	R—
POLYSTYRENES		U
POLYETHYLENE		U
POLYPROPYLENE		R—————
PVC—TYPE 1		RU
PVC—TYPE 2		U
POLYVINYLIDENE CHLORIDE		U
VINYLIDENE FLUORIDE		R———————
VINYL ESTER		U

RUBBERS		
BUTYL GR-1		R——
FLUORO ELASTOMERS VITON A		R——————— U
FLUORO ELASTOMERS KEL-F 3700		R—
HARD RUBBER		
HYPALON		U
KOROSEAL		U
NATURAL RUBBER (GRS)		U
NEOPRENE GR-M (CR)		U
NITRILE BUNA N (NBR)		U
NORDEL		
POLYBUTADIENE		R—
SILICONE		
POLYURETHANE		U

TEMP °F: 60 80 100 120 140 160 180 200 220 240 260 280 300 320 340 360 380 400 420 440 460

PHENOL, 10%

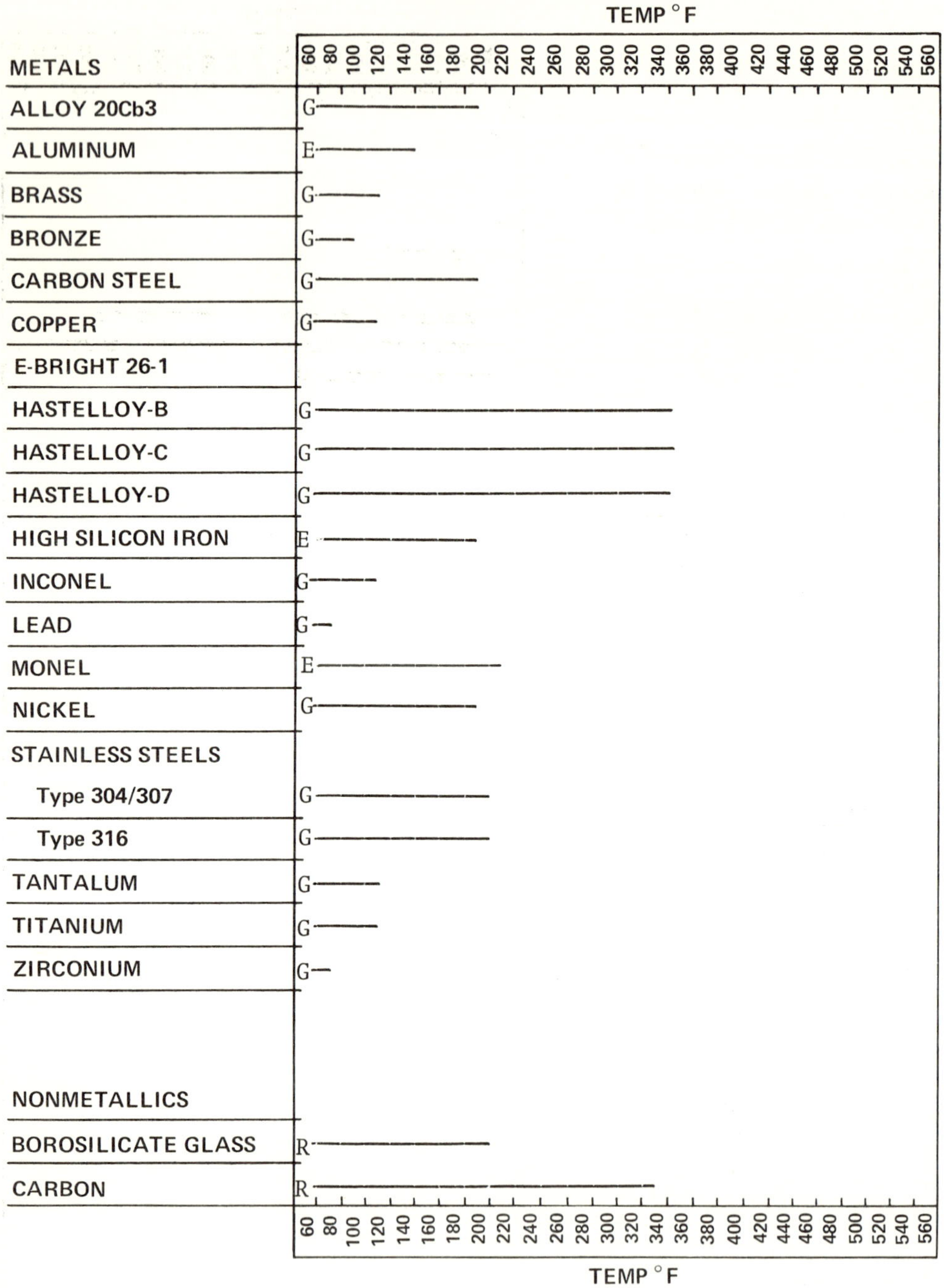

FOR METALS

E = < 2 Mils Penetration/Year; G = < 20 Mils Penetration/Year

S = < 50 Mils Penetration/Year; U = > 50 Mils Penetration/Year

FOR NONMETALLICS

R = Resistant

U = Unsatisfactory

PHENOL, 10%

TEMP °F: 60 80 100 120 140 160 180 200 220 240 260 280 300 320 340 360 380 400 420 440 460

PLASTICS	
ABS	U
ACRYLICS	R
ASBESTOS REINF., EPOXY	
ASBESTOS REINF., PHENOLIC	
CHLORINATED POLYETHER	R
CPVC	R
EPOXY	U
FLUOROCARBONS FEP	R
FLUOROCARBONS TFE	R
FURFURYL ALCOHOL	U
NORYL	
PHENOLIC	
POLYAMIDES–NYLON	U
POLYESTERS	
BISPHENOL A-FUMURATE	U
HYDROGENATED BISPHENOL-A-BISPHENOL-A	U
ISOPHTHALIC	U
CHLORINATED POLYESTERS	U
POLYSTYRENES	R
POLYETHYLENE	
POLYPROPYLENE	R
PVC–TYPE 1	
PVC–TYPE 2	
POLYVINYLIDENE CHLORIDE	R
VINYLIDENE FLUORIDE	R
VINYL ESTER	
RUBBERS	
BUTYL GR-1	R
FLUORO ELASTOMERS VITON A	R
FLUORO ELASTOMERS KEL-F 3700	
HARD RUBBER	
HYPALON	
KOROSEAL	U
NATURAL RUBBER (GRS)	R
NEOPRENE GR-M (CR)	U
NITRILE BUNA N (NBR)	U
NORDEL	
POLYBUTADIENE	
SILICONE	
POLYURETHANE	U

TEMP °F: 60 80 100 120 140 160 180 200 220 240 260 280 300 320 340 360 380 400 420 440 460

PHENYLHYDRAZINE

TEMP °F

METALS	60	80	100	120	140	160	180	200	220	240	260	280	300	320	340	360	380	400	420	440	460	480	500	520	540	560
ALLOY 20Cb3																										
ALUMINUM																										
BRASS																										
BRONZE																										
CARBON STEEL																										
COPPER																										
E-BRIGHT 26-1																										
HASTELLOY-B																										
HASTELLOY-C																										
HASTELLOY-D																										
HIGH SILICON IRON																										
INCONEL																										
LEAD																										
MONEL																										
NICKEL																										
STAINLESS STEELS																										
Type 304/307																										
Type 316																										
TANTALUM																										
TITANIUM																										
ZIRCONIUM																										
NONMETALLICS																										
BOROSILICATE GLASS																										
CARBON																										

TEMP °F

FOR METALS

E = < 2 Mils Penetration/Year; G = < 20 Mils Penetration/Year

S = < 50 Mils Penetration/Year; U = > 50 Mils Penetration/Year

FOR NONMETALLICS

R = Resistant

U = Unsatisfactory

PHENYLHYDRAZINE

TEMP °F: 60 80 100 120 140 160 180 200 220 240 260 280 300 320 340 360 380 400 420 440 460

PLASTICS	
ABS	U
ACRYLICS	
ASBESTOS REINF., EPOXY	
ASBESTOS REINF., PHENOLIC	
CHLORINATED POLYETHER	R
CPVC	U
EPOXY	
FLUOROCARBONS FEP	R
FLUOROCARBONS TFE	R
FURFURYL ALCOHOL	
NORYL	
PHENOLIC	
POLYAMIDES—NYLON	
POLYESTERS	
BISPHENOL A-FUMURATE	
HYDROGENATED BISPHENOL-A-BISPHENOL-A	
ISOPHTHALIC	
CHLORINATED POLYESTERS	
POLYSTYRENES	
POLYETHYLENE	U
POLYPROPYLENE	U
PVC—TYPE 1	U
PVC—TYPE 2	U
POLYVINYLIDENE CHLORIDE	U
VINYLIDENE FLUORIDE	R
VINYL ESTER	

RUBBERS	
BUTYL GR-1	U
FLUORO ELASTOMERS VITON A	R
FLUORO ELASTOMERS KEL-F 3700	
HARD RUBBER	
HYPALON	
KOROSEAL	
NATURAL RUBBER (GRS)	R
NEOPRENE GR-M (CR)	U
NITRILE BUNA N (NBR)	U
NORDEL	
POLYBUTADIENE	
SILICONE	
POLYURETHANE	

TEMP °F: 60 80 100 120 140 160 180 200 220 240 260 280 300 320 340 360 380 400 420 440 460

PHENYLHYDRAZINE HYDROCHLORIDE

TEMP °F

METALS	60	80	100	120	140	160	180	200	220	240	260	280	300	320	340	360	380	400	420	440	460	480	500	520	540	560
ALLOY 20Cb3																										
ALUMINUM																										
BRASS																										
BRONZE																										
CARBON STEEL																										
COPPER																										
E-BRIGHT 26-1																										
HASTELLOY-B																										
HASTELLOY-C																										
HASTELLOY-D																										
HIGH SILICON IRON																										
INCONEL																										
LEAD																										
MONEL																										
NICKEL																										
STAINLESS STEELS																										
Type 304/307																										
Type 316																										
TANTALUM																										
TITANIUM																										
ZIRCONIUM																										
NONMETALLICS																										
BOROSILICATE GLASS																										
CARBON	R	—	—	—	—	—	—	—	—	—	—	—	—	—	—											

TEMP °F

FOR METALS

E = < 2 Mils Penetration/Year; G = < 20 Mils Penetration/Year

S = < 50 Mils Penetration/Year; U = > 50 Mils Penetration/Year

FOR NONMETALLICS

R = Resistant

U = Unsatisfactory

PHENYLHYDRAZINE HYDROCHLORIDE

TEMP °F (60–460)

PLASTICS	Rating	Temperature range (°F)
ABS	U	60
ACRYLICS		
ASBESTOS REINF., EPOXY		
ASBESTOS REINF., PHENOLIC		
CHLORINATED POLYETHER	R	60–220
CPVC	U	60
EPOXY		
FLUOROCARBONS FEP	R	60–400
FLUOROCARBONS TFE	R	60–460
FURFURYL ALCOHOL		
NORYL		
PHENOLIC		
POLYAMIDES–NYLON		
POLYESTERS		
BISPHENOL A-FUMURATE		
HYDROGENATED BISPHENOL-A-BISPHENOL-A		
ISOPHTHALIC		
CHLORINATED POLYESTERS		
POLYSTYRENES		
POLYETHYLENE	U	60
POLYPROPYLENE	U	60
PVC–TYPE 1	RU	60
PVC–TYPE 2	U	60
POLYVINYLIDENE CHLORIDE		
VINYLIDENE FLUORIDE	R	60–120
VINYL ESTER		

RUBBERS	Rating	Temperature range (°F)
BUTYL GR-1		
FLUORO ELASTOMERS VITON A		
FLUORO ELASTOMERS KEL-F 3700		
HARD RUBBER		
HYPALON		
KOROSEAL		
NATURAL RUBBER (GRS)		
NEOPRENE GR-M (CR)		
NITRILE BUNA N (NBR)		
NORDEL		
POLYBUTADIENE		
SILICONE		
POLYURETHANE		

TEMP °F

PHOSGENE, GAS

METALS	TEMP °F (60 80 100 120 140 160 180 200 220 240 260 280 300 320 340 360 380 400 420 440 460 480 500 520 540 560)
ALLOY 20Cb3	
ALUMINUM	
BRASS	
BRONZE	
CARBON STEEL	
COPPER	
E-BRIGHT 26-1	
HASTELLOY-B	G ———
HASTELLOY-C	G ———
HASTELLOY-D	
HIGH SILICON IRON	
INCONEL	
LEAD	
MONEL	
NICKEL	
STAINLESS STEELS Type 304/307	
Type 316	
TANTALUM	
TITANIUM	
ZIRCONIUM	
NONMETALLICS	
BOROSILICATE GLASS	
CARBON	

FOR METALS

E = < 2 Mils Penetration/Year; G = < 20 Mils Penetration/Year

S = < 50 Mils Penetration/Year; U = > 50 Mils Penetration/Year

FOR NONMETALLICS

R = Resistant

U = Unsatisfactory

PHOSGENE, GAS

TEMP °F (60–460)

PLASTICS	
ABS	R
ACRYLICS	U
ASBESTOS REINF., EPOXY	
ASBESTOS REINF., PHENOLIC	
CHLORINATED POLYETHER	R
CPVC	U
EPOXY	
FLUOROCARBONS FEP	R
FLUOROCARBONS TFE	R
FURFURYL ALCOHOL	
NORYL	
PHENOLIC	
POLYAMIDES—NYLON	
POLYESTERS	
BISPHENOL A-FUMURATE	
HYDROGENATED BISPHENOL-A-BISPHENOL-A	
ISOPHTHALIC	
CHLORINATED POLYESTERS	
POLYSTYRENES	
POLYETHYLENE	U
POLYPROPYLENE	U
PVC—TYPE 1	U
PVC—TYPE 2	U
POLYVINYLIDENE CHLORIDE	
VINYLIDENE FLUORIDE	U
VINYL ESTER	

RUBBERS	
BUTYL GR-1	
FLUORO ELASTOMERS VITON A	U
FLUORO ELASTOMERS KEL-F 3700	
HARD RUBBER	
HYPALON	
KOROSEAL	
NATURAL RUBBER (GRS)	
NEOPRENE GR-M (CR)	
NITRILE BUNA N (NBR)	U
NORDEL	
POLYBUTADIENE	
SILICONE	
POLYURETHANE	

TEMP °F

PHOSGENE, LIQUID

TEMP °F: 60 80 100 120 140 160 180 200 220 240 260 280 300 320 340 360 380 400 420 440 460 480 500 520 540 560

METALS	TEMP °F (60–560)
ALLOY 20Cb3	
ALUMINUM	
BRASS	
BRONZE	
CARBON STEEL	
COPPER	
E-BRIGHT 26-1	
HASTELLOY-B	
HASTELLOY-C	
HASTELLOY-D	
HIGH SILICON IRON	
INCONEL	
LEAD	
MONEL	
NICKEL	
STAINLESS STEELS	
Type 304/307	G————————
Type 316	G————————
TANTALUM	G——————————————
TITANIUM	
ZIRCONIUM	

NONMETALLICS	TEMP °F (60–560)
BOROSILICATE GLASS	
CARBON	

FOR METALS

E = < 2 Mils Penetration/Year; G = < 20 Mils Penetration/Year

S = < 50 Mils Penetration/Year; U = > 50 Mils Penetration/Year

FOR NONMETALLICS

R = Resistant

U = Unsatisfactory

PHOSGENE, LIQUID

TEMP °F: 60 80 100 120 140 160 180 200 220 240 260 280 300 320 340 360 380 400 420 440 460

PLASTICS	
ABS	U
ACRYLICS	
ASBESTOS REINF., EPOXY	
ASBESTOS REINF., PHENOLIC	
CHLORINATED POLYETHER	R
CPVC	U
EPOXY	
FLUOROCARBONS FEP	R
FLUOROCARBONS TFE	R
FURFURYL ALCOHOL	
NORYL	
PHENOLIC	
POLYAMIDES—NYLON	
POLYESTERS	
BISPHENOL A-FUMURATE	
HYDROGENATED BISPHENOL-A-BISPHENOL-A	
ISOPHTHALIC	
CHLORINATED POLYESTERS	
POLYSTYRENES	
POLYETHYLENE	U
POLYPROPYLENE	U
PVC—TYPE 1	U
PVC—TYPE 2	U
POLYVINYLIDENE CHLORIDE	
VINYLIDENE FLUORIDE	
VINYL ESTER	

RUBBERS	
BUTYL GR-1	
FLUORO ELASTOMERS VITON A	U
FLUORO ELASTOMERS KEL-F 3700	
HARD RUBBER	
HYPALON	
KOROSEAL	
NATURAL RUBBER (GRS)	
NEOPRENE GR-M (CR)	
NITRILE BUNA N (NBR)	U
NORDEL	
POLYBUTADIENE	
SILICONE	
POLYURETHANE	

TEMP °F: 60 80 100 120 140 160 180 200 220 240 260 280 300 320 340 360 380 400 420 440 460

PHOSPHORIC ACID, 85%-50%

TEMP °F: 60, 80, 100, 120, 140, 160, 180, 200, 220, 240, 260, 280, 300, 320, 340, 360, 380, 400, 420, 440, 460, 480, 500, 520, 540, 560

METALS	Rating
ALLOY 20Cb3	E——G——
ALUMINUM	U
BRASS	U
BRONZE	U
CARBON STEEL	U
COPPER	U
E-BRIGHT 26-1	
HASTELLOY-B	E——G——
HASTELLOY-C	E——
HASTELLOY-D	E——G
HIGH SILICON IRON	G——
INCONEL	G——U
LEAD	G——
MONEL	S——
NICKEL	U
STAINLESS STEELS	
Type 304/307	E——
Type 316 9	G——
TANTALUM	E——
TITANIUM	S——
ZIRCONIUM	G——

NONMETALLICS	Rating
BOROSILICATE GLASS	R——
CARBON	R——

TEMP °F: 60, 80, 100, 120, 140, 160, 180, 200, 220, 240, 260, 280, 300, 320, 340, 360, 380, 400, 420, 440, 460, 480, 500, 520, 540, 560

FOR METALS

E = < 2 Mils Penetration/Year; G = < 20 Mils Penetration/Year

S = < 50 Mils Penetration/Year; U = > 50 Mils Penetration/Year

FOR NONMETALLICS

R = Resistant

U = Unsatisfactory

PHOSPHORIC ACID, 85%-50%

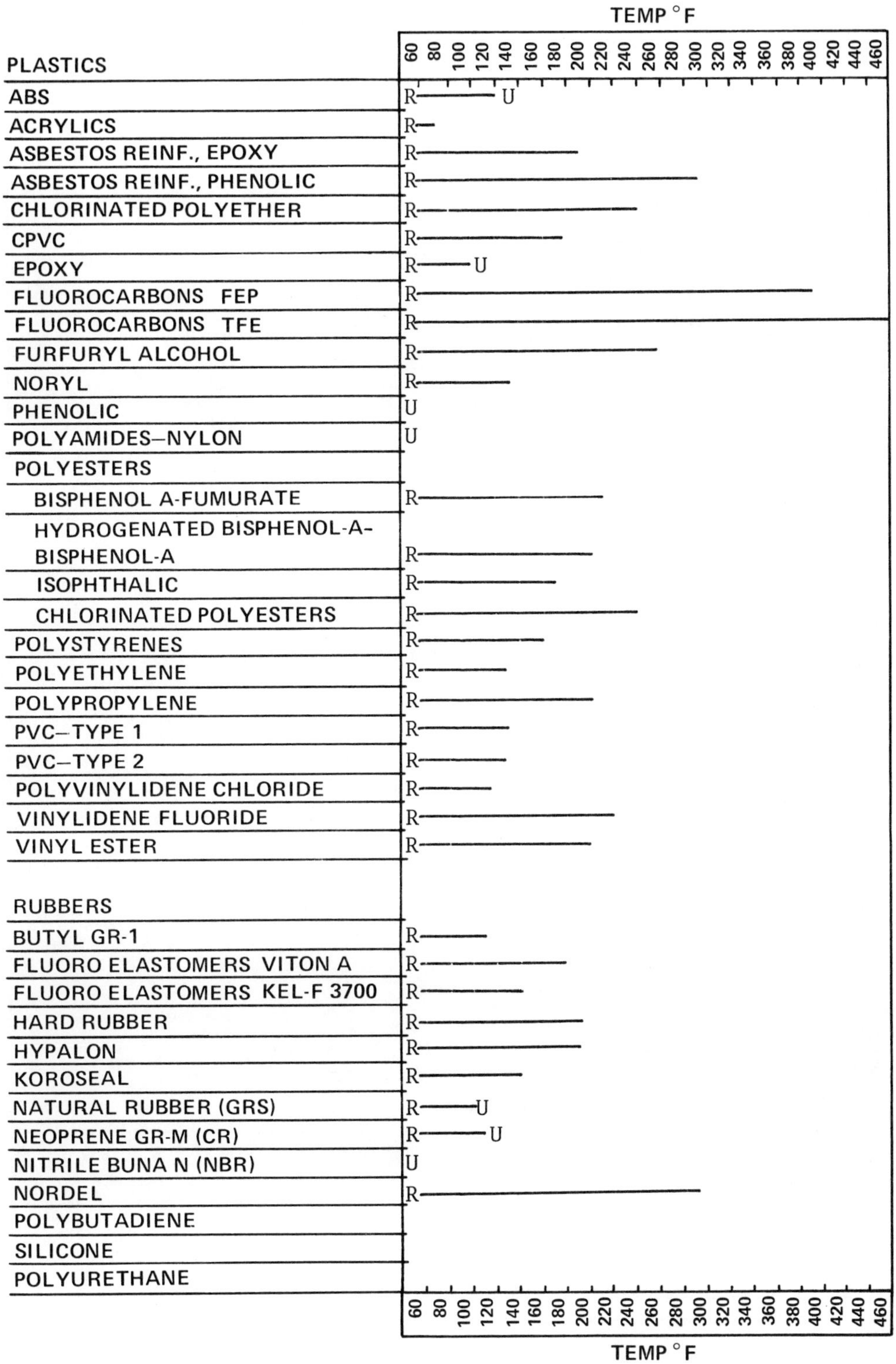

PHOSPHORIC ACID, 50%-25%

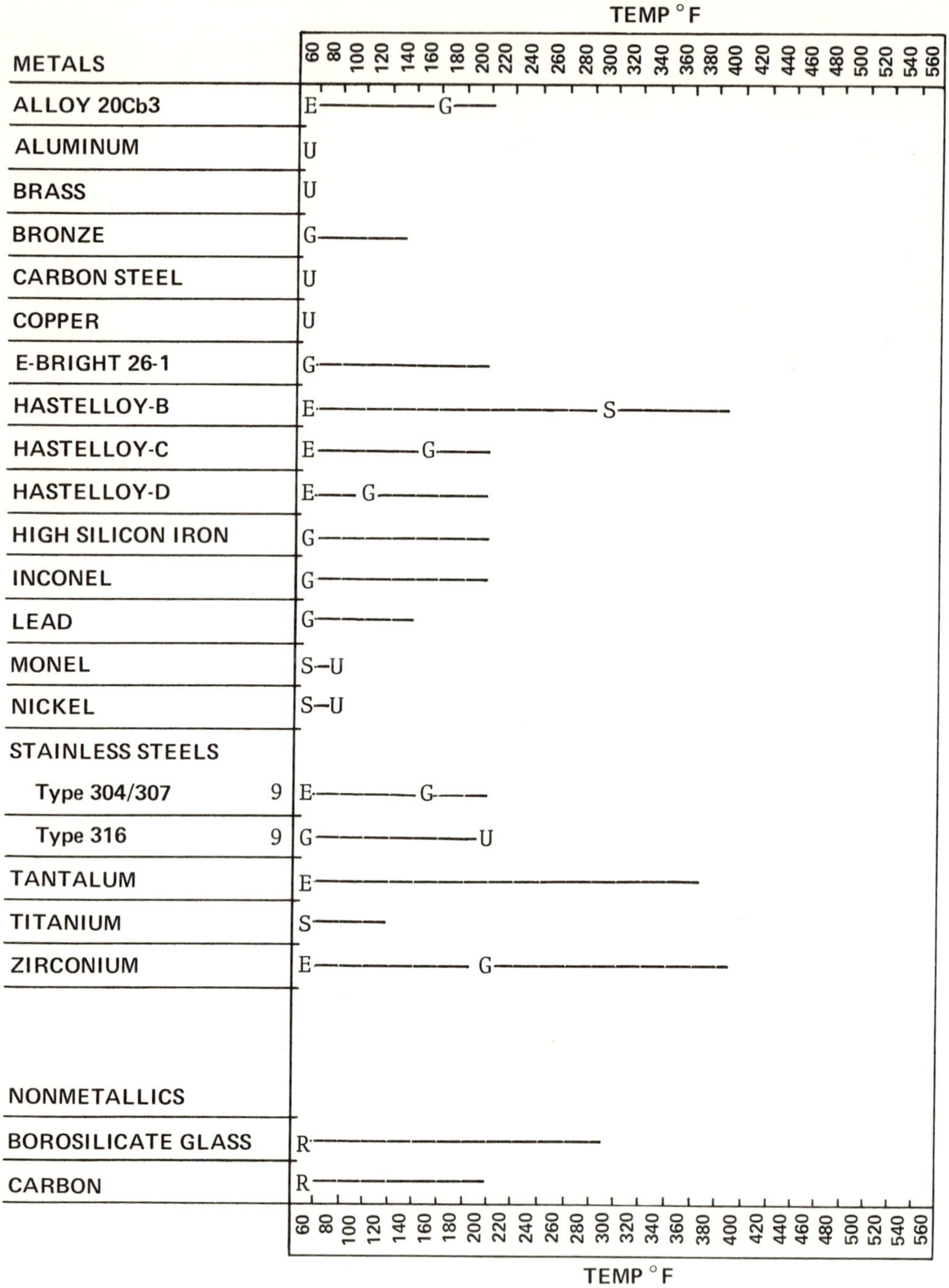

FOR METALS

E = < 2 Mils Penetration/Year; G = < 20 Mils Penetration/Year

S = < 50 Mils Penetration/Year; U = > 50 Mils Penetration/Year

FOR NONMETALLICS

R = Resistant

U = Unsatisfactory

PHOSPHORIC ACID, 50%-25%

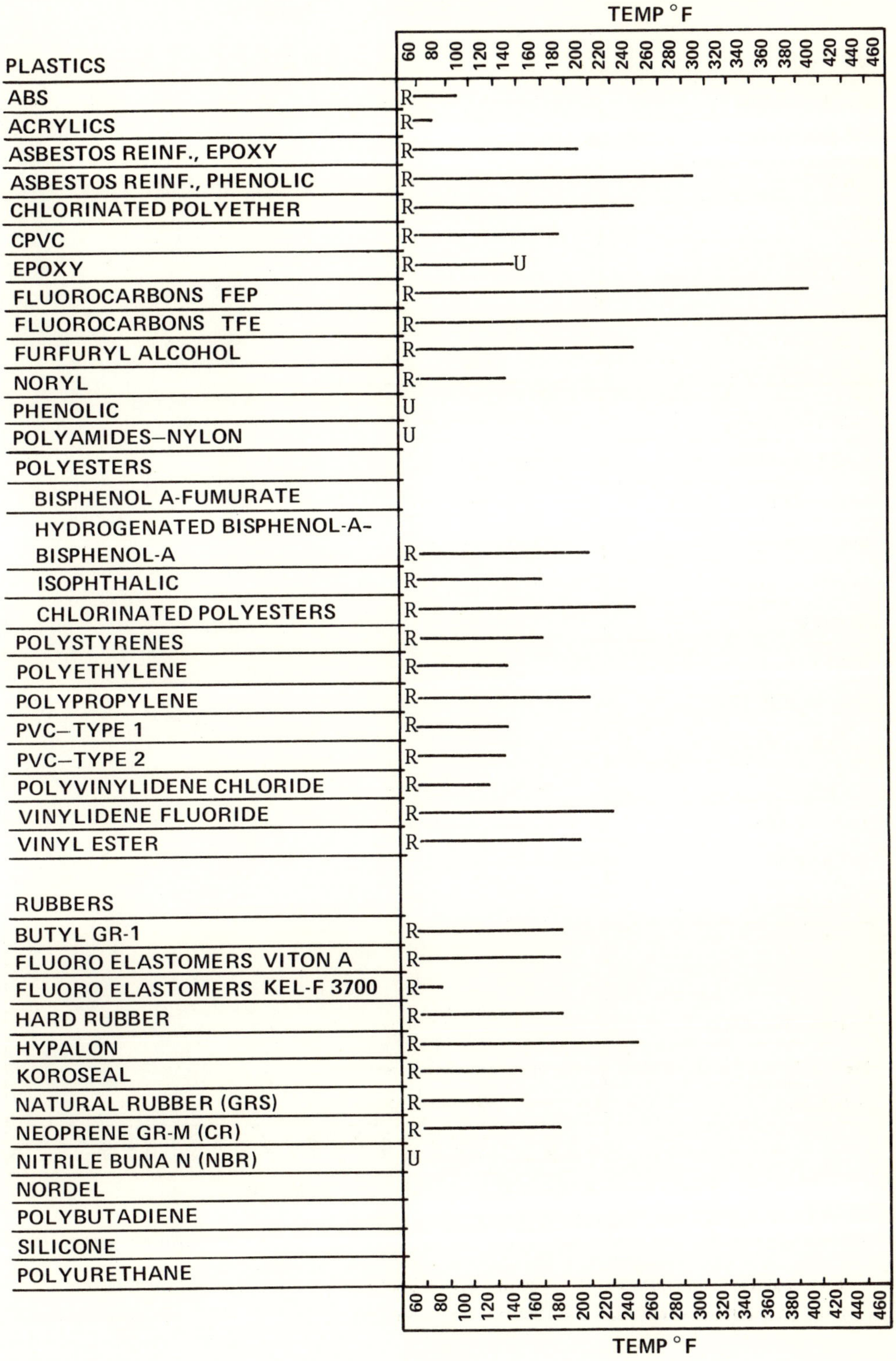

PHOSPHORIC ACID, 10%

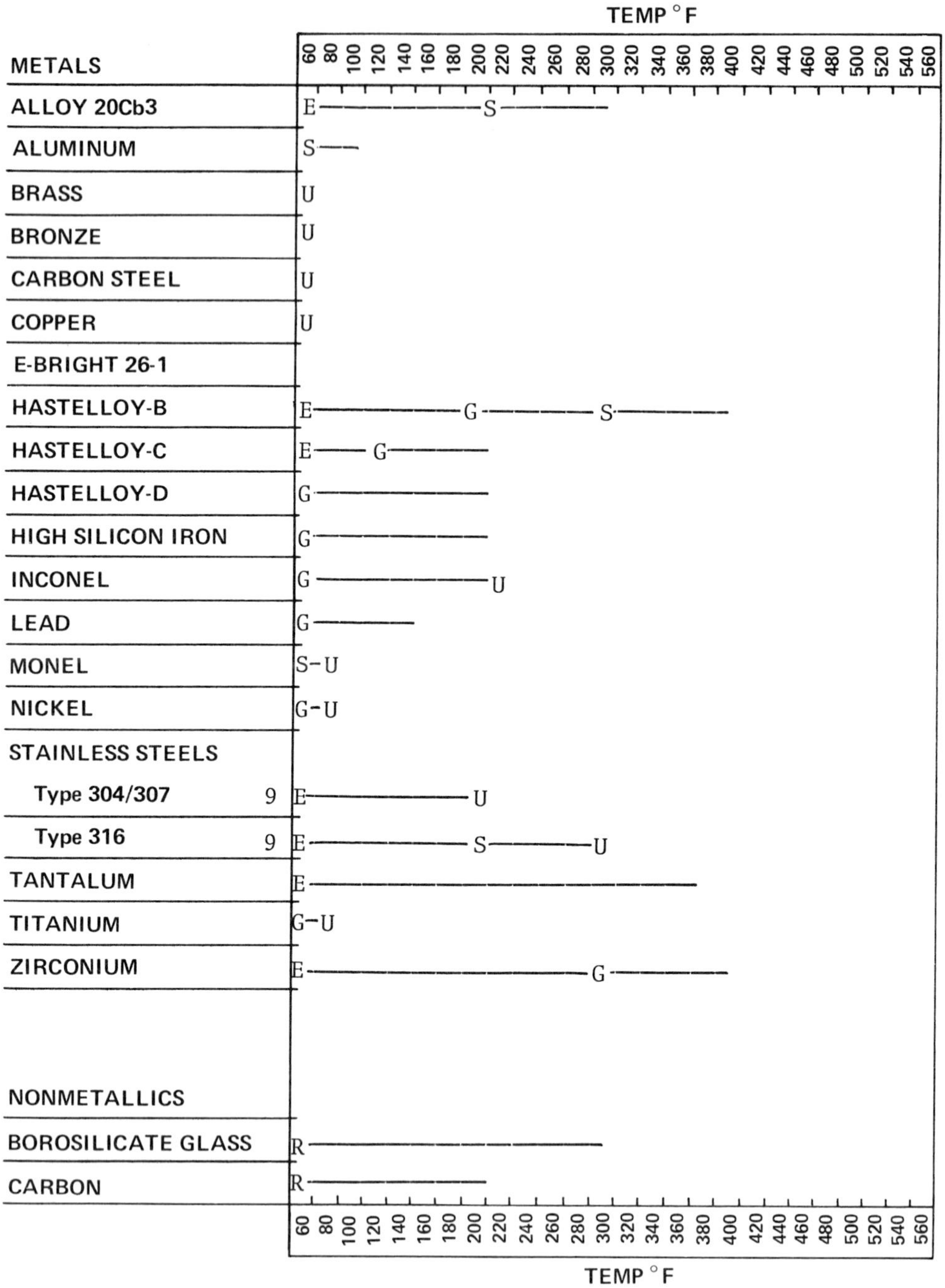

FOR METALS

E = < 2 Mils Penetration/Year; G = < 20 Mils Penetration/Year

S = < 50 Mils Penetration/Year; U = > 50 Mils Penetration/Year

FOR NONMETALLICS

R = Resistant

U = Unsatisfactory

PHOSPHORIC ACID, 10%

TEMP °F: 60, 80, 100, 120, 140, 160, 180, 200, 220, 240, 260, 280, 300, 320, 340, 360, 380, 400, 420, 440, 460

Material	Rating
PLASTICS	
ABS	R
ACRYLICS	R
ASBESTOS REINF., EPOXY	R
ASBESTOS REINF., PHENOLIC	R
CHLORINATED POLYETHER	R
CPVC	R
EPOXY	R
FLUOROCARBONS FEP	R
FLUOROCARBONS TFE	R
FURFURYL ALCOHOL	R
NORYL	R
PHENOLIC	
POLYAMIDES—NYLON	U
POLYESTERS	
BISPHENOL A-FUMURATE	
HYDROGENATED BISPHENOL-A-BISPHENOL-A	R
ISOPHTHALIC	R
CHLORINATED POLYESTERS	R
POLYSTYRENES	R
POLYETHYLENE	R
POLYPROPYLENE	R
PVC—TYPE 1	R
PVC—TYPE 2	R
POLYVINYLIDENE CHLORIDE	R
VINYLIDENE FLUORIDE	R
VINYL ESTER	R
RUBBERS	
BUTYL GR-1	R
FLUORO ELASTOMERS VITON A	R
FLUORO ELASTOMERS KEL-F 3700	R
HARD RUBBER	
HYPALON	R
KOROSEAL	R
NATURAL RUBBER (GRS)	R
NEOPRENE GR-M (CR)	R
NITRILE BUNA N (NBR)	U
NORDEL	
POLYBUTADIENE	
SILICONE	
POLYURETHANE	

PHOSPHORIC ANHYDRIDE (DRY OR MOIST)

TEMP °F (60, 80, 100, 120, 140, 160, 180, 200, 220, 240, 260, 280, 300, 320, 340, 360, 380, 400, 420, 440, 460, 480, 500, 520, 540, 560)

METALS	
ALLOY 20Cb3	G—
ALUMINUM	
BRASS	
BRONZE	
CARBON STEEL	U
COPPER	
E-BRIGHT 26-1	
HASTELLOY-B	
HASTELLOY-C	
HASTELLOY-D	
HIGH SILICON IRON	G—
INCONEL	
LEAD	
MONEL	
NICKEL	
STAINLESS STEELS	
Type 304/307	G————————
Type 316	G————————
TANTALUM	E—
TITANIUM	
ZIRCONIUM	
NONMETALLICS	
BOROSILICATE GLASS	
CARBON	

TEMP °F

FOR METALS

E = < 2 Mils Penetration/Year; G = < 20 Mils Penetration/Year
S = < 50 Mils Penetration/Year; U = > 50 Mils Penetration/Year

FOR NONMETALLICS

R = Resistant
U = Unsatisfactory

PHOSPHORIC ANHYDRIDE (DRY OR MOIST)

TEMP °F (60–460)

PLASTICS	TEMP °F
ABS	
ACRYLICS	
ASBESTOS REINF., EPOXY	
ASBESTOS REINF., PHENOLIC	
CHLORINATED POLYETHER	
CPVC	
EPOXY	
FLUOROCARBONS FEP	R (60–400)
FLUOROCARBONS TFE	R (60–460)
FURFURYL ALCOHOL	
NORYL	
PHENOLIC	
POLYAMIDES—NYLON	
POLYESTERS	
BISPHENOL A-FUMURATE	
HYDROGENATED BISPHENOL-A-BISPHENOL-A	
ISOPHTHALIC	
CHLORINATED POLYESTERS	
POLYSTYRENES	
POLYETHYLENE	R (60–80)
POLYPROPYLENE	
PVC—TYPE 1	
PVC—TYPE 2	
POLYVINYLIDENE CHLORIDE	
VINYLIDENE FLUORIDE	
VINYL ESTER	

RUBBERS	TEMP °F
BUTYL GR-1	
FLUORO ELASTOMERS VITON A	
FLUORO ELASTOMERS KEL-F 3700	
HARD RUBBER	
HYPALON	
KOROSEAL	
NATURAL RUBBER (GRS)	
NEOPRENE GR-M (CR)	
NITRILE BUNA N (NBR)	
NORDEL	
POLYBUTADIENE	
SILICONE	
POLYURETHANE	

TEMP °F: 60 80 100 120 140 160 180 200 220 240 260 280 300 320 340 360 380 400 420 440 460

PHOSPHORUS

TEMP °F: 60, 80, 100, 120, 140, 160, 180, 200, 220, 240, 260, 280, 300, 320, 340, 360, 380, 400, 420, 440, 460, 480, 500, 520, 540, 560

METALS	Rating
ALLOY 20Cb3	E
ALUMINUM	G
BRASS	E- U
BRONZE	
CARBON STEEL	
COPPER	G
E-BRIGHT 26-1	
HASTELLOY-B	E
HASTELLOY-C	E
HASTELLOY-D	E
HIGH SILICON IRON	G
INCONEL	
LEAD	
MONEL	E
NICKEL	E
STAINLESS STEELS	
Type 304/307	E
Type 316	E
TANTALUM	G
TITANIUM	
ZIRCONIUM	

NONMETALLICS	Rating
BOROSILICATE GLASS	
CARBON	

FOR METALS

E = < 2 Mils Penetration/Year; G = < 20 Mils Penetration/Year

S = < 50 Mils Penetration/Year; U = > 50 Mils Penetration/Year

FOR NONMETALLICS

R = Resistant

U = Unsatisfactory

PHOSPHORUS

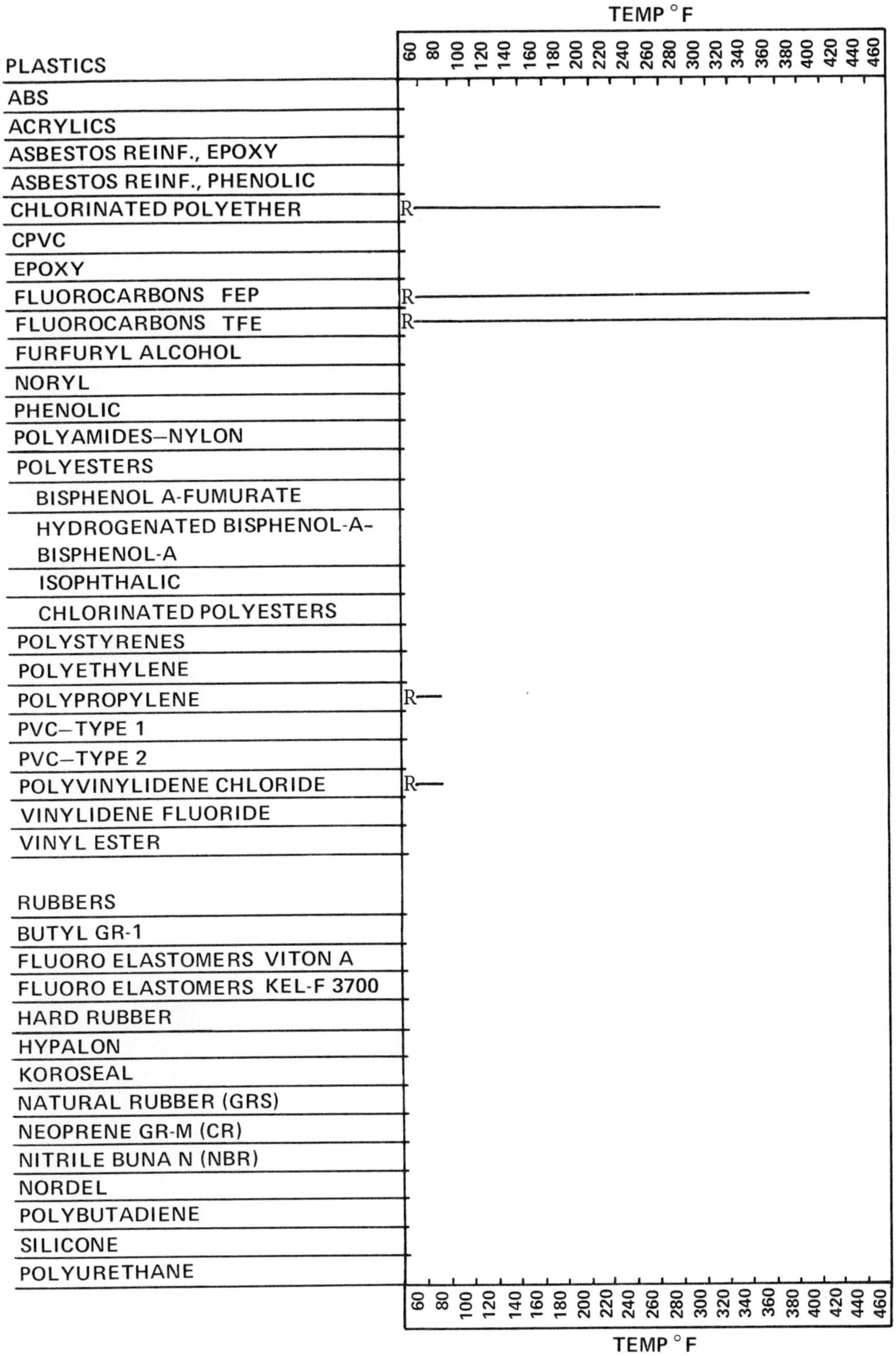
TEMP °F
60 80 100 120 140 160 180 200 220 240 260 280 300 320 340 360 380 400 420 440 460
PLASTICS
ABS
ACRYLICS
ASBESTOS REINF., EPOXY
ASBESTOS REINF., PHENOLIC
CHLORINATED POLYETHER
R
CPVC
EPOXY
FLUOROCARBONS FEP
R
FLUOROCARBONS TFE
R
FURFURYL ALCOHOL
NORYL
PHENOLIC
POLYAMIDES–NYLON
POLYESTERS
BISPHENOL A-FUMURATE
HYDROGENATED BISPHENOL-A-BISPHENOL-A
ISOPHTHALIC
CHLORINATED POLYESTERS
POLYSTYRENES
POLYETHYLENE
POLYPROPYLENE
R
PVC–TYPE 1
PVC–TYPE 2
POLYVINYLIDENE CHLORIDE
R
VINYLIDENE FLUORIDE
VINYL ESTER
RUBBERS
BUTYL GR-1
FLUORO ELASTOMERS VITON A
FLUORO ELASTOMERS KEL-F 3700
HARD RUBBER
HYPALON
KOROSEAL
NATURAL RUBBER (GRS)
NEOPRENE GR-M (CR)
NITRILE BUNA N (NBR)
NORDEL
POLYBUTADIENE
SILICONE
POLYURETHANE
60 80 100 120 140 160 180 200 220 240 260 280 300 320 340 360 380 400 420 440 460
TEMP °F

PHOSPHORUS OXYCHLORIDE

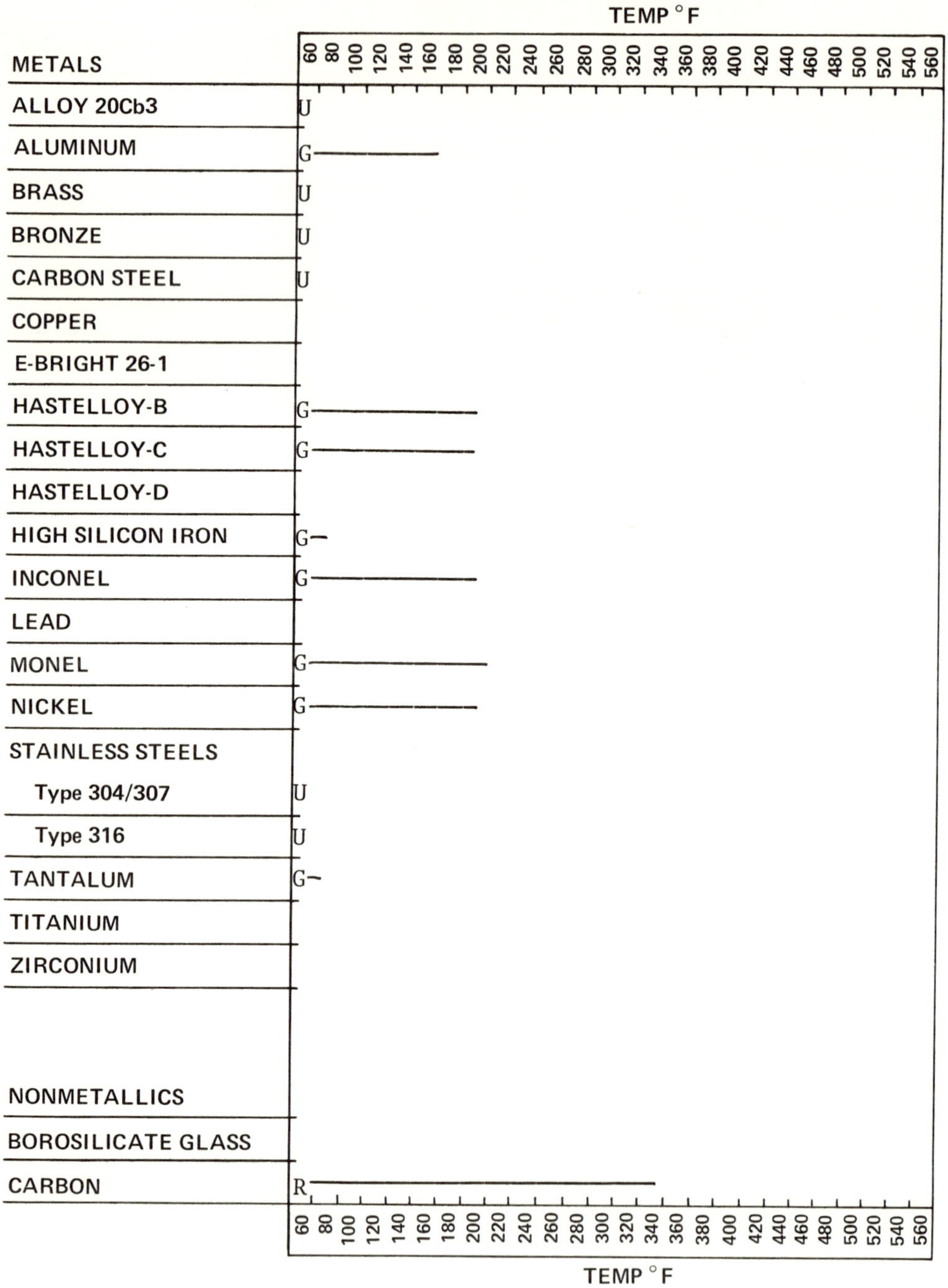

FOR METALS

E = < 2 Mils Penetration/Year; G = < 20 Mils Penetration/Year

S = < 50 Mils Penetration/Year; U = > 50 Mils Penetration/Year

FOR NONMETALLICS

R = Resistant

U = Unsatisfactory

PHOSPHORUS OXYCHLORIDE

TEMP °F: 60 80 100 120 140 160 180 200 220 240 260 280 300 320 340 360 380 400 420 440 460

PLASTICS	Rating
ABS	
ACRYLICS	
ASBESTOS REINF., EPOXY	
ASBESTOS REINF., PHENOLIC	
CHLORINATED POLYETHER	U
CPVC	
EPOXY	U
FLUOROCARBONS FEP	R
FLUOROCARBONS TFE	R
FURFURYL ALCOHOL	R
NORYL	
PHENOLIC	
POLYAMIDES–NYLON	
POLYESTERS	
BISPHENOL A-FUMURATE 75%	U
HYDROGENATED BISPHENOL-A-BISPHENOL-A	U
ISOPHTHALIC	U
CHLORINATED POLYESTERS 75%	U
POLYSTYRENES	
POLYETHYLENE	U
POLYPROPYLENE	U
PVC–TYPE 1	
PVC–TYPE 2	
POLYVINYLIDENE CHLORIDE	U
VINYLIDENE FLUORIDE	R
VINYL ESTER	

RUBBERS	Rating
BUTYL GR-1	
FLUORO ELASTOMERS VITON A	
FLUORO ELASTOMERS KEL-F 3700	
HARD RUBBER	
HYPALON	
KOROSEAL	
NATURAL RUBBER (GRS)	
NEOPRENE GR-M (CR)	
NITRILE BUNA N (NBR)	
NORDEL	
POLYBUTADIENE	
SILICONE	
POLYURETHANE	

TEMP °F: 60 80 100 120 140 160 180 200 220 240 260 280 300 320 340 360 380 400 420 440 460

PHOSPHORUS, PENTOXIDE

METALS		TEMP °F (60–560)
ALLOY 20Cb3		
ALUMINUM		G— (60)
BRASS		
BRONZE		
CARBON STEEL		
COPPER		G— (60)
E-BRIGHT 26-1		
HASTELLOY-B	18%	G (60–300)
HASTELLOY-C		U (60)
HASTELLOY-D		S (60–300)
HIGH SILICON IRON		
INCONEL		
LEAD		
MONEL		G— (60)
NICKEL		
STAINLESS STEELS		
Type 304/307		G— (60)
Type 316		G— (60)
TANTALUM		G— (60)
TITANIUM		
ZIRCONIUM		
NONMETALLICS		
BOROSILICATE GLASS		
CARBON		R (60–340)

FOR METALS

E = < 2 Mils Penetration/Year; G = < 20 Mils Penetration/Year

S = < 50 Mils Penetration/Year; U = > 50 Mils Penetration/Year

FOR NONMETALLICS

R = Resistant

U = Unsatisfactory

PHOSPHORUS, PENTOXIDE

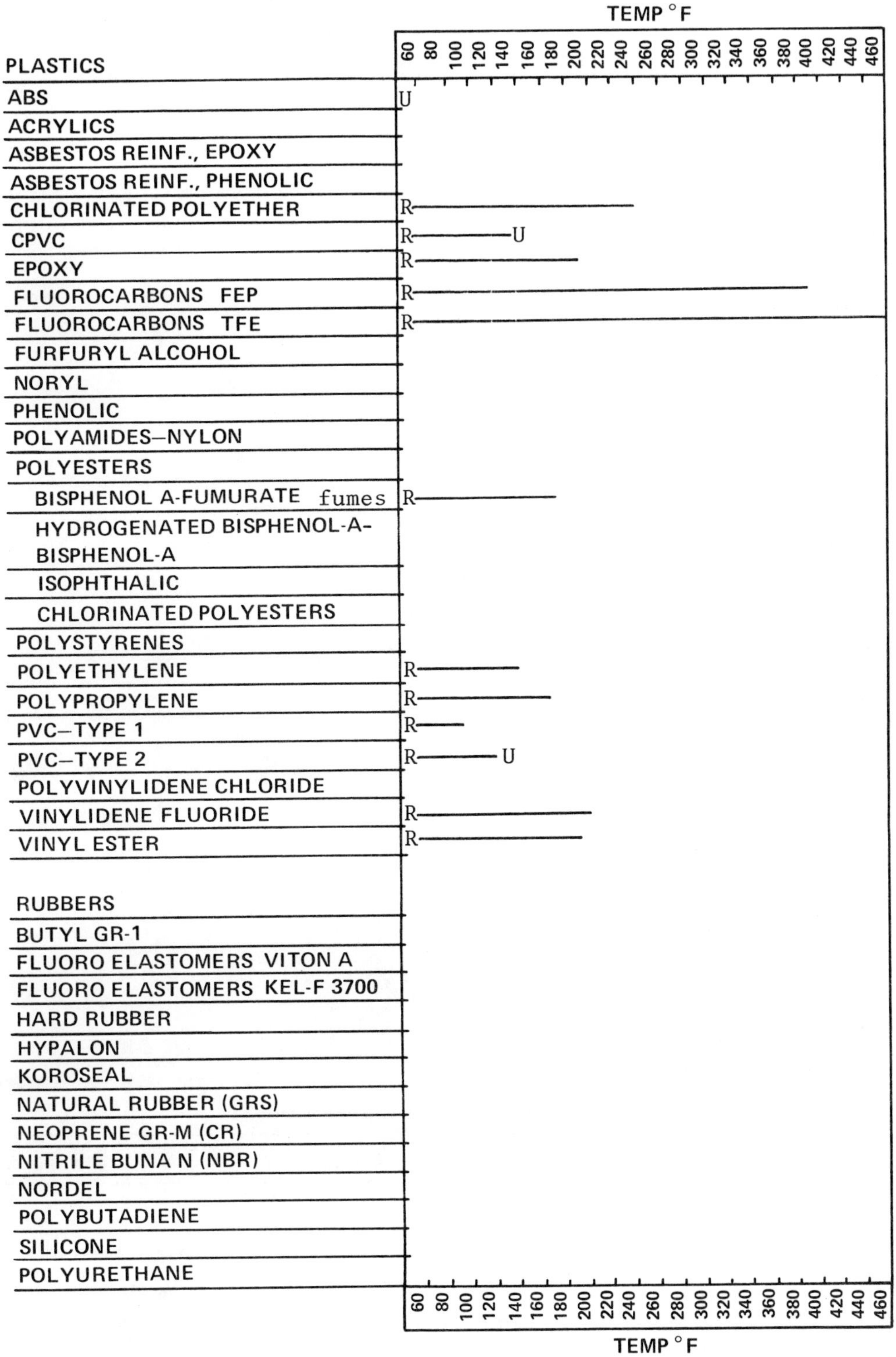

TEMP °F (60, 80, 100, 120, 140, 160, 180, 200, 220, 240, 260, 280, 300, 320, 340, 360, 380, 400, 420, 440, 460, 480, 500, 520, 540, 560)

METALS	TEMP °F
ALLOY 20Cb3	
ALUMINUM	
BRASS	
BRONZE	
CARBON STEEL	
COPPER	
E-BRIGHT 26-1	
HASTELLOY-B	
HASTELLOY-C	
HASTELLOY-D	
HIGH SILICON IRON	
INCONEL	
LEAD	
MONEL	
NICKEL	
STAINLESS STEELS	
Type 304/307	
Type 316	
TANTALUM	
TITANIUM	
ZIRCONIUM	
NONMETALLICS	
BOROSILICATE GLASS	
CARBON	R (60–340)

TEMP °F

FOR METALS

E = < 2 Mils Penetration/Year; G = < 20 Mils Penetration/Year

S = < 50 Mils Penetration/Year; U = > 50 Mils Penetration/Year

FOR NONMETALLICS

R = Resistant

U = Unsatisfactory

PHOSPHORUS, RED

PLASTICS	TEMP °F (60–460)
ABS	
ACRYLICS	
ASBESTOS REINF., EPOXY	
ASBESTOS REINF., PHENOLIC	
CHLORINATED POLYETHER	R (60–220)
CPVC	
EPOXY	
FLUOROCARBONS FEP	R (60–400)
FLUOROCARBONS TFE	R (60–460)
FURFURYL ALCOHOL	
NORYL	
PHENOLIC	
POLYAMIDES–NYLON	
POLYESTERS	
BISPHENOL A-FUMURATE	
HYDROGENATED BISPHENOL-A-BISPHENOL-A	
ISOPHTHALIC	
CHLORINATED POLYESTERS	
POLYSTYRENES	
POLYETHYLENE	
POLYPROPYLENE	R (60–80)
PVC–TYPE 1	
PVC–TYPE 2	
POLYVINYLIDENE CHLORIDE	R (60–80)
VINYLIDENE FLUORIDE	
VINYL ESTER	

RUBBERS	TEMP °F (60–460)
BUTYL GR-1	
FLUORO ELASTOMERS VITON A	
FLUORO ELASTOMERS KEL-F 3700	
HARD RUBBER	
HYPALON	
KOROSEAL	
NATURAL RUBBER (GRS)	
NEOPRENE GR-M (CR)	
NITRILE BUNA N (NBR)	
NORDEL	
POLYBUTADIENE	
SILICONE	
POLYURETHANE	

PHOSPHORUS TRICHLORIDE

TEMP °F (60, 80, 100, 120, 140, 160, 180, 200, 220, 240, 260, 280, 300, 320, 340, 360, 380, 400, 420, 440, 460, 480, 500, 520, 540, 560)

METALS	TEMP °F 60 – 560
ALLOY 20Cb3	
ALUMINUM	
BRASS	
BRONZE	
CARBON STEEL	
COPPER	
E-BRIGHT 26-1	
HASTELLOY-B	
HASTELLOY-C	
HASTELLOY-D	
HIGH SILICON IRON	
INCONEL	
LEAD	
MONEL	
NICKEL	
STAINLESS STEELS Type 304/307	
Type 316	
TANTALUM	E— (60)
TITANIUM	E— (60)
ZIRCONIUM	

NONMETALLICS	TEMP °F 60 – 560
BOROSILICATE GLASS	
CARBON	R———— (60 to 340)

FOR METALS

E = < 2 Mils Penetration/Year; G = < 20 Mils Penetration/Year

S = < 50 Mils Penetration/Year; U = > 50 Mils Penetration/Year

FOR NONMETALLICS

R = Resistant

U = Unsatisfactory

TEMP °F: 60 80 100 120 140 160 180 200 220 240 260 280 300 320 340 360 380 400 420 440 460

PLASTICS	
ABS	U
ACRYLICS	U
ASBESTOS REINF., EPOXY	
ASBESTOS REINF., PHENOLIC	
CHLORINATED POLYETHER	R
CPVC	U
EPOXY	
FLUOROCARBONS FEP	R
FLUOROCARBONS TFE	R
FURFURYL ALCOHOL	R
NORYL	
PHENOLIC	
POLYAMIDES—NYLON	
POLYESTERS	
BISPHENOL A-FUMURATE	U
HYDROGENATED BISPHENOL-A-BISPHENOL-A	
ISOPHTHALIC	
CHLORINATED POLYESTERS 80%	U
POLYSTYRENES	
POLYETHYLENE	R
POLYPROPYLENE	U
PVC—TYPE 1	U
PVC—TYPE 2	U
POLYVINYLIDENE CHLORIDE	U
VINYLIDENE FLUORIDE	R
VINYL ESTER	U

RUBBERS	
BUTYL GR-1	
FLUORO ELASTOMERS VITON A	R
FLUORO ELASTOMERS KEL-F 3700	
HARD RUBBER	
HYPALON	U
KOROSEAL	R
NATURAL RUBBER (GRS)	
NEOPRENE GR-M (CR)	U
NITRILE BUNA N (NBR)	U
NORDEL	
POLYBUTADIENE	
SILICONE	
POLYURETHANE	

TEMP °F: 60 80 100 120 140 160 180 200 220 240 260 280 300 320 340 360 380 400 420 440 460

PHOSPHORUS, YELLOW

TEMP °F

METALS	60	80	100	120	140	160	180	200	220	240	260	280	300	320	340	360	380	400	420	440	460	480	500	520	540	560
ALLOY 20Cb3																										
ALUMINUM																										
BRASS																										
BRONZE																										
CARBON STEEL																										
COPPER																										
E-BRIGHT 26-1																										
HASTELLOY-B																										
HASTELLOY-C																										
HASTELLOY-D																										
HIGH SILICON IRON																										
INCONEL																										
LEAD																										
MONEL																										
NICKEL																										
STAINLESS STEELS Type 304/307																										
Type 316																										
TANTALUM																										
TITANIUM																										
ZIRCONIUM																										
NONMETALLICS																										
BOROSILICATE GLASS																										
CARBON	R	—	—	—	—	—	—	—	—	—	—	—	—	—	—											

TEMP °F

FOR METALS

E = < 2 Mils Penetration/Year; G = < 20 Mils Penetration/Year

S = < 50 Mils Penetration/Year; U = > 50 Mils Penetration/Year

FOR NONMETALLICS

R = Resistant

U = Unsatisfactory

PHOSPHORUS, YELLOW

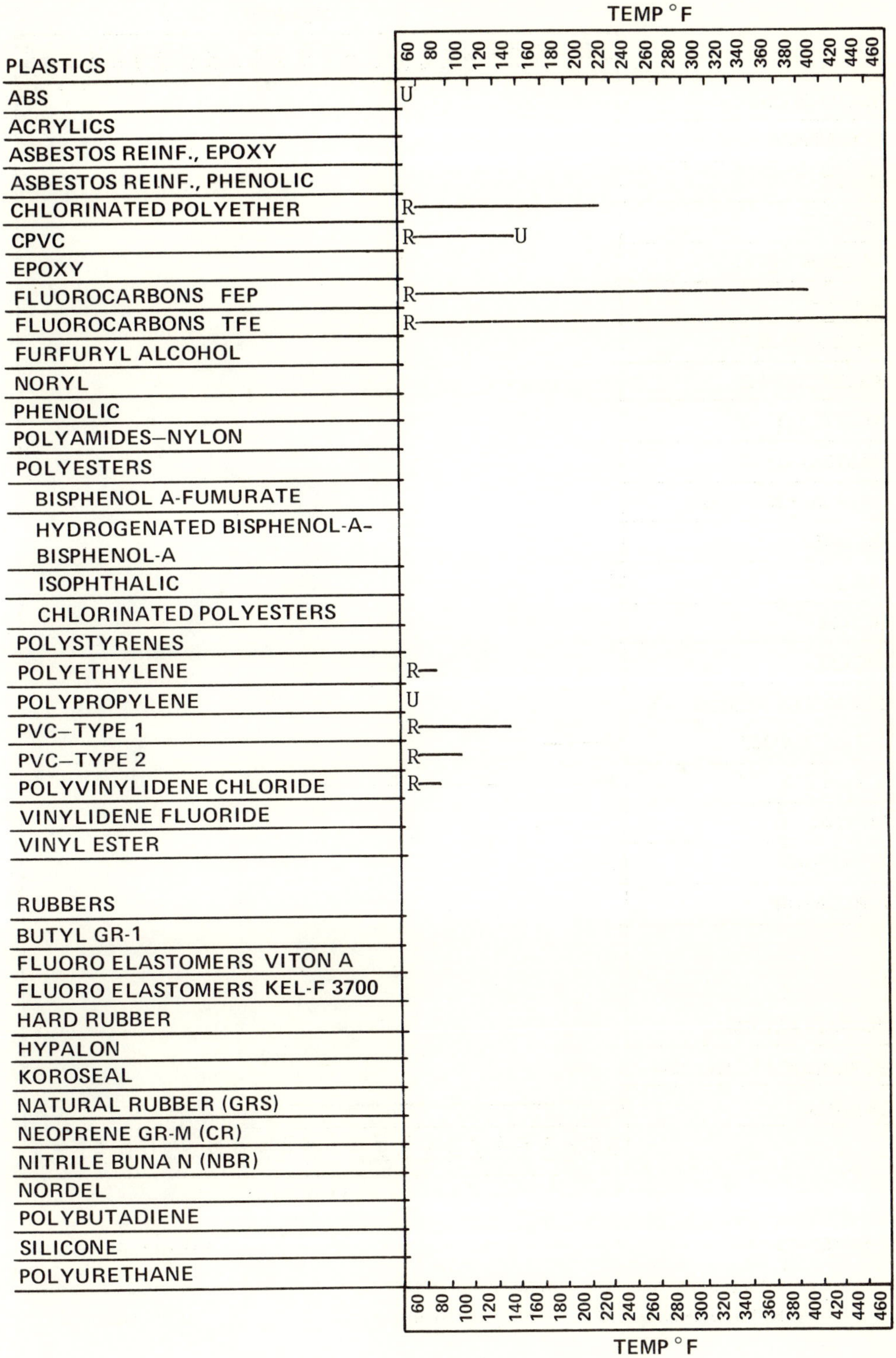

TEMP °F

METALS	60 80 100 120 140 160 180 200 220 240 260 280 300 320 340 360 380 400 420 440 460 480 500 520 540 560
ALLOY 20Cb3	
ALUMINUM	
BRASS	
BRONZE	
CARBON STEEL	U
COPPER	
E-BRIGHT 26-1	
HASTELLOY-B	
HASTELLOY-C	
HASTELLOY-D	
HIGH SILICON IRON	
INCONEL	
LEAD	G
MONEL	
NICKEL	
STAINLESS STEELS	
Type 304/307	E
Type 316	E
TANTALUM	
TITANIUM	G
ZIRCONIUM	
NONMETALLICS	
BOROSILICATE GLASS	
CARBON	R

TEMP °F

FOR METALS

E = < 2 Mils Penetration/Year; G = < 20 Mils Penetration/Year

S = < 50 Mils Penetration/Year; U = > 50 Mils Penetration/Year

FOR NONMETALLICS

R = Resistant

U = Unsatisfactory

PLASTICS	TEMP °F (60–460)
ABS	R
ACRYLICS	R
ASBESTOS REINF., EPOXY	
ASBESTOS REINF., PHENOLIC	
CHLORINATED POLYETHER	R
CPVC	R
EPOXY	
FLUOROCARBONS FEP	R
FLUOROCARBONS TFE	R
FURFURYL ALCOHOL	
NORYL	R
PHENOLIC	
POLYAMIDES—NYLON	
POLYESTERS	
BISPHENOL A-FUMURATE	
HYDROGENATED BISPHENOL-A-BISPHENOL-A	
ISOPHTHALIC	
CHLORINATED POLYESTERS	
POLYSTYRENES	
POLYETHYLENE	R
POLYPROPYLENE	R
PVC—TYPE 1	R
PVC—TYPE 2	R
POLYVINYLIDENE CHLORIDE	R
VINYLIDENE FLUORIDE	
VINYL ESTER	

RUBBERS	TEMP °F (60–460)
BUTYL GR-1	R
FLUORO ELASTOMERS VITON A	R
FLUORO ELASTOMERS KEL-F 3700	
HARD RUBBER	
HYPALON	R
KOROSEAL	
NATURAL RUBBER (GRS)	R
NEOPRENE GR-M (CR)	R
NITRILE BUNA N (NBR)	R
NORDEL	
POLYBUTADIENE	
SILICONE	
POLYURETHANE	

PHTHALIC ACID

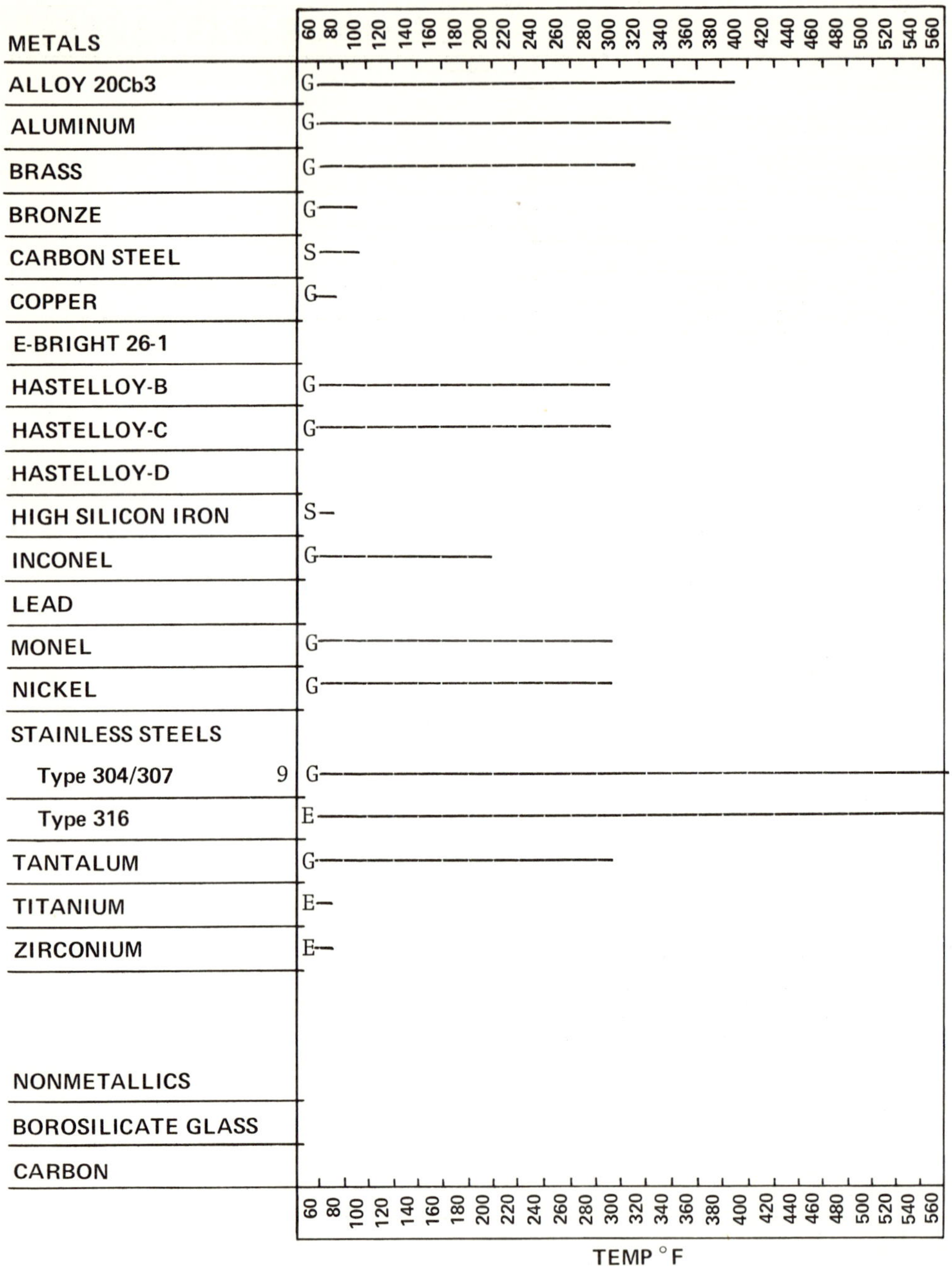

FOR METALS

E = < 2 Mils Penetration/Year; G = < 20 Mils Penetration/Year

S = < 50 Mils Penetration/Year; U = > 50 Mils Penetration/Year

FOR NONMETALLICS

R = Resistant

U = Unsatisfactory

PHTHALIC ACID

TEMP °F (scale: 60, 80, 100, 120, 140, 160, 180, 200, 220, 240, 260, 280, 300, 320, 340, 360, 380, 400, 420, 440, 460)

PLASTICS	
ABS	
ACRYLICS	
ASBESTOS REINF., EPOXY	R
ASBESTOS REINF., PHENOLIC	R
CHLORINATED POLYETHER	
CPVC	
EPOXY	
FLUOROCARBONS FEP	R
FLUOROCARBONS TFE	R
FURFURYL ALCOHOL	R
NORYL	
PHENOLIC	
POLYAMIDES—NYLON	
POLYESTERS	
BISPHENOL A-FUMURATE	R
HYDROGENATED BISPHENOL-A-BISPHENOL-A	R
ISOPHTHALIC	
CHLORINATED POLYESTERS	
POLYSTYRENES	
POLYETHYLENE 50%	R
POLYPROPYLENE	R
PVC—TYPE 1	
PVC—TYPE 2	
POLYVINYLIDENE CHLORIDE	
VINYLIDENE FLUORIDE	R
VINYL ESTER	R
RUBBERS	
BUTYL GR-1	
FLUORO ELASTOMERS VITON A	
FLUORO ELASTOMERS KEL-F 3700	
HARD RUBBER	
HYPALON	
KOROSEAL	R
NATURAL RUBBER (GRS)	
NEOPRENE GR-M (CR)	R
NITRILE BUNA N (NBR)	
NORDEL	
POLYBUTADIENE	
SILICONE	
POLYURETHANE	

TEMP °F

PHTHALIC ANHYDRIDE

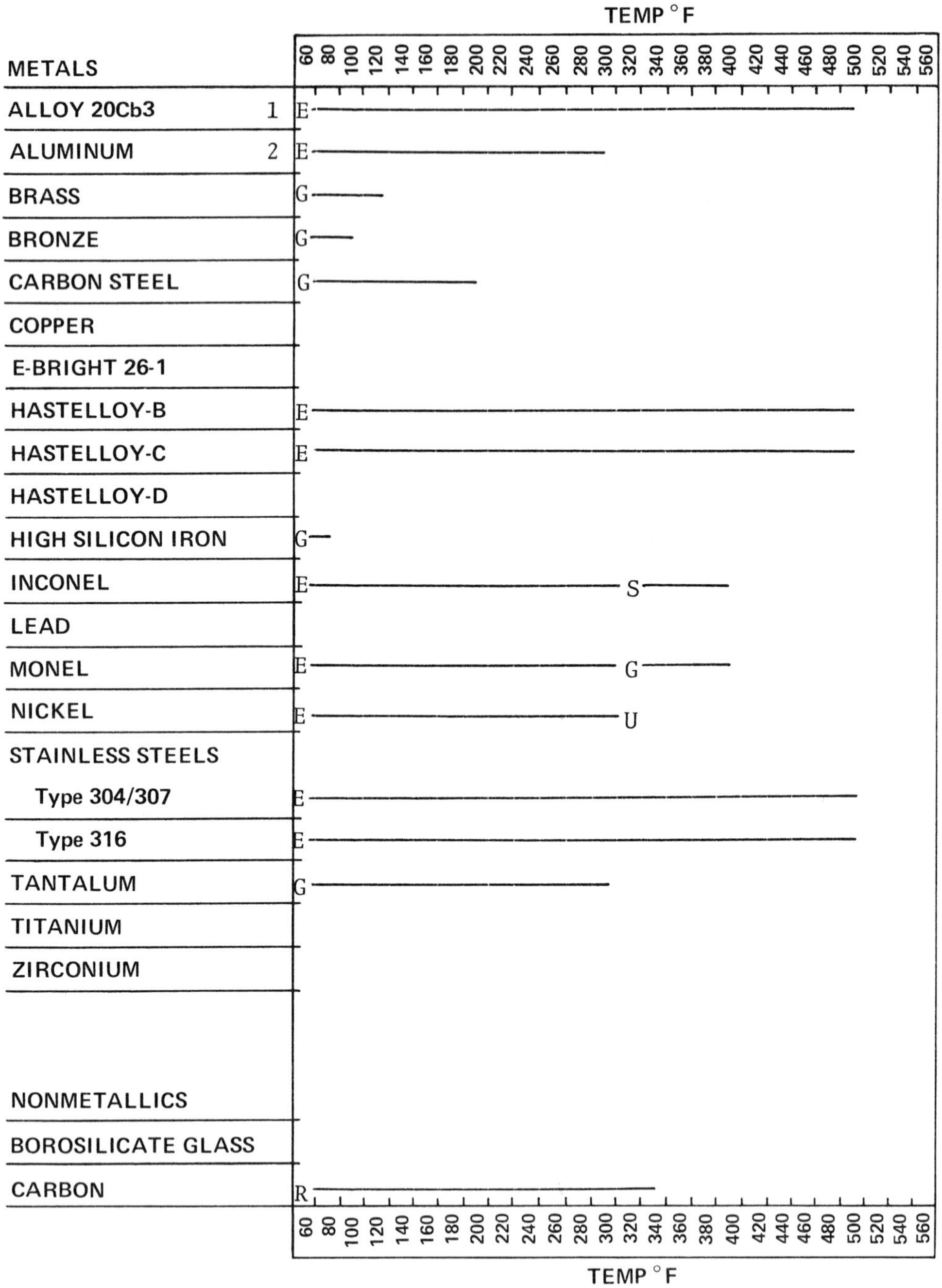

FOR METALS

E = < 2 Mils Penetration/Year; G = < 20 Mils Penetration/Year

S = < 50 Mils Penetration/Year; U = > 50 Mils Penetration/Year

FOR NONMETALLICS

R = Resistant

U = Unsatisfactory

PHTHALIC ANHYDRIDE

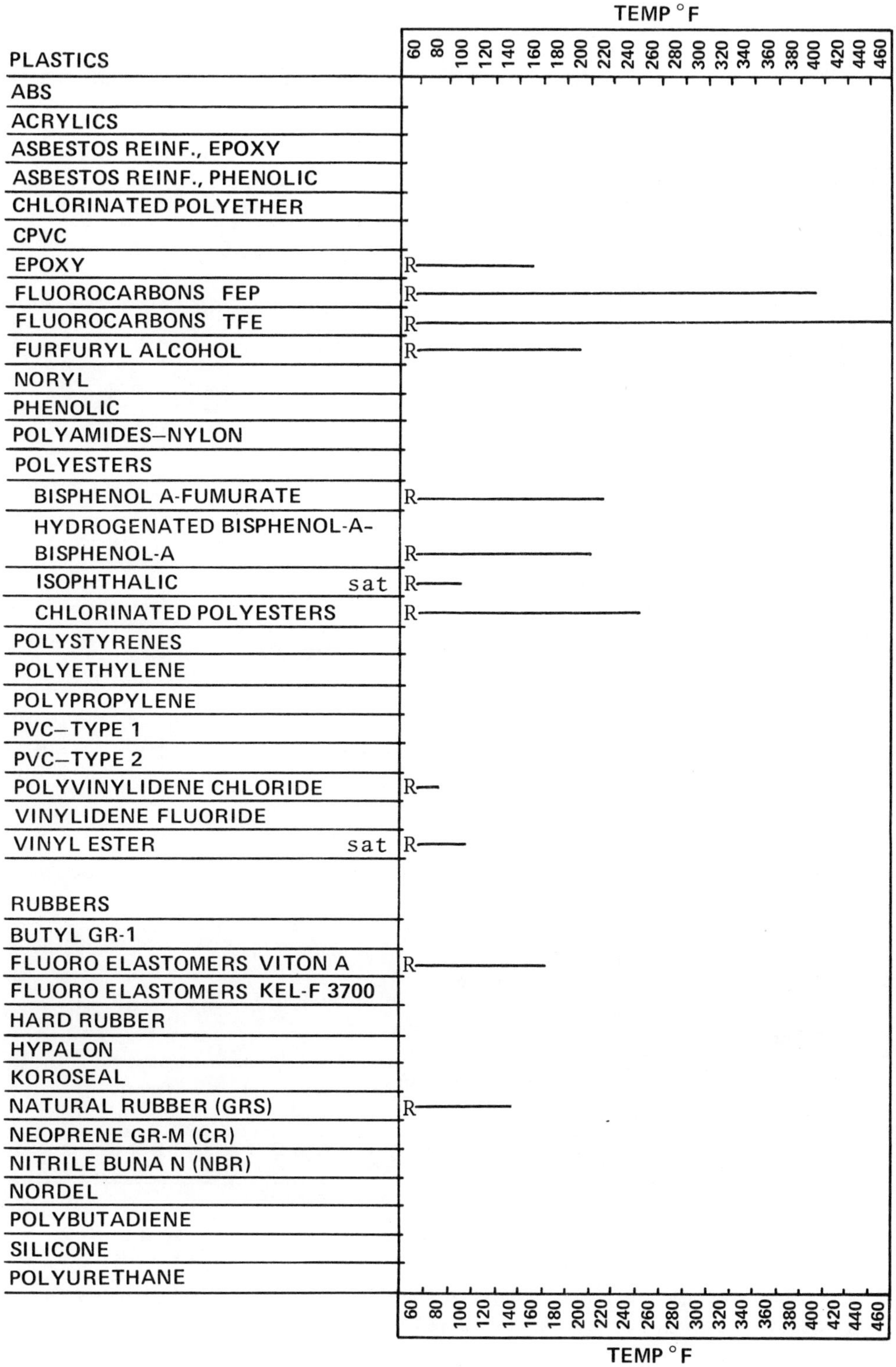

PICRIC ACID

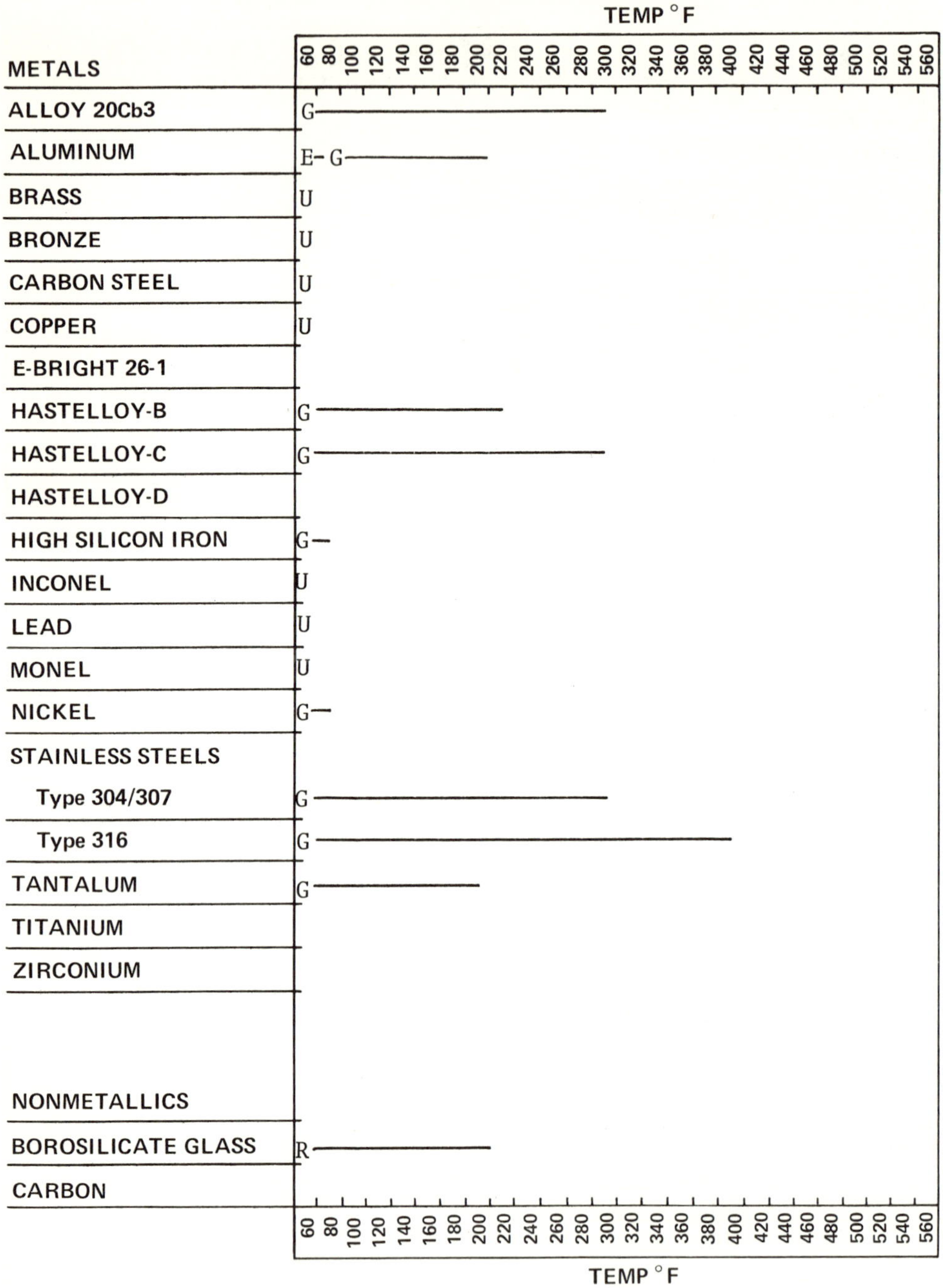

FOR METALS

E = < 2 Mils Penetration/Year; G = < 20 Mils Penetration/Year

S = < 50 Mils Penetration/Year; U = > 50 Mils Penetration/Year

FOR NONMETALLICS

R = Resistant

U = Unsatisfactory

PICRIC ACID

TEMP °F: 60 80 100 120 140 160 180 200 220 240 260 280 300 320 340 360 380 400 420 440 460

PLASTICS		TEMP °F
ABS		U
ACRYLICS		R
ASBESTOS REINF., EPOXY		
ASBESTOS REINF., PHENOLIC		
CHLORINATED POLYETHER		R
CPVC		U
EPOXY		R
FLUOROCARBONS FEP		R
FLUOROCARBONS TFE		R
FURFURYL ALCOHOL	sat	R
NORYL		
PHENOLIC		
POLYAMIDES—NYLON		U
POLYESTERS		
BISPHENOL A-FUMURATE		
HYDROGENATED BISPHENOL-A-BISPHENOL-A		
ISOPHTHALIC		U
CHLORINATED POLYESTERS		
POLYSTYRENES		
POLYETHYLENE		R
POLYPROPYLENE		R
PVC—TYPE 1		U
PVC—TYPE 2		U
POLYVINYLIDENE CHLORIDE		R
VINYLIDENE FLUORIDE		R
VINYL ESTER	10%	R

RUBBERS	TEMP °F
BUTYL GR-1	
FLUORO ELASTOMERS VITON A	R
FLUORO ELASTOMERS KEL-F 3700	
HARD RUBBER	
HYPALON	R
KOROSEAL	U
NATURAL RUBBER (GRS)	U
NEOPRENE GR-M (CR)	R
NITRILE BUNA N (NBR)	R — U
NORDEL	R
POLYBUTADIENE	
SILICONE	
POLYURETHANE	

TEMP °F: 60 80 100 120 140 160 180 200 220 240 260 280 300 320 340 360 380 400 420 440 460

PLATING SOLUTIONS, BRASS

TEMP °F: 60 80 100 120 140 160 180 200 220 240 260 280 300 320 340 360 380 400 420 440 460 480 500 520 540 560

METALS	
ALLOY 20Cb3	
ALUMINUM	
BRASS	
BRONZE	
CARBON STEEL	
COPPER	
E-BRIGHT 26-1	
HASTELLOY-B	
HASTELLOY-C	
HASTELLOY-D	
HIGH SILICON IRON	G
INCONEL	
LEAD	
MONEL	
NICKEL	
STAINLESS STEELS	
Type 304/307	
Type 316	G
TANTALUM	
TITANIUM	
ZIRCONIUM	
NONMETALLICS	
BOROSILICATE GLASS	
CARBON	R

TEMP °F: 60 80 100 120 140 160 180 200 220 240 260 280 300 320 340 360 380 400 420 440 460 480 500 520 540 560

FOR METALS

E = < 2 Mils Penetration/Year; G = < 20 Mils Penetration/Year

S = < 50 Mils Penetration/Year; U = > 50 Mils Penetration/Year

FOR NONMETALLICS

R = Resistant

U = Unsatisfactory

PLATING SOLUTIONS, BRASS

TEMP °F: 60 80 100 120 140 160 180 200 220 240 260 280 300 320 340 360 380 400 420 440 460

PLASTICS	
ABS	U
ACRYLICS	R
ASBESTOS REINF., EPOXY	
ASBESTOS REINF., PHENOLIC	
CHLORINATED POLYETHER	R
CPVC	R
EPOXY	
FLUOROCARBONS FEP	R
FLUOROCARBONS TFE	R
FURFURYL ALCOHOL	
NORYL	
PHENOLIC	
POLYAMIDES–NYLON	
POLYESTERS	
BISPHENOL A-FUMURATE	
HYDROGENATED BISPHENOL-A-BISPHENOL-A	
ISOPHTHALIC	
CHLORINATED POLYESTERS	R
POLYSTYRENES	
POLYETHYLENE	R
POLYPROPYLENE	R
PVC–TYPE 1	R
PVC–TYPE 2	R
POLYVINYLIDENE CHLORIDE	R
VINYLIDENE FLUORIDE	R
VINYL ESTER	

RUBBERS	
BUTYL GR-1	R
FLUORO ELASTOMERS VITON A	R
FLUORO ELASTOMERS KEL-F 3700	
HARD RUBBER	R
HYPALON	R
KOROSEAL	
NATURAL RUBBER (GRS)	R
NEOPRENE GR-M (CR)	R
NITRILE BUNA N (NBR)	R
NORDEL	R
POLYBUTADIENE	
SILICONE	
POLYURETHANE	

TEMP °F: 60 80 100 120 140 160 180 200 220 240 260 280 300 320 340 360 380 400 420 440 460

METALS	TEMP °F (60–560)
ALLOY 20Cb3	
ALUMINUM	
BRASS	
BRONZE	
CARBON STEEL	
COPPER	
E-BRIGHT 26-1	
HASTELLOY-B	
HASTELLOY-C	
HASTELLOY-D	
HIGH SILICON IRON	G—
INCONEL	
LEAD	
MONEL	
NICKEL	
STAINLESS STEELS Type 304/307	
Type 316	G——
TANTALUM	
TITANIUM	
ZIRCONIUM	
NONMETALLICS	
BOROSILICATE GLASS	
CARBON	R——

FOR METALS

E = < 2 Mils Penetration/Year; G = < 20 Mils Penetration/Year

S = < 50 Mils Penetration/Year; U = > 50 Mils Penetration/Year

FOR NONMETALLICS

R = Resistant

U = Unsatisfactory

TEMP °F

PLASTICS	60 80 100 120 140 160 180 200 220 240 260 280 300 320 340 360 380 400 420 440 460
ABS	U
ACRYLICS	R
ASBESTOS REINF., EPOXY	
ASBESTOS REINF., PHENOLIC	
CHLORINATED POLYETHER	R
CPVC	R
EPOXY	
FLUOROCARBONS FEP	R
FLUOROCARBONS TFE	R
FURFURYL ALCOHOL	
NORYL	
PHENOLIC	
POLYAMIDES–NYLON	
POLYESTERS	
BISPHENOL A-FUMURATE	R
HYDROGENATED BISPHENOL-A-BISPHENOL-A	
ISOPHTHALIC	
CHLORINATED POLYESTERS	U
POLYSTYRENES	
POLYETHYLENE	R
POLYPROPYLENE	R
PVC–TYPE 1	R
PVC–TYPE 2	R
POLYVINYLIDENE CHLORIDE	R
VINYLIDENE FLUORIDE	R
VINYL ESTER	
RUBBERS	
BUTYL GR-1	R
FLUORO ELASTOMERS VITON A	R
FLUORO ELASTOMERS KEL-F 3700	
HARD RUBBER	R
HYPALON	R
KOROSEAL	
NATURAL RUBBER (GRS)	R
NEOPRENE GR-M (CR)	R
NITRILE BUNA N (NBR)	R
NORDEL	R
POLYBUTADIENE	
SILICONE	
POLYURETHANE	

60 80 100 120 140 160 180 200 220 240 260 280 300 320 340 360 380 400 420 440 460

TEMP °F

PLATING SOLUTIONS, CHROME, 25%

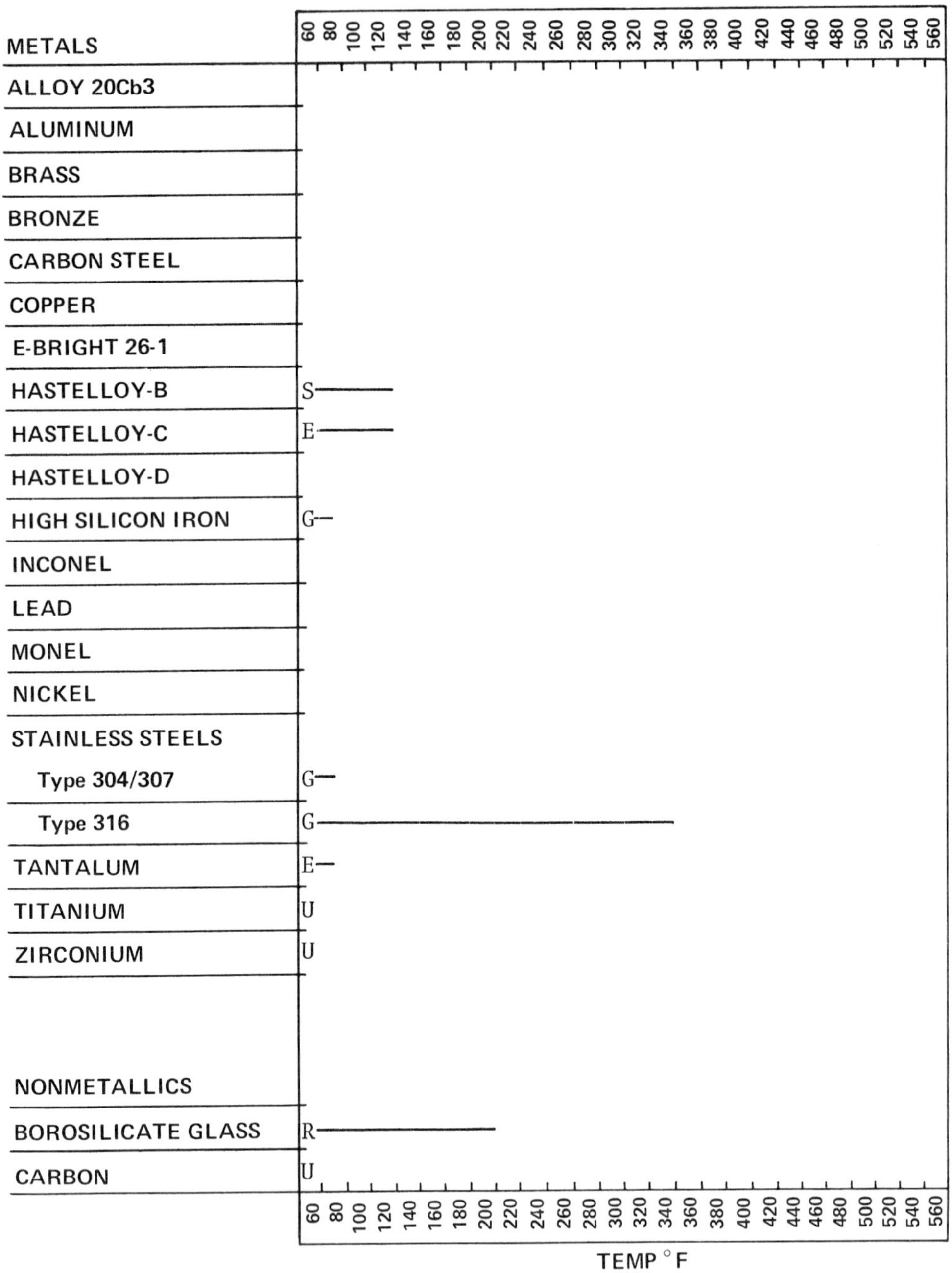

FOR METALS

E = < 2 Mils Penetration/Year; G = < 20 Mils Penetration/Year

S = < 50 Mils Penetration/Year; U = > 50 Mils Penetration/Year

FOR NONMETALLICS

R = Resistant

U = Unsatisfactory

PLATING SOLUTIONS, CHROME, 25%

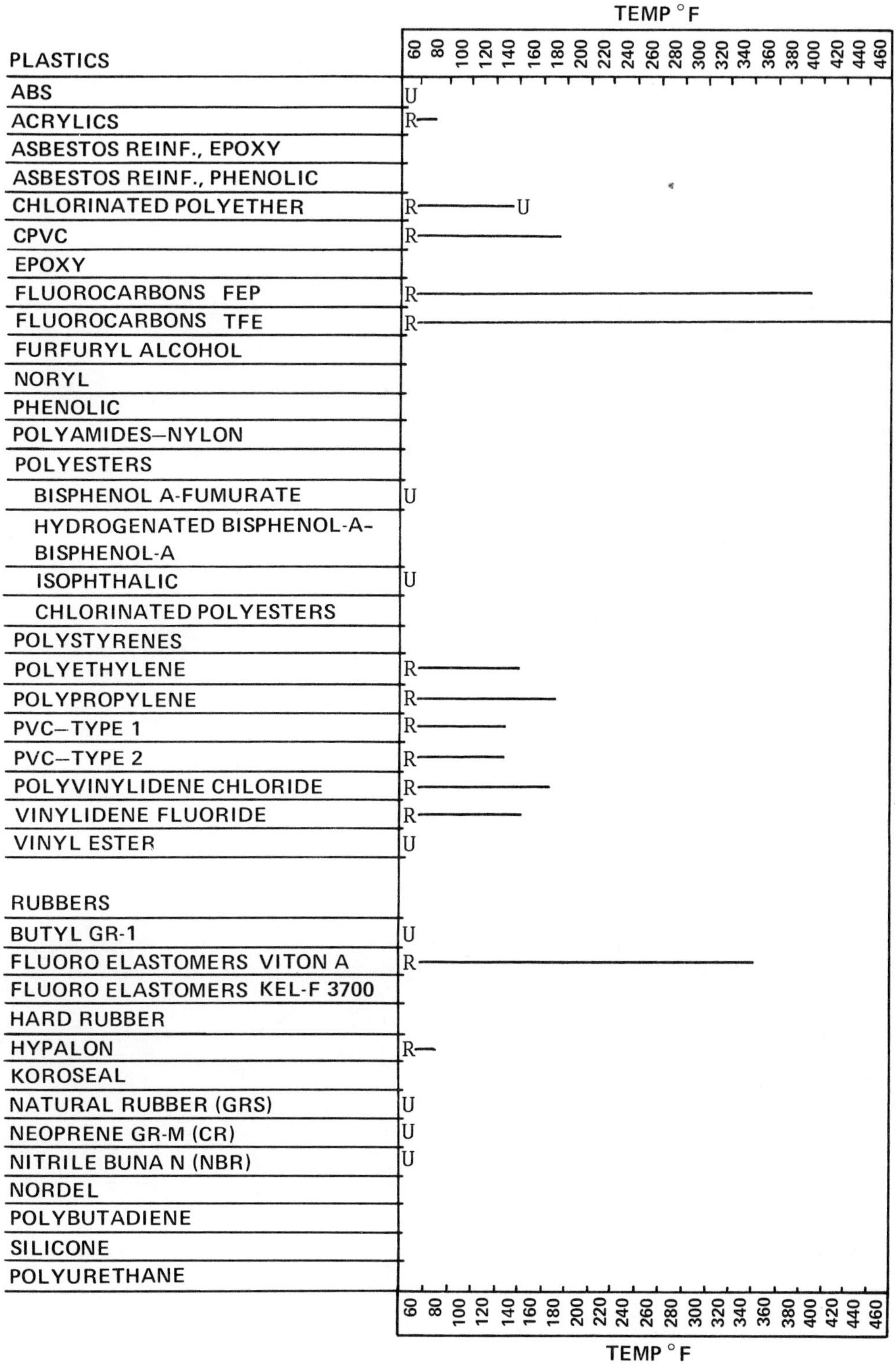

PLATING SOLUTIONS, CHROME, 40%

TEMP °F (60–560)

METALS	Rating
ALLOY 20Cb3	
ALUMINUM	
BRASS	
BRONZE	
CARBON STEEL	
COPPER	
E-BRIGHT 26-1	
HASTELLOY-B	
HASTELLOY-C	
HASTELLOY-D	
HIGH SILICON IRON	G—
INCONEL	
LEAD	
MONEL	
NICKEL	
STAINLESS STEELS	
Type 304/307	
Type 316	G————————
TANTALUM	E—
TITANIUM	U
ZIRCONIUM	U

NONMETALLICS	Rating
BOROSILICATE GLASS	R————
CARBON	U

TEMP °F (60–560)

FOR METALS

E = < 2 Mils Penetration/Year; G = < 20 Mils Penetration/Year

S = < 50 Mils Penetration/Year; U = > 50 Mils Penetration/Year

FOR NONMETALLICS

R = Resistant

U = Unsatisfactory

TEMP °F: 60 80 100 120 140 160 180 200 220 240 260 280 300 320 340 360 380 400 420 440 460

PLASTICS	
ABS	U
ACRYLICS	R
ASBESTOS REINF., EPOXY	
ASBESTOS REINF., PHENOLIC	
CHLORINATED POLYETHER	R
CPVC	R
EPOXY	
FLUOROCARBONS FEP	R
FLUOROCARBONS TFE	R
FURFURYL ALCOHOL	
NORYL	
PHENOLIC	
POLYAMIDES—NYLON	
POLYESTERS	
BISPHENOL A-FUMURATE	U
HYDROGENATED BISPHENOL-A-BISPHENOL-A	
ISOPHTHALIC	U
CHLORINATED POLYESTERS	
POLYSTYRENES	
POLYETHYLENE	U
POLYPROPYLENE	R
PVC—TYPE 1	R
PVC—TYPE 2	R
POLYVINYLIDENE CHLORIDE	R
VINYLIDENE FLUORIDE	R
VINYL ESTER	U
RUBBERS	
BUTYL GR-1	U
FLUORO ELASTOMERS VITON A	R
FLUORO ELASTOMERS KEL-F 3700	
HARD RUBBER	
HYPALON	R
KOROSEAL	
NATURAL RUBBER (GRS)	U
NEOPRENE GR-M (CR)	U
NITRILE BUNA N (NBR)	U
NORDEL	
POLYBUTADIENE	
SILICONE	
POLYURETHANE	

TEMP °F: 60 80 100 120 140 160 180 200 220 240 260 280 300 320 340 360 380 400 420 440 460

PLATING SOLUTIONS, COPPER (CYANIDE)

TEMP °F (60–560)

METALS	
ALLOY 20Cb3	
ALUMINUM	
BRASS	
BRONZE	
CARBON STEEL	
COPPER	
E-BRIGHT 26-1	
HASTELLOY-B	
HASTELLOY-C	
HASTELLOY-D	
HIGH SILICON IRON	G
INCONEL	
LEAD	
MONEL	
NICKEL	
STAINLESS STEELS Type 304/307	
Type 316	
TANTALUM	E
TITANIUM	
ZIRCONIUM	

NONMETALLICS	
BOROSILICATE GLASS	R
CARBON	R

TEMP °F

FOR METALS

E = < 2 Mils Penetration/Year; G = < 20 Mils Penetration/Year

S = < 50 Mils Penetration/Year; U = > 50 Mils Penetration/Year

FOR NONMETALLICS

R = Resistant

U = Unsatisfactory

TEMP °F (60–460)

PLASTICS	Rating
ABS	U
ACRYLICS	R
ASBESTOS REINF., EPOXY	
ASBESTOS REINF., PHENOLIC	
CHLORINATED POLYETHER	R
CPVC	R
EPOXY	
FLUOROCARBONS FEP	R
FLUOROCARBONS TFE	R
FURFURYL ALCOHOL	
NORYL	
PHENOLIC	
POLYAMIDES–NYLON	
POLYESTERS	
BISPHENOL A-FUMURATE	
HYDROGENATED BISPHENOL-A-BISPHENOL-A	
ISOPHTHALIC	
CHLORINATED POLYESTERS 10	R
POLYSTYRENES	
POLYETHYLENE	R
POLYPROPYLENE	R
PVC–TYPE 1	R
PVC–TYPE 2	R
POLYVINYLIDENE CHLORIDE	R
VINYLIDENE FLUORIDE	R
VINYL ESTER	

RUBBERS	Rating
BUTYL GR-1	R
FLUORO ELASTOMERS VITON A	R
FLUORO ELASTOMERS KEL-F 3700	
HARD RUBBER	R
HYPALON	R
KOROSEAL	
NATURAL RUBBER (GRS)	R
NEOPRENE GR-M (CR)	R
NITRILE BUNA N (NBR)	R
NORDEL	R
POLYBUTADIENE	
SILICONE	
POLYURETHANE	

TEMP °F (60–460)

TEMP °F

METALS	60	80	100	120	140	160	180	200	220	240	260	280	300	320	340	360	380	400	420	440	460	480	500	520	540	560
ALLOY 20Cb3																										
ALUMINUM																										
BRASS																										
BRONZE																										
CARBON STEEL																										
COPPER																										
E-BRIGHT 26-1																										
HASTELLOY-B																										
HASTELLOY-C																										
HASTELLOY-D																										
HIGH SILICON IRON	G																									
INCONEL																										
LEAD																										
MONEL																										
NICKEL																										
STAINLESS STEELS Type 304/307																										
Type 316																										
TANTALUM																										
TITANIUM																										
ZIRCONIUM																										
NONMETALLICS																										
BOROSILICATE GLASS																										
CARBON	R	—	—	—	—	—	—	—	—	—	—	—	—	—	—											

TEMP °F

FOR METALS

E = < 2 Mils Penetration/Year; G = < 20 Mils Penetration/Year

S = < 50 Mils Penetration/Year; U = > 50 Mils Penetration/Year

FOR NONMETALLICS

R = Resistant

U = Unsatisfactory

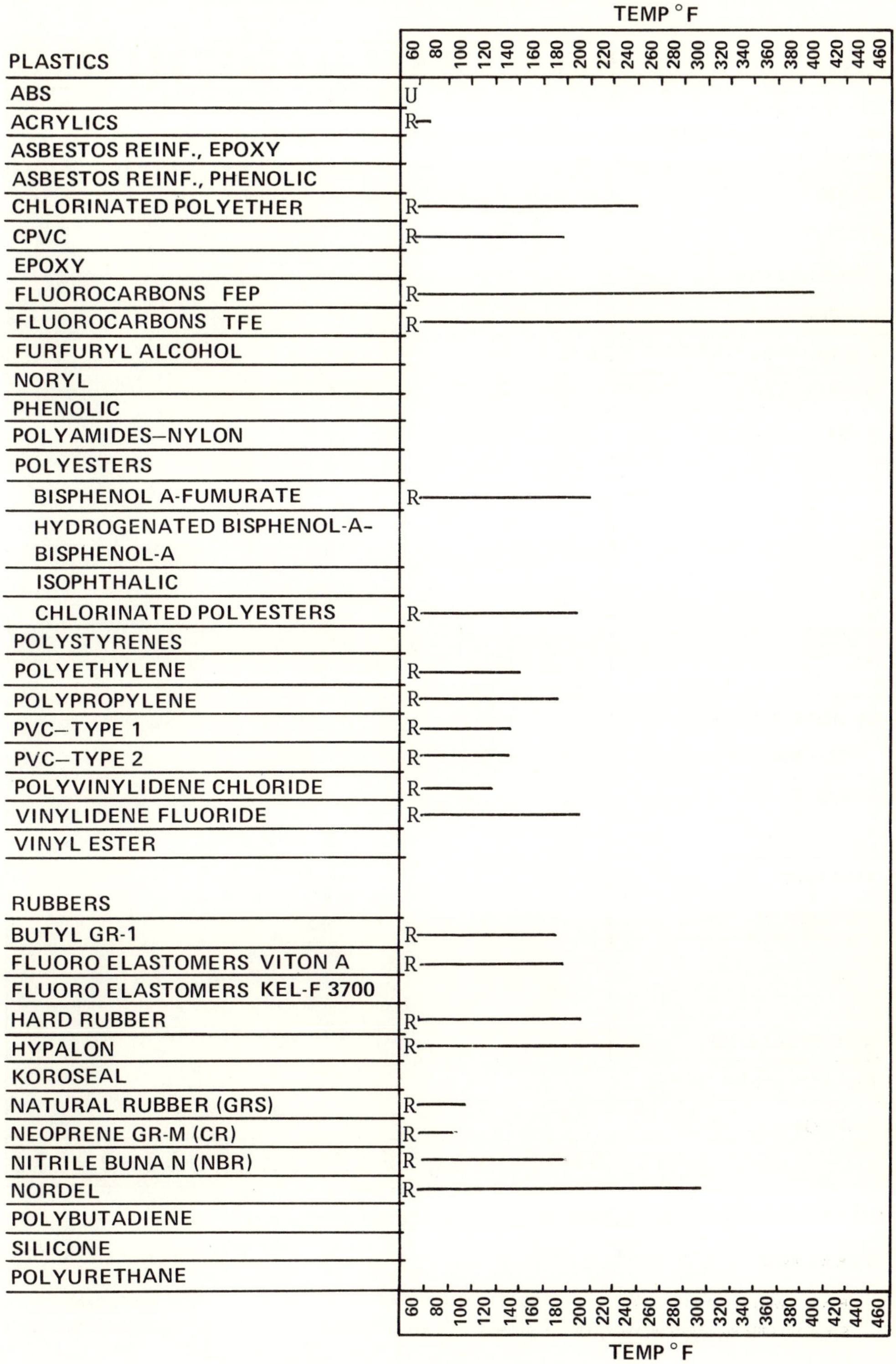
TEMP °F
60 80 100 120 140 160 180 200 220 240 260 280 300 320 340 360 380 400 420 440 460
PLASTICS
ABS U
ACRYLICS R
ASBESTOS REINF., EPOXY
ASBESTOS REINF., PHENOLIC
CHLORINATED POLYETHER R
CPVC R
EPOXY
FLUOROCARBONS FEP R
FLUOROCARBONS TFE R
FURFURYL ALCOHOL
NORYL
PHENOLIC
POLYAMIDES—NYLON
POLYESTERS
BISPHENOL A-FUMURATE R
HYDROGENATED BISPHENOL-A-BISPHENOL-A
ISOPHTHALIC
CHLORINATED POLYESTERS R
POLYSTYRENES
POLYETHYLENE R
POLYPROPYLENE R
PVC—TYPE 1 R
PVC—TYPE 2 R
POLYVINYLIDENE CHLORIDE R
VINYLIDENE FLUORIDE R
VINYL ESTER
RUBBERS
BUTYL GR-1 R
FLUORO ELASTOMERS VITON A R
FLUORO ELASTOMERS KEL-F 3700
HARD RUBBER R
HYPALON R
KOROSEAL
NATURAL RUBBER (GRS) R
NEOPRENE GR-M (CR) R
NITRILE BUNA N (NBR) R
NORDEL R
POLYBUTADIENE
SILICONE
POLYURETHANE
60 80 100 120 140 160 180 200 220 240 260 280 300 320 340 360 380 400 420 440 460
TEMP °F

PLATING SOLUTIONS, IRON

TEMP °F

METALS	60	80	100	120	140	160	180	200	220	240	260	280	300	320	340	360	380	400	420	440	460	480	500	520	540	560
ALLOY 20Cb3																										
ALUMINUM																										
BRASS																										
BRONZE																										
CARBON STEEL																										
COPPER																										
E-BRIGHT 26-1																										
HASTELLOY-B																										
HASTELLOY-C																										
HASTELLOY-D																										
HIGH SILICON IRON	G—																									
INCONEL																										
LEAD																										
MONEL																										
NICKEL																										
STAINLESS STEELS Type 304/307																										
Type 316																										
TANTALUM																										
TITANIUM																										
ZIRCONIUM																										
NONMETALLICS																										
BOROSILICATE GLASS																										
CARBON																										

TEMP °F

FOR METALS

E = < 2 Mils Penetration/Year; G = < 20 Mils Penetration/Year

S = < 50 Mils Penetration/Year; U = > 50 Mils Penetration/Year

FOR NONMETALLICS

R = Resistant

U = Unsatisfactory

PLATING SOLUTIONS, IRON

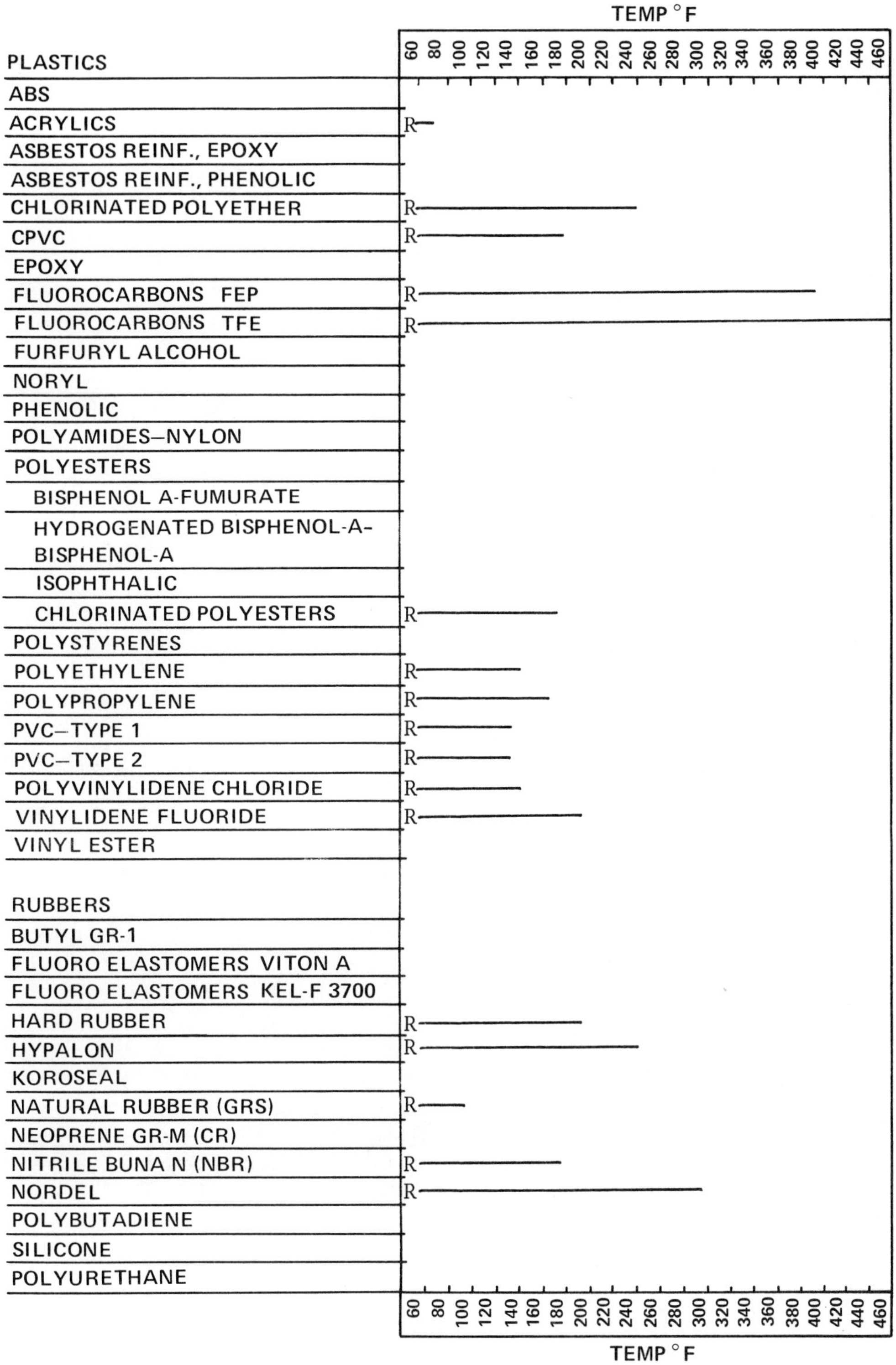

TEMP °F (60–560)

METALS	Rating	Temperature range (°F)
ALLOY 20Cb3		
ALUMINUM		
BRASS		
BRONZE		
CARBON STEEL		
COPPER		
E-BRIGHT 26-1		
HASTELLOY-B	E	60
HASTELLOY-C		
HASTELLOY-D		
HIGH SILICON IRON	G	60
INCONEL		
LEAD		
MONEL		
NICKEL		
STAINLESS STEELS Type 304/307		
Type 316		
TANTALUM		
TITANIUM		
ZIRCONIUM		
NONMETALLICS		
BOROSILICATE GLASS		
CARBON	R	60–340

FOR METALS

E = < 2 Mils Penetration/Year; G = < 20 Mils Penetration/Year

S = < 50 Mils Penetration/Year; U = > 50 Mils Penetration/Year

FOR NONMETALLICS

R = Resistant

U = Unsatisfactory

TEMP °F (60–460)

PLASTICS	Note	Rating
ABS		U
ACRYLICS		R
ASBESTOS REINF., EPOXY		
ASBESTOS REINF., PHENOLIC		
CHLORINATED POLYETHER		R
CPVC		R
EPOXY		
FLUOROCARBONS FEP		R
FLUOROCARBONS TFE		R
FURFURYL ALCOHOL		
NORYL		
PHENOLIC		
POLYAMIDES–NYLON		
POLYESTERS		
BISPHENOL A-FUMURATE	10	R
HYDROGENATED BISPHENOL-A-BISPHENOL-A		
ISOPHTHALIC		
CHLORINATED POLYESTERS	10	R
POLYSTYRENES		
POLYETHYLENE		R
POLYPROPYLENE		R
PVC–TYPE 1		R
PVC–TYPE 2		R
POLYVINYLIDENE CHLORIDE		R
VINYLIDENE FLUORIDE		R
VINYL ESTER		

RUBBERS	Rating
BUTYL GR-1	R
FLUORO ELASTOMERS VITON A	R
FLUORO ELASTOMERS KEL-F 3700	
HARD RUBBER	R
HYPALON	R
KOROSEAL	
NATURAL RUBBER (GRS)	R
NEOPRENE GR-M (CR)	R
NITRILE BUNA N (NBR)	R
NORDEL	R
POLYBUTADIENE	
SILICONE	
POLYURETHANE	

TEMP °F (60–460)

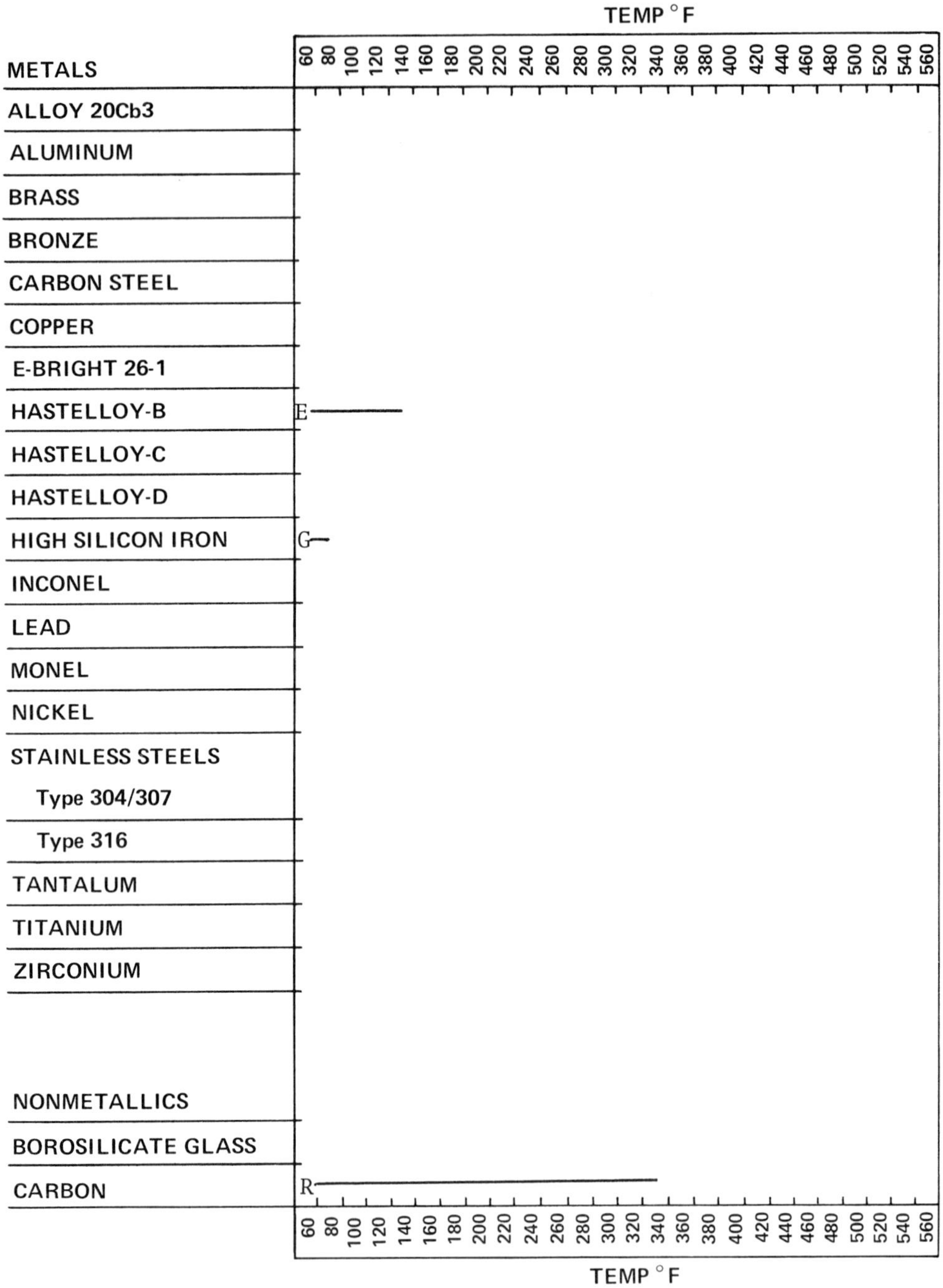

FOR METALS

E = < 2 Mils Penetration/Year; G = < 20 Mils Penetration/Year

S = < 50 Mils Penetration/Year; U = > 50 Mils Penetration/Year

FOR NONMETALLICS

R = Resistant

U = Unsatisfactory

TEMP °F: 60 80 100 120 140 160 180 200 220 240 260 280 300 320 340 360 380 400 420 440 460

PLASTICS	
ABS	U
ACRYLICS	R
ASBESTOS REINF., EPOXY	
ASBESTOS REINF., PHENOLIC	
CHLORINATED POLYETHER	R
CPVC	R
EPOXY	R U
FLUOROCARBONS FEP	R
FLUOROCARBONS TFE	R
FURFURYL ALCOHOL	
NORYL	
PHENOLIC	
POLYAMIDES–NYLON	
POLYESTERS	
BISPHENOL A-FUMURATE	R
HYDROGENATED BISPHENOL-A-BISPHENOL-A	
ISOPHTHALIC	
CHLORINATED POLYESTERS	R
POLYSTYRENES	
POLYETHYLENE	R
POLYPROPYLENE	R
PVC–TYPE 1	R
PVC–TYPE 2	R
POLYVINYLIDENE CHLORIDE	R
VINYLIDENE FLUORIDE	R
VINYL ESTER	

RUBBERS	
BUTYL GR-1	
FLUORO ELASTOMERS VITON A	R
FLUORO ELASTOMERS KEL-F 3700	
HARD RUBBER	R
HYPALON	R
KOROSEAL	
NATURAL RUBBER (GRS)	R
NEOPRENE GR-M (CR)	R
NITRILE BUNA N (NBR)	R
NORDEL	R
POLYBUTADIENE	
SILICONE	
POLYURETHANE	

TEMP °F: 60 80 100 120 140 160 180 200 220 240 260 280 300 320 340 360 380 400 420 440 460

TEMP °F (60–560)

METALS	
ALLOY 20Cb3	
ALUMINUM	
BRASS	
BRONZE	
CARBON STEEL	
COPPER	
E-BRIGHT 26-1	
HASTELLOY-B	
HASTELLOY-C	
HASTELLOY-D	
HIGH SILICON IRON	
INCONEL	
LEAD	
MONEL	
NICKEL	
STAINLESS STEELS Type 304/307	
Type 316	
TANTALUM	
TITANIUM	
ZIRCONIUM	

NONMETALLICS	Rating (60–560 °F)
BOROSILICATE GLASS	R (60–200)
CARBON	R (60–340)

TEMP °F

FOR METALS

E = < 2 Mils Penetration/Year; G = < 20 Mils Penetration/Year

S = < 50 Mils Penetration/Year; U = > 50 Mils Penetration/Year

FOR NONMETALLICS

R = Resistant

U = Unsatisfactory

TEMP °F (60, 80, 100, 120, 140, 160, 180, 200, 220, 240, 260, 280, 300, 320, 340, 360, 380, 400, 420, 440, 460)

PLASTICS	
ABS	U
ACRYLICS	R
ASBESTOS REINF., EPOXY	
ASBESTOS REINF., PHENOLIC	
CHLORINATED POLYETHER	R
CPVC	R
EPOXY	
FLUOROCARBONS FEP	R
FLUOROCARBONS TFE	R
FURFURYL ALCOHOL	
NORYL	
PHENOLIC	
POLYAMIDES—NYLON	
POLYESTERS	
BISPHENOL A-FUMURATE	
HYDROGENATED BISPHENOL-A-BISPHENOL-A	
ISOPHTHALIC	
CHLORINATED POLYESTERS	
POLYSTYRENES	
POLYETHYLENE	R
POLYPROPYLENE	R
PVC—TYPE 1	R
PVC—TYPE 2	R
POLYVINYLIDENE CHLORIDE	R
VINYLIDENE FLUORIDE	R
VINYL ESTER	

RUBBERS	
BUTYL GR-1	
FLUORO ELASTOMERS VITON A	R
FLUORO ELASTOMERS KEL-F 3700	
HARD RUBBER	
HYPALON	R
KOROSEAL	
NATURAL RUBBER (GRS)	
NEOPRENE GR-M (CR)	R
NITRILE BUNA N (NBR)	R
NORDEL	
POLYBUTADIENE	
SILICONE	
POLYURETHANE	

TEMP °F (60, 80, 100, 120, 140, 160, 180, 200, 220, 240, 260, 280, 300, 320, 340, 360, 380, 400, 420, 440, 460)

TEMP °F (60–560)

METALS	TEMP °F
ALLOY 20Cb3	
ALUMINUM	
BRASS	
BRONZE	
CARBON STEEL	
COPPER	
E-BRIGHT 26-1	
HASTELLOY-B	
HASTELLOY-C	
HASTELLOY-D	
HIGH SILICON IRON	G (60)
INCONEL	
LEAD	
MONEL	
NICKEL	
STAINLESS STEELS	
Type 304/307	
Type 316	
TANTALUM	
TITANIUM	
ZIRCONIUM	

NONMETALLICS	TEMP °F
BOROSILICATE GLASS	
CARBON	R (60–340)

FOR METALS

E = < 2 Mils Penetration/Year; G = < 20 Mils Penetration/Year
S = < 50 Mils Penetration/Year; U = > 50 Mils Penetration/Year

FOR NONMETALLICS

R = Resistant
U = Unsatisfactory

PLATING SOLUTIONS, SILVER

TEMP °F: 60 80 100 120 140 160 180 200 220 240 260 280 300 320 340 360 380 400 420 440 460

PLASTICS	TEMP °F
ABS	U
ACRYLICS	R—
ASBESTOS REINF., EPOXY	
ASBESTOS REINF., PHENOLIC	
CHLORINATED POLYETHER	R—
CPVC	R—
EPOXY	R—U
FLUOROCARBONS FEP	R—
FLUOROCARBONS TFE	R—
FURFURYL ALCOHOL	
NORYL	
PHENOLIC	
POLYAMIDES–NYLON	
POLYESTERS	
BISPHENOL A-FUMURATE	R—
HYDROGENATED BISPHENOL-A-BISPHENOL-A	
ISOPHTHALIC	
CHLORINATED POLYESTERS	U
POLYSTYRENES	
POLYETHYLENE	R—
POLYPROPYLENE	R—
PVC–TYPE 1	R—
PVC–TYPE 2	R—
POLYVINYLIDENE CHLORIDE	R—
VINYLIDENE FLUORIDE	R—
VINYL ESTER	

RUBBERS	TEMP °F
BUTYL GR-1	R—
FLUORO ELASTOMERS VITON A	R—
FLUORO ELASTOMERS KEL-F 3700	
HARD RUBBER	R—
HYPALON	R—
KOROSEAL	
NATURAL RUBBER (GRS)	R—
NEOPRENE GR-M (CR)	R—
NITRILE BUNA N (NBR)	R—
NORDEL	R—
POLYBUTADIENE	
SILICONE	
POLYURETHANE	

TEMP °F: 60 80 100 120 140 160 180 200 220 240 260 280 300 320 340 360 380 400 420 440 460

PLATING SOLUTIONS, TIN

TEMP °F: 60 80 100 120 140 160 180 200 220 240 260 280 300 320 340 360 380 400 420 440 460 480 500 520 540 560

METALS	TEMP °F
ALLOY 20Cb3	
ALUMINUM	
BRASS	
BRONZE	
CARBON STEEL	
COPPER	
E-BRIGHT 26-1	
HASTELLOY-B	E—
HASTELLOY-C	
HASTELLOY-D	
HIGH SILICON IRON	G—
INCONEL	
LEAD	
MONEL	
NICKEL	
STAINLESS STEELS Type 304/307	
Type 316	
TANTALUM	
TITANIUM	
ZIRCONIUM	
NONMETALLICS	
BOROSILICATE GLASS	
CARBON	R——————————— (60–340)

TEMP °F: 60 80 100 120 140 160 180 200 220 240 260 280 300 320 340 360 380 400 420 440 460 480 500 520 540 560

FOR METALS

E = < 2 Mils Penetration/Year; G = < 20 Mils Penetration/Year

S = < 50 Mils Penetration/Year; U = > 50 Mils Penetration/Year

FOR NONMETALLICS

R = Resistant

U = Unsatisfactory

PLATING SOLUTIONS, TIN

TEMP °F (60–460)

PLASTICS		Rating
ABS		U
ACRYLICS		R
ASBESTOS REINF., EPOXY		
ASBESTOS REINF., PHENOLIC		
CHLORINATED POLYETHER		R
CPVC		R
EPOXY		
FLUOROCARBONS FEP		R
FLUOROCARBONS TFE		R
FURFURYL ALCOHOL		
NORYL		
PHENOLIC		
POLYAMIDES–NYLON		
POLYESTERS		
BISPHENOL A-FUMURATE	10	R
HYDROGENATED BISPHENOL-A-BISPHENOL-A		
ISOPHTHALIC		
CHLORINATED POLYESTERS	10	R
POLYSTYRENES		
POLYETHYLENE		R
POLYPROPYLENE		R
PVC–TYPE 1		R
PVC–TYPE 2		R
POLYVINYLIDENE CHLORIDE		R
VINYLIDENE FLUORIDE		R
VINYL ESTER		

RUBBERS	Rating
BUTYL GR-1	R
FLUORO ELASTOMERS VITON A	R
FLUORO ELASTOMERS KEL-F 3700	
HARD RUBBER	R
HYPALON	R
KOROSEAL	
NATURAL RUBBER (GRS)	
NEOPRENE GR-M (CR)	R
NITRILE BUNA N (NBR)	R
NORDEL	R
POLYBUTADIENE	
SILICONE	
POLYURETHANE	

TEMP °F (60–460)

METALS	TEMP °F (60 80 100 120 140 160 180 200 220 240 260 280 300 320 340 360 380 400 420 440 460 480 500 520 540 560)
ALLOY 20Cb3	
ALUMINUM	
BRASS	
BRONZE	
CARBON STEEL	
COPPER	
E-BRIGHT 26-1	
HASTELLOY-B	
HASTELLOY-C	
HASTELLOY-D	
HIGH SILICON IRON	G
INCONEL	
LEAD	
MONEL	
NICKEL	
STAINLESS STEELS Type 304/307	
Type 316	
TANTALUM	
TITANIUM	
ZIRCONIUM	
NONMETALLICS	
BOROSILICATE GLASS	
CARBON	R

TEMP °F

FOR METALS

E = < 2 Mils Penetration/Year; G = < 20 Mils Penetration/Year

S = < 50 Mils Penetration/Year; U = > 50 Mils Penetration/Year

FOR NONMETALLICS

R = Resistant

U = Unsatisfactory

TEMP °F (60–460)

PLASTICS	Rating
ABS	U
ACRYLICS	R
ASBESTOS REINF., EPOXY	
ASBESTOS REINF., PHENOLIC	
CHLORINATED POLYETHER	R
CPVC	R
EPOXY	
FLUOROCARBONS FEP	R
FLUOROCARBONS TFE	R
FURFURYL ALCOHOL	
NORYL	
PHENOLIC	
POLYAMIDES–NYLON	
POLYESTERS	
BISPHENOL A-FUMURATE	
HYDROGENATED BISPHENOL-A-BISPHENOL-A	
ISOPHTHALIC	
CHLORINATED POLYESTERS	
POLYSTYRENES	
POLYETHYLENE	R
POLYPROPYLENE	R
PVC–TYPE 1	R
PVC–TYPE 2	R
POLYVINYLIDENE CHLORIDE	R
VINYLIDENE FLUORIDE	R
VINYL ESTER	

RUBBERS	Rating
BUTYL GR-1	R
FLUORO ELASTOMERS VITON A	R
FLUORO ELASTOMERS KEL-F 3700	
HARD RUBBER	R
HYPALON	R
KOROSEAL	
NATURAL RUBBER (GRS)	R
NEOPRENE GR-M (CR)	R
NITRILE BUNA N (NBR)	R
NORDEL	R
POLYBUTADIENE	
SILICONE	
POLYURETHANE	

TEMP °F (60–460)

POLYVINYL ACETATE

METALS	TEMP °F
ALLOY 20Cb3	G (60–120)
ALUMINUM	
BRASS	
BRONZE	
CARBON STEEL	
COPPER	
E-BRIGHT 26-1	
HASTELLOY-B	
HASTELLOY-C	
HASTELLOY-D	
HIGH SILICON IRON	
INCONEL	
LEAD	
MONEL	
NICKEL	
STAINLESS STEELS	
Type 304/307	E (60–180)
Type 316	E (60–180)
TANTALUM	G (60)
TITANIUM	
ZIRCONIUM	
NONMETALLICS	
BOROSILICATE GLASS	
CARBON	

TEMP °F: 60, 80, 100, 120, 140, 160, 180, 200, 220, 240, 260, 280, 300, 320, 340, 360, 380, 400, 420, 440, 460, 480, 500, 520, 540, 560

FOR METALS

E = < 2 Mils Penetration/Year; G = < 20 Mils Penetration/Year

S = < 50 Mils Penetration/Year; U = > 50 Mils Penetration/Year

FOR NONMETALLICS

R = Resistant

U = Unsatisfactory

POLYVINYL ACETATE

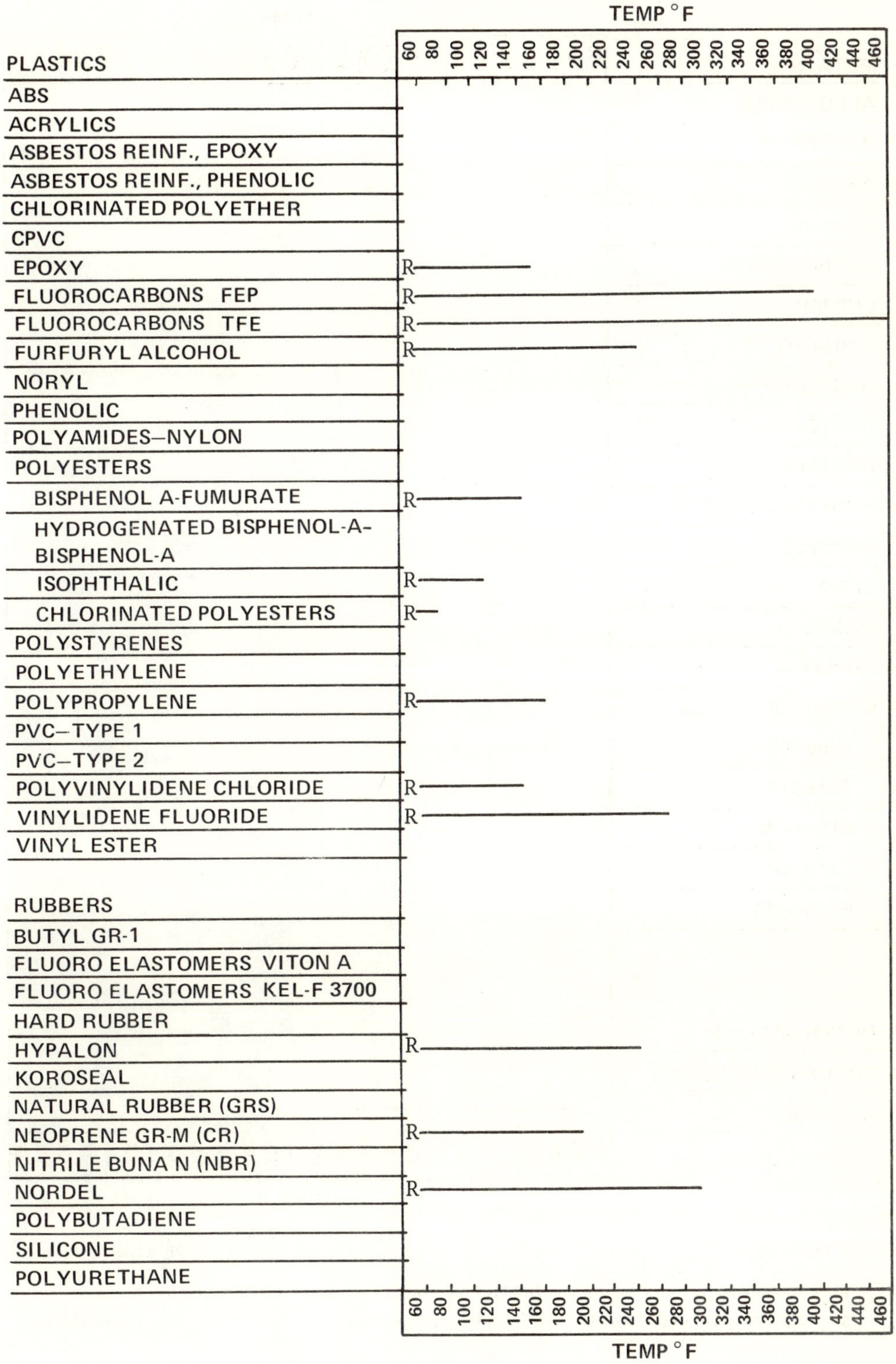

POTASSIUM ALUMINUM SULFATE

TEMP °F

METALS	60	80	100	120	140	160	180	200	220	240	260	280	300	320	340	360	380	400	420	440	460	480	500	520	540	560
ALLOY 20Cb3																										
ALUMINUM																										
BRASS																										
BRONZE																										
CARBON STEEL																										
COPPER																										
E-BRIGHT 26-1																										
HASTELLOY-B																										
HASTELLOY-C																										
HASTELLOY-D																										
HIGH SILICON IRON																										
INCONEL																										
LEAD																										
MONEL																										
NICKEL																										
STAINLESS STEELS																										
Type 304/307																										
Type 316																										
TANTALUM																										
TITANIUM																										
ZIRCONIUM																										
NONMETALLICS																										
BOROSILICATE GLASS																										
CARBON	R	—	—	—	—	—	—	—	—	—	—	—	—	—	—											

TEMP °F

FOR METALS

E = < 2 Mils Penetration/Year; G = < 20 Mils Penetration/Year

S = < 50 Mils Penetration/Year; U = > 50 Mils Penetration/Year

FOR NONMETALLICS

R = Resistant

U = Unsatisfactory

POTASSIUM ALUMINUM SULFATE

PLASTICS	TEMP °F (60–460)
ABS	
ACRYLICS	
ASBESTOS REINF., EPOXY	
ASBESTOS REINF., PHENOLIC	
CHLORINATED POLYETHER	R (to ~150)
CPVC	
EPOXY	
FLUOROCARBONS FEP	R (to ~400)
FLUOROCARBONS TFE	R (to 460)
FURFURYL ALCOHOL	
NORYL	R (to ~140)
PHENOLIC	
POLYAMIDES–NYLON	
POLYESTERS	
BISPHENOL A-FUMURATE	R (to ~200)
HYDROGENATED BISPHENOL-A-BISPHENOL-A	
ISOPHTHALIC	R (to ~180)
CHLORINATED POLYESTERS	R (to ~120)
POLYSTYRENES	
POLYETHYLENE	
POLYPROPYLENE	R (to ~220)
PVC–TYPE 1	R (to ~140)
PVC–TYPE 2	
POLYVINYLIDENE CHLORIDE	R (to ~180)
VINYLIDENE FLUORIDE	R (to ~280)
VINYL ESTER	R (to ~250)
RUBBERS	
BUTYL GR-1	
FLUORO ELASTOMERS VITON A	
FLUORO ELASTOMERS KEL-F 3700	
HARD RUBBER	R (to ~200)
HYPALON	R (to ~250)
KOROSEAL	
NATURAL RUBBER (GRS)	
NEOPRENE GR-M (CR)	
NITRILE BUNA N (NBR)	
NORDEL	R (to ~300)
POLYBUTADIENE	
SILICONE	
POLYURETHANE	

POTASSIUM BICARBONATE, 30%

TEMP °F

METALS	
ALLOY 20Cb3	G
ALUMINUM	U
BRASS	G
BRONZE	
CARBON STEEL	
COPPER	
E-BRIGHT 26-1	
HASTELLOY-B	G
HASTELLOY-C	G
HASTELLOY-D	G
HIGH SILICON IRON	
INCONEL	G
LEAD	
MONEL	G
NICKEL	G
STAINLESS STEELS Type 304/307	E
Type 316	E
TANTALUM	G
TITANIUM	E
ZIRCONIUM	E

NONMETALLICS	
BOROSILICATE GLASS	
CARBON	R

TEMP °F: 60 80 100 120 140 160 180 200 220 240 260 280 300 320 340 360 380 400 420 440 460 480 500 520 540 560

FOR METALS

E = < 2 Mils Penetration/Year; G = < 20 Mils Penetration/Year

S = < 50 Mils Penetration/Year; U = > 50 Mils Penetration/Year

FOR NONMETALLICS

R = Resistant

U = Unsatisfactory

POTASSIUM BICARBONATE, 30%

TEMP °F (scale 60–460)

PLASTICS	Conc.	Rating / range (°F)
ABS		R 60–140
ACRYLICS		
ASBESTOS REINF., EPOXY		
ASBESTOS REINF., PHENOLIC		
CHLORINATED POLYETHER		R 60–250
CPVC		R 60–185
EPOXY		R 60–300
FLUOROCARBONS FEP		R 60–400
FLUOROCARBONS TFE		R 60–460
FURFURYL ALCOHOL		R 60–250
NORYL		R 60–140
PHENOLIC		
POLYAMIDES–NYLON		
POLYESTERS		
BISPHENOL A-FUMURATE	10%	R 60–160 U
HYDROGENATED BISPHENOL-A-BISPHENOL-A		
ISOPHTHALIC	sat	R 60–140
CHLORINATED POLYESTERS		R 60–200
POLYSTYRENES		
POLYETHYLENE		R 60–150
POLYPROPYLENE		R 60–220
PVC–TYPE 1		R 60–140
PVC–TYPE 2		R 60–140
POLYVINYLIDENE CHLORIDE		R 60–180
VINYLIDENE FLUORIDE		R 60–280
VINYL ESTER	sat	R 60–170

RUBBERS	Conc.	Rating / range (°F)
BUTYL GR-1		
FLUORO ELASTOMERS VITON A		R 60–180
FLUORO ELASTOMERS KEL-F 3700		
HARD RUBBER		
HYPALON		
KOROSEAL		R 60–150
NATURAL RUBBER (GRS)		
NEOPRENE GR-M (CR)		R 60–220
NITRILE BUNA N (NBR)		R 60–180
NORDEL		
POLYBUTADIENE		
SILICONE		
POLYURETHANE		

TEMP °F

POTASSIUM BICHROMATE

TEMP °F (scale: 60, 80, 100, 120, 140, 160, 180, 200, 220, 240, 260, 280, 300, 320, 340, 360, 380, 400, 420, 440, 460, 480, 500, 520, 540, 560)

METALS	Concentration	Rating	Temp °F
ALLOY 20Cb3			
ALUMINUM			
BRASS			
BRONZE			
CARBON STEEL			
COPPER			
E-BRIGHT 26-1			
HASTELLOY-B			
HASTELLOY-C			
HASTELLOY-D			
HIGH SILICON IRON			
INCONEL		G	60
LEAD			
MONEL		G	60
NICKEL		G	60
STAINLESS STEELS			
Type 304/307	20%	G	60–200
Type 316	20%	G	60–200
TANTALUM		E	60
TITANIUM			
ZIRCONIUM			
NONMETALLICS			
BOROSILICATE GLASS			
CARBON		R	60–340

FOR METALS

E = < 2 Mils Penetration/Year; G = < 20 Mils Penetration/Year

S = < 50 Mils Penetration/Year; U = > 50 Mils Penetration/Year

FOR NONMETALLICS

R = Resistant

U = Unsatisfactory

POTASSIUM BICHROMATE

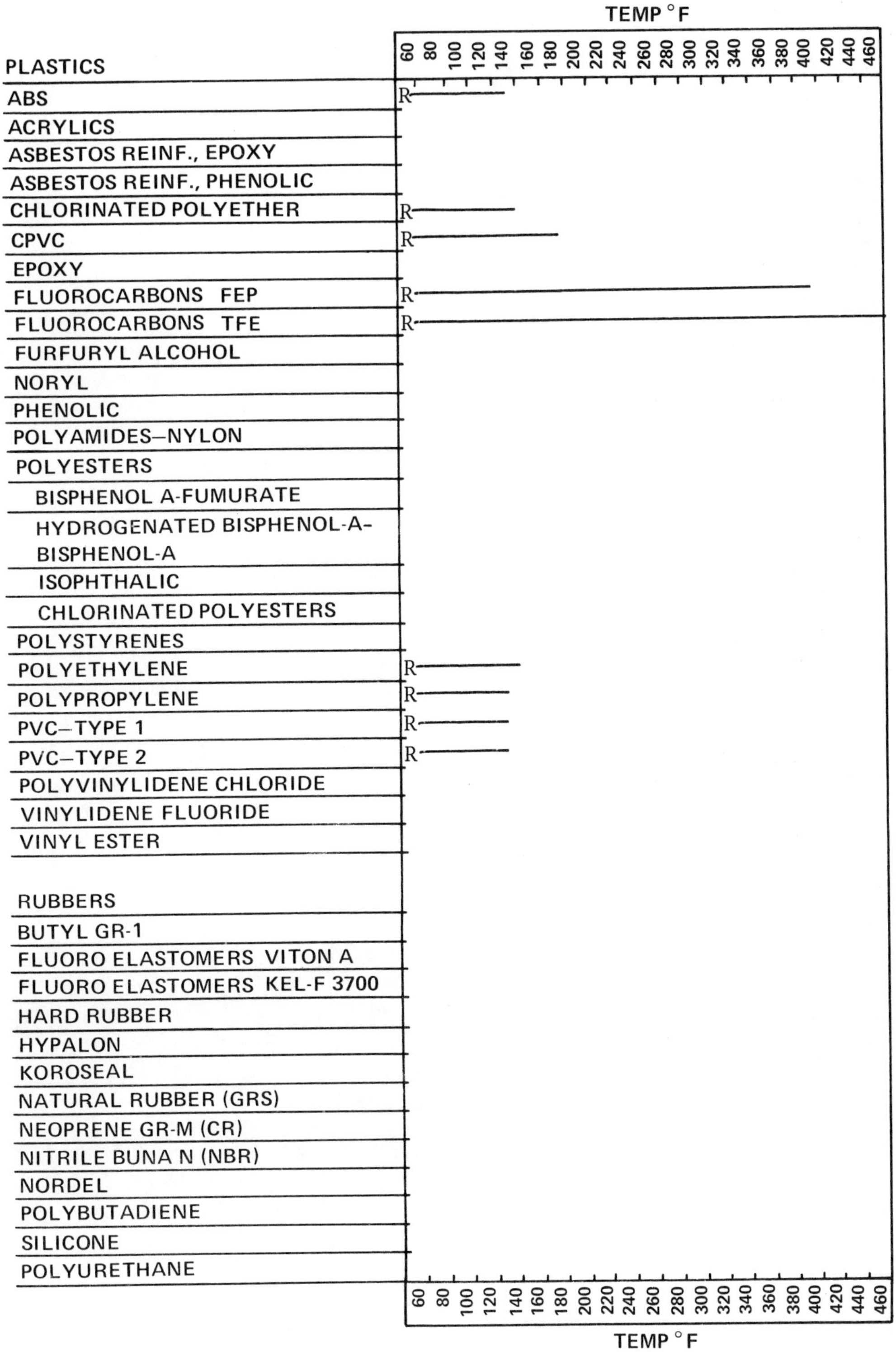

POTASSIUM BORATE

METALS	TEMP °F (60 80 100 120 140 160 180 200 220 240 260 280 300 320 340 360 380 400 420 440 460 480 500 520 540 560)
ALLOY 20Cb3	
ALUMINUM	
BRASS	
BRONZE	
CARBON STEEL	
COPPER	
E-BRIGHT 26-1	
HASTELLOY-B	
HASTELLOY-C	
HASTELLOY-D	
HIGH SILICON IRON	
INCONEL	
LEAD	
MONEL	
NICKEL	
STAINLESS STEELS	
Type 304/307	
Type 316	
TANTALUM	
TITANIUM	
ZIRCONIUM	
NONMETALLICS	
BOROSILICATE GLASS	
CARBON	R (60–340)

FOR METALS

E = < 2 Mils Penetration/Year; G = < 20 Mils Penetration/Year

S = < 50 Mils Penetration/Year; U = > 50 Mils Penetration/Year

FOR NONMETALLICS

R = Resistant

U = Unsatisfactory

POTASSIUM BORATE

PLASTICS	TEMP °F
ABS	U
ACRYLICS	
ASBESTOS REINF., EPOXY	
ASBESTOS REINF., PHENOLIC	
CHLORINATED POLYETHER	R
CPVC	R
EPOXY	
FLUOROCARBONS FEP	R
FLUOROCARBONS TFE	R
FURFURYL ALCOHOL	
NORYL	
PHENOLIC	
POLYAMIDES–NYLON	
POLYESTERS	
BISPHENOL A-FUMURATE	
HYDROGENATED BISPHENOL-A-BISPHENOL-A	
ISOPHTHALIC	
CHLORINATED POLYESTERS	
POLYSTYRENES	
POLYETHYLENE	R
POLYPROPYLENE	R
PVC–TYPE 1	R
PVC–TYPE 2	R
POLYVINYLIDENE CHLORIDE	R
VINYLIDENE FLUORIDE	R
VINYL ESTER	

RUBBERS	TEMP °F
BUTYL GR-1	
FLUORO ELASTOMERS VITON A	
FLUORO ELASTOMERS KEL-F 3700	
HARD RUBBER	
HYPALON	
KOROSEAL	
NATURAL RUBBER (GRS)	
NEOPRENE GR-M (CR)	
NITRILE BUNA N (NBR)	
NORDEL	
POLYBUTADIENE	
SILICONE	
POLYURETHANE	

TEMP °F: 60 80 100 120 140 160 180 200 220 240 260 280 300 320 340 360 380 400 420 440 460

POTASSIUM BROMATE

TEMP °F (60–560)

METALS	Rating
ALLOY 20Cb3	
ALUMINUM	
BRASS	
BRONZE	
CARBON STEEL	
COPPER	
E-BRIGHT 26-1	
HASTELLOY-B	
HASTELLOY-C	
HASTELLOY-D	
HIGH SILICON IRON	
INCONEL	
LEAD	
MONEL	
NICKEL	
STAINLESS STEELS Type 304/307	
Type 316	
TANTALUM	
TITANIUM	
ZIRCONIUM	

NONMETALLICS	Rating
BOROSILICATE GLASS	
CARBON	R (60–340 °F)

TEMP °F

FOR METALS

E = < 2 Mils Penetration/Year; G = < 20 Mils Penetration/Year

S = < 50 Mils Penetration/Year; U = > 50 Mils Penetration/Year

FOR NONMETALLICS

R = Resistant

U = Unsatisfactory

POTASSIUM BROMATE

TEMP °F (60–460)

PLASTICS		
ABS		
ACRYLICS		
ASBESTOS REINF., EPOXY		
ASBESTOS REINF., PHENOLIC		
CHLORINATED POLYETHER		R
CPVC		R
EPOXY		
FLUOROCARBONS FEP		R
FLUOROCARBONS TFE		R
FURFURYL ALCOHOL	10%	R
NORYL		
PHENOLIC		
POLYAMIDES—NYLON		
POLYESTERS		
BISPHENOL A-FUMURATE		
HYDROGENATED BISPHENOL-A-BISPHENOL-A		
ISOPHTHALIC		
CHLORINATED POLYESTERS		
POLYSTYRENES		
POLYETHYLENE		R
POLYPROPYLENE		R
PVC—TYPE 1		R
PVC—TYPE 2		R
POLYVINYLIDENE CHLORIDE		R
VINYLIDENE FLUORIDE		R
VINYL ESTER		
RUBBERS		
BUTYL GR-1		
FLUORO ELASTOMERS VITON A		R
FLUORO ELASTOMERS KEL-F 3700		
HARD RUBBER		
HYPALON		
KOROSEAL		
NATURAL RUBBER (GRS)		
NEOPRENE GR-M (CR)		
NITRILE BUNA N (NBR)		R
NORDEL		
POLYBUTADIENE		
SILICONE		
POLYURETHANE		

TEMP °F (60–460)

POTASSIUM BROMIDE, 30%

TEMP °F: 60 80 100 120 140 160 180 200 220 240 260 280 300 320 340 360 380 400 420 440 460 480 500 520 540 560

METALS		
ALLOY 20Cb3		G
ALUMINUM	1	G
BRASS	air free	G
BRONZE		G
CARBON STEEL		U
COPPER		G
E-BRIGHT 26-1		
HASTELLOY-B		E–G
HASTELLOY-C		E–G
HASTELLOY-D		E–G
HIGH SILICON IRON		G
INCONEL		G
LEAD		
MONEL	air free	G
NICKEL		G
STAINLESS STEELS		
Type 304/307	1	G
Type 316	1	G
TANTALUM		E
TITANIUM		E
ZIRCONIUM		E

NONMETALLICS	
BOROSILICATE GLASS	
CARBON	R

TEMP °F: 60 80 100 120 140 160 180 200 220 240 260 280 300 320 340 360 380 400 420 440 460 480 500 520 540 560

FOR METALS

E = < 2 Mils Penetration/Year; G = < 20 Mils Penetration/Year

S = < 50 Mils Penetration/Year; U = > 50 Mils Penetration/Year

FOR NONMETALLICS

R = Resistant

U = Unsatisfactory

POTASSIUM BROMIDE, 30%

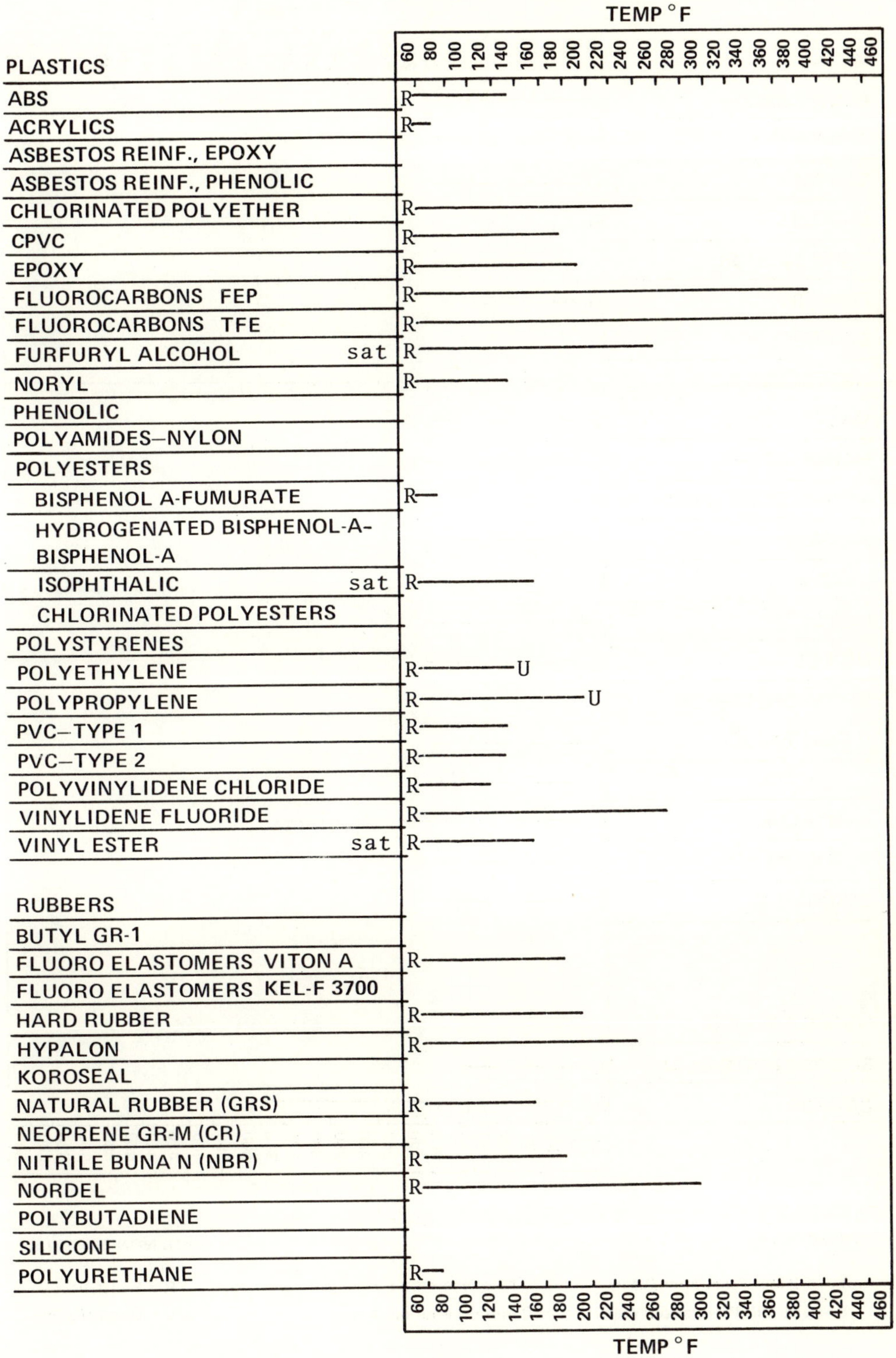

POTASSIUM CARBONATE, 50%

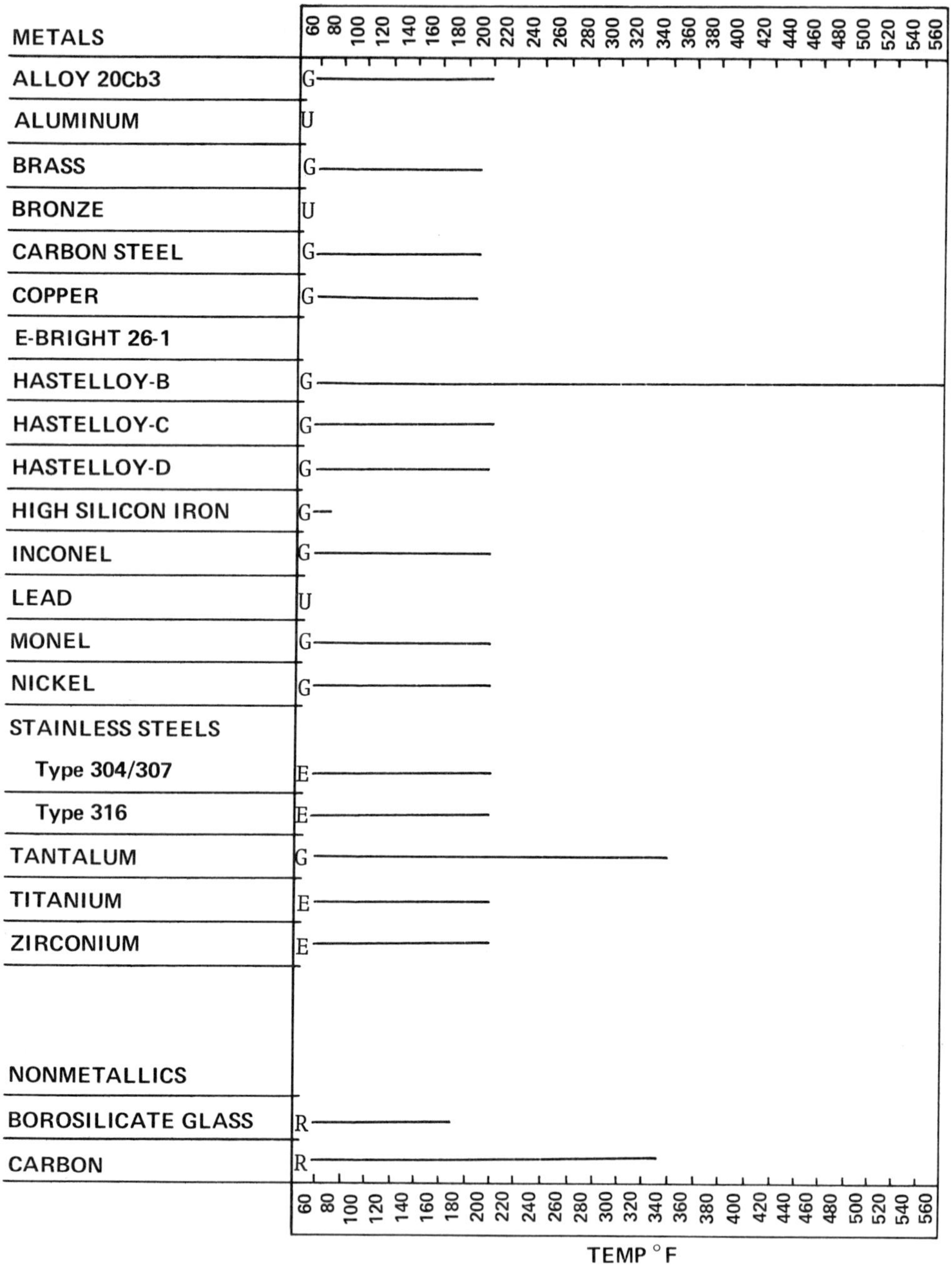

FOR METALS

E = < 2 Mils Penetration/Year; G = < 20 Mils Penetration/Year

S = < 50 Mils Penetration/Year; U = > 50 Mils Penetration/Year

FOR NONMETALLICS

R = Resistant

U = Unsatisfactory

POTASSIUM CARBONATE, 50%

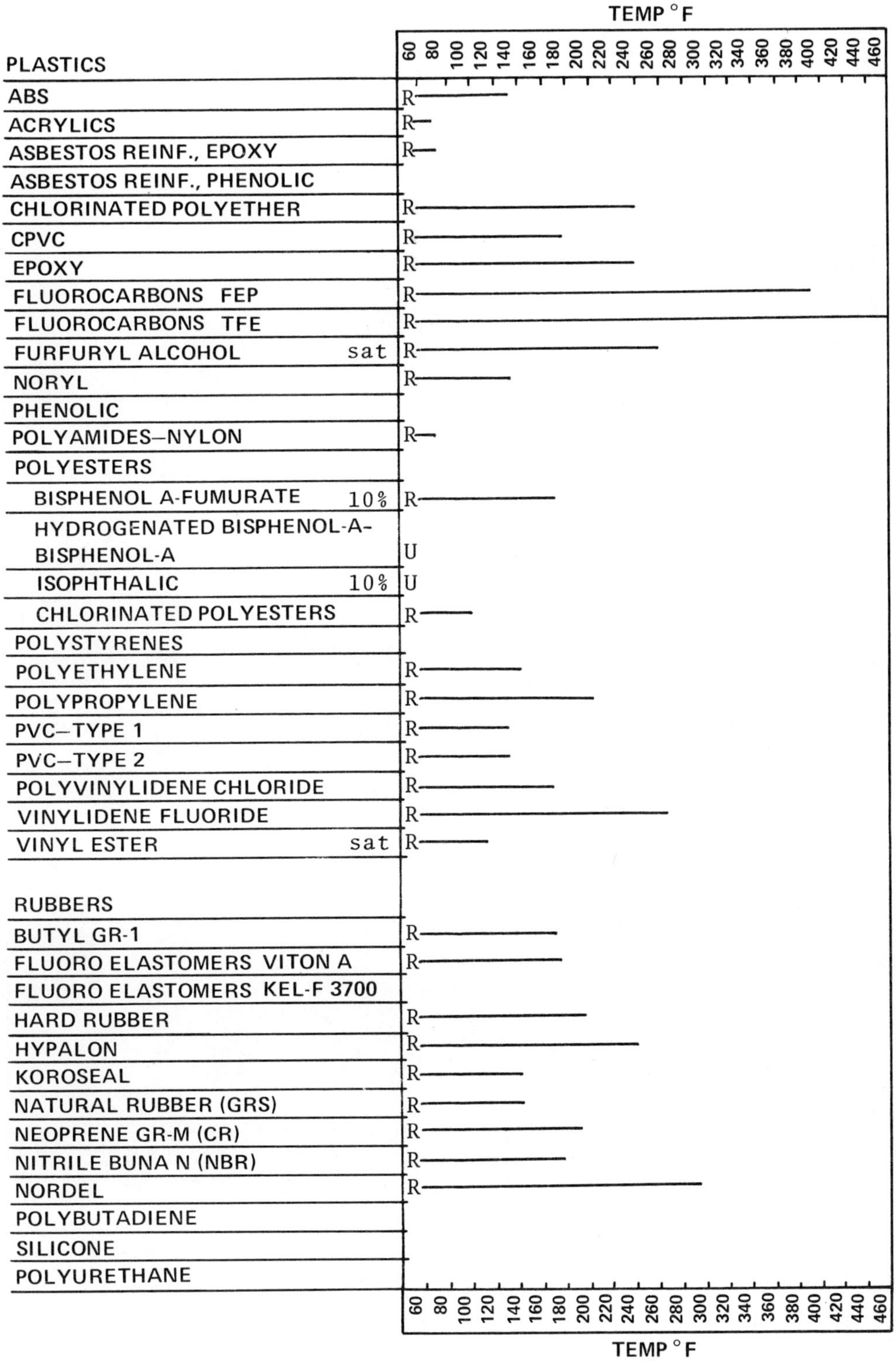

POTASSIUM CHLORATE, AQUEOUS, 30%

TEMP °F (60–560)

METALS	Rating
ALLOY 20Cb3	G
ALUMINUM	G
BRASS	U
BRONZE	U
CARBON STEEL	G
COPPER	
E-BRIGHT 26-1	
HASTELLOY-B	U
HASTELLOY-C	G
HASTELLOY-D	G
HIGH SILICON IRON	G
INCONEL	G
LEAD	
MONEL	G
NICKEL	G
STAINLESS STEELS	
Type 304/307	G
Type 316	E
TANTALUM	G
TITANIUM	E
ZIRCONIUM	E

NONMETALLICS	Rating
BOROSILICATE GLASS	R
CARBON	

FOR METALS

E = < 2 Mils Penetration/Year; G = < 20 Mils Penetration/Year

S = < 50 Mils Penetration/Year; U = > 50 Mils Penetration/Year

FOR NONMETALLICS

R = Resistant

U = Unsatisfactory

POTASSIUM CHLORATE, AQUEOUS, 30%

TEMP °F: 60, 80, 100, 120, 140, 160, 180, 200, 220, 240, 260, 280, 300, 320, 340, 360, 380, 400, 420, 440, 460

PLASTICS	
ABS	R
ACRYLICS	R
ASBESTOS REINF., EPOXY	
ASBESTOS REINF., PHENOLIC	
CHLORINATED POLYETHER	R
CPVC	R
EPOXY	
FLUOROCARBONS FEP	R
FLUOROCARBONS TFE	R
FURFURYL ALCOHOL sat	R
NORYL	R
PHENOLIC	
POLYAMIDES–NYLON	U
POLYESTERS	
BISPHENOL A-FUMURATE	
HYDROGENATED BISPHENOL-A-BISPHENOL-A	
ISOPHTHALIC	
CHLORINATED POLYESTERS	
POLYSTYRENES	
POLYETHYLENE	R
POLYPROPYLENE	R
PVC–TYPE 1	R
PVC–TYPE 2	R
POLYVINYLIDENE CHLORIDE	R
VINYLIDENE FLUORIDE	R
VINYL ESTER	

RUBBERS	
BUTYL GR-1	
FLUORO ELASTOMERS VITON A	R
FLUORO ELASTOMERS KEL-F 3700	
HARD RUBBER	R
HYPALON	R
KOROSEAL	
NATURAL RUBBER (GRS)	
NEOPRENE GR-M (CR)	
NITRILE BUNA N (NBR)	R
NORDEL	R
POLYBUTADIENE	
SILICONE	
POLYURETHANE	

TEMP °F: 60, 80, 100, 120, 140, 160, 180, 200, 220, 240, 260, 280, 300, 320, 340, 360, 380, 400, 420, 440, 460

POTASSIUM CHLORIDE, 30%

TEMP °F: 60 80 100 120 140 160 180 200 220 240 260 280 300 320 340 360 380 400 420 440 460 480 500 520 540 560

METALS		
ALLOY 20Cb3	1	E——— G———
ALUMINUM	1	U
BRASS		S—
BRONZE		G—
CARBON STEEL		G———
COPPER		S—
E-BRIGHT 26-1		
HASTELLOY-B	10-28%	G———
HASTELLOY-C	10-20%	G———
HASTELLOY-D		G—
HIGH SILICON IRON		G—
INCONEL	1	E——— G—
LEAD		G—
MONEL		E——— G—
NICKEL		E——— G—
STAINLESS STEELS		
Type 304/307	1,2	E———
Type 316	1	E———
TANTALUM		E———
TITANIUM		E———
ZIRCONIUM		E———

NONMETALLICS	
BOROSILICATE GLASS	
CARBON	R———

TEMP °F: 60 80 100 120 140 160 180 200 220 240 260 280 300 320 340 360 380 400 420 440 460 480 500 520 540 560

FOR METALS

E = < 2 Mils Penetration/Year; G = < 20 Mils Penetration/Year

S = < 50 Mils Penetration/Year; U = > 50 Mils Penetration/Year

FOR NONMETALLICS

R = Resistant

U = Unsatisfactory

POTASSIUM CHLORIDE, 30%

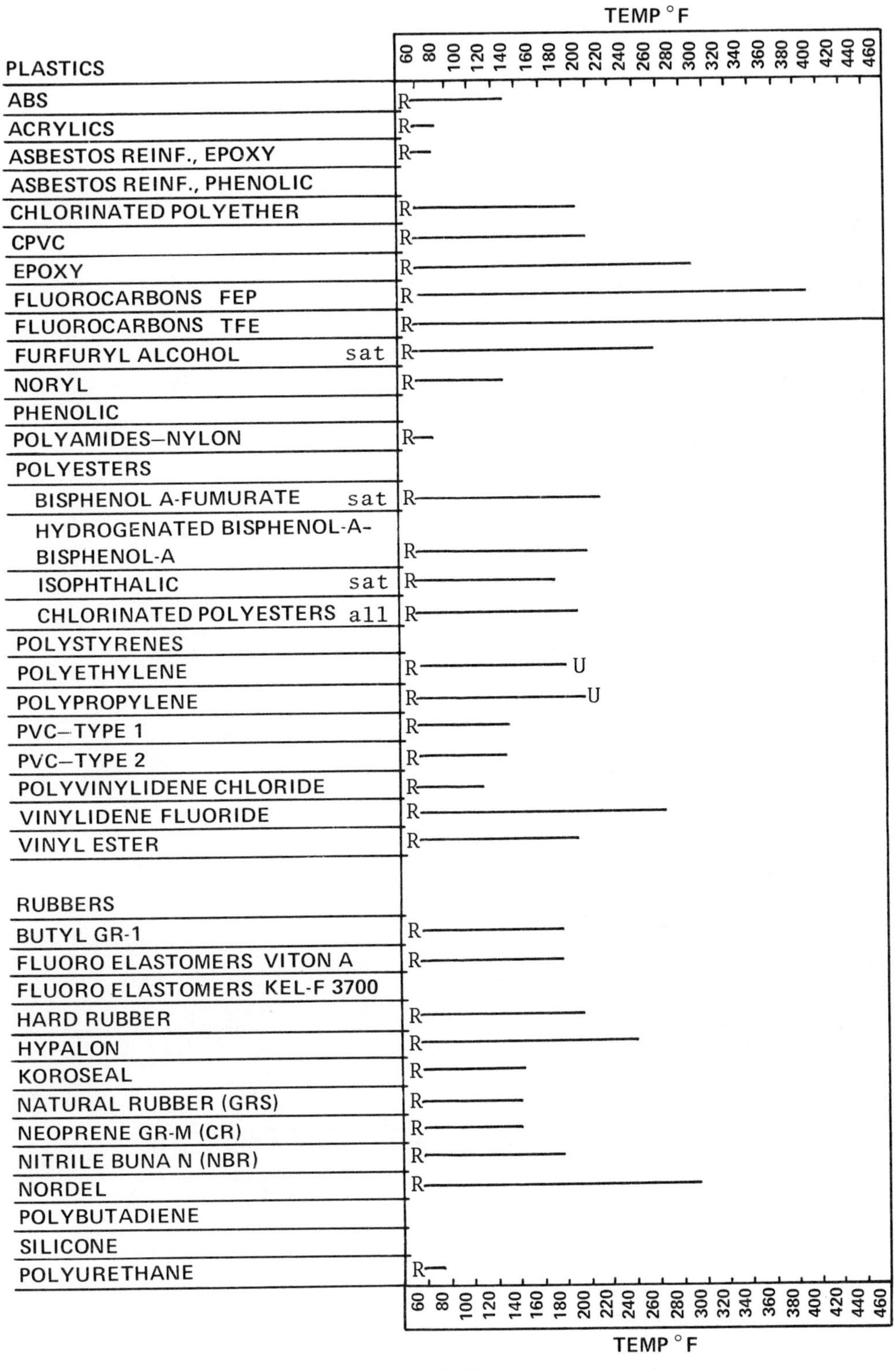

POTASSIUM CHROMATE, 30%

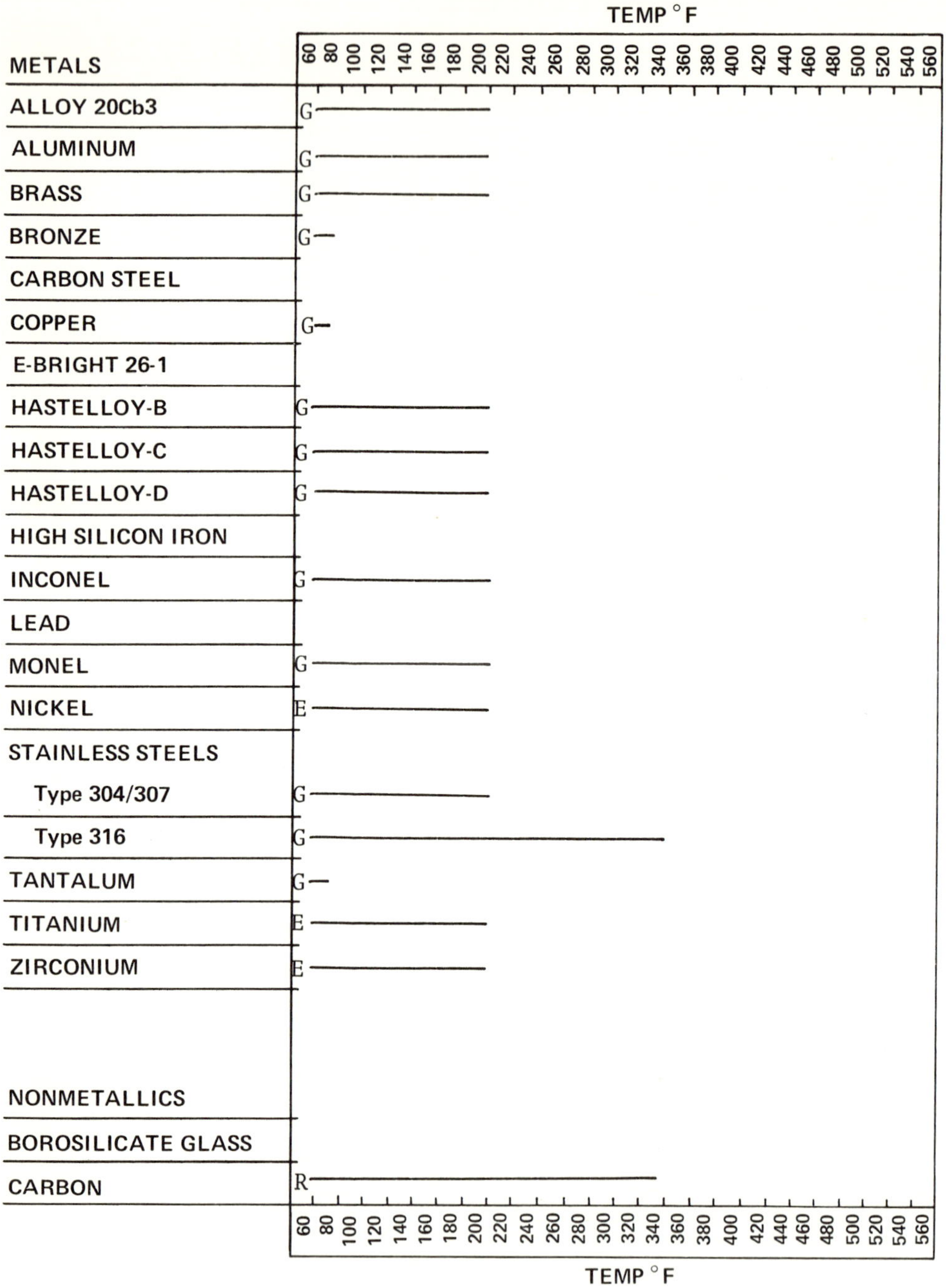

FOR METALS

E = < 2 Mils Penetration/Year; G = < 20 Mils Penetration/Year

S = < 50 Mils Penetration/Year; U = > 50 Mils Penetration/Year

FOR NONMETALLICS

R = Resistant

U = Unsatisfactory

POTASSIUM CHROMATE, 30%

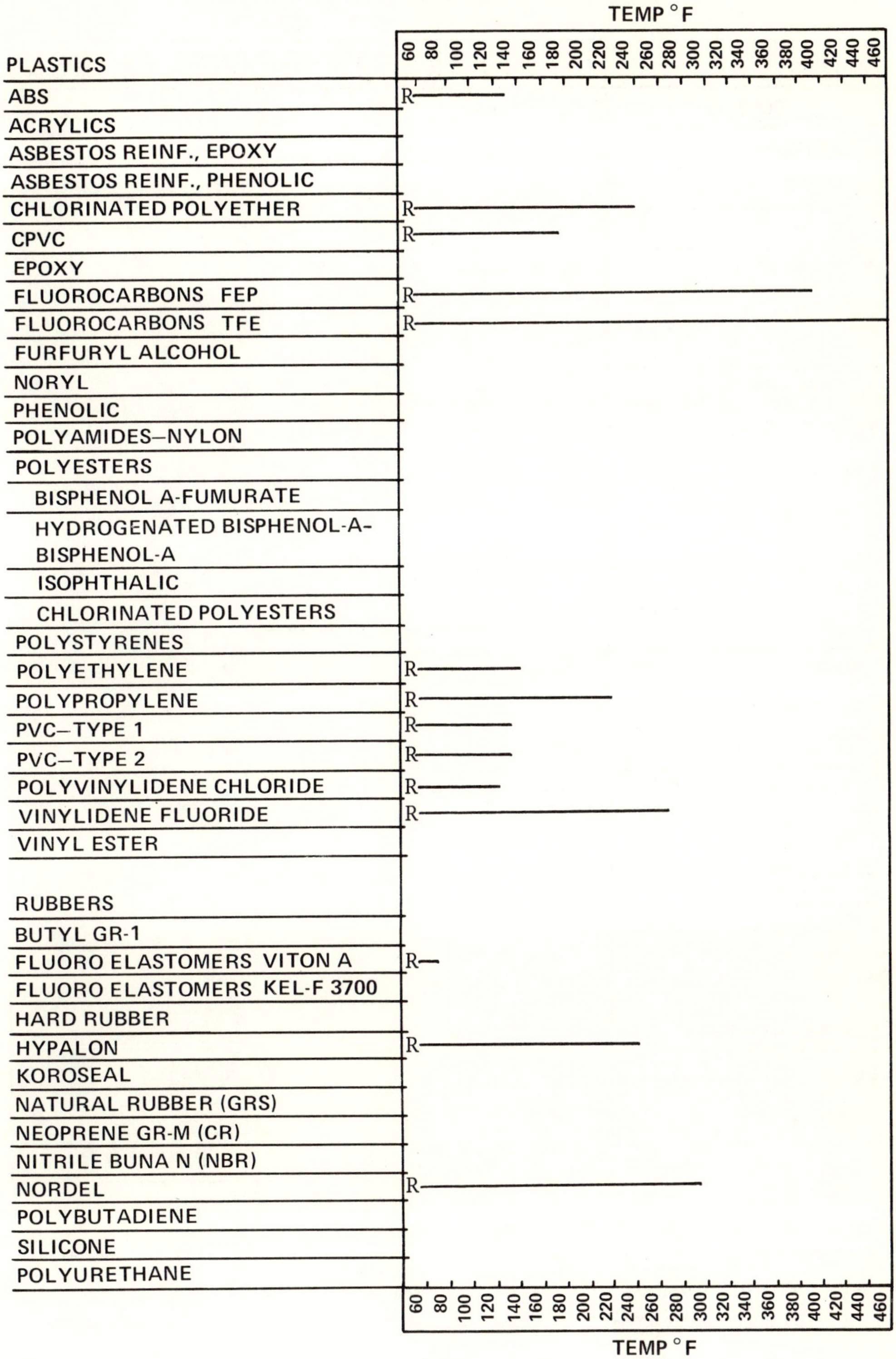

POTASSIUM CYANIDE, 30%

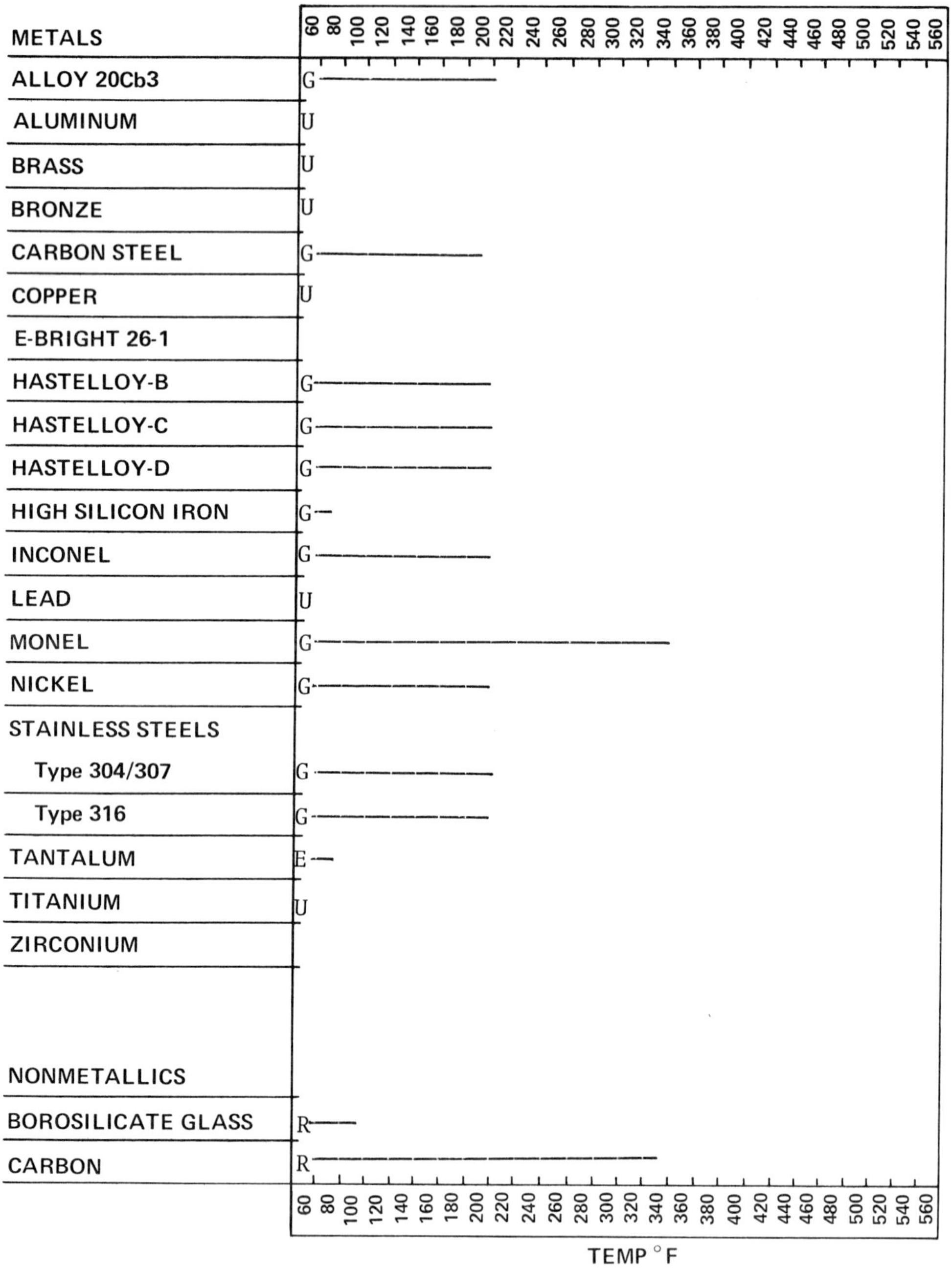

FOR METALS

E = < 2 Mils Penetration/Year; G = < 20 Mils Penetration/Year

S = < 50 Mils Penetration/Year; U = > 50 Mils Penetration/Year

FOR NONMETALLICS

R = Resistant

U = Unsatisfactory

POTASSIUM CYANIDE, 30%

TEMP °F (60–460)

PLASTICS	Conc.	Rating
ABS		R
ACRYLICS		R
ASBESTOS REINF., EPOXY		
ASBESTOS REINF., PHENOLIC		
CHLORINATED POLYETHER		R
CPVC		R
EPOXY	5%	R—U
FLUOROCARBONS FEP		R
FLUOROCARBONS TFE		R
FURFURYL ALCOHOL	sat	R
NORYL		R
PHENOLIC		
POLYAMIDES–NYLON		R
POLYESTERS		
BISPHENOL A-FUMURATE		
HYDROGENATED BISPHENOL-A-BISPHENOL-A		R
ISOPHTHALIC	sat	U
CHLORINATED POLYESTERS		
POLYSTYRENES		
POLYETHYLENE		R—U
POLYPROPYLENE		R—U
PVC–TYPE 1		R
PVC–TYPE 2		R
POLYVINYLIDENE CHLORIDE		R
VINYLIDENE FLUORIDE		R
VINYL ESTER	sat	U

RUBBERS	Rating
BUTYL GR-1	R
FLUORO ELASTOMERS VITON A	R
FLUORO ELASTOMERS KEL-F 3700	
HARD RUBBER	
HYPALON	R
KOROSEAL	R
NATURAL RUBBER (GRS)	R
NEOPRENE GR-M (CR)	R
NITRILE BUNA N (NBR)	R
NORDEL	R
POLYBUTADIENE	
SILICONE	
POLYURETHANE	R

TEMP °F

POTASSIUM DICHROMATE, 30%

TEMP °F: 60 80 100 120 140 160 180 200 220 240 260 280 300 320 340 360 380 400 420 440 460 480 500 520 540 560

METALS	
ALLOY 20Cb3	E
ALUMINUM	E G
BRASS	G
BRONZE	G
CARBON STEEL	G
COPPER	U
E-BRIGHT 26-1	
HASTELLOY-B	U
HASTELLOY-C 10-25%	G
HASTELLOY-D	
HIGH SILICON IRON	
INCONEL	G
LEAD	G
MONEL	G
NICKEL	G
STAINLESS STEELS	
Type 304/307	E
Type 316	E
TANTALUM	E
TITANIUM	E
ZIRCONIUM	E

NONMETALLICS	
BOROSILICATE GLASS	R
CARBON	R

TEMP °F: 60 80 100 120 140 160 180 200 220 240 260 280 300 320 340 360 380 400 420 440 460 480 500 520 540 560

FOR METALS

E = < 2 Mils Penetration/Year; G = < 20 Mils Penetration/Year

S = < 50 Mils Penetration/Year; U = > 50 Mils Penetration/Year

FOR NONMETALLICS

R = Resistant

U = Unsatisfactory

POTASSIUM DICHROMATE, 30%

TEMP °F

PLASTICS	Conc.	Rating	Bar extends to (°F, approx.)
ABS		R	100
ACRYLICS		R	80
ASBESTOS REINF., EPOXY		R	80
ASBESTOS REINF., PHENOLIC		R	80
CHLORINATED POLYETHER		R	225
CPVC		R	190
EPOXY		R	250
FLUOROCARBONS FEP		R	400
FLUOROCARBONS TFE		R	460+
FURFURYL ALCOHOL	10%	R	270
NORYL	40%	R	140
PHENOLIC			
POLYAMIDES–NYLON			
POLYESTERS			
BISPHENOL A-FUMURATE	sat	R	205
HYDROGENATED BISPHENOL-A-BISPHENOL-A			
ISOPHTHALIC	sat	R	180
CHLORINATED POLYESTERS	all	R	205
POLYSTYRENES			
POLYETHYLENE		R	150
POLYPROPYLENE		R	210
PVC–TYPE 1		R	140
PVC–TYPE 2		R	140
POLYVINYLIDENE CHLORIDE		R	155
VINYLIDENE FLUORIDE		R	280
VINYL ESTER		R	215

RUBBERS	Conc.	Rating	Bar extends to (°F, approx.)
BUTYL GR-1		R	150
FLUORO ELASTOMERS VITON A		R	190
FLUORO ELASTOMERS KEL-F 3700			
HARD RUBBER		U	
HYPALON		R	250
KOROSEAL			
NATURAL RUBBER (GRS)		U	
NEOPRENE GR-M (CR)		U	
NITRILE BUNA N (NBR)		R	190
NORDEL		R	300
POLYBUTADIENE			
SILICONE			
POLYURETHANE		R	80

Temperature scale (°F): 60 80 100 120 140 160 180 200 220 240 260 280 300 320 340 360 380 400 420 440 460

POTASSIUM FERRICYANIDE, 30%

TEMP °F: 60, 80, 100, 120, 140, 160, 180, 200, 220, 240, 260, 280, 300, 320, 340, 360, 380, 400, 420, 440, 460, 480, 500, 520, 540, 560

METALS		Rating
ALLOY 20Cb3		G
ALUMINUM	neutral	G
BRASS		G
BRONZE		G
CARBON STEEL		U
COPPER		G
E-BRIGHT 26-1		
HASTELLOY-B		G
HASTELLOY-C		G
HASTELLOY-D		G
HIGH SILICON IRON		G
INCONEL		G
LEAD		
MONEL		G
NICKEL		G
STAINLESS STEELS		
Type 304/307	1	G
Type 316		G
TANTALUM		E
TITANIUM		E
ZIRCONIUM		E

NONMETALLICS	Rating
BOROSILICATE GLASS	R
CARBON	R

TEMP °F

FOR METALS

E = < 2 Mils Penetration/Year; G = < 20 Mils Penetration/Year

S = < 50 Mils Penetration/Year; U = > 50 Mils Penetration/Year

FOR NONMETALLICS

R = Resistant

U = Unsatisfactory

POTASSIUM FERRICYANIDE, 30%

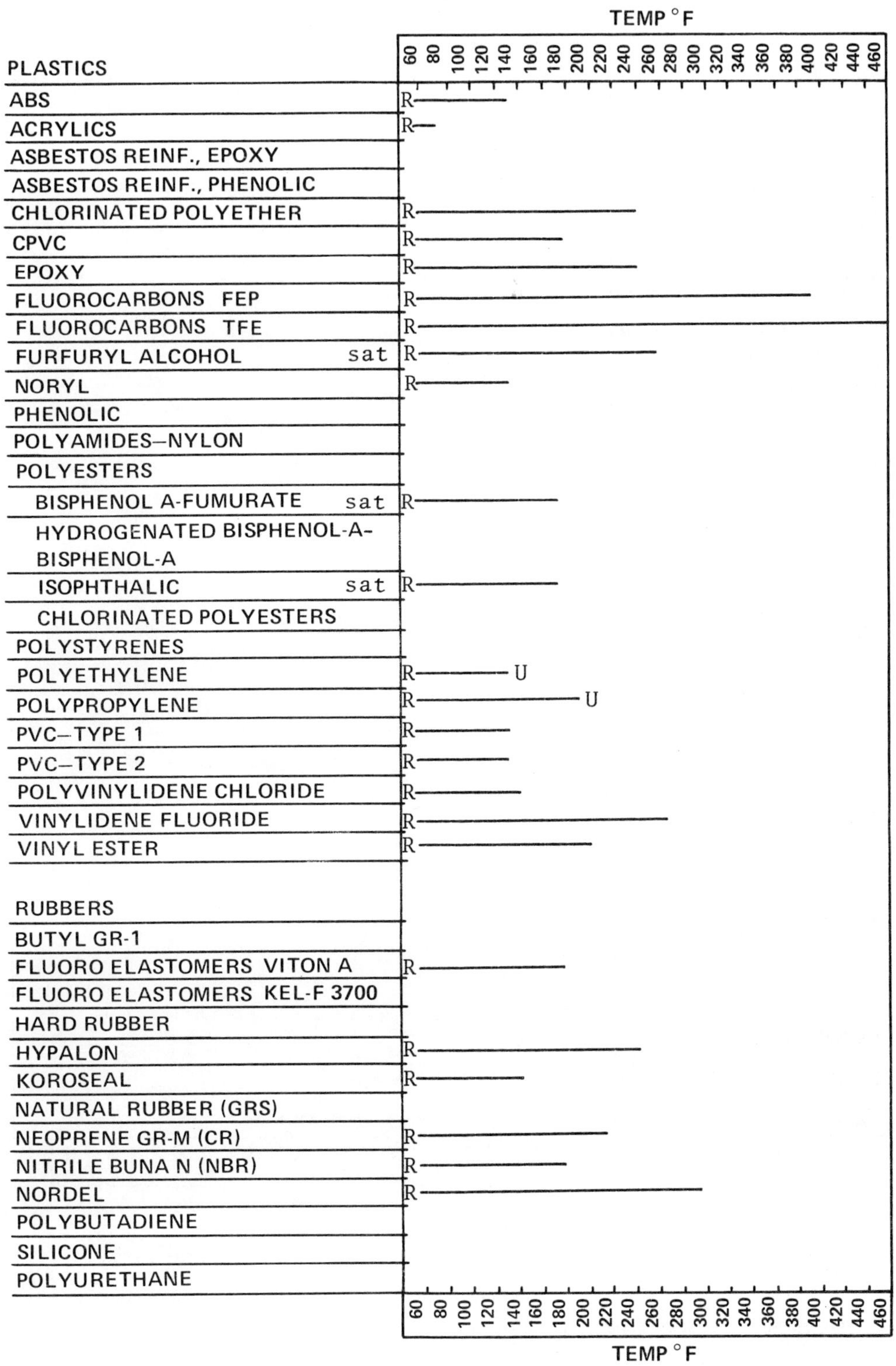

POTASSIUM FERROCYANIDE, 30%

METALS	Concentration	Rating	TEMP °F
ALLOY 20Cb3		G	60–210
ALUMINUM	neutral 10%	E	60–210
BRASS			
BRONZE		G	60–100
CARBON STEEL		G	60–100
COPPER			
E-BRIGHT 26-1			
HASTELLOY-B	20%	G	60–210
HASTELLOY-C	20%	G	60–210
HASTELLOY-D	20%	G	60–210
HIGH SILICON IRON		G	60–80
INCONEL	20%	G	60–210
LEAD			
MONEL	20%	G	60–210
NICKEL	20%	G	60–210
STAINLESS STEELS			
Type 304/307	1	G	60–210
Type 316		G	60–210
TANTALUM		G	60–210
TITANIUM			
ZIRCONIUM	10%	E	60–210

NONMETALLICS	Rating	TEMP °F
BOROSILICATE GLASS		
CARBON	R	60–340

TEMP °F: 60 80 100 120 140 160 180 200 220 240 260 280 300 320 340 360 380 400 420 440 460 480 500 520 540 560

FOR METALS

E = < 2 Mils Penetration/Year; G = < 20 Mils Penetration/Year

S = < 50 Mils Penetration/Year; U = > 50 Mils Penetration/Year

FOR NONMETALLICS

R = Resistant

U = Unsatisfactory

POTASSIUM FERROCYANIDE, 30%

TEMP °F (60–460)

PLASTICS		Rating
ABS		
ACRYLICS		
ASBESTOS REINF., EPOXY		
ASBESTOS REINF., PHENOLIC		
CHLORINATED POLYETHER		R
CPVC		R
EPOXY		R
FLUOROCARBONS FEP		R
FLUOROCARBONS TFE		R
FURFURYL ALCOHOL	sat	R
NORYL		R
PHENOLIC		
POLYAMIDES–NYLON		
POLYESTERS		
BISPHENOL A-FUMURATE	sat	R
HYDROGENATED BISPHENOL-A-BISPHENOL-A		
ISOPHTHALIC	sat	R
CHLORINATED POLYESTERS	all	R
POLYSTYRENES		
POLYETHYLENE		R — U
POLYPROPYLENE		R — U
PVC–TYPE 1		R
PVC–TYPE 2		R
POLYVINYLIDENE CHLORIDE		R
VINYLIDENE FLUORIDE		R
VINYL ESTER		R

RUBBERS	Rating
BUTYL GR-1	
FLUORO ELASTOMERS VITON A	R
FLUORO ELASTOMERS KEL-F 3700	
HARD RUBBER	
HYPALON	R
KOROSEAL	R
NATURAL RUBBER (GRS)	
NEOPRENE GR-M (CR)	R
NITRILE BUNA N (NBR)	R
NORDEL	
POLYBUTADIENE	
SILICONE	
POLYURETHANE	

TEMP °F (60–460)

POTASSIUM FLUORIDE

TEMP °F: 60, 80, 100, 120, 140, 160, 180, 200, 220, 240, 260, 280, 300, 320, 340, 360, 380, 400, 420, 440, 460, 480, 500, 520, 540, 560

METALS	
ALLOY 20Cb3	G
ALUMINUM	
BRASS	G
BRONZE	G
CARBON STEEL	
COPPER	G
E-BRIGHT 26-1	
HASTELLOY-B	
HASTELLOY-C	
HASTELLOY-D	
HIGH SILICON IRON	
INCONEL	G
LEAD	
MONEL	G
NICKEL	G
STAINLESS STEELS	
Type 304/307	G
Type 316	G
TANTALUM	G
TITANIUM	
ZIRCONIUM	

NONMETALLICS	
BOROSILICATE GLASS	
CARBON	R

TEMP °F

FOR METALS

E = < 2 Mils Penetration/Year; G = < 20 Mils Penetration/Year

S = < 50 Mils Penetration/Year; U = > 50 Mils Penetration/Year

FOR NONMETALLICS

R = Resistant

U = Unsatisfactory

POTASSIUM FLUORIDE

TEMP °F (60–460)

PLASTICS	
ABS	R
ACRYLICS	
ASBESTOS REINF., EPOXY	
ASBESTOS REINF., PHENOLIC	
CHLORINATED POLYETHER	R
CPVC	R
EPOXY	
FLUOROCARBONS FEP	R
FLUOROCARBONS TFE	R
FURFURYL ALCOHOL	
NORYL	R
PHENOLIC	
POLYAMIDES–NYLON	
POLYESTERS	
BISPHENOL A-FUMURATE	
HYDROGENATED BISPHENOL-A-BISPHENOL-A	
ISOPHTHALIC	
CHLORINATED POLYESTERS	
POLYSTYRENES	
POLYETHYLENE	R
POLYPROPYLENE	R
PVC–TYPE 1	R
PVC–TYPE 2	R
POLYVINYLIDENE CHLORIDE	R
VINYLIDENE FLUORIDE	R
VINYL ESTER	

RUBBERS	
BUTYL GR-1	
FLUORO ELASTOMERS VITON A	R
FLUORO ELASTOMERS KEL-F 3700	
HARD RUBBER	
HYPALON	
KOROSEAL	
NATURAL RUBBER (GRS)	
NEOPRENE GR-M (CR)	
NITRILE BUNA N (NBR)	R
NORDEL	
POLYBUTADIENE	
SILICONE	
POLYURETHANE	

TEMP °F (60–460)

POTASSIUM HYDROXIDE, 90%

METALS		TEMP °F
ALLOY 20Cb3		U 60
ALUMINUM		U 60
BRASS		U 60
BRONZE		S 60–80
CARBON STEEL	2	G 60–80
COPPER		S 60–80
E-BRIGHT 26-1		
HASTELLOY-B		G 60–120
HASTELLOY-C		G 60–140
HASTELLOY-D		G 60–120
HIGH SILICON IRON		U 60
INCONEL		G 60–80
LEAD		U 60
MONEL		G 60–200
NICKEL		G 60–200
STAINLESS STEELS		
Type 304/307		U 60
Type 316	2	G 60–340
TANTALUM		U 60
TITANIUM		U 60
ZIRCONIUM		G 60–80

NONMETALLICS		TEMP °F
BOROSILICATE GLASS		U 60
CARBON		R 60–340

FOR METALS

E = < 2 Mils Penetration/Year; G = < 20 Mils Penetration/Year
S = < 50 Mils Penetration/Year; U = > 50 Mils Penetration/Year

FOR NONMETALLICS

R = Resistant
U = Unsatisfactory

POTASSIUM HYDROXIDE, 90%

PLASTICS		Rating
ABS	sat	R
ACRYLICS		R
ASBESTOS REINF., EPOXY		R
ASBESTOS REINF., PHENOLIC		U
CHLORINATED POLYETHER		R
CPVC	sat	R
EPOXY	sat	R
FLUOROCARBONS FEP		R
FLUOROCARBONS TFE		R
FURFURYL ALCOHOL	sat	R
NORYL		
PHENOLIC		
POLYAMIDES—NYLON		R
POLYESTERS		
BISPHENOL A-FUMURATE		
HYDROGENATED BISPHENOL-A-BISPHENOL-A		U
ISOPHTHALIC		U
CHLORINATED POLYESTERS	10%	U
POLYSTYRENES		R
POLYETHYLENE		R
POLYPROPYLENE		R
PVC—TYPE 1		R
PVC—TYPE 2		R
POLYVINYLIDENE CHLORIDE		R
VINYLIDENE FLUORIDE		R
VINYL ESTER		
RUBBERS		
BUTYL GR-1		R
FLUORO ELASTOMERS VITON A		U
FLUORO ELASTOMERS KEL-F 3700		
HARD RUBBER		R
HYPALON		R
KOROSEAL		R
NATURAL RUBBER (GRS)		R
NEOPRENE GR-M (CR)		R
NITRILE BUNA N (NBR)		R
NORDEL		R
POLYBUTADIENE		
SILICONE		
POLYURETHANE		R

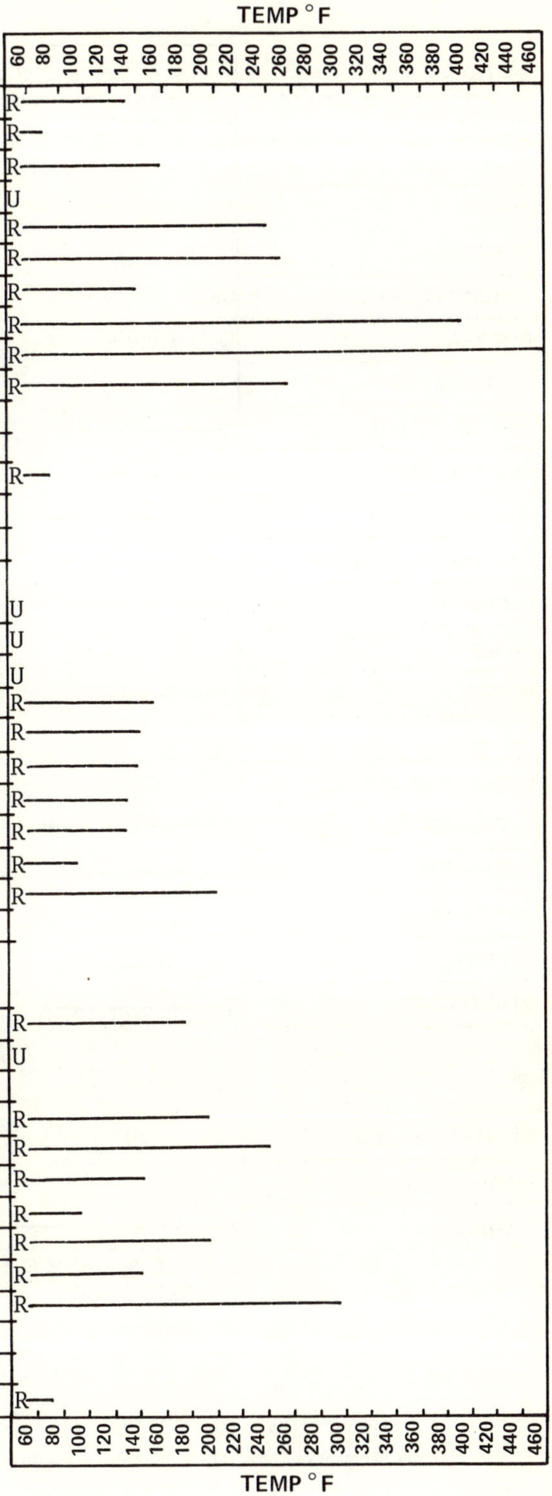

POTASSIUM HYDROXIDE, 50%

TEMP °F: 60, 80, 100, 120, 140, 160, 180, 200, 220, 240, 260, 280, 300, 320, 340, 360, 380, 400, 420, 440, 460, 480, 500, 520, 540, 560

METALS		
ALLOY 20Cb3		G —— U
ALUMINUM		U
BRASS		U
BRONZE		S–U
CARBON STEEL	2	G–U
COPPER		S–U
E-BRIGHT 26-1		
HASTELLOY-B		G ——
HASTELLOY-C		G ——
HASTELLOY-D		G ——
HIGH SILICON IRON		S ——
INCONEL		G ——
LEAD		U
MONEL	2	E ——
NICKEL		E ——
STAINLESS STEELS		
Type 304/307	2	G ——
Type 316	2	G —— U
TANTALUM		U
TITANIUM		E —— U
ZIRCONIUM		G ——

NONMETALLICS	
BOROSILICATE GLASS	U
CARBON	R ——

TEMP °F: 60, 80, 100, 120, 140, 160, 180, 200, 220, 240, 260, 280, 300, 320, 340, 360, 380, 400, 420, 440, 460, 480, 500, 520, 540, 560

FOR METALS

E = < 2 Mils Penetration/Year; G = < 20 Mils Penetration/Year

S = < 50 Mils Penetration/Year; U = > 50 Mils Penetration/Year

FOR NONMETALLICS

R = Resistant

U = Unsatisfactory

POTASSIUM HYDROXIDE, 50%

TEMP °F

60 80 100 120 140 160 180 200 220 240 260 280 300 320 340 360 380 400 420 440 460

PLASTICS	
ABS	R
ACRYLICS	R
ASBESTOS REINF., EPOXY	
ASBESTOS REINF., PHENOLIC	
CHLORINATED POLYETHER	R
CPVC	R
EPOXY	R U
FLUOROCARBONS FEP	R
FLUOROCARBONS TFE	R
FURFURYL ALCOHOL	R
NORYL	R
PHENOLIC	
POLYAMIDES—NYLON	R
POLYESTERS	
BISPHENOL A-FUMURATE	
HYDROGENATED BISPHENOL-A-BISPHENOL-A	U
ISOPHTHALIC	U
CHLORINATED POLYESTERS	
POLYSTYRENES	R
POLYETHYLENE	R
POLYPROPYLENE	R
PVC—TYPE 1	R
PVC—TYPE 2	R
POLYVINYLIDENE CHLORIDE	R
VINYLIDENE FLUORIDE	R
VINYL ESTER	

RUBBERS	
BUTYL GR-1	R
FLUORO ELASTOMERS VITON A	U
FLUORO ELASTOMERS KEL-F 3700	
HARD RUBBER	R
HYPALON	R
KOROSEAL	R
NATURAL RUBBER (GRS)	R
NEOPRENE GR-M (CR)	R
NITRILE BUNA N (NBR)	R
NORDEL	R
POLYBUTADIENE	
SILICONE	
POLYURETHANE	R

60 80 100 120 140 160 180 200 220 240 260 280 300 320 340 360 380 400 420 440 460

TEMP °F

POTASSIUM HYDROXIDE, 27%

TEMP °F: 60 80 100 120 140 160 180 200 220 240 260 280 300 320 340 360 380 400 420 440 460 480 500 520 540 560

METALS		
ALLOY 20Cb3		E ————————
ALUMINUM		U
BRASS		G — U
BRONZE		G ——— U
CARBON STEEL	2	G ———————
COPPER		G —— U
E-BRIGHT 26-1		
HASTELLOY-B		G ————————
HASTELLOY-C		G ————————
HASTELLOY-D		G ————————
HIGH SILICON IRON		G ——————— U
INCONEL		G ————————
LEAD		U
MONEL	2	E ————————
NICKEL		E ————————
STAINLESS STEELS		
Type 304/307	2	G ————————
Type 316	2	G —————————————————
TANTALUM		U
TITANIUM		U
ZIRCONIUM		E ————————

NONMETALLICS	
BOROSILICATE GLASS	U
CARBON	R —————————————————

TEMP °F: 60 80 100 120 140 160 180 200 220 240 260 280 300 320 340 360 380 400 420 440 460 480 500 520 540 560

FOR METALS

E = < 2 Mils Penetration/Year; G = < 20 Mils Penetration/Year

S = < 50 Mils Penetration/Year; U = > 50 Mils Penetration/Year

FOR NONMETALLICS

R = Resistant

U = Unsatisfactory

POTASSIUM HYDROXIDE, 27%

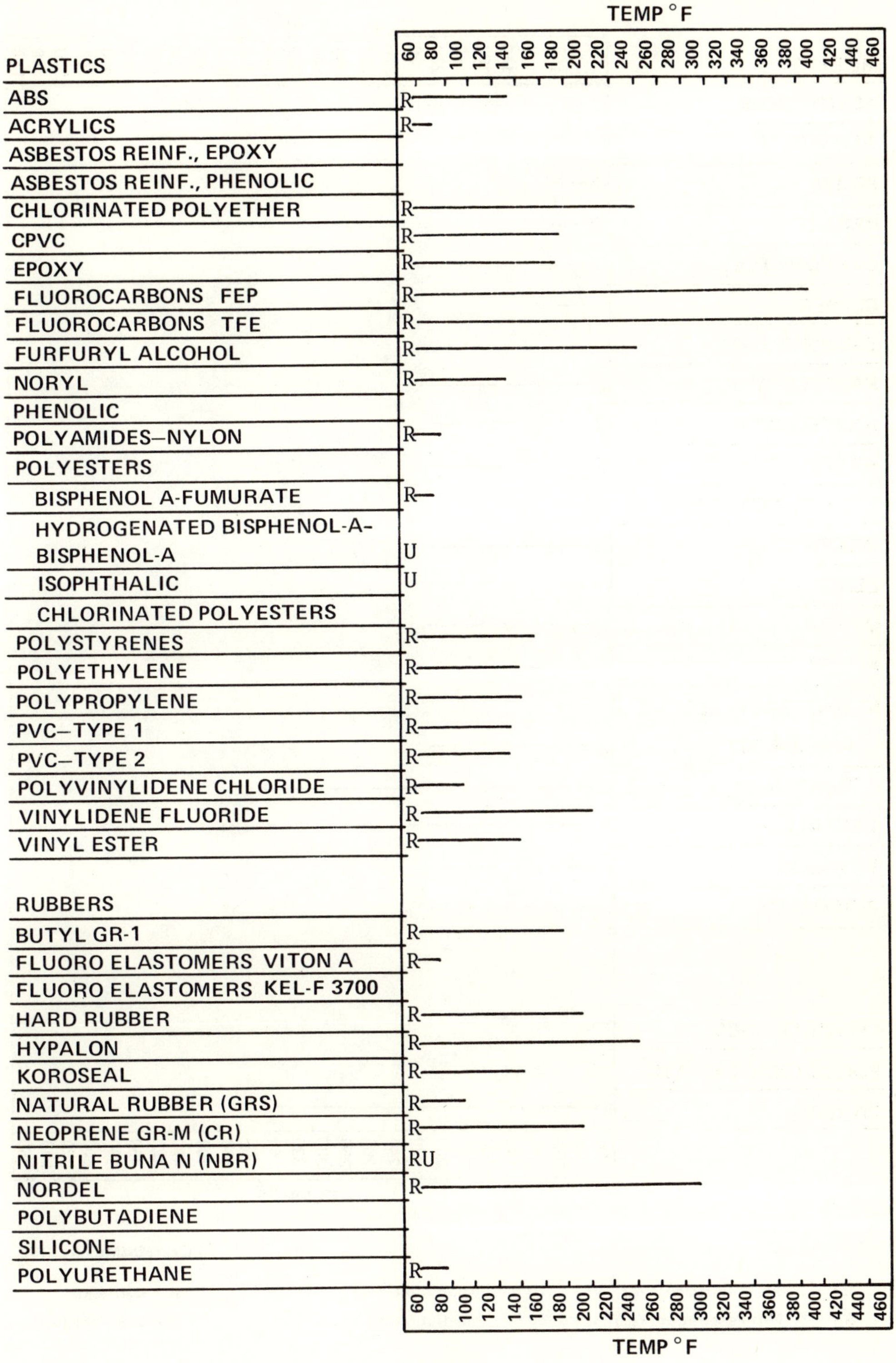

POTASSIUM HYDROXIDE, 5%

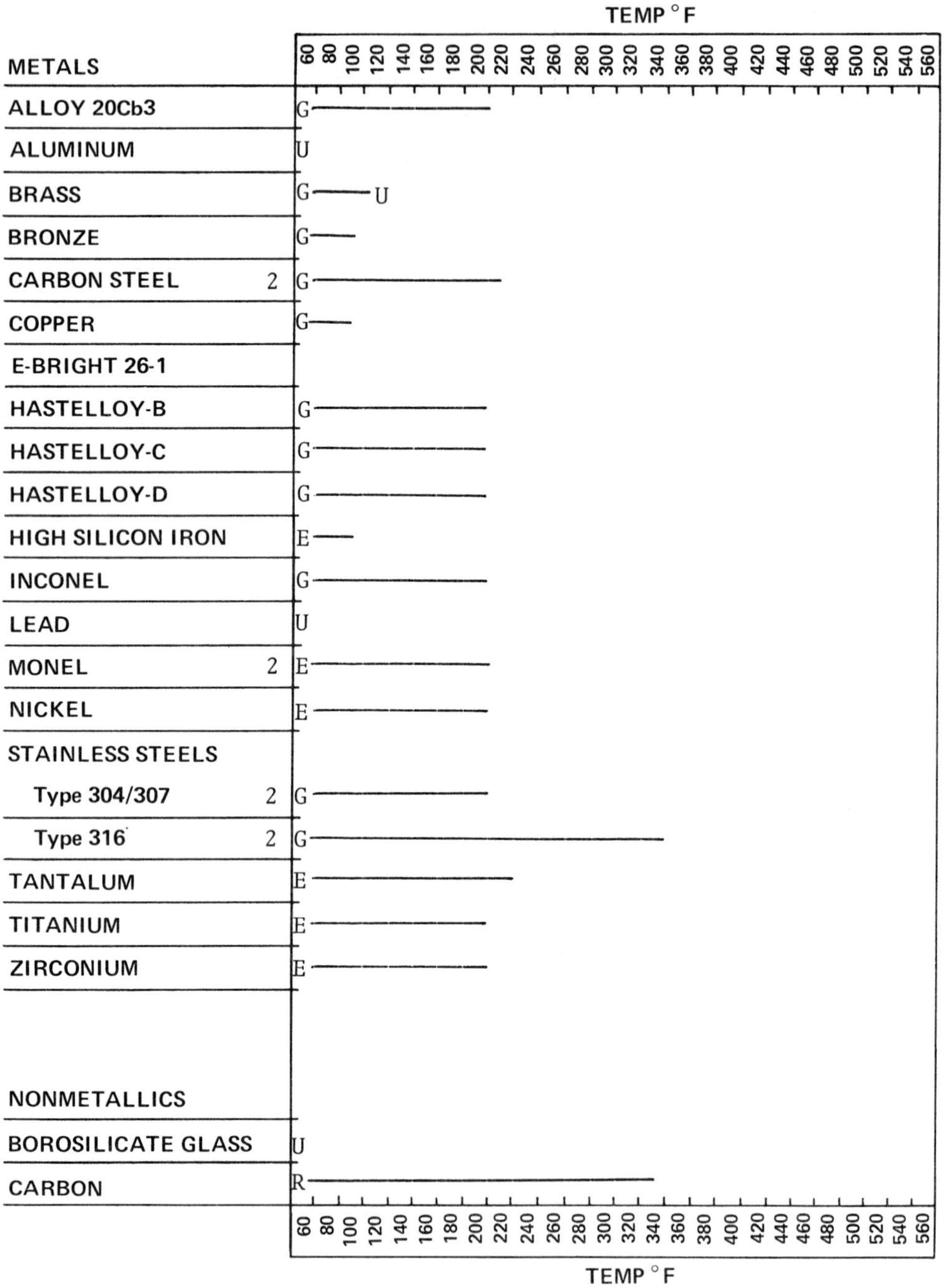

FOR METALS

E = < 2 Mils Penetration/Year; G = < 20 Mils Penetration/Year

S = < 50 Mils Penetration/Year; U = > 50 Mils Penetration/Year

FOR NONMETALLICS

R = Resistant

U = Unsatisfactory

POTASSIUM HYDROXIDE, 5%

TEMP °F: 60 80 100 120 140 160 180 200 220 240 260 280 300 320 340 360 380 400 420 440 460

PLASTICS	
ABS	R
ACRYLICS	R
ASBESTOS REINF., EPOXY	R — U
ASBESTOS REINF., PHENOLIC	
CHLORINATED POLYETHER	R
CPVC	R
EPOXY	R
FLUOROCARBONS FEP	R
FLUOROCARBONS TFE	R
FURFURYL ALCOHOL	R
NORYL	R
PHENOLIC	
POLYAMIDES—NYLON	R
POLYESTERS	
BISPHENOL A-FUMURATE	R
HYDROGENATED BISPHENOL-A-BISPHENOL-A	U
ISOPHTHALIC	U
CHLORINATED POLYESTERS	
POLYSTYRENES	R
POLYETHYLENE	R
POLYPROPYLENE	R
PVC—TYPE 1	R
PVC—TYPE 2	R
POLYVINYLIDENE CHLORIDE	R
VINYLIDENE FLUORIDE	R
VINYL ESTER	R
RUBBERS	
BUTYL GR-1	R
FLUORO ELASTOMERS VITON A	R
FLUORO ELASTOMERS KEL-F 3700	
HARD RUBBER	R
HYPALON	R
KOROSEAL	R
NATURAL RUBBER (GRS)	R
NEOPRENE GR-M (CR)	R
NITRILE BUNA N (NBR)	R
NORDEL	R
POLYBUTADIENE	
SILICONE	
POLYURETHANE	R

TEMP °F: 60 80 100 120 140 160 180 200 220 240 260 280 300 320 340 360 380 400 420 440 460

POTASSIUM HYPOCHLORITE

TEMP °F

METALS		60 80 100 120 140 160 180 200 220 240 260 280 300 320 340 360 380 400 420 440 460 480 500 520 540 560
ALLOY 20Cb3		G—
ALUMINUM		U
BRASS		G—
BRONZE		
CARBON STEEL		
COPPER		
E-BRIGHT 26-1		
HASTELLOY-B		
HASTELLOY-C		G————————
HASTELLOY-D		G————————
HIGH SILICON IRON		
INCONEL		U
LEAD		
MONEL		U
NICKEL		U
STAINLESS STEELS		
Type 304/307		U
Type 316	1	G—
TANTALUM		G————————
TITANIUM	40%	E————————
ZIRCONIUM		
NONMETALLICS		
BOROSILICATE GLASS		
CARBON		

TEMP °F

FOR METALS

E = < 2 Mils Penetration/Year; G = < 20 Mils Penetration/Year

S = < 50 Mils Penetration/Year; U = > 50 Mils Penetration/Year

FOR NONMETALLICS

R = Resistant

U = Unsatisfactory

POTASSIUM HYPOCHLORITE

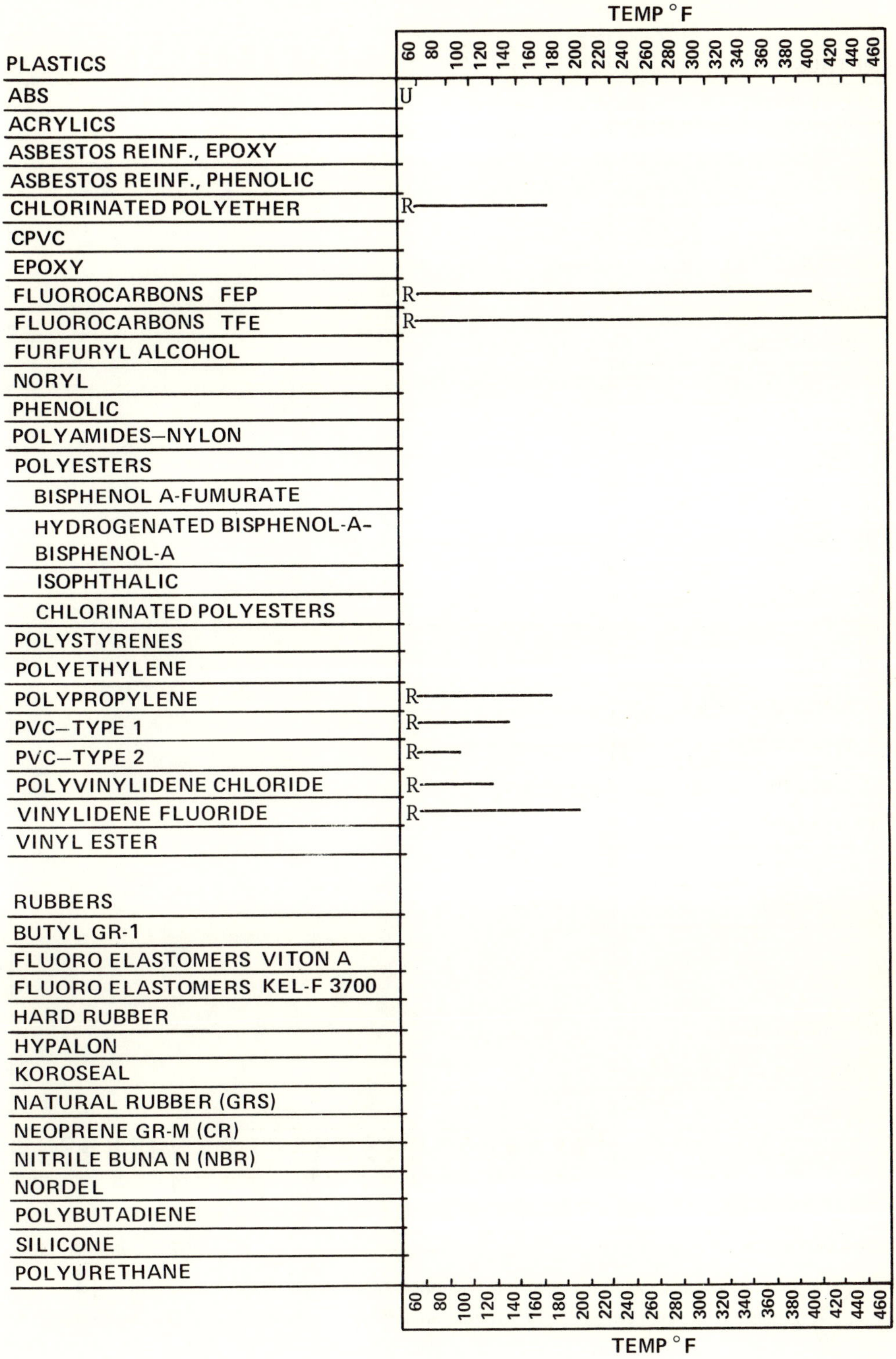

POTASSIUM IODIDE, 70%

TEMP °F: 60, 80, 100, 120, 140, 160, 180, 200, 220, 240, 260, 280, 300, 320, 340, 360, 380, 400, 420, 440, 460, 480, 500, 520, 540, 560

METALS	Rating
ALLOY 20Cb3	G
ALUMINUM	S
BRASS	
BRONZE	G
CARBON STEEL	S
COPPER	G
E-BRIGHT 26-1	
HASTELLOY-B	G
HASTELLOY-C	G
HASTELLOY-D	G
HIGH SILICON IRON	U
INCONEL	G
LEAD	
MONEL	G
NICKEL	G
STAINLESS STEELS	
Type 304/307	G
Type 316	G
TANTALUM	E
TITANIUM	E
ZIRCONIUM	E

NONMETALLICS	Rating
BOROSILICATE GLASS	
CARBON	

TEMP °F

FOR METALS

E = < 2 Mils Penetration/Year; G = < 20 Mils Penetration/Year

S = < 50 Mils Penetration/Year; U = > 50 Mils Penetration/Year

FOR NONMETALLICS

R = Resistant

U = Unsatisfactory

POTASSIUM IODIDE, 70%

TEMP °F: 60 80 100 120 140 160 180 200 220 240 260 280 300 320 340 360 380 400 420 440 460

PLASTICS	Rating	Line extent (°F)
ABS		
ACRYLICS		
ASBESTOS REINF., EPOXY		
ASBESTOS REINF., PHENOLIC		
CHLORINATED POLYETHER	R	60–250
CPVC		
EPOXY		
FLUOROCARBONS FEP	R	60–400
FLUOROCARBONS TFE	R	60–460
FURFURYL ALCOHOL		
NORYL		
PHENOLIC		
POLYAMIDES—NYLON		
POLYESTERS		
BISPHENOL A-FUMURATE	R	60–220
HYDROGENATED BISPHENOL-A-BISPHENOL-A		
ISOPHTHALIC		
CHLORINATED POLYESTERS		
POLYSTYRENES		
POLYETHYLENE	R	60–140
POLYPROPYLENE	R	60–140
PVC—TYPE 1	R	60–140
PVC—TYPE 2		
POLYVINYLIDENE CHLORIDE	R	60–150
VINYLIDENE FLUORIDE		
VINYL ESTER		

RUBBERS	Rating	Line extent (°F)
BUTYL GR-1		
FLUORO ELASTOMERS VITON A	R	60–75
FLUORO ELASTOMERS KEL-F 3700		
HARD RUBBER	R	60–180
HYPALON		
KOROSEAL		
NATURAL RUBBER (GRS)		
NEOPRENE GR-M (CR)	R	60–180
NITRILE BUNA N (NBR)		
NORDEL	R	60–300
POLYBUTADIENE		
SILICONE		
POLYURETHANE		

TEMP °F: 60 80 100 120 140 160 180 200 220 240 260 280 300 320 340 360 380 400 420 440 460

POTASSIUM NITRATE, 80%

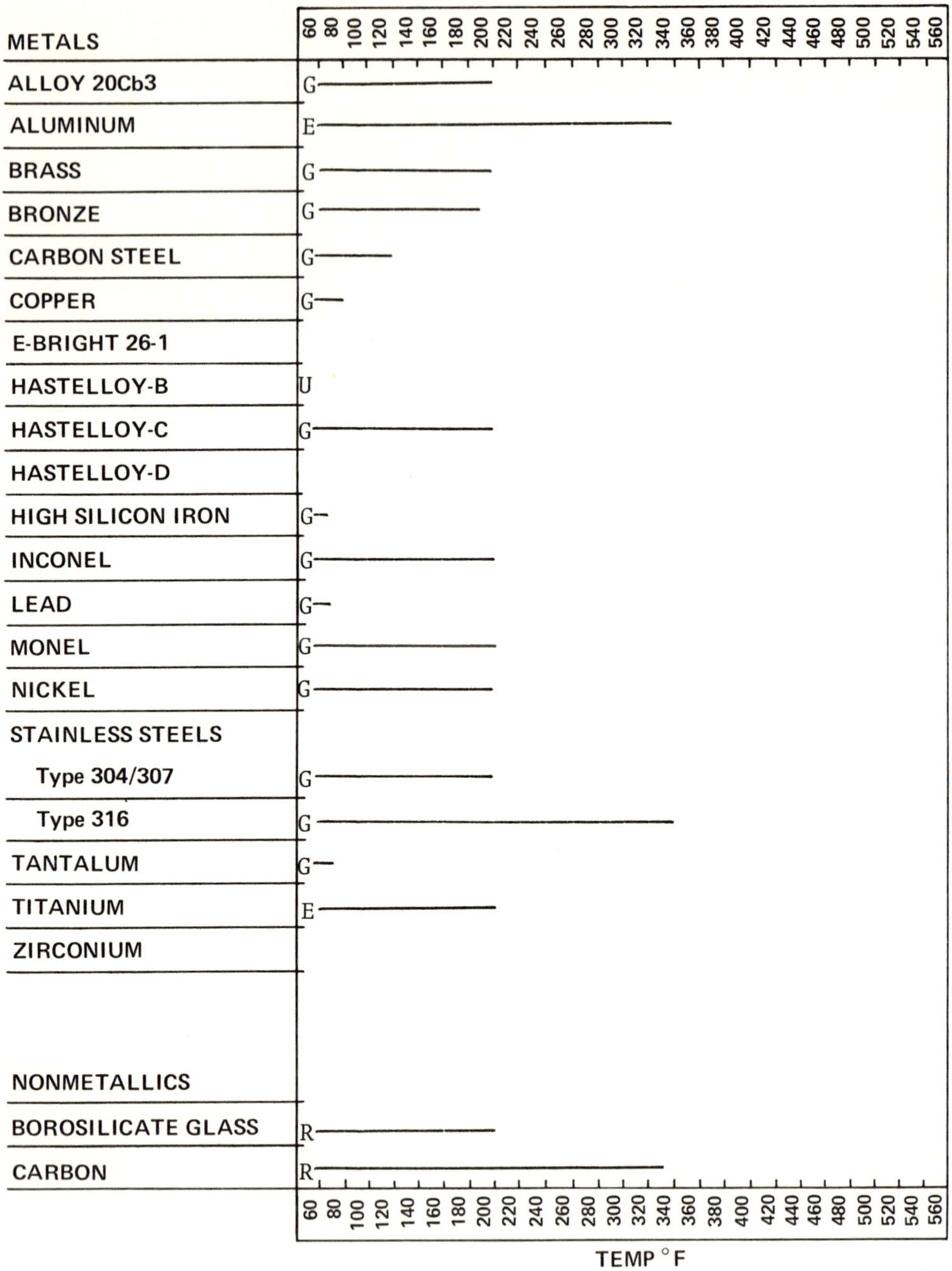

FOR METALS

E = < 2 Mils Penetration/Year; G = < 20 Mils Penetration/Year

S = < 50 Mils Penetration/Year; U = > 50 Mils Penetration/Year

FOR NONMETALLICS

R = Resistant

U = Unsatisfactory

POTASSIUM NITRATE, 80%

TEMP °F (scale: 60, 80, 100, 120, 140, 160, 180, 200, 220, 240, 260, 280, 300, 320, 340, 360, 380, 400, 420, 440, 460)

PLASTICS		
ABS		R
ACRYLICS		R
ASBESTOS REINF., EPOXY		R
ASBESTOS REINF., PHENOLIC		R
CHLORINATED POLYETHER		R
CPVC		R
EPOXY		R
FLUOROCARBONS FEP		R
FLUOROCARBONS TFE		R
FURFURYL ALCOHOL	sat	R
NORYL		R
PHENOLIC		
POLYAMIDES—NYLON		R
POLYESTERS		
BISPHENOL A-FUMURATE	sat	R
HYDROGENATED BISPHENOL-A-BISPHENOL-A		R
ISOPHTHALIC	sat	R
CHLORINATED POLYESTERS	all	R
POLYSTYRENES		
POLYETHYLENE		R
POLYPROPYLENE		R
PVC—TYPE 1		R
PVC—TYPE 2		R
POLYVINYLIDENE CHLORIDE		R
VINYLIDENE FLUORIDE		R
VINYL ESTER		R

RUBBERS		
BUTYL GR-1		R
FLUORO ELASTOMERS VITON A		R
FLUORO ELASTOMERS KEL-F 3700		
HARD RUBBER		R
HYPALON		R
KOROSEAL		R
NATURAL RUBBER (GRS)		R
NEOPRENE GR-M (CR)		R
NITRILE BUNA N (NBR)		R
NORDEL		R
POLYBUTADIENE		
SILICONE		
POLYURETHANE		

TEMP °F (scale: 60, 80, 100, 120, 140, 160, 180, 200, 220, 240, 260, 280, 300, 320, 340, 360, 380, 400, 420, 440, 460)

POTASSIUM NITRATE, 5%-1%

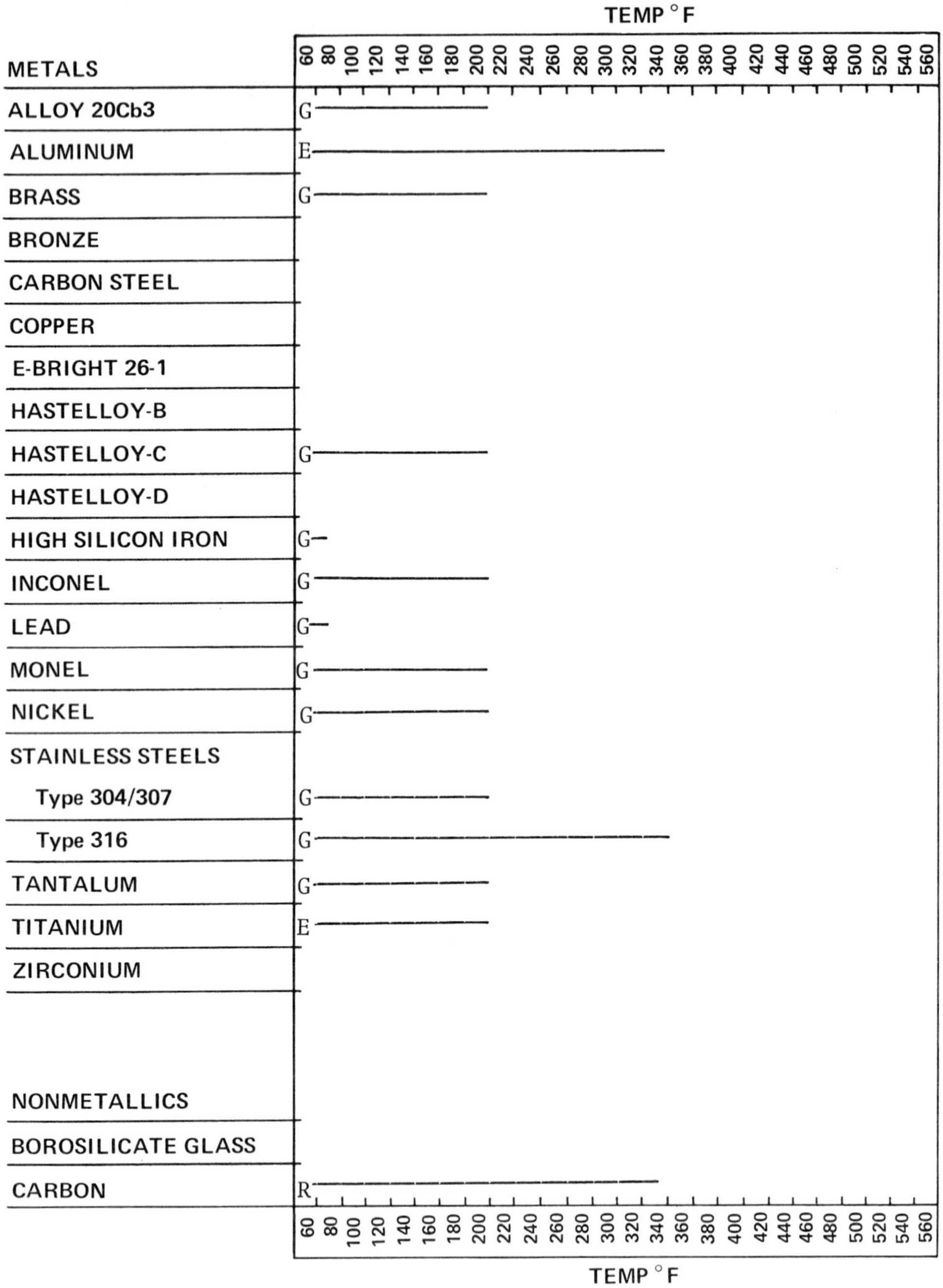

FOR METALS

E = < 2 Mils Penetration/Year; G = < 20 Mils Penetration/Year

S = < 50 Mils Penetration/Year; U = > 50 Mils Penetration/Year

FOR NONMETALLICS

R = Resistant

U = Unsatisfactory

POTASSIUM NITRATE, 5%-1%

TEMP °F (60–460)

PLASTICS	
ABS	R
ACRYLICS	R
ASBESTOS REINF., EPOXY	R
ASBESTOS REINF., PHENOLIC	R
CHLORINATED POLYETHER	R
CPVC	R
EPOXY	R
FLUOROCARBONS FEP	R
FLUOROCARBONS TFE	R
FURFURYL ALCOHOL	
NORYL	R
PHENOLIC	
POLYAMIDES–NYLON	R
POLYESTERS	
BISPHENOL A-FUMURATE	R
HYDROGENATED BISPHENOL-A-BISPHENOL-A	R
ISOPHTHALIC	R
CHLORINATED POLYESTERS	
POLYSTYRENES	
POLYETHYLENE	R
POLYPROPYLENE	R
PVC–TYPE 1	R
PVC–TYPE 2	R
POLYVINYLIDENE CHLORIDE	R
VINYLIDENE FLUORIDE	R
VINYL ESTER	R

RUBBERS	
BUTYL GR-1	
FLUORO ELASTOMERS VITON A	R
FLUORO ELASTOMERS KEL-F 3700	
HARD RUBBER	
HYPALON	R
KOROSEAL	R
NATURAL RUBBER (GRS)	
NEOPRENE GR-M (CR)	R
NITRILE BUNA N (NBR)	R
NORDEL	R
POLYBUTADIENE	
SILICONE	
POLYURETHANE	

TEMP °F (60–460)

POTASSIUM OXALATE, 20%

TEMP °F: 60, 80, 100, 120, 140, 160, 180, 200, 220, 240, 260, 280, 300, 320, 340, 360, 380, 400, 420, 440, 460, 480, 500, 520, 540, 560

METALS		
ALLOY 20Cb3	20%	G
ALUMINUM		
BRASS		
BRONZE		
CARBON STEEL		
COPPER		
E-BRIGHT 26-1		
HASTELLOY-B		
HASTELLOY-C		
HASTELLOY-D		
HIGH SILICON IRON		
INCONEL		G
LEAD		
MONEL		G
NICKEL		G
STAINLESS STEELS		
Type 304/307	20%	G
Type 316	20%	G
TANTALUM		E
TITANIUM		
ZIRCONIUM		

NONMETALLICS		
BOROSILICATE GLASS		
CARBON		

TEMP °F: 60, 80, 100, 120, 140, 160, 180, 200, 220, 240, 260, 280, 300, 320, 340, 360, 380, 400, 420, 440, 460, 480, 500, 520, 540, 560

FOR METALS

E = < 2 Mils Penetration/Year; G = < 20 Mils Penetration/Year

S = < 50 Mils Penetration/Year; U = > 50 Mils Penetration/Year

FOR NONMETALLICS

R = Resistant

U = Unsatisfactory

POTASSIUM OXALATE, 20%

PLASTICS		TEMP °F 60	80	100	120	140	160	180	200	220	240	260	280	300	320	340	360	380	400	420	440	460
ABS																						
ACRYLICS																						
ASBESTOS REINF., EPOXY																						
ASBESTOS REINF., PHENOLIC																						
CHLORINATED POLYETHER																						
CPVC																						
EPOXY																						
FLUOROCARBONS FEP		R																				
FLUOROCARBONS TFE		R																				
FURFURYL ALCOHOL	sat	R																				
NORYL																						
PHENOLIC																						
POLYAMIDES—NYLON																						
POLYESTERS																						
BISPHENOL A-FUMURATE																						
HYDROGENATED BISPHENOL-A-BISPHENOL-A																						
ISOPHTHALIC																						
CHLORINATED POLYESTERS																						
POLYSTYRENES																						
POLYETHYLENE																						
POLYPROPYLENE																						
PVC—TYPE 1																						
PVC—TYPE 2																						
POLYVINYLIDENE CHLORIDE																						
VINYLIDENE FLUORIDE																						
VINYL ESTER																						
RUBBERS																						
BUTYL GR-1																						
FLUORO ELASTOMERS VITON A																						
FLUORO ELASTOMERS KEL-F 3700																						
HARD RUBBER																						
HYPALON																						
KOROSEAL																						
NATURAL RUBBER (GRS)																						
NEOPRENE GR-M (CR)																						
NITRILE BUNA N (NBR)																						
NORDEL																						
POLYBUTADIENE																						
SILICONE																						
POLYURETHANE																						
		60	80	100	120	140	160	180	200	220	240	260	280	300	320	340	360	380	400	420	440	460

TEMP °F

POTASSIUM PERBORATE

TEMP °F (60 80 100 120 140 160 180 200 220 240 260 280 300 320 340 360 380 400 420 440 460 480 500 520 540 560)

METALS	TEMP °F
ALLOY 20Cb3	
ALUMINUM	
BRASS	
BRONZE	
CARBON STEEL	
COPPER	
E-BRIGHT 26-1	
HASTELLOY-B	
HASTELLOY-C	
HASTELLOY-D	
HIGH SILICON IRON	
INCONEL	
LEAD	
MONEL	
NICKEL	
STAINLESS STEELS	
Type 304/307	
Type 316	
TANTALUM	
TITANIUM	
ZIRCONIUM	

NONMETALLICS	TEMP °F
BOROSILICATE GLASS	
CARBON	R (60–340)

TEMP °F

FOR METALS

E = < 2 Mils Penetration/Year; G = < 20 Mils Penetration/Year

S = < 50 Mils Penetration/Year; U = > 50 Mils Penetration/Year

FOR NONMETALLICS

R = Resistant

U = Unsatisfactory

POTASSIUM PERBORATE

PLASTICS	TEMP °F (60–460)
ABS	R (60–140)
ACRYLICS	
ASBESTOS REINF., EPOXY	
ASBESTOS REINF., PHENOLIC	
CHLORINATED POLYETHER	R (60–220)
CPVC	R (60–180)
EPOXY	
FLUOROCARBONS FEP	R (60–400)
FLUOROCARBONS TFE	R (60–460)
FURFURYL ALCOHOL	
NORYL	
PHENOLIC	
POLYAMIDES–NYLON	
POLYESTERS	
BISPHENOL A-FUMURATE	
HYDROGENATED BISPHENOL-A-BISPHENOL-A	
ISOPHTHALIC	
CHLORINATED POLYESTERS	
POLYSTYRENES	
POLYETHYLENE	R (60–150)
POLYPROPYLENE	R (60–140)
PVC–TYPE 1	R (60–140)
PVC–TYPE 2	R (60–140)
POLYVINYLIDENE CHLORIDE	R (60–150)
VINYLIDENE FLUORIDE	R (60–280)
VINYL ESTER	

RUBBERS	TEMP °F (60–460)
BUTYL GR-1	
FLUORO ELASTOMERS VITON A	
FLUORO ELASTOMERS KEL-F 3700	
HARD RUBBER	
HYPALON	
KOROSEAL	
NATURAL RUBBER (GRS)	
NEOPRENE GR-M (CR)	
NITRILE BUNA N (NBR)	
NORDEL	
POLYBUTADIENE	
SILICONE	
POLYURETHANE	

POTASSIUM PERCHLORATE

METALS	Concentration	Rating	TEMP °F (bar extent)
ALLOY 20Cb3	30%	E	60–120
ALUMINUM			
BRASS			
BRONZE			
CARBON STEEL			
COPPER			
E-BRIGHT 26-1			
HASTELLOY-B			
HASTELLOY-C	25-75%	G	60–80
HASTELLOY-D			
HIGH SILICON IRON			
INCONEL			
LEAD			
MONEL			
NICKEL			
STAINLESS STEELS			
Type 304/307			
Type 316		G	60–340
TANTALUM			
TITANIUM	20%	E	60–80
ZIRCONIUM			

NONMETALLICS	Concentration	Rating	TEMP °F (bar extent)
BOROSILICATE GLASS			
CARBON		R	60–340

TEMP °F: 60 80 100 120 140 160 180 200 220 240 260 280 300 320 340 360 380 400 420 440 460 480 500 520 540 560

FOR METALS

E = < 2 Mils Penetration/Year; G = < 20 Mils Penetration/Year

S = < 50 Mils Penetration/Year; U = > 50 Mils Penetration/Year

FOR NONMETALLICS

R = Resistant

U = Unsatisfactory

POTASSIUM PERCHLORATE

TEMP °F

PLASTICS		60–460 °F
ABS		
ACRYLICS		
ASBESTOS REINF., EPOXY		
ASBESTOS REINF., PHENOLIC		
CHLORINATED POLYETHER		R
CPVC		R
EPOXY		
FLUOROCARBONS FEP		R
FLUOROCARBONS TFE		R
FURFURYL ALCOHOL		
NORYL	10%	R
PHENOLIC		
POLYAMIDES–NYLON		
POLYESTERS		
BISPHENOL A-FUMURATE		
HYDROGENATED BISPHENOL-A-BISPHENOL-A		
ISOPHTHALIC		
CHLORINATED POLYESTERS		
POLYSTYRENES		
POLYETHYLENE		R
POLYPROPYLENE		R
PVC–TYPE 1		R
PVC–TYPE 2		R
POLYVINYLIDENE CHLORIDE		R
VINYLIDENE FLUORIDE		R
VINYL ESTER		

RUBBERS		
BUTYL GR-1		
FLUORO ELASTOMERS VITON A		
FLUORO ELASTOMERS KEL-F 3700		
HARD RUBBER		
HYPALON		
KOROSEAL		
NATURAL RUBBER (GRS)		
NEOPRENE GR-M (CR)		
NITRILE BUNA N (NBR)		
NORDEL		
POLYBUTADIENE		
SILICONE		
POLYURETHANE		

TEMP °F

POTASSIUM PERMANGANATE, 20%

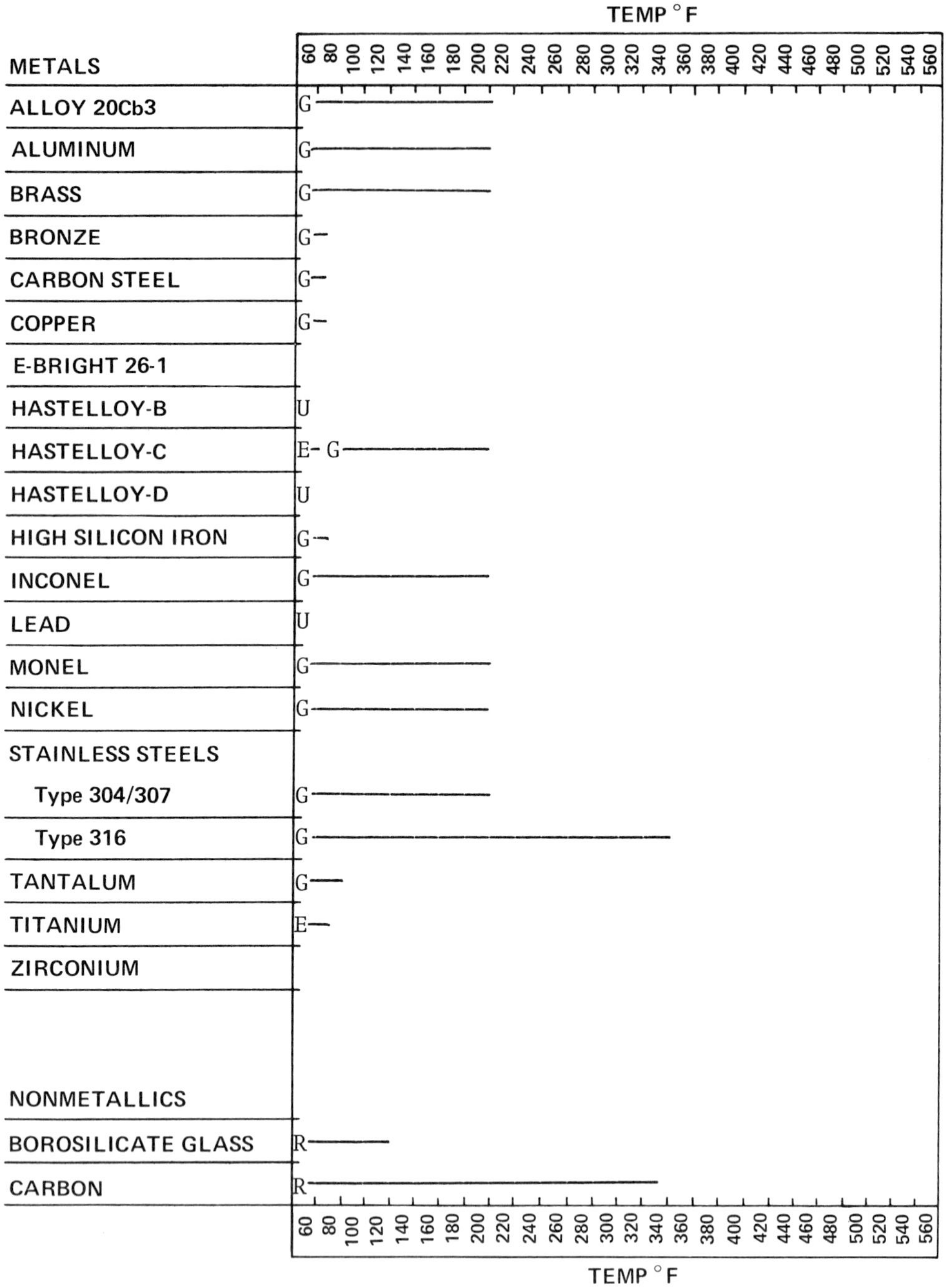

FOR METALS

E = < 2 Mils Penetration/Year; G = < 20 Mils Penetration/Year

S = < 50 Mils Penetration/Year; U = > 50 Mils Penetration/Year

FOR NONMETALLICS

R = Resistant

U = Unsatisfactory

POTASSIUM PERMANGANATE, 20%

TEMP °F: 60 80 100 120 140 160 180 200 220 240 260 280 300 320 340 360 380 400 420 440 460

PLASTICS		Rating
ABS	sat	R
ACRYLICS		U
ASBESTOS REINF., EPOXY		R
ASBESTOS REINF., PHENOLIC		R
CHLORINATED POLYETHER		R
CPVC		R U
EPOXY		R
FLUOROCARBONS FEP		R
FLUOROCARBONS TFE		R
FURFURYL ALCOHOL		R
NORYL		R
PHENOLIC		R
POLYAMIDES–NYLON		U
POLYESTERS		
BISPHENOL A-FUMURATE		R
HYDROGENATED BISPHENOL-A-BISPHENOL-A		
ISOPHTHALIC		R
CHLORINATED POLYESTERS		R
POLYSTYRENES		R
POLYETHYLENE		R
POLYPROPYLENE		R
PVC–TYPE 1		R
PVC–TYPE 2		R
POLYVINYLIDENE CHLORIDE		R
VINYLIDENE FLUORIDE		R
VINYL ESTER		R

RUBBERS	Rating
BUTYL GR-1	R
FLUORO ELASTOMERS VITON A	R
FLUORO ELASTOMERS KEL-F 3700	
HARD RUBBER	R
HYPALON	R
KOROSEAL	
NATURAL RUBBER (GRS)	U
NEOPRENE GR-M (CR)	U
NITRILE BUNA N (NBR)	RU
NORDEL	R
POLYBUTADIENE	
SILICONE	
POLYURETHANE	

TEMP °F: 60 80 100 120 140 160 180 200 220 240 260 280 300 320 340 360 380 400 420 440 460

POTASSIUM PERMANGANATE, 10%

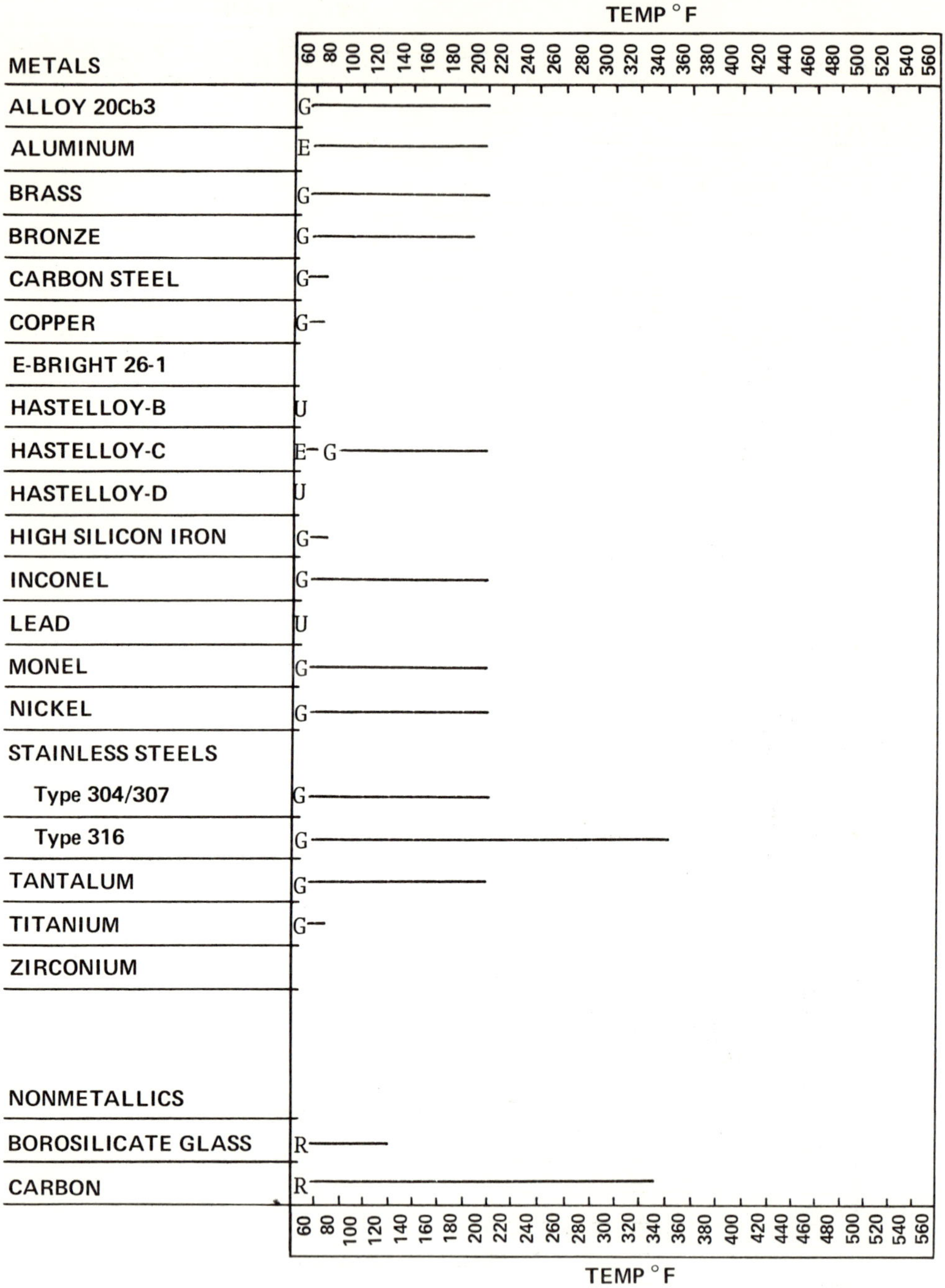

FOR METALS

E = < 2 Mils Penetration/Year; G = < 20 Mils Penetration/Year

S = < 50 Mils Penetration/Year; U = > 50 Mils Penetration/Year

FOR NONMETALLICS

R = Resistant

U = Unsatisfactory

POTASSIUM PERMANGANATE, 10%

TEMP °F: 60 80 100 120 140 160 180 200 220 240 260 280 300 320 340 360 380 400 420 440 460

PLASTICS		
ABS		U
ACRYLICS		U
ASBESTOS REINF., EPOXY		R
ASBESTOS REINF., PHENOLIC		R
CHLORINATED POLYETHER		R
CPVC		R
EPOXY		R
FLUOROCARBONS FEP		R
FLUOROCARBONS TFE		R
FURFURYL ALCOHOL	sat	R
NORYL		R
PHENOLIC		R
POLYAMIDES–NYLON		U
POLYESTERS		
BISPHENOL A-FUMURATE	sat	R
HYDROGENATED BISPHENOL-A-BISPHENOL-A		
ISOPHTHALIC	sat	U
CHLORINATED POLYESTERS		R
POLYSTYRENES		R
POLYETHYLENE		R
POLYPROPYLENE		R
PVC–TYPE 1		R
PVC–TYPE 2		R
POLYVINYLIDENE CHLORIDE		R
VINYLIDENE FLUORIDE		R
VINYL ESTER		R

RUBBERS	
BUTYL GR-1	R
FLUORO ELASTOMERS VITON A	R
FLUORO ELASTOMERS KEL-F 3700	
HARD RUBBER	R
HYPALON	R
KOROSEAL	
NATURAL RUBBER (GRS)	U
NEOPRENE GR-M (CR)	U
NITRILE BUNA N (NBR)	R
NORDEL	R
POLYBUTADIENE	
SILICONE	
POLYURETHANE	

TEMP °F: 60 80 100 120 140 160 180 200 220 240 260 280 300 320 340 360 380 400 420 440 460

POTASSIUM PERSULFATE

TEMP °F: 60, 80, 100, 120, 140, 160, 180, 200, 220, 240, 260, 280, 300, 320, 340, 360, 380, 400, 420, 440, 460, 480, 500, 520, 540, 560

METALS		TEMP °F
ALLOY 20Cb3		
ALUMINUM		
BRASS		
BRONZE		
CARBON STEEL		
COPPER		
E-BRIGHT 26-1		
HASTELLOY-B		
HASTELLOY-C	4%	E (short line from 60)
HASTELLOY-D		
HIGH SILICON IRON		
INCONEL		
LEAD		
MONEL		
NICKEL		
STAINLESS STEELS		
Type 304/307		
Type 316		G (line from 60 to about 340–360)
TANTALUM		
TITANIUM		
ZIRCONIUM		
NONMETALLICS		
BOROSILICATE GLASS		
CARBON		

TEMP °F: 60, 80, 100, 120, 140, 160, 180, 200, 220, 240, 260, 280, 300, 320, 340, 360, 380, 400, 420, 440, 460, 480, 500, 520, 540, 560

FOR METALS

E = < 2 Mils Penetration/Year; G = < 20 Mils Penetration/Year

S = < 50 Mils Penetration/Year; U = > 50 Mils Penetration/Year

FOR NONMETALLICS

R = Resistant

U = Unsatisfactory

POTASSIUM PERSULFATE

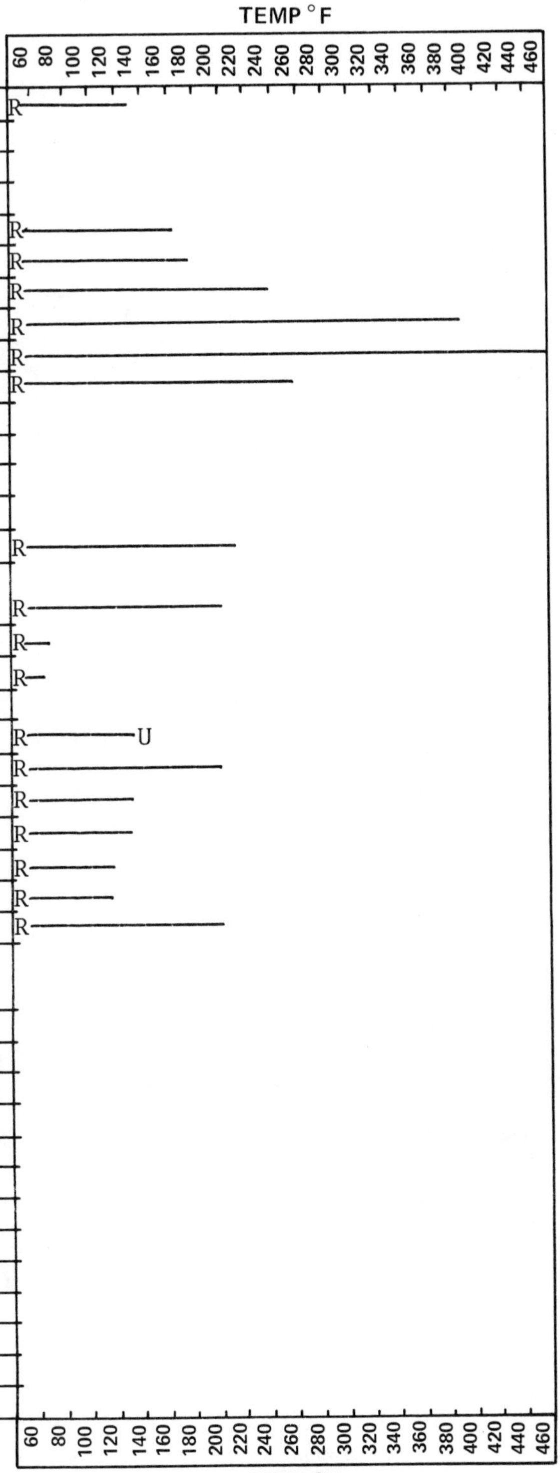

PLASTICS		Rating
ABS		R
ACRYLICS		
ASBESTOS REINF., EPOXY		
ASBESTOS REINF., PHENOLIC		
CHLORINATED POLYETHER		R
CPVC		R
EPOXY		R
FLUOROCARBONS FEP		R
FLUOROCARBONS TFE		R
FURFURYL ALCOHOL	sat	R
NORYL		
PHENOLIC		
POLYAMIDES–NYLON		
POLYESTERS		
BISPHENOL A-FUMURATE	sat	R
HYDROGENATED BISPHENOL-A-BISPHENOL-A		R
ISOPHTHALIC	sat	R
CHLORINATED POLYESTERS	all	R
POLYSTYRENES		
POLYETHYLENE		R … U
POLYPROPYLENE		R
PVC–TYPE 1		R
PVC–TYPE 2		R
POLYVINYLIDENE CHLORIDE		R
VINYLIDENE FLUORIDE		R
VINYL ESTER		R

RUBBERS		
BUTYL GR-1		
FLUORO ELASTOMERS VITON A		
FLUORO ELASTOMERS KEL-F 3700		
HARD RUBBER		
HYPALON		
KOROSEAL		
NATURAL RUBBER (GRS)		
NEOPRENE GR-M (CR)		
NITRILE BUNA N (NBR)		
NORDEL		
POLYBUTADIENE		
SILICONE		
POLYURETHANE		

POTASSIUM SULFATE, 10%

TEMP °F: 60 80 100 120 140 160 180 200 220 240 260 280 300 320 340 360 380 400 420 440 460 480 500 520 540 560

METALS		
ALLOY 20Cb3	10%	E
ALUMINUM	10%	E
BRASS		G
BRONZE		G
CARBON STEEL		G
COPPER		G
E-BRIGHT 26-1		
HASTELLOY-B	10%	E
HASTELLOY-C	10%	E
HASTELLOY-D		
HIGH SILICON IRON		E
INCONEL	10%	E–G
LEAD		G
MONEL	10%	E
NICKEL	10%	E
STAINLESS STEELS		
Type 304/307	10%	E
Type 316	10%	E
TANTALUM		E
TITANIUM	10%	E
ZIRCONIUM		

NONMETALLICS	
BOROSILICATE GLASS	R
CARBON	R

FOR METALS

E = < 2 Mils Penetration/Year; G = < 20 Mils Penetration/Year

S = < 50 Mils Penetration/Year; U = > 50 Mils Penetration/Year

FOR NONMETALLICS

R = Resistant

U = Unsatisfactory

POTASSIUM SULFATE, 10%

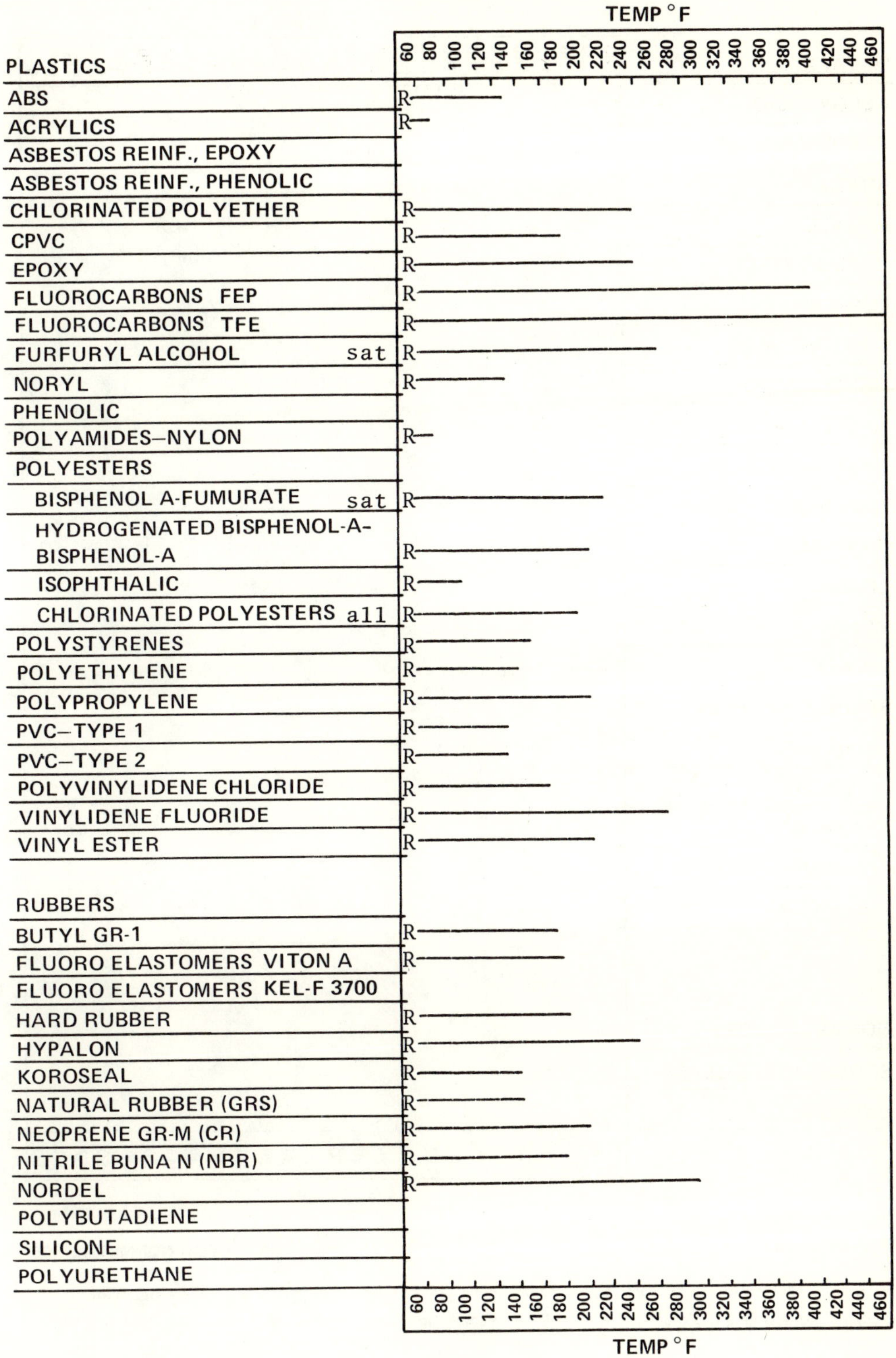

POTASSIUM SULFIDE

TEMP °F

METALS	60 80 100 120 140 160 180 200 220 240 260 280 300 320 340 360 380 400 420 440 460 480 500 520 540 560
ALLOY 20Cb3	G——
ALUMINUM	U
BRASS	U
BRONZE	U
CARBON STEEL	U
COPPER	
E-BRIGHT 26-1	
HASTELLOY-B	G——
HASTELLOY-C	G——
HASTELLOY-D	
HIGH SILICON IRON	G—
INCONEL	G—
LEAD	
MONEL	U
NICKEL	G——
STAINLESS STEELS	
Type 304/307	G————————
Type 316	G————————
TANTALUM	E—
TITANIUM	
ZIRCONIUM	
NONMETALLICS	
BOROSILICATE GLASS	
CARBON	

60 80 100 120 140 160 180 200 220 240 260 280 300 320 340 360 380 400 420 440 460 480 500 520 540 560

TEMP °F

FOR METALS

E = < 2 Mils Penetration/Year; G = < 20 Mils Penetration/Year

S = < 50 Mils Penetration/Year; U = > 50 Mils Penetration/Year

FOR NONMETALLICS

R = Resistant

U = Unsatisfactory

POTASSIUM SULFIDE

PLASTICS	TEMP °F
ABS	R 60–140
ACRYLICS	
ASBESTOS REINF., EPOXY	
ASBESTOS REINF., PHENOLIC	
CHLORINATED POLYETHER	R 60–220
CPVC	
EPOXY	
FLUOROCARBONS FEP	R 60–400
FLUOROCARBONS TFE	R 60–460
FURFURYL ALCOHOL	
NORYL	R 60–140
PHENOLIC	
POLYAMIDES—NYLON	
POLYESTERS	
BISPHENOL A-FUMURATE	
HYDROGENATED BISPHENOL-A-BISPHENOL-A	
ISOPHTHALIC	
CHLORINATED POLYESTERS	
POLYSTYRENES	
POLYETHYLENE	R 60–140
POLYPROPYLENE	R 60–200
PVC—TYPE 1	R 60–140
PVC—TYPE 2	R 60–140
POLYVINYLIDENE CHLORIDE	R 60–150
VINYLIDENE FLUORIDE	R 60–280
VINYL ESTER	

RUBBERS	TEMP °F
BUTYL GR-1	
FLUORO ELASTOMERS VITON A	
FLUORO ELASTOMERS KEL-F 3700	
HARD RUBBER	
HYPALON	
KOROSEAL	
NATURAL RUBBER (GRS)	
NEOPRENE GR-M (CR)	
NITRILE BUNA N (NBR)	
NORDEL	
POLYBUTADIENE	
SILICONE	
POLYURETHANE	

TEMP °F: 60 80 100 120 140 160 180 200 220 240 260 280 300 320 340 360 380 400 420 440 460

PROPANE

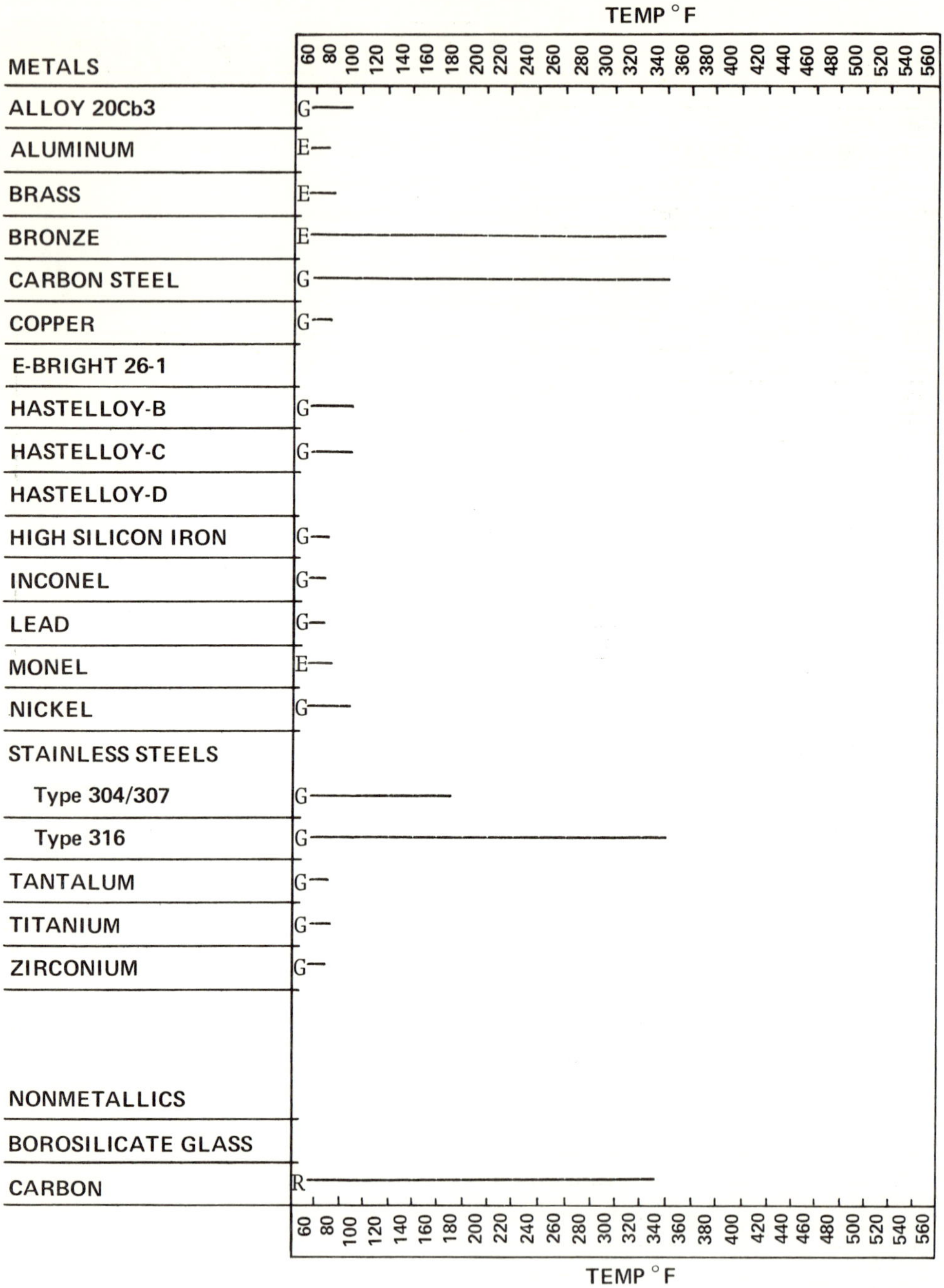

FOR METALS

E = < 2 Mils Penetration/Year; G = < 20 Mils Penetration/Year

S = < 50 Mils Penetration/Year; U = > 50 Mils Penetration/Year

FOR NONMETALLICS

R = Resistant

U = Unsatisfactory

TEMP °F

PLASTICS	Rating	Line extends to (°F, read from chart scale 60–460)
ABS	R	140
ACRYLICS		
ASBESTOS REINF., EPOXY		
ASBESTOS REINF., PHENOLIC		
CHLORINATED POLYETHER	R	250
CPVC	R	180
EPOXY	R	140
FLUOROCARBONS FEP	R	400
FLUOROCARBONS TFE	R	460
FURFURYL ALCOHOL		
NORYL		
PHENOLIC		
POLYAMIDES–NYLON	R	80
POLYESTERS		
BISPHENOL A-FUMURATE		
HYDROGENATED BISPHENOL-A-BISPHENOL-A	R	80
ISOPHTHALIC		
CHLORINATED POLYESTERS		
POLYSTYRENES		
POLYETHYLENE	R	100
POLYPROPYLENE	R	150
PVC–TYPE 1	R	140
PVC–TYPE 2	R	140
POLYVINYLIDENE CHLORIDE	R	150
VINYLIDENE FLUORIDE	R	280
VINYL ESTER		

RUBBERS	Rating	Line extends to (°F, read from chart scale 60–460)
BUTYL GR-1	U	
FLUORO ELASTOMERS VITON A	R	350
FLUORO ELASTOMERS KEL-F 3700		
HARD RUBBER		
HYPALON		
KOROSEAL		
NATURAL RUBBER (GRS)	U	
NEOPRENE GR-M (CR)	U	
NITRILE BUNA N (NBR)	R	200
NORDEL	R	300
POLYBUTADIENE		
SILICONE		
POLYURETHANE		

TEMP °F

PROPARGYL ALCOHOL

TEMP °F

METALS	60	80	100	120	140	160	180	200	220	240	260	280	300	320	340	360	380	400	420	440	460	480	500	520	540	560
ALLOY 20Cb3																										
ALUMINUM	G																									
BRASS																										
BRONZE																										
CARBON STEEL																										
COPPER																										
E-BRIGHT 26-1																										
HASTELLOY-B																										
HASTELLOY-C																										
HASTELLOY-D																										
HIGH SILICON IRON																										
INCONEL																										
LEAD																										
MONEL																										
NICKEL																										
STAINLESS STEELS Type 304/307																										
Type 316																										
TANTALUM	G																									
TITANIUM																										
ZIRCONIUM																										
NONMETALLICS																										
BOROSILICATE GLASS																										
CARBON																										

TEMP °F

FOR METALS

E = < 2 Mils Penetration/Year; G = < 20 Mils Penetration/Year

S = < 50 Mils Penetration/Year; U = > 50 Mils Penetration/Year

FOR NONMETALLICS

R = Resistant

U = Unsatisfactory

PLASTICS	TEMP °F (60–460)
ABS	U
ACRYLICS	
ASBESTOS REINF., EPOXY	
ASBESTOS REINF., PHENOLIC	
CHLORINATED POLYETHER	
CPVC	R
EPOXY	
FLUOROCARBONS FEP	R
FLUOROCARBONS TFE	R
FURFURYL ALCOHOL	
NORYL	
PHENOLIC	
POLYAMIDES–NYLON	
POLYESTERS	
BISPHENOL A-FUMURATE	
HYDROGENATED BISPHENOL-A-BISPHENOL-A	
ISOPHTHALIC	
CHLORINATED POLYESTERS	
POLYSTYRENES	
POLYETHYLENE	R
POLYPROPYLENE	U
PVC–TYPE 1	R
PVC–TYPE 2	R
POLYVINYLIDENE CHLORIDE	
VINYLIDENE FLUORIDE	
VINYL ESTER	
RUBBERS	
BUTYL GR-1	
FLUORO ELASTOMERS VITON A	
FLUORO ELASTOMERS KEL-F 3700	
HARD RUBBER	
HYPALON	
KOROSEAL	
NATURAL RUBBER (GRS)	
NEOPRENE GR-M (CR)	
NITRILE BUNA N (NBR)	
NORDEL	
POLYBUTADIENE	
SILICONE	
POLYURETHANE	

TEMP °F: 60 80 100 120 140 160 180 200 220 240 260 280 300 320 340 360 380 400 420 440 460

PROPYL ALCOHOL

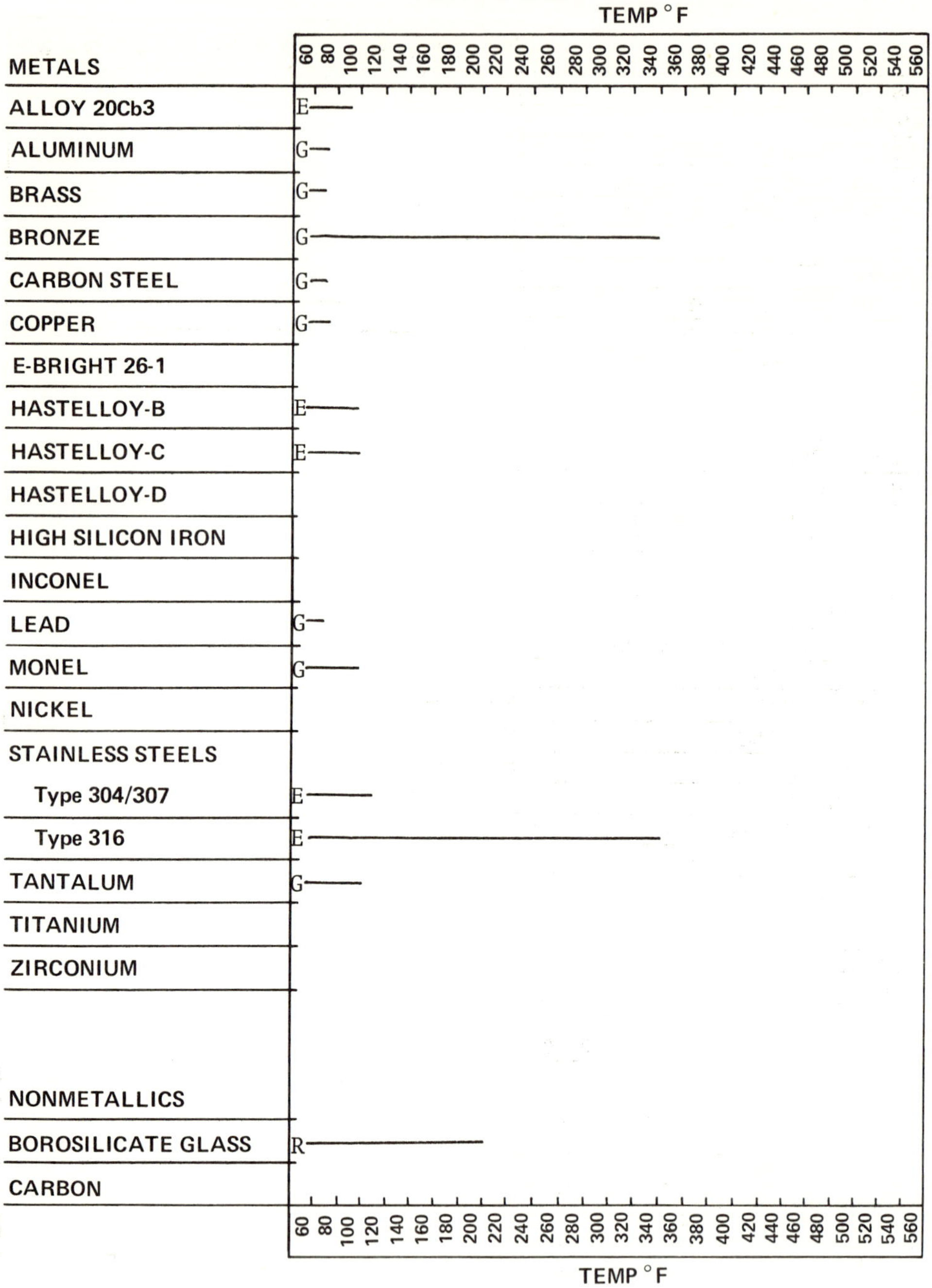

FOR METALS

E = < 2 Mils Penetration/Year; G = < 20 Mils Penetration/Year

S = < 50 Mils Penetration/Year; U = > 50 Mils Penetration/Year

FOR NONMETALLICS

R = Resistant

U = Unsatisfactory

PROPYL ALCOHOL

TEMP °F (60, 80, 100, 120, 140, 160, 180, 200, 220, 240, 260, 280, 300, 320, 340, 360, 380, 400, 420, 440, 460)

PLASTICS	
ABS	U
ACRYLICS	
ASBESTOS REINF., EPOXY	
ASBESTOS REINF., PHENOLIC	
CHLORINATED POLYETHER	R
CPVC	R
EPOXY	
FLUOROCARBONS FEP	R
FLUOROCARBONS TFE	R
FURFURYL ALCOHOL	
NORYL	
PHENOLIC	
POLYAMIDES–NYLON	
POLYESTERS	
BISPHENOL A-FUMURATE	
HYDROGENATED BISPHENOL-A-BISPHENOL-A	
ISOPHTHALIC	
CHLORINATED POLYESTERS	
POLYSTYRENES	
POLYETHYLENE	R
POLYPROPYLENE	R
PVC–TYPE 1	R
PVC–TYPE 2	R
POLYVINYLIDENE CHLORIDE	R
VINYLIDENE FLUORIDE	R
VINYL ESTER	

RUBBERS	
BUTYL GR-1	R
FLUORO ELASTOMERS VITON A	R
FLUORO ELASTOMERS KEL-F 3700	
HARD RUBBER	
HYPALON	R
KOROSEAL	
NATURAL RUBBER (GRS)	R
NEOPRENE GR-M (CR)	R
NITRILE BUNA N (NBR)	R
NORDEL	
POLYBUTADIENE	
SILICONE	
POLYURETHANE	

TEMP °F (60, 80, 100, 120, 140, 160, 180, 200, 220, 240, 260, 280, 300, 320, 340, 360, 380, 400, 420, 440, 460)

PROPYLENE DICHLORIDE

TEMP °F

METALS	60	80	100	120	140	160	180	200	220	240	260	280	300	320	340	360	380	400	420	440	460	480	500	520	540	560
ALLOY 20Cb3																										
ALUMINUM																										
BRASS																										
BRONZE																										
CARBON STEEL																										
COPPER																										
E-BRIGHT 26-1																										
HASTELLOY-B																										
HASTELLOY-C																										
HASTELLOY-D																										
HIGH SILICON IRON																										
INCONEL																										
LEAD																										
MONEL																										
NICKEL																										
STAINLESS STEELS																										
Type 304/307																										
Type 316																										
TANTALUM																										
TITANIUM																										
ZIRCONIUM																										
NONMETALLICS																										
BOROSILICATE GLASS																										
CARBON																										

60 80 100 120 140 160 180 200 220 240 260 280 300 320 340 360 380 400 420 440 460 480 500 520 540 560

TEMP °F

FOR METALS

E = < 2 Mils Penetration/Year; G = < 20 Mils Penetration/Year

S = < 50 Mils Penetration/Year; U = > 50 Mils Penetration/Year

FOR NONMETALLICS

R = Resistant

U = Unsatisfactory

PROPYLENE DICHLORIDE

PLASTICS	TEMP °F
ABS	U
ACRYLICS	U
ASBESTOS REINF., EPOXY	
ASBESTOS REINF., PHENOLIC	
CHLORINATED POLYETHER	R
CPVC	U
EPOXY	
FLUOROCARBONS FEP	R
FLUOROCARBONS TFE	R
FURFURYL ALCOHOL	
NORYL	
PHENOLIC	
POLYAMIDES—NYLON	
POLYESTERS	
BISPHENOL A-FUMURATE	
HYDROGENATED BISPHENOL-A-BISPHENOL-A	
ISOPHTHALIC	
CHLORINATED POLYESTERS	
POLYSTYRENES	
POLYETHYLENE	U
POLYPROPYLENE	U
PVC—TYPE 1	U
PVC—TYPE 2	U
POLYVINYLIDENE CHLORIDE	R
VINYLIDENE FLUORIDE	R
VINYL ESTER	

RUBBERS	TEMP °F
BUTYL GR-1	
FLUORO ELASTOMERS VITON A	
FLUORO ELASTOMERS KEL-F 3700	
HARD RUBBER	
HYPALON	
KOROSEAL	
NATURAL RUBBER (GRS)	
NEOPRENE GR-M (CR)	
NITRILE BUNA N (NBR)	
NORDEL	
POLYBUTADIENE	
SILICONE	
POLYURETHANE	

Temperature scale (°F): 60, 80, 100, 120, 140, 160, 180, 200, 220, 240, 260, 280, 300, 320, 340, 360, 380, 400, 420, 440, 460

PROPYLENE GLYCOL

METALS	TEMP °F (60–560)
ALLOY 20Cb3	
ALUMINUM	G
BRASS	
BRONZE	G
CARBON STEEL	G
COPPER	
E-BRIGHT 26-1	
HASTELLOY-B	
HASTELLOY-C	
HASTELLOY-D	
HIGH SILICON IRON	
INCONEL	
LEAD	
MONEL	
NICKEL	
STAINLESS STEELS	
Type 304/307	
Type 316	G
TANTALUM	
TITANIUM	
ZIRCONIUM	

NONMETALLICS	TEMP °F (60–560)
BOROSILICATE GLASS	R
CARBON	

FOR METALS

E = < 2 Mils Penetration/Year; G = < 20 Mils Penetration/Year

S = < 50 Mils Penetration/Year; U = > 50 Mils Penetration/Year

FOR NONMETALLICS

R = Resistant

U = Unsatisfactory

PLASTICS	TEMP °F (60–460)
ABS	R—
ACRYLICS	
ASBESTOS REINF., EPOXY	
ASBESTOS REINF., PHENOLIC	
CHLORINATED POLYETHER	
CPVC	
EPOXY	R———U
FLUOROCARBONS FEP	R———
FLUOROCARBONS TFE	R———
FURFURYL ALCOHOL	R———
NORYL	
PHENOLIC	
POLYAMIDES–NYLON	
POLYESTERS	
BISPHENOL A-FUMURATE	R———
HYDROGENATED BISPHENOL-A-BISPHENOL-A	R———
ISOPHTHALIC	R———
CHLORINATED POLYESTERS	R—
POLYSTYRENES	
POLYETHYLENE	R———
POLYPROPYLENE	R———
PVC–TYPE 1	
PVC–TYPE 2	
POLYVINYLIDENE CHLORIDE	
VINYLIDENE FLUORIDE	
VINYL ESTER	R———
RUBBERS	
BUTYL GR-1	
FLUORO ELASTOMERS VITON A	
FLUORO ELASTOMERS KEL-F 3700	
HARD RUBBER	
HYPALON	
KOROSEAL	
NATURAL RUBBER (GRS)	
NEOPRENE GR-M (CR)	R—
NITRILE BUNA N (NBR)	R—
NORDEL	
POLYBUTADIENE	
SILICONE	
POLYURETHANE	

TEMP °F: 60 80 100 120 140 160 180 200 220 240 260 280 300 320 340 360 380 400 420 440 460

PROPYLENE OXIDE

TEMP °F (60–560)

METALS		
ALLOY 20Cb3		
ALUMINUM		
BRASS		
BRONZE		
CARBON STEEL		
COPPER		
E-BRIGHT 26-1		
HASTELLOY-B		
HASTELLOY-C		
HASTELLOY-D		
HIGH SILICON IRON		
INCONEL		
LEAD		
MONEL		
NICKEL		
STAINLESS STEELS		
Type 304/307	90%	E
Type 316	90%	E
TANTALUM		G
TITANIUM		
ZIRCONIUM		

NONMETALLICS	
BOROSILICATE GLASS	
CARBON	

TEMP °F

FOR METALS

E = < 2 Mils Penetration/Year; G = < 20 Mils Penetration/Year

S = < 50 Mils Penetration/Year; U = > 50 Mils Penetration/Year

FOR NONMETALLICS

R = Resistant

U = Unsatisfactory

PROPYLENE OXIDE

PLASTICS	TEMP °F 60–460
ABS	
ACRYLICS	
ASBESTOS REINF., EPOXY	
ASBESTOS REINF., PHENOLIC	
CHLORINATED POLYETHER	
CPVC	
EPOXY	
FLUOROCARBONS FEP	R
FLUOROCARBONS TFE	R
FURFURYL ALCOHOL	
NORYL	
PHENOLIC	
POLYAMIDES—NYLON	
POLYESTERS	
BISPHENOL A-FUMURATE	
HYDROGENATED BISPHENOL-A-BISPHENOL-A	
ISOPHTHALIC	
CHLORINATED POLYESTERS	
POLYSTYRENES	
POLYETHYLENE	
POLYPROPYLENE	
PVC—TYPE 1	U
PVC—TYPE 2	U
POLYVINYLIDENE CHLORIDE	R
VINYLIDENE FLUORIDE	U
VINYL ESTER	U

RUBBERS	TEMP °F 60–460
BUTYL GR-1	
FLUORO ELASTOMERS VITON A	
FLUORO ELASTOMERS KEL-F 3700	
HARD RUBBER	
HYPALON	
KOROSEAL	
NATURAL RUBBER (GRS)	
NEOPRENE GR-M (CR)	
NITRILE BUNA N (NBR)	U
NORDEL	
POLYBUTADIENE	
SILICONE	
POLYURETHANE	

PYRIDINE

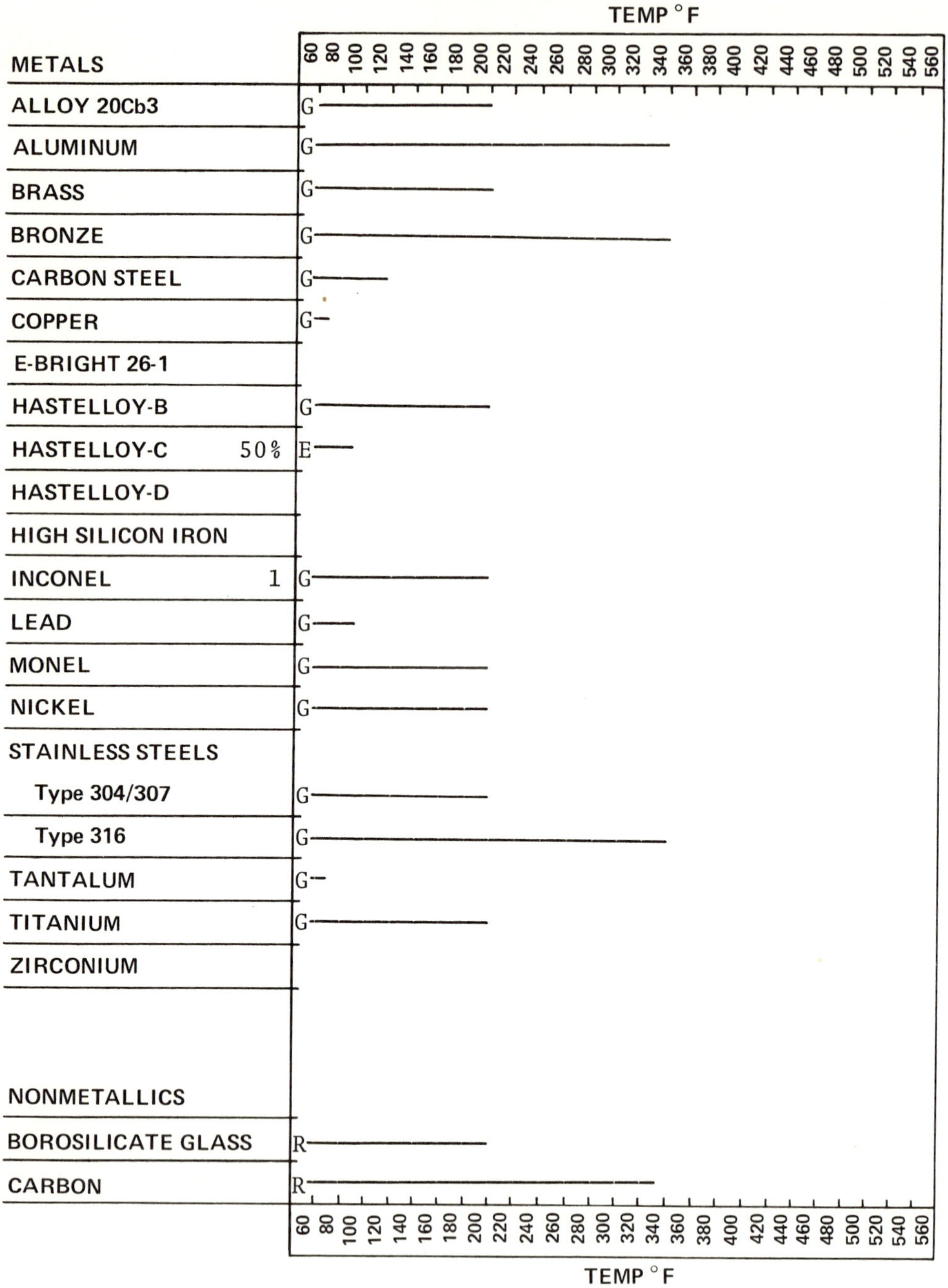

FOR METALS

E = < 2 Mils Penetration/Year; G = < 20 Mils Penetration/Year

S = < 50 Mils Penetration/Year; U = > 50 Mils Penetration/Year

FOR NONMETALLICS

R = Resistant

U = Unsatisfactory

PYRIDINE

TEMP °F (scale: 60, 80, 100, 120, 140, 160, 180, 200, 220, 240, 260, 280, 300, 320, 340, 360, 380, 400, 420, 440, 460)

PLASTICS	TEMP °F
ABS	
ACRYLICS	
ASBESTOS REINF., EPOXY	
ASBESTOS REINF., PHENOLIC	
CHLORINATED POLYETHER	R (60–250)
CPVC	
EPOXY	U
FLUOROCARBONS FEP	R (60–400)
FLUOROCARBONS TFE	R (60–460)
FURFURYL ALCOHOL	U
NORYL	
PHENOLIC	
POLYAMIDES—NYLON	U
POLYESTERS	
BISPHENOL A-FUMURATE	U
HYDROGENATED BISPHENOL-A-BISPHENOL-A	U
ISOPHTHALIC	U
CHLORINATED POLYESTERS	
POLYSTYRENES	
POLYETHYLENE	R (60–120) U (140)
POLYPROPYLENE	R (60–180)
PVC—TYPE 1	U
PVC—TYPE 2	
POLYVINYLIDENE CHLORIDE	
VINYLIDENE FLUORIDE	U
VINYL ESTER	U
RUBBERS	
BUTYL GR-1	R (60–100)
FLUORO ELASTOMERS VITON A	U
FLUORO ELASTOMERS KEL-F 3700	
HARD RUBBER	
HYPALON	
KOROSEAL	U
NATURAL RUBBER (GRS)	U
NEOPRENE GR-M (CR)	U
NITRILE BUNA N (NBR)	U
NORDEL	
POLYBUTADIENE	
SILICONE	
POLYURETHANE	

PYROGALLIC ACID

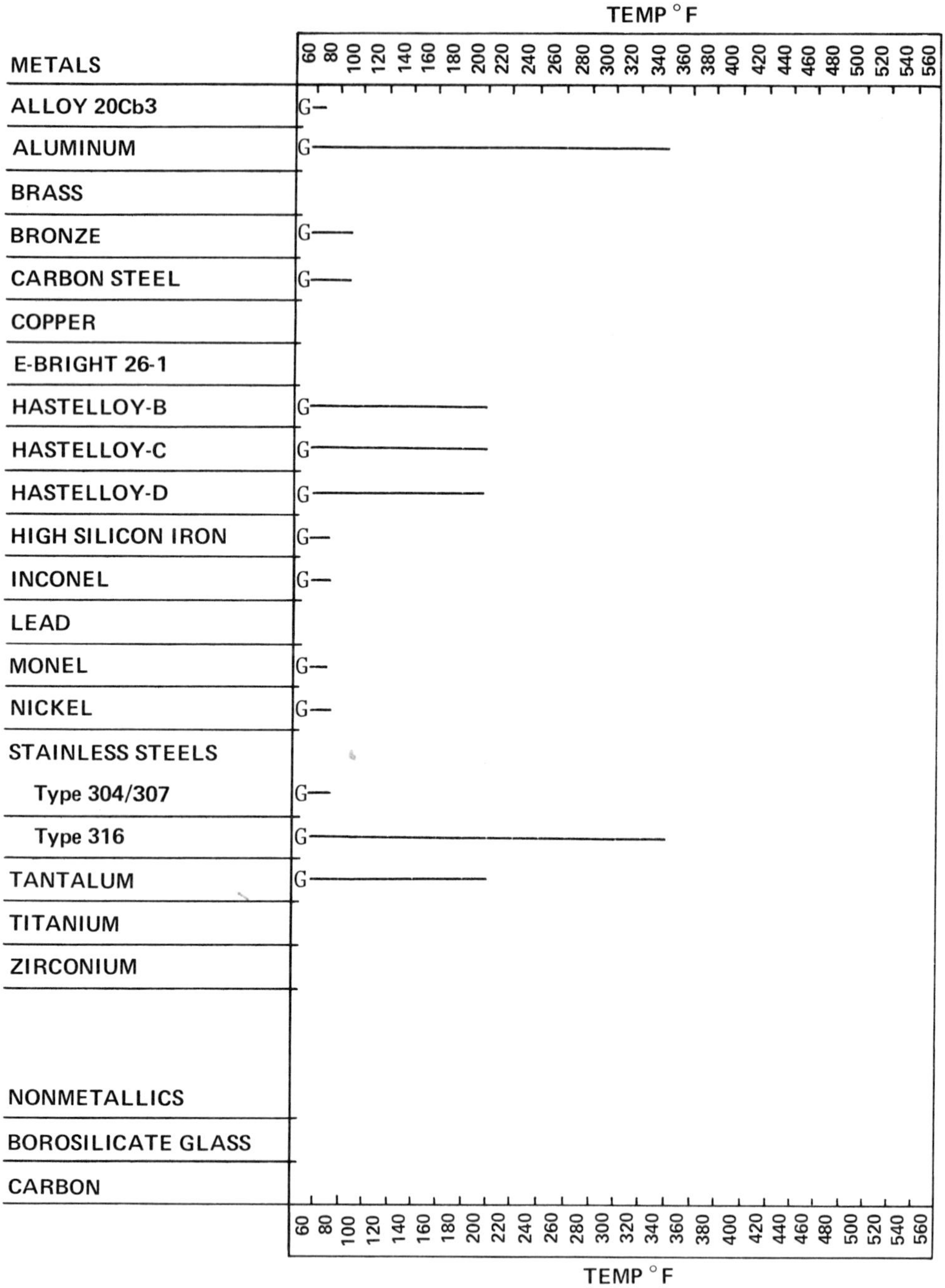

FOR METALS

E = < 2 Mils Penetration/Year; G = < 20 Mils Penetration/Year
S = < 50 Mils Penetration/Year; U = > 50 Mils Penetration/Year

FOR NONMETALLICS

R = Resistant
U = Unsatisfactory

PYROGALLIC ACID

TEMP °F (60–460)

PLASTICS	
ABS	
ACRYLICS	
ASBESTOS REINF., EPOXY	
ASBESTOS REINF., PHENOLIC	
CHLORINATED POLYETHER	R
CPVC	
EPOXY	
FLUOROCARBONS FEP	R
FLUOROCARBONS TFE	R
FURFURYL ALCOHOL	
NORYL	
PHENOLIC	
POLYAMIDES–NYLON	
POLYESTERS	
BISPHENOL A-FUMURATE	
HYDROGENATED BISPHENOL-A-BISPHENOL-A	
ISOPHTHALIC	
CHLORINATED POLYESTERS	
POLYSTYRENES	
POLYETHYLENE	
POLYPROPYLENE	
PVC–TYPE 1	R
PVC–TYPE 2	
POLYVINYLIDENE CHLORIDE	
VINYLIDENE FLUORIDE	
VINYL ESTER	

RUBBERS	
BUTYL GR-1	
FLUORO ELASTOMERS VITON A	
FLUORO ELASTOMERS KEL-F 3700	
HARD RUBBER	
HYPALON	R
KOROSEAL	
NATURAL RUBBER (GRS)	
NEOPRENE GR-M (CR)	R
NITRILE BUNA N (NBR)	
NORDEL	
POLYBUTADIENE	
SILICONE	
POLYURETHANE	

TEMP °F (60–460)

PYROGALLOL

TEMP °F

60 80 100 120 140 160 180 200 220 240 260 280 300 320 340 360 380 400 420 440 460 480 500 520 540 560

METALS	
ALLOY 20Cb3	
ALUMINUM	
BRASS	
BRONZE	G
CARBON STEEL	
COPPER	G
E-BRIGHT 26-1	
HASTELLOY-B	G
HASTELLOY-C	G
HASTELLOY-D	
HIGH SILICON IRON	G
INCONEL	
LEAD	
MONEL	
NICKEL	G
STAINLESS STEELS	
Type 304/307	G
Type 316	G
TANTALUM	
TITANIUM	
ZIRCONIUM	

NONMETALLICS	
BOROSILICATE GLASS	
CARBON	

TEMP °F

FOR METALS

E = < 2 Mils Penetration/Year; G = < 20 Mils Penetration/Year
S = < 50 Mils Penetration/Year; U = > 50 Mils Penetration/Year

FOR NONMETALLICS

R = Resistant
U = Unsatisfactory

TEMP °F (60, 80, 100, 120, 140, 160, 180, 200, 220, 240, 260, 280, 300, 320, 340, 360, 380, 400, 420, 440, 460)

PLASTICS	
ABS	
ACRYLICS	
ASBESTOS REINF., EPOXY	
ASBESTOS REINF., PHENOLIC	
CHLORINATED POLYETHER	
CPVC	
EPOXY	
FLUOROCARBONS FEP	R
FLUOROCARBONS TFE	R
FURFURYL ALCOHOL	
NORYL	
PHENOLIC	
POLYAMIDES–NYLON	
POLYESTERS	
BISPHENOL A-FUMURATE	
HYDROGENATED BISPHENOL-A-BISPHENOL-A	
ISOPHTHALIC	
CHLORINATED POLYESTERS	
POLYSTYRENES	
POLYETHYLENE	
POLYPROPYLENE	
PVC–TYPE 1	
PVC–TYPE 2	
POLYVINYLIDENE CHLORIDE	R
VINYLIDENE FLUORIDE	R
VINYL ESTER	

RUBBERS	
BUTYL GR-1	
FLUORO ELASTOMERS VITON A	
FLUORO ELASTOMERS KEL-F 3700	
HARD RUBBER	
HYPALON	
KOROSEAL	
NATURAL RUBBER (GRS)	
NEOPRENE GR-M (CR)	
NITRILE BUNA N (NBR)	
NORDEL	
POLYBUTADIENE	
SILICONE	
POLYURETHANE	

TEMP °F (60, 80, 100, 120, 140, 160, 180, 200, 220, 240, 260, 280, 300, 320, 340, 360, 380, 400, 420, 440, 460)

RAYON COAGULATING BATH

METALS		TEMP °F (60–560)
ALLOY 20Cb3		
ALUMINUM		
BRASS		
BRONZE		
CARBON STEEL		
COPPER		
E-BRIGHT 26-1		
HASTELLOY-B		G— (60)
HASTELLOY-C		G— (60)
HASTELLOY-D		
HIGH SILICON IRON		
INCONEL		
LEAD		
MONEL		
NICKEL		
STAINLESS STEELS		
Type 304/307		U (60)
Type 316	1	G— (60)
TANTALUM		
TITANIUM		
ZIRCONIUM		

NONMETALLICS	TEMP °F (60–560)
BOROSILICATE GLASS	
CARBON	

FOR METALS

E = < 2 Mils Penetration/Year; G = < 20 Mils Penetration/Year
S = < 50 Mils Penetration/Year; U = > 50 Mils Penetration/Year

FOR NONMETALLICS

R = Resistant
U = Unsatisfactory

RAYON COAGULATING BATH

Please see page 1 for notes on calibration.

TEMP °F: 60 80 100 120 140 160 180 200 220 240 260 280 300 320 340 360 380 400 420 440 460

PLASTICS		
ABS		R
ACRYLICS		
ASBESTOS REINF., EPOXY		
ASBESTOS REINF., PHENOLIC		
CHLORINATED POLYETHER		
CPVC		R
EPOXY		
FLUOROCARBONS FEP		R
FLUOROCARBONS TFE		R
FURFURYL ALCOHOL		
NORYL		
PHENOLIC		
POLYAMIDES—NYLON		
POLYESTERS		
BISPHENOL A-FUMURATE		R
HYDROGENATED BISPHENOL-A-BISPHENOL-A		
ISOPHTHALIC		
CHLORINATED POLYESTERS		
POLYSTYRENES		
POLYETHYLENE	11	R
POLYPROPYLENE		R
PVC—TYPE 1		R
PVC—TYPE 2		R
POLYVINYLIDENE CHLORIDE		
VINYLIDENE FLUORIDE		
VINYL ESTER		

RUBBERS		
BUTYL GR-1		
FLUORO ELASTOMERS VITON A		
FLUORO ELASTOMERS KEL-F 3700		
HARD RUBBER		
HYPALON		
KOROSEAL		
NATURAL RUBBER (GRS)		
NEOPRENE GR-M (CR)		
NITRILE BUNA N (NBR)		
NORDEL		
POLYBUTADIENE		
SILICONE		
POLYURETHANE		

TEMP °F: 60 80 100 120 140 160 180 200 220 240 260 280 300 320 340 360 380 400 420 440 460

SALENIC ACID, AQUEOUS

TEMP °F

METALS	60	80	100	120	140	160	180	200	220	240	260	280	300	320	340	360	380	400	420	440	460	480	500	520	540	560
ALLOY 20Cb3																										
ALUMINUM																										
BRASS																										
BRONZE																										
CARBON STEEL																										
COPPER																										
E-BRIGHT 26-1																										
HASTELLOY-B																										
HASTELLOY-C																										
HASTELLOY-D																										
HIGH SILICON IRON																										
INCONEL																										
LEAD																										
MONEL																										
NICKEL																										
STAINLESS STEELS Type 304/307																										
Type 316																										
TANTALUM																										
TITANIUM																										
ZIRCONIUM																										
NONMETALLICS																										
BOROSILICATE GLASS																										
CARBON																										

60 80 100 120 140 160 180 200 220 240 260 280 300 320 340 360 380 400 420 440 460 480 500 520 540 560

TEMP °F

FOR METALS

E = < 2 Mils Penetration/Year; G = < 20 Mils Penetration/Year
S = < 50 Mils Penetration/Year; U = > 50 Mils Penetration/Year

FOR NONMETALLICS

R = Resistant
U = Unsatisfactory

Please see page 1 for notes on calibration.

SALENIC ACID, AQUEOUS

TEMP °F: 60 80 100 120 140 160 180 200 220 240 260 280 300 320 340 360 380 400 420 440 460

PLASTICS	
ABS	U
ACRYLICS	
ASBESTOS REINF., EPOXY	
ASBESTOS REINF., PHENOLIC	
CHLORINATED POLYETHER	R
CPVC	R
EPOXY	
FLUOROCARBONS FEP	R
FLUOROCARBONS TFE	R
FURFURYL ALCOHOL	
NORYL	
PHENOLIC	
POLYAMIDES–NYLON	
POLYESTERS	
BISPHENOL A-FUMURATE	
HYDROGENATED BISPHENOL-A-BISPHENOL-A	
ISOPHTHALIC	
CHLORINATED POLYESTERS	
POLYSTYRENES	
POLYETHYLENE	R
POLYPROPYLENE	R
PVC–TYPE 1	R
PVC–TYPE 2	R
POLYVINYLIDENE CHLORIDE	
VINYLIDENE FLUORIDE	R
VINYL ESTER	

RUBBERS	
BUTYL GR-1	
FLUORO ELASTOMERS VITON A	
FLUORO ELASTOMERS KEL-F 3700	
HARD RUBBER	
HYPALON	
KOROSEAL	
NATURAL RUBBER (GRS)	
NEOPRENE GR-M (CR)	
NITRILE BUNA N (NBR)	
NORDEL	
POLYBUTADIENE	
SILICONE	
POLYURETHANE	

TEMP °F: 60 80 100 120 140 160 180 200 220 240 260 280 300 320 340 360 380 400 420 440 460

SALICIC ACID

	TEMP °F																									
METALS	60	80	100	120	140	160	180	200	220	240	260	280	300	320	340	360	380	400	420	440	460	480	500	520	540	560
ALLOY 20Cb3																										
ALUMINUM																										
BRASS																										
BRONZE																										
CARBON STEEL																										
COPPER																										
E-BRIGHT 26-1																										
HASTELLOY-B																										
HASTELLOY-C																										
HASTELLOY-D																										
HIGH SILICON IRON																										
INCONEL																										
LEAD																										
MONEL																										
NICKEL																										
STAINLESS STEELS																										
Type 304/307																										
Type 316																										
TANTALUM																										
TITANIUM																										
ZIRCONIUM																										
NONMETALLICS																										
BOROSILICATE GLASS	R	—	—	—	—	—	—	—																		
CARBON																										
	60	80	100	120	140	160	180	200	220	240	260	280	300	320	340	360	380	400	420	440	460	480	500	520	540	560
	TEMP °F																									

FOR METALS

E = < 2 Mils Penetration/Year; G = < 20 Mils Penetration/Year

S = < 50 Mils Penetration/Year; U = > 50 Mils Penetration/Year

FOR NONMETALLICS

R = Resistant

U = Unsatisfactory

SALICIC ACID

PLASTICS	TEMP °F (60–460)
ABS	R 60–140
ACRYLICS	
ASBESTOS REINF., EPOXY	
ASBESTOS REINF., PHENOLIC	
CHLORINATED POLYETHER	R 60–220, U
CPVC	
EPOXY	R 60–200
FLUOROCARBONS FEP	R 60–400
FLUOROCARBONS TFE	R 60–460
FURFURYL ALCOHOL	
NORYL	
PHENOLIC	
POLYAMIDES–NYLON	
POLYESTERS	
BISPHENOL A-FUMURATE	
HYDROGENATED BISPHENOL-A-BISPHENOL-A	
ISOPHTHALIC	R 60–180
CHLORINATED POLYESTERS	
POLYSTYRENES	
POLYETHYLENE	R 60–140
POLYPROPYLENE	R 60–140
PVC–TYPE 1	R 60–140
PVC–TYPE 2	R 60–140
POLYVINYLIDENE CHLORIDE	
VINYLIDENE FLUORIDE	
VINYL ESTER	
RUBBERS	
BUTYL GR-1	
FLUORO ELASTOMERS VITON A	
FLUORO ELASTOMERS KEL-F 3700	
HARD RUBBER	
HYPALON	
KOROSEAL	
NATURAL RUBBER (GRS)	
NEOPRENE GR-M (CR)	
NITRILE BUNA N (NBR)	
NORDEL	
POLYBUTADIENE	
SILICONE	
POLYURETHANE	

SALICYLADEHYDE

METALS	TEMP °F (60–560)
ALLOY 20Cb3	
ALUMINUM	
BRASS	
BRONZE	
CARBON STEEL	
COPPER	
E-BRIGHT 26-1	
HASTELLOY-B	
HASTELLOY-C	
HASTELLOY-D	
HIGH SILICON IRON	
INCONEL	
LEAD	
MONEL	
NICKEL	
STAINLESS STEELS	
Type 304/307	
Type 316	
TANTALUM	
TITANIUM	
ZIRCONIUM	
NONMETALLICS	
BOROSILICATE GLASS	
CARBON	R (60–340)

TEMP °F: 60, 80, 100, 120, 140, 160, 180, 200, 220, 240, 260, 280, 300, 320, 340, 360, 380, 400, 420, 440, 460, 480, 500, 520, 540, 560

FOR METALS

E = < 2 Mils Penetration/Year; G = < 20 Mils Penetration/Year
S = < 50 Mils Penetration/Year; U = > 50 Mils Penetration/Year

FOR NONMETALLICS

R = Resistant
U = Unsatisfactory

SALICYLADEHYDE

PLASTICS	TEMP °F (60–460)
ABS	U
ACRYLICS	
ASBESTOS REINF., EPOXY	
ASBESTOS REINF., PHENOLIC	
CHLORINATED POLYETHER	R—
CPVC	
EPOXY	
FLUOROCARBONS FEP	R————————————
FLUOROCARBONS TFE	R——————————————
FURFURYL ALCOHOL	
NORYL	
PHENOLIC	
POLYAMIDES–NYLON	
POLYESTERS	
BISPHENOL A-FUMURATE	
HYDROGENATED BISPHENOL-A-BISPHENOL-A	
ISOPHTHALIC	
CHLORINATED POLYESTERS	
POLYSTYRENES	
POLYETHYLENE	
POLYPROPYLENE	
PVC–TYPE 1	
PVC–TYPE 2	
POLYVINYLIDENE CHLORIDE	U
VINYLIDENE FLUORIDE	R———U
VINYL ESTER	

RUBBERS	TEMP °F (60–460)
BUTYL GR-1	
FLUORO ELASTOMERS VITON A	
FLUORO ELASTOMERS KEL-F 3700	
HARD RUBBER	
HYPALON	
KOROSEAL	
NATURAL RUBBER (GRS)	
NEOPRENE GR-M (CR)	
NITRILE BUNA N (NBR)	
NORDEL	
POLYBUTADIENE	
SILICONE	
POLYURETHANE	

SALICYLIC ACID

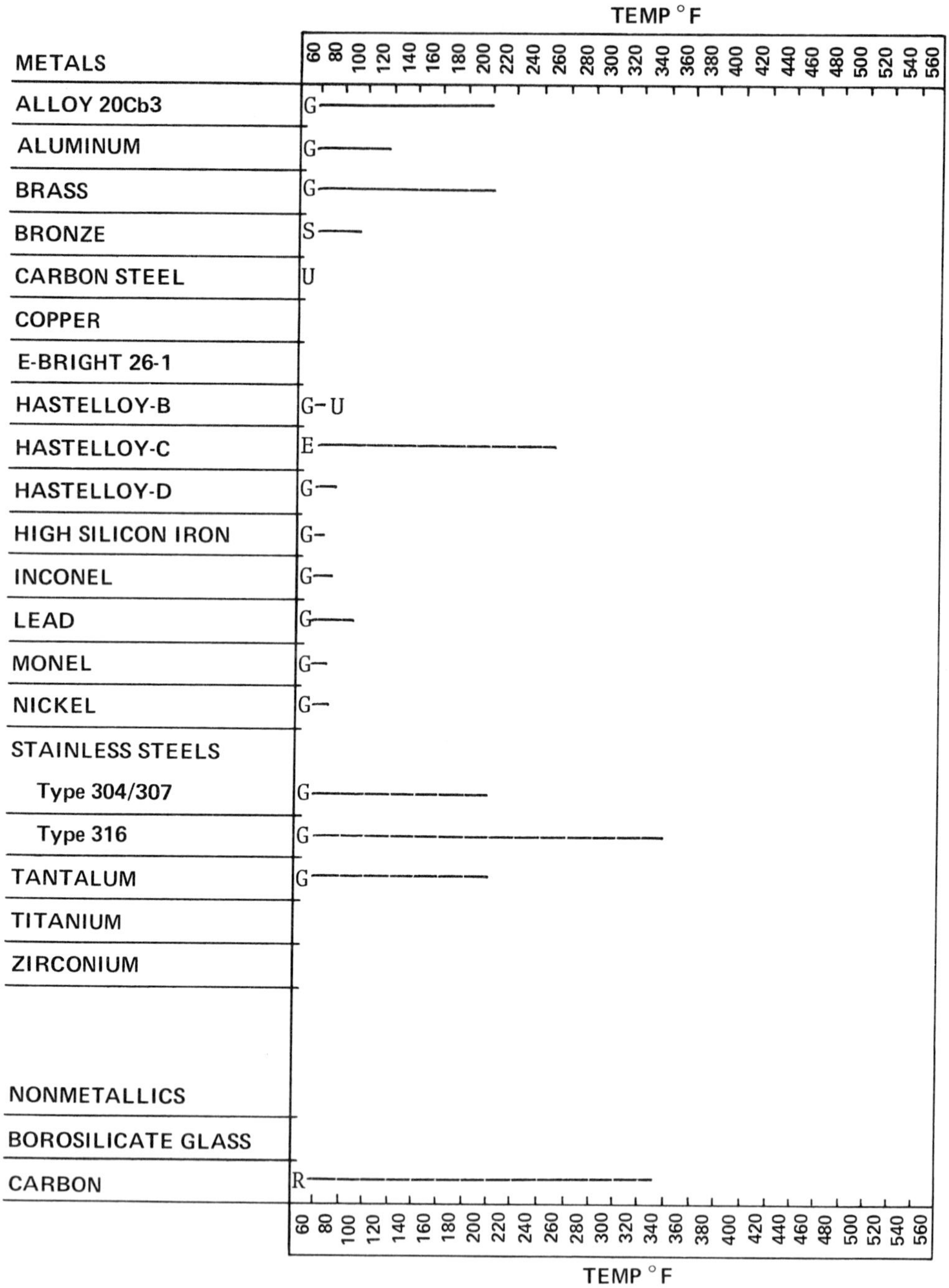

FOR METALS

E = < 2 Mils Penetration/Year; G = < 20 Mils Penetration/Year

S = < 50 Mils Penetration/Year; U = > 50 Mils Penetration/Year

FOR NONMETALLICS

R = Resistant

U = Unsatisfactory

SALICYLIC ACID

TEMP °F: 60 80 100 120 140 160 180 200 220 240 260 280 300 320 340 360 380 400 420 440 460

PLASTICS		Rating
ABS		
ACRYLICS		
ASBESTOS REINF., EPOXY		
ASBESTOS REINF., PHENOLIC		
CHLORINATED POLYETHER		R
CPVC		
EPOXY		
FLUOROCARBONS FEP		R
FLUOROCARBONS TFE		R
FURFURYL ALCOHOL	sat	R
NORYL		
PHENOLIC		
POLYAMIDES–NYLON		
POLYESTERS		
BISPHENOL A-FUMURATE		R
HYDROGENATED BISPHENOL-A-BISPHENOL-A		
ISOPHTHALIC		
CHLORINATED POLYESTERS		
POLYSTYRENES		
POLYETHYLENE		
POLYPROPYLENE		R
PVC–TYPE 1		R
PVC–TYPE 2		
POLYVINYLIDENE CHLORIDE		R
VINYLIDENE FLUORIDE		R
VINYL ESTER		

RUBBERS	Rating
BUTYL GR-1	R
FLUORO ELASTOMERS VITON A	R
FLUORO ELASTOMERS KEL-F 3700	
HARD RUBBER	
HYPALON	
KOROSEAL	R
NATURAL RUBBER (GRS)	R
NEOPRENE GR-M (CR)	U
NITRILE BUNA N (NBR)	RU
NORDEL	
POLYBUTADIENE	
SILICONE	
POLYURETHANE	

TEMP °F: 60 80 100 120 140 160 180 200 220 240 260 280 300 320 340 360 380 400 420 440 460

SILICONE OIL

TEMP °F: 60 80 100 120 140 160 180 200 220 240 260 280 300 320 340 360 380 400 420 440 460 480 500 520 540 560

METALS	
ALLOY 20Cb3	
ALUMINUM	
BRASS	
BRONZE	G ————————
CARBON STEEL	
COPPER	
E-BRIGHT 26-1	
HASTELLOY-B	
HASTELLOY-C	
HASTELLOY-D	
HIGH SILICON IRON	
INCONEL	
LEAD	
MONEL	
NICKEL	
STAINLESS STEELS	
Type 304/307	
Type 316	
TANTALUM	
TITANIUM	
ZIRCONIUM	

NONMETALLICS	
BOROSILICATE GLASS	
CARBON	

TEMP °F: 60 80 100 120 140 160 180 200 220 240 260 280 300 320 340 360 380 400 420 440 460 480 500 520 540 560

FOR METALS

E = $<$ 2 Mils Penetration/Year; G = $<$ 20 Mils Penetration/Year

S = $<$ 50 Mils Penetration/Year; U = $>$ 50 Mils Penetration/Year

FOR NONMETALLICS

R = Resistant

U = Unsatisfactory

SILICONE OIL

TEMP °F (60, 80, 100, 120, 140, 160, 180, 200, 220, 240, 260, 280, 300, 320, 340, 360, 380, 400, 420, 440, 460)

PLASTICS	SILICONE OIL
ABS	
ACRYLICS	
ASBESTOS REINF., EPOXY	
ASBESTOS REINF., PHENOLIC	
CHLORINATED POLYETHER	
CPVC	
EPOXY	
FLUOROCARBONS FEP	R———
FLUOROCARBONS TFE	R———
FURFURYL ALCOHOL	
NORYL	
PHENOLIC	
POLYAMIDES–NYLON	
POLYESTERS	
BISPHENOL A-FUMURATE	
HYDROGENATED BISPHENOL-A-BISPHENOL-A	
ISOPHTHALIC	
CHLORINATED POLYESTERS	
POLYSTYRENES	
POLYETHYLENE	R———U
POLYPROPYLENE	R———
PVC–TYPE 1	
PVC–TYPE 2	
POLYVINYLIDENE CHLORIDE	
VINYLIDENE FLUORIDE	
VINYL ESTER	
RUBBERS	
BUTYL GR-1	U
FLUORO ELASTOMERS VITON A	R———
FLUORO ELASTOMERS KEL-F 3700	
HARD RUBBER	
HYPALON	R———
KOROSEAL	
NATURAL RUBBER (GRS)	U
NEOPRENE GR-M (CR)	U
NITRILE BUNA N (NBR)	R———
NORDEL	
POLYBUTADIENE	
SILICONE	
POLYURETHANE	

TEMP °F (60, 80, 100, 120, 140, 160, 180, 200, 220, 240, 260, 280, 300, 320, 340, 360, 380, 400, 420, 440, 460)

SILVER CHLORIDE

TEMP °F

METALS		60	80	100	120	140	160	180	200	220	240	260	280	300	320	340	360	380	400	420	440	460	480	500	520	540	560
ALLOY 20Cb3		G—																									
ALUMINUM																											
BRASS		U																									
BRONZE		U																									
CARBON STEEL																											
COPPER																											
E-BRIGHT 26-1																											
HASTELLOY-B																											
HASTELLOY-C	10%	G—																									
HASTELLOY-D																											
HIGH SILICON IRON																											
INCONEL																											
LEAD																											
MONEL																											
NICKEL																											
STAINLESS STEELS																											
Type 304/307		U																									
Type 316		U																									
TANTALUM		E—																									
TITANIUM																											
ZIRCONIUM																											
NONMETALLICS																											
BOROSILICATE GLASS																											
CARBON																											

TEMP °F

FOR METALS

E = < 2 Mils Penetration/Year; G = < 20 Mils Penetration/Year

S = < 50 Mils Penetration/Year; U = > 50 Mils Penetration/Year

FOR NONMETALLICS

R = Resistant

U = Unsatisfactory

SILVER CHLORIDE

PLASTICS	TEMP °F (60–460)
ABS	
ACRYLICS	
ASBESTOS REINF., EPOXY	
ASBESTOS REINF., PHENOLIC	
CHLORINATED POLYETHER	
CPVC	
EPOXY	
FLUOROCARBONS FEP	R
FLUOROCARBONS TFE	R
FURFURYL ALCOHOL	
NORYL	
PHENOLIC	
POLYAMIDES–NYLON	R
POLYESTERS	
BISPHENOL A-FUMURATE	
HYDROGENATED BISPHENOL-A-BISPHENOL-A	
ISOPHTHALIC	
CHLORINATED POLYESTERS	
POLYSTYRENES	
POLYETHYLENE	R
POLYPROPYLENE	
PVC–TYPE 1	
PVC–TYPE 2	
POLYVINYLIDENE CHLORIDE	
VINYLIDENE FLUORIDE	
VINYL ESTER	
RUBBERS	
BUTYL GR-1	
FLUORO ELASTOMERS VITON A	
FLUORO ELASTOMERS KEL-F 3700	
HARD RUBBER	
HYPALON	
KOROSEAL	
NATURAL RUBBER (GRS)	
NEOPRENE GR-M (CR)	
NITRILE BUNA N (NBR)	
NORDEL	
POLYBUTADIENE	
SILICONE	
POLYURETHANE	

SILVER CYANIDE

TEMP °F (60, 80, 100, 120, 140, 160, 180, 200, 220, 240, 260, 280, 300, 320, 340, 360, 380, 400, 420, 440, 460, 480, 500, 520, 540, 560)

METALS	TEMP °F
ALLOY 20Cb3	
ALUMINUM	
BRASS	
BRONZE	
CARBON STEEL	
COPPER	
E-BRIGHT 26-1	
HASTELLOY-B	
HASTELLOY-C	
HASTELLOY-D	
HIGH SILICON IRON	
INCONEL	
LEAD	
MONEL	
NICKEL	
STAINLESS STEELS	
Type 304/307	
Type 316	
TANTALUM	E— (60)
TITANIUM	
ZIRCONIUM	
NONMETALLICS	
BOROSILICATE GLASS	
CARBON	R (60–340)

TEMP °F

FOR METALS

E = < 2 Mils Penetration/Year; G = < 20 Mils Penetration/Year

S = < 50 Mils Penetration/Year; U = > 50 Mils Penetration/Year

FOR NONMETALLICS

R = Resistant

U = Unsatisfactory

SILVER CYANIDE

PLASTICS	Rating	TEMP °F (end of bar, 60–460 scale)
ABS	R	140
ACRYLICS	R	80
ASBESTOS REINF., EPOXY		
ASBESTOS REINF., PHENOLIC		
CHLORINATED POLYETHER	R	250
CPVC	R	210
EPOXY	R	160
FLUOROCARBONS FEP	R	400
FLUOROCARBONS TFE	R	460
FURFURYL ALCOHOL	R	250
NORYL		
PHENOLIC		
POLYAMIDES—NYLON		
POLYESTERS		
BISPHENOL A-FUMURATE	R	220
HYDROGENATED BISPHENOL-A-BISPHENOL-A	R	210
ISOPHTHALIC	R	160
CHLORINATED POLYESTERS		
POLYSTYRENES		
POLYETHYLENE	R	150
POLYPROPYLENE	R	210
PVC—TYPE 1	R	140
PVC—TYPE 2	R	140
POLYVINYLIDENE CHLORIDE	R	80
VINYLIDENE FLUORIDE	R	280
VINYL ESTER		

RUBBERS	Rating	TEMP °F (end of bar, 60–460 scale)
BUTYL GR-1		
FLUORO ELASTOMERS VITON A	R	80
FLUORO ELASTOMERS KEL-F 3700		
HARD RUBBER	R	200
HYPALON	R	250
KOROSEAL		
NATURAL RUBBER (GRS)		
NEOPRENE GR-M (CR)	R	200
NITRILE BUNA N (NBR)		
NORDEL	R	300
POLYBUTADIENE		
SILICONE		
POLYURETHANE		

SILVER NITRATE

TEMP °F (60, 80, 100, 120, 140, 160, 180, 200, 220, 240, 260, 280, 300, 320, 340, 360, 380, 400, 420, 440, 460, 480, 500, 520, 540, 560)

METALS		Rating
ALLOY 20Cb3		G——
ALUMINUM		U
BRASS		U
BRONZE		U
CARBON STEEL		U
COPPER		U
E-BRIGHT 26-1		
HASTELLOY-B	10-40%	E—
HASTELLOY-C	10-40%	E—
HASTELLOY-D	10-40%	E—
HIGH SILICON IRON		G—
INCONEL		G—
LEAD		U
MONEL		U
NICKEL		U
STAINLESS STEELS		
Type 304/307		G——
Type 316		E— G——
TANTALUM	50%	E—
TITANIUM	50%	E—
ZIRCONIUM	50%	G—

NONMETALLICS	Rating
BOROSILICATE GLASS	R——
CARBON	R——

TEMP °F

FOR METALS

E = < 2 Mils Penetration/Year; G = < 20 Mils Penetration/Year

S = < 50 Mils Penetration/Year; U = > 50 Mils Penetration/Year

FOR NONMETALLICS

R = Resistant

U = Unsatisfactory

SILVER NITRATE

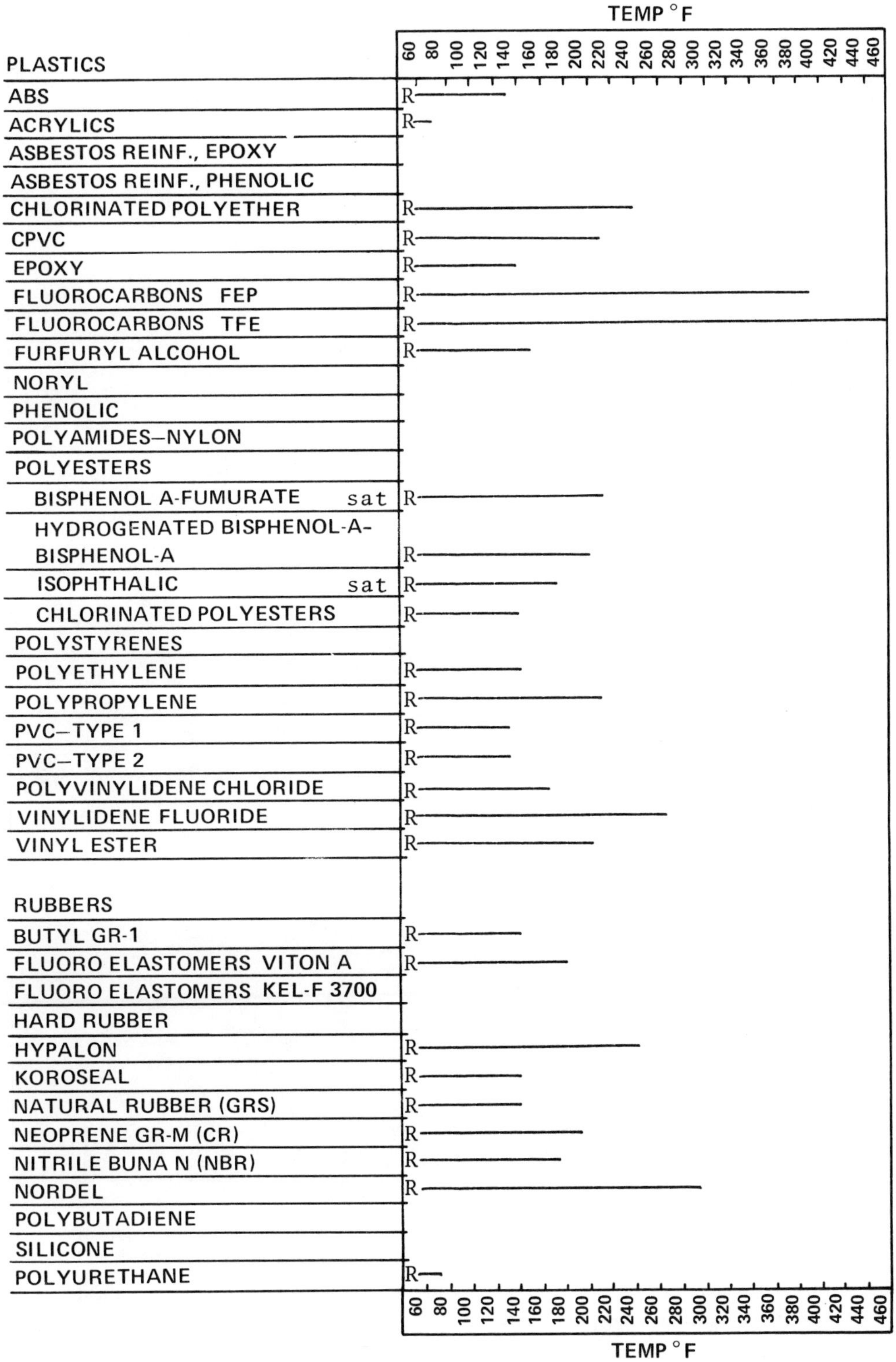

SOAPS

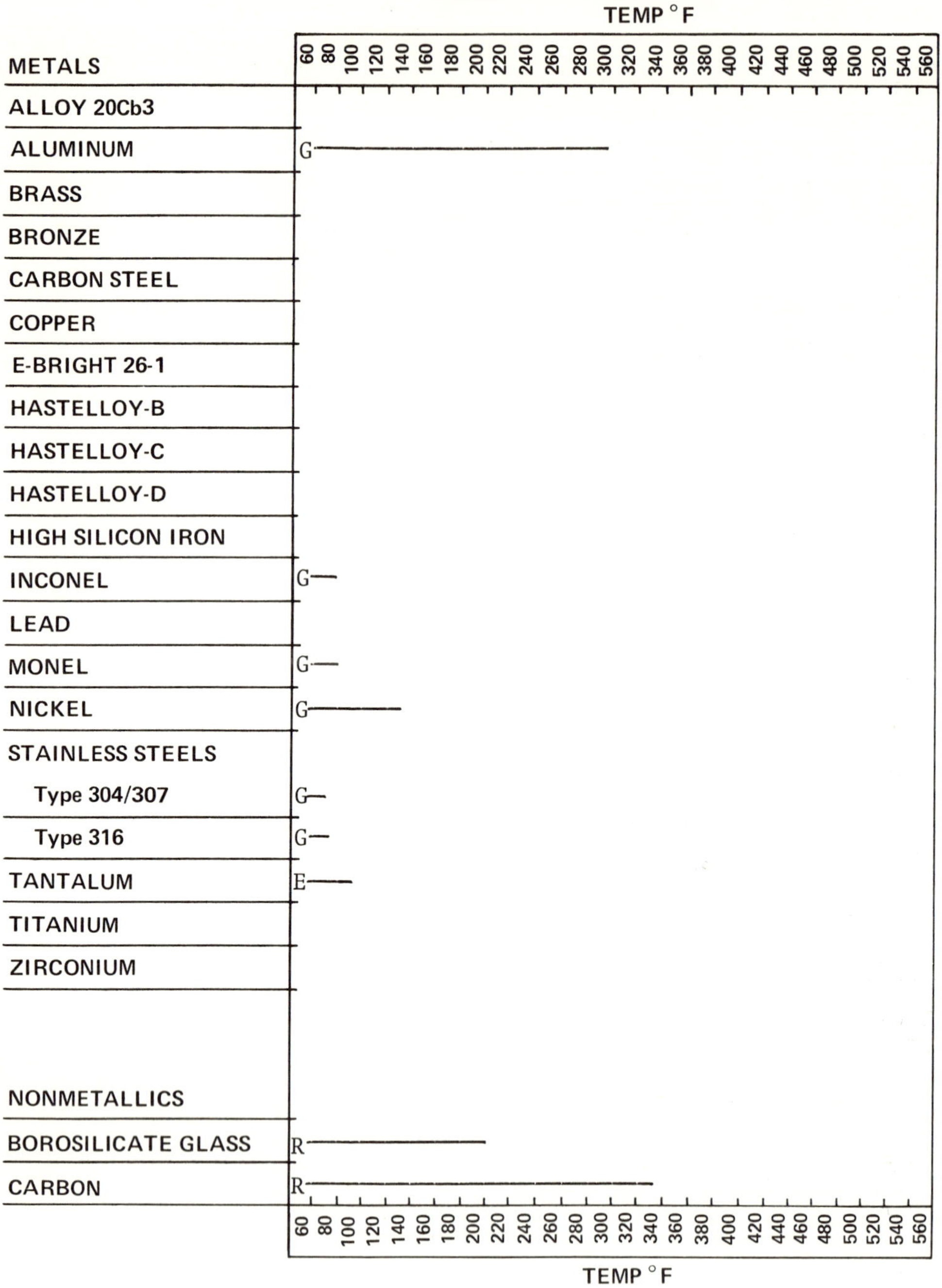

FOR METALS

E = < 2 Mils Penetration/Year; G = < 20 Mils Penetration/Year

S = < 50 Mils Penetration/Year; U = > 50 Mils Penetration/Year

FOR NONMETALLICS

R = Resistant

U = Unsatisfactory

PLASTICS	TEMP °F
ABS	R
ACRYLICS	R
ASBESTOS REINF., EPOXY	
ASBESTOS REINF., PHENOLIC	
CHLORINATED POLYETHER	R
CPVC	R
EPOXY	R
FLUOROCARBONS FEP	R
FLUOROCARBONS TFE	R
FURFURYL ALCOHOL	
NORYL	
PHENOLIC	
POLYAMIDES–NYLON	
POLYESTERS	
BISPHENOL A-FUMURATE	
HYDROGENATED BISPHENOL-A-BISPHENOL-A	
ISOPHTHALIC	
CHLORINATED POLYESTERS	
POLYSTYRENES	
POLYETHYLENE 11	R
POLYPROPYLENE	R
PVC–TYPE 1	R
PVC–TYPE 2	R
POLYVINYLIDENE CHLORIDE	
VINYLIDENE FLUORIDE	
VINYL ESTER	
RUBBERS	
BUTYL GR-1	
FLUORO ELASTOMERS VITON A	R
FLUORO ELASTOMERS KEL-F 3700	
HARD RUBBER	
HYPALON	R
KOROSEAL	
NATURAL RUBBER (GRS)	
NEOPRENE GR-M (CR)	R
NITRILE BUNA N (NBR)	R
NORDEL	
POLYBUTADIENE	
SILICONE	
POLYURETHANE	U

TEMP °F: 60 80 100 120 140 160 180 200 220 240 260 280 300 320 340 360 380 400 420 440 460

SOAP SOLUTIONS

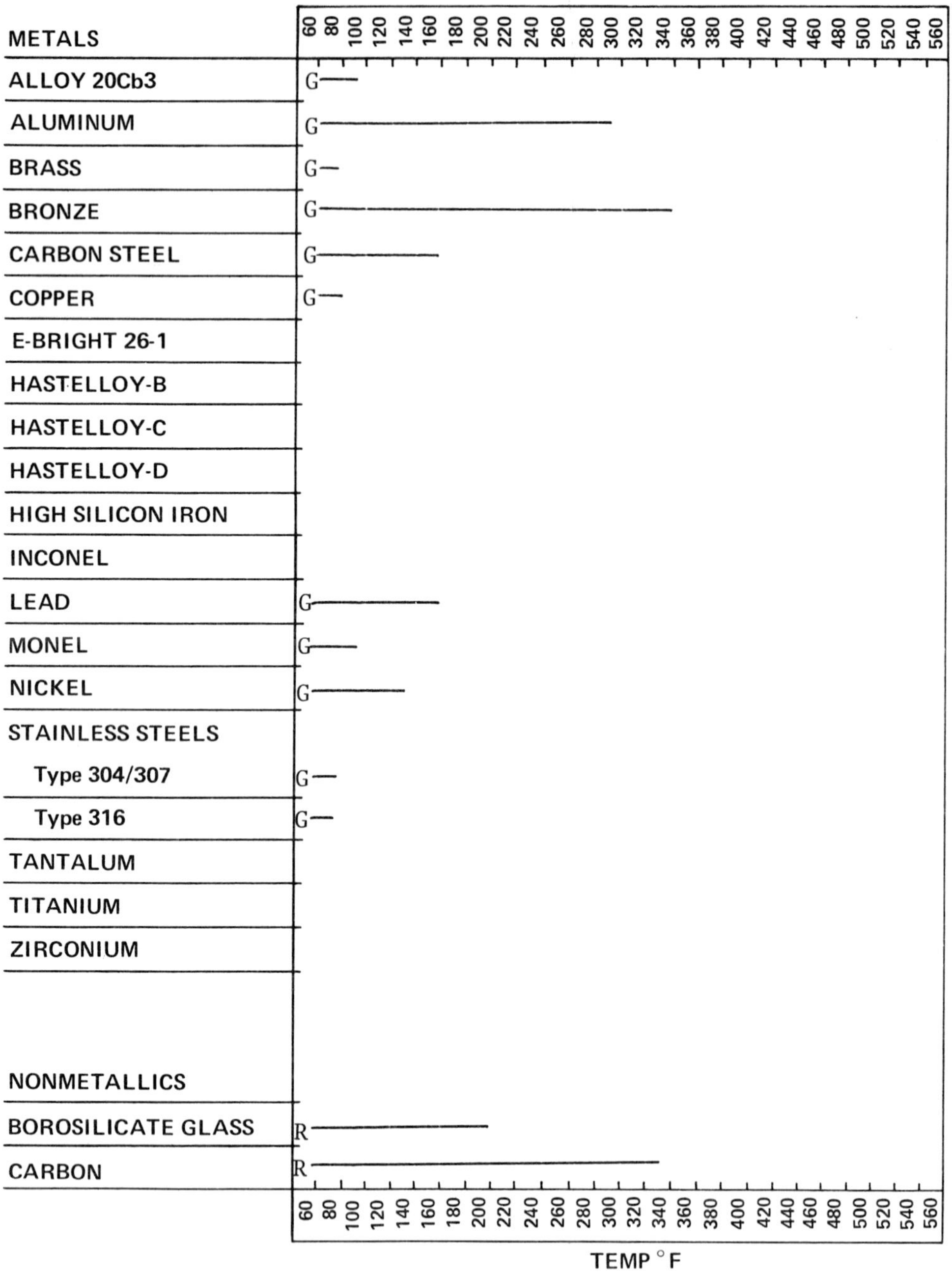

FOR METALS

E = < 2 Mils Penetration/Year; G = < 20 Mils Penetration/Year

S = < 50 Mils Penetration/Year; U = > 50 Mils Penetration/Year

FOR NONMETALLICS

R = Resistant

U = Unsatisfactory

SOAP SOLUTIONS

TEMP °F: 60 80 100 120 140 160 180 200 220 240 260 280 300 320 340 360 380 400 420 440 460

PLASTICS		TEMP °F
ABS		
ACRYLICS		R—
ASBESTOS REINF., EPOXY		
ASBESTOS REINF., PHENOLIC		
CHLORINATED POLYETHER		R—
CPVC		
EPOXY		
FLUOROCARBONS FEP		R—
FLUOROCARBONS TFE		R—
FURFURYL ALCOHOL		
NORYL		
PHENOLIC		
POLYAMIDES—NYLON		R—
POLYESTERS		
BISPHENOL A-FUMURATE	1%	R—
HYDROGENATED BISPHENOL-A-BISPHENOL-A		
ISOPHTHALIC	1%	R—
CHLORINATED POLYESTERS	1%	R—
POLYSTYRENES		
POLYETHYLENE		R—
POLYPROPYLENE		R—
PVC—TYPE 1		R—
PVC—TYPE 2		R—
POLYVINYLIDENE CHLORIDE		R—
VINYLIDENE FLUORIDE		
VINYL ESTER		

RUBBERS	TEMP °F
BUTYL GR-1	R—
FLUORO ELASTOMERS VITON A	R—
FLUORO ELASTOMERS KEL-F 3700	
HARD RUBBER	R—
HYPALON	R—
KOROSEAL	
NATURAL RUBBER (GRS)	R—
NEOPRENE GR-M (CR)	R—
NITRILE BUNA N (NBR)	R—
NORDEL	R—
POLYBUTADIENE	
SILICONE	
POLYURETHANE	R—

TEMP °F: 60 80 100 120 140 160 180 200 220 240 260 280 300 320 340 360 380 400 420 440 460

SOAP SOLUTION, 5%

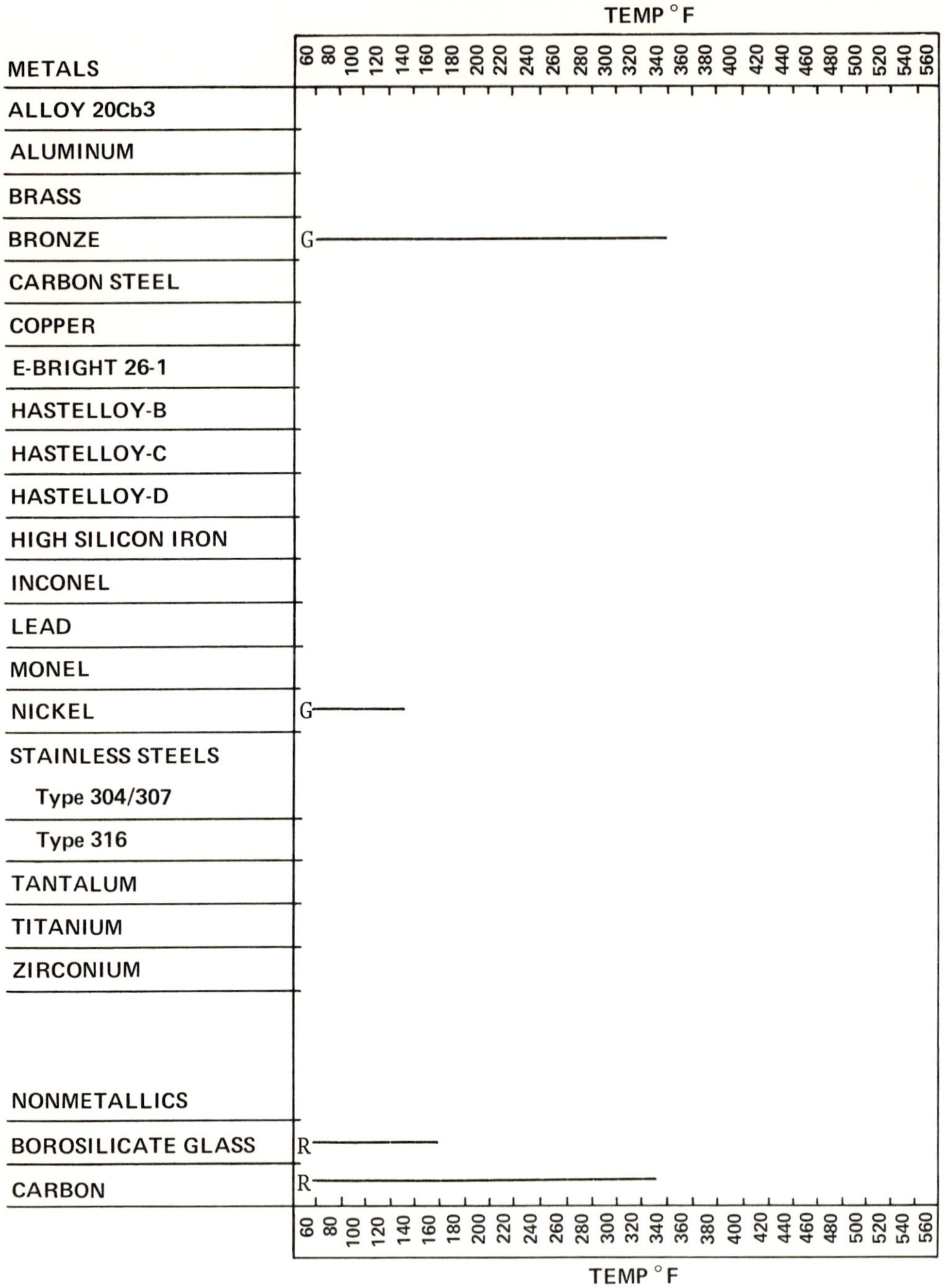

FOR METALS

E = < 2 Mils Penetration/Year; G = < 20 Mils Penetration/Year

S = < 50 Mils Penetration/Year; U = > 50 Mils Penetration/Year

FOR NONMETALLICS

R = Resistant

U = Unsatisfactory

SOAP SOLUTION, 5%

TEMP °F: 60 80 100 120 140 160 180 200 220 240 260 280 300 320 340 360 380 400 420 440 460

PLASTICS	TEMP °F
ABS	
ACRYLICS	
ASBESTOS REINF., EPOXY	
ASBESTOS REINF., PHENOLIC	
CHLORINATED POLYETHER	
CPVC	
EPOXY	
FLUOROCARBONS FEP	R
FLUOROCARBONS TFE	R
FURFURYL ALCOHOL	
NORYL	
PHENOLIC	
POLYAMIDES—NYLON	R
POLYESTERS	
BISPHENOL A-FUMURATE	
HYDROGENATED BISPHENOL-A-BISPHENOL-A	
ISOPHTHALIC	
CHLORINATED POLYESTERS	
POLYSTYRENES	
POLYETHYLENE	R
POLYPROPYLENE	R
PVC—TYPE 1	
PVC—TYPE 2	
POLYVINYLIDENE CHLORIDE	
VINYLIDENE FLUORIDE	
VINYL ESTER	
RUBBERS	
BUTYL GR-1	
FLUORO ELASTOMERS VITON A	
FLUORO ELASTOMERS KEL-F 3700	
HARD RUBBER	
HYPALON	R
KOROSEAL	
NATURAL RUBBER (GRS)	
NEOPRENE GR-M (CR)	R
NITRILE BUNA N (NBR)	
NORDEL	
POLYBUTADIENE	
SILICONE	
POLYURETHANE	

TEMP °F: 60 80 100 120 140 160 180 200 220 240 260 280 300 320 340 360 380 400 420 440 460

SODIUM ACETATE

TEMP °F

METALS	Conc.	Rating	Temp °F (range of line)
ALLOY 20Cb3		G	60–560
ALUMINUM		E	60–80
BRASS		G	60–200
BRONZE		G	60–200
CARBON STEEL	10%	U	60
COPPER		G	60–80
E-BRIGHT 26-1			
HASTELLOY-B	60%	G	60–200
HASTELLOY-C	60%	G	60–200
HASTELLOY-D	60%	G	60–200
HIGH SILICON IRON		G	60–80
INCONEL	60%	G	60–200
LEAD			
MONEL	60%	G	60–200
NICKEL	60%	G	60–200
STAINLESS STEELS			
Type 304/307		G	60–560
Type 316		G	60–560
TANTALUM		E	60–80
TITANIUM	sat	E	60–200
ZIRCONIUM	60%	E	60–200

NONMETALLICS	Conc.	Rating	Temp °F (range of line)
BOROSILICATE GLASS		R	60–200
CARBON		R	60–340

TEMP °F

FOR METALS

E = < 2 Mils Penetration/Year; G = < 20 Mils Penetration/Year

S = < 50 Mils Penetration/Year; U = > 50 Mils Penetration/Year

FOR NONMETALLICS

R = Resistant

U = Unsatisfactory

SODIUM ACETATE

TEMP °F

PLASTICS	60 80 100 120 140 160 180 200 220 240 260 280 300 320 340 360 380 400 420 440 460
ABS	R
ACRYLICS	
ASBESTOS REINF., EPOXY	
ASBESTOS REINF., PHENOLIC	
CHLORINATED POLYETHER	R
CPVC	R
EPOXY	R
FLUOROCARBONS FEP	R
FLUOROCARBONS TFE	R
FURFURYL ALCOHOL	R
NORYL	R
PHENOLIC	
POLYAMIDES—NYLON	
POLYESTERS	
BISPHENOL A-FUMURATE all	R
HYDROGENATED BISPHENOL-A-BISPHENOL-A	R
ISOPHTHALIC sat	R
CHLORINATED POLYESTERS all	R
POLYSTYRENES	
POLYETHYLENE	R
POLYPROPYLENE	R
PVC—TYPE 1	R
PVC—TYPE 2	R
POLYVINYLIDENE CHLORIDE	R
VINYLIDENE FLUORIDE	R
VINYL ESTER	R
RUBBERS	
BUTYL GR-1	
FLUORO ELASTOMERS VITON A	U
FLUORO ELASTOMERS KEL-F 3700	
HARD RUBBER	R
HYPALON	U
KOROSEAL	R
NATURAL RUBBER (GRS)	
NEOPRENE GR-M (CR)	R
NITRILE BUNA N (NBR)	RU
NORDEL	R
POLYBUTADIENE	
SILICONE	
POLYURETHANE	

60 80 100 120 140 160 180 200 220 240 260 280 300 320 340 360 380 400 420 440 460

TEMP °F

SODIUM BENZOATE

TEMP °F (60–560)

METALS	Rating
ALLOY 20Cb3	
ALUMINUM	
BRASS	
BRONZE	
CARBON STEEL	
COPPER	
E-BRIGHT 26-1	
HASTELLOY-B	G
HASTELLOY-C	G
HASTELLOY-D	
HIGH SILICON IRON	
INCONEL	
LEAD	
MONEL	G
NICKEL	
STAINLESS STEELS Type 304/307	
Type 316	
TANTALUM	E
TITANIUM	
ZIRCONIUM	

NONMETALLICS	Rating
BOROSILICATE GLASS	R
CARBON	R

TEMP °F

FOR METALS

E = < 2 Mils Penetration/Year; G = < 20 Mils Penetration/Year
S = < 50 Mils Penetration/Year; U = > 50 Mils Penetration/Year

FOR NONMETALLICS

R = Resistant
U = Unsatisfactory

SODIUM BENZOATE

TEMP °F (60–460)

PLASTICS	Rating	Bar extends to approx. (°F)
ABS	R	140
ACRYLICS		
ASBESTOS REINF., EPOXY		
ASBESTOS REINF., PHENOLIC		
CHLORINATED POLYETHER	R	250
CPVC	R	180
EPOXY	R	250
FLUOROCARBONS FEP	R	400
FLUOROCARBONS TFE	R	460
FURFURYL ALCOHOL	R	250
NORYL		
PHENOLIC		
POLYAMIDES–NYLON		
POLYESTERS		
BISPHENOL A-FUMURATE	R	180
HYDROGENATED BISPHENOL-A-BISPHENOL-A	R	210
ISOPHTHALIC sat	R	180
CHLORINATED POLYESTERS	R	180
POLYSTYRENES		
POLYETHYLENE	R	150
POLYPROPYLENE	R	210
PVC–TYPE 1	R	140
PVC–TYPE 2	R	140
POLYVINYLIDENE CHLORIDE	R	150
VINYLIDENE FLUORIDE	R	280
VINYL ESTER	R	180

RUBBERS	Rating
BUTYL GR-1	
FLUORO ELASTOMERS VITON A	
FLUORO ELASTOMERS KEL-F 3700	
HARD RUBBER	
HYPALON	
KOROSEAL	
NATURAL RUBBER (GRS)	
NEOPRENE GR-M (CR)	
NITRILE BUNA N (NBR)	
NORDEL	
POLYBUTADIENE	
SILICONE	
POLYURETHANE	

TEMP °F

SODIUM BICARBONATE, 20%

TEMP °F (60, 80, 100, 120, 140, 160, 180, 200, 220, 240, 260, 280, 300, 320, 340, 360, 380, 400, 420, 440, 460, 480, 500, 520, 540, 560)

METALS		
ALLOY 20Cb3		E
ALUMINUM		G
BRASS		G
BRONZE		G
CARBON STEEL		S
COPPER		S
E-BRIGHT 26-1		
HASTELLOY-B	10-20%	E
HASTELLOY-C	10-20%	E
HASTELLOY-D	10-20%	E
HIGH SILICON IRON		G
INCONEL		E
LEAD		G
MONEL		E
NICKEL		E
STAINLESS STEELS		
Type 304/307		E
Type 316		E
TANTALUM	all	E
TITANIUM		E
ZIRCONIUM		E

NONMETALLICS	
BOROSILICATE GLASS	R
CARBON	R

TEMP °F

FOR METALS

E = < 2 Mils Penetration/Year; G = < 20 Mils Penetration/Year

S = < 50 Mils Penetration/Year; U = > 50 Mils Penetration/Year

FOR NONMETALLICS

R = Resistant

U = Unsatisfactory

SODIUM BICARBONATE, 20%

TEMP °F: 60 80 100 120 140 160 180 200 220 240 260 280 300 320 340 360 380 400 420 440 460

PLASTICS		
ABS		R
ACRYLICS		R
ASBESTOS REINF., EPOXY		R
ASBESTOS REINF., PHENOLIC		R
CHLORINATED POLYETHER		R
CPVC		R
EPOXY		R
FLUOROCARBONS FEP		R
FLUOROCARBONS TFE		R
FURFURYL ALCOHOL	sat	R
NORYL		R
PHENOLIC		R
POLYAMIDES–NYLON		R
POLYESTERS		
BISPHENOL A-FUMURATE	sat	R
HYDROGENATED BISPHENOL-A-BISPHENOL-A		
ISOPHTHALIC	10%	R
CHLORINATED POLYESTERS	sat	R
POLYSTYRENES		R
POLYETHYLENE		R
POLYPROPYLENE		R
PVC–TYPE 1		R
PVC–TYPE 2		R
POLYVINYLIDENE CHLORIDE		R
VINYLIDENE FLUORIDE		R
VINYL ESTER		R

RUBBERS	
BUTYL GR-1	R
FLUORO ELASTOMERS VITON A	R
FLUORO ELASTOMERS KEL-F 3700	
HARD RUBBER	R
HYPALON	R
KOROSEAL	R
NATURAL RUBBER (GRS)	R
NEOPRENE GR-M (CR)	R
NITRILE BUNA N (NBR)	R
NORDEL	R
POLYBUTADIENE	R
SILICONE	
POLYURETHANE	

TEMP °F: 60 80 100 120 140 160 180 200 220 240 260 280 300 320 340 360 380 400 420 440 460

SODIUM BISULFATE

TEMP °F (scale: 60, 80, 100, 120, 140, 160, 180, 200, 220, 240, 260, 280, 300, 320, 340, 360, 380, 400, 420, 440, 460, 480, 500, 520, 540, 560)

METALS	Concentration	Rating
ALLOY 20Cb3	10-40%	E
ALUMINUM		U
BRASS	10%	G
BRONZE		G
CARBON STEEL		G
COPPER		S
E-BRIGHT 26-1		
HASTELLOY-B	10-30%	G
HASTELLOY-C	10-30%	G
HASTELLOY-D		G
HIGH SILICON IRON		G
INCONEL	10-30%	G
LEAD		G
MONEL	10-30%	G
NICKEL	10-50%	G
STAINLESS STEELS		
Type 304/307	10% 1,9	E — U
Type 316	10%	E — U
TANTALUM	to 25%	E
TITANIUM	10-30%	E — U
ZIRCONIUM	10-50%	E

NONMETALLICS		Rating
BOROSILICATE GLASS		R
CARBON		

TEMP °F

FOR METALS

E = < 2 Mils Penetration/Year; G = < 20 Mils Penetration/Year

S = < 50 Mils Penetration/Year; U = > 50 Mils Penetration/Year

FOR NONMETALLICS

R = Resistant

U = Unsatisfactory

SODIUM BISULFATE

PLASTICS		TEMP °F (60 80 100 120 140 160 180 200 220 240 260 280 300 320 340 360 380 400 420 440 460)
ABS	6.5%	R
ACRYLICS		R
ASBESTOS REINF., EPOXY		R
ASBESTOS REINF., PHENOLIC		R
CHLORINATED POLYETHER		R
CPVC		R
EPOXY		R
FLUOROCARBONS FEP		R
FLUOROCARBONS TFE		R
FURFURYL ALCOHOL	sat	R
NORYL		R
PHENOLIC		R
POLYAMIDES–NYLON		
POLYESTERS		
BISPHENOL A-FUMURATE	sat	R
HYDROGENATED BISPHENOL-A-BISPHENOL-A		
ISOPHTHALIC	sat	R
CHLORINATED POLYESTERS	all	R
POLYSTYRENES		R
POLYETHYLENE		R
POLYPROPYLENE		R
PVC–TYPE 1		R
PVC–TYPE 2		R
POLYVINYLIDENE CHLORIDE		R
VINYLIDENE FLUORIDE		R
VINYL ESTER		R
RUBBERS		
BUTYL GR-1		R
FLUORO ELASTOMERS VITON A		R
FLUORO ELASTOMERS KEL-F 3700		
HARD RUBBER		R
HYPALON		R
KOROSEAL		
NATURAL RUBBER (GRS)		R
NEOPRENE GR-M (CR)		R
NITRILE BUNA N (NBR)		R
NORDEL		
POLYBUTADIENE		R
SILICONE		
POLYURETHANE		

SODIUM BISULFITE

TEMP °F: 60 80 100 120 140 160 180 200 220 240 260 280 300 320 340 360 380 400 420 440 460 480 500 520 540 560

METALS		
ALLOY 20Cb3	10-40%	G
ALUMINUM		U
BRASS		G
BRONZE		U
CARBON STEEL		U
COPPER		G
E-BRIGHT 26-1		
HASTELLOY-B	10-40%	G
HASTELLOY-C		
HASTELLOY-D		
HIGH SILICON IRON		S
INCONEL	20%	G
LEAD		G
MONEL	10-40%	G
NICKEL	10-40%	G
STAINLESS STEELS		
Type 304/307	20% 2	G
Type 316	10-40%	G
TANTALUM		G
TITANIUM	10-40%	G
ZIRCONIUM		

NONMETALLICS	
BOROSILICATE GLASS	U
CARBON	

TEMP °F: 60 80 100 120 140 160 180 200 220 240 260 280 300 320 340 360 380 400 420 440 460 480 500 520 540 560

FOR METALS

E = < 2 Mils Penetration/Year; G = < 20 Mils Penetration/Year
S = < 50 Mils Penetration/Year; U = > 50 Mils Penetration/Year

FOR NONMETALLICS

R = Resistant
U = Unsatisfactory

SODIUM BISULFITE

TEMP °F (60–460)

PLASTICS		Rating
ABS		R
ACRYLICS		R
ASBESTOS REINF., EPOXY		R
ASBESTOS REINF., PHENOLIC		R
CHLORINATED POLYETHER		R
CPVC		R
EPOXY		R
FLUOROCARBONS FEP		R
FLUOROCARBONS TFE		R
FURFURYL ALCOHOL	sat	R
NORYL		R
PHENOLIC		R
POLYAMIDES–NYLON		U
POLYESTERS		
BISPHENOL A-FUMURATE	sat	R
HYDROGENATED BISPHENOL-A-BISPHENOL-A		R
ISOPHTHALIC	sat	R
CHLORINATED POLYESTERS	all	R
POLYSTYRENES		
POLYETHYLENE		R
POLYPROPYLENE		R ——— U
PVC–TYPE 1		R
PVC–TYPE 2		R
POLYVINYLIDENE CHLORIDE		R
VINYLIDENE FLUORIDE		R
VINYL ESTER	sat	R

RUBBERS	Rating
BUTYL GR-1	R
FLUORO ELASTOMERS VITON A	R
FLUORO ELASTOMERS KEL-F 3700	
HARD RUBBER	R
HYPALON	R
KOROSEAL	
NATURAL RUBBER (GRS)	R
NEOPRENE GR-M (CR)	R
NITRILE BUNA N (NBR)	R
NORDEL	R
POLYBUTADIENE	R
SILICONE	
POLYURETHANE	

TEMP °F (60–460)

SODIUM BORATE

TEMP °F: 60 80 100 120 140 160 180 200 220 240 260 280 300 320 340 360 380 400 420 440 460 480 500 520 540 560

METALS	
ALLOY 20Cb3	G
ALUMINUM	G
BRASS	
BRONZE	G
CARBON STEEL	S
COPPER	G
E-BRIGHT 26-1	
HASTELLOY-B	G
HASTELLOY-C	G
HASTELLOY-D	
HIGH SILICON IRON	
INCONEL	
LEAD	
MONEL	G
NICKEL	
STAINLESS STEELS	
Type 304/307	G
Type 316	G
TANTALUM	E
TITANIUM	
ZIRCONIUM	

NONMETALLICS	
BOROSILICATE GLASS	R
CARBON	

TEMP °F: 60 80 100 120 140 160 180 200 220 240 260 280 300 320 340 360 380 400 420 440 460 480 500 520 540 560

FOR METALS

E = < 2 Mils Penetration/Year; G = < 20 Mils Penetration/Year

S = < 50 Mils Penetration/Year; U = > 50 Mils Penetration/Year

FOR NONMETALLICS

R = Resistant

U = Unsatisfactory

SODIUM BORATE

TEMP °F (60–460)

PLASTICS		
ABS		
ACRYLICS		R
ASBESTOS REINF., EPOXY		
ASBESTOS REINF., PHENOLIC		
CHLORINATED POLYETHER		R
CPVC		R
EPOXY		R
FLUOROCARBONS FEP		R
FLUOROCARBONS TFE		R
FURFURYL ALCOHOL		R
NORYL		R
PHENOLIC		
POLYAMIDES–NYLON		R
POLYESTERS		
BISPHENOL A-FUMURATE		R
HYDROGENATED BISPHENOL-A-BISPHENOL-A		
ISOPHTHALIC	sat	R
CHLORINATED POLYESTERS	sat	R
POLYSTYRENES		
POLYETHYLENE		R
POLYPROPYLENE		R
PVC–TYPE 1		R
PVC–TYPE 2		R
POLYVINYLIDENE CHLORIDE		R
VINYLIDENE FLUORIDE		R
VINYL ESTER	sat	R

RUBBERS	
BUTYL GR-1	R
FLUORO ELASTOMERS VITON A	R
FLUORO ELASTOMERS KEL-F 3700	
HARD RUBBER	
HYPALON	R
KOROSEAL	
NATURAL RUBBER (GRS)	R
NEOPRENE GR-M (CR)	R
NITRILE BUNA N (NBR)	R
NORDEL	
POLYBUTADIENE	
SILICONE	
POLYURETHANE	

TEMP °F

SODIUM BROMIDE

TEMP °F (60–560)

METALS	Concentration	Rating
ALLOY 20Cb3		G
ALUMINUM	10%	S
BRASS	10%	G
BRONZE	10%	G
CARBON STEEL	10%	S
COPPER		
E-BRIGHT 26-1		
HASTELLOY-B	10-50%	G
HASTELLOY-C	10-50%	G
HASTELLOY-D	10-50%	G
HIGH SILICON IRON		
INCONEL	10-50%	G
LEAD		
MONEL	10-40%	G
NICKEL	10-50%	G
STAINLESS STEELS		
Type 304/307	10%	G
Type 316	10-50%	G
TANTALUM		G
TITANIUM		G
ZIRCONIUM		G

NONMETALLICS	Concentration	Rating
BOROSILICATE GLASS		
CARBON		R

TEMP °F

FOR METALS

E = < 2 Mils Penetration/Year; G = < 20 Mils Penetration/Year

S = < 50 Mils Penetration/Year; U = > 50 Mils Penetration/Year

FOR NONMETALLICS

R = Resistant

U = Unsatisfactory

SODIUM BROMIDE

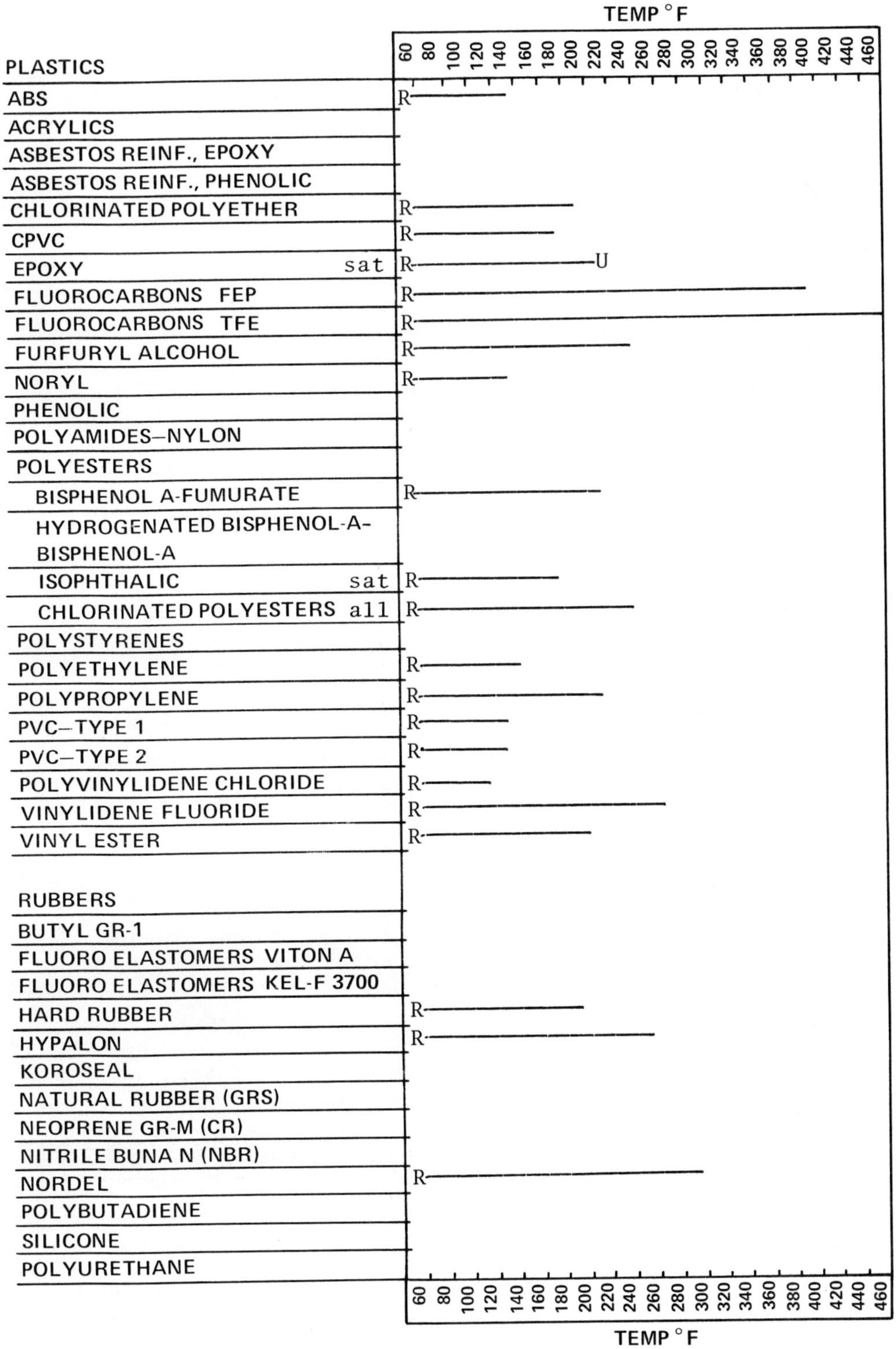

SODIUM CARBONATE

TEMP °F

METALS	Concentration	Rating / Temperature range (°F)
ALLOY 20Cb3	30%	E 60–200; G 200–560
ALUMINUM	30%	U
BRASS		G 60–80
BRONZE	0-25%	G 60–100
CARBON STEEL		G 60–120
COPPER	0-25%	G 60–100
E-BRIGHT 26-1		
HASTELLOY-B	10-100%	G 60–560
HASTELLOY-C	30%	G 60–200
HASTELLOY-D	30%	G 60–200
HIGH SILICON IRON		E 60–100
INCONEL	30%	E 60–200
LEAD		U
MONEL	30%	E 60–200
NICKEL	30%	E 60–200
STAINLESS STEELS		
Type 304/307	30%	G 60–200
Type 316	30%	G 60–340
TANTALUM	to 25%	E 60–200
TITANIUM		E 60–200
ZIRCONIUM		E 60–200

NONMETALLICS		
BOROSILICATE GLASS		R 60–140
CARBON		R 60–340

TEMP °F (scale: 60 80 100 120 140 160 180 200 220 240 260 280 300 320 340 360 380 400 420 440 460 480 500 520 540 560)

FOR METALS

E = < 2 Mils Penetration/Year; G = < 20 Mils Penetration/Year

S = < 50 Mils Penetration/Year; U = > 50 Mils Penetration/Year

FOR NONMETALLICS

R = Resistant

U = Unsatisfactory

TEMP °F (60–460)

PLASTICS		
ABS	25%	R
ACRYLICS		U
ASBESTOS REINF., EPOXY		R
ASBESTOS REINF., PHENOLIC		R
CHLORINATED POLYETHER		R
CPVC		R
EPOXY		R
FLUOROCARBONS FEP		R
FLUOROCARBONS TFE		R
FURFURYL ALCOHOL	50%	R
NORYL		
PHENOLIC		R
POLYAMIDES–NYLON		R
POLYESTERS		
BISPHENOL A-FUMURATE	sat	R
HYDROGENATED BISPHENOL-A-BISPHENOL-A	10%	R
ISOPHTHALIC	20%	R
CHLORINATED POLYESTERS	10%	R U
POLYSTYRENES		R
POLYETHYLENE		R
POLYPROPYLENE		R
PVC–TYPE 1		R
PVC–TYPE 2		R
POLYVINYLIDENE CHLORIDE		R
VINYLIDENE FLUORIDE		R
VINYL ESTER	sat	R

RUBBERS	
BUTYL GR-1	R
FLUORO ELASTOMERS VITON A	R
FLUORO ELASTOMERS KEL-F 3700	R
HARD RUBBER	R
HYPALON	R
KOROSEAL	R
NATURAL RUBBER (GRS)	R
NEOPRENE GR-M (CR)	R
NITRILE BUNA N (NBR)	R
NORDEL	R
POLYBUTADIENE	R
SILICONE	R
POLYURETHANE	

TEMP °F (60–460)

SODIUM CHLORATE

TEMP °F: 60 – 560

METALS	Concentration	Rating
ALLOY 20Cb3	to 70%	G
ALUMINUM	to 70%	G
BRASS		G
BRONZE		G
CARBON STEEL		S
COPPER		G
E-BRIGHT 26-1		
HASTELLOY-B		U
HASTELLOY-C	to 70%	E
HASTELLOY-D		U
HIGH SILICON IRON		G
INCONEL	10-70%	E
LEAD		
MONEL	10-70%	E
NICKEL	10-70%	E
STAINLESS STEELS		
Type 304/307	50%	G
Type 316	to 70%	G
TANTALUM		E
TITANIUM	10%	E
ZIRCONIUM		

NONMETALLICS	Rating
BOROSILICATE GLASS	
CARBON	

TEMP °F

FOR METALS

E = < 2 Mils Penetration/Year; G = < 20 Mils Penetration/Year

S = < 50 Mils Penetration/Year; U = > 50 Mils Penetration/Year

FOR NONMETALLICS

R = Resistant

U = Unsatisfactory

SODIUM CHLORATE

TEMP °F (60–460)

PLASTICS	Conc.	Rating
ABS		U
ACRYLICS		R
ASBESTOS REINF., EPOXY		
ASBESTOS REINF., PHENOLIC		R
CHLORINATED POLYETHER		R
CPVC		R
EPOXY		R
FLUOROCARBONS FEP		R
FLUOROCARBONS TFE		R
FURFURYL ALCOHOL		
NORYL		
PHENOLIC		R
POLYAMIDES—NYLON		U
POLYESTERS		
BISPHENOL A-FUMURATE	sat	R
HYDROGENATED BISPHENOL-A-BISPHENOL-A		R
ISOPHTHALIC	sat	U
CHLORINATED POLYESTERS	48%	R
POLYSTYRENES		R
POLYETHYLENE		R
POLYPROPYLENE		R
PVC—TYPE 1		R
PVC—TYPE 2		R
POLYVINYLIDENE CHLORIDE		R
VINYLIDENE FLUORIDE		R
VINYL ESTER	sat	R

RUBBERS	Conc.	Rating
BUTYL GR-1		
FLUORO ELASTOMERS VITON A		R
FLUORO ELASTOMERS KEL-F 3700		
HARD RUBBER		
HYPALON		R
KOROSEAL		
NATURAL RUBBER (GRS)		R
NEOPRENE GR-M (CR)		R
NITRILE BUNA N (NBR)		R
NORDEL		R
POLYBUTADIENE		R
SILICONE		
POLYURETHANE		

TEMP °F (60–460)

SODIUM CHLORIDE

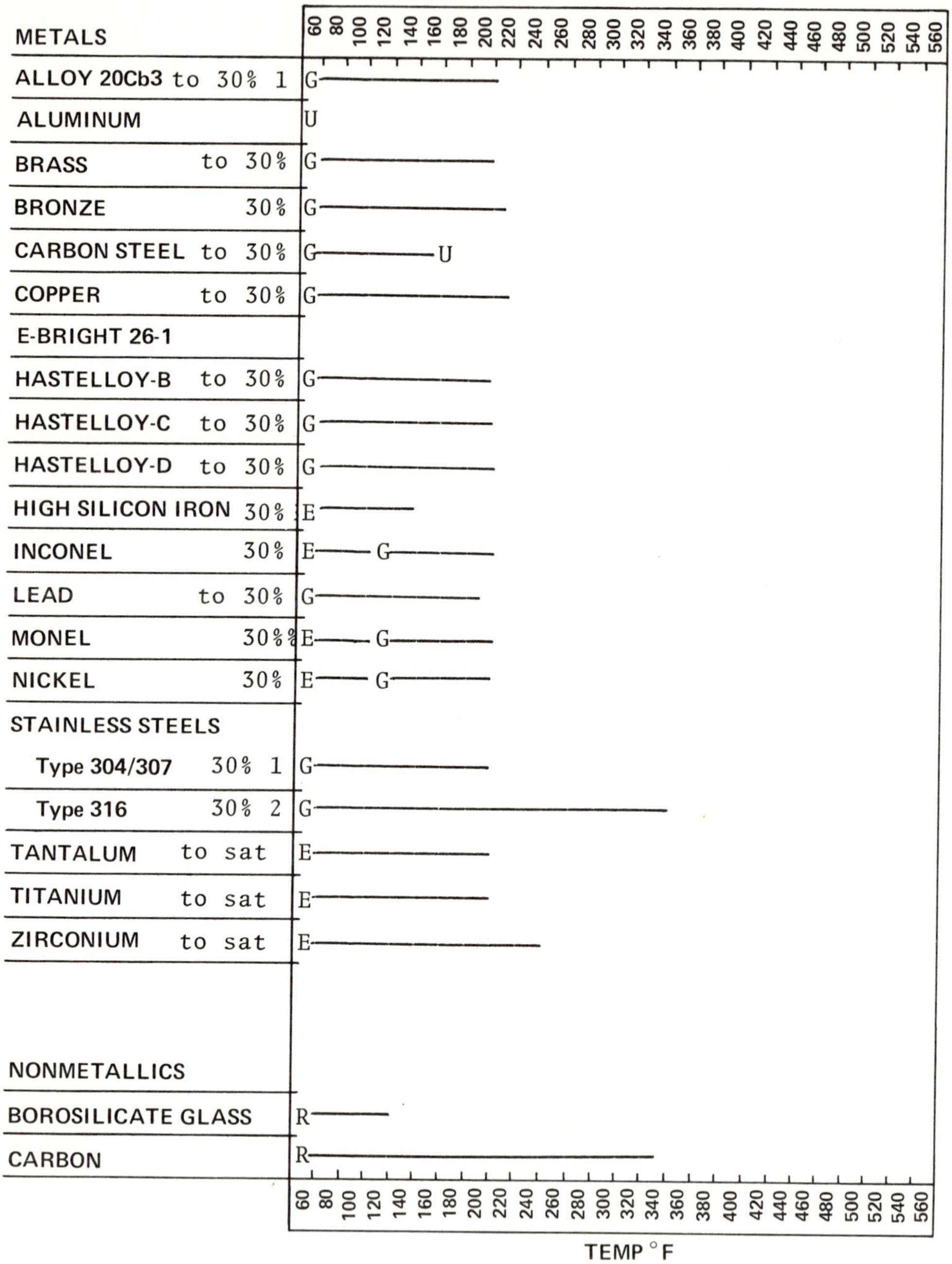

FOR METALS

E = < 2 Mils Penetration/Year; G = < 20 Mils Penetration/Year

S = < 50 Mils Penetration/Year; U = > 50 Mils Penetration/Year

FOR NONMETALLICS

R = Resistant

U = Unsatisfactory

SODIUM CHLORIDE

TEMP °F

PLASTICS		60 80 100 120 140 160 180 200 220 240 260 280 300 320 340 360 380 400 420 440 460
ABS	sat	R
ACRYLICS		R
ASBESTOS REINF., EPOXY		R
ASBESTOS REINF., PHENOLIC		R
CHLORINATED POLYETHER		R
CPVC		R
EPOXY	sat	R — U
FLUOROCARBONS FEP		R
FLUOROCARBONS TFE		R
FURFURYL ALCOHOL	sat	R
NORYL		
PHENOLIC		R
POLYAMIDES–NYLON	10%	R
POLYESTERS		
BISPHENOL A-FUMURATE		R
HYDROGENATED BISPHENOL-A-BISPHENOL-A		R
ISOPHTHALIC	sat	R
CHLORINATED POLYESTERS	all	R
POLYSTYRENES		R
POLYETHYLENE		R
POLYPROPYLENE		R
PVC–TYPE 1		R
PVC–TYPE 2		R
POLYVINYLIDENE CHLORIDE		R
VINYLIDENE FLUORIDE		R
VINYL ESTER	sat	R

RUBBERS		
BUTYL GR-1		R
FLUORO ELASTOMERS VITON A		R
FLUORO ELASTOMERS KEL-F 3700		R
HARD RUBBER		R
HYPALON		R
KOROSEAL		R
NATURAL RUBBER (GRS)		R
NEOPRENE GR-M (CR)		R
NITRILE BUNA N (NBR)		R
NORDEL		R
POLYBUTADIENE		
SILICONE	10%	R
POLYURETHANE		R

60 80 100 120 140 160 180 200 220 240 260 280 300 320 340 360 380 400 420 440 460

TEMP °F

SODIUM CHLORITE

TEMP °F: 60 80 100 120 140 160 180 200 220 240 260 280 300 320 340 360 380 400 420 440 460 480 500 520 540 560

METALS		TEMP °F
ALLOY 20Cb3		
ALUMINUM		
BRASS		
BRONZE		
CARBON STEEL		
COPPER		
E-BRIGHT 26-1		E—
HASTELLOY-B		
HASTELLOY-C	10%	G—
HASTELLOY-D		
HIGH SILICON IRON		
INCONEL		
LEAD		
MONEL		
NICKEL		
STAINLESS STEELS		
Type 304/307	5% 1	G————————
Type 316	5%	E————————
TANTALUM		
TITANIUM		
ZIRCONIUM		

NONMETALLICS		TEMP °F
BOROSILICATE GLASS		
CARBON		R——————————————————————

TEMP °F: 60 80 100 120 140 160 180 200 220 240 260 280 300 320 340 360 380 400 420 440 460 480 500 520 540 560

FOR METALS

E = < 2 Mils Penetration/Year; G = < 20 Mils Penetration/Year

S = < 50 Mils Penetration/Year; U = > 50 Mils Penetration/Year

FOR NONMETALLICS

R = Resistant

U = Unsatisfactory

SODIUM CHLORITE

TEMP °F (60–460)

PLASTICS	Conc.	Rating	Bar extends to (°F)
ABS			
ACRYLICS			
ASBESTOS REINF., EPOXY			
ASBESTOS REINF., PHENOLIC			
CHLORINATED POLYETHER		R	150
CPVC			
EPOXY	25%	R	150
FLUOROCARBONS FEP		R	400
FLUOROCARBONS TFE		R	460
FURFURYL ALCOHOL			
NORYL	20%	R	140
PHENOLIC			
POLYAMIDES—NYLON			
POLYESTERS			
BISPHENOL A-FUMURATE	50%	R	220
HYDROGENATED BISPHENOL-A-BISPHENOL-A			
ISOPHTHALIC	sat	U	
CHLORINATED POLYESTERS	sat	R	150
POLYSTYRENES			
POLYETHYLENE		U	
POLYPROPYLENE		U	
PVC—TYPE 1		R	140
PVC—TYPE 2		U	
POLYVINYLIDENE CHLORIDE		R	120
VINYLIDENE FLUORIDE		R	250
VINYL ESTER	50%	R	100

RUBBERS	Conc.	Rating	Bar extends to (°F)
BUTYL GR-1			
FLUORO ELASTOMERS VITON A			
FLUORO ELASTOMERS KEL-F 3700			
HARD RUBBER		R	200
HYPALON		R	250
KOROSEAL			
NATURAL RUBBER (GRS)			
NEOPRENE GR-M (CR)			
NITRILE BUNA N (NBR)			
NORDEL			
POLYBUTADIENE			
SILICONE			
POLYURETHANE			

TEMP °F

SODIUM CYANIDE

TEMP °F (60–560)

METALS	Concentration	Rating	Range shown (°F)
ALLOY 20Cb3	10%	E	60–200
ALUMINUM		U	
BRASS		U	
BRONZE		U	
CARBON STEEL		G	60–100
COPPER		U	
E-BRIGHT 26-1			
HASTELLOY-B		G	60–100
HASTELLOY-C		G	60–100
HASTELLOY-D			
HIGH SILICON IRON		G	60–80
INCONEL		G	60–200
LEAD		U	
MONEL		U	
NICKEL		U	
STAINLESS STEELS			
Type 304/307	10%	E	60–200
Type 316	10%	E	60–340
TANTALUM		G	60–200
TITANIUM		E	60–80
ZIRCONIUM			

NONMETALLICS	Rating	Range shown (°F)
BOROSILICATE GLASS	R	60–200
CARBON	R	60–340

TEMP °F

FOR METALS

E = < 2 Mils Penetration/Year; G = < 20 Mils Penetration/Year

S = < 50 Mils Penetration/Year; U = > 50 Mils Penetration/Year

FOR NONMETALLICS

R = Resistant

U = Unsatisfactory

SODIUM CYANIDE

TEMP °F (60, 80, 100, 120, 140, 160, 180, 200, 220, 240, 260, 280, 300, 320, 340, 360, 380, 400, 420, 440, 460)

PLASTICS		
ABS		R
ACRYLICS		R
ASBESTOS REINF., EPOXY		
ASBESTOS REINF., PHENOLIC		
CHLORINATED POLYETHER		R
CPVC		R
EPOXY		R
FLUOROCARBONS FEP		R
FLUOROCARBONS TFE		R
FURFURYL ALCOHOL	sat	R
NORYL		R
PHENOLIC		
POLYAMIDES–NYLON		R
POLYESTERS		
BISPHENOL A-FUMURATE	sat	R
HYDROGENATED BISPHENOL-A-BISPHENOL-A		
ISOPHTHALIC	sat	R
CHLORINATED POLYESTERS	50%	R—U
POLYSTYRENES		
POLYETHYLENE		R
POLYPROPYLENE		R
PVC–TYPE 1		R
PVC–TYPE 2		R
POLYVINYLIDENE CHLORIDE		R
VINYLIDENE FLUORIDE		R
VINYL ESTER		R

RUBBERS	
BUTYL GR-1	R
FLUORO ELASTOMERS VITON A	R
FLUORO ELASTOMERS KEL-F 3700	
HARD RUBBER	R
HYPALON	R
KOROSEAL	R
NATURAL RUBBER (GRS)	R
NEOPRENE GR-M (CR)	R
NITRILE BUNA N (NBR)	R
NORDEL	R
POLYBUTADIENE	
SILICONE	
POLYURETHANE	

TEMP °F (60, 80, 100, 120, 140, 160, 180, 200, 220, 240, 260, 280, 300, 320, 340, 360, 380, 400, 420, 440, 460)

SODIUM DICHROMATE

METALS		TEMP °F (60–560)
ALLOY 20Cb3		
ALUMINUM		G (60–220)
BRASS		U
BRONZE		U
CARBON STEEL		
COPPER		U
E-BRIGHT 26-1		
HASTELLOY-B		
HASTELLOY-C		
HASTELLOY-D		
HIGH SILICON IRON		
INCONEL		
LEAD		
MONEL		
NICKEL		
STAINLESS STEELS Type 304/307		
Type 316		
TANTALUM		E—
TITANIUM	sat	E—
ZIRCONIUM		
NONMETALLICS		
BOROSILICATE GLASS		
CARBON		R (60–340)

TEMP °F

FOR METALS

E = < 2 Mils Penetration/Year; G = < 20 Mils Penetration/Year

S = < 50 Mils Penetration/Year; U = > 50 Mils Penetration/Year

FOR NONMETALLICS

R = Resistant

U = Unsatisfactory

SODIUM DICHROMATE

TEMP °F (60–460)

PLASTICS	Concentration	Rating
ABS		R
ACRYLICS		
ASBESTOS REINF., EPOXY		R
ASBESTOS REINF., PHENOLIC		R
CHLORINATED POLYETHER		R
CPVC		R
EPOXY		R
FLUOROCARBONS FEP		R
FLUOROCARBONS TFE		R
FURFURYL ALCOHOL	sat	R
NORYL		
PHENOLIC		
POLYAMIDES–NYLON		
POLYESTERS		
BISPHENOL A-FUMURATE	sat	R
HYDROGENATED BISPHENOL-A-BISPHENOL-A		
ISOPHTHALIC	sat	R
CHLORINATED POLYESTERS		
POLYSTYRENES		
POLYETHYLENE		R
POLYPROPYLENE		R
PVC–TYPE 1		R
PVC–TYPE 2		R
POLYVINYLIDENE CHLORIDE		R
VINYLIDENE FLUORIDE		R
VINYL ESTER		R

RUBBERS	Concentration	Rating
BUTYL GR-1		
FLUORO ELASTOMERS VITON A		
FLUORO ELASTOMERS KEL-F 3700		
HARD RUBBER		
HYPALON	20%	R
KOROSEAL		
NATURAL RUBBER (GRS)		
NEOPRENE GR-M (CR)		U
NITRILE BUNA N (NBR)	10%	R
NORDEL		
POLYBUTADIENE		
SILICONE		
POLYURETHANE		

TEMP °F (60–460)

SODIUM FERRICYANIDE

TEMP °F: 60, 80, 100, 120, 140, 160, 180, 200, 220, 240, 260, 280, 300, 320, 340, 360, 380, 400, 420, 440, 460, 480, 500, 520, 540, 560

METALS	Rating
ALLOY 20Cb3	G—
ALUMINUM	
BRASS	U
BRONZE	
CARBON STEEL	
COPPER	
E-BRIGHT 26-1	
HASTELLOY-B	E————
HASTELLOY-C	E————
HASTELLOY-D	E————
HIGH SILICON IRON	G–
INCONEL	
LEAD	
MONEL	
NICKEL	
STAINLESS STEELS	
Type 304/307 1	G—
Type 316	G————————
TANTALUM	E–
TITANIUM	
ZIRCONIUM	

NONMETALLICS	Rating
BOROSILICATE GLASS	R—————
CARBON	R————————

TEMP °F: 60, 80, 100, 120, 140, 160, 180, 200, 220, 240, 260, 280, 300, 320, 340, 360, 380, 400, 420, 440, 460, 480, 500, 520, 540, 560

FOR METALS

E = < 2 Mils Penetration/Year; G = < 20 Mils Penetration/Year

S = < 50 Mils Penetration/Year; U = > 50 Mils Penetration/Year

FOR NONMETALLICS

R = Resistant

U = Unsatisfactory

SODIUM FERRICYANIDE

TEMP °F: 60 80 100 120 140 160 180 200 220 240 260 280 300 320 340 360 380 400 420 440 460

PLASTICS		
ABS		R
ACRYLICS		
ASBESTOS REINF., EPOXY		
ASBESTOS REINF., PHENOLIC		
CHLORINATED POLYETHER		R
CPVC		R
EPOXY		R
FLUOROCARBONS FEP		R
FLUOROCARBONS TFE		R
FURFURYL ALCOHOL		R
NORYL		R
PHENOLIC		
POLYAMIDES—NYLON		
POLYESTERS		
BISPHENOL A-FUMURATE	sat	R
HYDROGENATED BISPHENOL-A-BISPHENOL-A		
ISOPHTHALIC	sat	R
CHLORINATED POLYESTERS	all	R
POLYSTYRENES		
POLYETHYLENE		R
POLYPROPYLENE		R
PVC—TYPE 1		R
PVC—TYPE 2		R
POLYVINYLIDENE CHLORIDE		R
VINYLIDENE FLUORIDE		R
VINYL ESTER		R

RUBBERS	
BUTYL GR-1	
FLUORO ELASTOMERS VITON A	
FLUORO ELASTOMERS KEL-F 3700	
HARD RUBBER	R
HYPALON	R
KOROSEAL	
NATURAL RUBBER (GRS)	
NEOPRENE GR-M (CR)	
NITRILE BUNA N (NBR)	
NORDEL	R
POLYBUTADIENE	
SILICONE	
POLYURETHANE	

TEMP °F: 60 80 100 120 140 160 180 200 220 240 260 280 300 320 340 360 380 400 420 440 460

SODIUM FERROCYANIDE

TEMP °F: 60 80 100 120 140 160 180 200 220 240 260 280 300 320 340 360 380 400 420 440 460 480 500 520 540 560

METALS	TEMP °F
ALLOY 20Cb3	
ALUMINUM	
BRASS	
BRONZE	
CARBON STEEL	
COPPER	
E-BRIGHT 26-1	
HASTELLOY-B	
HASTELLOY-C	
HASTELLOY-D	
HIGH SILICON IRON	
INCONEL	
LEAD	
MONEL	
NICKEL	
STAINLESS STEELS Type 304/307	
Type 316	
TANTALUM	E — (60)
TITANIUM	
ZIRCONIUM	

NONMETALLICS	TEMP °F
BOROSILICATE GLASS	
CARBON	R (60–340)

TEMP °F

FOR METALS

E = < 2 Mils Penetration/Year; G = < 20 Mils Penetration/Year

S = < 50 Mils Penetration/Year; U = > 50 Mils Penetration/Year

FOR NONMETALLICS

R = Resistant

U = Unsatisfactory

SODIUM FERROCYANIDE

TEMP °F: 60 80 100 120 140 160 180 200 220 240 260 280 300 320 340 360 380 400 420 440 460

PLASTICS	
ABS	R
ACRYLICS	R
ASBESTOS REINF., EPOXY	
ASBESTOS REINF., PHENOLIC	
CHLORINATED POLYETHER	R
CPVC	R
EPOXY	R
FLUOROCARBONS FEP	R
FLUOROCARBONS TFE	R
FURFURYL ALCOHOL	
NORYL	R
PHENOLIC	
POLYAMIDES—NYLON	
POLYESTERS	
BISPHENOL A-FUMURATE	R
HYDROGENATED BISPHENOL-A-BISPHENOL-A	
ISOPHTHALIC sat	R
CHLORINATED POLYESTERS	
POLYSTYRENES	
POLYETHYLENE	R
POLYPROPYLENE	R
PVC—TYPE 1	R
PVC—TYPE 2	R
POLYVINYLIDENE CHLORIDE	R
VINYLIDENE FLUORIDE	R
VINYL ESTER	R

RUBBERS	
BUTYL GR-1	
FLUORO ELASTOMERS VITON A	
FLUORO ELASTOMERS KEL-F 3700	
HARD RUBBER	
HYPALON	
KOROSEAL	
NATURAL RUBBER (GRS)	
NEOPRENE GR-M (CR)	
NITRILE BUNA N (NBR)	
NORDEL	
POLYBUTADIENE	
SILICONE	
POLYURETHANE	

TEMP °F: 60 80 100 120 140 160 180 200 220 240 260 280 300 320 340 360 380 400 420 440 460

SODIUM FLUORIDE

METALS		TEMP °F (60–560)
ALLOY 20Cb3		G———
ALUMINUM		G—U
BRASS		
BRONZE		U
CARBON STEEL		U
COPPER		
E-BRIGHT 26-1		
HASTELLOY-B		G————
HASTELLOY-C		G————
HASTELLOY-D		
HIGH SILICON IRON		
INCONEL		E—
LEAD		
MONEL		E———
NICKEL		G———
STAINLESS STEELS		
Type 304/307	10%	S-U
Type 316	10%	G————————
TANTALUM		U
TITANIUM		E—
ZIRCONIUM	20%	E—

NONMETALLICS	TEMP °F (60–560)
BOROSILICATE GLASS	U
CARBON	R————————

TEMP °F: 60, 80, 100, 120, 140, 160, 180, 200, 220, 240, 260, 280, 300, 320, 340, 360, 380, 400, 420, 440, 460, 480, 500, 520, 540, 560

FOR METALS

E = < 2 Mils Penetration/Year; G = < 20 Mils Penetration/Year

S = < 50 Mils Penetration/Year; U = > 50 Mils Penetration/Year

FOR NONMETALLICS

R = Resistant

U = Unsatisfactory

SODIUM FLUORIDE

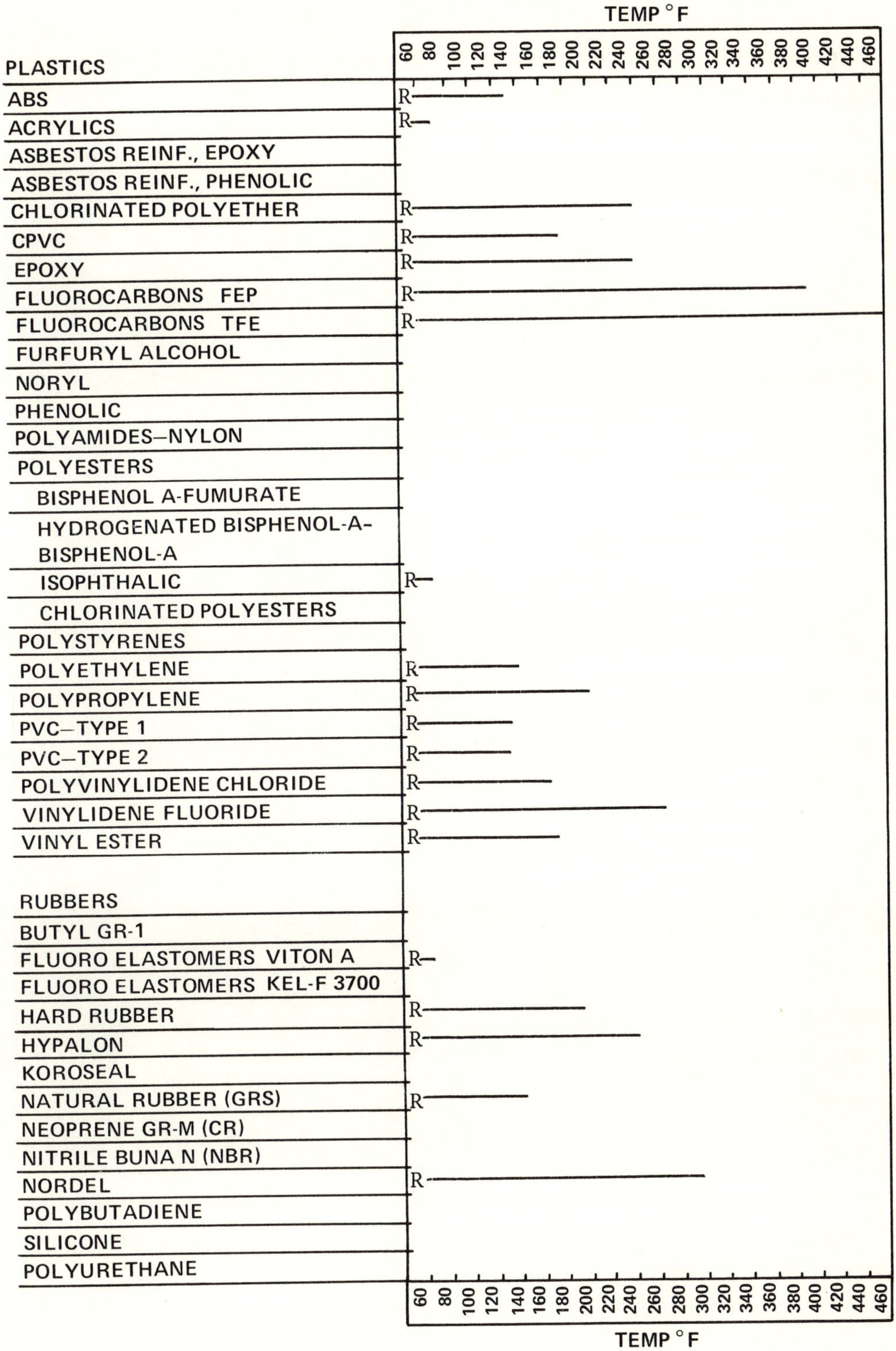

SODIUM HYDROXIDE SOLUTION

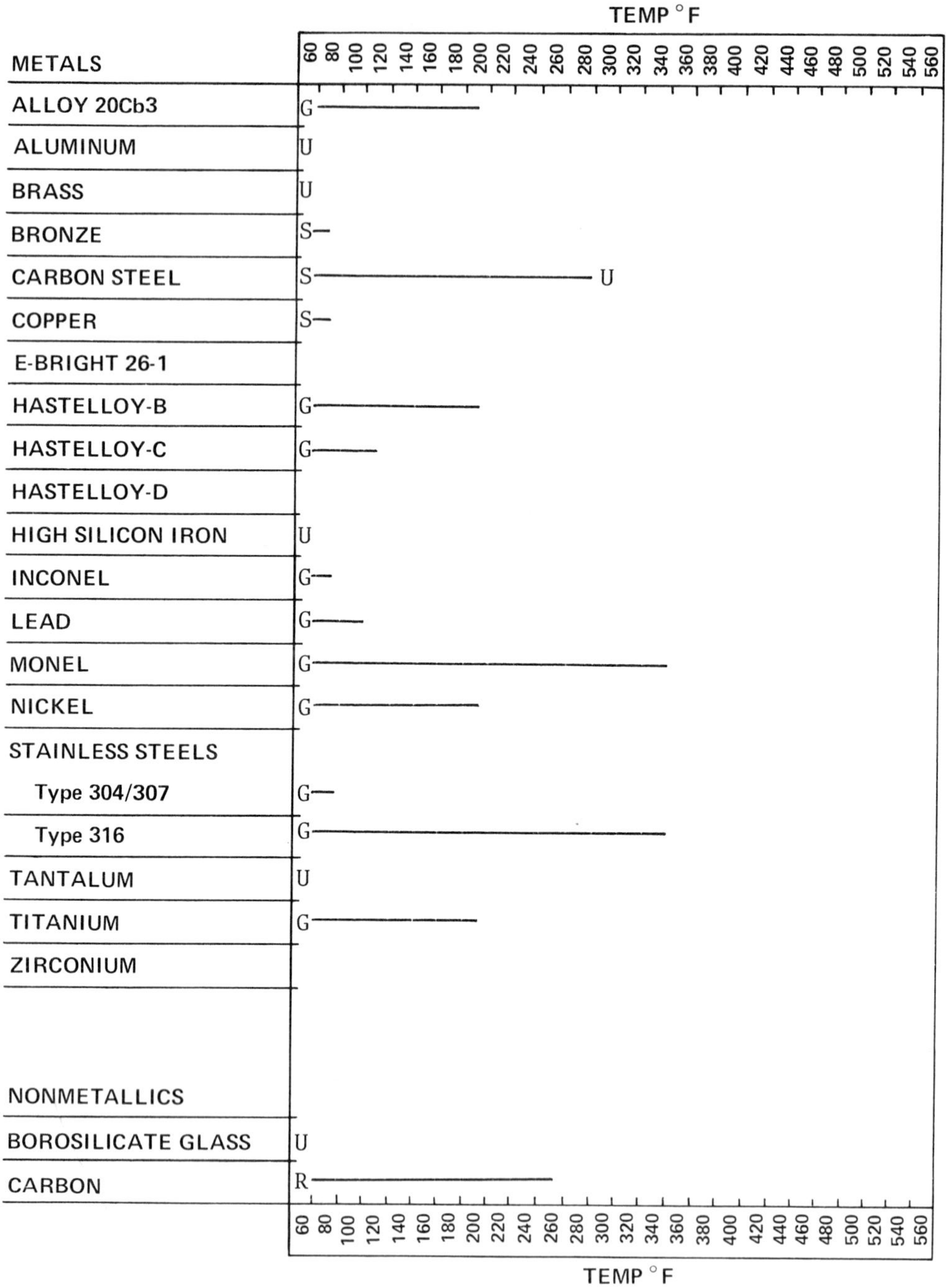

FOR METALS

E = < 2 Mils Penetration/Year; G = < 20 Mils Penetration/Year

S = < 50 Mils Penetration/Year; U = > 50 Mils Penetration/Year

FOR NONMETALLICS

R = Resistant

U = Unsatisfactory

SODIUM HYDROXIDE SOLUTION

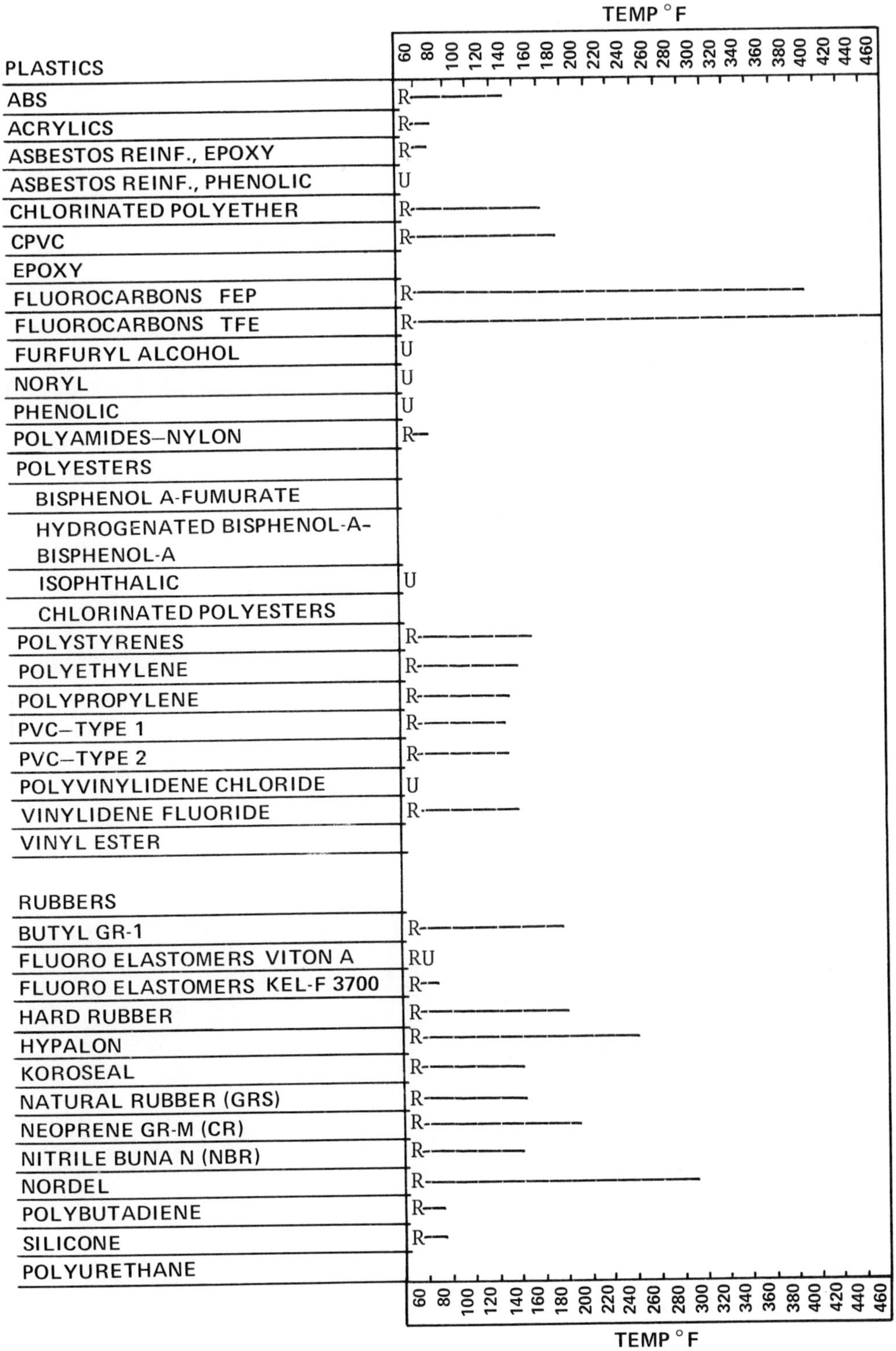

SODIUM HYDROXIDE, 70%

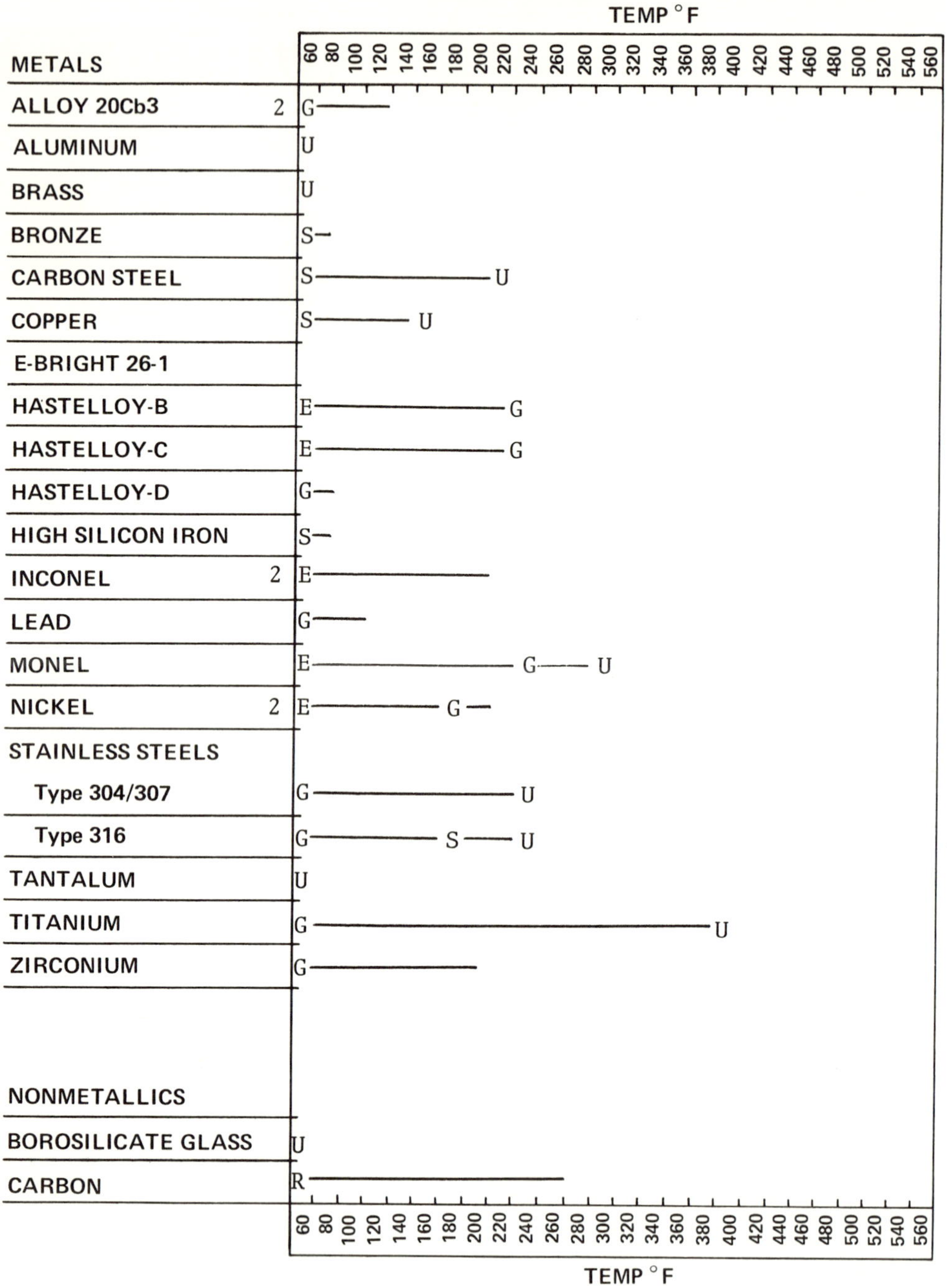

FOR METALS

E = < 2 Mils Penetration/Year; G = < 20 Mils Penetration/Year
S = < 50 Mils Penetration/Year; U = > 50 Mils Penetration/Year

FOR NONMETALLICS

R = Resistant
U = Unsatisfactory

SODIUM HYDROXIDE, 70%

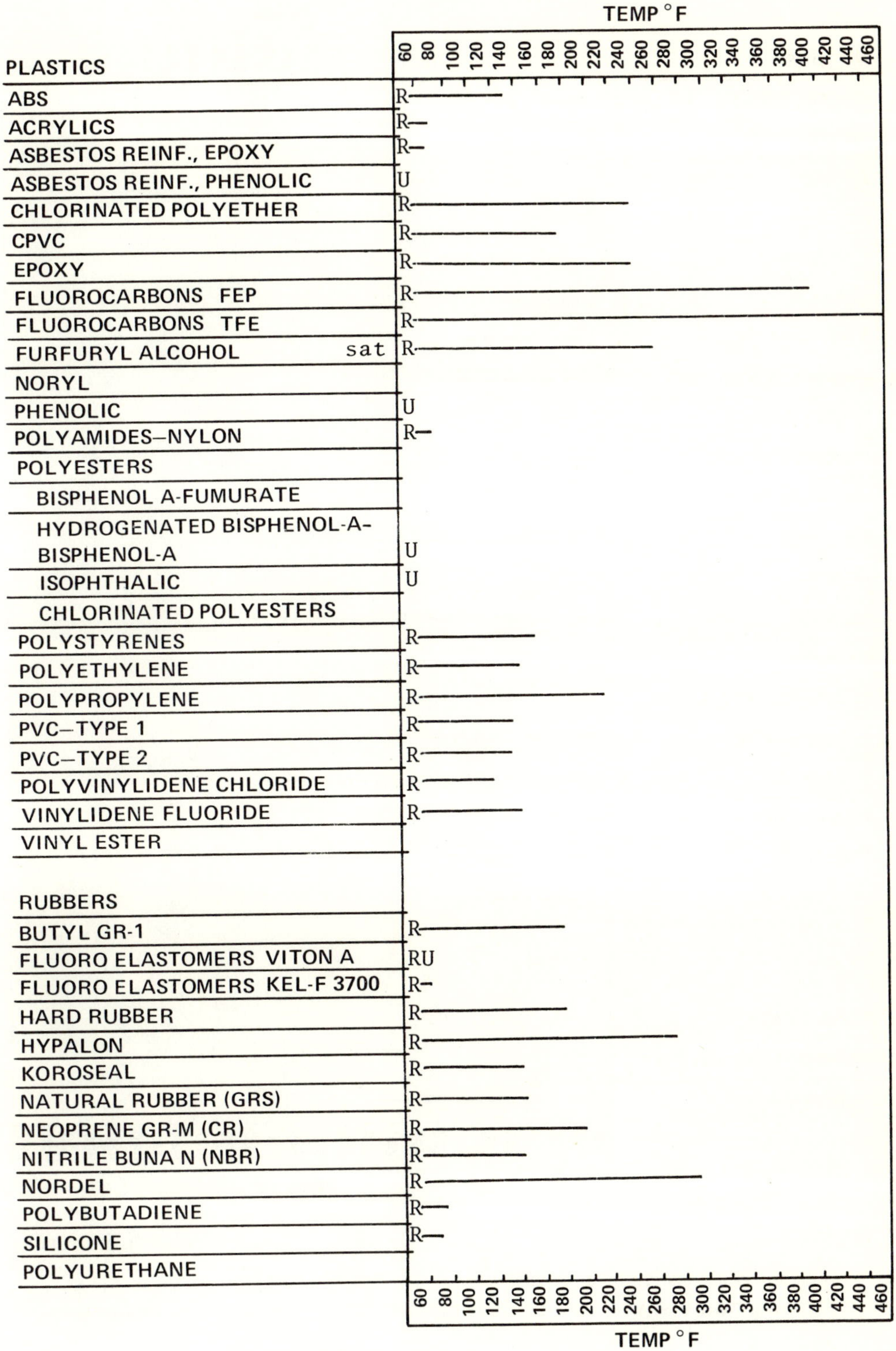

SODIUM HYDROXIDE, 50%

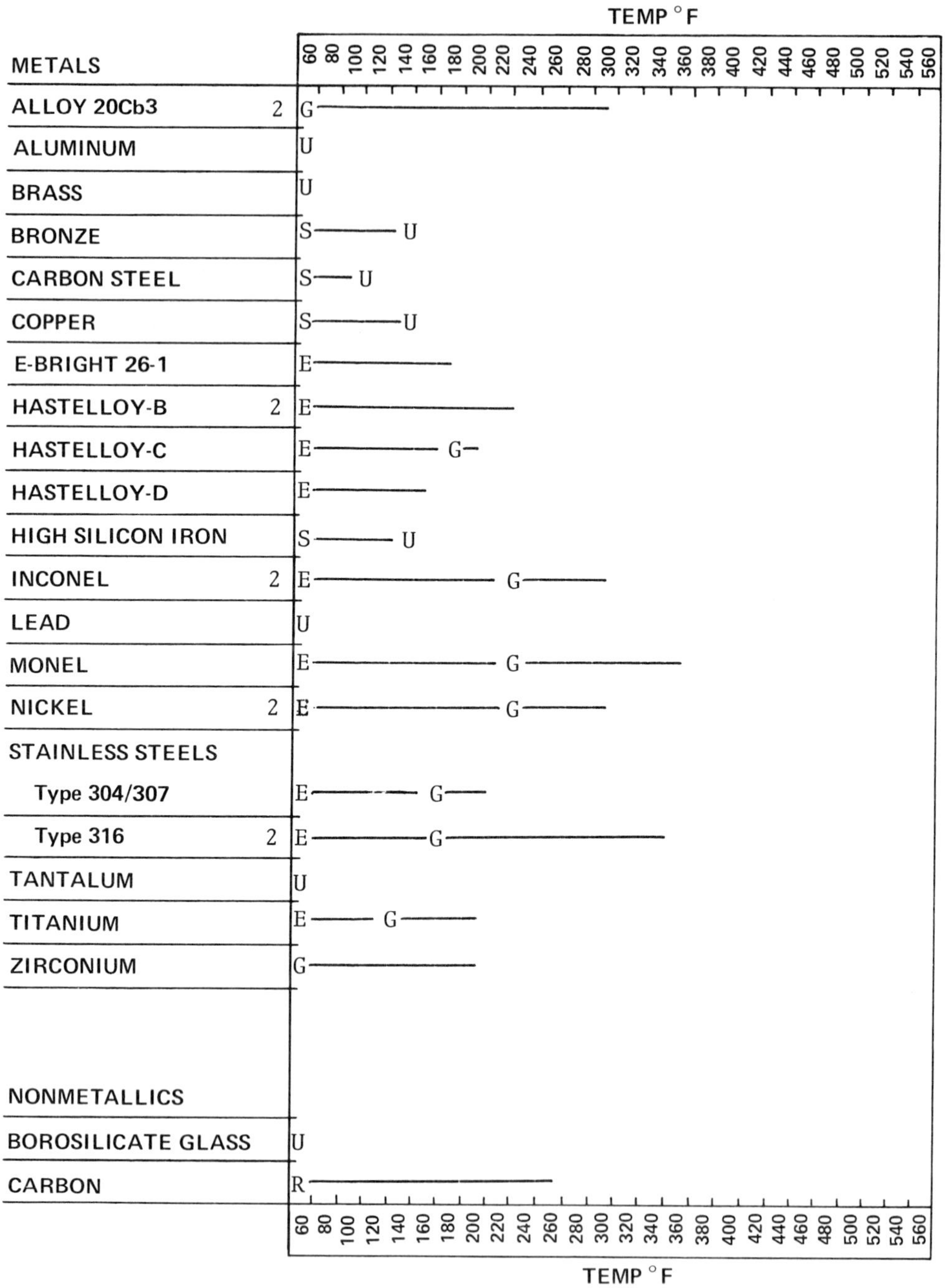

FOR METALS

E = < 2 Mils Penetration/Year; G = < 20 Mils Penetration/Year

S = < 50 Mils Penetration/Year; U = > 50 Mils Penetration/Year

FOR NONMETALLICS

R = Resistant

U = Unsatisfactory

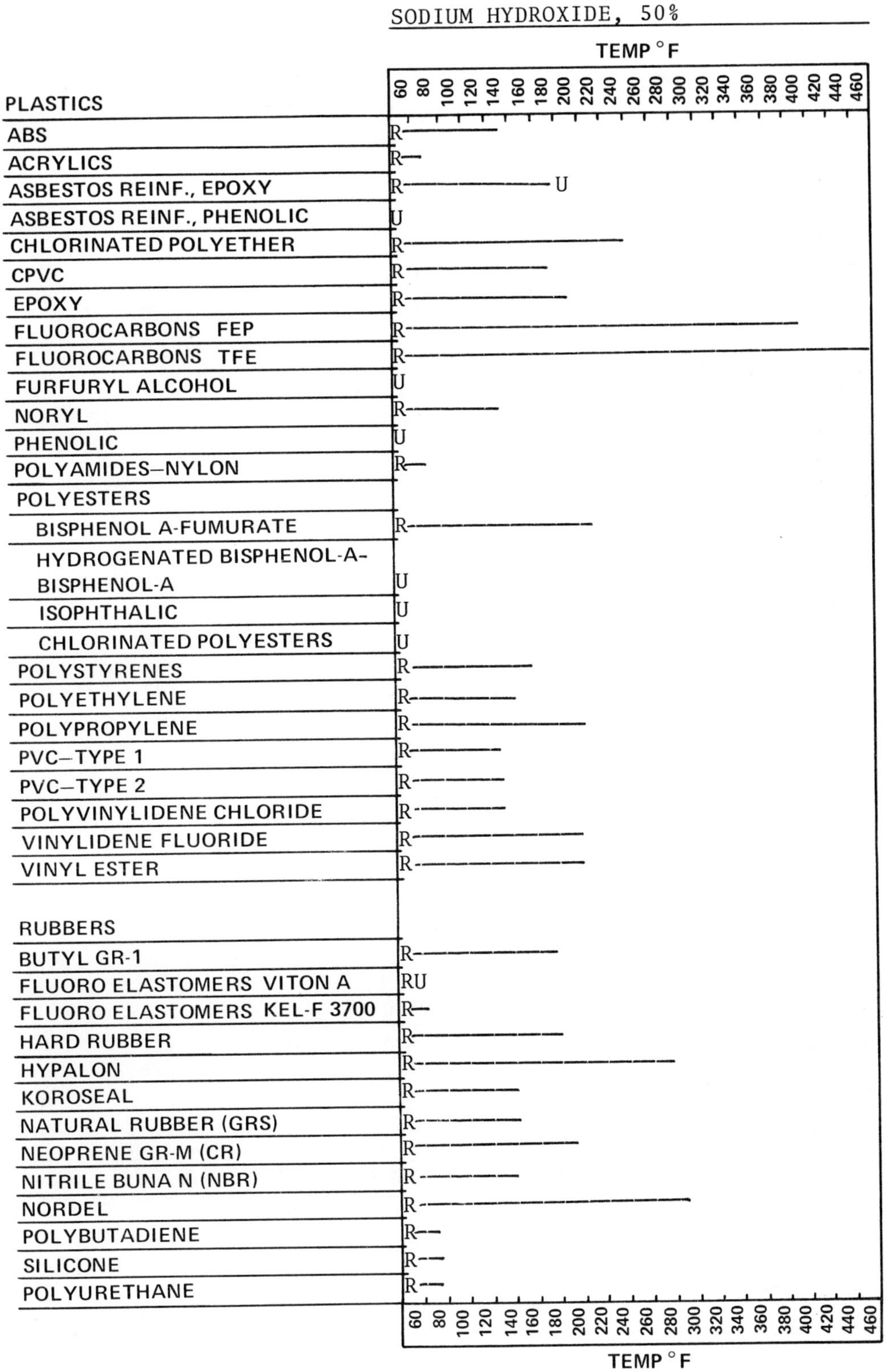
SODIUM HYDROXIDE, 50%
TEMP °F
60 80 100 120 140 160 180 200 220 240 260 280 300 320 340 360 380 400 420 440 460
PLASTICS
ABS R
ACRYLICS R
ASBESTOS REINF., EPOXY R U
ASBESTOS REINF., PHENOLIC U
CHLORINATED POLYETHER R
CPVC R
EPOXY R
FLUOROCARBONS FEP R
FLUOROCARBONS TFE R
FURFURYL ALCOHOL U
NORYL R
PHENOLIC U
POLYAMIDES–NYLON R
POLYESTERS
BISPHENOL A-FUMURATE R
HYDROGENATED BISPHENOL-A-BISPHENOL-A U
ISOPHTHALIC U
CHLORINATED POLYESTERS U
POLYSTYRENES R
POLYETHYLENE R
POLYPROPYLENE R
PVC–TYPE 1 R
PVC–TYPE 2 R
POLYVINYLIDENE CHLORIDE R
VINYLIDENE FLUORIDE R
VINYL ESTER R
RUBBERS
BUTYL GR-1 R
FLUORO ELASTOMERS VITON A RU
FLUORO ELASTOMERS KEL-F 3700 R
HARD RUBBER R
HYPALON R
KOROSEAL R
NATURAL RUBBER (GRS) R
NEOPRENE GR-M (CR) R
NITRILE BUNA N (NBR) R
NORDEL R
POLYBUTADIENE R
SILICONE R
POLYURETHANE R
60 80 100 120 140 160 180 200 220 240 260 280 300 320 340 360 380 400 420 440 460
TEMP °F

SODIUM HYDROXIDE, 30%

TEMP °F: 60 80 100 120 140 160 180 200 220 240 260 280 300 320 340 360 380 400 420 440 460 480 500 520 540 560

METALS		Rating
ALLOY 20Cb3	2	E—— G——————
ALUMINUM		U
BRASS		G——————
BRONZE		S——
CARBON STEEL		G——————
COPPER		S— U
E-BRIGHT 26-1		E——————
HASTELLOY-B	2	E——————————
HASTELLOY-C		G——————
HASTELLOY-D		G—
HIGH SILICON IRON		S———— U
INCONEL		E—————— G————
LEAD		U
MONEL		E————— G— U
NICKEL	2	E————————————
STAINLESS STEELS		
Type 304/307		E————— G—— U
Type 316		E————— G—— U
TANTALUM		U
TITANIUM		E——————
ZIRCONIUM		E——————

NONMETALLICS	Rating
BOROSILICATE GLASS	U
CARBON	R——————

TEMP °F: 60 80 100 120 140 160 180 200 220 240 260 280 300 320 340 360 380 400 420 440 460 480 500 520 540 560

FOR METALS

E = < 2 Mils Penetration/Year; G = < 20 Mils Penetration/Year

S = < 50 Mils Penetration/Year; U = > 50 Mils Penetration/Year

FOR NONMETALLICS

R = Resistant

U = Unsatisfactory

SODIUM HYDROXIDE, 30%

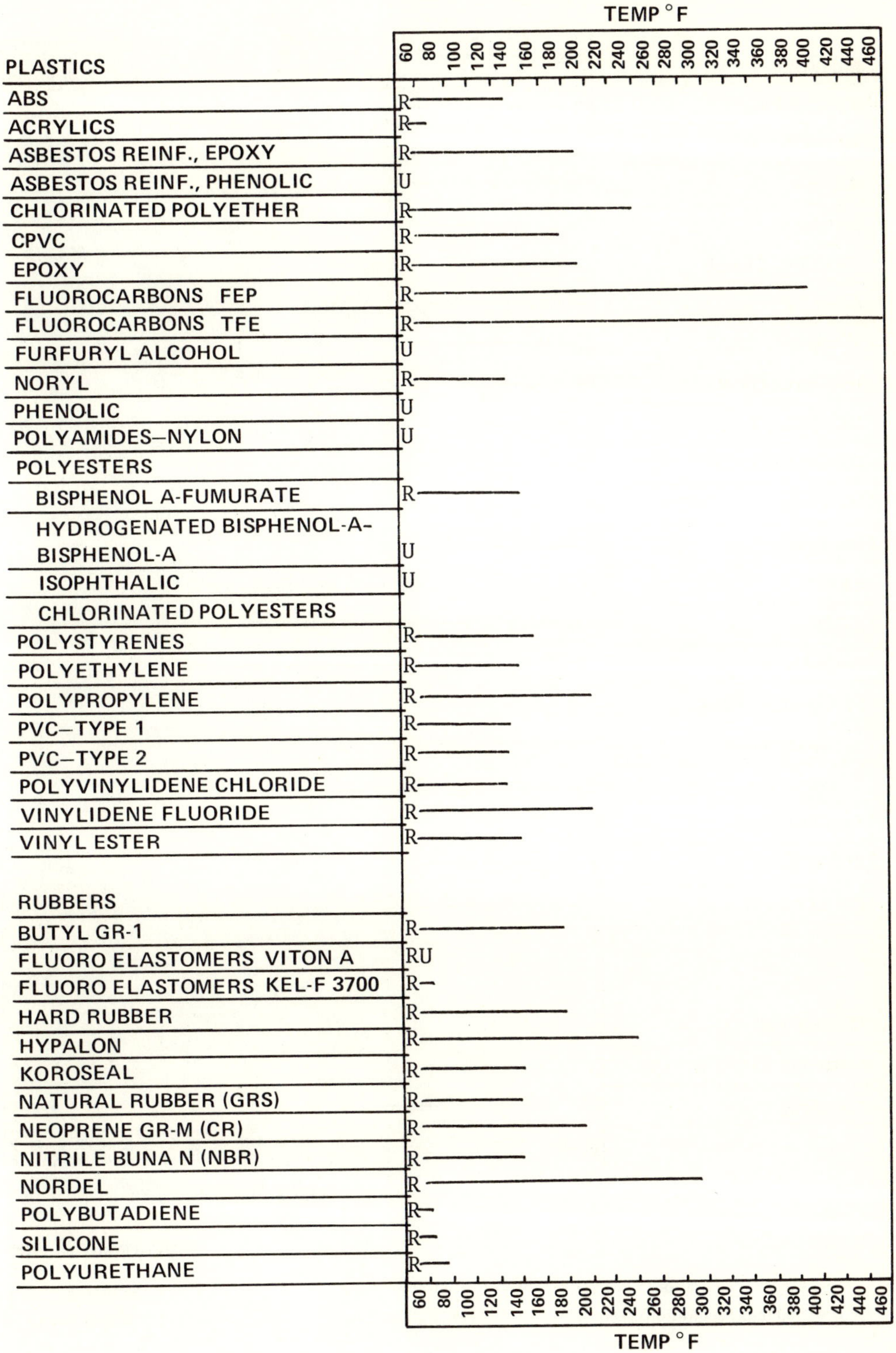

SODIUM HYDROXIDE, 15%

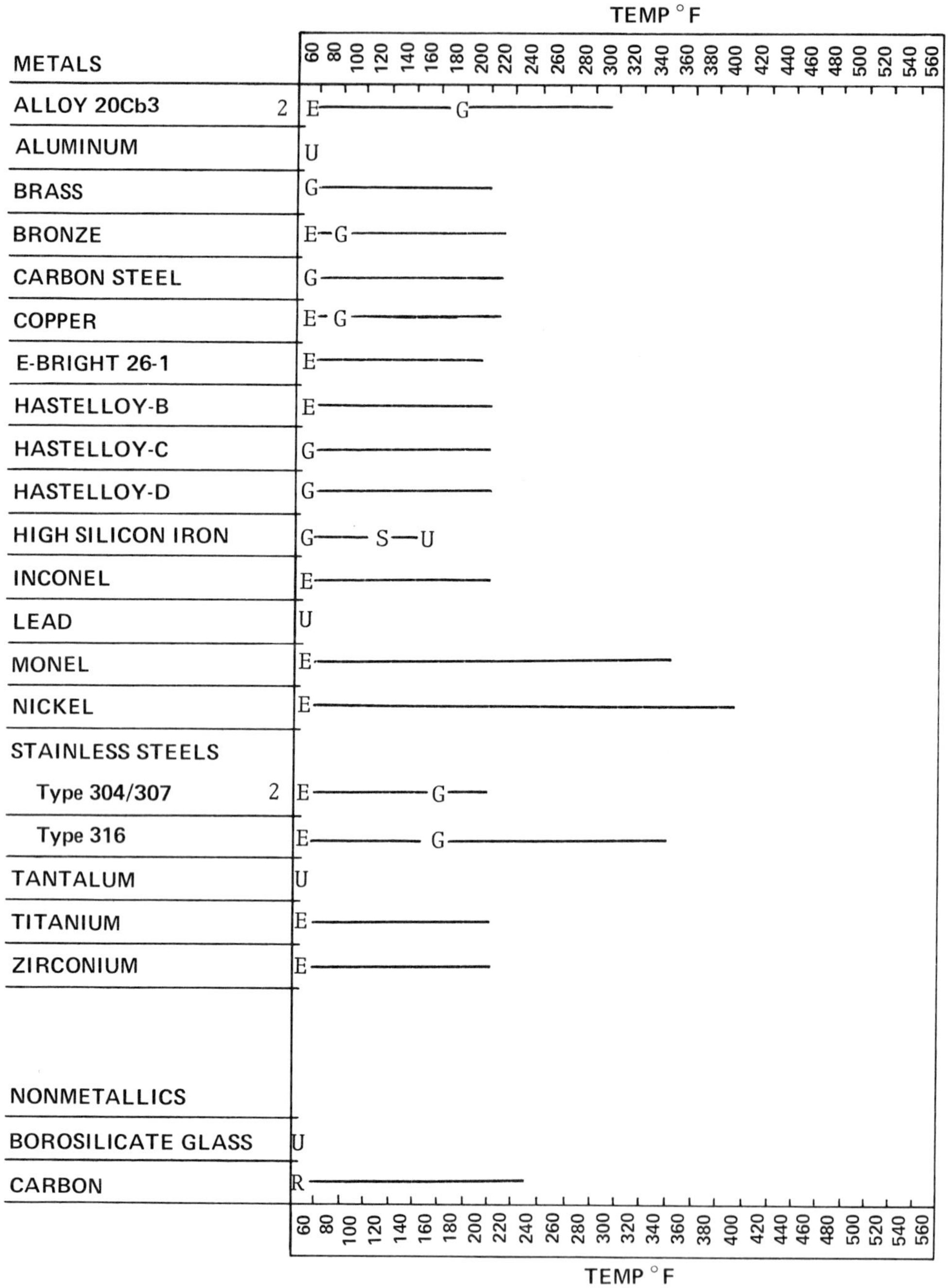

FOR METALS

E = < 2 Mils Penetration/Year; G = < 20 Mils Penetration/Year

S = < 50 Mils Penetration/Year; U = > 50 Mils Penetration/Year

FOR NONMETALLICS

R = Resistant

U = Unsatisfactory

SODIUM HYDROXIDE, 15%

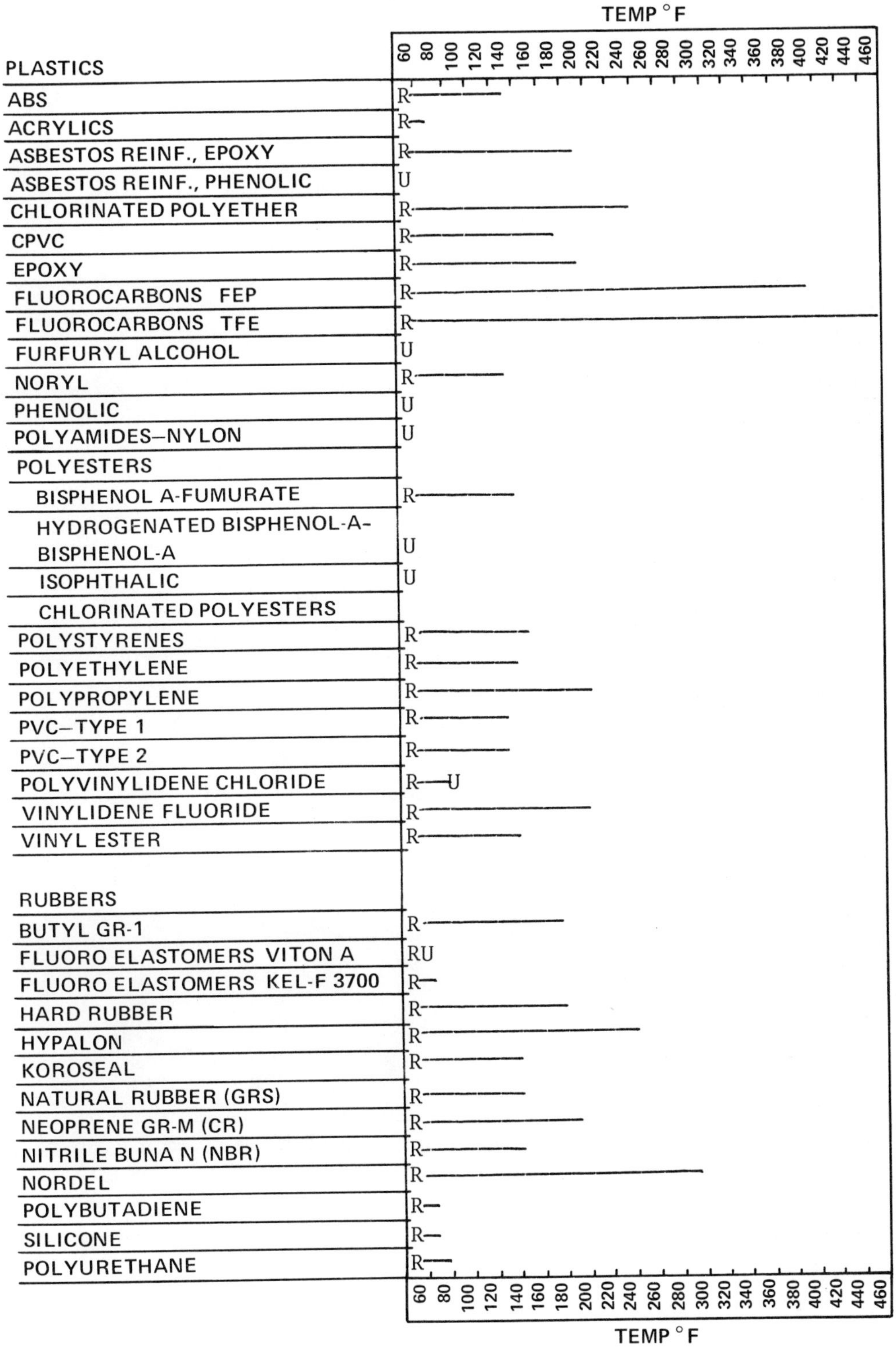

SODIUM HYDROXIDE, 10%

TEMP °F: 60, 80, 100, 120, 140, 160, 180, 200, 220, 240, 260, 280, 300, 320, 340, 360, 380, 400, 420, 440, 460, 480, 500, 520, 540, 560

METALS		Rating
ALLOY 20Cb3		E — G —
ALUMINUM		U
BRASS		G —
BRONZE		E- G —
CARBON STEEL		G —
COPPER		E- G —
E-BRIGHT 26-1		E —
HASTELLOY-B	2	E — G
HASTELLOY-C		G —
HASTELLOY-D		G —
HIGH SILICON IRON		G — S — U
INCONEL		E — G —
LEAD		U
MONEL	2	E — G —
NICKEL	2	E —
STAINLESS STEELS		
Type 304/307		E —
Type 316		E —
TANTALUM		U
TITANIUM		E —
ZIRCONIUM		E —
NONMETALLICS		
BOROSILICATE GLASS		U
CARBON		R —

TEMP °F: 60, 80, 100, 120, 140, 160, 180, 200, 220, 240, 260, 280, 300, 320, 340, 360, 380, 400, 420, 440, 460, 480, 500, 520, 540, 560

FOR METALS

E = < 2 Mils Penetration/Year; G = < 20 Mils Penetration/Year
S = < 50 Mils Penetration/Year; U = > 50 Mils Penetration/Year

FOR NONMETALLICS

R = Resistant
U = Unsatisfactory

SODIUM HYDROXIDE, 10%

TEMP °F (60, 80, 100, 120, 140, 160, 180, 200, 220, 240, 260, 280, 300, 320, 340, 360, 380, 400, 420, 440, 460)

PLASTICS	
ABS	R—
ACRYLICS	R—
ASBESTOS REINF., EPOXY	R—
ASBESTOS REINF., PHENOLIC	U
CHLORINATED POLYETHER	R—
CPVC	R—
EPOXY	R— U
FLUOROCARBONS FEP	R—
FLUOROCARBONS TFE	R—
FURFURYL ALCOHOL	U
NORYL	R—
PHENOLIC	U
POLYAMIDES–NYLON	R—
POLYESTERS	
BISPHENOL A-FUMURATE	R— U
HYDROGENATED BISPHENOL-A-BISPHENOL-A	R—
ISOPHTHALIC	U
CHLORINATED POLYESTERS	R— U
POLYSTYRENES	R—
POLYETHYLENE	R—
POLYPROPYLENE	R—
PVC–TYPE 1	R—
PVC–TYPE 2	R—
POLYVINYLIDENE CHLORIDE	R— U
VINYLIDENE FLUORIDE	R—
VINYL ESTER	R—
RUBBERS	
BUTYL GR-1	R—
FLUORO ELASTOMERS VITON A	RU
FLUORO ELASTOMERS KEL-F 3700	R—
HARD RUBBER	R—
HYPALON	R—
KOROSEAL	R—
NATURAL RUBBER (GRS)	R—
NEOPRENE GR-M (CR)	R—
NITRILE BUNA N (NBR)	R—
NORDEL	R—
POLYBUTADIENE	R—
SILICONE	R—
POLYURETHANE	

TEMP °F (60, 80, 100, 120, 140, 160, 180, 200, 220, 240, 260, 280, 300, 320, 340, 360, 380, 400, 420, 440, 460)

SODIUM HYPOCHLORITE (CONC.)

TEMP °F: 60 80 100 120 140 160 180 200 220 240 260 280 300 320 340 360 380 400 420 440 460 480 500 520 540 560

METALS	
ALLOY 20Cb3	
ALUMINUM	G—
BRASS	G—
BRONZE	S—
CARBON STEEL	U
COPPER	S—
E-BRIGHT 26-1	
HASTELLOY-B	U
HASTELLOY-C	U
HASTELLOY-D	
HIGH SILICON IRON	G—
INCONEL	U
LEAD	U
MONEL	S—
NICKEL	U
STAINLESS STEELS	
Type 304/307	U
Type 316	U
TANTALUM	G————————
TITANIUM	G———————
ZIRCONIUM	

NONMETALLICS	
BOROSILICATE GLASS	R————
CARBON	U

TEMP °F: 60 80 100 120 140 160 180 200 220 240 260 280 300 320 340 360 380 400 420 440 460 480 500 520 540 560

FOR METALS

E = < 2 Mils Penetration/Year; G = < 20 Mils Penetration/Year

S = < 50 Mils Penetration/Year; U = > 50 Mils Penetration/Year

FOR NONMETALLICS

R = Resistant

U = Unsatisfactory

SODIUM HYPOCHLORITE (CONC.)

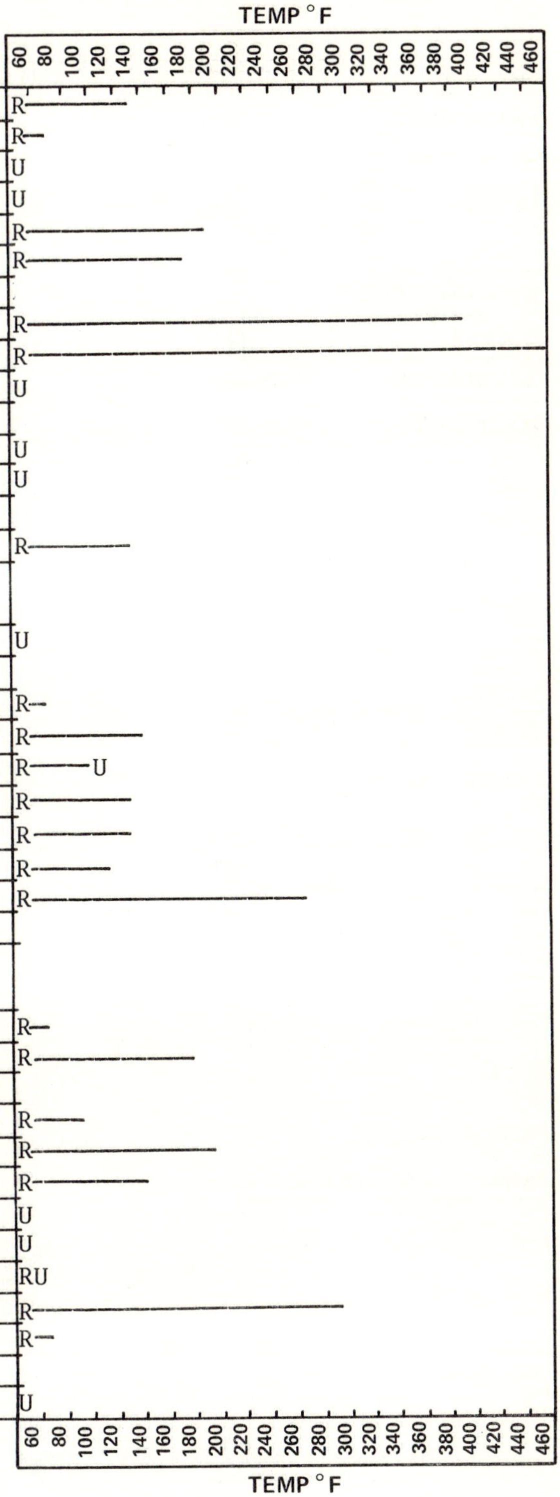

PLASTICS	
ABS	R
ACRYLICS	R
ASBESTOS REINF., EPOXY	U
ASBESTOS REINF., PHENOLIC	U
CHLORINATED POLYETHER	R
CPVC	R
EPOXY	
FLUOROCARBONS FEP	R
FLUOROCARBONS TFE	R
FURFURYL ALCOHOL	U
NORYL	
PHENOLIC	U
POLYAMIDES–NYLON	U
POLYESTERS	
BISPHENOL A-FUMURATE	R
HYDROGENATED BISPHENOL-A-BISPHENOL-A	
ISOPHTHALIC	U
CHLORINATED POLYESTERS	
POLYSTYRENES	R
POLYETHYLENE	R
POLYPROPYLENE	R U
PVC–TYPE 1	R
PVC–TYPE 2	R
POLYVINYLIDENE CHLORIDE	R
VINYLIDENE FLUORIDE	R
VINYL ESTER	

RUBBERS	
BUTYL GR-1	R
FLUORO ELASTOMERS VITON A	R
FLUORO ELASTOMERS KEL-F 3700	
HARD RUBBER	R
HYPALON	R
KOROSEAL	R
NATURAL RUBBER (GRS)	U
NEOPRENE GR-M (CR)	U
NITRILE BUNA N (NBR)	RU
NORDEL	R
POLYBUTADIENE	R
SILICONE	
POLYURETHANE	U

SODIUM HYPOCHLORITE

METALS	Concentration	TEMP °F (60–560)
ALLOY 20Cb3		S (60–80)
ALUMINUM		U
BRASS		U
BRONZE		U
CARBON STEEL		U
COPPER		U
E-BRIGHT 26-1	30%	G (60–80)
HASTELLOY-B		U
HASTELLOY-C	10%	E (60–100) — G (100–120)
HASTELLOY-D		U
HIGH SILICON IRON		G (60–80)
INCONEL		U
LEAD		U
MONEL		U
NICKEL		U
STAINLESS STEELS		
Type 304/307	10%	S (60–80)
Type 316	10%	S (60–80)
TANTALUM		G (60–200)
TITANIUM	6%	E (60–80)
ZIRCONIUM	6%	G (60–200)

NONMETALLICS	TEMP °F (60–560)
BOROSILICATE GLASS	R (60–140)
CARBON	U

FOR METALS

E = < 2 Mils Penetration/Year; G = < 20 Mils Penetration/Year

S = < 50 Mils Penetration/Year; U = > 50 Mils Penetration/Year

FOR NONMETALLICS

R = Resistant

U = Unsatisfactory

SODIUM HYPOCHLORITE

TEMP °F (60–460)

PLASTICS	Conc.	Rating
ABS		R
ACRYLICS		R
ASBESTOS REINF., EPOXY		
ASBESTOS REINF., PHENOLIC		
CHLORINATED POLYETHER		R
CPVC		R
EPOXY	5%	R–U
FLUOROCARBONS FEP		R
FLUOROCARBONS TFE		R
FURFURYL ALCOHOL	5%	U
NORYL	20%	R
PHENOLIC		
POLYAMIDES—NYLON	10%	R
POLYESTERS		
BISPHENOL A-FUMURATE	15%	U
HYDROGENATED BISPHENOL-A-BISPHENOL-A	10%	R
ISOPHTHALIC		U
CHLORINATED POLYESTERS		
POLYSTYRENES		
POLYETHYLENE		R
POLYPROPYLENE		R
PVC—TYPE 1		R
PVC—TYPE 2		R
POLYVINYLIDENE CHLORIDE	10%	R
VINYLIDENE FLUORIDE		R
VINYL ESTER	15%	R

RUBBERS	Conc.	Rating
BUTYL GR-1		R
FLUORO ELASTOMERS VITON A		R
FLUORO ELASTOMERS KEL-F 3700		
HARD RUBBER		R
HYPALON		R
KOROSEAL		R
NATURAL RUBBER (GRS)		R
NEOPRENE GR-M (CR)		R
NITRILE BUNA N (NBR)		RU
NORDEL		R
POLYBUTADIENE		R
SILICONE		
POLYURETHANE		

SODIUM IODIDE

TEMP °F

METALS	60	80	100	120	140	160	180	200	220	240	260	280	300	320	340	360	380	400	420	440	460	480	500	520	540	560
ALLOY 20Cb3																										
ALUMINUM																										
BRASS																										
BRONZE																										
CARBON STEEL																										
COPPER																										
E-BRIGHT 26-1																										
HASTELLOY-B																										
HASTELLOY-C																										
HASTELLOY-D																										
HIGH SILICON IRON																										
INCONEL																										
LEAD																										
MONEL																										
NICKEL																										
STAINLESS STEELS																										
Type 304/307																										
Type 316																										
TANTALUM	G	—																								
TITANIUM																										
ZIRCONIUM																										

NONMETALLICS	60	80	100	120	140	160	180	200	220	240	260	280	300	320	340	360	380	400	420	440	460	480	500	520	540	560
BOROSILICATE GLASS																										
CARBON	R	—	—	—	—	—	—	—	—	—	—	—	—	—	—											

TEMP °F

FOR METALS

E = < 2 Mils Penetration/Year; G = < 20 Mils Penetration/Year
S = < 50 Mils Penetration/Year; U = > 50 Mils Penetration/Year

FOR NONMETALLICS

R = Resistant
U = Unsatisfactory

SODIUM IODIDE

TEMP °F: 60 80 100 120 140 160 180 200 220 240 260 280 300 320 340 360 380 400 420 440 460

PLASTICS	TEMP °F
ABS	
ACRYLICS	
ASBESTOS REINF., EPOXY	
ASBESTOS REINF., PHENOLIC	
CHLORINATED POLYETHER	R
CPVC	
EPOXY	
FLUOROCARBONS FEP	R
FLUOROCARBONS TFE	R
FURFURYL ALCOHOL	
NORYL	
PHENOLIC	
POLYAMIDES–NYLON	
POLYESTERS	
BISPHENOL A-FUMURATE	
HYDROGENATED BISPHENOL-A-BISPHENOL-A	
ISOPHTHALIC	
CHLORINATED POLYESTERS	
POLYSTYRENES	
POLYETHYLENE	
POLYPROPYLENE	
PVC–TYPE 1	
PVC–TYPE 2	
POLYVINYLIDENE CHLORIDE	
VINYLIDENE FLUORIDE	R
VINYL ESTER	

RUBBERS	TEMP °F
BUTYL GR-1	
FLUORO ELASTOMERS VITON A	
FLUORO ELASTOMERS KEL-F 3700	
HARD RUBBER	
HYPALON	
KOROSEAL	
NATURAL RUBBER (GRS)	
NEOPRENE GR-M (CR)	
NITRILE BUNA N (NBR)	
NORDEL	
POLYBUTADIENE	
SILICONE	
POLYURETHANE	

60 80 100 120 140 160 180 200 220 240 260 280 300 320 340 360 380 400 420 440 460

TEMP °F

TEMP °F

METALS	Rating
ALLOY 20Cb3	G
ALUMINUM	U
BRASS	S
BRONZE	G
CARBON STEEL	U
COPPER	G
E-BRIGHT 26-1	
HASTELLOY-B	
HASTELLOY-C	
HASTELLOY-D	
HIGH SILICON IRON	G
INCONEL	
LEAD	
MONEL	G
NICKEL	G
STAINLESS STEELS	
Type 304/307	G
Type 316	G
TANTALUM	G
TITANIUM	
ZIRCONIUM	

NONMETALLICS	Rating
BOROSILICATE GLASS	R
CARBON	

Temperature scale (°F): 60, 80, 100, 120, 140, 160, 180, 200, 220, 240, 260, 280, 300, 320, 340, 360, 380, 400, 420, 440, 460, 480, 500, 520, 540, 560

TEMP °F

FOR METALS

E = < 2 Mils Penetration/Year; G = < 20 Mils Penetration/Year

S = < 50 Mils Penetration/Year; U = > 50 Mils Penetration/Year

FOR NONMETALLICS

R = Resistant

U = Unsatisfactory

SODIUM METAPHOSPHATE

PLASTICS	TEMP °F (60–460)
ABS	
ACRYLICS	
ASBESTOS REINF., EPOXY	
ASBESTOS REINF., PHENOLIC	
CHLORINATED POLYETHER	
CPVC	
EPOXY	
FLUOROCARBONS FEP	R
FLUOROCARBONS TFE	R
FURFURYL ALCOHOL	
NORYL	
PHENOLIC	
POLYAMIDES—NYLON	R
POLYESTERS	
BISPHENOL A-FUMURATE	
HYDROGENATED BISPHENOL-A-BISPHENOL-A	
ISOPHTHALIC	
CHLORINATED POLYESTERS	
POLYSTYRENES	
POLYETHYLENE	
POLYPROPYLENE	R
PVC—TYPE 1	
PVC—TYPE 2	
POLYVINYLIDENE CHLORIDE	
VINYLIDENE FLUORIDE	
VINYL ESTER	

RUBBERS	TEMP °F (60–460)
BUTYL GR-1	R
FLUORO ELASTOMERS VITON A	R
FLUORO ELASTOMERS KEL-F 3700	
HARD RUBBER	R
HYPALON	R
KOROSEAL	
NATURAL RUBBER (GRS)	R
NEOPRENE GR-M (CR)	R
NITRILE BUNA N (NBR)	R
NORDEL	
POLYBUTADIENE	
SILICONE	
POLYURETHANE	

SODIUM NITRATE

TEMP °F: 60, 80, 100, 120, 140, 160, 180, 200, 220, 240, 260, 280, 300, 320, 340, 360, 380, 400, 420, 440, 460, 480, 500, 520, 540, 560

METALS		Rating
ALLOY 20Cb3	to 60%	E
ALUMINUM	to 40%	E
BRASS	to 40%	G
BRONZE	to 30%	G
CARBON STEEL		G
COPPER	to 30%	G
E-BRIGHT 26-1		
HASTELLOY-B		U
HASTELLOY-C	10-30%	G
HASTELLOY-D		U
HIGH SILICON IRON	50%	E
INCONEL	to 40%	E
LEAD		U
MONEL	to 40%	G
NICKEL	to 40%	G
STAINLESS STEELS		
Type 304/307	to 50%	E
Type 316	to 60%	E
TANTALUM		G
TITANIUM	sat	E
ZIRCONIUM		G

NONMETALLICS	Rating
BOROSILICATE GLASS	R
CARBON	R

TEMP °F

FOR METALS

E = < 2 Mils Penetration/Year; G = < 20 Mils Penetration/Year

S = < 50 Mils Penetration/Year; U = > 50 Mils Penetration/Year

FOR NONMETALLICS

R = Resistant

U = Unsatisfactory

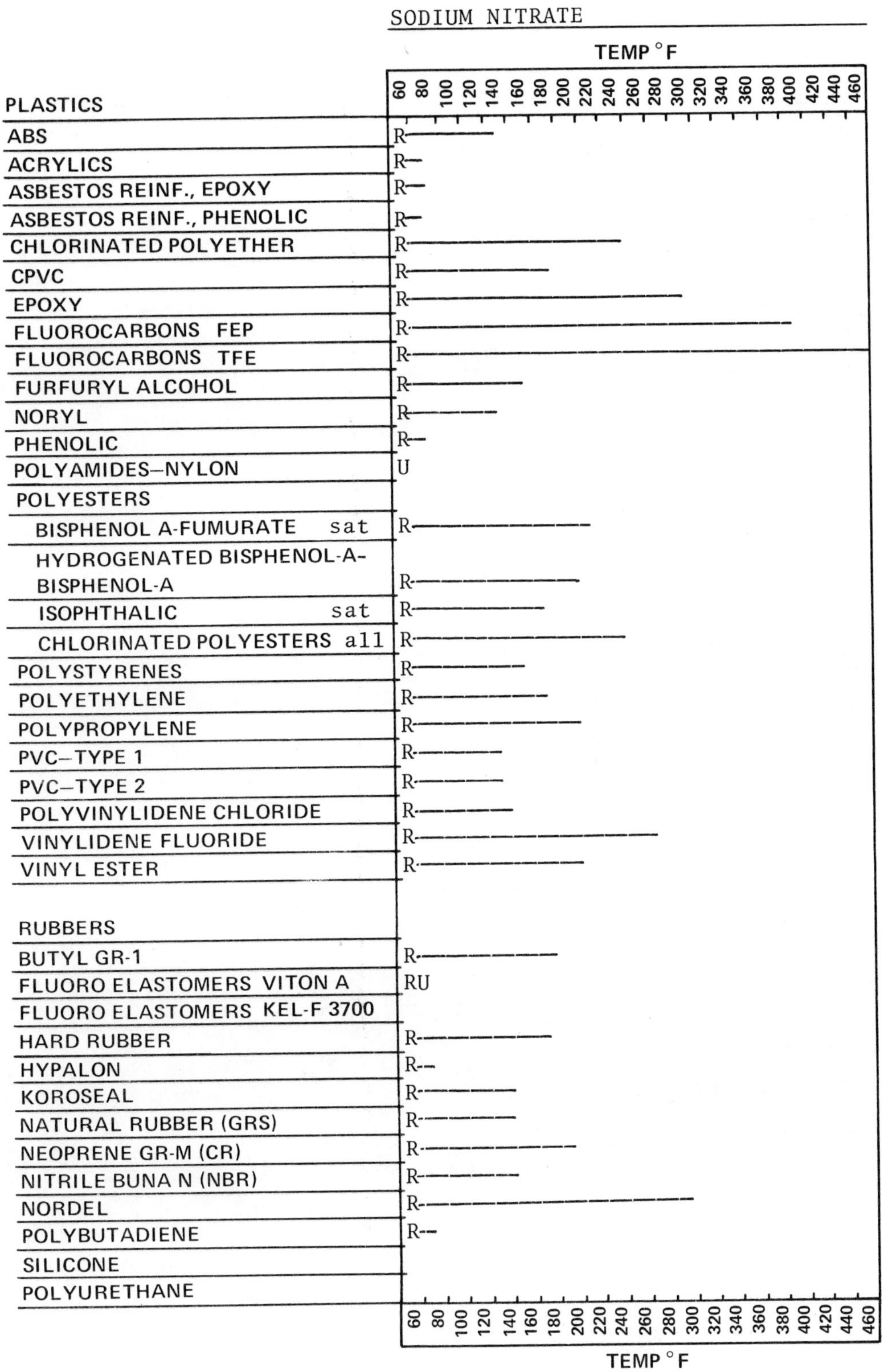
SODIUM NITRATE
TEMP °F
60 80 100 120 140 160 180 200 220 240 260 280 300 320 340 360 380 400 420 440 460
PLASTICS
ABS R
ACRYLICS R
ASBESTOS REINF., EPOXY R
ASBESTOS REINF., PHENOLIC R
CHLORINATED POLYETHER R
CPVC R
EPOXY R
FLUOROCARBONS FEP R
FLUOROCARBONS TFE R
FURFURYL ALCOHOL R
NORYL R
PHENOLIC R
POLYAMIDES–NYLON U
POLYESTERS
BISPHENOL A-FUMURATE sat R
HYDROGENATED BISPHENOL-A-BISPHENOL-A R
ISOPHTHALIC sat R
CHLORINATED POLYESTERS all R
POLYSTYRENES R
POLYETHYLENE R
POLYPROPYLENE R
PVC–TYPE 1 R
PVC–TYPE 2 R
POLYVINYLIDENE CHLORIDE R
VINYLIDENE FLUORIDE R
VINYL ESTER R
RUBBERS
BUTYL GR-1 R
FLUORO ELASTOMERS VITON A RU
FLUORO ELASTOMERS KEL-F 3700
HARD RUBBER R
HYPALON R
KOROSEAL R
NATURAL RUBBER (GRS) R
NEOPRENE GR-M (CR) R
NITRILE BUNA N (NBR) R
NORDEL R
POLYBUTADIENE R
SILICONE
POLYURETHANE
60 80 100 120 140 160 180 200 220 240 260 280 300 320 340 360 380 400 420 440 460
TEMP °F

SODIUM NITRITE

TEMP °F

METALS	Concentration	60 80 100 120 140 160 180 200 220 240 260 280 300 320 340 360 380 400 420 440 460 480 500 520 540 560
ALLOY 20Cb3	to 40%	G——————
ALUMINUM	to 40% 1	G——————
BRASS		
BRONZE		
CARBON STEEL		
COPPER		
E-BRIGHT 26-1		
HASTELLOY-B	to 40%	G——————
HASTELLOY-C	to 40%	G——————
HASTELLOY-D	to 40%	G——————
HIGH SILICON IRON		
INCONEL	to 40%	G——————
LEAD		
MONEL	to 40%	G——————
NICKEL	to 40%	G——————
STAINLESS STEELS		
Type 304/307	to 40%	G——————
Type 316	to 40%	G——————
TANTALUM		G——————
TITANIUM	sat	E——————
ZIRCONIUM	to 40%	E——————

NONMETALLICS		60 80 100 120 140 160 180 200 220 240 260 280 300 320 340 360 380 400 420 440 460 480 500 520 540 560
BOROSILICATE GLASS		
CARBON		

TEMP °F

FOR METALS

E = < 2 Mils Penetration/Year; G = < 20 Mils Penetration/Year

S = < 50 Mils Penetration/Year; U = > 50 Mils Penetration/Year

FOR NONMETALLICS

R = Resistant

U = Unsatisfactory

SODIUM NITRITE

TEMP °F: 60 80 100 120 140 160 180 200 220 240 260 280 300 320 340 360 380 400 420 440 460

PLASTICS		
ABS		R
ACRYLICS		
ASBESTOS REINF., EPOXY		
ASBESTOS REINF., PHENOLIC		
CHLORINATED POLYETHER		R
CPVC		R
EPOXY		R
FLUOROCARBONS FEP		R
FLUOROCARBONS TFE		R
FURFURYL ALCOHOL		
NORYL		
PHENOLIC		
POLYAMIDES–NYLON		
POLYESTERS		
BISPHENOL A-FUMURATE		R
HYDROGENATED BISPHENOL-A-BISPHENOL-A		R
ISOPHTHALIC	sat	R
CHLORINATED POLYESTERS		
POLYSTYRENES		
POLYETHYLENE		R
POLYPROPYLENE		R
PVC–TYPE 1		R
PVC–TYPE 2		R
POLYVINYLIDENE CHLORIDE		R
VINYLIDENE FLUORIDE		R
VINYL ESTER	all	R

RUBBERS	
BUTYL GR-1	
FLUORO ELASTOMERS VITON A	
FLUORO ELASTOMERS KEL-F 3700	
HARD RUBBER	
HYPALON	
KOROSEAL	
NATURAL RUBBER (GRS)	
NEOPRENE GR-M (CR)	
NITRILE BUNA N (NBR)	
NORDEL	
POLYBUTADIENE	
SILICONE	
POLYURETHANE	

TEMP °F: 60 80 100 120 140 160 180 200 220 240 260 280 300 320 340 360 380 400 420 440 460

SODIUM PERBORATE, 10%

TEMP °F: 60, 80, 100, 120, 140, 160, 180, 200, 220, 240, 260, 280, 300, 320, 340, 360, 380, 400, 420, 440, 460, 480, 500, 520, 540, 560

METALS	
ALLOY 20Cb3	G
ALUMINUM	G
BRASS	S
BRONZE	G
CARBON STEEL	G
COPPER	G
E-BRIGHT 26-1	
HASTELLOY-B 10%	G
HASTELLOY-C	G
HASTELLOY-D	G
HIGH SILICON IRON	
INCONEL	G
LEAD	U
MONEL	G
NICKEL	G
STAINLESS STEELS	
Type 304/307	G
Type 316	G
TANTALUM	
TITANIUM	
ZIRCONIUM	

NONMETALLICS	
BOROSILICATE GLASS	R
CARBON	

TEMP °F

FOR METALS

E = < 2 Mils Penetration/Year; G = < 20 Mils Penetration/Year

S = < 50 Mils Penetration/Year; U = > 50 Mils Penetration/Year

FOR NONMETALLICS

R = Resistant

U = Unsatisfactory

SODIUM PERBORATE, 10%

TEMP °F (60–460)

PLASTICS	
ABS	
ACRYLICS	
ASBESTOS REINF., EPOXY	
ASBESTOS REINF., PHENOLIC	
CHLORINATED POLYETHER	R
CPVC	
EPOXY	
FLUOROCARBONS FEP	R
FLUOROCARBONS TFE	R
FURFURYL ALCOHOL	
NORYL	
PHENOLIC	
POLYAMIDES–NYLON	
POLYESTERS	
BISPHENOL A-FUMURATE	
HYDROGENATED BISPHENOL-A-BISPHENOL-A	
ISOPHTHALIC	
CHLORINATED POLYESTERS	
POLYSTYRENES	
POLYETHYLENE	R
POLYPROPYLENE	R
PVC–TYPE 1	R
PVC–TYPE 2	
POLYVINYLIDENE CHLORIDE	R
VINYLIDENE FLUORIDE	
VINYL ESTER	

RUBBERS	
BUTYL GR-1	R
FLUORO ELASTOMERS VITON A	R
FLUORO ELASTOMERS KEL-F 3700	
HARD RUBBER	
HYPALON	R
KOROSEAL	
NATURAL RUBBER (GRS)	R
NEOPRENE GR-M (CR)	R
NITRILE BUNA N (NBR)	R
NORDEL	
POLYBUTADIENE	
SILICONE	
POLYURETHANE	

TEMP °F (60–460)

SODIUM PEROXIDE, 10%

TEMP °F: 60 80 100 120 140 160 180 200 220 240 260 280 300 320 340 360 380 400 420 440 460 480 500 520 540 560

METALS		
ALLOY 20Cb3		G
ALUMINUM		G
BRASS		U
BRONZE		U
CARBON STEEL		G
COPPER		U
E-BRIGHT 26-1		
HASTELLOY-B	10%	G
HASTELLOY-C	10%	G
HASTELLOY-D		G
HIGH SILICON IRON		G
INCONEL		G
LEAD		U
MONEL		G
NICKEL		G
STAINLESS STEELS		
Type 304/307		G
Type 316		G
TANTALUM		
TITANIUM		
ZIRCONIUM		

NONMETALLICS	
BOROSILICATE GLASS	R
CARBON	

TEMP °F: 60 80 100 120 140 160 180 200 220 240 260 280 300 320 340 360 380 400 420 440 460 480 500 520 540 560

FOR METALS

E = < 2 Mils Penetration/Year; G = < 20 Mils Penetration/Year

S = < 50 Mils Penetration/Year; U = > 50 Mils Penetration/Year

FOR NONMETALLICS

R = Resistant

U = Unsatisfactory

SODIUM PEROXIDE, 10%

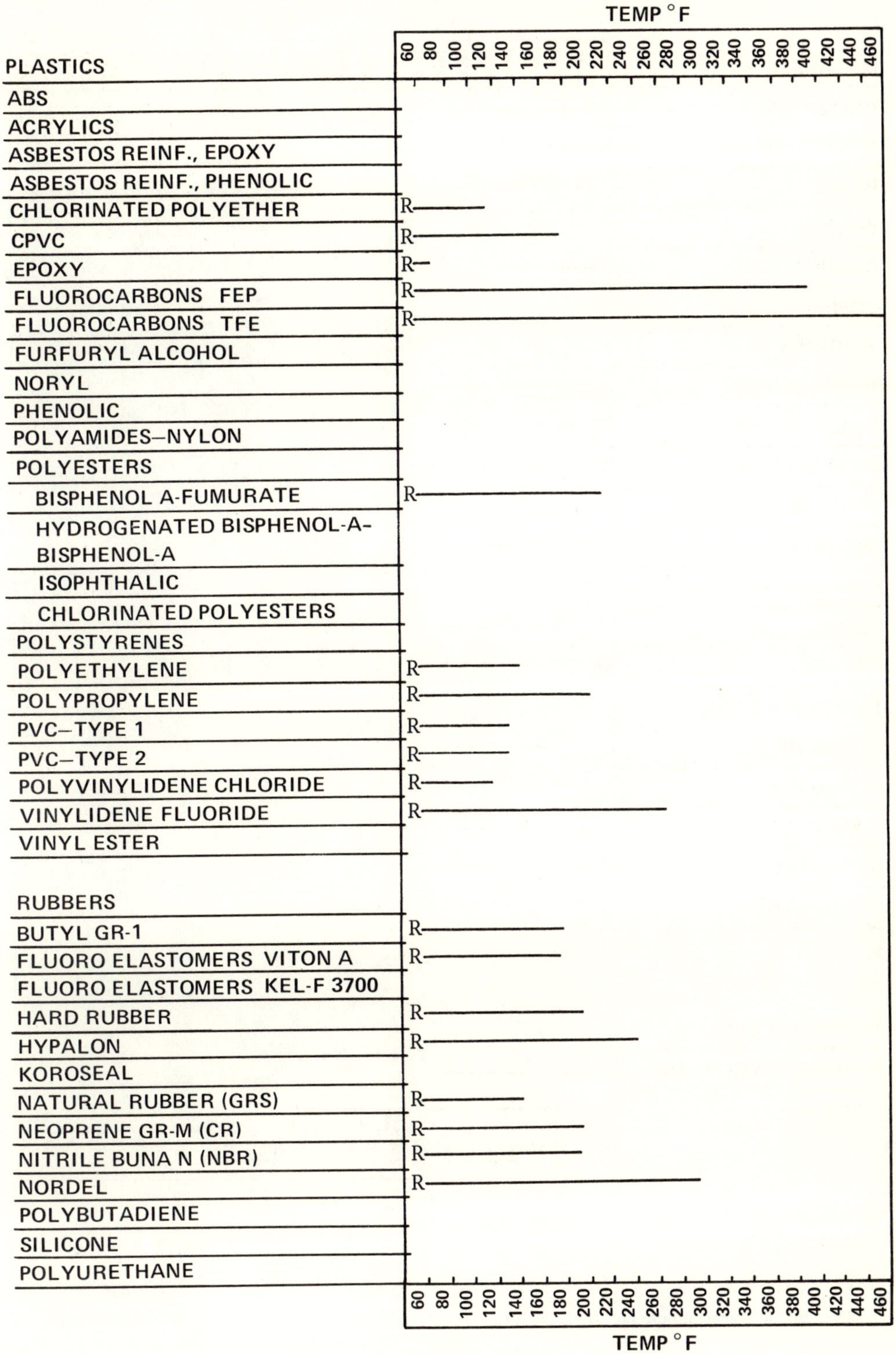

SODIUM PHOSPHATE ALKALINE

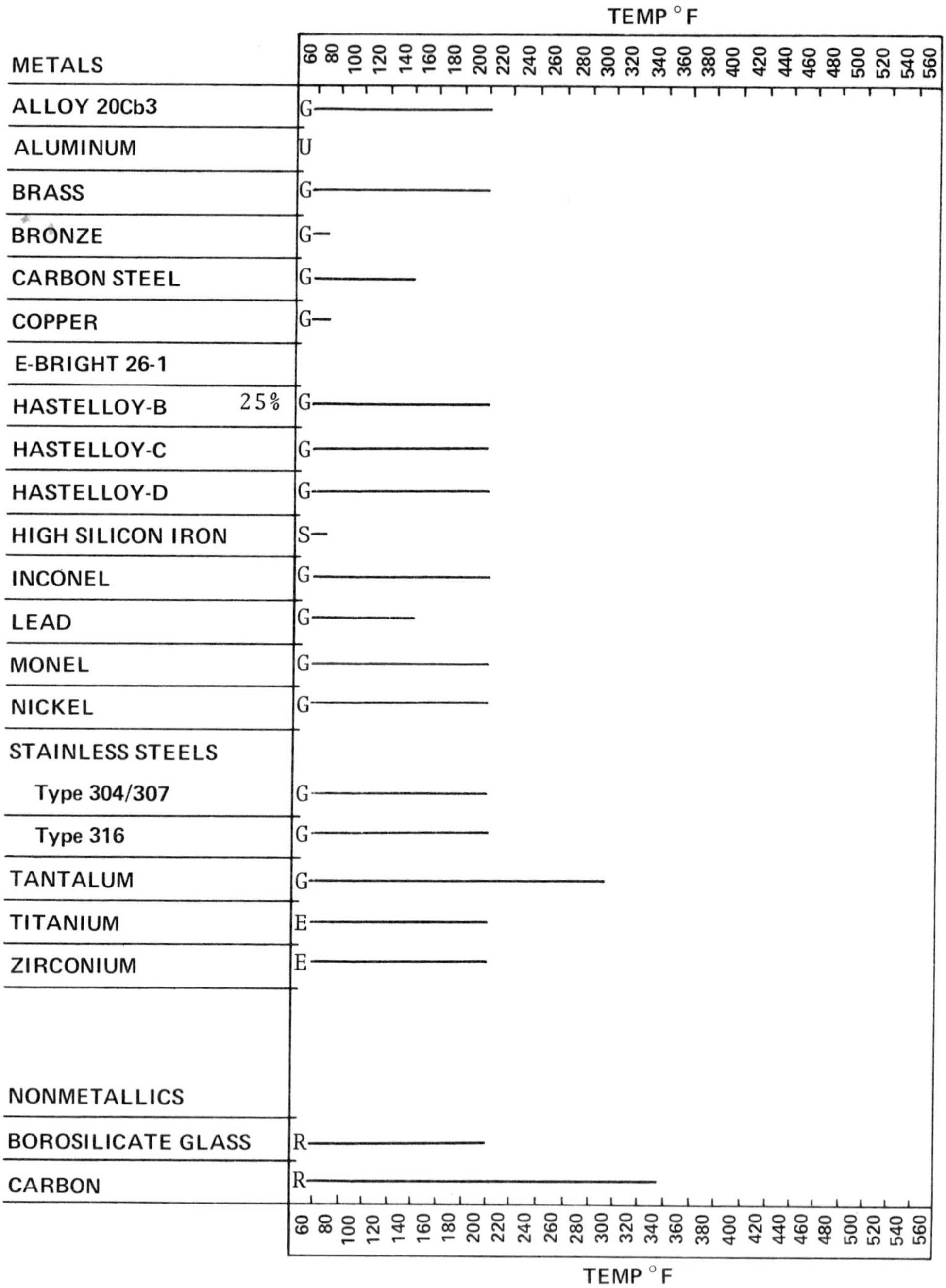

FOR METALS

E = < 2 Mils Penetration/Year; G = < 20 Mils Penetration/Year

S = < 50 Mils Penetration/Year; U = > 50 Mils Penetration/Year

FOR NONMETALLICS

R = Resistant

U = Unsatisfactory

SODIUM PHOSPHATE ALKALINE

TEMP °F: 60 80 100 120 140 160 180 200 220 240 260 280 300 320 340 360 380 400 420 440 460

PLASTICS	
ABS	
ACRYLICS	R
ASBESTOS REINF., EPOXY	
ASBESTOS REINF., PHENOLIC	
CHLORINATED POLYETHER	R
CPVC	
EPOXY	U
FLUOROCARBONS FEP	R
FLUOROCARBONS TFE	R
FURFURYL ALCOHOL	
NORYL	
PHENOLIC	
POLYAMIDES–NYLON	R
POLYESTERS	
BISPHENOL A-FUMURATE	
HYDROGENATED BISPHENOL-A-BISPHENOL-A	
ISOPHTHALIC	
CHLORINATED POLYESTERS	
POLYSTYRENES	
POLYETHYLENE	R
POLYPROPYLENE	R
PVC–TYPE 1	
PVC–TYPE 2	
POLYVINYLIDENE CHLORIDE	R
VINYLIDENE FLUORIDE	R
VINYL ESTER	

RUBBERS	
BUTYL GR-1	R
FLUORO ELASTOMERS VITON A	R
FLUORO ELASTOMERS KEL-F 3700	
HARD RUBBER	R
HYPALON	R
KOROSEAL	
NATURAL RUBBER (GRS)	R
NEOPRENE GR-M (CR)	R
NITRILE BUNA N (NBR)	R
NORDEL	
POLYBUTADIENE	R
SILICONE	
POLYURETHANE	

TEMP °F: 60 80 100 120 140 160 180 200 220 240 260 280 300 320 340 360 380 400 420 440 460

SODIUM PHOSPHATE ACID

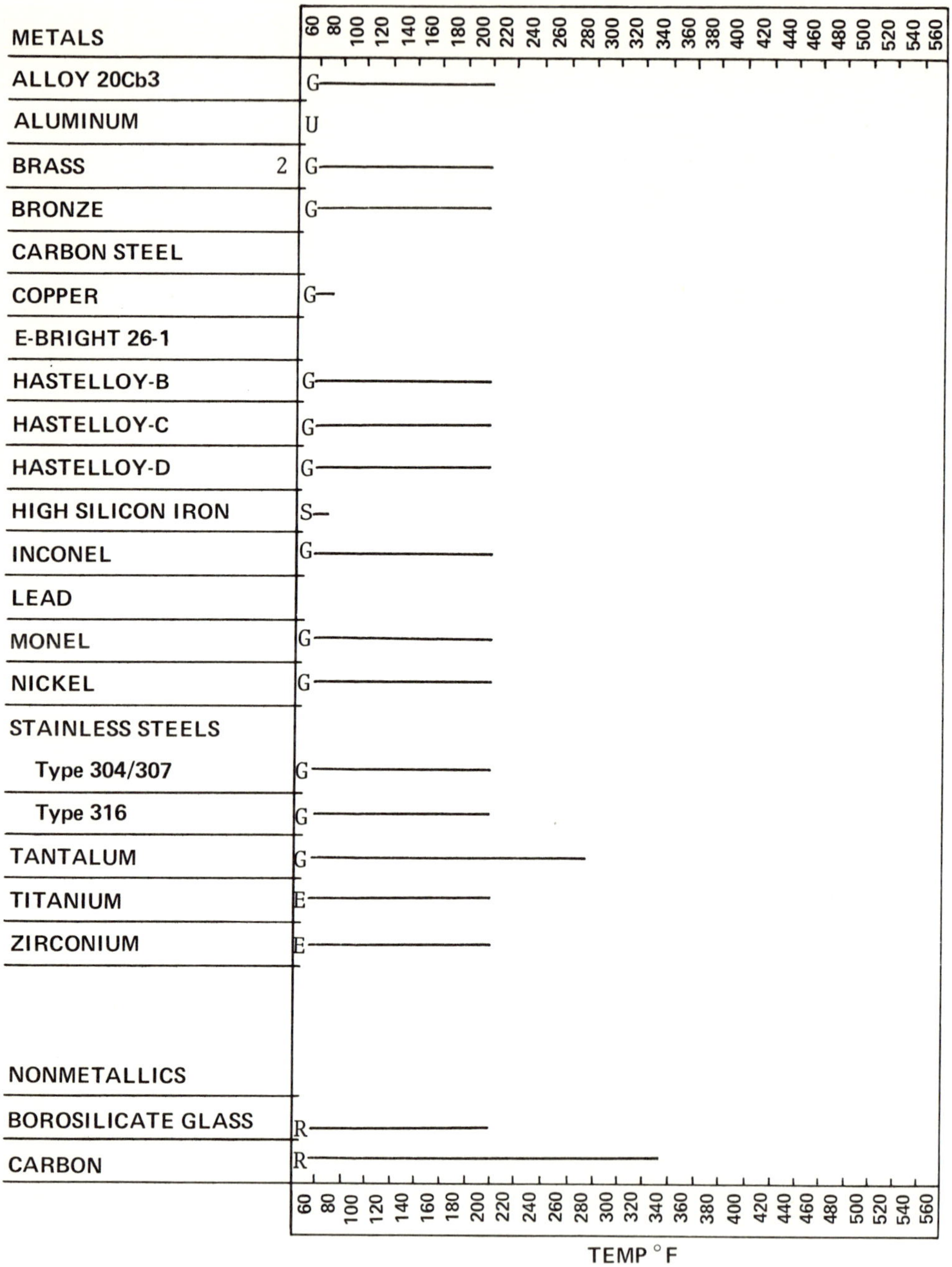

FOR METALS

E = < 2 Mils Penetration/Year; G = < 20 Mils Penetration/Year

S = < 50 Mils Penetration/Year; U = > 50 Mils Penetration/Year

FOR NONMETALLICS

R = Resistant

U = Unsatisfactory

SODIUM PHOSPHATE ACID

TEMP °F (60–460)

PLASTICS	Rating
ABS	R
ACRYLICS	R
ASBESTOS REINF., EPOXY	
ASBESTOS REINF., PHENOLIC	
CHLORINATED POLYETHER	R
CPVC	
EPOXY	U
FLUOROCARBONS FEP	R
FLUOROCARBONS TFE	R
FURFURYL ALCOHOL	
NORYL	
PHENOLIC	
POLYAMIDES–NYLON	R
POLYESTERS	
BISPHENOL A-FUMURATE	
HYDROGENATED BISPHENOL-A-BISPHENOL-A	
ISOPHTHALIC	
CHLORINATED POLYESTERS	
POLYSTYRENES	
POLYETHYLENE	R
POLYPROPYLENE	R
PVC–TYPE 1	R
PVC–TYPE 2	R
POLYVINYLIDENE CHLORIDE	R
VINYLIDENE FLUORIDE	R
VINYL ESTER	

RUBBERS	Rating
BUTYL GR-1	
FLUORO ELASTOMERS VITON A	R
FLUORO ELASTOMERS KEL-F 3700	
HARD RUBBER	
HYPALON	R
KOROSEAL	
NATURAL RUBBER (GRS)	R
NEOPRENE GR-M (CR)	
NITRILE BUNA N (NBR)	R
NORDEL	
POLYBUTADIENE	R
SILICONE	
POLYURETHANE	

TEMP °F

SODIUM PHOSPHATE NEUTRAL

TEMP °F (scale 60–560)

METALS	Rating	Up to (°F)
ALLOY 20Cb3	G	200
ALUMINUM	U	
BRASS	G	200
BRONZE	G	200
CARBON STEEL		
COPPER	G	80
E-BRIGHT 26-1		
HASTELLOY-B	G	200
HASTELLOY-C	G	200
HASTELLOY-D	G	200
HIGH SILICON IRON	S	80
INCONEL	G	200
LEAD		
MONEL	G	200
NICKEL	G	200
STAINLESS STEELS		
Type 304/307	G	200
Type 316	G	200
TANTALUM	G	300
TITANIUM	G	200
ZIRCONIUM	G	200

NONMETALLICS	Rating	Up to (°F)
BOROSILICATE GLASS	R	200
CARBON	R	340

TEMP °F

FOR METALS

E = < 2 Mils Penetration/Year; G = < 20 Mils Penetration/Year

S = < 50 Mils Penetration/Year; U = > 50 Mils Penetration/Year

FOR NONMETALLICS

R = Resistant

U = Unsatisfactory

SODIUM PHOSPHATE NEUTRAL

TEMP °F: 60, 80, 100, 120, 140, 160, 180, 200, 220, 240, 260, 280, 300, 320, 340, 360, 380, 400, 420, 440, 460

PLASTICS	
ABS	
ACRYLICS	R
ASBESTOS REINF., EPOXY	R
ASBESTOS REINF., PHENOLIC	R
CHLORINATED POLYETHER	R
CPVC	
EPOXY	U
FLUOROCARBONS FEP	R
FLUOROCARBONS TFE	R
FURFURYL ALCOHOL	U
NORYL	
PHENOLIC	R
POLYAMIDES–NYLON	R
POLYESTERS	
BISPHENOL A-FUMURATE	
HYDROGENATED BISPHENOL-A-BISPHENOL-A	
ISOPHTHALIC	
CHLORINATED POLYESTERS	
POLYSTYRENES	U
POLYETHYLENE	R
POLYPROPYLENE	R
PVC–TYPE 1	R
PVC–TYPE 2	
POLYVINYLIDENE CHLORIDE	R
VINYLIDENE FLUORIDE	R
VINYL ESTER	

RUBBERS	
BUTYL GR-1	
FLUORO ELASTOMERS VITON A	R
FLUORO ELASTOMERS KEL-F 3700	
HARD RUBBER	
HYPALON	R
KOROSEAL	
NATURAL RUBBER (GRS)	R
NEOPRENE GR-M (CR)	
NITRILE BUNA N (NBR)	R
NORDEL	
POLYBUTADIENE	R
SILICONE	
POLYURETHANE	

TEMP °F

SODIUM SILICATE

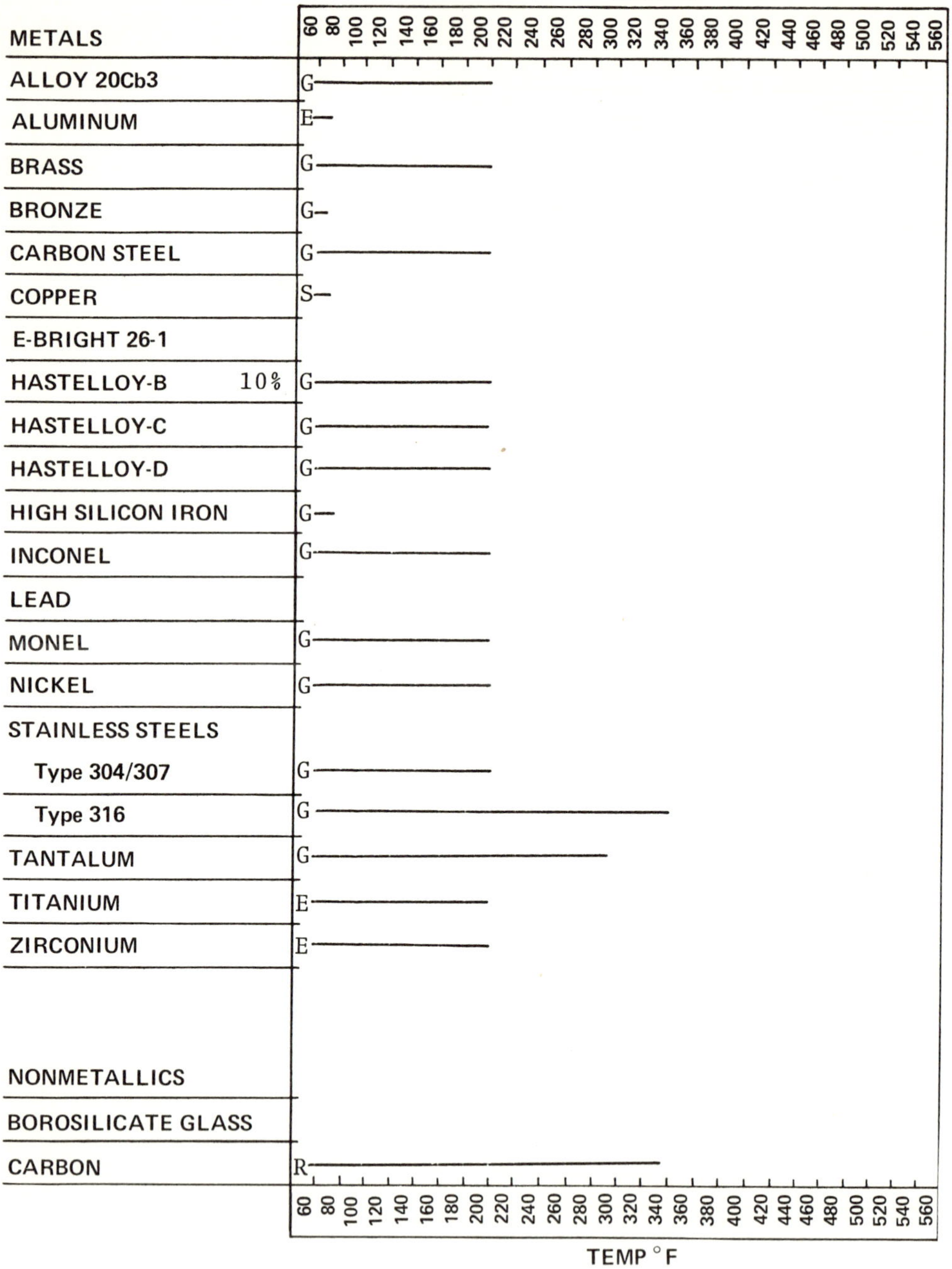

FOR METALS

E = < 2 Mils Penetration/Year; G = < 20 Mils Penetration/Year

S = < 50 Mils Penetration/Year; U = > 50 Mils Penetration/Year

FOR NONMETALLICS

R = Resistant

U = Unsatisfactory

SODIUM SILICATE

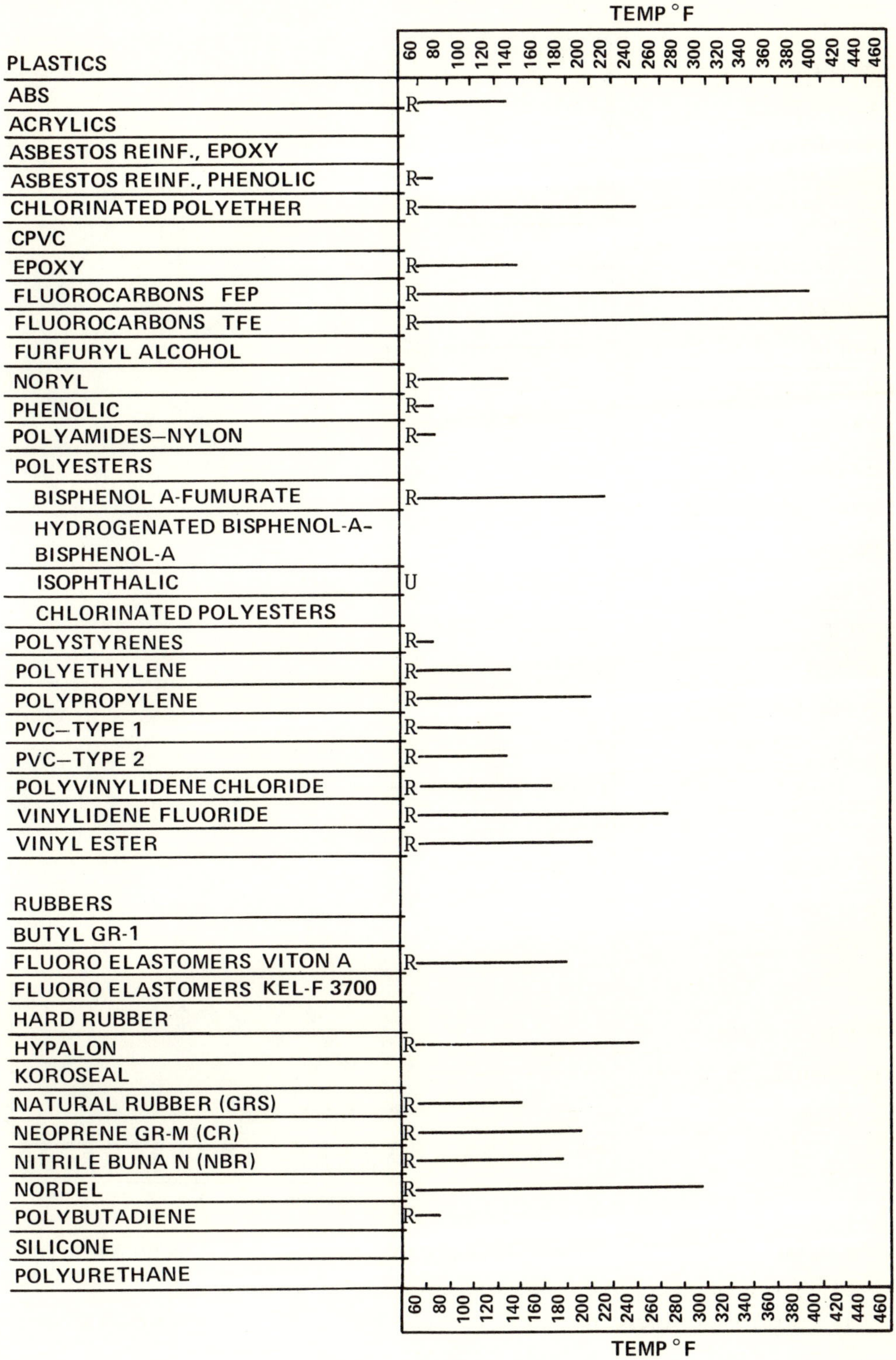

SODIUM SULFATE

METALS		TEMP °F (60–560)
ALLOY 20Cb3		E
ALUMINUM	30%	E
BRASS		G
BRONZE		G
CARBON STEEL		G
COPPER		G
E-BRIGHT 26-1	10%	E
HASTELLOY-B		G
HASTELLOY-C		G
HASTELLOY-D		G
HIGH SILICON IRON		G
INCONEL		G
LEAD		S
MONEL	30%	G
NICKEL		G
STAINLESS STEELS		
Type 304/307	2	E
Type 316		E
TANTALUM	10-20%	E
TITANIUM	10-20%	E
ZIRCONIUM		
NONMETALLICS		
BOROSILICATE GLASS		R
CARBON		R

TEMP °F: 60 80 100 120 140 160 180 200 220 240 260 280 300 320 340 360 380 400 420 440 460 480 500 520 540 560

FOR METALS

E = < 2 Mils Penetration/Year; G = < 20 Mils Penetration/Year

S = < 50 Mils Penetration/Year; U = > 50 Mils Penetration/Year

FOR NONMETALLICS

R = Resistant

U = Unsatisfactory

SODIUM SULFATE

PLASTICS		
ABS		R
ACRYLICS		R
ASBESTOS REINF., EPOXY		R
ASBESTOS REINF., PHENOLIC		R
CHLORINATED POLYETHER		R
CPVC		R
EPOXY		R
FLUOROCARBONS FEP		R
FLUOROCARBONS TFE		R
FURFURYL ALCOHOL	sat	R
NORYL		R
PHENOLIC		
POLYAMIDES–NYLON		R
POLYESTERS		
BISPHENOL A-FUMURATE	sat	R
HYDROGENATED BISPHENOL-A-BISPHENOL-A		R
ISOPHTHALIC		R
CHLORINATED POLYESTERS	sat	R — U
POLYSTYRENES		
POLYETHYLENE		R
POLYPROPYLENE		R
PVC–TYPE 1		R
PVC–TYPE 2		R
POLYVINYLIDENE CHLORIDE		R
VINYLIDENE FLUORIDE		R
VINYL ESTER	all	R
RUBBERS		
BUTYL GR-1		R
FLUORO ELASTOMERS VITON A		R
FLUORO ELASTOMERS KEL-F 3700		
HARD RUBBER		R
HYPALON		R
KOROSEAL		R
NATURAL RUBBER (GRS)		R
NEOPRENE GR-M (CR)		R
NITRILE BUNA N (NBR)		R
NORDEL		R
POLYBUTADIENE		
SILICONE		
POLYURETHANE		R

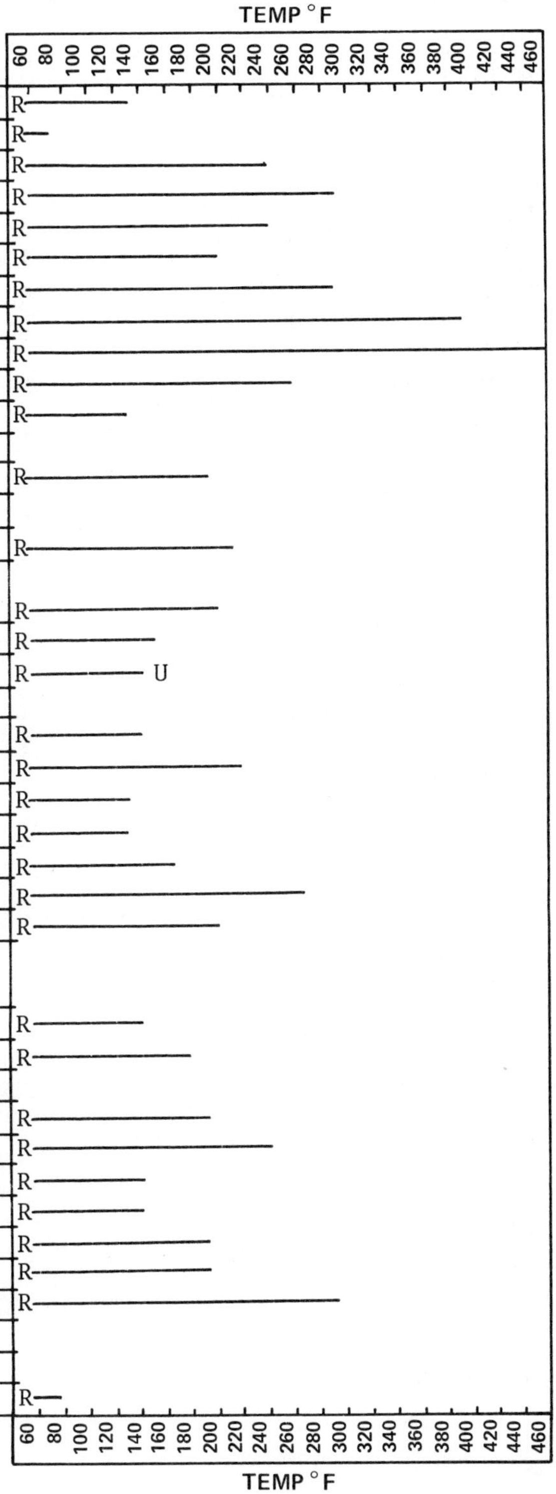

SODIUM SULFIDE TO 50%

TEMP °F: 60 80 100 120 140 160 180 200 220 240 260 280 300 320 340 360 380 400 420 440 460 480 500 520 540 560

METALS		Rating (TEMP °F)
ALLOY 20Cb3		G 60–200
ALUMINUM		U
BRASS		U
BRONZE		U
CARBON STEEL		G–
COPPER		U
E-BRIGHT 26-1		
HASTELLOY-B		G 60–200
HASTELLOY-C		G 60–200
HASTELLOY-D		G 60–200
HIGH SILICON IRON		G–
INCONEL		G 60–200
LEAD		S–
MONEL		G 60–200
NICKEL		G 60–200; S 220–300
STAINLESS STEELS		
Type 304/307	1	G 60–200; U 220
Type 316		G 60–200; U 220
TANTALUM	to 10%	G 60–200
TITANIUM	10%	E 60–200
ZIRCONIUM		U

NONMETALLICS	
BOROSILICATE GLASS	
CARBON	

TEMP °F: 60 80 100 120 140 160 180 200 220 240 260 280 300 320 340 360 380 400 420 440 460 480 500 520 540 560

FOR METALS

E = < 2 Mils Penetration/Year; G = < 20 Mils Penetration/Year

S = < 50 Mils Penetration/Year; U = > 50 Mils Penetration/Year

FOR NONMETALLICS

R = Resistant

U = Unsatisfactory

SODIUM SULFIDE TO 50%

TEMP °F: 60 80 100 120 140 160 180 200 220 240 260 280 300 320 340 360 380 400 420 440 460

PLASTICS		
ABS		R
ACRYLICS		
ASBESTOS REINF., EPOXY		R
ASBESTOS REINF., PHENOLIC		
CHLORINATED POLYETHER		R
CPVC		R
EPOXY	to 10%	R
FLUOROCARBONS FEP		R
FLUOROCARBONS TFE		R
FURFURYL ALCOHOL	10%	R
NORYL	25%	R
PHENOLIC		
POLYAMIDES—NYLON		R
POLYESTERS		
BISPHENOL A-FUMURATE	sat	R — U
HYDROGENATED BISPHENOL-A-BISPHENOL-A		
ISOPHTHALIC	sat	U
CHLORINATED POLYESTERS		U
POLYSTYRENES		R
POLYETHYLENE		R — U
POLYPROPYLENE		R — U
PVC—TYPE 1		R
PVC—TYPE 2		R
POLYVINYLIDENE CHLORIDE		R
VINYLIDENE FLUORIDE		R
VINYL ESTER	all	R
RUBBERS		
BUTYL GR-1		
FLUORO ELASTOMERS VITON A		R
FLUORO ELASTOMERS KEL-F 3700		
HARD RUBBER		R
HYPALON		R
KOROSEAL		R
NATURAL RUBBER (GRS)		R
NEOPRENE GR-M (CR)		R
NITRILE BUNA N (NBR)		R
NORDEL		R
POLYBUTADIENE		R
SILICONE		
POLYURETHANE		

TEMP °F: 60 80 100 120 140 160 180 200 220 240 260 280 300 320 340 360 380 400 420 440 460

SODIUM SULFITE, 10%

TEMP °F: 60 80 100 120 140 160 180 200 220 240 260 280 300 320 340 360 380 400 420 440 460 480 500 520 540 560

METALS		Rating
ALLOY 20Cb3		E
ALUMINUM		G
BRASS		G
BRONZE		U
CARBON STEEL		G
COPPER		S
E-BRIGHT 26-1		
HASTELLOY-B		U
HASTELLOY-C		G
HASTELLOY-D		U
HIGH SILICON IRON		U
INCONEL		G
LEAD		G
MONEL		G
NICKEL		G
STAINLESS STEELS		
Type 304/307		E
Type 316		E
TANTALUM	to sat	G
TITANIUM	sat	E
ZIRCONIUM		

NONMETALLICS	Rating
BOROSILICATE GLASS	U
CARBON	R

TEMP °F: 60 80 100 120 140 160 180 200 220 240 260 280 300 320 340 360 380 400 420 440 460 480 500 520 540 560

FOR METALS

E = < 2 Mils Penetration/Year; G = < 20 Mils Penetration/Year

S = < 50 Mils Penetration/Year; U = > 50 Mils Penetration/Year

FOR NONMETALLICS

R = Resistant

U = Unsatisfactory

SODIUM SULFITE, 10%

PLASTICS		TEMP °F (60–460)
ABS		R
ACRYLICS		R
ASBESTOS REINF., EPOXY		R
ASBESTOS REINF., PHENOLIC		R
CHLORINATED POLYETHER		R
CPVC		R
EPOXY		R
FLUOROCARBONS FEP		R
FLUOROCARBONS TFE		R
FURFURYL ALCOHOL		R
NORYL		
PHENOLIC		R
POLYAMIDES—NYLON		R
POLYESTERS		
BISPHENOL A-FUMURATE	sat	R
HYDROGENATED BISPHENOL-A-BISPHENOL-A		
ISOPHTHALIC	sat	R
CHLORINATED POLYESTERS	all	R
POLYSTYRENES		
POLYETHYLENE		R
POLYPROPYLENE		R
PVC—TYPE 1		R
PVC—TYPE 2		R
POLYVINYLIDENE CHLORIDE		R
VINYLIDENE FLUORIDE		R
VINYL ESTER	all	R
RUBBERS		
BUTYL GR-1		R
FLUORO ELASTOMERS VITON A		R
FLUORO ELASTOMERS KEL-F 3700		
HARD RUBBER		R
HYPALON		R
KOROSEAL		R
NATURAL RUBBER (GRS)		R
NEOPRENE GR-M (CR)		R
NITRILE BUNA N (NBR)		R
NORDEL		R
POLYBUTADIENE		R
SILICONE		
POLYURETHANE		R

TEMP °F: 60 80 100 120 140 160 180 200 220 240 260 280 300 320 340 360 380 400 420 440 460

SODIUM THIOSULFATE (HYPO.)

TEMP °F (60–560)

METALS		Rating
ALLOY 20Cb3		G—
ALUMINUM		G–
BRASS		U
BRONZE		U
CARBON STEEL		U
COPPER		U
E-BRIGHT 26-1		
HASTELLOY-B		G—
HASTELLOY-C		G—
HASTELLOY-D		
HIGH SILICON IRON		G–
INCONEL		
LEAD		G–
MONEL		G–
NICKEL		
STAINLESS STEELS		
Type 304/307		G–
Type 316		G——
TANTALUM	to 25%	E——
TITANIUM	25%	E——
ZIRCONIUM		
NONMETALLICS		
BOROSILICATE GLASS		R——
CARBON		R——

TEMP °F

FOR METALS

E = < 2 Mils Penetration/Year; G = < 20 Mils Penetration/Year
S = < 50 Mils Penetration/Year; U = > 50 Mils Penetration/Year

FOR NONMETALLICS

R = Resistant
U = Unsatisfactory

SODIUM THIOSULFATE (HYPO.)

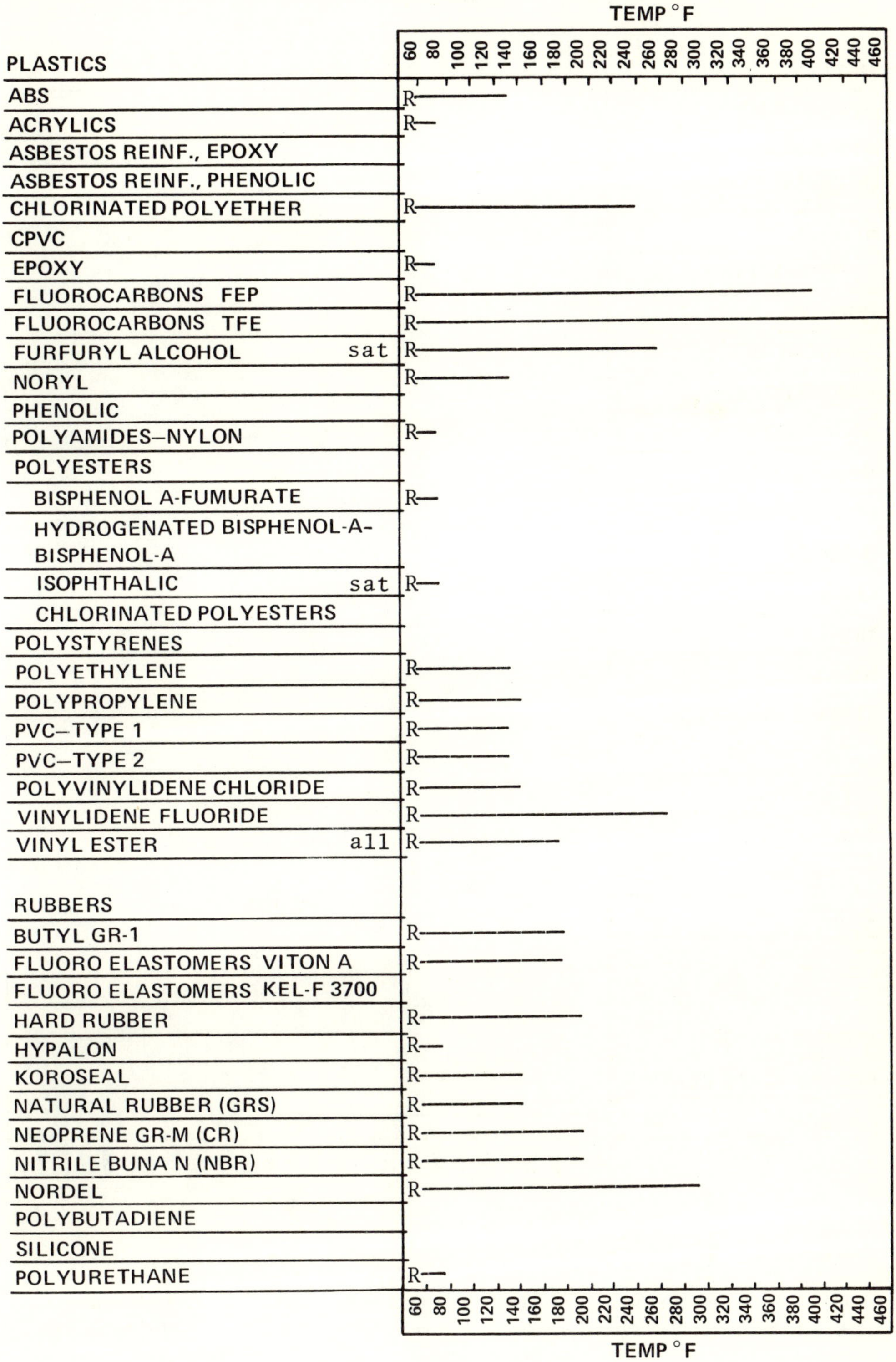

SOUR CRUDE OIL

TEMP °F: 60, 80, 100, 120, 140, 160, 180, 200, 220, 240, 260, 280, 300, 320, 340, 360, 380, 400, 420, 440, 460, 480, 500, 520, 540, 560

METALS		Rating
ALLOY 20Cb3		
ALUMINUM		
BRASS		
BRONZE		
CARBON STEEL		
COPPER		
E-BRIGHT 26-1		
HASTELLOY-B		
HASTELLOY-C		E
HASTELLOY-D	58%	G
HIGH SILICON IRON		
INCONEL		
LEAD		
MONEL		
NICKEL		
STAINLESS STEELS Type 304/307		
Type 316		
TANTALUM		
TITANIUM		
ZIRCONIUM		
NONMETALLICS		
BOROSILICATE GLASS		
CARBON		R

FOR METALS

E = < 2 Mils Penetration/Year; G = < 20 Mils Penetration/Year

S = < 50 Mils Penetration/Year; U = > 50 Mils Penetration/Year

FOR NONMETALLICS

R = Resistant

U = Unsatisfactory

SOUR CRUDE OIL

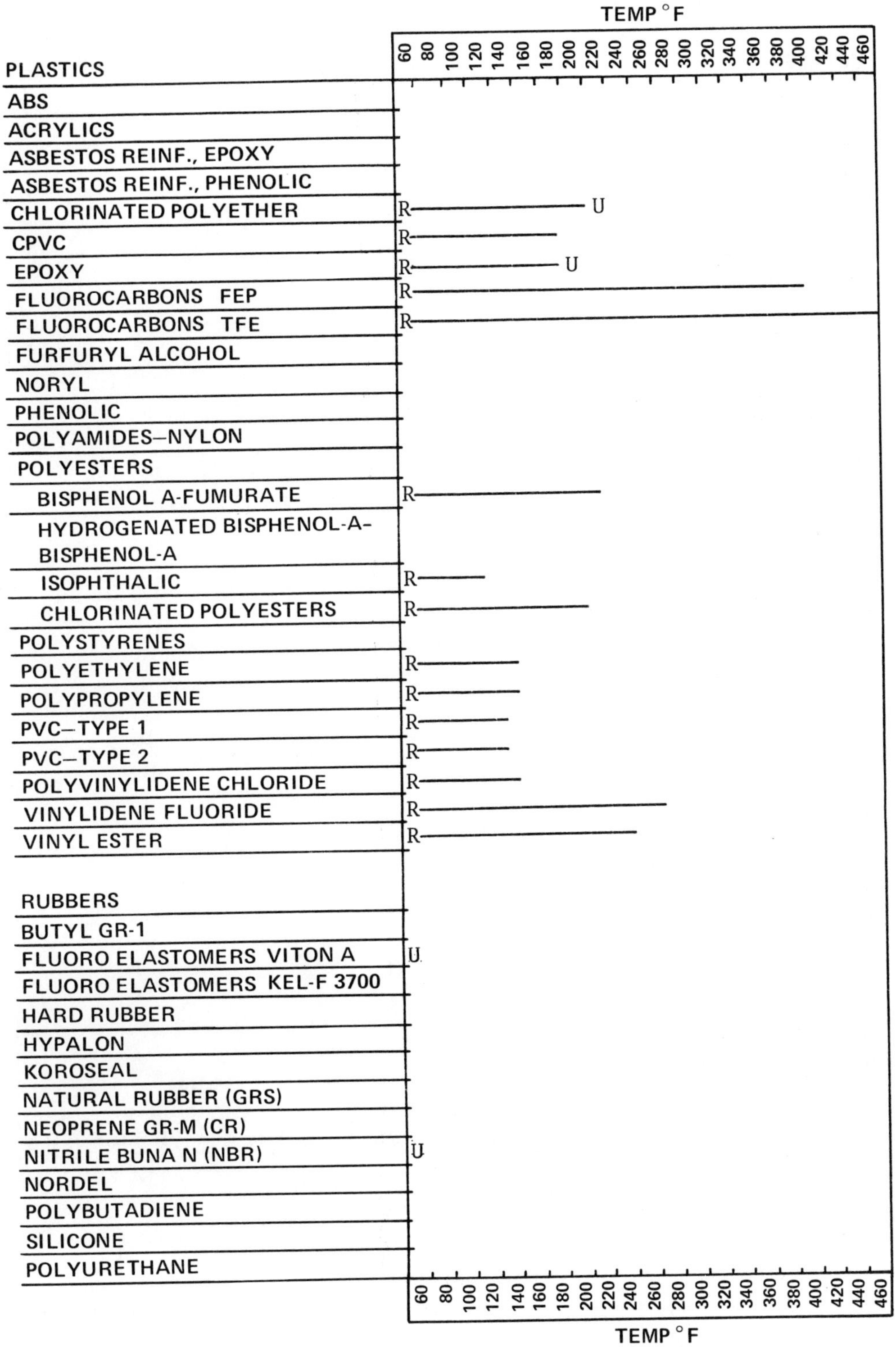

SPECULUM PLATING SOLUTION

TEMP °F

METALS	60	80	100	120	140	160	180	200	220	240	260	280	300	320	340	360	380	400	420	440	460	480	500	520	540	560
ALLOY 20Cb3																										
ALUMINUM																										
BRASS																										
BRONZE																										
CARBON STEEL																										
COPPER																										
E-BRIGHT 26-1																										
HASTELLOY-B																										
HASTELLOY-C																										
HASTELLOY-D																										
HIGH SILICON IRON																										
INCONEL																										
LEAD																										
MONEL																										
NICKEL																										
STAINLESS STEELS Type 304/307																										
Type 316																										
TANTALUM																										
TITANIUM																										
ZIRCONIUM																										
NONMETALLICS																										
BOROSILICATE GLASS																										
CARBON																										

TEMP °F

FOR METALS

E = < 2 Mils Penetration/Year; G = < 20 Mils Penetration/Year
S = < 50 Mils Penetration/Year; U = > 50 Mils Penetration/Year

FOR NONMETALLICS

R = Resistant
U = Unsatisfactory

SPECULUM PLATING SOLUTION

TEMP °F (60–460)

PLASTICS	
ABS	
ACRYLICS	
ASBESTOS REINF., EPOXY	
ASBESTOS REINF., PHENOLIC	
CHLORINATED POLYETHER	R
CPVC	
EPOXY	
FLUOROCARBONS FEP	R
FLUOROCARBONS TFE	R
FURFURYL ALCOHOL	
NORYL	
PHENOLIC	
POLYAMIDES–NYLON	
POLYESTERS	
BISPHENOL A-FUMURATE	
HYDROGENATED BISPHENOL-A-BISPHENOL-A	
ISOPHTHALIC	
CHLORINATED POLYESTERS	
POLYSTYRENES	
POLYETHYLENE	
POLYPROPYLENE	R
PVC–TYPE 1	
PVC–TYPE 2	
POLYVINYLIDENE CHLORIDE	R
VINYLIDENE FLUORIDE	R
VINYL ESTER	

RUBBERS	
BUTYL GR-1	
FLUORO ELASTOMERS VITON A	
FLUORO ELASTOMERS KEL-F 3700	
HARD RUBBER	
HYPALON	
KOROSEAL	
NATURAL RUBBER (GRS)	
NEOPRENE GR-M (CR)	
NITRILE BUNA N (NBR)	
NORDEL	
POLYBUTADIENE	
SILICONE	
POLYURETHANE	

TEMP °F (60–460)

STANNIC CHLORIDE

METALS	Concentration	Rating	TEMP °F (range)
ALLOY 20Cb3		U	60
ALUMINUM		U	60
BRASS		U	60
BRONZE		U	60
CARBON STEEL		U	60
COPPER		U	60
E-BRIGHT 26-1			
HASTELLOY-B	10-50%	G	60–200
HASTELLOY-C	to 50%	G	60–200
HASTELLOY-D			
HIGH SILICON IRON		U	60
INCONEL		U	60
LEAD		U	60
MONEL		U	60
NICKEL		U	60
STAINLESS STEELS			
Type 304/307		U	60
Type 316		U	60
TANTALUM		G	60–300
TITANIUM	20%	E	60–200
ZIRCONIUM	24%	E	60–200

NONMETALLICS	Rating	TEMP °F (range)
BOROSILICATE GLASS	R	60–200
CARBON	R	60–340

TEMP °F: 60 80 100 120 140 160 180 200 220 240 260 280 300 320 340 360 380 400 420 440 460 480 500 520 540 560

FOR METALS

E = < 2 Mils Penetration/Year; G = < 20 Mils Penetration/Year

S = < 50 Mils Penetration/Year; U = > 50 Mils Penetration/Year

FOR NONMETALLICS

R = Resistant

U = Unsatisfactory

STANNIC CHLORIDE

TEMP °F

PLASTICS		60 80 100 120 140 160 180 200 220 240 260 280 300 320 340 360 380 400 420 440 460
ABS		R
ACRYLICS		
ASBESTOS REINF., EPOXY		
ASBESTOS REINF., PHENOLIC		
CHLORINATED POLYETHER		R
CPVC		R
EPOXY		R
FLUOROCARBONS FEP		R
FLUOROCARBONS TFE		R
FURFURYL ALCOHOL	sat	R
NORYL		
PHENOLIC		
POLYAMIDES–NYLON		
POLYESTERS		
BISPHENOL A-FUMURATE	sat	R
HYDROGENATED BISPHENOL-A-BISPHENOL-A		
ISOPHTHALIC	sat	R
CHLORINATED POLYESTERS		R
POLYSTYRENES		
POLYETHYLENE		R
POLYPROPYLENE		R
PVC–TYPE 1		R
PVC–TYPE 2		R
POLYVINYLIDENE CHLORIDE		R
VINYLIDENE FLUORIDE		R
VINYL ESTER	all	R

RUBBERS		
BUTYL GR-1		R
FLUORO ELASTOMERS VITON A		R
FLUORO ELASTOMERS KEL-F 3700		
HARD RUBBER		R
HYPALON		R
KOROSEAL		
NATURAL RUBBER (GRS)		R
NEOPRENE GR-M (CR)		U
NITRILE BUNA N (NBR)		R
NORDEL		R
POLYBUTADIENE		
SILICONE		
POLYURETHANE		

60 80 100 120 140 160 180 200 220 240 260 280 300 320 340 360 380 400 420 440 460

TEMP °F

STANNOUS CHLORIDE

TEMP °F: 60 80 100 120 140 160 180 200 220 240 260 280 300 320 340 360 380 400 420 440 460 480 500 520 540 560

METALS		
ALLOY 20Cb3	10%	E
ALUMINUM		U
BRASS		U
BRONZE		U
CARBON STEEL		U
COPPER		U
E-BRIGHT 26-1	10%	G
HASTELLOY-B	1	G
HASTELLOY-C		G
HASTELLOY-D		
HIGH SILICON IRON		S
INCONEL	dry	G
LEAD		U
MONEL	dry	G
NICKEL	dry	G
STAINLESS STEELS		
Type 304/307		U
Type 316	10%	E
TANTALUM		G
TITANIUM	sat	E
ZIRCONIUM	to 70%	G

NONMETALLICS		
BOROSILICATE GLASS		R
CARBON		

TEMP °F: 60 80 100 120 140 160 180 200 220 240 260 280 300 320 340 360 380 400 420 440 460 480 500 520 540 560

FOR METALS

E = < 2 Mils Penetration/Year; G = < 20 Mils Penetration/Year

S = < 50 Mils Penetration/Year; U = > 50 Mils Penetration/Year

FOR NONMETALLICS

R = Resistant

U = Unsatisfactory

STANNOUS CHLORIDE

TEMP °F (60–460)

PLASTICS		
ABS		R
ACRYLICS		R
ASBESTOS REINF., EPOXY		
ASBESTOS REINF., PHENOLIC		
CHLORINATED POLYETHER		R
CPVC		R
EPOXY		R
FLUOROCARBONS FEP		R
FLUOROCARBONS TFE		R
FURFURYL ALCOHOL		R
NORYL		R
PHENOLIC		
POLYAMIDES–NYLON		U
POLYESTERS		
BISPHENOL A-FUMURATE	sat	R
HYDROGENATED BISPHENOL-A-BISPHENOL-A		
ISOPHTHALIC	sat	R
CHLORINATED POLYESTERS	all	R
POLYSTYRENES		
POLYETHYLENE		R
POLYPROPYLENE		R
PVC–TYPE 1		R
PVC–TYPE 2		R
POLYVINYLIDENE CHLORIDE		R
VINYLIDENE FLUORIDE		R
VINYL ESTER	all	R

RUBBERS		
BUTYL GR-1		R
FLUORO ELASTOMERS VITON A		R
FLUORO ELASTOMERS KEL-F 3700		
HARD RUBBER		
HYPALON	15%	R
KOROSEAL		
NATURAL RUBBER (GRS)		R
NEOPRENE GR-M (CR)		U
NITRILE BUNA N (NBR)		R
NORDEL		R
POLYBUTADIENE		
SILICONE		
POLYURETHANE		

TEMP °F (60–460)

STEARIC ACID

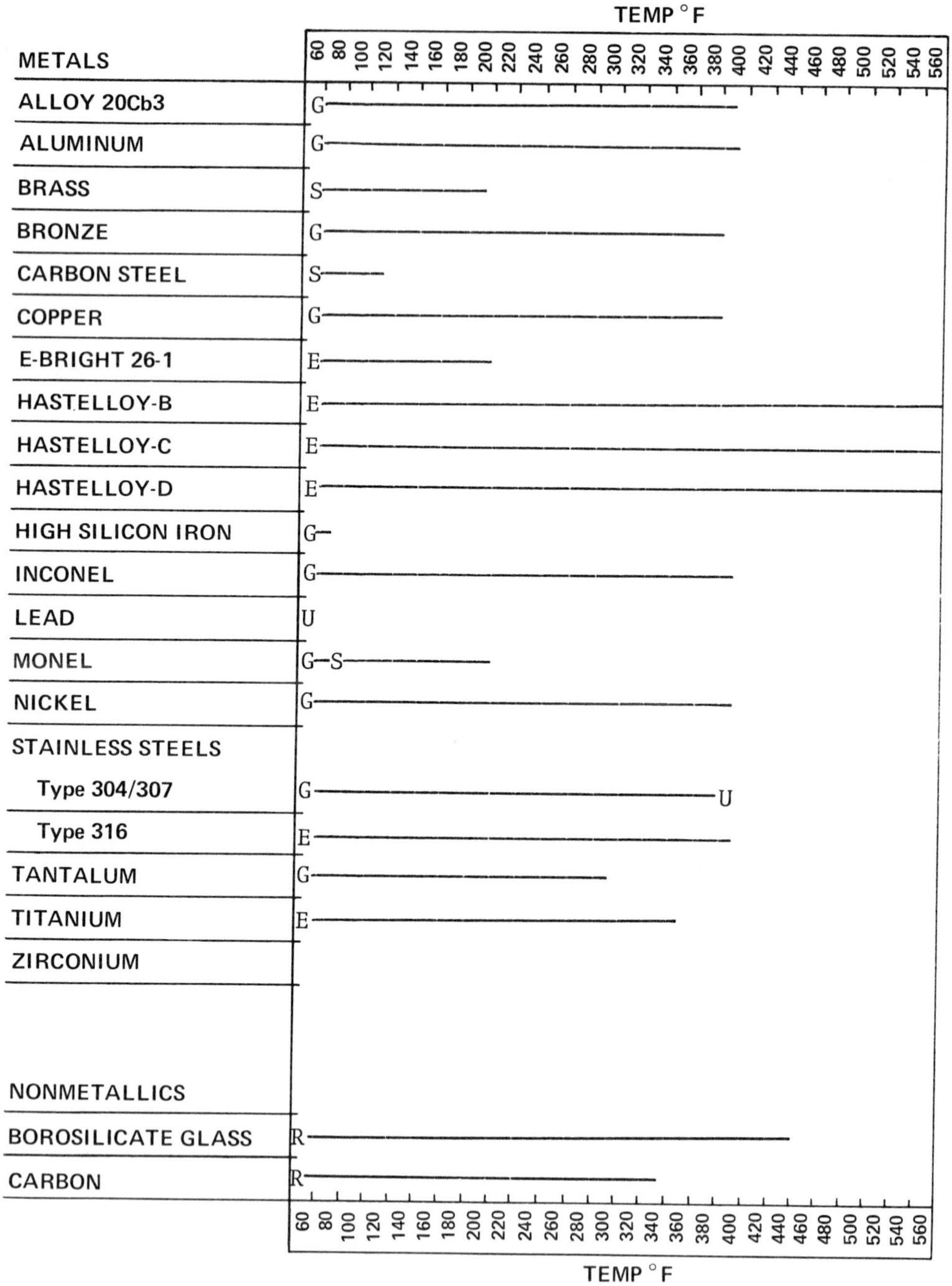

FOR METALS

E = < 2 Mils Penetration/Year; G = < 20 Mils Penetration/Year
S = < 50 Mils Penetration/Year; U = > 50 Mils Penetration/Year

FOR NONMETALLICS

R = Resistant
U = Unsatisfactory

STEARIC ACID

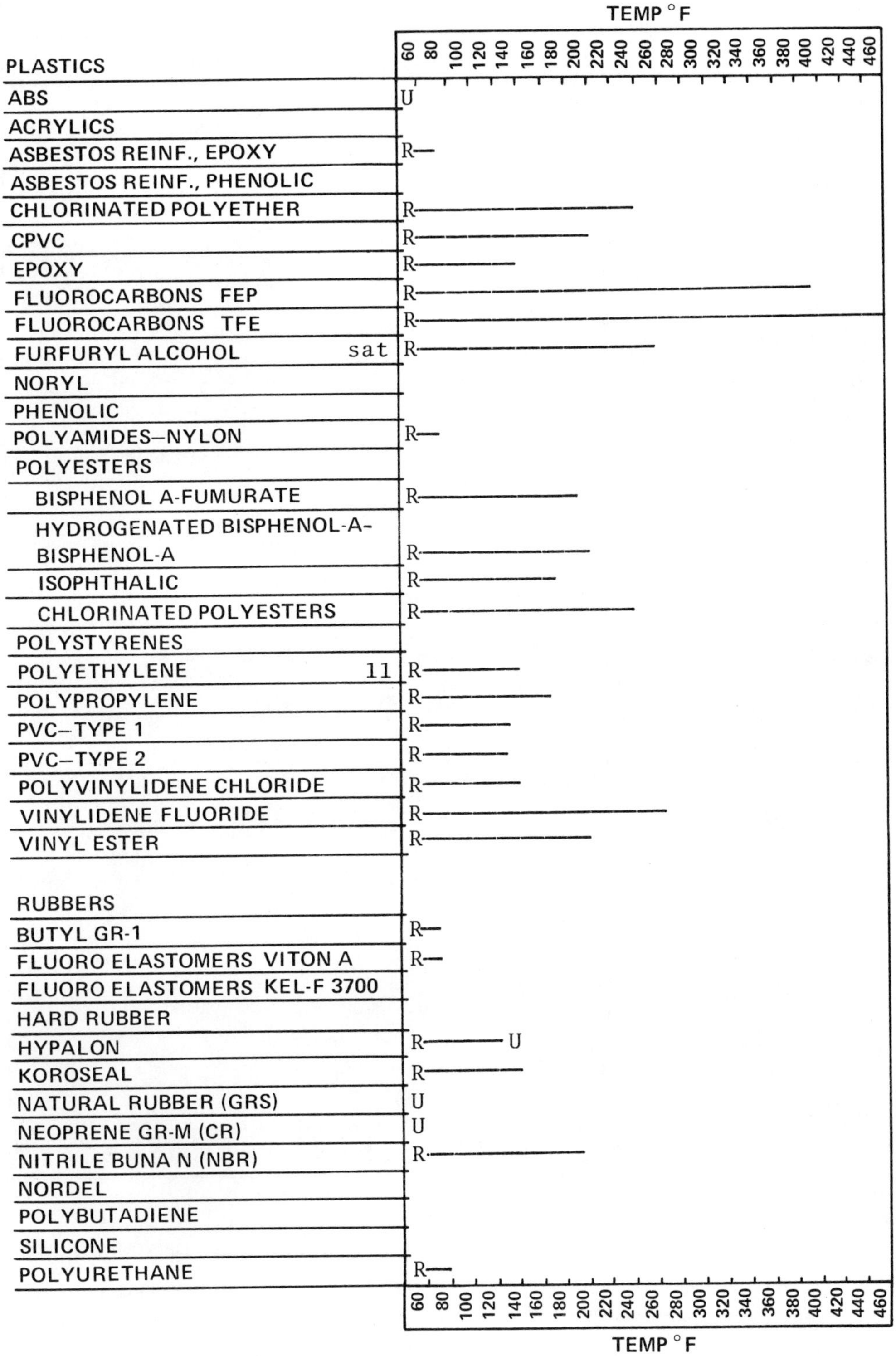

STODDARD'S SOLVENT

TEMP °F (60–560)

METALS	Rating
ALLOY 20Cb3	G
ALUMINUM	G
BRASS	
BRONZE	G
CARBON STEEL	G
COPPER	
E-BRIGHT 26-1	
HASTELLOY-B	
HASTELLOY-C	
HASTELLOY-D	
HIGH SILICON IRON	
INCONEL	
LEAD	
MONEL	G
NICKEL	G
STAINLESS STEELS	
Type 304/307	G
Type 316	G
TANTALUM	
TITANIUM	
ZIRCONIUM	

NONMETALLICS	Rating
BOROSILICATE GLASS	R
CARBON	

TEMP °F

FOR METALS

E = < 2 Mils Penetration/Year; G = < 20 Mils Penetration/Year
S = < 50 Mils Penetration/Year; U = > 50 Mils Penetration/Year

FOR NONMETALLICS

R = Resistant
U = Unsatisfactory

STODDARD'S SOLVENT

TEMP °F

PLASTICS	60 – 460 °F
ABS	R
ACRYLICS	
ASBESTOS REINF., EPOXY	
ASBESTOS REINF., PHENOLIC	
CHLORINATED POLYETHER	
CPVC	U
EPOXY	
FLUOROCARBONS FEP	R
FLUOROCARBONS TFE	R
FURFURYL ALCOHOL	
NORYL	
PHENOLIC	
POLYAMIDES–NYLON	R
POLYESTERS	
BISPHENOL A-FUMURATE	
HYDROGENATED BISPHENOL-A-BISPHENOL-A	
ISOPHTHALIC	
CHLORINATED POLYESTERS	
POLYSTYRENES	
POLYETHYLENE	U
POLYPROPYLENE	R
PVC–TYPE 1	R
PVC–TYPE 2	U
POLYVINYLIDENE CHLORIDE	
VINYLIDENE FLUORIDE	R
VINYL ESTER	
RUBBERS	
BUTYL GR-1	U
FLUORO ELASTOMERS VITON A	R
FLUORO ELASTOMERS KEL-F 3700	
HARD RUBBER	
HYPALON	
KOROSEAL	
NATURAL RUBBER (GRS)	U
NEOPRENE GR-M (CR)	U
NITRILE BUNA N (NBR)	R
NORDEL	
POLYBUTADIENE	
SILICONE	U
POLYURETHANE	

TEMP °F: 60 80 100 120 140 160 180 200 220 240 260 280 300 320 340 360 380 400 420 440 460

SUCCINIC ACID

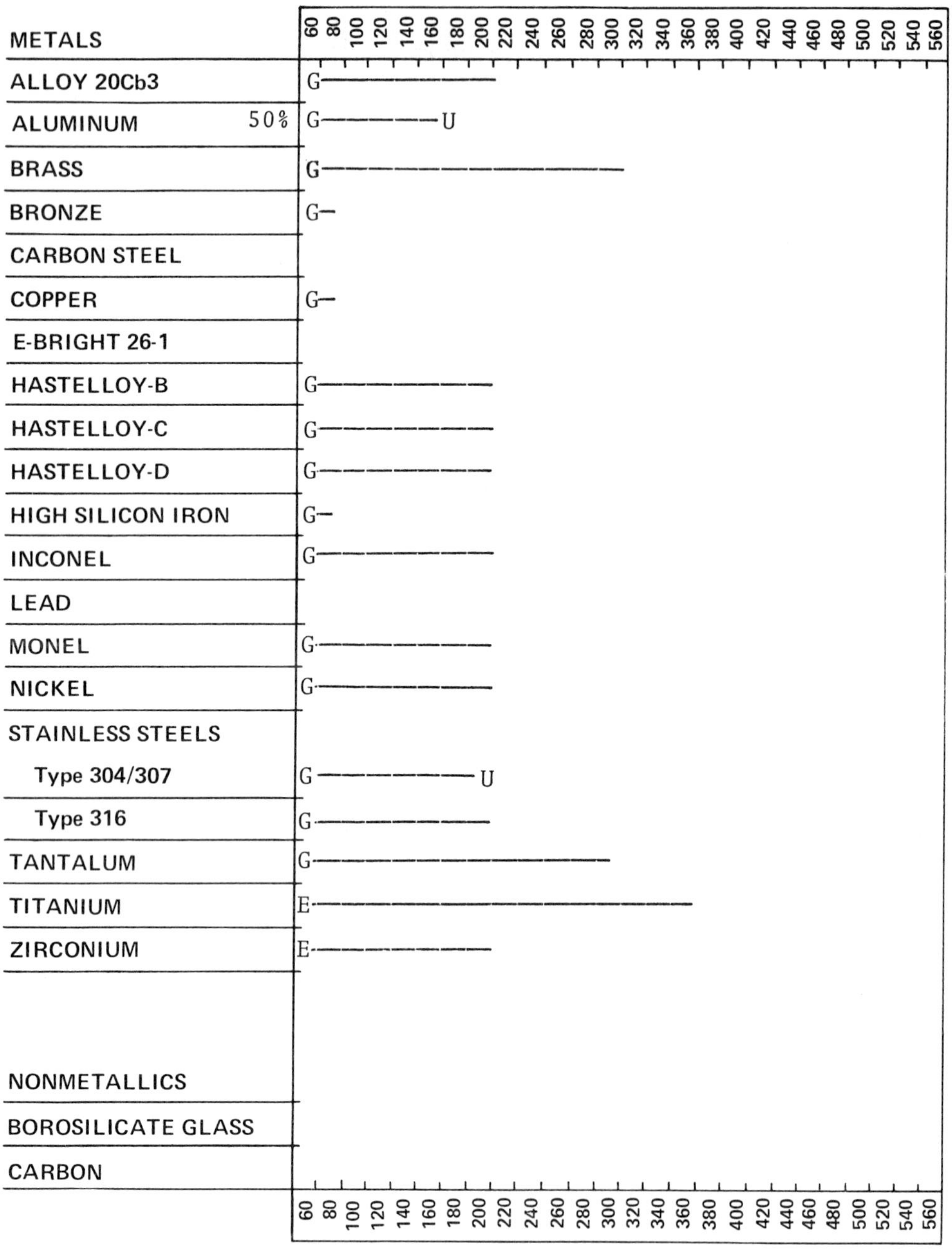

FOR METALS

E = < 2 Mils Penetration/Year; G = < 20 Mils Penetration/Year

S = < 50 Mils Penetration/Year; U = > 50 Mils Penetration/Year

FOR NONMETALLICS

R = Resistant

U = Unsatisfactory

SUCCINIC ACID

TEMP °F (60–460)

PLASTICS	
ABS	
ACRYLICS	
ASBESTOS REINF., EPOXY	
ASBESTOS REINF., PHENOLIC	
CHLORINATED POLYETHER	
CPVC	
EPOXY	
FLUOROCARBONS FEP	R
FLUOROCARBONS TFE	R
FURFURYL ALCOHOL	
NORYL	
PHENOLIC	
POLYAMIDES–NYLON	
POLYESTERS	
BISPHENOL A-FUMURATE	
HYDROGENATED BISPHENOL-A-BISPHENOL-A	
ISOPHTHALIC	
CHLORINATED POLYESTERS	
POLYSTYRENES	
POLYETHYLENE	
POLYPROPYLENE	R
PVC–TYPE 1	R
PVC–TYPE 2	
POLYVINYLIDENE CHLORIDE	R
VINYLIDENE FLUORIDE	R
VINYL ESTER	

RUBBERS	
BUTYL GR-1	
FLUORO ELASTOMERS VITON A	
FLUORO ELASTOMERS KEL-F 3700	
HARD RUBBER	
HYPALON	
KOROSEAL	
NATURAL RUBBER (GRS)	
NEOPRENE GR-M (CR)	
NITRILE BUNA N (NBR)	
NORDEL	
POLYBUTADIENE	
SILICONE	
POLYURETHANE	

TEMP °F (60–460)

SULFAMIC ACID

TEMP °F

METALS	60	80	100	120	140	160	180	200	220	240	260	280	300	320	340	360	380	400	420	440	460	480	500	520	540	560
ALLOY 20Cb3																										
ALUMINUM																										
BRASS																										
BRONZE																										
CARBON STEEL																										
COPPER																										
E-BRIGHT 26-1	E																									
HASTELLOY-B																										
HASTELLOY-C																										
HASTELLOY-D																										
HIGH SILICON IRON																										
INCONEL																										
LEAD																										
MONEL																										
NICKEL																										
STAINLESS STEELS Type 304/307																										
Type 316																										
TANTALUM	G																									
TITANIUM																										
ZIRCONIUM																										
NONMETALLICS																										
BOROSILICATE GLASS																										
CARBON																										

TEMP °F

FOR METALS

E = < 2 Mils Penetration/Year; G = < 20 Mils Penetration/Year

S = < 50 Mils Penetration/Year; U = > 50 Mils Penetration/Year

FOR NONMETALLICS

R = Resistant

U = Unsatisfactory

SULFAMIC ACID

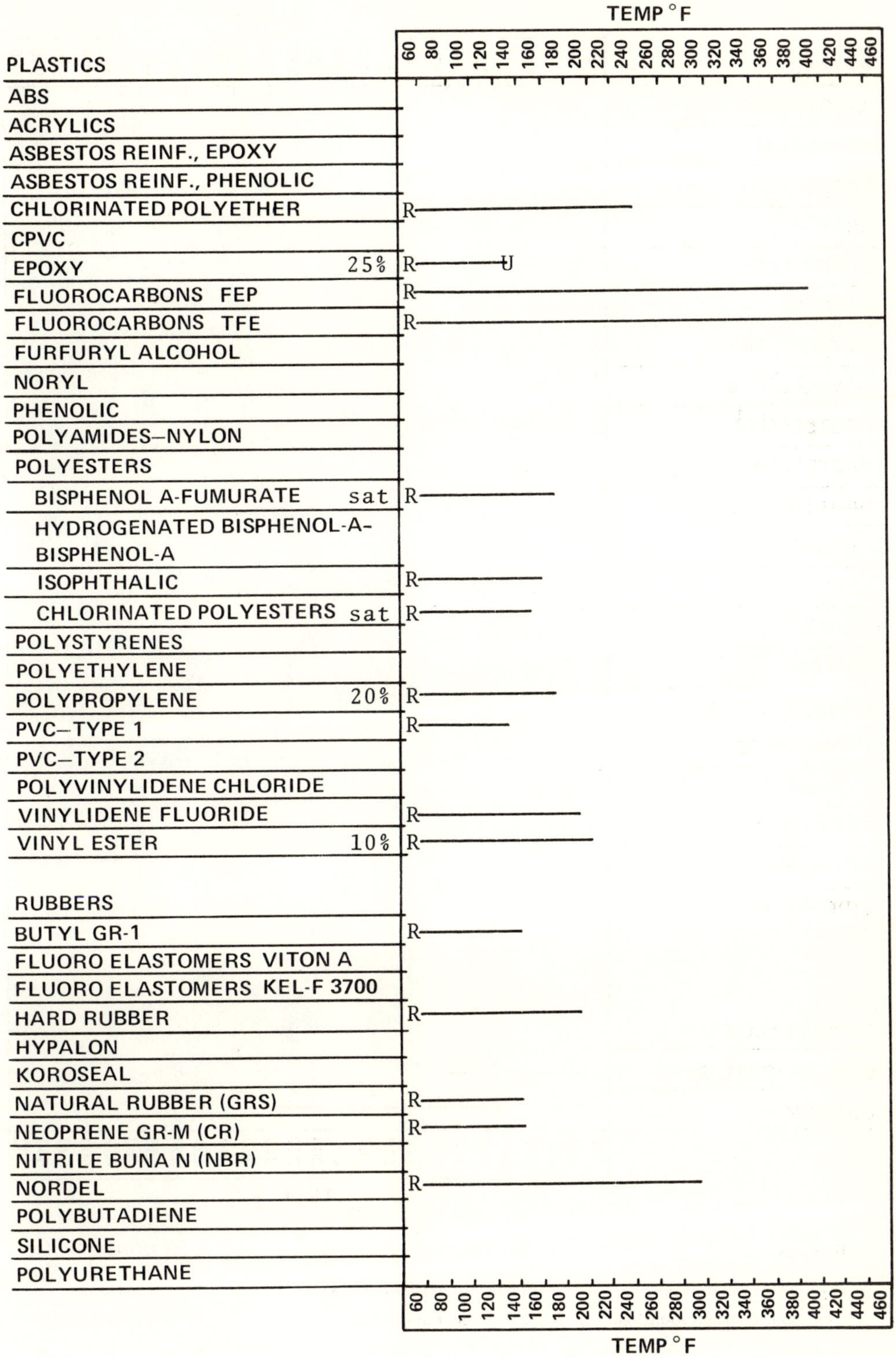

SULFONATED DETERGENTS

TEMP °F

METALS	Concentration	Rating	TEMP °F (60–560)
ALLOY 20Cb3			
ALUMINUM			
BRASS			
BRONZE			
CARBON STEEL			
COPPER			
E-BRIGHT 26-1			
HASTELLOY-B			
HASTELLOY-C	50%	E	60–150
HASTELLOY-D			
HIGH SILICON IRON			
INCONEL			
LEAD			
MONEL			
NICKEL			
STAINLESS STEELS Type 304/307			
Type 316			
TANTALUM			
TITANIUM			
ZIRCONIUM			

NONMETALLICS	Concentration	Rating	TEMP °F (60–560)
BOROSILICATE GLASS		R	60–200
CARBON		R	60–340

TEMP °F

FOR METALS

E = < 2 Mils Penetration/Year; G = < 20 Mils Penetration/Year

S = < 50 Mils Penetration/Year; U = > 50 Mils Penetration/Year

FOR NONMETALLICS

R = Resistant

U = Unsatisfactory

SULFONATED DETERGENTS

TEMP °F: 60 80 100 120 140 160 180 200 220 240 260 280 300 320 340 360 380 400 420 440 460

PLASTICS	SULFONATED DETERGENTS
ABS	
ACRYLICS	
ASBESTOS REINF., EPOXY	
ASBESTOS REINF., PHENOLIC	
CHLORINATED POLYETHER	R——U
CPVC	
EPOXY	R——
FLUOROCARBONS FEP	R——
FLUOROCARBONS TFE	R——
FURFURYL ALCOHOL	R——
NORYL	
PHENOLIC	
POLYAMIDES–NYLON	
POLYESTERS	
BISPHENOL A-FUMURATE	R——
HYDROGENATED BISPHENOL-A-BISPHENOL-A	
ISOPHTHALIC	R——
CHLORINATED POLYESTERS 50%	R——
POLYSTYRENES	
POLYETHYLENE	
POLYPROPYLENE	
PVC–TYPE 1	
PVC–TYPE 2	
POLYVINYLIDENE CHLORIDE	
VINYLIDENE FLUORIDE	
VINYL ESTER	R——

RUBBERS	SULFONATED DETERGENTS
BUTYL GR-1	
FLUORO ELASTOMERS VITON A	
FLUORO ELASTOMERS KEL-F 3700	
HARD RUBBER	
HYPALON	
KOROSEAL	
NATURAL RUBBER (GRS)	
NEOPRENE GR-M (CR)	
NITRILE BUNA N (NBR)	
NORDEL	
POLYBUTADIENE	
SILICONE	
POLYURETHANE	

TEMP °F: 60 80 100 120 140 160 180 200 220 240 260 280 300 320 340 360 380 400 420 440 460

SULFATE LIQUORS

TEMP °F: 60 80 100 120 140 160 180 200 220 240 260 280 300 320 340 360 380 400 420 440 460 480 500 520 540 560

METALS		
ALLOY 20Cb3		G
ALUMINUM		G
BRASS		U
BRONZE		U
CARBON STEEL		S
COPPER		G
E-BRIGHT 26-1		
HASTELLOY-B		G
HASTELLOY-C		G
HASTELLOY-D		
HIGH SILICON IRON		
INCONEL		
LEAD		
MONEL		G
NICKEL		
STAINLESS STEELS		
Type 304/307	2,9	G
Type 316		G
TANTALUM		G
TITANIUM		
ZIRCONIUM		

NONMETALLICS	
BOROSILICATE GLASS	
CARBON	

TEMP °F: 60 80 100 120 140 160 180 200 220 240 260 280 300 320 340 360 380 400 420 440 460 480 500 520 540 560

FOR METALS

E = < 2 Mils Penetration/Year; G = < 20 Mils Penetration/Year

S = < 50 Mils Penetration/Year; U = > 50 Mils Penetration/Year

FOR NONMETALLICS

R = Resistant

U = Unsatisfactory

SULFATE LIQUORS

TEMP °F (60–460)

PLASTICS	TEMP °F 60–460
ABS	
ACRYLICS	
ASBESTOS REINF., EPOXY	
ASBESTOS REINF., PHENOLIC	
CHLORINATED POLYETHER	
CPVC	
EPOXY	
FLUOROCARBONS FEP	R——————
FLUOROCARBONS TFE	R———————
FURFURYL ALCOHOL	
NORYL	
PHENOLIC	
POLYAMIDES—NYLON	
POLYESTERS	
BISPHENOL A-FUMURATE	R———
HYDROGENATED BISPHENOL-A-BISPHENOL-A	
ISOPHTHALIC	
CHLORINATED POLYESTERS	
POLYSTYRENES	
POLYETHYLENE	
POLYPROPYLENE	R—
PVC—TYPE 1	
PVC—TYPE 2	
POLYVINYLIDENE CHLORIDE	
VINYLIDENE FLUORIDE	
VINYL ESTER	R———
RUBBERS	
BUTYL GR-1	
FLUORO ELASTOMERS VITON A	R—
FLUORO ELASTOMERS KEL-F 3700	
HARD RUBBER	
HYPALON	R—
KOROSEAL	
NATURAL RUBBER (GRS)	
NEOPRENE GR-M (CR)	R—
NITRILE BUNA N (NBR)	
NORDEL	
POLYBUTADIENE	
SILICONE	
POLYURETHANE	

TEMP °F (60–460)

SULFITE LIQUORS

METALS	TEMP °F (60–560)
ALLOY 20Cb3	G
ALUMINUM	U
BRASS	U
BRONZE	
CARBON STEEL	U
COPPER	
E-BRIGHT 26-1	
HASTELLOY-B	
HASTELLOY-C	E
HASTELLOY-D	
HIGH SILICON IRON	U
INCONEL	U
LEAD	E
MONEL	
NICKEL	
STAINLESS STEELS	
Type 304/307	G U
Type 316	G
TANTALUM	
TITANIUM	E
ZIRCONIUM	

NONMETALLICS	TEMP °F (60–560)
BOROSILICATE GLASS	R
CARBON	

FOR METALS

E = < 2 Mils Penetration/Year; G = < 20 Mils Penetration/Year

S = < 50 Mils Penetration/Year; U = > 50 Mils Penetration/Year

FOR NONMETALLICS

R = Resistant

U = Unsatisfactory

SULFITE LIQUORS

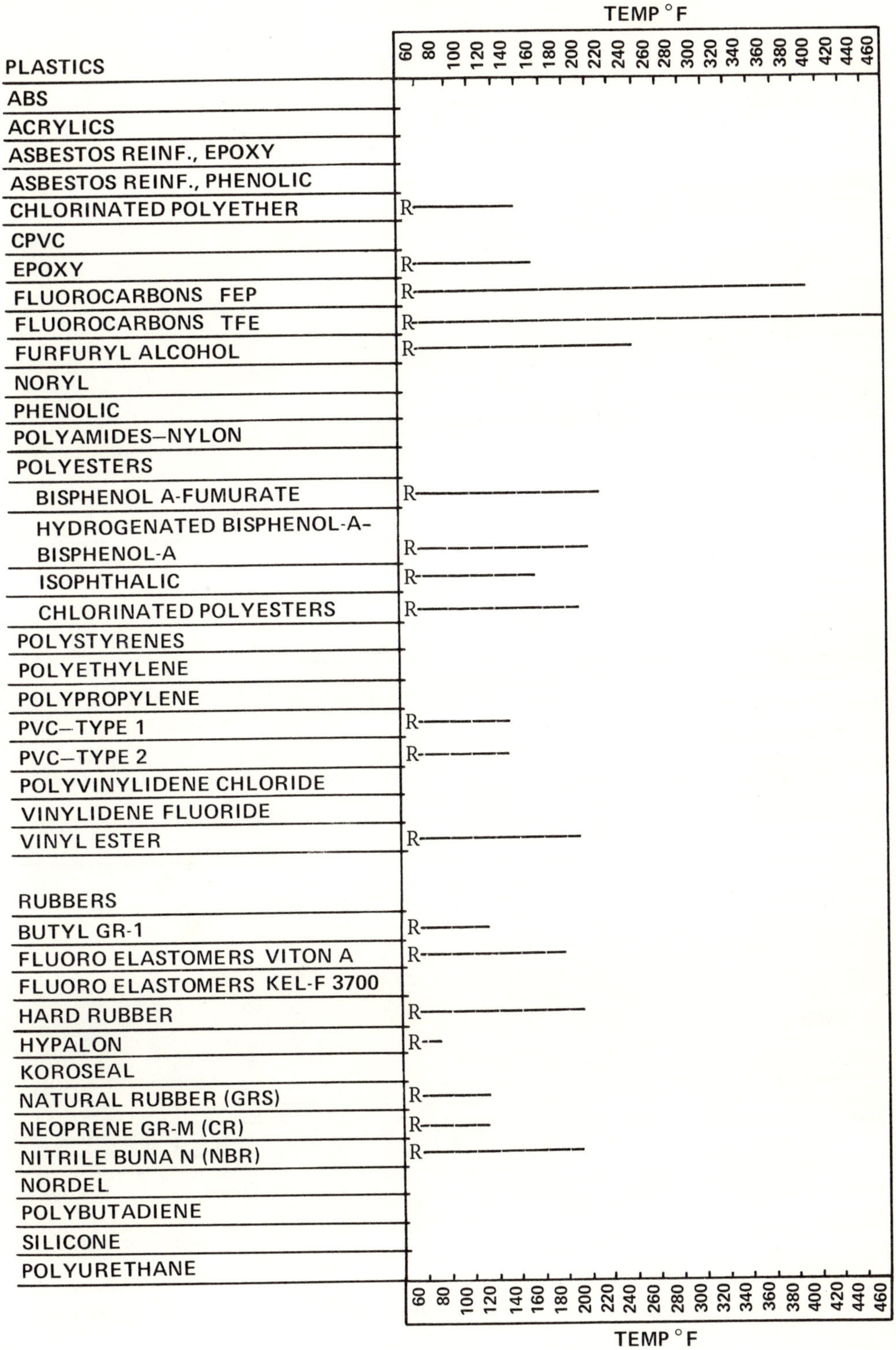

SULFUR

TEMP °F: 60, 80, 100, 120, 140, 160, 180, 200, 220, 240, 260, 280, 300, 320, 340, 360, 380, 400, 420, 440, 460, 480, 500, 520, 540, 560

METALS	Rating
ALLOY 20Cb3	E ———
ALUMINUM	E ———
BRASS	G—U
BRONZE	S—
CARBON STEEL	U
COPPER	S—
E-BRIGHT 26-1	
HASTELLOY-B	U
HASTELLOY-C	E ———
HASTELLOY-D	E ———
HIGH SILICON IRON	S—
INCONEL	E ———
LEAD	
MONEL	E ——— G ——— U
NICKEL	E ——— G ——— U
STAINLESS STEELS	
Type 304/307	E ———
Type 316	E ———
TANTALUM	G ———
TITANIUM	E ———
ZIRCONIUM	G ———

NONMETALLICS	Rating
BOROSILICATE GLASS	
CARBON	R ———

TEMP °F: 60, 80, 100, 120, 140, 160, 180, 200, 220, 240, 260, 280, 300, 320, 340, 360, 380, 400, 420, 440, 460, 480, 500, 520, 540, 560

FOR METALS

E = < 2 Mils Penetration/Year; G = < 20 Mils Penetration/Year

S = < 50 Mils Penetration/Year; U = > 50 Mils Penetration/Year

FOR NONMETALLICS

R = Resistant

U = Unsatisfactory

SULFUR

TEMP °F: 60 80 100 120 140 160 180 200 220 240 260 280 300 320 340 360 380 400 420 440 460

PLASTICS	
ABS	U
ACRYLICS	U
ASBESTOS REINF., EPOXY	
ASBESTOS REINF., PHENOLIC	
CHLORINATED POLYETHER	R
CPVC	R
EPOXY	R
FLUOROCARBONS FEP	R
FLUOROCARBONS TFE	R
FURFURYL ALCOHOL	R
NORYL	
PHENOLIC	
POLYAMIDES–NYLON	R
POLYESTERS	
BISPHENOL A-FUMURATE	
HYDROGENATED BISPHENOL-A-BISPHENOL-A	
ISOPHTHALIC	
CHLORINATED POLYESTERS	R
POLYSTYRENES	
POLYETHYLENE	R
POLYPROPYLENE	R
PVC–TYPE 1	R
PVC–TYPE 2	R
POLYVINYLIDENE CHLORIDE	R
VINYLIDENE FLUORIDE	R
VINYL ESTER	R

RUBBERS	
BUTYL GR-1	
FLUORO ELASTOMERS VITON A	U
FLUORO ELASTOMERS KEL-F 3700	
HARD RUBBER	
HYPALON	R
KOROSEAL	
NATURAL RUBBER (GRS)	
NEOPRENE GR-M (CR)	
NITRILE BUNA N (NBR)	U
NORDEL	
POLYBUTADIENE	
SILICONE	
POLYURETHANE	

TEMP °F: 60 80 100 120 140 160 180 200 220 240 260 280 300 320 340 360 380 400 420 440 460

SULFUR CHLORIDE

TEMP °F: 60 80 100 120 140 160 180 200 220 240 260 280 300 320 340 360 380 400 420 440 460 480 500 520 540 560

METALS		
ALLOY 20Cb3	dry	G—
ALUMINUM		S—
BRASS	dry	G—
BRONZE		U
CARBON STEEL		U
COPPER		U
E-BRIGHT 26-1		
HASTELLOY-B		
HASTELLOY-C		G————
HASTELLOY-D		
HIGH SILICON IRON		U
INCONEL		G—
LEAD		G—
MONEL		S—
NICKEL		G————————————
STAINLESS STEELS		
Type 304/307	dry	G————
Type 316	dry	G——U
TANTALUM	dry	E—
TITANIUM		
ZIRCONIUM		

NONMETALLICS	
BOROSILICATE GLASS	R———
CARBON	

TEMP °F: 60 80 100 120 140 160 180 200 220 240 260 280 300 320 340 360 380 400 420 440 460 480 500 520 540 560

FOR METALS

E = < 2 Mils Penetration/Year; G = < 20 Mils Penetration/Year
S = < 50 Mils Penetration/Year; U = > 50 Mils Penetration/Year

FOR NONMETALLICS

R = Resistant
U = Unsatisfactory

TEMP °F: 60 80 100 120 140 160 180 200 220 240 260 280 300 320 340 360 380 400 420 440 460

PLASTICS	
ABS	
ACRYLICS	
ASBESTOS REINF., EPOXY	
ASBESTOS REINF., PHENOLIC	
CHLORINATED POLYETHER	U
CPVC	
EPOXY	U
FLUOROCARBONS FEP	R
FLUOROCARBONS TFE	R
FURFURYL ALCOHOL	R
NORYL	
PHENOLIC	
POLYAMIDES–NYLON	
POLYESTERS	
BISPHENOL A-FUMURATE	
HYDROGENATED BISPHENOL-A-BISPHENOL-A vapor	U
ISOPHTHALIC	U
CHLORINATED POLYESTERS	U
POLYSTYRENES	
POLYETHYLENE	U
POLYPROPYLENE	U
PVC–TYPE 1	
PVC–TYPE 2	
POLYVINYLIDENE CHLORIDE	U
VINYLIDENE FLUORIDE	R
VINYL ESTER vapor	U

RUBBERS	
BUTYL GR-1	U
FLUORO ELASTOMERS VITON A	R
FLUORO ELASTOMERS KEL-F 3700	
HARD RUBBER	
HYPALON	
KOROSEAL	
NATURAL RUBBER (GRS)	U
NEOPRENE GR-M (CR)	U
NITRILE BUNA N (NBR)	U
NORDEL	
POLYBUTADIENE	
SILICONE	
POLYURETHANE	

TEMP °F: 60 80 100 120 140 160 180 200 220 240 260 280 300 320 340 360 380 400 420 440 460

SULFUR DICHLORIDE

TEMP °F

METALS	60	80	100	120	140	160	180	200	220	240	260	280	300	320	340	360	380	400	420	440	460	480	500	520	540	560
ALLOY 20Cb3																										
ALUMINUM																										
BRASS																										
BRONZE																										
CARBON STEEL																										
COPPER																										
E-BRIGHT 26-1																										
HASTELLOY-B																										
HASTELLOY-C																										
HASTELLOY-D																										
HIGH SILICON IRON																										
INCONEL																										
LEAD																										
MONEL																										
NICKEL																										
STAINLESS STEELS Type 304/307																										
Type 316																										
TANTALUM	G																									
TITANIUM																										
ZIRCONIUM																										
NONMETALLICS																										
BOROSILICATE GLASS																										
CARBON																										

TEMP °F

FOR METALS

E = < 2 Mils Penetration/Year; G = < 20 Mils Penetration/Year

S = < 50 Mils Penetration/Year; U = > 50 Mils Penetration/Year

FOR NONMETALLICS

R = Resistant

U = Unsatisfactory

SULFUR DICHLORIDE

TEMP °F: 60 80 100 120 140 160 180 200 220 240 260 280 300 320 340 360 380 400 420 440 460

PLASTICS		
ABS		
ACRYLICS		
ASBESTOS REINF., EPOXY		
ASBESTOS REINF., PHENOLIC		
CHLORINATED POLYETHER		
CPVC		
EPOXY		
FLUOROCARBONS FEP		R
FLUOROCARBONS TFE		R
FURFURYL ALCOHOL		
NORYL		
PHENOLIC		
POLYAMIDES–NYLON		
POLYESTERS		
BISPHENOL A-FUMURATE		U
HYDROGENATED BISPHENOL-A-BISPHENOL-A	vapor	U
ISOPHTHALIC	vapor	U
CHLORINATED POLYESTERS		U
POLYSTYRENES		
POLYETHYLENE		
POLYPROPYLENE		
PVC–TYPE 1		
PVC–TYPE 2		
POLYVINYLIDENE CHLORIDE		U
VINYLIDENE FLUORIDE		R
VINYL ESTER	vapor	U

RUBBERS	
BUTYL GR-1	
FLUORO ELASTOMERS VITON A	R
FLUORO ELASTOMERS KEL-F 3700	
HARD RUBBER	
HYPALON	
KOROSEAL	
NATURAL RUBBER (GRS)	
NEOPRENE GR-M (CR)	
NITRILE BUNA N (NBR)	
NORDEL	
POLYBUTADIENE	
SILICONE	
POLYURETHANE	

TEMP °F: 60 80 100 120 140 160 180 200 220 240 260 280 300 320 340 360 380 400 420 440 460

SULFUR DIOXIDE, DRY

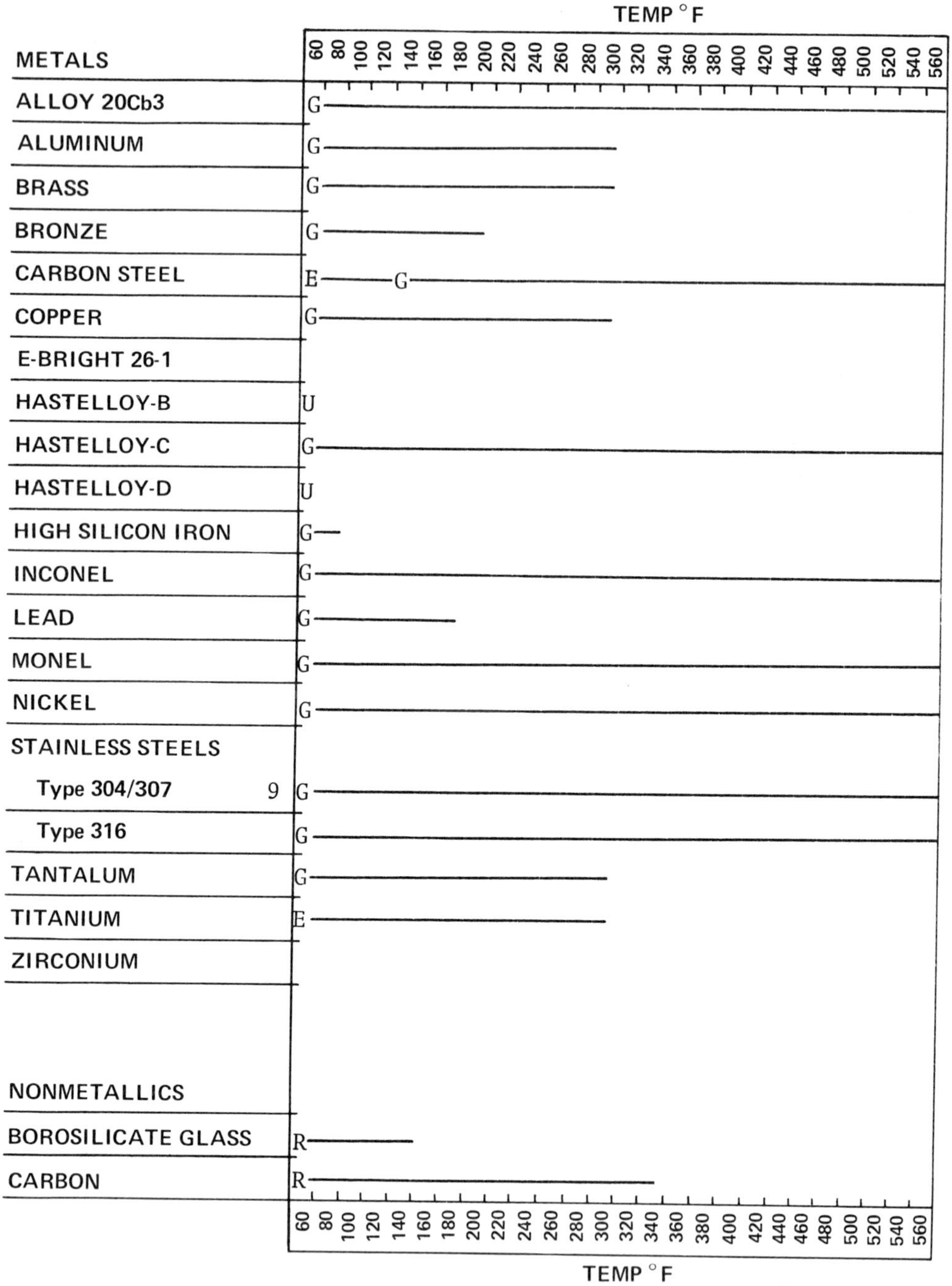

FOR METALS

E = < 2 Mils Penetration/Year; G = < 20 Mils Penetration/Year
S = < 50 Mils Penetration/Year; U = > 50 Mils Penetration/Year

FOR NONMETALLICS

R = Resistant
U = Unsatisfactory

SULFUR DIOXIDE, DRY

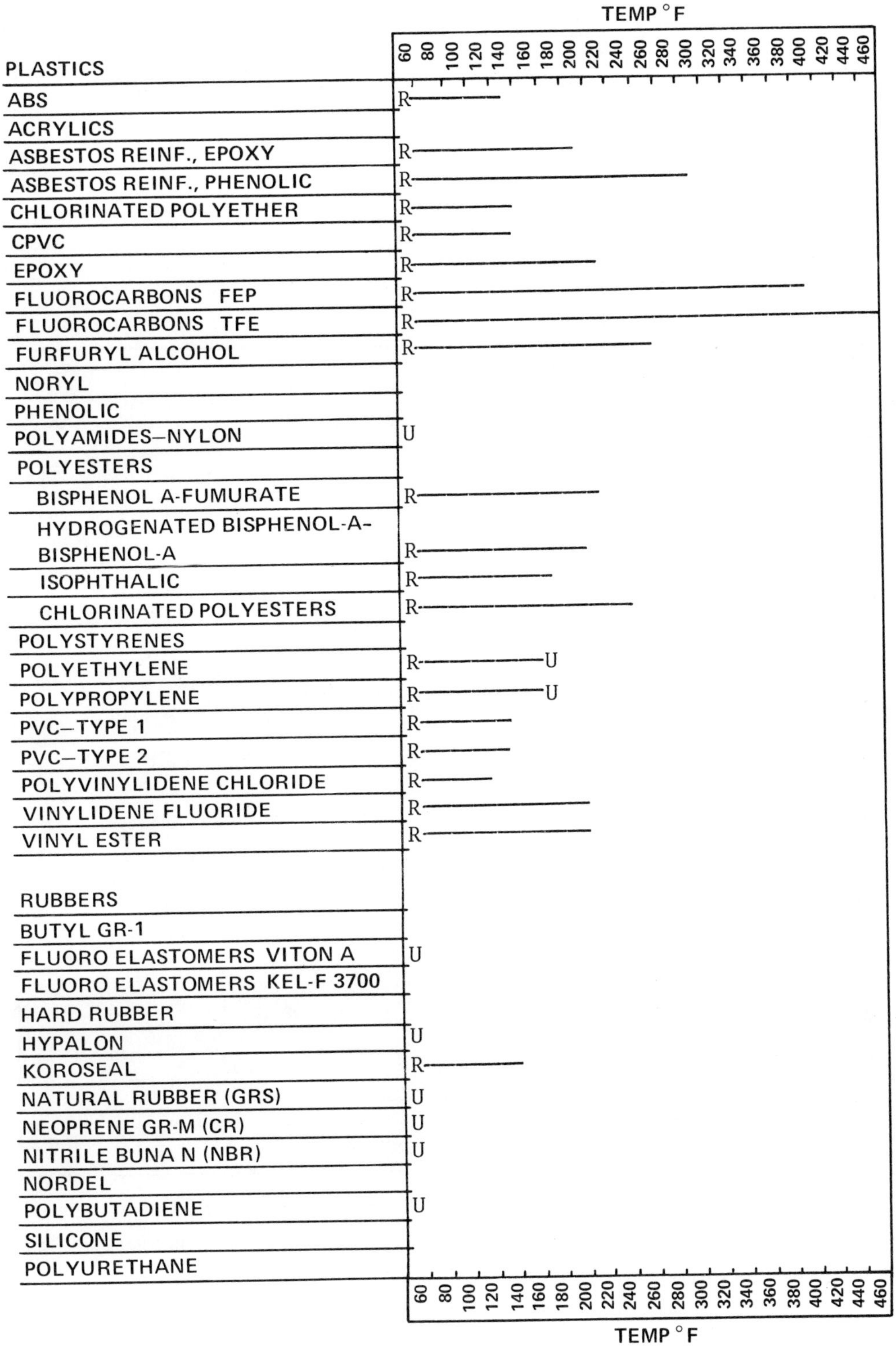

SULFUR DIOXIDE, WET

TEMP °F: 60 80 100 120 140 160 180 200 220 240 260 280 300 320 340 360 380 400 420 440 460 480 500 520 540 560

METALS	Rating
ALLOY 20Cb3	G
ALUMINUM	U
BRASS	U
BRONZE	U
CARBON STEEL	U
COPPER	U
E-BRIGHT 26-1	E
HASTELLOY-B	U
HASTELLOY-C	E — G
HASTELLOY-D	U
HIGH SILICON IRON	S
INCONEL	U
LEAD	G
MONEL	U
NICKEL	U
STAINLESS STEELS	
Type 304/307	U
Type 316	G
TANTALUM	G
TITANIUM	U
ZIRCONIUM	

NONMETALLICS	Rating
BOROSILICATE GLASS	R
CARBON	R

TEMP °F: 60 80 100 120 140 160 180 200 220 240 260 280 300 320 340 360 380 400 420 440 460 480 500 520 540 560

FOR METALS

E = < 2 Mils Penetration/Year; G = < 20 Mils Penetration/Year

S = < 50 Mils Penetration/Year; U = > 50 Mils Penetration/Year

FOR NONMETALLICS

R = Resistant

U = Unsatisfactory

SULFUR DIOXIDE, WET

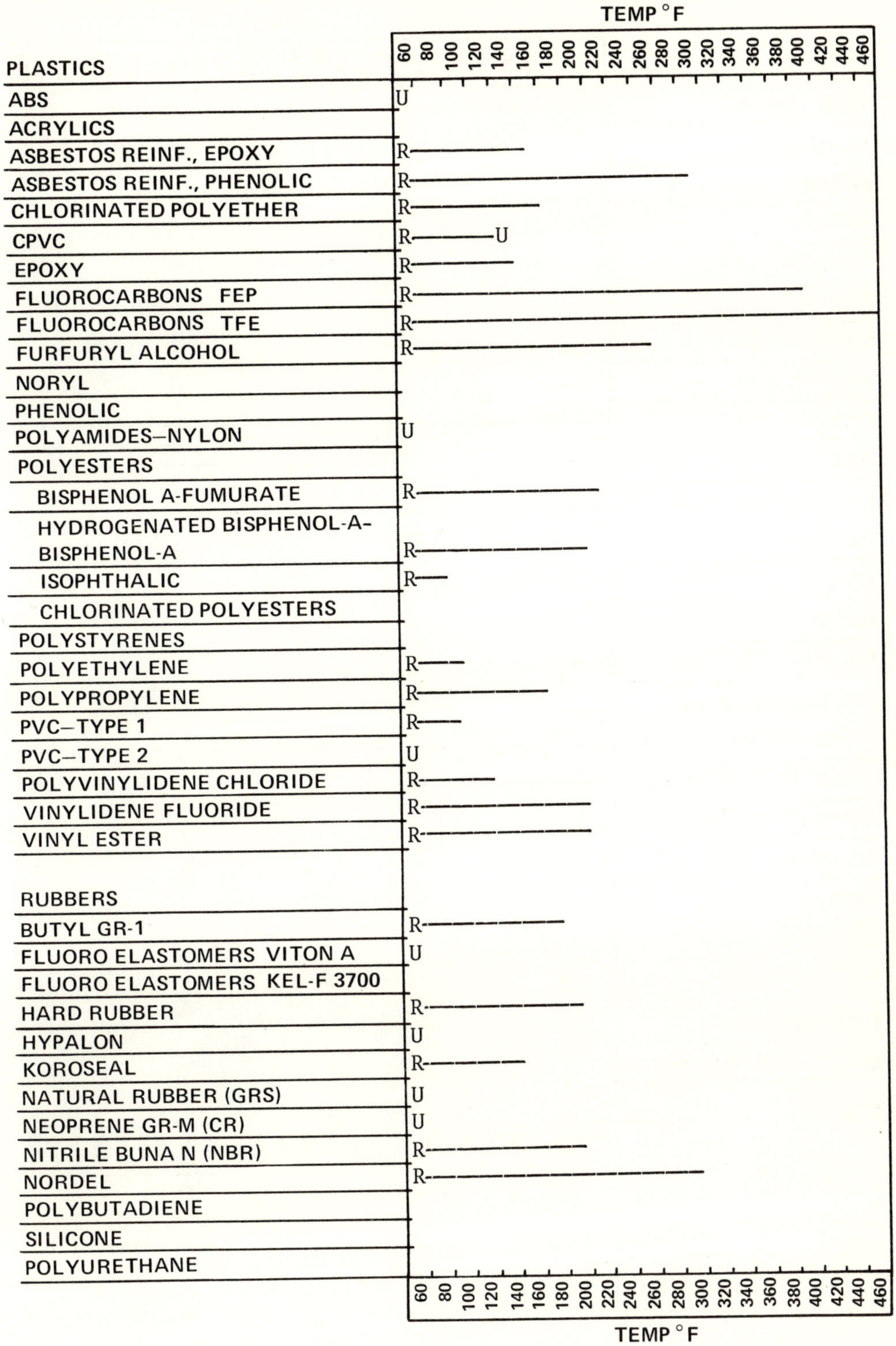

SULFUR TRIOXIDE

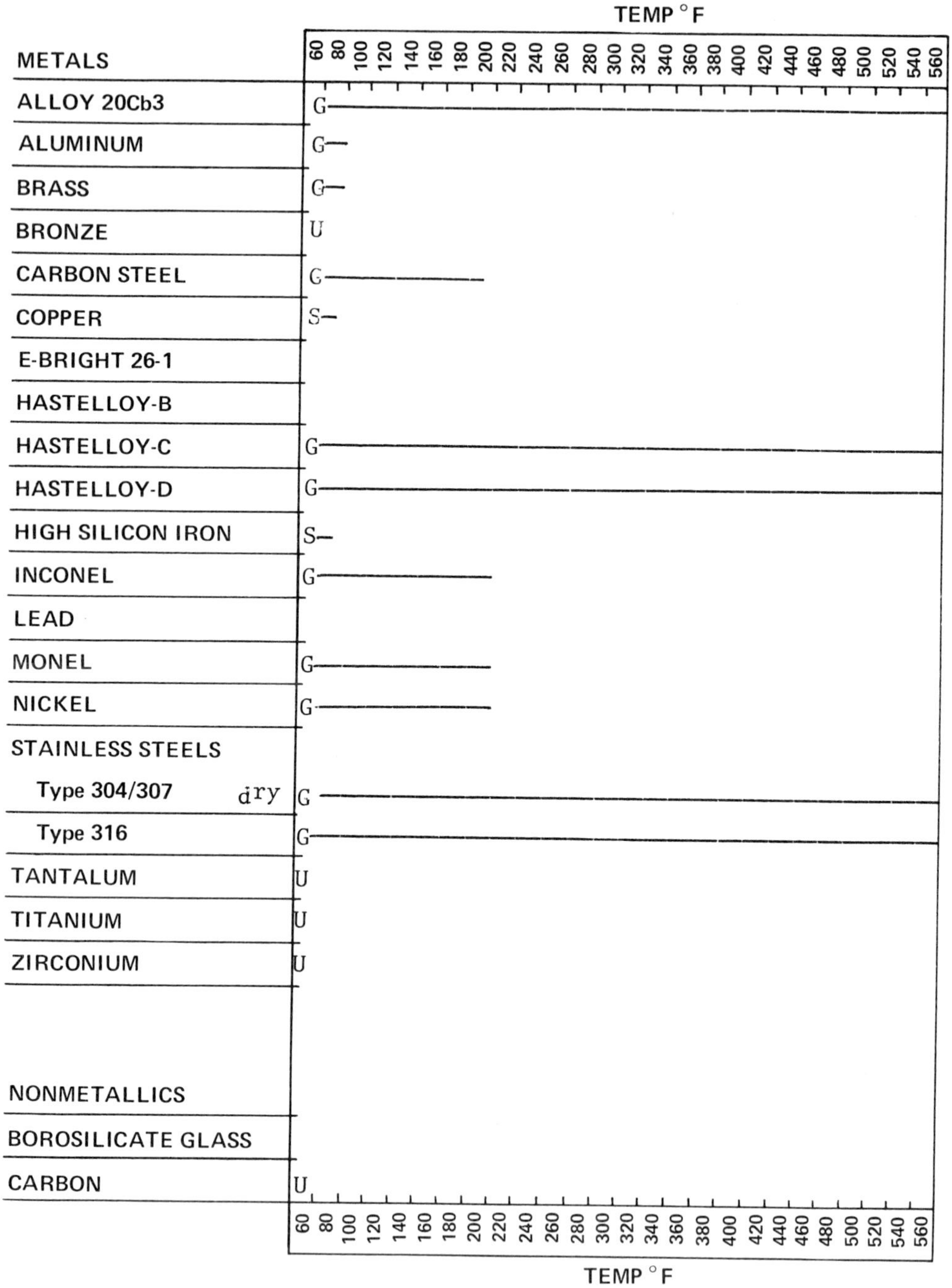

FOR METALS

E = < 2 Mils Penetration/Year; G = < 20 Mils Penetration/Year

S = < 50 Mils Penetration/Year; U = > 50 Mils Penetration/Year

FOR NONMETALLICS

R = Resistant

U = Unsatisfactory

SULFUR TRIOXIDE

TEMP °F (60–460)

PLASTICS	
ABS	U
ACRYLICS	
ASBESTOS REINF., EPOXY	R—
ASBESTOS REINF., PHENOLIC	
CHLORINATED POLYETHER	R—
CPVC	R—
EPOXY	U
FLUOROCARBONS FEP	R—
FLUOROCARBONS TFE	R—
FURFURYL ALCOHOL	U
NORYL	
PHENOLIC	
POLYAMIDES—NYLON	R—
POLYESTERS	
BISPHENOL A-FUMURATE	R—
HYDROGENATED BISPHENOL-A-BISPHENOL-A	
ISOPHTHALIC	R—
CHLORINATED POLYESTERS	R—U
POLYSTYRENES	
POLYETHYLENE	R—
POLYPROPYLENE	U
PVC—TYPE 1	R—
PVC—TYPE 2	R—
POLYVINYLIDENE CHLORIDE	R—
VINYLIDENE FLUORIDE	U
VINYL ESTER	R—

RUBBERS	
BUTYL GR-1	
FLUORO ELASTOMERS VITON A	R—
FLUORO ELASTOMERS KEL-F 3700	
HARD RUBBER	R—
HYPALON	U
KOROSEAL	U
NATURAL RUBBER (GRS)	R—
NEOPRENE GR-M (CR)	U
NITRILE BUNA N (NBR)	RU
NORDEL	R—
POLYBUTADIENE	R—
SILICONE	
POLYURETHANE	

TEMP °F (60–460)

SULFURIC ACID, 103%

TEMP °F

METALS	60 80 100 120 140 160 180 200 220 240 260 280 300 320 340 360 380 400 420 440 460 480 500 520 540 560
ALLOY 20Cb3	
ALUMINUM	
BRASS	
BRONZE	
CARBON STEEL	
COPPER	
E-BRIGHT 26-1	
HASTELLOY-B	
HASTELLOY-C	
HASTELLOY-D	
HIGH SILICON IRON	U
INCONEL	U
LEAD	
MONEL	U
NICKEL	U
STAINLESS STEELS	
Type 304/307	U
Type 316	G—
TANTALUM	U
TITANIUM	
ZIRCONIUM	
NONMETALLICS	
BOROSILICATE GLASS	R———————————————————
CARBON	

TEMP °F

FOR METALS

E = < 2 Mils Penetration/Year; G = < 20 Mils Penetration/Year

S = < 50 Mils Penetration/Year; U = > 50 Mils Penetration/Year

FOR NONMETALLICS

R = Resistant

U = Unsatisfactory

SULFURIC ACID, 103%

TEMP °F (60–460)

PLASTICS	TEMP °F
ABS	U
ACRYLICS	U
ASBESTOS REINF., EPOXY	
ASBESTOS REINF., PHENOLIC	U
CHLORINATED POLYETHER	U
CPVC	U
EPOXY	U
FLUOROCARBONS FEP	R (60–400)
FLUOROCARBONS TFE	R (60–460)
FURFURYL ALCOHOL	
NORYL	
PHENOLIC	U
POLYAMIDES–NYLON	U
POLYESTERS	
BISPHENOL A-FUMURATE	
HYDROGENATED BISPHENOL-A-BISPHENOL-A	U
ISOPHTHALIC	U
CHLORINATED POLYESTERS	
POLYSTYRENES	U
POLYETHYLENE	
POLYPROPYLENE	U
PVC–TYPE 1	U
PVC–TYPE 2	U
POLYVINYLIDENE CHLORIDE	U
VINYLIDENE FLUORIDE	U
VINYL ESTER	U
RUBBERS	
BUTYL GR-1	
FLUORO ELASTOMERS VITON A	
FLUORO ELASTOMERS KEL-F 3700	
HARD RUBBER	
HYPALON	
KOROSEAL	U
NATURAL RUBBER (GRS)	
NEOPRENE GR-M (CR)	U
NITRILE BUNA N (NBR)	
NORDEL	
POLYBUTADIENE	
SILICONE	
POLYURETHANE	U

TEMP °F (60–460)

SULFURIC ACID, 100%

TEMP °F: 60 80 100 120 140 160 180 200 220 240 260 280 300 320 340 360 380 400 420 440 460 480 500 520 540 560

METALS		TEMP °F
ALLOY 20Cb3		E— G—
ALUMINUM		U
BRASS		U
BRONZE		U
CARBON STEEL		G—
COPPER		U
E-BRIGHT 26-1		
HASTELLOY-B		G— U
HASTELLOY-C		E— G— U
HASTELLOY-D		G— U
HIGH SILICON IRON		E—
INCONEL		U
LEAD		G—
MONEL		U
NICKEL		U
STAINLESS STEELS		
Type 304/307	9	G—
Type 316	9	G—
TANTALUM		G—
TITANIUM		U
ZIRCONIUM		U

NONMETALLICS	TEMP °F
BOROSILICATE GLASS	R—
CARBON	U

TEMP °F: 60 80 100 120 140 160 180 200 220 240 260 280 300 320 340 360 380 400 420 440 460 480 500 520 540 560

FOR METALS

E = < 2 Mils Penetration/Year; G = < 20 Mils Penetration/Year
S = < 50 Mils Penetration/Year; U = > 50 Mils Penetration/Year

FOR NONMETALLICS

R = Resistant
U = Unsatisfactory

SULFURIC ACID, 100%

TEMP °F (60–460)

PLASTICS	Rating
ABS	U
ACRYLICS	U
ASBESTOS REINF., EPOXY	U
ASBESTOS REINF., PHENOLIC	
CHLORINATED POLYETHER	U
CPVC	U
EPOXY	U
FLUOROCARBONS FEP	R (60–400 °F)
FLUOROCARBONS TFE	R (60–460 °F)
FURFURYL ALCOHOL	U
NORYL	
PHENOLIC	
POLYAMIDES–NYLON	U
POLYESTERS	
BISPHENOL A-FUMURATE	
HYDROGENATED BISPHENOL-A-BISPHENOL-A	U
ISOPHTHALIC	U
CHLORINATED POLYESTERS	
POLYSTYRENES	U
POLYETHYLENE	U
POLYPROPYLENE	U
PVC–TYPE 1	U
PVC–TYPE 2	U
POLYVINYLIDENE CHLORIDE	U
VINYLIDENE FLUORIDE	
VINYL ESTER	U

RUBBERS	Rating
BUTYL GR-1	U
FLUORO ELASTOMERS VITON A	R (60–200 °F)
FLUORO ELASTOMERS KEL-F 3700	
HARD RUBBER	
HYPALON	
KOROSEAL	U
NATURAL RUBBER (GRS)	U
NEOPRENE GR-M (CR)	U
NITRILE BUNA N (NBR)	U
NORDEL	
POLYBUTADIENE	
SILICONE	U
POLYURETHANE	U

TEMP °F

SULFURIC ACID, 98%

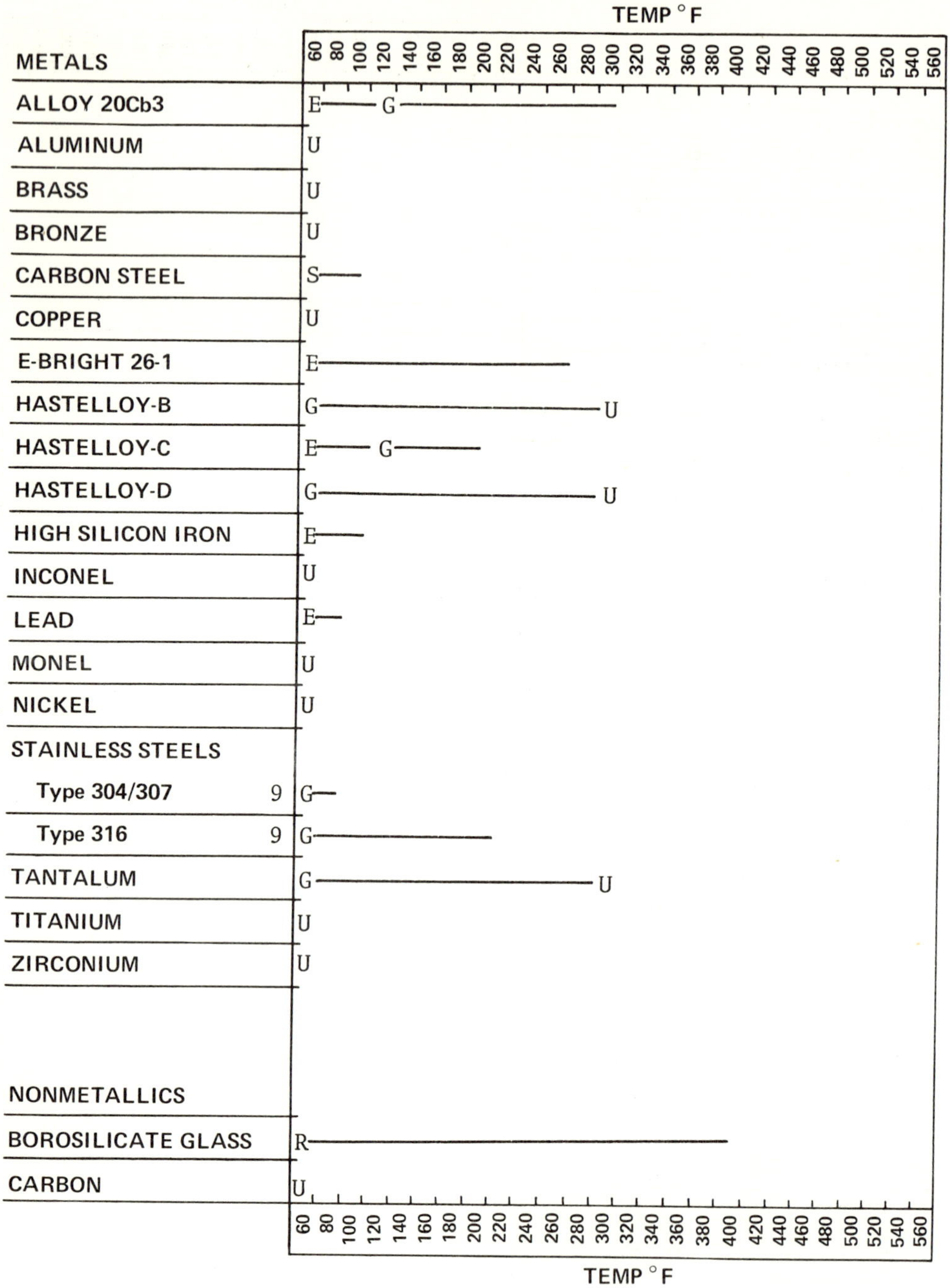

FOR METALS

E = < 2 Mils Penetration/Year; G = < 20 Mils Penetration/Year

S = < 50 Mils Penetration/Year; U = > 50 Mils Penetration/Year

FOR NONMETALLICS

R = Resistant

U = Unsatisfactory

SULFURIC ACID, 98%

TEMP °F: 60 80 100 120 140 160 180 200 220 240 260 280 300 320 340 360 380 400 420 440 460

PLASTICS	
ABS	U
ACRYLICS	U
ASBESTOS REINF., EPOXY	U
ASBESTOS REINF., PHENOLIC	U
CHLORINATED POLYETHER	U
CPVC	U
EPOXY	U
FLUOROCARBONS FEP	R (60–400 °F)
FLUOROCARBONS TFE	R (60–460 °F)
FURFURYL ALCOHOL	U
NORYL	
PHENOLIC	U
POLYAMIDES–NYLON	U
POLYESTERS	
BISPHENOL A-FUMURATE	U
HYDROGENATED BISPHENOL-A-BISPHENOL-A	U
ISOPHTHALIC	U
CHLORINATED POLYESTERS	U
POLYSTYRENES	U
POLYETHYLENE	U
POLYPROPYLENE	R (60–140 °F)
PVC–TYPE 1	U
PVC–TYPE 2	U
POLYVINYLIDENE CHLORIDE	U
VINYLIDENE FLUORIDE	R (60–160 °F)
VINYL ESTER	U
RUBBERS	
BUTYL GR-1	U
FLUORO ELASTOMERS VITON A	R (60–350 °F)
FLUORO ELASTOMERS KEL-F 3700	
HARD RUBBER	
HYPALON	R (60–120 °F) U
KOROSEAL	U
NATURAL RUBBER (GRS)	U
NEOPRENE GR-M (CR)	U
NITRILE BUNA N (NBR)	U
NORDEL	
POLYBUTADIENE	
SILICONE	U
POLYURETHANE	U

TEMP °F: 60 80 100 120 140 160 180 200 220 240 260 280 300 320 340 360 380 400 420 440 460

SULFURIC ACID, 95%

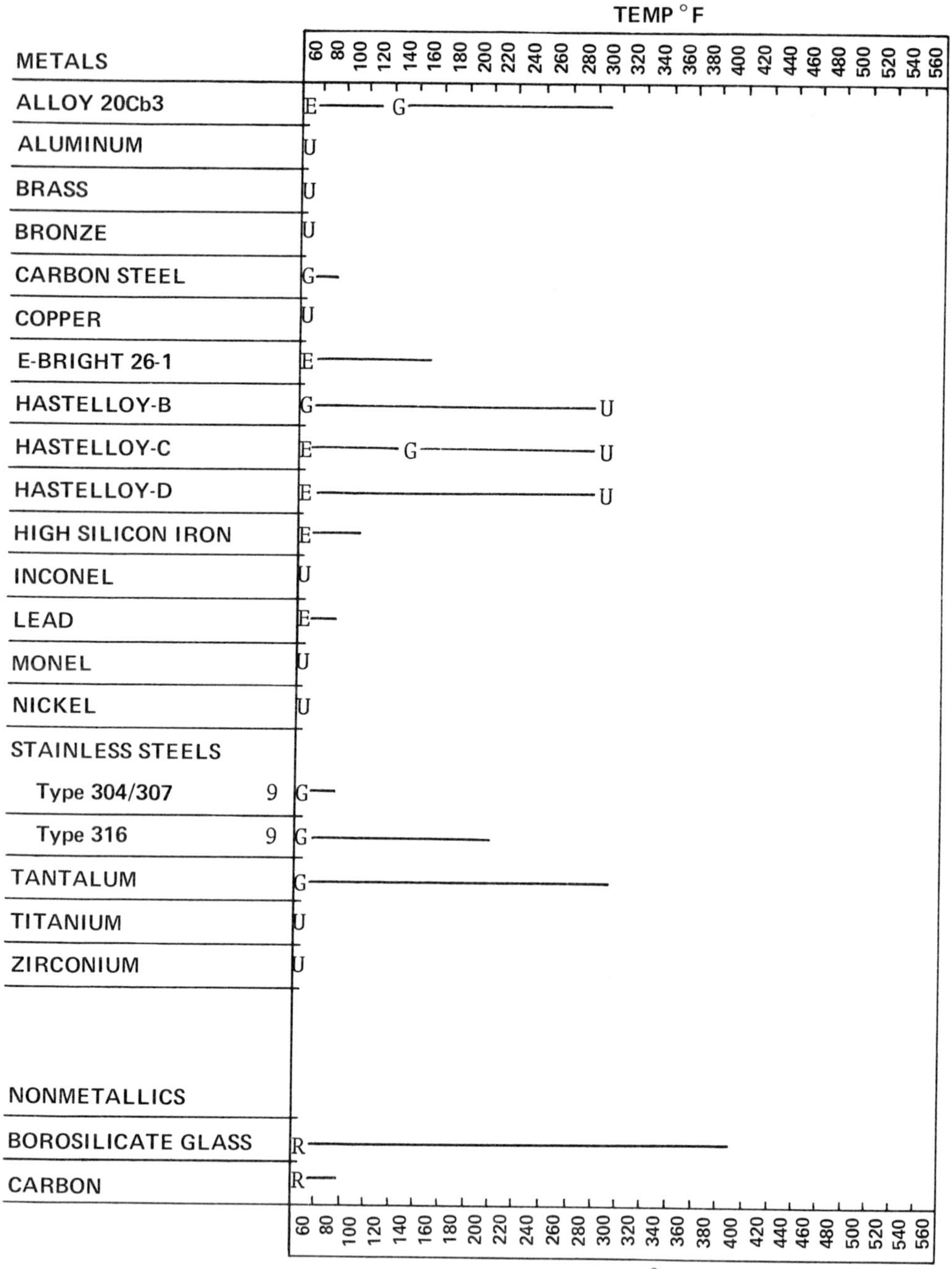

FOR METALS

E = < 2 Mils Penetration/Year; G = < 20 Mils Penetration/Year

S = < 50 Mils Penetration/Year; U = > 50 Mils Penetration/Year

FOR NONMETALLICS

R = Resistant

U = Unsatisfactory

SULFURIC ACID, 95%

TEMP °F: 60 80 100 120 140 160 180 200 220 240 260 280 300 320 340 360 380 400 420 440 460

PLASTICS	
ABS	U
ACRYLICS	U
ASBESTOS REINF., EPOXY	U
ASBESTOS REINF., PHENOLIC	U
CHLORINATED POLYETHER	R
CPVC	U
EPOXY	U
FLUOROCARBONS FEP	R
FLUOROCARBONS TFE	R
FURFURYL ALCOHOL	U
NORYL	R
PHENOLIC	U
POLYAMIDES–NYLON	U
POLYESTERS	
BISPHENOL A-FUMURATE	
HYDROGENATED BISPHENOL-A-BISPHENOL-A	U
ISOPHTHALIC	U
CHLORINATED POLYESTERS	U
POLYSTYRENES	U
POLYETHYLENE	U
POLYPROPYLENE	RU
PVC–TYPE 1	R
PVC–TYPE 2	U
POLYVINYLIDENE CHLORIDE	U
VINYLIDENE FLUORIDE	R
VINYL ESTER	U
RUBBERS	
BUTYL GR-1	U
FLUORO ELASTOMERS VITON A	R
FLUORO ELASTOMERS KEL-F 3700	
HARD RUBBER	U
HYPALON	R
KOROSEAL	U
NATURAL RUBBER (GRS)	U
NEOPRENE GR-M (CR)	U
NITRILE BUNA N (NBR)	U
NORDEL	
POLYBUTADIENE	
SILICONE	U
POLYURETHANE	U

TEMP °F: 60 80 100 120 140 160 180 200 220 240 260 280 300 320 340 360 380 400 420 440 460

SULFURIC ACID, 90%

TEMP °F: 60, 80, 100, 120, 140, 160, 180, 200, 220, 240, 260, 280, 300, 320, 340, 360, 380, 400, 420, 440, 460, 480, 500, 520, 540, 560

METALS	Note	Rating
ALLOY 20Cb3		E—S—U
ALUMINUM		U
BRASS		U
BRONZE		U
CARBON STEEL		U
COPPER		U
E-BRIGHT 26-1		U
HASTELLOY-B		E—G—U
HASTELLOY-C		E—G—U
HASTELLOY-D		E-G—U
HIGH SILICON IRON		E—
INCONEL		U
LEAD		E—
MONEL		U
NICKEL		U
STAINLESS STEELS		
Type 304/307	9	G–U
Type 316	9	G–U
TANTALUM		G—
TITANIUM		U
ZIRCONIUM		U

NONMETALLICS	Rating
BOROSILICATE GLASS	R—
CARBON	R—

FOR METALS

E = < 2 Mils Penetration/Year; G = < 20 Mils Penetration/Year

S = < 50 Mils Penetration/Year; U = > 50 Mils Penetration/Year

FOR NONMETALLICS

R = Resistant

U = Unsatisfactory

SULFURIC ACID, 90%

TEMP °F: 60 80 100 120 140 160 180 200 220 240 260 280 300 320 340 360 380 400 420 440 460

PLASTICS	
ABS	U
ACRYLICS	
ASBESTOS REINF., EPOXY	U
ASBESTOS REINF., PHENOLIC	R—U
CHLORINATED POLYETHER	R—
CPVC	U
EPOXY	U
FLUOROCARBONS FEP	R—
FLUOROCARBONS TFE	R—
FURFURYL ALCOHOL	U
NORYL	R—
PHENOLIC	
POLYAMIDES–NYLON	U
POLYESTERS	
BISPHENOL A-FUMURATE	
HYDROGENATED BISPHENOL-A-BISPHENOL-A	U
ISOPHTHALIC	U
CHLORINATED POLYESTERS	
POLYSTYRENES	U
POLYETHYLENE	U
POLYPROPYLENE	R—
PVC–TYPE 1	R—
PVC–TYPE 2	U
POLYVINYLIDENE CHLORIDE	U
VINYLIDENE FLUORIDE	R—
VINYL ESTER	U
RUBBERS	
BUTYL GR-1	U
FLUORO ELASTOMERS VITON A	R—
FLUORO ELASTOMERS KEL-F 3700	
HARD RUBBER	U
HYPALON	R—
KOROSEAL	U
NATURAL RUBBER (GRS)	U
NEOPRENE GR-M (CR)	U
NITRILE BUNA N (NBR)	U
NORDEL	
POLYBUTADIENE	
SILICONE	U
POLYURETHANE	U

TEMP °F: 60 80 100 120 140 160 180 200 220 240 260 280 300 320 340 360 380 400 420 440 460

SULFURIC ACID, 80%

TEMP °F: 60, 80, 100, 120, 140, 160, 180, 200, 220, 240, 260, 280, 300, 320, 340, 360, 380, 400, 420, 440, 460, 480, 500, 520, 540, 560

METALS		
ALLOY 20Cb3		E—S—U
ALUMINUM		U
BRASS		U
BRONZE		U
CARBON STEEL		U
COPPER		U
E-BRIGHT 26-1		U
HASTELLOY-B		E—G—U
HASTELLOY-C		E—G—U
HASTELLOY-D		E—S—U
HIGH SILICON IRON		E—
INCONEL		U
LEAD		G—
MONEL		G—U
NICKEL		U
STAINLESS STEELS		
Type 304/307	9	G—
Type 316	9	G—S—U
TANTALUM		G—
TITANIUM		U
ZIRCONIUM		G—S—U

NONMETALLICS	
BOROSILICATE GLASS	R—
CARBON	R—

TEMP °F

FOR METALS

E = < 2 Mils Penetration/Year; G = < 20 Mils Penetration/Year
S = < 50 Mils Penetration/Year; U = > 50 Mils Penetration/Year

FOR NONMETALLICS

R = Resistant
U = Unsatisfactory

SULFURIC ACID, 80%

TEMP °F (scale 60–460)

PLASTICS	Rating
ABS	U
ACRYLICS	
ASBESTOS REINF., EPOXY	U
ASBESTOS REINF., PHENOLIC	R (60–120) U
CHLORINATED POLYETHER	R (60–220)
CPVC	R (60–240)
EPOXY	U
FLUOROCARBONS FEP	R (60–400)
FLUOROCARBONS TFE	R (60–460)
FURFURYL ALCOHOL	U
NORYL	R (60–80)
PHENOLIC	U
POLYAMIDES–NYLON	U
POLYESTERS	
BISPHENOL A-FUMURATE	U
HYDROGENATED BISPHENOL-A-BISPHENOL-A	U
ISOPHTHALIC	U
CHLORINATED POLYESTERS	R (60–80)
POLYSTYRENES	U
POLYETHYLENE	U
POLYPROPYLENE	R (60–180)
PVC–TYPE 1	R (60–140)
PVC–TYPE 2	U
POLYVINYLIDENE CHLORIDE	U
VINYLIDENE FLUORIDE	R (60–220)
VINYL ESTER	U

RUBBERS	Rating
BUTYL GR-1	R (60–100)
FLUORO ELASTOMERS VITON A	R (60–340)
FLUORO ELASTOMERS KEL-F 3700	
HARD RUBBER	U
HYPALON	R (60–80)
KOROSEAL	U
NATURAL RUBBER (GRS)	U
NEOPRENE GR-M (CR)	U
NITRILE BUNA N (NBR)	U
NORDEL	
POLYBUTADIENE	
SILICONE	U
POLYURETHANE	U

TEMP °F

SULFURIC ACID, 70%

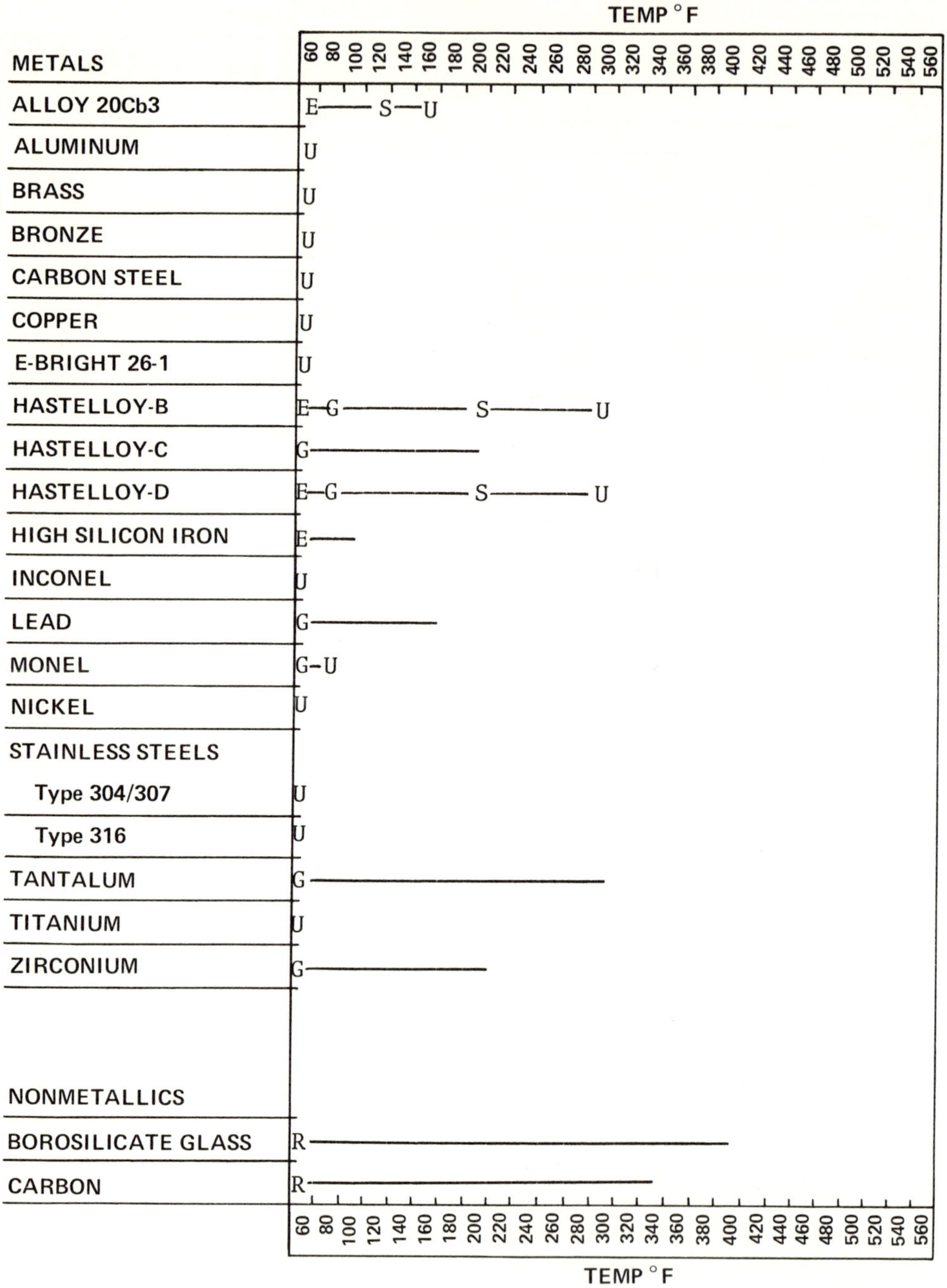

FOR METALS

E = < 2 Mils Penetration/Year; G = < 20 Mils Penetration/Year

S = < 50 Mils Penetration/Year; U = > 50 Mils Penetration/Year

FOR NONMETALLICS

R = Resistant

U = Unsatisfactory

SULFURIC ACID, 70%

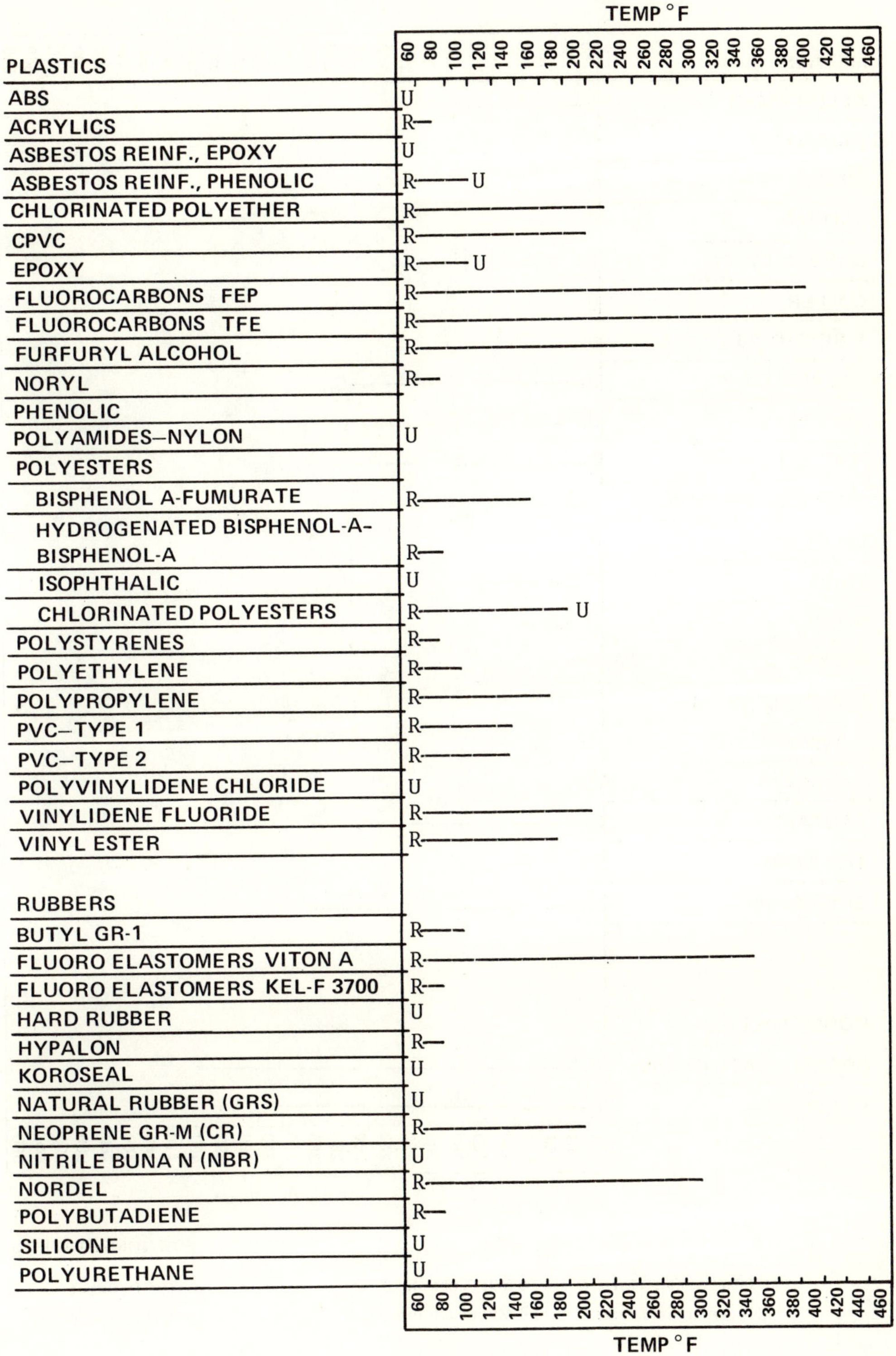

SULFURIC ACID, 60%

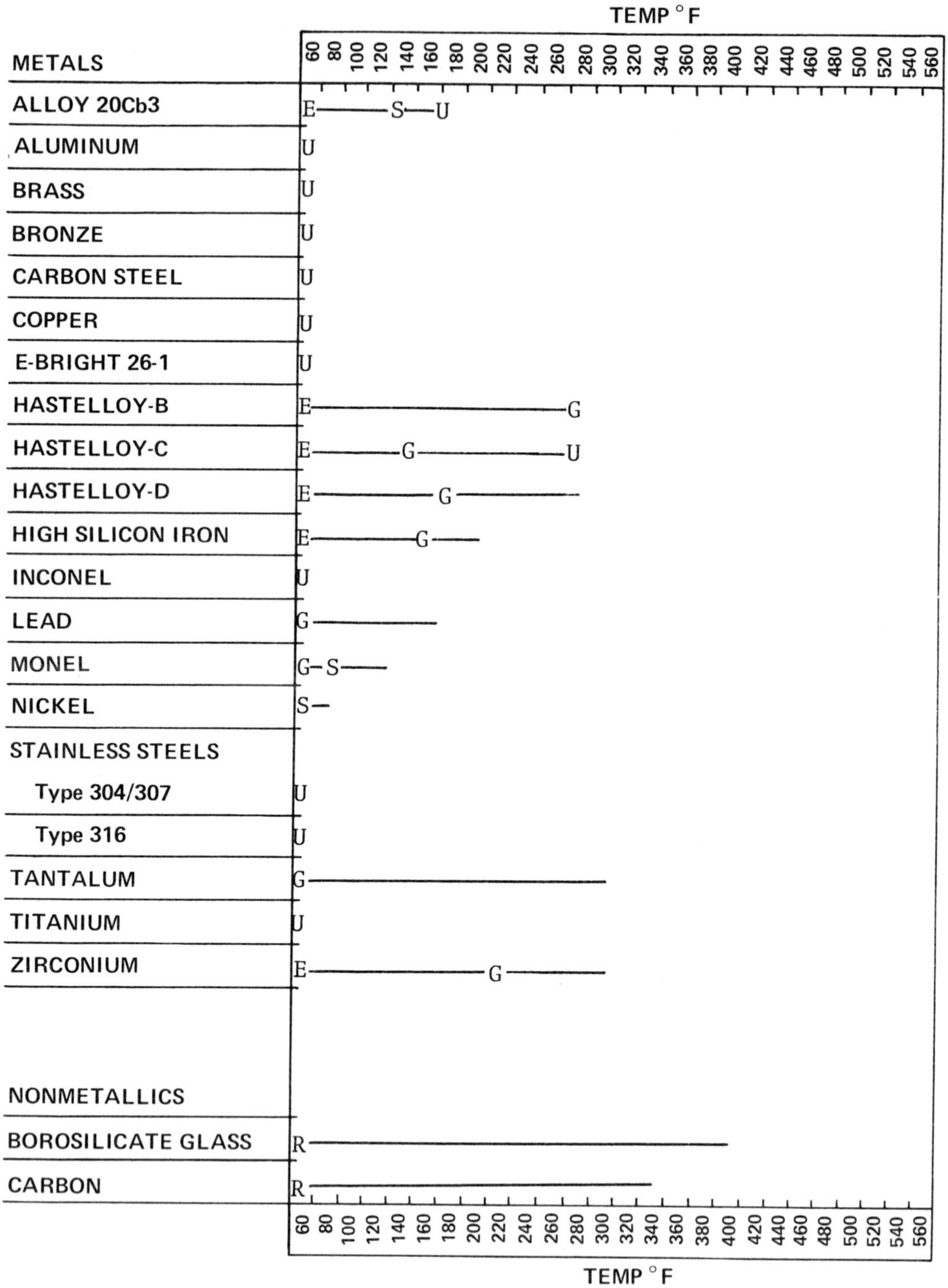

FOR METALS

E = < 2 Mils Penetration/Year; G = < 20 Mils Penetration/Year
S = < 50 Mils Penetration/Year; U = > 50 Mils Penetration/Year

FOR NONMETALLICS

R = Resistant
U = Unsatisfactory

SULFURIC ACID, 60%

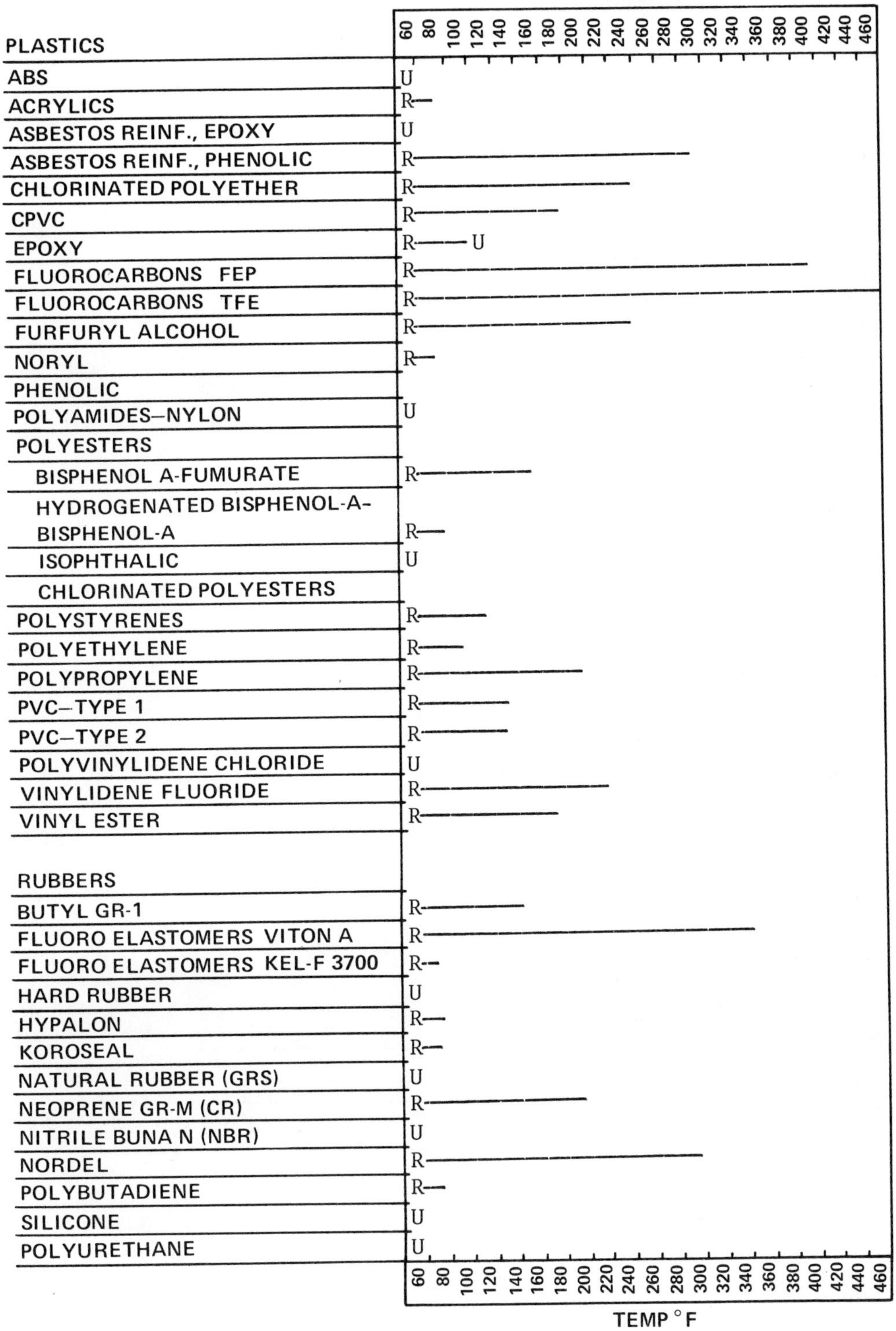

SULFURIC ACID, 50%

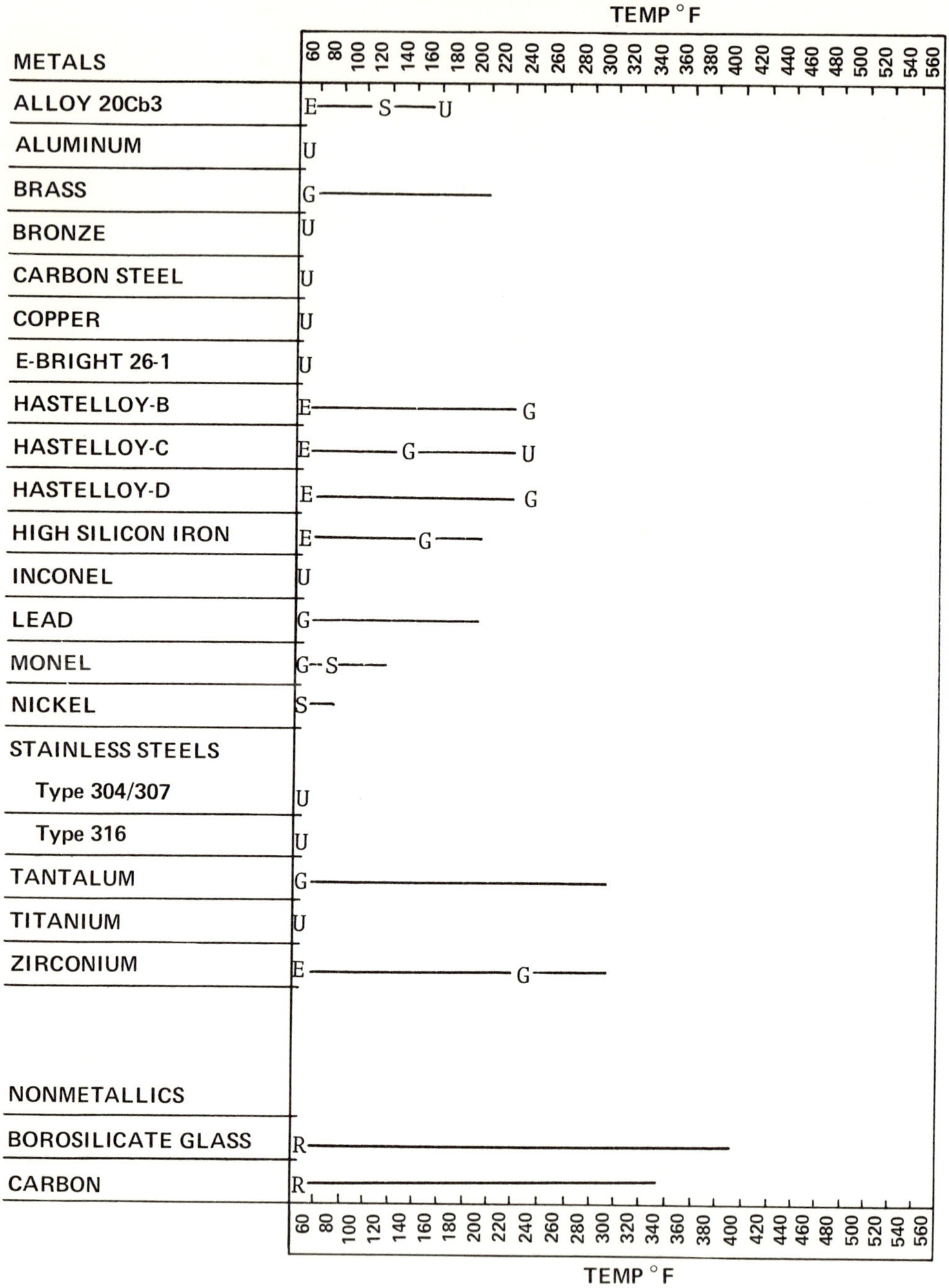

FOR METALS

E = < 2 Mils Penetration/Year; G = < 20 Mils Penetration/Year

S = < 50 Mils Penetration/Year; U = > 50 Mils Penetration/Year

FOR NONMETALLICS

R = Resistant

U = Unsatisfactory

SULFURIC ACID, 50%

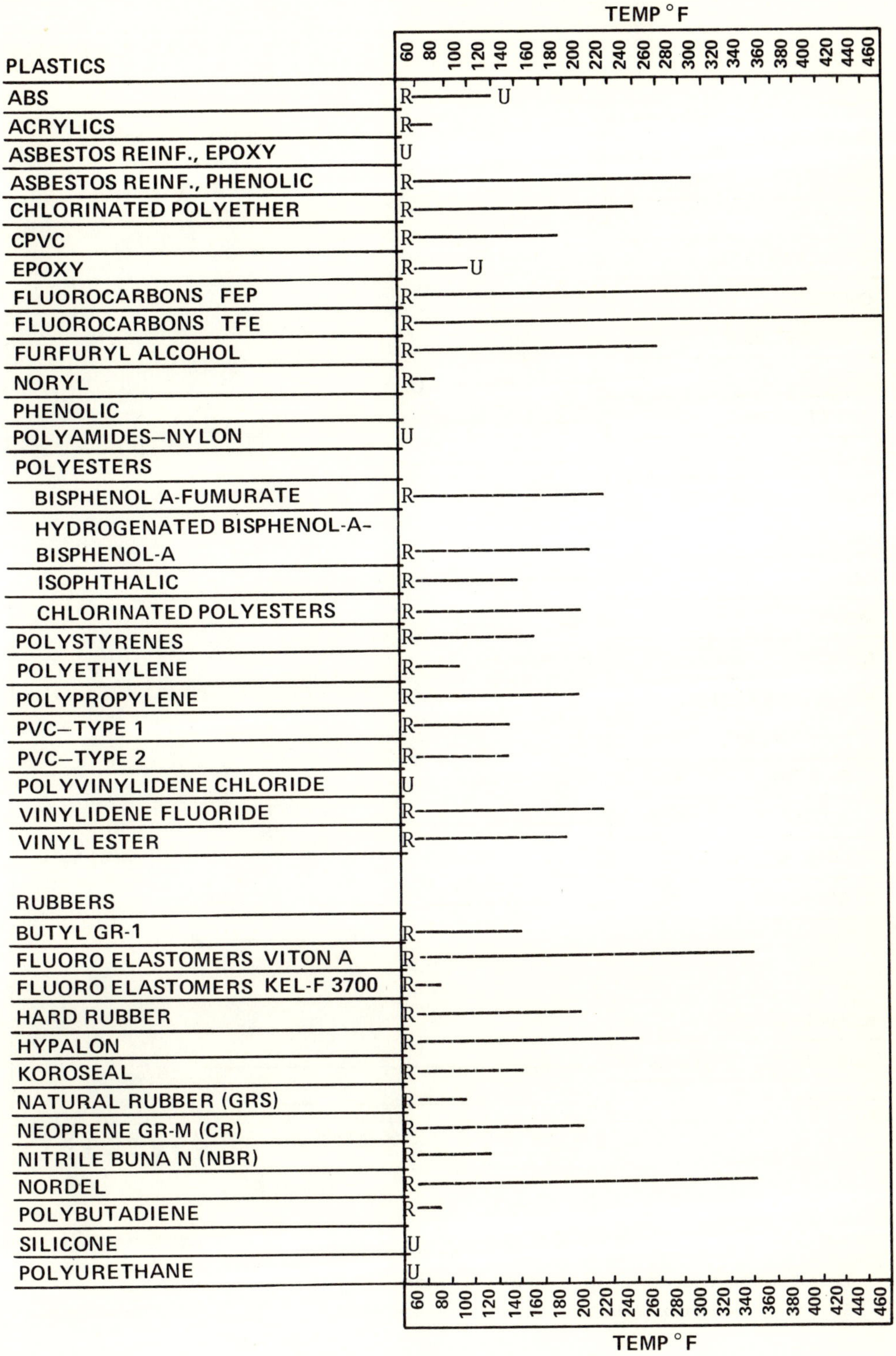

SULFURIC ACID, 30%

METALS	TEMP °F (60–560)
ALLOY 20Cb3	E—— G——
ALUMINUM	U
BRASS	G——
BRONZE	U
CARBON STEEL	U
COPPER	U
E-BRIGHT 26-1	U
HASTELLOY-B	E——G
HASTELLOY-C	E—— G——U
HASTELLOY-D	E—— ——G
HIGH SILICON IRON	E—— G—
INCONEL	U
LEAD	G——
MONEL	S—U
NICKEL	S—U
STAINLESS STEELS	
Type 304/307	U
Type 316	U
TANTALUM	G——
TITANIUM	U
ZIRCONIUM	E——G——

NONMETALLICS	TEMP °F (60–560)
BOROSILICATE GLASS	R——
CARBON	R——

FOR METALS

E = < 2 Mils Penetration/Year; G = < 20 Mils Penetration/Year

S = < 50 Mils Penetration/Year; U = > 50 Mils Penetration/Year

FOR NONMETALLICS

R = Resistant

U = Unsatisfactory

SULFURIC ACID, 30%

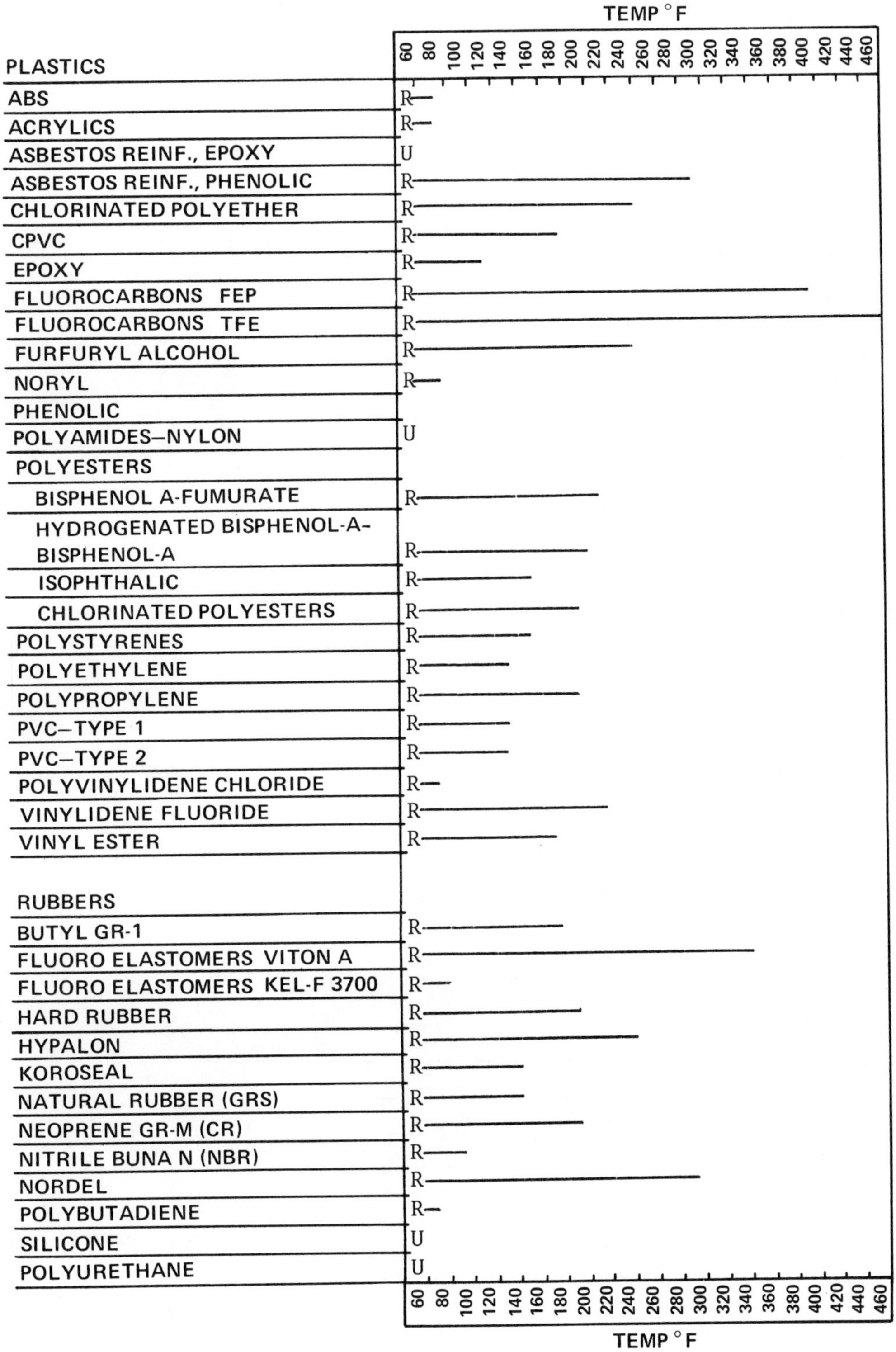

SULFURIC ACID, 10%

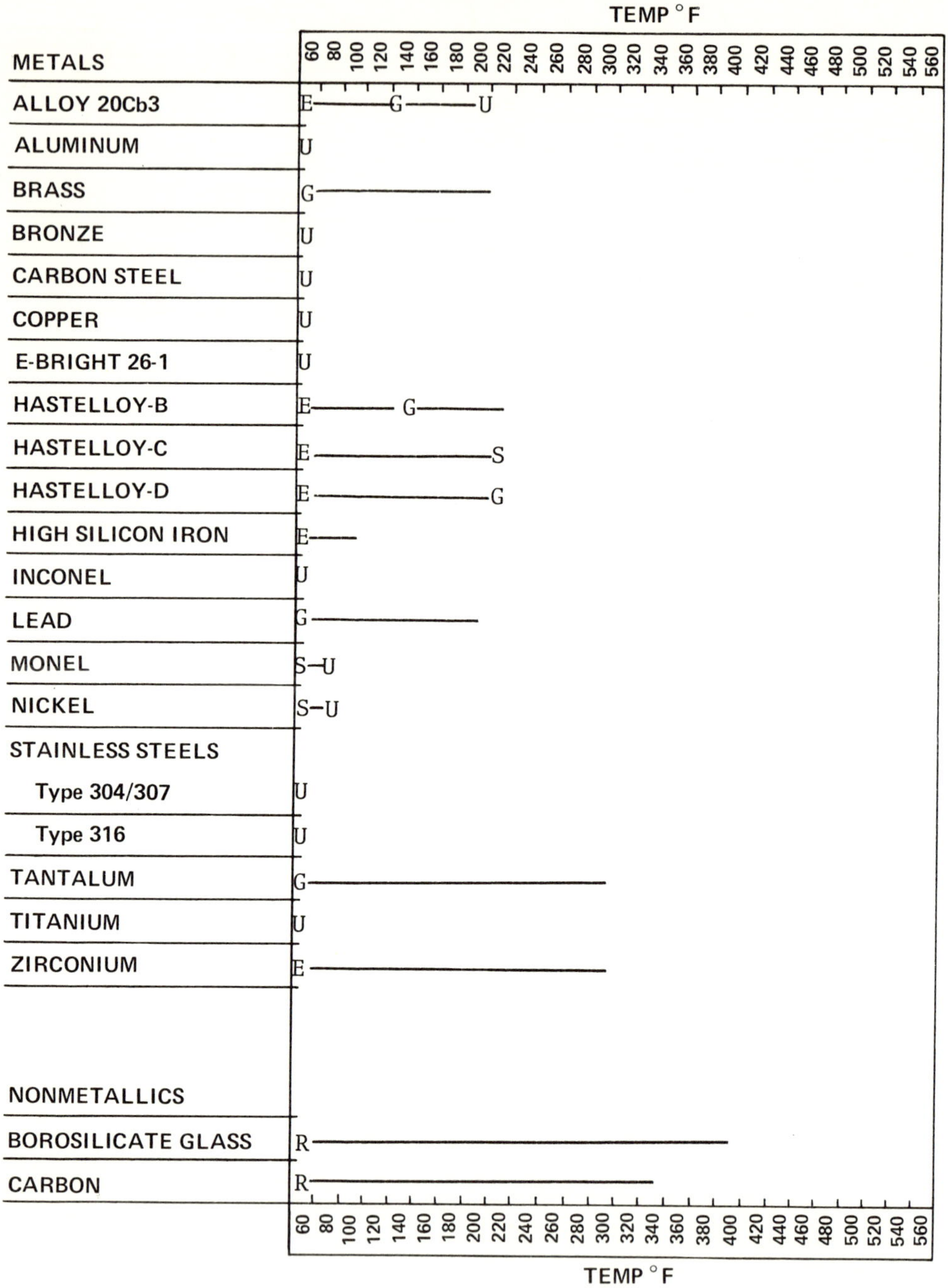

FOR METALS

E = < 2 Mils Penetration/Year; G = < 20 Mils Penetration/Year

S = < 50 Mils Penetration/Year; U = > 50 Mils Penetration/Year

FOR NONMETALLICS

R = Resistant

U = Unsatisfactory

SULFURIC ACID, 10%

TEMP °F: 60 80 100 120 140 160 180 200 220 240 260 280 300 320 340 360 380 400 420 440 460

PLASTICS	
ABS	R
ACRYLICS	R
ASBESTOS REINF., EPOXY	R
ASBESTOS REINF., PHENOLIC	R
CHLORINATED POLYETHER	R
CPVC	R
EPOXY	R — U
FLUOROCARBONS FEP	R
FLUOROCARBONS TFE	R
FURFURYL ALCOHOL	R
NORYL	R
PHENOLIC	R
POLYAMIDES—NYLON	U
POLYESTERS	
BISPHENOL A-FUMURATE	R
HYDROGENATED BISPHENOL-A-BISPHENOL-A	R
ISOPHTHALIC	R
CHLORINATED POLYESTERS	R
POLYSTYRENES	R
POLYETHYLENE	R
POLYPROPYLENE	R
PVC—TYPE 1	R
PVC—TYPE 2	R
POLYVINYLIDENE CHLORIDE	R
VINYLIDENE FLUORIDE	R
VINYL ESTER	R
RUBBERS	
BUTYL GR-1	R
FLUORO ELASTOMERS VITON A	R
FLUORO ELASTOMERS KEL-F 3700	R
HARD RUBBER	R
HYPALON	R
KOROSEAL	R
NATURAL RUBBER (GRS)	R
NEOPRENE GR-M (CR)	R
NITRILE BUNA N (NBR)	R
NORDEL	R
POLYBUTADIENE	R
SILICONE	U
POLYURETHANE	U

TEMP °F: 60 80 100 120 140 160 180 200 220 240 260 280 300 320 340 360 380 400 420 440 460

SULFUROUS ACID

TEMP °F: 60 80 100 120 140 160 180 200 220 240 260 280 300 320 340 360 380 400 420 440 460 480 500 520 540 560

METALS		Rating
ALLOY 20Cb3		G——
ALUMINUM		G—
BRASS		G—
BRONZE		U
CARBON STEEL		U
COPPER		S——
E-BRIGHT 26-1	5%	G——
HASTELLOY-B		G——
HASTELLOY-C		G——
HASTELLOY-D		G——
HIGH SILICON IRON		U
INCONEL		G—
LEAD		G——
MONEL		U
NICKEL		U
STAINLESS STEELS		
Type 304/307		U
Type 316	9	E–G—— U
TANTALUM		E——
TITANIUM		E——
ZIRCONIUM		S——

NONMETALLICS	Rating
BOROSILICATE GLASS	R——
CARBON	R—

TEMP °F: 60 80 100 120 140 160 180 200 220 240 260 280 300 320 340 360 380 400 420 440 460 480 500 520 540 560

FOR METALS

E = < 2 Mils Penetration/Year; G = < 20 Mils Penetration/Year

S = < 50 Mils Penetration/Year; U = > 50 Mils Penetration/Year

FOR NONMETALLICS

R = Resistant

U = Unsatisfactory

SULFUROUS ACID

TEMP °F (60 – 460)

PLASTICS	Conc.	Rating
ABS		R
ACRYLICS		R
ASBESTOS REINF., EPOXY		R
ASBESTOS REINF., PHENOLIC		R
CHLORINATED POLYETHER		R
CPVC		R
EPOXY	to 20%	R
FLUOROCARBONS FEP		R
FLUOROCARBONS TFE		R
FURFURYL ALCOHOL		R
NORYL		
PHENOLIC		R
POLYAMIDES–NYLON		U
POLYESTERS		
BISPHENOL A-FUMURATE		R
HYDROGENATED BISPHENOL-A-BISPHENOL-A	25%	R
ISOPHTHALIC	10%	U
CHLORINATED POLYESTERS	10%	R
POLYSTYRENES	10%	R
POLYETHYLENE		R
POLYPROPYLENE		R
PVC–TYPE 1		R
PVC–TYPE 2		R
POLYVINYLIDENE CHLORIDE		R
VINYLIDENE FLUORIDE		R
VINYL ESTER	10%	R

RUBBERS	Rating
BUTYL GR-1	R
FLUORO ELASTOMERS VITON A	R
FLUORO ELASTOMERS KEL-F 3700	
HARD RUBBER	R
HYPALON	R
KOROSEAL	R
NATURAL RUBBER (GRS)	U
NEOPRENE GR-M (CR)	R
NITRILE BUNA N (NBR)	RU
NORDEL	
POLYBUTADIENE	
SILICONE	
POLYURETHANE	

TEMP °F (60 – 460)

SULFURYL CHLORIDE

TEMP °F: 60 80 100 120 140 160 180 200 220 240 260 280 300 320 340 360 380 400 420 440 460 480 500 520 540 560

METALS	
ALLOY 20Cb3	
ALUMINUM	G
BRASS	
BRONZE	
CARBON STEEL	
COPPER	
E-BRIGHT 26-1	
HASTELLOY-B	
HASTELLOY-C	E
HASTELLOY-D	
HIGH SILICON IRON	G
INCONEL	
LEAD	
MONEL	G
NICKEL	G
STAINLESS STEELS Type 304/307	
Type 316	
TANTALUM	E
TITANIUM	
ZIRCONIUM	

NONMETALLICS	
BOROSILICATE GLASS	
CARBON	

TEMP °F: 60 80 100 120 140 160 180 200 220 240 260 280 300 320 340 360 380 400 420 440 460 480 500 520 540 560

FOR METALS

E = < 2 Mils Penetration/Year; G = < 20 Mils Penetration/Year

S = < 50 Mils Penetration/Year; U = > 50 Mils Penetration/Year

FOR NONMETALLICS

R = Resistant

U = Unsatisfactory

SULFURYL CHLORIDE

PLASTICS	TEMP °F (60 80 100 120 140 160 180 200 220 240 260 280 300 320 340 360 380 400 420 440 460)
ABS	
ACRYLICS	
ASBESTOS REINF., EPOXY	
ASBESTOS REINF., PHENOLIC	
CHLORINATED POLYETHER	
CPVC	
EPOXY	
FLUOROCARBONS FEP	R (60–400)
FLUOROCARBONS TFE	R (60–460)
FURFURYL ALCOHOL	
NORYL	
PHENOLIC	
POLYAMIDES—NYLON	
POLYESTERS	
BISPHENOL A-FUMURATE	U
HYDROGENATED BISPHENOL-A-BISPHENOL-A	
ISOPHTHALIC	
CHLORINATED POLYESTERS	U
POLYSTYRENES	
POLYETHYLENE	U
POLYPROPYLENE	U
PVC—TYPE 1	
PVC—TYPE 2	
POLYVINYLIDENE CHLORIDE	
VINYLIDENE FLUORIDE	
VINYL ESTER	

RUBBERS	TEMP °F (60 80 100 120 140 160 180 200 220 240 260 280 300 320 340 360 380 400 420 440 460)
BUTYL GR-1	
FLUORO ELASTOMERS VITON A	U
FLUORO ELASTOMERS KEL-F 3700	
HARD RUBBER	
HYPALON	
KOROSEAL	
NATURAL RUBBER (GRS)	
NEOPRENE GR-M (CR)	
NITRILE BUNA N (NBR)	
NORDEL	
POLYBUTADIENE	
SILICONE	
POLYURETHANE	

TALL OIL

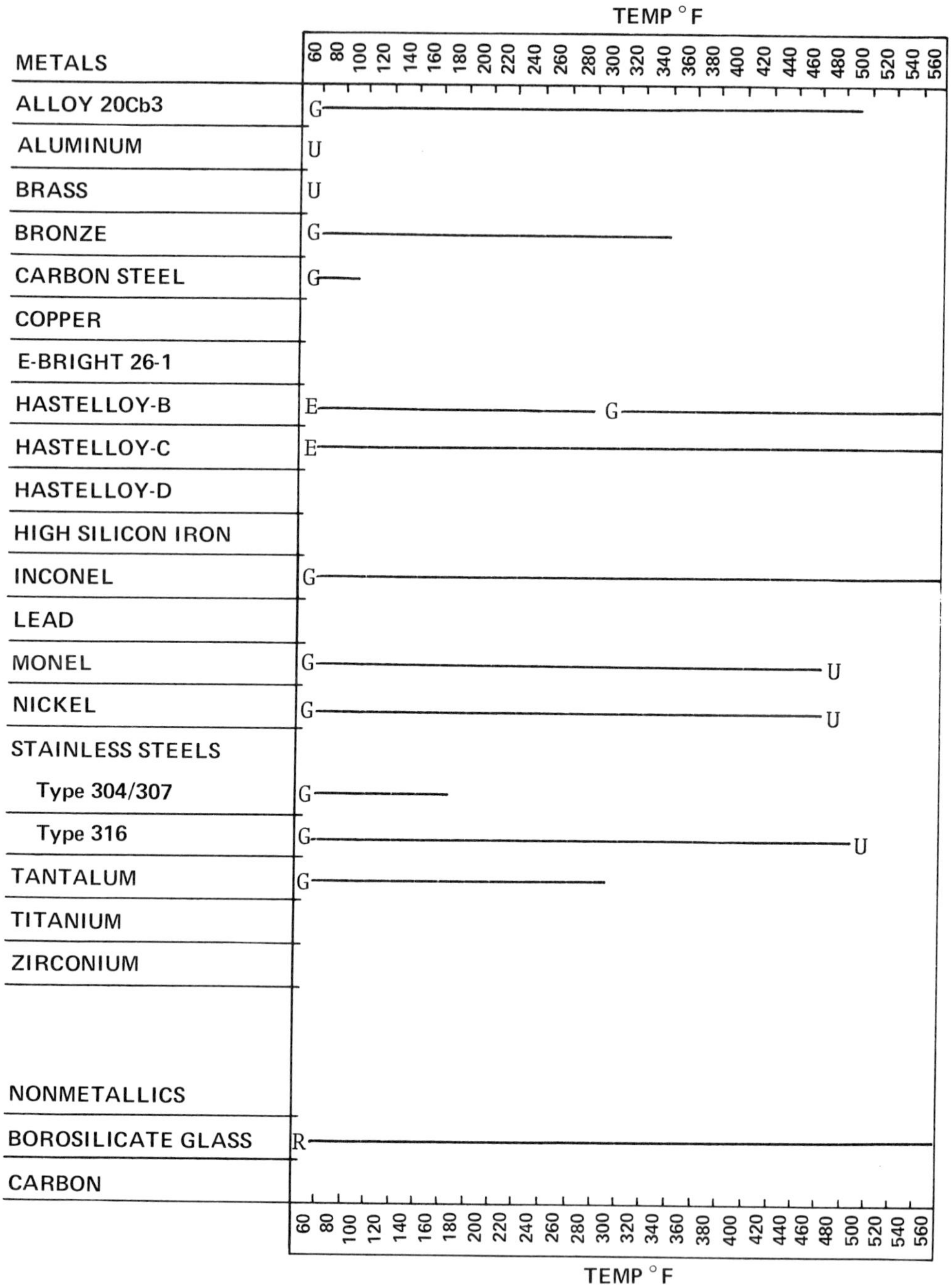

FOR METALS

E = < 2 Mils Penetration/Year; G = < 20 Mils Penetration/Year
S = < 50 Mils Penetration/Year; U = > 50 Mils Penetration/Year

FOR NONMETALLICS

R = Resistant
U = Unsatisfactory

Please see page 1 for notes on calibration.

TALL OIL

TEMP °F: 60, 80, 100, 120, 140, 160, 180, 200, 220, 240, 260, 280, 300, 320, 340, 360, 380, 400, 420, 440, 460

PLASTICS	
ABS	
ACRYLICS	
ASBESTOS REINF., EPOXY	
ASBESTOS REINF., PHENOLIC	
CHLORINATED POLYETHER	R
CPVC	R
EPOXY	
FLUOROCARBONS FEP	R
FLUOROCARBONS TFE	R
FURFURYL ALCOHOL	
NORYL	
PHENOLIC	
POLYAMIDES–NYLON	
POLYESTERS	
BISPHENOL A-FUMURATE	R
HYDROGENATED BISPHENOL-A-BISPHENOL-A	
ISOPHTHALIC	
CHLORINATED POLYESTERS	
POLYSTYRENES	
POLYETHYLENE	R
POLYPROPYLENE	R
PVC–TYPE 1	R
PVC–TYPE 2	R
POLYVINYLIDENE CHLORIDE	
VINYLIDENE FLUORIDE	R
VINYL ESTER	

RUBBERS	
BUTYL GR-1	U
FLUORO ELASTOMERS VITON A	R
FLUORO ELASTOMERS KEL-F 3700	
HARD RUBBER	
HYPALON	
KOROSEAL	
NATURAL RUBBER (GRS)	U
NEOPRENE GR-M (CR)	U
NITRILE BUNA N (NBR)	R
NORDEL	
POLYBUTADIENE	
SILICONE	
POLYURETHANE	

TEMP °F: 60, 80, 100, 120, 140, 160, 180, 200, 220, 240, 260, 280, 300, 320, 340, 360, 380, 400, 420, 440, 460

TANNIC ACID

TEMP °F: 60 80 100 120 140 160 180 200 220 240 260 280 300 320 340 360 380 400 420 440 460 480 500 520 540 560

METALS		
ALLOY 20Cb3		G
ALUMINUM		U
BRASS	air free	G
BRONZE		U
CARBON STEEL		U
COPPER		G
E-BRIGHT 26-1		
HASTELLOY-B	10%	G
HASTELLOY-C	10%	G
HASTELLOY-D		
HIGH SILICON IRON		G
INCONEL		G
LEAD		U
MONEL	10%	G
NICKEL		G
STAINLESS STEELS		
Type 304/307		G
Type 316		G
TANTALUM		G
TITANIUM	10-20%	E
ZIRCONIUM	10-25%	E

NONMETALLICS	
BOROSILICATE GLASS	R
CARBON	R

TEMP °F: 60 80 100 120 140 160 180 200 220 240 260 280 300 320 340 360 380 400 420 440 460 480 500 520 540 560

FOR METALS

E = < 2 Mils Penetration/Year; G = < 20 Mils Penetration/Year

S = < 50 Mils Penetration/Year; U = > 50 Mils Penetration/Year

FOR NONMETALLICS

R = Resistant

U = Unsatisfactory

TANNIC ACID

TEMP °F: 60 80 100 120 140 160 180 200 220 240 260 280 300 320 340 360 380 400 420 440 460

PLASTICS		
ABS		R
ACRYLICS		R
ASBESTOS REINF., EPOXY		
ASBESTOS REINF., PHENOLIC		
CHLORINATED POLYETHER		R
CPVC		R
EPOXY		R — U
FLUOROCARBONS FEP		R
FLUOROCARBONS TFE		R
FURFURYL ALCOHOL		R
NORYL	10%	R
PHENOLIC		
POLYAMIDES–NYLON		R
POLYESTERS		
BISPHENOL A-FUMURATE	sat	R
HYDROGENATED BISPHENOL-A-BISPHENOL-A		R
ISOPHTHALIC	sat	R
CHLORINATED POLYESTERS	all	R
POLYSTYRENES		
POLYETHYLENE		R
POLYPROPYLENE		R
PVC–TYPE 1		R
PVC–TYPE 2		R
POLYVINYLIDENE CHLORIDE		R
VINYLIDENE FLUORIDE		R
VINYL ESTER	all	R

RUBBERS		
BUTYL GR-1		R
FLUORO ELASTOMERS VITON A		R
FLUORO ELASTOMERS KEL-F 3700		
HARD RUBBER		R
HYPALON		R
KOROSEAL		R
NATURAL RUBBER (GRS)		R
NEOPRENE GR-M (CR)		U
NITRILE BUNA N (NBR)		U
NORDEL		
POLYBUTADIENE		
SILICONE		
POLYURETHANE		R

TEMP °F: 60 80 100 120 140 160 180 200 220 240 260 280 300 320 340 360 380 400 420 440 460

TANNING LIQUORS

TEMP °F: 60 80 100 120 140 160 180 200 220 240 260 280 300 320 340 360 380 400 420 440 460 480 500 520 540 560

METALS	Rating
ALLOY 20Cb3	E——G——S
ALUMINUM	
BRASS	
BRONZE	
CARBON STEEL	
COPPER	
E-BRIGHT 26-1	
HASTELLOY-B	
HASTELLOY-C	E—
HASTELLOY-D	
HIGH SILICON IRON	G—
INCONEL	
LEAD	G—
MONEL	
NICKEL	
STAINLESS STEELS	
Type 304/307	E—G— S—U
Type 316	E—G— S—U
TANTALUM	
TITANIUM	
ZIRCONIUM	

NONMETALLICS	Rating
BOROSILICATE GLASS	R————————
CARBON	R————————————————

TEMP °F: 60 80 100 120 140 160 180 200 220 240 260 280 300 320 340 360 380 400 420 440 460 480 500 520 540 560

FOR METALS

E = < 2 Mils Penetration/Year; G = < 20 Mils Penetration/Year

S = < 50 Mils Penetration/Year; U = > 50 Mils Penetration/Year

FOR NONMETALLICS

R = Resistant

U = Unsatisfactory

TANNING LIQUORS

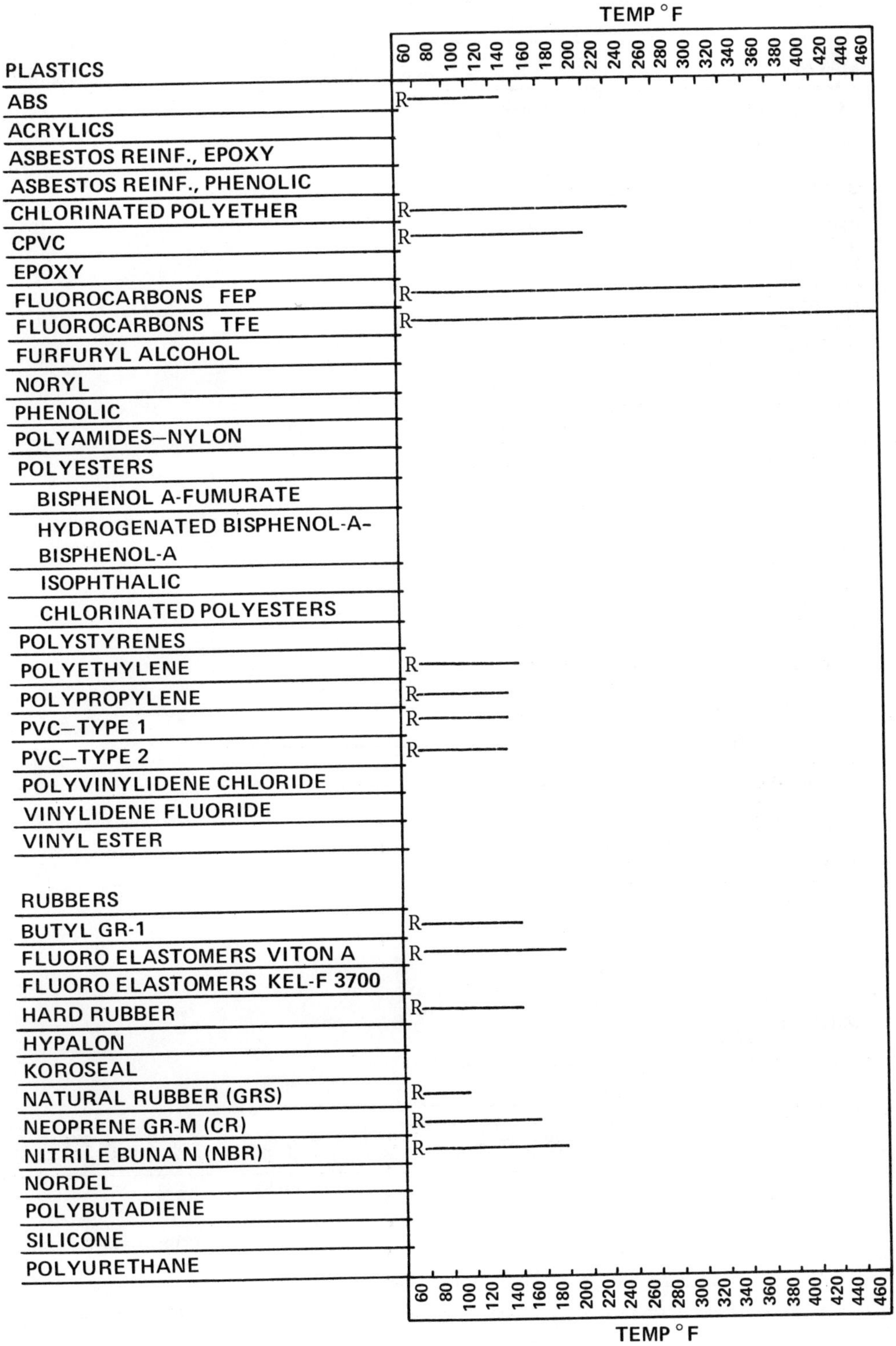

TARTARIC ACID

TEMP °F (60–560)

METALS		
ALLOY 20Cb3	to 50%	E
ALUMINUM	to 50%	G-U
BRASS	2	S
BRONZE		G
CARBON STEEL		U
COPPER		G
E-BRIGHT 26-1	50%	E
HASTELLOY-B	all	G
HASTELLOY-C	all	G
HASTELLOY-D	all	G
HIGH SILICON IRON		G
INCONEL	to 60%	G
LEAD		U
MONEL	to 50%	G
NICKEL	to 50%	G
STAINLESS STEELS		
Type 304/307		E
Type 316		E
TANTALUM	to 50%	E
TITANIUM	to 50%	E
ZIRCONIUM	all	E

NONMETALLICS	
BOROSILICATE GLASS	R
CARBON	R

TEMP °F

FOR METALS

E = < 2 Mils Penetration/Year; G = < 20 Mils Penetration/Year

S = < 50 Mils Penetration/Year; U = > 50 Mils Penetration/Year

FOR NONMETALLICS

R = Resistant

U = Unsatisfactory

PLASTICS	Conc.	Rating / TEMP °F
ABS		R to 140
ACRYLICS		
ASBESTOS REINF., EPOXY		
ASBESTOS REINF., PHENOLIC		
CHLORINATED POLYETHER		R to 250
CPVC		R to 180
EPOXY	11%	R to 150, U at 160
FLUOROCARBONS FEP		R to 400
FLUOROCARBONS TFE		R to 460
FURFURYL ALCOHOL	sat	R to 270
NORYL		
PHENOLIC		
POLYAMIDES–NYLON		R to 80
POLYESTERS		
BISPHENOL A-FUMURATE	sat	R to 220
HYDROGENATED BISPHENOL-A-BISPHENOL-A		R to 210
ISOPHTHALIC	sat	R to 180
CHLORINATED POLYESTERS	all	R to 250
POLYSTYRENES		
POLYETHYLENE		R to 150
POLYPROPYLENE		R to 150
PVC–TYPE 1		R to 140
PVC–TYPE 2		R to 140
POLYVINYLIDENE CHLORIDE		R to 150
VINYLIDENE FLUORIDE		R to 250
VINYL ESTER	all	R to 210

RUBBERS	Conc.	Rating / TEMP °F
BUTYL GR-1		R to 190
FLUORO ELASTOMERS VITON A		R to 190
FLUORO ELASTOMERS KEL-F 3700		
HARD RUBBER		R to 200
HYPALON		R to 200
KOROSEAL		R to 150
NATURAL RUBBER (GRS)		R to 150
NEOPRENE GR-M (CR)		R to 200
NITRILE BUNA N (NBR)		R to 190
NORDEL		
POLYBUTADIENE		
SILICONE		
POLYURETHANE		R to 90

TEMP °F: 60 80 100 120 140 160 180 200 220 240 260 280 300 320 340 360 380 400 420 440 460

TETRAETHYL LEAD

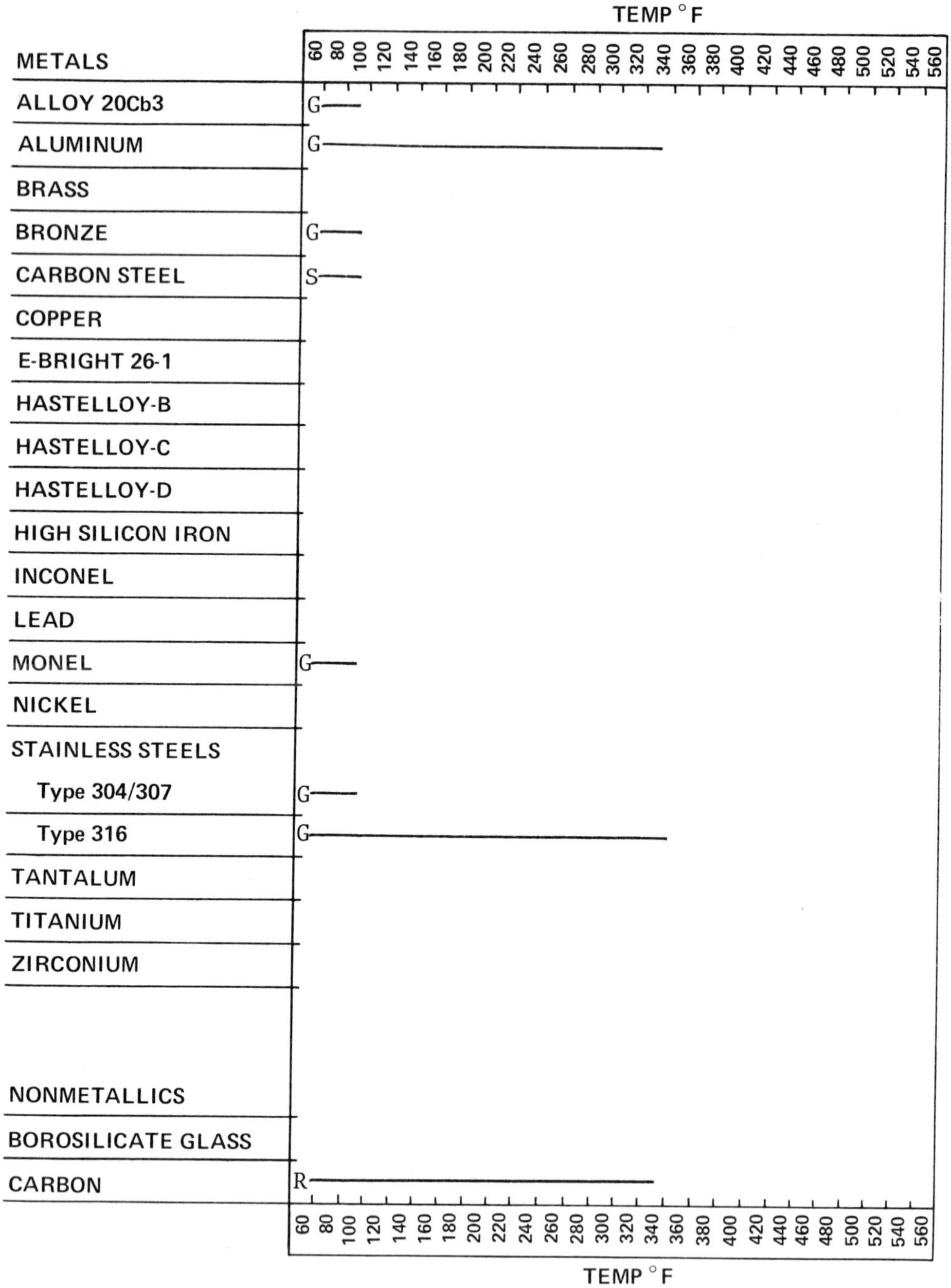

FOR METALS

E = < 2 Mils Penetration/Year; G = < 20 Mils Penetration/Year

S = < 50 Mils Penetration/Year; U = > 50 Mils Penetration/Year

FOR NONMETALLICS

R = Resistant

U = Unsatisfactory

TETRAETHYL LEAD

TEMP °F: 60 80 100 120 140 160 180 200 220 240 260 280 300 320 340 360 380 400 420 440 460

PLASTICS	
ABS	U
ACRYLICS	
ASBESTOS REINF., EPOXY	
ASBESTOS REINF., PHENOLIC	
CHLORINATED POLYETHER	R
CPVC	R U
EPOXY	R
FLUOROCARBONS FEP	R
FLUOROCARBONS TFE	R
FURFURYL ALCOHOL	
NORYL	
PHENOLIC	
POLYAMIDES–NYLON	
POLYESTERS	
BISPHENOL A-FUMURATE	
HYDROGENATED BISPHENOL-A-BISPHENOL-A	
ISOPHTHALIC	
CHLORINATED POLYESTERS	
POLYSTYRENES	
POLYETHYLENE	R
POLYPROPYLENE	R
PVC–TYPE 1	R
PVC–TYPE 2	R
POLYVINYLIDENE CHLORIDE	
VINYLIDENE FLUORIDE	R
VINYL ESTER	

RUBBERS	
BUTYL GR-1	
FLUORO ELASTOMERS VITON A	R
FLUORO ELASTOMERS KEL-F 3700	
HARD RUBBER	
HYPALON	U
KOROSEAL	
NATURAL RUBBER (GRS)	
NEOPRENE GR-M (CR)	
NITRILE BUNA N (NBR)	RU
NORDEL	
POLYBUTADIENE	
SILICONE	
POLYURETHANE	

TEMP °F: 60 80 100 120 140 160 180 200 220 240 260 280 300 320 340 360 380 400 420 440 460

TETRAHYDROFURAN

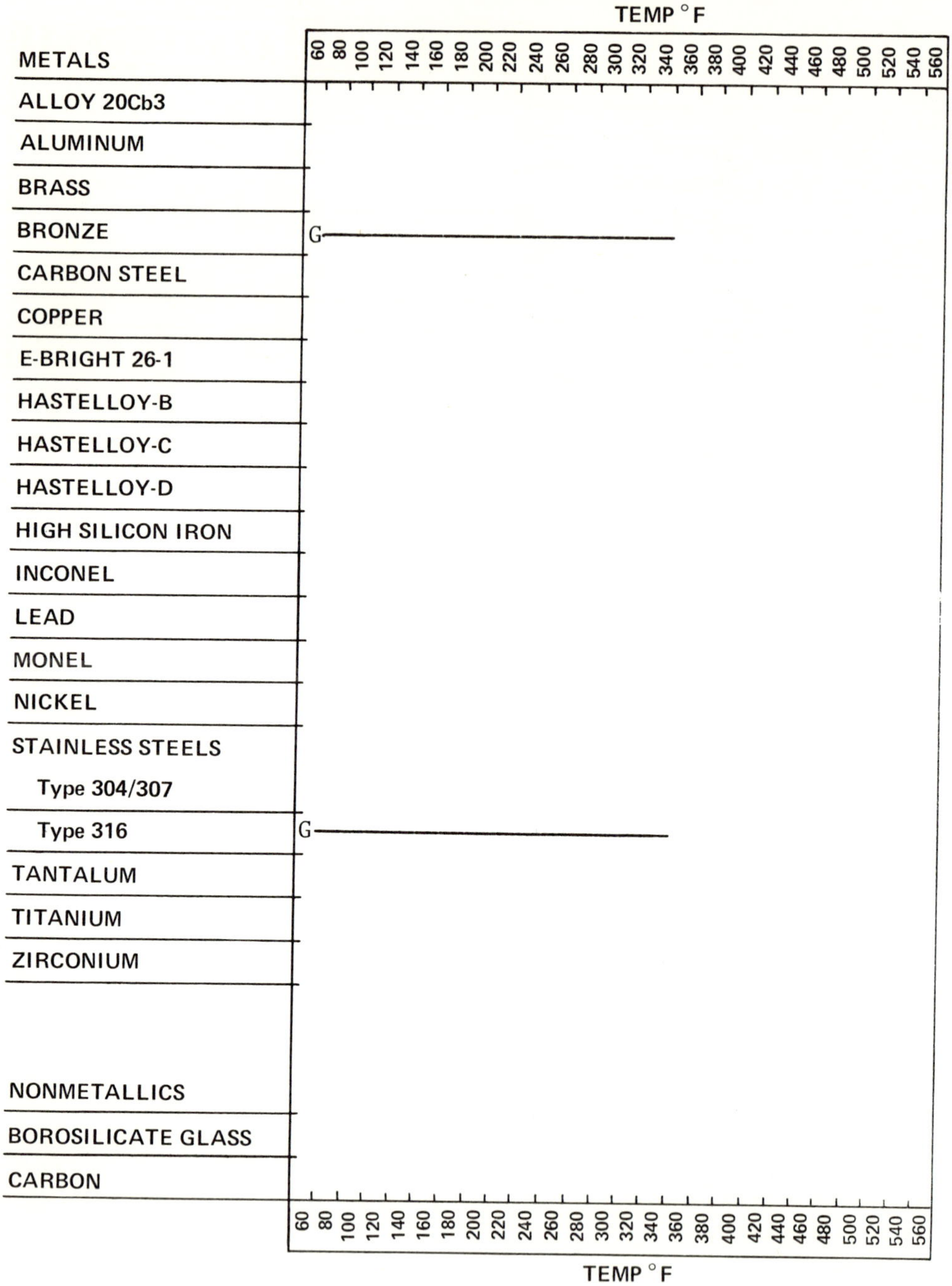

FOR METALS

E = < 2 Mils Penetration/Year; G = < 20 Mils Penetration/Year

S = < 50 Mils Penetration/Year; U = > 50 Mils Penetration/Year

FOR NONMETALLICS

R = Resistant

U = Unsatisfactory

TETRAHYDROFURAN

PLASTICS	60	80	100	120	140	160	180	200	220	240	260	280	300	320	340	360	380	400	420	440	460
ABS	U																				
ACRYLICS																					
ASBESTOS REINF., EPOXY																					
ASBESTOS REINF., PHENOLIC																					
CHLORINATED POLYETHER	R	—																			
CPVC	U																				
EPOXY																					
FLUOROCARBONS FEP	R	—	—	—	—	—	—	—	—	—	—	—	—	—	—	—	—	—			
FLUOROCARBONS TFE	R	—	—	—	—	—	—	—	—	—	—	—	—	—	—	—	—	—	—	—	—
FURFURYL ALCOHOL	R	—	—																		
NORYL																					
PHENOLIC																					
POLYAMIDES—NYLON	R	—	—	—	—	—	—	—													
POLYESTERS																					
BISPHENOL A-FUMURATE																					
HYDROGENATED BISPHENOL-A-BISPHENOL-A																					
ISOPHTHALIC																					
CHLORINATED POLYESTERS																					
POLYSTYRENES																					
POLYETHYLENE	U																				
POLYPROPYLENE	RU																				
PVC—TYPE 1	U																				
PVC—TYPE 2	U																				
POLYVINYLIDENE CHLORIDE																					
VINYLIDENE FLUORIDE	RU																				
VINYL ESTER																					
RUBBERS																					
BUTYL GR-1																					
FLUORO ELASTOMERS VITON A	RU																				
FLUORO ELASTOMERS KEL-F 3700																					
HARD RUBBER																					
HYPALON	U																				
KOROSEAL																					
NATURAL RUBBER (GRS)																					
NEOPRENE GR-M (CR)																					
NITRILE BUNA N (NBR)	U																				
NORDEL																					
POLYBUTADIENE																					
SILICONE																					
POLYURETHANE																					

TEMP °F

TETRAMETHYL AMMONIUM HYDROXIDE, 50%

TEMP °F

METALS	60	80	100	120	140	160	180	200	220	240	260	280	300	320	340	360	380	400	420	440	460	480	500	520	540	560
ALLOY 20Cb3																										
ALUMINUM																										
BRASS																										
BRONZE																										
CARBON STEEL																										
COPPER																										
E-BRIGHT 26-1																										
HASTELLOY-B																										
HASTELLOY-C																										
HASTELLOY-D																										
HIGH SILICON IRON																										
INCONEL																										
LEAD																										
MONEL																										
NICKEL																										
STAINLESS STEELS																										
Type 304/307																										
Type 316																										
TANTALUM																										
TITANIUM																										
ZIRCONIUM																										
NONMETALLICS																										
BOROSILICATE GLASS																										
CARBON																										

60 80 100 120 140 160 180 200 220 240 260 280 300 320 340 360 380 400 420 440 460 480 500 520 540 560

TEMP °F

FOR METALS

E = < 2 Mils Penetration/Year; G = < 20 Mils Penetration/Year
S = < 50 Mils Penetration/Year; U = > 50 Mils Penetration/Year

FOR NONMETALLICS

R = Resistant
U = Unsatisfactory

TETRAMETHYL AMMONIUM HYDROXIDE, 50%

TEMP °F

60 80 100 120 140 160 180 200 220 240 260 280 300 320 340 360 380 400 420 440 460

PLASTICS	
ABS	
ACRYLICS	
ASBESTOS REINF., EPOXY	
ASBESTOS REINF., PHENOLIC	
CHLORINATED POLYETHER	R
CPVC	
EPOXY	
FLUOROCARBONS FEP	R
FLUOROCARBONS TFE	R
FURFURYL ALCOHOL	
NORYL	
PHENOLIC	
POLYAMIDES–NYLON	
POLYESTERS	
BISPHENOL A-FUMURATE	
HYDROGENATED BISPHENOL-A-BISPHENOL-A	
ISOPHTHALIC	
CHLORINATED POLYESTERS	
POLYSTYRENES	
POLYETHYLENE	
POLYPROPYLENE	R
PVC–TYPE 1	
PVC–TYPE 2	
POLYVINYLIDENE CHLORIDE	R
VINYLIDENE FLUORIDE	R
VINYL ESTER	
RUBBERS	
BUTYL GR-1	
FLUORO ELASTOMERS VITON A	
FLUORO ELASTOMERS KEL-F 3700	
HARD RUBBER	
HYPALON	
KOROSEAL	
NATURAL RUBBER (GRS)	
NEOPRENE GR-M (CR)	
NITRILE BUNA N (NBR)	
NORDEL	
POLYBUTADIENE	
SILICONE	
POLYURETHANE	

60 80 100 120 140 160 180 200 220 240 260 280 300 320 340 360 380 400 420 440 460

TEMP °F

TETRALIN

TEMP °F: 60 80 100 120 140 160 180 200 220 240 260 280 300 320 340 360 380 400 420 440 460 480 500 520 540 560

METALS	TEMP °F (60–560)
ALLOY 20Cb3	
ALUMINUM	G—
BRASS	
BRONZE	
CARBON STEEL	
COPPER	
E-BRIGHT 26-1	
HASTELLOY-B	
HASTELLOY-C	
HASTELLOY-D	
HIGH SILICON IRON	
INCONEL	
LEAD	
MONEL	
NICKEL	
STAINLESS STEELS	
Type 304/307	
Type 316	
TANTALUM	G—
TITANIUM	
ZIRCONIUM	

NONMETALLICS	TEMP °F (60–560)
BOROSILICATE GLASS	R———
CARBON	

TEMP °F

FOR METALS

E = < 2 Mils Penetration/Year; G = < 20 Mils Penetration/Year

S = < 50 Mils Penetration/Year; U = > 50 Mils Penetration/Year

FOR NONMETALLICS

R = Resistant

U = Unsatisfactory

TETRALIN

PLASTICS	TEMP °F (60–460)
ABS	
ACRYLICS	
ASBESTOS REINF., EPOXY	
ASBESTOS REINF., PHENOLIC	
CHLORINATED POLYETHER	
CPVC	
EPOXY	
FLUOROCARBONS FEP	R————————————
FLUOROCARBONS TFE	R———————————————
FURFURYL ALCOHOL	
NORYL	
PHENOLIC	
POLYAMIDES—NYLON	
POLYESTERS	
BISPHENOL A-FUMURATE	
HYDROGENATED BISPHENOL-A-BISPHENOL-A	
ISOPHTHALIC	
CHLORINATED POLYESTERS	
POLYSTYRENES	
POLYETHYLENE	R——
POLYPROPYLENE	R———U
PVC—TYPE 1	
PVC—TYPE 2	
POLYVINYLIDENE CHLORIDE	
VINYLIDENE FLUORIDE	
VINYL ESTER	

RUBBERS	TEMP °F (60–460)
BUTYL GR-1	U
FLUORO ELASTOMERS VITON A	
FLUORO ELASTOMERS KEL-F 3700	
HARD RUBBER	
HYPALON	
KOROSEAL	
NATURAL RUBBER (GRS)	U
NEOPRENE GR-M (CR)	U
NITRILE BUNA N (NBR)	U
NORDEL	
POLYBUTADIENE	
SILICONE	
POLYURETHANE	

TEMP °F: 60 80 100 120 140 160 180 200 220 240 260 280 300 320 340 360 380 400 420 440 460

TETRAPHOSPHORIC ACID

TEMP °F (scale: 60, 80, 100, 120, 140, 160, 180, 200, 220, 240, 260, 280, 300, 320, 340, 360, 380, 400, 420, 440, 460, 480, 500, 520, 540, 560)

METALS	Rating	Line extends to (°F)
ALLOY 20Cb3	G	220
ALUMINUM	U	
BRASS	S	120
BRONZE		
CARBON STEEL	U	
COPPER		
E-BRIGHT 26-1		
HASTELLOY-B	G	300
HASTELLOY-C	G	300
HASTELLOY-D		
HIGH SILICON IRON		
INCONEL	G	210
LEAD		
MONEL	S	80
NICKEL	U	
STAINLESS STEELS		
Type 304/307	U	
Type 316	U	
TANTALUM		
TITANIUM		
ZIRCONIUM		

NONMETALLICS	Rating
BOROSILICATE GLASS	
CARBON	

TEMP °F

FOR METALS

E = < 2 Mils Penetration/Year; G = < 20 Mils Penetration/Year

S = < 50 Mils Penetration/Year; U = > 50 Mils Penetration/Year

FOR NONMETALLICS

R = Resistant

U = Unsatisfactory

TETRAPHOSPHORIC ACID

PLASTICS	TEMP °F (60–460)
ABS	
ACRYLICS	
ASBESTOS REINF., EPOXY	
ASBESTOS REINF., PHENOLIC	
CHLORINATED POLYETHER	
CPVC	
EPOXY	
FLUOROCARBONS FEP	R (to about 400 °F)
FLUOROCARBONS TFE	R (to 460 °F)
FURFURYL ALCOHOL	
NORYL	
PHENOLIC	
POLYAMIDES–NYLON	
POLYESTERS	
BISPHENOL A-FUMURATE	
HYDROGENATED BISPHENOL-A-BISPHENOL-A	
ISOPHTHALIC	
CHLORINATED POLYESTERS	
POLYSTYRENES	
POLYETHYLENE	
POLYPROPYLENE	
PVC–TYPE 1	
PVC–TYPE 2	
POLYVINYLIDENE CHLORIDE	
VINYLIDENE FLUORIDE	
VINYL ESTER	

RUBBERS	TEMP °F (60–460)
BUTYL GR-1	
FLUORO ELASTOMERS VITON A	
FLUORO ELASTOMERS KEL-F 3700	
HARD RUBBER	
HYPALON	
KOROSEAL	
NATURAL RUBBER (GRS)	
NEOPRENE GR-M (CR)	
NITRILE BUNA N (NBR)	
NORDEL	
POLYBUTADIENE	
SILICONE	
POLYURETHANE	

THIONYL CHLORIDE

TEMP °F: 60, 80, 100, 120, 140, 160, 180, 200, 220, 240, 260, 280, 300, 320, 340, 360, 380, 400, 420, 440, 460, 480, 500, 520, 540, 560

METALS	
ALLOY 20Cb3	
ALUMINUM	U
BRASS	
BRONZE	
CARBON STEEL	U
COPPER	
E-BRIGHT 26-1	
HASTELLOY-B	
HASTELLOY-C	
HASTELLOY-D	
HIGH SILICON IRON	
INCONEL	
LEAD	G
MONEL	G
NICKEL	G
STAINLESS STEELS	
Type 304/307	
Type 316	
TANTALUM	G
TITANIUM	
ZIRCONIUM	

NONMETALLICS	
BOROSILICATE GLASS	R
CARBON	

TEMP °F: 60, 80, 100, 120, 140, 160, 180, 200, 220, 240, 260, 280, 300, 320, 340, 360, 380, 400, 420, 440, 460, 480, 500, 520, 540, 560

FOR METALS

E = < 2 Mils Penetration/Year; G = < 20 Mils Penetration/Year

S = < 50 Mils Penetration/Year; U = > 50 Mils Penetration/Year

FOR NONMETALLICS

R = Resistant

U = Unsatisfactory

THIONYL CHLORIDE

TEMP °F

PLASTICS	60	80	100	120	140	160	180	200	220	240	260	280	300	320	340	360	380	400	420	440	460
ABS	U																				
ACRYLICS																					
ASBESTOS REINF., EPOXY																					
ASBESTOS REINF., PHENOLIC																					
CHLORINATED POLYETHER	U																				
CPVC	U																				
EPOXY																					
FLUOROCARBONS FEP	R	—	—	—	—	—	—	—	—	—	—	—	—	—	—	—	—	—			
FLUOROCARBONS TFE	R	—	—	—	—	—	—	—	—	—	—	—	—	—	—	—	—	—	—	—	—
FURFURYL ALCOHOL	U																				
NORYL																					
PHENOLIC																					
POLYAMIDES–NYLON																					
POLYESTERS																					
BISPHENOL A-FUMURATE	U																				
HYDROGENATED BISPHENOL-A-BISPHENOL-A																					
ISOPHTHALIC																					
CHLORINATED POLYESTERS	U																				
POLYSTYRENES																					
POLYETHYLENE	R	—	—	—	—																
POLYPROPYLENE	R	—	—																		
PVC–TYPE 1	U																				
PVC–TYPE 2	U																				
POLYVINYLIDENE CHLORIDE	U																				
VINYLIDENE FLUORIDE																					
VINYL ESTER	U																				
RUBBERS																					
BUTYL GR-1	U																				
FLUORO ELASTOMERS VITON A	U																				
FLUORO ELASTOMERS KEL-F 3700																					
HARD RUBBER																					
HYPALON																					
KOROSEAL																					
NATURAL RUBBER (GRS)	U																				
NEOPRENE GR-M (CR)	U																				
NITRILE BUNA N (NBR)	U																				
NORDEL																					
POLYBUTADIENE																					
SILICONE																					
POLYURETHANE																					

TEMP °F

THREAD CUTTING OILS

TEMP °F: 60 80 100 120 140 160 180 200 220 240 260 280 300 320 340 360 380 400 420 440 460 480 500 520 540 560

METALS	
ALLOY 20Cb3	
ALUMINUM	
BRASS	
BRONZE	
CARBON STEEL	G
COPPER	
E-BRIGHT 26-1	
HASTELLOY-B	
HASTELLOY-C	
HASTELLOY-D	
HIGH SILICON IRON	
INCONEL	
LEAD	
MONEL	
NICKEL	
STAINLESS STEELS	
Type 304/307	G
Type 316	G
TANTALUM	
TITANIUM	
ZIRCONIUM	

NONMETALLICS	
BOROSILICATE GLASS	R
CARBON	R

TEMP °F: 60 80 100 120 140 160 180 200 220 240 260 280 300 320 340 360 380 400 420 440 460 480 500 520 540 560

FOR METALS

E = < 2 Mils Penetration/Year; G = < 20 Mils Penetration/Year

S = < 50 Mils Penetration/Year; U = > 50 Mils Penetration/Year

FOR NONMETALLICS

R = Resistant

U = Unsatisfactory

THREAD CUTTING OILS

TEMP °F (60–460)

PLASTICS	TEMP °F 60 80 100 120 140 160 180 200 220 240 260 280 300 320 340 360 380 400 420 440 460
ABS	
ACRYLICS	
ASBESTOS REINF., EPOXY	
ASBESTOS REINF., PHENOLIC	
CHLORINATED POLYETHER	R
CPVC	
EPOXY	
FLUOROCARBONS FEP	R
FLUOROCARBONS TFE	R
FURFURYL ALCOHOL	
NORYL	
PHENOLIC	
POLYAMIDES–NYLON	
POLYESTERS	
BISPHENOL A-FUMURATE	
HYDROGENATED BISPHENOL-A-BISPHENOL-A	
ISOPHTHALIC	
CHLORINATED POLYESTERS	
POLYSTYRENES	
POLYETHYLENE	
POLYPROPYLENE	R
PVC–TYPE 1	R
PVC–TYPE 2	
POLYVINYLIDENE CHLORIDE	R
VINYLIDENE FLUORIDE	R
VINYL ESTER	
RUBBERS	
BUTYL GR-1	
FLUORO ELASTOMERS VITON A	
FLUORO ELASTOMERS KEL-F 3700	
HARD RUBBER	
HYPALON	
KOROSEAL	
NATURAL RUBBER (GRS)	
NEOPRENE GR-M (CR)	
NITRILE BUNA N (NBR)	
NORDEL	
POLYBUTADIENE	
SILICONE	
POLYURETHANE	

TEMP °F (60–460)

TIN CHLORIDE

TEMP °F: 60 80 100 120 140 160 180 200 220 240 260 280 300 320 340 360 380 400 420 440 460 480 500 520 540 560

METALS	Rating (from 60 °F)
ALLOY 20Cb3	
ALUMINUM	
BRASS	
BRONZE	
CARBON STEEL	
COPPER	
E-BRIGHT 26-1	
HASTELLOY-B	G—
HASTELLOY-C	
HASTELLOY-D	
HIGH SILICON IRON	
INCONEL	
LEAD	
MONEL	
NICKEL	
STAINLESS STEELS	
Type 304/307	U
Type 316	G————
TANTALUM	E—
TITANIUM	
ZIRCONIUM	

NONMETALLICS	Rating (from 60 °F)
BOROSILICATE GLASS	R——————
CARBON	

TEMP °F: 60 80 100 120 140 160 180 200 220 240 260 280 300 320 340 360 380 400 420 440 460 480 500 520 540 560

FOR METALS

E = < 2 Mils Penetration/Year; G = < 20 Mils Penetration/Year

S = < 50 Mils Penetration/Year; U = > 50 Mils Penetration/Year

FOR NONMETALLICS

R = Resistant

U = Unsatisfactory

TIN CHLORIDE

TEMP °F (60 80 100 120 140 160 180 200 220 240 260 280 300 320 340 360 380 400 420 440 460)

PLASTICS		
ABS		R
ACRYLICS		
ASBESTOS REINF., EPOXY		
ASBESTOS REINF., PHENOLIC		
CHLORINATED POLYETHER		
CPVC		R
EPOXY		
FLUOROCARBONS FEP		R
FLUOROCARBONS TFE		R
FURFURYL ALCOHOL	sat	R
NORYL		
PHENOLIC		
POLYAMIDES–NYLON		
POLYESTERS		
BISPHENOL A-FUMURATE		
HYDROGENATED BISPHENOL-A-BISPHENOL-A		
ISOPHTHALIC		
CHLORINATED POLYESTERS		
POLYSTYRENES		
POLYETHYLENE		R
POLYPROPYLENE		R
PVC–TYPE 1		R
PVC–TYPE 2		R
POLYVINYLIDENE CHLORIDE		
VINYLIDENE FLUORIDE		
VINYL ESTER		
RUBBERS		
BUTYL GR-1		R
FLUORO ELASTOMERS VITON A		
FLUORO ELASTOMERS KEL-F 3700		
HARD RUBBER		
HYPALON		
KOROSEAL		R
NATURAL RUBBER (GRS)		R
NEOPRENE GR-M (CR)		R
NITRILE BUNA N (NBR)		R
NORDEL		
POLYBUTADIENE		
SILICONE		
POLYURETHANE		

TEMP °F (60 80 100 120 140 160 180 200 220 240 260 280 300 320 340 360 380 400 420 440 460)

TEMP °F (60–560)

METALS		Rating
ALLOY 20Cb3		E—
ALUMINUM		U
BRASS		U
BRONZE		
CARBON STEEL		
COPPER		
E-BRIGHT 26-1		
HASTELLOY-B		
HASTELLOY-C		G—
HASTELLOY-D		
HIGH SILICON IRON		
INCONEL		G—
LEAD		
MONEL		G—
NICKEL		
STAINLESS STEELS		
Type 304/307	dry	G—
Type 316		G—
TANTALUM		E—
TITANIUM		
ZIRCONIUM		

NONMETALLICS	Rating
BOROSILICATE GLASS	R————————
CARBON	

TEMP °F

FOR METALS

E = < 2 Mils Penetration/Year; G = < 20 Mils Penetration/Year

S = < 50 Mils Penetration/Year; U = > 50 Mils Penetration/Year

FOR NONMETALLICS

R = Resistant

U = Unsatisfactory

TITANIUM TETRACHLORIDE

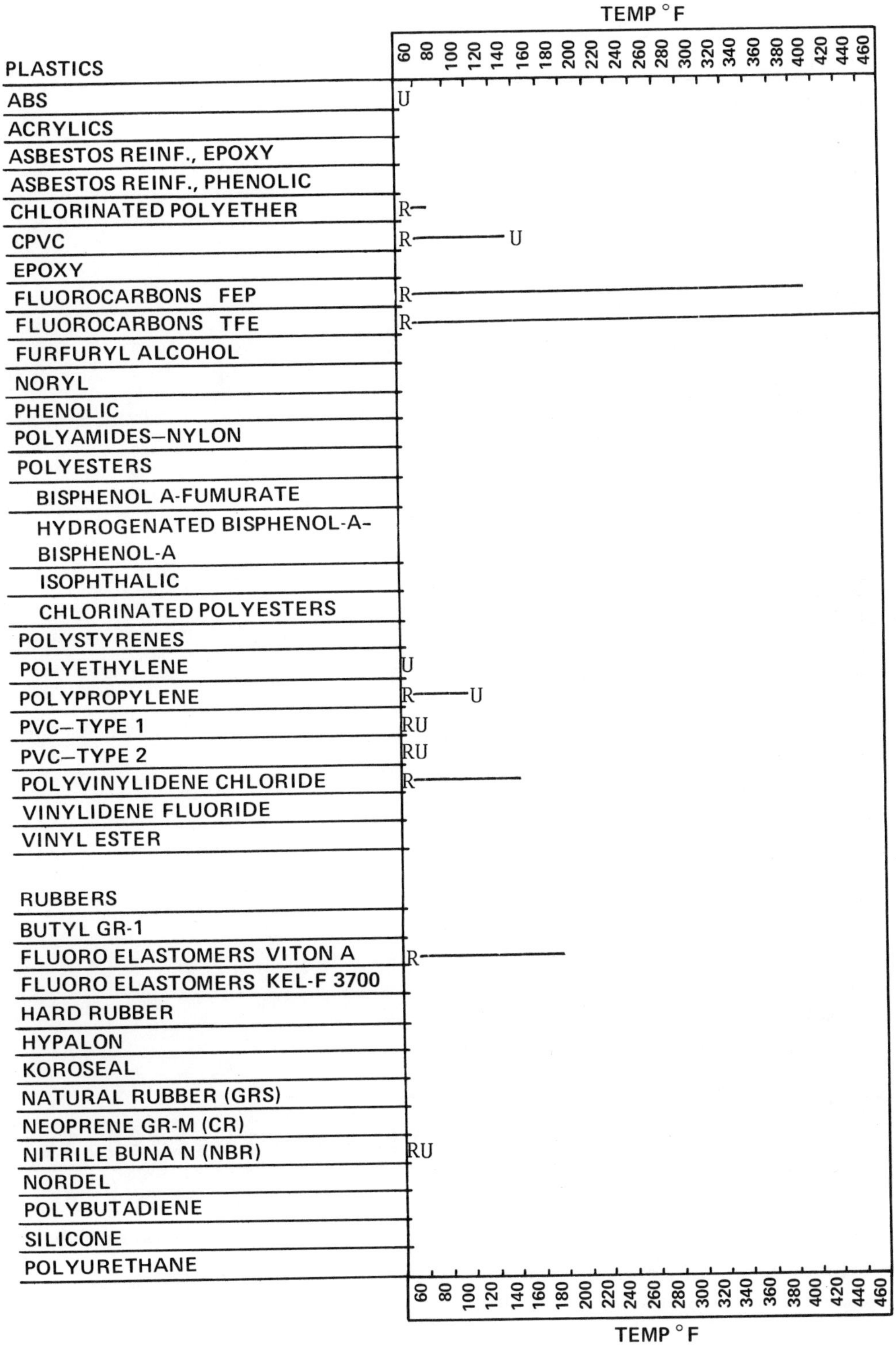

TOLUENE, TOLUOL

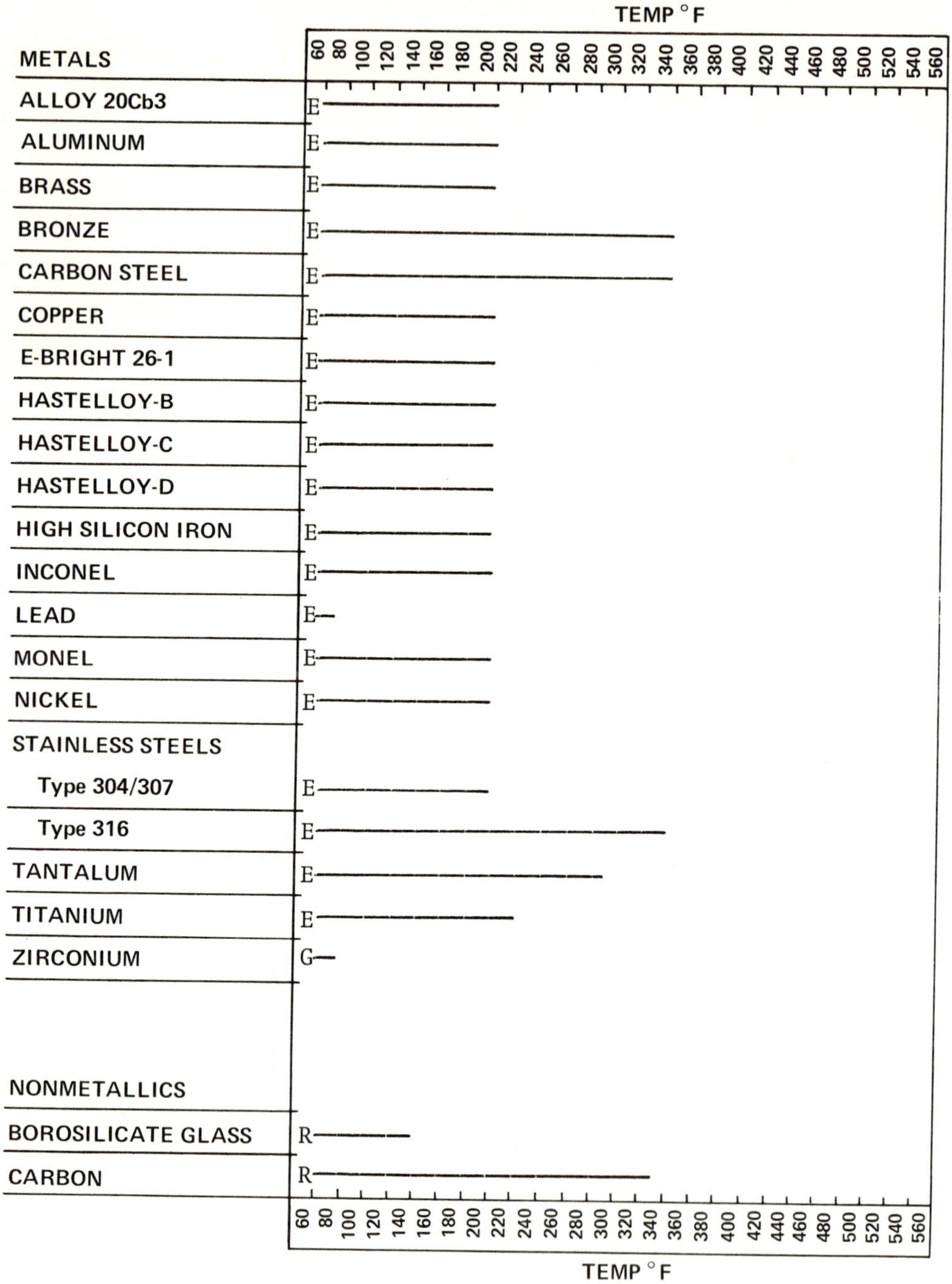

FOR METALS

E = < 2 Mils Penetration/Year; G = < 20 Mils Penetration/Year
S = < 50 Mils Penetration/Year; U = > 50 Mils Penetration/Year

FOR NONMETALLICS

R = Resistant
U = Unsatisfactory

TOLUENE, TOLUOL

TEMP °F: 60 80 100 120 140 160 180 200 220 240 260 280 300 320 340 360 380 400 420 440 460

PLASTICS	
ABS	U
ACRYLICS	U
ASBESTOS REINF., EPOXY	R—
ASBESTOS REINF., PHENOLIC	R——
CHLORINATED POLYETHER	R—
CPVC	U
EPOXY	R—— U
FLUOROCARBONS FEP	R——
FLUOROCARBONS TFE	R——
FURFURYL ALCOHOL	R——
NORYL	
PHENOLIC	
POLYAMIDES—NYLON	R——
POLYESTERS	
BISPHENOL A-FUMURATE	U
HYDROGENATED BISPHENOL-A-BISPHENOL-A	R—U
ISOPHTHALIC	R—— U
CHLORINATED POLYESTERS	R—— U
POLYSTYRENES	U
POLYETHYLENE	U
POLYPROPYLENE	RU
PVC—TYPE 1	U
PVC—TYPE 2	U
POLYVINYLIDENE CHLORIDE	R—
VINYLIDENE FLUORIDE	R——
VINYL ESTER	R——

RUBBERS	
BUTYL GR-1	U
FLUORO ELASTOMERS VITON A	R——
FLUORO ELASTOMERS KEL-F 3700	U
HARD RUBBER	
HYPALON	U
KOROSEAL	U
NATURAL RUBBER (GRS)	U
NEOPRENE GR-M (CR)	U
NITRILE BUNA N (NBR)	U
NORDEL	U
POLYBUTADIENE	U
SILICONE	U
POLYURETHANE	U

TEMP °F: 60 80 100 120 140 160 180 200 220 240 260 280 300 320 340 360 380 400 420 440 460

TOLUENE-KEROSENE, 75%-25%

TEMP °F

METALS	60	80	100	120	140	160	180	200	220	240	260	280	300	320	340	360	380	400	420	440	460	480	500	520	540	560
ALLOY 20Cb3																										
ALUMINUM																										
BRASS																										
BRONZE																										
CARBON STEEL																										
COPPER																										
E-BRIGHT 26-1																										
HASTELLOY-B																										
HASTELLOY-C																										
HASTELLOY-D																										
HIGH SILICON IRON																										
INCONEL																										
LEAD																										
MONEL																										
NICKEL																										
STAINLESS STEELS Type 304/307																										
Type 316																										
TANTALUM																										
TITANIUM																										
ZIRCONIUM																										
NONMETALLICS																										
BOROSILICATE GLASS																										
CARBON																										

60 80 100 120 140 160 180 200 220 240 260 280 300 320 340 360 380 400 420 440 460 480 500 520 540 560

TEMP °F

FOR METALS

E = < 2 Mils Penetration/Year; G = < 20 Mils Penetration/Year
S = < 50 Mils Penetration/Year; U = > 50 Mils Penetration/Year

FOR NONMETALLICS

R = Resistant
U = Unsatisfactory

TOLUENE-KEROSENE, 75%-25%

TEMP °F

PLASTICS	60	80	100	120	140	160	180	200	220	240	260	280	300	320	340	360	380	400	420	440	460
ABS																					
ACRYLICS																					
ASBESTOS REINF., EPOXY																					
ASBESTOS REINF., PHENOLIC																					
CHLORINATED POLYETHER	R	—	U																		
CPVC																					
EPOXY																					
FLUOROCARBONS FEP	R	—	—	—	—	—	—	—	—	—	—	—	—	—	—	—	—	—			
FLUOROCARBONS TFE	R	—	—	—	—	—	—	—	—	—	—	—	—	—	—	—	—	—	—	—	—
FURFURYL ALCOHOL																					
NORYL																					
PHENOLIC																					
POLYAMIDES–NYLON																					
POLYESTERS																					
BISPHENOL A-FUMURATE																					
HYDROGENATED BISPHENOL-A-BISPHENOL-A																					
ISOPHTHALIC																					
CHLORINATED POLYESTERS																					
POLYSTYRENES																					
POLYETHYLENE																					
POLYPROPYLENE	R	—																			
PVC–TYPE 1																					
PVC–TYPE 2																					
POLYVINYLIDENE CHLORIDE	R	—																			
VINYLIDENE FLUORIDE																					
VINYL ESTER																					
RUBBERS																					
BUTYL GR-1																					
FLUORO ELASTOMERS VITON A	R	—																			
FLUORO ELASTOMERS KEL-F 3700																					
HARD RUBBER																					
HYPALON																					
KOROSEAL																					
NATURAL RUBBER (GRS)																					
NEOPRENE GR-M (CR)																					
NITRILE BUNA N (NBR)																					
NORDEL																					
POLYBUTADIENE																					
SILICONE																					
POLYURETHANE																					

TEMP °F

TOMATO JUICE

METALS	TEMP °F (60–560)
ALLOY 20Cb3	G
ALUMINUM	G
BRASS	
BRONZE	S
CARBON STEEL	S
COPPER	
E-BRIGHT 26-1	
HASTELLOY-B	G
HASTELLOY-C	G
HASTELLOY-D	
HIGH SILICON IRON	
INCONEL	
LEAD	
MONEL	G
NICKEL	G
STAINLESS STEELS	
Type 304/307	G
Type 316	G
TANTALUM	E
TITANIUM	
ZIRCONIUM	

NONMETALLICS	TEMP °F (60–560)
BOROSILICATE GLASS	R
CARBON	R

FOR METALS

E = < 2 Mils Penetration/Year; G = < 20 Mils Penetration/Year

S = < 50 Mils Penetration/Year; U = > 50 Mils Penetration/Year

FOR NONMETALLICS

R = Resistant

U = Unsatisfactory

TOMATO JUICE

TEMP °F: 60 80 100 120 140 160 180 200 220 240 260 280 300 320 340 360 380 400 420 440 460

PLASTICS	
ABS	R
ACRYLICS	
ASBESTOS REINF., EPOXY	
ASBESTOS REINF., PHENOLIC	
CHLORINATED POLYETHER	R
CPVC	
EPOXY	
FLUOROCARBONS FEP	R
FLUOROCARBONS TFE	R
FURFURYL ALCOHOL	
NORYL	
PHENOLIC	
POLYAMIDES–NYLON	
POLYESTERS	
BISPHENOL A-FUMURATE	
HYDROGENATED BISPHENOL-A-BISPHENOL-A	
ISOPHTHALIC	
CHLORINATED POLYESTERS	
POLYSTYRENES	
POLYETHYLENE	
POLYPROPYLENE	R
PVC–TYPE 1	
PVC–TYPE 2	
POLYVINYLIDENE CHLORIDE	R
VINYLIDENE FLUORIDE	R
VINYL ESTER	

RUBBERS	
BUTYL GR-1	
FLUORO ELASTOMERS VITON A	R
FLUORO ELASTOMERS KEL-F 3700	
HARD RUBBER	
HYPALON	
KOROSEAL	
NATURAL RUBBER (GRS)	
NEOPRENE GR-M (CR)	R
NITRILE BUNA N (NBR)	R
NORDEL	
POLYBUTADIENE	
SILICONE	
POLYURETHANE	

TEMP °F: 60 80 100 120 140 160 180 200 220 240 260 280 300 320 340 360 380 400 420 440 460

TEMP °F: 60 80 100 120 140 160 180 200 220 240 260 280 300 320 340 360 380 400 420 440 460 480 500 520 540 560

METALS	
ALLOY 20Cb3	
ALUMINUM	
BRASS	
BRONZE	
CARBON STEEL	
COPPER	
E-BRIGHT 26-1	
HASTELLOY-B	
HASTELLOY-C	
HASTELLOY-D	
HIGH SILICON IRON	
INCONEL	
LEAD	
MONEL	
NICKEL	
STAINLESS STEELS	
Type 304/307	
Type 316	
TANTALUM	
TITANIUM	
ZIRCONIUM	

NONMETALLICS	
BOROSILICATE GLASS	
CARBON	R

TEMP °F: 60 80 100 120 140 160 180 200 220 240 260 280 300 320 340 360 380 400 420 440 460 480 500 520 540 560

FOR METALS

E = < 2 Mils Penetration/Year; G = < 20 Mils Penetration/Year

S = < 50 Mils Penetration/Year; U = > 50 Mils Penetration/Year

FOR NONMETALLICS

R = Resistant

U = Unsatisfactory

TOXAPHENE-XYLENE, 90%-10%

TEMP °F (60, 80, 100, 120, 140, 160, 180, 200, 220, 240, 260, 280, 300, 320, 340, 360, 380, 400, 420, 440, 460)

PLASTICS	
ABS	
ACRYLICS	
ASBESTOS REINF., EPOXY	
ASBESTOS REINF., PHENOLIC	
CHLORINATED POLYETHER	R———U
CPVC	
EPOXY	
FLUOROCARBONS FEP	R———————
FLUOROCARBONS TFE	R—————————
FURFURYL ALCOHOL	
NORYL	
PHENOLIC	
POLYAMIDES–NYLON	
POLYESTERS	
BISPHENOL A-FUMURATE	
HYDROGENATED BISPHENOL-A-BISPHENOL-A	
ISOPHTHALIC	
CHLORINATED POLYESTERS	
POLYSTYRENES	
POLYETHYLENE	
POLYPROPYLENE	R—
PVC–TYPE 1	
PVC–TYPE 2	
POLYVINYLIDENE CHLORIDE	R—
VINYLIDENE FLUORIDE	
VINYL ESTER	

RUBBERS	
BUTYL GR-1	
FLUORO ELASTOMERS VITON A	
FLUORO ELASTOMERS KEL-F 3700	
HARD RUBBER	
HYPALON	
KOROSEAL	
NATURAL RUBBER (GRS)	
NEOPRENE GR-M (CR)	
NITRILE BUNA N (NBR)	
NORDEL	
POLYBUTADIENE	
SILICONE	
POLYURETHANE	

TEMP °F (60, 80, 100, 120, 140, 160, 180, 200, 220, 240, 260, 280, 300, 320, 340, 360, 380, 400, 420, 440, 460)

TEMP °F (60–560)

METALS	60 – 560 °F
ALLOY 20Cb3	G—
ALUMINUM	G—
BRASS	
BRONZE	G—
CARBON STEEL	G—
COPPER	
E-BRIGHT 26-1	
HASTELLOY-B	G—
HASTELLOY-C	G—
HASTELLOY-D	
HIGH SILICON IRON	
INCONEL	
LEAD	
MONEL	G—
NICKEL	G—
STAINLESS STEELS	
Type 304/307	G—
Type 316	G—
TANTALUM	
TITANIUM	
ZIRCONIUM	
NONMETALLICS	
BOROSILICATE GLASS	R—
CARBON	

TEMP °F: 60, 80, 100, 120, 140, 160, 180, 200, 220, 240, 260, 280, 300, 320, 340, 360, 380, 400, 420, 440, 460, 480, 500, 520, 540, 560

FOR METALS

E = < 2 Mils Penetration/Year; G = < 20 Mils Penetration/Year
S = < 50 Mils Penetration/Year; U = > 50 Mils Penetration/Year

FOR NONMETALLICS

R = Resistant
U = Unsatisfactory

TRANSFORMER OIL

TEMP °F: 60 80 100 120 140 160 180 200 220 240 260 280 300 320 340 360 380 400 420 440 460

PLASTICS	
ABS	
ACRYLICS	
ASBESTOS REINF., EPOXY	
ASBESTOS REINF., PHENOLIC	
CHLORINATED POLYETHER	
CPVC	
EPOXY	R——U
FLUOROCARBONS FEP	R——
FLUOROCARBONS TFE	R——
FURFURYL ALCOHOL	
NORYL	
PHENOLIC	
POLYAMIDES—NYLON	
POLYESTERS	
BISPHENOL A-FUMURATE	R——
HYDROGENATED BISPHENOL-A-BISPHENOL-A	
ISOPHTHALIC	R—
CHLORINATED POLYESTERS	R—
POLYSTYRENES	
POLYETHYLENE	R—
POLYPROPYLENE	R—U
PVC—TYPE 1	R—
PVC—TYPE 2	
POLYVINYLIDENE CHLORIDE	
VINYLIDENE FLUORIDE	
VINYL ESTER	R——
RUBBERS	
BUTYL GR-1	U
FLUORO ELASTOMERS VITON A	R——
FLUORO ELASTOMERS KEL-F 3700	
HARD RUBBER	U
HYPALON	
KOROSEAL	
NATURAL RUBBER (GRS)	U
NEOPRENE GR-M (CR)	R—
NITRILE BUNA N (NBR)	R——
NORDEL	
POLYBUTADIENE	
SILICONE	
POLYURETHANE	

TEMP °F: 60 80 100 120 140 160 180 200 220 240 260 280 300 320 340 360 380 400 420 440 460

TRIBUTYL PHOSPHATE

TEMP °F: 60 80 100 120 140 160 180 200 220 240 260 280 300 320 340 360 380 400 420 440 460 480 500 520 540 560

METALS	
ALLOY 20Cb3	G
ALUMINUM	G
BRASS	
BRONZE	G
CARBON STEEL	G
COPPER	
E-BRIGHT 26-1	
HASTELLOY-B	G
HASTELLOY-C	G
HASTELLOY-D	
HIGH SILICON IRON	
INCONEL	
LEAD	
MONEL	G
NICKEL	G
STAINLESS STEELS	
Type 304/307	G
Type 316	G
TANTALUM	
TITANIUM	
ZIRCONIUM	

NONMETALLICS	
BOROSILICATE GLASS	R
CARBON	

TEMP °F: 60 80 100 120 140 160 180 200 220 240 260 280 300 320 340 360 380 400 420 440 460 480 500 520 540 560

FOR METALS

E = < 2 Mils Penetration/Year; G = < 20 Mils Penetration/Year
S = < 50 Mils Penetration/Year; U = > 50 Mils Penetration/Year

FOR NONMETALLICS

R = Resistant
U = Unsatisfactory

TRIBUTYL PHOSPHATE

PLASTICS	TEMP °F
ABS	U (60)
ACRYLICS	
ASBESTOS REINF., EPOXY	
ASBESTOS REINF., PHENOLIC	
CHLORINATED POLYETHER	R (60–220)
CPVC	U (60)
EPOXY	
FLUOROCARBONS FEP	R (60–400)
FLUOROCARBONS TFE	R (60–460)
FURFURYL ALCOHOL	
NORYL	
PHENOLIC	
POLYAMIDES–NYLON	R (60–80)
POLYESTERS	
BISPHENOL A-FUMURATE	
HYDROGENATED BISPHENOL-A-BISPHENOL-A	
ISOPHTHALIC	
CHLORINATED POLYESTERS	
POLYSTYRENES	
POLYETHYLENE	R (60–140)
POLYPROPYLENE	R (60–120), U (140)
PVC–TYPE 1	U (60)
PVC–TYPE 2	U (60)
POLYVINYLIDENE CHLORIDE	R (60–80)
VINYLIDENE FLUORIDE	R (60–120), U (140)
VINYL ESTER	R (60–140)

RUBBERS	TEMP °F
BUTYL GR-1	U (60)
FLUORO ELASTOMERS VITON A	U (60)
FLUORO ELASTOMERS KEL-F 3700	
HARD RUBBER	
HYPALON	U (60)
KOROSEAL	
NATURAL RUBBER (GRS)	U (60)
NEOPRENE GR-M (CR)	U (60)
NITRILE BUNA N (NBR)	U (60)
NORDEL	
POLYBUTADIENE	
SILICONE	
POLYURETHANE	U (60)

TEMP °F

METALS		Rating
ALLOY 20Cb3		
ALUMINUM		U
BRASS		G
BRONZE		G
CARBON STEEL		U
COPPER		G
E-BRIGHT 26-1		
HASTELLOY-B	all	G
HASTELLOY-C	all	G
HASTELLOY-D	all	G
HIGH SILICON IRON		G
INCONEL		G
LEAD		U
MONEL		G
NICKEL		G
STAINLESS STEELS		
Type 304/307		U
Type 316		U
TANTALUM		G
TITANIUM		U
ZIRCONIUM		U

NONMETALLICS	Rating
BOROSILICATE GLASS	R
CARBON	R

TEMP °F: 60 80 100 120 140 160 180 200 220 240 260 280 300 320 340 360 380 400 420 440 460 480 500 520 540 560

FOR METALS

E = < 2 Mils Penetration/Year; G = < 20 Mils Penetration/Year

S = < 50 Mils Penetration/Year; U = > 50 Mils Penetration/Year

FOR NONMETALLICS

R = Resistant

U = Unsatisfactory

TRICHLOROACETIC ACID

TEMP °F: 60 80 100 120 140 160 180 200 220 240 260 280 300 320 340 360 380 400 420 440 460

PLASTICS		
ABS		
ACRYLICS		
ASBESTOS REINF., EPOXY		
ASBESTOS REINF., PHENOLIC		
CHLORINATED POLYETHER		U
CPVC		
EPOXY		U
FLUOROCARBONS FEP		R
FLUOROCARBONS TFE		R
FURFURYL ALCOHOL		R
NORYL	10%	R
PHENOLIC		
POLYAMIDES–NYLON		U
POLYESTERS		
BISPHENOL A-FUMURATE	50%	R
HYDROGENATED BISPHENOL-A-BISPHENOL-A		R
ISOPHTHALIC	50%	R U
CHLORINATED POLYESTERS	50%	R
POLYSTYRENES		
POLYETHYLENE		R
POLYPROPYLENE		R U
PVC–TYPE 1		R
PVC–TYPE 2		
POLYVINYLIDENE CHLORIDE		
VINYLIDENE FLUORIDE		R U
VINYL ESTER	50%	R

RUBBERS		
BUTYL GR-1		U
FLUORO ELASTOMERS VITON A		R
FLUORO ELASTOMERS KEL-F 3700		
HARD RUBBER		
HYPALON		
KOROSEAL		
NATURAL RUBBER (GRS)		U
NEOPRENE GR-M (CR)		U
NITRILE BUNA N (NBR)		RU
NORDEL		
POLYBUTADIENE		
SILICONE		
POLYURETHANE		

TEMP °F: 60 80 100 120 140 160 180 200 220 240 260 280 300 320 340 360 380 400 420 440 460

TRICHLOROACETIC ACID, 2N

TEMP °F: 60 80 100 120 140 160 180 200 220 240 260 280 300 320 340 360 380 400 420 440 460 480 500 520 540 560

METALS	
ALLOY 20Cb3	
ALUMINUM	U
BRASS	
BRONZE	
CARBON STEEL	U
COPPER	
E-BRIGHT 26-1	
HASTELLOY-B	G
HASTELLOY-C	G
HASTELLOY-D	G
HIGH SILICON IRON	
INCONEL	
LEAD	
MONEL	
NICKEL	
STAINLESS STEELS	
Type 304/307	U
Type 316	U
TANTALUM	
TITANIUM	
ZIRCONIUM	E

NONMETALLICS	
BOROSILICATE GLASS	
CARBON	R

TEMP °F: 60 80 100 120 140 160 180 200 220 240 260 280 300 320 340 360 380 400 420 440 460 480 500 520 540 560

FOR METALS

E = < 2 Mils Penetration/Year; G = < 20 Mils Penetration/Year
S = < 50 Mils Penetration/Year; U = > 50 Mils Penetration/Year

FOR NONMETALLICS

R = Resistant
U = Unsatisfactory

TRICHLOROACETIC ACID, 2N

TEMP °F (60–460)

PLASTICS	Rating (from 60°F)
ABS	
ACRYLICS	
ASBESTOS REINF., EPOXY	
ASBESTOS REINF., PHENOLIC	
CHLORINATED POLYETHER	U
CPVC	
EPOXY	U
FLUOROCARBONS FEP	R (line to 400)
FLUOROCARBONS TFE	R (line to 460)
FURFURYL ALCOHOL	
NORYL	R (line to 140)
PHENOLIC	U
POLYAMIDES–NYLON	U
POLYESTERS	
BISPHENOL A-FUMURATE	
HYDROGENATED BISPHENOL-A-BISPHENOL-A	
ISOPHTHALIC	U
CHLORINATED POLYESTERS	
POLYSTYRENES	
POLYETHYLENE	
POLYPROPYLENE	R (line to 160), U (180)
PVC–TYPE 1	
PVC–TYPE 2	
POLYVINYLIDENE CHLORIDE	
VINYLIDENE FLUORIDE	
VINYL ESTER	

RUBBERS	Rating (from 60°F)
BUTYL GR-1	U
FLUORO ELASTOMERS VITON A	
FLUORO ELASTOMERS KEL-F 3700	
HARD RUBBER	
HYPALON	
KOROSEAL	
NATURAL RUBBER (GRS)	U
NEOPRENE GR-M (CR)	U
NITRILE BUNA N (NBR)	RU
NORDEL	
POLYBUTADIENE	
SILICONE	
POLYURETHANE	

TEMP °F

TRICHLOROETHYLENE

TEMP °F: 60 80 100 120 140 160 180 200 220 240 260 280 300 320 340 360 380 400 420 440 460 480 500 520 540 560

METALS		
ALLOY 20Cb3		G
ALUMINUM		E — G
BRASS	dry	G
BRONZE		G
CARBON STEEL		G
COPPER		G
E-BRIGHT 26-1		
HASTELLOY-B		E
HASTELLOY-C		E
HASTELLOY-D		
HIGH SILICON IRON		G
INCONEL		E — G
LEAD		U
MONEL	1	E
NICKEL	1	E
STAINLESS STEELS		
Type 304/307	1	G
Type 316	1	G
TANTALUM		G
TITANIUM	all	E
ZIRCONIUM		G

NONMETALLICS	
BOROSILICATE GLASS	R
CARBON	R

TEMP °F: 60 80 100 120 140 160 180 200 220 240 260 280 300 320 340 360 380 400 420 440 460 480 500 520 540 560

FOR METALS

E = < 2 Mils Penetration/Year; G = < 20 Mils Penetration/Year

S = < 50 Mils Penetration/Year; U = > 50 Mils Penetration/Year

FOR NONMETALLICS

R = Resistant

U = Unsatisfactory

TRICHLOROETHYLENE

TEMP °F: 60 80 100 120 140 160 180 200 220 240 260 280 300 320 340 360 380 400 420 440 460

PLASTICS	
ABS	U
ACRYLICS	U
ASBESTOS REINF., EPOXY	U
ASBESTOS REINF., PHENOLIC	R (60–300)
CHLORINATED POLYETHER	R (60–150)
CPVC	U
EPOXY	R (60–150)
FLUOROCARBONS FEP	R (60–400)
FLUOROCARBONS TFE	R (60–460)
FURFURYL ALCOHOL	R (60–180)
NORYL	
PHENOLIC	
POLYAMIDES–NYLON	U
POLYESTERS	
BISPHENOL A-FUMURATE	U
HYDROGENATED BISPHENOL-A-BISPHENOL-A	
ISOPHTHALIC	U
CHLORINATED POLYESTERS	R (60–140) U
POLYSTYRENES	
POLYETHYLENE	U
POLYPROPYLENE	RU
PVC–TYPE 1	U
PVC–TYPE 2	U
POLYVINYLIDENE CHLORIDE	R (60–80)
VINYLIDENE FLUORIDE	R (60–280)
VINYL ESTER	U

RUBBERS	
BUTYL GR-1	U
FLUORO ELASTOMERS VITON A	R (60–350)
FLUORO ELASTOMERS KEL-F 3700	
HARD RUBBER	U
HYPALON	U
KOROSEAL	U
NATURAL RUBBER (GRS)	U
NEOPRENE GR-M (CR)	U
NITRILE BUNA N (NBR)	U
NORDEL	
POLYBUTADIENE	
SILICONE	
POLYURETHANE	U

TEMP °F: 60 80 100 120 140 160 180 200 220 240 260 280 300 320 340 360 380 400 420 440 460

TRIETHANOLAMINE

TEMP °F: 60, 80, 100, 120, 140, 160, 180, 200, 220, 240, 260, 280, 300, 320, 340, 360, 380, 400, 420, 440, 460, 480, 500, 520, 540, 560

METALS	
ALLOY 20Cb3	G
ALUMINUM	G
BRASS	U
BRONZE	G
CARBON STEEL	G
COPPER	
E-BRIGHT 26-1	
HASTELLOY-B	
HASTELLOY-C	
HASTELLOY-D	
HIGH SILICON IRON	G
INCONEL	
LEAD	
MONEL	G
NICKEL	
STAINLESS STEELS	
Type 304/307	G
Type 316	G
TANTALUM	G
TITANIUM	
ZIRCONIUM	

NONMETALLICS	
BOROSILICATE GLASS	R
CARBON	R

TEMP °F: 60, 80, 100, 120, 140, 160, 180, 200, 220, 240, 260, 280, 300, 320, 340, 360, 380, 400, 420, 440, 460, 480, 500, 520, 540, 560

FOR METALS

E = < 2 Mils Penetration/Year; G = < 20 Mils Penetration/Year
S = < 50 Mils Penetration/Year; U = > 50 Mils Penetration/Year

FOR NONMETALLICS

R = Resistant
U = Unsatisfactory

TRIETHANOLAMINE

TEMP °F: 60, 80, 100, 120, 140, 160, 180, 200, 220, 240, 260, 280, 300, 320, 340, 360, 380, 400, 420, 440, 460

PLASTICS	
ABS	R—
ACRYLICS	R—
ASBESTOS REINF., EPOXY	
ASBESTOS REINF., PHENOLIC	
CHLORINATED POLYETHER	R—
CPVC	R—U
EPOXY	R—
FLUOROCARBONS FEP	R—
FLUOROCARBONS TFE	R—
FURFURYL ALCOHOL	
NORYL	
PHENOLIC	
POLYAMIDES—NYLON	
POLYESTERS	
BISPHENOL A-FUMURATE	R—
HYDROGENATED BISPHENOL-A-BISPHENOL-A	
ISOPHTHALIC	U
CHLORINATED POLYESTERS	R—
POLYSTYRENES	
POLYETHYLENE	RU
POLYPROPYLENE	R—
PVC—TYPE 1	RU
PVC—TYPE 2	RU
POLYVINYLIDENE CHLORIDE	R—
VINYLIDENE FLUORIDE	R—
VINYL ESTER	
RUBBERS	
BUTYL GR-1	R—
FLUORO ELASTOMERS VITON A	R—
FLUORO ELASTOMERS KEL-F 3700	
HARD RUBBER	
HYPALON	R—
KOROSEAL	
NATURAL RUBBER (GRS)	R—
NEOPRENE GR-M (CR)	R—
NITRILE BUNA N (NBR)	U
NORDEL	R—
POLYBUTADIENE	
SILICONE	
POLYURETHANE	U

TEMP °F: 60, 80, 100, 120, 140, 160, 180, 200, 220, 240, 260, 280, 300, 320, 340, 360, 380, 400, 420, 440, 460

TRIETHYLAMINE

TEMP °F: 60 80 100 120 140 160 180 200 220 240 260 280 300 320 340 360 380 400 420 440 460 480 500 520 540 560

METALS	Rating
ALLOY 20Cb3	G
ALUMINUM	
BRASS	
BRONZE	
CARBON STEEL	
COPPER	
E-BRIGHT 26-1	
HASTELLOY-B	
HASTELLOY-C	
HASTELLOY-D	
HIGH SILICON IRON	G
INCONEL	
LEAD	
MONEL	G
NICKEL	
STAINLESS STEELS	
Type 304/307	G
Type 316	G
TANTALUM	E
TITANIUM	
ZIRCONIUM	

NONMETALLICS	Rating
BOROSILICATE GLASS	
CARBON	R

TEMP °F: 60 80 100 120 140 160 180 200 220 240 260 280 300 320 340 360 380 400 420 440 460 480 500 520 540 560

FOR METALS

E = < 2 Mils Penetration/Year; G = < 20 Mils Penetration/Year

S = < 50 Mils Penetration/Year; U = > 50 Mils Penetration/Year

FOR NONMETALLICS

R = Resistant

U = Unsatisfactory

TRIETHYLAMINE

PLASTICS	TEMP °F (60–460)
ABS	U
ACRYLICS	
ASBESTOS REINF., EPOXY	
ASBESTOS REINF., PHENOLIC	
CHLORINATED POLYETHER	R———————
CPVC	R——— U
EPOXY	R—
FLUOROCARBONS FEP	R———————————————
FLUOROCARBONS TFE	R——————————————————
FURFURYL ALCOHOL	
NORYL	
PHENOLIC	
POLYAMIDES–NYLON	
POLYESTERS	
BISPHENOL A-FUMURATE	R———
HYDROGENATED BISPHENOL-A-BISPHENOL-A	
ISOPHTHALIC	
CHLORINATED POLYESTERS	
POLYSTYRENES	
POLYETHYLENE	R—
POLYPROPYLENE	U
PVC–TYPE 1	R————
PVC–TYPE 2	RU
POLYVINYLIDENE CHLORIDE	R—
VINYLIDENE FLUORIDE	R———— U
VINYL ESTER	

RUBBERS	TEMP °F (60–460)
BUTYL GR-1	
FLUORO ELASTOMERS VITON A	R——————
FLUORO ELASTOMERS KEL-F 3700	
HARD RUBBER	
HYPALON	
KOROSEAL	
NATURAL RUBBER (GRS)	
NEOPRENE GR-M (CR)	
NITRILE BUNA N (NBR)	R——————
NORDEL	
POLYBUTADIENE	
SILICONE	
POLYURETHANE	

TRIMETHYL PROPANE

TEMP °F

METALS	60	80	100	120	140	160	180	200	220	240	260	280	300	320	340	360	380	400	420	440	460	480	500	520	540	560
ALLOY 20Cb3																										
ALUMINUM																										
BRASS																										
BRONZE																										
CARBON STEEL																										
COPPER																										
E-BRIGHT 26-1																										
HASTELLOY-B																										
HASTELLOY-C																										
HASTELLOY-D																										
HIGH SILICON IRON																										
INCONEL																										
LEAD																										
MONEL																										
NICKEL																										
STAINLESS STEELS Type 304/307																										
Type 316																										
TANTALUM																										
TITANIUM																										
ZIRCONIUM																										
NONMETALLICS																										
BOROSILICATE GLASS																										
CARBON																										
	60	80	100	120	140	160	180	200	220	240	260	280	300	320	340	360	380	400	420	440	460	480	500	520	540	560

TEMP °F

FOR METALS

E = < 2 Mils Penetration/Year; G = < 20 Mils Penetration/Year

S = < 50 Mils Penetration/Year; U = > 50 Mils Penetration/Year

FOR NONMETALLICS

R = Resistant

U = Unsatisfactory

TRIMETHYL PROPANE

TEMP °F (60–460)

PLASTICS	60 → 460 °F
ABS	U
ACRYLICS	
ASBESTOS REINF., EPOXY	
ASBESTOS REINF., PHENOLIC	
CHLORINATED POLYETHER	R————————
CPVC	R——U
EPOXY	
FLUOROCARBONS FEP	R————————————————
FLUOROCARBONS TFE	R——————————————————
FURFURYL ALCOHOL	
NORYL	
PHENOLIC	
POLYAMIDES–NYLON	
POLYESTERS	
BISPHENOL A-FUMURATE	
HYDROGENATED BISPHENOL-A-BISPHENOL-A	
ISOPHTHALIC	
CHLORINATED POLYESTERS	
POLYSTYRENES	
POLYETHYLENE	
POLYPROPYLENE	U
PVC–TYPE 1	R———
PVC–TYPE 2	R———
POLYVINYLIDENE CHLORIDE	R———
VINYLIDENE FLUORIDE	R——————
VINYL ESTER	

RUBBERS	60 → 460 °F
BUTYL GR-1	
FLUORO ELASTOMERS VITON A	
FLUORO ELASTOMERS KEL-F 3700	
HARD RUBBER	
HYPALON	
KOROSEAL	
NATURAL RUBBER (GRS)	
NEOPRENE GR-M (CR)	
NITRILE BUNA N (NBR)	
NORDEL	
POLYBUTADIENE	
SILICONE	
POLYURETHANE	

TEMP °F: 60 80 100 120 140 160 180 200 220 240 260 280 300 320 340 360 380 400 420 440 460

TRISODIUM PHOSPHATE

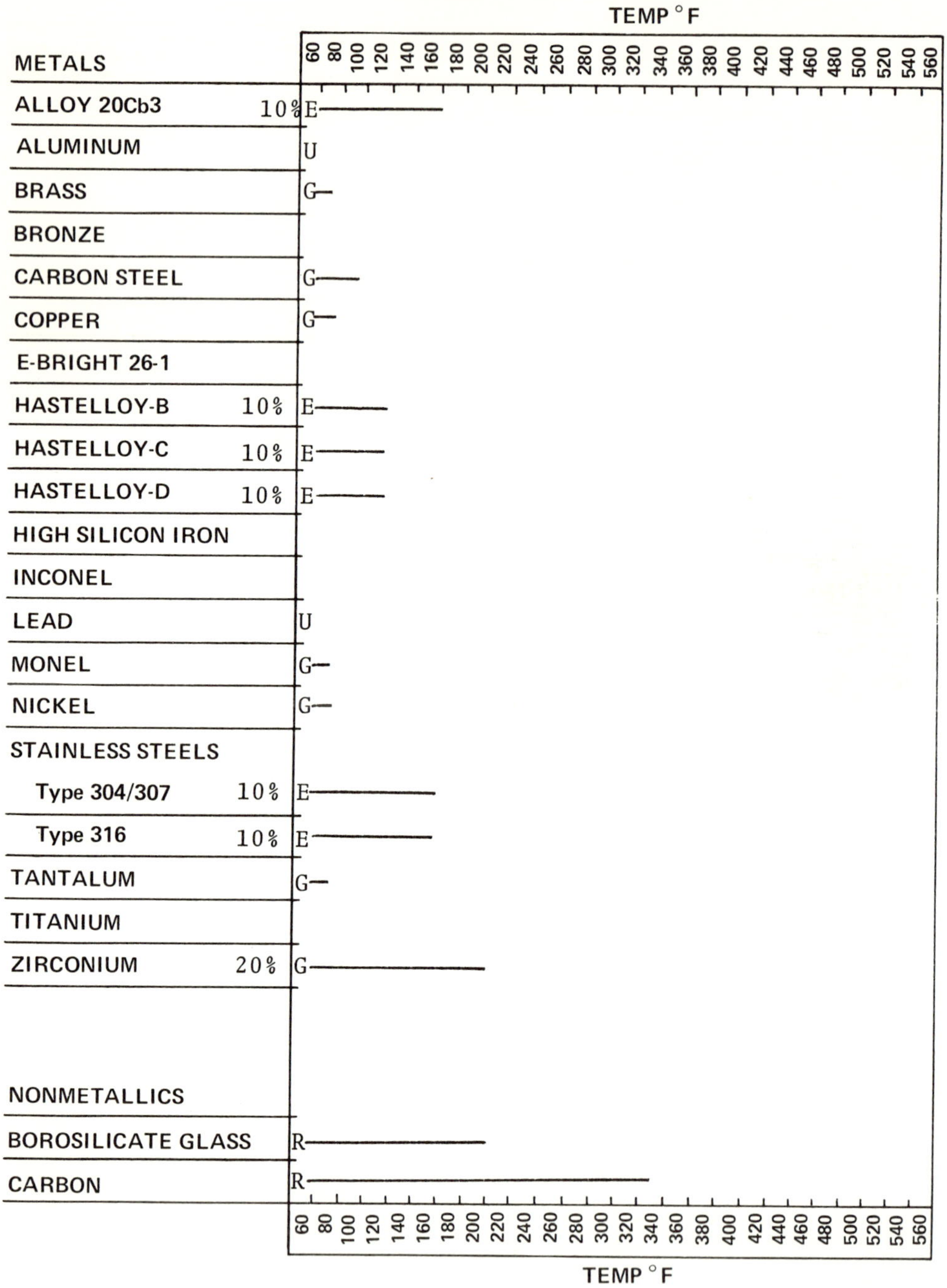

FOR METALS

E = < 2 Mils Penetration/Year; G = < 20 Mils Penetration/Year

S = < 50 Mils Penetration/Year; U = > 50 Mils Penetration/Year

FOR NONMETALLICS

R = Resistant

U = Unsatisfactory

TRISODIUM PHOSPHATE

TEMP °F: 60 80 100 120 140 160 180 200 220 240 260 280 300 320 340 360 380 400 420 440 460

PLASTICS	Conc.	Rating
ABS		R
ACRYLICS		
ASBESTOS REINF., EPOXY		R
ASBESTOS REINF., PHENOLIC		U
CHLORINATED POLYETHER		R
CPVC		R
EPOXY		R
FLUOROCARBONS FEP		R
FLUOROCARBONS TFE		R
FURFURYL ALCOHOL	25%	R
NORYL		
PHENOLIC		
POLYAMIDES–NYLON		R
POLYESTERS		
BISPHENOL A-FUMURATE	50%	R
HYDROGENATED BISPHENOL-A-BISPHENOL-A	10%	R
ISOPHTHALIC	sat	U
CHLORINATED POLYESTERS		R—U
POLYSTYRENES		
POLYETHYLENE		R
POLYPROPYLENE		R
PVC–TYPE 1		R
PVC–TYPE 2		R
POLYVINYLIDENE CHLORIDE		R
VINYLIDENE FLUORIDE		R
VINYL ESTER	all	R

RUBBERS	Conc.	Rating
BUTYL GR-1		R
FLUORO ELASTOMERS VITON A		R
FLUORO ELASTOMERS KEL-F 3700		
HARD RUBBER		R
HYPALON		R
KOROSEAL		R
NATURAL RUBBER (GRS)		R
NEOPRENE GR-M (CR)		R
NITRILE BUNA N (NBR)		R
NORDEL		
POLYBUTADIENE		
SILICONE		
POLYURETHANE		R

TEMP °F: 60 80 100 120 140 160 180 200 220 240 260 280 300 320 340 360 380 400 420 440 460

TURPENTINE

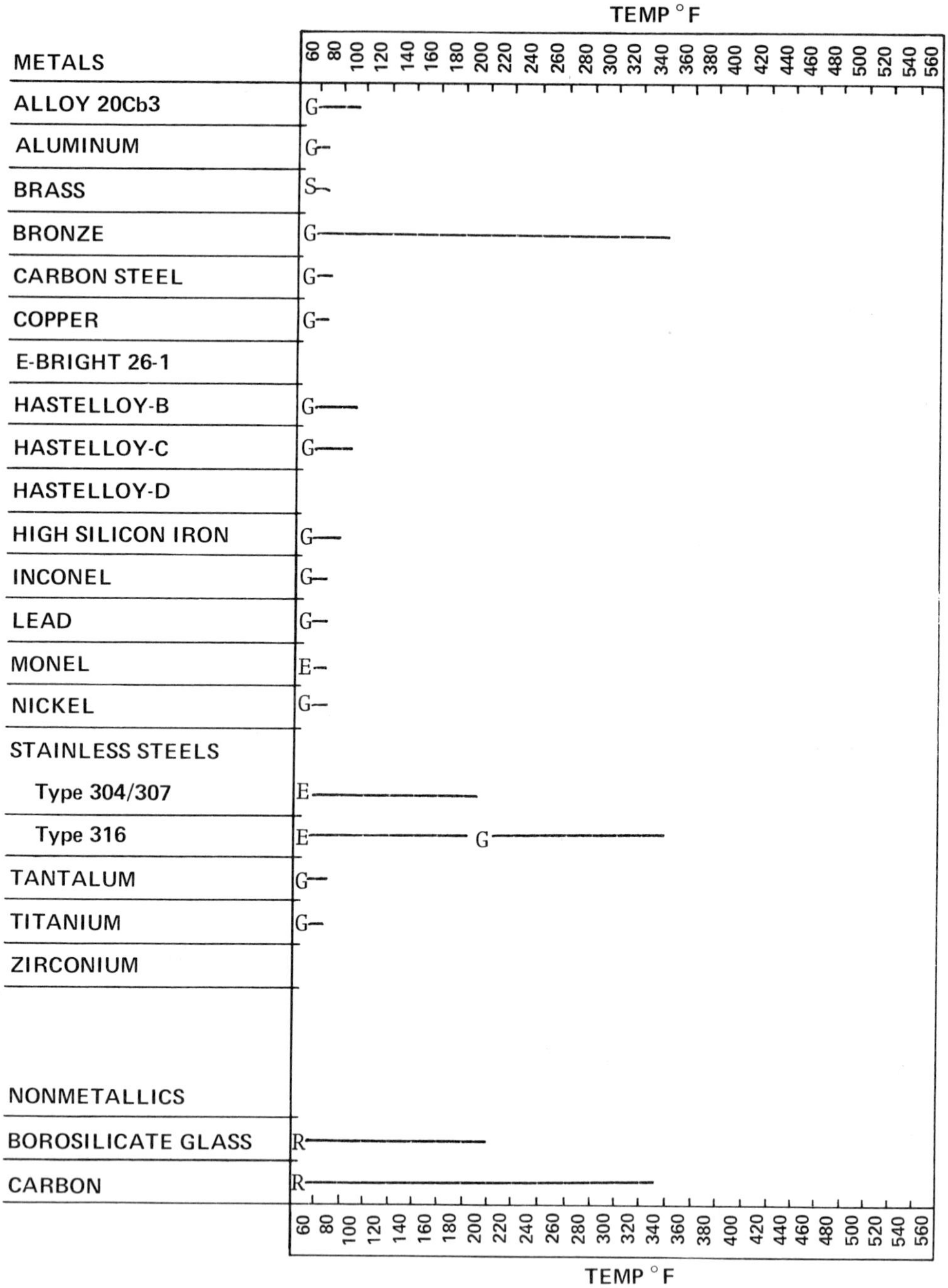

FOR METALS

E = < 2 Mils Penetration/Year; G = < 20 Mils Penetration/Year
S = < 50 Mils Penetration/Year; U = > 50 Mils Penetration/Year

FOR NONMETALLICS

R = Resistant
U = Unsatisfactory

TURPENTINE

PLASTICS	TEMP °F (60–460)
ABS	U
ACRYLICS	R—
ASBESTOS REINF., EPOXY	R—
ASBESTOS REINF., PHENOLIC	R—
CHLORINATED POLYETHER	R—
CPVC	R—U
EPOXY	R—
FLUOROCARBONS FEP	R—
FLUOROCARBONS TFE	R—
FURFURYL ALCOHOL	
NORYL	
PHENOLIC	
POLYAMIDES–NYLON	R—
POLYESTERS	
BISPHENOL A-FUMURATE	R—
HYDROGENATED BISPHENOL-A-BISPHENOL-A	
ISOPHTHALIC	R—
CHLORINATED POLYESTERS	R—
POLYSTYRENES	
POLYETHYLENE	U
POLYPROPYLENE	R—U
PVC–TYPE 1	R—
PVC–TYPE 2	U
POLYVINYLIDENE CHLORIDE	R—
VINYLIDENE FLUORIDE	R—
VINYL ESTER	R—
RUBBERS	
BUTYL GR-1	U
FLUORO ELASTOMERS VITON A	R—
FLUORO ELASTOMERS KEL-F 3700	
HARD RUBBER	
HYPALON	U
KOROSEAL	
NATURAL RUBBER (GRS)	U
NEOPRENE GR-M (CR)	U
NITRILE BUNA N (NBR)	R—
NORDEL	U
POLYBUTADIENE	
SILICONE	
POLYURETHANE	U

TEMP °F: 60 80 100 120 140 160 180 200 220 240 260 280 300 320 340 360 380 400 420 440 460

UREA

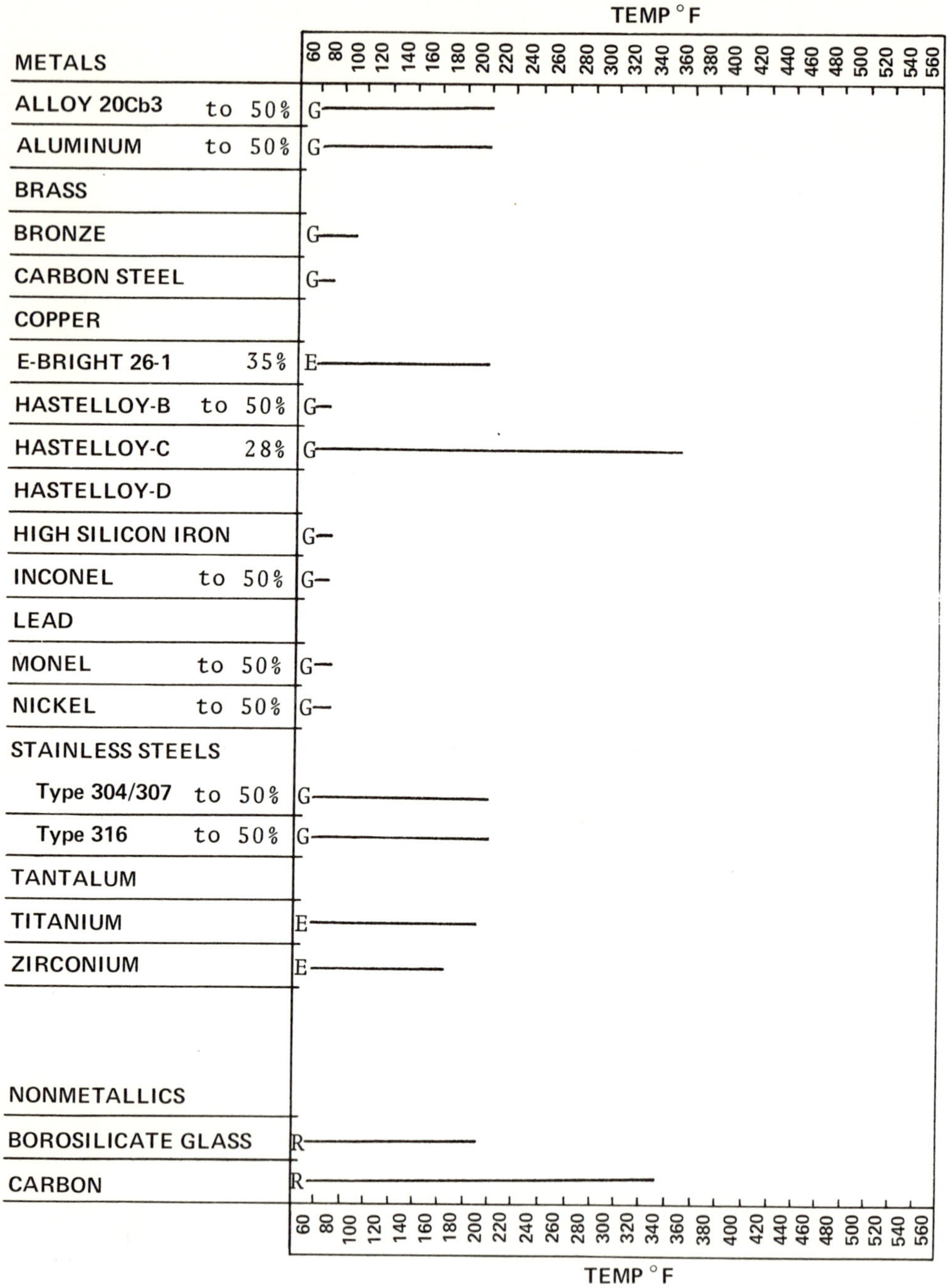

FOR METALS

E = < 2 Mils Penetration/Year; G = < 20 Mils Penetration/Year
S = < 50 Mils Penetration/Year; U = > 50 Mils Penetration/Year

FOR NONMETALLICS

R = Resistant
U = Unsatisfactory

UREA

Please see page 1 for notes on calibration.

TEMP °F: 60 80 100 120 140 160 180 200 220 240 260 280 300 320 340 360 380 400 420 440 460

PLASTICS		Rating
ABS		R
ACRYLICS		R
ASBESTOS REINF., EPOXY		
ASBESTOS REINF., PHENOLIC		
CHLORINATED POLYETHER		R
CPVC		R
EPOXY		R
FLUOROCARBONS FEP		R
FLUOROCARBONS TFE		R
FURFURYL ALCOHOL		R
NORYL		
PHENOLIC		
POLYAMIDES–NYLON		R
POLYESTERS		
BISPHENOL A-FUMURATE		R
HYDROGENATED BISPHENOL-A-BISPHENOL-A		
ISOPHTHALIC	sat	R
CHLORINATED POLYESTERS	50%	R
POLYSTYRENES		
POLYETHYLENE		R
POLYPROPYLENE		R
PVC–TYPE 1		R
PVC–TYPE 2		R
POLYVINYLIDENE CHLORIDE		R
VINYLIDENE FLUORIDE		R
VINYL ESTER	sat	R

RUBBERS		Rating
BUTYL GR-1		R
FLUORO ELASTOMERS VITON A		R
FLUORO ELASTOMERS KEL-F 3700		
HARD RUBBER		R
HYPALON		R
KOROSEAL		R
NATURAL RUBBER (GRS)		R
NEOPRENE GR-M (CR)		R
NITRILE BUNA N (NBR)		R
NORDEL		
POLYBUTADIENE		
SILICONE		
POLYURETHANE		

TEMP °F: 60 80 100 120 140 160 180 200 220 240 260 280 300 320 340 360 380 400 420 440 460

URANYL SULFATE, 5%

	TEMP °F 60	80	100	120	140	160	180	200	220	240	260	280	300	320	340	360	380	400	420	440	460	480	500	520	540	560
METALS																										
ALLOY 20Cb3																										
ALUMINUM																										
BRASS																										
BRONZE																										
CARBON STEEL																										
COPPER																										
E-BRIGHT 26-1																										
HASTELLOY-B																										
HASTELLOY-C																										
HASTELLOY-D																										
HIGH SILICON IRON																										
INCONEL																										
LEAD																										
MONEL																										
NICKEL																										
STAINLESS STEELS Type 304/307																										
Type 316																										
TANTALUM																										
TITANIUM																										
ZIRCONIUM	G	—	—	—	—	—	—	—	—	—	—	—	—													
NONMETALLICS																										
BOROSILICATE GLASS																										
CARBON																										

FOR METALS

E = < 2 Mils Penetration/Year; G = < 20 Mils Penetration/Year

S = < 50 Mils Penetration/Year; U = > 50 Mils Penetration/Year

FOR NONMETALLICS

R = Resistant

U = Unsatisfactory

URANYL SULFATE, 5%

TEMP °F: 60, 80, 100, 120, 140, 160, 180, 200, 220, 240, 260, 280, 300, 320, 340, 360, 380, 400, 420, 440, 460

PLASTICS	
ABS	
ACRYLICS	
ASBESTOS REINF., EPOXY	
ASBESTOS REINF., PHENOLIC	
CHLORINATED POLYETHER	
CPVC	
EPOXY	
FLUOROCARBONS FEP	R
FLUOROCARBONS TFE	R
FURFURYL ALCOHOL	
NORYL	
PHENOLIC	
POLYAMIDES—NYLON	
POLYESTERS	
BISPHENOL A-FUMURATE	
HYDROGENATED BISPHENOL-A-BISPHENOL-A	
ISOPHTHALIC	
CHLORINATED POLYESTERS	
POLYSTYRENES	
POLYETHYLENE	R
POLYPROPYLENE	R
PVC—TYPE 1	
PVC—TYPE 2	
POLYVINYLIDENE CHLORIDE	
VINYLIDENE FLUORIDE	R
VINYL ESTER	

RUBBERS	
BUTYL GR-1	
FLUORO ELASTOMERS VITON A	
FLUORO ELASTOMERS KEL-F 3700	
HARD RUBBER	
HYPALON	
KOROSEAL	
NATURAL RUBBER (GRS)	
NEOPRENE GR-M (CR)	
NITRILE BUNA N (NBR)	
NORDEL	
POLYBUTADIENE	
SILICONE	
POLYURETHANE	

TEMP °F: 60, 80, 100, 120, 140, 160, 180, 200, 220, 240, 260, 280, 300, 320, 340, 360, 380, 400, 420, 440, 460

URINE

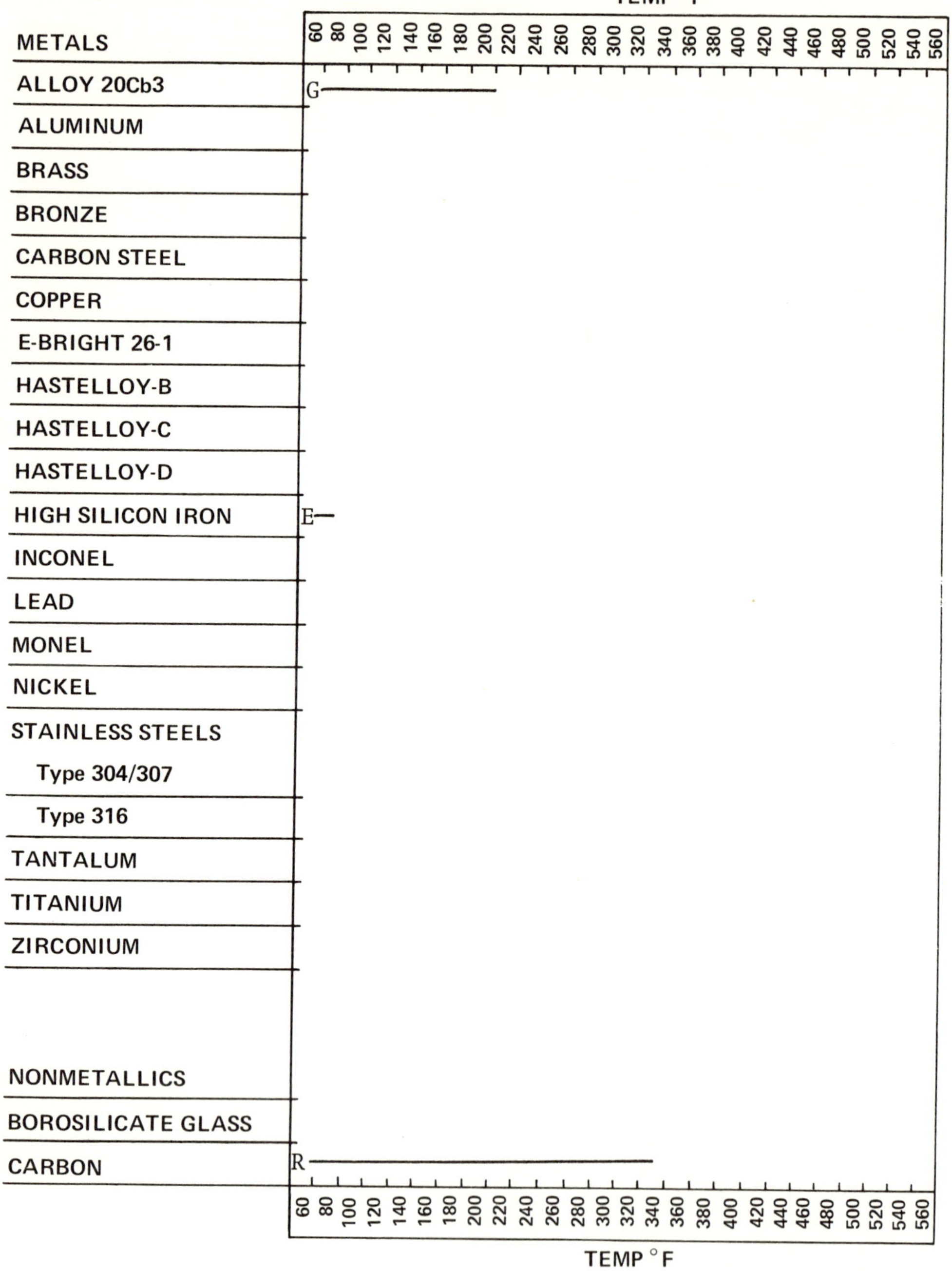

FOR METALS

E = < 2 Mils Penetration/Year; G = < 20 Mils Penetration/Year

S = < 50 Mils Penetration/Year; U = > 50 Mils Penetration/Year

FOR NONMETALLICS

R = Resistant

U = Unsatisfactory

URINE

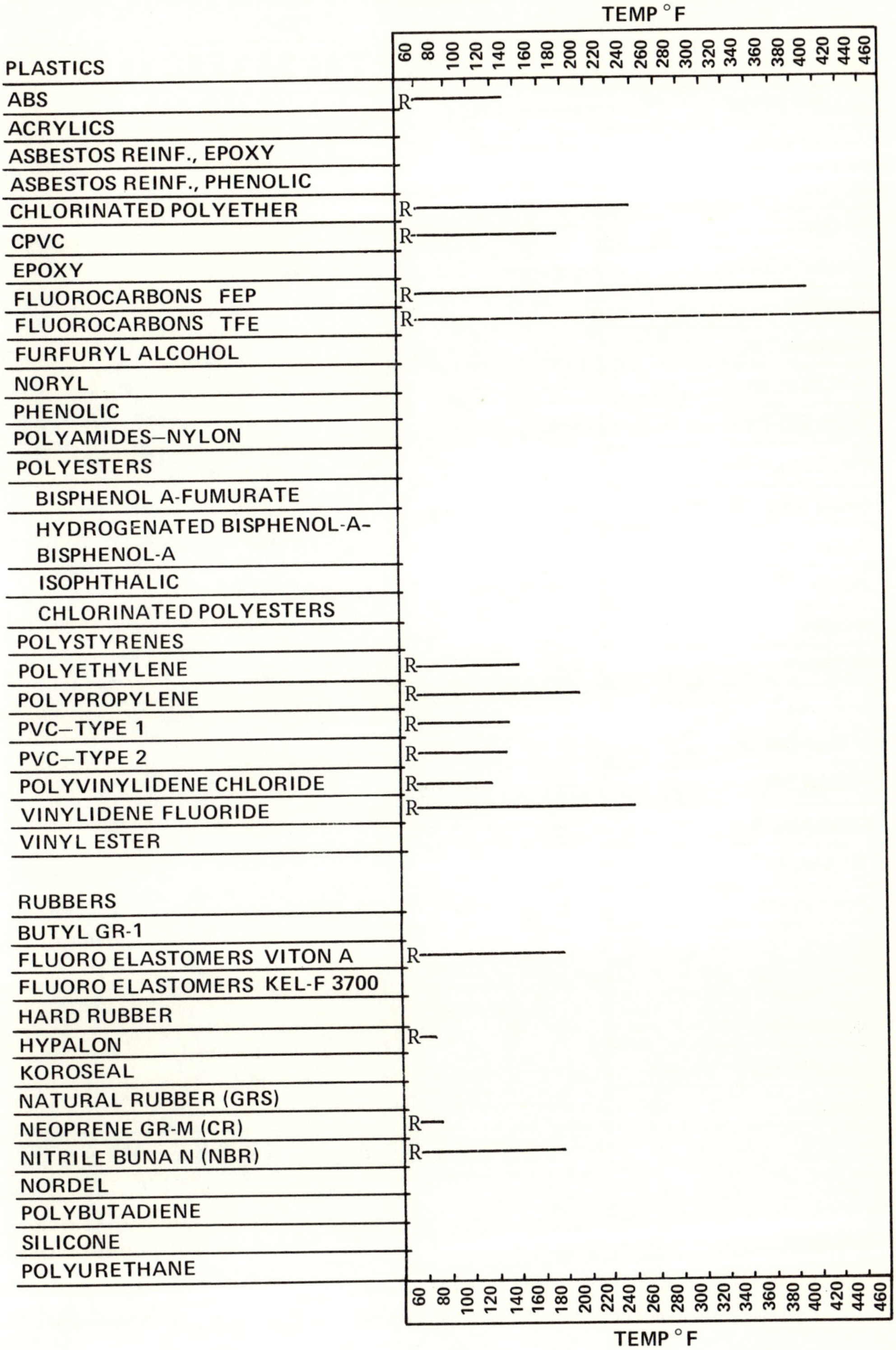
TEMP °F
60 80 100 120 140 160 180 200 220 240 260 280 300 320 340 360 380 400 420 440 460
PLASTICS
ABS R
ACRYLICS
ASBESTOS REINF., EPOXY
ASBESTOS REINF., PHENOLIC
CHLORINATED POLYETHER R
CPVC R
EPOXY
FLUOROCARBONS FEP R
FLUOROCARBONS TFE R
FURFURYL ALCOHOL
NORYL
PHENOLIC
POLYAMIDES—NYLON
POLYESTERS
BISPHENOL A-FUMURATE
HYDROGENATED BISPHENOL-A-BISPHENOL-A
ISOPHTHALIC
CHLORINATED POLYESTERS
POLYSTYRENES
POLYETHYLENE R
POLYPROPYLENE R
PVC—TYPE 1 R
PVC—TYPE 2 R
POLYVINYLIDENE CHLORIDE R
VINYLIDENE FLUORIDE R
VINYL ESTER
RUBBERS
BUTYL GR-1
FLUORO ELASTOMERS VITON A R
FLUORO ELASTOMERS KEL-F 3700
HARD RUBBER
HYPALON R
KOROSEAL
NATURAL RUBBER (GRS)
NEOPRENE GR-M (CR) R
NITRILE BUNA N (NBR) R
NORDEL
POLYBUTADIENE
SILICONE
POLYURETHANE
60 80 100 120 140 160 180 200 220 240 260 280 300 320 340 360 380 400 420 440 460
TEMP °F

VEGETABLE OIL

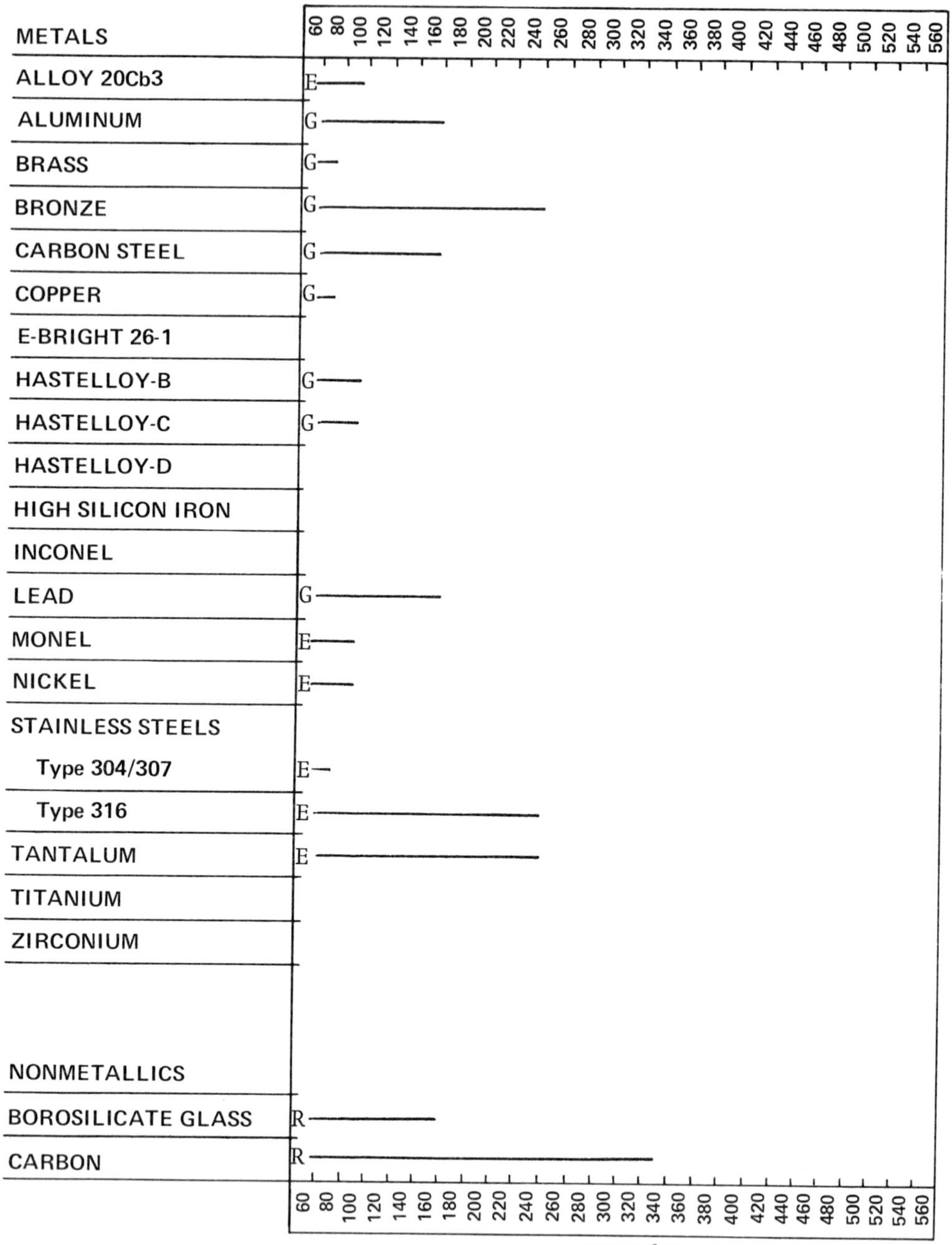

FOR METALS

E = < 2 Mils Penetration/Year; G = < 20 Mils Penetration/Year

S = < 50 Mils Penetration/Year; U = > 50 Mils Penetration/Year

FOR NONMETALLICS

R = Resistant

U = Unsatisfactory

VEGETABLE OIL

Please see page 1 for notes on calibration.

TEMP °F: 60 80 100 120 140 160 180 200 220 240 260 280 300 320 340 360 380 400 420 440 460

PLASTICS	
ABS	
ACRYLICS	
ASBESTOS REINF., EPOXY	
ASBESTOS REINF., PHENOLIC	
CHLORINATED POLYETHER	R
CPVC	R
EPOXY	
FLUOROCARBONS FEP	R
FLUOROCARBONS TFE	R
FURFURYL ALCOHOL	
NORYL	
PHENOLIC	
POLYAMIDES–NYLON	
POLYESTERS	
BISPHENOL A-FUMURATE	
HYDROGENATED BISPHENOL-A-BISPHENOL-A	
ISOPHTHALIC	
CHLORINATED POLYESTERS	
POLYSTYRENES	
POLYETHYLENE	
POLYPROPYLENE	
PVC–TYPE 1	
PVC–TYPE 2	
POLYVINYLIDENE CHLORIDE	R
VINYLIDENE FLUORIDE	R
VINYL ESTER	
RUBBERS	
BUTYL GR-1	R
FLUORO ELASTOMERS VITON A	R
FLUORO ELASTOMERS KEL-F 3700	
HARD RUBBER	U
HYPALON	
KOROSEAL	
NATURAL RUBBER (GRS)	U
NEOPRENE GR-M (CR)	R
NITRILE BUNA N (NBR)	R
NORDEL	
POLYBUTADIENE	
SILICONE	
POLYURETHANE	

TEMP °F: 60 80 100 120 140 160 180 200 220 240 260 280 300 320 340 360 380 400 420 440 460

VINEGAR

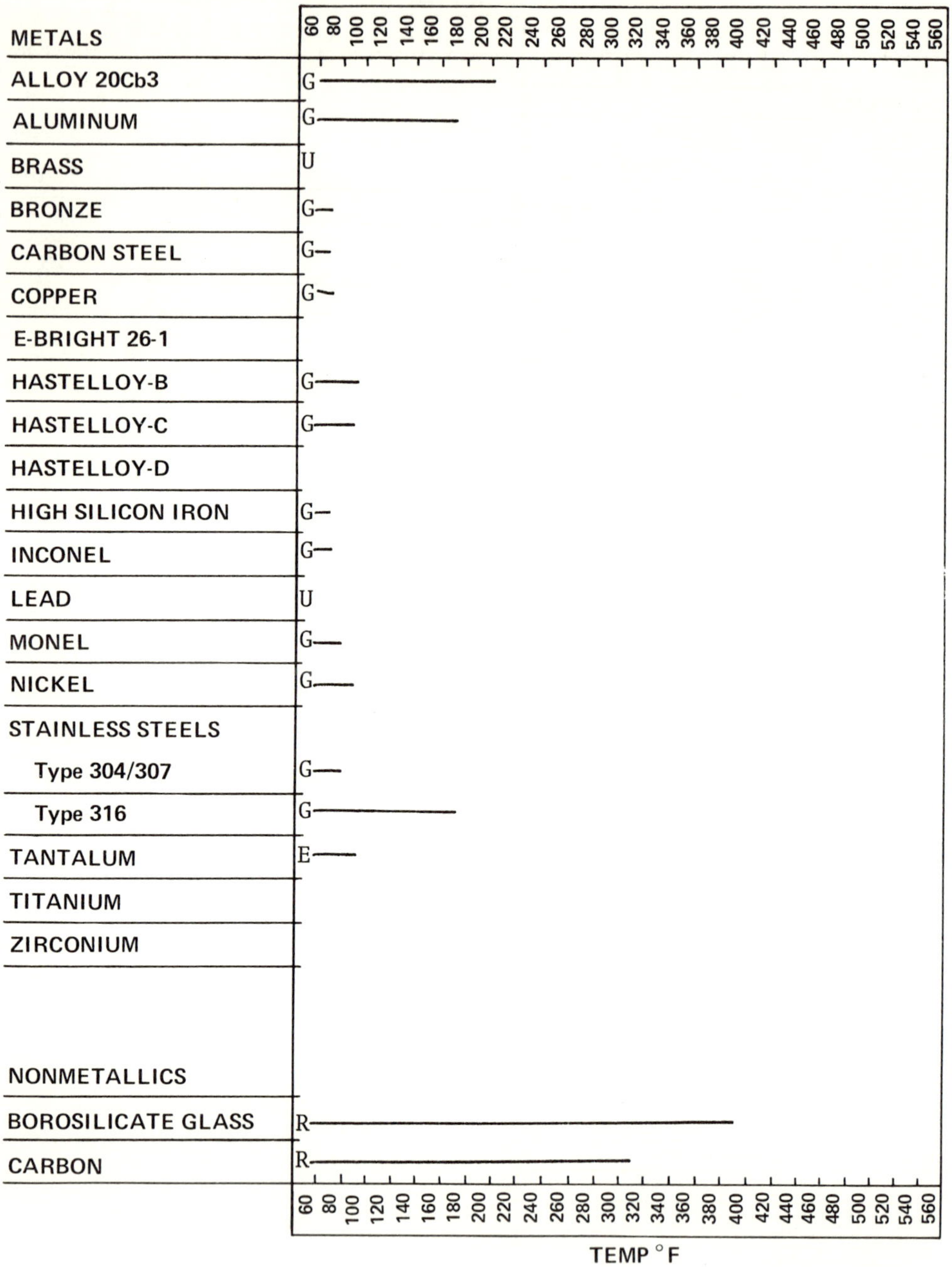

FOR METALS

E = < 2 Mils Penetration/Year; G = < 20 Mils Penetration/Year

S = < 50 Mils Penetration/Year; U = > 50 Mils Penetration/Year

FOR NONMETALLICS

R = Resistant

U = Unsatisfactory

VINEGAR

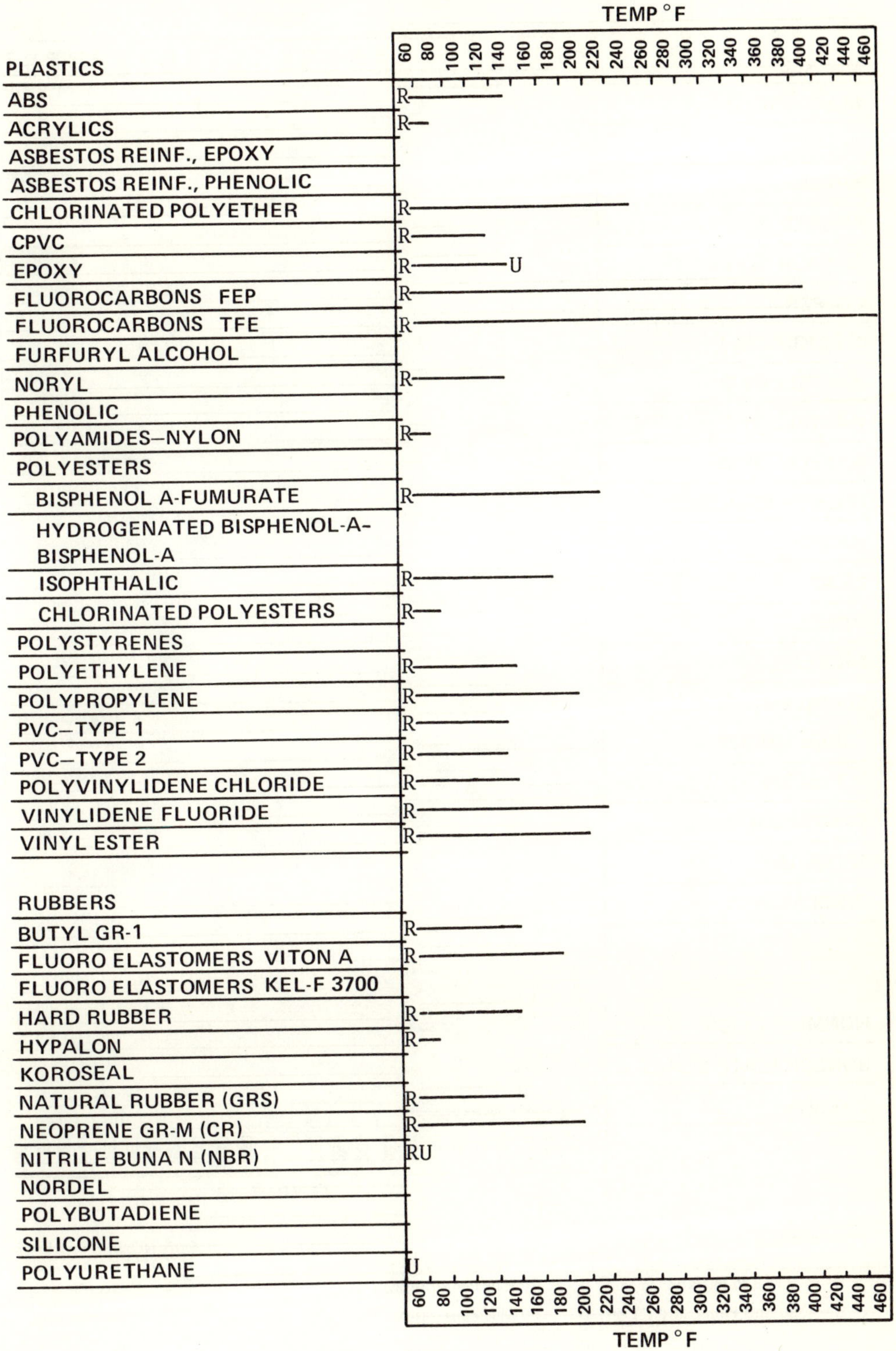

TEMP °F
60 80 100 120 140 160 180 200 220 240 260 280 300 320 340 360 380 400 420 440 460
PLASTICS
ABS
ACRYLICS
ASBESTOS REINF., EPOXY
ASBESTOS REINF., PHENOLIC
CHLORINATED POLYETHER
CPVC
EPOXY
FLUOROCARBONS FEP
FLUOROCARBONS TFE
FURFURYL ALCOHOL
NORYL
PHENOLIC
POLYAMIDES–NYLON
POLYESTERS
BISPHENOL A-FUMURATE
HYDROGENATED BISPHENOL-A-BISPHENOL-A
ISOPHTHALIC
CHLORINATED POLYESTERS
POLYSTYRENES
POLYETHYLENE
POLYPROPYLENE
PVC–TYPE 1
PVC–TYPE 2
POLYVINYLIDENE CHLORIDE
VINYLIDENE FLUORIDE
VINYL ESTER
RUBBERS
BUTYL GR-1
FLUORO ELASTOMERS VITON A
FLUORO ELASTOMERS KEL-F 3700
HARD RUBBER
HYPALON
KOROSEAL
NATURAL RUBBER (GRS)
NEOPRENE GR-M (CR)
NITRILE BUNA N (NBR)
NORDEL
POLYBUTADIENE
SILICONE
POLYURETHANE
R
U
RU
TEMP °F

VINYL ACETATE

METALS	TEMP °F (60–560)
ALLOY 20Cb3	
ALUMINUM	
BRASS	
BRONZE	
CARBON STEEL	
COPPER	
E-BRIGHT 26-1	
HASTELLOY-B	
HASTELLOY-C	
HASTELLOY-D	
HIGH SILICON IRON	
INCONEL	
LEAD	
MONEL	
NICKEL	
STAINLESS STEELS	
Type 304/307	E (60–160)
Type 316	E (60–160) G (160–300)
TANTALUM	
TITANIUM	
ZIRCONIUM	
NONMETALLICS	
BOROSILICATE GLASS	
CARBON	R (60–340)

FOR METALS

E = < 2 Mils Penetration/Year; G = < 20 Mils Penetration/Year

S = < 50 Mils Penetration/Year; U = > 50 Mils Penetration/Year

FOR NONMETALLICS

R = Resistant

U = Unsatisfactory

VINYL ACETATE

PLASTICS	TEMP °F (60–460)
ABS	U (60)
ACRYLICS	
ASBESTOS REINF., EPOXY	
ASBESTOS REINF., PHENOLIC	
CHLORINATED POLYETHER	R (60–140)
CPVC	U (60)
EPOXY	R (60–80); U (100)
FLUOROCARBONS FEP	R (60–400)
FLUOROCARBONS TFE	R (60–460)
FURFURYL ALCOHOL	
NORYL	
PHENOLIC	
POLYAMIDES–NYLON	
POLYESTERS	
BISPHENOL A-FUMURATE	U (60)
HYDROGENATED BISPHENOL-A-BISPHENOL-A	U (60)
ISOPHTHALIC	U (60)
CHLORINATED POLYESTERS	
POLYSTYRENES	
POLYETHYLENE	R (60–140)
POLYPROPYLENE	R (60–140)
PVC–TYPE 1	U (60)
PVC–TYPE 2	U (60)
POLYVINYLIDENE CHLORIDE	
VINYLIDENE FLUORIDE	R (60–240)
VINYL ESTER	

RUBBERS	TEMP °F (60–460)
BUTYL GR-1	
FLUORO ELASTOMERS VITON A	R (60–180)
FLUORO ELASTOMERS KEL-F 3700	
HARD RUBBER	
HYPALON	R (60–240)
KOROSEAL	
NATURAL RUBBER (GRS)	
NEOPRENE GR-M (CR)	
NITRILE BUNA N (NBR)	RU (60)
NORDEL	
POLYBUTADIENE	
SILICONE	
POLYURETHANE	

WATER, ACID MINE

METALS	Rating	TEMP °F
ALLOY 20Cb3		
ALUMINUM		
BRASS	U	60
BRONZE	U	60
CARBON STEEL		
COPPER	U	60
E-BRIGHT 26-1		
HASTELLOY-B		
HASTELLOY-C		
HASTELLOY-D		
HIGH SILICON IRON	G	60
INCONEL		
LEAD		
MONEL		
NICKEL		
STAINLESS STEELS		
Type 304/307	E	60–120
Type 316	E	60–120
TANTALUM	E	60–120
TITANIUM		
ZIRCONIUM		

NONMETALLICS	Rating	TEMP °F
BOROSILICATE GLASS	R	60–200
CARBON		

FOR METALS

E = < 2 Mils Penetration/Year; G = < 20 Mils Penetration/Year

S = < 50 Mils Penetration/Year; U = > 50 Mils Penetration/Year

FOR NONMETALLICS

R = Resistant

U = Unsatisfactory

WATER, ACID MINE

Please see page 1 for notes on calibration.

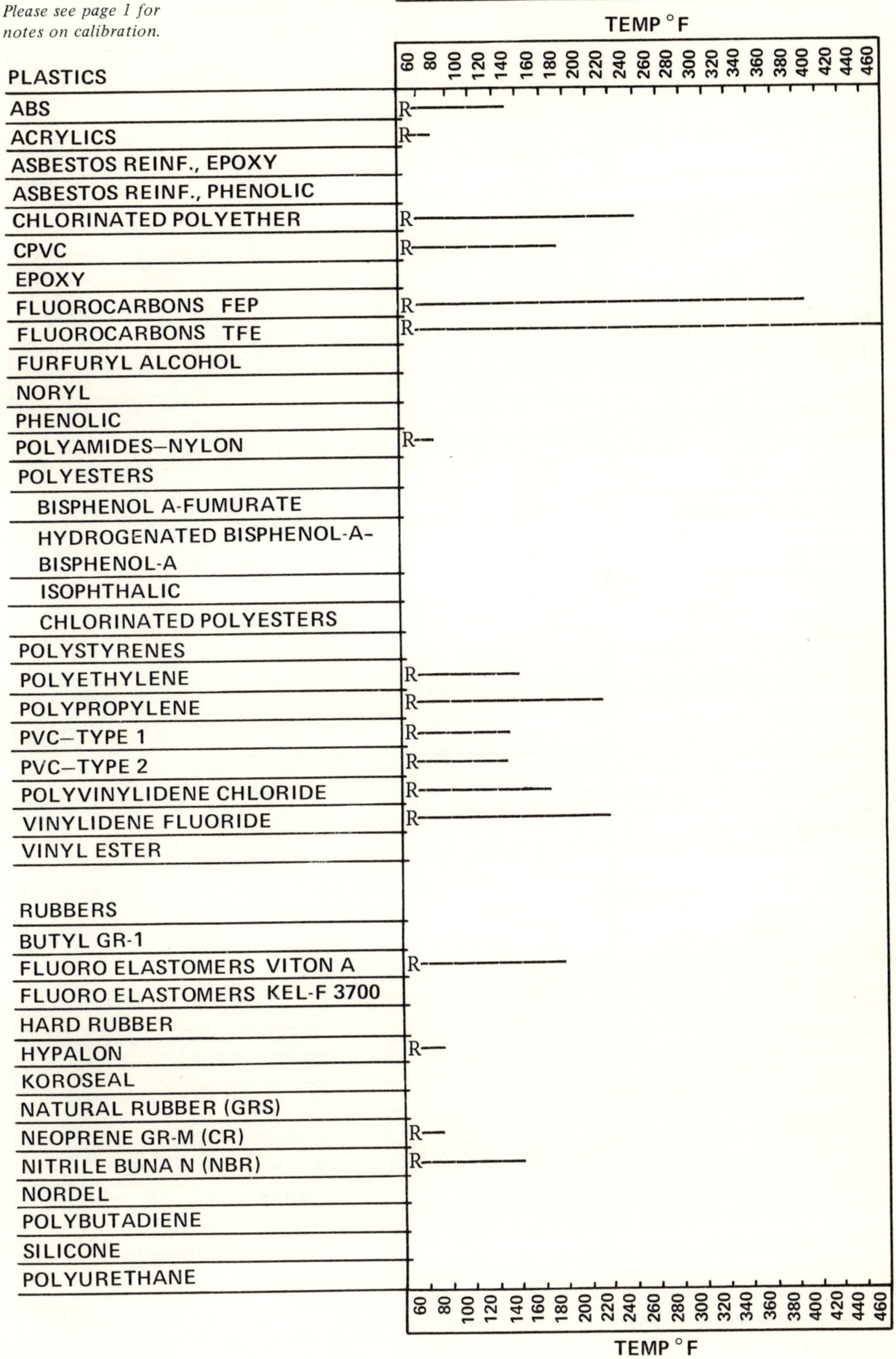

WATER, DEMINERALIZED

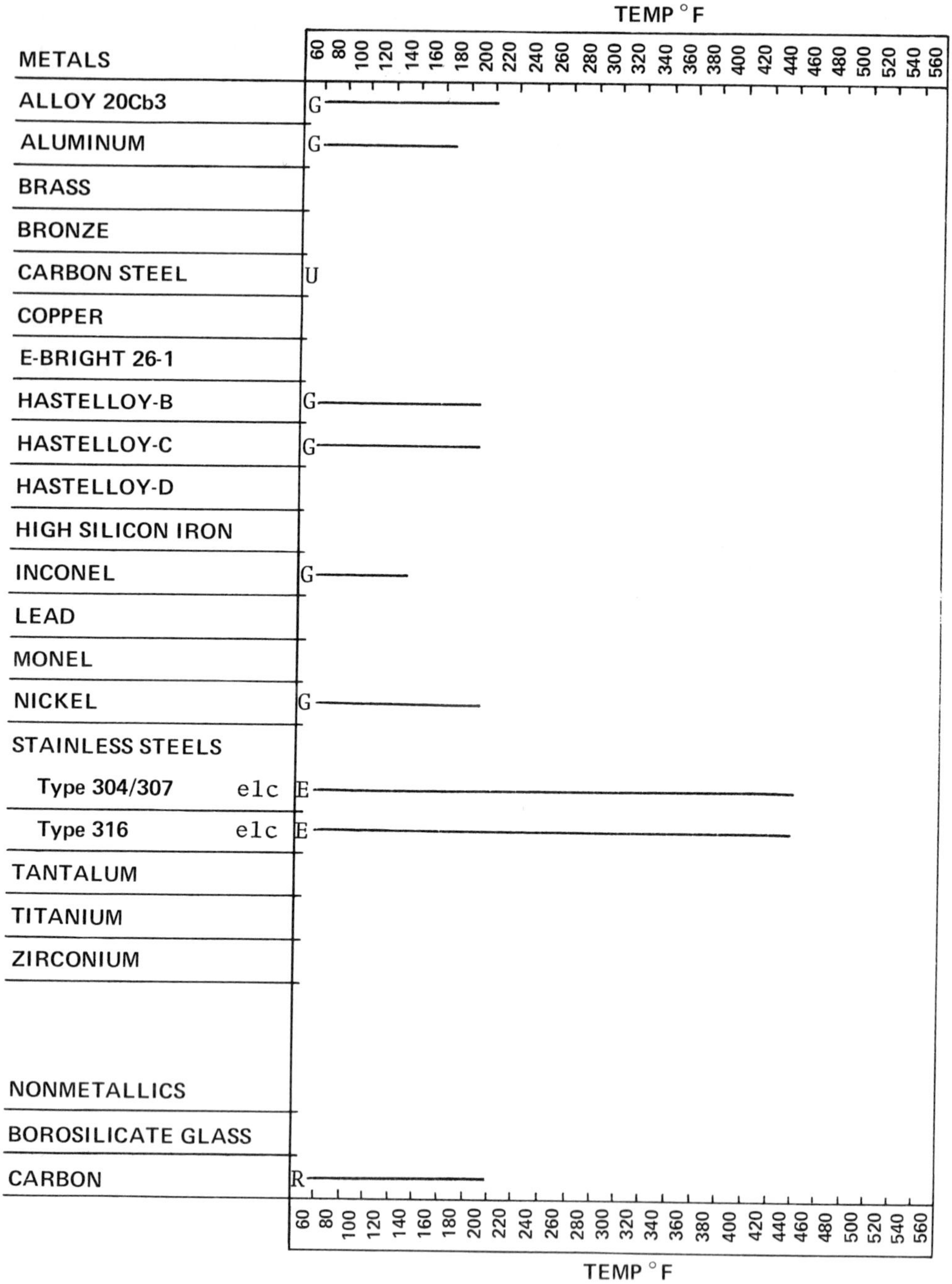

FOR METALS

E = < 2 Mils Penetration/Year; G = < 20 Mils Penetration/Year

S = < 50 Mils Penetration/Year; U = > 50 Mils Penetration/Year

FOR NONMETALLICS

R = Resistant

U = Unsatisfactory

WATER, DEMINERALIZED

PLASTICS	TEMP °F (60–460)
ABS	R
ACRYLICS	
ASBESTOS REINF., EPOXY	
ASBESTOS REINF., PHENOLIC	
CHLORINATED POLYETHER	R
CPVC	R
EPOXY	R
FLUOROCARBONS FEP	R
FLUOROCARBONS TFE	R
FURFURYL ALCOHOL	R
NORYL	R
PHENOLIC	
POLYAMIDES–NYLON	
POLYESTERS	
BISPHENOL A-FUMURATE	
HYDROGENATED BISPHENOL-A-BISPHENOL-A	R
ISOPHTHALIC	R
CHLORINATED POLYESTERS	R
POLYSTYRENES	
POLYETHYLENE	R
POLYPROPYLENE	R
PVC–TYPE 1	R
PVC–TYPE 2	R
POLYVINYLIDENE CHLORIDE	R
VINYLIDENE FLUORIDE	R
VINYL ESTER	R

RUBBERS	TEMP °F (60–460)
BUTYL GR-1	R
FLUORO ELASTOMERS VITON A	R
FLUORO ELASTOMERS KEL-F 3700	
HARD RUBBER	
HYPALON	R
KOROSEAL	
NATURAL RUBBER (GRS)	R
NEOPRENE GR-M (CR)	R
NITRILE BUNA N (NBR)	R
NORDEL	R
POLYBUTADIENE	
SILICONE	
POLYURETHANE	

WATER, DISTILLED

TEMP °F: 60 80 100 120 140 160 180 200 220 240 260 280 300 320 340 360 380 400 420 440 460 480 500 520 540 560

METALS	Rating
ALLOY 20Cb3	G
ALUMINUM	U
BRASS	G
BRONZE	G
CARBON STEEL	U
COPPER	G
E-BRIGHT 26-1	
HASTELLOY-B	G
HASTELLOY-C	E
HASTELLOY-D	G
HIGH SILICON IRON	G
INCONEL	
LEAD	U
MONEL	U
NICKEL	S
STAINLESS STEELS	
Type 304/307	G
Type 316	G
TANTALUM	
TITANIUM	
ZIRCONIUM	

NONMETALLICS	Rating
BOROSILICATE GLASS	R
CARBON	R

TEMP °F: 60 80 100 120 140 160 180 200 220 240 260 280 300 320 340 360 380 400 420 440 460 480 500 520 540 560

FOR METALS

E = < 2 Mils Penetration/Year; G = < 20 Mils Penetration/Year

S = < 50 Mils Penetration/Year; U = > 50 Mils Penetration/Year

FOR NONMETALLICS

R = Resistant

U = Unsatisfactory

WATER, DISTILLED

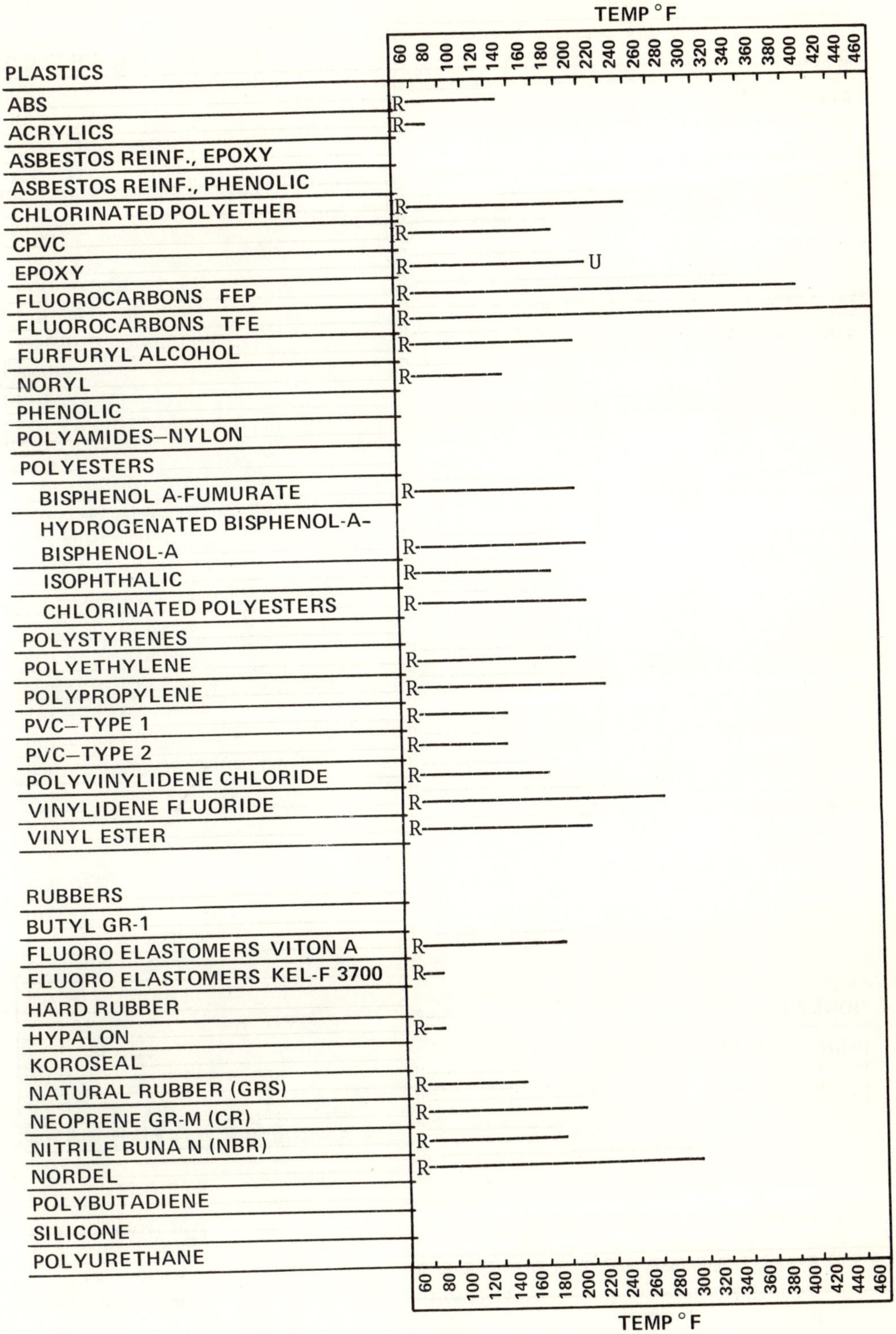

WATER, SALT

TEMP °F: 60, 80, 100, 120, 140, 160, 180, 200, 220, 240, 260, 280, 300, 320, 340, 360, 380, 400, 420, 440, 460, 480, 500, 520, 540, 560

METALS		Rating
ALLOY 20Cb3		
ALUMINUM		U
BRASS		S
BRONZE		G
CARBON STEEL		G
COPPER		G
E-BRIGHT 26-1		
HASTELLOY-B		
HASTELLOY-C		
HASTELLOY-D		
HIGH SILICON IRON		
INCONEL		
LEAD		G
MONEL		G
NICKEL		
STAINLESS STEELS		
Type 304/307	1	G
Type 316	1	G
TANTALUM		E
TITANIUM		
ZIRCONIUM		

NONMETALLICS	Rating
BOROSILICATE GLASS	R
CARBON	R

TEMP °F: 60, 80, 100, 120, 140, 160, 180, 200, 220, 240, 260, 280, 300, 320, 340, 360, 380, 400, 420, 440, 460, 480, 500, 520, 540, 560

FOR METALS

E = < 2 Mils Penetration/Year; G = < 20 Mils Penetration/Year
S = < 50 Mils Penetration/Year; U = > 50 Mils Penetration/Year

FOR NONMETALLICS

R = Resistant
U = Unsatisfactory

WATER, SALT

PLASTICS	TEMP °F (60–460)
ABS	R
ACRYLICS	R
ASBESTOS REINF., EPOXY	
ASBESTOS REINF., PHENOLIC	
CHLORINATED POLYETHER	R
CPVC	R
EPOXY 10%	R ——— U
FLUOROCARBONS FEP	R
FLUOROCARBONS TFE	R
FURFURYL ALCOHOL	
NORYL	
PHENOLIC	
POLYAMIDES—NYLON	R
POLYESTERS	
BISPHENOL A-FUMURATE	
HYDROGENATED BISPHENOL-A-BISPHENOL-A	R
ISOPHTHALIC	R
CHLORINATED POLYESTERS	
POLYSTYRENES	
POLYETHYLENE	R
POLYPROPYLENE	R
PVC—TYPE 1	R
PVC—TYPE 2	R
POLYVINYLIDENE CHLORIDE	R
VINYLIDENE FLUORIDE	R
VINYL ESTER	R
RUBBERS	
BUTYL GR-1	R
FLUORO ELASTOMERS VITON A	R
FLUORO ELASTOMERS KEL-F 3700	
HARD RUBBER	R
HYPALON	R
KOROSEAL	
NATURAL RUBBER (GRS)	R
NEOPRENE GR-M (CR)	R
NITRILE BUNA N (NBR)	R
NORDEL	R
POLYBUTADIENE	
SILICONE	
POLYURETHANE	U

WATER, SEA

TEMP °F: 60 80 100 120 140 160 180 200 220 240 260 280 300 320 340 360 380 400 420 440 460 480 500 520 540 560

METALS	
ALLOY 20Cb3	G
ALUMINUM	G
BRASS	G
BRONZE	G
CARBON STEEL	S
COPPER	G
E-BRIGHT 26-1	
HASTELLOY-B	G — U
HASTELLOY-C	E — G
HASTELLOY-D	
HIGH SILICON IRON	U
INCONEL	G
LEAD	
MONEL	G
NICKEL	G
STAINLESS STEELS	
Type 304/307	G
Type 316	G
TANTALUM	E
TITANIUM	E
ZIRCONIUM	E

NONMETALLICS	
BOROSILICATE GLASS	R
CARBON	R

TEMP °F: 60 80 100 120 140 160 180 200 220 240 260 280 300 320 340 360 380 400 420 440 460 480 500 520 540 560

FOR METALS

E = < 2 Mils Penetration/Year; G = < 20 Mils Penetration/Year

S = < 50 Mils Penetration/Year; U = > 50 Mils Penetration/Year

FOR NONMETALLICS

R = Resistant

U = Unsatisfactory

WATER, SEA

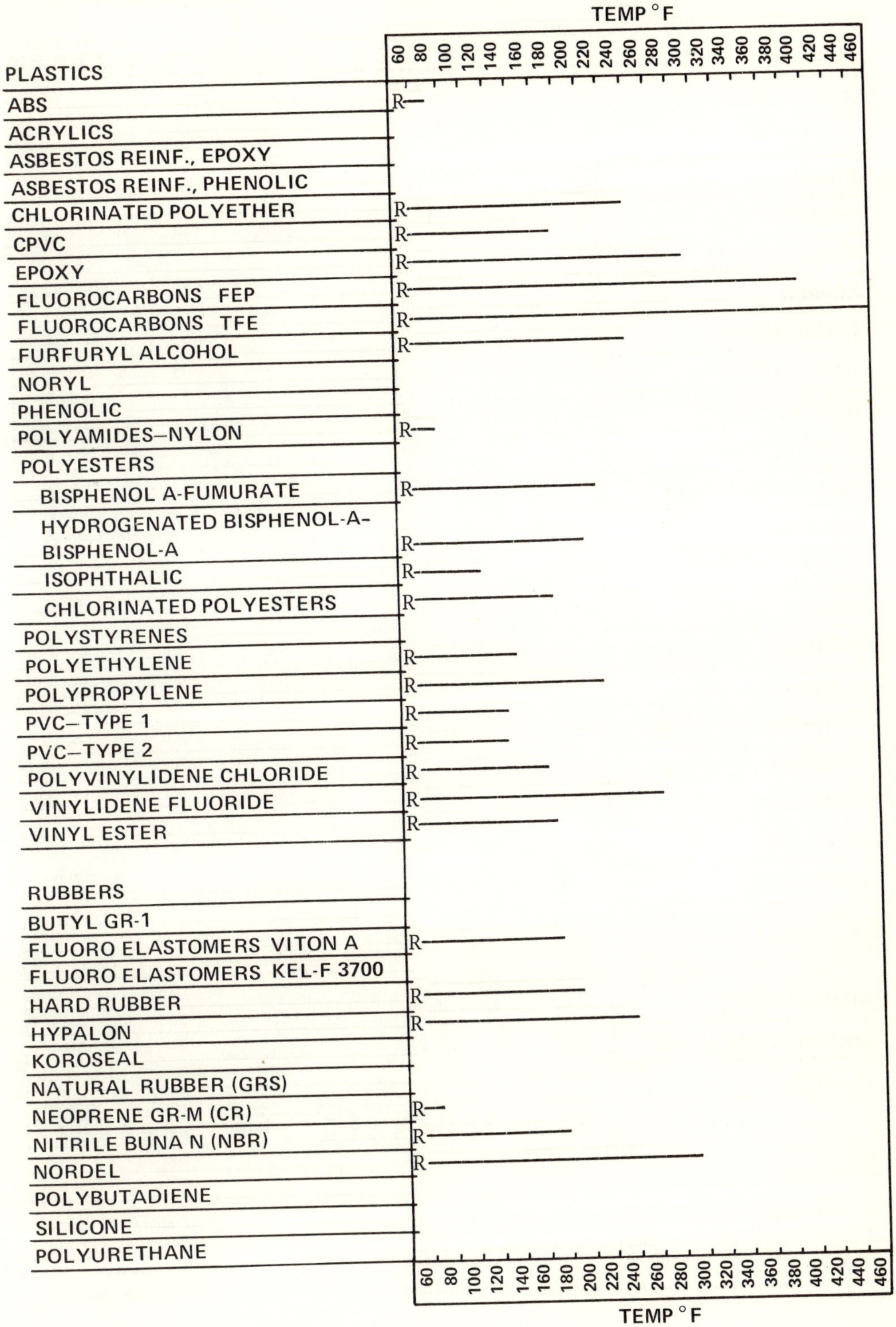
TEMP °F
60 80 100 120 140 160 180 200 220 240 260 280 300 320 340 360 380 400 420 440 460
PLASTICS
ABS
ACRYLICS
ASBESTOS REINF., EPOXY
ASBESTOS REINF., PHENOLIC
CHLORINATED POLYETHER
CPVC
EPOXY
FLUOROCARBONS FEP
FLUOROCARBONS TFE
FURFURYL ALCOHOL
NORYL
PHENOLIC
POLYAMIDES—NYLON
POLYESTERS
BISPHENOL A-FUMURATE
HYDROGENATED BISPHENOL-A-BISPHENOL-A
ISOPHTHALIC
CHLORINATED POLYESTERS
POLYSTYRENES
POLYETHYLENE
POLYPROPYLENE
PVC—TYPE 1
PVC—TYPE 2
POLYVINYLIDENE CHLORIDE
VINYLIDENE FLUORIDE
VINYL ESTER
RUBBERS
BUTYL GR-1
FLUORO ELASTOMERS VITON A
FLUORO ELASTOMERS KEL-F 3700
HARD RUBBER
HYPALON
KOROSEAL
NATURAL RUBBER (GRS)
NEOPRENE GR-M (CR)
NITRILE BUNA N (NBR)
NORDEL
POLYBUTADIENE
SILICONE
POLYURETHANE
R
TEMP °F

WATER, SEWAGE

TEMP °F (60–560)

METALS	Rating
ALLOY 20Cb3	
ALUMINUM	
BRASS	G—
BRONZE	G—
CARBON STEEL	G—
COPPER	G—
E-BRIGHT 26-1	
HASTELLOY-B	
HASTELLOY-C	
HASTELLOY-D	
HIGH SILICON IRON	
INCONEL	
LEAD	
MONEL	
NICKEL	
STAINLESS STEELS	
Type 304/307	G—
Type 316	G—
TANTALUM	E——
TITANIUM	
ZIRCONIUM	

NONMETALLICS	Rating
BOROSILICATE GLASS	
CARBON	

TEMP °F (60–560)

FOR METALS

E = < 2 Mils Penetration/Year; G = < 20 Mils Penetration/Year

S = < 50 Mils Penetration/Year; U = > 50 Mils Penetration/Year

FOR NONMETALLICS

R = Resistant

U = Unsatisfactory

WATER, SEWAGE

PLASTICS	TEMP °F (60–460)
ABS	
ACRYLICS	
ASBESTOS REINF., EPOXY	
ASBESTOS REINF., PHENOLIC	
CHLORINATED POLYETHER	R
CPVC	R
EPOXY	
FLUOROCARBONS FEP	R
FLUOROCARBONS TFE	R
FURFURYL ALCOHOL	
NORYL	
PHENOLIC	
POLYAMIDES–NYLON	
POLYESTERS	
BISPHENOL A-FUMURATE	
HYDROGENATED BISPHENOL-A-BISPHENOL-A	
ISOPHTHALIC	
CHLORINATED POLYESTERS	
POLYSTYRENES	
POLYETHYLENE	R
POLYPROPYLENE	R
PVC–TYPE 1	R
PVC–TYPE 2	R
POLYVINYLIDENE CHLORIDE	R
VINYLIDENE FLUORIDE	R
VINYL ESTER	
RUBBERS	
BUTYL GR-1	
FLUORO ELASTOMERS VITON A	R
FLUORO ELASTOMERS KEL-F 3700	
HARD RUBBER	
HYPALON	
KOROSEAL	
NATURAL RUBBER (GRS)	
NEOPRENE GR-M (CR)	
NITRILE BUNA N (NBR)	R
NORDEL	
POLYBUTADIENE	
SILICONE	
POLYURETHANE	

WHISKEY

TEMP °F: 60 80 100 120 140 160 180 200 220 240 260 280 300 320 340 360 380 400 420 440 460 480 500 520 540 560

METALS	
ALLOY 20Cb3	G
ALUMINUM	S
BRASS	G
BRONZE	G
CARBON STEEL	G
COPPER	G
E-BRIGHT 26-1	
HASTELLOY-B	E
HASTELLOY-C	E
HASTELLOY-D	
HIGH SILICON IRON	
INCONEL	
LEAD	
MONEL	G
NICKEL	G
STAINLESS STEELS	
Type 304/307	E
Type 316	E
TANTALUM	E
TITANIUM	
ZIRCONIUM	

NONMETALLICS	
BOROSILICATE GLASS	R
CARBON	R

TEMP °F: 60 80 100 120 140 160 180 200 220 240 260 280 300 320 340 360 380 400 420 440 460 480 500 520 540 560

FOR METALS

E = < 2 Mils Penetration/Year; G = < 20 Mils Penetration/Year

S = < 50 Mils Penetration/Year; U = > 50 Mils Penetration/Year

FOR NONMETALLICS

R = Resistant

U = Unsatisfactory

WHISKEY

TEMP °F (60–460)

PLASTICS	
ABS	R
ACRYLICS	U
ASBESTOS REINF., EPOXY	
ASBESTOS REINF., PHENOLIC	
CHLORINATED POLYETHER	R
CPVC	R
EPOXY	
FLUOROCARBONS FEP	R
FLUOROCARBONS TFE	R
FURFURYL ALCOHOL	
NORYL	R
PHENOLIC	
POLYAMIDES–NYLON	R
POLYESTERS	
BISPHENOL A-FUMURATE	R
HYDROGENATED BISPHENOL-A-BISPHENOL-A	
ISOPHTHALIC	
CHLORINATED POLYESTERS	
POLYSTYRENES	
POLYETHYLENE 11	R
POLYPROPYLENE	R
PVC–TYPE 1	R
PVC–TYPE 2	R
POLYVINYLIDENE CHLORIDE	R
VINYLIDENE FLUORIDE	R
VINYL ESTER	
RUBBERS	
BUTYL GR-1	R
FLUORO ELASTOMERS VITON A	R
FLUORO ELASTOMERS KEL-F 3700	
HARD RUBBER	
HYPALON	R
KOROSEAL	
NATURAL RUBBER (GRS)	R
NEOPRENE GR-M (CR)	R
NITRILE BUNA N (NBR)	R
NORDEL	
POLYBUTADIENE	
SILICONE	
POLYURETHANE	U

TEMP °F (60–460)

WHITE LIQUOR

TEMP °F: 60 80 100 120 140 160 180 200 220 240 260 280 300 320 340 360 380 400 420 440 460 480 500 520 540 560

METALS	
ALLOY 20Cb3	G
ALUMINUM	G
BRASS	
BRONZE	S
CARBON STEEL	S
COPPER	
E-BRIGHT 26-1	
HASTELLOY-B	G
HASTELLOY-C	G
HASTELLOY-D	
HIGH SILICON IRON	
INCONEL	
LEAD	
MONEL	S
NICKEL	
STAINLESS STEELS	
Type 304/307	G
Type 316	G
TANTALUM	
TITANIUM	
ZIRCONIUM	

NONMETALLICS	
BOROSILICATE GLASS	R
CARBON	

TEMP °F: 60 80 100 120 140 160 180 200 220 240 260 280 300 320 340 360 380 400 420 440 460 480 500 520 540 560

FOR METALS

E = < 2 Mils Penetration/Year; G = < 20 Mils Penetration/Year

S = < 50 Mils Penetration/Year; U = > 50 Mils Penetration/Year

FOR NONMETALLICS

R = Resistant

U = Unsatisfactory

WHITE LIQUOR

TEMP °F: 60 80 100 120 140 160 180 200 220 240 260 280 300 320 340 360 380 400 420 440 460

PLASTICS	
ABS	R
ACRYLICS	
ASBESTOS REINF., EPOXY	
ASBESTOS REINF., PHENOLIC	
CHLORINATED POLYETHER	R
CPVC	R
EPOXY	
FLUOROCARBONS FEP	R
FLUOROCARBONS TFE	R
FURFURYL ALCOHOL	
NORYL	
PHENOLIC	
POLYAMIDES–NYLON	R
POLYESTERS	
BISPHENOL A-FUMURATE	R
HYDROGENATED BISPHENOL-A-BISPHENOL-A	
ISOPHTHALIC	
CHLORINATED POLYESTERS	
POLYSTYRENES	
POLYETHYLENE	R
POLYPROPYLENE	R
PVC–TYPE 1	R
PVC–TYPE 2	R
POLYVINYLIDENE CHLORIDE	
VINYLIDENE FLUORIDE	
VINYL ESTER	R

RUBBERS	
BUTYL GR-1	
FLUORO ELASTOMERS VITON A	R
FLUORO ELASTOMERS KEL-F 3700	
HARD RUBBER	
HYPALON	
KOROSEAL	
NATURAL RUBBER (GRS)	
NEOPRENE GR-M (CR)	
NITRILE BUNA N (NBR)	RU
NORDEL	R
POLYBUTADIENE	
SILICONE	
POLYURETHANE	

TEMP °F: 60 80 100 120 140 160 180 200 220 240 260 280 300 320 340 360 380 400 420 440 460

WINES

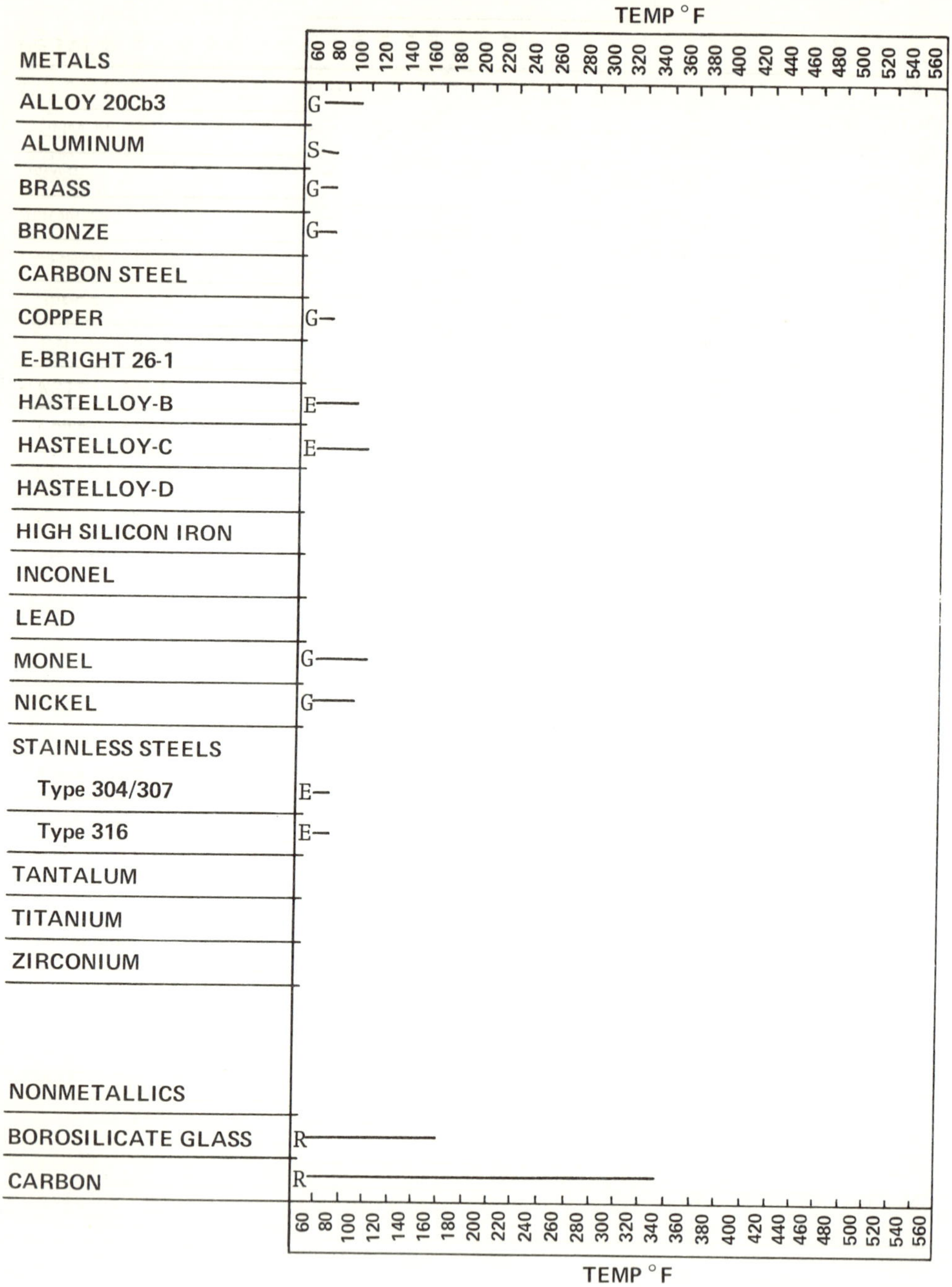

FOR METALS

E = < 2 Mils Penetration/Year; G = < 20 Mils Penetration/Year

S = < 50 Mils Penetration/Year; U = > 50 Mils Penetration/Year

FOR NONMETALLICS

R = Resistant

U = Unsatisfactory

WINES

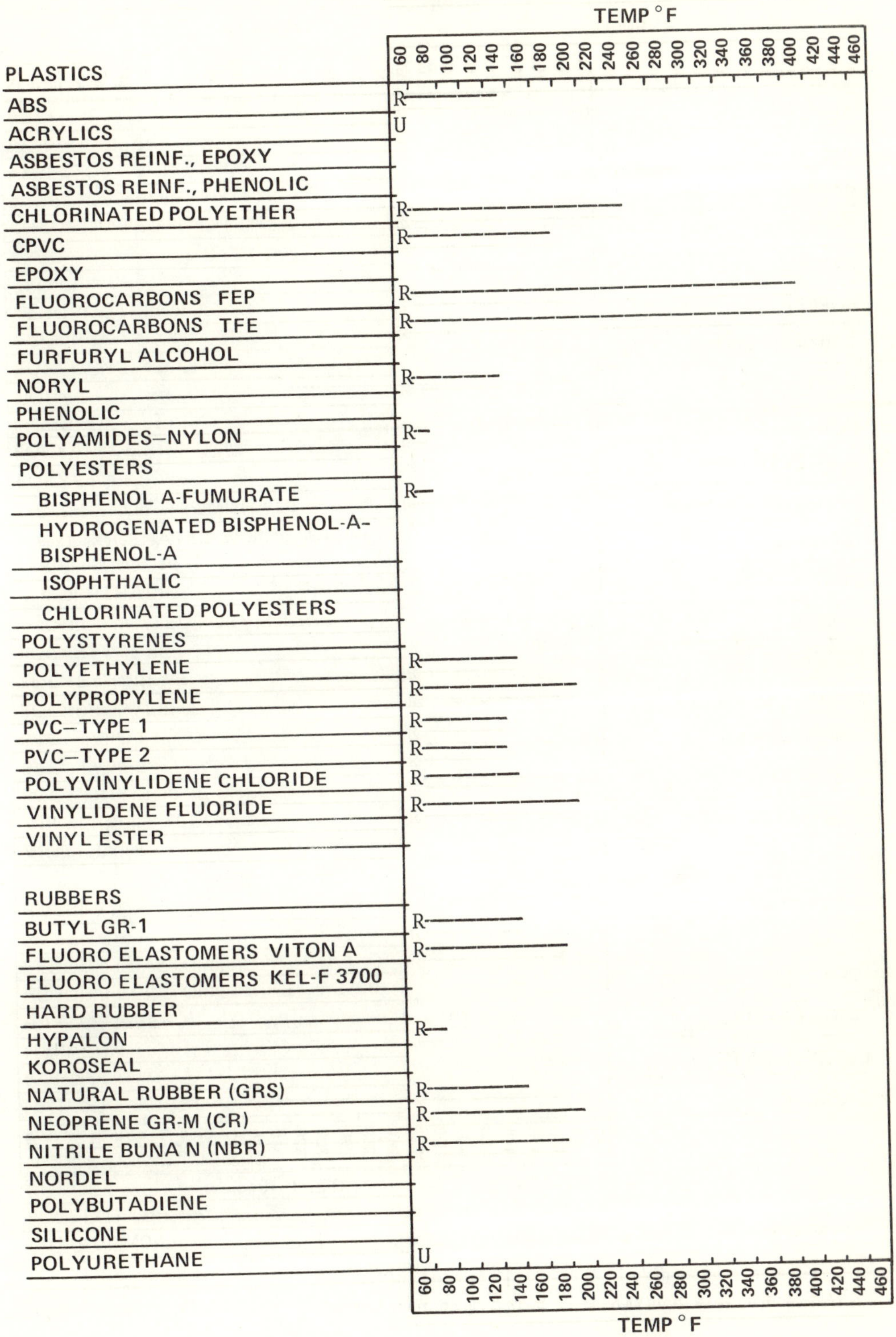
TEMP °F
60 80 100 120 140 160 180 200 220 240 260 280 300 320 340 360 380 400 420 440 460
PLASTICS
ABS R
ACRYLICS U
ASBESTOS REINF., EPOXY
ASBESTOS REINF., PHENOLIC
CHLORINATED POLYETHER R
CPVC R
EPOXY
FLUOROCARBONS FEP R
FLUOROCARBONS TFE R
FURFURYL ALCOHOL
NORYL R
PHENOLIC
POLYAMIDES–NYLON R
POLYESTERS
BISPHENOL A-FUMURATE R
HYDROGENATED BISPHENOL-A-BISPHENOL-A
ISOPHTHALIC
CHLORINATED POLYESTERS
POLYSTYRENES
POLYETHYLENE R
POLYPROPYLENE R
PVC–TYPE 1 R
PVC–TYPE 2 R
POLYVINYLIDENE CHLORIDE R
VINYLIDENE FLUORIDE R
VINYL ESTER
RUBBERS
BUTYL GR-1 R
FLUORO ELASTOMERS VITON A R
FLUORO ELASTOMERS KEL-F 3700
HARD RUBBER
HYPALON R
KOROSEAL
NATURAL RUBBER (GRS) R
NEOPRENE GR-M (CR) R
NITRILE BUNA N (NBR) R
NORDEL
POLYBUTADIENE
SILICONE
POLYURETHANE U
60 80 100 120 140 160 180 200 220 240 260 280 300 320 340 360 380 400 420 440 460
TEMP °F

XYLENE (XYLOL)

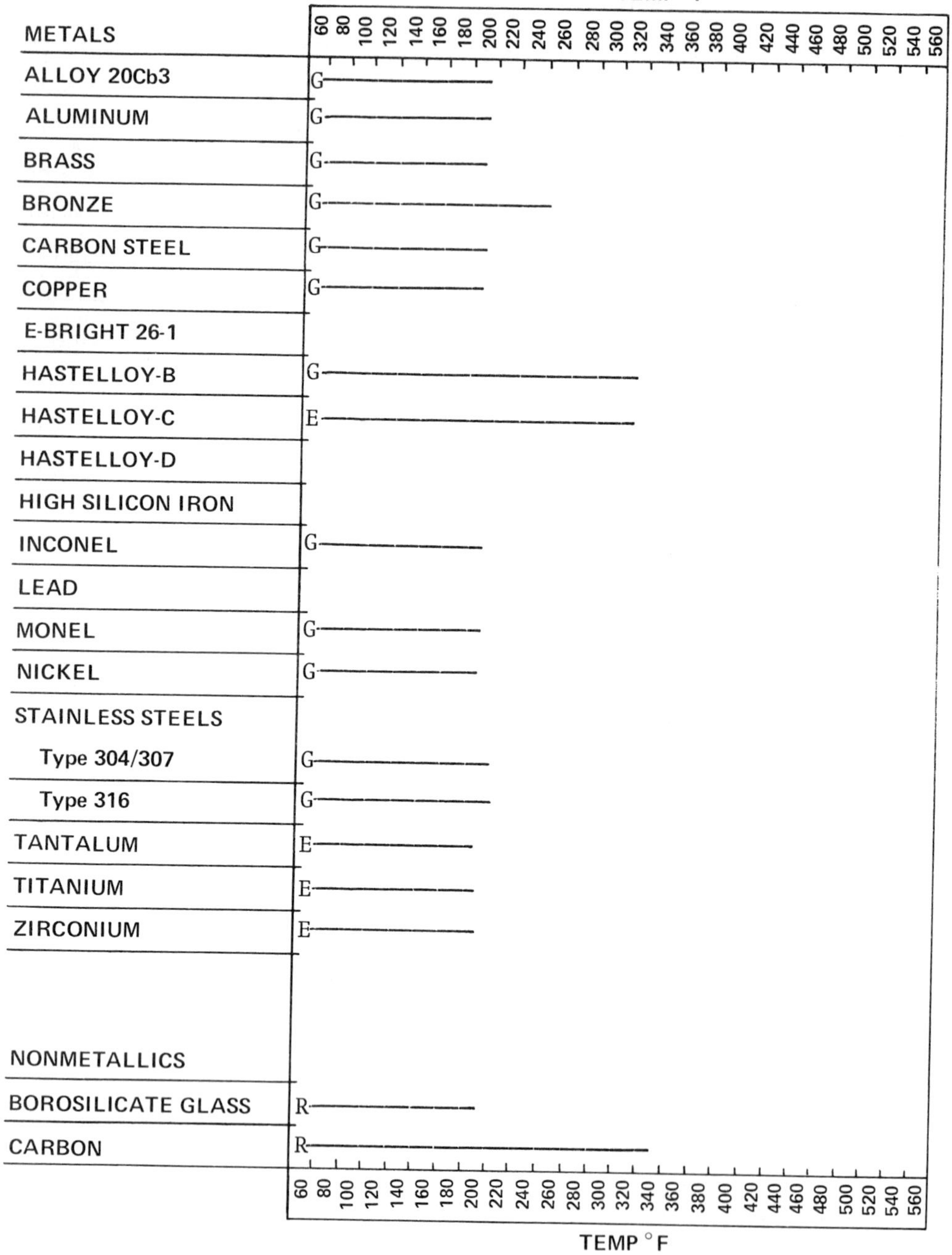

FOR METALS

E = < 2 Mils Penetration/Year; G = < 20 Mils Penetration/Year
S = < 50 Mils Penetration/Year; U = > 50 Mils Penetration/Year

FOR NONMETALLICS

R = Resistant
U = Unsatisfactory

XYLENE (XYLOL)

Please see page 1 for notes on calibration.

TEMP °F (60–460)

PLASTICS	60 → 460 °F
ABS	U
ACRYLICS	
ASBESTOS REINF., EPOXY	RU
ASBESTOS REINF., PHENOLIC	R—
CHLORINATED POLYETHER	R—
CPVC	U
EPOXY	R—U
FLUOROCARBONS FEP	R—
FLUOROCARBONS TFE	R—
FURFURYL ALCOHOL	R—
NORYL	
PHENOLIC	
POLYAMIDES—NYLON	R—
POLYESTERS	
BISPHENOL A-FUMURATE	R—
HYDROGENATED BISPHENOL-A-BISPHENOL-A	R—
ISOPHTHALIC	U
CHLORINATED POLYESTERS	R—
POLYSTYRENES	
POLYETHYLENE	U
POLYPROPYLENE	RU
PVC—TYPE 1	U
PVC—TYPE 2	U
POLYVINYLIDENE CHLORIDE	U
VINYLIDENE FLUORIDE	R—
VINYL ESTER	R—

RUBBERS	60 → 460 °F
BUTYL GR-1	U
FLUORO ELASTOMERS VITON A	R—
FLUORO ELASTOMERS KEL-F 3700	
HARD RUBBER	U
HYPALON	U
KOROSEAL	U
NATURAL RUBBER (GRS)	U
NEOPRENE GR-M (CR)	U
NITRILE BUNA N (NBR)	U
NORDEL	
POLYBUTADIENE	
SILICONE	
POLYURETHANE	U

TEMP °F

ZINC CHLORIDE

TEMP °F: 60 80 100 120 140 160 180 200 220 240 260 280 300 320 340 360 380 400 420 440 460 480 500 520 540 560

METALS		
ALLOY 20Cb3	to 80%	G
ALUMINUM		U
BRASS		U
BRONZE		U
CARBON STEEL		U
COPPER		U
E-BRIGHT 26-1		E
HASTELLOY-B		G
HASTELLOY-C		G
HASTELLOY-D		G
HIGH SILICON IRON		E
INCONEL	dry	G
LEAD		S
MONEL	to 80%	G
NICKEL	to 80%	G
STAINLESS STEELS		
Type 304/307		U
Type 316		G
TANTALUM		E
TITANIUM	to 70%	E — G — U
ZIRCONIUM	10-20%	G

NONMETALLICS	
BOROSILICATE GLASS	R
CARBON	R

TEMP °F

FOR METALS

E = < 2 Mils Penetration/Year; G = < 20 Mils Penetration/Year

S = < 50 Mils Penetration/Year; U = > 50 Mils Penetration/Year

FOR NONMETALLICS

R = Resistant

U = Unsatisfactory

ZINC CHLORIDE

Please see page 1 for notes on calibration.

TEMP °F: 60, 80, 100, 120, 140, 160, 180, 200, 220, 240, 260, 280, 300, 320, 340, 360, 380, 400, 420, 440, 460

Material	Conc.	Rating
PLASTICS		
ABS		R
ACRYLICS		R
ASBESTOS REINF., EPOXY		R
ASBESTOS REINF., PHENOLIC		R
CHLORINATED POLYETHER		R
CPVC		R
EPOXY		R
FLUOROCARBONS FEP		R
FLUOROCARBONS TFE		R
FURFURYL ALCOHOL	sat	R
NORYL		R
PHENOLIC		R
POLYAMIDES–NYLON		U
POLYESTERS		
BISPHENOL A-FUMURATE		R
HYDROGENATED BISPHENOL-A-BISPHENOL-A		R
ISOPHTHALIC	sat	R
CHLORINATED POLYESTERS	all	R
POLYSTYRENES		R
POLYETHYLENE		R
POLYPROPYLENE		R
PVC–TYPE 1		R
PVC–TYPE 2		R
POLYVINYLIDENE CHLORIDE		R
VINYLIDENE FLUORIDE		R
VINYL ESTER	sat	R
RUBBERS		
BUTYL GR-1		R
FLUORO ELASTOMERS VITON A		R
FLUORO ELASTOMERS KEL-F 3700		
HARD RUBBER		R
HYPALON		R
KOROSEAL		R
NATURAL RUBBER (GRS)		R
NEOPRENE GR-M (CR)		R
NITRILE BUNA N (NBR)		R
NORDEL		R
POLYBUTADIENE		R
SILICONE		
POLYURETHANE		

TEMP °F: 60, 80, 100, 120, 140, 160, 180, 200, 220, 240, 260, 280, 300, 320, 340, 360, 380, 400, 420, 440, 460

ZINC NITRATE

METALS	TEMP °F (60–560)
ALLOY 20Cb3	
ALUMINUM	
BRASS	
BRONZE	
CARBON STEEL	
COPPER	
E-BRIGHT 26-1	
HASTELLOY-B	
HASTELLOY-C	
HASTELLOY-D	
HIGH SILICON IRON	
INCONEL	
LEAD	
MONEL	
NICKEL	
STAINLESS STEELS	
Type 304/307	G (60–140)
Type 316	G (60–140)
TANTALUM	
TITANIUM	
ZIRCONIUM	

NONMETALLICS	TEMP °F (60–560)
BOROSILICATE GLASS	
CARBON	R (60–340)

FOR METALS

E = < 2 Mils Penetration/Year; G = < 20 Mils Penetration/Year
S = < 50 Mils Penetration/Year; U = > 50 Mils Penetration/Year

FOR NONMETALLICS

R = Resistant
U = Unsatisfactory

ZINC NITRATE

TEMP °F: 60, 80, 100, 120, 140, 160, 180, 200, 220, 240, 260, 280, 300, 320, 340, 360, 380, 400, 420, 440, 460

PLASTICS		Rating
ABS		R
ACRYLICS		
ASBESTOS REINF., EPOXY		
ASBESTOS REINF., PHENOLIC		
CHLORINATED POLYETHER		R
CPVC		R
EPOXY		U
FLUOROCARBONS FEP		R
FLUOROCARBONS TFE		R
FURFURYL ALCOHOL		U
NORYL		
PHENOLIC		
POLYAMIDES—NYLON		
POLYESTERS		
BISPHENOL A-FUMURATE		R
HYDROGENATED BISPHENOL-A-BISPHENOL-A		R
ISOPHTHALIC	sat	R
CHLORINATED POLYESTERS		
POLYSTYRENES		
POLYETHYLENE		R
POLYPROPYLENE		R
PVC—TYPE 1		R
PVC—TYPE 2		R
POLYVINYLIDENE CHLORIDE		R
VINYLIDENE FLUORIDE		R
VINYL ESTER	sat	R
RUBBERS		
BUTYL GR-1		
FLUORO ELASTOMERS VITON A		R
FLUORO ELASTOMERS KEL-F 3700		
HARD RUBBER		
HYPALON		
KOROSEAL		R
NATURAL RUBBER (GRS)		R
NEOPRENE GR-M (CR)		R
NITRILE BUNA N (NBR)		
NORDEL		
POLYBUTADIENE		
SILICONE		
POLYURETHANE		

TEMP °F: 60, 80, 100, 120, 140, 160, 180, 200, 220, 240, 260, 280, 300, 320, 340, 360, 380, 400, 420, 440, 460

ZINC SULFATE, 30%

TEMP °F: 60, 80, 100, 120, 140, 160, 180, 200, 220, 240, 260, 280, 300, 320, 340, 360, 380, 400, 420, 440, 460, 480, 500, 520, 540, 560

METALS	
ALLOY 20Cb3	E
ALUMINUM	U
BRASS	G
BRONZE	G
CARBON STEEL	U
COPPER	G
E-BRIGHT 26-1	
HASTELLOY-B	G
HASTELLOY-C	G
HASTELLOY-D	G
HIGH SILICON IRON	E
INCONEL	E
LEAD	G
MONEL	G
NICKEL	E
STAINLESS STEELS	
Type 304/307	E
Type 316	E
TANTALUM	E
TITANIUM	E
ZIRCONIUM	E

NONMETALLICS	
BOROSILICATE GLASS	R
CARBON	R

TEMP °F: 60, 80, 100, 120, 140, 160, 180, 200, 220, 240, 260, 280, 300, 320, 340, 360, 380, 400, 420, 440, 460, 480, 500, 520, 540, 560

FOR METALS

E = < 2 Mils Penetration/Year; G = < 20 Mils Penetration/Year

S = < 50 Mils Penetration/Year; U = > 50 Mils Penetration/Year

FOR NONMETALLICS

R = Resistant

U = Unsatisfactory

ZINC SULFATE, 30%

TEMP °F: 60 80 100 120 140 160 180 200 220 240 260 280 300 320 340 360 380 400 420 440 460

Material	Conc.	Rating
PLASTICS		
ABS		R
ACRYLICS		R
ASBESTOS REINF., EPOXY		R
ASBESTOS REINF., PHENOLIC		R
CHLORINATED POLYETHER		R
CPVC		R
EPOXY		R
FLUOROCARBONS FEP		R
FLUOROCARBONS TFE		R
FURFURYL ALCOHOL	sat	R
NORYL		R
PHENOLIC		R
POLYAMIDES–NYLON		R
POLYESTERS		
BISPHENOL A-FUMURATE	sat	R
HYDROGENATED BISPHENOL-A-BISPHENOL-A		R
ISOPHTHALIC	sat	R
CHLORINATED POLYESTERS	all	R
POLYSTYRENES		R
POLYETHYLENE		R
POLYPROPYLENE		R
PVC–TYPE 1		R
PVC–TYPE 2		R
POLYVINYLIDENE CHLORIDE		R
VINYLIDENE FLUORIDE		R
VINYL ESTER	all	R
RUBBERS		
BUTYL GR-1		R
FLUORO ELASTOMERS VITON A		R
FLUORO ELASTOMERS KEL-F 3700		
HARD RUBBER		R
HYPALON		R
KOROSEAL		R
NATURAL RUBBER (GRS)		R
NEOPRENE GR-M (CR)		R
NITRILE BUNA N (NBR)		R
NORDEL		R
POLYBUTADIENE		
SILICONE		
POLYURETHANE		

TEMP °F: 60 80 100 120 140 160 180 200 220 240 260 280 300 320 340 360 380 400 420 440 460